MW00838284

# UHMWPE Biomaterials Handbook

Ultra-High Molecular Weight Polyethylene in Total Joint Replacement and Medical Devices

# UHMWPE Biomaterials Handbook

## Ultra-High Molecular Weight Polyethylene in Total Joint Replacement and Medical Devices

Second Edition

Editor
Steven M. Kurtz, PhD

AMSTERDAM • BOSTON • HEIDELBERG • LONDON • NEW YORK • OXFORD • PARIS • SAN DIEGO • SAN FRANCISCO • SINGAPORE • SYDNEY • TOKYO

Academic Press is an Imprint of Elsevier

Academic Press is an imprint of Elsevier
30 Corporate Drive, Suite 400, Burlington, MA 01803, USA
525 B Street, Suite 1900, San Diego, California 92101-4495, USA
84 Theobald's Road, London WC1X 8RR, UK



**Library of Congress Cataloging-in-Publication Data**
Application submitted

**British Library Cataloguing-in-Publication Data**
A catalogue record for this book is available from the British Library.

ISBN: 978- 0-12-374721-1



Printed in China

09 10 11 12 10 9 8 7 6 5 4 3 2 1

To Karen, for listening to all my stories.

# Contents

# Foreword

The clinical history of UHMWPE biomaterials continues to expand at a surprising pace. Five years ago, when the first edition of the handbook was published, radiation crosslinked materials had been recently introduced. Today, a second generation of radiation crosslinked materials are in clinical use, and Vitamin E stabilized UHMWPE has emerged as a new, internationally standardized biomaterial. In retrospect, the speed of advancement in this field is astonishing, considering that for the first three decades of the clinical history of UHMWPE biomaterials (1962–1997), few clinically relevant innovations occurred beyond the removal of calcium stearate and changes in sterilization practice. Because of the ongoing developments in this field, I felt that it was appropriate to undertake a major revision of the first edition, culminating in the preparation of this volume.

The second edition differs from its predecessor in several notable respects. The first major change is the involvement of many additional UHMWPE experts, to provide an expanded and diversified perspective on advancements in particular subjects. For example, I have asked pioneers in the use of remelted and annealed materials to explain the development history and the research findings for their respective families of crosslinked polyethylenes. A second major change for this volume is its breadth: the first edition contained 15 chapters, whereas the 2nd edition includes 35. The major expansion of the current edition not only reflects feedback I received from readers in both industry and academia, but also better captures my personal wishes for a more comprehensive treatment of the subject, which simply wasn't practical five years ago. For example, although I wanted to include a chapter on in vivo oxidation of UHMWPE five years ago in the first edition, I felt that consensus in the field and my own research on the topic had not yet advanced to the point where a dedicated chapter was appropriate. In other instances, I simply did not have the time to research the background of certain "pet" projects, such as composites and fibers in UHMWPE. In other cases, such as the fields of biotribology and the biological response to wear debris, I felt that other authors would be better able to explain the background in this discipline than I could. As a result, this edition of the *Handbook* captures, to a much greater extent than its predecessor, the incredible breadth and depth of scientific inquiry in the field of UHMWPE biomaterials. I am endebted to my collaborators for dedicating their personal time and expertise to assist me in this undertaking.

The first edition of the *Handbook* has its origins in a review article, "Advances in the sterilization, processing, and crosslinking of ultra-high molecular weight polyethylene for total joint arthroplasty" (*Biomaterials* 1999; 20: 1659–88), which I co-authored with Orhun Muratoglu, Mark Evans, and Av Edidin. Our review was written between 1997 and 1998, at a time when highly crosslinked and thermally treated UHMWPE materials were about to be clinically introduced for total hip replacement. Several other important milestones in the clinical use of UHMWPE had occurred, including the abandonment of gamma sterilization in air (at least in the United States), and the trend to reduce calcium stearate (a processing additive) in the UHMWPE powder.

If we include the alternative bearing solutions (e.g., metal-on-metal and ceramic-on-ceramic) for total joint arthroplasty, with their myriad of design options, the number of choices available to the orthopedic surgeon continues to expand far beyond what was available in 2004 when the first edition of the *Handbook* was published. Yet, with all the advanced technologies and design options available today, we should still be mindful that for very elderly patients, artificial joints incorporating conventional UHMWPE will continue to afford long-term clinical benefits that could potentially last the rest of these patients' natural lives. On the other hand, the more recently-introduced alternative bearing technologies, including crosslinked UHMWPE, should provide the greatest benefit to young patients (less than 60 years in age) who lead an active lifestyle and who need a total hip replacement. For patients in need of knee arthroplasty, shoulder arthroplasty, or total disc replacement, conventional UHMWPE continues to prevail as the polymeric bearing material of choice.

The latest generation of new processing technology has contributed to some confusion, among surgeons and researchers alike, regarding the technical details associated with first- and second-generation highly crosslinked UHMWPE. I am also frequently asked about the differences between blending and doping UHMWPE with Vitamin E. My goal in writing the *UHMWPE Handbook* is to explain the common concepts beneath all UHMWPE biomaterials, as well as the technical differences in specific formulations for an audience of surgeons, researchers, and students who may be starting work in this field. Some of the early chapters in this book may also be of interest to current or prospective patients who are motivated to learn more about current treatment options. Because all of the alternative bearing solutions (both highly crosslinked UHMWPE as well as hard-on-hard bearings) have their origins in the 1950s, 60s, and 70s, it is important to understand the historical context in which the first and second generation of highly crosslinked UHMWPEs were originally developed. With this in mind, I have also taken care to review the historical development of UHMWPE bearings for joint replacement.

The story of UHMWPE in orthopedics, seemingly immutable and static, still continues to evolve. Early in 2002, I began to expand my website, the UHMWPE Lexicon (www.uhmwpe.org), with an online monograph of six introductory chapters covering the basic scientific principles and clinical performance of UHMWPE in hip replacement. The response to these online chapters was overwhelmingly positive, and encouraged me to revise the first six chapters for hardcopy publication, and to develop the additional nine chapters for first edition of the *UHMWPE Handbook*. This website continues today as a repository of UHMWPE knowledge accumulating in "real time."

Despite my diligent efforts to summarize the state of the art in this field, I appreciate that current trends related to UHMWPE in orthopedics will give way to new ideas as further clinical data becomes available in the future. For this reason, I hope that the UHMWPE Lexicon website (www.uhmwpe.org) will continue to disseminate new research findings to the orthopedic and polymer research communities. The Lexicon website and this *Handbook* play complementary roles. When the accumulation of new ideas and findings has diffused sufficiently into the orthopedic clinical and research practice, it will be time once again to update this written work. I look forward to your comments and suggestions for future expansion of the UHMWPE Lexicon website and this *UHMWPE Handbook*.

The present edition would not have been possible without the suggestions and advice of many supportive colleagues. I am especially grateful to my co-authors, experts in the fields of joint arthroplasty and spine, as well as in the testing and evaluation of UHMWPE, who have cheerfully contributed chapters to this book on relatively short notice. I have also included acknowledgements at the ends of chapters to thank my many friends and associates for their contributions.

—*Steven Kurtz*, PhD
Philadelphia, PA
February 22, 2009
skurtz@drexel.edu

# Contributors


**Steven M. Kurtz, Ph.D** Corporate Vice President, Exponent, Inc. Research Professor, Drexel University, Philadelphia PA

**Ryan M. Baxter, B.S., Ph.D Candidate** Drexel University, Philadelphia, PA, USA

**Anuj Bellare, Ph.D** Orthopaedic Nanotechnology Laboratory, Brigham & Women's Hospital, Boston, MA, USA

**Jorgen Bergström, Ph.D** Veryst Engineering, Needham, MA, USA

**Pierangiola Bracco, Ph.D** Dipartimento di Chimica IFM, Università di Torino, Torino, Italy

**Charles R. Bragdon, Ph.D** Massachusetts General Hospital, Boston, MA, USA

**Luigi Costa, Ph.D** Dipartimento di Chimica IFM, Università di Torino, Torino, Italy

**Judd S. Day, Ph.D** Exponent Inc., Philadelphia, PA, USA

**John H. Dumbleton Ph.D DSc** Consultancy in Medical Devices, Ridgewood, NJ

**Avram A. Edidin, Ph.D** Drexel University, Philadelphia, PA, USA

**Aaron Essner, M.S.** Stryker Orthopaedics, Inc., Mahwah, NJ

**John Fisher, Ph.D, DEng, CEng, CSci, FIMechE, FIPEM** University of Leeds, Leeds, UK

**Theresa A. Freeman, Ph.D** Thomas Jefferson University, Philadelphia, PA, USA

**Jevan Furmanski, Ph.D** University of California, Berkeley, CA, USA

**Stefan M. Gabriel, Ph.D, PE** Moximed, Hayward, CA

**Jeremy L. Gilbert, Ph.D** Syracuse University, Syracuse, NY, USA

**Hani Haider, Ph.D** University of Nebraska Medical Center, Omaha, NE, USA

**Allyson Ianuzzi, Ph.D** Exponent Inc., Philadelphia, PA, USA

**Eileen Ingham, Ph.D** University of Leeds, Leeds, UK

**M. Shah Jahan, Ph.D** University of Memphis, Memphis, TN, USA

**Susan P. James, Ph.D** Colorado State University, Fort Collins, CO, USA

**Daniel MacDonald, B.S., Ph.D. Candidate** Implant Research Center, Drexel University, Philadelphia, PA, USA

**Michael T. Manley, Ph.D** Stryker Orthopedics, Mahwah, NJ, USA

**Francisco J. Medel, Ph.D** Drexel University, Philadelphia, PA, USA

**Chimba Mkandawire, Ph.D**, CAISS Exponent Inc., Philadelphia, PA, USA

**Orhun K. Muratoglu, Ph.D** Massachusetts General Hospital, Boston, MA, USA

**Rachael (Kurkowski) Oldinski, Ph.D** Colorado State University, Fort Collins, CO, USA

**Erin Oneida, B.S., Ph.D. Candidate** Cornell University, Ithaca, NY, USA

**Kevin L. Ong, Ph.D** Exponent Inc., Philadelphia, PA, USA

**Ebru Oral, Ph.D** Massachusetts General Hospital, Boston, MA, USA

**Laura Richards, Ph.D** University of Leeds, Leeds, UK

**Clare M. Rimnac, Ph.D** Case Western Reserve University, Cleveland, OH, USA

**Herb Schwartz, Ph.D** Schwartz Biomedical, Fort Wayne, IN, USA

**Michael C. Sobieraj, Ph.D.** Case Western Reserve University, Cleveland, OH, USA

**Stephen Spiegelberg, Ph.D** Cambridge Polymer Group, Boston, MA, USA

**Marla J. Steinbeck, Ph.D** Drexel University, Philadelphia, PA, USA

**Kate Sutton, MA** ELS Homer Stryker Center, Mahwah, NJ

**Joanne L. Tipper, Ph.D** University of Leeds, Leeds, UK

**Marta L. Villarraga, Ph.D** Exponent Inc., Philadelphia, PA, USA

**Aiguo Wang Ph.D** Stryker Orthopaedics, Inc., Mahwah, NJ

**James D. Wernle** Syracuse University, Syracuse, NY, USA

**Min Zhang, Ph.D** University of Washington, Seattle, WA, USA

Chapter 1

# A Primer on UHMWPE

Steven M. Kurtz, PhD



## 1.1 INTRODUCTION

Ultra-high molecular weight polyethylene (UHMWPE) is a unique polymer with outstanding physical and mechanical properties. Most notable are its chemical inertness, lubricity, impact resistance, and abrasion resistance. These characteristics of UHMWPE have been exploited since the 1950s in a wide range of industrial applications (Figure 1.1), including pickers for textile machinery, lining for coal chutes and dump trucks, runners for bottling production lines, as well as bumpers and siding for ships and harbors. Over 90% of the UHMWPE produced in the world is used by industry.

For the past 45 years, UHMWPE has also been used in orthopedics as a bearing material in artificial joints. Each year, about 2 million joint replacement procedures are performed around the world, and the majority of these joint replacements incorporate UHMWPE. Despite the success of these restorative procedures, orthopedic and spine implants have only a finite lifetime. Wear and damage of the UHMWPE components has historically been one of the factors limiting implant longevity. In the past 10 years, highly crosslinked UHMWPE biomaterials have shown dramatic reductions in wear in clinical use around the world. The orthopedic community awaits confirmation that these reductions in wear will be associated with improved long-term survival, as expected.

UHMWPE comes from a family of polymers with a deceptively simple chemical composition, consisting of only hydrogen and carbon. However, the simplicity inherent in its chemical composition belies a more complex hierarchy of organizational structures at the molecular and supermolecular length scales. At a molecular level, the carbon backbone of polyethylene can twist, rotate, and fold into ordered crystalline regions. At a supermolecular level, the UHMWPE consists of powder (also known as resin or flake) that must be consolidated at elevated temperatures and pressures to form a bulk material. Further layers of complexity are introduced by chemical changes that arise in UHMWPE due to radiation sterilization and processing.

The purpose of this *Handbook* is to explore the complexities inherent in UHMWPE and an increasingly diverse field of UHMWPE biomaterials that include radiation

**FIGURE 1.1** Dump truck liner of UHMWPE, an example of an industrial application for the polymer.

crosslinking, composites, and antioxidants such as Vitamin E. This book is intended to provide the reader with a background in the terminology, history, and recent advances related to its use in orthopedics. A monograph such as this is helpful in several respects. First, it is important that members of the surgical community have access to up-to-date knowledge about the properties of UHMWPE so that this information can be more accurately communicated to their patients. Second, members of the orthopedic research community need access to timely synthesis of the existing literature so that future studies are more effectively planned to fill in existing gaps in our current understanding. Finally, this *Handbook* may also serve as a resource for university students at both the undergraduate and graduate levels.

This introductory chapter starts with the basics, assuming the reader is not familiar with polymers, let alone polyethylene. The chapter provides basic information about polymers in general, describes the structure and composition of polyethylene, and explains how UHMWPE differs from other polymers (including high density polyethylene [HDPE]) and from other materials (e.g., metals and ceramics). The concepts of crystallinity and thermal transitions are introduced at a basic level. Readers familiar with these basic polymer concepts may want to consider skipping ahead to the next chapter.

## 1.2 WHAT IS A POLYMER?

The *ultra-high molecular weight polyethylene* (UHMWPE) used in orthopedic applications is a type of *polymer* generally classified as a *linear homopolymer*. Our first task is to explain what is meant by all of these terms. Before proceeding to a definition of UHMWPE, one needs to first understand what constitutes a linear homopolymer.

A *polymer* is a molecule consisting of many (*poly-*) parts (*-mer*) linked together by chemical covalent bonds. The individual parts, or *monomer* segments, of a polymer can all be the same. In such a case, we have a *homopolymer* as illustrated in Figure 1.2. If the parts of a polymer are different, it is termed a *copolymer*. These differences in chemical structure are also illustrated in Figure 1.2, with generic symbols (A, B) for the monomers.

Polymers can be either *linear* or *branched* as illustrated in Figure 1.3. The tendency for a polymer to exhibit branching is governed by its synthesis conditions.

Keep in mind that the conceptual models of polymer structure illustrated in Figures 1.2 and 1.3 have been highly simplified. For example, it is possible for a copolymer to have a wide range of substructural elements giving rise to an impressive range of possibilities. In industrial practice, polyethylenes, including UHMWPE, are frequently copolymerized with other monomers (e.g., polypropylene) to achieve improved processing characteristics

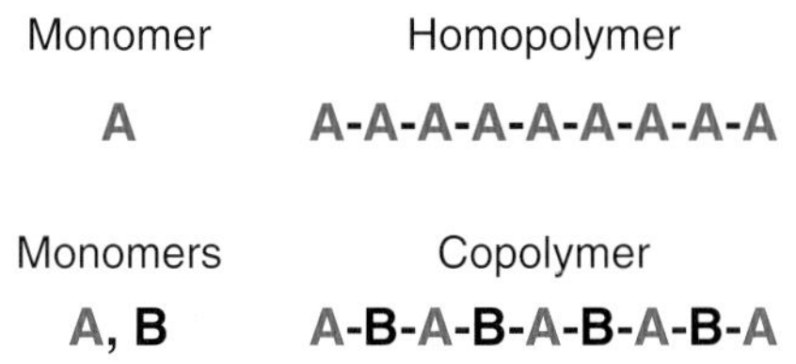

**FIGURE 1.2** Schematics of homopolymer and copolymer structure.

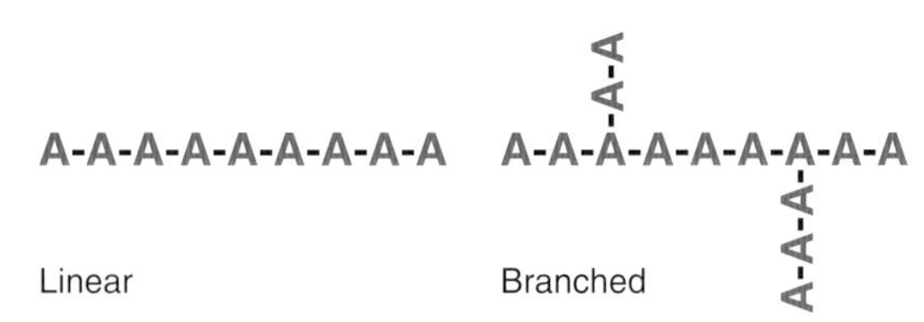


**FIGURE 1.3** Schematics of linear and branched polymer structures.

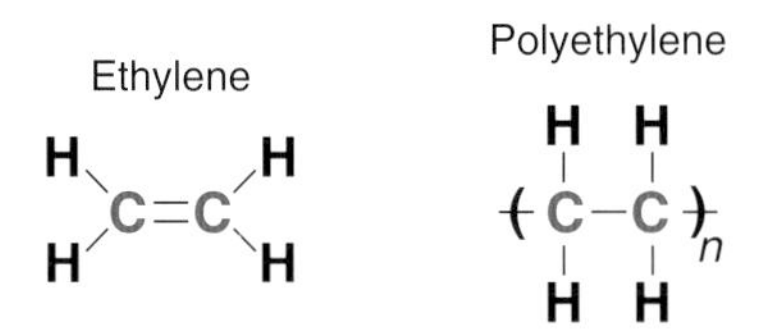


**FIGURE 1.4** Schematic of the chemical structures of ethylene and polyethylene.

or to alter the physical and mechanical properties of the polymer. For example, according to ISO 11542, which is the industrial standard for UHMWPE, the polymer can contain a large concentration of copolymer (up to 50%) and still be referred to as "UHMWPE." However, most of the UHMWPEs used to fabricate orthopedic implants are homopolymers, and so we will restrict our further discussion to polymers with only a single type of monomer.

The principal feature of a polymer that distinguishes it from other materials, such as metals and ceramics, is its molecular size. In a metallic alloy or ceramic, the elemental building blocks are individual metal atoms (e.g., Co, Cr, Mo) or relatively small molecules (e.g., metal carbides or oxides). In a polymer, however, the molecular size can comprise over 100,000 monomer units, with molecular weights of up to millions of g/mol.

The molecular chain architecture of a polymer also imparts many unique attributes, including temperature and rate dependence. Some of these unique properties are further illustrated in the specific case of UHMWPE in subsequent sections of this chapter. For further background on general polymer concepts, the reader is referred to texts by Rodriguez [1] and Young [2].

## 1.3 WHAT IS POLYETHYLENE?

Polyethylene is a polymer formed from ethylene ($C_2H_4$), which is a gas having a molecular weight of 28g/mol. The generic chemical formula for polyethylene is $-(C_2H_4)_n-$, where *n* is the degree of polymerization. A schematic of the chemical structures for ethylene and polyethylene are shown in Figure 1.4.

**TABLE 1.1** Typical Average Physical Properties of High Density Polyethylene (HDPE), Ultra-high Molecular Weight Polyethylene (UHMWPE), Adapted from [3]

| Property | HDPE | UHMWPE |
|---|---|---|
| Molecular weight ($10^6$ g/mol) | 0.05–0.25 | 3.5–7.5 |
| Melting temperature (°C) | 130–137 | 132–138 |
| Poisson's ratio | 0.40 | 0.46 |
| Specific gravity | 0.952–0.965 | 0.925–0.945 |
| Tensile modulus of elasticity* (GPa) | 0.4–4.0 | 0.5–0.8 |
| Tensile yield strength* (MPa) | 26–33 | 21–28 |
| Tensile ultimate strength* (MPa) | 22–31 | 39–48 |
| Tensile ultimate elongation* (%) | 10–1200 | 350–525 |
| Impact strength, Izod* (J/m of notch; 3.175 mm thick specimen) | 21–214 | >1070 (No Break) |
| Degree of crystallinity (%) | 60–80 | 39–75 |

**Testing conducted at 23°C.*

For an ultra-high molecular weight polyethylene, the molecular chain can consist of as many as 200,000 ethylene repeat units. Put another way, the molecular chain of UHMWPE contains up to 400,000 carbon atoms.

There are several kinds of polyethylene (LDPE, LLDPE, HDPE, UHMWPE), which are synthesized with different molecular weights and chain architectures. LDPE and LLDPE refer to low density polyethylene and linear low density polyethylene, respectively. These polyethylenes generally have branched and linear chain architectures, respectively, each with a molecular weight of typically less than 50,000 g/mol.

High density polyethylene (HDPE) is a linear polymer with a molecular weight of up to 200,000 g/mol. UHMWPE, in comparison, has a viscosity average molecular weight of 6 million g/mol. In fact, the molecular weight is so ultra-high that it cannot be measured directly by conventional means and must instead be inferred by its intrinsic viscosity (IV).

Table 1.1 summarizes the physical and mechanical properties of HDPE and UHMWPE. As shown in the table, UHMWPE has a higher ultimate strength and impact strength than HDPE.

Perhaps more relevant from a clinical perspective, UHMWPE is significantly more abrasion- and wear-resistant than HDPE. The following wear data for UHMWPE and HDPE was collected using a contemporary, multidirectional hip simulator [3]. Based on hip simulator data, shown in Figure 1.5, the volumetric wear rate for HDPE is 4.3 times greater than that of UHMWPE.

In the early 1960s, UHMWPE was classified as a form of high-density polyethylene (HDPE) among members of the polymer industry [4]. Thus, Charnley's earlier references to UHMWPE as HDPE are technically accurate for his time [5] but have contributed to some confusion over the years as to exactly what kinds of polyethylenes have been used clinically. From a close reading of Charnley's works, it is clear that HDPE is used synonymously with RCH-1000, the trade name for UHMWPE produced by Hoechst in Germany [6]. With the exception of a small series of 22 patients who were implanted with silane-crosslinked HDPE at Wrighington [7], there is no evidence in the literature that lower molecular weight polyethylenes have been used clinically.

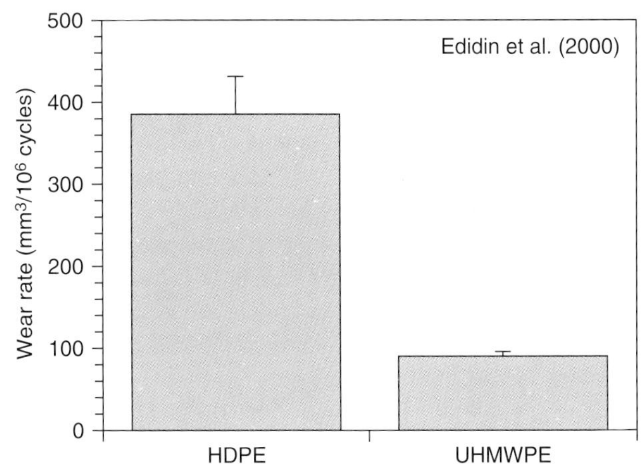


**FIGURE 1.5** Comparison of wear rates of HDPE and UHMWPE in a multidirectional hip simulator [3].

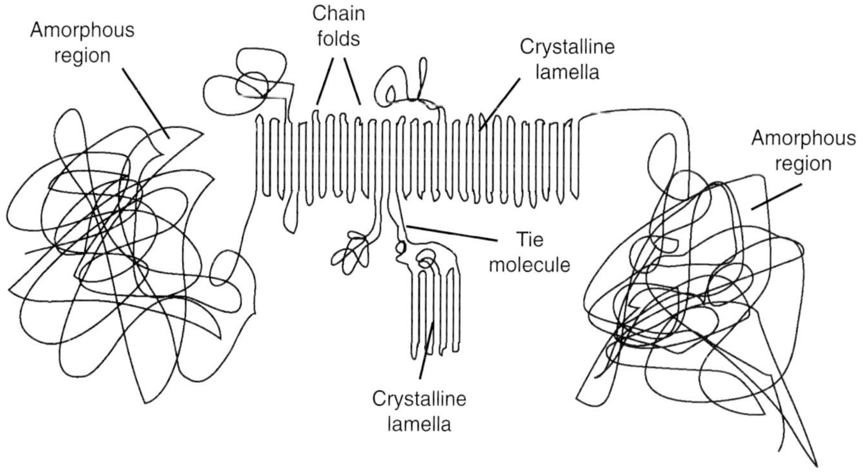

**FIGURE 1.6** Morphological features of UHMWPE.

## 1.4 CRYSTALLINITY

One can visualize the molecular chain of UHMWPE as a tangled string of spaghetti over a kilometer long. Because the chain is not static, but imbued with internal (thermal) energy, the molecular chain can become mobile at elevated temperatures. When cooled below the melt temperature, the molecular chain of polyethylene has the tendency to rotate about the C-C bonds and create chain folds. This chain folding, in turn, enables the molecule to form local ordered, sheetlike regions known as crystalline lamellae. These lamellae are embedded within amorphous (disordered) regions and may communicate with surrounding lamellae by tie molecules. All of these morphological features of UHMWPE are shown schematically in Figure 1.6.

The degree and orientation of crystalline regions within a polyethylene depends upon a variety of factors, including its molecular weight, processing conditions, and environmental conditions (such as loading), and will be discussed in later chapters of this work.

The crystalline lamellae are microscopic and invisible to the naked eye. The lamellae diffract visible light, giving UHMWPE a white, opaque appearance at room temperature. At temperatures above the melt temperature of the lamellae, around 137°C, UHMWPE becomes translucent. The lamellae are on the order of 10–50 nm in thickness and 10–50 μm in length [8]. The average spacing between lamellae is on the order of 50 nm [9].

The crystalline morphology of UHMWPE can be visualized using transmission electron microscopy (TEM), which can magnify the polymer by up to 16,000 times. An ultramicrotomed slice of the polymer is typically stained with uranyl acetate to improve contrast in the TEM. The staining procedure makes the amorphous regions turn gray in the micrograph. The lamellae, which are impervious to the contrast agent, appear as white lines with a dark outline. From the TEM micrograph in Figure 1.7, one can appreciate the composite nature of UHMWPE as an interconnected network of amorphous and crystalline regions.

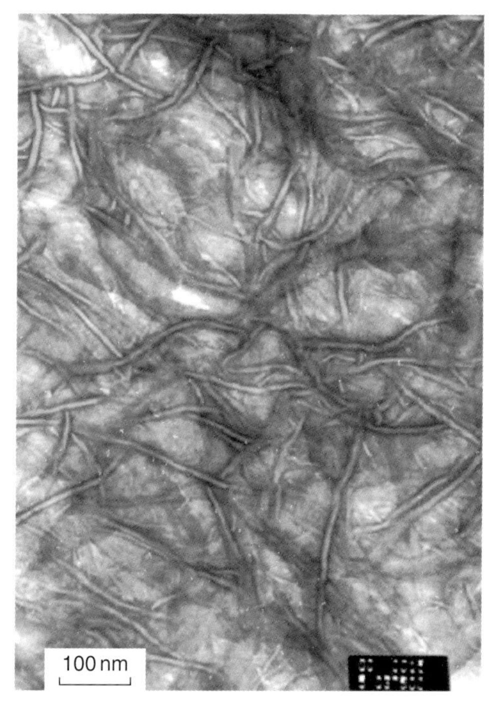


**FIGURE 1.7** TEM micrograph of UHMWPE showing amorphous and crystalline regions (lamellae).

## 1.5 THERMAL TRANSITIONS

As already indicated, one of the distinguishing characteristics of polymeric materials is the temperature dependence of their properties. Returning to our conceptual model of an UHMWPE molecule as a mass of incredibly long spaghetti, one must also imagine it to be jiggling and writhing with thermal energy. Generally speaking, many polymers undergo three major thermal transitions: the glass transition temperature ($T_g$), the melt temperature ($T_m$), and the flow temperature ($T_f$).

The glass transition ($T_g$) is the temperature below which the polymer chains behave like a brittle glass.

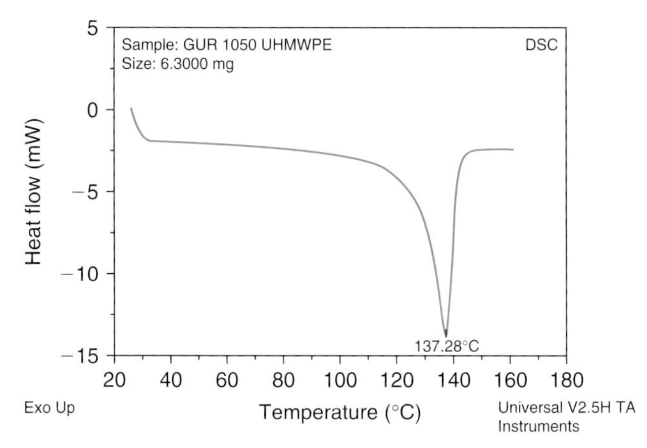


**FIGURE 1.8** Differential scanning calorimetry (DSC) trace for UHMWPE.

Below $T_g$, the polymer chains have insufficient thermal energy to slide past one another, and the only way for the material to respond to mechanical stress is by stretching (or rupture) of the bonds constituting the molecular chain. In UHMWPE, the glass transition occurs around −120°C.

As we raise the temperature above $T_g$, the amorphous regions within the polymer gain increased mobility. When the temperature of UHMWPE rises above 60–90°C, the smaller crystallites in the polymer begin to melt. The melting behavior of semicrystalline polymers, including UHMWPE, is typically measured using differential scanning calorimetry (DSC). DSC measures the amount of heat needed to increase the temperature of a polymer sample. Some representative DSC data for UHMWPE is shown in Figure 1.8.

The DSC trace for UHMWPE shows two key features. The first feature of the DSC curve is the peak melting temperature ($T_m$), which occurs around 137°C and corresponds to the point at which the majority of the crystalline regions have melted. The melt temperature reflects the thickness of the crystals as well as their perfection. Thicker and more perfect polyethylene crystals will tend to melt at a higher temperature than smaller crystals.

In addition, the area underneath the melting peak is proportional to the crystallinity of the UHMWPE. DSC provides a measure of the total heat energy per unit mass (also referred to as the change in enthalpy, $\Delta H$) required to melt the crystalline regions within the sample. By comparing the change in enthalpy of a UHMWPE sample to that of a perfect 100% crystal, one can calculate the degree of crystallinity of the UHMWPE. Most bulk UHMWPEs are about 50% crystalline.

As the temperature of a semicrystalline polymer is raised above the melt temperature, it may undergo a flow transition and become liquid. Polyethylenes with a molecular weight of less than 500,000 g/mol can be observed to undergo such a flow transition ($T_f$). However, when the molecular weight of polyethylene increases above 500,000 g/mol, the entanglement of the immense polymer chains prevents it from flowing. UHMWPE does not exhibit a flow transition for this reason.

## 1.6 OVERVIEW OF THE *HANDBOOK*

This *Handbook* is organized into three main sections. The first section, which consists of three chapters, reviews the basic scientific, engineering, and clinical foundations for UHMWPE. For example, in Chapter 2, we explain how UHMWPE must be formed into bulk components from the resin powder using extrusion or compression molding techniques. In Chapter 3, we review the techniques associated with sterilization and packaging of UHMWPE implants. Chapters 5–12 cover the basic clinical applications of UHMWPE in the lower extremities, upper extremities, and the spine.

The second part of the *Handbook* is focused on the development of UHMWPE biomaterials technologies for orthopedics. Chapters 13 and 14 summarize the state of the art as it pertains to remelted and annealed highly crosslinked UHMWPE, and Chapters 15 and 16 describe the advances in stabilization of highly crosslinked and conventional UHMWPE using doped and blended vitamin E. Chapter 17 covers advances in UHMWPE composites, including the history of Poly II, and Chapter 18 describes advances in creating microcomposites of UHMWPE and hyaluronan, a biomolecule that promotes lubrication in cartilage. Chapter 19 summarizes advancements in high pressure crystallization and crosslinking of UHMWPE, including the history of Hylamer. The last chapter in this section, Chapter 20, is a compendium of the processing, packaging, and sterilization information for first- and second-generation highly crosslinked UHMWPE materials that are currently used in hip and knee arthroplasty.

The topics outlined in this *Handbook* may be used as a resource in undergraduate, as well as graduate, courses in biomaterials and orthopedic biomechanics. Students in these disciplines can learn a great deal from exposure to the historical development of total joint replacements within the context of UHMWPE. The first main sections of this book, which cover the fundamentals of UHMWPE and clinical applications in the spine and upper and lower extremities, are intended as a resource for undergraduate instruction.

The third section of this book contains state-of-the-art reviews on cutting-edge topics of ongoing UHMWPE research, including mechanisms of crosslinking (Chapter 21); *in vivo* oxidation (Chapter 22); tissue response to wear debris (Chapter 23); general characterization methods of UHMWPE (Chapter 24); hip and knee simulation (Chapters 25 and 26); wear particle analysis (Chapter 27); clinical surveillance of UHMWPE implants (Chapter 28); and detailed perspectives of characterizing UHMWPE, including electron spin resonance (Chapter 29); fatigue and fracture behavior (Chapter 30); notch sensitivity (Chapter 31); small punch testing (Chapter 32); nanoindentation (Chapter 33); microCT wear analysis (Chapter 34); and constitutive modeling (Chapter 35). The third section of this book, which covers these specialized topics related to UHMWPE, is intended for an audience of graduate students and orthopedic researchers.

Understanding basic chemical structure and morphology is an important starting point for appreciating the unique and outstanding properties of UHMWPE. The chapters that follow and describe the processing, as well as the sterilization, of UHMWPE will continue to build upon the conceptual foundation established in this introduction.

## REFERENCES

1. Rodriguez F. *Principles of polymer systems*. New York: Hemisphere; 1989.
2. Young RJ. *Introduction to polymers*. London: Chapman and Hall Ltd.; 1983.
3. Edidin AA, Kurtz SM. The influence of mechanical behavior on the wear of four clinically relevant polymeric biomaterials in a hip simulator. *J Arthroplasty* 2000;**15**:321–31.
4. Chubberley AH. Ultra-high molecular weight polyethylenes. In: *Modern plastics encyclopaedia*. New York: McGraw-Hill Publications; 1965.
5. Charnley J. Tissue reaction to the polytetraflourocthyene. *Lancet* 1963;**II**:1379.
6. Charnley J. Low friction principle. In: *Low friction arthroplasty of the hip: theory and practice*. Berlin: Springer-Verlag; 1979.
7. Wroblewski BM, Siney PD, Dowson D, Collins SN. Prospective clinical and joint simulator studies of a new total hip arthroplasty using alumina ceramic heads and cross-linked polyethylene cups. *J Bone Joint Surg* 1996;**78B**:280–5.
8. Kurtz SM, Muratoglu OK, Evans M, Edidin AA. Advances in the processing, sterilization, and crosslinking of ultra-high molecular weight polyethylene for total joint arthroplasty. *Biomaterials* 1999;**20**:1659–88.
9. Bellare A, Schnablegger H, Cohen RE. A small-angle x-ray scattering study of high-density polyethyelene and ultra-high molecular weight polyethylene. *Macromolecules* 1995;**17**:2325–33.

Chapter 2

# From Ethylene Gas to UHMWPE Component: The Process of Producing Orthopedic Implants

Steven M. Kurtz, PhD



## 2.1 INTRODUCTION

At a conceptual level, polyethylene consists only of carbon and hydrogen, as was described in the previous chapter. However, if the discussion of polyethylene is to proceed from ideal abstractions to actual physical implants, three real-world steps need to occur. First, the UHMWPE must be polymerized from ethylene gas. Second, the polymerized UHMWPE, in the form of resin powder, needs to be consolidated into a sheet, rod, or near-net shaped implant (Figure 2.1). Finally, in most instances, the UHMWPE implant needs to be machined into its final shape (Figure 2.1). A small subset of implants are consolidated into their final form directly, in a process known as direct compression molding, without need of additional machining.

Each of these three principal steps produces a subtle alteration of the properties of UHMWPE. In some cases, such as machining, the change in the material may only occur in the topography and appearance of the surface. On the other hand, changes in the polymerization and conversion of the UHMWPE can impact the physical and mechanical properties of the entire implant.

Because many of the details used in the polymerization, conversion, and machining of UHMWPE are proprietary, very little is written in the public domain literature about the techniques used to make actual implants. The little that has been written about implant production can be confusing to the uninitiated because of the abundance of trade names, which evolve over time. Not only do industrial trade names change over time, but the actual processing techniques have also improved. Surgeons and researchers are strongly cautioned against overgeneralizing about the properties or techniques of any single resin or process, which has likely changed over the past 50 years.

In recent years, there has been increasing interest in incorporating antioxidants, such as vitamin E, into UHMWPE. This chapter covers the resin characteristics of UHMWPE without antioxidants and methods of consolidation into rods, sheets, and direct molded forms. The reader will find vitamin E blended UHMWPE resins (e.g., GUR 1020-E and GUR 1050-E) covered in Chapter 16; these new antioxidant and UHMWPE resin blends can readily be consolidated using the methods described in the current chapter.

**FIGURE 2.1** Typical processing steps in the manufacture of UHMWPE implants, starting with the resin powder (A). (B) Semifinished rods that have been consolidated from the resin powder; (C) Machining of the UHMWPE rods on a lathe; (D) UHMWPE acetabular components after machining. Pictures provided courtesy of David Schroeder (Biomet, Inc., Warsaw, Indiana, USA).

## 2.2 POLYMERIZATION: FROM ETHYLENE GAS TO UHMWPE POWDER

The polymerization of UHMWPE was commercialized by Ruhrchemie AG, based in northern Germany, during the 1950s. Ruhrchemie AG itself evolved in 1928 with the mission of developing useful chemicals from coal (carbon); its shareholders consisted of 28 coal mining companies. In 1953, chemists from the nearby Max Planck Institute approached scientists at Ruhrchemie AG in Oberhausen with a brown, wet (not fully dried) mass that they claimed was a new form of polyethylene, produced in a new low-pressure process. Convinced of the commercial utility of such a material (the dangers of high-pressure polymerization, such as in the production of LDPE, were widely appreciated), development on UHMWPE began shortly thereafter at Ruhrchemie AG. In 1955, the first commercial polymerization of UHMWPE began, and during that same year, the material was first introduced at the K55, a polymer trade show.

Since the 1950s, the UHMWPE powders have been produced by Ruhrchemie (currently known as Ticona) using the Ziegler process, which has been described by Birnkraut [1]. The main ingredients for producing UHMWPE are ethylene (a reactive gas), hydrogen, and titanium tetra chloride (the catalyst). The polymerization takes place in a solvent used for mass and heat transfer. The ingredients require that the polymerization be conducted in specialized production plants capable of handling these volatile and potentially dangerous chemicals. The last ingredient (catalyst) has been improved consistently since the 1950s because it is the key to producing white UHMWPE powder with reduced impurities.

The requirements for medical grade UHMWPE powder are specified in ASTM standard F648 and ISO standard 5834-1. In the standards, medical grade resins are described as Types 1, 2, or 3, depending upon their molecular weight and producer (Table 2.1). The trace impurities of titanium, aluminum, and chlorine are residuals from the catalyst, whereas the trace levels of calcium, as well as the ash content, depend upon the storage and handling of the powder after polymerization.

Currently Ticona (Oberhausen, Germany) produces a Type 1 and Type 2 resin with the trade names of GUR 1020 and 1050, respectively. Prior to 2002, Basell Polyolefins (Wilmington, Delaware, USA) produced Type 3 resin with the 1900 trade name. This resin was discontinued in January 2002 and is no longer produced. Two orthopedic

**TABLE 2.1 Requirements for Medical Grade UHMWPE Powders (per ASTM 648 and ISO 5834-1)**

| Property | Requirements | |
|---|---|---|
| Resin type | Types 1–2 | Type 3 |
| Trade name | GUR 1020 and 1050 | 1900H |
| Producer | Ticona, Inc. | Basell Polyolefins (now discontinued) |
| Ash, mg/kg, (maximum) | 125 | 300 |
| Titanium, ppm, (maximum) | 40 | 150 |
| Aluminum, ppm, (maximum) | 20 | 100 |
| Calcium, ppm, (maximum) | 5 | 50 |
| Chlorine, ppm, (maximum) | 30 | 90 |

**TABLE 2.2 Nomenclature of Ticona and Basell UHMWPE Resins (The average molecular weight is calculated based on intrinsic viscosity.)**

| Resin designation | Producer | Average molecular weight, ASTM calculation ($10^6$ g/mol) | Calcium stearate added? |
|---|---|---|---|
| GUR 1020 | Ticona | 3.5 | No |
| GUR 1050 | Ticona | 5.5–6 | No |
| 1900H* | Basell | >4.9 | No |

**Note: Production of this resin was discontinued in 2002.*

manufacturers have maintained large stockpiles of this resin, so orthopedic implants will continue to be fabricated from this resin, at least in the near future.

The orthopedics literature contains numerous references to different trade names for UHMWPE, which fall into two categories: GUR resins currently produced by Ticona and the 1900 resins, previously produced by Basell. The UHMWPE grades currently used in the orthopedic industry are summarized in Table 2.2.

## 2.2.1 GUR Resins

Ticona (formerly known as Hoechst, and before that, as Ruhrchemie AG) currently supplies premium grade UHMWPE for orthopedic applications. Resin is currently manufactured in Bishop, Texas, USA, and Oberhausen, Germany. The resin grades produced at Bishop and Oberhausen use the same catalyst technology and undergo similar resin synthesis processes. When tested in accordance with ASTM F 648 and ISO 5834, the physical and mechanical properties of the Bishop and Oberhausen resins are indistinguishable. Ticona uses the designation GUR for its UHMWPE grades worldwide; the acronym GUR stands for "Granular," "UHMWPE," and "Ruhrchemie."

Hoechst changed its resin nomenclature between 1992 and 1998. Resins distributed to both the general and orthopedic marketplaces prior to 1992 were designated as, for example, GUR 415 and GUR 412 in the United States or CHIRULEN P in Europe. In October 1992, a fourth digit was added to all general and orthopedic products to allow inclusion of additional resins to the product lines. For example, GUR 415 became GUR 4150. In 1994, after the sale of the CHIRULEN trade name, Hoechst in Europe decided to exclusively designate all resins for the surgical implant market with a first digit of "1." By comparison, Hoechst in North America renamed GUR 4150 as GUR 4150 HP (to designate "high purity") for this application in August 1993. In 1998, all of the nomenclature was consolidated with the availability of four grades for the worldwide orthopedic market: GUR 1150, 1050, 1120, and 1020 resins (Table 2.2).

The first digit of the grade name was originally the loose bulk density of the resin, that is, the weight measurement of a fixed volume of loose, unconsolidated powder; the "4" corresponded to a bulk density of over 400 g/l for standard grades. The second digit indicates the presence ("1") or absence ("0") of calcium stearate, and the third digit is correlated to the average molecular weight of the resin. The fourth digit is an internal code designation. In 2002, Ticona discontinued the production of GUR 1120 and 1150, which contained added calcium stearate.

Most recently, in 2009, Ticona introduced two new vitamin E-blended resins to the orthopedic community to improve the oxidation resistance of UHMWPE after irradiation. GUR 1020-E and 1050-E are blended with 1,000 ppm of the biocompatible antioxidant, vitamin E, but otherwise conform to the properties and processability of GUR 1020 and 1050. Further details about vitamin-E blended UHMWPE are summarized in Chapter 16.

## 2.2.2 1900 Resins

Although Basell, prior to 2002, produced six grades of 1900 resin, only one of these grades (1900H) is currently used orthopedic applications (Table 2.2). Since its introduction by the Hercules Powder Company (Wilmington, Delaware, USA), 1900 resin has been through transitions in nomenclature and production. During the 1960s, Hercules produced several grades of Hi-Fax 1900 UHMWPE with molecular weights ranging from 2 million to 4 million g/mol [2]. In 1983 Hercules formed a joint venture with Montedison in Italy to become Himont. In April 1995, Montedison and Shell Oil (Netherlands) formed Montell Polyolefins; the company was merged into BASF and Shell, forming Basell Polyolefins in 2001. In January 2002, Basell divested itself of its UHMWPE production facilities and sold its technology to Polialden, a Brazilian UHMWPE manufacturer.

The resin designation of 1900 has stayed the same through the past 3 decades of transitions while Montell changed the manufacturing facilities. Over this time, three different reactors were used to produce the different grades of 1900. Originally the A-line reactor was used, but in 1989–1990 Himont began producing the resin on their F-line. The latter line was a semicontinuous process as compared to the batch processing used earlier. In 1996–1997 the F-line was replaced with the G-line. In early 2002, the G-line production facility was dismantled by Basell. The current owner of the 1900 resin technology, Polialden, has not yet announced whether they intend to resume production for orthopedic use in Brazil.

## 2.2.3 Molecular Weight

The mechanical behavior of UHMWPE is related to its average molecular weight, which is routinely inferred from intrinsic viscosity measurements [3, 4]. There are two commonly-used methods for calculating the viscosity average molecular weight ($M\nu$) for UHMWPE based on the intrinsic viscosity (IV) using the Mark-Houwink equation:

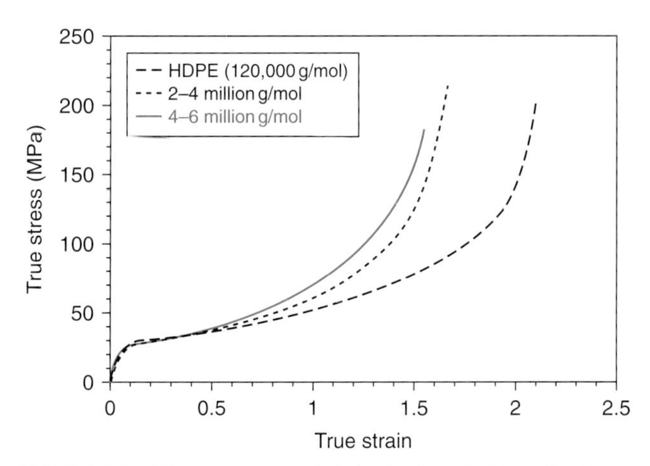


**FIGURE 2.2** The true-stress strain behavior in uniaxial tension (room temperature, 30 mm/min) for two grades of UHMWPE, in comparison with HDPE.

1. ASTM D4020-05:

$$M_\nu = 53,700\ \mathrm{IV}^{1.37}$$

2. Margolies equation, used outside North America:

$$M_\nu = 53,700\ \mathrm{IV}^{1.49}$$

In the preceding equations, $M\nu$ has units of g/mol and IV has units of dL/g. Alternative methods for molecular weight characterization of UHMWPE include sequential extraction [5] and gel permeation chromatography (GPC), a type of size exclusion chromatography [6].

The intrinsic viscosity (IV) of UHMWPE is related to the bulk impact strength and abrasive wear resistance after conversion to bulk form, although the relationships are nonlinear. The maximum impact strength of both Ticona and Basell is found between 16 and 20 IV, which is the equivalent to a molecular weight range of 2.4 million to 3.3 million using the ASTM calculation for molecular weight. As the IV increases, abrasive wear resistance increases, as measured by sand slurry testing, reaching a plateau for IV greater than 20.

Molecular weight also influences the static fracture response as well as the mechanical behavior of UHMWPE at large strains [7]. For example, beyond the polymer yield point, the hardening or cold drawing portion behavior in uniaxial tension is sensitive to the molecular weight. Figure 2.2 illustrates the true-stress strain curve in uniaxial tension (room temperature, 30 mm/min) for two grades of UHMWPE, in comparison with HDPE.

Under biaxial drawing conditions of the small punch test [8], the large-deformation mechanical behavior of polyethylene also displays strong molecular weight dependence. Representative small punch test data, shown in Figure 2.3, were conducted at room temperature at a rate of 0.5 mm/min.

## 2.2.4 GUR versus 1900 Resin

Variations between the material properties of converted GUR and 1900 resins have been explained by differences

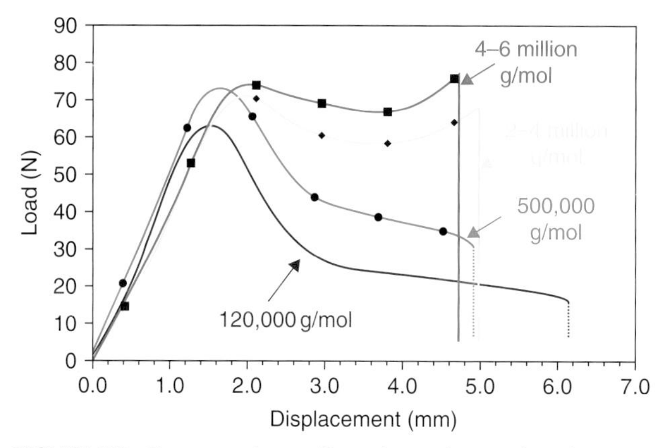


FIGURE 2.3 Representative small punch test data conducted at room temperature at a rate of 0.5 mm/min.

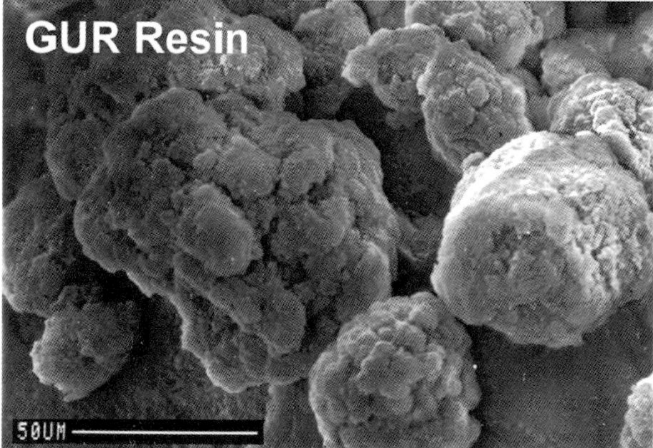


FIGURE 2.4 The scanning electron micrographs illustrate the subtle differences in nascent particle morphology between Basell and Ticona resins.

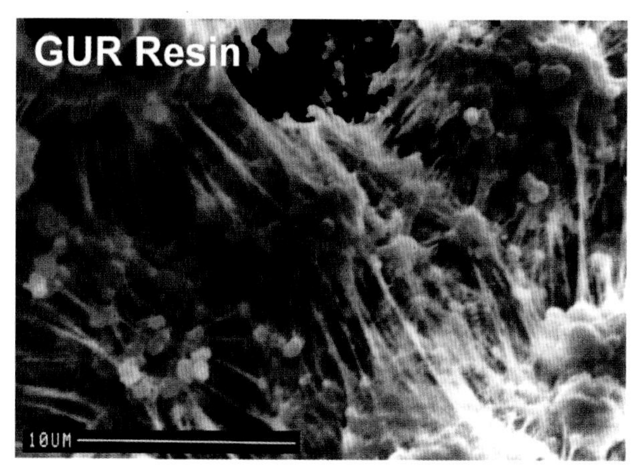


FIGURE 2.5 SEM image of submicron size fibrils in GUR resins.

in the average resin particle size, the size distribution, and morphology of the resin particles [9, 10]. Ticona resins have a mean particle size of about 140 μm [9, 10], whereas the mean particle size of 1900 resin is around 300 μm [10]. The distribution of particle sizes also varies between Ticona and Basell resins. When studied under a scanning electron microscope, GUR and 1900 resin powders consist of numerous spheroidal particles [9, 10]. The scanning electron micrographs seen in Figure 2.4 (provided courtesy of Rizwan Gul [9]) illustrate the subtle differences in nascent particle morphology between Basell and Ticona resins, which are likely related to the catalyst package and polymerization conditions [1].

The most notable finding from studies of powder morphology is that the Ticona resins are characterized by a fine network of submicron-sized fibrils, which interconnect the microscopic spheroids. The fibrils are illustrated in the SEM micrograph in Figure 2.5, provided courtesy of Rizwan Gul [9].

There is some evidence that Basell resins have a different molecular weight distribution than the Ticona resins [9]. For example, depending upon the processing conditions, Basell 1900 exhibits spherulitic crystalline morphology [9, 11], typically associated with slightly lower molecular weight polyethylenes [12]. Spherulites are not observed in Ticona resins, which have a lamellar crystalline morphology [9, 13, 14]. Differences in the molecular weight distribution between the Basell and Ticona produced resins may account for the variation in impact strength observed experimentally at the same average molecular weight. However, despite clues in the literature, the differences in molecular weight distribution between Basell and Ticona resins have yet to be explicitly quantified.

### 2.2.5 Calcium Stearate

Calcium stearate has been used by Hoechst since 1955 and by Montell since the late 1960s. The additive acts as a scavenger for residual catalyst components that can potentially corrode conversion equipment; calcium stearate also acts as a lubricant and a release agent [15]. The catalyst used by Ticona today has a higher activity than the one originally used for UHMWPE production. Thus, a larger quantity of UHMWPE can be synthesized with the same amount of catalyst, resulting in a lower residual catalyst concentration [1]. The trace element level of calcium in UHMWPE is directly proportional to the addition of calcium stearate, which is currently certified as food grade [16]. When calcium stearate is added to any UHMWPE resin, regardless of the manufacturer or catalyst technology used, the polymer particles are surface coated by the calcium stearate.

During the 1980s, the influence of calcium stearate on the properties and performance of UHMWPE total joint replacements was considered a controversial issue [17]. In several studies, the presence of trace calcium levels in UHMWPE has been associated with fusion defects and oxidation of UHMWPE [15, 18–25]. Using high-resolution synchrotron infrared spectroscopy, Muratoglu et al. [26] resolved the molecular vibrations associated with calcium stearate in the grain boundary layers of fusion defects in GUR 4150HP; examination of converted GUR 4050 and HIMONT 1900, in contrast, showed no evidence of calcium stearate. In an accelerated aging study by Swarts and colleagues [20], GUR 4150HP exhibited more oxidation than reduced-calcium stearate resins (GUR 4050 and Montell 1900H).

However, the mere presence of calcium stearate does not automatically imply poor consolidation and decreased fracture resistance in UHMWPE. For example, using the J-integral method, Baldini and colleagues [27] reported that the fracture resistance of GUR 1020 and 1120 were "comparable." Lykins and Evans [28] have suggested that fusion defects may result from inappropriate control of processing variables (e.g., temperature, pressure, time, heating rate) during conversion of resin to stock material. Thus, it remains to be established whether calcium stearate plays a role in the mechanical behavior of well-consolidated UHMWPE except at sufficiently high enough concentrations to interfere with the sintering of the powder.

Thus, research conducted during the 1990s indicated that calcium stearate may be present at the boundaries of fusion defects in UHMWPE and that fusion defects may in turn deleteriously affect the fatigue and fracture behavior of UHMWPE. However, in the absence of fusion defects, the deleterious effects of trace levels of calcium stearate were not conclusively established. Furthermore, polymerization and processing technology had, by the late 1990s, evolved to the point that the additive was no longer necessary. Consequently, orthopedic manufacturers began switching to UHMWPE resins without added calcium stearate (e.g., GUR 1020 and 1050). By 2002, demand for the calcium stearate-containing resins (GUR 1120 and 1150) dropped to the point that Ticona discontinued their production.

### 2.2.6 DSM Resin

As of 2008, DSM (Netherlands) has declared their capability of producing a Type 3 resin without added calcium stearate. However, there are currently no reports that this resin has been evaluated in scientific studies or employed in orthopedic implants. Regardless, it appears that Type 3 resin will remain in the ASTM standards for UHMWPE at the present time, and further details about the performance of DSM material are anticipated.

## 2.3 CONVERSION: FROM UHMWPE POWDER TO CONSOLIDATED FORM

UHMWPE is produced as powder and must be consolidated under elevated temperatures and pressures because of its high melt viscosity. As already discussed (in Section 1.4), UHMWPE does not flow like lower molecular weight polyethylenes when raised above its melting temperature. For this reason, many thermoplastic processing techniques, such as injection molding, screw extrusion, or blow molding, are not practical for UHMWPE. Instead, semifinished UHMWPE is typically produced by compression molding and ram extrusion.

The process of consolidation in UHMWPE requires the proper combination of temperature, pressure, and time. The precise combinations of these variables used to produce commercially available molded and extruded stock materials remain proprietary, but the scientific principles underlying consolidation of UHMWPE are generally well understood [9, 10, 29–41]. The governing mechanism of consolidation is self-diffusion, whereby the UHMWPE chains (or chain segments) in adjacent resin particles intermingle at a molecular level. The kinetics of intergranular diffusion are promoted by close proximity of the interfaces (at elevated pressures) and thermally-activated mobility of the polymer chains (at elevated temperatures). As a diffusion-limited process, consolidation of UHMWPE requires sufficient time at elevated temperature and pressure for the molecular chains to migrate across grain boundaries.

Consolidated UHMWPE retains a memory of its prior granular structure, which is especially evident when calcium stearate is added to the resin [9, 29, 38, 41]. The ultrastructure of UHMWPE may be visualized by either optical or scanning electron microscopy, but special preparation methods are needed in either case. A standard technique, described in ASTM F648, involves microtoming thin films and observing them under optical microscopy, preferably under dark field conditions [38]. However, in today's calcium-stearate-free resins, the presence of grain boundaries in well-consolidated material is usually difficult to detect, as illustrated in Figure 2.6, which is an optical micrograph of a 100 μm-thick section of GUR 1020 (provided courtesy of Rolf Kaldeweier, Ticona, Inc.).

Grain boundaries can also be visualized by freezing a UHMWPE sample in liquid nitrogen and then fracturing the sample while frozen [9]. Etching of fracture surfaces can further highlight the intergranular regions, allowing for visualization of the UHMWPE ultrastructure [41]. Scanning electron microscopy is then used to inspect the surface of the freeze-fractured or etched UHMWPE surfaces.

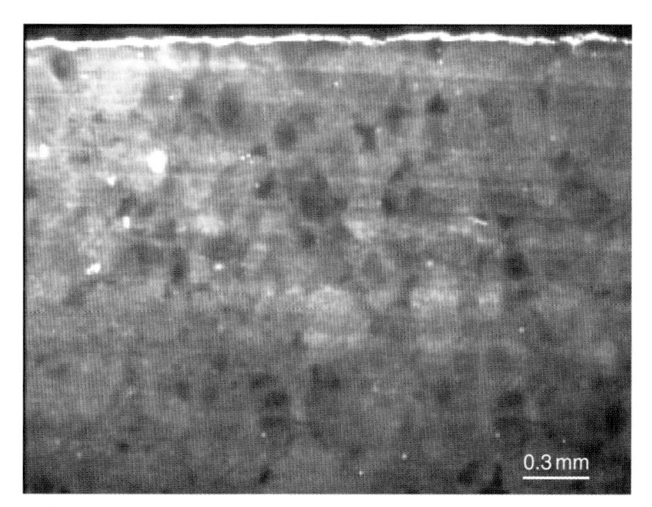


**FIGURE 2.6** An optical micrograph of a 100 μm-thick section of GUR 1020.

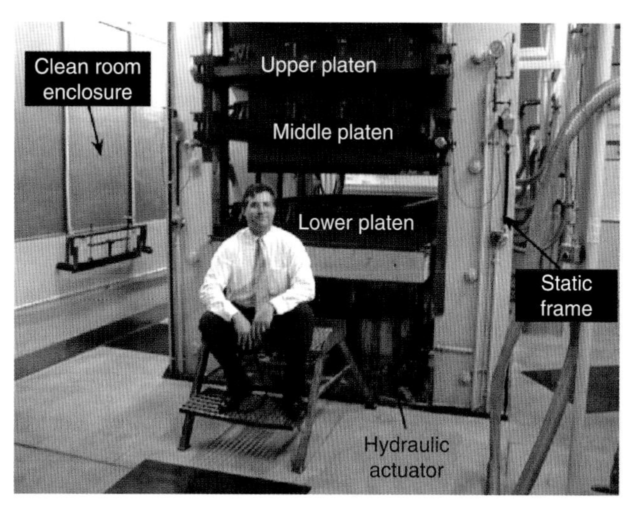


**FIGURE 2.7** Compression Molding Press (along with the author, for scale) for production of 1 m by 2 m sheets of UHMWPE. This press is located at Poly Hi Solidur MediTECH in Vreden, Germany, and was originally used by Ticona in the production of Chirulen sheets of UHMWPE. The press is still used today in the production of medical grade UHMWPE.

In contrast with grain boundaries, which reflect the normal ultrastructure of UHMWPE, consolidation *defects* may arise when the proper combination of pressure, temperature, and time are not used [28]. More typically, consolidation defects represent a single resin particle, or a highly localized region, which has not fully fused with its neighbors. A standard technique (specified in ASTM F648), involving optical microscopy of thin sections, has been developed to quantify the presence of fusion defects in UHMWPE. Nondestructive methods, including laser candling and ultrasound, are also used in industry for inspection of medical grade UHMWPE, but the use of these techniques has not been standardized.

## 2.3.1 Compression Molding of UHMWPE

Historically, the UHMWPE powder has been converted by compression molding since the 1950s because the industries in the area around Ruhrchemie already had experience with this processing technique. At first, the semifinished material was distributed under the trade name RCH-1000/Hostalen GUR 412. During the 1970s, the compression molded UHMWPE was manufactured and distributed by Ruhrchemie/Hoechst and later distributed by the company Europlast specifically for orthopedic applications under the trade name CHIRULEN, which denoted that the material was produced in Germany by a dedicated press using established processing parameters. Thus, references in the orthopedic literature to RCH-1000 and CHIRULEN apply to compression molded UHMWPE, which was similar to contemporary GUR 1020. Note, however, that GUR 412 would have contained calcium stearate, whereas GUR 1020 does not.

Today, compression molded sheets of GUR 1020 and 1050 are produced commercially by two companies (Orthoplastics [formerly known as Perplas Medical] and MediTECH, a division of Quadrant [formerly known as Poly Hi Solidur Meditech]). Ticona stopped producing compression molded UHMWPE in 1994. Orthoplastics' molding facility is in England, whereas MediTECH produces medical grade UHMWPE sheets in the United States and in Germany.

An example of a compression molding press, currently installed at MediTECH in Vreden, Germany, is shown in Figure 2.7. This particular press, originally designed by Hoechst in the 1970s for production of CHIRULEN, molds two 1 m by 2 m sheets in a single press cycle. One UHMWPE sheet is pressed between the upper and middle platens, and the second is produced between the middle and lower platens. The platens are oil heated and hydraulically actuated from below. The heating and loading systems are all computer controlled. Finally, the entire press is contained in a clean room to reduce the introduction of extraneous matter into the sheet.

Depending upon the size of a press, the UHMWPE sheet may range in size between 1 m by 2 m (shown in Figure 2.7) to 2 m by 4 m, with thicknesses of 30 mm to 80 mm. However, the facilities necessary to mold a 2 m by 4 m sheet of UHMWPE are considerably larger than that shown in the photo (Figure 2.7). The press operated by Orthoplastics, for example, is over three stories tall and enclosed within its own clean room structure.

Due to the relatively low thermal conductivity of UHMWPE, the duration of the molding cycle will depend upon the particular geometry of the press and the size of the sheet to be produced, but the processing time can last up to 24 h. The long molding times are necessary to maintain the slow, uniform heating and cooling rates throughout the entire sheet during the molding process.

After molding, the sheet is typically turned into rods or other preform shapes to facilitate subsequent machining operations by orthopedic manufacturers. Thus, the final shape of UHMWPE stock material today is not necessarily dictated by the conversion method.

### 2.3.2 Ram Extrusion of UHMWPE

In contrast with compression molding, which originated in Germany in the 1950s, ram extrusion of UHMWPE was developed by converters in the United States during the 1970s. Historically, a wide range of UHMWPE resins have been extruded over the past 3 decades. From the early 1970s to early 1980s, ram-extruded Hi-Fax 1900 resin from Hercules Powder Company (Wilmington, Delaware, USA) was commonly supplied to bulk form converters. In the early 1960s Hoechst's GUR 412 resin was made available to converters. Although GUR 412 tended to have a lower extraneous particle count than 1900, it was also more difficult to process using ram extrusion due to its lower average molecular weight and lower melt viscosity. In the mid 1980s, converters began ram extruding using the higher average molecular weight Hoechst GUR 415 resin, which also had a lower extraneous particle count relative to 1900 and was easier to extrude than GUR 412. During the 1980s and early 1990s, extruded rods of GUR 412, 415, and (more rarely) 1900 CM (a calcium stearate containing grade of 1900 resin) were all used in orthopedics. Thus, it is difficult to generalize about the resin types used in implants without tracing the lot numbers used by a particular manufacturer.

Today only a few converters supply medical grade GUR 1020 and 1050 ram extruded UHMWPE to the orthopedic industry. Medical grade extrusion facilities for MediTECH and Westlake Plastics are based in the United States, whereas Orthoplastics' medical grade extrusion is in England.

Like compression molding facilities, a medical grade extruder is typically maintained in a clean room environment to reduce the introduction of extraneous matter into the UHMWPE. The generic schematic of a ram extruder is illustrated in Figure 2.8 (the clean room is not shown).

UHMWPE powder is fed continuously into an extruder. The extruder itself consists essentially of a hopper that allows powder to enter a heated receiving chamber, a horizontal reciprocating ram, a heated die, and an outlet. Within the extruder, the UHMWPE is maintained under pressure by the ram, as well as by the back pressure of the molten UHMWPE. The back pressure is caused by frictional forces of the molten resin against the heated die wall surface as it is forced horizontally through the outlet. Beyond the outlet, the rod of UHMWPE is slowly cooled in a series of electric heating mantles.

A wide range of rod diameters is achievable (up to 12 inches in diameter) using extrusion, but the rate of production depends on the rod size due to the increased cooling times needed with the larger rod diameters. Rod sizes ranging from 20 to 80 mm in diameter are most often used in orthopedic applications. Typical production rates are on the order of mm/minute.

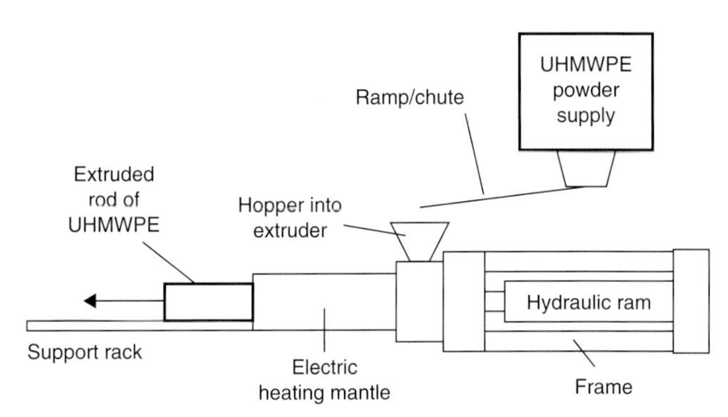


**FIGURE 2.8** Schematic of a ram extruder.

### 2.3.3 Hot Isostatic Pressing of ArCom UHMWPE

A third conversion method, known as hot isostatic pressing, is used by one orthopedic manufacturer (Biomet, Inc., Warsaw, Indiana, USA) for conversion of resin to stock material [9]. This multistep conversion process, referred to as ArCom by the manufacturer, begins with the manufacture of a cylindrical compact through cold isostatic pressing, which expels most of the air (Figure 2.9). Subsequently, the compacted green rods are sintered in a hot isostatic pressure (HIP) furnace in an argon-filled pouch to prevent degradation of the UHMWPE (Figure 2.9). The resulting ArCom rod stock is essentially isotropic due to the hydrostatic sintering process and may be considered a compression molded form of the resin. Finished implants are then made by from the iso-molded rod stock by either turning or milling operations [42], as illustrated in Figure 2.1.

### 2.3.4 Direct Compression Molding of UHMWPE

In direct compression molding (DCM), sometimes also called net shape compression molding, the manufacturer of the polyethylene insert effectively converts the resin to a finished or semifinished part using individual molds. One advantage of DCM is the extremely smooth surface finish obtained with a complete absence of machining marks at the articulating surface. In addition, higher processing pressures may be attained, if desired, because the projected surface area of each individual part mold is relatively small compared to the area of large molds used to compression mold sheets.

Direct compression molding has been used for over 25 years to produce tibial and acetabular inserts. Historically, the process may have been adopted because the cutting and

**FIGURE 2.9** Steps in the processing of ArCom UHMWPE (Biomet, Inc.: Warsaw, Indiana, USA). (A) Resin powder is first poured into a polymeric can with a compressible lid; (B) The can is inserted into a cold isostatic press for compaction of the powder; (C) Cold pressed (green) rod of UHMWPE after cold compaction; (D) The green rod is sealed in an argon-filled foil pouch; (E) The foil-wrapped green rods are loaded in a metallic mesh rack and lowered into a hot isostatic press; (F) The foil-wrapped rod of consolidated UHMWPE following hot isostatic pressing. The foil is peeled off the rod and machined into components as shown in Figure 2.1. Pictures provided courtesy of David Schroeder, Biomet, Inc.

milling machinery of the day was not numerically controlled and thus less able to accurately produce complicated curves as required to make knee inserts. The physical and mechanical properties of the finished product can be tailored to some degree by varying the DCM molding cycle, as detailed by Chen, Truss, and others [10, 32, 35]. DCM can also be used to produce UHMWPE with properties indistinguishable from stock material produced by closely monitored conversion of compression molded sheet and extruded rod.

Overall, the literature suggests that the DCM process can produce a specimen or implant with mechanical

**TABLE 2.3 Summary of Conversion Methods and Resins Used in the Production of ArCom During Both Historical and Contemporary Time Frames**

| ArCom conversion method | Historical UHMWPE resin (1993–2002) | Contemporary UHMWPE resin (2002–present) |
|---|---|---|
| Isostatically molding | 1900H | GUR 1050 |
| Direct compression molding | 1900H | 1900H* |

*This resin continues to be used because it has been stockpiled by the manufacturer.*

properties similar to those exhibited by high-quality bulk converted UHMWPE stock. The variations in final properties seen as functions of temperature, pressure, density, and cooling rate would presumably affect the production of bulk converted sheet and rod. However, bulk conversion needs to be performed much less often to convert a given mass of resin, and thus issues of monitoring and maintaining quality need to be performed less often as well. From a manufacturability standpoint, the difficulties associated with process control of DCM has led most implant producers to obtain bulk converted sheet or rod.

**TABLE 2.4 Breakdown of 680 Individual UHMWPE Lots Compared in ASTM Survey**

| | GUR 1020 (Type 1) | GUR 1050 (Type 2) | Total |
|---|---|---|---|
| Ram extruded | 113 | 218 | 331 |
| Compression molded | 186 | 163 | 349 |
| Total | 299 | 381 | 680 |

## 2.3.5 ArCom

ArCom is the trade name of a proprietary conversion process, which denotes that the resin has been compression molded in the presence of argon (Biomet, Inc., Warsaw, Indiana, USA). ArCom is produced by hot isostatic pressing (see Section 2.3.3) as well as by direct compression molding (see Section 2.3.4).

The type of resins used with ArCom are summarized in Table 2.3. Both 1900H and GUR resins have been fabricated using the ArCom iso-molding process. Between 1993 and 2002, iso-molded ArCom was produced with 1900H resin. In 2002, after Basell discontinued the production of 1900 resin (see Section 2.2.2), Biomet switched production of iso-molded rod stock to GUR resins produced by Ticona (Table 2.3). On the other hand, Biomet has stockpiled sufficient 1900H resin to continue production of direct compression molded ArCom for the foreseeable future (Table 2.3).

## 2.3.6 Properties of Extruded versus Molded UHMWPE

As might be expected from our previous discussion of compression molding and extrusion, the conversion method could have an effect on the properties of UHMWPE. In practice, however, the differences between extrusion and compression molding are slight in the hands of a skilled converter. In 2001, a survey was conducted by the author under the auspices of ASTM to evaluate the variation in physical and mechanical properties of UHMWPE as a function of resin and conversion method. Three commercial suppliers provided physical and mechanical property data collected during certification of 680 individual lots of medical grade UHMWPE, which were produced between 1998 and 2001. The breakdown of the individual lots in the total data set, according to resin and conversion method, is summarized in Table 2.4 and indicates a relatively even distribution between resins and conversion methods.

Not surprisingly, the data reported by the converters for contemporary UHMWPE was found to exceed the requirements for Type 1 and 2 resins, as set forth in ASTM F648. For density, resin but not conversion method was found to be a significant factor (Table 2.5). For the impact strength and the tensile mechanical properties, resin as well as conversion method were both found to be significant factors (Table 2.5). Although statistically significant, differences in the density and tensile properties of UHMWPE were, in general, not substantial (less than 21% difference in means between the Extruded GUR 1020 and Molded GUR 1050, Table 2.5).

Note that this study did not evaluate differences between converters, which could be another source of variation in material properties of UHMWPE. Because the sintering conditions can significantly influence the density, as well as the mechanical properties of UHMWPE, one should not make conclusions of preferred resin and processing method without having all relevant data available from a particular converter.

Studies have reported subtle differences in the morphology and fatigue crack propagation behavior of extruded versus compression molded UHMWPE. For example,

**TABLE 2.5** **Summary of Mean (± Standard Deviation) Physical and Tensile Mechanical Properties of Extruded and Molded UHMWPE**

| Material | Density (kg/m$^3$) | Tensile yield (MPa) | Ultimate tensile strength (MPa) | Elongation to failure (%) |
|---|---|---|---|---|
| Extruded GUR 1020 | 935 ± 1 | 22.3 ± 0.5 | 53.7 ± 4.4 | 452 ± 19 |
| Molded GUR 1020 | 935 ± 1 | 21.9 ± 0.7 | 51.1 ± 7.7 | 440 ± 32 |
| Extruded GUR 1050 | 931 ± 1 | 21.5 ± 0.5 | 50.7 ± 4.2 | 395 ± 23 |
| Molded GUR 1050 | 930 ± 2 | 21.0 ± 0.7 | 46.8 ± 6.4 | 373 ± 29 |

investigations of UHMWPE morphology suggest that compression molded material has an isotropic crystalline orientation, whereas in ram extruded UHMWPE the morphology varies slightly as a function of the distance from the centerline [31]. Similarly, crack propagation studies have found more isotropic crack propagation behavior in compression molded sheets as opposed to ram extruded rods of UHMWPE [13].

## 2.4 MACHINING: FROM CONSOLIDATED FORM TO IMPLANT

Orthopedic manufacturers generally machine UHMWPE components into their final form. Even components that are direct molded may be machined on the back surface to accommodate a locking mechanism. However, the actual morphology of machining marks depends upon the manufacturing conditions as well as the type of UHMWPE material (e.g., conventional versus highly crosslinked) [43]. An example of machining marks in an as-machined (never implanted) GUR 1050 UHMWPE component is shown in Figure 2.10.

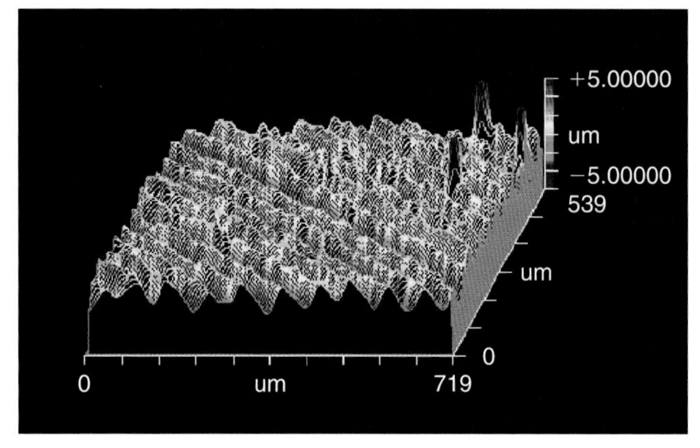


**FIGURE 2.10** An example of machining marks in an as-machined (never implanted) GUR 1050 UHMWPE component.

Machining of UHMWPE components consists of milling and turning operations for both roughing and finishing steps. In some cases, the resin converter may supply the stock in a shape that approximates the cross-section of the finished implant. Such preshaping or preforming offers advantages of efficiency and speed to the manufacturer. Because UHMWPE can be damaged by excessive heat, the feed rate, tool cutting force, and spindle speed must be closely monitored during manufacture. Close mechanical tolerances generally require the manufacturing environment temperature to be carefully controlled. Milling machine spindles in the early to mid-1980s could develop up to 4000 rpm, while newer machinery can develop up to 8000 rpm; the latest machines develop speeds of 12,000 rpm.Overheating of the UHMWPE is avoided by proper optimization of the feed rate and tool cutting force during machining operations.

The actual cutting speeds, tool feed rates, and depths of cut used to machine UHMWPE components are proprietary, and hence little information is available in the literature about the effect of machining parameters on the tribological properties of UHMWPE [44]. Song et al. [44] have proposed an idealized model for the surface morphology of as-machined UHMWPE as a triangular wave, in which the peak to peak distance, d, is given by the following equation:

$$d = \frac{f}{2s}$$

where $f$ is the tool feed rate and $s$ is the cutting speed. Song's model also proposes that the base angle of the machining marks is related to the geometry of the cutting tool. Given the numerous (potentially interrelated) factors influencing the surface topology of as-manufactured components, it is difficult (based on measurements of the surface alone) to quantify the machining conditions used to produce the surface finishes of a randomly sampled component.

Cooper and colleagues have classified abrasive wear in UHMWPE as involving macroscopic or microscopic asperity contact between the sliding surfaces [45]. Because of the initial difference in surface roughness at the UHMWPE and metal counterfaces, the initial wear rate involves the removal of the larger macroscopic asperities on the UHMWPE surface, whereas the long-term wear rate is governed by the microscopic asperity size of the metal counterface [45]. Thus, changes to the surface roughness of UHMWPE components may be expected to affect the initial wear rate because removal of machining marks will occur within the contact zone during the first stages of wear in an orthopedic bearing [46].

## 2.5 CONCLUSION

As we shall see in Chapter 4, the first UHMWPE hip components were produced by Charnley himself in his home workshop or in the machine shop at Wrightington starting in 1962. The sophistication of the implant manufacturing process has changed considerably since that time. Because of the complexities inherent in producing UHMWPE implants, three highly specialized industries have developed over the past 4 decades to address different stages of the production pipeline. Polymer resin producers, such as Ticona, focus on polymerization of medical grade UHMWPE resin, whereas a separate industry of converters have evolved to address consolidation of UHMWPE into semifinished stock materials, catering to the needs of orthopedic manufacturers. The task of implant design, UHMWPE selection, implant machining, packaging, sterilization, and distribution all typically fall within the purview of orthopedic manufacturers.

The choice of conversion method (and converter) is at least as important a decision as the choice of resin for an UHMWPE component because both factors introduce a subtle change in the morphology and material properties of the consolidated polymer. However, there continues to be limited consensus as to which resin and conversion method would be universally superior for all orthopedic applications. Consequently, it is left to each orthopedic manufacturer to determine which conversion method (and resin) is most appropriate for applications in hip, knee, shoulder, and spine implants.

## REFERENCES

1. Birnkraut HW. Synthesis of UHMWPE. In: Willert HG, Buchtorn GH, Eyerer P, editors. *Ultra-high molecular weight polyethylene as a biomaterial in orthopedic surgery*: Hogrefe & Huber Publishers; 1991.
2. Chubberley AH. Ultra-high molecular weight polyethylenes. In: *Modern plastics encyclopaedia*. New York: McGraw-Hill Publications; 1965.
3. Li S, Burstein AH. Ultra-high molecular weight polyethylene. The material and its use in total joint implants. *J Bone Joint Surg Am* 1994;**76**:1080–90.
4. Eyerer P, Frank A, Jin R. Characterization of ultrahigh molecular weight polyethylene (UHMWPE): extraction and viscometry of UHMWPE. *Plastverarbeiter* 1985;**36**:46–54.
5. Kusy RP, Whitley JQ. Use of a sequential extraction technique to determine the MWD of bulk UHMWPE. *J Appl Poly Sci* 1986;**32**:4263–9.
6. Wagner HL, Dillon JG. Viscosity and molecular weight distribution of ultra-high molecular weight polyethylene. *J Appl Poly Sci* 1988;**36**:567–82.
7. Kurtz SM, Pruitt L, Jewett CW, Crawford RP, Crane DJ, Edidin AA. The yielding, plastic flow, and fracture behavior of ultra-high molecular weight polyethylene used in total joint replacements. *Biomaterials* 1998;**19**:1989–2003.
8. Edidin AA, Kurtz SM. Development and validation of the small punch test for UHMWPE used in total joint replacements. In: Katsube N, Soboyejo W, Sacks M, editors. *Functional biomaterials*. Winterthur, Switzerland: Trans Tech Publications Ltd.; 2001.
9. Gul R. *Improved UHMWPE for use in total joint replacement*. Ph.D. Dissertation. Boston: Massachusetts Institute of Technology; 1997.
10. Han KS, Wallace JF, Truss RW, Geil PH. Powder compaction, sintering, and rolling of ultra-high molecular weight polyethylene and its composites. *J Macromol Sci-Phys* 1981;**B19**:313–49.
11. Weightman B, Light D. A comparison of RCH 1000 and Hi-Fax 1900 ultra-high molecular weight polyethylenes. *Biomaterials* 1985;**6**:177–83.
12. Voigt-Martin IG, Fisher EW, Mandelkern L. Morphology of melt-crystallized linear polyethylene fractions and its dependence on molecular weight and crystallization temperature. *J Poly Sci (Poly Phys)* 1980;**18**:2347–67.
13. Pruitt L, Bailey L. Factors affecting the near-threshold fatigue behavior of surgical grade ultra high molecular weight polyethylene. *Polymer* 1998;**39**:1545–53.
14. Goldman M, Gronsky R, Ranganathan R, Pruitt L. The effects of gamma radiation sterilization and ageing on the structure and morphology of medical grade ultra-high molecular weight polyethylene. *Polymer* 1996;**37**:2909–13.
15. Eyerer P, Ellwanger R, Federolf H-A, Kurth M, Madler H. Polyethylene. In: Williams D, Cahn R, editors. *Concise encyclopaedia of medical and dental materials*. Oxford: Pergamon; 1990.
16. Stein H. Personal Communication. July 8, 1997.
17. Kurtz SM, Muratoglu OK, Evans M, Edidin AA. Advances in the processing, sterilization, and crosslinking of ultra- high molecular weight polyethylene for total joint arthroplasty. *Biomaterials* 1999;**20**:1659–88.
18. Resin consolidation issues with UHMWPE. *Report No. Y-BEM-069*. Warsaw: Biomet, Inc.; 1995.
19. Gsell R, King R, Swarts D. Quality indicators of high-performance UHMWPE. *Report No*. Warsaw: Zimmer, Inc.; 1997.
20. Swarts D, Gsell R, King R, Devanathan D, Wallace S, Lin S, Rohr W. Aging of calcium stearate-free polyethylene. *Trans 5th World Biomater Conf* 1996;**2**:196.
21. Schmidt MB, Hamilton JV. The effects of calcium stearate on the properties of UHMWPE. *42nd Orthop Res Soc* 1996;**21**:22.
22. Hamilton JV, Wang HC, Sung C. The effect of fusion defects on the mechanical properties of UHMWPE. *Trans 5th World Biomater Conf* 1996;**2**:511.
23. Blunn GW, Joshi AB, Minns RJ, Lidgren L, Lilley P, Ryd L, Engelbrecht E, Walker PS. Wear in retrieved condylar knee arthroplastics. A comparison of wear in different designs of 280 retrieved condylar knee prostheses. *J Arthroplasty* 1997;**12**:281–90.
24. Walker PS, Blunn GW, Lilley PA. Wear testing of materials and surfaces for total knee replacement. *J Biomed Mater Res* 1996;**33**:159–75.
25. Wrona M, Mayor MB, Collier JP, Jensen RE. The correlation between fusion defects and damage in tibial polyethylene bearings. *Clin Orthop* 1994;**299**:92–103.

26. Muratoglu OK, Jasty M, Harris WH. High resolution synchrotron infra-red microscopy of the structure of fusion defects in UHMWPE. *Trans of the 43rd Orthop Res Soc* 1997;**22**:773.
27. Baldini TH, Rimnac CM, Wright TM. The effect of resin type and sterilization method on the static (J-integral) fracture resistance of UHMW polyethylene. *Orthop Res Soc* 1997;**43**:780.
28. Lykins MD, Evans MA. A comparison of extruded and molded UHMWPE. *Trans 21st Soc Biomater* 1995;**18**:385.
29. Bastiaansen CWM, Meyer HEH, Lemstra PJ. Memory effects in polyethylenes: influence of processing and crystallization history. *Polymer* 1990;**31**:1435–40.
30. Barnetson A, Hornsby PR. Observations on the sintering of ultra-high molecular weight polyethylene (UHMWPE) powders. *J Materials Sci Letters* 1995;**14**:80–4.
31. Bellare A, Cohen RE. Morphology of rod stock and compression-moulded sheets of ultra-high- molecular-weight polyethylene used in orthopaedic implants. *Biomaterials* 1996;**17**:2325–33.
32. Truss RW, Han KS, Wallace JF, Geil PH. Cold compaction molding and sintering of ultra-high molecular weight polyethylene. *Poly Engr Sci* 1980;**20**:747–55.
33. Shenoy AV, Saini DR. Compression moulding of ultra-high molecular weight polyethylene. *Plast Rubber Proc Appl* 1985;**5**:313–17.
34. Wang X-Y, Li S-Y, Salovey R. Processing of ultra-high molecular weight polyethylene. *J Appl Poly Sci* 1988;**35**:2165–71.
35. Chen K-C, Ellis EJ, Crugnola A. Effects of molding cycle on the molecular structure and abrasion resistance of ultra-high molecular weight polyethylene. *ANTEC '81* 1981;**39**:270–2.
36. McKenna GB, Crissman JM, Khoury F. Deformation and failure of ultra-high molecular weight polyethylene. *ANTEC '81* 1981;**39**:82–4.
37. Zachariades AE. The effect of powder particle fusion on the mechanical properties of ultra-high molecular weight polyethylene. *Poly Engr Sci* 1985;**25**:747–50.
38. Farrar DF, Brain AA. The microstructure of ultra-high molecular weight polyethylene used in total joint replacements. *Biomaterials* 1997;**18**:1677–85.
39. Halldin GW, Kamel IL. Powder processing of ultra-high molecular weight polyethylene. I. Powder characterization and compaction. *Poly Eng Sci* 1977;**17**:21–6.
40. Halldin GW, Kamel IL. Powder processing of ultra-high molecular weight polyethylene. II. Sintering. *ANTEC 77* 1977;**35**:298–300.
41. Olley RH, Hosier IL, Bassett DC, Smith NG. On morphology of consolidated UHMWPE resin in hip cups. *Biomaterials* 1999;**20**:2037–46.
42. ArCom processed polyethylene. *Report No. Y-BMT-503*. Warsaw: Biomet, Inc.; 1997.
43. Kurtz SM, Turner J, Herr M, Edidin AA. Deconvolution of surface topology for quantification of initial wear in highly crosslinked acetabular components for THA. *JBMR (Appl Biomater)* 2002. In Press.
44. Song J, Liu P, Cremens M, Bonutti P. Effects of machining on tribological behavior of ultra-high molecular weight polyethylene (UHMWPE) under dry reciprocating sliding. *Wear* 1999;**225–229**:716–23.
45. Cooper JR, Dowson D, Fisher J. Macroscopic and microscopic wear mechanisms in ultra-high molecular weight polyethylene. *Wear* 1993;**162–164**:378–84.
46. Wang A, Stark C, Dumbleton JH. Role of cyclic plastic deformation in the wear of UHMWPE acetabular cups. *J Biomed Mater Res* 1995;**29**:619–26.

Chapter 3

# Packaging and Sterilization of UHMWPE

Steven M. Kurtz, PhD



## 3.1 INTRODUCTION

After fabrication, either by machining or by direct compression molding, UHMWPE components for total joint replacement are packaged and sterilized prior to external distribution to the clinic. Unlike polymerization and resin conversion, which are typically under the direct control of specialized vendors, the choice of packaging and sterilization method falls within the purview of the implant designer. Although the implant designer is responsible for selecting the packaging and sterilization methods for a particular type of orthopedic component, the actual packaging and sterilization processes themselves may be amenable to outsourcing. For example, gas plasma sterilizers are commercially distributed as standalone units and can be incorporated directly into a manufacturer's facility (e.g., STERRAD System: Advanced Sterilization Products, Irvine, California, USA), whereas gamma and ethylene oxide sterilization requires specialized facilities and is typically performed by an outside vendor (e.g., Isomedix Services, Steris Corporation, Mentor, Ohio, USA).

During the 1990s, the packaging and sterilization of UHMWPE was a controversial topic. As recently as 1995, UHMWPE was typically sterilized with a nominal dose of 25 to 40 kGy of gamma radiation in the presence of air. By 1998, all of the major orthopedic manufacturers in the United States were either sterilizing UHMWPE using gamma radiation in a reduced oxygen environment or sterilizing without ionizing radiation, using ethylene oxide or gas plasma. The shift in sterilization practice was catalyzed by mounting evidence that gamma sterilization in air, followed by long-term shelf storage, promoted oxidative chain scission and degradation of desirable physical, chemical, and mechanical properties of UHMWPE. Whether packaged in air-permeable or barrier packaging, gamma sterilized UHMWPE contains macroradicals that will react with available oxygen in air or dissolved in body fluids.

A wide range of choices is currently available to the implant designer for packaging and sterilization of UHMWPE implants, as summarized in Table 3.1. Implants can be sterilized with or without ionizing radiation. When sterilized using gamma radiation, UHMWPE components are contained in a barrier package with a reduced oxygen environment. Gas plasma and ethylene oxide-sterilized implants are packaged in gas-permeable packaging to allow access of the sterilizing medium to the UHMWPE surface. Gas sterilization is now widely used as the method of choice for highly crosslinked and stabilized UHMWPE materials.

The purpose of this chapter is to review historical and contemporary packaging and sterilization methods for UHMWPE. Obviously, all of the sterilization methods currently employed by the orthopedic community fulfill their intended purpose, namely the eradication of bacterial agents, which may result in sepsis and premature revision.

**TABLE 3.1** Summary of Sterilization Processes for UHMWPE Implants (Note that gamma air sterilization is listed as a historical reference* for comparison purposes only.)

| Sterilization process | Packaging type | Gamma radiation dose | Contemporary method? |
|---|---|---|---|
| Gamma air | Gas permeable | 25–40 kGy | No* (historical) |
| Gamma inert | Barrier packaging, reduced oxygen atmosphere | 25–40 kGy | Yes |
| Gas plasma | Gas permeable | None | Yes |
| Ethylene oxide | Gas permeable | None | Yes |

**Certain small manufacturers in Europe may still use gamma irradiation in air [1].*

The diverse sterilization methods in current use reflect the lack of scientific consensus as to which of the currently favored sterilization methods provides the most advantageous long-term UHMWPE product for the ultimate user, namely the patient.

In the past 5 years, new evidence has emerged that certain first-generation, polymeric barrier packaging systems may not have been as effective at preventing UHMWPE oxidation on the shelf as originally thought [1, 2]. Another current concern with gamma-sterilized components is that they can be expected to oxidize *in vivo* during long-term exposure to the environment in the human body [2]. The clinical significance of these recent findings is not fully appreciated at the present time and is the subject of ongoing research by the scientific community. The findings from a recent survey of contemporary orthopedic implant packaging are detailed in the present chapter, whereas the topic of *in vivo* oxidation is discussed in Chapter 22.

**FIGURE 3.1** An example of historical air-permeable packaging used with gamma sterilization, consisting of a box, two nested, polymeric packages, and an inner foam insert (Osteonics, Allendale, New Jersey, USA). The UHMWPE component is not shown.

## 3.2 GAMMA STERILIZATION IN AIR

Starting in the 1960s, UHMWPE components for joint replacement have been stored in air-permeable packaging and gamma sterilized with a nominal dose of 25 kGy (2.5 Mrad) [3]. An example of air-permeable packaging is shown in Figure 3.1. It consists of a box, two nested, polymeric packages, and an inner foam insert (the UHMWPE component is not shown). The outer box also typically contains a booklet of information for the surgeon, as well as stickers identifying the catalog and lot numbers of the implant, for affixing to the patient's medical records (not shown in the Figure 3.1). More examples of air permeable packaging from different manufacturers are included in Figure 3.2 and Figure 3.3. Although the packaging of orthopedic implants has evolved during the past 45 years, gamma sterilization of UHMWPE components remains an industry standard today [4].

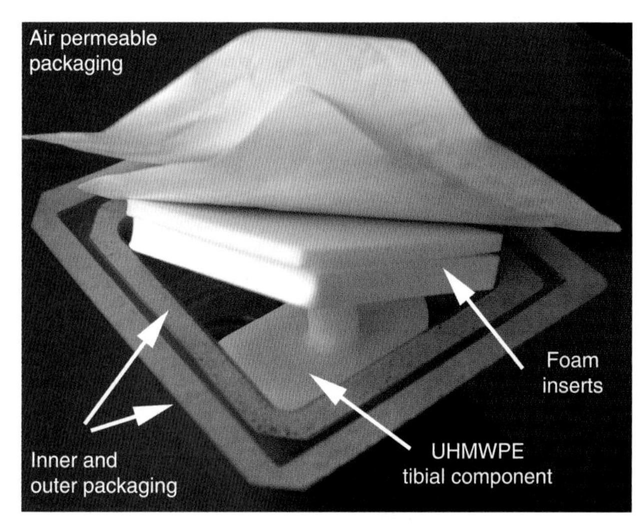


**FIGURE 3.2** Historical air permeable packaging used with gamma sterilization (Wright Medical, Arlington, Tennessee, USA).

**FIGURE 3.3** Historical air permeable packaging used with gamma sterilization (DePuy, Warsaw, Indiana, USA).

Historically, UHMWPE components have received a dose ranging between 25 and 40 kGy during sterilization. Charnley's pioneering invention of the total hip replacement, detailed in Chapter 4, was achieved with UHMWPE that was gamma sterilized in air. Ethylene oxide sterilization was considered and rejected by Charnley, as documented in his personal correspondence from October 19, 1971. In this correspondence, Charnley outlined seven reasons for the superiority of gamma sterilization over ethylene oxide. He concludes, "In view of the fact that I have personally over ten years experience on nearly six thousand patients using this plastic in the hip, I would be very unhappy to change this for something which might possibly expose the tissues to a new and unknown source of toxicity" (Charnley, J. Personal Communication, October 19, 1971). The complete text of this letter is reproduced in Chapter 4. In light of Charnley's rejection of ethylene oxide for sterilizing hip implants, it is hardly surprising that gamma radiation became the sterilization method of choice for UHMWPE.

The effect of sterilization of UHMWPE was not extensively researched in orthopedics until the mid-1990s, when clinical studies implicated wear particles, produced by *in vivo* damage to implant components, as one of multiple potential mechanisms for osteolysis. This realization led to a number of discoveries related to the science of producing UHMWPE for implant applications, including the dual role that gamma sterilization plays in both preservation of implant properties (with beneficial crosslinking) and its role in initiating a complex cascade of chemical reactions in the polymer, which ultimately result in oxidation and degradation of material properties.

The motivation for changing packaging techniques was to prevent oxidation of macroradicals in the UHMWPE, which persist for years after irradiation [5], and can be replenished by the complex, cyclic cascade of chemical reactions that follows oxidation (see Chapter 21). During shelf storage, UHMWPE components that are gamma sterilized in air-permeable packaging undergo oxidative degradation, resulting in an increase in density and crystallinity [6–10], and more importantly, in a loss of mechanical properties associated with progressive embrittlement [11–13]. The loss of mechanical properties during long-term shelf aging in air frequently manifests most severely in a subsurface band, located 1–2 mm below the articulating surface. The development of this subsurface embrittled region has been associated with fatigue damage, including delamination, of tricompartmental and unicondylar knee replacements [13–15]. The key concept is that an excessive duration of shelf aging in air, rather than the fact that an implant was gamma sterilized, was identified in the 1990s as the cause of degradation.

As anticipated one decade ago [4], gamma sterilization of UHMWPE components in the presence of air has continued as a clinically relevant issue into the 21st century. If we assume that the frequency of total hip and knee procedures performed annually in the United States increased at a rate of 10% in the 1980s and at a rate of 5% in the 1990s, reaching an estimated 500,000 annual procedures in 1995 [16], then at least 2 million US patients may have been implanted with an UHMWPE component that was sterilized in air during the period of 1980 to 1989, when gamma irradiation in air was the standard sterilization practice. From the same calculation, an additional 2 million patients are estimated to have been implanted with air-sterilized UHMWPE components in the United States between 1990 and 1995. Furthermore, McGovern et al. [15] reported that there was the potential for components that were gamma sterilized in air and stored in air to remain undetected in clinical inventory as late as January 2000. Therefore, based on the population of patients already implanted with air-sterilized UHMWPE components during the 1980s, 1990s, and possibly the early 2000s, clinical interest in the long-term clinical outcomes of patients implanted with gamma-air irradiated UHMWPE has continued.

## 3.3 GAMMA STERILIZATION IN BARRIER PACKAGING

There are many types of barrier packaging currently in use by the orthopedic industry (Table 3.2). The details about current packaging systems for orthopedic components are proprietary but basically consist of evacuating the air from the packaging and backfilling with an inert gas (e.g., nitrogen or argon). The barrier in the package consists of polymer laminates or metallic foils to block gas diffusion (see the photo of barrier packaged tibial components in Figure 3.4, provided in January 2003, courtesy of David Schroeder, Biomet, Inc.). Three further examples of contemporary

**TABLE 3.2 Summary of Contemporary Barrier Packaging Techniques Used for Gamma Sterilization of UHMWPE among Major Orthopedic Manufacturers (As of October 2008)**

| Company | Package environment | Sterilization/packaging trade name |
|---|---|---|
| Biomet | Argon flushed, near-vacuum sealed | ArCom |
| DePuy, Inc. | Near vacuum | GVF (gamma vacuum foil) |
| Stryker Howmedica Osteonics | Nitrogen | N2-Vac; Duration |
| Zimmer, Inc. | Nitrogen | |

**FIGURE 3.4** Contemporary barrier packaging used with the fabrication of ArCom (Biomet, Inc., Warsaw, Indiana, USA). ArCom packaging currently uses a glass film interposed between polymer sheets for an oxygen barrier. The packaging is argon-flushed and then vacuum-sealed.

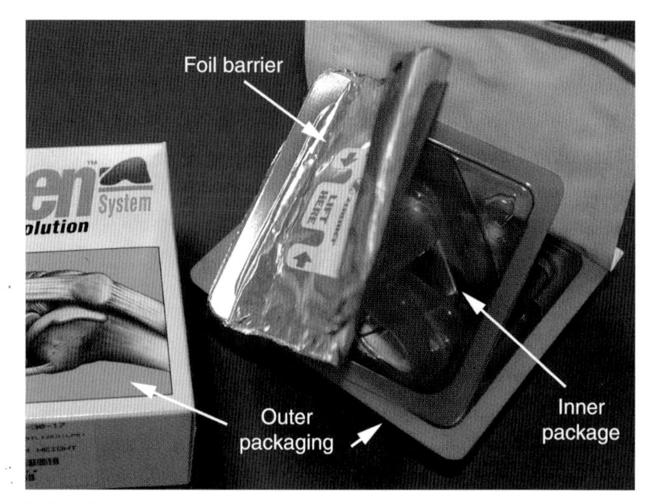


**FIGURE 3.5** Contemporary nitrogen-filled barrier packaging for gamma sterilization of UHMWPE components used by Zimmer, Inc. (Warsaw, Indiana, USA).

barrier packaging for UHMWPE components from different manufacturers are shown in Figures 3.5–3.7.

As shown in Table 3.2, many major orthopedic manufacturers continue to distribute UHMWPE components that are barrier packaged and gamma sterilized in an inert environment. The goal of barrier packaging is to minimize oxidative degradation during long-term shelf storage. Because gamma sterilized UHMWPE in a low oxygen environment still contains macroradicals, the successful mitigation of oxidation during shelf storage depends upon its ability to limit access of oxygen to the polymer. As highlighted in a recent survey [1] and covered in a subsequent section of this chapter, polymeric barriers may not be successful at preventing UHMWPE oxidation during shelf aging if they do not also incorporate an impermeable glass or metallic layer into the laminate.

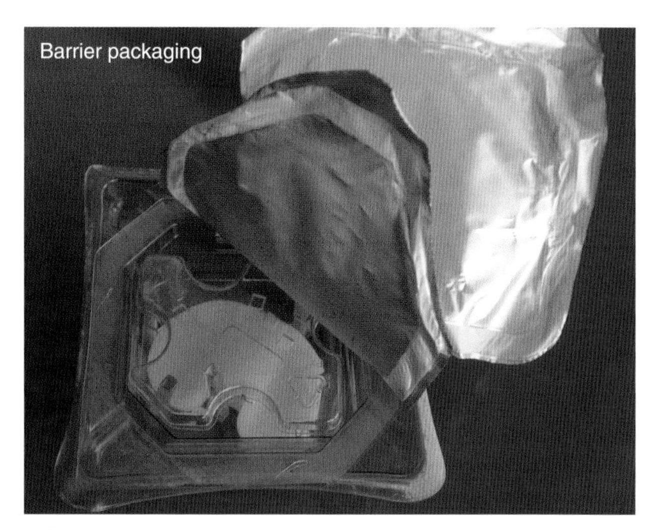


**FIGURE 3.6** Contemporary N2-Vac barrier packaging for gamma sterilization of UHMWPE components used by Stryker Howmedica Osteonics (Mahwah, New Jersey, USA).

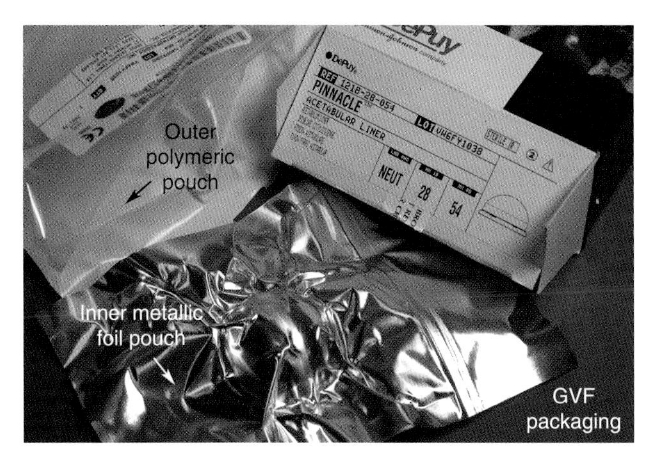


**FIGURE 3.7** Contemporary GVF barrier packaging for gamma sterilization of UHMWPE components, used by DePuy Orthopedics (Warsaw, Indiana, USA).

Concern about degradation during shelf aging prompted some orthopedic manufacturers in the 1990s to adopt sterilization of conventional UHMWPE using gas plasma or ethylene oxide. These sterilization methods admittedly generate no free radicals that can subsequently oxidize during shelf storage. However, UHMWPE sterilized in this manner also does not receive a tribological benefit associated with radiation-induced crosslinking. McKellop et al. [17] reported on the wear performance of UHMWPE in a contemporary hip simulator following gamma irradiation in air, gamma irradiation in an inert gas, ethylene oxide, or gas plasma. Between 2 and 5 million cycles (each million of cycles corresponds to about a year of use *in vivo* for an average patient), the wear rate of the gamma sterilized UHMWPE was significantly lower than UHMWPE sterilized by either gas plasma or ethylene oxide. For example, the wear rate from 3.5 to 5 million cycles for ethylene oxide-sterilized UHMWPE was reported as $40 \pm 0.6\,mm^3$/million cycles; in contrast, the wear rate for UHMWPE that was gamma irradiated in an air-permeable package was found to be $18.5 \pm 0.9\,mm^3$/million cycles [17]. A similar trend has been reported by Wang et al. [18], who observed over a 50% drop in the hip simulator wear rate following a single 25 kGy dose of gamma sterilization. When oxygen is excluded from the package during sterilization, further crosslinking, and additional improvement in wear performance, may be achieved in a hip simulator relative to gamma sterilization in air [17].

The effect of gamma inert sterilization on wear has been evaluated in clinical studies. The clinical wear rate of gamma air and gamma inert sterilized components is comparable. However, gas plasma sterilized implants have a clinical wear rate that is approximately two times greater than gamma air sterilized liners [19]. This is because wear processes in hip replacements are dominated by crosslinking. Even following gamma sterilization in air, the beneficial effects of crosslinking dominate over the deleterious effects of oxidative degradation, helping to explain the decades of successful use of gamma-air sterilized components starting in the 1960s:

> Perhaps surprisingly, the mean true wear rate for the conventional polyethylene liners that had been sterilized with gamma radiation in air was not significantly different from that for the liners that had been sterilized with gamma radiation in vacuum-barrier packaging.
>
> Sychterz et al., JBJS, Vol. 86-A, 2004 [20]

Because gamma sterilization, whether in air or in a low oxygen environment, produces a comparable extent of radiation crosslinking, one would not expect a substantial difference between gamma-air and gamma-inert sterilized polyethylene in a hip replacement. To achieve a clinically meaningful reduction in wear for total hip replacements, a higher level of radiation treatment is necessary than can be achieved using a standard sterilization dose (25–40 kGy).

## 3.4 ETHYLENE OXIDE GAS STERILIZATION

Ethylene oxide gas (EtO) has been a commercially available sterilization method since the 1970s. Highly toxic, EtO neutralizes bacteria, spores, and viruses. UHMWPE is a good candidate for EtO sterilization because it contains no constituents that will react with or bind to the toxic gas. The efficacy of EtO sterilization depends upon stringent control of process conditions, including humidity, duration, and temperature [21, 22]. When properly performed, EtO has been shown to effectively sterilize the interface of assembled modular components, such as metal-backed tibial and patellar devices [22]. However, because of its toxicity and hazardous residues, ethylene oxide sterilization is conducted in accordance with domestic and international standards [22, 23]. Laboratory studies suggest that sterilization using ethylene oxide gas does not substantially influence the physical, chemical, and mechanical properties of UHMWPE [24–27].

Sterilization of UHMWPE is accomplished by diffusion of EtO into the near-surface regions during several hours [21, 22]. After sterilization, residual EtO is then allowed to diffuse out of the UHMWPE to reduce the likelihood of an adverse topical reaction [21]. In a study by Ries et al. [22], which provided details of a validated EtO sterilization cycle for UHMWPE, the following protocol was employed: 18 hours of preconditioning at 65% relative humidity, followed by 5 hours of exposure to 100% ethylene oxide gas at 0.04 MPa, followed by 18 hours of forced air aeration. Thus, the entire sterilization cycle with EtO took a total of 41 hours. Preconditioning, exposure, and aeration were all conducted at 46°C.

Based on a limited number of retrieval studies, the clinical experience with EtO-sterilized UHMWPE components has thus far been favorable [13, 28]. In a study by White et al. [28], researchers compared the surface damage

**FIGURE 3.8** Contemporary gas-permeable packaging for ethylene sterilization of Durasul highly crosslinked UHMWPE components, used by Zimmer, Inc. (Warsaw, Indiana, USA).

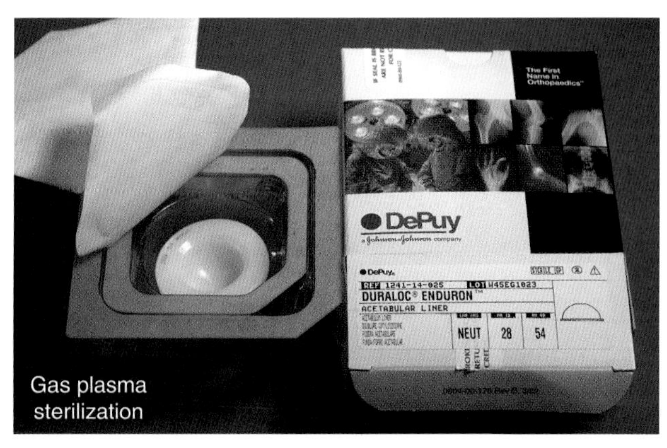


**FIGURE 3.9** Contemporary packaging for gamma sterilization of Enduron conventional UHMWPE components, used by DePuy Orthopedics (Warsaw, Indiana, USA).

and physical and mechanical properties of 26 retrieved UHMWPE tibial components of identical design that had been either sterilized with EtO or with gamma radiation (presumably in air). Components that were sterilized using EtO showed significantly less surface damage and delamination than gamma radiation sterilized components. The radiation sterilized components, on the other hand, had decreased ductility (elongation to failure), decreased toughness, and increased crystallinity. In a study of 150 retrieved acetabular components, Sutula and colleagues observed no evidence of rim cracking or delamination in any of the 17 ethylene oxide sterilized components [13]. In contrast, rim cracking was found in 19% of the acetabular components that were gamma radiation sterilized in air. Although the cited retrieval studies are necessarily limited by their retrospective nature, EtO sterilization appears not to induce certain modes of surface damage, which may be exacerbated by oxidative embrittlement of UHMWPE following gamma radiation sterilization in air.

Today, ethylene oxide continues to be used as a contemporary sterilization method for UHMWPE by major orthopedic manufacturers. Smith & Nephew, Inc. (Memphis, Tennessee, USA) and Wright Medical Technology, Inc. (Arlington, Tennessee, USA) have employed ethylene oxide since the 1990s [4]. More recently, however, Zimmer, Inc. (Warsaw, Indiana, USA) chose ethylene oxide for sterilization of their highly crosslinked UHMWPE (Durasul) hip and knee products [29] (Figure 3.8).

## 3.5 GAS PLASMA STERILIZATION

Low-temperature gas plasma is a relatively new commercially-available sterilization method that was applied to UHMWPE in the 1990s [17, 25, 30]. Gas plasma is a surface sterilization method that relies upon ionized gas for deactivation of biological organisms [21]. Two examples of commercially-available gas plasma sterilization methods that have been evaluated for compatibility with UHMWPE include Plazlyte (Abtox, Inc., Mundelein, Illinois, USA), which involves low-temperature peracetic acid gas plasma, and Sterrad (Advanced Sterilization Products, Irvine, California, USA), which uses low-temperature hydrogen peroxide gas plasma. Gas plasma sterilization is accomplished at temperatures lower than 50°C [31, 32]. Recent laboratory investigations suggest that low temperature gas plasma does not substantially affect the physical, chemical, or mechanical properties of UHMWPE [17, 25, 30, 33–35].

Gas plasma is an attractive sterilization method because it does not leave toxic residues or involve environmentally hazardous byproducts [31, 32]. Based on product literature available from the manufacturer, the sterilization cycle time for the Plazlyte system is 3–4 hours; the Sterrad system has an even shorter sterilization cycle of 75 minutes [31, 32]. Because no lengthy aeration period is required after gas plasma sterilization, it potentially offers substantial time and cost savings over EtO sterilization [31]. Because of its recent introduction, retrieval data from *in vivo* gas plasma sterilized UHMWPE components are not yet available.

Gas plasma has gained increased acceptance as a method for sterilizing UHMWPE components for total joint replacement. Gas plasma (Plazlyte system) has been used by DePuy Orthopaedics, Inc., (Warsaw, Indiana, USA) since the 1990s to routinely sterilize UHMWPE components [4] (Figure 3.9). Zimmer, Inc., (Warsaw, Indiana, USA) and Stryker (Mahwah, New Jersey, USA) currently employ gas plasma for sterilization of highly crosslinked UHMWPE components (e.g., Longevity, Prolong, X3).

## 3.6 THE TORINO SURVEY OF CONTEMPORARY ORTHOPEDIC PACKAGING

The contemporary packaging of both conventional and highly crosslinked UHMWPE components was surveyed

**TABLE 3.3** Seven Types of Packaging for UHMWPE and Proposed Classification System, Adapted from [1]

| Type | Packaging components | Packaging classification |
|---|---|---|
| 1 | Double PET blister with Tyvek cover | Class I: Air permeable |
| 2 | Two bags in PE/polyamide multilayer | Class II: Multilayer polymer film barriers |
| 3 | Two bags in PE/PET/PE multilayer | |
| 4 | Two bags in PE/PET/PVOH multilayer | |
| 5 | Double PET blister with Al foil cover | Class III: Aluminum foil barriers and multilayer polymer film |
| 6 | PET blister with Al foil cover, inside PET blister/Tyvek cover | |
| 7 | Al foil pouch inside PET blister with Tyvek cover or PE/PET/PE multilayer pouch | |

by researchers from the University of Torino, Italy [1]. This survey, conducted between 2003 and 2004, included not only an assessment of the packaging but also an evaluation of the chemical reactions within the packaged components. Key aspects of the study with respect to the packaging are summarized herein.

The survey was cleverly organized as a supplier competition by a major orthopedic center (CTO) in northern Italy. A total of 100 sterilized, never-implanted UHMWPE hip, knee, and shoulder components were submitted by 19 orthopedic manufacturers to researchers at the University of Torino for analysis. Six of the manufacturers were Italian (Citieffe, Gruppo Bioimpianti, Hit Medica, Lima Lto, Permedica, SAMO), and the remaining manufacturers were based in Europe (e.g., Aesculap, Centerpulse, Endoplus, Euros, Link, Mathys, Tornier) or the United States (Biomet, DePuy, Smith & Nephew, Stryker, Zimmer), representing the main world producers. Both conventional and highly crosslinked UHMWPE implants were included in this study.

The outer packaging was evaluated for the presence of an expiration date, suggesting a remaining shelf life, and the implants were classified as either expired or nonexpired components. It was unexpected that certain manufacturers would enter components with expired packaging into the survey, especially when it was explicitly framed as a competition for business with a high-volume regional orthopedic hospital. Perhaps these manufacturers considered that their expired implants were expendable and thus only suitable for scientific study.

The inner implant packaging materials were characterized into three general classes, including gas permeable packaging (Class I); multilayer polymer film barrier packaging (Class II); and barrier packaging using a combination of metallic foils and polymeric materials (Class III). The polymeric materials used in the packaging of the UHMWPE were specifically identified (e.g., polyethylene terephthalate [PET], polyethylene [PE], polyamide, polyvinyl alcohol [PVOH]) by declaration of the manufacturer and by using Fourier transform infrared spectroscopy (FTIR) (System 2000, Perkin-Elmer, Shelton, Connecticut, USA). The packaging materials were identified as one of seven types, as summarized in Table 3.3.

The oxidation levels in the sterilized components were quantified using the ASTM oxidation index and found to vary considerably among radiation-sterilized polyethylene inserts, but not the gas sterilized inserts. None of the gas-sterilized components were found to contain detectable macroradicals, hydroperoxide content, or oxidation, regardless of whether or not they were highly crosslinked.

Surprisingly, the survey documented the continued use of air permeable (Type 1) packaging for contemporary gamma sterilized components fabricated by certain smaller European orthopedic manufacturers. 16/65 radiation sterilized liners submitted to the Torino Survey as "ready to implant" as recently as 2004 were gamma irradiated in air. Among radiation sterilized inserts, the highest oxidation levels were associated with gas permeable (Class I) and polymer barrier packaging (Class II) (Figure 3.10, Table 3.4). For gas permeable packaging (Class I), significantly higher levels of oxidation were associated with expired packaging (Figure 3.10). Relatively less variation in oxidation was observed in polyethylene stored within polymeric and foil barrier packaging (Class III, Figure 3.10). The oxidation index (OI) values were very similar for polyethylene whether it was stored in expired or unexpired Class III packaging (Figure 3.10).

Gamma irradiation in air has long been considered to be a historical sterilization process for orthopedic manufacturers in the United States, but the results of the Torino survey quantified the ongoing prevalence of this practice for a major European surgical center. This survey represents the first comprehensive study documenting elevated oxidation content in gamma sterilized UHMWPE components stored within contemporary [36], multilayer polymer film barrier packaging (Class II). Little evidence of oxidation was found in UHMWPE stored within foil and polymer-based barrier packaging (Class III).

Manufacturers participated in the Torino survey as a requirement for consideration as a supplier by our orthopedic hospital (CTO). Consequently, this study was limited as a survey of implants offered competitively by manufacturers for analysis. No sampling criteria were imposed on the manufacturers. Therefore, the packaging evaluated in the survey should be interpreted as illustrative, rather than as a statistically representative sample for Italy or any other European country.

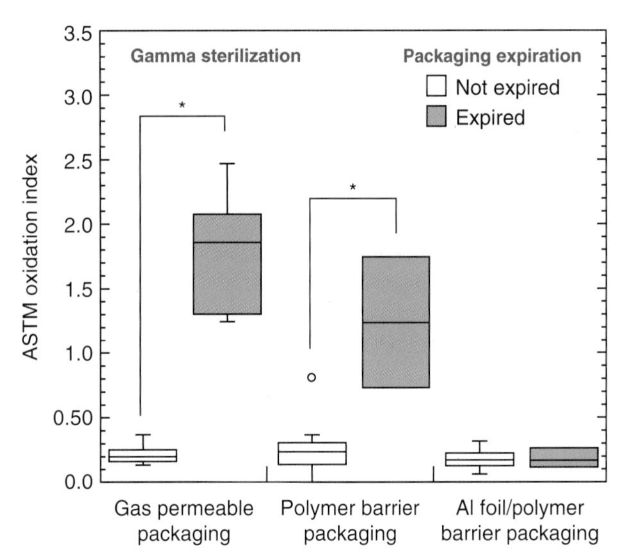


**FIGURE 3.10** Oxidation levels in gamma sterilized polyethylene components, adapted from [1]. Packaging was classified as gas permeable, multilayer polymer film barrier, or aluminum foil and polymer film barrier (see Table 3.4). Gamma sterilization in gas permeable packaging is the equivalent to gamma sterilization in air. The expiration date was obtained directly from the packaging. (The * symbol corresponds to $p < 0.05$ for Mann-Whitney tests.)

## 3.7 SHELF LIFE OF UHMWPE COMPONENTS FOR TJR

The shelf life of UHMWPE components was initially predicated on the ability of the package to maintain sterility, and a consensus practice of 5-year shelf life was adopted in Europe for medical implants so that sterility could be assured. In 1998, the Medical Devices Agency (MDA) warned that implants gamma sterilized in air and stored in air should be removed from circulation in England after 5 years or the manufactures' shelf life [37]. To date, no similar warning has ever been issued by the Food and Drug Administration in the United States or national regulatory bodies in other countries. The European Committee for Standardization (CEN) attempted, but was ultimately not successful, in establishing standards that limit the shelf life of UHMWPE components to 5 years. Because implants are distributed multinationally, packaging for components sold in the United States may stipulate a 5-year shelf life. However, it should be noted that, even today, there is no US standard for shelf life of UHMWPE components after sterilization by gamma radiation, ethylene oxide, or gas plasma.

The formation and propagation of chemical reactions to produce oxidation in UHMWPE requires the initial presence of alkyl macroradicals from irradiation [1]. Data from the Torino packaging survey [1] indicates that when an UHMWPE implant contains no alkyl macroradicals, gas sterilization—whether by gas plasma or ethylene oxide—prevents oxidation during shelf aging, regardless of the packaging expiration status. During the Torino survey, it was inferred based on the timing of the study and the expiration dates on the packages, that the manufacturer's shelf life was greater than 5 years in gas-sterilized implants provided by one manufacturer. A shelf life of longer than 5 years appears to be justifiable for gas-sterilized UHMWPE implants with undetectable macroradicals, providing the packaging sterility can be assured for durations longer than 5 years.

The situation is different with radiation-sterilized implants, especially if the packaging system is new and

**TABLE 3.4** Oxidation and Hydroperoxide Content in Unexpired, Gamma Sterilized Orthopedic UHMWPE Implants from the Torino Packaging Survey, Adapted from [1]

| Packaging classification | Number of implants tested | Median hydroperoxide index (range) | Median ASTM oxidation index (range) |
|---|---|---|---|
| Class I: Gas permeable | 10 | 0.52 (0.43–1.29) | 0.20 (0.13–0.37) |
| Class II: Multilayer polymer film barriers | 20 | 0.78 (0–1.17) | 0.23 (0–0.81) |
| Class III: Aluminum foil barriers and multilayer polymer film | 21 | 0.37 (0.20–0.63) | 0.17 (0.06–0.32) |

its long-term integrity has not been validated by the manufacturer. Accelerated aging methods have been developed for evaluating the integrity of medical packaging [38]. Nevertheless, due to proprietary variations in industry packaging practices, regulatory agencies such as the US Food and Drug Administration (FDA) currently require orthopedic manufacturers to submit real-time shelf aging data to validate accelerated aging tests and to establish the maximum shelf life for a particular UHMWPE product. Efforts at the American Society for Testing and Materials (ASTM) to develop protocols for establishing the shelf life of UHMWPE components used in total joint replacement have not reached consensus, and the task force established for this purpose was disbanded.

## 3.8 OVERVIEW OF CURRENT TRENDS

Fifteen years ago, gamma sterilization of UHMWPE in air-permeable packaging was the norm in orthopedics. Today, a broad range of sterilization and packaging methods are employed by orthopedic companies for UHMWPE components. For gamma sterilization, barrier packaging methods have been widely adopted and, until recently, were thought to satisfactorily address the historical problem of oxidation during shelf aging. Certain first-generation polymeric barrier packaging systems used for orthopedic implants have now been shown to be permeable to oxidation during 5 years of shelf life. Aside from conference abstracts [39, 40], little international data is available in the peer-reviewed literature to judge the efficacy of the numerous current packaging and shelf life strategies across the orthopedic industry from the perspective of reducing UHMWPE oxidation during shelf storage [41, 42]. The limited available data for gamma sterilization in a low oxygen environment suggests that, with the appropriate choice of packaging materials and conditions, oxidation of UHMWPE components may be limited for 5 to 10 years [1, 39–42]. However, as also demonstrated by the findings of the Torino survey [1], there is no guarantee that a manufacturer's new barrier packaging system will, in fact, exclude oxygen throughout its anticipated shelf life. Therefore, real time shelf aging studies are considered necessary to validate the effectiveness of every manufacturers' contemporary barrier packaging systems. It is now better appreciated that packaging incorporating impermeable barriers, such as metallic or glass foils, appear to be necessary to effectively reduce oxidation during shelf aging. The generality of this finding across different orthopedic manufacturers using radiation sterilization, particularly in Europe, remains poorly understood.

Gas-permeable packaging continues to be used for ethylene oxide and gas plasma, which emerged at the turn of the century as industrially viable, alternative methods for sterilizing UHMWPE, especially following radiation crosslinking and thermal stabilization. It is also noteworthy that gas sterilization, whether by EtO or GP, does not generate macroradicals in UHMWPE, and therefore the oxidation cascade reaction associated with gamma irradiation in air or in a low oxygen environment is thereby prevented from initiating. The proliferation of sterilization and proprietary packaging techniques has thus far made it difficult to establish broad industry standards in this area. Standards organizations, such as ASTM, have thus far been unable to harmonize the methods used to establish the shelf life of contemporary UHMWPE packaging and sterilization techniques.

## 3.9 ACKNOWLEDGMENTS

The author is grateful for the collaboration with researchers from Torino, including Prof. Luigi Costa and Dr. Pierangiola Bracco, on the survey of orthopedic packaging conducted at their institution. Special thanks are also extended to Janet Krevolin (Zimmer, Inc.), Ray Gsell (Zimmer, Inc.), Shi-Shen Yao and Paul Serekian (Stryker Orthopedics, Inc.), Jorge Ochoa and Mark Haynes (DePuy Orthopaedics, Inc.), and David Schroeder (Biomet, Inc.) for their helpful discussions and editorial assistance with the first edition of this chapter.

## REFERENCES

1. Costa L, Bracco P, Brach del Prever EM, Kurtz SM, Gallinaro P. Oxidation and oxidation potential in contemporary packaging for polyethylene total joint replacement components. *J Biomed Mater Res B Appl Biomater* 2006 July;**78**(1):20–6.
2. Kurtz SM, Hozack WJ, Purtill JJ, Marcolongo M, Kraay MJ, Goldberg VM, et al. Significance of in vivo degradation for polyethylene in total hip arthroplasty. *Clin Orthop Relat Res* 2006;**453**:47–57.
3. Isaac GH, Dowson D, Wroblewski BM. An investigation into the origins of time-dependent variation in penetration rates with Charnley acetabular cups—wear, creep or degradation?. *Proc Inst Mech Eng* 1996;**210**(3):209–16.
4. Kurtz SM, Muratoglu OK, Evans M, Edidin AA. Advances in the processing, sterilization, and crosslinking of ultra-high molecular weight polyethylene for total joint arthroplasty. *Biomaterials* 1999;**20**(18):1659–88.
5. Jahan MS, Wang C. Combined chemical and mechanical effects on free radicals in UHMWPE joints during implantation. *J Biomed Mater Res* 1991;**25**:1005–17.
6. Kurth M, Eyerer P, Ascherl R, Dittel K, Holz U. An evaluation of retrieved UHMWPE hip joint cups. *J Biomater Appl* 1988;**3**(1):33–51.
7. Rimnac CM, Klein RW, Betts F, Wright TM. Post-irradiation aging of ultra-high molecular weight polyethylene. *J Bone Joint Surg* 1994;**76A**(7):1052–6.
8. Bostrom MP, Bennett AP, Rimnac CM, Wright TM. The natural history of ultra-high molecular weight polyethylene. *Clin Orthop* 1994;**309**(309):20–8.

9. Kurtz SM, Rimnac CM, Bartel DL. Degradation rate of ultra-high molecular weight polyethylene. *J Orthop Res* 1997;**15**(1):57–61.
10. Kurtz SM, Bartel DL, Rimnac CM. Post-irradiation aging affects the stresses and strains in UHMWPE components for total joint replacement. *Clin Orthop* 1998;**350**:209–20.
11. Edidin AA, Jewett CW, Kwarteng K, Kalinowski A, Kurtz SM. Degradation of mechanical behavior in UHMWPE after natural and accelerated aging. *Biomaterials* 2000;**21**(14):1451–60.
12. Collier JP, Sperling DK, Currier JH, Sutula LC, Saum KA, Mayor MB. Impact of gamma sterilization on clinical performance of polyethylene in the knee. *J Arthroplasty* 1996;**11**(4):377–89.
13. Sutula LC, Collier JP, Saum KA, Currier BH, Currier JH, Sanford WM, et al. Impact of gamma sterilization on clinical performance of polyethylene in the hip. *Clin Orthop* 1995;**319**:28–40.
14. Kennedy FE, Currier JH, Plumet S, Duda JL, Gestwick DP, Collier JP, et al. Contact fatigue failure of ultra-high molecular weight polyethylene bearing components of knee prostheses. *J Tribol* 2000;**122**:332–9.
15. McGovern TF, Ammeen DJ, Collier JP, Currier BH, Engh GA. Rapid polyethylene failure of unicondylar tibial components sterilized with gamma irradiation in air and implanted after a long shelf life. *J Bone Joint Surg Am* 2002 June;**84-A**(6):901–6.
16. Orthopaedic Products. In: Smith RC, Geier MA, Reno J, Sarasohn-Kahn J, editors. *Medical & Healthcare Marketplace Guide*. New York: IDD Enterprises, L.P.; 1996. p. 265–72.
17. McKellop HA, Shen F-W, Campbell P, Ota T. Effect of molecular weight, calcium stearate, and sterilization methods on the wear of ultra-high molecular weight polyethylene acetabular cups in a hip simulator. *J Orthop Res* 1999;**17**(3):329–39.
18. Wang A, Essner A, Polineni VK, Stark C, Dumbleton JH. Lubrication and wear of ultra-high molecular weight polyethylene in total joint replacements. *Tribol Int* 1998;**31**(1–3):17–33.
19. Hopper Jr. RH, Young, AM, Orishimo KF, Engh Jr. CA. Effect of terminal sterilization with gas plasma or gamma radiation on wear of polyethylene liners. *J Bone Joint Surg Am* 2003 March;**85-A**(3):464–8.
20. Sychterz CJ, Orishimo KF, Engh CA. Sterilization and polyethylene wear: clinical studies to support laboratory data. *J Bone Joint Surg Am* 2004 May;**86-A**(5):1017–22.
21. Bruck SD, Mueller EP. Radiation sterilization of polymeric implant materials. *J Biomed Mater Res* 1988;**22**(A2 Suppl.):133–44.
22. Ries MD, Weaver K, Beals N. Safety and efficacy of ethylene oxide sterilized polyethylene in total knee arthroplasty. *Clin Orthop* 1996;**331**:159–63.
23. Page BFJ, Cyr H. A guide to AAMI's TIR for EtO-sterilized medical devices. *Med Dev Diag Indust* 1998;**20**(2):73–8.
24. Goldman M, Lee M, Gronsky R, Pruitt L. Oxidation of ultrahigh molecular weight polyethylene characterized by Fourier Transform Infrared Spectrometry. *J Biomed Mater Res* 1997;**37**(1):43–50.
25. Collier JP, Sutula LC, Currier BH, Currier JH, Wooding RE, Williams IR, et al. Overview of polyethylene as a bearing material: comparison of sterilization methods. *Clin Orthop* 1996;**333**(333): 76–86.
26. Ries MD, Weaver K, Rose RM, Gunther J, Sauer W, Beals N. Fatigue strength of polyethylene after sterilization by gamma irradiation or ethylene oxide. *Clin Orthop* 1996;**333**:87–95.
27. Wang A, Sun DC, Yau S-S, Edwards B, Sokol M, Essner A, et al. Orientation softening in the deformation and wear of ultra-high molecular weight polyethylene. *Wear* 1997;**203–204**:230–41.
28. White SE, Paxson RD, Tanner MG, Whiteside LA. Effects of sterilization on wear in total knee arthroplasty. *Clin Orthop* 1996;**331**(331):164 71.
29. Muratoglu OK, Kurtz SM. Alternative bearing surfaces in hip replacement. In: Sinha R, editor. *Hip Replacement: current Trends and Controversies*. New York: Marcel Dekker; 2002. p. 1–46.
30. Goldman M, Pruitt L. Comparison of the effects of gamma radiation and low temperature hydrogen peroxide gas plasma sterilization on the molecular structure, fatigue resistance, and wear behavior of UHMWPE. *J Biomed Mater Res* 1998 June 5;**40**(3):378–84.
31. Feldman LA, Hui HK. Compatibility of medical devices and materials with low-temperature hydrogen peroxide gas plasma. *Med Dev Diag Indust* 1997;**19**(12):57–62.
32. Kyi MS, Holton J, Ridgway GL. Assessment of the efficacy of a low temperature hydrogen peroxide gas plasma sterilization system. *J Hosp Infect* 1995;**31**(4):275–84.
33. Reeves EA, Barton DC, FitzPatrick DP, Fisher J. Comparison of gas plasma and gamma irradiation in air sterilization on the delamination wear of the ultra-high molecular weight polyethylene used in knee replacements. *Proceedings of the Institution of Mechanical Engineers* 2000;**214**(3):249–55.
34. Charlebois SJ, Daniels AU, Lewis G. Isothermal microcalorimetry: an analytical technique for assessing the dynamic chemical stability of UHMWPE. *Biomaterials* 2003 January;**24**(2):291–6.
35. McKellop H, Shen FW, Lu B, Campbell P, Salovey R. Effect of sterilization method and other modifications on the wear resistance of acetabular cups made of ultra-high molecular weight polyethylene. A hip-simulator study. *J Bone Joint Surg Am* 2000 December;**82-A**(12):1708–25.
36. Costa L, Luda MP, Trossarelli L, Brach del Prever EM, Crova M, Gallinaro P. Oxidation in orthopaedic UHMWPE sterilized by gamma-radiation and ethylene oxide. *Biomaterials* 1998 April–May;**19**(7–9):659–68.
37. Ultra-High Molecular Weight Polyethylene (UHMWPE) components of joint replacement implants. Medical Devices Agency Safety Notice MDA SN 9816; 1998.
38. ASTM F 1980–02. *Standard guide for accelerated aging of sterile medical device packages*. West Conshohocken, PA: American Society for Testing and Materials; 2002.
39. Edidin AA, Muth J, Spiegelberg S, Schaffner SR. Sterilization of UHMWPE in nitrogen prevents oxidative degradation for more than ten years. *Transactions of the 46th Orthopedic Research Society*. Orlando, FL; 2000 February 9–13. p. 1.
40. Greer K. A prospective study of the effects of five years shelf aging on the properties of UHMWPE following gamma sterilization in air or in a vacuum foil pouch. *Transactions of the 28th Society for Biomaterials*; 2003. p. 378.
41. Streicher RM. Ionizing irradiation for sterilization and modification of high molecular weight polyethylenes. *Plast Rubber Proc Appl* 1988;**10**(4):221–9.
42. Lu S, Orr JF, Buchanan FJ. The influence of inert packaging on the shelf ageing of gamma-irradiation sterilised ultra-high molecular weight polyethylene. *Biomaterials* 2003 January;**24**(1):139–45.

Chapter 4

# The Origins of UHMWPE in Total Hip Arthroplasty

Steven M. Kurtz, PhD



## 4.1 INTRODUCTION AND TIMELINE

Introduced clinically in November 1962 by Sir John Charnley, UHMWPE articulating against a metallic femoral head remains the gold standard bearing surface combination for total hip arthroplasty. Considering how rapidly technology can change in the field of orthopedics, the long-term role that ultrahigh molecular weight polyethylene (UHMWPE) has played in joint arthroplasty over the past 45 years is fairly remarkable. Starting in the 1970s, researchers have attempted to improve UHMWPE for orthopedic applications, starting with the introduction of a carbon-fiber reinforced material (Poly II, described in Chapter 17) [1]. However, this composite UHMWPE was not found to exhibit consistent and improved clinical results relative to the conventional UHMWPE introduced by Charnley.

In Japan during the 1970s, two important technological advancements occurred: one was the introduction of alumina ceramic as an alternative bearing surface for UHMWPE [2]. The second advancement involved the clinical introduction of a highly crosslinked UHMWPE by over 1000 kGy of gamma irradiation in air [3]. A similar advancement in highly crosslinked UHMWPE also occurred in South Africa during the 1970s, where researchers in Praetoria clinically introduced a UHMWPE that was gamma irradiated with up to 700 kGy in the presence of acetylene [4].

During the 1980s, two other noteworthy developments occurred relative to polyethylene in joint replacements. In the early 1980s, Char F. Thackray, Ltd. of Leeds began development on an injection molded HDPE that could be crosslinked by silane chemistry (the same type of crosslinking used in silicone polymers) [5, 6]. Only 22 of these implants were produced and implanted by Dr. Wroblewski starting in 1986 [7]. After an initial bedding in period, these crosslinked HDPE components have been found to exhibit very low clinical wear rates.

In the late 1980s, a joint venture between DePuy Orthopedics and DuPont developed a highly crystalline form of UHMWPE distributed under the trade name of Hylamer [8, 9]. The clinical history of Hylamer, which unfolded during the 1990s, has been mixed and therefore controversial [1]. Although several orthopedic centers have reported worse clinical performance using Hylamer than with conventional UHMWPE, other surgeons have experienced satisfactory or even improved performance.

Thus, it would be misleading to think that the use of UHMWPE in the field of orthopedics has stood still for the past 5 decades. As summarized in Table 4.1, researchers and clinicians have tried on several occasions to modify

**TABLE 4.1 Timeline of Early UHMWPE Development for Joint Replacement**

| Date | Comment |
| --- | --- |
| 1958 | Charnley develops the technique of low friction arthroplasty (LFA). Using PTFE as the bearing material [10], implants were fabricated either by Charnley in his home workshop or in the machine shop at Wrightington and chemically sterilized. |
| 1962 | Charnley adopts UHMWPE for use in his LFA. Components were chemically sterilized. |
| 1968 | Start of Leeds production of the Charnley LFA by Chas F. Thackray, Ltd. of Leeds. The UHMWPE was gamma irradiated. |
| 1969 | General commercial release of the Charnley LFA by Chas F. Thackray, Ltd. of Leeds. UHMWPE were marketed as gamma irradiated (in air) with a minimum dose of 2.5 Mrad. |
| 1970s | Commercial release of Poly II—Carbon Fiber Reinforced UHMWPE for THA/TKA by Zimmer, Inc. |
| 1972 | Use of alumina ceramic heads articulating against UHMWPE in Japan. |
| 1980 to 1984 | Codevelopment of silane-crosslinked HDPE by University of Leeds, Wrightington Hospital, and Thackray. |
| 1980s | Commercial release of Hylamer (Extended Chain Recrystalized UHMWPE) for THA/TKA/TSA by DePuy Orthopedics. |
| 1998 | Clinical introduction of first-generation highly crosslinked and thermally stabilized UHMWPEs for THA. |

UHMWPE for use in joint replacements. Nevertheless, the UHMWPE introduced by Charnley remains the gold standard for artificial hips and now other artificial joints, including the knee and shoulder.

This chapter will focus on the historical development of UHMWPE for use in hip replacements by John Charnley. The main sources of this chapter have been Charnley's journal publications and books, as well as an outstanding biography of Charnley written by William Waugh [11]. Preparation of this chapter also entailed interviews and the review of archives and implants at Wrightington Hospital, DePuy International (Leeds), and the Thackray Museum of Leeds.

This chapter addresses the following questions: (1) How was UHMWPE introduced into hip arthroplasty? (2) What methods were used to sterilize the first UHMWPE implants? (3) What methods were used to test and evaluate UHMWPE (as well as other candidate biomaterials) prior to implantation? The goal of this chapter is to review the historical origins of UHMWPE as an orthopedic biomaterial.

## 4.2 THE ORIGINS OF A GOLD STANDARD (1958 TO 1982)

The rationale for John Charnley's design of an artificial joint started with a series of frictional experiments on animal and human joints in the 1950s [12, 13]. From these experiments, Charnley concluded that natural joints functioned well because of their low coefficient of friction, which result from the unique properties of cartilage tissue that promote lubrication. When the natural joint is compressed, the cartilage tissue expels water between the contacting surfaces, which become separated, at least in part, by a thin film of pressurized synovial fluid. In addition to water, the synovial fluid contains proteins and other biological constituents that facilitate lubrication. The pressurized synovial fluid film carries the joint force that protects cartilage tissue from wear during walking or other load-bearing activities. Because of arthritis and other joint diseases, cartilage can lose its unique lubricious characteristics.

Charnley realized that an artificial joint fabricated from synthetic materials like metal or plastic could not operate with purely hydrodynamic lubrication. In 1959, Charnley wrote that "attempts to lubricate any artificial joint must be based on the idea of using boundary lubrication. The substance which seems ideally suited for this purpose is PTFE because not only has this a low coefficient of friction (0.04–0.05) but it is a substance which is readily tolerated by animal tissues by virtue of its chemical inertness." Therefore, friction, not wear, was Charnley's primary reason for selecting PTFE. From the onset, Charnley's artificial joint design was based on the principles of boundary lubrication, during which pressure in the synovial fluid is not sufficient to fully separate the joint surfaces for load bearing. Thus, the artificial joint surfaces are expected to contact each other during boundary lubrication. This regime of partial lubrication is in contrast with a situation (common for some industrial applications) in which bearing surfaces must articulate dry, or without any lubrication to assist with reducing friction.

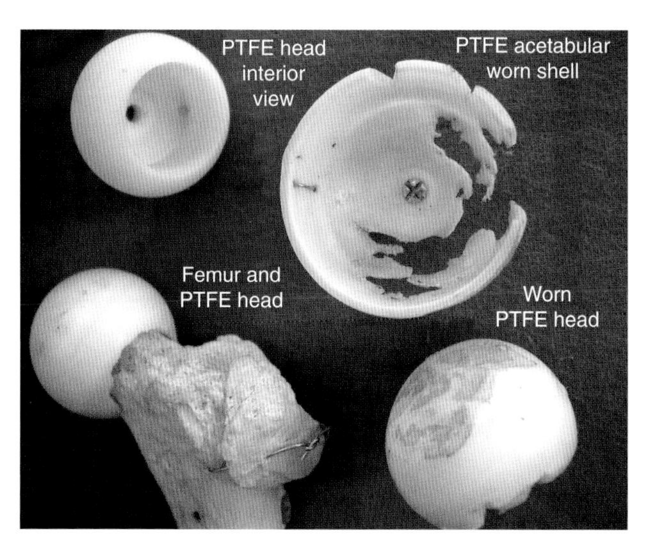


**FIGURE 4.1** Acetabulum was replaced with a thin shell of PTFE and femoral head surface was replaced with a PTFE ball.

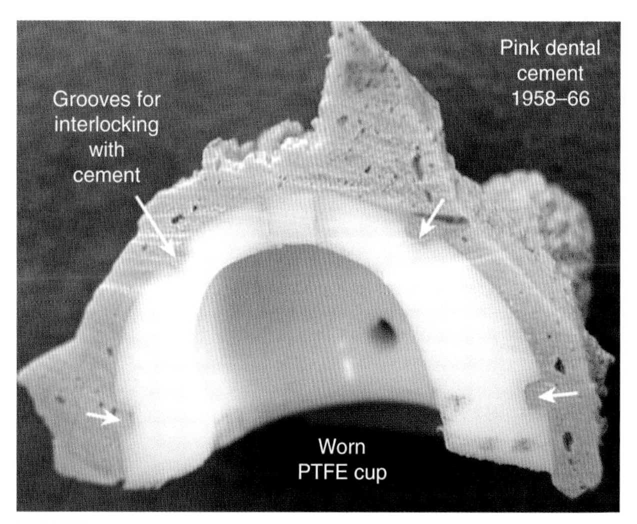


**FIGURE 4.2** The cross-sectional view of a worn PTFE implant, retrieved with its cement mantle.

## 4.3 CHARNLEY'S FIRST HIP ARTHROPLASTY DESIGN WITH PTFE (1958)

Charnley's design of an artificial hip joint was the product of an evolutionary process between 1958 and 1960 [14]. Five design iterations occurred. Charnley's initial "double cup" design in 1958 mimicked the natural joint. The acetabulum was replaced with a thin shell of PTFE and the femoral head surface was replaced with a PTFE ball, as shown in Figure 4.1, taken from the collection at the Charnley Museum at Wrightington.

For this initial design, the femoral head and the shell were press fit into the bone, with the idea that the friction between the bone and the PTFE would prevent relative motion. Unfortunately, the tolerances between the head and shell were such that the articulation tended to occur mainly between the PTFE shell and the acetabulum, resulting in abrasion of the shell and destruction of the underlying bone. Wear also occurred of the PTFE femoral head prosthesis, but more concerning from a clinical standpoint was the concomitant loss of blood supply to the femoral bone inside the PTFE cavity, which lead to necrosis [10].

## 4.4 IMPLANT FIXATION WITH PINK DENTAL ACRYLIC CEMENT (1958 TO 1966)

In subsequent designs of the hip arthroplasty, Charnley made use of pink dental acrylic cement, circulated under the trade name of Nu-Life [15], to fix the stem to the bone. After the fixation failures of the PTFE shell were observed, Charnley also started to fix the PTFE acetabular cups as well as the stems. Starting in 1958, the cement used by Charnley consisted of acrylic powder which was mixed with liquid monomer. The acrylic powder was chemically sterilized by formaldehyde vapor [15], but it was not necessary to sterilize the monomer, because the liquid (methyl methacrylate) itself is bactericidal. As shown in Figure 4.2, the cross-sectional view of a worn implant, retrieved with its cement mantle, the cup was designed with grooves in the back surface to facilitate interlocking at the PTFE-cement interface.

Although Charnley popularized the use of bone cement for hip arthroplasty, Haboush previously reported the use of acrylic cement to fix a femoral component in hip arthroplasty in 1953. In Haboush's previous experience, however, the fixation failed because the femoral prosthesis did not extend into the femur and was attached to the cut end of the bone. In the case of Charnley, the femoral stems extended deep into the femoral canal. The bone cement acted to effectively transfer the compressive forces and twisting (i.e., moments) that are imparted to the artificial hip during regular daily activities from Charnley's prosthesis to the bone. Like the grout that holds bathroom tiles to a shower wall, bone cement acted as a void filler and an interlocking agent between the implant and the bone. Bone cement is a poor adhesive to bone or PTFE. According to Charnley's biography [15] and records at Wrightington, Charnley used the pink Nu-Life dental cement until 1966, when he switched to the CMW formulation (CMW is an acronym for calculated molecular weight).

## 4.5 INTERIM HIP ARTHROPLASTY DESIGNS WITH PTFE (1958 TO 1960)

Charnley's second, interim design for a total hip replacement included an acetabular component of PTFE articulating against a metallic femoral head from a cemented Austin Moore or Thompson prosthesis [12]. The femoral head was 41.5 mm (1 5/8 inches) in diameter. He initially employed the largest feasible femoral head, which would result in the lowest contact stress. However, his engineering colleagues soon persuaded Charnley to reduce the femoral head diameter, based on the rationale that it would decrease the frictional torque. In other words, Charnley theorized that the frictional resistance during twisting of the joint would be reduced by a smaller femoral head. The concern was that the friction imparted to the joint during normal walking could lead to loosening of the cup from the acetabulum. This theory was judged to be valid because the joint surfaces were not intended to be lubricated hydronamically. In a hydrodynamic bearing, having a large femoral head would be an asset, because the surface sliding speeds would be greater, facilitating the development of a fluid film to separate the articulating surfaces. In an artificial joint, on the other hand, in which the surfaces would always be in contact, reducing the frictional resistance of the joint was a major concern for Charnley. Consequently, third and fourth iterations of this design, between 1958 and 1960 [16], employed cemented femoral components with diameters of 28.5 mm and 25.25 mm, respectively.

These three interim implant designs, preserved in the collection at Wrightington Hospital, are shown in Figure 4.3. The PTFE components in Figure 4.3 were retrieved at revision surgery and are severely worn. The wear is most evident in the sectioned 25.3-mm diameter acetabular component. The femoral components, on the other hand, appear pristine. The femoral heads are polished to a mirror finish (note the reflection of my hands holding the camera).

## 4.6 FINAL HIP ARTHROPLASTY DESIGN WITH PTFE (1960 TO 1962)

The fifth and final design iteration, which was clinically introduced in January 1960, included an acetabular component of PTFE, articulating against a cemented femoral component having a 22.225 mm (7/8 inch) diameter head [10]. In May 1961, Charnley published the early results of his low friction arthroplasty (LFA) in *Lancet* [10]. The early results of the new operation were extremely encouraging, with patients, previously handicapped by joint disease, returning to pain-free mobility within weeks of the operation. He wrote that "negligible wear" was observed on "close scrutiny" of radiographs for patients implanted for 10 months [10].

An example of a radiograph from a short-term implanted PTFE low friction arthroplasty is shown in Figure 4.4. This radiograph shows the initial orientation of the PTFE cup with respect to the pelvis. Initially, the femoral head is centered in the acetabular cup.

Within only a few years, patients began reporting to Wrightington with pain and inflammation associated with their PTFE artificial joints. Radiographic examination revealed severe wear, and upon revision, the joints were

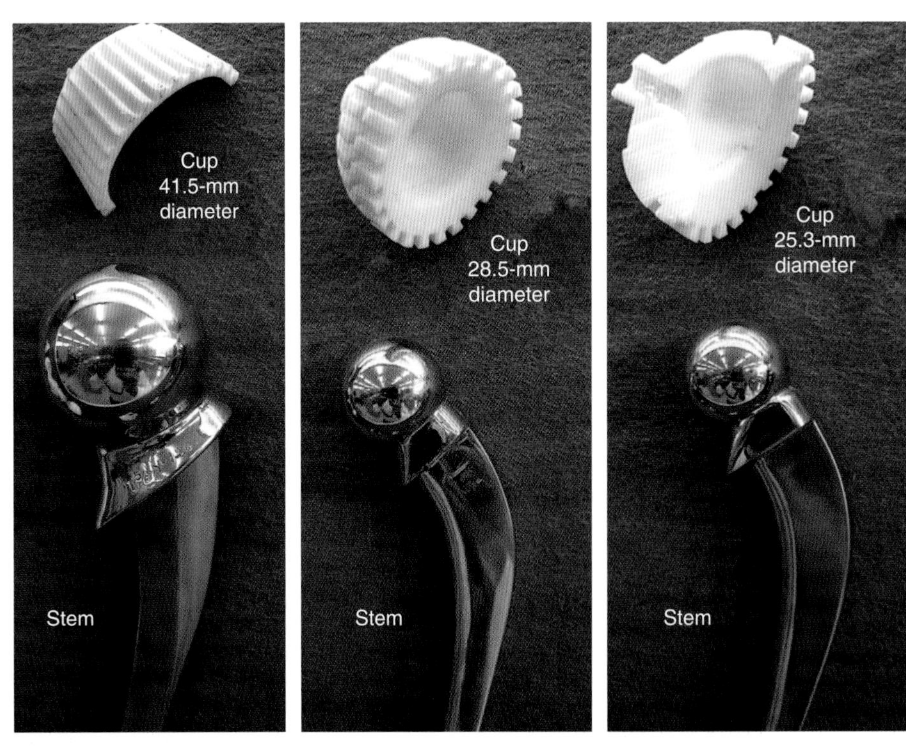


**FIGURE 4.3** Three interim PTFE implant designs by Charnley.

found to be surrounded by 100–200 ml of caseous, purulent tissue (examples of which were thoughtfully saved and can still be viewed, preserved at the Charnley museum at Wrightington). Although the joint articulation was successful in the first 1 to 2 years after surgery, over 99% had to be revised within 2 to 3 years of implantation due to severe wear and the inflammatory response provoked by the PTFE wear debris.

The radiograph in Figure 4.5 shows the PTFE artificial hip joint prior to revision (same patient as in the previous radiograph). Note that the femoral head has now translated vertically with respect to the center of acetabular cup, indicative of severe wear. The photograph to the right (not the same component as in the radiographs) shows how the femoral head tunnels into the PTFE acetabular cup during wear. The cup has been sectioned for illustrative purposes.

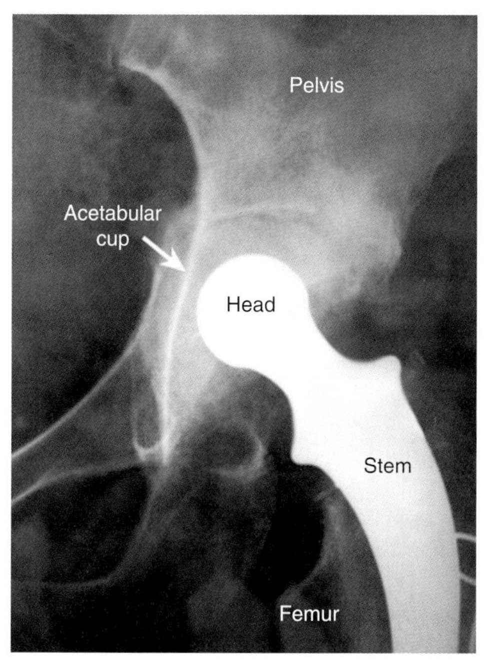


**FIGURE 4.4** The initial orientation of the PTFE cup with respect to the pelvis.

Charnley was faced with an ethical dilemma. On one hand, he had many elderly patients demanding relief from their debilitating pain, and some were willing to receive a PTFE component even if it would require revision in a few years. On the other hand, the short-term benefit was balanced by the risks associated with revision surgery and loss of bone stock caused by the PTFE debris.

## 4.7 IMPLANT FABRICATION AT WRIGHTINGTON

Up until 1963, the acetabular cups used at Wrightington were fabricated either by Charnley himself at his home workshop or by technicians at the hospital workshop in Wrightington [17, 18] (Figure 4.6). Charnley's home workshop was used to develop implant prototypes, as well as instruments. His 1/2 hp lathe, purchased in 1946, is now on display at the Thackray museum in Leeds. Charnley's biography recounts that "throughout his life, he was able to 'mock up' any surgical device in his workshop, and try it out himself before having it manufactured."

Charnley's 1961 publication indicates that "the research committee of the Manchester Regional Hospital Board has built and equipped a research workshop in the hospital and has provided the salary for a fitter and turner who will make the surgical implants." Charnley's research workshop and laboratory still remain at Wrightington Hospital, and they have been partly converted into a museum.

Between 1958 and 1966, H. Craven was Charnley's technician and was responsible for machining PTFE as well as UHMWPE cups for use at Wrightington [18]. He remembers ordering the PTFE in billets, which were predrilled by

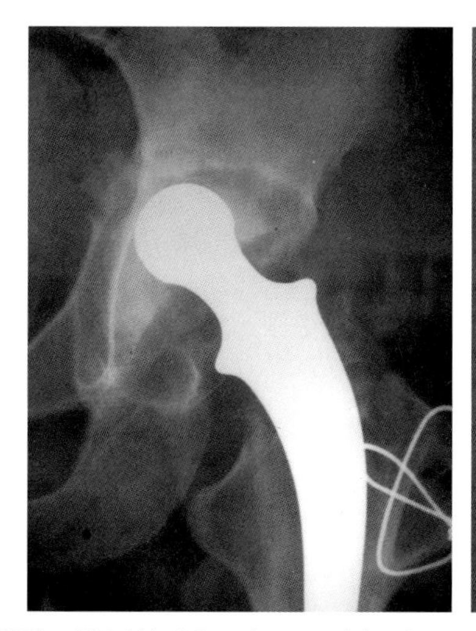

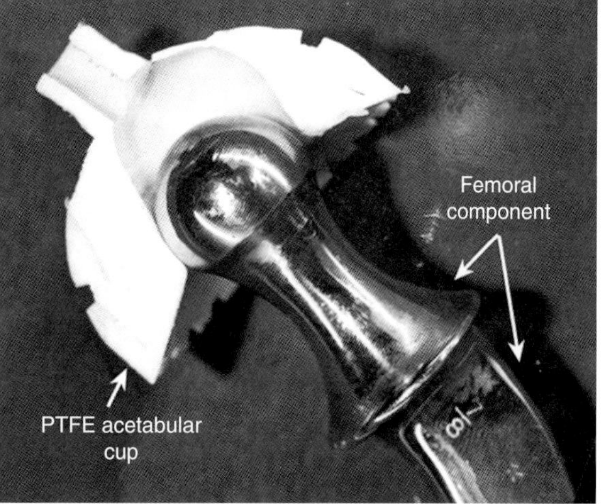


**FIGURE 4.5** PTFE artificial hip joint prior to revision (same patient as in the previous radiograph).

**FIGURE 4.6** The first UHMWPE hip components were fabricated either at the green lathe in his home workshop (the photo behind the lathe was taken of Charnley), or at the machine shop in Wrightington Hospital.

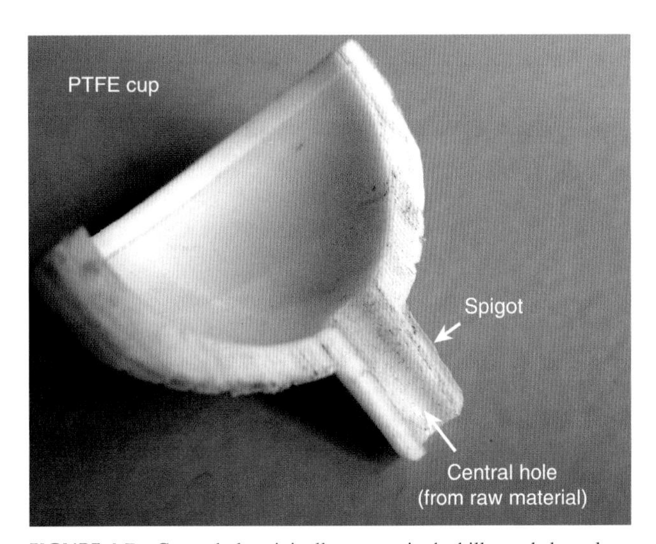


**FIGURE 4.7** Center hole originally present in the billet ended up along the central axis of the cup.

the supplier with a hole to improve the dimensional stability of the polymer. Because the cups were turned on a lathe, the center hole originally present in the billet ended up along the central axis of the cup, as shown in the section of the PTFE cup shown in Figure 4.7. The spigot was a design feature, which was used to orient the cup with respect to a locator hole drilled in the acetabulum.

The cups originally took about 45 minutes to machine. However, as the number of procedures increased, Craven developed a machine in 1962 which could produce a UHMWPE cup in less than 5 minutes. Craven's machine was sold to Thackray in 1963 after it was decided that the cups would be produced commercially [18].

## 4.8 THE FIRST WEAR TESTER

Craven built the first wear testing rig at Wrightington for Charnley between 1959 and 1960. With little money available to support research, the stainless steel parts of the wear tester were scrounged from the local scrap yard, where the proprietor was accustomed to reserving raw materials for the hospital. The rig, shown in a black and white photo in Figure 4.8 on display in Wrightington Hospital, consisted of four stations.

Within each test chamber, a stationary 1/4-inch diameter polymer pin was mounted superiorly with respect to a sliding, stainless steel plate. The height of each pin was recorded continuously during the test by a height gauge, each having a sensitivity of 0.001 inch. The test chambers were mounted on a reciprocating, sliding base. The sliding distance was approximately 1 inch, and the contact pressure was 700 psi (5 MPa). One full day of testing PTFE in this rig (24 hrs) was considered by Craven to be equivalent to 18 months of *in vivo* wear. Thus, Craven's testing rig subjected polymer pins of candidate biomaterials to unidirectional sliding conditions in an aqueous (saline) environment, using PTFE as the control.

Charnley was certainly aware of the effects of testing joint surfaces in synovial fluid as opposed to aqueous environment. His animal and human joint frictional testing, for example, which were conducted in the 1950s, included a range of lubricating fluids [12, 13]. However, for wear testing of artificial joint materials, Charnley advocated saline as the medium, because he felt that synovial fluid would not be a harsh enough test medium for the joint materials.

Still on display in Wrightington, the rig is only partially assembled today. Inspection reveals a very sturdily constructed apparatus. Craven's pride in his craftsmanship is evident because his name is stamped into the top of the rig.

In the 1970s, more complex testing rigs were built at Wrightington, which would subject polymer pins to multidirectional sliding (still using saline as a lubricant). However, from a historical perspective, the first test rig built by Craven is the most significant of the wear testers

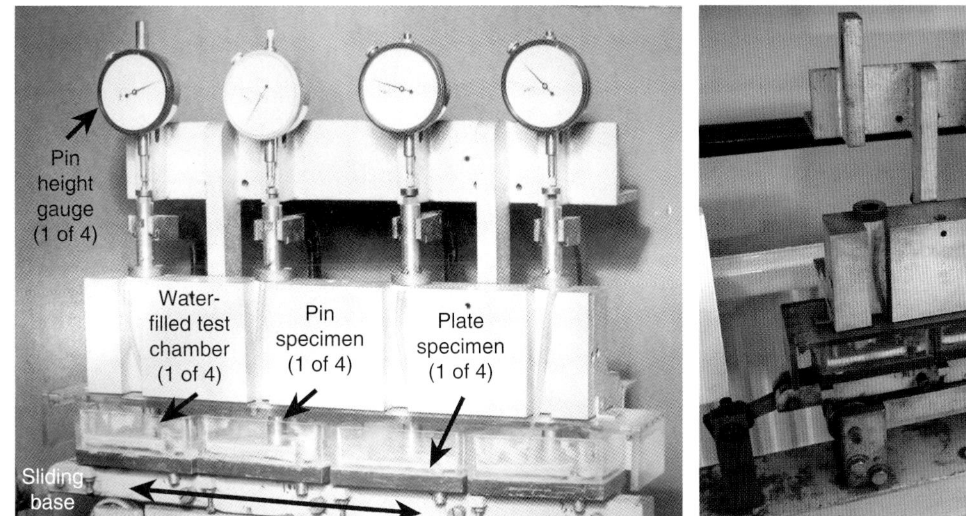

**FIGURE 4.8** Wear testing rig built at Wrightington, constructed by Harry Craven, which was used to test pins of PTFE and the first UHMWPE.

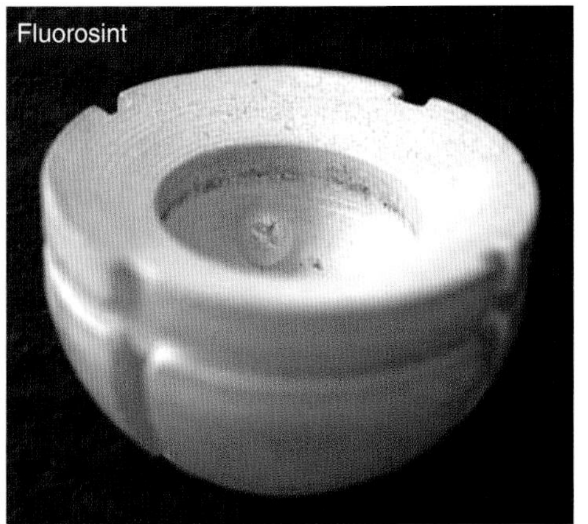


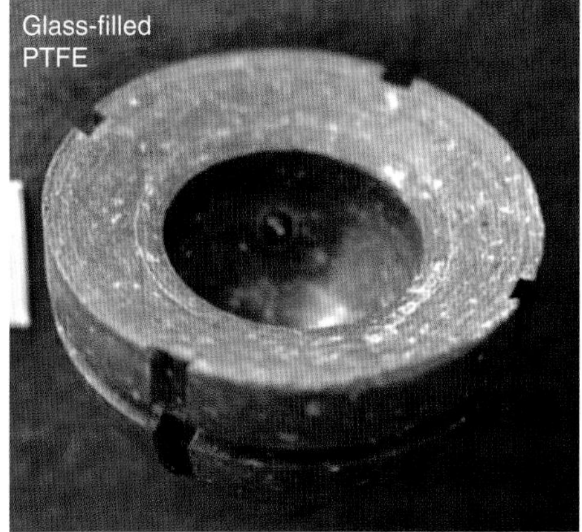


**FIGURE 4.9** Proprietary filled PTFE with trade name “Fluorosint” and glass-filled PTFE.

at Wrightington because it was used by Charnley to screen materials for the artificial hip after PTFE began to show evidence of severe wear *in vivo*. As we shall see, data collected using Craven’s wear tester ultimately established the superiority of UHMWPE over other polymeric materials for use in joint replacement.

## 4.9 SEARCHING TO REPLACE PTFE

After the failure of PTFE cups *in vivo*, Charnley experimented with filled PTFE [14]. Charnley tested and ultimately implanted glass-filled PTFE as well as a proprietary filled PTFE produced by Polypenco under the trade name “Fluorosint” (Figure 4.9).

In the wear tester, Craven showed that both materials showed evidence of scratching the stainless steel counterface (presumably because of the glass fibers). On the other hand, the wear rates were improved. For instance, the wear tester predicted that the wear rate of Fluorosint would be 10 times greater than PTFE. Charnley ordered that cups be fashioned and implanted immediately.

In the body, both types of filled PTFE exhibited severe wear, similar to PTFE. With the Fluorosint components, the wear debris presented an added complication of darkened tissue around the joint, which needed to be excised. The photograph in Figure 4.10, taken of a retrieved Fluorosint component (encased in pink acrylic dental cement) at Wrightington, illustrates the severe wear exhibited by such devices.

After the clinical failures of filled PTFE, Charnley abandoned polymers entirely in early 1962 and started to implant Thompson prostheses. However, the gloom that pervaded Wrightington at that time was short-lived because a promising new material soon arrived at the hospital through a series of most fortunate events.

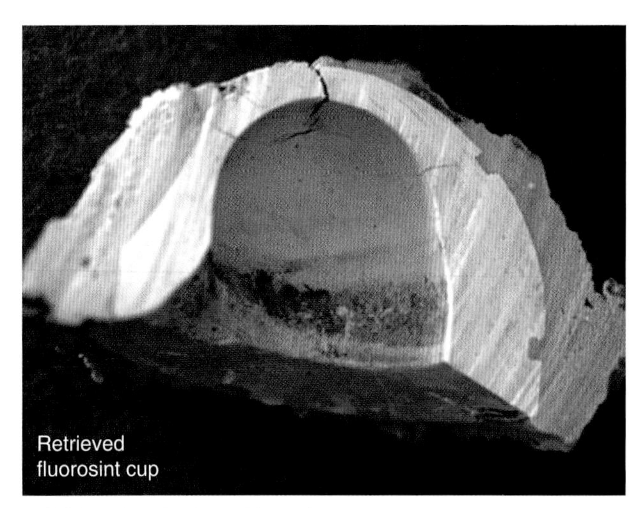


**FIGURE 4.10** Retrieved Fluorosint component (encased in pink acrylic dental cement).

AD.6352
† CHARNLEY ACETABULAR CUP, high density polyethylene, LARGE SIZE, with stainless steel* radiopaque marker and
wire mesh Cement Restrictor, stainless steel.*
(STERILE)

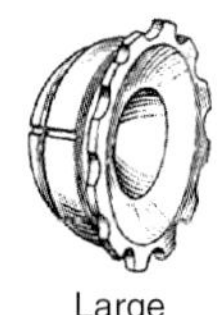


AD.6353
† CHARNLEY ACETABULAR CUP, as above but SMALL SIZE, and
wire mesh Cement Restrictor, stainless steel.*
(STERILE)

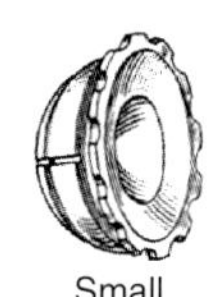


AD.6354
SPARE CEMENT RESTRICTOR.

Patent applied for

AD.6355 CHARNLEY INTRODUCER for Cement Restrictor, stainless steel.

* EN58J Stainless Steel to British Standard Specification, 3531: Part 1: 1968.

† Sterilised by gamma radiation at 2.5 M. Rads.

**FIGURE 4.11** Chas F. Thackray, Ltd. catalog page from the 1969 product release brochure.

## 4.10 UHMWPE ARRIVES AT WRIGHTINGTON

The story of how UHMWPE was introduced to orthopedics is recounted in considerable detail by Waugh in Charnley's biography [19]. Interviews with Craven, who was 32 years old at the time, as well as Charnley's writings, provide the basis for what is known about the circumstances at Wrightington in 1962.

In May of that year, a salesman (Mr. V. C. Binns) arrived at Wrightington and was introduced to Craven by the hospital's supply officer. Binns' products, small gears and bushings, were fabricated from UHMWPE, and he provided Craven with a sample of the material. Charnley's first reaction to UHMWPE is described in his biography as follows: "When shown the material, Charnley dug his thumb-nail into it and walked out, telling Craven he was wasting his time" [20].

Despite Charnley's negative initial remarks, Craven decided to test UHMWPE on his wear testing rig anyway while the surgeon was away in Copenhagen for a meeting. When Charnley returned to Wrightington in June, Craven showed him the 3 weeks of continuous testing data he had collected in his absence. Charnley then wrote,

> After running day and night for three weeks, this new material, which very few people even in engineering circles had heard about at that time, had not worn as much as PTFE would have worn in 24 hours under the same conditions. There was no doubt about it, "we were on." [20].

Although the wear results were encouraging, Charnley did not implant UHMWPE into patients until he was convinced of its biocompatibility. He first wrote to the producers of the UHMWPE, which was marketed under the trade name of RCH-1000. The German company that produced UHMWPE was known as the Ruhrchemie, although it was later merged into Hoechst. From his correspondence,

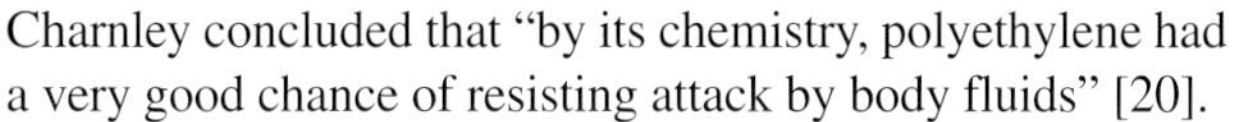

Charnley concluded that "by its chemistry, polyethylene had a very good chance of resisting attack by body fluids" [20].

Prior to implanting cups in patients, however, he implanted the UHMWPE in his own thigh, both in the bulk and finely divided (particulate) state. He also implanted PTFE in his thigh, to serve as a historical control. The results of this personal biocompatibility test, published in *Lancet* in 1963 [19], convinced Charnley that UHMWPE wear debris was biocompatible, whereas the particulate PTFE was not. Armed with this information, Charnley began implanting UHMWPE in patients during November 1962.

## 4.11 IMPLANT STERILIZATION PROCEDURES AT WRIGHTINGTON

Because the sterilization of UHMWPE continues to be a controversial topic, even today, it is helpful to review the early methods in use at Wrightington for joint replacement components. The stainless steel femoral components were produced by Thackray [17]. Prior to 1968, only the final polishing of the femoral heads would be performed at Wrightington. After 1968, the stems were fabricated entirely by Thackray. The femoral components were provided nonsterile by Thackray and sterilized at the hospital by autoclaving. This sterilization method was not appropriate for UHMWPE, which would distort after prolonged steam exposure.

According to Dr. Wroblewski, who started his service at Wrightington in 1967, the UHMWPE cups were chemically sterilized by wrapping them in gauze and soaking them in Cidex (glutaraldehyde) overnight. Although UHMWPE cups produced commercially by Thackray were gamma irradiated starting in 1968 [21], it remains unclear based upon the documents available today, when, prior to 1968, gamma irradiation was initiated for cups that were implanted at Wrightington [21]. Figure 4.11 shows a Chas F. Thackray, Ltd. catalog page from the 1969 product release brochure of the Charnley acetabular cup, indicating that the components were provided after sterilization with 2.5 Mrad of gamma radiation.

Charnley personally held strong views on the subject of UHMWPE sterilization. In personal correspondence with Thackrays, dated October 17, 1971, Charnley discusses seven reasons why gamma sterilization is preferable to ethylene oxide. The context of this letter is apparently in response to the suggestion by an official from the Scottish Home and Health Department that UHMWPE hip sockets be sterilized with ethylene oxide rather than gamma irradiation. Charnley's letter, reproduced here in its entirety, is fascinating because it not only shows his personal style and scientific opinions regarding sterilization, but the first point in the correspondence also documents his results from wear testing highly crosslinked UHMWPE with up to 10 Mrad (100 kGy) of gamma radiation. One can only imagine how the history of UHMWPE in orthopedics would have been rewritten had the wear testing apparatus at Wrightington been more biofidelic, and thus able to detect the substantial improvement in multidirectional wear behavior with elevated doses of ionizing radiation. Hip arthroplasty patients would have to wait over 2 decades, until the late 1990s, for highly crosslinked UHMWPE to be rediscovered and widely accepted by members of the orthopedic research community.

Wrightington Hospital Management Committee
Centre for Hip Surgery
Wrightington Hospital
Near Wigan

19th October, 1971

Mr. G. Robinson ,
Chas. F. Thackray Limited,
P.O. Box 171
Park Street,
LEEDS LS1 1RQ.

Dear Mr. Robinson,

Sterilisation of Charnley Total Hip Sockets of High Density Polyethylene
Report of Dr. Weymes, Sterilisation Research Unit,
Scottish Home and Health Department

I have read this report sent to Mr. Bury and dated 7th October, 1971, and feel it incumbent upon me, as advisor to Thackrays in the Charnley total hip programme, to make some comments. I will send a copy of this letter direct to Dr. Weymes.

I think it is important first of all to emphasise that this report is, of course, a personal opinion of Dr. Weymes and this should not be lost sight of in the categoric manner in which the statements are made.

1. In our own Biomechanical Laboratories here where we have apparatus for the testing of wear second to none in the world, and we have shown that gamma radiation up to 10 Mrad does not increase the rate of wear of high density polyethylene. There are some slight changes in colour from white to slightly cream colour but absolutely no alteration in the rate of wear. This amount of radiation is four times the medical sterilisation dose.

2. In the categories into which the U.K. Atomic Energy Authority at Wantage groups different plastics in their sensitivity to radiation, it is notable that high density polyethylene is rated as next to the most resistant group of plastics.

3. Wright of Petrochemicals published one of the first studies of high energy irradiation of high density polyethylene (Journal of Applied Polymer Science, Vol. 7 1963. pp. 1905–1918) and stated "ultimate tensile strength and impact strength increased with electron radiation, but very large doses (of the order of 25 Mrad) are required to effect a worthwhile improvement".

4. There is a paper from Japan (Matsubara Watanabe, Wear, 10, (1967) 214–222) indicating an improvement in the wear properties of high density polyethylene after gamma irradiation.

5. Since I consider sterilisation by irradiation a very suitable method there seems no point in using ethylene oxide which must surely be soluble in this plastic. Ethylene oxide freely penetrates two layers of low density polyethylene 250 micron thick and until proved otherwise, I think one has to accept that it will be very soluble in high density polyethylene. I fail to see how one can be sure of getting out all the ethylene oxide when double wrapped, even with the precautions mentioned in this letter.

It is to be noted that the possibility of harm to living tissues by implanted plastics is now foremost in peoples' minds so that there seems no point in adding an additional source of tissue damage no matter how remote this possibility may be considered when it is not absolutely necessary.

6. To say that ethylene oxide sterilisation is the method of choice is also not universally accepted in England. I have made special representations to Dr. Kelsey of the Central Public Health Laboralories at Colindale on two occasions over the last four years. On both occasions he has produced more objections to ethylene oxide sterilisation than attractions. Dr. Kelsey has sent me considerable literature and references in his own work against the efficacy of ethylene oxide sterilisation. Rather than support the use of ethylene oxide sterilisation Dr. Kelsey apparently would prefer to popularise sterilisation by low temperature steam combined with formaldehyde vapour for plastics materials and other substances likely to be damaged by dry heat or conventional autoclaving.

7. It is not irrelevant to mention that though the acrylic cement powder used with this socket is sterilised in the U.S.A. by ethylene oxide, a working party in the Deportment of Health and Social Security has rejected this method in favour of gamma radiation as an instruction to British manufacturers in the future.

You will therefore see that this problem of sterilisation by ethylene oxide is by no means as simple as the very sweeping statement of Dr. Weymes might suggest. In view of the fact that I have personally over ten years experience on nearly six thousand patients using this plastic in the hip joint sterilised by gamma radiation, I would be very unhappy to have to change this for something which might possibly expose the tissues to a new and unknown source of toxicity.

Yours sincerely,

John Charnley, C.B.E., D.Sc., F.R.C.S.
Consultant Orthopaedic Surgeon and
Director of Hip Surgery

Copy to: Dr. C. Weymes,
Sterilisation Research Unit,
Scottish Home and Health Department.

## 4.12 SUMMARY

In summary, UHMWPE was subjected to wear testing and biocompatibility testing prior to its first use in patients, during November 1962. After its clinical introduction, Charnley continued his research on the wear properties of polymers, but he never found a material better suited for joint replacements than UHMWPE. While it is clear that he evaluated different polymers, including Hi-Fax UHMWPE material from the United States, his cups were always fabricated from the RCH-1000 compression molded UHMWPE material produced in Germany by Ruhrchemie (now Ticona).

Understanding the origins of UHMWPE in orthopedics is didactic in several respects. The first lesson is that the clinical introduction of UHMWPE was made possible by the close collaboration of surgical and engineering talent. That is not to suggest that the collaboration was necessarily harmonious because the available evidence would suggest that Charnley and Craven were both stubborn and opinionated. However, there was clearly a respect for the talents of the surgeon for the engineer, and vice versa. Mutual respect, collaboration, as well as genuine talent, were the necessary foundations to achieve the breakthrough for joint arthroplasty which was desperately needed. The remarkable clinical performance of UHMWPE following its introduction is recounted in the following chapter.

## 4.13 ACKNOWLEDGMENTS

This chapter would not have been possible without the assistance of many supportive colleagues in England. Professor John Fisher (University of Leeds), Dr. Mike Wroblewski (Wrightington Hospital), and Harry Craven (Lancashire) were all instrumental in providing details regarding the undocumented history of Wrightington during the time of Charnley. Drs. Ken Brummit and

Graham Isaac (DePuy International) and Alan Humphries (Thackray Museum) were all extremely helpful in tracing the historical background of Thackray during the early development period of the artificial hip. The author is particularly thankful to Dr. Brummit for providing a copy of Charnley's correspondence.

## REFERENCES

1. Kurtz SM, Muratoglu OK, Evans M, Edidin AA. Advances in the processing, sterilization, and crosslinking of ultra- high molecular weight polyethylene for total joint arthroplasty. *Biomaterials* 1999;**20**:1659–88.
2. Shikata T, Oonishi H, Hashimato Y, et al. Wear resistance of irradiated UHMW polyethylenes to $Al_2O_3$ ceramics in total hip prostheses. *Trans of the 3rd Annu Meet of the Soc for Biomater* 1977:118.
3. Oonishi H. Long term clinical results of THR. Clinical results of THR of an alumina head with a cross-linked UHMWPE cup. *Orthop Surg Traumatol* 1995;**38**:1255–64.
4. Grobbelaar CJ, Du Plessis TA, Marais F. The radiation improvement of polyethylene prostheses: A preliminary study. *J Bone Joint Surg* 1978;**60-B**:370–4.
5. Atkinson JR, Cicek RZ. Silane cross-linked polyethylene for prosthetic applications. Certain physical and mechanical properties related to the nature of the material. *Biomaterials* 1983;**4**:267–75.
6. Atkinson JR, Cicek RZ. Silane crosslinked polyethylene for prosthetic applications. Part II. Creep and wear behavior and a preliminary moulding test. *Biomaterials* 1984;**5**:326–35.
7. Wroblewski BM, Siney PD, Dowson D, Collins SN. Prospective clinical and joint simulator studies of a new total hip arthroplasty using alumina ceramic heads and cross-linked polyethylene cups. *J Bone & Joint Surg* 1996;**78B**:280–5.
8. Champion AR, Li S, Saum K, Howard E, Simmons W. The effect of crystallinity on the physical properties of UHMWPE. *Trans of the 40th Orthop Res Soc* 1994;**19**:585.
9. Li S, Burstein AH. Ultra-high molecular weight polyethylene. The material and its use in total joint implants. *J Bone Joint Surg Am* 1994;**76**:1080–90.
10. Charnley J. Arthroplasty of the hip: a new operation. *Lancet* 1961;**I**:1129–32.
11. Waugh W. *John charnley: the man and the hip*. London: Springer-Verlag; 1990.
12. Waugh W. The growth of an idea: 1951–1961. In: *John Charnley: the man and the hip*. London: Springer-Verlag; 1990.
13. Charnley J. The lubrication of animal joints. *Inst Mech Eng: symp Biomech* 1959;**17**:12–22.
14. Charnley J. Low Friction Principle. In: *Low friction arthroplasty of the hip: theory and practice*. Berlin: Springer-Verlag; 1979.
15. Waugh W. Bone cement: grout not glue. In: *John Charnley: the man and the hip*. London: Springer-Verlag; 1990.
16. Charnley J, Kamangar A, Longfield MD. The optimum size of prosthetic heads in relation to the wear of plastic sockets in total replacement of the hip. *Med Biol Eng* 1969;**7**:31–9.
17. Waugh W. The Charnley-Thackray Relationship 1947–1982. In: *John Charnley: the man and the hip*. London: Springer-Verlag; 1990.
18. Craven H. Personal Communication. July, 2002.
19. Charnley J. Tissue reaction to the polytetrafluoroethylene. *Lancet* 1963;**II**:1379.
20. Waugh W. The plan fulfilled 1959–1969. In: *John Charnley: the man and the hip*. London: Springer-Verlag; 1990.
21. Isaac GH, Dowson D, Wroblewski BM. An investigation into the origins of time-dependent variation in penetration rates with Charnley acetabular cups—wear, creep or degradation? *Proc Inst Mech Eng [H]* 1996;**210**:209–16.

# Chapter 5

# The Clinical Performance of UHMWPE in Hip Replacements

Steven M. Kurtz, PhD



## 5.1 INTRODUCTION

Although UHMWPE has a clinical track record spanning the past 45 years, the first decade of clinical implementation occurred at Wrightington under the exclusive direction of John Charnley. Following the PTFE debacle, Charnley withheld from publishing his experience with UHMWPE low friction arthroplasties until the 1970s. Thus, in the 1960s, the proliferation of total joint replacements beyond Wrightington was strictly controlled by Charnley [1]. However, copies of Charnley's designs began to appear in the United States during the mid-1970s, and after the details of the operation became widely known, total hip replacement underwent a period of steady growth, which lasted through the 1980s. Starting in the 1990s, joint arthroplasty entered a period of steady, exponential growth, which has been projected to last throughout the next 2 decades, barring unforeseen changes in the availability of funds, the emergence of new disruptive technologies, and/or the size of the orthopedic surgeon workforce [2]. The American Academy of Orthopedic Surgeons (AAOS) has proclaimed the first decade of our century (2002 to 2011) as "The United States Bone and Joint Decade."

The prevailing optimism regarding joint replacement is founded not only on the historically successful performance of these surgical procedures in the United States but also upon the demographics of our gentrifying population; increases in the prevalence of obesity; and the dramatic rise in direct-to-consumer advertising of joint replacements via mass media and the Internet [2–4]. The chart in Figure 5.1 shows the age distribution of patients who received a primary hip replacement in 2006 [5]. As a result of this combination of factors, future demand for primary hip replacement procedures is projected to increase by 171% by 2030 [2, 5]. The majority of these future total hip arthroplasty (THA) patients are anticipated be less than 65 years in age [5].

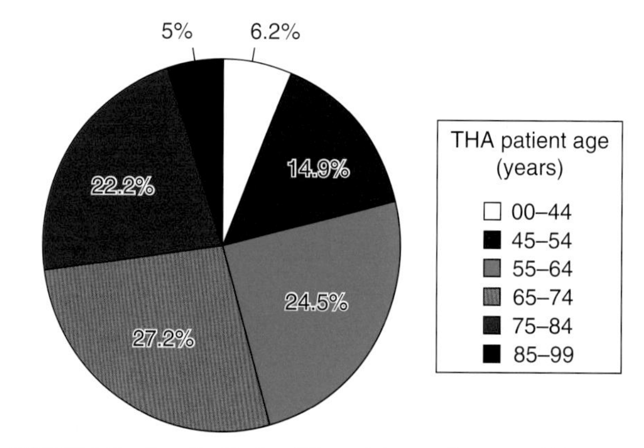


**FIGURE 5.1** Age distribution (%) receiving a primary total hip replacement in 2006. Note that 46% of hip arthroplasty patients in the United States were less than 65 years of age. (Source: Nationwide Inpatient Sample). Data analysis courtesy of Edmund Lau, Exponent, Inc.

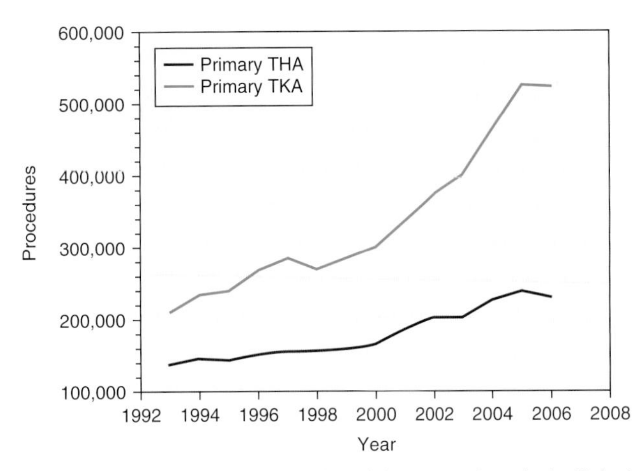


**FIGURE 5.2** Growth of primary hip and knee procedures in the United States between 1993 and 2006 (Source: Nationwide Inpatient Sample). Data analysis courtesy of Edmund Lau, Exponent, Inc.

The graph in Figure 5.2 charts the historical growth of primary total hip and knee procedures in the United States between 1993 and 2006. The data plotted in Figure 5.2 indicate that fewer primary total joint procedures are performed for the hip and opposed to the knee. This trend is not expected to change substantially by 2030 [2, 5]. Although THA is performed less frequently than TKA, the clinical performance of the hip has been studied to a much greater extent than that of the knee. Consequently, the focus of the current chapter is on the clinical performance of THA.

Due to the role that wear has historically played in promoting osteolysis, aseptic loosening, and ultimately implant revision, this chapter reviews the historical development of concepts related to quantifying clinical wear performance of gamma sterilized UHMWPE in hip replacements. The foundations of our current understanding of this topic were established in the publications of Charnley, published in the 1970s. Additional background material about the biological, design, and implant factors related to implant wear may be found in previous reviews [6, 7]. Several excellent book chapters and review articles also have been published that describe the range of factors contributing to wear of UHMWPE materials in the laboratory [8–11]. Our recent thinking on the subject of wear in joint replacements can be found in a special issue of the *Journal of the AAOS* in July 2008 dedicated to implant wear (http://www.jaaos.org/content/vol16/suppl_1/index.dtl). Readers who are interested in the performance of contemporary, highly crosslinked UHMWPE and hard-on-hard bearings in the hip may wish to skip ahead to Chapter 6.

## 5.2 JOINT REPLACEMENTS DO NOT LAST FOREVER

Joint replacements are highly successful, especially during their first decade of use. Nevertheless, joint replacements may still be revised at a rate of 1% per year or less during the first 10 years of implantation for a variety of implant, patient, and surgeon-related factors [12–16]. Scandinavian outcome studies for THA suggest that the long-term survivorship decreases markedly after 10 years of implantation, especially for patients less than 55 years old [13].

The revision rate for total joint replacements remains a significant burden to the health care economies of Western countries and varies from 10–20% depending upon the nationality [13]. In the United States, for example, analysis of epidemiological data from the National Hospital Discharge Survey (NHDS) suggests that revision procedures have occurred at a ratio of about 15.2% to 20.5% relative to the number of primary surgeries between 1990 and 2002 [17]. According to data for 11,543 revisions in Sweden from 1979 to 1998, 75.7% of hip revision surgeries occurred due to aseptic loosening [12]. However, analysis of 2005 data from the United States suggests that dislocation, rather than aseptic loosening, is currently the most frequently reported reason for revision in total hip arthroplasty [18].

Wear of conventional UHMWPE has historically been recognized as the primary culprit responsible for inflammatory bone loss and late revision of hip replacements. Researchers have estimated that for each day of patient activity, around 100 million microscopic UHMWPE wear particles are released into the tissues surrounding the hip joint [19]. This particulate wear debris can initiate a cascade of adverse tissue response leading to osteolysis (bone death) and ultimately aseptic loosening of the components [20–23]. The radiograph in Figure 5.3 (provided courtesy of Av Edidin, PhD, Drexel University) shows an example of an osteolytic lesion in the pelvis located superior to the historical, gamma-sterilized acetabular component.

Based on a review of the literature, Dumbleton et al. [24] suggest that radiographic wear rates of less than 0.05 mm/y are below an "osteolysis threshold" below which patients are not expected to be at risk of developing osteolysis. Osteolysis, in turn, may be associated with the need for revision, depending upon the location (i.e., in the pelvis or femur) and rate of progression. As noted by Hozack et al. [25]:

> Polyethylene wear alone (stage I) is an indication of impending failure, and when symptoms develop (stage IIA), revision should be undertaken. The development of radiographic lysis is a critical event, and as soon as osteolysis develops (stage IIB or III), revision should be undertaken immediately. From the perspective of the revision surgeon, there is great value in the early intervention for polyethylene wear and pelvic osteolysis.

Therefore, understanding the natural history of UHMWPE wear in a clinical setting, starting with the pioneering work of Charnley, is an important first step to improving the longevity of THA.

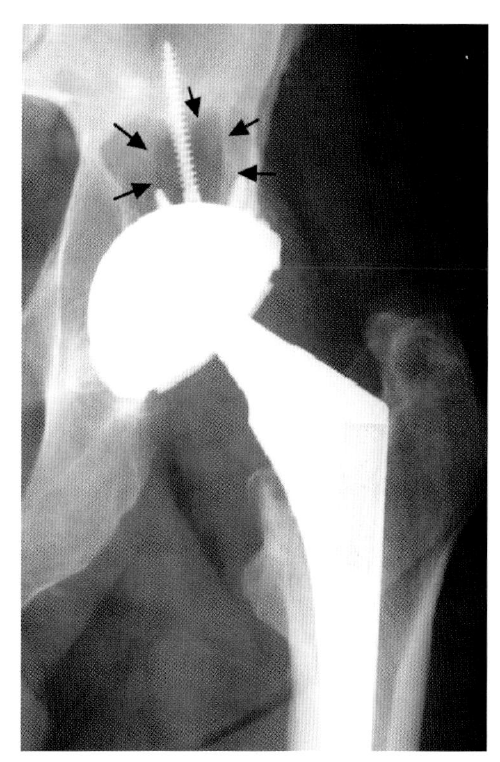

FIGURE 5.3 Example of an osteolytic lesion in the pelvis, located superior to the metal-backed acetabular component.

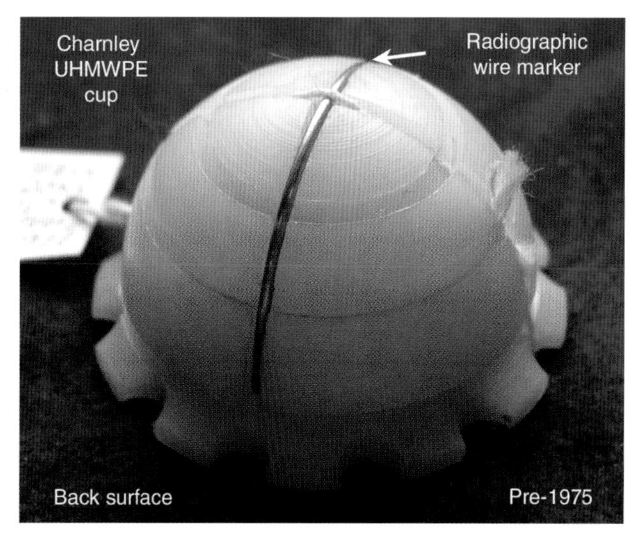


FIGURE 5.4 Produced between 1968 and 1975, the configuration of the wire marker was similar to that used in Charnley and Cupic's study.

## 5.3 RANGE OF CLINICAL WEAR PERFORMANCE IN CEMENTED ACETABULAR COMPONENTS

Charnley and coworkers first developed radiographic techniques for evaluating the wear rate of UHMWPE acetabular components in patients. In 1973, Charnley and Cupic reported on the long-term wear performance the first cohort of patients to receive a UHMWPE component between November 1962 and December 1963 [26]. During this time period, 170 patients received a cemented low friction arthroplasty with an UHMWPE component; a total of 185 acetabular cups were implanted. Because of the elderly population originally implanted with the components, many had died or were too infirm to travel to the clinic for follow-up examinations (over 2/3 of the patients were over 60 years of age at the time of implantation). Thus, only 106 out of the original 185 UHMWPE cups could still be evaluated after 9 or 10 years of implantation. The complications for this series included a 4–6% rate of infection, 1–2% rate of mechanical loosening, and a 2% incidence of late dislocation.

Early in 1963, Charnley introduced the use of a semicircular wire marker on the back of the acetabular component to assist with the measurement of wear using radiographs. Figure 5.4, a photograph taken of an unused Charnley cup from the collection at the Thackray Museum at Leeds, shows the wire marker inserted into the cement groove. Although the implant shown in Figure 5.4 was produced between 1968 and 1975, the configuration of the wire marker was similar to that used in Charnley and Cupic's study.

Charnley and Cupic [26] calculated wear from the radiographs by measuring the narrowest distance between the femoral head and the back of the liner in the weight-bearing and non-weight-bearing regions of the cup. The difference in thickness in the weight-bearing and non-weight-bearing regions was then divided by two to estimate the radial penetration of the head, which was attributed to wear of the cup. These measurements only had a precision of 0.5 mm, so that wear of less than 0.5 mm could not be detected using this technique. In contrast, contemporary computer-based radiographic wear analysis techniques have a precision of 0.1 mm or less [27], as we shall see in Section 5.8.

A total of 72 components out of Charnley's 9- to 10-year series were measured for wear. The other hips in the series either did not have a wire marker or did not have radiographs. Charnley reported the average wear rate was 0.15 mm/y, but analysis of his data (presented in Table 1 of his paper) indicates that this average excluded patients with undetectable (0 mm) wear. The distribution of wear rates in Charnley's 72 patients is shown as a histogram in Figure 5.5, based on the data from his Table 1, and includes patients with undetectable (0 mm) wear.

The average (mean) wear, including those patients with undetectable wear, was 1.18 mm (range: 0 to 4 mm). A standard normal (Gaussian) distribution is also superimposed on these data, showing that the distribution is skewed to the left, with a greater number of patients exhibiting wear below the mean than above it. Also, there is a small number of patients exhibiting wear much greater than the mean. This skewed distribution in UHMWPE wear performance in a patient population was well recognized by

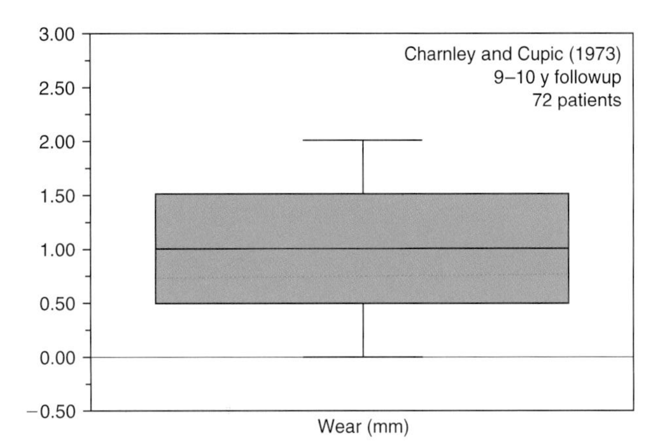


**FIGURE 5.5** Histogram showing the distribution of wear measured radiographically in Charnley and Cupic's study [26].

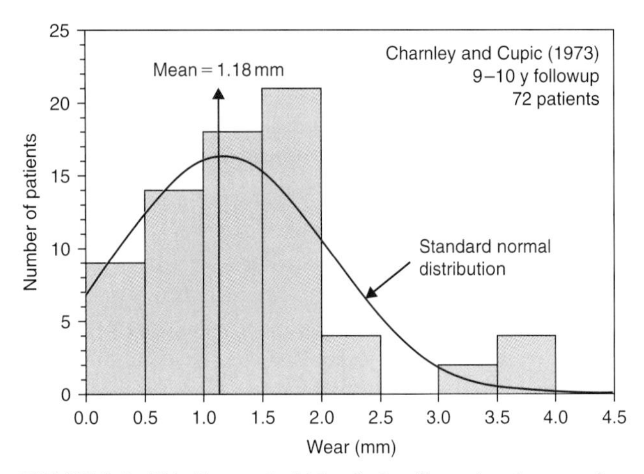


**FIGURE 5.6** This "box and whiskers" plot, illustrating the same data shown in Figure 5.5, is helpful to summarize the distribution of wear results.

Charnley [26, 28] and has been observed in subsequent studies using contemporary modular hip designs [7]. For this reason, it is sometimes more useful to characterize clinical wear performance using nonparametric statistical methods, which make no assumptions about the underlying distribution of the data. A "box and whiskers" plot (or more simply, a box plot), shown in Figure 5.6 for the same wear data as in the histogram derived from Charnley and Cupic's study [26], is helpful to summarize the distribution of the wear data.

For readers who may not be familiar with a box plot, it can be interpreted as follows: The top and bottom faces of the box correspond to the 25% and 75% percentiles of the data (e.g., 0.5 mm and 1.5 mm of wear, respectively). The median (50% percentile) is the horizontal line inside the box (at 1.0 mm of wear). Thus, the box itself represents the interquartile range, or the central tendency of the distribution corresponding to the "middle half" of the population. The median is the center of the distribution, with half the population lying to either side. Note that the median for the wear data (1 mm) is a bit lower than the arithmetic mean or average (1.18 mm). Finally, the "whiskers" of the plot correspond to the 10% and 90% of the distribution (e.g., 0.0 mm and 2.0 mm of wear, respectively), providing an indication of breadth of the central 80% of the population. Some statisticians like to plot the data outside the 10% and 90%iles as individual data points or outliers, but I have not adopted that convention here.

## 5.4 WEAR VERSUS WEAR RATE OF HIP REPLACEMENTS

The absolute amount of wear observed in a group of patients is going to depend on the time interval between radiographic examinations. Consequently, clinical researchers starting with Charnley have been interested in the linear wear rate (LWR), or the change in apparent femoral head penetration (P) over time (t):

$$LWR = \frac{\Delta P}{\Delta t} \tag{1}$$

Charnley also suggested that the volumetric wear rate, as opposed to the linear wear rate, may also be a clinically relevant metric for wear because the biological stimulus may be related to the volume of wear debris [29]. If the femoral head penetrates the cup following a linear trajectory (an assumption that was verified by Charnley's clinical experience with PTFE), the wear volume will be approximated as a cylinder having the projected circular area ($A$) of the femoral head and a height equal to the depth of penetration. Under this assumption, the volumetric wear rate (VWR) can be calculated as follows:

$$VWR = A \times LWR = \frac{\pi}{4} D^2 \times \frac{\Delta P}{\Delta t} \tag{2}$$

In a study published in 1975, Charnley and Halley [30] calculated the wear rate for the 72 patients with 9–10 years of follow-up by analyzing the serial radiographs. Charnley and Halley found that, after an initial bedding-in period during the first several years, the rate of wear progressed linearly with time, as shown in Figure 5.7 (adapted from Charnley and Halley's paper).

Figure 5.7 plots the *average* radiographic wear (again, for patients with measurable wear) for 72 patients, so one must understand that some variability would be superimposed on this picture if all of the individual data points had been plotted. Overall, the average wear for the entire series was observed to decrease over time. In the first 5 years, the wear rate was 0.18 mm/y for all 72 patients, and during the 5- to 10-year period, the average wear dropped to 0.10 mm/y.

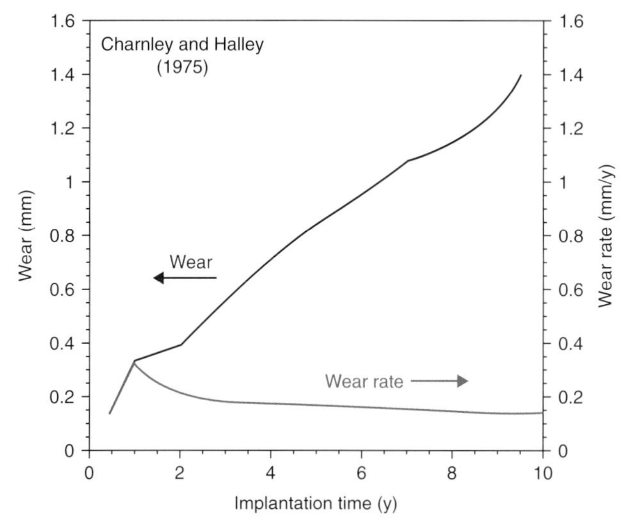


**FIGURE 5.7** After an initial bedding-in period during the first several years, the rate of wear progressed quasi-linearly with time. The overall wear rate, however, appeared to slowly decrease over time.

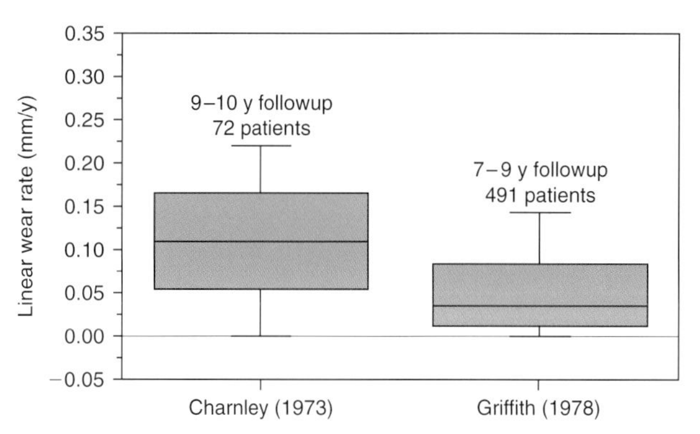


**FIGURE 5.8** Distribution of THA wear rates observed in two studies from Wrightington Hospital.

## 5.5 COMPARING WEAR RATES BETWEEN DIFFERENT CLINICAL STUDIES

If two studies have different follow-up periods, it should be possible to compare the wear rates. In practice, however, differences between patient groups, surgeon groups, and implant systems greatly complicate the task of reconciling differences in wear behavior observed in clinical studies. For example, consider studies published by Charnley and coworkers from Wrightington that describe the wear behavior of UHMWPE in two cohorts. The first cohort, which we have already discussed, consisted of 72 arthroplasties that were implanted between 1962 and 1963 and followed for 9–10 years [26]. The second cohort, described by Griffith et al., consisted of 493 arthroplasties implanted between 1967 and 1968 and followed for 7–9 years (mean, 8.3y) [28].

Both cohorts of patients were implanted at Wrightington using the same cemented design using an UHMWPE cup fabricated from RCH-1000 (Ruhrchemie, Germany). Presumably, these two studies should be the easiest to compare because they were directed by the same surgeons and drew patients from the same elderly populations. However, there are significant differences in the resulting wear rates for both studies: The average wear rate for patients in Charnley's study was 0.15 mm/y, whereas the wear rate in Griffith's study was 0.07 mm/y. The average wear rates reported by both studies did not include patients exhibiting undetectable wear, which is an important omission if we consider that 110 of Griffith's 491 hips (i.e., 22%) were measured with 0 mm of wear after 8.3 years of implantation.

When we include the patients with zero wear, the distribution of wear rates in both studies is summarized as a box plot in Figure 5.8, taking into account the calculation as described in Equation (1).

**TABLE 5.1 Linear Wear Rate (mm/y) Based on Radiographs for Charnley Components Implanted between 1962 and 1968**

| Percentiles | Wear rate (mm/y) Charnley (1973) | Wear rate (mm/y) Griffith (1978) |
|---|---|---|
| 10 | 0.00 | 0.00 |
| 25 | 0.06 | 0.01 |
| 50 | 0.11 | 0.04 |
| 75 | 0.17 | 0.09 |
| 90 | 0.22 | 0.15 |

The data indicate that the distribution of wear rates is more skewed by low-wearing patients in Griffith's (1978) study than in Charnley's (1973) study. The distribution of linear wear data for the two studies, in terms of percentiles, is tabulated in Table 5.1.

Thus, over 60% of the patients in both studies exhibited wear lower than the average values of 0.15 and 0.07 mm/y reported by Charnley et al. and Griffith et al., respectively. Also, in both studies, 10% of the patients (i.e., in the 90% ile) exhibited wear at 2–3 times the median rate!

Of particular interest to Griffith, Charnley and coworkers were able to explain the variability in clinical wear performance, not only within the context of a single study, but also between the two different studies. For example, Griffith et al. examined whether age, gender, postoperative activity level, and the involvement of other joints with hip disease influenced the magnitude of wear. The researchers observed that male patients younger than 60 years of age at the time of implantation appeared to be associated with the incidence of heavy wear (defined as greater than 1.4 mm). Griffith et al. also tried to explain the difference between their results and Charnley and Cupic's previous study. The researchers suggest that an improvement in surface finish of the femoral head may have contributed to the apparent improvement clinical wear performance. They also suggest that the UHMWPE used in Charnley and Cupic's earlier study may have been "of inferior quality."

Over the years, the difference in wear rates observed between the two studies has continued to intrigue researchers [31]. Griffith's paper strongly suggests that differences in manufacturing of the implants, as opposed to the patients, is most likely responsible for the difference in wear rates as compared with Charnley and Cupic's earlier study. Unfortunately, it is no longer possible to obtain written documentation describing the manufacture of components in the early 1960s at Wrightington. Many of the details about implant manufacturing during that time period have been provided by Harry Craven, who served as Charnley's engineer and technician between 1958 and 1966 when he was 30–36 years old [32].

On one hand, commercial production of the Charnley prosthesis by Thackray, Ltd. is reported to have started between 1968 and 1969. The production dates have been confirmed from several sources [31, 33], including a commercial release pamphlet from 1969 produced by Thackray that is available at the Thackray Museum in Leeds (see Figure 4.11 in Chapter 4).

Thackray's documentation clearly indicates that the UHMWPE cups were gamma irradiated by a dose of 2.5 Mrad. It has been reported that no sterilization records from that time survive today [33].

On the other hand, documents obtained at Wrightington suggest that 1968 to 1969 may only have applied to distribution of Charnley implants to the outside world. According to an internal Wrightington publication [34]:

> The total hip prosthesis which I call "low friction arthroplasty," in which a plastic socket is an integral part of the design, is being released (November 1966) under pressure from surgical colleagues to a restricted number of surgeons (in particular those who have worked with me over six months) prior to considering a general release in January 1968.
>
> John Charnley, November 1966

The release described in Charnley's internal report is at least consistent with Griffith's statement (on p. 46 of his paper) that Wrightington shifted to commercial production of femoral heads in 1966 [28]. Craven, who was himself responsible for manufacturing some of the implants and instruments for Charnley between 1958 and 1966, recalls that Thackray was producing components for Wrightington as early as 1963 [32]. Specifically, Craven recalls fabricating a specialized machine for stamping out the scalloping for the rim in the standard cup and selling the machine to Thackray between 1963 and 1964 for the sum of 360 £. Taken together, the available evidence suggests that Thackray was producing implants for Wrightington prior to 1969. Because Thackray only sterilized UHMWPE using gamma irradiation, the inference is that prior to 1969, perhaps as early as 1963, UHMWPE components were used at Wrightington that had been gamma irradiated. Therefore, there is convincing evidence to suggest that the components in Griffith's study, which were implanted at Wrightington between 1967 and 1968, were also gamma irradiated.

The crosslinking produced by a single dose of gamma radiation (even in air) has the beneficial result of increasing the resistance to adhesive/abrasive wear and could explain, at least in part, the lower wear rates in Griffith's study. According to wear testing by Wang et al. using a contemporary multidirectional hip simulator [35], changing from 0 to 2.5 Mrad of irradiation (in air) drops the wear rate from 140 to 90 mm3/million cycles (using 32 mm diameter heads), corresponding to a reduction of about 36%.

A substantial increase in the molecular weight of the UHMWPE could also explain the improved wear resistance observed in Griffith's study. The wear resistance of UHMWPE is imparted by its molecular weight, which ranges today, on average, between 2 and 6 million. Griffith et al. suggest that the quality of the UHMWPE may have been inferior in 1961 to 1962 [28], even though the source of the material remained the same (Ruhrchemie AG, currently known as Ticona, distributed locally in Lancashire by a company known as High Density Plastics [32]). In wear testing of UHMWPE and HDPE (molecular weight 200,000g/mol) using a contemporary multidirectional hip simulator [36], differences in the molecular weight of the polyethylene can affect the wear rate by a factor of 4.

The surface finish of the heads, suggested by Griffith [28], is a possible but unlikely reason for the change in clinical wear rates. A wide range in surface roughness values were observed in Isaac's analysis of retrieved Charnley cups [31, 37], with no association to the clinical wear rate. Similarly, the retrieval work of Hall et al. also showed no significant relationship between surface roughness of the femoral head and clinical wear rate in a group of retrieved metal-backed acetabular components that were implanted without cement [38]. Thus, in light of recently published studies, the roughness of the femoral head seems an unlikely explanation for the change in wear behavior for the entire cohort of Griffith's patients.

In summary, the following three factors, either alone or in combination, could explain the differences in wear rates between Charnley's and Griffith's studies:

1. Radiation-induced crosslinking;
2. Molecular weight of the UHMWPE; and
3. Surface finish of the femoral head.

Our inability to conclusively identify the explanation for the difference in wear rates between these two studies underscores the difficulty in comparing two retrospective series of patients, even when performed at the same institution and by the same group of investigators.

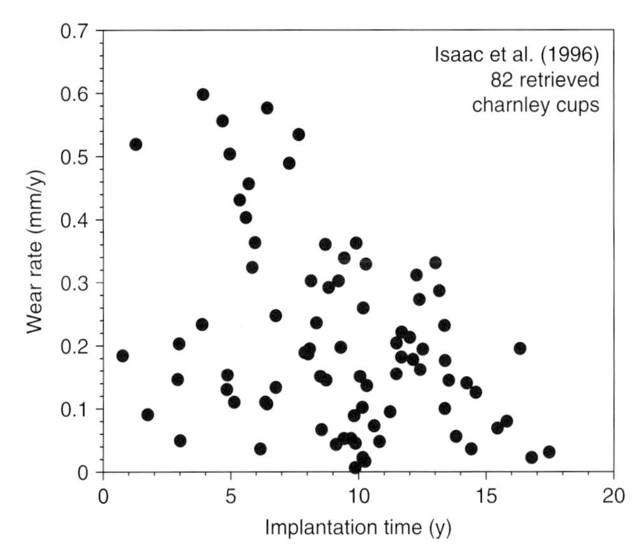


**FIGURE 5.9** Wear rate versus implantation time. Adapted from [31].

## 5.6 COMPARISON OF WEAR RATES IN CLINICAL AND RETRIEVAL STUDIES

Two additional clinical studies of interest related to the Charnley prosthesis were performed by Isaac et al. [31, 37] on a group of 100 explanted UHMWPE cups that were retrieved at revision surgery after an average of 9 years of implantation, with an implantation time of up to 17.5y. Unlike previous studies conducted by Charnley and his research colleagues, who focused on the wear behavior of UHMWPE functioning arthroplasties, the retrieved components represented a group of patients in which the surgery had failed. Thus, the study of retrieved implants provides insight into factors leading to the revision (i.e., clinical failure) of total joint arthroplasty. In the study of retrieved Charnley components, wear of the cups was measured radiographically, as well as directly from the actual explants, in 86 out of 100 cases (in 14 components the wear was too severe to be accurately measured). The average wear rate for this series of explanted cups was 0.2 mm/y, but again a wide variation in rates was observed. The wear rates for individual cups are plotted in Figure 5.9 as a function of implantation time (redrawn from Isaac et al. [31]).

## 5.7 CURRENT METHODS FOR MEASURING CLINICAL WEAR IN THA

Wear can be measured from radiographs using manual or computer-assisted techniques. The Charnley method, developed in the 1970s for measuring radiographic wear, has been reviewed in Section 5.3. Livermore and colleagues [39] later improved on the Charnley wear measurement technique with the use of circular templates. Despite the widespread use of Charnley and Livermore techniques throughout Europe and the United States, these manual methods have poor intra- and interobserver repeatability and are unsuitable for measuring wear of less than 0.3 to 0.5 mm [24, 27], as would presumably be the case for newly developed highly crosslinked UHMWPE materials for hip replacement.

For detecting low levels of initial wear in UHMWPE hip components, researchers have developed both 2-D and 3-D computer assisted radiographic wear measurement methods, as summarized in a recent review [24]. In addition, radiostereometric analysis (RSA) in conjunction with digital radiography promises to have improved accuracy for measuring initial wear in UHMWPE [40]. However, a number of practical limitations preclude the use of RSA and 3-D computer-assisted techniques in a generalized clinical setting. For example, the use of RSA requires the addition of tantalum markers to the acetabular and femoral components and is limited to only a few, specialized radiographic suites worldwide. Another advanced 3-D wear assessment technique, developed by Devane et al. [41], requires that a manufacturer provide a CAD model of an implant design to reconstruct wear from multiple radiographic views.

In contrast, a 2-D computer-assisted technique utilizing edge detection and computer vision, developed by Martell and Berdia, improves intra- and interobserver repeatability tenfold over manual techniques [27]. Figure 5.10 illustrates the Martell technique for computer-assisted measurement of radiographic wear.

The Martell technique has been adapted to allow 3-D wear assessment. Comparing 2-D and 3-D analysis on a large clinical population, Martell has shown that 87–90% of wear is detected using 2-D on AP pelvis radiographs alone, with 2-D and 3-D wear values being highly correlated ($r^2 = 0.993$). Additionally, due to poor lateral pelvic radiographic quality, repeatability of the 3-D analysis is up to four times worse than the 2-D technique [40, 42]. Two-dimensional analysis is therefore the preferred method for wear detection in THA using the Martell technique. Using a validated dynamic phantom total hip wear model, the accuracy of the Martell 2-D analysis has been shown to be ±0.073 mm, while the precision is ±0.023 mm [43]. Finally, the Martell technique is implant independent, requiring only digitized A-P radiographs and knowledge of the head size.

The technology for interpreting low wear rates in modern, highly crosslinked UHMWPE bearings is complex. Further details about radiographic wear measurement and RSA can be found in Chapter 28.

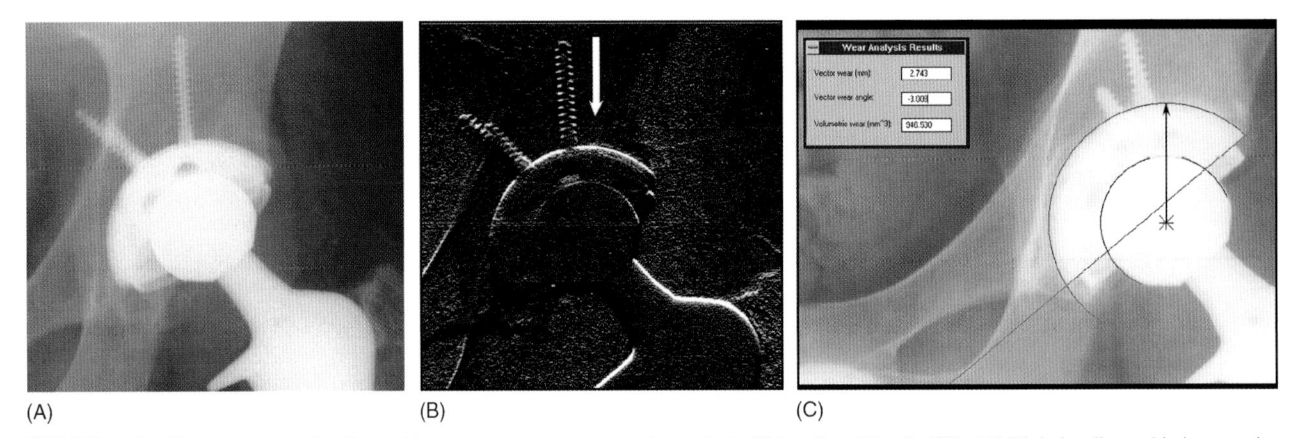


**FIGURE 5.10** Computer-assisted radiographic wear measurement using the method of Martell and Berdia [27]. (A) Digital radiographic image prior to wear analysis. (B) The results of processing the digital image using edge detectors in the direction of the white arrow. Each white dot represents a potential point on the edge of the prosthesis. (C) The computer has chosen the best circle fits for the acetabular shell and femoral head using the edge points created in the previous image. The change in position of the femoral head with respect to the acetabular center over time is reported as wear and is displayed on the screen. Images provided courtesy of J. Martell, MD (University of Chicago, Chicago, Illinois, USA).

**TABLE 5.2** Linear Wear Rate (mm/y) Based on Radiographs for the Modular HG I Acetabular Component as a Function of Implantation Time

| | Implanted 1 – 6y N = 35 patients | | Implanted 7 – 10y N = 43 patients | | Implanted 11 + y N = 29 patients | |
|---|---|---|---|---|---|---|
| Percentiles | 2-D | 3-D | 2-D | 3-D | 2-D | 3-D |
| 10 | −0.070 | 0.080 | 0.040 | 0.058 | 0.054 | 0.064 |
| 25 | 0.080 | 0.122 | 0.060 | 0.070 | 0.077 | 0.090 |
| 50 | 0.150 | 0.200 | 0.090 | 0.110 | 0.120 | 0.130 |
| 75 | 0.228 | 0.287 | 0.170 | 0.210 | 0.182 | 0.185 |
| 90 | 0.340 | 0.480 | 0.272 | 0.272 | 0.218 | 0.232 |

*Previously published data [42] provided courtesy of J. Martell, MD (University of Chicago, Chicago, Illinois, USA). See also Figure 5.11.*

**TABLE 5.3** Volumetric Wear Rate ($mm^3$/y) Based on Radiographs for the Modular HG I Acetabular Component as a Function of Implantation Time

| | Implanted 1 – 6y N = 35 patients | | Implanted 7 – 10y N = 43 patients | | Implanted 11 + y N = 29 patients | |
|---|---|---|---|---|---|---|
| Percentiles | 2-D | 3-D | 2-D | 3-D | 2-D | 3-D |
| 10 | 2.2 | 26.6 | 17.4 | 20.9 | 28.9 | 31.4 |
| 25 | 44.3 | 48.9 | 28.0 | 31.1 | 37.3 | 41.6 |
| 50 | 75.0 | 88.3 | 42.2 | 51.7 | 60.2 | 56.1 |
| 75 | 127.5 | 149.3 | 90.1 | 91.8 | 82.0 | 80.9 |
| 90 | 182.8 | 226.4 | 135.2 | 132.3 | 113.7 | 108.0 |

*Previously published data [42] provided courtesy of J. Martell, MD (University of Chicago, Chicago, Illinois, USA). See also Figure 5.11.*

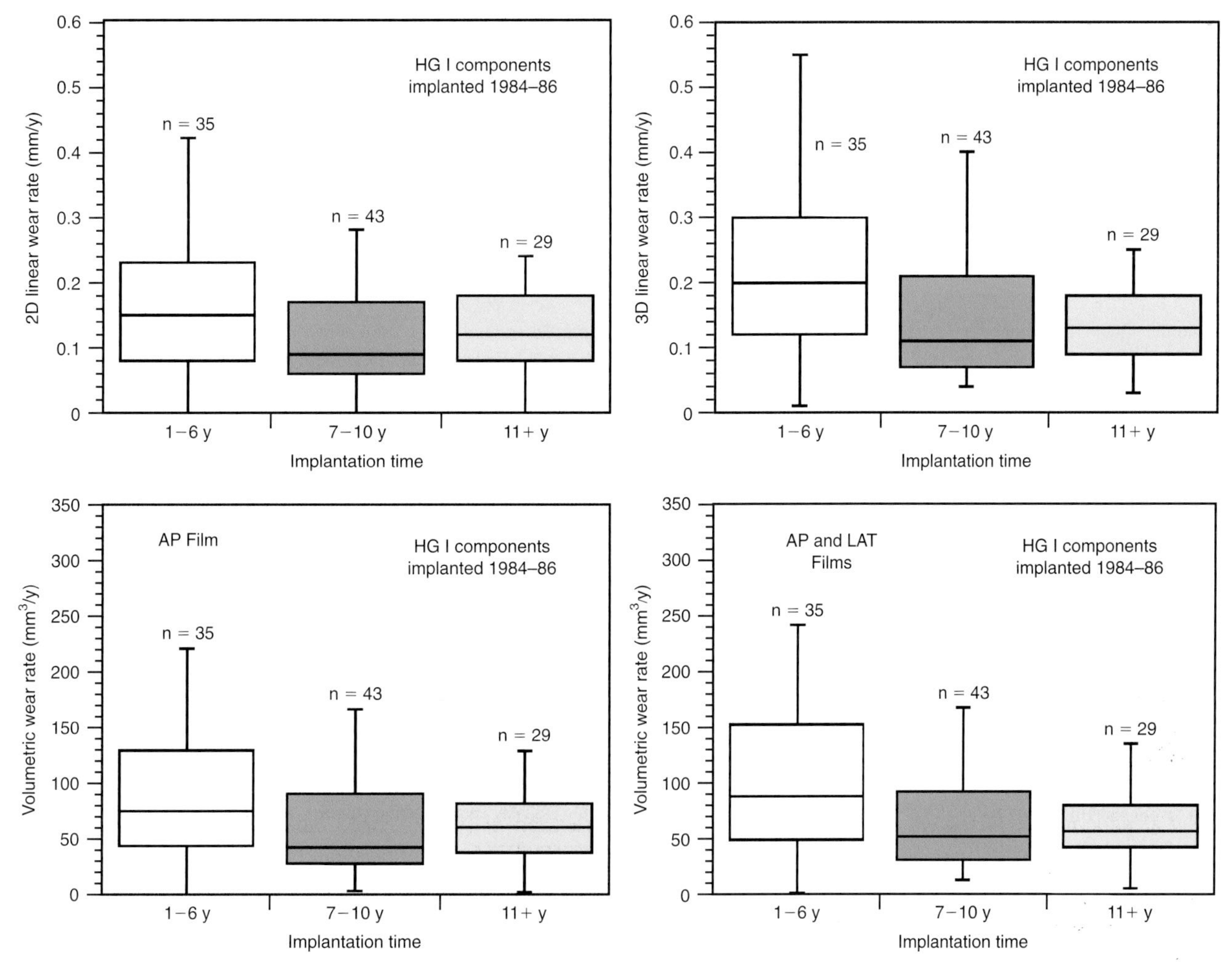


**FIGURE 5.11** Distribution of linear and volumetric wear rates estimated radiographs in a first-generation modular acetabular component design (Harris-Gallante: Zimmer: Warsaw, Indiana, USA). These graphs show the tendency for the wear rates to decrease with implantation time. Results are compared based on analysis of anterior-posterior radiographs alone (2-D), as well as based on the combined analysis of anterior-posterior and lateral radiographs (3-D). Based on data provided courtesy of J. Martell, MD (University of Chicago, Chicago, Illinois, USA).

## 5.8 RANGE OF CLINICAL WEAR PERFORMANCE IN MODULAR ACETABULAR COMPONENTS

Qualitatively, the trends observed for cemented components also appear to be applicable to modular acetabular components, evaluated using current computer-assisted wear measurement techniques (Tables 5.2 and 5.3, Figure 5.11). The "box and whisker" plots shown in Figure 5.11 compare the linear wear (penetration) of UHMWPE for a first-generation modular acetabular component design. Data previously published by Martell and colleagues [42] has compared radiographs using planar (AP) radiographs alone, as compared with a 3-D analysis that includes planar and lateral radiographs (Figure 5.11). Note that, as already indicated, the lateral radiographs capture about 90% of the total wear of the hip joint.

## 5.9 CONCLUSION

UHMWPE has a long, successful clinical track record in total hip replacements dating back to the 1960s. However, the *in vivo* wear rate of conventional UHMWPE for some THA patients falls above the threshold for osteolysis and may ultimately lead to the need for long-term revision. Although accurate computer-assisted methods have now been developed to track the progression of *in vivo* wear of UHMWPE, these techniques reveal a range of performance within the human body. Even when implant- and surgeon-related factors are held constant, patient factors (e.g., activity level, obesity, etc.) play major roles in the clinical wear rate [6,44–46]. Due to the importance of reducing the broad distribution of wear rates observed *in vivo*, major research efforts were directed to developing improved UHMWPE materials for hip replacements in the 1990s, leading to the clinical introduction of highly crosslinked UHMWPE in 1998. Although highly crosslinked UHMWPE continues to be widely used around the world for total hip arthroplasty, metal-on-metal and ceramic-on-ceramic alternatives have become increasingly popular, as detailed in the following chapter.

## 5.10 ACKNOWLEDGMENTS

The author is indebted to John Martell, MD, from the University of Chicago, and to Av Edidin, PhD, Drexel University, for their helpful comments and editorial assistance with this chapter.

## REFERENCES

1. Waugh W. *The plan fulfilled 1959–1969. John Charnley: the man and the hip*. London: Springer-Verlag; 1990. p. 113–38.
2. Kurtz S, Ong K, Lau E, Mowat F, Halpern M. Projections of primary and revision hip and knee arthroplasty in the United States from 2005 to 2030. *J Bone Joint Surg Am* 2007;**89**(4):780–5.
3. Bozic KJ, Smith AR, Hariri S, Adeoye S, Gourville J, Maloney WJ, et al. The 2007 ABJS Marshall Urist Award: the impact of direct-to-consumer advertising in orthopaedics. *Clin Orthop Relat Res* 2007;**458**:202–19.
4. Adeoye S, Bozic KJ. Direct to consumer advertising in healthcare: history, benefits, and concerns. *Clin Orthop Relat Res* 2007;**457**:96–104.
5. Kurtz S, Lau E, Ong K, Zhao K, Kelly M, Bozic K. Future young patient demand for primary and revision joint replacement: National projections from 2010 to 2030. *Trans of the 54th Orthop Res Soc* 2008;**33**:1784.
6. Schmalzried TP, Dorey FJ, McKellop H. The multifactorial nature of polyethylene wear in vivo. *J Bone Joint Surg Am* 1998;**80**(8): 1234–42. discussion 1242-1233.
7. Chapter 5: How should wear-related implant surveillance be carried out and what methods are indicated to diagnose wear-related problems. In: Wright TM, Goodman SB, editors. *Implant Wear*. Rosemont, IL: American Academy of Orthopedic Surgeons; 2001.
8. Clarke IC, Kabo JM. Wear in total hip replacement. In: Amstutz HC, editor. *Hip arthroplasty*. New York: Churchill Livingstone; 1991. p. 535–49.
9. McKellop HA. Wear assessment. In: Callaghan JJ, Rosenberg AG, Rubash HE, editors *The adult hip*. Philadelphia: Lippincott-Raven Publishers; 1998. p. 231–46.
10. Sauer WL, Anthony ME. Predicting the clinical wear performance of orthopaedic bearing surfaces. In: Jacobs JJ, Craig TL, editors *Alternative bearing surfaces in total joint replacement*. West Conshohoken: American Society for Testing and Materials; 1998. p. 1–29.
11. Wang A, Essner A, Polineni VK, Stark C, Dumbleton JH. Lubrication and wear of ultra-high molecular weight polyethylene in total joint replacements. *Tribol Int* 1998;**31**(1–3):17–33.
12. Malchau H, Herberts P, Söderman P, Odén A. Prognosis of total hip replacement: update and validation of results from the Swedish National Hip Arthroplasty Registry, 1979–1998. *67th Annual Meeting of the American Academy of Orthopaedic Surgeons, Scientific Exhibition*; 2000.
13. Malchau H, Herberts P, Söderman P, Odén A. Prognosis of total hip replacement: update of results and risk-ratio analysis for revision and re-revision from the Swedish National Hip Arthroplasty Registry, 1979–2000. *69th Annual Meeting of the American Academy of Orthopaedic Surgeons, Scientific Exhibition*; 2002.
14. Fitzpatrick R, Shortall E, Sculpher M, Murray D, Morris R, Lodge M, et al. Primary total hip replacement surgery: a systematic review of outcomes and modelling of cost-effectiveness associated with different prostheses. *Health Technol Assess* 1998;**2**(20):1–64.
15. Barrack RL. Concerns with cementless modular acetabular components. *Orthopedics* 1996;**19**(9):741–3.
16. Kurtz SM, Ong KL, Schmier J, Mowat F, Saleh K, Dybvik E, et al. Future clinical and economic impact of revision total hip and knee arthroplasty. *J Bone Joint Surg Am* 2007;**89**(Suppl. 3):144–51.
17. Kurtz SM, Mowat F, Ong K, Chan N, Lau E, Halpern M. Prevalence of primary and revision total hip and knee arthroplasty in the United States from 1990 through 2002. *J Bone Joint Surg Am* 2005;**87**(7):1487–97.
18. Bozic K, Kurtz SM, Lau E, Ong K, Vail T, Berry D. Revision total joint arthroplasty: early insights into cause of failure, type of revision from newly implemented diagnosis and procedure codes. *Trans of the 54th Orthop Res Soc* 2008;**33**:235.
19. Muratoglu OK, Kurtz SM. Alternative bearing surfaces in hip replacement. In: Sinha R, editor. *Hip replacement: current trends and controversies*. New York: Marcel Dekker; 2002. p. 1–46.
20. Willert HG, Bertram H, Buchhorn GH. Osteolysis in alloarthroplasty of the hip. The role of bone cement fragmentation. *Clin Orthop Relat Res* 1990(258):108–21.
21. Willert HG. Reactions of the articular capsule to wear products of artificial joint prostheses. *J Biomed Mater Res* 1977;**11**(2): 157–64.
22. Goldring SR, Jasty M, Roueke CM, Bringhurst FR, Harris WH. Formation of a synovial-like membrane at the bone-cement interface. Its role in bone resorption and implant loosening after total hip replacement. *Arthritis Rheum* 1986;**29**(7):836–42.
23. Jasty M, Floyd WEI, Schiller AL, Goldring SR, Harris WH. Localized osteolysis in stable, non-septic total hip replacement. *J Bone and Joint Surg* 1986;**68A**:912–19.
24. Dumbleton JH, Manley MT, Edidin AA. A literature review of the association between wear rate and osteolysis in total hip arthroplasty. *J Arthroplasty* 2002;**17**(5):649–61.
25. Hozack WJ, Mesa JJ, Carey C, Rothman RH. Relationship between polyethylene wear, pelvic osteolysis, and clinical symptomatology in patients with cementless acetabular components. A framework for decision making. *J Arthroplasty* 1996;**11**(7):769–72.
26. Charnley J, Cupic Z. The nine and ten year results of the low-friction arthroplasty of the hip. *Clin Orthop Rel Res* 1973;**95**:9–25.
27. Martell JM, Berdia S. Determination of polyethylene wear in total hip replacements with use of digital radiographs. *J Bone Joint Surg Am* 1997;**79**(11):1635–41.
28. Griffith MJ, Seidenstein MK, Williams D, Charnley J. Socket wear in Charnley low friction arthroplasty of the hip. *Clin Orthop* 1978;**137**(137):37–47.
29. Charnley J, Kamangar A, Longfield MD. The optimum size of prosthetic heads in relation to the wear of plastic sockets in total replacement of the hip. *Med Biol Eng* 1969;**7**(1):31–9.
30. Charnley J, Halley DK. Rate of wear in total hip replacement. *Clin Orthop* 1975;**112**(112):170–9.
31. Isaac GH, Dowson D, Wroblewski BM. An investigation into the origins of time-dependent variation in penetration rates with

Charnley acetabular cups—wear, creep or degradation? *Proc Inst Mech Eng [H]* 1996;**210**(3):209–16.
32. Craven H. Personal communication. Lancashire; 2002.
33. Brummit K. Personal communication. Leeds; 2002.
34. Charnley J. Total prosthetic replacement of the hip joint using a socket of high density polyethylene. *Report No.: 1*. Wigan: Center for Hip Surgery, Wrightington Hospital; 1966 November.
35. Wang A, Essner A, Polineni VK, Sun DC, Stark C, Dumbleton JH. Wear mechanisms and wear testing of ultra-high molecular weight polyethylene in total joint replacements. *Polyethylene Wear in Orthopaedic Implants Workshop, Society for Biomaterials*. New Orleans, LA; 1997 April 30–May 5. p. 4–18.
36. Edidin AA, Kurtz SM. The influence of mechanical behavior on the wear of four clinically relevant polymeric biomaterials in a hip simulator. *J Arthroplasty* 2000;**15**(3):321–31.
37. Isaac GH, Wroblewski BM, Atkinson JR, Dowson D. A tribological study of retrieved hip prostheses. *Clin Orthop* 1992;**276**(276): 115–25.
38. Hall RM, Siney P, Unsworth A, Wroblewski BM. The effect of surface topography of retrieved femoral heads on the wear of UHMWPE sockets. *Med Eng Phys* 1997;**19**(8):711–19.
39. Livermore J, Ilstrup D, Morrey B. Effect of femoral head size on wear of the polyethylene acetabular component. *J Bone Joint Surg [Am]* 1990;**72**(4):518–28.
40. Bragdon CR, Malchau H, Larson SL, Borlin N, Karrholm J, Harris WH. Validation of digital radiography, for use with radiosteriometric analysis (RSA) using a dynamic phantom wear model. *Transactions of the 48th Orthopedic Research Society*. Dallas, TX; 2002. p. 1020.
41. Devane PA, Bourne RB, Rorabeck CH, Hardie RM, Horne JG. Measurement of polyethylene wear in metal-backed acetabular cups. I. Three-dimensional technique. *Clin Orthop* 1995;**319** (319):303–16.
42. Martell J, Berkson E, Jacobs JJ. The performance of 2D vs. 3D computerized wear analysis in the Harris Galante acetabular component. *Orthop Trans* 2000;**25**:564.
43. Bragdon CR, Yuan X, Perinchief R, Malchau H, Karrholm J, Harris WH. Precision and reproducibility of radiostereometric analysis (RSA) to determine polyethylene wear in a total hip replacement model. *Trans ORS* 2001;**47**:1005.
44. Zahiri CA, Schmalzried TP, Szuszczewicz ES, Amstutz HC. Assessing activity in joint replacement patients. *J Arthroplasty* 1998;**13**(8):890–5.
45. Schmalzried TP, Shepherd EF, Dorey FJ, Jackson WO, dela Rosa M, Fa'vae F, et al. The John Charnley award. Wear is a function of use, not time. *Clin Orthop* 2000;**381**(381):36–46.
46. McClung CD, Zahiri CA, Higa JK, Amstutz HC, Schmalzried TP. Relationship between body mass index and activity in hip or knee arthroplasty patients. *J Orthop Res* 2000;**18**(1):35–9.

Chapter 6

# Contemporary Total Hip Arthroplasty: Hard-on-Hard Bearings and Highly Crosslinked UHMWPE

Steven M. Kurtz, PhD and Kevin Ong, PhD



## 6.1 INTRODUCTION

During the 1980s and early 1990s, aseptic loosening and osteolysis emerged as major problems in orthopedics that were perceived to limit the longevity of joint replacements [1]. As discussed in Chapter 5, it was not until the early 1990s that the production of UHMWPE debris at the articulating surface of joint replacements was widely recognized to play a central role in initiating osteolysis [1–6]. Since that time, orthopedic research efforts have focused increasingly on improving UHMWPE for joint replacements, with the goals of reducing wear and, by implication, improving implant survival, especially for young, active patients. Throughout this chapter, we will refer to UHMWPE that has been irradiated with the historical sterilization dose of 25 to 40 kGy as "conventional" material, whereas UHMWPE that has been irradiated with a dose of greater than 40 kGy will be termed "highly crosslinked."

Contemporary hip bearings may be generally classified as either hard-on-hard or hard-on-soft. In the 1990s, growing concern over long-term osteolysis provoked by UHMWPE wear debris from hard-on-soft bearings led to the concurrent development of highly crosslinked UHMWPEs, as well as hard-on-hard bearing alternatives. Over the past decade, hip bearings incorporating metal-on-metal (MOM) or ceramic-on-ceramic (COC) articulations were widely adopted in orthopedics due to their "ultra-low" wear rates in hip simulator studies [7], even when clinical factors such as subluxation are taken into account. In recent years, a novel ceramic-on-metal bearing (COM) combination has also been clinically introduced [8]. In the laboratory, current MOM, COC, and COM bearings have been reported to reduce the production of wear debris by *two to three orders of magnitude* relative to conventional UHMWPE. When compared against highly crosslinked UHMWPE, COC and COM (but not necessarily MOM) exhibit substantially lower wear in a laboratory setting.

For hard-on-soft hip bearings incorporating conventional or highly crosslinked UHMWPE, the femoral head may be fabricated from a metallic alloy (usually CoCr alloy) or a ceramic biomaterial. Contemporary ceramic femoral heads may be produced from alumina, zirconia-toughened alumina matrix composite (e.g., Biolox Delta, CeramTec), or oxidized zirconium composites (e.g., Oxinium, Smith & Nephew). Throughout this chapter, we will refer to metal-on-UHMWPE bearings as M-PE, and the variety of contemporary ceramic-on-UHMWPE bearings, in general, will be referred to as C-PE. An example of a contemporary M-PE hip bearing is illustrated in Figure 6.1.

The first generation of hard-on-hard bearing technologies has been available since the 1950s and 1970s, but MOM and COC joints were superseded in orthopedics for many decades by Charnley's low-friction M-PE design. Initially, the proliferation of modern MOM and COC alternative bearing technologies for hip replacement was limited due, in part, to stringent manufacturing requirements, contributing to substantially higher cost relative to designs incorporating UHMWPE. Regulatory hurdles posed by the FDA also precluded substantial volumes of hard-on-hard bearings from the US market until 2003. Today, however, manufacturing of high-tolerance MOM and COC bearings has now become common across the orthopedic industry. MOM and COC are widely used in orthopedics worldwide because they can substantially reduce the wear rate of arthroplasties. Furthermore, because of the large head sizes currently achievable with MOM and COM bearings, these implants currently offer improved stability and reduced risk for dislocation in comparison with today's modular M-PE, C-PE, and COC implants, which typically must sacrifice head size to maintain a greater insert thickness than MOM implants.

The patterns of bearing usage vary considerably from country to country around the world. In a survey of 125 university hospitals across the European Union conducted between January 2002 and 2003, Scheerlinck and coworkers found substantial differences in bearing preference between centers in Scandinavian and Anglo Saxon countries, Benelux and Germanic countries, and Southern Europe [9]. In the United States, United Kingdom, Scandinavia, and Australia, hard-on-soft bearings have historically dominated orthopedic practice in total hip replacement, whereas in certain western European countries, such as France and Germany, hard-on-hard bearings have been more highly utilized. According to the Australian Registry, between 2001 and 2007, 57% of primary conventional total hip replacements implanted in their country were M-PE, 13% were C-PE, 19% were COC, and 11% were MOM [10]. In the United States, hard-on-soft bearings are currently used in the majority of hip replacements; however, hard-on-hard bearings, and in particular MOM, has grown substantially over the past 5 years, as illustrated by a 35-hospital network that was surveyed in 2007 (Figure 6.2) [11]. These available data illustrate that surgeons' bearing preferences differ not only by country but also as a function of time. In the absence of validated arthroplasty registries, it continues to be difficult to obtain nationally

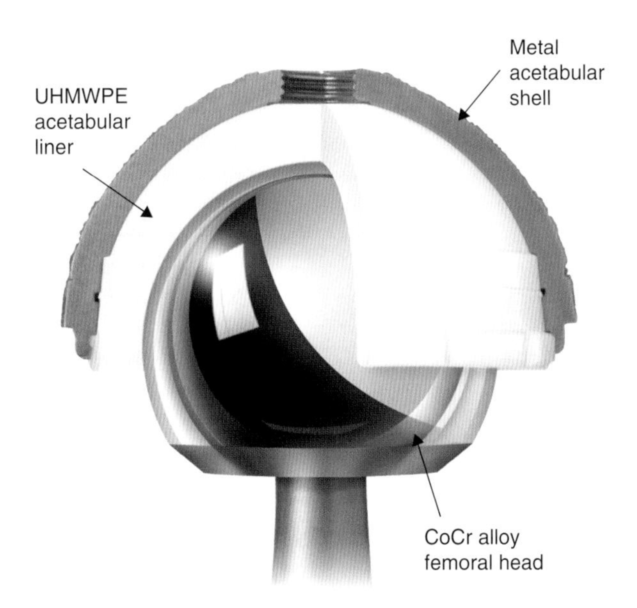


**FIGURE 6.1** Cutaway section of a contemporary metal-on-UHMWPE (M-PE) hip bearing, consisting of a large-diameter CoCr femoral head articulating against a UHMWPE acetabular liner. The liner fits within a metallic acetabular shell. Image provided courtesy of Stryker Orthopedics (Mahwah, New Jersey, USA).

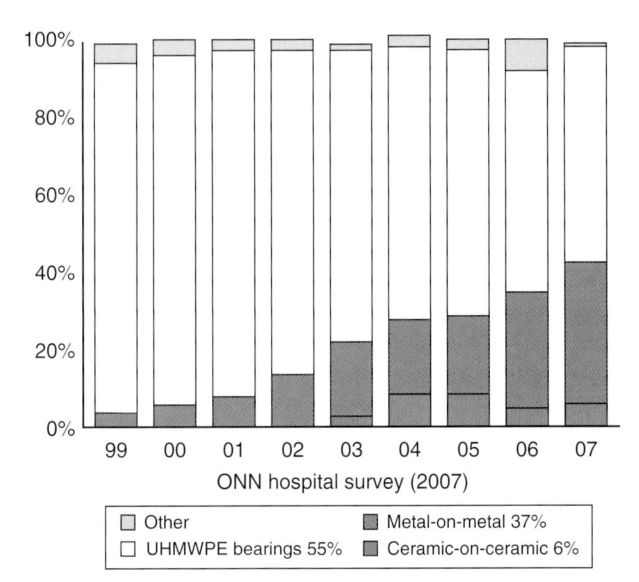


**FIGURE 6.2** Alternative bearing prevalence for THA surveyed in 35 hospitals in the United States (1999 to 2007). UHMWPE bearings include metal-on-UHMWPE and ceramic-on-UHMWPE combinations. Adapted from [11].

representative data for bearing usage in the United States, for many countries in Europe, and in Asia.

Despite their widespread use, especially among younger and more highly active patients, MOM and COC bearings are not expected to supplant UHMWPE bearings in the future. Complete conversion of hip arthroplasties to hard-on-hard bearings is unlikely, not only because of their increased cost, but also due to additional unique risks associated with these implants that may not be warranted for elderly or low-demand patient populations. Some of the unique risks associated with alternative bearings include noise and squeaking (in both MOM and COC bearings); tissue necrosis, hypersensitivity, and potential long-term health risks from metal ions (in MOM bearings); and implant fracture (in COC and C-PE bearings), which are generally not encountered with hip replacements incorporating M-PE bearing surfaces. The clinical success of hard-on-hard bearings is also especially sensitive to implantation technique, and thus these alternatives are less forgiving than UHMWPE bearings. On the other hand, due to the expanding size of the young patient population needing hip replacements [12] (see Figure 5.1 in Chapter 5), hard-on-hard bearings will continue to play an important role in the hip treatment options for the foreseeable future.

This chapter first summarizes MOM, COC, and COM alternative bearing designs and some of the unique trade-offs associated with their use in comparison with highly crosslinked UHMWPE. The history of MOM bearings is particularly noteworthy because it predates the use of UHMWPE in artificial hip joints. We also review the use of ceramics as a counter face in articulations with UHMWPE. Highly crosslinked UHMWPE remains the most widely used alternative to conventional UHMWPE in orthopedics today. Thus, this chapter also provides an introduction to highly crosslinked and thermally stabilized UHMWPE and describes the characteristics of the most prevalent alternative to conventional UHMWPE in joint arthroplasty. Over the past 10 years, an extensive body of knowledge has emerged on different types of highly crosslinked UHMWPE materials, and we will explore first- and second-generation materials in greater detail in later chapters of this book. Consequently, we introduce highly crosslinked materials in this chapter to provide a context with which to compare modern hard-on-hard bearing choices confronting contemporary orthopedic practitioners.

## 6.2 METAL-ON-METAL HIP BEARINGS

Although a variety of metallic biomaterials have been employed in joint arthroplasty, alloys of cobalt, 28% chromium, and 6% molybdenum (hereafter CoCr) are viewed as the gold standard for use in MOM bearings. CoCr alloy is also considered the gold standard as a femoral head material for articulations against conventional as well as highly crosslinked UHMWPE [13, 14]. CoCr alloys (e.g., Vitallium) have been used for hip replacement since 1938, when the biomaterial was employed in the Smith-Petersen mold arthroplasty [15].

Interestingly, earlier (unsuccessful) iterations of the mold arthroplasty, dating from 1923 to 1937, employed glass, Pyrex, viscaloid, and bakelite [15], prior to the use of CoCr alloy. An example of a metallic Smith-Petersen mold arthroplasty (with its original packaging) is shown in Figure 6.3 (photographed at the Thackray Museum, Leeds, England).

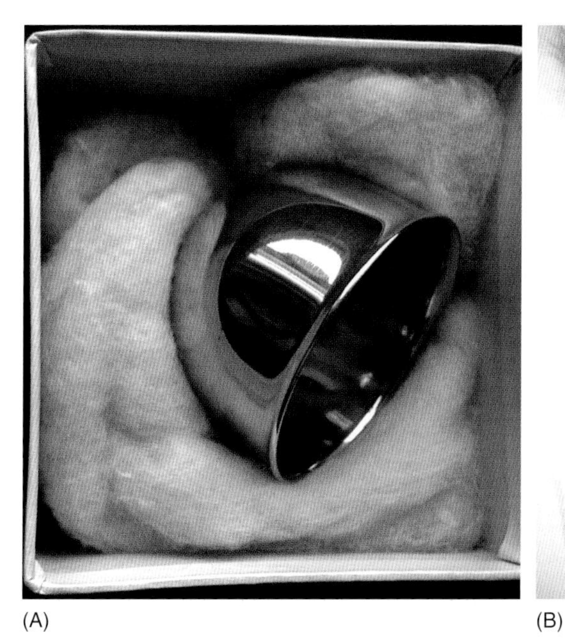

(A)

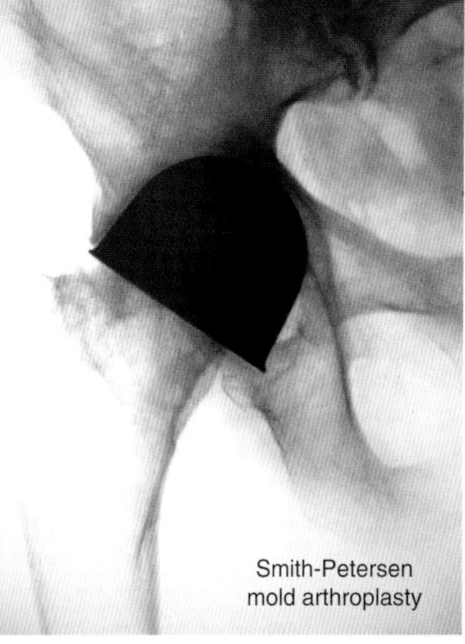


(B)

**FIGURE 6.3** An example of a metallic Smith-Petersen mold arthroplasty (with its original packaging).

**TABLE 6.1 Minimum ASTM Specifications for Mechanical Properties of Cast, Forged, and Wrought CoCr Alloys for Use in Implants**

| Condition | ASTM standard | Yield strength (MPa) | Ultimate tensile strength (MPa) | Ultimate elongation (%) | Hardness, HRC, typical |
|---|---|---|---|---|---|
| Casting | F75 | 450 | 655 | 8 | NA |
| Forged | F799 | 827 | 1172 | 12 | 35 |
| Wrought and annealed | F1537 | 517 | 897 | 20 | 25 |
| Wrought and hot worked | F1537 | 700 | 1000 | 12 | 28 |
| Wrought and warm worked | F1537 | 827 | 1172 | 12 | 35 |

The Smith-Petersen implant is an early example of a hemiarthroplasty, in which only the femoral side of the hip joint is replaced. By contrast, in a total hip joint replacement, both the femoral and acetabular surfaces are replaced. The radiograph, shown in Figure 6.3, was obtained at the Charnley museum in Wrightington (England) and illustrates this early design.

The production of CoCr for use in implants has changed considerably since the days of the Smith-Petersen. CoCr implants were originally produced solely by investment casting, whereas higher strength and ductility CoCr can by achieved today by forging or by use of wrought CoCr (see Table 6.1). In particular, wrought CoCr alloys have superior hardness, yield strength, and ultimate strength relative to cast CoCr due to the more uniform carbide microstructure [14].

Today, MOM implants are fabricated from wrought CoCr alloys with either a high carbon (>0.20%) or low carbon (<0.05%) composition. The specifications for CoCr medical-grade alloys (cast, wrought, and forged) are covered in ASTM standards F75, F799, and F1537 (Table 6.1). For more details about the composition and microstructure of CoCr alloys used in orthopedics, the reader is referred to several reviews and book chapters [14, 16–19]. Although the material specifications may differ for these alloys, no significant difference in the wear performance of MOM implants with various fabrication conditions have been observed [20–22].

MOM bearings have been documented with very low long-term clinical wear rates. At a consensus development meeting convened in 1996, orthopedic researchers generally agreed that MOM joints wear clinically at a rate of 1–5 $mm^3$/y [23], which is 10–100 times lower than the typical wear rate for conventional CoCr/UHMWPE bearing couple. Generally speaking, the *in vivo* wear rates are too low to be detected using the radiographic techniques currently employed for CoCr/UHMWPE hip replacements, as were described in Chapter 5. Instead, clinical wear rates for MOM joints have generally been obtained from geometric measurements of retrieved historical and contemporary components, performed using high-precision coordinate measurement machines (CMM) [24–28].

Further details about first-generation (historical), contemporary MOM, and MOM hip resurfacing designs are provided in the following three sections. In a fourth section, we also address the potential biological risks associated with long-term implantation of MOM hip replacements.

## 6.2.1 Historical Overview of Metal-on-Metal

In 1938, Wiles was reported to have performed the first hip arthroplasty, consisting of stainless-steel femoral and metal components that were fixed to the bone without cement [29]. Although many clinical records of Wiles' patients were lost in World War II, some radiographs of Wiles' prosthesis can still be found at the Charnley museum in Wrightington (Wigan, England), as shown in Figure 6.4.

Between the 1950s and 1970s, pioneering British surgeons like McKee [30] and Ring [31] developed MOM joint replacements fabricated from cast cobalt chrome molybdenum (CoCr) alloy. McKee's first designs of a metal-on-metal joint from the 1950s employed screw fixation. The following radiograph (obtained at Wrightington's museum) shown in Figure 6.5 is an early McKee design from 1957.

Later versions of McKee's design, referred to as the McKee-Farrar prosthesis, were clinically introduced in the 1960s. The McKee-Farrar prosthesis employed cement fixation, whereas the Ring prosthesis, developed in the 1960s, employed screw fixation. The photographs of the McKee-Farrar and the Ring prostheses, shown in Figure 6.6, were taken from implants in the collection at the Thackray Museum (Leeds, England).

In the case of first-generation MOM designs, problems with design tolerances and manufacturing of the articulations have been cited as contributing to bearing seizure and early loosening [26]. Surgical factors may also have contributed to poor short-term survivorship of first-generation MOM designs [32]. Issues related to the potential for metallic hypersensitivity, as well as the (unknown) carcinogenicity associated with the long-term exposure to metal wear debris, has been consistently raised with early MOM joint replacements [33–36].

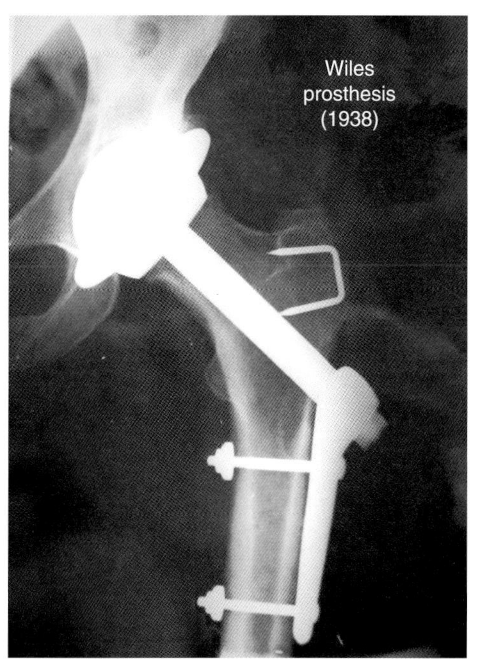


**FIGURE 6.4** Radiograph of Wiles' prosthesis found at the Charnley museum in Wrightington (Wigan, England).

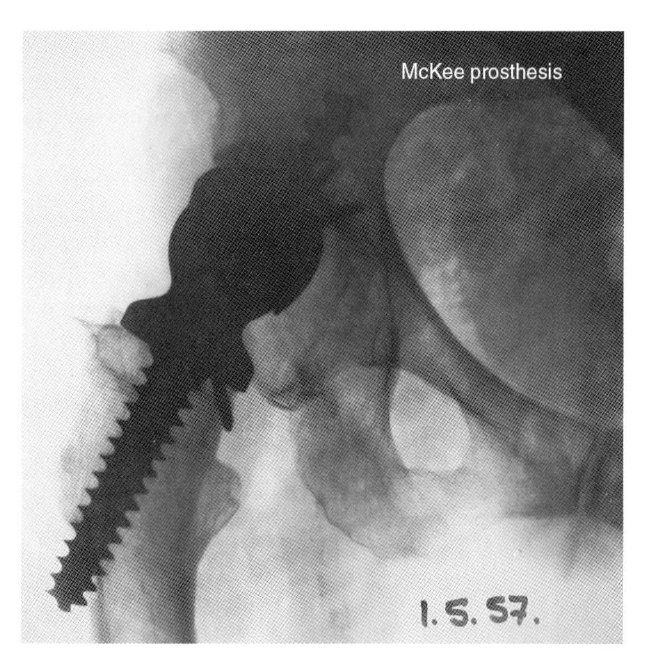


**FIGURE 6.5** Radiograph is an early McKee design from 1957 obtained at Wrightington's museum.

## 6.2.2 Contemporary (Second Generation) Metal-on-Metal Hip Designs

Starting in the 1980s, members of the clinical community in Europe became intrigued by observations of long-term successful survivorship among some first-generation MOM designs [37]. Studies in England and Scandinavia, published in the 1980s and 1990s, suggested that the long-term survivorship of McKee-Farrar prostheses is comparable to the Charnley designs [38–40].

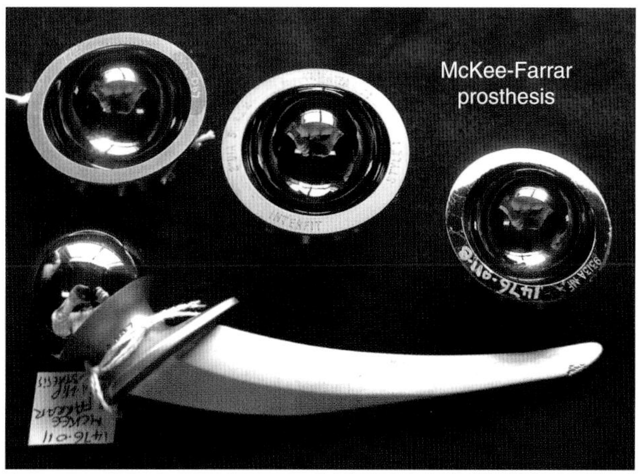


(A)

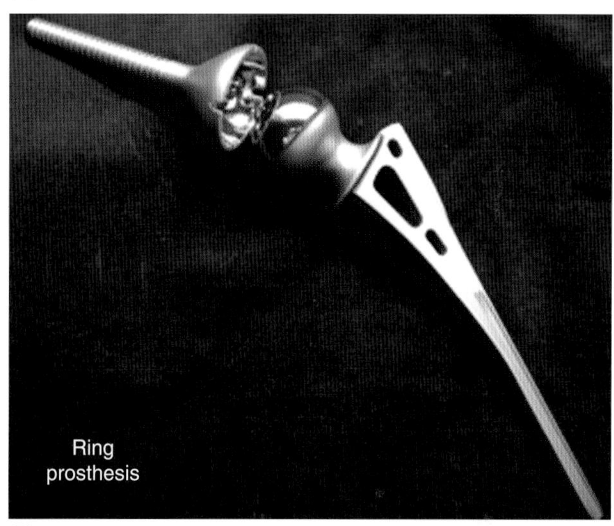


(B)

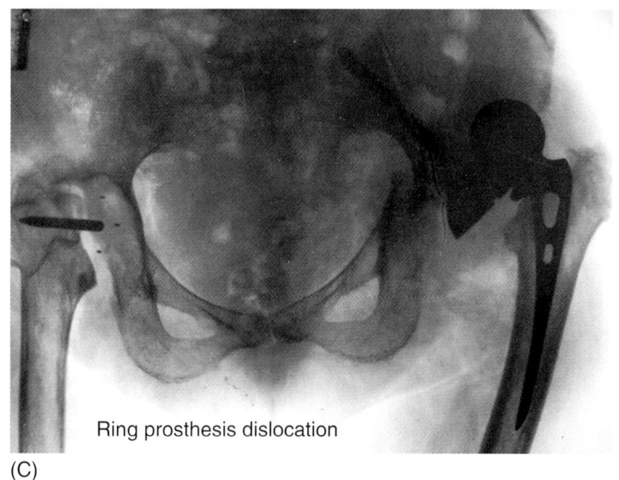


(C)

**FIGURE 6.6** McKee-Farrar and the Ring prostheses taken from implants in the collection at the Thackray Museum, along with a radiograph of a dislocated Ring prosthesis obtained from the Charnley Museum at Wrightington.

Second-generation MOM designs were clinically introduced during 1988 by Sulzer Orthopedics (currently Centerpulse Orthopedics, Winterthur, Switzerland) [37, 41–43]. Sulzer's design was approved for marketing in the

United States by the FDA in August 1999. Between 1988 and 2000, it is estimated that 125,000 of these second-generation MOM components have been implanted worldwide [14].

Sulzer's second-generation MOM designs incorporated a CoCr articulating surface, but the acetabular component consisted of a modular shell and a UHMWPE liner embedded with a CoCr insert. A cross-section of the contemporary "sandwich" type design, distributed under the trade name METASUL (Centerpulse Orthopedics, Winterthur, Switzerland), is illustrated in Figure 6.7.

UHMWPE continues to be used in second-generation MOM and, as we shall see, in certain COC designs as well, primarily as a means for achieving implant fixation. However, UHMWPE is also used in these alternative bearing designs for the objective of preserving intraoperative modularity [44]. According to Rieker, "This embedded solution was chosen to assure complete compatibility with the shells already commercially available (same operative technique and instruments)" [45]. Also, by incorporating UHMWPE into the bearing design, the same acetabular shell could be used for a wider range of liner designs, both conventional and alternative.

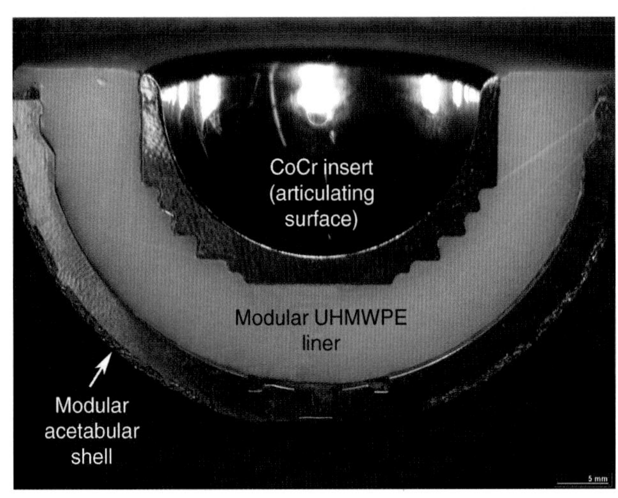


**FIGURE 6.7** A cross-section of the contemporary sandwich-type design, distributed under the trade name METASUL (Centerpulse Orthopedics, Winterthur, Switzerland).

Other contemporary MOM designs, employing a modular taper-fit connection between the CoCr insert and the metal shell, have also been clinically introduced by companies such as Biomet (Warsaw, Indiana, USA), DePuy Orthopaedics (Warsaw, Indiana, USA), Wright Medical (Arlington, Tennessee, USA), and Smith & Nephew (Memphis, Tennessee, USA). Unlike the METASUL design, these other taper-lock modular MOM designs do not incorporate an interpositional UHMWPE layer.

## 6.2.3 Metal-on-Metal Hip Resurfacing

Hip resurfacing arthroplasty, as shown in Figure 6.8, involves treatment of the diseased hip by removing or resurfacing the diseased surfaces of the femoral head and the acetabulum, without full resection of the femoral head and neck. Hip resurfacing evolved from the early Smith-Petersen mold arthroplasty in the early 1920s but was largely abandoned in the late 1960s due to poor clinical outcomes and less than optimal wear characteristics of the bearing materials. Some of the early MOM hip resurfacing prostheses included those developed by Haboush (USA), Muller (Switzerland), Gerard (France), and Nishio (Japan). Over the next decade or so, there was another resurgence in the use of hip resurfacing, but most of the designs were metal-on-polyethylene bearings. The interest waned in the mid-1980s following poor outcomes, primarily due to excessive polyethylene wear and particle-induced osteolysis. However, the development of better-engineered, low-wear metallic alloys stimulated the reintroduction of the next generation of MOM hip resurfacing in the early 1990s, behind the work of surgeons such as McMinn (Cormet and Birmingham Hip Resurfacing) and Amstutz (Conserve Plus). For more details about the evolution of MOM

(A)

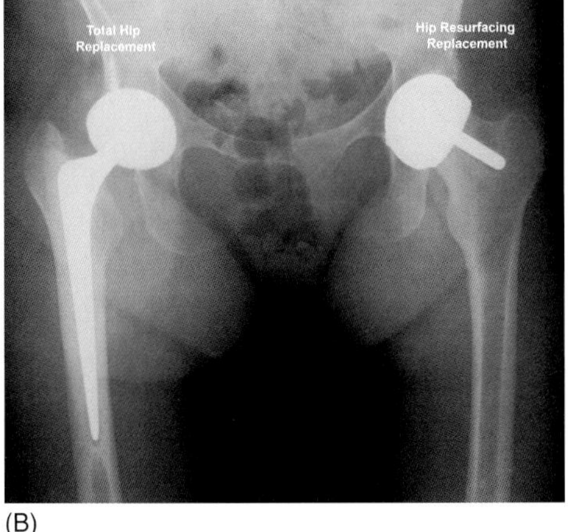


(B)

**FIGURE 6.8** Contemporary metal-on-metal hip resurfacing prosthesis, along with a radiograph of a total hip replacement (left) and a hip resurfacing replacement (right).

hip resurfacing, the reader is referred to several recent reviews [46–48].

Hip resurfacing arthroplasty provides an attractive alternative to conventional total hip replacement, particularly for younger, more active patients, because the risk of total hip replacement failure is greater in this patient population [49]. Furthermore, hip resurfacing utilizes larger diameter femoral heads than total hip replacements, thus providing increased range of motion and improvement in the stability of the joint, which is one of the most common complications associated with total hip replacements [50, 51]. Any potential revision surgery of a hip resurfacing is also theoretically an easier conversion than revision of a conventional long stem THA [52, 53]. Firstly, the femoral canal is left intact during the hip resurfacing, which leaves more bone stock available for the revision stem [52, 53]. In addition, when the cup is well fixed, it may not be necessary to remove the cup because it can be left in place to mate with a large diameter modular femoral head with a standard long stem [52, 53].

According to hip registry data [10, 54], the primary reasons for short-term revision of hip resurfacing arthroplasty are periprosthetic (femoral neck) fracture and implant loosening, which account for up to 41% and 24% of the revisions, respectively. In particular, older female patients are at greater risk for revision. Intraoperative notching of the neck is a common finding for patients who encounter neck fractures, while dissection around the neck region may also cause damage to the blood vessels and lead to altered vascularity to the femoral head or avascular necrosis [55, 56]. Surgeons have emphasized the importance of careful patient selection for hip resurfacing [57, 58], while the surgical procedure has also been associated with a learning curve [58]. The procedures may also be complicated by the abnormal bone anatomy commonly found in hip resurfacing patients [59, 60], which provide additional challenges in restoring hip mechanics and could induce bone remodeling changes around the implant [61, 62]. The reader is referred to recent reviews [63, 64] for more information about hip resurfacing.

### 6.2.4 Potential Biological Risks Associated with MOM Joints

Despite the ultralow wear rates afforded by second-generation MOM hip implants, concerns remain about the potential health risks associated with long-term metal ion exposure. The wear particles in MOM articulations range between 6 nm and 5 mm [65–68]. Due to their smaller wear particle size, MOM hip implants have been estimated to release about a 100 times greater number of wear particles than conventional UHMWPE hip implants [65, 67]. In particular, the nanometer-sized metallic wear particles are more easily digested by cells, bound into proteins, and/or dissolved into body fluids than the larger UHMWPE wear particles [36, 68–71].

The wear products of MOM joint articulation are transported systemically and are manifested in elevated chromium and cobalt levels in a patient's serum and urine, raising the potential risk for carcinogenesis [36, 69, 70]. However, epidemiological studies of cancer risk in patients with MOM remain inconclusive, due to the relatively small patient populations evaluated, the Scandinavian basis of the studies, and the typically rare incidence of the disease [35, 72]. There have also been reports of metal hypersensitivity associated with the implantation of MOM prostheses, but the incidence of this complication is reported to be extremely rare [73]. Hypersensitivity may manifest itself in the form of an allergic response on the patient's skin; however, there is no accepted clinical test to reliably test a patent's sensitivity to metals [74]. It remains unclear whether metal hypersensitivity is a contributing factor to implant performance or survivorship.

Although MOM bearings generally produce very low wear rates, recent clinical studies have reported elevated concentration of metal ions and the presence of metallosis, particularly when the cups are positioned in high inclination angles [75, 76]. Malpositioning of the cup may cause edge loading of the cup, resulting in disruption of the bearing lubrication and increased wear [76, 77]. More recently, the risks of lymphocytic response, pseudotumors, and osteolysis in MOM hips have also been reported [78–82]. In some instances, lesions may develop secondary to particulate debris generation or elevated chromium and/or cobalt levels [78, 79, 82], but the etiology is still unknown. Osteolysis in MOM hips is thought to be a rare complication [83,84], and according to limited intermediate clinical follow-up, the prevalence of osteolysis in MOM hips is substantially lower compared to that of conventional UHMWPE. Recent reports have also linked early osteolysis in MOM hips to a delayed hypersensitivity to metal [80, 84, 85], but this association is not completely understood.

In summary, for the reasons previously outlined, the orthopedics community continues to study the biological and carcinogenic implications of metallic wear debris, which are not fully understood at the present time. Due to ongoing clinical concern, researchers are continuing to monitor the long-term health effects associated with MOM bearings.

## 6.3 CERAMICS IN HIP ARTHROPLASTY

Alumina ceramics have been used successfully, as both femoral heads and acetabular components, in certain hip replacement designs since the 1970s [86–88]. Alumina, therefore, is a widely recognized, well-validated ceramic biomaterial that has evolved though gradual improvements in manufacturing over the past 30 years and thus has a long-term track record for biostability and biocompatibility [89]. However, limited design options, catastrophic fracture risk and, more recently, squeaking have all been well documented in alumina bearings over the years, thus there continues to be motivation in certain circles to develop more advanced ceramic materials for contemporary total hip arthroplasty.

In this section, we outline the history of ceramics in orthopedics and provide an introductory overview of ceramic materials relevant to hip replacement. This section also discusses the use of ceramic femoral heads as bearing surfaces with UHMWPE and covers current designs of ceramic-on-ceramic (COC) alternative bearings. The final part of this section contains an overview of wear mechanisms and ceramic fracture risk in historical, as well as in current, ceramic materials.

## 6.3.1 Historical Overview of Ceramics in THA

The application of ceramic materials in hip arthroplasty has its origins in Europe and Japan. Pierre Boutin, in collaboration with Ceraver Inc. from France, first reported on the clinical results of COC hip arthroplasty in 1971 and 1972 [86, 87]. In 1977, Shikata in Japan introduced the concept of using alumina femoral heads with UHMWPE acetabular components.

The first-generation of COC components consisted of monolithic acetabular component, fabricated entirely from alumina ceramic ($Al_2O_3$), articulating against an alumina ceramic femoral head, fitted using a tapered interlock to the metallic femoral stem. COC designs developed by Sedel in Paris [90], and by Mittelmeier in Germany [91], also employed $Al_2O_3/Al_2O_3$ bearing surfaces. These early designs incorporated tapered screw threads and/or spikes on the back side of the acetabular component for initial fixation to the pelvis. These early COC implants were at first implanted without cement. Although a wide range of COC designs were investigated in Europe during the 1970s and 1980s, only the Mittelmeier design was clinically released (as the "Autophor" design) in significant numbers within the United States during the 1980s.

Monolithic alumina acetabular components, including those of the Autophor or Mittelmeier design, were generally associated with unacceptable levels of loosening [92, 93]. Due to the bioinertness of the ceramic material, it proved to be more difficult to achieve long-term stable fixation with uncemented monolithic ceramic cups as compared with metal-backed UHMWPE acetabular component designs. Migration or tilting of the monolithic ceramic acetabular component in some cases contributed to edge loading and a variety of undesirable late complications, including edge loading, accelerated wear, and impingement [92]. *In vivo* fracture of the ceramic components was also a significant problem associated with certain first-generation COC components [92].

However, those early COC components that survived the hurdles of fracture, migration, and loosening exhibited extremely low clinical wear rates [94–97]. The *in vivo* wear rates for first-generation COC components has been found to range between 3–9 μm/y for linear wear [95, 96] and 1–5 $mm^3$/y for volumetric wear [97]. As with MOM designs, the radiographic wear assessments for COC components are complicated by the very low wear rates and poor edge detection on X-rays, and thus direct measurements from retrievals has generally been reported in the literature [95–97].

## 6.3.2 Ceramic Biomaterials for Hip Arthroplasty

The potential for extremely low clinical wear rates, necessary to reduce the risk of osteolysis, inspired increased interest in developing new COC designs for hip arthroplasty during the 1990s [98]. Today, the motivation for new designs is best illustrated by the shift toward femoral heads greater than 28 mm in diameter, resulting in the demand for thinner acetabular liners than can be accommodated safely using alumina. In addition, as noted previously, the desire to reduce the fracture risk and acoustic emissions has led to continuous improvement of ceramic materials for orthopedic load bearing applications over the past 30 years (Table 6.2).

**TABLE 6.2** Various Properties of Alumina and Zirconia Ceramics Used in the Total Hip Reconstruction [99–101]

| Property | Alumina in 1970s | Alumina in 1980s | Alumina in 1990s | Zirconia | ZTA* |
|---|---|---|---|---|---|
| Bending strength (MPa) | >450 | >500 | >550 | >900 | >1000 |
| Fracture toughness (MPa.m$^{1/2}$) | 4 | 4 | 4 | 8 | 5.7 |
| Vickers hardness (0.1) | 1800 | 1900 | 2000 | 1250 | 1975 |
| Grain size (micron) | 4.5 | 3.2 | 1.8 | <0.5 | <1.5 (alumina matrix) |
| Young's modulus (GPa) | 380 | 380 | 380 | 210 | 350 |

*Note: ZTA: Zirconia toughened alumina microcomposite.*

The ceramic materials used in orthopedics today are formed by fusing or sintering microscopic grains of alumina ($Al_2O_3$) and/or zirconia ($ZrO_2$) ceramic powder into a consolidated product. The processing of ceramic parts for hip replacement is now performed in accordance with a wide range of international quality standards and stringent FDA regulatory requirements [102]. The manufacturing stages for ceramic components generally include an isostatic pressing step and machining into near net (or green) shape; a firing step in a furnace for final sintering of the powder; hot isostatic pressing (hipping) to reduce porosity; tempering to improve toughness; and several high-precision machining and/or polishing steps to achieve the desired surface finish [103]. The strength of the ceramic component depends upon the purity and size of the granular powder particles, as well as upon the powder composition (i.e., alumina versus zirconia). The micrograph in Figure 6.9 (provided courtesy of CeramTec AG, Plochingen, Germany) shows an example of sintered alumina microstructure in which the average grain size is 2 μm or less.

Ceramics are brittle, polycrystalline solids. For this family of materials, the inherent fracture resistance is intimately related to the size and distribution of internal flaws or defects, which may occur near grain boundaries. Thus, reducing the grain size of alumina ceramics over the past 30 years (Table 6.2) has resulted in an overall improvement in its fracture strength.

Despite brittle characteristics, ceramic components enjoy several outstanding tribological properties, including their hardness (Table 6.2), which contributes to wear and scratch resistance. Ceramic surfaces are also more hydrophilic than the CoCr surfaces of a femoral head, as illustrated in Figure 6.10 by the water droplets shown in the pictures (provided courtesy of CeramTec AG, Plochingen, Germany). The improved wettability of ceramics contributes to lower friction than CoCr when articulated against UHMWPE under physiologic loading and lubrication conditions [104].

There are three types of ceramics that are clinically relevant to contemporary total hip replacement, including alumina, zirconia, and zirconia-toughened alumina composites. A fourth ceramic biomaterial, oxidized zirconium, is a ceramic-metal composite with a surface layer of zirconia. A fifth biomaterial, silicon nitride, is in advanced stages of commercialization for hip arthroplasty.

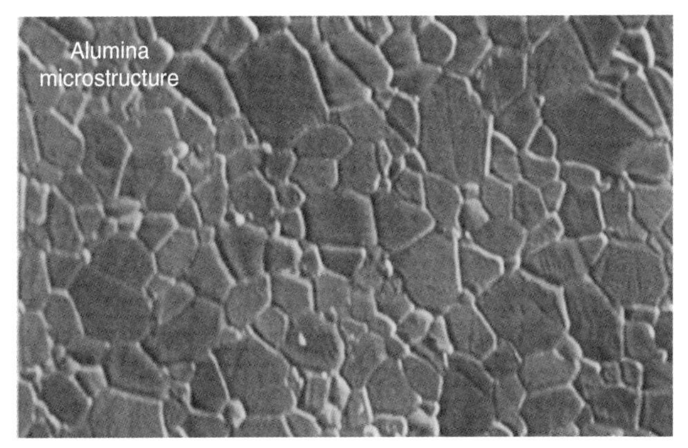


**FIGURE 6.9** The micrograph (provided courtesy of CeramTec AG, Plochingen, Germany) shows an example of sintered alumina microstructure, in which the average grain size is 2 μm or less.

### 6.3.2.1 *Alumina*

Alumina has the longest history of successful use in articulations with UHMWPE and against itself in COC bearings. For detailed reviews of alumina materials used in orthopedics, the reader is referred to previous review articles [89, 99]. Currently, CeramTec AG (Plochingen, Germany) is the world's largest supplier of medical-grade alumina ceramic, which is distributed under the trade name of BIOLOX Forte. CeramTec has reported that 1.3 million BIOLOX Forte femoral heads have been sold for hip arthroplasty between 1995 and 2002 [105]. An additional 350,000 ceramic inserts for COC components have been sold by CeramTec during the same time period [105]. As of mid-2008, the number of BIOLOX Forte components in hip arthroplasty was reported by the manufacturer to reach 5 million [106].

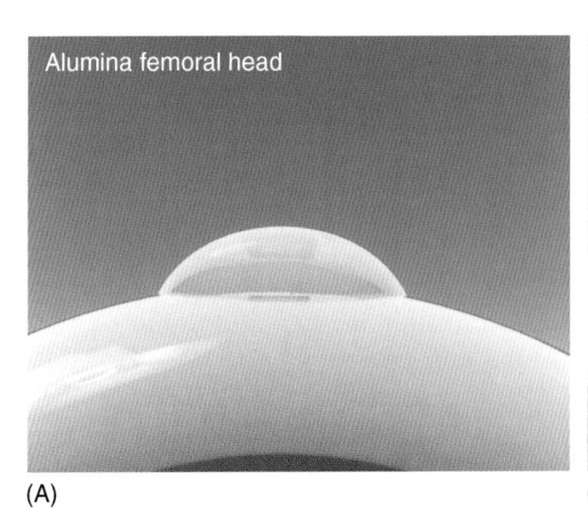


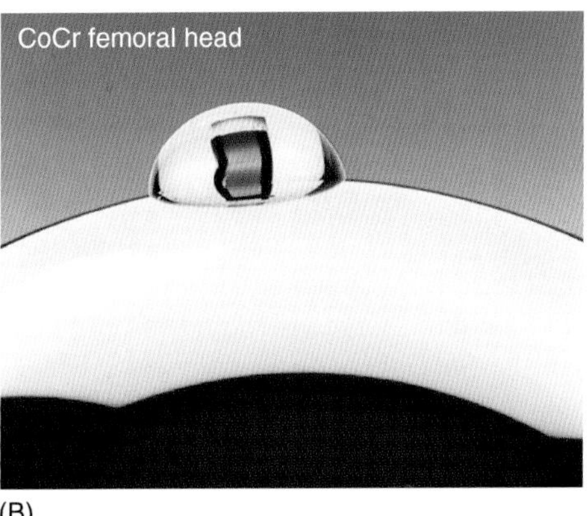


**FIGURE 6.10** Ceramic surfaces are also more hydrophilic than the CoCr surfaces of a femoral head, shown by the water droplets in the pictures (provided courtesy of CeramTec AG, Plochingen, Germany).

#### 6.3.2.2 Zirconia

Zirconia ceramic was widely used in orthopedic applications between 1985 and 2001, until manufacturing difficulties at the world's largest supplier resulted in its abrupt decline. Zirconia was initially chosen for commercialization due largely to its higher strength relative to alumina (Table 6.2). However, zirconia is a considerably more complex material than alumina. The reader is referred to several excellent reviews summarizing the properties of zirconia ceramic biomaterials that have recently been published [107–109]. Unlike alumina, zirconia is a metastable ceramic, consisting of monoclinic, tetragonal, and cubic phases [108]. Under a combination of temperature, humidity, and stress, zirconia can undergo a phase transformation from tetragonal to monoclinic, which results in a volume change [99, 108]. This phase transformation can have desirable consequences, such as the generation of a compressive stress field at the tip of a propagating crack, resulting in crack growth resistance. This material characteristic is referred to as "phase transformation toughening" [108]. However, the phase-induced volumetric transformation can also have disastrous consequences if not properly controlled. For this reason, the phase transformation property of zirconia is typically stabilized with the addition of magnesia or yttria [99]. The most common type of zirconia used in orthopedics is termed Y-TZP, corresponding to yttria stabilized-tetragonal phase, polycrystalline zirconia [99]. The chemical composition of Y-TZP is about 5.1% yttria ($Y_2O_3$) and 93–94% zirconia ($ZrO_2$) [108].

Until recently, St. Gobain Desmarquest (Vincennes, France) was the world's largest supplier of medical-grade stabilized zirconia ceramic. The company reported that between 1985 and 2001 it sold a total of 500,000 components fabricated from this material under the trade name of Prozyr [110]. Starting in 2000, Desmarquest received reports of an unusually large number of fractures from components that were fabricated in early 1998, when the company implanted a change in manufacturing processes from use of a batch furnace to a continuous kiln or tunnel furnace [110]. These zirconia fractures were particularly troubling because they involved components that complied with applicable technical standards. During August 2001, Desmarquest announced a voluntary recall of nine batches of femoral heads, all fabricated with a tunnel furnace during the late 1990s [110]. The recall had global implications because the femoral heads were distributed to over 51 companies worldwide [111]. Recall announcements soon followed from national regulatory agencies. On September 13, 2001, the FDA warned that "surgeons should not implant artificial hips with the St. Gobain zirconia ceramic heads manufactured since the processing change in 1998" [111]. The company suspended distribution of zirconia ceramics for orthopedic applications in August 2001 [110].

Since the recall, zirconia has fallen out of favor as a bulk orthopedic bearing biomaterial. However, zirconia continues to be the subject of ongoing research by the orthopedic community in an effort to understand the clinical consequences of manufacturing variations on the patients implanted with these materials. Retrieval studies have refined the methodologies of detecting phase transformation in explanted zirconia components. X-ray diffraction was initially used to track phase transformation, but more recently Raman spectroscopy has emerged as an alternative technique. These analytical techniques, in combination with steam autoclaving studies, are now the preferred accelerated aging method to screen zirconia biomaterials for their biostability.

In addition to improvements in experimental techniques to detect and simulate phase transformation in zirconia, clinical studies and retrieval studies have noted the tendency of certain components to exhibit surface roughening *in vivo*, presumably as a consequence of local phase transformation. *In vivo* changes in zirconia surface roughness are by no means universal, and observations of phase transformation are highly variable in zirconia from retrieval to retrieval. However, in several population-based studies of zirconia articulating against UHMWPE, a gradual increase in wear rates has been observed over time.

The 2001 recall, and the ongoing research on zirconia ceramics, has not dampened the general enthusiasm for ceramic materials in orthopedics. Although alumina is currently the historic reference material for orthopedic applications, either for articulations with UHMWPE, or for use in COC alternate bearings, a newer and more complex alumina composite incorporating zirconia has emerged as the current state-of-the-art biomaterial.

#### 6.3.2.3 Zirconia-Toughened Alumina Matrix Composite

Starting in June 2000, a new alumina matrix nanocomposite material (BIOLOX Delta, CeramTec, Plochingen, Germany) has been available as a femoral head material [101]. This ceramic biomaterial is now broadly used across the orthopedic industry in both femoral heads and acetabular liners. According to the manufacturer, more than 320,000 femoral heads and 160,000 acetabular inserts have been implanted on a worldwide basis as of 2008 [106]. The primary advantage of this alumina matrix composite is its increased strength, fracture toughness, and wear resistance relative to alumina (Table 6.2) [112].

Zirconia-toughened alumina matrix composite (ZTA) is engineered to manage tetragonal-to-monoclinic phase transformation in zirconia to improve the overall strength of the alumina matrix composite [113]. BIOLOX Delta, the ZTA commercialized by CeramTec, consists of an alumina matrix (82% by volume) reinforced by zirconia (17 w/w%), strontium aluminate (0.5 w/w%), and chromium oxide (0.5 w/w%) (Figure 6.11). The additives are incorporated into the alumina matrix to provide crack tip blunting and

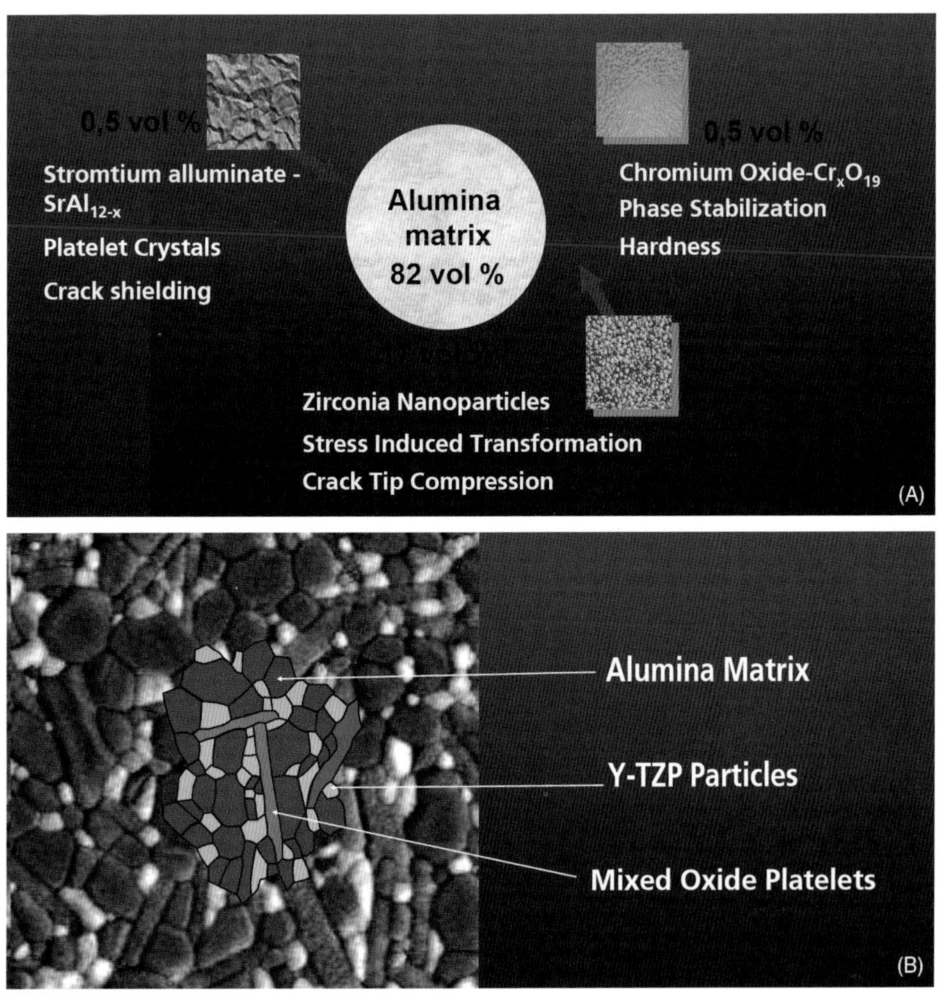


**FIGURE 6.11** Chemical composition (A) and microstructure (B) of zirconia-toughened alumina composite, BIOLOX Delta (provided courtesy of CeramTec AG, Plochingen, Germany).

toughening mechanisms, as well as to increase the hardness of the composite. Clinical studies are underway to determine the effectiveness and reliability of this new biomaterial in C-PE and COC articulations. In clinical use, BIOLOX Delta is now combined with alumina or highly crosslinked UHMWPE as shown in Figure 6.12.

### 6.3.2.4 Oxidized Zirconium

Oxidized zirconium is a proprietary implant material of Smith & Nephew Orthopedics (Memphis, Tennessee, USA) marketed under the trade name OXINIUM [114]. OXINIUM material is fabricated by heating the zirconium alloy device in the presence of air, which converts the surface to a black zirconium oxide ceramic (~5 μm thick) (Figure 6.13). *In vivo* phase transformation from metastable tetragonal to stable monoclinic is not an issue with oxidized zirconium because its phase content is over 95% stable monoclinic. Clinically introduced for the knee in December 1997 and for the hip in October 2002, oxidized zirconium is intended only for hard-on-soft bearings both in total hip and total knee arthroplasty [115–118]. OXINIUM material is currently not used in hard-on-hard bearings.

The ceramic surface of OXINIUM material provides increased scratch resistance and is more wettable and provides lower UHMWPE wear in a hip simulator than that provided by a CoCr alloy femoral head, without the negative risks associated with ceramic fracture because of the ductile zirconium alloy substrate [115]. Early (12-month) radiographic wear data for OXINIUM material were reported in a conference abstract [117], but to the authors' knowledge longer-term follow-up data have not yet been reported. Retrieval studies suggest that the oxidized zirconium surface layer may be damaged by a dislocation, during which the femoral head rides over the metallic edge of the metal shell [119, 120].

### 6.3.2.5 Silicon Nitride

Silicon nitride ($Si_3N_4$) is a recent entry into the ceramic biomaterials arena for hard-on-hard hip bearings [121–124]. With an elastic modulus of 300 GPa and a fracture toughness

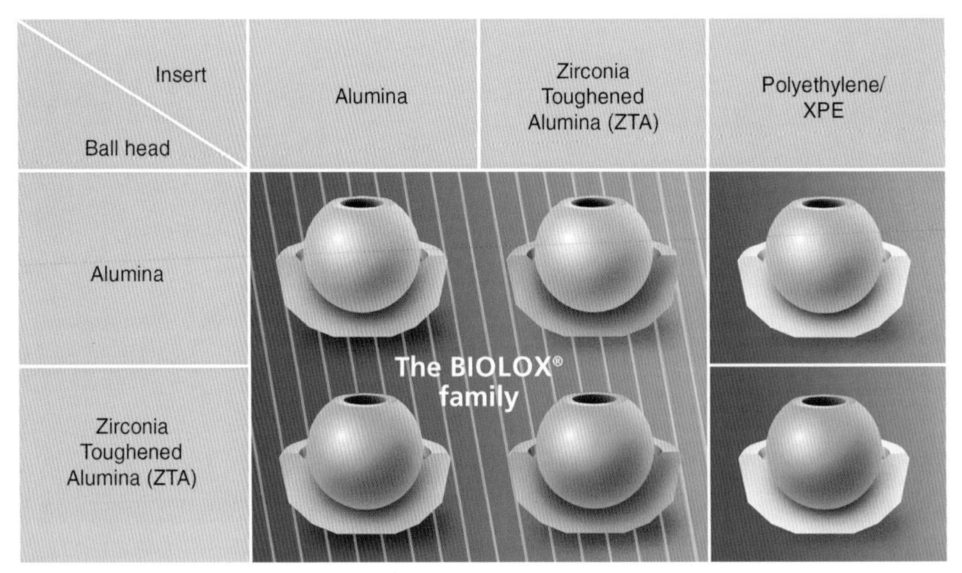


**FIGURE 6.12** Hard-on-hard and hard-on-soft bearing combinations currently used with contemporary medical-grade ceramics (provided courtesy of CeramTec AG, Plochingen, Germany).

**FIGURE 6.13** Hip bearing incorporating an oxidized zirconium femoral head articulating against a UHMWPE acetabular liner in a metal shell, provided courtesy of Smith & Nephew (Memphis, Tennessee, USA).

of $10\,MPam^{1/2}$, silicon nitride provides higher strength than alumina. Historically, the biocompatibility of $Si_3N_4$ has been a controversial topic in the literature, but recent data would suggest that these concerns may be ceramic-formulation dependent [124]. Recently, a $Si_3N_4$ ceramic formulation was commercialized by Amedica (Salt Lake City, Utah, USA) for C-PE, COC, and COM hip bearing applications. The ceramic reportedly consists of 90% $Si_3N_4$ powder (having a mean particle size of $0.5\,\mu m$), sintered with 6 wt% yttrium oxide ($Y_2O_3$) and 4 wt% alumina ($Al_2O_3$) [121]. Wear testing of COM and COC bearings in a hip simulator demonstrated ultralow wear rates, comparable to or lower than alumina-alumina [121]. According to the manufacturer, human implantation of silicon nitride is anticipated to begin in 2009.

### 6.3.3 Ceramic-on-UHMWPE

As summarized in the previous section, laboratory tests of ceramic biomaterials have demonstrated substantial advantages over CoCr for articulation against UHMWPE. The harder ceramic surfaces should theoretically be more scratch resistant than CoCr femoral heads [125, 126]. Unfortunately, it has thus far proven difficult to conclusively establish, as a general premise, the clinical benefit of ceramic femoral heads in patients. For reasons noted previously associated with *in vivo* phase transformation and surface roughening of zirconia, clinical wear studies published in the past 5 years with zirconia heads have reported mixed results in the literature, ranging from no significant difference to inferior performance relative to CoCr heads.

With alumina femoral heads, the clinical history is longer but no less inconclusive. Sugano et al. [127] reported on the use of first-generation alumina ceramic femoral heads (Bioceram, Kyocera Corporation, Kyoto, Japan) in a series of 57 hips in 50 patients, who were implanted between 1981 and 1983. After 10 years of follow-up, they observed an average wear rate of 0.1 mm/y. Because a control group was not included in this study, the authors did not compare with the wear observed using CoCr heads in a comparable patient population. A similar limitation has been noted in previous retrospective clinical studies reporting wear rates for zirconia against UHMWPE, which show, overall, an average wear rate of 0.1 mm/y [108]. However, a wear rate of 0.1 mm/y is typically observed in the literature as an average clinical wear rate for CoCr when used with

UHMWPE [128]. In prospective randomized trials, a significant improvement in wear for zirconia femoral head patients could be detected between CoCr and zirconia femoral heads implanted bilaterally in the same patient for 7.1 years on average [129], whereas a prospective study of two matched cohorts with 4.3 years of follow-up found similar wear rates in both CoCr and zirconia femoral head groups [130]. Other studies have found increased wear associated with zirconia heads, attributed to phase transformation [131].

Sychterz et al. [132] compared the *in vivo* wear of a cohort of 81 patients implanted with alumina femoral heads, with a well-matched control group of 43 patients implanted CoCr femoral heads. At 7 years follow-up, the radiographic wear rate for patients with alumina was $0.09 \pm 0.07$ mm/y; the wear rate in the control group was $0.07 \pm 0.04$ mm/y. The authors concluded that the wear rates of UHMWPE hip replacements using alumina and CoCr femoral heads were similar.

Hendrich et al. [133] reported on the largest matched series of patients with 28-mm diameter alumina ($n = 100$) and CoCr ($n = 109$) femoral heads using the Harris-Galante acetabular component design. The alumina components in this case were identified as BIOLOX (CeramTec). Despite the identical acetabular component designs and femoral head sizes, there were significant differences in follow-up period, which averaged 8y for the CoCr group and 5y for the alumina group. The average wear rates were $0.14 \pm 0.11$ mm/y for components with CoCr femoral heads and $0.13 \pm 0.08$ mm/y for components with the alumina femoral heads. This difference was not statistically significant ($p = 0.46$). Furthermore, a power analysis showed that the clinical study was powered sufficiently to detect a 0.038 mm/y difference (28%) in wear rates with 80% power.

Hopper et al. [134] performed a retrospective analysis of 512 hips (137 ceramic, 430 CoCr heads) that were followed radiographically for femoral head penetration with 3-year minimum follow-up. Due to difficulties encountered in radiographic assessment of zirconia heads, this study focused almost exclusively on alumina femoral heads [135]. Twenty-five factors were examined for their hypothesized influence on wear, including age, gender, preoperative diagnosis, cup design, head diameter, head material, UHMWPE material (Enduron versus Hylamer), liner geometry, and UHMWPE sterilization method (gas sterilized versus gamma sterilized). A ceramic femoral head was found to significantly lower the *in vivo* femoral head penetration rate by 0.03 mm/y; however, head material was only the sixth most important factor in the multivariate statistical model and explained approximately 1% of the variability in the penetration rate data. In comparison, changing the UHMWPE material from uncrosslinked, gas sterilized to crosslinked, gamma sterilized was the single most important factor, explaining 12% of the variability in clinical penetration rates.

Wroblewski et al. from Wrightington hospital have performed a long-term prospective observational cohort study on a group of 17 patients implanted with 19 experimental, silane-crosslinked HDPE acetabular cups and 22 mm diameter alumina femoral heads [167, 177]. These patients were relatively young, with an average age of 47 years (range: 26 to 58 years) at the beginning of the study. The cemented cups were similar in design to the Low Friction Arthroplasty developed by Charnley and described in Chapters 4 and 5. The most recent follow-up of this cohort after an average of 17 years of implantation (range: 15 to 18.1 years) was completed for nine patients with 11 hips [177]. The clinical results for this group was judged to be excellent, and the mean radiographic penetration rate was 0.019 mm/y. None of the patients showed evidence of osteolysis on radiographs. Although the young patients in this study showed evidence of much lower wear than patients implanted with conventional M-PE bearings at this institution, it is unclear whether the wear reduction can be attributed to the novel chemically crosslinked acetabular cups, the alumina femoral heads, or the C-PE bearing combination.

Urban et al. have also reported long-term clinical performance of alumina C-PE bearings in a retrospective, single-surgeon series with no matching M-PE control group [178]. The 56 patients (64 hips) included in this retrospective study were implanted with a first-generation cemented UHMWPE cup (gamma irradiated in air), a modular 32-mm alumina femoral head, and a stainless steel cemented femoral stem between 1978 and 1981. The patients were originally 69 years old, on average, when they were implanted with a C-PE bearing. At the time of the study, 19 patients with 22 hips were still alive and available for follow-up after a mean implantation time of 18 years (range: 17 to 21 years). The mean penetration rate for this group was 0.034 mm/y in 18 hips with available radiographs (range: 0.000 to 0.077 mm/y), and there was no evidence of osteolysis around any of the acetabular or femoral components. The penetration rates for the C-PE patients surviving long-term implantation were lower than values previously reported for conventional M-PE bearings (typically 0.1 mm/y, as discussed in Chapter 5). However, with only 32% (18/56) of the patients with radiographs available for analysis at the time of the study, the penetration rates for the majority of the cohort could not be reported.

The clinical history for C-PE bearings in the literature includes alumina and, more recently, zirconia femoral heads. It remains difficult to generalize about C-PE bearings because the clinical performance over the years has been ceramic-material specific, especially when zirconia is concerned. Indeed, the most recently published report by the Australian Hip Registry found no significant difference in survivorship up to 7 years when comparing modern M-PE and C-PE bearings between 2001 and 2007 [10]. However, the Australian hip registry study did not include data on the type of ceramic femoral head material and considered all ceramic femoral heads grouped in a single category. Although alumina has the longest successful clinical history of any of the medical ceramics used thus far in C-PE bearings, the clinical studies reported to date in the literature have been retrospective in nature and limited to single centers. For all of these reasons, and despite the fact that ceramic heads show clear benefits in the laboratory, it remains controversial whether ceramic heads alone substantially improve the wear rate or survivorship of UHMWPE acetabular components. Multicenter, prospective randomized trials would be useful to more conclusively demonstrate the differences in clinical performance between modern ceramic and CoCr femoral heads in THA.

### 6.3.4 Contemporary Ceramic-on-Ceramic Hip Implants

Today's manufacturers of second-generation COC alternative bearings claim to have addressed the issue of implant migration in contemporary COC designs with the use of modular acetabular components incorporating a metal-backing having some type of biocompatible coating [136]. Short-term studies of migration in contemporary COC designs have yielded mixed results. Bohler et al. [136] reported on a series of 73 modular hip arthroplasties with identical stems and metal-backed acetabular components, except for a ceramic or UHMWPE insert. Radiographic analysis of Bohler's patients during the first 3 years of implantation revealed significantly higher vertical migration of the acetabular components with ceramic inlays in older (>60y) and osteoporotic patients. However, a more recent randomized study with 53 patients, used radiostereoanalysis (RSA) techniques to measure changes in acetabular lilt and migration after 2 years of follow-up [137]. The researchers concluded that there was "no marked difference in outcomes" between acetabular components with a UHMWPE or ceramic liner. Thus, the potential for migration in current COC designs remains a somewhat controversial topic, and longer-term studies are still needed to address this issue.

Modern COC implants are modular, and contemporary articulating surfaces are fabricated from high-purity alumina or ZTA (e.g., BIOLOX Forte or BIOLOX Delta) or a mixed combination of the two materials (Figure 6.12). The acetabular insert may be fabricated entirely from alumina or ZTA and fitted to the metal shell with a taper junction. Alternatively, the alumina or ZTA insert may be embedded within an UHMWPE liner. Currently, the majority of COC bearings are fabricated with a tapered insert design, rather than using the sandwich design [105]. Both types of COC designs are depicted in Figure 6.14.

Although the sandwich design was initially considered a viable design option for COC implants, these designs have fallen out of favor due to a series of failures reported worldwide [138–141]. As a result of sandwich design failures, which are discussed in greater detail in Section 6.3.7, the modular tapered design has emerged as the current paradigm for COC bearings. This design configuration imposes limits on head size because of the minimum thickness of ceramic liner that can be accommodated without risk of fracture, at least with alumina liners. To improve liner strength, one orthopedic manufacturer has devised a shrink fit titanium shell around the alumina liner as a variation on the modular taper design (Trident, Stryker Orthopedics, Mahwah, New Jersey, USA) [142].

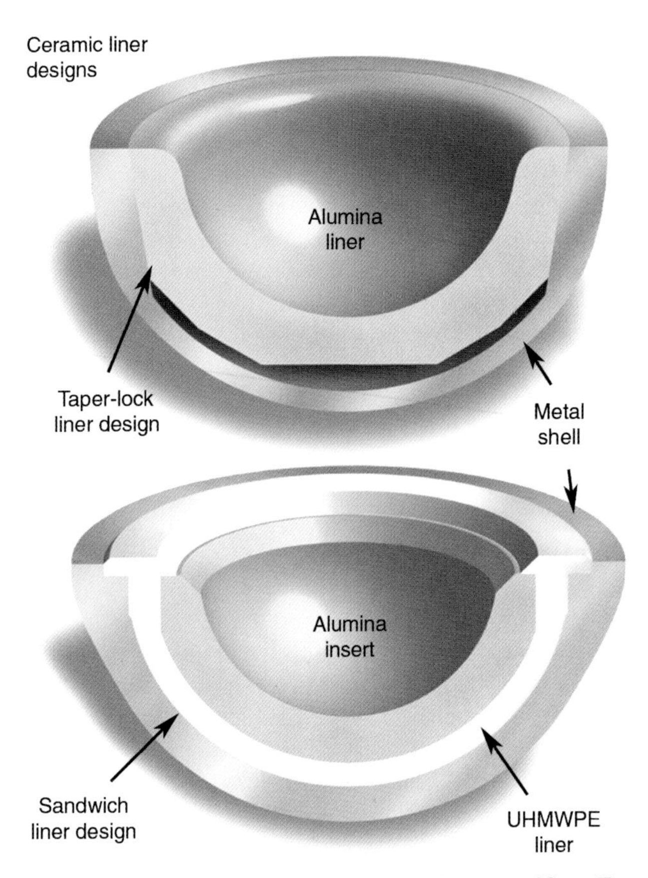


**FIGURE 6.14** Two types of COC designs, provided courtesy of CeramTec (Plochingen, Germany).

Alumina-alumina COC hip designs are clinically available throughout Europe and Asia but were approved for marketing in the United States starting in 2003 due to restrictions by the Food and Drug Administration (FDA). As of 2008, COC bearings incorporating ZTA are not yet approved by the FDA for use in the United States. Starting during the 1990s, the FDA required implant manufacturers to conduct clinical studies to demonstrate the safety of their second-generation COC designs. The FDA has required CeramTec (currently the main supplier of ceramic components to the US market) to maintain the same quality systems as orthopedic manufacturers [102], due in part to the manufacturing issues that were associated with the 2001 recall of zirconia femoral heads. In February 2003, two orthopedic companies (Howmedica Osteonics, Mahwah New Jersey, USA and Wright Medical Technology, Arlington Tennessee, USA) were granted permission by the FDA to market COC designs in the United States. Since

**FIGURE 6.15** Ceramic-on-metal hip bearing, consisting of a ZTA (BIOLOX-Delta) femoral head articulating against a CoCr alloy insert in a Ti alloy metal shell (Depuy International, Leeds). Image provided courtesy of Prof. John Fisher, University of Leeds.

that time, several other manufacturers have received regulatory clearance for COC bearings. Thus, COC implants have undergone particularly rigorous regulatory scrutiny during the past decade, prior to their release by the FDA and the general clinical introduction to the American orthopedic community.

### 6.3.5 Differential Hardness Bearings: Ceramic-on-Metal

Differential hardness bearings borrow a concept from industrial tribology applications, in which it is customary for the counterfaces in a bearing to be fabricated from different materials. Using this philosophy, researchers from Leeds University tested ceramic femoral heads articulating on CoCr alloy cups and found a marked reduction in wear rates when compared with MOM and COC bearings under both idealized and adverse rim loading conditions [8]. Furthermore, the COM bearings exhibited less metal ion release than the MOM bearings under the same *in vitro* test conditions. The concept of differential hardness bearings has since been confirmed in hip simulator studies using CoCr alloys with mismatched hardness [143], as well as by testing silicon nitride articulating on CoCr [121].

COM technology was commercialized in 2005 by DePuy International (Leeds, United Kingdom), using a BIOLOX Delta ceramic head and a CoCr acetabular liner (Figure 6.15). Short-term follow-up at 6 months confirmed lower metal-ion release for COM *in vivo* compared with MOM bearings, as predicted by the hip simulator studies [8]. COM bearings are available for the international market but not yet approved by the FDA for use in the United States.

It must be emphasized that differential hardness has been verified in hip simulator studies only for combinations in which a ceramic head articulates against a CoCr liner. Conversely, it is not recommended that metal heads be used in combination with a ceramic liner. Two case studies of metal-on-ceramic combination bearings have been associated with severe wear and, for one revision case in which a metallic head was used to replace a fractured ceramic component, the head wear that ensued resulted in cobalt poisoning and blindness [144, 145].

### 6.3.6 Wear Mechanisms in Ceramic Bearings

From the perspective of reducing wear-debris induced osteolysis, modern ceramic bearings are associated with ultralow wear rates in contemporary hip simulator studies [7]. In clinical studies, the magnitude of wear in COC bearings is too low to be measured using current radiographic techniques, including RSA (Chapter 28). For this reason, most of our knowledge of *in vivo* wear in COC bearings is based on retrieval studies and comparisons between retrievals and hip simulator studies.

Microscopically, the wear mechanism in alumina ceramics is recognized to initiate with intergranular erosion, followed by isolated grain pull-out. As this process evolves, an entire layer of the ceramic surface becomes disrupted, resulting in a higher surface roughness and increased local friction. The mechanisms of wear in a retrieved alumina hip socket are illustrated in Figure 6.16. Because of its microscopic length scale, worn regions of a ceramic bearing are often only distinguishable by a change in optical reflectivity from shiny (in highly polished, as-manufactured surfaces) to matte and nonreflective (in worn regions). For this reason, researchers often highlight the wear surface of a ceramic bearing by rubbing it with a marker or pencil lead so that its overall shape can be documented in photographs (Figure 6.17).

Many retrieved ceramic bearings also show a macroscopic pattern of wear known as "stripe wear" or "edge loading wear" (Figure 6.17) [28, 146]. This wear pattern was first explained by Nevelos and colleagues [28], who found that microseparation of the ceramic bearing during gait resulted in head-liner contact at the equator of the cup at heel strike. Under these conditions, there is a line of contact under peak loading conditions of the joint, which produces elevated stresses and focal wear. In the early stages of edge-loading, the wear scar has the appearance of a narrow, elliptical stripe on the cup (hence early references to this mechanism as "stripe wear" [28]). Although edge loading increases the wear rate of ceramic bearings [146], even under these severe circumstances the wear is still less than C-PE and M-PE bearings incorporating highly crosslinked UHMWPE.

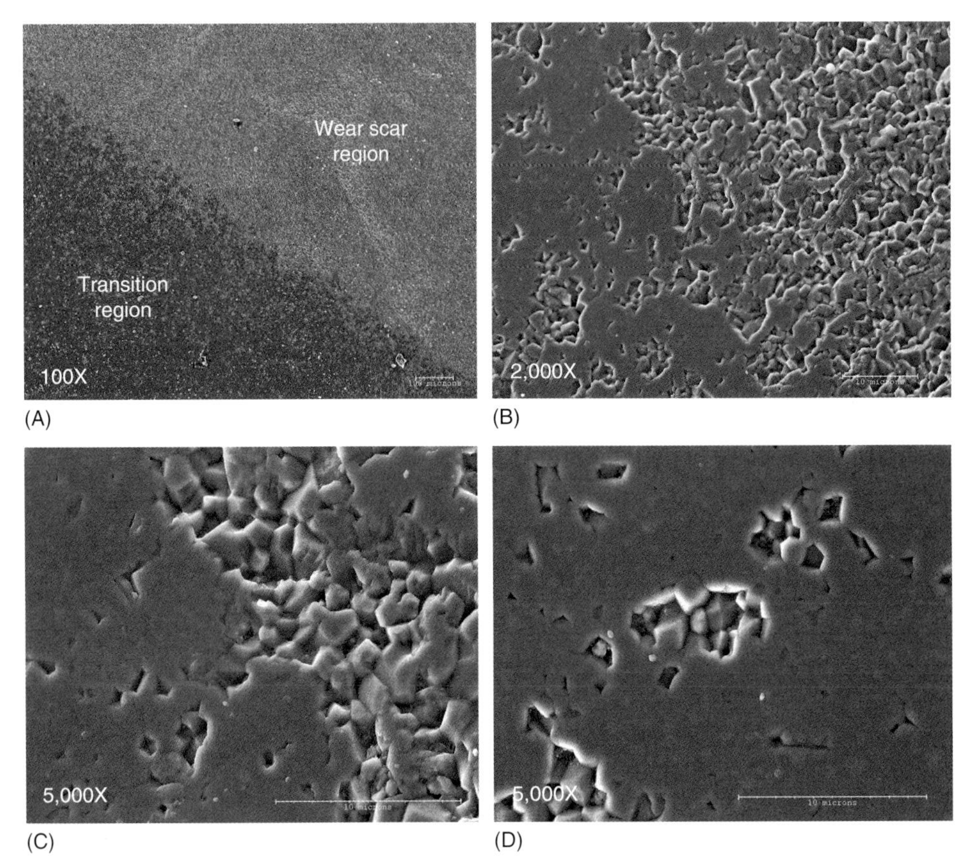


**FIGURE 6.16** Microscopic wear mechanisms in a retrieved alumina COC bearing revised after 17 months for aseptic loosening. Field emission scanning electron micrographs were taken in the vicinity of a wear scar located superiorly, near the equator, and the surrounding transition region (A) at 100X. The wear mechanism underwent a gradual transition from isolated grain boundary erosion to a progressively intercommunicating network of intergranular disruption at 2,000X (B) and 5,000X (C). Within a few millimeters away from the transition region, isolated grain boundary erosion and pullout can also be observed at 5,000X (D).

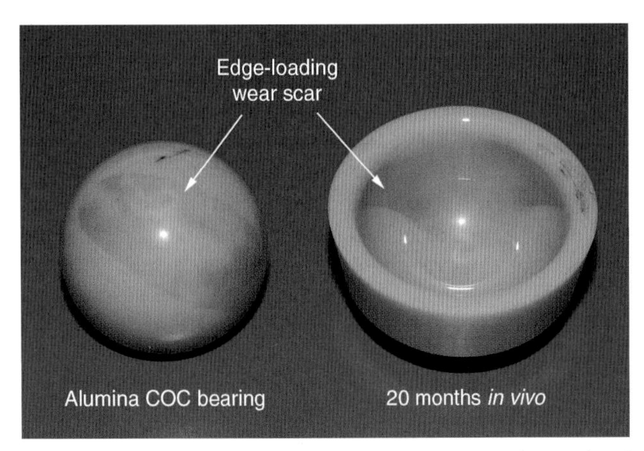


**FIGURE 6.17** Macroscopic pattern of edge-loading wear in a retrieved COC bearing. The wear scar was colored with a marker to facilitate photodocumentation.

Although generally not troubling from an osteolysis perspective, edge-loading wear has been implicated in increasing surface friction and may be one of the factors contributing to squeaking in certain COC bearings. We will return to edge loading and squeaking in Section 6.4. Edge loading wear, in combination with severe impingement, has been documented in two case studies to result in osteolysis from a combination of ceramic and metal particles [147].

### 6.3.7 *In Vivo* Fracture Risk of Ceramic Components for THA

The defining issue with ceramic components for hip replacement remains their *in vivo* fracture risk. A search of PubMed in October 2008 using the keywords "fracture," "ceramics," and "hip" yielded 180 articles, with 56 of these papers published between 2005 and 2008. Thus, ceramic fracture is not merely a historical issue but is a clinically relevant topic for ceramic hip components used in the 21st century. The *in vivo* fracture of a ceramic component is a serious complication requiring immediate revision. The revision of a fractured femoral head may be complicated. As shown in Figure 6.18 (provided courtesy of Prof. Clare Rimnac, Case Western University), a ceramic femoral head typically fractures into multiple fragments, which may be difficult for the revising surgeon to completely clean from the surrounding tissues.

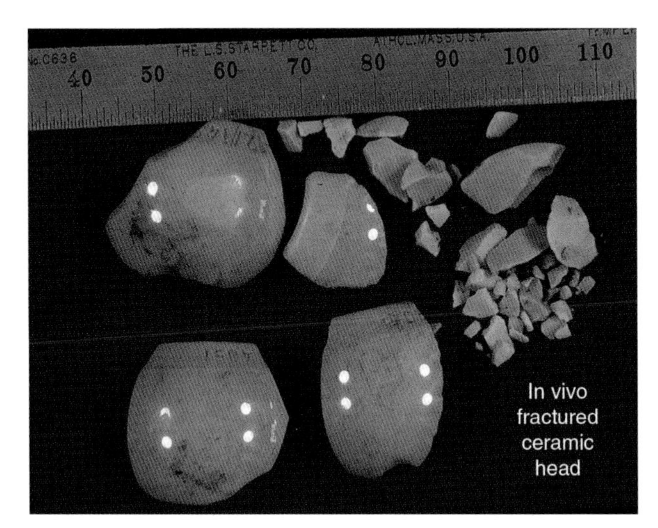


**FIGURE 6.18** A ceramic femoral head typically fractures into multiple fragments.

Furthermore, the ceramic fragments typically damage the Morse taper of the femoral component, requiring the use of a CoCr head for revision. In some cases, the taper is severely damaged by *in vivo* abrasion by ceramic fracture fragments that the entire stem must be revised. Due to the potential for trunnion damage, surgeons were once cautioned not to implant a new primary ceramic head during a revision if the stem is not also going to be revised [148]. Today, if the trunnion is not severely damaged, metallic sleeves are now available to "regenerate" the taper in a revision surgery so that a new ceramic femoral head can be used. An example of a "revision" ceramic head with a metallic sleeve is illustrated in Figure 6.19.

The early clinical experience with COC in Europe was associated with high catastrophic fracture rates. The historical fracture rates ranged between 0% to as high as 13% in some studies [105]. Although the overall fracture rates of modern ceramics in orthopedics is thought to be extremely low, precise quantification of the rates remains problematic, in part due to their rarity. The *in vivo* fracture rates within the overall patient population for contemporary alumina hip components have been reported by CeramTec to range between 1 and 3 in 10,000 [105, 149]. On the other hand, a higher *in vivo* fracture rate has been reported in clinical studies of alumina COC bearings. In a recent survey of the literature, 24 fracture cases were identified out of 35,000 implantations, a fracture rate on the order of 1 in 1,000 [149]. However, these overall rates paint an incomplete picture because an individual surgeon's experience with ceramic bearings may vary from zero fractures in a 5-year period [150] to a 10% overall fracture rate [138] depending upon patient, implant, and clinical factors, even with third-generation alumina components and modern implant designs. Consequently, it remains problematic and perhaps overly simplistic to specify an overall *in vivo* fracture rate of contemporary alumina COC implants. As of yet, there are

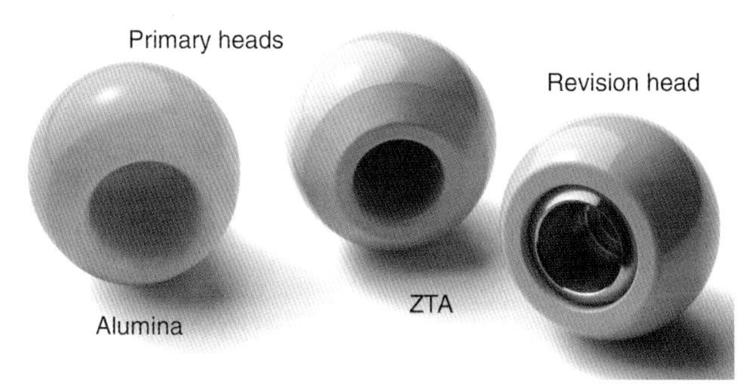


**FIGURE 6.19** Comparison of a ceramic revision femoral head fabricated from zirconia-toughened alumina (ZTA) with primary ceramic heads. The revision head has a metal sleeve to "regenerate" the trunnion of a femoral stem, which may become damaged *in vivo*.

no data available regarding the *in vivo* fracture rate of alumina matrix composite (ZTA) ceramics in hip arthroplasty.

Design factors play a role in the fracture risk of contemporary alumina COC bearings, as illustrated by the recent clinical performance of sandwich-type bearings mentioned earlier (Figure 6.14) [138–141]. In the clinical experience of Kircher et al. in Germany, the fracture rate in a cohort of 50 sandwich-type ceramic liners was 18% and was largely responsible for their overall fracture rate of 10% in their COC series [138]. A clinical trial in the United States was suspended due to high failure rates of the alumina insert in a COC sandwich design (14/315, 4.4%) [140]. The failure mechanism of this design was attributed to disruption of the ceramic-UHMWPE interface, and fracture of the ceramic inlay was shown to occur after it became dislodged or displaced from the UHMWPE liner. Sandwich-type COC designs have been reported with high fracture rates in clinical studies conducted in Europe [138], the United States [140], and in South Korea [141].

Fracture of ceramic components may also occur during implantation. Intraoperative component fractures ("chipping") were documented to occur in 2% to 3% of early cases performed in clinical trials in the United States [142, 149]. These intraoperative fractures were produced by misalignment of the alumina liner in the metal shell, which results in a small fragment of the ceramic chipping off the rim when the insert is seated in the shell by impaction. Although rim chipping may be noticeable by the surgeon and corrected by substituting a new, undamaged liner prior to completing the procedure, the incidence of intraoperative chipping was sufficiently high to motivate one orthopedic manufacturer to redesign the alumina liner in a protective titanium sleeve [142].

Zirconia initially enjoyed a low overall failure rate, similar to alumina, until the events of 2001 forced St. Gobain Desmarquest to suspend its international sales of orthopedic ceramic products. For example, in one of the groups of recalled zirconia femoral heads, a total of 227 *in vivo* fractures were reported for a series of 683 manufactured heads, corresponding to a fracture rate of 33%, according to the data made publicly available by St. Gobain [151].

Due to the possibility that an unforeseen manufacturing change may significantly influence the clinical performance of ceramics, the FDA has taken the unprecedented step of requiring alumina ceramic component suppliers to comply with the strict manufacturing and design controls that historically have only been applied to orthopedic companies. Thus far, only one ceramic manufacturer (i.e., CeramTec) has produced COC components in compliance with the stringent quality requirements of the FDA.

## 6.4 NOISE AND SQUEAKING FROM HARD-ON-HARD BEARINGS

Noise and squeaking have recently emerged as new complications for hard-on-hard bearings. In some cases, the noise is sufficiently loud as to be considered unacceptable to the patient, necessitating revision. Although acoustic emissions have been documented in both MOM and COC bearings, the phenomenon has been most frequently studied in contemporary alumina-alumina bearings [152–160]. The drop in the adoption of COC bearings in the United States between 2005 and 2007 is attributed to concerns by surgeons about bearing noise [106].

Our understanding of acoustic emissions in hard-on-hard bearings remains in the early stages. Squeaking is thought to result from a breakdown of lubrication and high friction at the head-liner interface, producing a forced vibration of the entire arthroplasty construct [160]. Analysis of retrieved components have documented edge loading wear and rim impingement in the majority of patients that were revised with complaints of squeaking [160]. On the other hand, squeaking has been observed in components with no evidence of edge loading wear and/or impingement upon retrieval. Edge loading is frequently noted in retrieved silent bearings [146]. Furthermore, the observation that squeaking is transient in some patients suggests that factors other than edge loading and impingement can also be responsible for acoustic emissions.

Squeaking is considered a new complication because, despite 3 decades of alumina-alumina bearing usage in Europe, reports of clinically significant squeaking requiring revision have only recently surfaced in the literature [153, 156, 160]. The prevalence of noise or squeaking in clinical series varies [153, 154], with rates ranging from 2.7% to 20%, depending upon whether patients are surveyed directly or surgeons are asked about the unsolicited complaints of noise from their patients. Many aspects about the prevalence, significance, and natural history of bearing noise have not yet been fully elucidated.

Research is underway to develop test methods to screen hard-on-hard bearings for their propensity to generate noise [157]. Methods are also being developed to accurately and reproducibly measure bearing noise in a clinical setting. Until further research is conducted to elucidate the clinical, implant, and patient factors associated with the generation of chronic bearing noise, it is still too early to identify potential countermeasures.

## 6.5 HIGHLY CROSSLINKED UHMWPE

Highly crosslinked and thermally treated UHMWPE was clinically introduced for hip bearings in the United States during the late 1990s. Since that time, highly crosslinked UHMWPE has steadily replaced conventional polyethylene for hip arthroplasty [161]. Over 2 million patients worldwide have been implanted with highly crosslinked UHMWPE (Chapters 13 and 14). Therefore, highly crosslinked UHMWPE is now widely accepted by surgeons as the main alternative to hard-on-hard bearings for improving the wear resistance of hip arthroplasty [162].

In this section, we first briefly review the historical experience of highly crosslinked UHMWPE for hip replacement and summarize the general characteristics of first-generation materials in current clinical use. This section also describes the effect of thermal treatment on the properties of this family of materials, as well as the motivation to develop second-generation materials. The reader will find more detailed information, including clinical data, for first- and second-generation highly crosslinked UHMWPE in later chapters of this book (Chapters 13–15 and 20).

### 6.5.1 Historical Clinical Experience with Highly Crosslinked UHMWPE

During the 1970s, Oonishi in Japan [163–165] and Grobbelaar in South Africa [166] implanted highly irradiated UHMWPE hip cups. These formulations of UHMWPE were produced using doses of up to 1000 kGy in air, by Oonishi, and up to 100 kGy in the presence of acetylene, by Grobbelaar. During the 1980s, a small clinical study of 22 patients was conducted by Wroblewski using an experimental, chemically crosslinked high density polyethylene (HDPE) [167]. Due to the small numbers of patients, lack of matched controls, and a substantial number of patients lost to follow-up, these three clinical studies of highly crosslinked polyethylene provide only anecdotal confirmation of the benefits of increased crosslinking. These isolated clinical experiments from around the world cannot really be considered a generation of highly crosslinked UHMWPE but are noteworthy for their innovation nonetheless.

### 6.5.2 First-Generation Highly Crosslinked UHMWPE

Starting in 1998, orthopedic manufacturers introduced the so-called first generation of highly crosslinked UHMWPEs for total hip replacement [168]. These materials are processed with a total dose ranging from 50 to 105 kGy, depending on the manufacturer (Chapter 20). Saturation of crosslinking is achieved at around 100 kGy of absorbed dose. Besides choice of dose, each manufacturer adopted a different route for production that includes a proprietary combination of three important factors: an irradiation step, a postirradiation thermal processing step, and thirdly, a sterilization step (see Figure 6.20). It is now well established in the literature that first generation highly crosslinked UHMWPE materials exhibit reduced wear compared to conventional UHMWPE, both experimentally [168] and clinically [169].

Under ideal (*in vitro*) circumstances, the magnitude of the wear reduction for first-generation highly crosslinked UHMWPEs is proportional to the absorbed dose, but in practical terms the perceived wear reduction is also influenced by experimental study design, including such factors as the radiographic wear measurement technique (Chapter 28) and the choice of control material used as a comparison (Chapter 20). The choice of irradiation source (gamma or e-beam) does not substantially influence wear resistance

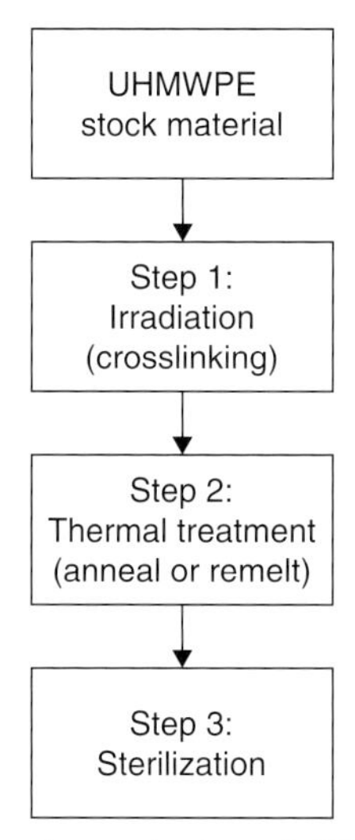


**FIGURE 6.20** Route for production of "first generation" highly crosslinked and thermally treated UHMWPE that includes a proprietary combination of three important factors.

of the resulting polymer, and commercial materials are made using either methodology.

Although crosslinking is highly beneficial from the perspective of wear resistance, it comes at the price of less molecular mobility, lower material ductility, and reduced fatigue and fracture resistance due to the crosslinking of the molecular chains [170]. Thus crosslinking improves certain properties at the expense of others, so developers of orthopedic implants must balance the amount of crosslinking achieved with the maintenance of mechanical properties and/or oxidation resistance.

The final tuning step used in the production of first-generation highly crosslinked UHMWPE is thermal processing of the irradiated material. Thermal processing below the melt transition (~137°C) is known as annealing, whereas remelting refers to thermal processing above the melt transition. The choice of thermal treatment has a significant impact on the crystallinity and mechanical properties of highly crosslinked UHMWPE [171]. It can also influence the resistance of a material to *in vivo* oxidation (Chapters 14 and 22).

At a dose of 100 kGy, the elastic modulus, yield stress, and ultimate stress of a remelted material is significantly lower than the respective properties for an annealed material (Table 6.3).

Figure 6.21 compares the uniaxial tensile behavior of unirradiated UHMWPE material with conventionally sterilized (30 kGy, in N2) polyethylene and with both annealed and remelted highly crosslinked polyethylenes (100 kGy).

For the two highly crosslinked UHMWPEs shown in Figure 6.21, the annealed material has an average degree of crystallinity of 60%, whereas the remelted material has a crystallinity of 43%. Throughout the entire stress-strain curve, the higher crystallinity of the annealed material results in a greater resistance to plastic deformation when compared to remelted material. Therefore, the selection of postirradiation thermal treatment is the second most important decision for an implant designer because it will influence not only its oxidative stability but also the crystallinity, yield strength, and ultimate tensile strength of the highly crosslinked polyethylene. These reduced mechanical properties may not influence wear but will certainly influence the resistance of

**TABLE 6.3** Effect of Postirradiation Thermal Treatment on Uniaxial Mechanical Properties [172]

| Dose (gamma) | Heat treatment | Yield stress (MPa) | Ultimate stress (MPa) | Elongation to failure (%) |
|---|---|---|---|---|
| 100 kGy | None | 23.2 ± 0.2 | 47.6 ± 2.0 | 238 ± 13 |
| 100 kGy | 110°C anneal | 23.0 ± 0.3 | 47.3 ± 1.5 | 230 ± 12 |
| 100 kGy | 130°C anneal | 22.6 ± 0.2 | 48.5 ± 1.5 | 231 ± 13 |
| 100 kGy | 150°C remelt | 19.5 ± 0.3 | 43.9 ± 3.9 | 246 ± 12 |

*Note that these irradiation treatments were achieved with a single dose. Properties were determined from treated rods of GUR 1050.*

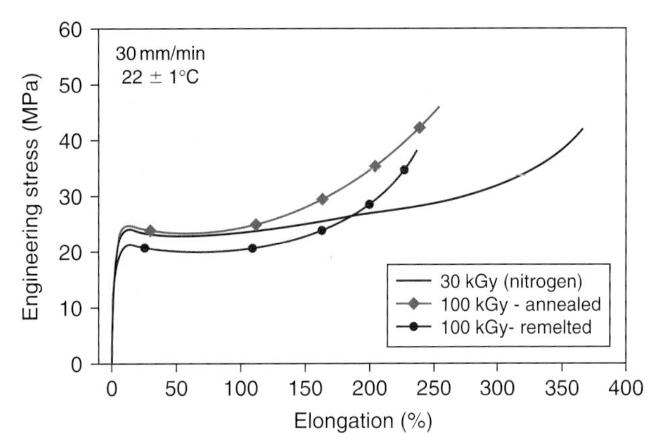


**FIGURE 6.21** Uniaxial stress-strain behavior of two highly crosslinked UHMWPEs. The annealed material has an average degree of crystallinity of 60%, whereas the remelted material has a crystallinity of 43%.

the material to damage caused by impingement or bearing lift off. Although the virtues of annealing and remelting have been debated in the literature [14, 173], as well as in Chapters 13 and 14 of the present work, ultimately the overall clinical performance of highly crosslinked UHMWPE has been generally successful for both types of materials (Chapter 20).

Over the past 8 years, a few reports of rim fracture of remelted crosslinked liners have surfaced in the literature [174, 175], as well as in FDA Maude reports (see http://www.accessdata.fda.gov/scripts/cdrh/cfdocs/cfMAUDE/search.CFM). These case reports are usually associated with thin liners associated with acetabular shells implanted with a high abduction angle. Due to the small numbers reported thus far in the literature, the incidence of rim fracture necessitating revision of crosslinked polyethylenes remains poorly understood. None of the numerous clinical studies of remelted polyethylenes have reported rim fracture as a clinically relevant failure mode. However, the few rare cases of rim fracture have provided motivation for improving the mechanical behavior of crosslinked polyethylenes, especially for thin liners.

### 6.5.3 Second-Generation Highly Crosslinked UHMWPE

Highly crosslinked UHMWPE was clinically introduced at a time when 28 mm femoral heads were widely used, whereas today there is a demand for larger diameter articulations for hip arthroplasty. With the incorporation of highly crosslinked polyethylene into new large-diameter cup designs (e.g., Figure 6.1), modes of clinical failure other than wear may become new limiting factors for the long-term clinical performance of crosslinked polyethylene. These factors include component fracture associated with rim loading, thin liners, and impingement-related damage due to component malpositioning. The clinical introduction of thin acetabular liners incorporating highly crosslinked UHMWPE raises new questions regarding the mechanical properties of these thin liner designs being able to withstand structural fatigue loading.

Given the broad range of highly crosslinked materials currently available for hip arthroplasty, controversies remain regarding the relevance of *in vivo* oxidation [176] and the risk of rim impingement damage and fracture [174, 175]. The controversies surrounding the importance of *in vivo* oxidation and rim fracture continue to be debated in scientific circles and have led to the development of additional second generation highly crosslinked UHMWPE. Examples of these materials include X3 (Stryker Orthopedics: Mahwah, New Jersey, USA), as well as ArCom XL and E-Poly (Biomet Orthopedics, Warsaw, Indiana, USA) (Chapter 20). The intent of these second generation materials is to reduce the potential for material oxidation in the long term while preserving the bulk mechanical properties necessary to use crosslinked polyethylene in higher stress applications, such as thin acetabular liners and stabilized knee designs. Laboratory testing suggests that material design goals have been satisfied in these new polymer formulations. However, due to clinical introduction within the past 2 years, clinical data regarding the performance of these new formulations is not yet available.

## 6.6 SUMMARY

Today, orthopedic surgeons and patients have even more alternative bearings for hip replacement to choose from than were available 5 years ago. Among hard-on-soft bearings, highly crosslinked UHMWPE has now superseded conventional UHMWPE for hip arthroplasty, especially in the United States. MOM, COC, and highly crosslinked acetabular liners—used in conjunction with either a CoCr or ceramic femoral head—all have the potential to significantly reduce the clinical wear rates relative to the existing gold standard of conventional UHMWPE. All three of these mainstream alternative bearings incorporate the successful elements of historical precedents.

The use of hard-on-hard bearings does entail potential risks for the patient. With MOM bearings, the concern is the potential for hypersensitivity and the biological risks associated with elevated metal ion exposure. MOM hip resurfacing also introduces the risk of femoral neck fracture because the femoral neck is retained during the primary procedure, unlike in conventional total hip replacement. With COC, the concern is the risk of fracture for the femoral head and/or the acetabular liner. For all hard-on-hard bearings, there is the potential for noise and squeaking, which in some cases has required revision. Over the past 5 years, new hard-on-hard bearing combinations, including different ceramic biomaterials, as well as COM bearings, have also been introduced clinically; however, it is too early to judge their long-term risks relative to other bearing couples.

The alternative bearings that have been reviewed in this chapter represent the current state of the art in orthopedics, and all may be expected to reduce osteolysis when properly implanted. Although all of these new alternative bearings have the potential to reduce wear, none may claim to be as forgiving of malpositioning and impingement as conventional UHMWPE for hip arthroplasty. The ultimate goal for all of these alternative bearings is to reduce the overall incidence of hip revision surgery. Many years of clinical follow-up are still needed to verify the attainment of this long-term objective.

## REFERENCES


1. NIH Consensus Statement: Total Hip Replacement. *National Institutes of Health Technology Assessment Conference,* 1994 September 12–14.
2. Harris WH. Aseptic loosening in total hip arthroplasty secondary to osteolysis induced by wear debris from titanium-alloy modular femoral heads. *J Bone Joint Surg Am* 1991;**73**(3):470–2.
3. Harris WH. Osteolysis and particle disease in hip replacement. *Acta Orthop Scand* 1994;**65**(1):113–23.
4. Peters Jr. PC, Engh GA, Dwyer KA, Vinh TN. Osteolysis after total knee arthroplasty without cement. *J Bone Joint Surg Am* 1992;**74**(6):864–76.
5. Schmalzried TP, Guttmann D, Grecula M, Amstutz HC. The relationship between the design, position, and articular wear of acetabular components inserted without cement and the development of pelvic osteolysis. *J Bone Joint Surg Am* 1994;**76**(5):677–88.
6. Willert HG, Bertram H, Buchhorn GH. Osteolysis in alloarthroplasty of the hip. The role of bone cement fragmentation. *Clin Orthop Relat Res* 1990(258):108–21.
7. Clarke IC, Good V, Williams P, Schroeder D, Anissian L, Stark A, et al. Ultra-low wear rates for rigid-on-rigid bearings in total hip replacements. *Proc Inst Mech Eng [H]* 2000;**214**(4):331–47.
8. Williams S, Schepers A, Isaac G, Hardaker C, Ingham E, van der Jagt D, et al. The 2007 Otto Aufranc Award. Ceramic-on-metal hip arthroplasties: a comparative in vitro and in vivo study. *Clin Orthop Relat Res* 2007;**465**:23–32.
9. Scheerlinck T, Druyts P, Casteleyn PP. The use of primary total hip arthroplasty in university hospitals of the European Union. *Acta Orthop Belg* 2004;**70**(3):231–9.
10. Australian Orthopaedic Association National Joint Replacement Registry. *Annual report*. 2008 [cited September 21, 2008]. Available from: http://www.dmac.adelaide.edu.au/aoanjrr/publications.jsp
11. Mendenhall S. Hospital resources and implant cost management–a 2007 update. *Orthop Network News* 2008;**19**(3):13–19.
12. Kurtz S, Lau E, Ong K, Zhao K, Kelly M, Bozic K. Future young patient demand for primary and revision joint replacement: National projections from 2010 to 2030. *Trans of the 54th Orthop Res Soc* 2008;**33**:1784.
13. Sauer WL, Anthony ME. Predicting the clinical wear performance of orthopaedic bearing surfaces. In: Jacobs JJ, Craig TL, editors. *Alternative bearing surfaces in total joint replacement.* West Conshohoken: American Society for Testing and Materials; 1998. p. 1–29.
14. Muratoglu OK, Kurtz SM. Alternative bearing surfaces in hip replacement. In: Sinha R, editor. *Hip replacement: current trends and controversies*. New York: Marcel Dekker; 2002. p. 1–46.
15. Smith-Petersen MN. Evolution of mould arthroplasty of the hip joint. *J Bone Joint Br* 1948;**30-B**(1):59–75.
16. Lemons JE. Metals and Alloys. In: Petty W, editor. *Total joint replacement*. Philaldelphia: W.B. Saunders Company; 1991. p. 21–7.
17. Dearnley PA. A review of metallic, ceramic and surface-treated metals used for bearing surfaces in human joint replacements. *Proc Inst Mech Eng [H]* 1999;**213**(2):107–35.
18. Park JB. Metallic biomaterials. In: Bronzino JD, editor. *The biomedical engineering handbook*. Boca Raton, FL: CRC Press LLC; 1995. p. 537–51.
19. Varano R, Yue S, Bobyn JD, Medley J. Co-Cr-Mo alloys used in metal-metal bearing surfaces. In: Jacobs JJ, Craig TL, editors. *Alternative bearing surfaces in total joint replacement*. West Conshohoken: American Society for Testing and Materials; 1998. p. 55–68.
20. Bowsher JG, Nevelos J, Williams PA, Shelton JC. 'Severe' wear challenge to 'as-cast' and 'double heat-treated' large-diameter metal-on-metal hip bearings. *Proc Inst Mech Eng [H]* 2006;**220**:135–43.
21. Dowson D, Hardaker C, Flett M, Isaac GH. A hip joint simulator study of the performance of metal-on-metal joints. Part I: the role of materials. *J Arthroplasty* 2004;**19**(Suppl 3):118–23.
22. Isaac GH, Thompson J, Williams S, Fisher J. Metal-on-metal bearing surfaces: materials, manufacture, design, optimization, and alternatives. *Proc Inst Mech Eng [H]* 2006;**220**:119–33.
23. Amstutz HC, Campbell P, McKellop H, Schmalzreid TP, Gillespie WJ, Howie D, et al. Metal on metal total hip replacement workshop consensus document. *Clin Orthop* 1996;**329**(329 Suppl):S297–303.
24. Sieber HP, Rieker CB, Kottig P. Analysis of 118 second-generation metal-on-metal retrieved hip implants. *J Bone Joint Surg Br* 1999;**81**(1):46–50.
25. Willert HG, Buchhorn GH. Retrieval studies on classic cemented metal-on-metal hip endoprostheses. In: Rieker CB, Windler M, Wyss U, editors. *Metasul: a metal-on-metal bearing*. Bern: Hans Huber; 1999. p. 39–60.
26. Kothari M, Bartel DL, Booker JF. Surface geometry of retrieved McKee-Farrar total hip replacements. *Clin Orthop* 1996;**329**(329 Suppl):S141–7.
27. McKellop H, Park SH, Chiesa R, Doorn P, Lu B, Normand P, et al. In vivo wear of three types of metal on metal hip prostheses during two decades of use. *Clin Orthop* 1996;**329**(329 Suppl):S128–40.
28. Nevelos JE, Ingham E, Doyle C, Fisher J, Nevelos AB. Analysis of retrieved alumina ceramic components from Mittelmeier total hip prostheses. *Biomaterials* 1999;**20**(19):1833–40.
29. Wiles P. The surgery of the osteo-arthritic hip. *Br J Surg* 1957; **45**:488–97.
30. McKee GK, Watson-Farrar J. Replacement of arthritic hips by the McKee-Farrar prosthesis. *J Bone Joint Surg [Br]* 1966;**48**(2):245–59.
31. Ring PA. Complete replacement arthroplasty of the hip by the ring prosthesis. *J Bone Joint Surg Br* 1968;**50**(4):720–31.
32. Zahiri CA, Schmalzried TP, Ebramzadeh E, Szuszczewicz ES, Salib D, Kim C, et al. Lessons learned from loosening of the McKee-Farrar metal-on-metal total hip replacement. *J Arthroplasty* 1999;**14**(3):326–32.
33. Willert HG. Reactions of the articular capsule to wear products of artificial joint prostheses. *J Biomed Mater Res* 1977;**11**(2):157–64.
34. Willert HG, Buchhorn GH, Gobel D, Koster G, Schaffner S, Schenk R, et al. Wear behavior and histopathology of classic cemented metal on metal hip endoprostheses. *Clin Orthop Relat Res* 1996(329 Suppl):S160–86.

35. Tharani R, Dorey FJ, Schmalzried TP. The risk of cancer following total hip or knee arthroplasty. *J Bone Joint Surg Am* 2001;**83-A**(5):774–80.
36. Jacobs JJ, Skipor AK, Doorn PF, Campbell P, Schmalzried TP, Black J, et al. Cobalt and chromium concentrations in patients with metal on metal total hip replacements. *Clin Orthop* 1996;**329**(329 Suppl):S256–63.
37. Muller ME. The benefits of metal-on-metal total hip replacements. *Clin Orthop* 1995;**311**(311):54–9.
38. August AC, Aldam CH, Pynsent PB. The McKee-Farrar hip arthroplasty. A long-term study. *J Bone Joint Surg Br* 1986;**68**(4):520–7.
39. Jacobsson SA, Djerf K, Wahlstrom O. A comparative study between McKee-Farrar and Charnley arthroplasty with long-term follow-up periods. *J Arthroplasty* 1990;**5**(1):9–14.
40. Jacobsson SA, Djerf K, Wahlstrom O. Twenty-year results of McKee-Farrar versus Charnley prosthesis. *Clin Orthop* 1996(329 Suppl):S60–8.
41. Wagner M, Wagner H. Medium-term results of a modern metal-on-metal system in total hip replacement. *Clin Orthop* 2000(379):123–33.
42. Wagner M, Wagner H. Preliminary results of uncemented metal on metal stemmed and resurfacing hip replacement arthroplasty. *Clin Orthop* 1996;**329** (329 Suppl):S78–88.
43. Weber BG. METASUL from 1988 to today. In: Rieker CB, Windler M, Wyss U, editors *Metasul: a metal-on-metal bearing*. Bern: Hans Huber; 1999. p. 23–8.
44. Schmidt M, Weber H, Schon R. Cobalt chromium molybdenum metal combination for modular hip prostheses. *Clin Orthop* 1996;**329**(329 Suppl):S35–47.
45. Rieker CB, Weber H, Schön R, Windler M, Wyss U. Development of the METASUL articulations. In: Rieker CB, Windler M, Wyss U, editors. *Metasul: a metal-on-metal bearing*. Bern: Hans Huber; 1999. p. 15–21.
46. Amstutz HC, Le Duff MJ. Background of metal-on-metal resurfacing. *Proc Inst Mech Eng [H]* 2006;**220**(2):85–94.
47. Clarke IC, Donaldson T, Bowsher JG, Nasser S, Takahashi T. Current concepts of metal-on-metal hip resurfacing. *Orthop Clin North Am* 2005;**36**(2):143–62. viii.
48. McMinn D, Daniel J. History and modern concepts in surface replacement. *Proc Inst Mech Eng [H]* 2006;**220**(2):239–51.
49. Malchau H, Herberts P, Eisler T, Garellick G, Soderman P. The swedish total hip replacement register. *J Bone Joint Surg Am* 2002;**84-A**(Suppl 2):2–20.
50. Bozic KJ, Kurtz S, Lau E, Ong K, Vail T, Berry D. The epidemiology of revision total hip arthroplasty in the United States. *18th annual meeting of the American association of hip and knee surgeons*, Dallas, TX: 2008.
51. Phillips CB, Barrett JA, Losina E, Mahomed NN, Lingard EA, Guadagnoli E, et al. Incidence rates of dislocation, pulmonary embolism, and deep infection during the first six months after elective total hip replacement. *J Bone Joint Surg Am* 2003;**85-A**(1):20–6.
52. Amstutz HC, Beaule PE, Dorey FJ, Le Duff MJ, Campbell PA, Gruen TA. Metal-on-metal hybrid surface arthroplasty: two to six-year follow-up study. *J Bone Joint Surg Am* 2004;**86-A**(1):28–39.
53. Mont MA, Ragland PS, Etienne G, Seyler TM, Schmalzried TP. Hip resurfacing arthroplasty. *J Am Acad Orthop Surg* 2006;**14**(8):454–63.
54. National Joint Registry. *National Joint Registry for England and Wales: 4th Annual report*, 2007.
55. Beaule PE, Campbell PA, Hoke R, Dorey F. Notching of the femoral neck during resurfacing arthroplasty of the hip: a vascular study. *J Bone Joint Surg Br* 2006;**88**(1):35–9.
56. Shimmin AJ, Back D. Femoral neck fractures following Birmingham hip resurfacing: a national review of 50 cases. *J Bone Joint Surg Br* 2005;**87**(4):463–4.
57. Schmalzried TP, Silva M, de la Rosa MA, Choi ES, Fowble VA. Optimizing patient selection and outcomes with total hip resurfacing. *Clin Orthop Relat Res* 2005;**441**:200–4.
58. Siebel T, Maubach S, Morlock MM. Lessons learned from early clinical experience and results of 300 ASR hip resurfacing implantations. *Proc Inst Mech Eng [H]* 2006;**220**(2):345–53.
59. Beaule PE, Harvey N, Zaragoza E, Le Duff MJ, Dorey FJ. The femoral head/neck offset and hip resurfacing. *J Bone Joint Surg Br* 2007;**89**(1):9–15.
60. Silva M, Lee KH, Heisel C, Dela Rosa MA, Schmalzried TP. The biomechanical results of total hip resurfacing arthroplasty. *J Bone Joint Surg Am* 2004;**86-A**(1):40–6.
61. Long JP, Bartel DL. Surgical variables affect the mechanics of a hip resurfacing system. *Clin Orthop Relat Res* 2006;**453**:115–22.
62. Ong KL, Day JS, Kurtz SM, Field RE, Manley MT. Role of surgical position on interface stress and initial bone remodeling stimulus around hip resurfacing arthroplasty. *J Arthroplasty* 2008.; *in press*.
63. Ong KL, Manley MT, Kurtz SM. Have contemporary hip resurfacing designs reached maturity? A review. *J Bone Joint Surg Am* 2008;**90**(Suppl 3):81–8.
64. Shimmin A, Beaule PE, Campbell P. Metal-on-metal hip resurfacing arthroplasty. *J Bone Joint Surg Am* 2008;**90**(3):637–54.
65. Firkins PJ, Tipper JL, Saadatzadeh MR, Ingham E, Stone MH, Farrar R, et al. Quantitative analysis of wear and wear debris from metal-on-metal hip prostheses tested in a physiological hip joint simulator. *Biomed Mater Eng* 2001;**11**(2):143–57.
66. Doorn PF, Campbell PA, Worrall J, Benya PD, McKellop HA, Amstutz HC. Metal wear particle characterization from metal on metal total hip replacements: transmission electron microscopy study of periprosthetic tissues and isolated particles. *J Biomed Mater Res* 1998;**42**(1):103–11.
67. Doorn PF, Campbell P, Amstutz H. Particle disease in metal-on-metal total hip replacements. In: Rieker CB, Windler M, Wyss U, editors *Metasul: a metal-on-metal bearing*. Bern: Hans Huber; 1999. p. 113–19.
68. Fisher J, Ingham E, Stone M, Wroblewski BM, Besong AA, Tipper JL, et al. Wear particle morphologies in artificial hip joints: Particle size is critical to the response of macrophages. In: Rieker CB, Windler M, Wyss U, editors *Metasul: a metal-on-metal bearing*. Bern: Hans Huber; 1999. p. 121–4.
69. Jacobs JJ, Skipor AK, Patterson LM, Hallab NJ, Paprosky WG, Black J, et al. Metal release in patients who have had a primary total hip arthroplasty. A prospective, controlled, longitudinal study. *J Bone Joint Surg Am* 1998;**80**(10):1447–58.
70. Jacobs JJ, Hallab NJ, Skipor AK, Urban RM, Mikecz K, Glant TT. Metallic wear and corrosion products: biological implications. In: Rieker CB, Windler M, Wyss U, editors. *Metasul: a metal-on-metal bearing*. Bern: Hans Huber; 1999. p. 125–32.
71. Brodner W, Bitzan P, Meisinger V, Kaider A, Gottsauner-Wolf F, Kotz R. Elevated serum cobalt with metal-on-metal articulating surfaces. *J Bone Joint Surg Br* 1997;**79**(2):316–21.
72. Visuri T, Pukkala E, Paavolainen P, Pulkkinen P, Riska EB. Cancer risk after metal on metal and polyethylene on metal total hip arthroplasty. *Clin Orthop* 1996;**329** (329 Suppl):S280–9.
73. Willert HG, Buchhorn GH, Fayyazi A. Hypersensitivity to wear products in metal-on-metal articulation. In: Zippel H, Dietrich M, editors. *Bioceramics in joint arthroplasty, 8th Biolox symposium proceedings*. Darmstadt: Steinkopff Verlag; 2003. p. 65–72.

74. Hallab N, Merritt K, Jacobs JJ. Metal sensitivity in patients with orthopaedic implants. *J Bone Joint Surg Am* 2001;**83-A**(3):428–36.
75. De Haan R, Campbell PA, Su EP, De Smet KA. Revision of metal-on-metal resurfacing arthroplasty of the hip: the influence of malpositioning of the components. *J Bone Joint Surg Br* 2008;**90**(9):1158–63.
76. De Haan R, Pattyn C, Gill HS, Murray DW, Campbell PA, De Smet KA. Correlation between inclination of the acetabular component and metal ion levels in metal-on-metal hip resurfacing replacement. *J Bone Joint Surg Br* 2008;**90**(10):1291–7.
77. Williams S, Leslie I, Isaac G, Jin Z, Ingham E, Fisher J. Tribology and wear of metal-on-metal hip prostheses: influence of cup angle and head position. *J Bone Joint Surg Am* 2008;**90**(Suppl 3):111–17.
78. Boardman DR, Middleton FR, Kavanagh TG. A benign psoas mass following metal-on-metal resurfacing of the hip. *J Bone Joint Surg Br* 2006;**88**(3):402–4.
79. Clayton RA, Beggs I, Salter DM, Grant MH, Patton JT, Porter DE. Inflammatory pseudotumor associated with femoral nerve palsy following metal-on-metal resurfacing of the hip. A case report. *J Bone Joint Surg Am* 2008;**90**(9):1988–93.
80. Davies AP, Willert HG, Campbell PA, Learmonth ID, Case CP. An unusual lymphocytic perivascular infiltration in tissues around contemporary metal-on-metal joint replacements. *J Bone Joint Surg Am* 2005;**87**(1):18–27.
81. Gruber FW, Bock A, Trattnig S, Lintner F, Ritschl P. Cystic lesion of the groin due to metallosis: a rare long-term complication of metal-on-metal total hip arthroplasty. *J Arthroplasty* 2007;**22**(6):923–7.
82. Pandit H, Glyn-Jones S, McLardy-Smith P, Gundle R, Whitwell D, Gibbons CL, et al. Pseudotumours associated with metal-on-metal hip resurfacings. *J Bone Joint Surg Br* 2008;**90**(7):847–51.
83. Silva M, Heisel C, Schmalzried TP. Metal-on-metal total hip replacement. *Clin Orthop Relat Res* 2005(430):53–61.
84. Willert HG, Buchhorn GH, Fayyazi A, Flury R, Windler M, Koster G, et al. Metal-on-metal bearings and hypersensitivity in patients with artificial hip joints. A clinical and histomorphological study. *J Bone Joint Surg Am* 2005;**87**(1):28–36.
85. Park YS, Moon YW, Lim SJ, Yang JM, Ahn G, Choi YL. Early osteolysis following second-generation metal-on-metal hip replacement. *J Bone Joint Surg Am* 2005;**87**(7):1515–21.
86. Boutin P. [Alumina and its use in surgery of the hip. (Experimental study)]. *Presse Med* 1971;**79**(14):639–40.
87. Boutin P. [Total arthroplasty of the hip by fritted aluminum prosthesis. Experimental study and 1st clinical applications]. *Rev Chir Orthop Reparatrice Appar Mot* 1972;**58**(3):229–46.
88. Shikata T, Oonishi H, Hashimato Y, et al. Wear resistance of irradiated UHMW polyethylenes to Al2O3 ceramics in total hip prostheses. *Trans of the 3rd Annu Meet of the Soc for Biomater* 1977:118.
89. Hannouche D, Hamadouche M, Nizard R, Bizot P, Meunier A, Sedel L. Ceramics in total hip replacement. *Clin Orthop Relat Res* 2005(430):62–71.
90. Sedel L, Kerboull L, Christel P, Meunier A, Witvoet J. Alumina-on-alumina hip replacement. Results and survivorship in young patients. *J Bone Joint Surg Br* 1990;**72**(4):658–63.
91. Mittelmeier H. Ceramic prosthetic devices. *Hip* 1984:146–60.
92. Winter M, Griss P, Scheller G, Moser T. Ten- to 14-year results of a ceramic hip prosthesis. *Clin Orthop* 1992(282):73–80.
93. Mahoney OM, Dimon JH, 3rd. Unsatisfactory results with a ceramic total hip prosthesis. *J Bone Joint Surg Am* 1990;**72**(5):663–71.
94. Jazrawi LM, Bogner E, Della Valle CJ, Chen FS, Pak KI, Stuchin SA, et al. Wear rates of ceramic-on-ceramic bearing surfaces in total hip implants: a 12-year follow-up study. *J Arthroplasty* 1999;**14**(7):781–7.
95. Dorlot JM, Christel P, Meunier A. Wear analysis of retrieved alumina heads and sockets of hip prostheses. *J Biomed Mater Res* 1989;**23**(A3 Suppl):299–310.
96. Bizot P, Nizard R, Lerouge S, Prudhommeaux F, Sedel L. Ceramic/ceramic total hip arthroplasty. *J Orthop Sci* 2000;**5**(6):622–7.
97. Nevelos AB, Evans PA, Harrison P, Rainforth M. Examination of alumina ceramic components from total hip arthroplasties. *Proc Inst Mech Eng [H]* 1993;**207**(3):155–62.
98. Boehler M, Plenk Jr. H., Salzer M. Alumina ceramic bearings for hip endoprostheses: the Austrian experiences. *Clin Orthop* 2000;**379**(379):85–93.
99. Willmann G. Ceramics for total hip replacement—what a surgeon should know. *Orthopedics* 1998;**21**(2):173–7.
100. Willmann G. Ceramic femoral head retrieval data. *Clin Orthop* 2000;**379**(379):22–8.
101. Merkert P. Next generation ceramic bearings. In: Zippel H, Dietrich M, editors. *Bioceramics in joint arthroplasty, 8th Biolox symposium proceedings*. Darmstadt: Steinkopff Verlag; 2003. p. 123–5.
102. Dobbs H. Quality improvement resulting from legal and regulatory developments. In: Zippel H, Dietrich M, editors, *Bioceramics in joint arthroplasty, 8th Biolox symposium proceedings*. Darmstadt: Steinkopff Verlag; 2003. p. 205–8.
103. Griesmayr G, Dietrich M, Kasprowitsch J, Dobbs H. Improvements in processing and manufacturing at CeramTec. In: Zippel H, Dietrich M, editors, *Bioceramics in joint arthroplasty, 8th Biolox symposium proceedings*. Darmstadt: Steinkopff Verlag; 2003. p. 209–14.
104. Morlock M, Nassutt R, Wimmer MA, Schneider E. Influence of resting periods on friction in artificial hip joint articulations. In: Garino JP, Willmann G, editors. *Bioceramics in joint arthroplasty, Proceedings of the 7th international BIOLOX symposium*. Stuttgart: Thieme; 2002. p. 6–20.
105. Willmann G. Fiction and facts concerning the reliability of ceramics in THR. In: Zippel H, Dietrich M, editors. *Bioceramics in joint arthroplasty, 8th Biolox symposium proceedings*. Darmstadt: Steinkopff Verlag; 2003. p. 193–6.
106. Heros R. Personal communication. CeramTec; 2008.
107. Piconi C, Maccauro G. Zirconia as a ceramic biomaterial. *Biomaterials* 1999;**20**(1):1–25.
108. Cales B. Zirconia as a sliding material: histologic, laboratory, and clinical data. *Clin Orthop* 2000;**379**(379):94–112.
109. Clarke IC, Manaka M, Green DD, Williams P, Pezzotti G, Kim YH, et al. Current status of zirconia used in total hip implants. *J Bone Joint Surg Am* 2003;**85-A**(Suppl 4):73–84.
110. *Information on breakages reported on Prozyr femoral heads: Key dates*. http://wwwprozyrcom/PAGES_UK/Biomedical/historichtm: St. Gobain, Desmarquest; 2002 (accessed: 28.04.03).
111. *Recall of Zirconia Ceramic Femoral Heads for Hip Implants*. http://wwwfdagov/cdrh/recalls/zirconiahiphtml: United States Food and Drug Administration; 2001 (accessed: 28.04.03).
112. Green DD, Williams PA, Donaldson TK. C. CI. BIOLOX-Forte vs. BIOLOX Delta under microseparation test mode in the USA. *Trans of the 51st Orthop Res Soc* 2005;**30**:239.
113. Affatato S, Torrecillas R, Taddei P, Rocchi M, Fagnano C, Ciapetti G, et al. Advanced nanocomposite materials for orthopaedic applications. I. A long-term in vitro wear study of zirconia-toughened alumina. *J Biomed Mater Res B Appl Biomater* 2006; **78**(1): 76–82.
114. Sheth NP, Lementowski P, Hunter G, Garino JP. Clinical applications of oxidized zirconium. *J Surg Orthop Adv* 2008;**17**(1):17–26.

115. Bourne RB, Barrack R, Rorabeck CH, Salehi A, Good V. Arthroplasty options for the young patient: Oxinium on cross-linked polyethylene. *Clin Orthop Relat Res* 2005;**441**:159–67.
116. Good V, Ries M, Barrack RL, Widding K, Hunter G, Heuer D. Reduced wear with oxidized zirconium femoral heads. *J Bone Joint Surg Am* 2003;**85-A**(Suppl 4):105–10.
117. Li MG, Zhou ZK, Wood DJ, Rohrl SM, Ioppolo JL, Nivbrant B. Low wear with high-crosslinked polyethylene especially in combination with oxinium heads. A RSA evaluation. *Trans of the 52nd Orthop Res Soc* 2006;**31**:643.
118. Ezzet KA, Hermida JC, Colwell Jr. CW, D'Lima DD. Oxidized zirconium femoral components reduce polyethylene wear in a knee wwear simulator. *Clin Orthop Relat Res* 2004(428):120–4.
119. Evangelista GT, Fulkerson E, Kummer F, Di Cesare PE. Surface damage to an Oxinium femoral head prosthesis after dislocation. *J Bone Joint Surg Br* 2007;**89**(4):535–7.
120. Kop AM, Whitewood C, Johnston DJ. Damage of oxinium femoral heads subsequent to hip arthroplasty dislocation three retrieval case studies. *J Arthroplasty* 2007;**22**(5):775–9.
121. Bal BS, Khandkar A, Lakshminarayanan R, Clarke I, Hoffman AA, Rahaman MN. Fabrication and testing of silicon nitride bearings in total hip arthroplasty winner of the 2007 "HAP" PAUL award. *J Arthroplasty* 2008.
122. Bal BS, Khandkar A, Lakshminarayanan R, Clarke I, Hoffman AA, Rahaman MN. Testing of silicon nitride ceramic bearings for total hip arthroplasty. *J Biomed Mater Res B Appl Biomater* 2008.
123. Mazzocchi M, Gardini D, Traverso PL, Faga MG, Bellosi A. On the possibility of silicon nitride as a ceramic for structural orthopaedic implants. Part II: chemical stability and wear resistance in body environment. *J Mater Sci Mater Med* 2008;**19**(8):2889–901.
124. Mazzocchi M, Bellosi A. On the possibility of silicon nitride as a ceramic for structural orthopaedic implants. Part I: processing, microstructure, mechanical properties, cytotoxicity. *J Mater Sci Mater Med* 2008;**19**(8):2881–7.
125. Lancaster JG, Dowson D, Isaac GH, Fisher J. The wear of ultra-high molecular weight polyethylene sliding on metallic and ceramic counterfaces representative of current femoral surfaces in joint replacement. *Proc Inst Mech Eng [H]* 1997;**211**(1):17–24.
126. Cuckler JM, Bearcroft J, Asgian CM. Femoral head technologies to reduce polyethylene wear in total hip arthroplasty. *Clin Orthop* 1995;**317**(317):57–63.
127. Sugano N, Nishii T, Nakata K, Masuhara K, Takaoka K. Polyethylene sockets and alumina ceramic heads in cemented total hip arthroplasty. A ten-year study. *J Bone Joint Surg Br* 1995;**77**(4):548–56.
128. Dumbleton JH, Manley MT, Edidin AA. A literature review of the association between wear rate and osteolysis in total hip arthroplasty. *J Arthroplasty* 2002;**17**(5):649–61.
129. Kim YH. Comparison of polyethylene wear associated with cobalt-chromium and zirconia heads after total hip replacement. A prospective, randomized study. *J Bone Joint Surg Am* 2005;**87**(8):1769–76.
130. Kraay MJ, Thomas RD, Rimnac CM, Fitzgerald SJ, Goldberg VM. Zirconia versus Co-Cr femoral heads in total hip arthroplasty: early assessment of wear. *Clin Orthop Relat Res* 2006;**453**:86–90.
131. Hernigou P, Bahrami T. Zirconia and alumina ceramics in comparison with stainless-steel heads. Polyethylene wear after a minimum ten-year follow-up. *J Bone Joint Surg Br* 2003;**85**(4):504–9.
132. Sychterz CJ, Engh Jr. CA, Young AM, Hopper Jr. RH, Engh CA Comparison of in vivo wear between polyethylene liners articulating with ceramic and cobalt-chrome femoral heads. *J Bone Joint Surg Br* 2000;**82**(7):948–51.
133. Hendrich C, Goebel S, Roller C, Kirschner S, Martell JM. Wear performance of 28 millimeter femoral heads with the Harris-Galante cup: Comparison of alumina and cobalt chrome. In: Zippel H, Dietrich M, editors. *Bioceramics in joint arthroplasty, 8th Biolox symposium proceedings*. Darmstadt: Steinkopff Verlag; 2003. p. 177–82.
134. Hopper Jr. RH, Engh Jr. CA, Fowlkes LB, Engh CA. The pros and cons of polyethylene sterilization with gamma irradiation. *Clin Orthop Relat Res* 2004(429):54–62.
135. Hopper Jr RH. Personal communication. AORI; 2008.
136. Bohler M, Schachinger W, Wolfl G, Krismer M, Mayr G, Salzer M. Comparison of migration in modular sockets with ceramic and polyethylene inlays. *Orthopedics* 2000;**23**(12):1261–6.
137. Schwämmlein D, Schmidt R, Schikora N, Graef B, Willmann G, Holzwarth U, et al. Migration patterns of press-fit cups with polyethylene or alumina liner-A randomized clinical trial using radiostereoanalysis. In: Garino JP, Willmann G, editors. *Bioceramics in joint arthroplasty, Proceedings of the 7th international BIOLOX symposium*. Stuttgart: Thieme; 2002. p. 81–5.
138. Kircher J, Bader R, Schroeder B, Mittelmeier W. Extremely high fracture rate of a modular acetabular component with a sandwich polyethylene ceramic insertion for THA: a preliminary report. *Arch Orthop Trauma Surg* 2008.
139. Popescu D, Gallart X, Garcia S, Bori G, Tomas X, Riba J. Fracture of a ceramic liner in a total hip arthroplasty with a sandwich cup. *Arch Orthop Trauma Surg* 2008;**128**(8):783–5.
140. Poggie RA, Turgeon TR, Coutts RD. Failure analysis of a ceramic bearing acetabular component. *J Bone Joint Surg Am* 2007;**89**(2):367–75.
141. Ha YC, Kim SY, Kim HJ, Yoo JJ, Koo KH. Ceramic liner fracture after cementless alumina-on-alumina total hip arthroplasty. *Clin Orthop Relat Res* 2007;**458**:106–10.
142. D'Antonio JA, Capello WN, Manley MT, Naughton M, Sutton K. A titanium-encased alumina ceramic bearing for total hip arthroplasty: 3- to 5-year results. *Clin Orthop Relat Res* 2005;**441**:151–8.
143. Barnes CL, DeBoer D, Corpe RS, Nambu S, Carroll M, Timmerman I. Wear performance of large-diameter differential-hardness hip bearings. *J Arthroplasty* 2008;**23**(6 Suppl 1):56–60.
144. Steens W, von Foerster G, Katzer A. Severe cobalt poisoning with loss of sight after ceramic-metal pairing in a hip—a case report. *Acta Orthop* 2006;**77**(5):830–2.
145. Valenti JR, Del Rio J, Amillo S. Catastrophic wear in a metal-on-ceramic total hip arthroplasty. *J Arthroplasty* 2007;**22**(6):920–2.
146. Walter WL, Insley GM, Walter WK, Tuke MA. Edge loading in third generation alumina ceramic-on-ceramic bearings: stripe wear. *J Arthroplasty* 2004;**19**(4):402–13.
147. Murali R, Bonar SF, Kirsh G, Walter WK, Walter WL. Osteolysis in Third-Generation Alumina Ceramic-on-Ceramic Hip Bearings With Severe Impingement and Titanium Metallosis. *J Arthroplasty* 2008.
148. Krikler S, Schatzker J. Ceramic head failure. *J Arthroplasty* 1995;**10**(6):860–2.
149. Tateiwa T, Clarke IC, Williams PA, Garino J, Manaka M, Shishido T, et al. Ceramic total hip arthroplasty in the United States: safety and risk issues revisited. *Am J Orthop* 2008;**37**(2):E26–31.
150. D'Antonio J, Capello W, Manley M, Naughton M, Sutton K. Alumina ceramic bearings for total hip arthroplasty: five-year results of a prospective randomized study. *Clin Orthop Relat Res* 2005(436):164–71.
151. *Information on breakages reported on Prozyr femoral heads: Batches & product configurations concerned by breakages*. http://wwwprozyrcom/PAGES_UK/Biomedical/breakageshtm: St. Gobain, Desmarquest; 2003 (accessed: 28.04.03).

152. Rosneck J, Klika A, Barsoum W. A rare complication of ceramic-on-ceramic bearings in total hip arthroplasty. *J Arthroplasty* 2008;**23**(2):311–13.
153. Restrepo C, Parvizi J, Kurtz SM, Sharkey PF, Hozack WJ, Rothman RH. The noisy ceramic hip: is component malpositioning the cause? *J Arthroplasty* 2008;**23**(5):643–9.
154. Baek SH, Kim SY. Cementless total hip arthroplasty with alumina bearings in patients younger than fifty with femoral head osteonecrosis. *J Bone Joint Surg Am* 2008;**90**(6):1314–20.
155. Yang CC, Kim RH, Dennis DA. The squeaking hip: a cause for concern-disagrees. *Orthopedics* 2007;**30**(9):739–42.
156. Walter WL, O'Toole GC, Walter WK, Ellis A, Zicat BA. Squeaking in ceramic-on-ceramic hips: the importance of acetabular component orientation. *J Arthroplasty* 2007;**22**(4):496–503.
157. Taylor S, Manley MT, Sutton K. The role of stripe wear in causing acoustic emissions from alumina ceramic-on-ceramic bearings. *J Arthroplasty* 2007;**22**(7 Suppl 3):47–51.
158. Ranawat AS, Ranawat CS. The squeaking hip: a cause for concern-agrees. *Orthopedics* 2007;**30**(9):738–43.
159. Eickmann TH, Clarke IC, Gustafson GA. Squeaking in a ceramic on ceramic total hip. In: Zippel H, Dietrich M, editors. *Bioceramics in joint arthroplasty, 8th Biolox symposium proceedings*. Darmstadt: Steinkopff Verlag; 2003. p. 187–91.
160. Walter WK, Waters TS, Gillies M, Donohoo S, Kurtz SM, Ranawat AS, et al. Squeaking hips. *J Bone Joint Surg [Am]* 2008;**90**(Suppl 4). In press.
161. Kurtz SM, Muratoglu OK, Evans M, Edidin AA. Advances in the processing, sterilization, and crosslinking of ultra- high molecular weight polyethylene for total joint arthroplasty. *Biomaterials* 1999;**20**(18):1659–88.
162. Jasty M, Rubash HE, Muratoglu O. Highly cross-linked polyethylene: the debate is over—in the affirmative. *J Arthroplasty* 2005;**20**(4 Suppl 2):55–8.
163. Oonishi H, Takayama Y, Tsuji E. Improvement of polyethylene by irradiation in artificial joints. *Radiation Physics and Chemistry* 1992;**39**(6):495–504.
164. Oonishi H. Long term clinical results of THR. Clinical results of THR of an alumina head with a cross-linked UHMWPE cup. *Orthop Surg and Traumatol* 1995;**38**(11):1255–64.
165. Oonishi H, Takayama Y, Tsuji E. The low wear of cross-linked polyethylene socket in total hip prostheses. In: Wise DL, Trantolo DJ, Altobelli DE, Yaszemski MJ, Gresser JD, Schwartz ER, editors. *Encyclopedic handbook of biomaterials and bioengineering Part A: Materials*. New York: Marcel Dekker; 1995. p. 1853–68.
166. Grobbelaar CJ, Du Plessis TA, Marais F. The radiation improvement of polyethylene prostheses: A preliminary study. *J Bone Joint Surg* 1978;**60-B**(3):370–4.
167. Wroblewski BM, Siney PD, Dowson D, Collins SN. Prospective clinical and joint simulator studies of a new total hip arthroplasty using alumina ceramic heads and cross-linked polyethylene cups. *J Bone & Joint Surg* 1996;**78B**:280–5.
168. Crowninshield RD, Muratoglu OK. How have new sterilization techniques and new forms of polyethylene influenced wear in total joint replacement? *J Am Acad Orthop Surg* 2008;**16**(Suppl 1):S80–5.
169. Callaghan JJ, Cuckler JM, Huddleston JI, Galante JO. How have alternative bearings (such as metal-on-metal, highly cross-linked polyethylene, and ceramic-on-ceramic) affected the prevention and treatment of osteolysis? *J Am Acad Orthop Surg* 2008;**16**(Suppl 1):S33–8.
170. Gencur SJ, Rimnac CM, Kurtz SM. Fatigue crack propagation resistance of virgin and highly crosslinked, thermally treated ultra-high molecular weight polyethylene. *Biomaterials* 2006;**27**(8):1550–7.
171. Kurtz SM, Villarraga ML, Herr MP, Bergstrom JS, Rimnac CM, Edidin AA. Thermomechanical behavior of virgin and highly cross-linked ultra-high molecular weight polyethylene used in total joint replacements. *Biomaterials* 2002;**23**(17):3681–97.
172. Kurtz SM, Cooper C, Siskey R, Hubbard N. Effects of dose rate and thermal treatment on the physical and mechanical properties of highly crosslinked UHMWPE used in total joint replacements. *Transactions of the 49th orthopedic research society.* New Orleans, LA; 2003 February 9-13.
173. Kurtz SM, Manley M, Wang A, Taylor S, Dumbleton J. Comparison of the properties of annealed crosslinked (Crossfire) and conventional polyethylene as hip bearing materials. *Bull Hosp Jt Dis* 2002-2003;**61**(1-2):17–26.
174. Halley D, Glassman A, Crowninshield RD. Recurrent dislocation after revision total hip replacement with a large prosthetic femoral head. A case report. *J Bone Joint Surg Am* 2004;**86-A**(4):827–30.
175. Tower SS, Currier JH, Currier BH, Lyford KA, Van Citters DW, Mayor MB. Rim cracking of the cross-linked longevity polyethylene acetabular liner after total hip arthroplasty. *J Bone Joint Surg Am* 2007;**89**(10):2212–17.
176. Kurtz SM, Hozack WJ, Purtill JJ, Marcolongo M, Kraay MJ, Goldberg VM, et al. Significance of in vivo degradation for polyethylene in total hip arthroplasty. *Clin Orthop Relat Res* 2006;**453**:47–57.
177. Wroblewski BM, Siney PD, Fleming PA. Low-friction arthroplasty of the hip using alumina ceramic and cross-linked polyethylene. A 17-year follow-up report. *J Bone Joint Surg Br* 2005 Sep;**87**(9):1220–1.
178. Urban JA, Garvin KL, Boese CK, Bryson L, Pedersen DR, Callaghan JJ, et al. Ceramic-on-polyethylene bearing surfaces in total hip arthroplasty. Seventeen to twenty-one-year results. *J Bone Joint Surg Am* Nov 2001;**83-A**(11):1688–94.

Chapter 7

# The Origins and Adaptations of UHMWPE for Knee Replacements

Steven M. Kurtz, PhD



## 7.1 INTRODUCTION

Knee arthroplasty, referring to surgical reconstruction of the knee joint, has its origins in the late 19th century as the treatment for severe joint degeneration resulting from tuberculosis [1, 2]. In 1890, Gluck from the Charité hospital in Berlin described his design of a fixed hinged knee replacement with components fashioned from ivory [3]. These overly constrained hinged knee replacements suffered from short-term failure, and Gluck later retracted his endorsement of this surgical procedure. In the early 20th century, attempts at knee arthroplasty also involved the implantation of autogenous tissue (such as muscle fascia), as well as chromicized pig's bladder, to serve as articulating surfaces of a reconstructed knee joint [4].

UHMWPE has been used in knee replacements since the late 1960s, when Frank Gunston developed a cemented implant design at Wrightington Hospital [5, 6]. This early knee replacement resurfaced the individual condyles of the femur and the tibia. Total knee arthroplasty (TKA), which replaces the articulation between the femur and tibia, as well as between the femur and the patella, was developed in the 1970s, primarily at surgical centers in North America. The basic anatomical landmarks and implant features of a typical total knee replacement are illustrated in Figure 7.1.

Contemporary knee arthroplasty comprises a broad range of surgical procedures, which are tailored by the physician for the specific needs of the patient. For patients with mild arthritis, which is confined to one of the condyles of the knee, the surgeon might decide to perform a unicondylar knee arthroplasty (UKA). If both condyles of the knee are diseased, but the patella remains intact, the surgeon may perform a bicondylar (also referred to as bicompartmental) total knee arthroplasty (TKA). When both condyles, as well as the patello-femoral joint are diseased, a tricompartmental total knee replacement (illustrated in Figure 7.1) is performed. If only the patella is diseased, a surgeon might opt to implant a patellar component (this procedure is also referred to as patellar resurfacing). Finally, in the case of extreme circumstances, such as a salvage revision operation or in the event of tumor resection, a semiconstrained hinged knee design might be employed. In all of these procedures, UHMWPE plays a primary role as a polymeric component, articulating either against a metallic component, or in some cases (such as in a patellar resurfacing) the UHMWPE may articulate against cartilage.

The manner in which UHMWPE was fundamentally adapted for TKA evolved most quickly during the 1970s. In 1975, Ewald [7] wrote that "the problem we are faced with today is to select the best design among the 300 total knee prostheses currently commercially available or in the process of development around the world." Although the main adaptations of UHMWPE for TKA were firmly established by the end of the 1970s, the 1980s and 1990s were still associated with continuous incremental improvements in the design of the femoral, tibial, and patellar components to

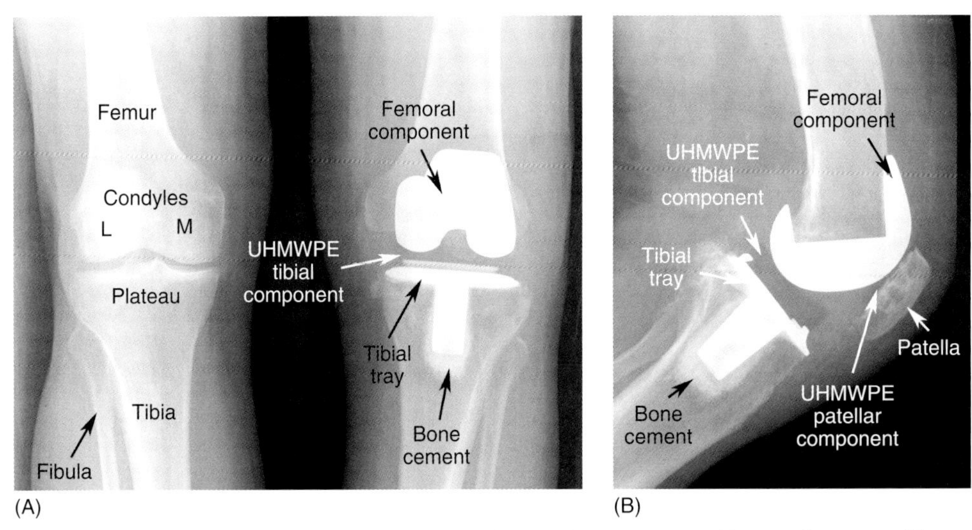


**FIGURE 7.1** (A) Anterior/posterior and (B) lateral radiographic views of an Insall/Burstein II total knee replacement, with associated anatomical landmarks and implant terminology. In 1 A, the femoral condyles are designated as medial (M) and lateral (L).

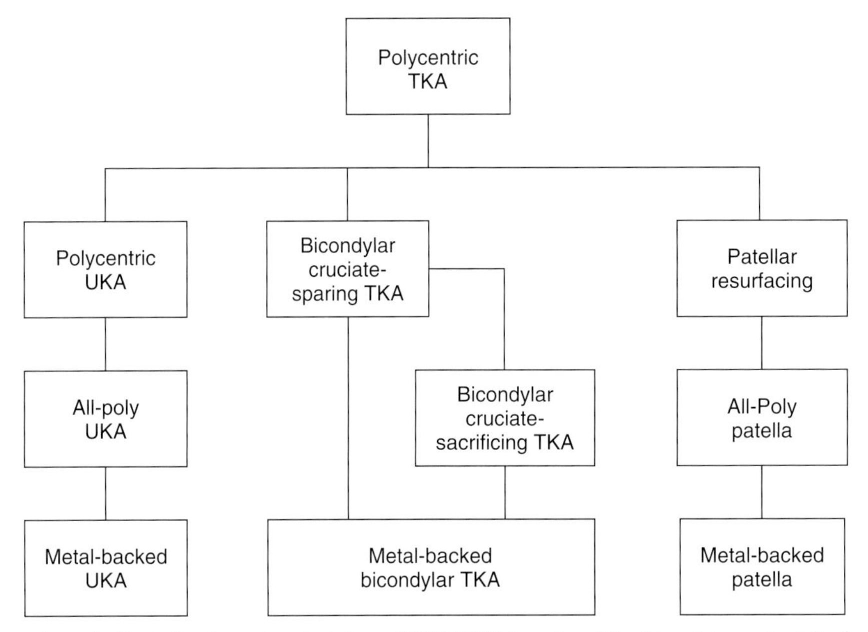


**FIGURE 7.2** Flowchart schematic illustrating the major adaptations of UHMWPE for unicondylar, bicondylar, and patello-femoral replacements.

address recurring problems with positioning and loosening. During the 1980s and 1990s, there were also major strides in the surgical instrumentation and the techniques used to implant the artificial knee components, which, when coupled with improved implant designs and fixation methods, have contributed to improved survivorship.

Because so many different surgeons and engineers have designed knee replacements over the past 3 decades, the history of TKA is much more complex than THA [8, 9]. In this chapter, we start with tracing the origins of how UHMWPE came to be used in knee arthroplasty when Frank Gunston worked at Wrightington. The remaining sections of this chapter focus specifically on five fundamental adaptations of UHMWPE for knee replacement since the 1970s (Figure 7.2). These five evolutionary stages for UHMWPE in TKA include: (1) Gunston's initial design concept for the Polycentric total knee replacement, which replaced both condyles of the femur individually; (2) the adaptation of Gunston's design to unicondylar knee arthroplasty for carefully selected groups of patients; (3) the evolution to a bicondylar total knee, in which the tibial and femoral components were joined for ease of insertion and anatomical positioning; (4) the resurfacing of the patello-femoral joint; and (5) the incorporation of metal backing in the design of UHMWPE components. It should be emphasized that all of these major evolutionary steps in the clinical application of UHMWPE for TKA were initiated in the 1970s, even if the final embodiments of these

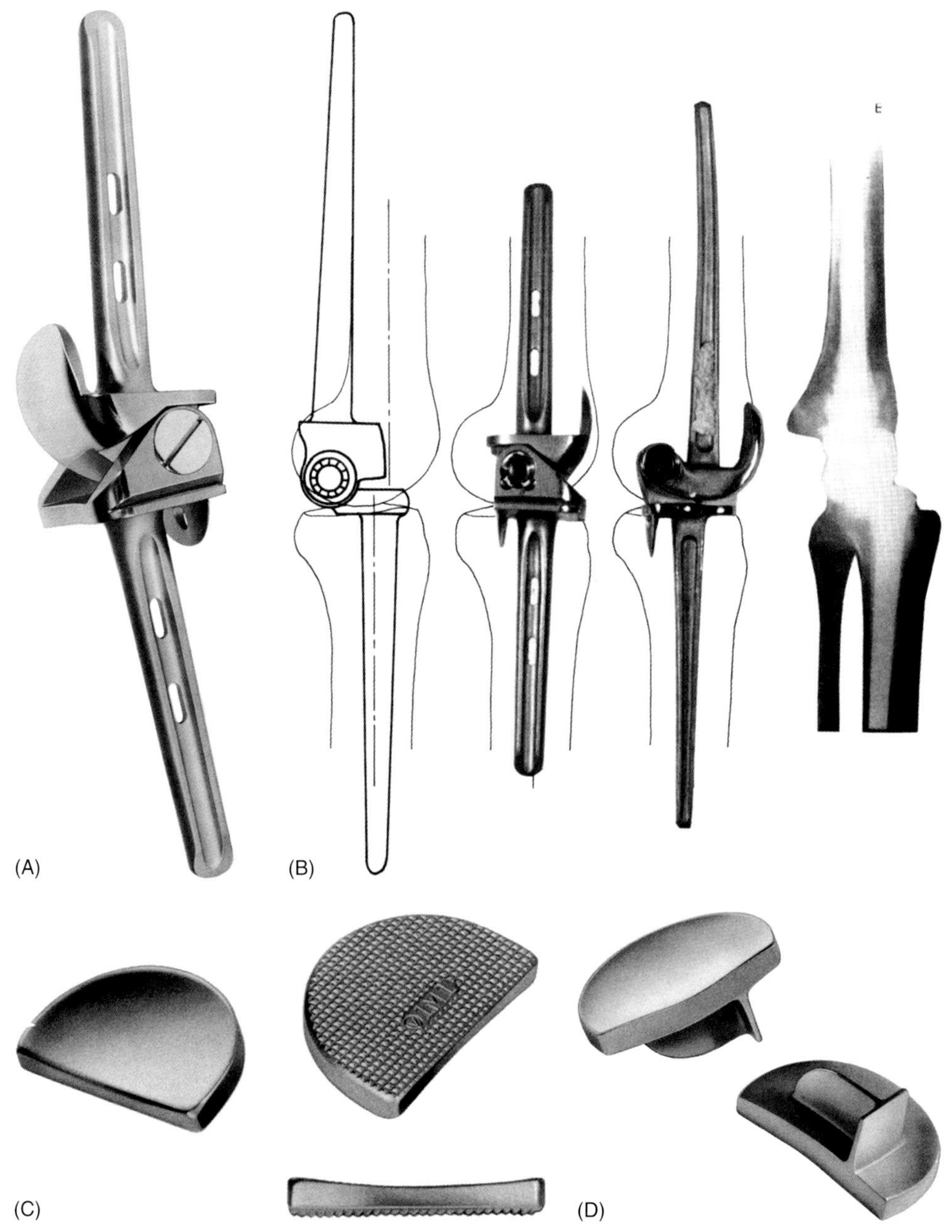


FIGURE 7.3 Examples of knee arthroplasty during the 1960s. (A) Walldius hinged knee replacement; (B) The Shiers, Walldius, and Guépar hinged knee replacements, superimposed over the anatomy of the knee. Reprinted with permission from Walker PS. *Human joints and their artificial replacements*. Springfield, IL: CC Thomas Publisher, 1977; (C) MacIntosh tibial plateau; and (D) McKeever tibial plateau.

design concepts did not reach fruition until the following decades. Even today, surgeons and biomechanical engineers continue to debate and refine their understanding of these fundamental adaptations of UHMWPE for knee arthroplasty.

## 7.2 FRANK GUNSTON AND THE WRIGHTINGTON CONNECTION TO TKA

In the 1960s, the mainstream treatments for knee arthritis included fusion, replacement with a metallic hinged prosthesis, or implantation of a metallic tibial resurfacing component (Figure 7.3). Hinged prostheses, like the Shiers or the Walldius knee for example, were state of the art at that time (Figure 7.3A, B). These fixed hinge knee prostheses required substantial bone resection and were associated with problems of loosening due to their overconstraint and their inability to accommodate internal-external rotation. Metallic tibial resurfacing components, like the McKeever and MacIntosh prosthesis (Figure 7.3C, D), as well as the Townley resurfacing prosthesis, were also employed during this time period [8]. Unfortunately, the knee implant solutions available in the mid-1960s were fraught with high complication rates and unacceptable long-term functionality.

UHMWPE was introduced for knee arthroplasty at the same place, and at around the same time, as it was introduced for hip replacement. As we have seen in Chapter 4, John Charnley introduced UHMWPE for hip arthroplasty in November 1962 at Wrightington Hospital, in Lancashire, England. By the mid-1960s, the hip arthroplasty using UHMWPE had become routine at Wrightington. Although Charnley's hip implants were not widely available during the 1960s, Wrightington nevertheless quickly evolved into a training center for orthopedic surgeons, who traveled from around the world to learn the latest techniques in hip arthroplasty.

In 1967, Frank H. Gunston, an orthopedic surgeon from Winnipeg, Manitoba, Canada was granted a traveling fellowship to study hip arthroplasty at Wrightington. Gunston was initially trained as an engineer, and he was especially attracted to Wrightington because of its machine shop and unique experimental facilities. Like all of the visiting registrars, Gunston learned hip arthroplasty by assisting with the hip surgeries being performed at Wrightington. He also helped with Charnley's ongoing projects related to *in vitro* testing of hip replacements.

During his fellowship at Wrightington, Gunston was struck by the problem of treating rheumatoid arthritis patients, who were afflicted both at the hip and knee. These patients continued to be debilitated after their hip replacement due to their ongoing knee arthritis. In addition, for these rheumatoid patients, the pain relief associated with their hip arthroplasty made them dissatisfied with the prevailing treatment options for knee arthritis.

In this context, Gunston developed a design for knee arthroplasty reflecting his exposure to UHMWPE and hip arthroplasty at Wrightington (Figure 7.4). Gunston's design incorporated two separate condylar replacements, each consisting of a convex metallic component (or "runner"), which was implanted on the posterior aspect of the femur, and a concave UHMWPE component (or "track"), which was implanted in the tibia. The implants were cemented into place without disturbing the cruciate ligaments.

In 1971, Gunston wrote that "the biomechanical principles and experience gained from total hip arthroplasty were combined with an analysis of normal knee movement to determine a solution [for knee replacement]." Gunston himself machined the first UHMWPE components for knee arthroplasty out of the RCH-1000 material that was available in the machine shop at Wrightington. In his 1971 paper, Gunston acknowledges Charnley "for his continued encouragement and the use of the facilities of the Center for Hip Surgery." Later, when Gunston returned from his fellowship to begin his orthopedic practice at Winnipeg, he continued to machine all of his own UHMWPE tibial components for his use, as well as for colleagues who requested them. Now retired from orthopedics, Gunston continues to operate on classic cars and on steam engines.

Although Gunston was clearly influenced by his fellowship experience at Wrightington, Charnley himself did not actively participate in the design of the first artificial knee. Charnley's interests at the time were firmly directed toward improving hip replacement surgery. In 1970, after Gunston had returned to Canada, Charnley did develop his own independent total knee replacement design, which was also intended for patients at Wrightington with rheumatoid

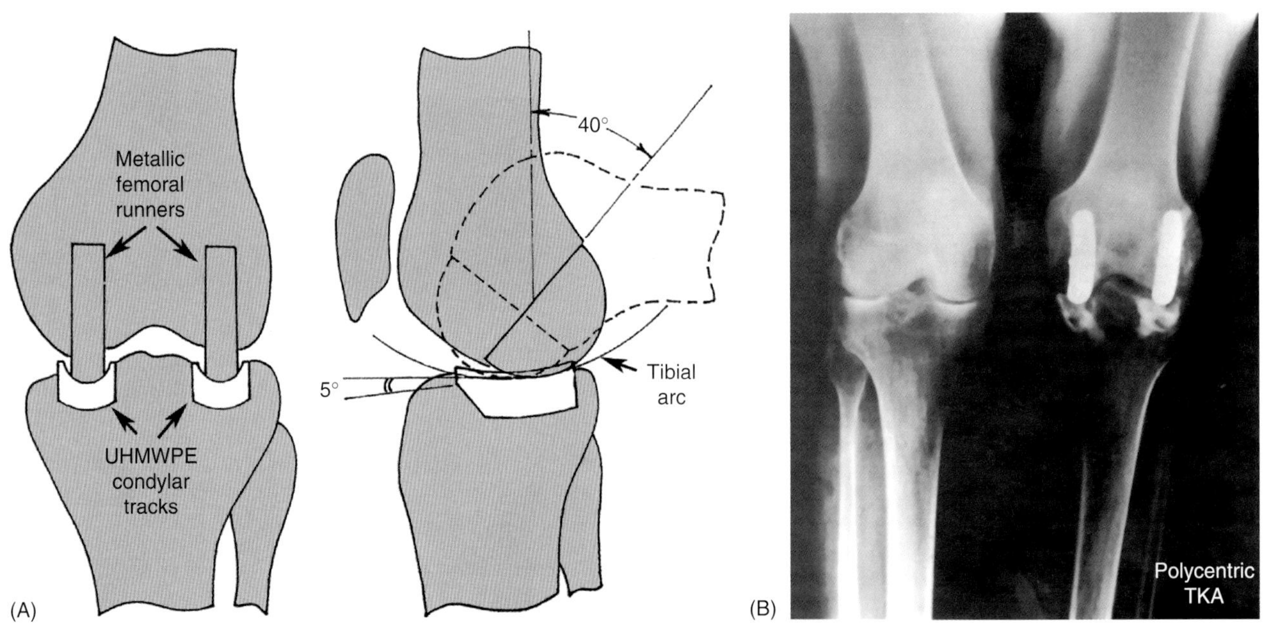


**FIGURE 7.4** Schematic of the Polycentric TKA design by Frank Gunston (A). Adapted from [5]. (B) Radiograph of a Polycentric TKA.

arthritis. Charnley's knee design, which was distributed by Thackrays in the 1970s as the Load Angle Inlay, had a convex UHMWPE component articulating against a flat metallic tibial plateau. However, Charnley's knee design, although unique, was not successful and never became widely adopted.

## 7.3 POLYCENTRIC KNEE ARTHROPLASTY

Unlike previous hinged prostheses, Gunston's design attempted to incorporate the complex kinematics of the knee joint. Gunston recognized that during flexion, the femur rolls and slides posteriorly back across the tibial condyles about successive instant centers of rotation (Figure 7.5). In his 1971 paper [5], Gunston described the motion of the knee during flexion as rotation about a "polycentric pathway." Consequently, Gunston's knee design was named the "Polycentric."

Gunston decided to publish his experience with the Polycentric rather than patent the design. In 1969, at a Canadian orthopedic meeting, Gunston met Lowell Peterson, MD, from the Mayo Clinic (Rochester, Minnesota, USA). Subsequently, Peterson's colleague, Richard Bryan, MD, traveled to Winnipeg in December 1969 to study the procedure and to further evaluate Gunston's clinical results [10].

Gunston freely shared his drawings with Bryan and Peterson, and after some design modifications, the first Polycentric TKA was performed at the Mayo Clinic by July 1970. The first 81 Polycentric knees at the Mayo Clinic were fabricated by the hospital [10]. However, up until 1978, the vast majority of the 1938 Polycentric components implanted at the Mayo Clinic were manufactured by orthopedic companies, like Howmedica (Rutherford, New Jersey, USA).

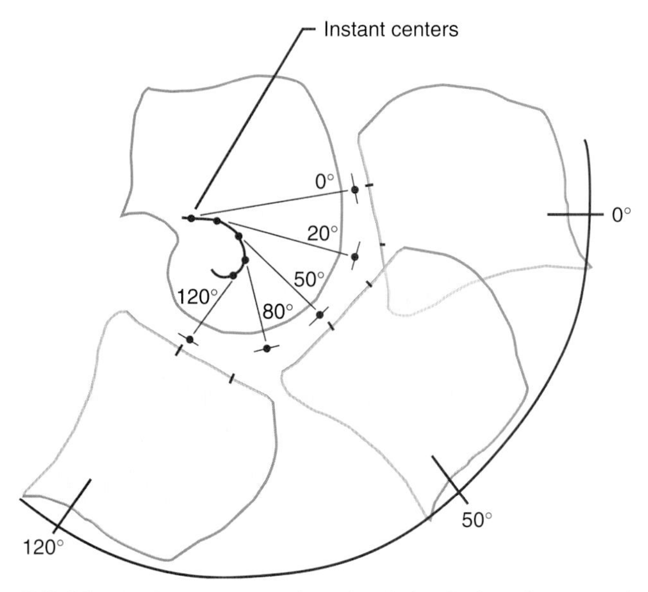


**FIGURE 7.5** Instant centers of rotation during flexion of the natural knee joint. Adapted from [5].

The Polycentric design was further modified by researchers at Mayo in 1973 to provide wider UHMWPE tracks [11]. Additional refinements of the Polycentric design, such as the introduction of UHMWPE runners with variable height, were proposed by Cracchiolo [12]. A review by Peterson et al. [13] in 1975 provides an instructive summary of the different types of Polycentric knee components, which were available from implant companies such as Howmedica and DePuy at the time.

The UHMWPE material used for Polycentric components was RCH-1000. Gunston sterilized the UHMWPE tibial components he personally manufactured using an autoclave. However, orthopedic companies like Howmedica, which fabricated most of Polycentric components implanted during the 1970s, sterilized their UHMWPE components with gamma radiation and used air-permeable packaging.

Proper placement of the four implant components was a major challenge for Polycentric TKA, and even with the specialized instruments that were developed in the 1970s, intraoperative alignment could be difficult (Figure 7.6). Polycentric TKA was initially performed on rheumatoid patients. With experience, the physicians at Mayo started implanting in patients with other forms of arthritis. Due to the shape of the components, polycentric TKA was contraindicated in patients with severe deformity of the femur or tibia.

Short-term results of the Polycentric were highly encouraging, especially considering the improved mobility of most patients, who had previously been debilitated by knee pain [5, 6, 14–16]. Unfortunately, loosening, infection, subsidence of the implant components, and instability all proved to be important complications for Polycentric TKA. In 1976, Gunston reported on 89 of his patients who were followed from 2 to 7.5y [17]. He reported nine cases of loosening (10%) and six cases of infection (6%). Although the Polycentric provided pain relief and improved mobility for 81% of his patients, the amount of flexion that could be achieved with this design was generally lower than what would be needed for stair climbing and getting out of a chair. For example, the average range of motion for Gunston's patients was 91° before the operation, but was 89.9° after Polycentric TKA.

Wear and fatigue damage were observed on UHMWPE tracks of the Polycentric TKA, but the observations of surface damage were typically associated with suboptimal placement of the components. Shoji et al. [18], for example, observed evidence up to 5.3 mm wear, fatigue cracks, and "crater like defects" in retrieved UHMWPE wear in Polycentric TKA. The four components analyzed by Shoji et al. [18] were implanted for 2 years or less. Noting the generally positive short-term outcomes of the Polycentric, the authors advised that proper surgical technique, not the performance of the UHMWPE per se, was critical to avoiding future failures of this design.

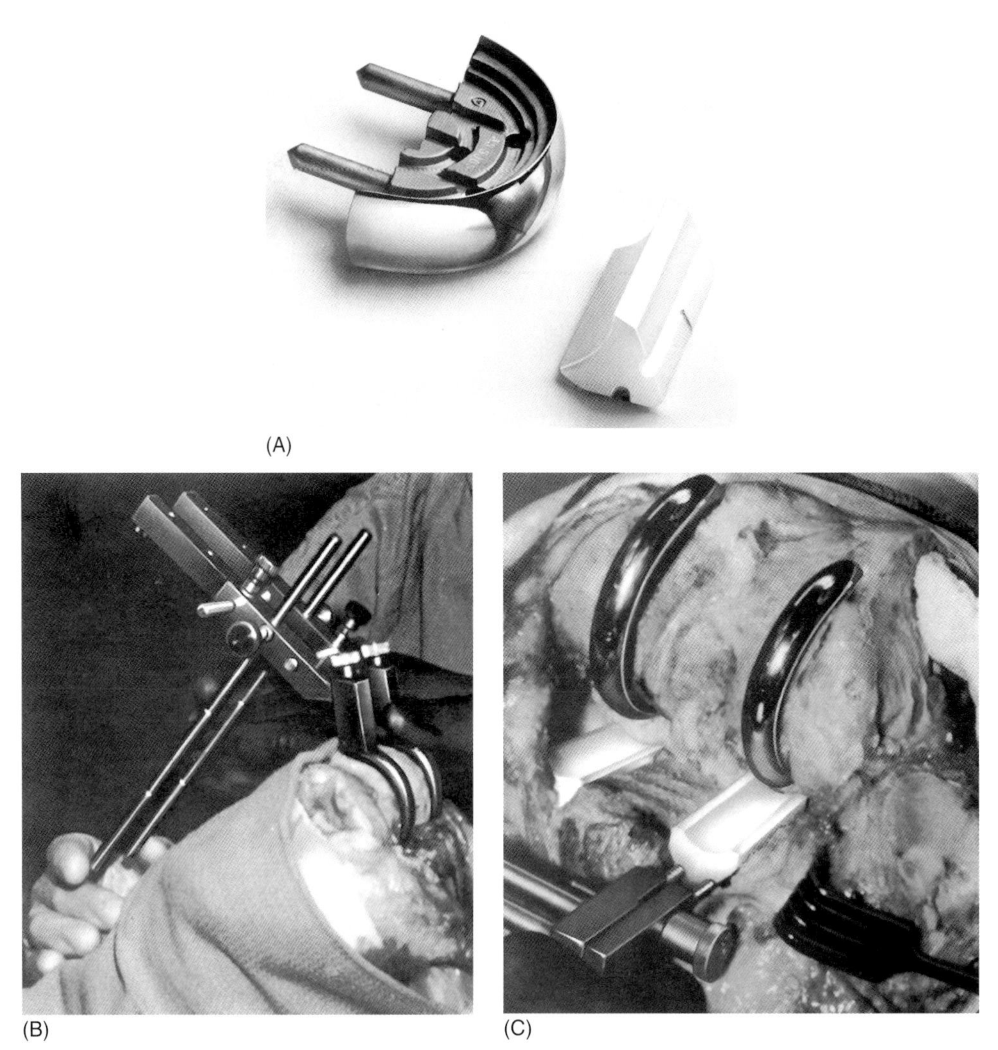

FIGURE 7.6 Geometry (A) and implantation technique (B, C) of Polycentric TKA components.

In 1984, Lewallen reported that the 10-year survivorship for the Polycentric TKA at the Mayo Clinic was 66% [19], with a 13% incidence of instability due to ligament laxity, 7% incidence of loosening, 3% incidence of infection, and 4% incidence of tibio-femoral joint pain. Although Polycentric TKA has not been clinically used since the 1970s, it can still be viewed in many respects as the great-grandfather of current knee designs. As we shall see, the fundamental applications for UHMWPE in contemporary TKA have their origins in Polycentric, if only in response to the problems encountered with this pioneering design.

## 7.4 UNICONDYLAR POLYCENTRIC KNEE ARTHROPLASTY

Surgeons at the Mayo Clinic started using Polycentric unicondylar knee arthroplasties (UKAs) starting in 1971 [20]. This procedure used identical implant components as the total knee replacement version of the Polycentric. Skolnick [20] described the 1-year results for 14 knees in 13 patients in 1975. By 1979, Bryan et al. [10] had reported that 338 polycentric UKAs had been performed at the Mayo Clinic.

At the Mayo Clinic, Polycentric UKA was only performed on patients with unicondylar disease (osteoarthritis) [20, 21]. An initial assessment was based on radiographic screening. However, the UKA implantation technique involved the same exposure as in TKA, so that both compartments could be directly examined and evidence of unicompartmental disease visually confirmed. Mallory also reported the utility of polycentric UKA in the treatment of posttraumatic arthritis due to fracture-induced deformity of the knee [22].

The main feature of Polycentric UKA was the relief of pain and restoration of mobility in the majority of patients [20, 22]. In Skolnick's study, the average range of motion was 119° prior to the procedure and 116° afterwards [20].

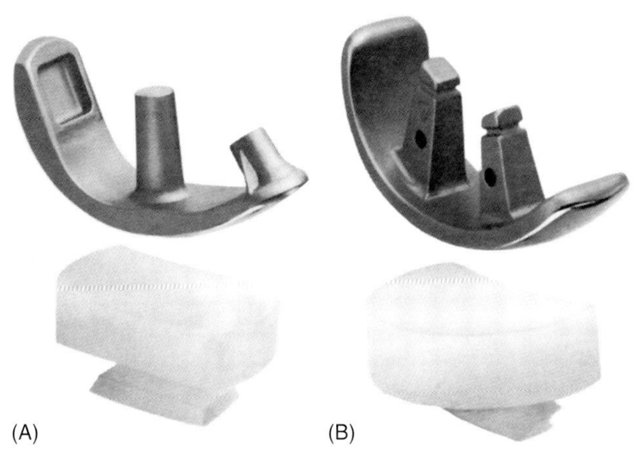


**FIGURE 7.7** First-generation, cemented unicondylar knee designs from the mid-1970s. (A) Geomedic; (B) Savastano unicondylar hemi-knee prostheses.

Because the implants and the procedure were identical, Polycentric UKA can be considered a special case of the Polycentric TKA. The primary difference is in the patient selection. Although superceded in the 1970s by other UKA designs, such as the Geomedic, Savastano, and Marmor prostheses [23], the Polycentric UKA is nonetheless the earliest example for the use of UHMWPE to treat unicompartmental disease in the knee.

Examples of first-generation cemented designs from the mid-1970s, intended specifically for UKA, are shown in Figures 7.7, 7.8, 7.10, and 7.12. These UKA designs differ from the Polycentric in several key respects. First, the tibial-femoral contact is less constrained than the rail-in-track geometry of the Polycentric. In addition, the tibial component is broader, providing greater coverage of the plateau during internal-external rotation. Finally, the femoral component also provides greater coverage of the femoral condyle than the Polycentric, enabling contact throughout a larger range of motion. Further details about the clinical performance of UHMWPE in UKA can be found in Chapter 8.

## 7.5 BICONDYLAR TOTAL KNEE ARTHROPLASTY

Bicondylar knee replacements evolved from difficulties with implanting two sets of condylar prostheses. Problems with surgical positioning of four individual femoral and tibial components prompted implant designers to physically join the separate compartments on the tibial and femoral side, respectively. Bicondylar knee replacements can be classified as "cruciate sparing" or "cruciate sacrificing" depending upon whether the PCL was excised during the installation of the UHMWPE tibial component. In this section, we highlight some of the features of early bicondylar knee designs, which were introduced in the 1970s.

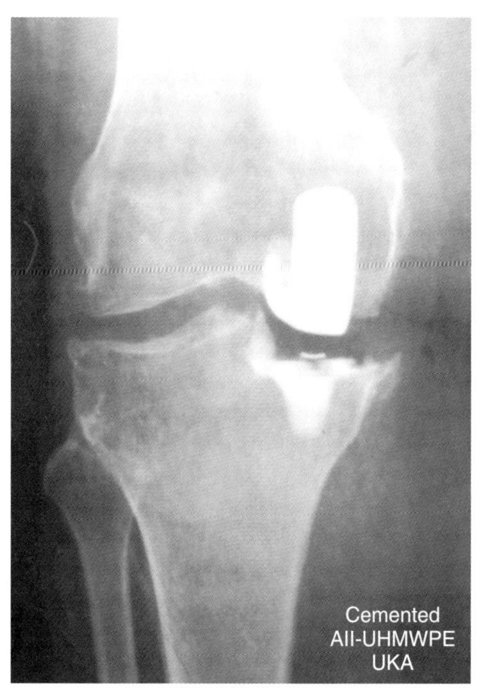


**FIGURE 7.8** Radiograph of a knee implanted with the Hospital for Special Surgery Uni-Condylar prosthesis, which included an all-UHMWPE tibial component. Image provided courtesy of Professor Clare Rimnac, Case Western Reserve University, Cleveland, Ohio, USA.

### 7.5.1 Cruciate Sparing Bicondylar Prostheses

The evolutionary step from the Polycentric knee to bicondylar knee arthroplasty is perhaps best appreciated in the "Geometric" knee design, developed by a group of five surgeons at the Mayo Clinic and clinically introduced in April 1971 [24]. The design was marketed under the Geomedic trade name (Howmedica, Rutherford, New Jersey, USA) (Figures 7.9A and 7.10A). Like the Polycentric, the Geomedic design preserved the PCL structures of the knee, with only an anterior UHMWPE bar joining the two condyles.

Although the two components of the Geometric knee were somewhat easier to implant than the four components of the Polycentric knee, long-term fixation continued to present an important problem for the Geometric design, as well as for other all-UHMWPE tibial components that were developed in this time period. Under radiographic examination, radiolucent lines of at least 1 mm were observed in the cement layer underneath the tibial component in 38% of long-term implanted Geometric knees [25]. Using revision or moderate-to-severe pain as the end point, Rand and Coventry [25] reported that the 10-year survivorship of the Geometric knee was 69%.

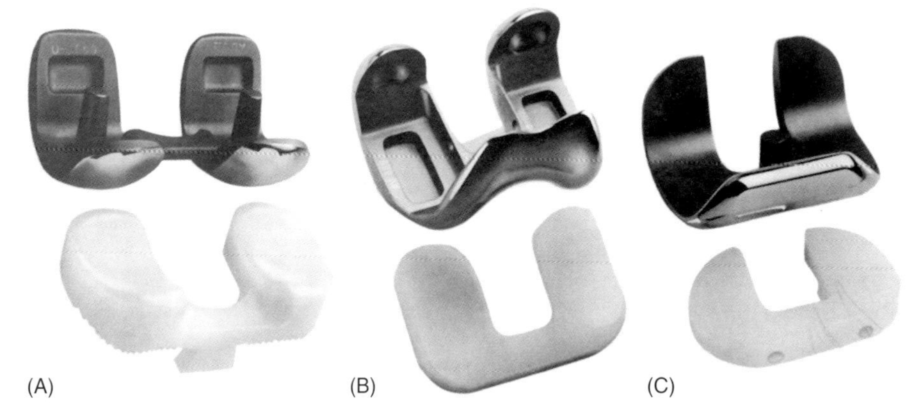


**FIGURE 7.9** Cruciate-sparing, bicondylar knee designs from the mid-1970s. (A) Geomedic; (B) Townley; and (C) Freeman-Swanson total knee prostheses.

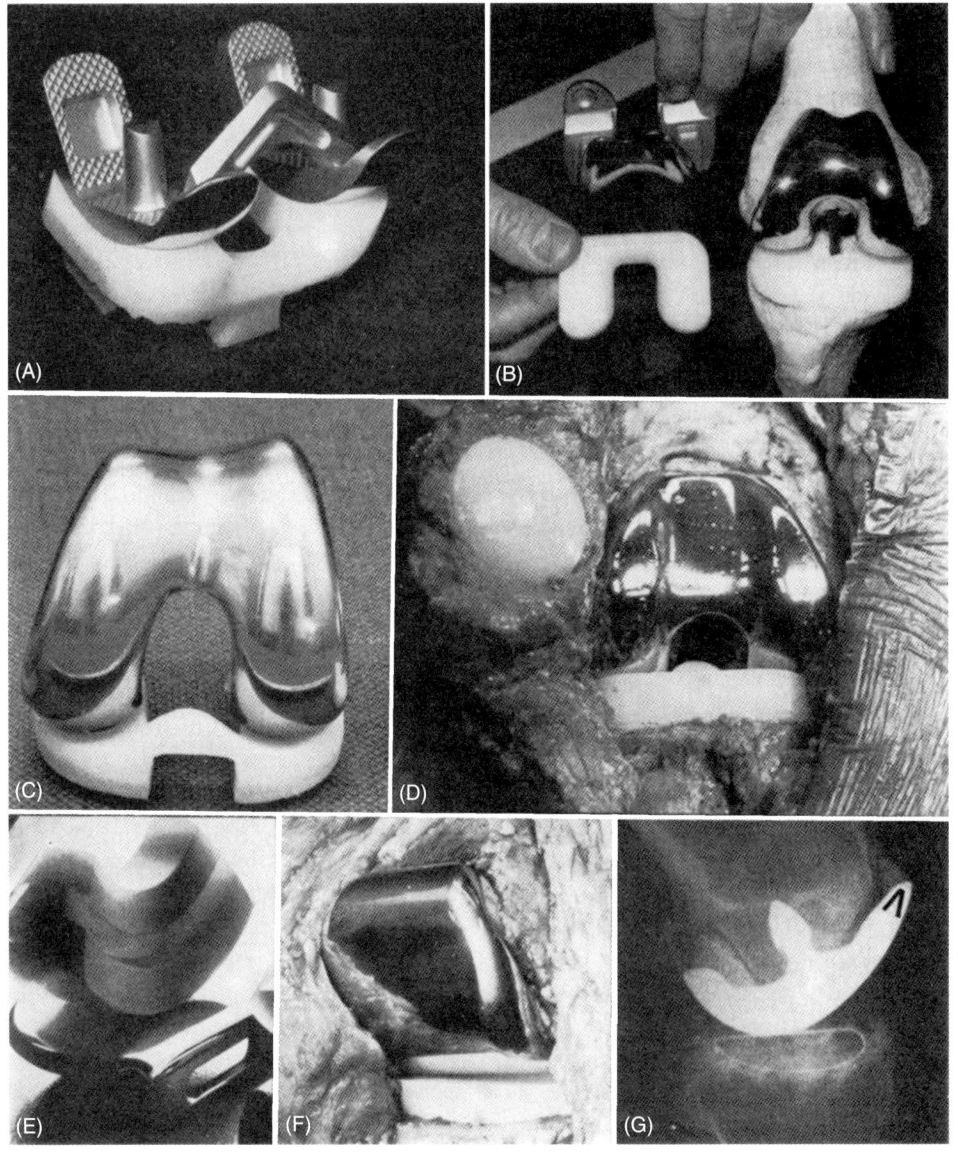


**FIGURE 7.10** Knee arthroplasty designs from the 1970s. (A) Modified geometric; (B) Townley; (C) Leeds; (D) Total Condylar; (E) Charnley load-angle inlay; (F) Freeman-Swanson; (G) Marmor modular. Reprinted with permission from Walker PS. *Human joints and their artificial replacements*. Springfield, IL: CC Thomas Publisher, 1977.

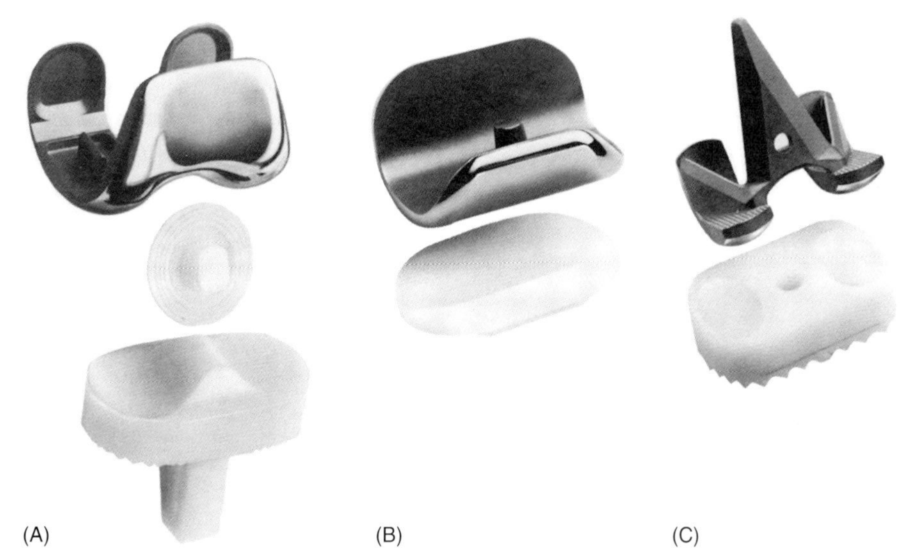


**FIGURE 7.11** Cruciate-sacrificing, bicondylar knee designs from the mid-1970s. (A) Total Condylar; (B) Freeman-Swanson; and (C) Fowler total knee prostheses.

During the 1970s, many other types of bicondylar cruciate retaining total knee replacement designs were developed, as reviewed by Walker [26]. The Townley knee, developed in 1972, is another example of a well-known first-generation, PCL-sparing knee design from this period [27]. Figure 7.9 provides examples of three cruciate retaining designs from a 1975 Howmedica catalog; Walker's book provides additional examples of designs from this time period (Figure 7.10). Although the geometry of these designs varied, they provided greater coverage of the condylar surfaces than the original Polycentric dual condylar components conceived by Gunston.

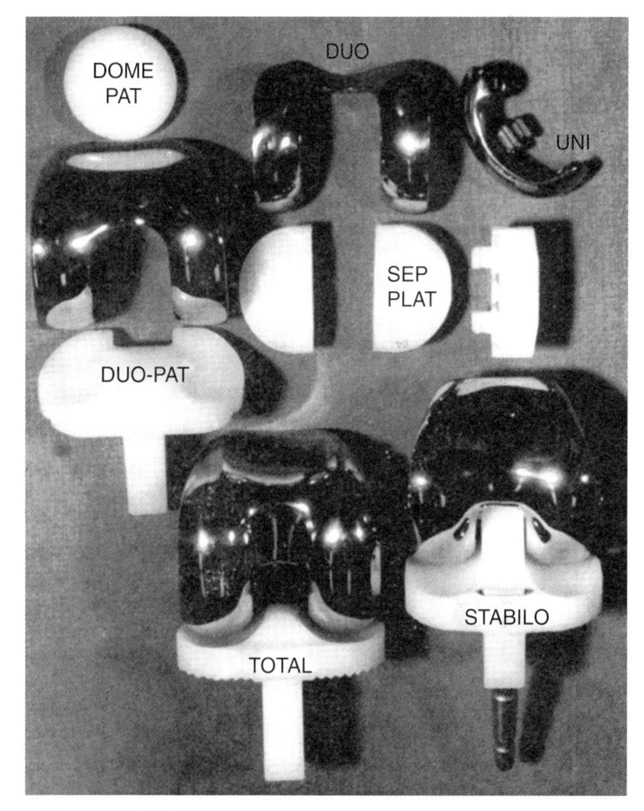


**FIGURE 7.12** Family of unicondylar and bicondylar knee prostheses designed at the Hospital for Special Surgery by Insall, Ranawat, and Walker during the 1970s. Reprinted with permission from Walker PS. *Human joints and their artificial replacements*. Springfield, IL: CC Thomas Publisher, 1977.

## 7.5.2 The Total Condylar Knee

Knee implant designers in the 1970s did not universally agree on the need to preserve the cruciate ligament, especially in light of problems encountered with tibial component fixation. In addition, many failures of bicondylar cruciate sparing designs, like the Polycentric, resulted from instability and ligament laxity during the progression of arthritis. By sacrificing the PCL, implant designers could not only geometrically resurface the ends of the tibia and femur, thereby gaining improved fixation, but they also had the potential to correct anatomical deformities produced by disease and old age. Also, it was not clear that the PCL continued to function competently in elderly patients.

A team of surgeons and engineers, including Peter Walker, PhD, John Insall, MD, and Chitranjan Ranawat, MD, from the Hospital for Special Surgery (HSS), developed the Total Condylar Knee arthroplasty in 1973 [8, 28, 29] (Figures 7.10D, 7.11A, 7.12, and 7.13). This tricompartmental design advanced the state of the art of knee design in several respects, most notably by developing the procedure, instrumentation, and components necessary to simultaneously replace both the tibiofemoral compartments and the patello-femoral compartment. All of the implant components in this design were cemented into place. The geometry of the UHMWPE tibial component

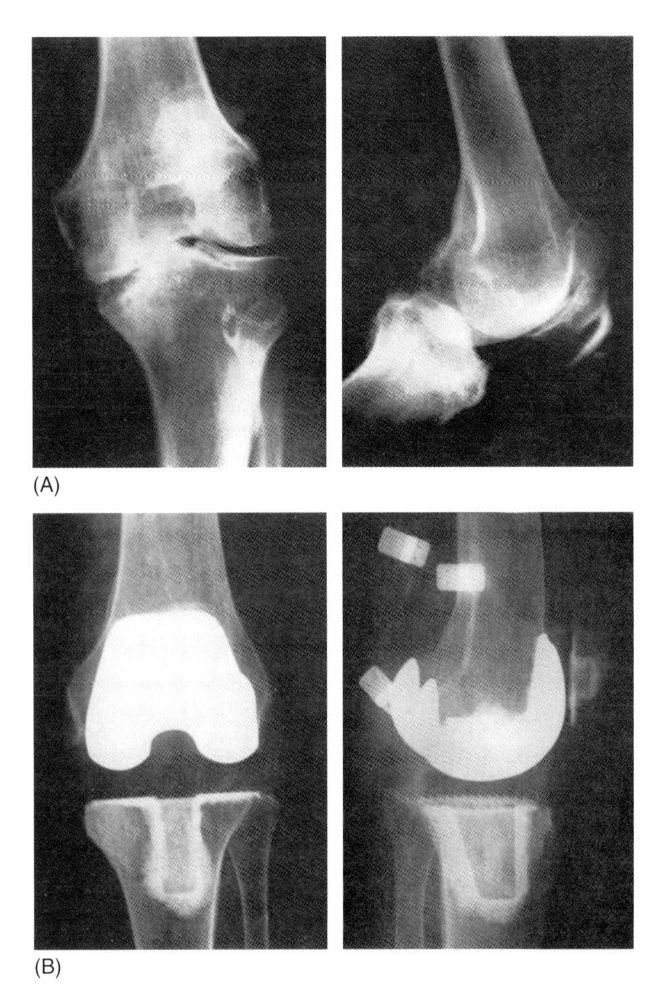


**FIGURE 7.13** Pre- (A) and postoperative (B) radiographs of a Total Condylar Knee Prosthesis. The components are well positioned. Reprinted with permission from Walker PS. *Human joints and their artificial replacements*. Springfield, IL: CC Thomas Publisher, 1977.

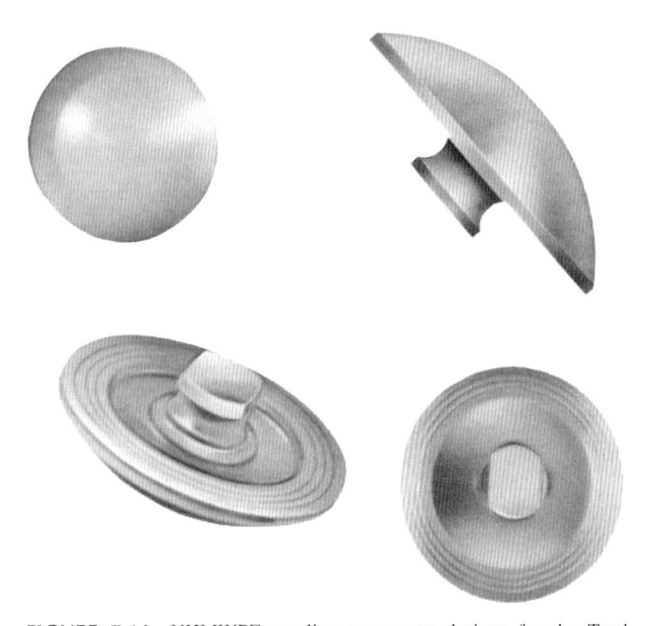

**FIGURE 7.14** UHMWPE patellar component designs for the Total Condylar Knee from the mid-1970s.

included an innovative stem to improve cemented fixation and to reduce subsidence into the femur. In addition, the femoral component included a wide anterior flange for articulation with the UHMWPE patellar button (Figures 7.10D, 7.14).

The surgeon-designers have reported impressive clinical results with the Total Condylar Prosthesis, especially when compared with the outcomes from previous knee designs [28–34]. The first cohort of patients implanted with the Total Condylar knee, between March 1974 and December 1977, have been followed closely for up to 20 years [30, 31, 35, 36]. Depending upon whether evidence of radiographic loosening or clinical failure was used for the endpoint, Ranawat et al. reported a 91–94% survivorship at 15 years [36].

The Total Condylar Prosthesis (TCP) design was modified several times in the 1970s by the surgeon inventors, giving rise to a family of Total Condylar prostheses (e.g., TCP I, II, III, etc.) [8, 28]. According to Insall, "The total condylar knee prosthesis II was developed because, at times, the articular geometry of the total condylar prosthesis did not provide sufficient anteriorposterior stability, and a few cases of posterior subluxation of the tibia occurred" [28]. For one of the designs (referred to by Insall as the TCP II), a special type of cruciate sacrificing knee replacement, incorporating a vertical UHMWPE post into the UHMWPE tibial component, was devised to provide greater joint stability, as well as to extend the range of motion during flexion activities (Figure 7.15).

The posterior-stabilized (PS) total condylar prosthesis (TCP II) was clinically introduced in 1978 (Figure 7.15) [37]. In this design, the vertical UHMWPE post of the tibial component makes contact with a horizontal cam when the knee was loaded and flexed, such as during stair climbing or rising from a chair. The PS design resulted in marked improvements in functional capabilities of joint replacement patients. The average postoperative range of motion for the PS design was 115° [37], whereas with the previous, unstabilized total condylar knee, an average range of motion of 90° had been reported [28]. With the PS knee, 76% of the patients could now climb stairs normally or walk an unlimited distance.

As discussed by Walker [26], many other types of bicondylar cruciate sacrificing knee designs were conceived in the 1970s (Figure 7.11). The Freeman-Swanson knee prosthesis [38], developed at London Hospital, is an another example of a well-known design from this time period (Figure 7.16). However, the Total Condylar Knee is considered by many orthopedic researchers and surgeons to be an archetype or "gold standard" among this first generation of bicondylar knee designs. The excellent clinical results reported by the designing surgeons of the Total Condylar Knee have been confirmed by orthopedic centers around the world [39, 40].

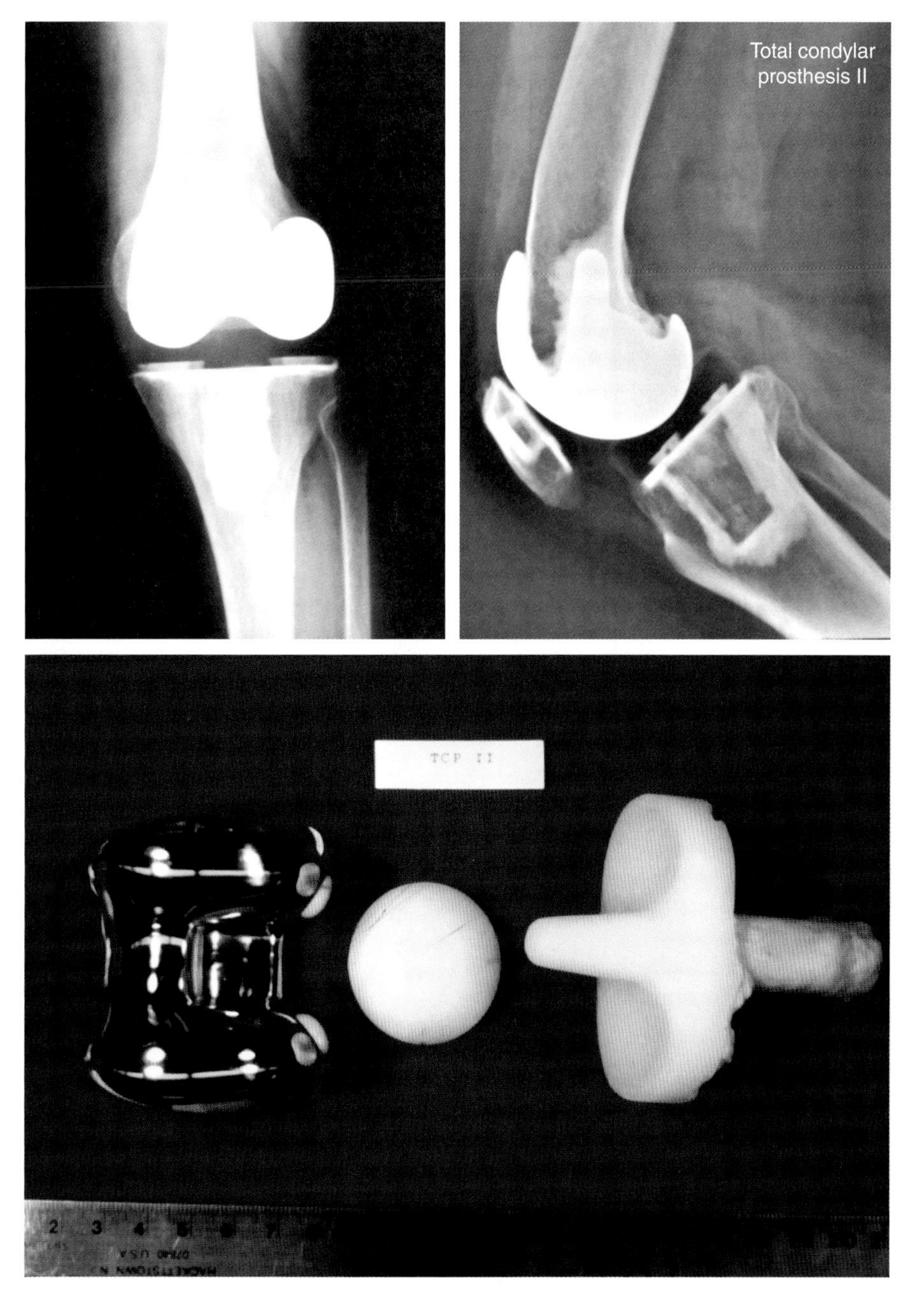


FIGURE 7.15 Radiographs and photographs of a retrieved Total Condylar Prosthesis II (TCP II), clinically introduced in 1978. Images provided courtesy of Professor Clare Rimnac, Case Western Reserve University, Cleveland, Ohio, USA.

## 7.6 PATELLO-FEMORAL ARTHROPLASTY

The Polycentric knee design, as originally conceived by Gunston, did not address the patello-femoral joint. Gunston mentions in his 1971 paper [5] that "the patella may be retained because no impingement of the patella on the prosthesis occurs." Over time, the limitation of this approach became increasingly apparent, as one of the major reasons for revising the Polycentric was due to patella pain [19]. Gunston later noted that "marked patello-femoral involvement in osteolysis may require trimming of the patella or patellectomy with usually unsatisfactory clinical result" [6].

Gunston's response to severe patello-femoral pain was to develop a separate patello-femoral arthroplasty, which consisted of a metallic patellar button articulating against an UHMWPE track implanted in the femur [17]. Because this solution required the implantation of two additional components, Gunston cautioned against using this solution for patello-femoral replacement "indiscriminately" [17].

Within this context, the patello-femoral arthroplasty offered by the Total Condylar Knee was a far more elegant

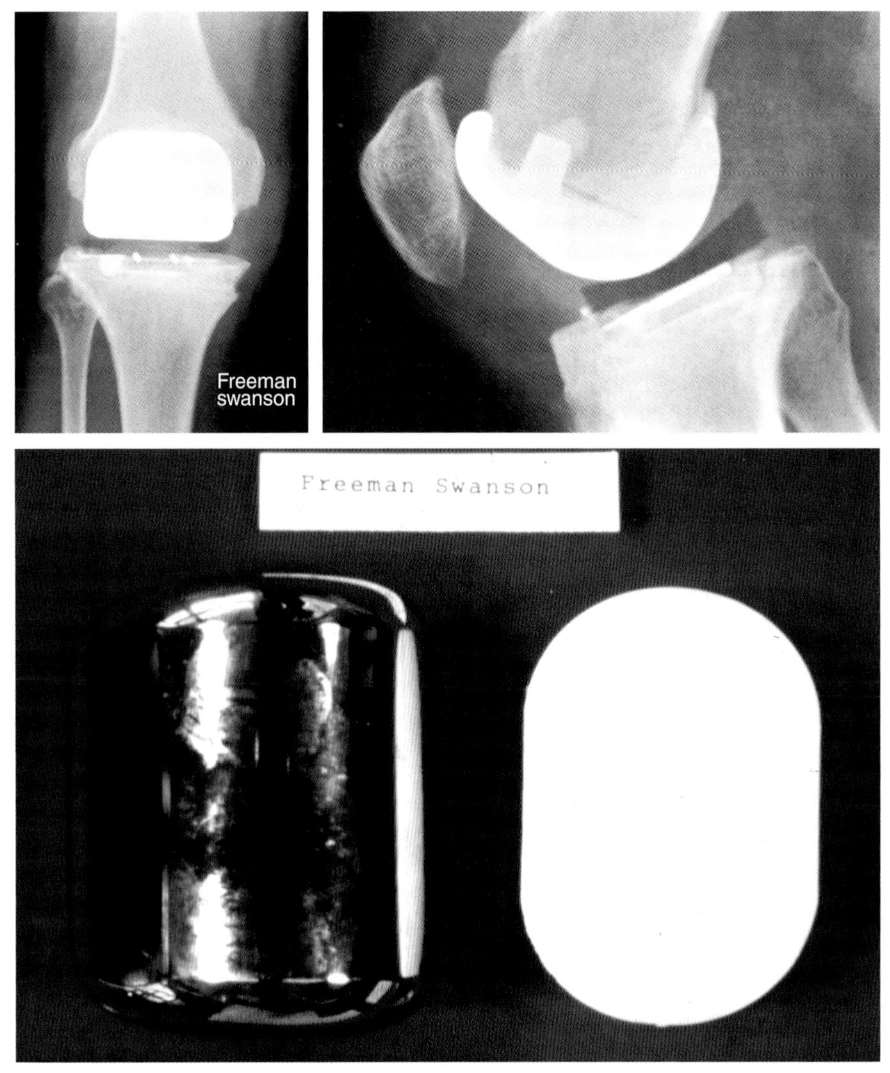


FIGURE 7.16 Radiographs and photographs of a retrieved Freeman Swanson total knee arthroplasty, clinically introduced in the 1970s. Images provided courtesy of Professor Clare Rimnac, Case Western Reserve University, Cleveland, Ohio, USA.

and simpler solution than previous designs. The general concept of the convex UHMWPE patellar component, originally developed for the Total Condylar Knee (Figure 7.14), remains relevant today. In the hands of the surgeons at the Hospital for Special Surgery, this patellar implant design initially provided good or excellent results in 95% of patients [32].

The design of the patello-femoral arthroplasty has undergone many evolutionary design changes since the days of the Total Condylar Knee. On the patellar side, the profile of the domed articulating surface, as well as the size and number of cement fixation pegs, has been tailored to improve conformity of contact, as well as to improve alignment of the patellar component with respect to the femoral condyle. On the femoral side, the design of the anterior flange of the femoral component has evolved to more closely reproduce the anatomic tracking of the patella during deep flexion activities such as stair climbing or squatting. Despite these evolutionary improvements in the design of the articulation, the dome-shaped patellar component included with the Total Condylar knee replacement is considered the foundation of current design concepts incorporating UHMWPE in patello-femoral arthroplasty.

## 7.7 UHMWPE WITH METAL BACKING

The fifth major evolutionary step in the use of UHMWPE for knee arthroplasty was the incorporation of metal backing into tibial component designs during the late 1970s (Figure 7.17). In the total condylar knee, and other similar knee designs, the UHMWPE tibial component was mechanically fixed to the metallic tibial tray. Designs of this type are currently referred to as "fixed bearing" designs because the UHMWPE tibial component remains fixed with respect to the metal backing. In addition to providing integral fixation

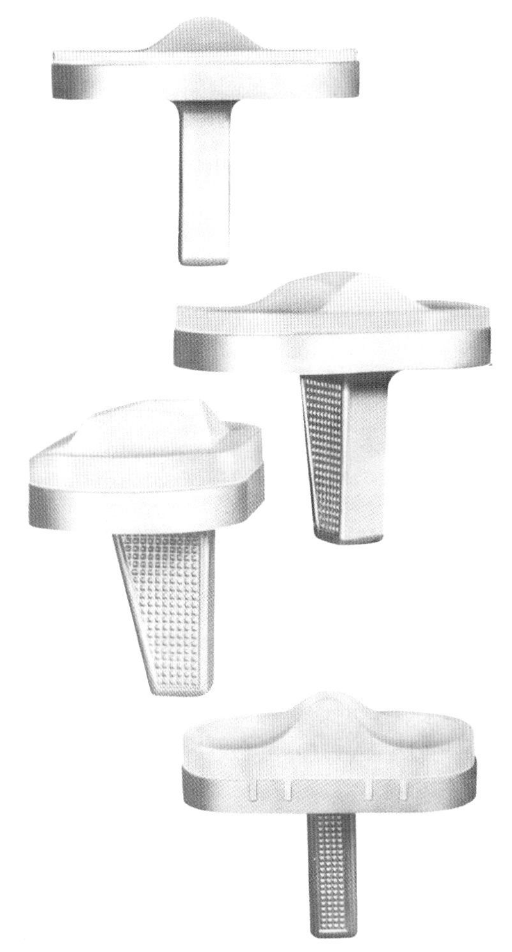

**FIGURE 7.17** Metal backing of the Total Condylar Knee Replacement from the late 1970s.

between the tibial tray and the underlying cement, the introduction of metal backing also made it possible to conceive of a unique family of "mobile bearing" knee designs, which also have their origins in the late 1970s.

## 7.7.1 Fixed Bearing TKA

Although the cemented all-polyethylene Total Condylar Prosthesis was intended to provide improved fixation over previous designs, troubling radiolucencies were nonetheless observed at the cement interface within the first 10 years of implantation [41]. To address the problem of implant fixation, metal tibial trays were initially used to improve the integrity of the cement-prosthesis interface. Finite element analyses later demonstrated that metal backing had the further theoretical benefit of lowering the stresses in the cement and in the subchondral bone [42, 43]. The use of metal backing was later adapted to patellar and unicondylar components for similar reasons.

Notwithstanding the potential advantages afforded by improved fixation and modularity, the use of metal backing with UHMWPE components for knee replacement has been controversial [44]. Human anatomy imposes geometric space constraints on the overall size of orthopedic components. Consequently, the inclusion of metal backing requires the use of a thinner UHMWPE insert than would otherwise be possible without a metallic tibial tray. Under comparable joint loading scenarios, the contact and subsurface stresses in UHMWPE tibial components increase as the thickness decreases [45]. Based on elasticity and finite element solutions of the UHMWPE tibial component in the total condylar knee, Bartel et al. [45] recommended in 1986 that "a thickness of more than eight to ten millimeters should be maintained when possible."

In the 1990s, clinical failures were reported with metal-backed tibial and patellar components due to the reduced thickness of UHMWPE that could be accommodated with such designs [46]. Consequently, all-UHMWPE patellar components are currently more widely used by orthopedic surgeons than metal-backed designs. Today, the Food and Drug Administration recommends a minimum UHMWPE thickness of 6 mm for metal-backed tibial components. Despite some early setbacks with first-generation thin inserts, fixed-bearing metal-backed UHMWPE tibial components currently represent the standard of care in TKA.

## 7.7.2 Mobile Bearing TKA

The mobile bearing is conceptually an interesting and unique adaptation of UHMWPE for metal backed components in TKA, which have their genesis in the late 1970s. In mobile bearing tibial designs, the UHMWPE insert articulates against both a polished femoral component and a polished metal tibial tray (Figure 7.18). Mobile knee bearings comprise two families of designs, including meniscal bearings and rotating platform knees. In meniscal bearings, the UHMWPE portion of the knee implant consists of two individual condylar components, which are constrained to slide in a polished A/P groove in the metal tray. In rotating platform bearings, the UHMWPE tibial component contains an inferior stabilization peg that articulates with a centralizing hole in the tibial tray. A mobile bearing patellar component, based on the rotating platform design, has also been clinically introduced.

One of the first meniscal bearings, known as the Oxford Knee, was developed as a unicondylar prosthesis in 1977 (Biomet, Warsaw, Indiana, USA). Starting in 1977, Michael J. Pappas, PhD, and Frederick F. Buechel, MD, from Newark, New Jersey, USA independently developed several designs of Low Contact Stress (LCS) mobile bearing total joint replacements (DePuy Orthopedics, Warsaw, Indiana, USA) [47]. The philosophy of these mobile bearing designs was to reduce contact stresses at the articulating surface by making the tibial component and femoral components highly conforming. To provide rotational range of motion for the knee, the designers incorporated a second articulation between the UHMWPE component and the polished metal

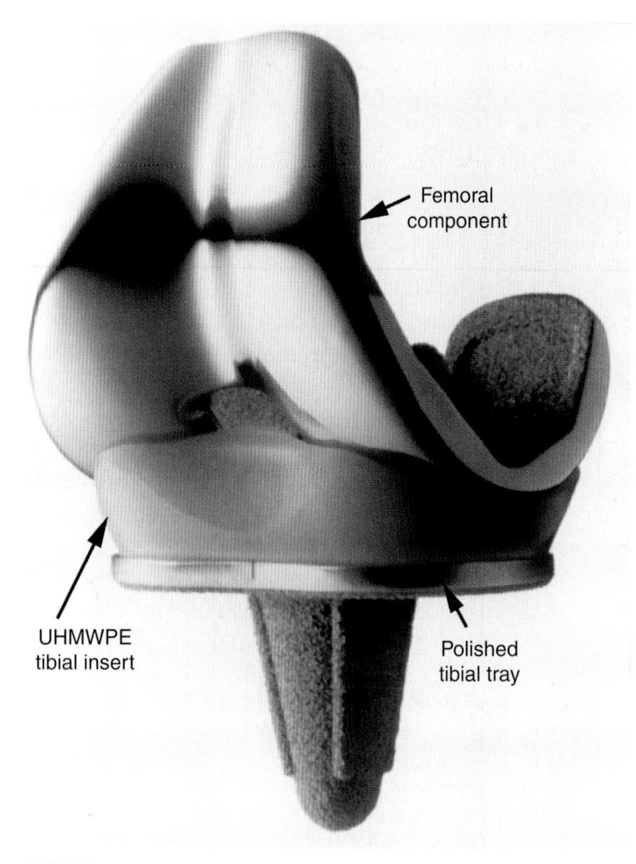


**FIGURE 7.18** Photograph of an LCS rotating platform mobile bearing total knee component. Images provided courtesy of J.B. VanMeter, DePuy Orthopedics, Warsaw, Indiana, USA.

tray. A unique mobile bearing patellar component, which enabled rotation between the UHMWPE patellar component and the metal backing, was also designed by Pappas and Buechel [47].

Mobile bearings were sufficiently different from fixed bearing devices that the FDA required a multicenter clinical study to be performed before they could be marketed in the United States. Following the successful conclusion of the clinical trials in the early 1980s, the LCS mobile bearing knees were clinically introduced by DePuy Orthopedics in 1985 [47]. The designers of the LCS implant system report excellent results after 20 years of follow-up [48–50]. For further information about mobile bearing knee designs, the reader may wish to consult two recent reviews [51, 52] and a book [53] based on the LCS experience.

## 7.8 CONCLUSION

Within a decade of incorporating UHMWPE into TKA, this new form of surgery reached the same level of consistency and success as THA. However, the introduction of UHMWPE influenced the historical development of THA and TKA in different ways. With THA, Charnley had already perfected the design of the Low Friction Arthroplasty, but the introduction of UHMWPE ensured the long-term durability of the prosthesis. With TKA, on the other hand, UHMWPE was accepted from the outset as the material of choice by implant designers, and the design concepts, not the material, were forced to evolve to achieve successful performance in the knee.

Starting with Gunston's pioneering knee design, the principal adaptations for UHMWPE used in TKA were firmly established in the 1970s. In 1986, Insall [33] opined that "very little future improvement can be expected by tinkering with the [total condylar] prosthesis itself, especially for routine cases. What is needed is better surgical training, better instruments, and wider availability of custom designs for special circumstances." Today, UHMWPE continues to serve as the only widely used bearing material for articulation with metallic components in TKA. In the following chapter, we review aspects related to the clinical performance of UHMWPE in the knee.

## 7.9 ACKNOWLEDGMENTS

Many thanks to Professor Clare Rimnac, Case Western Reserve University; Professor Donald Bartel, Cornell University; Frank Gunston, Brandon, Manitoba; and Professor David Lyttle, University of Manitoba, for helpful advice and discussions. Thanks also to Paul Serekian, Howmedica Osteonics, for providing access to the catalog archives for knee replacement during the 1960s, and to J.B. VanMeter and Donald McNulty, DePuy Orthopedics, for assistance with researching the background of the LCS design.

## REFERENCES

1. Robertsson O. The Swedish Knee Arthroplasty Register: validity and outcome. *Ph.D. Dissertation*. Lund, Sweden: Orthopedics, Lund University; 2000.
2. Robertsson O, Lewold S, Knutson K, Lidgren L. The Swedish Knee Arthroplasty Project. *Acta Orthop Scand* 2000;**71**:7–18.
3. Gluck T. Die Invaginationsmethods der Osteo- und Arthoplastik. *Berl Klin Wschr* 1890;**19**:732.
4. Speed JS, Smith H. Arthroplasty: a review of the past ten years. *Surg Gynec Obstet* 1940;**70**:224–30.
5. Gunston FH. Polycentric knee arthroplasty. Prosthetic simulation of normal knee movement. *J Bone Joint Surg Br* 1971;**53**:272–7.
6. Gunston FH. Polycentric knee arthroplasty. Prosthetic simulation of normal knee movement: interim report. *Clin Orthop* 1973;**94**:128–35.
7. Ewald FC. Metal to plastic total knee replacement. *Orthop Clin North Am* 1975;**6**:811–21.
8. Walker PS. Historical development of artificial joints. In: Walker PS, editor. *Human joints and their artificial replacements*. Springfield, IL: Charles C. Thomas, Publisher; 1977.
9. Vince KG. Evolution of total knee arthroplasty. In: Scott WN, editor. *The knee*. St. Louis, MO: Mosby-Year Book, Inc., 1994.
10. Bryan RS, Peterson LF. Polycentric total knee arthroplasty: a prognostic assessment. *Clin Orthop* 1979:23–8.

11. Bloom JD, Bryan RS. Wide-track polycentric total knee arthroplasty: one year follow-up study. *Clin Orthop* 1977:210–13.
12. Cracchiolo 3rd A. Polycentric knee arthroplasty using tibial prosthetic units of a variable height. A preliminary report of design characteristics and a concept of clinical use. *Clin Orthop* 1973;**94**:140–7.
13. Peterson LF, Bryan RS, Combs JJ Jr. Polycentric knee arthroplasty. *Curr Pract Orthop Surg* 1975;**6**:2–10.
14. Bryan RS, Peterson LF. Polycentric total knee arthroplasty. *Orthop Clin North Am* 1973;**4**:575–84.
15. Bryan RS, Peterson LF, Combs JJ Jr. Polycentric knee arthroplasty. A preliminary report of postoperative complications in 450 knees. *Clin Orthop* 1973;**94**:148–52.
16. Bryan RS, Peterson LF, Combs JJ Jr. Polycentric knee arthroplasty. A review of 84 patients with more than one year follow-up. *Clin Orthop* 1973;**94**:136–9.
17. Gunston FH, MacKenzie RI. Complications of polycentric knee arthroplasty. *Clin Orthop* 1976;**120**:11–7.
18. Shoji H, D'Ambrosia RD, Lipscomb PR. Failed polycentric total knee prostheses. *J Bone Joint Surg Am* 1976;**58**:773–7.
19. Lewallen DG, Bryan RS, Peterson LF. Polycentric total knee arthroplasty. A ten-year follow-up study. *J Bone Joint Surg Am* 1984;**66**:1211–8.
20. Skolnick MD, Bryan RS, Peterson LF. Unicompartmental polycentric knee arthroplasty: description and preliminary results. *Clin Orthop* 1975:208–14.
21. Jones WT, Bryan RS, Peterson LF, Ilstrup DM. Unicompartmental knee arthroplasty using polycentric and geometric hemicomponents. *J Bone Joint Surg Am* 1981;**63**:946–54.
22. Mallory TH. The use of polycentric knee arthroplasty in the treatment of fracture deformities of the knee. *Clin Orthop* 1973;**97**:114–6.
23. Marmor L. Unicompartmental knee arthroplasty. Ten- to 13-year follow-up study. *Clin Orthop* 1988:14–20.
24. Skollnick MD, Bryan RS, Peterson LF, Combs Jr. JJ, Ilstrup DM. Polycentric total knee arthroplasty. A two-year follow-up study. *J Bone Joint Surg Am* 1976;**58**:743–8.
25. Rand JA, Coventry MB. Ten-year evaluation of geometric total knee arthroplasty. *Clin Orthop* 1988:168–73.
26. Walker PS. Requirements for successful total knee replacements. Design considerations. *Orthop Clin North Am* 1989;**20**:15–29.
27. Townley CO. The anatomic total knee resurfacing arthroplasty. *Clin Orthop* 1985:82–96.
28. Insall J, Tria AJ, Scott WN. The total condylar knee prosthesis: the first 5 years. *Clin Orthop* 1979:68–77.
29. Insall J, Ranawat CS, Scott WN, Walker P. Total condylar knee replacment: preliminary report. *Clin Orthop* 1976;**120**:149–54.
30. Vince KG, Insall JN, Kelly MA. The total condylar prosthesis. 10- to 12-year results of a cemented knee replacement. *J Bone Joint Surg Br* 1989;**71**:793–7.
31. Scuderi GR, Insall JN, Windsor RE, Moran MC. Survivorship of cemented knee replacements. *J Bone Joint Surg Br* 1989; **71**:798–803.
32. Ranawat CS, Rose HA, Bryan WJ. Replacement of the patello-femoral joint with the total condylar knee arthroplasty. *Int Orthop* 1984;**8**:61–5.
33. Insall JN, Kelly M. The total condylar prosthesis. *Clin Orthop* 1986:43–8.
34. Insall J, Scott WN, Ranawat CS. The total condylar knee prosthesis. A report of two hundred and twenty cases. *J Bone Joint Surg Am* 1979;**61**:173–80.
35. Rodriguez JA, Bhende H, Ranawat CS. Total condylar knee replacement: a 20-year follow-up study. *Clin Orthop* 2001:10–7.
36. Ranawat CS, Flynn Jr. WF, Saddler S, Hansraj KK, Maynard MJ. Long-term results of the total condylar knee arthroplasty. A 15-year survivorship study. *Clin Orthop* 1993:94–102.
37. Insall JN, Lachiewicz PF, Burstein AH. The posterior stabilized condylar prosthesis: a modification of the total condylar design. Two to four-year clinical experience. *J Bone Joint Surg Am* 1982;**64**:1317–23.
38. Freeman MA, Swanson SA, Todd RC. Total replacement of the knee using the Freeman-Swanson knee prosthesis. *Clin Orthop* 1973;**94**:153–70.
39. Borden LS, Heyne T, Belhobek G, Marks KE, Stulberg BN, Wilde AH. Total condylar prosthesis. *Orthop Clin North Am* 1982;**13**:123–30.
40. Aglietti P, Rinonapoli E. Total condylar knee arthroplasty. A five-year follow-up study of 33 knees. *Clin Orthop* 1984:104–11.
41. Ecker ML, Lotke PA, Windsor RE, Cella JP. Long-term results after total condylar knee arthroplasty. Significance of radiolucent lines. *Clin Orthop* 1987:151–8.
42. Lewis JL, Askew MJ, Jaycox DP. A comparative evaluation of tibial component designs of total knee prostheses. *J Bone Joint Surg Am* 1982;**64**:129–35.
43. Bartel DL, Burstein AH, Santavicca EA, Insall JN. Performance of the tibial component in total knee replacement. *J Bone Joint Surg Am* 1982;**64**:1026–33.
44. Rodriguez JA, Baez N, Rasquinha V, Ranawat CS. Metal-backed and all-polyethylene tibial components in total knee replacement. *Clin Orthop* 2001;**392**:174–83.
45. Bartel DL, Bicknell VL, Wright TM. The effect of conformity, thickness, and material on stresses in ultra-high molecular weight components for total joint replacement. *J Bone Joint Surg [Am]* 1986;**68**:1041–51.
46. Collier JP, Mayor MB, Surprenant VA, Surprenant HP, Dauphinais LA, Jensen RE. The biomechanical problems of polyethylene as a bearing surface. *Clin Orthop* 1990;**261**:107–13.
47. Buechel Jr. FF. The LCS story. In: Hamelynck KJ, Stiehl JB, editors. *LCS mobile bearing knee arthroplasty: 25 years of worldwide experience*. Berlin: Springer; 2002.
48. Buechel Sr. FF. Long-term follow-up after mobile-bearing total knee replacement. *Clin Orthop* 2002:40–50.
49. Buechel Sr. FF, Buechel Jr. FF, Pappas MJ, Dalessio J. Twenty-year evaluation of the New Jersey LCS rotating platform knee replacement. *J Knee Surg* 2002;**15**:84–9.
50. Buechel Sr. FF, Buechel Jr. FF, Pappas MJ, D'Alessio J. Twenty-year evaluation of meniscal bearing and rotating platform knee replacements. *Clin Orthop* 2001:41–50.
51. Callaghan JJ, Insall JN, Greenwald AS, Dennis DA, Komistek RD, Murray DW, Bourne RB, Rorabeck CH, Dorr LD. Mobile-bearing knee replacement: concepts and results. *Instr Course Lect* 2001;**50**:431–49.
52. Stiehl JB. World experience with low contact stress mobile-bearing total knee arthroplasty: a literature review. *Orthopedics* 2002;**25**:s213–7.
53. Hamelynck KJ, Stiehl JB. *LCS mobile bearing knee arthroplasty: 25 years of worldwide experience*. Berlin: Springer; 2002.

Chapter 8

# The Clinical Performance of UHMWPE in Knee Replacements

Steven M. Kurtz, PhD



## 8.1 INTRODUCTION

Total and unicondylar knee arthroplasties (TKA and UKA) are highly successful surgical procedures. As we have seen in Chapter 7, the clinical track record for TKA was established during 1970s, when total knee arthroplasty (TKA) gained the same high level of reliability as hip arthroplasty. Since that time, most total knee arthroplasties have enjoyed a long-term survivorship of over 90% after 10 years of implantation. UKA was also established in the 1970s (see Chapter 7) and enjoyed a resurgence of clinical interest in the 1990s with the advance of minimally invasive surgical techniques.

Fixed bearing and mobile bearing total knee designs currently both share the clinical spotlight as safe and effective treatment options for patients requiring total knee replacement. Figure 8.1 shows examples of contemporary fixed-bearing knee replacement components that are commercially available today. Chapter 7 contains an example of a contemporary mobile bearing knee prosthesis (Figure 7.14). Despite decades of continuous advancement in design, instrumentation, and surgical technique for UKA and TKA, conventional UHMWPE remains the gold standard polymeric bearing material for use in both unicondylar and bicondylar knee arthroplasty, whether in fixed or mobile bearing designs.

Data from the Nationwide Inpatient Sample indicates that 525,000 primary knee procedures were performed in the United States in 2006, and the demand for knee arthroplasty is projected to exceed 4 million primary procedures by 2030 [1]. As already discussed in Chapter 5, the incidence of knee arthroplasty is greater than hip arthroplasty (see Figure 5.2). In the United States, over 60% of primary knee procedures are performed on women [2].

Although knee arthroplasty enjoys a remarkable clinical track record, problems with wear and fatigue damage of UHMWPE continue to limit the longevity of both unicondylar and bicondylar knee replacement components. Unlike in the hip, where radiographic techniques have been developed to quantify *in vivo* wear rates, there currently exist no standard and widely accepted techniques for

UHMWPE Biomaterials Handbook

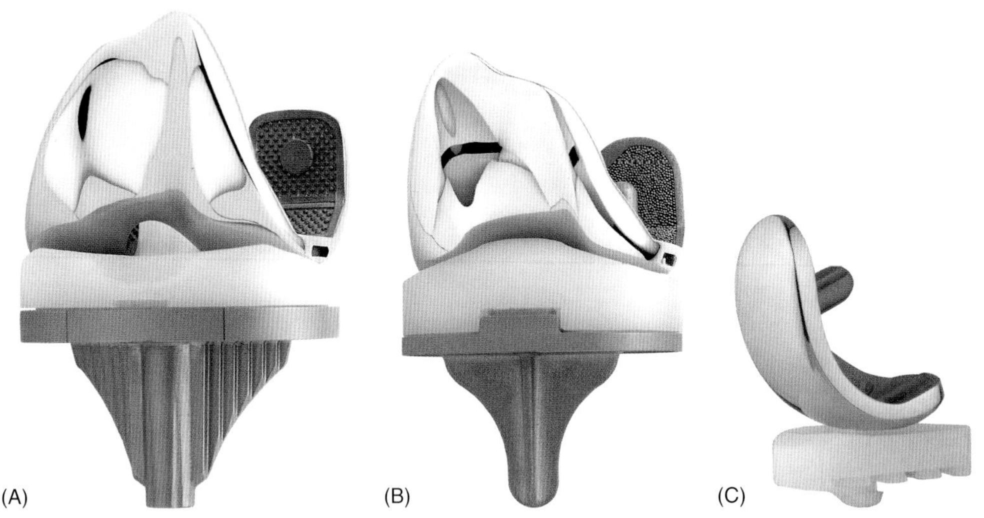


**FIGURE 8.1** Examples of contemporary fixed-bearing knee replacement components. (A) Scorpio posterior-stabilized total knee prostheses; (B) Duracon total knee prostheses; and (C) Eius unicondylar prostheses. Images provided courtesy of Stryker Orthopedics (Mahwah, New Jersey, USA).

tracking the clinical performance of UHMWPE in patients with knee replacement. Thus, today the most effective way to evaluate the *in vivo* performance of UHMWPE continues to be the analysis of retrieved components from revision surgery or from autopsy donations.

This chapter contains four main sections covering TKA and, where applicable, UKA as well. The first section reviews the biomechanical considerations of knee arthroplasty that distinguish it from hip replacement. The second section describes the survivorship of TKA and UKA and outlines measures of clinical performance for UHMWPE in knee arthroplasty. The third section is devoted to wear and osteolysis in TKA. In the fourth and final section of this chapter, we describe alternatives to metal-on-conventional UHMWPE articulation for knee arthroplasty.

## 8.2 BIOMECHANICS OF TOTAL KNEE ARTHROPLASTY

### 8.2.1 Anatomical Considerations

The knee is one of the most complex joints in the body. Unlike the hip, in which the joint surfaces are highly conforming, in the knee the articulating surfaces are more nonconforming. The geometry of the knee permits extreme flexion (up to 140°), which may be needed for activities such as getting out of a chair or for squatting [3]. On the other hand, when the joint is unloaded, there is over 5 mm of laxity in the soft tissue structures surrounding the knee [4] to enable complex rotations and relative sliding of the joint surfaces. The principal ligaments of the knee, including the collaterals and the cruciates, which play an important role in the function of knee replacements, are shown in Figure 8.2.

The importance and biomechanical role of the soft tissues surrounding the knee joint continue to be debated [5, 6]. The posterior cruciate ligament (PCL), in particular, has been an ongoing source of controversy in TKA. This ligament inserts on the posterior aspect of the tibia, and during activities such as stair climbing or rising from a chair, prevents the condyles of the femur from sliding forward off the anterior edge of the tibial plateau. The controversy over the PCL, which started in the 1970s and continues to this day, centers around whether or not it should be sacrificed during knee arthroplasty [5]. Proponents of PCL sacrifice maintain that the implant should provide the constraint and posterior stabilization of the anatomic knee. Such total knee designs are referred to as "posterior-stabilized" knees (or more simply, PS knees), and include a central post of the tibial component, which engages with a cam in the femoral component to constrain anterior relative motion. Opponents of posterior stabilization maintain that the role of knee arthroplasty is to mimic the anatomic knee as closely as possible, which includes preserving the original PCL. The controversy over conformity and constraint is reflected in the wide range of total knee replacement designs currently available to orthopedic surgeons. As we shall see in Section 8.3, both posterior stabilized and PCL-sparing designs of knee replacement have successful survivorships that are statistically indistinguishable. Consequently, today the decision to sacrifice the PCL remains a function of the patient (i.e., whether the PCL is too diseased to function) and the personal preference of the surgeon.

### 8.2.2 Knee Joint Loading

It is difficult to address the clinical performance of UHMWPE in TKA without reference to the demanding

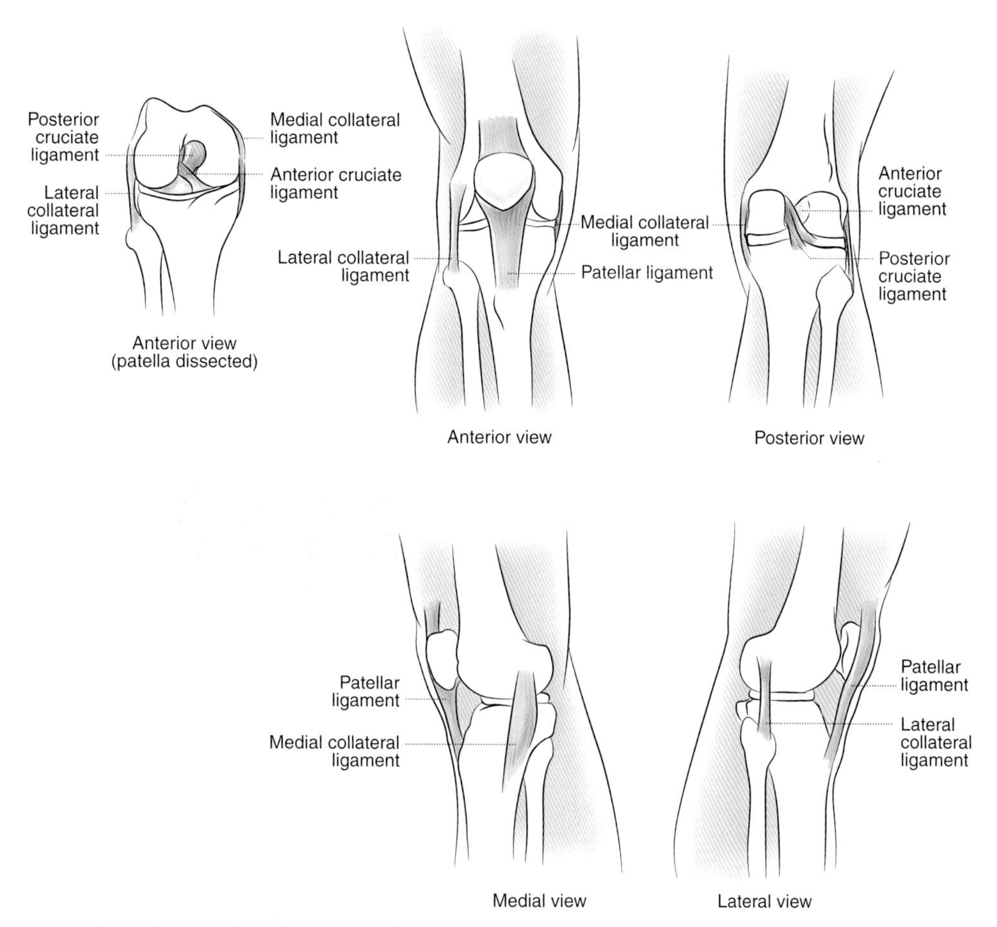


FIGURE 8.2 Anatomy of cruciate and collateral ligaments of the knee.

**TABLE 8.1 Summary of Average Knee Joint Loading for Activities of Daily Living**

| Activity | Reference | Patellofemoral joint | | Tibiofemoral joint (compression) | | Tibiofemoral joint (anterior shear) | |
|---|---|---|---|---|---|---|---|
| | | Knee angle (°) | Force (× BW) | Knee angle (°) | Force (× BW) | Knee angle (°) | Force (× BW) |
| Walking | [8–10] | 10 | 0.5 | 15 | 3.0–3.5 | 5 | 0.4 |
| Squatting | [3] | 140 | 6.0–7.6 | 140 | 5.0–5.6 | 140 | 2.9–3.5 |
| Rising from a chair | [11] | 120 | 3.1 | 120 | 3–7 | 120 | 2.3 |
| Stair climbing/descent | [9] | 60 | 3.3 | 45–60 | 3.8–4.3 | 5 | 0.6 |

*Force is expressed in units of body weight (BW).*

functional environment of the knee joint. Starting in the 1960s [7], researchers began to develop models to calculate the load transmission across the contacting surfaces of the knee joint for various activities. Early models to predict the forces across the knee joint are statically indeterminate because the number of independent soft tissue structures acting across the knee joint is greater than what can be solved for using the equations of static equilibrium. However, by making simplifying assumptions in the number of active muscle groups across the joint during a particular activity, it turns out that reasonable estimates of joint reaction forces may be obtained for activities such as walking, stair climbing/descent, squatting, and rising from a chair (Table 8.1). For example, the early models of Morrison suggested that the tibiofemoral contact force ranged from two to four times body weight (2–4 BW) during regular gait, with an average of three times body weight [7, 8]. For certain activities, such as rising from a chair or squatting, the forces across the knee can be up to 7.6 times body weight (Table 8.1).

Surprisingly, consistent conclusions are reached using more complex biomechanical models of the knee [12–16], as well as by measuring *in vivo* forces from telemeterized knee replacements [17, 18]. For further discussion of the biomechanics of the knee joint, the reader is referred to the works of Morrison [7], Paul [19], Walker [20], as well as to review articles [21, 22].

## 8.2.3 Stresses in UHMWPE Tibial and Patellar Components for TKR

Since the 1980s, researchers have associated the stress levels acting on UHMWPE components with the extent and severity of surface damage observed in retrieved components. However, the precise relationship between wear, damage, and stress acting on the UHMWPE has remained elusive, due to the numerous factors contributing to clinical performance. The stresses acting on UHMWPE components for total joint replacement are design specific. Today, finite element analysis is the primary method for calculating stress distributions in specific designs of UHMWPE components. Experimentally, pressure sensitive film may also be used to quantify contact stress, but not other stress components, which may be associated with surface damage.

Contemporary finite element models take into account the complex three-dimensional geometry of tibial, femoral, and patellar components (Figure 8.3). The models must incorporate relevant boundary conditions, including the appropriate joint loading, as was discussed in Section 8.2.1.

A finite element simulation of a total joint simulation must also incorporate a realistic material model for UHMWPE. Depending upon the scope and question posed by a finite element analyst, a simple linear elastic or isotropic plasticity material model for UHMWPE may be sufficient. These simple material models, however, are not suitable for simulation of cyclic loading, for determining the time-dependent response to loading, or for prediction of large deformation behavior leading up to failure in UHMWPE. Recently, a more advanced material model has been developed and validated for conventional and highly crosslinked UHMWPE (see Chapter 35). Based on the principles of polymer physics, this hybrid model accurately incorporates rate effects, viscoplasticity, and evolution of anisotropy during large deformations for a wide range of UHMWPE materials.

Beyond the complexities introduced by geometry, boundary conditions, and material properties, the surgical malpositioning of components, or geometric discontinuities for a particular hip or knee design, may also produce regional stress concentrations. However, in this section of the chapter we will focus on describing the macroscopic stress state of UHMWPE within the contact area of the articulating surface of well-aligned components.

The magnitude and distributions of stress in the knee are different from the hip. In the hip, the spherical contacting surfaces are highly conforming, and the effective (von Mises) stress levels are below yield, and, thus, below the onset of irrecoverable plastic deformation. Consequently, for hip components, UHMWPE can reasonably be considered to behave as an elastic material at the continuum level. Elasticity solutions have been developed to calculate

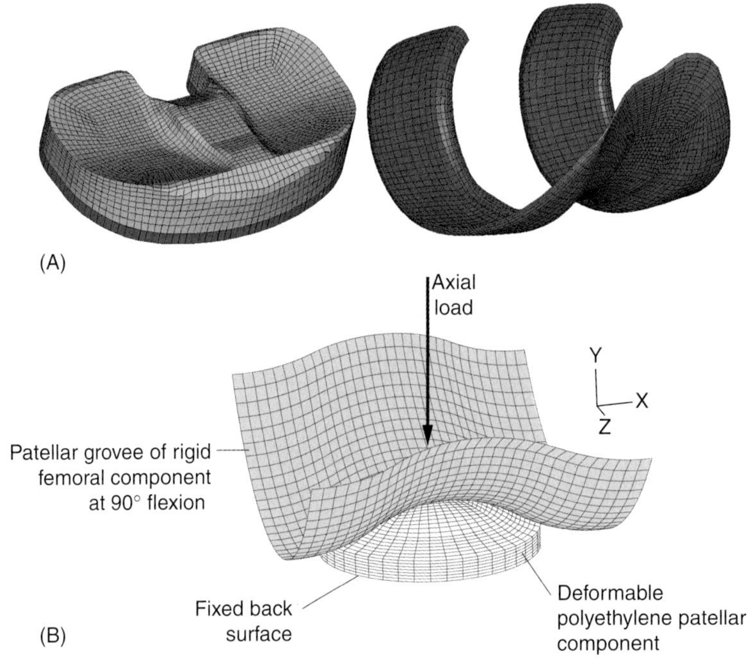


**FIGURE 8.3** Three-dimensional finite element models used to simulate tibio-femoral contact (A) and patello-femoral contact (B).

the contact stress distributions for metal-backed acetabular components [23]. In this regard, a specialized analytical solution is required for UHMWPE acetabular components because the more widely used Hertzian contact theory does not apply. Finite element analyses are nonetheless helpful to get a complete picture of the stress state in hip replacements and have proved to be necessary when addressing the effects of specific hip component designs. From finite element analysis, researchers have observed that the maximum shear stress occurs at or very near the surface of conforming UHMWPE acetabular bearings. This observation has been used to explain why certain forms of surface damage, such as pitting or delamination, are rarely observed in retrieved UHMWPE acetabular components. As we have seen, the primary damage mode in the hip is adhesive/abrasive wear.

The stress levels for UHMWPE tibial and patellar components are generally higher than in acetabular components. Knee joint forces are comparable to those at the hip. However, due to the nonconformity between the tibial and femoral components surfaces, the joint forces of the knee are distributed over a much smaller area than in the hip. The von Mises stresses in knee replacements are typically greater than the offset yield stress of UHMWPE and are thus of sufficient magnitude to result in irrecoverable plastic deformation of the component. Figure 8.4 shows the distribution of compressive stress (minimum principal stress) on the surface of a contemporary knee replacement (the corresponding mesh for the analysis is shown in Figure 8.3). If the component is sectioned perpendicular to the surface, one can visualize the stress state in the UHMWPE through the thickness of the tibial component using finite element analysis. Figure 8.5 shows the effective (von Mises) stress in a perpendicular section through the thickness of the tibial component under the center of contact. As shown by Bartel et al. [24, 25], the maximum von Mises stress occurs at a depth of 1 to 2 mm below the articulating surface. The location in maximum von Mises stress also coincides with the location in the maximum shear stress within the UHMWPE component. Thus, not only is the magnitude of stress greater in knee components, but the distribution through the thickness varies substantially from hip replacement components.

Although the computational tools for predicting the stress state within UHMWPE tibial components have advanced considerably since the 1980s, these sophisticated models remain limited to comparative assessment of different design geometries. The ability to extrapolate predicted stress results into the clinical environment remains extremely limited. As stated earlier, the magnitude of stresses predicted in finite element model of knee arthroplasty, while associated with surface damage in a general sense, have not yet been directly linked to long-term survivorship in patients.

It would be logical to suppose that designs that are subjected to a higher stress level would exhibit worse clinical performance. However, this has not necessarily been the case in actual clinical practice. The Insall-Burstein (IB) and the Miller-Gallante (MG) (Figure 8.6) are examples of knee prostheses designs that have functioned successfully under substantially different stress states. The IB knee is a cruciate-sacrificing design, and the tibial component is designed with concave condylar surfaces for conformity with the femoral component [26]. The MG knee, in contrast, is a cruciate-sparing design and includes a relatively flat tibial

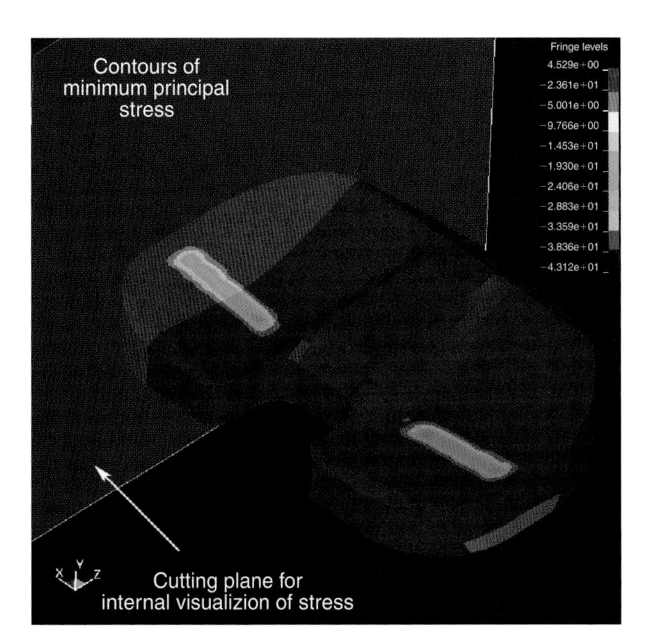


**FIGURE 8.4** Distribution of minimum principal stress on the surface of a tibial component at heel strike during normal gait, using the model shown in Figure 8.3A. Also shown is a cutting plane, which will be used to examine the internal stress distribution of the component (see Figure 8.5).

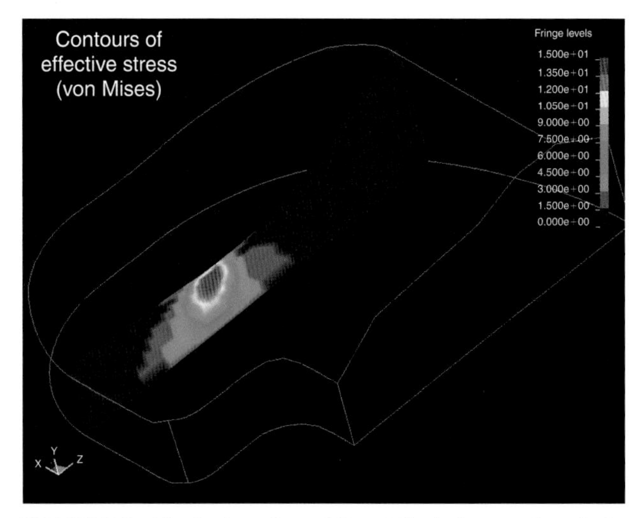


**FIGURE 8.5** Distribution of von Mises (effective) stress through the thickness of a tibial component. Note that the location of the maximum von Mises stress, which also corresponds to the location of maximum shear stress, is located at a depth of 1–2 mm from the articulating surface. The slice is taken perpendicular to the contact surface of a tibial component at heel strike during normal gait, using the model shown in Figures 4A and 4.

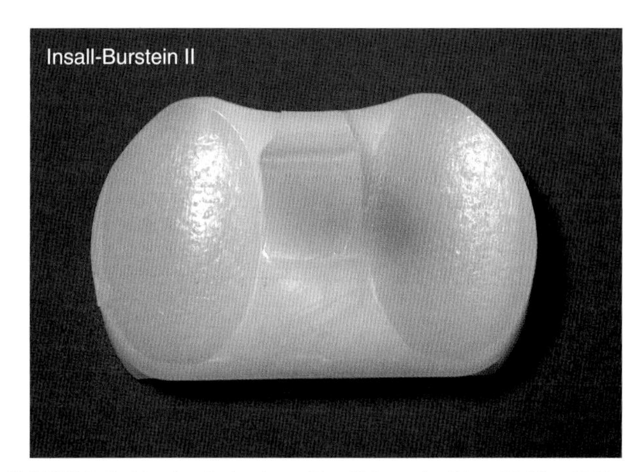


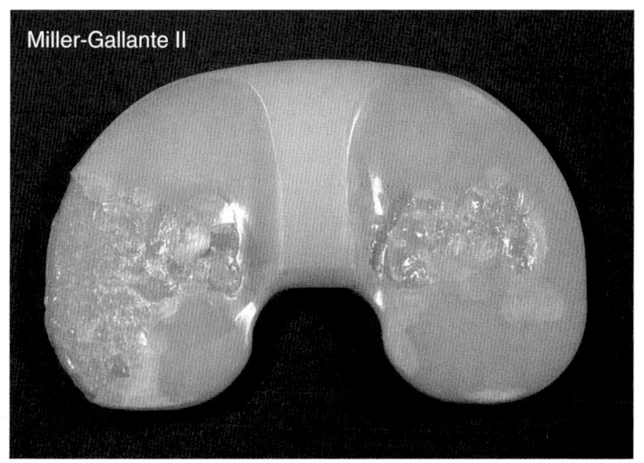


**FIGURE 8.6** Previously implanted Insall-Burstein (II) and Miller-Gallante (II) tibial inserts with evidence of pitting and delamination, respectively.

component to reduce constraint. In the MG knee, the tibial component is nonconforming with the femoral component, especially during activities that involve full flexion.

The contact stresses and von Mises (effective) stresses are up to 20% higher for the nonconforming MG design as compared with the more conforming IB design under identical loading conditions [27]. Depending upon the loading, the magnitude of the von Mises stresses for both designs may be sufficiently high to produce localized yielding and permanent deformation of the UHMWPE insert [27]. Despite differences in stress levels, the two designs both exhibit successful clinical performance in terms of long-term survivorship [26, 28]. Interestingly, the designs do exhibit different predominant modes of surface damage. As illustrated in Figure 8.6, IB knees tend to show greater evidence of pitting than MG knees, which tend to delaminate when oxidized.

The magnitude of stresses subjected to a UHMWPE component is not the sole variable governing wear and damage to total knee replacements. Additional factors, such as the amount of constraint provided by different knee designs [6], the presence of fusion defects in the UHMWPE [29], and the extent of oxidation in the UHMWPE [30] have all been identified as factors related to the clinical performance of tibial components. Due to the complexity of the multiple factors contributing to *in vivo* wear and surface damage of UHMWPE tibial components, researchers and implant designers in orthopedics have not yet developed the capability to accurately simulate the clinical performance of different knee designs on a computer in advance of human clinical trials.

## 8.3 CLINICAL PERFORMANCE OF UHMWPE IN KNEE ARTHROPLASTY

The orthopedic literature contains thousands of articles describing the clinical performance of various knee replacement designs. As in the hip, the clinical performance of knee replacement is most unambiguously defined in terms of survivorship. Clinicians may disagree as to the precise etiology of a TKA failure, but the date of a revision surgery is a precise endpoint for the procedure.

On the other hand, survivorship alone does not fully capture the clinical performance of UHMWPE in the knee. Surface damage and wear of the UHMWPE insert are also important metrics of clinical performance of knee arthroplasty. If a knee prosthesis survives the first 10 years of implantation, wear behavior of the insert plays an increasing role in the longevity of the joint replacement. In the following sections, we describe the survivorship of TKA, as well as the assessment of wear and surface damage in knee arthroplasty.

### 8.3.1 Survivorship of Knee Arthroplasty

Total knee replacements are, in general, highly successful implants. Knee arthroplasty is so successful, in fact, that is often difficult to discriminate between different knee designs, which generally appear to enjoy comparably high survival rates. The long-term survivorship of total knee arthroplasty, as reflected in the orthopedic literature, has been summarized in a recent meta-analysis performed by Forster [31]. In Forster's study, the knee designs were classified as either posterior stabilized or nonstabilized. The analysis also evaluated whether metal backing of the tibial component influenced implant survival. The 10–11y survivorship rates from Forster's analysis are summarized in Table 8.2. Although the survival rates for all of the design types considered was over 90%, the breadth of the confidence intervals precluded distinguishing among different designs. As discussed by Forster, the orthopedic literature may be limited by publication bias (the tendency to report only successful outcomes), and the vast majority of the published studies had to be excluded from the meta-analysis

**TABLE 8.2 Summary of Total Knee Replacement Survival at 10–11y, Based on a Meta-Analysis of the Orthopedic Literature Published by Forster [31], Covering 5950 TKAs**

| TKA design group | Original cohort size | Success rate at 10–11y | 95% confidence intervals |
|---|---|---|---|
| Posterior stabilized design | 1698 | 92.7% | 88.0%–95.4% |
| Nonstabilized design | 2218 | 92.4% | 90.3%–94.1% |
| Metal-backed tibial component | 1561 | 92.7% | 89.4%–94.6% |
| All-polyethylene tibial component | 473 | 95.9% | 92.7%–97.8% |
| Nonstabilized, metal-backed tibial component | 2034 | 91.4% | 88.8%–93.4% |
| Posterior stabilized, metal-backed tibial component | 1272 | 92.8% | 83.7%–94.2% |
| Posterior stabilized, all-polyethylene tibial component | 289 | 93.3% | 86.9%–96.9% |
| Nonstabilized, all-polyethylene, tibial component | 184 | 98.9% | 95.7%–99.7% |

due to inadequate reporting of survivorship tables in individual studies.

A more comprehensive evaluation of survivorship in contemporary knee arthroplasty (both UKA and TKA) can be derived from the national knee implant registries in Sweden and Norway [32–34]. The Swedish Knee Arthroplasty Register, for example, started in 1975 and, as of 1997, had registered 57,533 procedures [32]. A total of 6865 knee arthroplasties were registered in the Swedish Knee Register in 2001 (the year for which the most recent data are available), which corresponds to a 15% increase over the previous year. The system used to collect data for the national registry in Sweden has recently undergone extensive reverification and validation [34].

The Scandinavian national knee registries provide unique, population-based outcome data for TKA and UKA. The Swedish knee registry data, for example, shows a significant effect of patient age, gender, and disease (osteoarthritis [OA] versus rheumatoid arthritis [RA]) on the survivorship of TKA and UKA (Figures 8.7 and 8.8). The developers of the Scandinavian national registries argue that the early identification of inadequate designs has led to their withdrawal from the marketplace and resulted in national revision rates that have declined steadily over time (Figure 8.9).

Interestingly, the Swedish registry researchers have observed a "learning curve" phenomenon with UKA. For designs with technically demanding implantation procedures, the risk of UKA revision in Sweden was associated with the number of procedures performed at a surgical site [35]. The negative Swedish experience with UKA in patients with RA also underscores the importance of proper patient selection for this procedure (Figure 8.8).

## 8.3.2 Reasons for Knee Arthroplasty Revision Surgery

Today, a patient's knee replacement may be revised for aseptic loosening; infection; fracture; joint stiffness; tibio/femoral instability due to collateral ligament instability; patellar complications; extensor mechanism rupture; and/or wear or failure of the UHMWPE component [36–38]. According to Vince, the surgeon revising a knee arthroplasty must first identify and correct the root cause of the previous surgery, not merely treat the symptoms of the failed knee [36]. If the root cause of a revision is malalignment of the femoral component, simply exchanging a worn UHMWPE tibial insert will not address the central problem for the patient and, conversely, will predispose the new tibial component to early failure as well.

Infection, loosening, and patellar complications have been identified as prevalent reasons of TKA revision (Figure 8.10). In a study of 440 revision surgeries performed between 1982 and 1999, Fehring et al. reported that infection was the single largest cause of knee arthroplasty revision within the first 5 years of implantation [37]. In a study of 212 knee revisions performed between 1997 and 2000 by Sharkey et al., infection was responsible for 25% of revisions within the first 2 years of implantation, but only 7.8% of components implanted over 2 years were infected [38].

Sharkey et al. [38] also found that 25% of knee revisions were associated with wear or surface damage of the tibial or patellar insert. Interestingly, the timing of knee revisions associated with UHMWPE wear has two major peaks (Figure 8.11), the first occurring before 5y and the second peak in revisions occurring between 6–10y. For components revised for wear within the first 5y of implantation, patellar

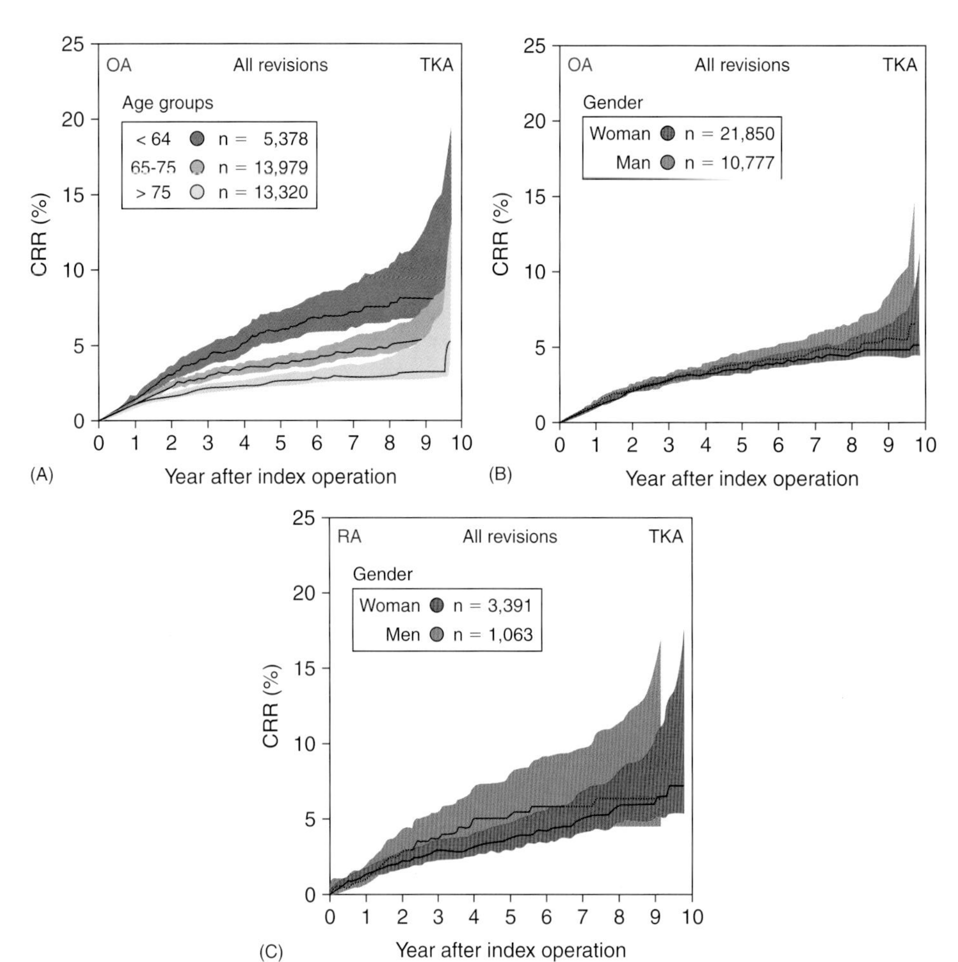


**FIGURE 8.7** Total knee arthroplasty (TKA) survivorship, expressed as the cumulative risk of revision (CRR), from the Swedish National Knee Arthroplasty Register [74]. Data were compiled for patients as a function of age (A) and for those with (B) osteoarthritis (OA) and (C) rheumatoid arthritis (RA). Reprinted with permission.

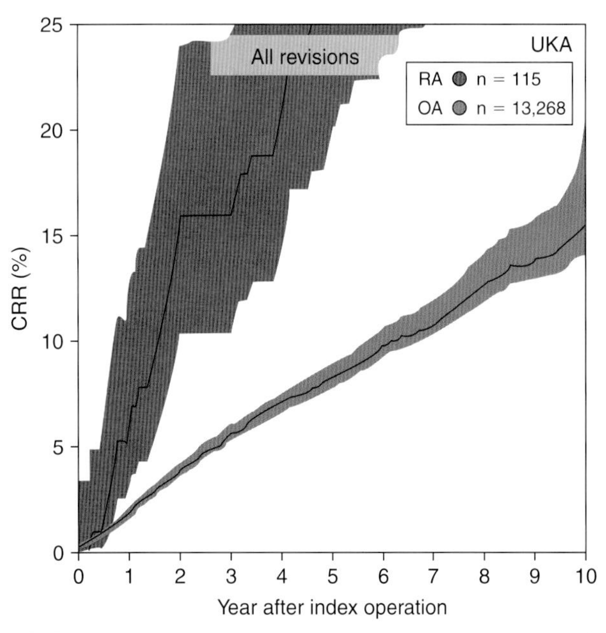


**FIGURE 8.8** Unicompartmental knee arthroplasty (UKA) survivorship, expressed as the cumulative risk of revision (CRR). Data were compiled for patients with osteoarthritis (OA) and rheumatoid arthritis (RA) by the Swedish National Knee Arthroplasty Register [74]. Reprinted with permission.

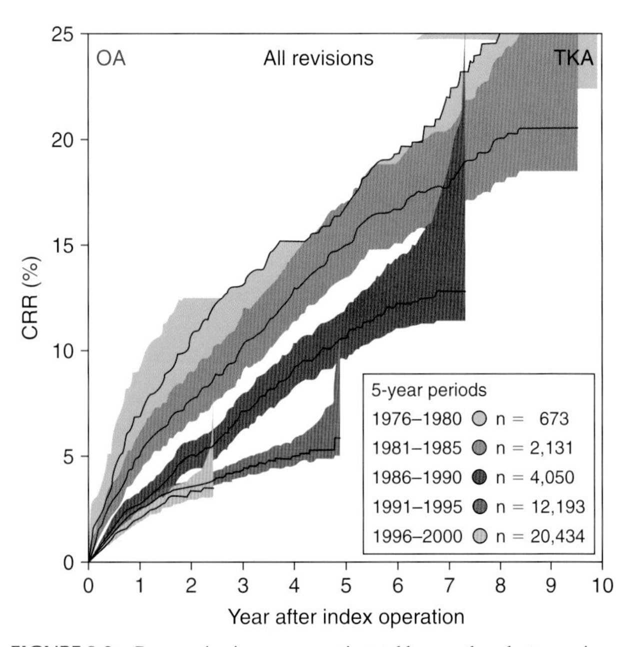


**FIGURE 8.9** Progressive improvement in total knee arthroplasty survivorship in 5y intervals from 1975 to 2000. Data were compiled by the Swedish National Knee Arthroplasty Register [74]. Reprinted with permission.

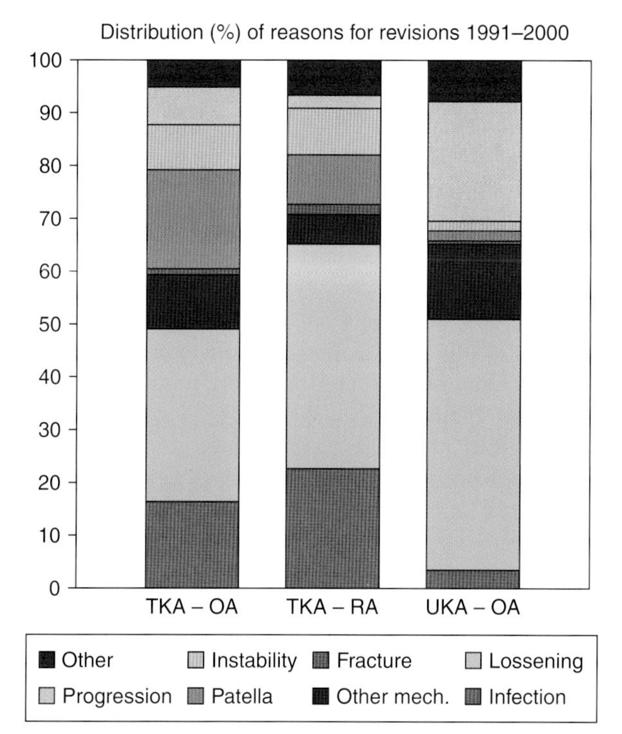


FIGURE 8.10 Reasons for knee arthroplasty revision from the Swedish National Knee Arthroplasty Register [74]. Reprinted with permission.

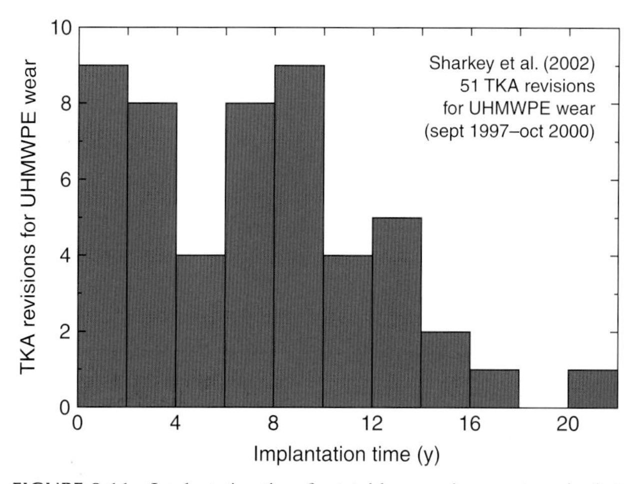


FIGURE 8.11 Implantation time for total knee replacements revised at one institution, in which UHMWPE wear was identified as a reason for revision. Adapted from [38].

problems, loosening, instability, and malalignment were typically also identified as reasons for revision [38]. For knee arthroplasties that survive the first 5 years of implantation, the clinical performance of the UHMWPE plays an increasing role in the longevity of the artificial knee.

### 8.3.3 Articulating Surface Damage Modes

Wear and damage to UHMWPE components for knee replacement is not a new phenomenon and has been clinically

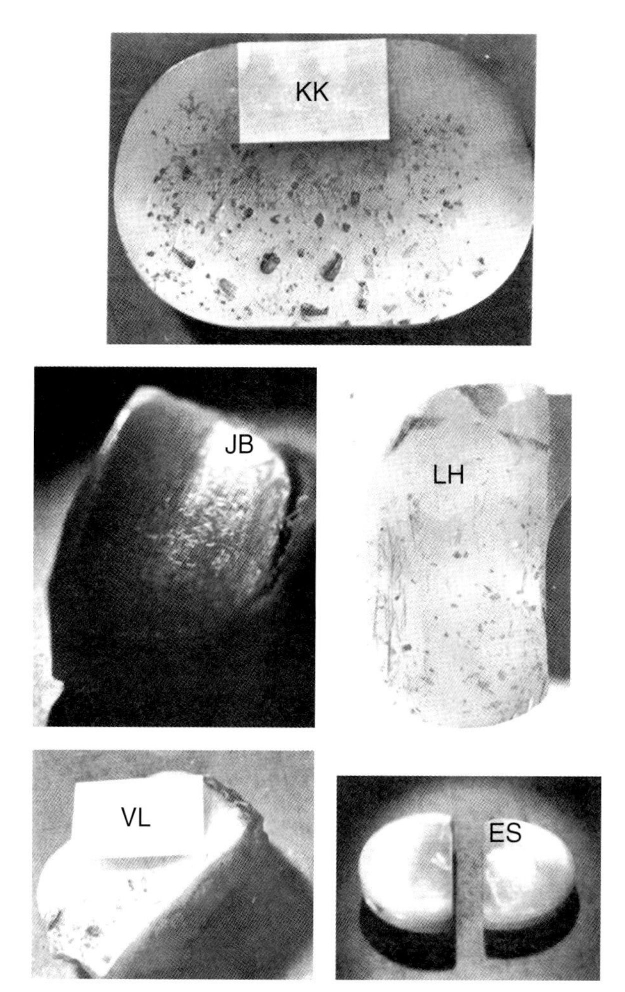


FIGURE 8.12 Examples of UHMWPE surface damage (e.g., scratching, burnishing, pitting) documented with tibial components during the 1970s. *KK* Freman-Swanson; *JB* Duo-Condylar; *LH* Geometric; *VL* Duo-Condylar; and *ES* Marmor modular knee prosthesis designs. From Walker, *Human Joints and Their Artificial Replacements*, 1977 [40]. Courtesy of Charles C. Thomas, Publisher, Ltd., Springfield, Illinois, USA.

observed since the 1970s. As noted in Chapter 7, in 1976 Shoji et al. [39] reported severe wear and fatigue cracks in a small series of Polycentric knee components that were implanted for 2 years or less, but the UHMWPE failures were attributed to surgical misalignment of the implant components. Walker's treatise on natural and artificial joints [40], published in 1977, also documented examples of pitting, scratching, and burnishing in UHMWPE knee replacement components (Figure 8.12). However, the focus on developing improved designs dominated knee arthroplasty during the 1970s. Consequently, the development of reliable methods to quantify surface damage in artificial knees was delayed until the 1980s.

In 1983, Hood et al. [41] published a seminal paper that established a reproducible and semiquantitative method for scoring the modes and prevalence of surface damage in UHMWPE components for knee arthroplasty. Hood and coworkers recognized the immediate need for a system to

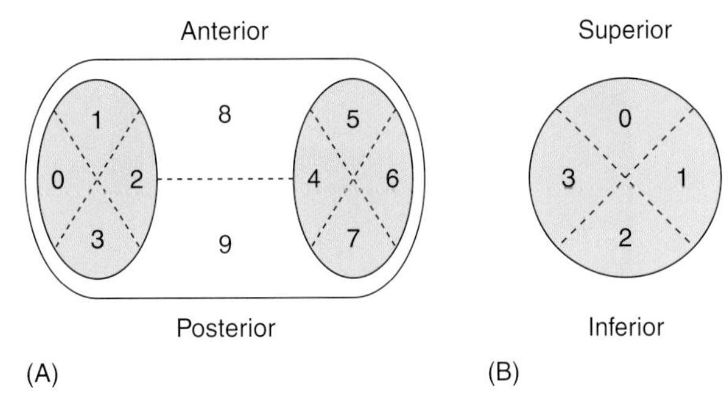


**FIGURE 8.13** The damage scoring method of Hood et al. [41] involved analyzing wear and surface damage within 10 surface regions of UHMWPE tibial components (A) and within four surface regions of UHMWPE patellar components (B).

quantify surface damage and wrote that, "Now that metal and polyethylene joint implants are in the second decade of common usage, it would be prudent to develop a method for characterizing the changes in surface and structural conditions of retrieved implants so as to allow the study of the effects of this mechanical degradation on implant performance" [41]. Although originally developed using Total Condylar Prostheses, Hood's general method of classifying and scoring damage modes for UHMWPE components has remained relevant to a surprisingly wide range of different knee replacement designs over the past 20 years.

Under a light microscope at 10× magnification, Hood studied 10 regions of the articulating surface of tibial components, and four regions of the surface of patellar components, for evidence of wear or surface damage (Figure 8.13). Seven modes of surface damage were identified:

Pitting: This damage mode is sometimes referred to as "cratering" and is characterized by Hood as surface defects 2–3 mm in diameter and 1–2 mm deep (Figure 8.14). Pitting is classified as a mode of fatigue wear and involves the liberation of millimeter-sized pieces of wear debris from the articulating surface. Because the wear debris produced by pitting is considered to be too large to provoke an osteolytic response, from a biological perspective pitting is a more benign wear mechanism than adhesive/abrasive wear, which produces micrometer-sized debris.

Embedded debris: Hood and coworkers initially restricted this damage mode to PMMA debris, but it is possible that bone chips or metallic beads or fragments from the back surface of metallic components could also become embedded in the UHMWPE (Figure 8.14). Embedded debris can result in third body wear of the UHMWPE and metallic surfaces. In addition, embedded debris can scratch the metallic surface, resulting in further abrasive wear of the UHMWPE.

Scratching: This damage mode is identified as linear features on the articulating surface, produced by plowing of microscopic asperities on the opposing metallic surface , or by third body debris (Figure 8.14). This is a mode of abrasive wear.

Delamination: This damage mode is a more severe manifestation of fatigue wear than pitting and involves the removal of sheets of UHMWPE from the articulating surface. If the tibial component is sufficiently thick, the UHMWPE underlying the delamination may continue to serve as a functional bearing surface (e.g., the Miller-Gallante knee shown in Figure 8.6). In other cases, when the tibial component is too thin and embrittled due to oxidation, delamination can result in catastrophic wear of the UHMWPE, necessitating revision (Figure 8.15).

Surface deformation: This damage mode corresponds to a permanent (irrecoverable) change in the surface geometry of the implant (Figure 8.16). It is sometimes referred to as "plastic deformation," "cold flow," or "creep." Unlike the other damage modes, plastic deformation does not result in material removal and thus does not strictly correspond to wear.

Burnishing: This damage mode is characterized as "wear polishing" and is characteristic of adhesive/abrasive wear (Figure 8.16). From a biological perspective, burnishing produces wear debris that is within the size range that can stimulate an osteolytic response.

Abrasion: Hood et al. describe abrasion as a shredding or "tufting" of the UHMWPE surface. This is classified as a mode of abrasive wear.

Within each surface region of the tibial or patellar component, Hood and colleagues graded the presence and extent of the seven damage modes on a 0 to 3 scale [41]. A score of 0 corresponded to the absence of the damage mode within the specified region. Scores of 1, 2, and 3 corresponded to observation of the damage mode over less than 10%, 10–50%, or over 50% of the specified region, respectively. An overall assessment of each damage mode was determined by summing the scores across all of the surface regions. For each damage mode, the maximum score was 30 (3 max score/region × 10 regions) for the tibial component and 12 (3 max score × 4 regions) for the patellar component. A total damage score was computed by adding up the total damage mode scores, corresponding to a maximum score of 210 (3 max score/region × 10 regions × 7 damage modes) for the tibial component and 84 (3 max score × 4 regions × 7 damage modes) for the patellar component.

The Hood method is semiquantitative and thus allows researchers to compare the location of damage (e.g., medial condyle versus lateral condyle; anterior versus posterior), the prevalence of damage modes within a single design, and also differences in damage between designs. For example, in Hood's assessment of retrieved Total Condylar Prostheses, Hood et al. noted that scratching was the most prevalent form of surface damage (observed in 90% of the tibial component retrievals), followed by pitting (81%),

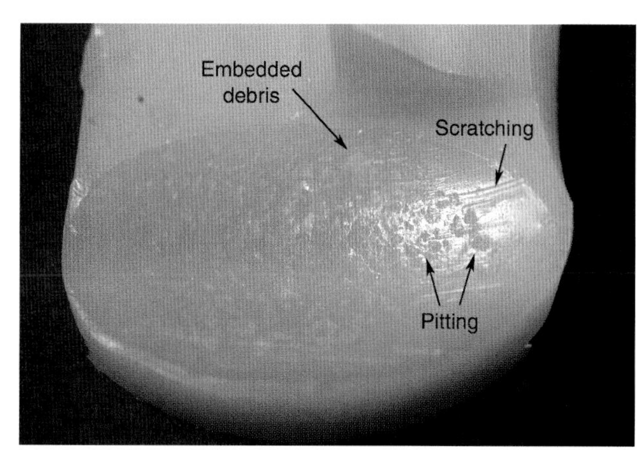


FIGURE 8.14 Examples of pitting, scratching, and embedded debris surface damage modes.

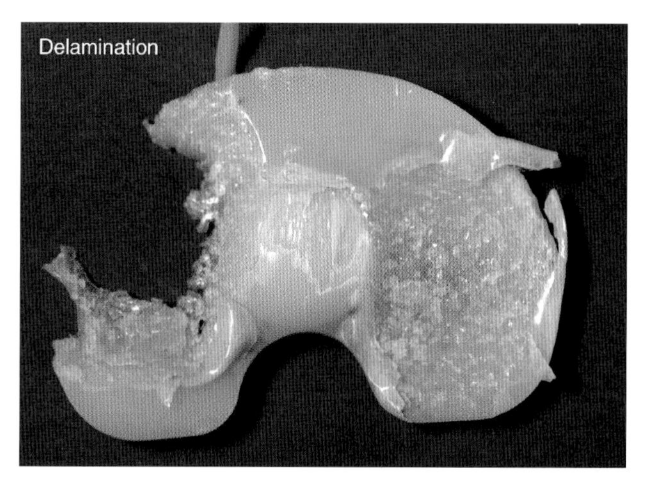


FIGURE 8.15 Severe delamination of an Ortholoc tibial component (Richards, Memphis, Tennessee, USA).

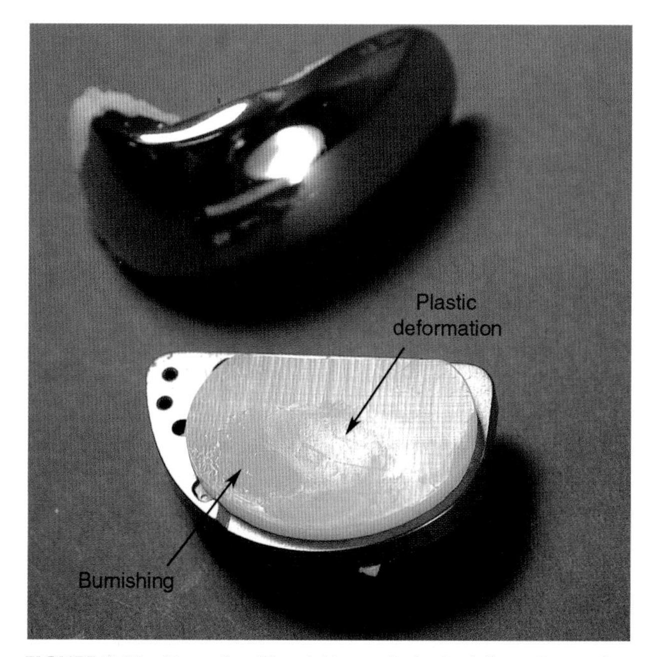


FIGURE 8.16 Example of burnishing and plastic deformation surface damage modes.

burnishing (75%), surface deformation (62%), cement debris (48%), abrasion (41%), and delamination (4%) [41]. Hood and coworkers also noted a significant correlation between patient weight and total damage score, as well as between implantation time and damage score, suggesting that fatigue mechanisms resulting from cyclic loading were likely involved in the generation of wear and damage to knee components.

Although originally developed for the Total Condylar knee, Hood's method has been applied to a wide range of cruciate sparing as well as cruciate sacrificing tibial component designs during the past 2 decades [42–44]. Hood's method has also been adapted to quantify damage to vertical stabilizing posts [45]. After 20 years, Hood's method continues to serve as the fundamental reference for damage modes in UHMWPE components for knee replacement.

## 8.4 OSTEOLYSIS AND WEAR IN TKA

Osteolysis, which may be provoked by exposure of bone to particles of UHMWPE, bone cement, or metallic debris, has been a major concern for hip replacement. However, only recently has osteolysis been regarded as an important complication of TKA. An example of severe osteolysis in TKA is shown in Figure 8.17.

Prior to 1992 [46], osteolysis was generally not noted in the orthopedic knee literature except as isolated case reports. Within the past 10 years, however, there has been increased interest in wear and osteolysis as it relates to knee replacement. Several clinical and retrieval studies related to osteolysis in TKA have been summarized in Table 8.3.

### 8.4.1 Incidence and Significance of Osteolysis in TKA

Whether in the hip or the knee, for osteolysis to expand, wear debris particles need access to the periprosthetic bone. From review of the orthopedic knee literature, it appears that cemented fixation of knee components is an effective barrier to particle access and explains the lower incidence of osteolysis in studies with cemented components (Table 8.3). In this regard, much of the knee osteolysis literature, including the 1992 study by Peters et al. [46], relates to this complication in cementless TKA (Table 8.3).

The incidence of osteolysis with cementless fixation has been reported as high as 30% within the first 5 years of implantation [47] (Table 8.3). In studies of cemented components, in contrast, the incidence of osteolysis is generally lower and has been reported to range between 0% and 20% (Table 8.3). In addition to the fixation method, implant factors (e.g., design [48–50]) and patient factors (e.g., obesity [51]) have been shown to have a significant influence on the incidence of osteolysis in TKA.

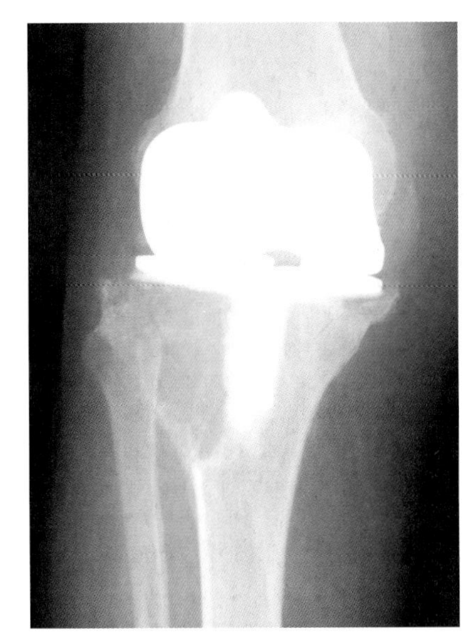
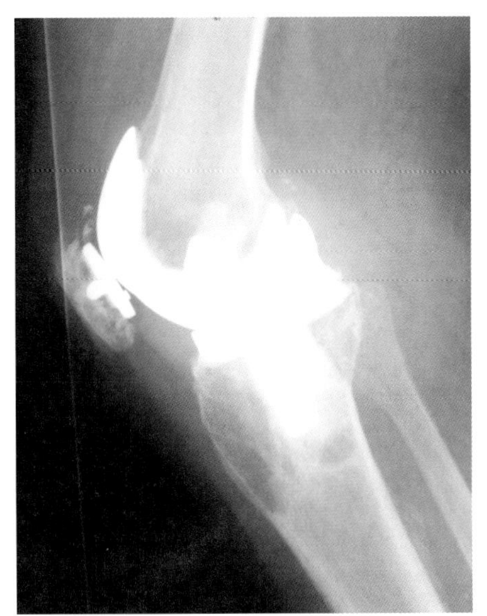

**FIGURE 8.17** Osteolysis in the knee on AP and lateral radiographs. A large osteolytic lesion is present in the lateral and anterior portions of the tibia.

Although osteolysis is now recognized as a clinically relevant problem for TKA, a mature understanding of the role of osteolysis in knee implant failure has not yet been reached. Osteolysis may be difficult to visualize using plane radiographs, especially in the femur where the anterior flange of the femoral component obscures the field of view [49]. Furthermore, osteolysis must be interpreted in the context of normal changes in bone density (bone remodeling) that occurs following TKA.

Not all researchers use the same definition of osteolysis in their radiographic analyses, complicating the comparison of results between studies. Some researchers identify radiolucencies around the margin of an implant as a "linear lytic defect" [52], whereas in other studies, only a focal lesion or cyst is classified as osteolysis [46]. Of greatest clinical concern are unstable or expansile lesions (so-called "balloon lesions") that grow over time and lead to aseptic loosening of a prosthetic component.

Because of difficulty in unambiguously diagnosing osteolysis in the knee, other complications (e.g., infection, instability, malalignment, patellar tracking) may result in implant revision before osteolysis has progressed to the point that it can be identified radiologically. Due to the number of unanswered questions regarding the etiology and significance of osteolysis in TKA, clinical research on this topic is likely to continue for the foreseeable future.

## 8.4.2 Methods to Assess *In Vivo* Wear in TKA

Clinical interest in osteolysis has led to research on improved clinical methods to quantify UHMWPE wear for TKA. The most widely used clinical measure of wear in the knee is based on analysis of AP radiographs [57] by estimating the minimum distance between the femoral condyles and the tibial baseplate (Figure 8.18). The separation between the tibial baseplate and the condyles of the femoral component gives an indication of the relative wear between the medial and lateral condyles. If the magnification of the radiograph is known or can be ascertained (based on the known width of the baseplate or tibial stem, for example), absolute distance measurements can be performed from the radiograph. By comparing changes in apparent component thickness in a series of radiographs over time, the *in vivo* wear rate for a particular TKA patient can be calculated using this "minimum distance" method. However, tilting of the tibial baseplate relative to the plane of the radiograph leads to "false shortening," as described by Fukuoka [57], and greatly complicates this radiographic wear assessment technique.

An example of radiographic false shortening produced by tibial tilt is illustrated in Figure 8.18. Without knowing the clinical history of these two radiographs, it is difficult to judge, based on direct comparison of the 2-D images alone, whether a difference in tibial component height has occurred. In fact, the two radiographs shown in this figure are taken on the same patient within 4 months of implantation, and hence the actual amount of wear in both Figures 8.18A and 8.18B is actually expected to be negligible, notwithstanding the false shortening captured in Figure 8.18A. Consequently, properly aligned prosthetic components, relative to the plane of the radiograph, are essential for direct quantitative measurement of radiographic wear in TKA.

Despite its limitations, the minimum distance method of estimating wear in TKA may be extremely useful clinically, especially in a case of severe wear, when quantitative radiographic measurements may not be necessary for a surgeon to ascertain that a clinical failure has occurred.

**TABLE 8.3** Summary of Osteolysis Studies for TKA

| Study | Prosthesis design | Fixation | Mean follow-up | Incidence of osteolysis in study population | Osteolysis diagnosis | Incidence of revisions for osteolysis |
|---|---|---|---|---|---|---|
| Peters [46] | Synatomic & Arizona (DePuy) | Cementless | 3.5y | 27/174 (16%) | Radiographs | 15/27 (56%) |
| Cambadi [53] | AMK (DePuy) & PCA (Howmedica) | Cementless | 2.6y | 30/271 (11%) | Radiographs | 18/30 (60%) |
| Ezzet [47] | AMK (DePuy) | Cemented tibia, Cemented and cementless femur | 4.7y | 17/83 (20%) overall; 0/12 (0%) cemented, no screws; 14/46 (30%) cemented tibia with screws and uncemented femur | Radiographs | NA |
| Kim [54] | PCA (Howmedica) | Cementless | >7y | 54/60 (90%) tibia; 48/60 (80%) patella; 0/60 (0%) femur | Radiographs | 6/48 patellae (13%) |
| Robinson [55] | Posterior stabilized (65%), constrained implant (30%) | Cemented and cementless | 4.7y | Not studied (revisions only) | Radiographs | 17/185 (9%) |
| Whiteside [48] | Ortholoc II and Ortholoc Modular, short and long stem (Wright Medical) | Cementless | 3–7y (Ortholoc II) and 2–4y (Ortholoc Modular) | 0/675 Ortholoc II; 28/124 (23%) Ortholoc Modular with long stem; 19/112 (17%) Ortholoc Molar with short stem | Radiographs | 2/47 Ortholoc Modular (4%) |
| Mikulak [56] | Posterior stabilized | Cemented | 4.7y | Not studied (revisions only) | Radiographs | 16/557 (2.9%) |
| Spicer [51] | PFC (Johnson & Johnson) | Cemented and cementless | 6y | 29/751 (3.9%) | Radiographs | 11/751 (1.5%) |
| Huang [49] | LCS (DePuy), PCA (Howmedica), Miller-Gallante (Zimmer), Tricon (Smith & Nephew) | Cemented and cementless | 8y | Not studied (only revisions due to wear and osteolysis studied) | Radiographs and intraoperative observations | 16/34 (47%) mobile bearing group; 6/46 (13%) fixed bearing group |
| O'Rourke [50] | IB II (Zimmer) | Cemented | 6.4y | 17/105 (16%) | Radiographs | 2/17 (12%) |
| Weber [52] | AGC (Biomet) | Cemented | 6.3y (Monoblock design), 5.5y (Modular design) | 40/698 (5.7%) Monoblock design; 73/353 (20.7%) Modular design | Radiographs | 1/40 (2.5%) Monoblock design; 6/73 (8.2%) Modular design |

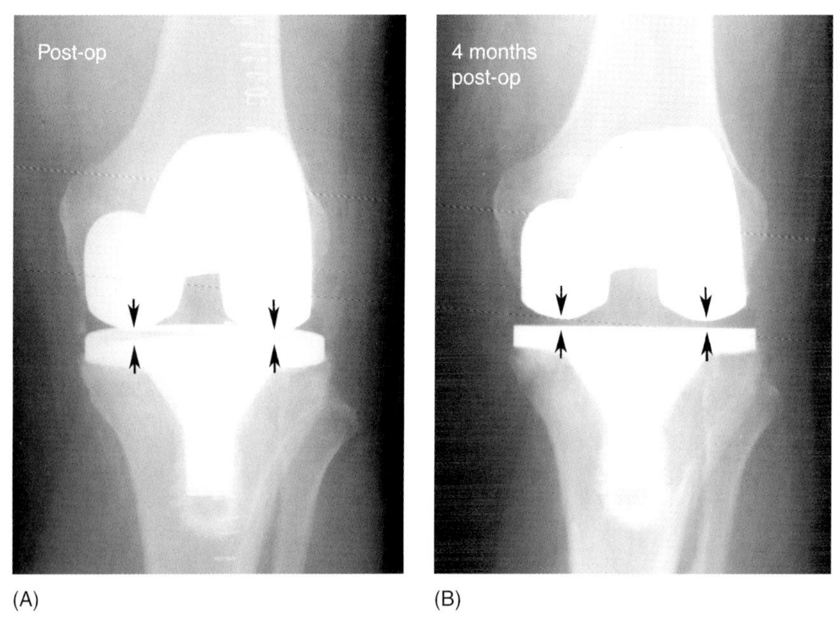


**FIGURE 8.18** Tilting of the femoral and/or tibial components relative to the plane of the X-ray complicates radiographic measurement of wear in TKA. Some studies have measured knee wear as the difference in minimum distance separating the condylar surface of the femur and the tibial tray (denoted with arrows). Note that the initial postoperative radiograph (A) shows substantially more tilt of the tibial tray than the radiograph taken 4 months later (B).

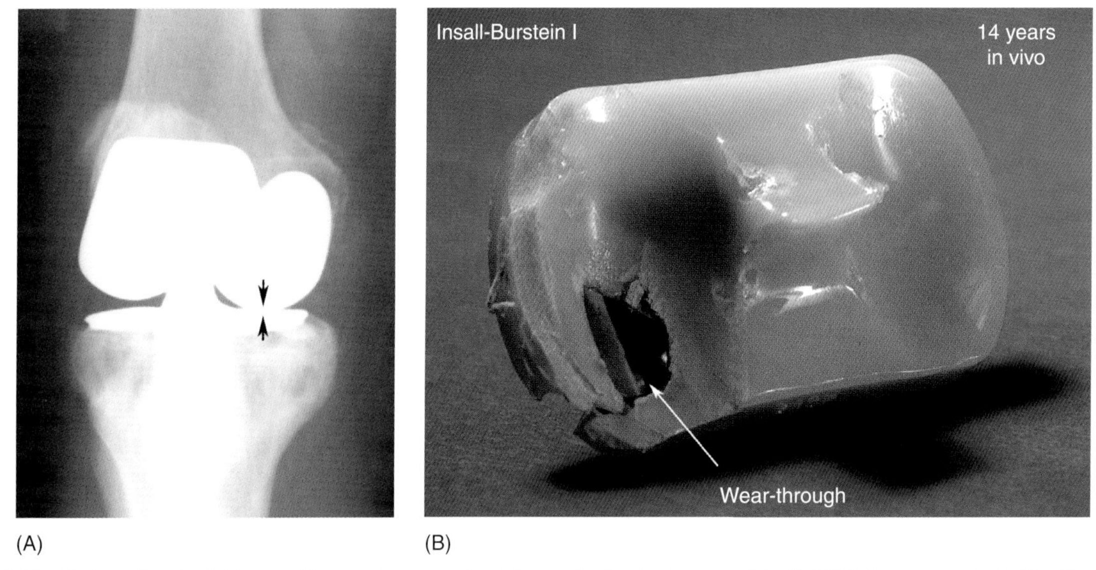


**FIGURE 8.19** Failure of an Insall-Burstein I knee replacement after 14 years *in vivo* due to wear of the UHMWPE component. (A) Standing AP radiograph shows narrowing of the joint space in the medial compartment (indicated by arrows). (B) The retrieved UHMWPE component was worn completely through on the medial side. The tibial tray was polished (i.e., burnished) from articulation with the femoral component. Severe metalosis of the surrounding tissues was noted upon revision.

For example, severe wear isolated to a single compartment can readily be perceived from a well-aligned radiograph, although it may not always be possible to judge if the component has completely worn through if the tibial component is even slightly tilted (Figure 8.19A). In the example illustrated in Figure 8.19, revision surgery of the TKA revealed extensive metalosis of the joint space, and examination of the retrieved component confirmed that the component had worn through in the posterior medial compartment (Figure 8.19B). However, except in cases of severe wear, as previously noted, either careful alignment or computer-aided image analysis techniques are necessary to quantify the minute and gradual changes in tibial component thickness that typically occur over time in TKA based on 2-D radiographs.

Sanzén et al. [58] have described a method to measure *in vivo* UHMWPE of tibial components using fluoroscopy-guided AP radiographs in a series of patients with the

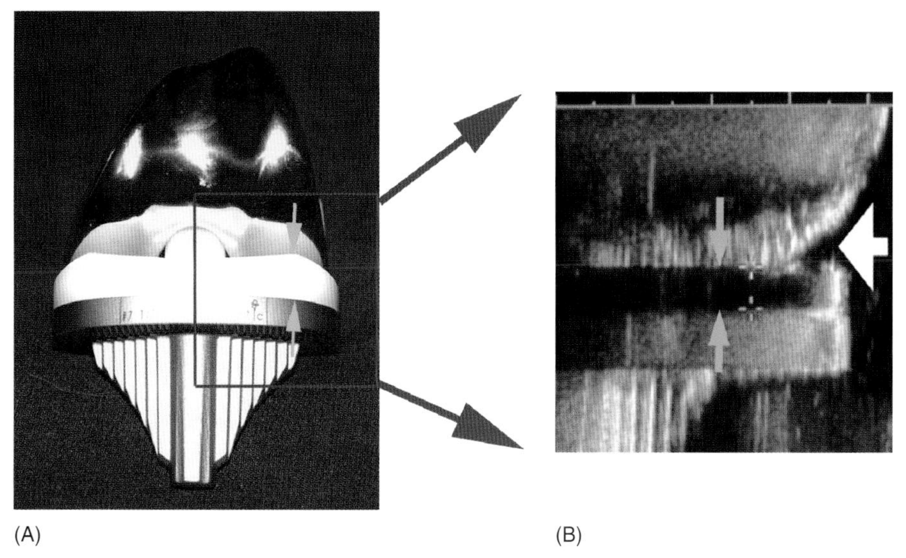


**FIGURE 8.20** Ultrasound as a method to measure UHMWPE wear in TKA. The figure shows a posterior stabilized total knee replacement (A) and an ultrasound of the same prosthesis taken in a water bath without soft tissues (B). The red inset represents the ultrasound region shown in (B). The UHMWPE tibial insert is visible between the green arrows. The large white arrow indicates the direction from which the ultrasound probe was applied. The images were provided courtesy of John Martell, MD, University of Chicago (Chicago, Illinois, USA).

same design of knee replacement (PCA, Howmedica, Rutherford, New Jersey, USA). In the first step of Sanzén's procedure, the plane of the AP radiograph was oriented perpendicular to the plane of the tibial baseplate using fluoroscopy. Image magnification was then corrected by the known width of the baseplate, and the perpendicular distance from the baseplate to the femoral condyles was measured and defined as the "femorotibial distance." In the PCA design, the tibial surface is flat and the component has constant thickness in the contact region of the condyles. Consequently, for this particular design, wear was computed by subtracting the two-dimensional femorotibial distance from the (known) initial thickness of the component. Using this technique, Sanzén reports absolute *in vivo* wear measurements with a precision of 0.1 mm.Sanzén's work does not include a description of the absolute accuracy of their technique, however.

Fukuoka and colleagues [57] have argued that a two-dimensional radiographic wear measurement technique "seems to include some degree of inaccuracy," even though Sanzén's method accounts for errors introduced by component rotation. In their study, Fukuoka [57] describes a "simple" method for computing three-dimensional wear from 2-D radiographs. Fukuoka's method is referred to in his work as "3-D/2-D matching," in which a 3-D computer model of a tibial baseplate and femoral component, along with contours extracted from the AP radiograph, are used as inputs into optimization software that matches the pose of the implant components necessary to reconstruct the 2-D view obtained in the AP radiograph. When the "optimal" pose has been found by the software, the 3-D wear vectors can be calculated. Fukuoka reports an accuracy (specifically, an RMS error) of 0.04 mm when comparing the tibiofemoral distance calculated using his technique with the physical measurements of actual components in a laboratory setting.

Although fluoroscope-assisted or computer-assisted radiographic techniques are widely employed among the orthopedic community, ultrasound has recently been introduced as a promising alternative for wear assessment in TKA [59]. Using an ultrasound probe, it is possible to visualize not only the femoral and tibial component surfaces but also identify the contours of the UHMWPE component (Figure 8.20). Sofka and associates, in a study of 24 TKA patients who were being screened for deep venous thrombosis, found that ultrasonic insert thickness measurements were highly correlated with radiographic measurements ($r^2 = 0.64$) [59]. However, the absolute accuracy of the technique was not reported. Consequently, ultrasound may be regarded as an experimental technique currently under development.

All of the *in vivo* wear measurement techniques described in this section related to TKA have certain drawbacks and unique limitations. Certainly all of these methods are predicated on the presence of a metallic tibial baseplate. In this respect, radiographic wear assessment in the hip and the knee share a common limitation, although the inclusion of a radiographic wire marker in all-UHMWPE cups has somewhat alleviated this concern. In addition, from a practical perspective, both fluoroscope-assisted or computer-assisted radiographic techniques for TKA are more cumbersome and nonstandard than the methods used to obtain radiographs for routine clinical diagnosis. Ultrasound as a wear measurement technique for TKA is in its infancy, insofar as widespread acceptance is concerned. Finally, the absolute accuracy of these *in vivo* wear measurement techniques, when applied to clinical radiographs, remains as yet unquantified.

For all of these reasons, the most reliable and accurate method of wear assessment in TKA continues to be inspection of UHMWPE components following revision surgery or autopsy removal. The dimensional measurements obtainable from retrieved components using a coordinate measuring machine, for example, have an accuracy at least an order of magnitude better than using radiographic techniques. Further research in knee imaging technology will be needed to close the gap in accuracy between *in vivo* and *ex vivo* wear measurement in TKA.

## 8.4.3 Backside Wear

We have thus far been concerned primarily with wear at the articulating surface between the femoral condyles and UHMWPE tibial insert. Recently, however, researchers have drawn attention to backside articulation (i.e., between the tibial insert and the metallic tray or baseplate) as potentially a clinically relevant source of wear debris [50, 60, 61]. Relative motion between the UHMWPE insert and metallic tray is resisted by the locking mechanism. These mechanisms are proprietary and design specific. Consequently, it is difficult to generalize about the integrity of locking mechanisms and propensity for backside wear.

One well-studied example of a first-generation modular knee replacement is the Insall-Burstein II (IB II) [50, 60]. A retrieved IB II component is illustrated in Figure 8.21 along with the locking mechanism. Backside wear is typically characterized as burnishing or scratching of the UHMWPE component, sometimes with removal of machining marks from the UHMWPE surface. There may also be evidence of extrusion of the UHMWPE into screw holes or recesses on the back surface. All of these features are evident in the retrieved component shown in Figure 8.21. Similar observations have been reported for this design by O'Rourke et al. [50]. The tibial baseplate also typically exhibits scratching, burnishing, or other evidence of articulation, especially when the component is fabricated from titanium alloy, as was the case with the IB II. More severe cases of backside wear, not shown in the referenced figure, include pitting of the UHMWPE surface.

Researchers have used damage scoring techniques, measured the height of UHMWPE extruded in screw holes, and quantified the relative motion between the insert and the tray [60, 61]. Engh and coworkers have recently measured the amount of insert-tray relative motion in 10 different designs of new, retrieved, and autopsy-retrieved tibial inserts [60]. The magnitude of relative motion in the new inserts ranged between 6 and 157 μm. In the retrieved and autopsy-retrieved implants, the magnitude of relative motion ranged between 104 and 718 μm. The researchers concluded that insert-tray relative motion increased after implantation under *in vivo* loading.

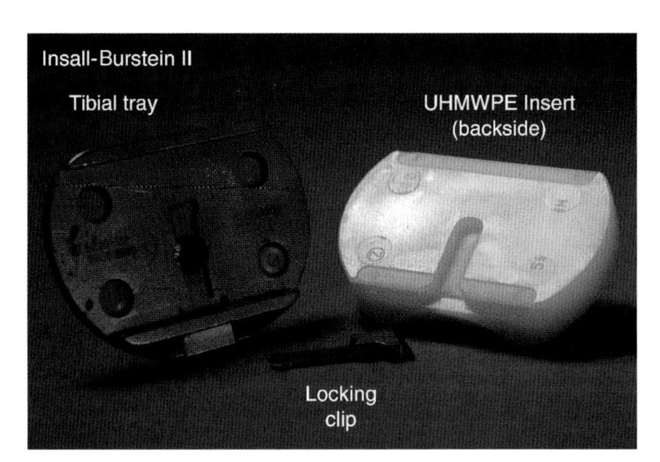


FIGURE 8.21 Backside wear of an Insall-Burstein II knee replacement. The inferior surface of the tibial tray shows is burnished from relative motion with respect to the tibial tray. Note the four impressions on the backside of the insert, which correspond to the four screw holes on the tibial tray. Between the screw hole impressions, the machined inscriptions on the backside of the insert have been worn away.

The clinical significance of backside wear remains very much open to scientific debate. Engh and coworkers [60] have postulated that "perhaps the combination of articulating surface wear and backside wear has produced a greater volume of debris, which has caused the increased occurrence of osteolysis observed with the use of modular implants." However, the magnitude of relative motion at the back surface is still orders of magnitude less than the sliding that occurs at the articulating surface.

Alternate hypotheses, involving "ease of access" and particle migration, have been suggested to account for differences in osteolysis incidence among different metal-backed tibial designs with otherwise identical locking mechanisms [48]. In other words, modularity may not necessarily increase the total magnitude of wear debris but may provide easier access of what debris is produced to the bone underlying the tibial baseplate. For these reasons, backside wear in TKA continues to be a controversial topic among members of the orthopedic research community.

## 8.4.4 Damage to Posts in PS Tibial Components

The post in posterior-stabilized (PS) condylar components has also been identified as a potential site of impingement and wear in TKA [45, 62, 63]. In studies thus far, post damage has been quantified using damage scoring techniques. Estimates of wear volume generated from the post in PS knees have not yet been reported.

A wide range of damage modes have been observed in posts; the most concerning include fatigue damage, fracture, and adhesive wear [45]. Although designed to articulate with a cam in the femoral component during flexion, the UHMWPE post may contact the femoral component

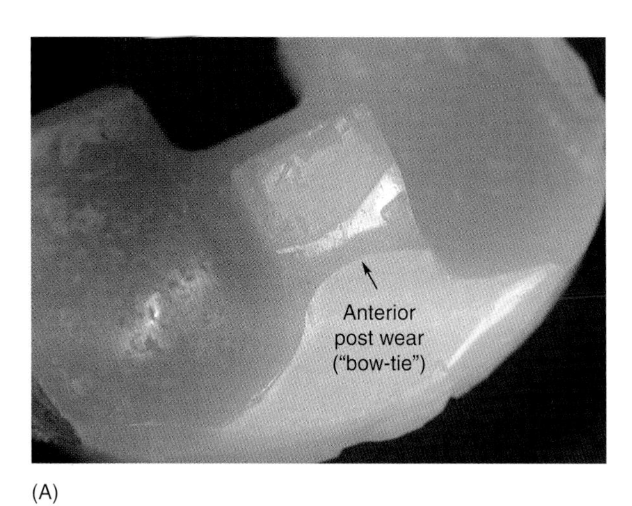


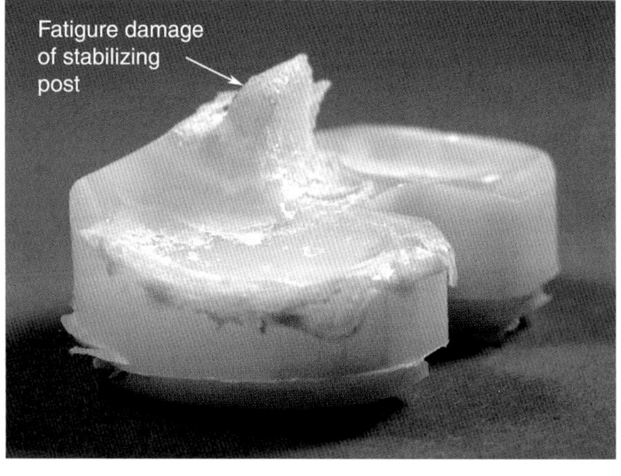


FIGURE 8.22 Examples of post wear in PFC total knee replacements (Johnson & Johnson). (A) Anterior post wear, producing a bow tie wear scar; (B) fatigue damage to the stabilizing post.

during hyperflexion, resulting in anterior wear. When axial torque is applied to the knee (from internal/external rotation of the femur with respect to the tibia), the sides and corners of the post may impinge against the sides of the femoral condyles, producing the characteristic bow tie wear scar (Figure 8.22). In more severe cases, fatigue damage or fracture of the post may occur [45].

However, it remains to be seen whether PS knees are prone to this type of damage collectively or if post wear is only limited to certain designs [62]. Based on a study of seven retrieved inserts (n = 2 IB/PS II, n = 5 PFC), Callaghan et al. suggested that the post served to transmit axial torque across the joint, contributing to backside wear of the inserts [63]. Thus, it is also unclear whether damage to the post is related to wear modes in remote regions of the insert.

## 8.5 ALTERNATIVES TO CONVENTIONAL UHMWPE IN TKA

Despite its successful track record, wear and damage of the UHMWPE insert compromises the longevity of knee arthroplasty. Osteolysis, a rare occurrence with all-UHMWPE tibial components, has been documented with increasing frequency in modular knee replacements. Clearly, improved wear behavior of the UHMWPE insert would be advantageous from the perspective of reducing the risk of osteolysis and aseptic loosening. One strategy for reducing wear is to improve the lubricity and scratch resistance of the femoral component using ceramic surfaces. Another tactic is to improve the wear resistance of UHMWPE by elevated crosslinking. Both approaches to improving wear resistance of knee bearings are currently under clinical investigation, but it remains too early to determine whether encouraging laboratory data will be translated to improved longevity in patients.

### 8.5.1 Ceramic Bearings in TKA

The majority of knee bearings employ a CoCr alloy femoral component articulating on UHMWPE. However, for the same reasons that researchers have explored ceramic femoral heads for hip replacement (detailed in Chapter 6), there has been longstanding curiosity about ceramic femoral components for total knee applications [64, 65]. Because of their complex geometry, relatively thin cross-section, and the risk of fracture, the manufacture of ceramic femoral components is considerably more challenging than the spherical components used for total hip bearings. Alumina ceramics represent a documented alternative bearing surface for TKA, with over 20 years of clinical experience in Japan pioneered by Kyocera [66]. Zirconia femoral components have also been developed in Japan and evaluated in short-term clinical trials [67, 68]. In addition to the use of alumina and zirconia in femoral components, ZTA alumina matrix composite (BIOLOX Delta, CeramTec) is now under clinical evaluation as a C-PE bearing for TKA in Europe (Figure 8.23).

Another alternative bearing for the knee, oxidized zirconium (OXINIUM, Smith & Nephew) is now widely used (see Chapter 6, Figure 8.24) [69–71]. Unlike monolithic alumina or zirconia femoral components, oxidized zirconium is produced by diffusion of oxygen into a zirconium-niobium alloy, resulting in a hardened ceramic surface. Thus, the fracture risk associated with monolithic ceramics such as alumina and zirconia is thought to be reduced with OXINIUM. However, because to date no clinical factures of ceramic femoral components for TKA have yet been reported in the literature [67], the ability to compare actual clinical fracture rates among alternative knee bearings is not yet possible.

Despite the theoretical appeal of ceramic femoral components based on encouraging laboratory data, the clinical utilization of ceramic-on-UHMWPE (C-PE) bearings in

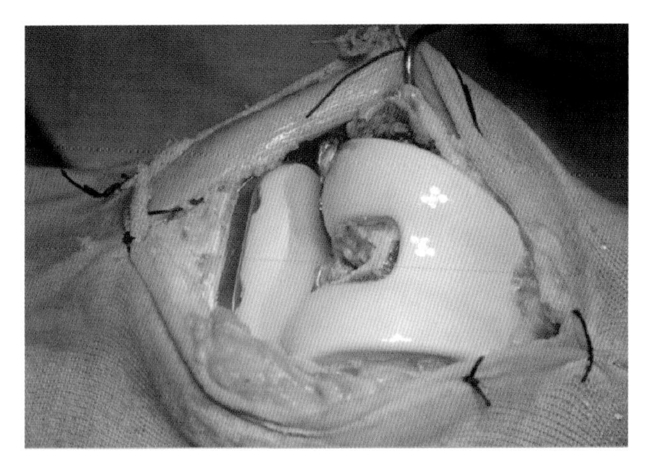

**FIGURE 8.23** Zirconia toughened alumina matrix composite femoral component (BIOLOX® Delta, CeramTec). Image provided courtesy of CeramTec.

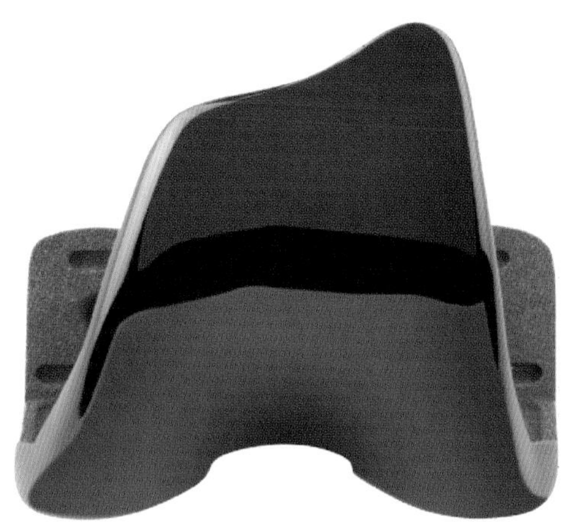

**FIGURE 8.24** Oxidized zirconium femoral component (OXINIUM, Smith & Nephew). Image provided courtesy of Smith & Nephew.

TKA remains limited at the present time. Because of the excellent clinical performance of M-PE bearings in TKA, proving long-term superiority of C-PE bearings for TKA is a challenging task that, to date, has not yet been shown. Overall, the use of materials other than CoCr articulating against UHMWPE in the knee literature is still in its early stages, and the clinical benefits of an alternative approach to this gold standard bearing couple have yet to be clearly demonstrated.

### 8.5.2 Highly Crosslinked UHMWPE in TKA

Highly crosslinked UHMWPE is now the most widely used alternative to conventional UHMWPE. Today, four out of the five major orthopedic manufacturers offer highly crosslinked UHMWPE for total knee arthroplasty (Prolong, Zimmer, Inc.; X3, Stryker Orthopaedics; XLK, DePuy Orthopedics; and EPoly, Biomet, Inc.). However, there is not universal agreement in the clinical community that highly crosslinked UHMWPE is either necessary or desirable for the knee. Due to the reduction in ductility and fracture resistance associated with radiation crosslinking (see Chapter 6), the introduction of highly crosslinked UHMWPE to TKA has been somewhat more gradual than what was observed in THA. Several *in vitro* studies using knee simulators have demonstrated that significant reduction of wear and surface damage can be achieved using highly crosslinked UHMWPE tibial inserts, even under aggressive loading conditions [72, 73].

Although these materials were clinically introduced for knee applications in 2001, it remains too early to ascertain whether the benefits observed *in vitro* will be translated to the clinical situation. Additional information about the types of highly crosslinked UHMWPE that are available for use in knee applications may be found in Chapters 13–15 and 20. Because highly crosslinked UHMWPE has not yet been clinically proven in TKA, its use remains controversial and the topic of ongoing orthopedic research.

## SUMMARY

Conventional UHMWPE continues to play a crucial role in the long-term success of knee arthroplasty around the world. The articulation of UHMWPE against a stainless steel or cobalt chrome alloy counterface has been the gold standard bearing couple in the knee starting in the 1960s, as we have seen in Chapter 7. Without UHMWPE, patients with debilitating knee arthritis would be faced with choosing either a metallic hinge (vintage 1950s technology), with its associated restrictions on activity, or permanent fusion of their joint. It remains a sober fact that there currently are few clinically proven alternatives to conventional UHMWPE as a bearing material in the knee. With the growing interest in alternative bearings for TKA, the orthopedic community awaits clinical data demonstrating their long-term performance relative to CoCr articulating against conventional UHMWPE.

## ACKNOWLEDGMENTS

I am grateful to my colleagues at the Rothman Institute, William Hozack, MD, and Peter Sharkey, MD, Gina Bissett, James Purtill, MD, and Jay Parvisi, MD. By generous donation of their time and research collaboration, they have deepened my insight into the clinical topics covered in this chapter.

Special thanks to Professor Otto Robertsson, University of Lund, for permission to reproduce the figures related to the Swedish Knee Registry. Thanks also to Clare Rimnac, PhD, for editorial assistance with this chapter, and to John Martell, MD, for many helpful discussions and for providing the figure related to ultrasound.

## REFERENCES


1. Kurtz S, Lau E, Ong K, Zhao K, Kelly M, Bozic K. Future young patient demand for primary and revision joint replacement: national projections from 2010 to 2030. *Trans of the 54th Orthop Res Soc* 2008 April;**33**:1784.
2. Kurtz SM, Mowat F, Ong K, Chan N, Lau E, Halpern M. Prevalence of primary and revision total hip and knee arthroplasty in the United States from 1990 through 2002. *J Bone Joint Surg Am* 2005 July;**87**(7):1487–97.
3. Dahlkvist NJ, Mayo P, Seedhom BB. Forces during squatting and rising from a deep squat. *Eng Med* 1982 April;**11**(2):69–76.
4. Piziali RL, Seering WP, Nagel DA, Schurman DJ. The function of the primary ligaments of the knee in anterior-posterior and medial-lateral motions. *J Biomech* 1980;**13**(9):777–84.
5. Trousdale RT, Pagnano MW. Fixed-bearing cruciate-retaining total knee arthroplasty. *Clin Orthop* 2002 November(404):58–61.
6. Sathasivam S, Walker PS. The conflicting requirements of laxity and conformity in total knee replacement. *J Biomech* 1999 March;**32**(3):239–47.
7. Morrison JB. Function of the knee joint in various activities. *Biomed Eng* 1969 December;**4**(12):573–80.
8. Morrison JB. The mechanics of the knee joint in relation to normal walking. *J Biomech* 1970 January;**3**(1):51–61.
9. Reilly DT, Martens M. Experimental analysis of the quadriceps muscle force and patello-femoral joint reaction force for various activities. *Acta Orthop Scand* 1972;**43**(2):126–37.
10. Harrington IJ. A bioengineering analysis of force actions at the knee in normal and pathological gait. *Biomed Eng* 1976 May;**11**(5):167–72.
11. Ellis MI, Seedhom BB, Wright V. Forces in the knee joint whilst rising from a seated position. *J Biomed Eng* 1984 April;**6**(2):113–20.
12. Patriarco AG, Mann RW, Simon SR, Mansour JM. An evaluation of the approaches of optimization models in the prediction of muscle forces during human gait. *J Biomech* 1981;**14**(8):513–25.
13. Rohrle H, Scholten R, Sigolotto C, Sollbach W, Kellner H. Joint forces in the human pelvis-leg skeleton during walking. *J Biomech* 1984;**17**(6):409–24.
14. Davy DT, Audu ML. A dynamic optimization technique for predicting muscle forces in the swing phase of gait. *J Biomech* 1987; **20**(2):187–201.
15. Olney SJ, Winter DA. Predictions of knee and ankle moments of force in walking from EMG and kinematic data. *J Biomech* 1985; **18**(1):9–20.
16. Amstutz HC, Grigoris P, Dorey FJ. Evolution and future of surface replacement of the hip. *J Orthop Sci* 1998;**3**(3):169–86.
17. Gsell RA, Taylor GK, Lin ST. Shelf aging of net shape compression molded 1900H ultra-high molecular weight polyethylene (UHMWPE) components. *Transactions of the 44st Orthopedic Research Society*; New Orleans, LA; 1998 February 13-16, p. 364.
18. Taylor SJ, Walker PS. Forces and moments telemetered from two distal femoral replacements during various activities. *J Biomech* 2001 July;**34**(7):839–48.
19. Paul JP. Force actions transmitted by joints in the human body. *Proc R Soc Lond B Biol Sci* 1976 January 20;**192**(1107):163–72.
20. Walker PS. Historical development of artificial joints. In: Walker PS, editor. *Human joints and their artificial replacements*. Springfield, IL: Charles C. Thomas, Publisher; 1977. p. 253–75.
21. Andriacchi TP, Mikosz KR. Musculoskeletal dynamics, locomotion, and clinical applications. In: Mow VC, Hayes WC, editors. *Basic orthopedic biomechanics*. New York: Raven Press; 1991.
22. Burstein AH, Wright TM. *Fundamentals of orthopedic biomechanics*. Baltimore, MD: 1994.
23. Bartel DL, Burstein AH, Toda MD, Edwards DL. The effect of conformity and plastic thickness on contact stresses in metal-backed plastic implants. *J Biomech Eng* 1985;**107**(3):193–9.
24. Bartel DL, Bicknell VL, Wright TM. The effect of conformity, thickness, and material on stresses in ultra-high molecular weight components for total joint replacement. *J Bone Joint Surg [Am]* 1986;**68**(7):1041–51.
25. Bartel DL, Rawlinson JJ, Burstein AH, Ranawat CS, Flynn Jr. WF Stresses in polyethylene components of contemporary total knee replacements. *Clin Orthop* 1995;**317**:76–82.
26. Thadani PJ, Vince KG, Ortaaslan SG, Blackburn DC, Cudiamat CV. Ten- to 12-year follow-up of the Insall-Burstein I total knee prosthesis. *Clin Orthop* 2000 November;**380**(380):17–29.
27. Kurtz SM, Bartel DL, Rimnac CM. Post-irradiation aging affects the stresses and strains in UHMWPE components for total joint replacement. *Clin Orthop* 1998;**350**:209–20.
28. Berger RA, Rosenberg AG, Barden RM, Sheinkop MB, Jacobs JJ, Galante JO. Long-term follow-up of the Miller-Galante total knee replacement. *Clin Orthop* 2001 July(388):58–67.
29. Blunn GW, Joshi AB, Minns RJ, Lidgren L, Lilley P, Ryd L, et al. Wear in retrieved condylar knee arthroplasties. A comparison of wear in different designs of 280 retrieved condylar knee prostheses. *J Arthroplasty* 1997;**12**(3):281–90.
30. Bell CJ, Walker PS, Abeysundera MR, Simmons JM, King PM, Blunn GW. Effect of oxidation on delamination of ultrahigh-molecular-weight polyethylene tibial components. *J Arthroplasty* 1998;**13**(3):280–90.
31. Forster MC. Survival analysis of primary cemented total knee arthroplasty: Which designs last?. *J Arthroplasty* 2003 April;**18**(3):265–70.
32. Robertsson O, Knutson K, Lewold S, Lidgren L. The Swedish Knee Arthroplasty Register 1975-1997: an update with special emphasis on 41,223 knees operated on in 1988-1997. *Acta Orthop Scand* 2001 October;**72**(5):503–13.
33. Robertsson O, Lewold S, Knutson K, Lidgren L. The Swedish Knee Arthroplasty Project. *Acta Orthop Scand* 2000 February;**71**(1):7–18.
34. Robertsson O. The Swedish Knee Arthroplasty Register: validity and outcome. *Ph.D. Dissertation*. Lund, Sweden: Lund University; 2000.
35. Robertsson O, Knutson K, Lewold S, Lidgren L. The routine of surgical management reduces failure after unicompartmental knee arthroplasty. *J Bone Joint Surg Br* 2001 January;**83**(1):45–9.
36. Vince KG. Why knees fail. *J Arthroplasty* 2003 April;**18**(3 Suppl 1): 39–44.
37. Fehring TK, Odum S, Griffin WL, Mason JB, Nadaud M. Early failures in total knee arthroplasty. *Clin Orthop* 2001 November;**392**(392):315–18.
38. Sharkey PF, Hozack WJ, Rothman RH, Shastri S, Jacoby SM. Insall Award paper. Why are total knee arthroplasties failing today?. *Clin Orthop* 2002 November(404):7–13.
39. Shoji H, D'Ambrosia RD, Lipscomb PR. Failed polycentric total knee prostheses. *J Bone Joint Surg Am* 1976 September;**58**(6):773–7.
40. Walker PS. *Human joints and their artificial replacements*. Springfield, Il: CC Thomas Publisher; 1977.
41. Hood RW, Wright TM, Burstein AH. Retrieval analysis of total knee prostheses: a method and its application to 48 total condylar prostheses. *J Biomed Mater Res* 1983 September;**17**(5):829–42.
42. Wright TM, Rimnac CM, Stulberg SD, Mintz L, Tsao AK, Klein RW, et al. Wear of polyethylene in total joint replacements: observations from retrieved PCA knee implants. *Clin Orthop* 1992;**276**(276):126–34.

43. Won CH, Rohatgi S, Kraay MJ, Goldberg VM, Rimnac CM. Effect of resin type and manufacturing method on wear of polyethylene tibial components. *Clin Orthop* 2000;**376**(376):161–71.
44. Kurtz SM, Rimnac CM, Pruitt L, Jewett CW, Goldberg V, Edidin AA. The relationship between the clinical performance and large deformation mechanical behavior of retrieved UHMWPE tibial inserts. *Biomaterials* 2000;**21**(3):283–91.
45. Furman BD, Mahmood F, Wright TM, Haas SB. Insall Burstein PS II has more severe anterior wear and fracture of the tibial post than the Insall Burstein I. *Transactions of the Orthopedic Research Society.* Anaheim, CA; 2003 April 28-May 2, 1999, p. 1404.
46. Peters Jr. PC, Engh GA, Dwyer KA, Vinh TN Osteolysis after total knee arthroplasty without cement. *J Bone Joint Surg Am* 1992;**74**(6):864–76.
47. Ezzet KA, Garcia R, Barrack RL. Effect of component fixation method on osteolysis in total knee arthroplasty. *Clin Orthop* 1995 December;**321**(321):86–91.
48. Whiteside LA. Effect of porous-coating configuration on tibial osteolysis after total knee arthroplasty. *Clin Orthop* 1995 December (321):92–7.
49. Huang CH, Ma HM, Liau JJ, Ho FY, Cheng CK. Osteolysis in failed total knee arthroplasty: a comparison of mobile-bearing and fixed-bearing knees. *J Bone Joint Surg Am* 2002 December; **84-A**(12):2224–9.
50. O'Rourke MR, Callaghan JJ, Goetz DD, Sullivan PM, Johnston RC. Osteolysis associated with a cemented modular posterior-cruciate-substituting total knee design: five to eight-year follow-up. *J Bone Joint Surg Am* 2002 August;**84-A**(8):1362–71.
51. Spicer DD, Pomeroy DL, Badenhausen WE, Schaper Jr. LA, Curry JI, Suthers KE, et al. Body mass index as a predictor of outcome in total knee replacement. *Int Orthop* 2001;**25**(4):246–9.
52. Weber AB, Worland RL, Keenan J, Van Bowen J. A study of polyethylene and modularity issues in >1000 posterior cruciate-retaining knees at 5 to 11 years. *J Arthroplasty* 2002 December;**17**(8):987–91.
53. Cadambi A, Engh GA, Dwyer KA, Vinh TN. Osteolysis of the distal femur after total knee arthroplasty. *J Arthroplasty* 1994 December; **9**(6):579–94.
54. Kim YH, Oh JH, Oh SH. Osteolysis around cementless porous-coated anatomic knee prostheses. *J Bone Joint Surg Br* 1995 March; **77**(2):236–41.
55. Robinson EJ, Mulliken BD, Bourne RB, Rorabeck CH, Alvarez C. Catastrophic osteolysis in total knee replacement. A report of 17 cases. *Clin Orthop* 1995;**321**(321):98–105.
56. Mikulak SA, Mahoney OM, dela Rosa MA, Schmalzried TP. Loosening and osteolysis with the press-fit condylar posterior-cruciate-substituting total knee replacement. *J Bone Joint Surg Am* 2001 March;**83-A**(3):398–403.
57. Fukuoka Y, Hoshino A, Ishida A. A simple radiographic measurement method for polyethylene wear in total knee arthroplasty. *IEEE Trans Rehabil Eng* 1999;**7**(2):228–33.
58. Sanzén L, Sahlstrom A, Gentz CF, Johnell IR. Radiographic wear assessment in a total knee prosthesis. 5- to 9-year follow-up study of 158 knees. *J Arthroplasty* 1996;**11**(6):738–42.
59. Sofka CM, Adler RS, Laskin R. Sonography of polyethylene liners used in total knee arthroplasty. *AJR Am J Roentgenol* 2003 May; **180**(5):1437–41.
60. Engh GA, Lounici S, Rao AR, Collier MB. In vivo deterioration of tibial baseplate locking mechanisms in contemporary modular total knee components. *J Bone Joint Surg Am* 2001 November;**83-A**(11):1660–5.
61. Wasielewski RC, Parks N, Williams I, Surprenant H, Collier JP, Engh G. Tibial insert undersurface as a contributing source of polyethylene wear debris. *Clin Orthop* 1997 December;**345**(345):53–9.
62. Callaghan JJ. O'Rourke MR. Picking your implant: all PS knees are not alike!. *Orthopedics* 2002 September;**25**(9):977–8.
63. Callaghan JJ, O'Rourke MR, Goetz DD, Schmalzried TP, Campbell PA, Johnston RC. Tibial post impingement in posterior-stabilized total knee arthroplasty. *Clin Orthop* 2002 November(404):83–8.
64. Hernigou P, Nogier A, Manicom O, Poignard A, De Abreu L, Filippini P. Alternative femoral bearing surface options for knee replacement in young patients. *The Knee* 2004 June;**11**(3):169–72.
65. Bal BS, Garino J, Ries M, Oonishi H. Ceramic bearings in total knee arthroplasty. *J Knee Surg* 2007 October;**20**(4):261–70.
66. Oonishi H, Kim SC, Kyomoto M, Masuda S, Asano T, Clarke IC. Change in UHMWPE properties of retrieved ceramic total knee prosthesis in clinical use for 23 years. *J Biomed Mater Res B Appl Biomater* 2005 August;**74**(2):754–9.
67. Bal BS, Greenberg DD, Buhrmester L, Aleto TJ. Primary TKA with a zirconia ceramic femoral component. *J Knee Surg* 2006 April; **19**(2):89–93.
68. Tsukamoto R, Chen S, Asano T, Ogino M, Shoji H, Nakamura T, et al. Improved wear performance with crosslinked UHMWPE and zirconia implants in knee simulation. *Acta Orthop* 2006 June;**77**(3):505–11.
69. White SE, Whiteside LA, McCarthy DS, Anthony M, Poggie RA. Simulated knee wear with cobalt chromium and oxidized zirconium knee femoral components. *Clin Orthop Relat Res* 1994 December (309):176–84.
70. Ezzet KA, Hermida JC, Colwell Jr. CW, D'Lima DD. Oxidized zirconium femoral components reduce polyethylene wear in a knee wear simulator. *Clin Orthop Relat Res* 2004 November(428):120–4.
71. Laskin RS. An oxidized Zr ceramic surfaced femoral component for total knee arthroplasty. *Clin Orthop Relat Res* 2003 November (416):191–6.
72. Muratoglu OK, Bragdon CR, O'Connor DO, Perinchief RS, Jasty M, Harris WH. Aggressive wear testing of a cross-linked polyethylene in total knee arthroplasty. *Clin Orthop* 2002 November(404):89–95.
73. Muratoglu OK, Mark A, Vittetoe DA, Harris WH, Rubash HE. Polyethylene damage in total knees and use of highly crosslinked polyethylene. *J Bone Joint Surg Am* 2003;**85-A**(Suppl 1):S7–S13.
74. Lindgren L. *Annual Report 2002 – The Swedish Knee Arthroplasty Register – Part II*. Lund, Sweden: University of Lund; 2002.

Chapter 9

# The Clinical Performance of UHMWPE in Shoulder Replacements

Stefan M. Gabriel, PhD, PE



## 9.1 INTRODUCTION

Shoulder replacement, although done much less frequently than hip and knee replacement, is the third most prevalent joint replacement procedure worldwide. Current shoulder replacement systems rely on ultrahigh molecular weight polyethylene (UHMWPE) components for motion and load bearing. Because of this critical role, the performance of UHMWPE components can determine the overall performance of the replacement system.

When considering the performance of UHMWPE in shoulder replacement components, one should have a basic understanding of the anatomical and biomechanical system into which they are placed, as well as the ways that system can be compromised by disease or trauma. It is also useful to gain a historical perspective on the origins and evolution of design and technique, as well as see a number of systems currently in use. It is especially critical to also examine overall measures of replacement success, as well as specific measures of wear or damage to directly assess the performance of UHMWPE components in particular. This chapter presents information covering these areas of consideration, as well as a discussion of alternatives to standard, contemporary device designs and materials to give an overview of shoulder replacement, and an assessment of UHMWPE performance in clinical use in the shoulder.

## 9.2 THE SHOULDER JOINT

The shoulder is made up of a number of bones, ligaments, and muscles (Figure 9.1). The partial ball defined by the head of the humerus and the partial socket defined by the glenoid of the scapula form the main articulating ball and socket geometry of the joint. Along with the passive and active stability provided by the surrounding joint capsule, ligaments, and the muscles of the rotator cuff, the geometry allows the normal shoulder joint to achieve the largest range of motion of any joint in the human body. Rotation of the humerus relative to the glenoid of the scapula allows positioning of the elbow at any of a number of points on an essentially spherical surface covering nearly a full hemisphere.

Things that can upset the structural balance of the shoulder include arthritis, tendon and ligament abrasions and ruptures, and deterioration and fracture of the bones. Rheumatoid arthritis (RA) affects the soft tissues and

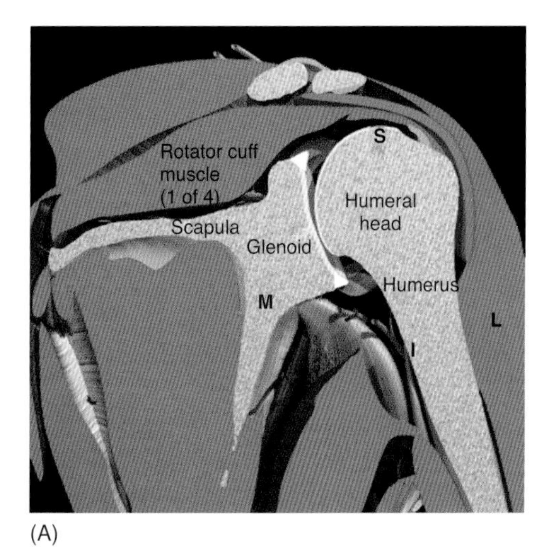


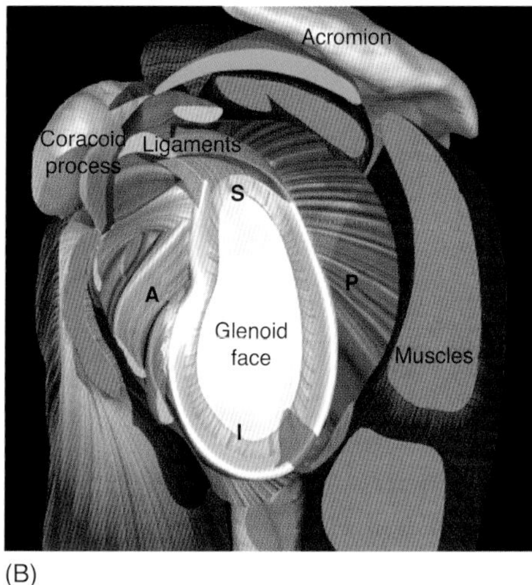


**FIGURE 9.1** Normal anatomy of the Glenohumeral joint showing the bones, muscles, tendons, ligaments, and capsule at the shoulder in frontal section (**A**) and glenoid face (**B**) views. The superior (S), inferior (I), medial (M), lateral (L), anterior (A) and posterior (P) directions are also denoted on each view (images used with permission of Primal Pictures, Ltd., London).

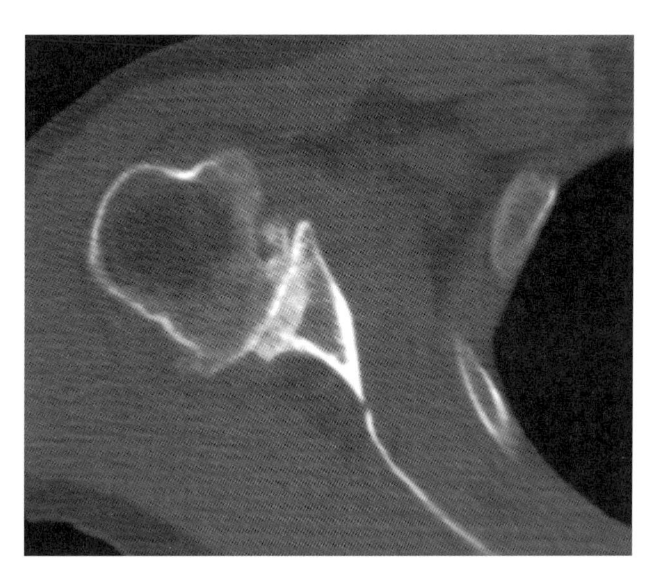

**FIGURE 9.2** Computed tomography image of a shoulder showing an axillary view (looking up under the arm). The lateral direction is to the left, and the anterior direction is up in this view. Note the damage to the humeral head and the severe erosion on the posterior aspect of the glenoid (image courtesy of Jon JP Warner, Massachusetts General Hospital, Boston).

cartilage around the joint, causing a loss of stiffness, strength, and integrity of the affected structures leading to instability of the shoulder. In the absence of other disease processes, RA commonly results in erosion of the humeral head and central erosion of the glenoid face with a corresponding loss of stabilization [1, 2].

Osteoarthritis (OA) affects the articular cartilage and underlying bones at the joint. Primary OA is associated with progressive wearing of the humeral head and the posterior aspect of the glenoid (Figure 9.2) [1, 2]. Dislocation caused by laxity or loss of stabilization in one direction or another due to a specific ligament or muscle's weakness or tearing can cause wear of the humeral head and glenoid in various locations, depending on the direction of instability. Fractures due to trauma or the presence of tumors can result in multiple bone fragments. This most often affects the proximal humerus, requiring repair of the bone and reconstruction of the muscle and ligament attachment sites.

## 9.3 SHOULDER REPLACEMENT

Disease and/or trauma can sometimes cause such pain and debilitation that surgical intervention is warranted. In such a case, replacement of all or part of the articulating bones at the shoulder can be undertaken in order to relieve pain and restore function.

### 9.3.1 Procedures

When disease, pathology, or trauma lead to debilitating pain or unacceptable loss of function, shoulder arthroplasty, the replacement of one or both of the articular surfaces of the glenohumeral joint, is often performed. In a hemiarthroplasty, only the humeral articular surface is replaced, and in a total shoulder arthroplasty (TSA), both humeral and glenoid articular surfaces are replaced (Figure 9.3). In the prosthesis design shown in Figure 9.3, a convex hemispherical humeral component attached to a stem is used to replace the humeral head, and a concave polyethylene component is used to replace the articular surface of the glenoid. In contrast to this, in a reverse or inverse prosthesis design, a concave polyethylene lined, stemmed component replaces the humeral head, and a convex hemispherical component fixed to the scapula is used to replace the articular surface of the glenoid (Figure 9.4).

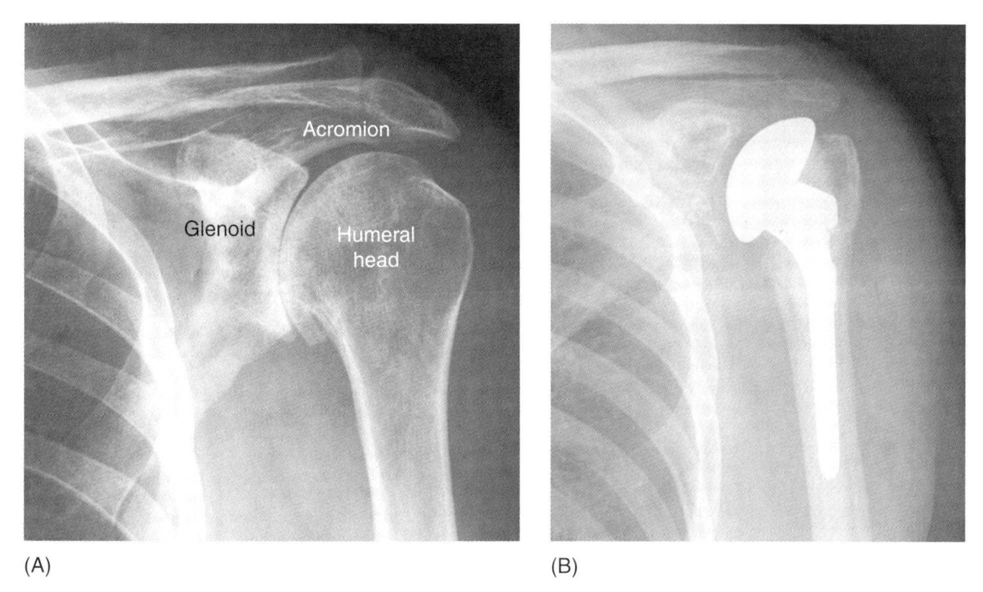


**FIGURE 9.3** Radiographic images of frontal views of shoulders before and after total shoulder arthroplasty. The preoperative anatomy in (**A**) the placement of the humeral component in (B) can be clearly seen. The UHMWPE glenoid component in (**B**) is radiolucent and is evident only as a space between the humeral component and bone of the glenoid (images courtesy of Jon JP Warner, Massachusetts General Hospital).

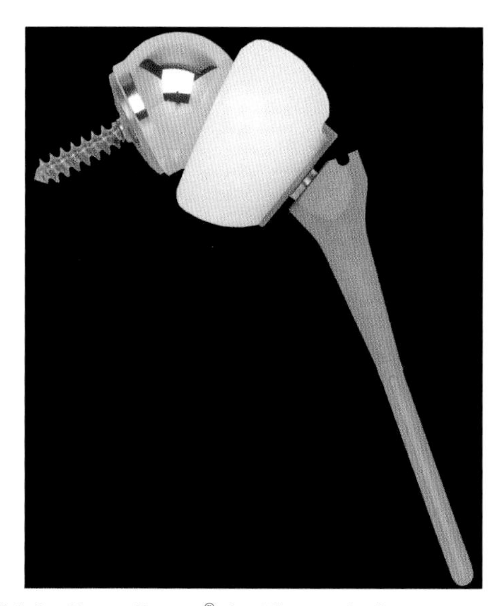

**FIGURE 9.4** Encore Reverse® shoulder prosthesis system components, circa 2003. The metal screw-fixed ball is implanted in the scapula to replace the glenoid, and the concave polyethylene component mounted on the stem is implanted into the proximal humerus to replace the humeral head (image courtesy of Encore Medical, Austin, TX).

In either case, to replace the humeral articular surface, the humeral head is resected at its base and the medullary cavity of the proximal humeral bone is prepared to accept the stem of a metal humeral prosthesis. In a fracture repair procedure, the bone fragments are reapproximated and held in place around the stem and under the head of a metal humeral component by wires. Many humeral components include holes in the stem to facilitate this reattachment. One type of humeral component used in hemiarthroplasty procedures is the bipolar design, in which a smaller head attached to the humeral stem articulates within a polyethylene cup, which is, in turn, within a polished metal shell that then articulates against the natural glenoid surface. Another type of humeral component is a metal shell used in resurfacing procedures in which the humeral head is shaped to accept such a metal cap. Biological fixation (bone ongrowth and/or ingrowth), fixation with polymethylmethacrylate (PMMA) bone cement, and/or fixation via press fit are relied on to hold the humeral component within or on the bone.

In a TSA, the glenoid articular surface is also removed and the underlying glenoid bone is shaped to receive a glenoid component. In some cases of glenoid erosion, fracture, or malformation, bone graft material is placed under one side or the other of a glenoid component to correct glenoid position and orientation. Glenoid components are held in place with biological fixation, screw fixation through holes in the metal backing of a component and into the bone, and/or with PMMA bone cement.

## 9.3.2 Patient Population

In the United States, the number of shoulder arthroplasty procedures done yearly has grown from around 13,900 per year in 1993 to around 34,500 per year in 2005 [3]. The percentage of these that are hemiarthroplasties has decreased from a high of 62.2% in 2001 to a low of 51.5% in 2005 [3]. Depending on the particular subgroup of patients, or country studied, the prevalence of hemiarthroplasties as a percentage of the total number of replacements ranges from 14% [4] in a multicenter study of OA, to 65% in a study of fracture repair [5], to 68% in a study of hemiarthroplasty and TSA for patients younger than 50 years old [6], and to 87% in the Swedish shoulder arthroplasty register started in 1999 [7]. In 2003, Dr. Charles Neer, II, the pioneer of shoulder replacement, estimated that overall, around 20% of shoulder replacements were hemiarthroplasties [8], which is generally consistent with the percentage of hemiarthroplasty versus TSA at high-volume centers [9]. It has been estimated that 5% of degenerative arthritis patients undergoing shoulder arthroplasty present with rotator cuff deficiency [10], the functional deficit for which the reverse prosthesis design was developed to specifically address. From this, the percentage of TSAs done using reverse prosthesis components could be estimated to be around 5% worldwide.

The average patient age for a series of shoulder replacements in most studies is in the range of 55 to 75 years old [2, 4, 5, 11–32]. Remarkably, this has remained relatively constant since the first published series of hemiarthroplasties [23, 24]. Relatively recent studies of groups of patients with reverse prosthesis components reported somewhat higher patient ages with averages from 65 to 78 years old [33–39]. TSA patients greater than or equal to 90 years old have also been studied as a separate group [40]. On the other end of the age spectrum, Sperling, Cofield, and Rowland [41], along with Hayes and Flatow [1], define young patients as being younger than 50 years old, and these patients also continue to be studied as a separate group [6, 42, 43].

Arthritis is noted as the reason for more than half of all shoulder replacements. A sample of published studies shows a range of 50–74% of all indications being arthritis [7, 25, 29, 44]. Within the arthritis group, the largest number of patients present with RA and the next largest number with OA. Sojbjerg and colleagues, as well as Sharma and Dreghorn, report this ratio of RA to OA at 2 to 1 [29, 45], both Rahme and colleagues [7] and Snyder [44] at about 3 to 2, Sanchez-Sotelo and coworkers at 4 to 3 [28], and Neer and colleagues [25] and Torchia and associates [31] at approximately 1 to 1.

Fracture/trauma is the next most common problem addressed by shoulder replacement. Snyder reports the rate as 9% [44], Torchia and associates as 12% [31], Neer and colleagues as 23% [25], Sojbjerg and colleagues as 30% [29], and Rahme and associates as 35% [7]. Fractures of the proximal humerus are unfortunately seen relatively often. They are reported to account for 5% of all fractures seen in the emergency room [46], and their incidence is expected to triple in upcoming decades [47]. In fact, repair of the humerus after fracture was the driving reason for the development of the first modern shoulder replacement component [8], and the first reported series of modern shoulder replacements were done to correct problems caused by humeral fractures [24].

All other indications leading to total shoulder replacement (TSR) are generally below approximately 5% each over a range of reported studies [44]. It is interesting to note, however, that replacement required for revision of prior surgery has been reported to be from 0% to about 4% [44] or from 6% [7] up to 11% [25] of all cases.

Reverse prosthesis designs were developed for use in cases of rotator cuff tear arthropathy, and although they continue to be used primarily in such cases, they have also been used in others. From a series of 240 reverse prosthesis cases, Wall and colleagues [39] report that 31% were used for rotator cuff tear arthropathy, 22.5% for revision arthroplasty, 17% for massive rotator cuff tears, 14% for primary OA, and 14% for posttraumatic arthritis. In marked contrast to other TSA studies, only 1 of the 240 cases reported by Wall and coworkers was done for RA [39].

### 9.3.3 History

The first recorded TSR was performed by a French surgeon named Péan in 1893 using a constrained design (connected glenoid and humeral components) to treat tuberculous arthritis of the shoulder [48]. The patient was a 37-year-old man, and the prosthesis components reportedly functioned relatively well until they were removed 2 years later because of infection [48].

The modern era of shoulder replacement was ushered in by Dr. Charles Neer, II. In 1953, in response to the relatively poor results of humeral head resection for patients with proximal humeral fractures, Dr. Neer implanted a Vitallium humeral component of his own design in a hemiarthroplasty procedure (Figure 9.5). In 1955, he reported on his first series of 12 patients treated in this way.

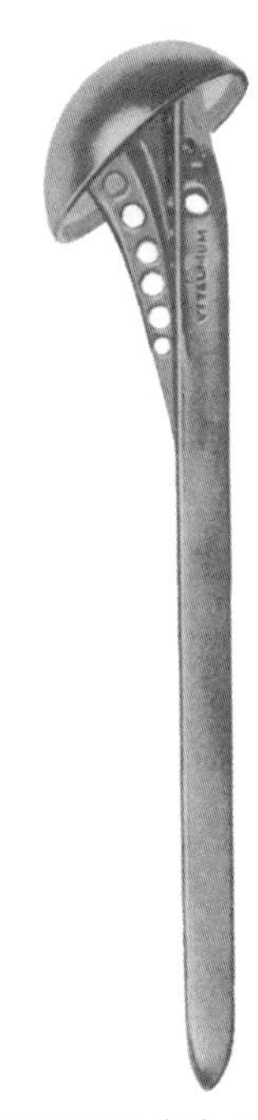

**FIGURE 9.5** Humeral component designed and used by Dr. Neer for his first shoulder hemiarthroplasties (image courtesy of Howmedica Osteonics, Allendale, NJ).

Constrained, fixed-fulcrum (captured ball in socket) devices for TSA were also considered by Dr. Neer and others. In 1972, Neer and Robert Averill tried three designs of fixed-fulcrum prostheses, but function was inadequate [25]. Successful TSRs were also performed in the early 1970s by Stillbrink (four patients), Kenmore (three patients), and Zipple using polyethylene glenoid components with Neer's original humeral component [24, 49]. This is the first recorded use of polyethylene in the shoulder, and the date identifies the polyethylene as RCH-1000, the trade name for UHMWPE produced by Hoechst in Germany.

In 1973, Dr. Neer also performed a TSR for the first time [24]. He used a cemented polyethylene glenoid component with his Neer humeral prosthesis that had been modified to more beneficially articulate with a glenoid component [24]. The polyethylene glenoid, an RCH-1000 device, was designed by Robert G. Averill, manufactured by Howmedica, and sterilized by ethylene oxide gas [8]. Between 1973 and 1982, Dr. Neer designed a number of glenoid components, both with and without metal backings [25].

The foundation that Neer laid for shoulder replacement can be seen today in many ways. The continued production and use of the 1973 version of his humeral component as the Neer II and the many other humeral and glenoid components based on this original design is testament to its stature in the surgical community. The basic concepts included in the Neer prosthesis, such as all-polyethylene and metal-backed keeled glenoid components and wire fixation holes and stabilization flanges on press-fit humeral stems, persist in current designs.

The reverse shoulder concept was first developed and implanted in Europe by Professor Paul Grammont in 1985 to allow treatment of patients with rotator cuff tear arthropathy [50]. Initial versions included a relatively constrained articulation between the ball and socket and were largely unsuccessful because of loosening [51]. In 1987, Grammont and Baulot designed a semiconstrained reverse prosthesis and recommended it be used prudently and in patients older than 70 years of age [51]. After years of use in Europe, the first reverse prosthesis case was done in North America in February 2003 [52]. In November 2003, the FDA first cleared a reverse prosthesis design for sale in the United States, and the first United States case was done in March 2004 [38].

## 9.4 BIOMECHANICS OF TOTAL SHOULDER REPLACEMENT

During the course of normal activities of arm use, the loads at the shoulder, and more specifically, the loads between the humeral head and the glenoid, can vary widely in both magnitude and direction. Anglin and colleagues give a review of the reported resultant glenohumeral forces from different studies and for different activities as a multiple of body weight (BW) [53]. These range in magnitude from $0.9 \times$ BW during arm abduction to more than $7 \times$ BW during push-ups [53]. For a realistic BW of 160 pounds for a shoulder patient, these would result in glenohumeral forces ranging from 144 to 1120 pounds. These are relatively large loads that could be a factor in damage to the UHMWPE of a glenoid component.

The direction of this force also varies with different activities. Anglin and associates report a variation of superiorly directed force on the glenoid from 10.9 to 23.6 degrees and a variation of anterior–posteriorly directed force from 2.2 degrees posterior to 17.5 degrees anterior for a variety of common activities [53]. Poppen and Walker report a variation of superior–inferiorly directed force on the glenoid in the plane of the scapula from around 60 degrees inferior to approximately 60 degrees superior during arm abduction [54]. The direction and variation also are shown to depend on the neutral, internal rotation, or external rotation position of the humerus during abduction with internal rotation resulting in more superiorly directed forces and external rotation resulting in more inferiorly and horizontally directed forces [54]. The changing direction of the load, both superior–inferiorly and anterior–posteriorly, is noteworthy. That is because of the clinically identified "rocking horse" glenoid, which describes a loosened glenoid component being alternately angled one way and then another by the loads applied to it.

Finite element studies have been conducted using the load magnitudes and directions mentioned earlier to examine predicted stresses within all-polyethylene and metal-backed glenoid components, within the cement mantle used for component fixation, and in the surrounding glenoid bone [55, 56]. Of particular interest are stresses in the cement mantle because they can influence the long-term fixation of the polyethylene in the bone. For all-UHMWPE glenoid component designs, increased cement stresses were predicted for keeled versus pegged designs [55]. Decreased cement stresses were predicted for metal-backed versus all-UHMWPE glenoid component designs [56].

Loads of nontrivial magnitude, such as those described previously, moving across the surface of the glenoid component have the potential to cause stress-related damage and wear to the polyethylene similar to that seen in UHMWPE knee and hip joint replacement components. Stresses in the UHMWPE of the glenoid component articular surface have been studied by Swieszkowski and colleagues [57] following the methods of Bartel and associates [58]. As shown by Bartel and coworkers and again by Swieszkowski and colleagues, stresses predicted in an UHMWPE component vary with thickness of the polyethylene, conformity between the metal and polyethylene surfaces, and the presence or absence of metal backing [57, 58]. For loads and geometries corresponding to glenoid and shoulder humeral components, the stresses in the UHMWPE were predicted to be in the range of 5 to 30 MPa [57].

Stresses were substantially higher for less conforming humeral–glenoid component pairs than for more conforming ones, and somewhat higher for glenoid components with metal backings and thinner UHMWPE layers [57].

The effects of load and conformity between the humeral and glenoid components on joint stability have also been studied. A study by Karduna and coworkers investigated the importance of prosthetic glenohumeral conformity in reproducing the force-displacement relationships of the natural joint as well as its implications for glenoid component stresses [59]. Relatively nonconforming components were shown to develop lower strains and better reproduce natural glenohumeral force-displacement relationships than more conforming components [59]. Studies by Warner and associates examined the relative importance of the structures at the shoulder in maintaining that force-displacement relationship for inferior humeral translation [60, 61]. It was shown that increased compressive force across the joint provided by the rotator cuff muscles was the most important factor in maintaining stability [60, 61].

It should be noted that the loads, motions, and stresses considered and reported in the studies assumed or implied normal kinematics and geometry of the shoulder joint and anatomically correct placement of humeral and glenoid components relative to the other structures at the shoulder. This is noteworthy because, in reality, the disease or trauma state that leads to a TSA most often affects one or more of the surrounding bony, ligamentous, tendinous, and muscular structures at the shoulder, leaving it in a deteriorated state. In fact, in some cases, it is a deterioration of one or more of the structures at the shoulder that is the underlying cause of the disease progression to the point of requiring shoulder replacement.

Rotator cuff tear arthropathy is one such condition for which standard TSA is not particularly successful. In recreating the geometry of the humeral head and glenoid face, standard TSA components generally may not provide adequate constraint against superior translation of the humerus in the absence of fully functional rotator cuff muscles [62]. In contrast, the reverse TSA prosthesis components interact to keep the humerus from translating superiorly when the deltoid muscle is activated to elevate the arm. In addition, the fulcrum for the rotation at the shoulder with a reverse TSA is more medial than with a standard TSA. This results in a greater mechanical advantage for the deltoid muscle and a functional improvement for the patient. In restoring function, the prosthesis components can be subjected to loads that can be relatively high, especially in younger, more active patients. This is one of the reasons that the reverse prosthesis components are recommended by some clinicians for use only in older (>70 years old) patients and only in cases of salvage procedures [34, 37, 38, 63].

In all cases, humeral and glenoid components are not placed into an otherwise well-functioning joint environment, but rather into a diseased or damaged and surgically disrupted (to implant the components) and surgically repaired joint environment. It is also possible that the implanted components will not interact properly with the surrounding shoulder structures. One of the effects of this is that, in addition to normal loads and motions, replacement components can be faced with potentially damaging exceptions to the normal shoulder loads and motions.

## 9.5 CONTEMPORARY TOTAL SHOULDER REPLACEMENTS

It has been estimated that more than 100 shoulder prosthesis designs are currently in use worldwide [49]. Where there is articulation between components, UHMWPE is used on one of the counterfaces, the other being polished metal, usually cobalt chromium alloy. Methods of differentiation between designs include the mechanical interaction between components (e.g., fixed fulcrum, semiconstrained, unconstrained, etc.), the intended indications to be addressed (e.g., primary, total, revision, fracture, etc.), and/or the overall nature of the design (e.g., monoblock, modular, bipolar, metal-backed glenoid, etc.). These categories can aid in the understanding of the classes of designs that exist, but they are not definitive because of the overlap between groups.

Some of the available prosthesis systems that include UHMWPE components are (alphabetically by manufacturer):

- Biomet (Warsaw, Indiana, USA) (Figure 9.6)
  - Bio-Modular: A modular cobalt chromium humeral component with keeled and three-pegged UHMWPE glenoid components developed with Drs. Russell Warren and David Dines and introduced in 1987. The glenoid components are sterilized by gamma in argon. It is intended for primary, hemiarthroplasty, total, and fracture use with cemented fixation.
  - Bi-Angular: A modular cobalt chromium or titanium stem humeral component with UHMWPE glenoid components with an angled keel or metal backed with an angled keel and three pegs sterilized by gamma in argon. The system was developed with Dr. Richard Worland and introduced in the mid-1990s. It is intended for primary, hemiarthroplasty, total, and fracture use with cemented fixation.
  - BiPolar: A modular head with a cobalt chromium shell and an inner UHMWPE bearing surface and retaining ring capturing an inner cobalt chromium head. The concept was originally developed in the 1970s and was intended for use in a hemiarthroplasty salvage procedure.

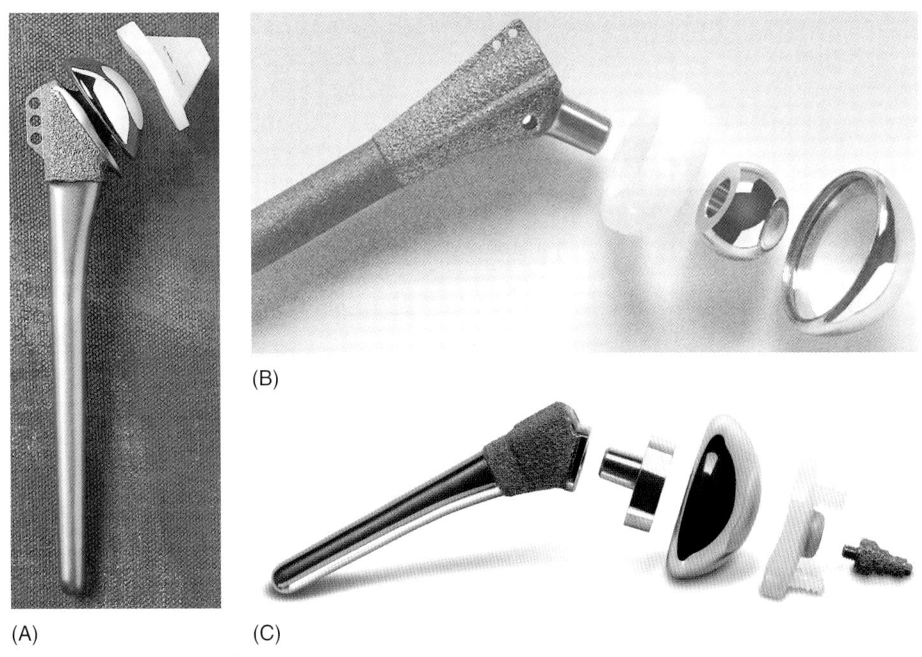


**FIGURE 9.6** Bio-Modular® and Bi-Angular®/BiPolar contemporary shoulder prosthesis system components. The Bio-Modular® components are shown in (**A**) an exploded view of the BiPolar assembly is shown in (**B**) and an exploded view of components of the Comprehensive® system, including the modular hybrid glenoid, is shown in (**C**) (images courtesy of Biomet, Warsaw, IN).

- Comprehensive: Prosthesis system including a Modular Hybrid Glenoid (which features an Arcom UHMWPE that is isostatically molded and gamma sterilized and packaged in an argon environment) base and optional central pegs of UHMWPE and Regenerex (a porous titanium construct to allow for biologic fixation to enhance glenoid fixation). The system was developed with Drs. David Dines, Russell Warren, Edward Craig, and Donald Lee and was introduced with the modular glenoid in 2007. Reverse components within the Comprehensive system were cleared for sale in the United States by the FDA in July 2008.
- Integrated: A collection of monoblock and modular cobalt chromium humeral prostheses within a single system. Included are the Kirschner II-C monoblock and the Atlas modular design. Glenoid components include all-poly UHMWPE keeled and pegged components as well as a metal-backed, screw-fixed design. Components of the system are intended for primary, hemiarthroplasty, total, and fracture use with cemented and press-fit fixation.

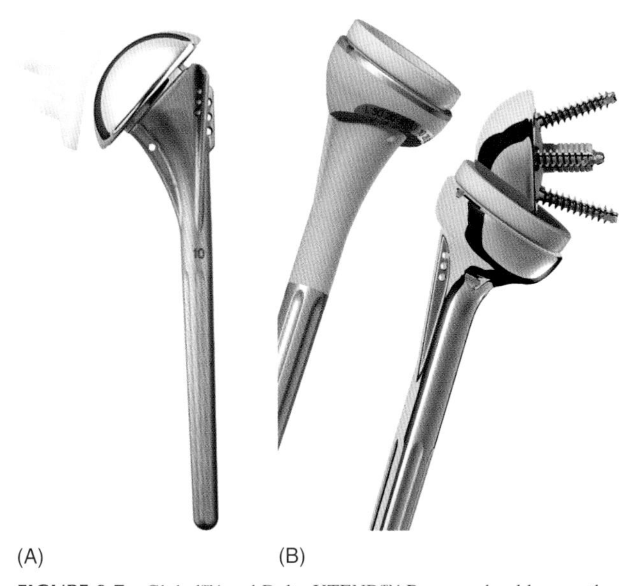


**FIGURE 9.7** Global™ and Delta XTEND™ Reverse shoulder prosthesis system components. Components of the Global system are shown in (**A**) and components of the Delta XTEND Reverse system are shown in (**B**). (images courtesy of DePuy Orthopaedics, Inc., Warsaw, IN).

- DePuy Orthopaedics (Warsaw, Indiana, USA) (Figure 9.7)
  - Global and Global Advantage and Global FX: Modular cobalt chromium humeral prostheses developed with Drs. Charles Rockwood and Fredrick Matsen, III, and first introduced in 1990. They include UHMWPE keeled and five-pegged glenoid components sterilized by gas plasma. They are intended for primary, hemiarthroplasty, total, and fracture use (FX specifically) with press-fit fixation.
  - Delta CTA Reverse and XTEND Reverse: Reverse design prosthesis component systems incorporating HA coating to enhance screw fixation of the glenoid component, and Enduron UHMWPE (gamma sterilized and packaged in vacuum foil). The Delta

CTA was cleared for sale in the United States by the FDA in November 2003.

- Encore (Austin, Texas, USA)
  - Foundation and Foundation fracture: Modular titanium stem humeral prostheses with keeled and pegged UHMWPE glenoid components sterilized by gamma in nitrogen. Developed with Drs. Richard Friedman and Mark Frankle, it is intended for primary, hemiarthroplasty, total, and fracture use with cemented and press-fit fixation.
  - Reverse: Reverse design prosthesis component system. The Encore Reverse system was cleared for sale in the United States by the FDA in March 2005.
- Smith & Nephew (Memphis, Tennessee, USA) (Figure 9.8)
  - Neer II: The Neer monoblock cobalt chromium prosthesis design was developed in 1973 by Dr. Charles Neer, II [25]. It includes a keeled UHMWPE glenoid component sterilized by ethylene oxide gas (EtO). The Neer II design is also marketed by Biomet. Dr. Neer originally presented five different UHMWPE glenoid components, two all poly and three metal backed, that were designed for use with the humeral component and allowed the choice of a range of different constraints [25]. Although still in use, it has been supplanted by the Neer 3 design in the United States.
  - Neer 3: The update to the Neer humeral component with modified fins and head position relative to the stem and added stem markings. It is intended for primary, hemiarthroplasty, total, and fracture use for cemented or press-fit fixation.
  - Cofield and Cofield2: Monoblock and modular cobalt chromium humeral prostheses developed with Dr. Robert Cofield. First introduced in 1983, the Cofield humeral components provide a surface designed for bone ingrowth. They include a keeled UHMWPE and a metal-backed, screw-fixed glenoid sterilized by EtO. They are intended for primary, hemiarthroplasty, total, and fracture use with cemented fixation.
  - Modular Shoulder System: A modular titanium stem humeral component with a keeled UHMWPE glenoid component sterilized by EtO. Cobalt chromium, titanium nitrite, and ceramic humeral heads are available as part of the system. It is intended for primary, hemiarthroplasty, total, and fracture use with press-fit fixation.
- Stryker Howmedica Osteonics (Mahwah, New Jersey, USA) (Figure 9.9)
  - Solar: Modular titanium stem humeral prostheses with UHMWPE angled-peg glenoid components sterilized by gamma in nitrogen and a vacuum. They are intended for primary, hemiarthroplasty, total, and fracture use with cemented and press-fit fixation. The system also includes a bipolar head that can be used with the stems.
- Tornier (Stafford, Texas, USA) (Figure 9.10)
  - Aequalis and Aequalis Fracture: Modular titanium or cobalt chromium stem, and humeral head prostheses with UHMWPE keeled and pegged glenoid components. Developed with Drs. Gilles Walch and Pascal Boileau, the system was introduced in the early 1990s. It was the first system to have multiple stem angles and offset head modularity, and its introduction marked the beginning of the

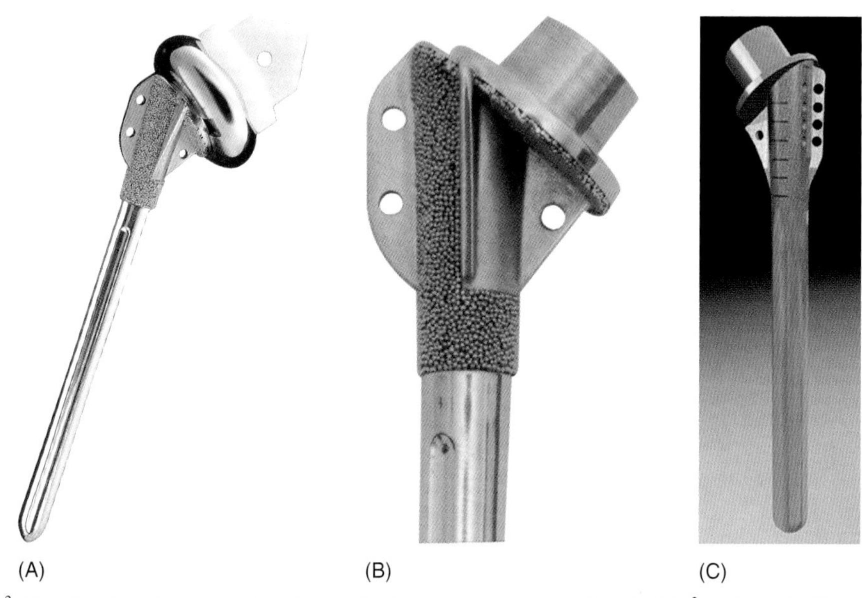


**FIGURE 9.8** Cofield², Neer 3 and contemporary shoulder prosthesis system components. The Cofield² is shown in (**A**) and (**B**). A close-up view of the ingrowth surface of the proximal humeral stem is shown in (**B**). The humeral stem of the modular Neer 3 is shown in (**C**). (images courtesy of Smith & Nephew, Memphis, TN).

third-generation, or adaptable, prosthesis designs, so-called because it was more possible than in previous designs to adapt the prosthesis to the anatomy of the patient. They are intended for primary, hemiarthroplasty, total, and fracture use with cemented and press-fit fixation.
  - Aequalis Reversed: Reverse design prosthesis component system. The Reversed system was cleared for sale in the United States by the FDA in August 2004.
- Zimmer (Warsaw, Indiana, USA) (includes Centerpulse prosthesis systems, which Zimmer acquired when it purchased Centerpulse in October 2003) (Figure 9.11)

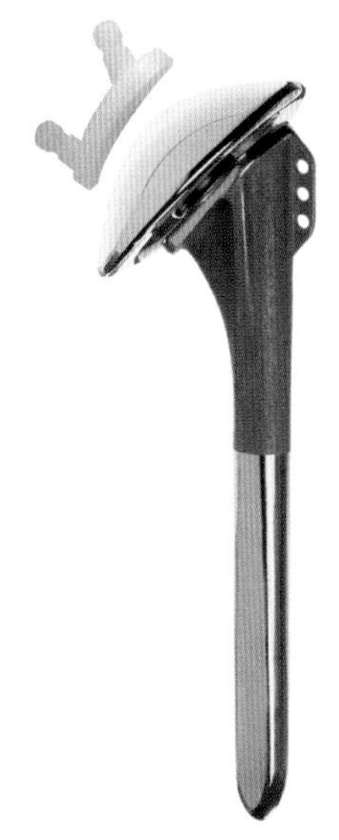

**FIGURE 9.9** Solar® contemporary shoulder prosthesis system components (image courtesy of Howmedica Osteonics, Mahwah, NJ).

  - Bigliani/Flatow: Modular cobalt chromium stem humeral prostheses with keeled and pegged UHMWPE glenoid components sterilized by gamma in nitrogen. Developed with Drs. Louis Bigliani and Evan Flatow, it was introduced in 1999. It is intended for primary, hemiarthroplasty, total, and fracture use with cemented and press-fit fixation.
  - Select: Modular titanium stem humeral prostheses with keeled and pegged UHMWPE glenoid components sterilized by gamma in an oxygen-free environment. Developed with Dr. Wayne Burkhead, it was introduced in 1987. It is intended for primary, hemiarthroplasty, total, and fracture use with cemented fixation.
  - Anatomical Shoulder: Modular titanium stem and cobalt chromium humeral head prostheses with UHMWPE four-pegged glenoid components sterilized by gamma in an oxygen-free environment. Developed with Drs. Christian Gerber and Jon Warner, the first implantation was performed in 1995. It is intended for primary, hemiarthroplasty, total, and fracture use with cemented and press-fit fixation.
  - Anatomical Shoulder Inverse/Reverse: Reverse design prosthesis component system. The Inverse/Reverse system was cleared for sale in the United States by the FDA in January 2006.

As can be seen from this partial listing of devices, the number of components and designs from which a surgeon can choose is relatively large. (Another list of shoulder prostheses with pictures is available at http://depts.washington.edu/shoulder/CommonUSShoulderProstheses.htm.)

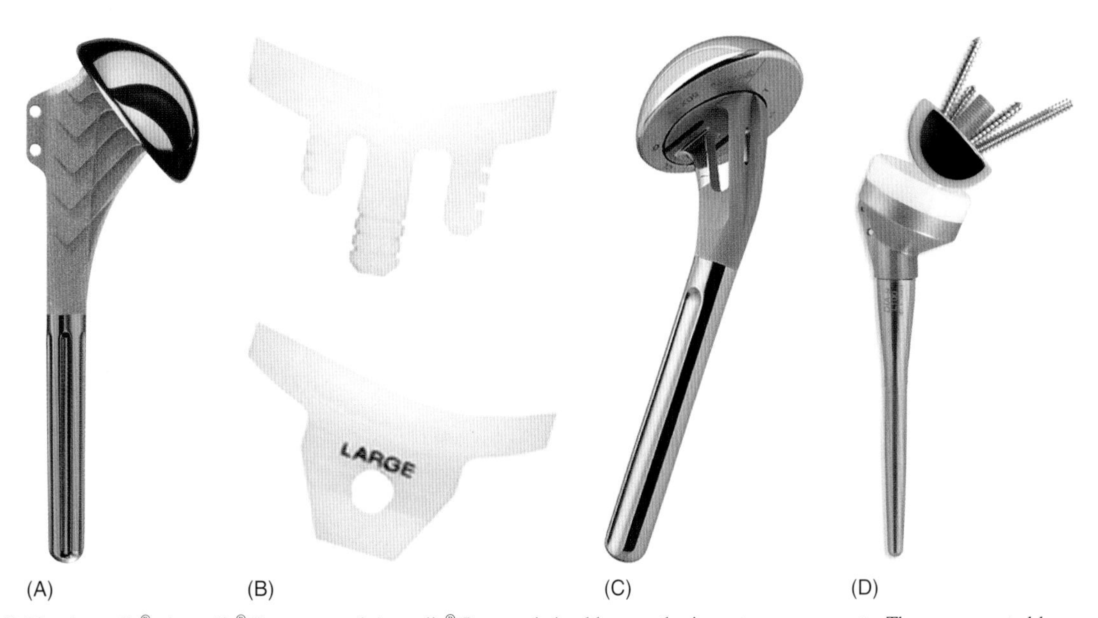


**FIGURE 9.10** Aequalis®, Aequalis® Fracture, and Aequalis® Reversed shoulder prosthesis system components. The noncemented humeral component assembly is shown in (**A**). Pegged (above) and keeled (below) all-polyethylene glenoid components are shown in (**B**). An Aequalis fracture prosthesis assembly showing the back of the modular indexing head is shown in (**C**). Aequalis Reversed system prosthesis components are shown in (**D**). (images courtesy of Tornier, Stafford, TX).

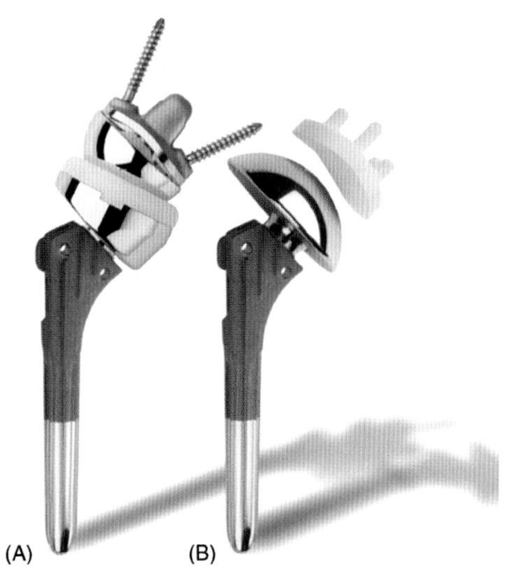


**FIGURE 9.11** Anatomical Shoulder™ contemporary shoulder prosthesis system components. Inverse / Reverse components are shown in (**A**) and stem and glenoid components and an offset head component are shown in (**B**). (Images courtesy of Zimmer Holdings, Inc, Warsaw, IN).

There is a full range of choices depending on indication and surgeon preference. First-generation (monoblock) humeral component designs, such as the Neer II, Cofield, and Kirschner-II-C, continue to be used along with second-generation (modular) designs and third-generation (adaptable) designs, such as the Aequalis and Anatomical prostheses, with comparable success [64, 65], and reverse prosthesis designs are offered in the United States by Biomet, DePuy, Encore, Tornier, and Zimmer.

The humeral heads or glenospheres are polished cobalt chromium, and humeral stems are cobalt chromium or titanium. The glenoid components or humeral liners in all the systems are made from UHMWPE. The standard TSA glenoid components are available in all-polyethylene keeled and pegged designs, metal-backed keeled and pegged designs, and metal-backed, screw-fixed designs. A range of sterilization methods from gas plasma to EtO to gamma in inert and vacuum environments are used by the manufacturers [66] with a goal of decreasing the wear of the components as compared with UHMWPE sterilized by gamma in air. The UHMWPE used in bipolar humeral heads as intermediate head bearing and retaining surfaces is manufactured and sterilized in the same way as glenoid components from those companies that offer them.

## 9.6 CLINICAL PERFORMANCE OF TOTAL SHOULDER ARTHROPLASTY

Total Shoulder Arthroplasty is a very successful procedure, relieving pain and restoring function to the majority of patients who receive it. Successful outcomes, however, are not uniform, and there are problem areas common to different devices and procedures, as well as different problem areas depending on a number of factors, including the type of arthroplasty performed, and the underlying disease or trauma leading to the arthroplasty.

### 9.6.1 Overall Clinical Success Rates

The goal of TSA is to reduce or remove pain, reconstruct the articular surfaces of the glenohumeral joint, and to restore function to the shoulder by restoring strength and movement. In general, contemporary TSR achieves these goals very well. This has been determined by objective functional measurement as well as subjective measures of patient satisfaction for postoperative follow-up times ranging from 2 years to more than 25 years. The success of TSA in restoring function has been reported to be from 42–95% [4, 14, 22, 24–27, 29, 32, 39, 44, 67–69]. Patient satisfaction with TSA has been reported to range from 42–94% [1, 2, 5, 12, 18, 21, 27, 28, 39, 70–73].

Taking a closer look at these results shows differences with respect to a number of variables. The indications for TSA are varied and therefore present unique challenges to the repair. In general, standard TSA performed to address problems caused by primary arthritis (OA or RA) are the most successful [18, 20, 22, 24, 26, 29, 74], and the success rates for standard TSA performed to address problems due to fracture are substantially lower [1, 5]. In younger patients (less than 50 years old), the results for TSA are also generally less successful than those performed in older patients [1, 6, 21]. Similarly, results for reverse TSA are better when it is used in cases of rotator cuff tear arthropathy than when it is used in cases of revision or posttraumatic arthritis [38, 39].

As with any mechanical system, time in service also affects the continued effectiveness of the procedure. For one group of studies, a range of 90–95% success rate was reported for short-term results (up to 5 years), and a much lower 55–88% success rate was reported for results up to and longer than 10 years [75]. Survivorship analyses for standard TSAs also show this expected drop off with time *in vivo* [30, 31, 34, 74, 76]. One study showed a 100% survivorship from 0 to 2 years followed by 95% from 4 to 6 years and 92% up to 9 years [74], while others show 93% survivorship up to 10 years and 87% up to 15 years [31], 92% up to 15 years and 87% up to 20 years [30], and 88% up to 15 years and 85% up to 20 years [71]. For bipolar prosthesis components, Diaz-Borjon and coworkers reported a 93.3% survivorship to 11 years, although the number of cases reviewed in their study was small (33 shoulders in 31 patients) [16].

In spite of this natural decline with increased postoperative time, the functional benefit from TSA is quantifiable,

**TABLE 9.1 Complications with Total Shoulder Arthroplasty Ranked According to Prevalence**

| Standard TSA | | Reverse TSA | |
|---|---|---|---|
| **Prevalence ranking (approx. % of all cases)[a]** | **Complication** | **Prevalence ranking (approx. % of all cases)[b]** | **Complication** |
| 1 (10% or more) | Glenoid loosening | 1 (around 10%) | Glenoid loosening |
| 2 (up to 4%) | Instability | 2 (around 5%) | Infection |
| 3 (up to 4%) | Postoperative rotator cuff tear/retear | 3 (up to 5%) | Dislocation |
| 4 (around 3%) | Periprosthetic bone fracture | 4 (around 4%) | Hematoma |
| 5 (less than 1%) | Infection | 5 (up to 4%) | Humeral loosening |
| 6 (less than 1%) | Neural injuries | 6 (up to 4%) | Fracture |
| 7 | Prosthesis dissociation/ fracture/wear | 7 (less than 1% each) | Prosthesis dissociation/ neural injuries |

[a]*Wirth and Rockwood, Jr. 1996 [79], Noble and Bell 1995 [72].*
[b]*Calculated from review data in McFarland 2006 [78].*

demonstrable, and long lasting [19]. From a study of 994 shoulder arthroplasties, 15,414 hip arthroplasties, and 34,471 knee arthroplasties from 1994 to 2001 in the Maryland Health Services Cost Review Commission database, Farmer and coworkers determined that shoulder arthroplasty had fewer complications, shorter hospital stays, and lower costs than either hip or knee procedures [77]. The fact that the success of TSR is so good overall and that its effectiveness persists to such a degree for so long is testament to the inherent longevity of the replacement components and the excellent surgical techniques used for implantation.

In spite of these good results, there are complications that adversely affect the performance of TSRs. Table 9.1 lists complications seen in TSA in the order of their prevalence [72, 78, 79]. One or more of these complications can cause failure of TSA, leading to revision. Revision rates range from 0–50% in reported TSA series [4, 25, 28, 29, 44, 67, 69, 71, 74, 80].

## 9.6.2 Loosening

Many other clinicians also note glenoid loosening as the primary complication and the major reason for failure of TSA [1, 4, 10, 12, 21, 26, 28, 29, 31, 34, 41, 44, 56, 74, 76, 81–94]. Among unsatisfactory arthroplasties, Franta and colleagues report that 63% of them included loose glenoid components [86].

Evidence of impending or eventual loosening is also reported frequently in the form of radiolucent lines at prosthesis component fixation locations. Reports of clinical experience note a relatively high incidence of glenoid radiolucencies [1, 10, 30, 31, 56, 74, 79, 89], although there is not always a correlation seen between radiolucencies and negative clinical results [10, 12, 18, 28, 30, 95]. On the other hand, roentgen stereophotogrammetric studies of glenoid migration indicate that components may be undergoing more motion away from their implanted positions and orientations than may be indicated by radiolucencies [91, 96, 97].

The complications listed in Table 9.1 are interrelated. Loosening of a glenoid component can cause instability, as can a preexisting or postoperative tear of the rotator cuff muscles. Instability due to lax or torn ligamentous or muscular structures can, on the other hand, lead to glenoid loosening. This instability is identified as the main culprit in TSA failure by some authors [17, 29, 98, 99]. Superior instability of the glenohumeral joint allows higher than normal superior motion of the humeral head, which can lead to intermittent superior edge loading of the glenoid component [1, 29, 70, 79, 98–100]. Loss of active stabilization of the joint has been shown to increase the variability of glenohumeral motion *in vivo* [62]. The increased translation has been estimated to be around 3 mm to around 11 mm [101]. This intermittent loading can lead to rocking of the glenoid component.

Fixation of the glenoid component within the bone of the glenoid is generally good, but as noted earlier, it seems to be the weak link of TSA. Primary fixation is achieved either with polymethylmethacrylate (PMMA) bone cement or via screws. Cemented all-polyethylene components are widely used, but metal-backed components with bone ingrowth surfaces are also used with cement or supplemental screw fixation. In spite of the radiolucencies seen, cemented all-polyethylene glenoids are seen as the gold standard of fixation [102] with pressurized cementing technique resulting in a relative decrease in glenoid loosening complications [89].

### 9.6.3 Wear

Another possible interrelationship between TSA complications is that between component loosening and wear or damage to the UHMWPE component. The association among wear, wear debris, and osteolysis leading to aseptic component loosening has been widely reported for total hip arthroplasty (THA) and total knee arthroplasty (TKA). In comparison, the number of reports of osteolysis or the possibility of osteolysis in TSA is relatively small [12, 90, 100, 103–106]. Although it is a relatively small problem overall in TSA, damage to UHMWPE components can be drastic and affect the longevity of the procedure.

Wear of UHMWPE in TSA leading to osteolysis or the possibility of osteolysis has been reported as part of a clinical series [12, 94] as part of a study of negative outcomes [21], as part of studies examining retrievals [103, 105, 107], and as part of studies examining wear debris [104, 106]. In a report of a series of implants, Boileau and colleagues noted three metal-backed glenoid components that showed complete UHMWPE wear-through and osteolysis at revision [12]. Weldon and associates noted changes in the UHMWPE surfaces of glenoid components that they attributed to wear and deformation [107]. In 60% of the components, these changes substantially altered the glenohumeral stability provided by the glenoid [107]. Hasan and coworkers studied unsuccessful arthroplasties after an average of 3.6 years follow-up [21]. They reported a 20% overall incidence of glenoid UHMWPE wear and a 59% overall incidence of loose glenoid components with 77% of the cases with loose glenoids also exhibiting stiffness of the shoulder and/or malpositioning of the implants [21]. Franta and coworkers also reported similar findings in a group of 232 unsatisfactory arthroplasties with 63% exhibiting glenoid loosening and 30% showing UHMWPE wear [86].

Observations of retrieved glenoid components show that wear and damage to the UHMWPE component can vary from being relatively subtle to being catastrophic [13, 81, 103, 105, 108–111]. Relatively severe wear and damage was seen in one group of Global (DePuy Orthopaedics, Inc.) glenoid components made of Hylamer UHMWPE (developed by a DePuy Orthopaedics and E. I. DuPont Company joint venture) and sterilized by gamma irradiation in air [111]. As more fully described in Chapter 19, Hylamer UHMWPE was developed to have a higher yield strength and tensile strength than conventional UHMWPE, and it also had a higher tensile modulus. Oxidation damage due to the gamma sterilization in air that becomes progressively worse during shelf storage was more severe in the Hylamer material than in conventional UHMWPE and rendered many of the Hylamer glenoids made and sterilized at the time prone to wear and damage clinically [111]. Hylamer UHMWPE glenoids made and sterilized from 1990 to 1993 were affected in this way. The clinical effect of these Hylamer glenoids can be seen in the reported data from at least one study following 124 shoulders done for primary OA out to 2.2 to 11.6 years (average 5.1 yrs) [20]. In this study, the results for 10-year survivorship without revision was 86% but improved to 90% when the adverse data associated with Hylamer glenoids sterilized in air were excluded [20]. Changes in sterilization methods with Hylamer from 1993 to 1998 and the elimination in 1998 of Hylamer from glenoid components have limited the possibilities of adverse clinical results resulting from this issue [111].

Retrieval studies of glenoid components identify the damage modes on UHMWPE glenoid components. Scarlat and colleagues reported on a group of 43 retrieved glenoid components that had been retrieved after an average of 2.5 years of service, 67% of which were keeled all-polyethylene designs, 21% of which were metal-backed designs, and the remainder of which were nonkeeled all-polyethylene designs [105]. Of the 21 cases where the reasons for the original arthroplasty and component removal were known, preimplantation instability was noted in 38% and all but three components were removed because of looseness (two were removed at autopsy and one was removed due to infection) [105]. Of the four damage modes identified, rim wear was the most prevalent (70%), followed by surface irregularity (67%), fracture (30%), and central wear (20%) [105]. Characterization of UHMWPE wear was also done by Braman and coworkers on 20 glenoid components retrieved at revision [108]. They found diffuse surface wear to be the most prevalent (90%) with rim wear also common (70%) [108].

A damage mode identification and grading scheme used for UHMWPE knee tibial components and hip acetabular components has also been applied to a small group of retrieved UHMWPE glenoid components (10 components) [103]. Scratching was noted as the most prevalent damage mode and was seen in 90% of the examined regions [103]. Abrasion (68%), pitting (60%), delamination (58%), deformation (40%), embedded debris (28%), and burnishing (8%) were also reported [103]. Component fractures of four glenoids, and complete wear-through of one glenoid component, were also noted [103]. The data shows a combination of abrasive and fatigue wear mechanisms at work in the shoulder. This is similar to the damage modes seen in knee components and in contrast to the predominant wear modes seen in total hip components.

Further analysis of the type of wear in UHMWPE glenoid components has also been conducted by examining the wear debris generated in the shoulder. Isolation and characterization of the UHMWPE wear debris generated in failed shoulder, hip, and knee arthroplasties gives added insight into wear in the shoulder and how it compares with that in the hip and knee. Wirth and coworkers [106] and Mabrey and associates [104] have applied American Society for Testing and Materials (ASTM) standard F1877 Standard Practice for Characterization of Particles to quantitatively describe the UHMWPE wear particles from shoulder, hip, and knee components. The presence of substantial numbers of UHMWPE wear particles in the tissues surrounding aseptically loose glenoid components is direct evidence corresponding to clinical observations that wear and loosening are related. UHMWPE wear particles isolated from total shoulders are comparable in size and form to those isolated from total knees and total hips [104, 106]. In general, the shoulder particles are more similar to knee particles, being larger and more oblong than hip particles [104, 106] (Figure 9.12). The data is consistent with the different articulation characteristics of hip, knee, and shoulder arthroplasties. In hip arthroplasty, the metal head rotates within the UHMWPE liner. In contrast, in knee and shoulder arthroplasties, both rotation and translation of

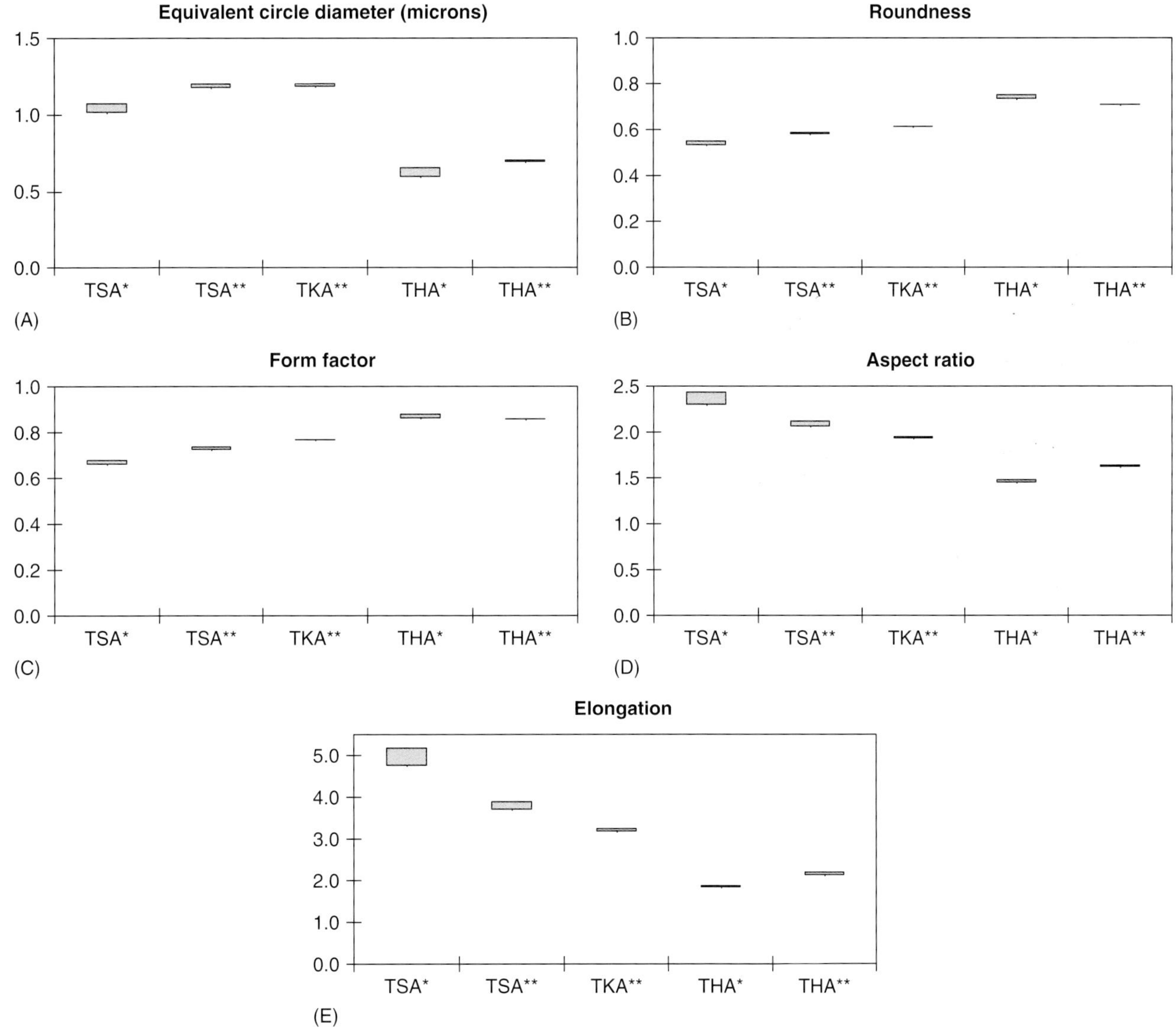


**FIGURE 9.12** Plots of data showing ASTM F1877-98 quantitative description characteristics of retrieved UHMWPE particles from TSA, TKA, and THA. Data from Mabrey et al. 2002 [104] is noted with a double asterisk (**) and data from Wirth et al. 1999 [106] is noted with a single asterisk (*). The boxes in each plot enclose the data values of the mean ± the standard error in the mean reported for each implant type. The equivalent circle diameter (the diameter, in microns, of the circle with the same area as the projected area of a particle) is plotted in (**A**). The roundness (how closely the projected shape of a particle resembles a circle on the basis of diameter [perfect match with a circle = 1.0]) is plotted in (**B**). The form factor (how closely the projected shape of a particle resembles a circle on the basis of perimeter [perfect match with a circle = 1.0]) is plotted in (**C**). The aspect ratio (maximum straight-line particle length divided by maximum straight-line particle width) is plotted in (**D**). The elongation (actual particle length divided by mean particle width) is plotted in (**E**). It is interesting to note that where differences are apparent, the TSA and TKA particles are similar to one another and generally larger (as shown by the equivalent circle diameter data) and more elongated (as shown by the aspect ratio and elongation data) than THA particles.

the metal components occur relative to the polyethylene components.

## 9.7 CONTROVERSIES IN SHOULDER REPLACEMENT

Although shoulder replacements generally achieve the goal of pain relief, repair of structure, and restoration of function, there are still many ways to go about it and many products and techniques from which to choose. The introduction of UHMWPE as a glenoid articular bearing surface allowed surgeons to contemplate and perform TSR for a broader range of indications than was possible with a monoblock humeral component, and introduction of the reverse TSA prosthesis further expanded the cases that could be successfully treated. As noted earlier, however, the addition of a glenoid component opens up the possibility for complications related to its fixation and durability. For some surgeons, this leaves open the question of whether to use a glenoid component.

Results and rates of TSA and hemiarthroplasty are varied, but both definitely seem to have a place in treatment of the shoulder. The Swedish Register of shoulder arthroplasty reports that 87% of shoulder replacements from 1999 to 2001 have been hemiarthroplasties because of the high rates of radiolucency seen in other studies and because of the relative difficulty of implanting glenoid components [7]. On the other hand, in a large, multicenter study (571 cases of a series of 1701 arthroplasties), 86% of the cases were TSAs performed on the basis of surgeon preference [4]. Overall in the United States from 1993 to 2005, TSA rates increased more rapidly than did hemiarthroplasty rates, but more hemiarthroplasties than TSAs were still done [3]. These data are consistent with the observation that high volume surgeons prefer TSA to hemiarthroplasty more than 3 to 1 [9, 112] and that the percentage of patients operated on in high volume urban centers increased [113].

In a study of results of shoulder arthroplasty from 1988 to 2000 in New York state, although no significant differences were seen, there was an indication of better results in the form of lower short-term readmission rates for procedures done at high-volume centers as compared to those done at lower-volume centers [114]. Better outcomes have also been shown for TSA compared with hemiarthroplasty in general [69, 115–117], with TSA yielding a significantly higher pain relief and abduction function and a significantly lower risk for revision than hemiarthroplasty [30]. TSA also showed better results in cases of OA in one study [118] and in cases of glenoid erosion in another [22], although glenoid bone loss has also been noted as a contraindication for TSA [2].

For OA with a functioning rotator cuff, the results of total and hemiarthroplasty have been shown to be comparable [11]. Short-term, prospective, randomized studies of TSA versus hemiarthroplasty (follow-up to 2 years) also show comparable results, although trends are more favorable for TSA [119, 120]. In cases of rotator cuff tear or arthropathy, however, hemiarthroplasty has been noted as a preference because of the association of glenoid loosening with rotator cuff problems [29, 32, 99]. The recent availability of the reverse TSA prostheses, however, is likely to impact that. It is also interesting to see reports showing half of the revised TSAs being converted to hemiarthroplasties [29, 70], although in one study a larger percentage of patients were satisfied after revision to another TSA than after revision to a hemiarthroplasty [70]. Long-term data (follow-up to >25 years) shows both TSA and hemiarthroplasty to be good options. Survivorship is reported to be 92% for TSA and 89% for hemiarthroplasty at 15 years, and 87% for TSA and 89% for hemiarthroplasty at 20 years [30].

UHMWPE can still play a vital role in shoulder arthroplasty, even when a glenoid component is not used. The bipolar prosthesis used in hemiarthroplasty has an UHMWPE internal liner and retaining ring that holds the inner head within the outer metal shell. In short-term follow-up (around 3 years), studies report 95% success and 92% satisfaction with a bipolar prosthesis used without a glenoid component, even when used in cases with massive rotator cuff tears [11, 32]. For longer-term follow-up (5 to 11 years, average of 6.7 years), the bipolar prosthesis also shows very good results with survivorship of 93% at 11 years, although the number of cases is relatively small (33 shoulders in 31 patients) [16]. In spite of the relatively few cases on which to draw conclusions, proponents of the bipolar design concept note that the dual articulation of the device is likely to result in less damage to an unresurfaced glenoid than a standard prosthetic humeral head [11, 32].

Even after the choice to use a glenoid component has been made, there are still a number of alternatives to consider. All-polyethylene components are available in both keeled and pegged variants. The keeled design goes back to the time of the first TSA in the 1970s, and the pegged design is a more recent development. Fixing the glenoid component in the bone is challenging in part because of the relatively small amount of bone available, especially in cases of erosion or fracture. Keels and pegs are different answers to this challenge. For comparable keeled and pegged designs, clinical results comparing radiolucency rates [74] and stability of the components *in vivo* [96, 121] indicate that the pegged design may be better than the keeled design. On the other hand, one study notes incidences of peg breakage [26]. Theoretical results predicted lower stresses beneath the pegged design compared with the keeled design in normal glenoid bone [55]. In rheumatoid bone, however, theoretical results predicted lower stresses beneath the keeled design [55].

The important glenoid component questions of all-polyethylene or metal-backed and cemented or noncemented are also not mutually exclusive. Fixation of all-polyethylene components is performed only with bone cement. The addition of a metal backing opens up the possibilities for glenoid fixation. The surfaces of some metal-backed glenoid components are textured to allow for bony ingrowth to achieve biological fixation. The metal backings of some glenoid components also allow for screw fixation. For the most challenging cases of glenoid repair requiring bone grafting to aid in component fixation and to reconstruct the proper location and orientation of the glenoid face, both all-polyethylene as well as metal-backed components have been cemented with some success out to an average of 5 years follow-up [2].

Arguments in favor of metal-backed glenoid components include theoretical tensile stresses that are predicted to be lower beneath metal-backed components than beneath corresponding all-polyethylene glenoid components [56]. Early results for contemporary modular metal-backed designs are also promising [94], and there is the potential for strong fixation via osseointegration offered by metal-backed glenoid components [70, 102]. Godeneche and colleagues note an 83% rate of radiolucency for all-polyethylene glenoids compared with a 20% rate of radiolucency for metal-backed glenoids [18]. Boileau and associates also note a similar discrepancy in radiolucency rates (85% for all polyethylene, 25% metal backed), but go on to report increased wear and loosening in the metal-backed glenoids compared with the all-polyethylene ones [12]. Radiolucency around the glenoid component may be associated with the use of PMMA bone cement for fixation. When PMMA is used in surgery, its exothermic curing may cause thermal damage to the adjacent bone [122].

The preponderance of the evidence is that better clinical results are seen for cemented, all-polyethylene components, especially with third-generation cementing technique, leading Ibarra and colleagues to call them the gold standard for fixation [102]. Clinical reports note a higher revision rate for metal-backed components (6.8%) than for all-polyethylene components (1.7%) [69], incidences of screw breakage with metal-backed components [44, 70], incidences of dissociation of the UHMWPE liner from the metal backing [18, 70, 123], and wear of the UHMWPE liner through to the metal backing [12, 103]. Theoretical results also show larger UHMWPE stresses in metal-backed glenoids than in all-polyethylene glenoids [57]. Also, a metal backing requires a thinner UHMWPE component than a corresponding all-polyethylene glenoid with the same overall component thickness. As a result of the poorer outcomes for some metal backed versus all-polyethylene components, some clinicians have abandoned use of metal-backed components [121].

Overall, good results are obtained by surgeons on both sides of each of these questions. Glenoid component use and choice of design, material, and fixation method of the glenoid component can be important. Of utmost importance, however, is the technical skill of the surgeon in assessing the joint, positioning the components correctly within the joint, and reconstructing and repairing the bony and soft-tissue structures.

## 9.8 FUTURE DIRECTIONS IN TOTAL SHOULDER ARTHROPLASTY

In spite of the success of Total Shoulder Arthroplasty, its shortcomings provide opportunities for improvement. Continuing investigation is focused on incremental improvement in areas of current strength as well as on decreasing or eliminating problem areas.

### 9.8.1 Design

Starting from three sizes of a monoblock humeral component without a glenoid component, TSA design has already evolved a great deal. Third-generation components with modularity, which includes head offsets and head-stem angulation, have been widely used to great success.

The issue of glenoid fixation, however, remains a concern. One approach that could prove successful in TSA is the creation of glenoid components featuring integrated bony ingrowth features. An example of this is the creation of a component with UHMWPE on the articular surface and a porous-structured metal layer on the fixation surface. One such porous-structured metal is Hedrocel (Zimmer Holdings, Inc.), an interconnected network of tantalum struts that can be combined with UHMWPE to create a monoblock component. Andreykiv and coworkers have studied such a concept in theoretical modeling of fixation with some positive results [124]. The potential for durable biological fixation of such a component may address current shortcomings of cemented and screw-based fixation.

### 9.8.2 Materials

Changes to the humeral head and to the glenoid liner are beginning to be made to address the potential for UHMWPE damage that exists in current TSA prosthesis systems. The standard humeral head material is a highly polished cobalt chromium alloy. Modular heads are also offered with titanium nitride-bearing surfaces as well as ceramic surfaces, but they are not widely used. The goal of these alternative surfaces to the standard cobalt chromium is to decrease the friction between the humeral and glenoid articular surfaces and thereby decrease the potential for damage to the UHMWPE. That addresses one side of the current wear couple.

On the other side of the wear couple is the UHMWPE bearing surface. With the exception of Hylamer (Chapter 19), conventional UHMWPE has been used and sterilization methods have been updated so that currently, all of the major manufacturers process the polyethylene in such a way as to reduce or inhibit oxidative degradation. These methods include gamma sterilization in nitrogen or other inert atmospheres, gas plasma and EtO sterilization, and storage in vacuum. Highly crosslinked UHMWPEs with increased abrasion resistance have been used in THA acetabular liners. Decreased mechanical properties of these compared with nonhighly crosslinked UHMWPEs, however, make them less attractive for use in shoulders where, as discussed earlier, the characteristics of the glenohumeral articulation are such that fatigue wear and damage causing pitting, delamination, and fracture are more prevalent. Enhancement or even elimination of UHMWPE, such as has been done for metal-on-metal and ceramic-on-ceramic hips, does not seem to be imminent for TSR components.

## 9.9 CONCLUSION

In general, the success rate of TSR exceeds 90% for a wide range of common shoulder problems. As in hip and knee arthroplasty, the aim of shoulder arthroplasty is to reduce pain, restore damaged anatomy, and restore function. As with hip and knee arthroplasty, there are continuing concerns with shoulder arthroplasty regarding component fixation and persisting instability of the joint.

The introduction of UHMWPE to shoulder arthroplasty via the addition of glenoid components has increased the applicability of the procedure and has been a great success, even though there remains a substantial subset of cases that are successfully addressed via hemiarthroplasty. The addition of UHMWPE has, however, brought along its associated issues. Glenoid fixation is the biggest of these and the continuing persistence of a variety of glenoid component designs and metal backings and fixation methods attests to the fact that a clearly superior answer has not yet been found. Wear and damage is the other issue specific to the UHMWPE glenoid, and, although it remains a relatively small percentage of all cases, the ramifications can be severe.

In spite of these issues, shoulder replacement with UHMWPE remains a successful procedure bringing lasting pain relief and restoration of function to tens of thousands of patients worldwide every year.

## 9.10 ACKNOWLEDGMENTS

Many thanks to Jon JP Warner of Massachusetts General Hospital and industry representatives Terry Armstrong, Jodelle Brosig, Masood Durkhshan, Monika Gibson, Jason Kirsch, Elaine Mattheus, Marly Moate, Brian Sauls, Kate Smith, Lori Stoneburner, and Melissa West for their assistance with figures of radiographs, anatomy, and current prostheses. Many thanks also to Steven Kurtz of Drexel University for his editorial assistance and to Avram Edidin of Drexel University for his discussions of historical information.

## REFERENCES

1. Hayes PR, Flatow EL. Total shoulder arthroplasty in the young patient. *Instr Course Lect* 2001;**50**:73–88.
2. Hill JM, Norris TR. Long-term results of total shoulder arthroplasty following bone-grafting of the glenoid. *J Bone Joint Surg Am* 2001;**83-A**(6):877–83.
3. NIS, *data on upper extremity arthroplasty from 1993 to 2005*; 2008.
4. Edwards TB, et al. The influence of rotator cuff disease on the results of shoulder arthroplasty for primary osteoarthritis: results of a multicenter study. *J Bone Joint Surg Am* 2002;**84-A**(12):2240–8.
5. Boileau P, et al. Shoulder arthroplasty for the treatment of the sequelae of fractures of the proximal humerus. *J Shoulder Elbow Surg* 2001;**10**(4):299–308.
6. Sperling JW, Cofield RH, Rowland CM. Minimum fifteen-year follow-up of Neer hemiarthroplasty and total shoulder arthroplasty in patients aged fifty years or younger. *J Shoulder Elbow Surg* 2004;**13**(6):604–13.
7. Rahme H, Jacobsen MB, Salomonsson B. The Swedish Elbow Arthroplasty Register and the Swedish Shoulder Arthroplasty Register: two new Swedish arthroplasty registers. *Acta Orthop Scand* 2001;**72**(2):107–12.
8. Neer CS II. Telephone conversation; 2003 June 19.
9. Jain NB, et al. Total arthroplasty versus hemiarthroplasty for glenohumeral osteoarthritis: role of provider volume. *J Shoulder Elbow Surg* 2005;**14**(4):361–7.
10. Keller J, et al. Glenoid replacement in total shoulder arthroplasty. *Orthopedics* 2006;**29**(3):221–6.
11. Arredondo J, Worland RL. Bipolar shoulder arthroplasty in patients with osteoarthritis: short-term clinical results and evaluation of birotational head motion. *J Shoulder Elbow Surg* 1999;**8**(5): 425–9.
12. Boileau P, et al. Cemented polyethylene versus uncemented metal-backed glenoid components in total shoulder arthroplasty: a prospective, double-blind, randomized study. *J Shoulder Elbow Surg* 2002;**11**(4):351–9.
13. Cheung EV, Sperling JW, Cofield RH. Polyethylene insert exchange for wear after total shoulder arthroplasty. *J Shoulder Elbow Surg* 2007;**16**(5):574–8.
14. Chin PY, et al. Complications of total shoulder arthroplasty: are they fewer or different? *J Shoulder Elbow Surg* 2006;**15**(1):19–22.
15. Clinton J, et al. Nonprosthetic glenoid arthroplasty with humeral hemiarthroplasty and total shoulder arthroplasty yield similar self-assessed outcomes in the management of comparable patients with glenohumeral arthritis. *J Shoulder Elbow Surg* 2007;**16**(5):534–8.
16. Diaz-Borjon E, et al. Shoulder replacement in end-stage rotator cuff tear arthropathy: 5- to 11-year follow-up analysis of the bi-polar shoulder prosthesis. *J Surg Orthop Adv* 2007;**16**(3):123–30.
17. Dines JS, et al. Outcomes analysis of revision total shoulder replacement. *J Bone Joint Surg Am* 2006;**88**(7):1494–500.
18. Godeneche A, et al. Prosthetic replacement in the treatment of osteoarthritis of the shoulder: early results of 268 cases. *J Shoulder Elbow Surg* 2002;**11**(1):11–18.

19. Goldberg BA, et al. The magnitude and durability of functional improvement after total shoulder arthroplasty for degenerative joint disease. *J Shoulder Elbow Surg* 2001;**10**(5):464–9.
20. Haines JF, et al. The results of arthroplasty in osteoarthritis of the shoulder. *J Bone Joint Surg Br* 2006;**88**(4):496–501.
21. Hasan SS, et al. Characteristics of unsatisfactory shoulder arthroplasties. *J Shoulder Elbow Surg* 2002;**11**(5):431–41.
22. Iannotti JP, Norris TR. Influence of preoperative factors on outcome of shoulder arthroplasty for glenohumeral osteoarthritis. *J Bone Joint Surg Am* 2003;**85-A**(2):251–8.
23. Neer CS II. Articular replacement for the humeral head. *J Bone Joint Surg Am* 1955;**37-A**(2):215–28.
24. Neer CS II. Replacement arthroplasty for glenohumeral osteoarthritis. *J Bone Joint Surg Am* 1974;**56**(1):1–13.
25. Neer CS II, Watson KC, Stanton FJ. Recent experience in total shoulder replacement. *J Bone Joint Surg Am* 1982;**64**(3):319–37.
26. Norris TR, Iannotti JP. Functional outcome after shoulder arthroplasty for primary osteoarthritis: a multicenter study. *J Shoulder Elbow Surg* 2002;**11**(2):130–5.
27. Rice RS, et al. Augmented glenoid component for bone deficiency in shoulder arthroplasty. *Clin Orthop Relat Res* 2008;**466**(3):579–83.
28. Sanchez-Sotelo J, et al. Radiographic assessment of uncemented humeral components in total shoulder arthroplasty. *J Arthroplasty* 2001;**16**:180–7.
29. Sojbjerg JO, et al. Late results of total shoulder replacement in patients with rheumatoid arthritis. *Clin Orthop Relat Res* 1999(366):39–45.
30. Sperling JW, et al. Total shoulder arthroplasty versus hemiarthroplasty for rheumatoid arthritis of the shoulder: results of 303 consecutive cases. *J Shoulder Elbow Surg* 2007;**16**(6):683–90.
31. Torchia ME, Cofield RH, Settergren CR. Total shoulder arthroplasty with the Neer prosthesis: long-term results. *J Shoulder Elbow Surg* 1997;**6**(6):495–505.
32. Worland RL, et al. Bipolar shoulder arthroplasty for rotator cuff arthropathy. *J Shoulder Elbow Surg* 1997;**6**(6):512–15.
33. Bufquin T, et al. Reverse shoulder arthroplasty for the treatment of three- and four-part fractures of the proximal humerus in the elderly: a prospective review of 43 cases with a short-term follow-up. *J Bone Joint Surg Br* 2007;**89**(4):516–20.
34. Guery J, et al. Reverse total shoulder arthroplasty. Survivorship analysis of eighty replacements followed for five to ten years. *J Bone Joint Surg Am* 2006;**88**(8):1742–7.
35. Katzer A, et al. Two-year results after exchange shoulder arthroplasty using inverse implants. *Orthopedics* 2004;**27**(11):1165–7.
36. Levy J, et al. The use of the reverse shoulder prosthesis for the treatment of failed hemiarthroplasty for proximal humeral fracture. *J Bone Joint Surg Am* 2007;**89**(2):292–300.
37. Matsen FA III, et al. The reverse total shoulder arthroplasty. *J Bone Joint Surg Am* 2007;**89**(3):660–7.
38. Rockwood Jr. CA The reverse total shoulder prosthesis. The new kid on the block. *J Bone Joint Surg Am* 2007;**89**(2):233–5.
39. Wall B, et al. Reverse total shoulder arthroplasty: a review of results according to etiology. *J Bone Joint Surg Am* 2007;**89**(7):1476–85.
40. Churchill RS. Elective shoulder arthroplasty in patients older than ninety years of age. *J Shoulder Elbow Surg* 2008;**17**(3):376–9.
41. Sperling JW, Cofield RH, Rowland CM. Neer hemiarthroplasty and Neer total shoulder arthroplasty in patients fifty years old or less. Long-term results. *J Bone Joint Surg Am* 1998;**80**(4):464–73.
42. Burroughs PL, et al. Shoulder arthroplasty in the young patient. *J Arthroplasty* 2003;**18**(6):792–8.
43. Levy JC, et al. Young patients with shoulder chondrolysis following arthroscopic shoulder surgery treated with total shoulder arthroplasty. *J Shoulder Elbow Surg* 2008;**17**(3):380–8.
44. Snyder G. Shoulder implant system. In: Jench L, Wilson J, editors *Clinical performance of skeletal prostheses*. London: Chapman & Hall; 1996.
45. Sharma S, Dreghorn CR. Registry of shoulder arthroplasty – the Scottish experience. *Ann R Coll Surg Engl* 2006;**88**(2):122–6.
46. Horak J, Nilsson BE. Epidemiology of fracture of the upper end of the humerus. *Clin Orthop Relat Res* 1975(112):250–3.
47. Palvaxen M, et al. Update in the epidemiology of proximal humeral fractures. *Clin Orthop Relat Res* 2006;**442**:87–92.
48. Lugli T. Artificial shoulder joint by Pean (1893): the facts of an exceptional intervention and the prosthetic method. *Clin Orthop Relat Res* 1978(133):215–18.
49. Rockwood Jr. CA The century in orthopedics-a century of shoulder arthroplasty innovations and discoveries. *Orthop Today* 2000 February. Slack, Inc.
50. Grammont PM, Baulot E. Delta shoulder prosthesis for rotator cuff rupture. *Orthopedics* 1993;**16**(1):65–8.
51. Boulahia A, et al. Early results of a reverse design prosthesis in the treatment of arthritis of the shoulder in elderly patients with a large rotator cuff tear. *Orthopedics* 2002;**25**(2):129–33.
52. *First North American 'reverse' shoulder replacement surgery performed at Toronto Western Hospital*. Canada Newswire; 2003 March 25.
53. Anglin C, Wyss UP, Pichora DR. Glenohumeral contact forces. *Proc Inst Mech Eng [H]* 2000;**214**(6):637–44.
54. Poppen NK, Walker PS. Forces at the glenohumeral joint in abduction. *Clin Orthop Relat Res* 1978(135):165–70.
55. Lacroix D, Murphy LA, Prendergast PJ. Three-dimensional finite element analysis of glenoid replacement prostheses: a comparison of keeled and pegged anchorage systems. *J Biomech Eng* 2000;**122**(4):430–6.
56. Lacroix D, Prendergast PJ. Stress analysis of glenoid component designs for shoulder arthroplasty. *Proc Inst Mech Eng [H]* 1997;**211**(6):467–74.
57. Swieszkowski W, Bednarz P, Prendergast PJ. Contact stresses in the glenoid component in total shoulder arthroplasty. *Proc Inst Mech Eng [H]* 2003;**217**(1):49–57.
58. Bartel DL, Bicknell VL, Wright TM. The effect of conformity, thickness, and material on stresses in ultra-high molecular weight components for total joint replacement. *J Bone Joint Surg Am* 1986;**68**(7):1041–51.
59. Karduna AR, et al. Total shoulder arthroplasty biomechanics: a study of the forces and strains at the glenoid component. *J Biomech Eng* 1998;**120**(1):92–9.
60. Warner JJ, et al. Effect of joint compression on inferior stability of the glenohumeral joint. *J Shoulder Elbow Surg* 1999;**8**(1):31–6.
61. Warner JJ, Warren RF. Quantitative determination of articular pressure in the human shoulder joint. *J Shoulder Elbow Surg* 2001;**10**(5):496–7.
62. Mahfouz M, et al. In vivo determination of the dynamics of normal, rotator cuff-deficient, total, and reverse replacement shoulders. *J Bone Joint Surg Am* 2005;**87**(Suppl. 2):107–13.
63. Carroll RM, et al. Conversion of painful hemiarthroplasty to total shoulder arthroplasty: long-term results. *J Shoulder Elbow Surg* 2004;**13**(6):599–603.
64. Churchill RS, et al. Humeral component modularity may not be an important factor in the outcome of shoulder arthroplasty for glenohumeral osteoarthritis. *Am J Orthop* 2005;**34**(4):173–6.

65. Mileti J, et al. Monoblock and modular total shoulder arthroplasty for osteoarthritis. *J Bone Joint Surg Br* 2005;**87**(4):496–500.
66. Guide to polyethylene in joint implants. *Orthop Today* 1999 October–November. Slack, Inc.
67. Frankle M, et al. The reverse shoulder prosthesis for glenohumeral arthritis associated with severe rotator cuff deficiency. a minimum two-year follow-up study of sixty patients surgical technique. *J Bone Joint Surg Am* 2006;**88**(Suppl. 1 Pt. 2):178–90.
68. Frankle M, et al. The reverse shoulder prosthesis for glenohumeral arthritis associated with severe rotator cuff deficiency. A minimum two-year follow-up study of sixty patients. *J Bone Joint Surg Am* 2005;**87**(8):1697–705.
69. Radnay CS, et al. Total shoulder replacement compared with humeral head replacement for the treatment of primary glenohumeral osteoarthritis: a systematic review. *J Shoulder Elbow Surg* 2007;**16**(4):396–402.
70. Antuna SA, et al. Glenoid revision surgery after total shoulder arthroplasty. *J Shoulder Elbow Surg* 2001;**10**(3):217–24.
71. Deshmukh AV, et al. Total shoulder arthroplasty: long-term survivorship, functional outcome, and quality of life. *J Shoulder Elbow Surg* 2005;**14**(5):471–9.
72. Noble JS, Bell RH. Failure of total shoulder arthroplasty: why does it occur? *Semin Arthroplasty* 1995;**6**(4):280–8.
73. Vanhove B, Beugnies A. Grammont's reverse shoulder prosthesis for rotator cuff arthropathy. A retrospective study of 32 cases. *Acta Orthop Belg* 2004;**70**(3):219–25.
74. Trail IA, Nuttall D. The results of shoulder arthroplasty in patients with rheumatoid arthritis. *J Bone Joint Surg Br* 2002;**84**(8):1121–5.
75. Mackay DC, Hudson B, Williams JR. Which primary shoulder and elbow replacement? A review of the results of prostheses available in the UK. *Ann R Coll Surg Engl* 2001;**83**(4):258–65.
76. Matsen FA, III Bicknell RT, Lippitt SB. Shoulder arthroplasty: the socket perspective. *J Shoulder Elbow Surg* 2007;**16**(5 Suppl.): S241–7.
77. Farmer KW, et al. Shoulder arthroplasty versus hip and knee arthroplasties: a comparison of outcomes. *Clin Orthop Relat Res* 2007;**455**:183–9.
78. McFarland EG, et al. The reverse shoulder prosthesis: a review of imaging features and complications. *Skeletal Radiol* 2006;**35**(7):488–96.
79. Wirth Jr. MA, Rockwood CA Complications of total shoulder-replacement arthroplasty. *J Bone Joint Surg Am* 1996;**78**(4):603–16.
80. Bohsali Jr. KI, Wirth MA, Rockwood CA Complications of total shoulder arthroplasty. *J Bone Joint Surg Am* 2006;**88**(10):2279–92.
81. Buckingham BP, et al. Patient functional self-assessment in late glenoid component failure at three to eleven years after total shoulder arthroplasty. *J Shoulder Elbow Surg* 2005;**14**(4):368–74.
82. Chebli C, et al. Factors affecting fixation of the glenoid component of a reverse total shoulder prosthesis. *J Shoulder Elbow Surg* 2008;**17**(2):323–7.
83. Cheung EV, Sperling JW, Cofield RH. Revision shoulder arthroplasty for glenoid component loosening. *J Shoulder Elbow Surg* 2008;**17**(3):371–5.
84. Deutsch A, et al. Clinical results of revision shoulder arthroplasty for glenoid component loosening. *J Shoulder Elbow Surg* 2007;**16**(6):706–16.
85. Elhassan B, et al. Glenoid reconstruction in revision shoulder arthroplasty. *Clin Orthop Relat Res* 2008;**466**(3):599–607.
86. Franta AK, et al. The complex characteristics of 282 unsatisfactory shoulder arthroplasties. *J Shoulder Elbow Surg* 2007;**16**(5):555–62.
87. Gagey O, Pourjamasb B, Court C. [Revision arthroplasty of the shoulder for painful glenoid loosening: a series of 14 cases with acromial prostheses reviewed at four year follow up]. *Rev Chir Orthop Reparatrice Appar Mot* 2001;**87**(3):221–8.
88. Gartsman GM, Hasan SS. What's new in shoulder and elbow surgery. *J Bone Joint Surg Am* 2006;**88**(1):230–43.
89. Klepps S, et al. Incidence of early radiolucent glenoid lines in patients having total shoulder replacements. *Clin Orthop Relat Res* 2005(435):118–25.
90. Matsen FA III, et al. Glenoid component failure in total shoulder arthroplasty. *J Bone Joint Surg Am* 2008;**90**(4):885–96.
91. Nagels J, et al. Patterns of loosening of the glenoid component. *J Bone Joint Surg Br* 2002;**84**(1):83–7.
92. Rosenberg N, et al. Improvements in survival of the uncemented Nottingham total shoulder prosthesis: a prospective comparative study. *BMC Musculoskelet Disord* 2007;**8**:76.
93. Scalise JJ, Iannotti JP. Bone grafting severe glenoid defects in revision shoulder arthroplasty. *Clin Orthop Relat Res* 2008; **466**(1):139–45.
94. Skirving AP. Total shoulder arthroplasty—current problems and possible solutions. *J Orthop Sci* 1999;**4**(1):42–53.
95. Mileti J, et al. Radiographic analysis of polyethylene glenoid components using modern cementing techniques. *J Shoulder Elbow Surg* 2004;**13**(5):492–8.
96. Nuttall D, Haines JF, Trail II. A study of the micromovement of pegged and keeled glenoid components compared using radiostereometric analysis. *J Shoulder Elbow Surg* 2007;**16**(3 Suppl.):S65–70.
97. Rahme H, Mattsson P, Larsson S. Stability of cemented all-polyethylene keeled glenoid components. A radiostereometric study with a two-year follow-up. *J Bone Joint Surg Br* 2004;**86**(6):856–60.
98. Gerber A, Ghalambor N, Warner JJ. Instability of shoulder arthroplasty: balancing mobility and stability. *Orthop Clin North Am* 2001;**32**(4):661–70.
99. Warren RF, Coleman SH, Dines JS. Instability after arthroplasty: the shoulder. *J Arthroplasty* 2002;**17**(4 Suppl. 1):28–31.
100. Zilber S, et al. Total shoulder arthroplasty using the superior approach: influence on glenoid loosening and superior migration in the long-term follow-up after Neer II prosthesis installation. *J Shoulder Elbow Surg* 2008;**17**(4):554–63.
101. Iannotti JP, Williams GR. Total shoulder arthroplasty. Factors influencing prosthetic design. *Orthop Clin North Am* 1998;**29**(3):377–91.
102. Ibarra C, Dines DM, McLaughlin JA. Glenoid replacement in total shoulder arthroplasty. *Orthop Clin North Am* 1998;**29**(3):403–13.
103. Gunther SB, et al. Retrieved glenoid components: a classification system for surface damage analysis. *J Arthroplasty* 2002;**17**(1):95–100.
104. Mabrey JD, et al. Standardized analysis of UHMWPE wear particles from failed total joint arthroplasties. *J Biomed Mater Res* 2002;**63**(5):475–83.
105. Scarlat III MM, Matsen FA. Observations on retrieved polyethylene glenoid components. *J Arthroplasty* 2001;**16**(6):795–801.
106. Wirth MA, et al. Isolation and characterization of polyethylene wear debris associated with osteolysis following total shoulder arthroplasty. *J Bone Joint Surg Am* 1999;**81**(1):29–37.
107. Weldon III EJ, et al. Intrinsic stability of unused and retrieved polyethylene glenoid components. *J Shoulder Elbow Surg* 2001;**10**(5):474–81.

108. Braman JP, et al. Alterations in surface geometry in retrieved polyethylene glenoid component. *J Orthop Res* 2006;**24**(6):1249–60.
109. Levy JC, et al. Use of the reverse shoulder prosthesis for the treatment of failed hemiarthroplasty in patients with glenohumeral arthritis and rotator cuff deficiency. *J Bone Joint Surg Br* 2007;**89**(2):189–95.
110. Nyffeler RW, et al. Analysis of a retrieved delta III total shoulder prosthesis. *J Bone Joint Surg Br* 2004;**86**(8):1187–91.
111. Rockwood Jr. CA, Wirth MA Observation on retrieved Hylamer glenoids in shoulder arthroplasty: problems associated with sterilization by gamma irradiation in air. *J Shoulder Elbow Surg* 2001;**11**(2):191–7.
112. Jain N, et al. The relationship between surgeon and hospital volume and outcomes for shoulder arthroplasty. *J Bone Joint Surg Am* 2004;**86-A**(3):496–505.
113. Jain NB, et al. Trends in the epidemiology of total shoulder arthroplasty in the United States from 1990–2000. *Arthritis Rheum* 2006;**55**(4):591–7.
114. Lyman S, et al. The association between hospital volume and total shoulder arthroplasty outcomes. *Clin Orthop Relat Res* 2005(432):132–7.
115. Bishop JY, Flatow EL. Humeral head replacement versus total shoulder arthroplasty: clinical outcomes—a review. *J Shoulder Elbow Surg* 2005;**14**(1 Suppl. S):141S–146S.
116. Pfahler M, et al. Hemiarthroplasty versus total shoulder prosthesis: results of cemented glenoid components. *J Shoulder Elbow Surg* 2006;**15**(2):154–163.
117. Rickert M, Loew M. [Hemiarthroplasty or total shoulder replacement in glenohumeral osteoarthritis?]. *Orthopade* 2007;**36**(11):1013–1016.
118. Edwards TB, et al. A comparison of hemiarthroplasty and total shoulder arthroplasty in the treatment of primary glenohumeral osteoarthritis: results of a multicenter study. *J Shoulder Elbow Surg* 2003;**12**(3):207–213.
119. Lo IK, et al. Quality-of-life outcome following hemiarthroplasty or total shoulder arthroplasty in patients with osteoarthritis. A prospective, randomized trial. *J Bone Joint Surg Am* 2005;**87**(10):2178–2185.
120. Marx RG. Quality of life after shoulder arthroplasty-total shoulder arthroplasty versus hemiarthroplasty. *Nat Clin Pract Rheumatol* 2006;**2**(5):250–251.
121. Williams GR, Abboud JA. Total shoulder arthroplasty: glenoid component design. *J Shoulder Elbow Surg* 2005;**14**(1 Suppl. S):122S–128S.
122. Churchill RS, et al. Glenoid cementing may generate sufficient heat to endanger the surrounding bone. *Clin Orthop Relat Res* 2004(419):76–79.
123. Levy O, Copeland SA. Rotational dissociation of glenoid components in a total shoulder prosthesis: an indication that sagittal torque forces may be important in glenoid component design. *J Shoulder Elbow Surg* 2001;**10**(2):197.
124. Andreykiv A, et al. Bone ingrowth simulation for a concept glenoid component design. *J Biomech* 2005;**38**(5):1023–1033.

Chapter 10

# The Clinical Performance of UHMWPE in Elbow Replacements

Judd S. Day, PhD



## 10.1 INTRODUCTION

Because of its important role in manipulating the hand, loss of elbow function can impair many activities of daily living leading to diminished quality of life. Although total elbow arthroplasty has been successfully used for the treatment of relatively inactive patients with rheumatoid arthritis, implant survival has been limited when used for treatment of osteoarthritis or trauma. Like the knee, the axis of rotation of the elbow shifts in both position and orientation during flexion. This makes the use of a fixed-hinge device unacceptable for total elbow arthroplasty (TEA). Early fixed-hinge TEA systems, although inherently stable, suffered from early loosening because of the increased forces induced at the cement interface because of this unnatural constraint. Compared to the knee, stability of the natural elbow depends to a greater degree on the conforming bony anatomy and to a lesser degree on the soft tissue structures. For this reason, it has been difficult to design a TEA system that relies purely on joint resurfacing for patients with compromised soft tissue function. As will be discussed in this chapter, modern implants are generally based on two types of designs: resurfacing-type implants or linked-hinge devices that depend on a "sloppy hinge" design to allow for some degree of joint laxity. Both types of design rely on a metal-on-UHMWPE bearing couple for function. As TEA designs have improved, so has the survival of the devices. This has led to use of TEA for increased indications rather than as a salvage procedure. According to the National Inpatient Survey, the number of primary TEA procedures performed has increased 2.5-fold between 1993 and 2006

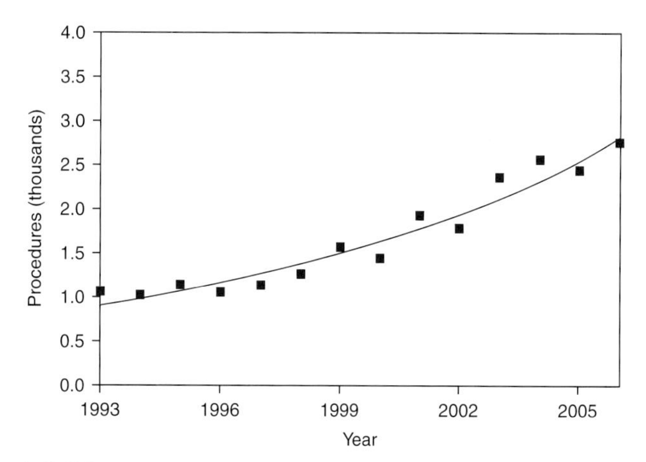


FIGURE 10.1 Primary arthroplasty procedures in the United States between 1993 and 2006.

(Figure 10.1). However, as the longevity of TEA has increased, UHMWPE wear and osteolysis has become an increased concern. This chapter will review the basic anatomy of the elbow joint, address contemporary and historic designs, as well as design-related issues for TEA.

## 10.2 ANATOMY OF THE ELBOW

### 10.2.1 Osteoarticular Anatomy

The elbow acts as a fulcrum for the forearm during lifting and together with the shoulder provides the ability to

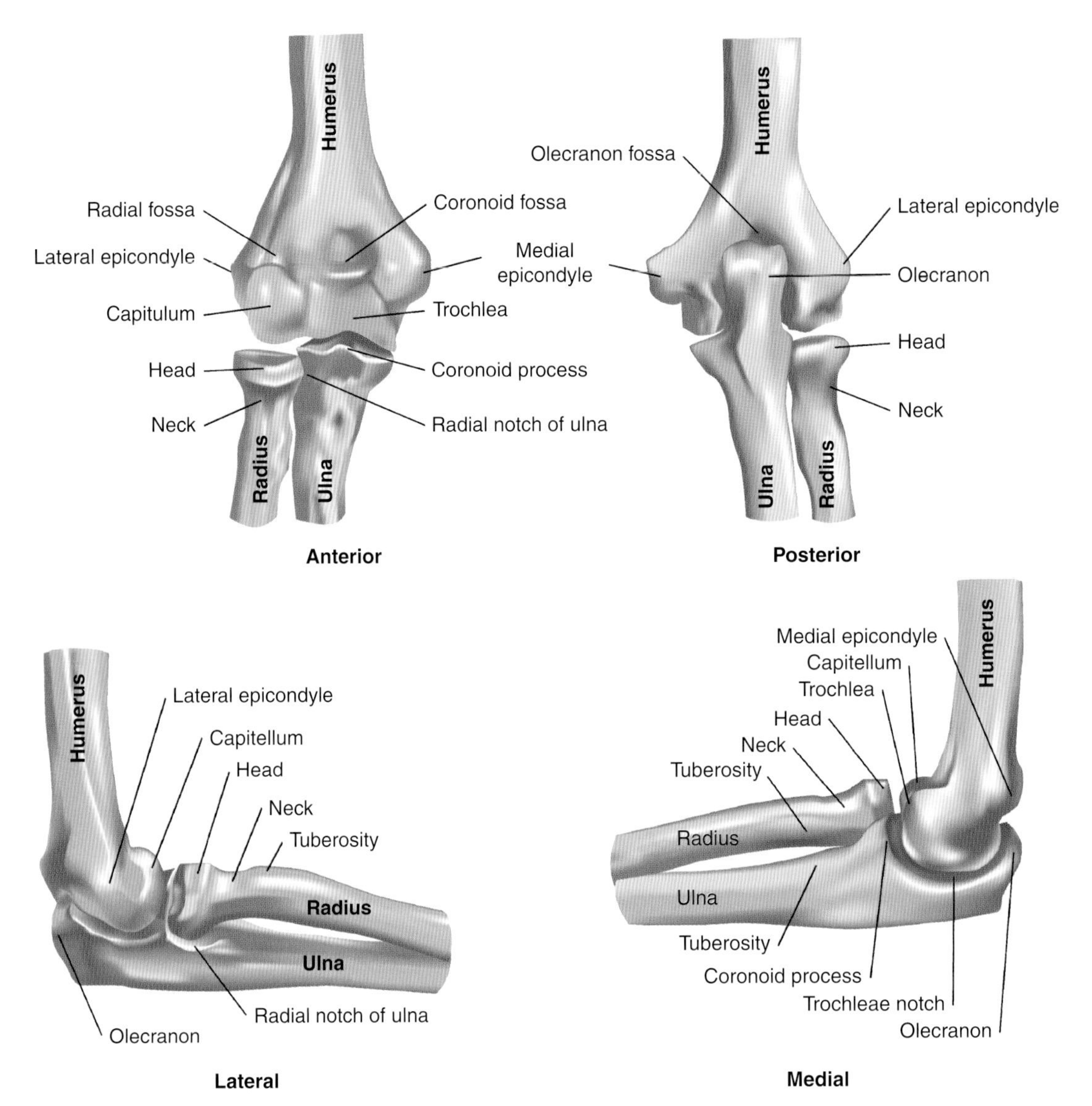


**FIGURE 10.2** Osteoarticular anatomy of the elbow.

locate the hand in space. Compared to many other joints in the human body, the elbow is highly conforming and inherently stable. The elbow joint complex is composed of three bones, the distal humerus, proximal radius, and proximal ulna. There are three articulating surfaces: the hingelike humeroulnar joint, which provides primary stability and mobility during elbow flexion/extension; the humeroradial joint, which allows gliding motion between the capitellum and the radial head; and the radioulnar joint, a pivot joint that enables pronation (palm facing down) and supination (palm facing up) of the wrist (Figure 10.2).

The congruent humeroulnar joint provides much of the stability of the elbow. The spool-shaped trochlea, located on the distal humerus, articulates against the greater sigmoid notch in a tongue-and-groove-like manner, such that medial and lateral gliding is constrained [1]. The trochlea is covered by articular cartilage over an arch of 300 degrees [2]. The medial ridge of the trochlea is more prominent than the lateral ridge, resulting in a 6 to 8 degree valgus tilt with respect to the axis of the humerus [3]. The articular condyles are also rotated 30 degrees anterior to the humeral axis [2]. In flexion the coronoid process of the ulna locks into the coronoid fossa of the distal humerus, further enhancing joint stability. In extension, stability is enhanced by the interlocking fit between the tip of the olecranon and the olecranon fossa [4].

Stability of the elbow is further enhanced by articulation of the concave radial head against the hemispherical capitellum and tracking of the rim of the radial head against the trochleocapitellar groove. The radial head provides resistance to valgus loading of the elbow. Cartilage covers the entire concave surface of the head as well as an arc of approximately 280 degrees of the rim where it bears against the ulna [3]. The radial head also contributes to stability, and radial shortening can result in varus/valgus laxity and excessive ulnar rotation [1]. However, contact

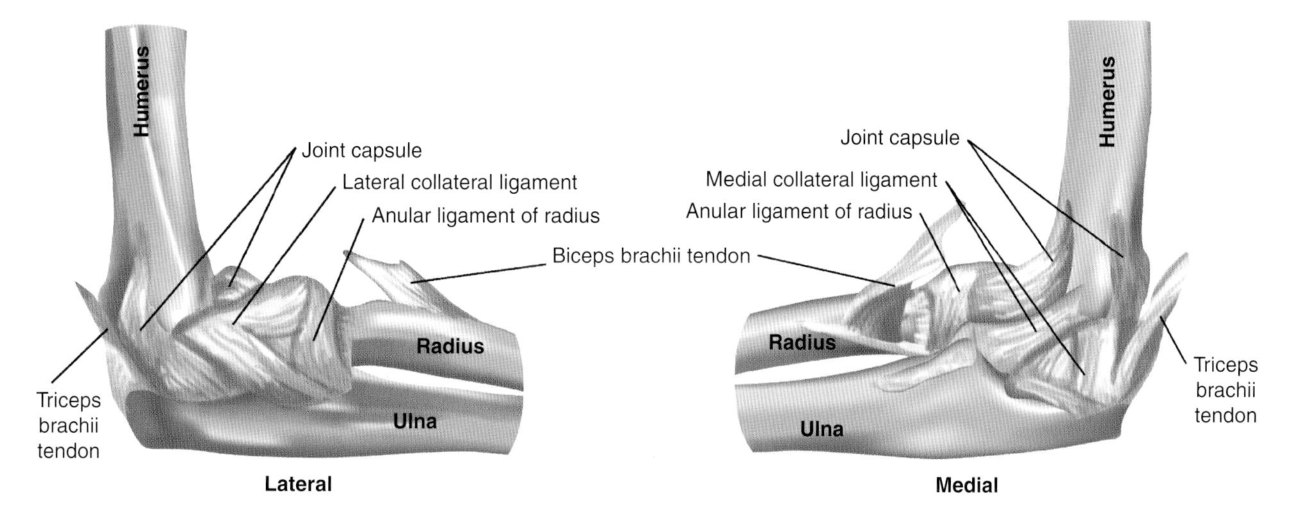


**FIGURE 10.3** Soft tissue anatomy of the elbow.

at the humeroradial joint varies during flexion/extension and forearm rotation. During flexion the radial head moves proximally, increasing contact with the capitellum. During supination humeroradial contact is reduced [1]. It has also been noted that the radial head is not perfectly circular and is variably offset from the axis of the radial neck. This could have important implications for the design of radial head implants [4].

## 10.2.2 Soft Tissue Anatomy

Static constraint of the elbow is the result of both bony structure and capsuloligamentous stabilizers. The soft tissue stabilizers include the medial and lateral collateral ligament complexes, the joint capsule, the annular ligament, and the interosseous membrane (Figure 10.3).

The joint capsule surrounds all three articulations of the elbow joint and is thickened medially and laterally to form the collateral ligament complexes [2]. The joint capsule is loose anteriorly and especially posterior to allow full flexion and extension of the elbow [1]. It is most lax in 80 degrees of elbow flexion. Transverse and obliquely directed bands in the anterior capsule provide significant stability when taut in extension [4].

The medial collateral ligament complex consists of three components: the anterior bundle, posterior bundle, and transverse ligament. The anterior bundle is the main stabilizer against valgus torques of the elbow [2]. The anterior bundle provides maximal constraint when the elbow is fully extended. The posterior bundle is taut, providing constraint when the elbow is flexed. The transverse bundle is not believed to contribute significantly to joint stability [1].

The lateral collateral ligament complex also consists of four components: the lateral ulnar collateral ligament, the fan-shaped radial collateral ligament, the accessory collateral ligament, and the annular ligament. This complex restrains the elbow from varus torques. The lateral ulnar collateral ligament provides both varus and posterolateral stability [3]. It is most taut in extreme flexion, but because its origin is near to the flexion axis of rotation, it provides stability throughout the range of motion [4]. The annular ligament holds the radial head against the radial notch of the ulna and prevents subluxation during traction on the forearm [1]. The radial head is further stabilized by the radial collateral ligament [3]. Both the medial and lateral collateral ligament complexes limit medial and lateral instability of the ulna on the humerus.

The interosseous membrane is a thin fibrous sheet of connective tissue that connects the radius to the ulna along their length. Although the interosseous membrane is primarily lax with the elbow in distraction and the forearm pronated (i.e., carrying an object), it becomes taut when a pushing force is applied across the wrist at the distal radius (i.e., pushing up from a chair). This results in transfer of force to the ulna and load sharing at the elbow [1] (Figure 10.4).

## 10.2.3 Muscular Anatomy

While a number of muscles crossing the elbow are involved with motion of the wrist and fingers, relatively few function to move the elbow joint. The biceps brachii, brachialis, and brachioradialis are the primary elbow flexors. The tricpes brachii is the primary elbow extensor. While the anconeus is thought to also contribute to elbow extension, it is also likely to provide dynamic constraint to varus and posterolateral instability of the elbow [4]. Pronation is powered by the pronator teres, pronator quadratus, and brachioradialis (from supinated position). Supination is the result of biceps brachii and supinator, as well as brachioradialis (from pronated position) (Figure 10.5). It is important to note that the muscles crossing the elbow can act as dynamic stabilizers, compressing the joint surfaces and enhancing the geometric stability of its conforming surfaces.

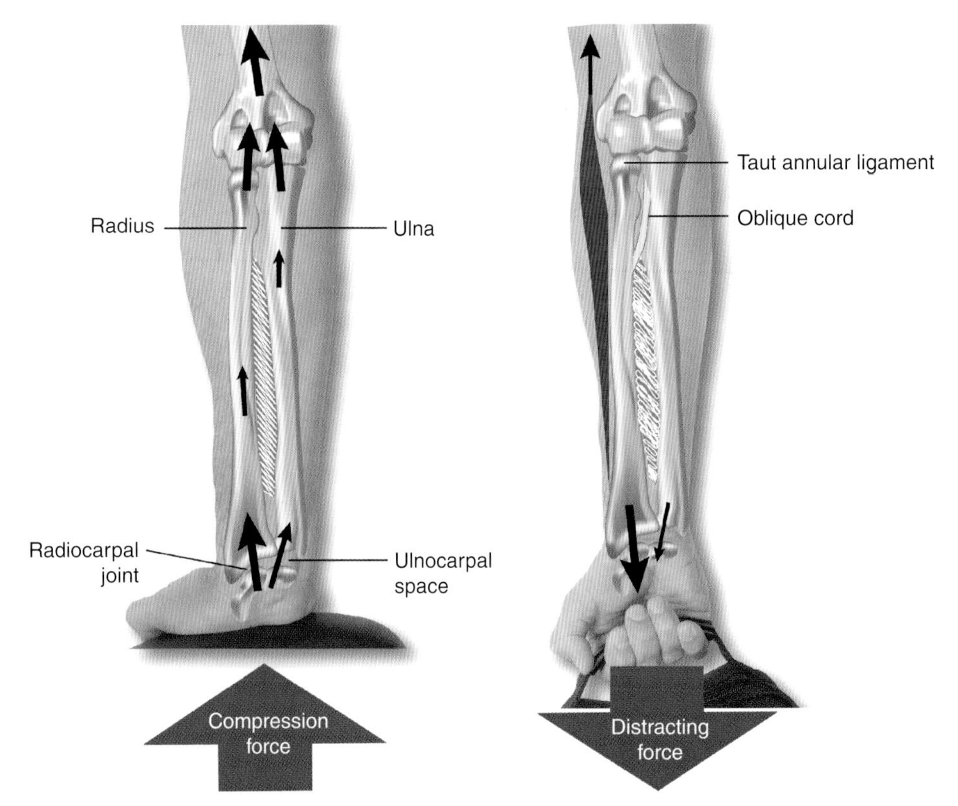


FIGURE 10.4 Role of the interosseus membrane.

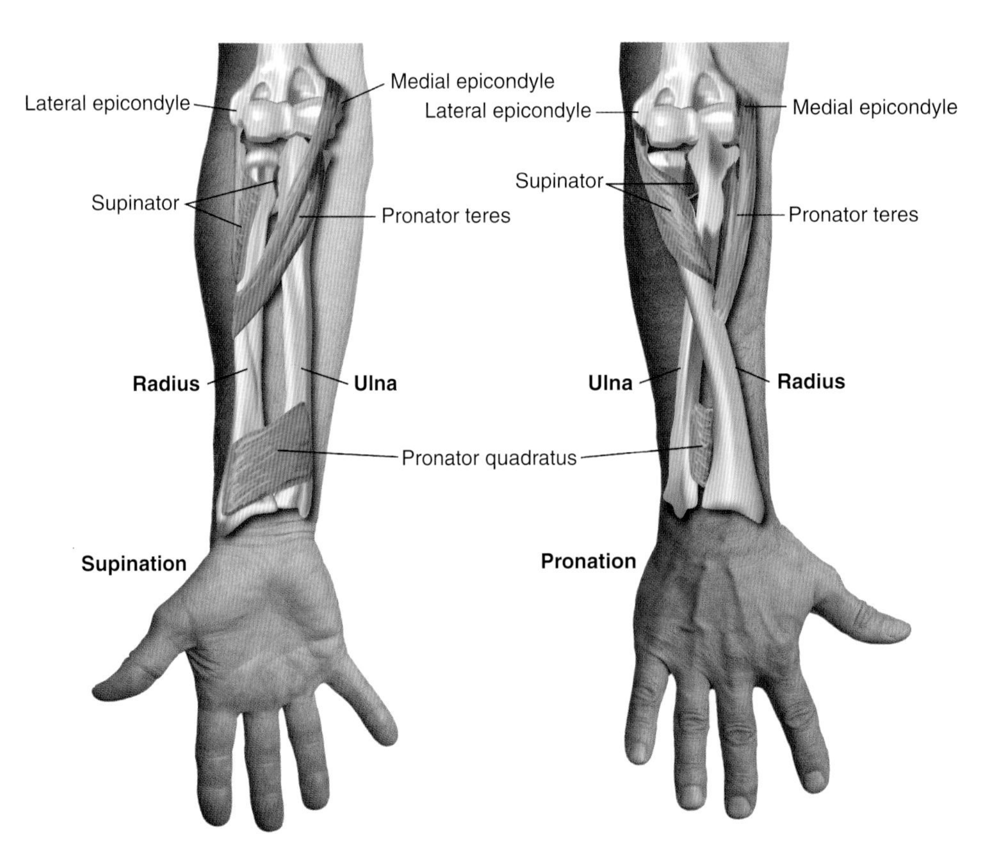


FIGURE 10.5 Forearm pronator and supinators.

## 10.3 ELBOW BIOMECHANICS

### 10.3.1 Kinematics

The humeroulnar and humeroradial joints provide flexion and extension of the elbow. The normal range of motion in flexion is from slight hyperextension to about 140 degrees of flexion. The radio-capitellar and proximal ulnar joints allow around 90 degrees of supination and 80 degrees of pronation. It has been reported that most activities of daily living can be accomplished with an arc of 30 to 130 degrees of flexion and 100 degrees of forearm rotation (50 degrees of pronation to 50 degrees of supination). While it is possible to compensate for loss of forearm pronation by shoulder abduction, there are no effective mechanisms to replace supination [3].

The flexion axis of the elbow joint is not constant throughout the arc of motion, but is instead considered to be described as a "sloppy hinge." The axis runs approximately from the center of the radial head to the center of the joint surface of the proximal ulna. The axis of rotation is approximately 3 to 5 degrees internally rotated relative to the plane of the medial and lateral epicondyles and is oriented in 4 to 8 degrees valgus with respect to the long axis of the humerus. It has been demonstrated that the orientation of the axis of rotation varies both throughout flexion and depending on forearm pronation/supination [4]. The axis of rotation in pronation/supination runs from the center of the radial head through the center of the distal ulna. This axis shifts slightly ulnar and volar (toward the palm) during supination and radial and dorsal during pronation. The radius also moves proximally with pronation of the forearm and distally with supination [3]. The ovoid shape of the radial head is thought to contribute to this camlike effect.

When the elbow is fully extended, the axes of the humerus and ulna are not aligned. The deviation of these axes from a straight line is called the carrying angle. The carrying angle averages 11 to 14 degrees in males and 13 to 16 degrees in females [2].

### 10.3.2 Joint Loading

Transmission of loads through the elbow joint to the humerus occurs mainly in compression at the radial head and coronoid process of the ulna. Peak forces occur between 0 and 30 degrees of flexion. At these joint angles there is a lack of mechanical advantage for the elbow flexors, thus necessitating higher forces [5]. Because of the relatively short moment arm of the biceps, joint reaction forces at the elbow are approximately eight to nine times the magnitude of a vertical force applied at the hand [6]. Resultant joint forces of 0.3 to 0.5 times body weight are routinely encountered during normal daily activities, and forces of approximately three times body weight are reported for strenuous activities [7, 8]. The direction of the joint compressive forces change during flexion from an anterior direction at full extension to a more posterior direction when the elbow is flexed (Figure 10.6). Mineralization and density of the subchondral bone, an indication of highly loaded areas, is greater on the anterior capitellum and distal and anteriorly on the trochlea [9]. In the ulna, mineralization is greater in the proximal and distal regions of the trochlear notch [10].

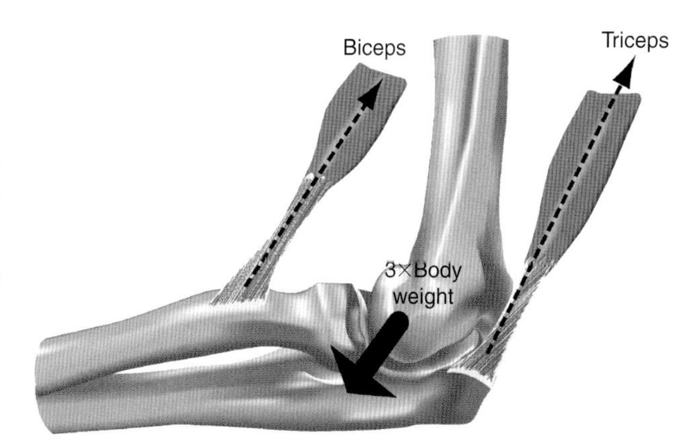


FIGURE 10.6 Because of the poor mechanical advantage of the biceps, loading of the joint reaction forces during strenuous activities can reach three times body weight.

Load sharing between the humeroulnar joint and the humeroradial joint varies with both varus/valgus orientation and with forearm pronation/supination. The proportion of force transmitted by the radial head increases as the forearm rotates from supination to pronation [5, 11]. In valgus alignment, contact of the radial head is increased at the capitellum, and in varus alignment the radial head is moved away from the capitellum and loading is decreased [12].

## 10.4 IMPLANT DESIGN

### 10.4.1 Historical Context

The history of elbow arthroplasty has been reviewed by Coonrad, who defined four eras [13]. The first era, from 1885 to 1947, was dominated by procedures such as joint resection, interposition, and anatomic arthroplasty. In the second era, from 1947 to 1970, surgeons experimented with hemireplacements made out of metal, acrylic, or silicone as well as hinged total joint arthroplasties. These straight-hinged arthroplasties did not have an anatomic center of rotation and had limited success clinically.

The third era of elbow arthroplasty design began in 1972 when Dee described a metal-on-metal constrained hinge design fixed with bone cement [14]. This was a cobalt chrome intramedullary hinged joint that articulated by an axis pin. The device was fit into the distal humerus and proximal radius (Figure 10.7). This device was reported to have excellent short-term results and was followed by other

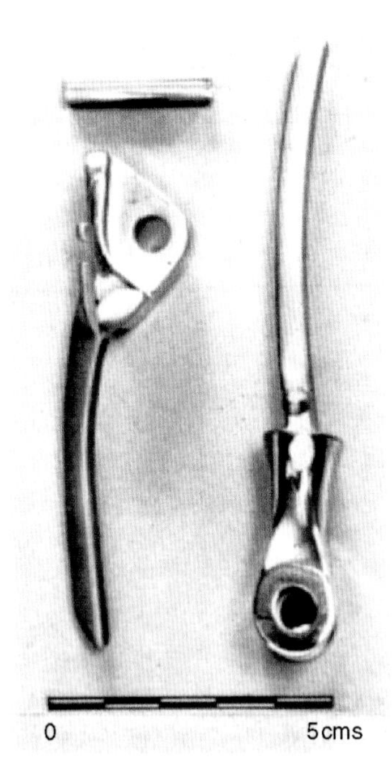


**FIGURE 10.7** Constrained hinge device by Dee. Reproduced with permission and copyright © of the British Editorial Society of Bone and Joint Surgery [14].

similar devices (McKee, Stanmore, Shiers, GSB, Mazas, Souter, Silva, Mayo, Coonrad). However, constrained devices had high rates of failure after 2 to 3 years primarily because of prosthetic loosening. The uniaxial fixed hinge of these devices was not able to reproduce the small amount of rotational and translation movement of the natural elbow during flexion. This resulted in high stresses at the interface between the stem of the component and the bone cement. Loosening was usually noted first at the humeral component, followed by the ulnar component [15].

## 10.4.2 Contemporary Designs

The modern era of elbow arthroplasty began in about 1975. With increased knowledge of elbow biomechanics, it was realized that the elbow did not rotate about a fixed hinge and that out-of-plane motion does occur during flexion and extension. Elbow designs must either reproduce or compensate for not reproducing the anatomic arc of motion and valgus carrying angle [16]. Over constraint and/or misalignment of the anatomic axis of motion results in loading of the cement mantle rather than the soft tissues. This results in loosening of the components and increased UHMWPE wear. However, when insufficient constraint is provided, either by the soft tissues or the device components, the resulting unstable joint is more likely to dislocate. Therefore, design of elbow arthroplasty systems requires a delicate balance between providing constraint through device design and the need for soft tissue balancing.

Modern implants are generally based on two design principles: unlinked-resurfacing-type implants or linked-hinge-type devices. In resurfacing type designs the joint surfaces are replaced by a metal-on-UHMWPE articulation that resembles (to varying extents) the original anatomy. While a degree of constraint can be achieved by increasing the conformity of the articulating surfaces, these devices normally depend on the integrity of the surrounding soft tissues for stability. Some historical examples of unlinked devices include the Capitellocondylar, the London design, the Stevens-Street design, the Kudo, and the Sorbie-Questor designs [13, 17–20]. Linked devices typically articulate through a hinge-type articulating surface. These devices are referred to as "sloppy hinges" because they are designed to allow a certain amount of laxity at the hinge and are therefore semiconstrained. When the degree of laxity of the device exceeds that of the soft tissues, forces are transferred through the soft tissues. However, the devices are inherently stable, even in an otherwise unstable elbow because of the physical linkage of the ulnar and humeral components. Historical examples of linked hinge-type devices include the Mayo Clinic design, Pritchard-Walker, the Coonrad/Coonrad-Morrey, Volz, Schlein, and the Triaxial design [13, 21–23]. In addition to resurfacing-type and semi-constrained linked devices, new hybrid designs have recently entered the market that can be implanted in a linked or unlinked configuration. A number of contemporary and historical designs will be reviewed in the following section.

The radial head plays an important role in resisting valgus and axial forces in the intact elbow and reducing the load on the humeroulnar joint. However, integrating the complex anatomy of the radial head into a TEA design has proved difficult. Although radial head replacement has been used successfully, TEA designs often rely on replacing the humeroulnar articulation and excise the radial head. Replacement of the humeroradial articulation is possible in designs like the Latitude and Sorbie-Questor [24]. Radial head replacement was possible with early versions of the Capitellocondylar elbow but was discontinued because of concern that the increased constraint applied by the radial head would contribute to loosening of the components. This proved unfounded, however, and the small number of elbows that were implanted with radial head components had reasonable survival rates [25].

Humeral loosening had been noted as a common complication of early, constrained designs [13]. While the move to semiconstrained and resurfacing designs led to a reduction in the incidence of humeral loosening, various design strategies have been employed to further ensure stability of the humeral component. Intramedullary stems were added to resurfacing-type designs, such as the Kudo, and were lengthened for the Souter-Strathclyde [26]. An anterior flange was added to the Coonrad-Morrey to resist posterior displacement and torsional stresses at the stem cement interface. A bone graft is interposed between

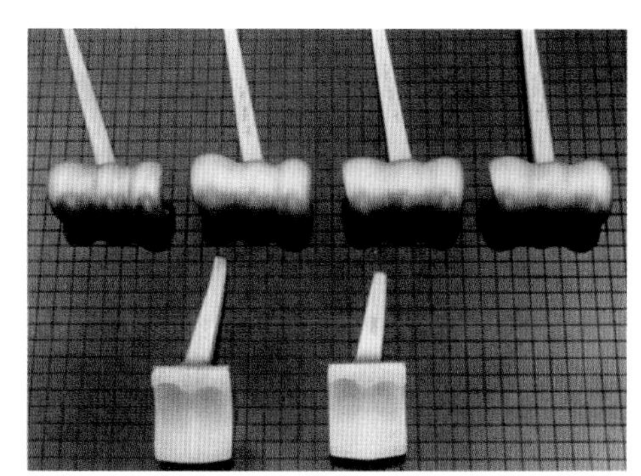
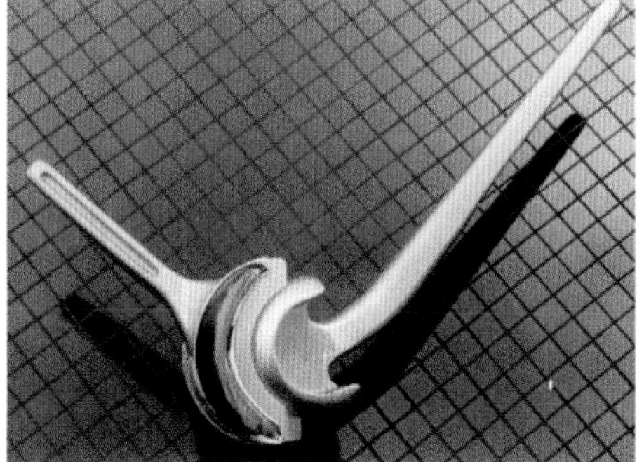

**FIGURE 10.8** Capitellocondylar Elbow. Reprinted with permission from *The Journal of Bone and Joint Surgery, Inc.* [33].

the flange and anterior humerus to ensure load transfer between the flange and the humerus. Although the reduction in load transfer through the cement interface has not yet been fully quantified, clinical studies indicate that the bone graft does not atrophy in the majority of cases, thus indicating that it is loaded [27]. The use of an anterior flange has also been incorporated into the Tornier Latitude, Biomet Discovery, and the Biomet Huene.

A number of influential historic and contemporary elbow designs are described in the following sections.

#### *10.4.2.1 Capitellocondylar Prosthesis (Johnson & Johnson Orthopaedics Inc., New Brunswick, New Jersey, USA)*

The Capitellocondylar Prosthesis, developed with Dr. F. J. Ewald, was first implanted in 1974 [27]. This unlinked and minimally constrained device was designed to anatomically resurface the articulating surfaces of the elbow with minimal bone resection. A cobalt chrome humeral component with an intramedullary stem is used. The humeral component has a semicircular shaped surface profile for articulation against an UHMWPE bearing surface on the ulnar component. The humeral surface has two shallow grooves, one for the trochlea and the other between the trochlea and capitellum. A ridge on the ulnar component articulates with the trochlear groove, similar to a natural elbow (Figure 10.8). The ulnar component was composed completely of UHMWPE from 1974 to 1977, after which a metal stem and backing was added to prevent breakage of the ulnar fixation stem [28]. Both components are intended for cemented use. This device provides little intrinsic constraint, and therefore joint laxity is higher than the intact elbow, even under joint compression. The device, therefore, depends almost completely on the soft tissue and muscles for stability. Although it was originally possible to use this device with a radial head replacement to improve joint stability, this option was discontinued due to early radiographic loosening of the radial head. However, the long-term results for this small group of patients did not indicate a high rate of loosening [24]. While the Capitellocondylar Prosthesis had a very low rate of clinical loosening, recurrent dislocation, due to its lack of inherent stability, was a common complication [29–33].

#### *10.4.2.2 Souter-Strathclyde Total Elbow (Stryker-Howmedica-Osteonics, Limerick, Ireland)*

The Souter-Strathclyde total elbow was designed in 1973 and used clinically beginning in 1977. It was marketed initially by Zimmer (London, England) and later by Stryker-Howmedica-Osteonics (Limerick, Ireland). This cemented device consists of a cobalt chrome alloy humeral component modeled on the anatomy of the trochlear surface and a keeled all UHMWPE ulnar component modeled on the surface of the olecranon (Figure 10.9). Metal-backed versions of the ulnar component are also available. Its anatomical surface design results in an 8 degree valgus angulation when fully extended and a 6 degree varus angulation in maximum flexion. Although the Souter-Strathclyde device is unlinked, it can be considered to be semiconstrained because of the high degree of conformity between the ulnar and humeral bearing surfaces. Schneeberger noted that this device was more constrained than other designs, with a functional varus/valgus laxity of 6.5 degrees [34]. This device has been associated with a high incidence of loosening, especially of the humeral component [35–38]. It is believed that the relatively high degree of constraint induced by the surface geometry leads to increased transmission of force directly to the cement-stem interface. It was also noted that this device is sensitive to surgical placement and that the surgical technique commonly resulted in lateralization of the device [4]. Longer stemmed versions of the humeral component and metal-backed ulnar components have been associated with a lower rate of loosening [35, 36, 39–41].

#### *10.4.2.3 Coonrad-Morrey Total Elbow Arthroplasty (Zimmer, Warsaw, Indiana, USA)*

The Coonrad/Coonrad-Morrey total elbow arthroplasty was initially designed in 1969 as a low friction metal on

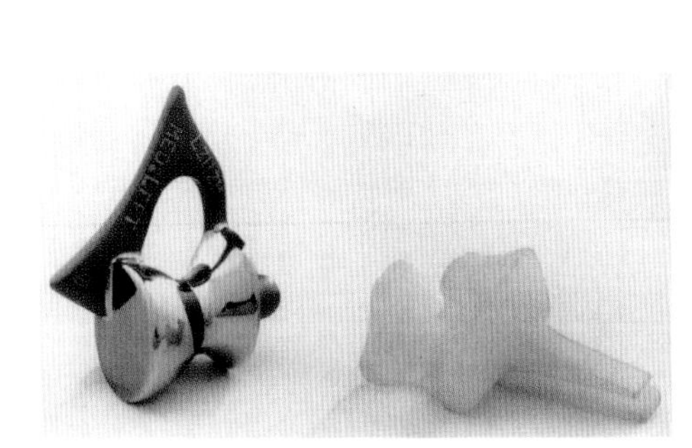
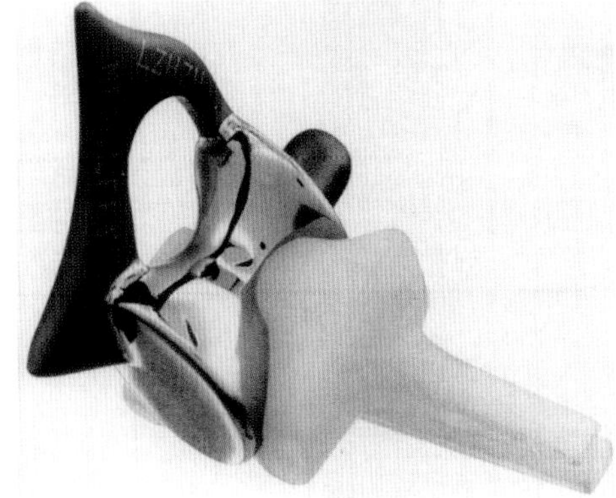

**FIGURE 10.9** Souter-Strathclyde Elbow. Reprinted with permission from *The Journal of Bone and Joint Surgery, Inc.* [26].

UHMWPE hinge. The initial design was modified in 1978 to increase the varus/valgus laxity from 2–3 degrees to 7 degrees. The device consists of cemented titanium humeral and ulnar stems. The ulnar component articulates though a press fit UHMWPE bushing onto a cobalt chrome axis pin. The humeral component is inset with two UHMWPE bushings that provide bearing surfaces for both the axis pin and for the ulnar component at its limits in varus and valgus. In early design iterations, the axis was locked in place by a C-ring. However, it was found that the C-ring could dissociate from the axis and was replaced by a snap-fit titanium pin that is inserted into the axis. An anterior flange was added to the humeral component in 1981 to reduce the transmission of forces through the cement mantle and enhance stability (Figure 10.10). This has resulted in reduced loosening of the humeral component [42, 43]. Ulnar loosening has been addressed by treating the proximal third of the ulnar component to enhance fixation. During the evolution of this device, titanium plasma spray, PMMA precoat, and a beaded surface have been used. UHMWPE type and sterilization methods have also changed during the life of this device. While increased rates of ulnar radiolucency have been noted in devices with a PMMA precoat [4], it is not clear what effect the evolution of UHMWPE type and packaging have had on *in vivo* wear rates.

### *10.4.2.4 GSB III Total Elbow (Zimmer, Warsaw, Indiana, USA)*

The first version of the Gschwend-Scheier-Bahler (GSB) prosthesis was introduced in 1972 as a metal-on-metal fully constrained hinge. The current semiconstrained version was introduced in 1978. The device consists of a cemented cobalt chrome alloy humeral and a cemented two-part ulnar component. Flanges on the humeral component are designed to transmit forces directly to the condyles, thereby reducing the stress at the cement-stem interface during torsional loading. A total of 180 degrees of flexion with 10 degrees of hyperextension are possible by articulation with UHMWPE bushings at a sloppy hinge. This also allows 12 to 14 degrees of valgus laxity [44]. An oval loose link connection between the proximal and distal portions of the ulnar component allows 5 degrees of ulnar axial rotation and axial pistoning of the ulnar stem within a UHMWPE bushing (Figure 10.11). The GSB III currently uses gamma sterilized Sulene PE packaged for a shelf life of 5 years. Because of its loose link design, this device depends on the soft tissues for stability. While this design may reduce loading of the cement mantle, a number of cases of disassembly of the two-part ulnar component has been reported in the literature [44, 45].

### *10.4.2.5 Acclaim Total Elbow (DePuy, Warsaw, Indiana, USA)*

The DePuy Acclaim Total Elbow System was designed to operate as either an unlinked, semiconstrained, or linked elbow. This device consists of cobalt chrome humeral and ulnar stems with a choice of modular bearing surfaces that are chosen intraoperatively. In the unlinked mode a spool shaped, polished cobalt chrome bearing surface is fastened to the humeral component, and a conforming UHMWPE surface is fastened to the ulnar component. In the linked mode the ulnar bearing surface is replaced by a metal block with a cylindrical passage. The metal component articulates directly against a UHMWPE axis bushing that is installed over the axis pin. Varus/valgus motion was limited by bearing of the metal ulnar component against a C-shaped UHMWPE bushing mounted on the humeral component (Figure 10.12).

DePuy recalled ulnar bearing components for the Acclaim from the market in 2005 after it was discovered that a minimal chamfer on the metal ulnar bearing component could cause premature wear of the UHMWPE sleeve, resulting in dissociation of the axis. A business decision was made to no longer market this elbow, and it was voluntarily withdrawn from the market.

(A)

(B)

**FIGURE 10.10** Evolution of the Coonrad-Morrey Total Elbow. (A) Original Coonrad. Image provided courtesy of Bernard Morrey, MD. (B) Contemporary Coonrad-Morrey.

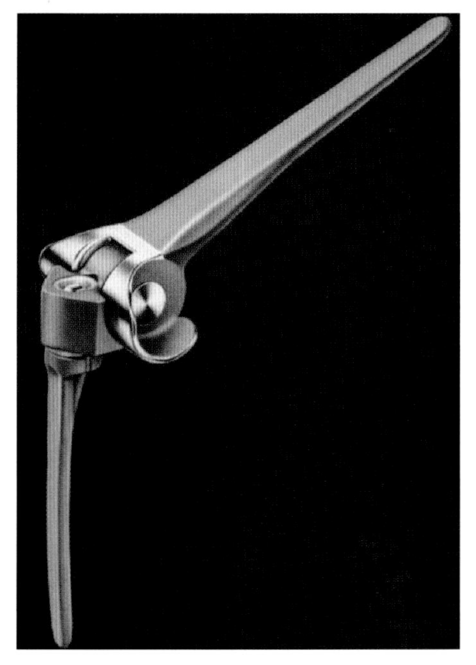

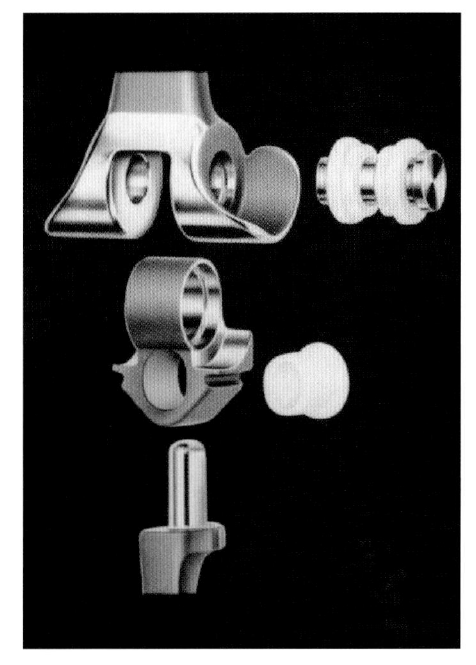

**FIGURE 10.11** GSB III Total Elbow. Reprinted with permission from Elsevier [44].

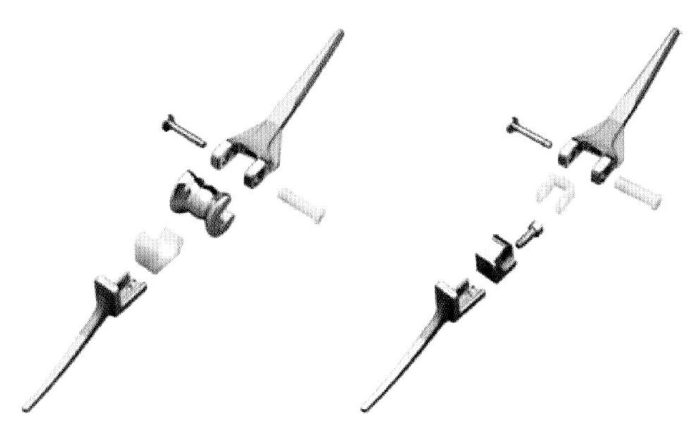

FIGURE 10.12 DePuy Acclaim Total Elbow in unlinked configuration (left) and linked configuration (right). Reprinted with permission of Spring Science + Business Media [61].

### 10.4.2.6 Latitude Total Elbow (Tornier, Saint-Ismier, France)

The Latitude Total Elbow consists of a cobalt chrome humeral stem, an ulnar component, and a radial component. The humeral component is modular with an interchangeable, highly polished bearing surface. The bearing surface consists of a spool-shaped humeroulnar surface and a spherical humeroradial surface. The ulnar component consists of a stem and a UHMWPE bearing surface that is complimentary to its mating humeral surface while permitting 7 degrees of varus/valgus laxity. The device can be converted from an unlinked to a linked design by installing a cap on the ulnar component. The radial component is a bipolar radial head with two separate articulating surfaces: a UHMWPE bearing that directly bears against the hemispherical capitellar surface and a metal on polyethelyene spherical bearing that provides a ±10 degree range of motion (Figure 10.13). The UHMWPE bearing is manufactured from GUR 1050 and gamma sterilized in an argon atmosphere. The Latitude was launched in 2001 in the United States and is intended for cemented use only.

### 10.4.2.7 Solar Elbow System (Stryker, Mahwah, New Jersey, USA)

Launched in 1998, the Solar Elbow System is similar to the Hospital for Special Surgery design that had been previously marketed as the Osteonics Linked Semi-Constrained Total Elbow. This device is a linked semiconstrained system with cemented titanium ulnar and humeral components. The articulation uses a UHMWPE bushing with a metal axle that locks into the lateral portion of the humeral component. A C-shaped UHMWPE bearing surface is also attached to the humeral component. The articulation is loaded centrally along a concave UHMWPE surface. The rest of the ulnar component articulates with the condylar portion of the humeral bushing at extremes of motion

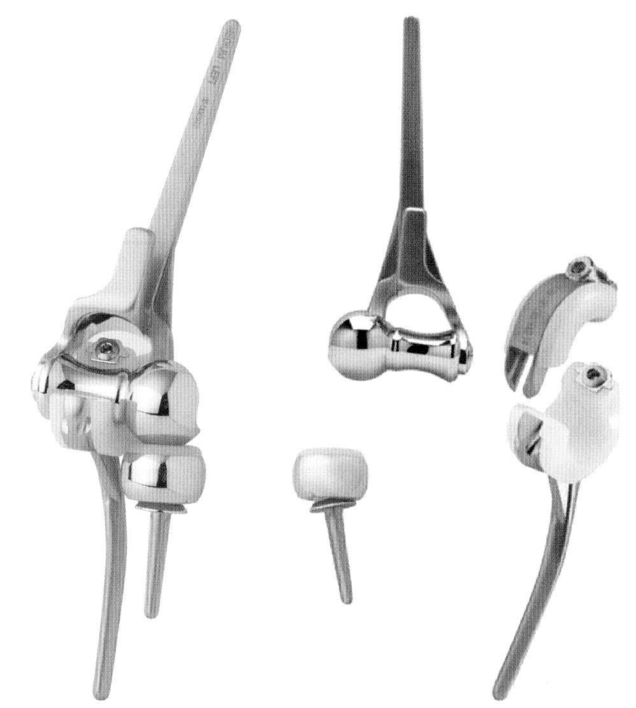

FIGURE 10.13 Tornier Latitude Total Elbow. Images provided courtesy of Tornier, Inc.

(Figure 10.14). This device allows full extension and flexion up to 140 degrees with approximately 7 degrees of laxity in the varus/valgus plane during supination and pronation.

### 10.4.2.8 Discovery Elbow System (Biomet, Warsaw, Indiana, USA)

Developed with Drs. Thomas J. Graham and Hill Hastings, II, this semiconstrained linked design uses two cobalt chrome spherical condyles that are fixed to the humeral component and articulates with a concave hemispherical UHMWPE surface on the ulnar component. The spherical design of the articulating surfaces provide a varus/valgus laxity of 7 degrees. This shape is designed to increase the area of contact between the cobalt chrome and UHMWPE to avoid edge loading and reduce stresses in the UHMWPE (Figure 10.15). The mechanical constraint of the articulation allows for 55 degrees of hyperextension and 175 degrees of flexion; however, range of motion is limited to the patient's anatomy. The device has also been designed to allow posterior access during bushing revision. The UHMWPE is direct compression molded and sterilized by gamma irradiation in argon. Both the humeral and ulnar components are plasma sprayed titanium alloy. A number of design considerations have been included to retain the anatomic axes with respect to both the humeral and ulnar components.

FIGURE 10.14 Disassembled Stryker Solar Elbow System. Reprinted with permission from Elsevier [59].

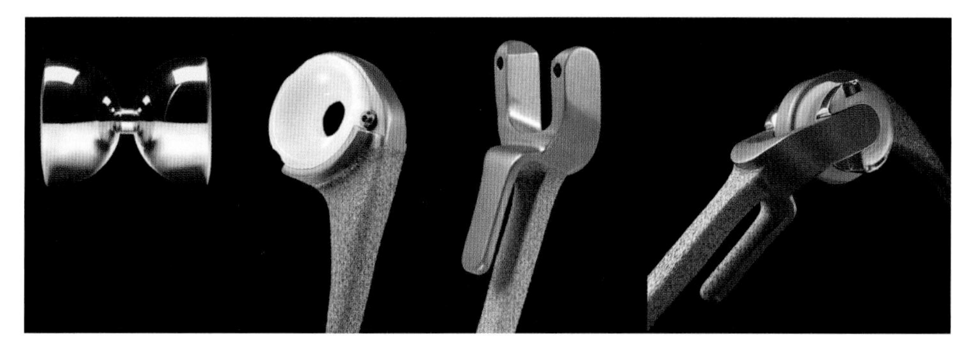

FIGURE 10.15 Biomet Discovery Elbow System. Images provided courtesy of Biomet, Inc.

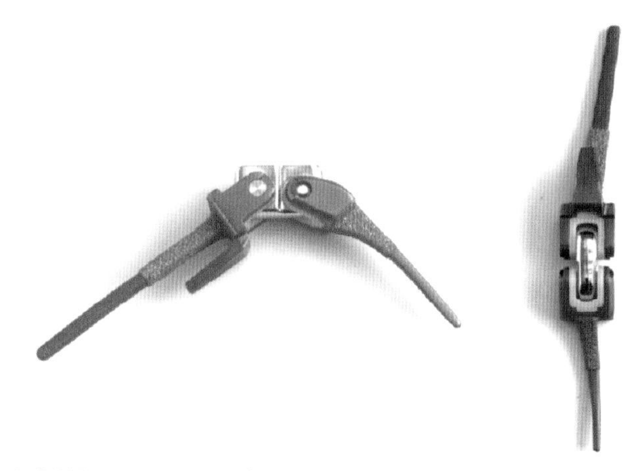

FIGURE 10.16 Huene Biaxial Elbow System. Images provided courtesy of Biomet, Inc.

### 10.4.2.9 Huene Biaxial Elbow System (Biomet, Warsaw, Indiana, USA)

The Huene Biaxial Elbow system is a constrained linked system that relies on two parallel axes to reproduce anterior-posterior translocation of the axis of rotation. The device consists of plasma-coated titanium humeral and ulnar components intended for cemented implantation. Flanges are included on the titanium components to reduce loading of the cement mantles. Saddle-shaped UHMWPE bearings on each component are connected through two axles to a polished cobalt chromium alloy linkage. The axles each run in a UHMWPE bushing (Figure 10.16). The UHMWPE is sterilized by gamma irradiation in argon. Designed with Dr. Donald R. Huene, this device allows for 16 degrees of varus-valgus laxity and 10 degrees of rotational laxity with a greater degree of laxity at the ulnar axis than the humeral axis. The mechanical constraint of the articulation allows 6 degrees of hyperextension and 188 degrees of flexion.

### 10.4.2.10 Kudo Elbow System (Biomet, Warsaw, Indiana, USA)

The Kudo system was developed in 1972 with Dr. Hiroshi Kudo (Type 1) as an unlinked, minimally constrained stainless-steel-on-UHMWPE resurfacing prosthesis. The original design was cylindrical in shape [18]. This design was slightly modified in 1975 to a more physiological saddle-like shape with a reciprocally shaped ulnar component (Type 2). The articulating surfaces were not congruent, and the radius of curvature of the ulnar component was slightly larger than that of the humeral component such that a slight degree of varus-valgus motion and axial rotation were possible. In 1983 the humeral component was modified by adding an intramedullary stem to increase fixation strength and reduce the occurrence of proximal migration of the humeral component (Type 3). In 1988 the design of the humeral component was again modified by switching to titanium alloy with a porous coated stem (Type 4). In 1989 a metal back support was added to the ulnar component to enable uncemented implantation. The use of a titanium-on-UHMWPE bearing surface was associated with metallosis and osteolysis in some patients [46]. In 1991, the material of the humeral component was changed to a cobalt-chromium alloy (Type 5).

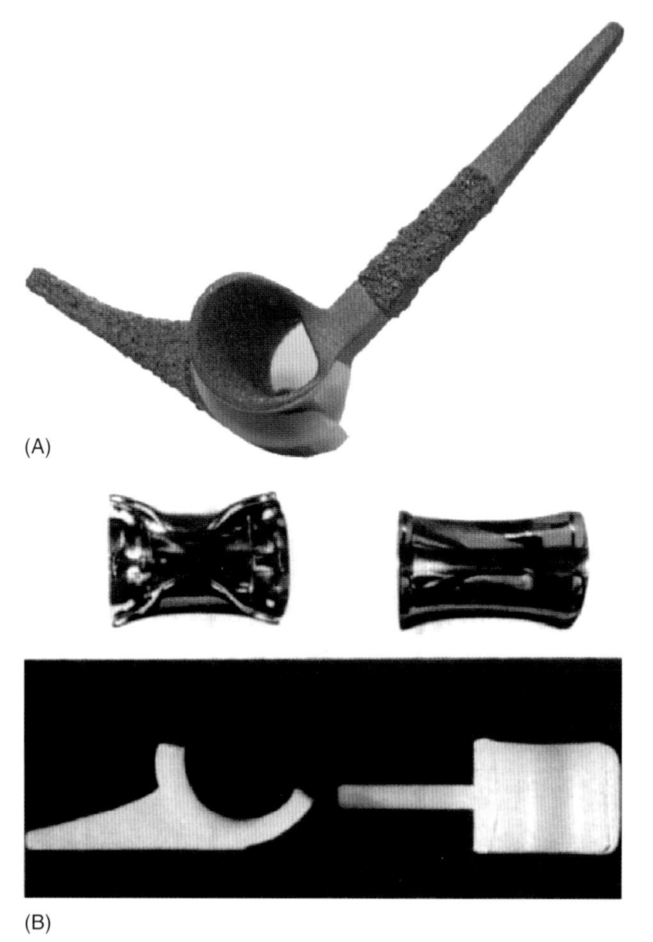


**FIGURE 10.17** Kudo Elbow System. (A) Assembled Kudo Elbow. Image provided courtesy of Biomet, Inc. (B) Disassembled Kudo humeral component and all UHMWPE ulnar component. Reprinted with permission from *The Journal of Bone and Joint Surgery, Inc.* [18].

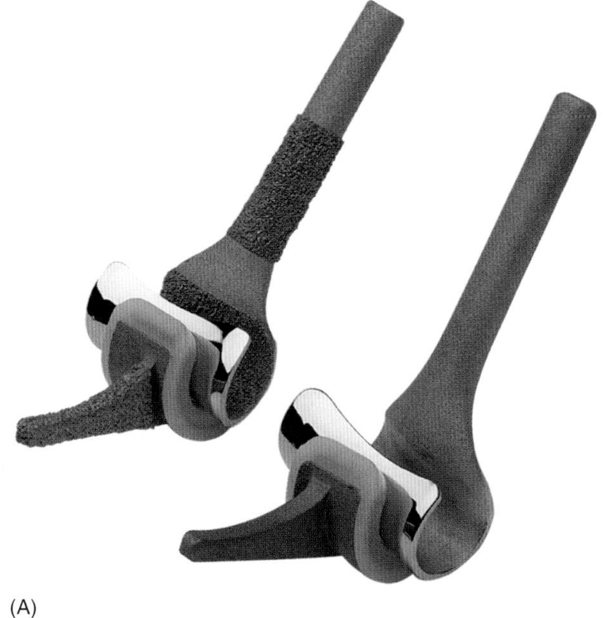

**FIGURE 10.18** Biomet i.B.P. Elbow System. (A) Uncemented and cemented versions of the i.B.P. Image provided courtesy of Biomet. (B) Disassembled i.B.P. showing bearing surfaces. Reprinted with permission from Elsevier [62].

The current system consists of cobalt chrome components with a titanium porous coating on the ulnar component and the distal half of the stem (Figure 10.17). The UHMWPE is direct compression molded and sterilized by gamma irradiation in argon. It may be used either cemented or uncemented. While the Kudo system has had high survival rates at some institutions, the rate of reported loosening of the humeral and ulnar components has been varied [18, 43, 46–53]. The Kudo Elbow System has been used extensively in Europe and Japan, but it is not currently marketed in the United States.

### 10.4.2.11 i.B.P. Elbow System (Biomet, Warsaw, Indiana, USA)

The instrumented Bone Preserving (i.B.P.) elbow was developed with Dr. Joseph Pooley, and although it is marketed in Europe, it is currently not available in the United States. It is an unlinked device with a noncongruent surface geometry. It is similar in form to the Kudo Elbow System, also offered by Biomet. The humeral component is composed of cobalt chrome alloy with a titanium porous coating. The ulnar component is titanium alloy with a UHMWPE bearing surface that is direct compression molded and sterilized by gamma irradiation in Argon (Figure 10.18).

### 10.4.2.12 UNI-Elbow and rHead (Small Bone Innovations, Morrisville, Pennsylvania, USA)

The UNI-Elbow Radio Capitellum System is an anatomic design for use in conjunction with the rHead to replace the humeroradial articulation while preserving the humeroulnar joint. The device is unconstrained and unlinked. A highly polished convex capitellar surface articulates against a concave UHMWPE radial component in a spherical relation (Figure 10.19). Alternatively, the rHead can be used as a radial head hemiarthroplasty. A variety of configurations of the rHead implant are available, including a bipolar implant with a metal-on-UHMWPE bearing couple for a 10 degree angle of rotation. The components are

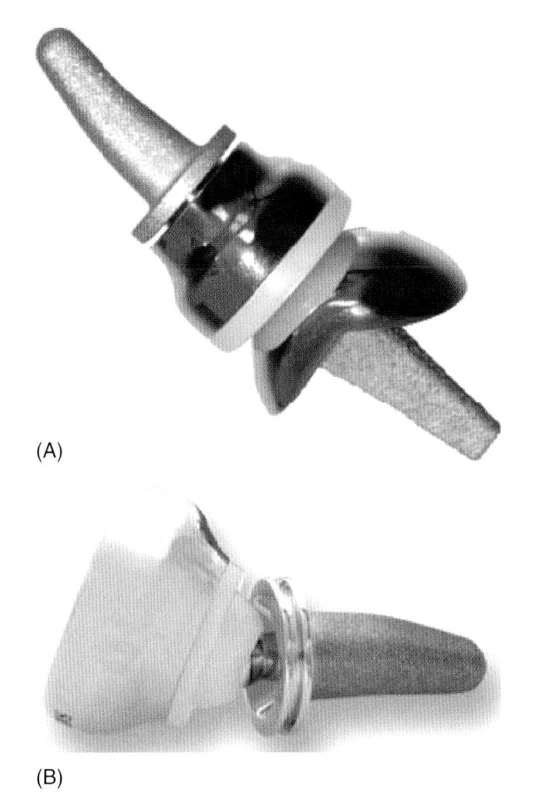


**FIGURE 10.19** (A) UNI-Elbow RadioCapitelum and (B) rHead RECON. Images provided courtesy of Small Bones Innovations, Inc.

composed of cobalt chrome alloy and plasma sprayed with commercially pure titanium. The implants are packaged and supplied sterile via a standard ethylene oxide (EO) sterilization cycle. The cycle consists of humidity and temperature controlled preconditioning, 100% EO exposure in an evacuated chamber, and environmentally controlled outgassing. This device is intended for uncemented use.

## 10.5 OSTEOLYSIS AND WEAR

Osteolysis has long been recognized as an important mode of failure for hip and knee arthroplasties. In the elbow, the predominant failure modes have historically been aseptic loosening of linked designs and instability of unlinked designs. As implant designs have improved, indications have expanded from an elderly, relatively inactive rheumatoid population to include younger and noninflammatory patients. It is increasingly being recognized that UHMWPE wear and osteolysis are important long-term complications.

Because of the relatively recent realization that UHMWPE wear could become an issue in TEA, there have been few studies of bushing wear, and there are no consensus standards for wear testing of elbow designs. For this reason, most of what is known about wear and TEA comes from case reports and a small number of retrieval studies. The majority of this material is reported for linked designs. There are few, if any, studies comparing implants produced with modern material types and packaging techniques to older materials.

Lee et al. reported a retrospective review of 919 replacements with the Coonrad-Morrey implant [54]. Patients were treated for rheumatoid conditions, trauma-related conditions, and revision of a previous TEA. Wear was evaluated radiographically by measuring the alignment of the ulnar component with respect to the humeral component in full elbow extension. Angles in excess of 7 degrees of total varus valgus laxity indicated displacement greater than that allowed by the design tolerances. Angles greater than 10 degrees indicated complete wear of the bushings. They found that 1.3% of those studied had received a bushing exchange for bushing wear at an average of 7.9 years postimplantation and that an additional 0.7% had measurable bushing wear. This was markedly lower than the results reported by Gill et al., who reported 7% completely worn bushings and 13% partially worn in a population of rheumatoid patients with an average of 7 years follow-up [42]. Wright and Hastings reported a mean time to revision of 60 months for 10 patients revised for bushing wear [55]. In treatment for posttraumatic TEA, Hasting and Theng reported bushing wear in more than 25% of 17 cases [56]. Osteolysis has been noted at the time of revision or bushing exchange [55, 57]. Lee reported that the patients who had bushing exchange were younger than the rest of the population, and there was a trend toward patients treated for posttraumatic arthritis rather than for inflammatory disease. It was noted that a number of patients in this group had moderate to strenuous occupations, including a contractor and a lumberjack. Bushing exchange was associated with deformity of the joint at the time of the initial arthroplasty, and the articulation that was at the limits of the designed angular tolerance in the immediate postoperative radiographs for 9 of the 12 patients. Osteolysis was not considered to have compromised the function or fixation of any of the devices that underwent bushing exchange. This was attributed to a presumably smaller absolute volume of wear debris generated by the smaller articular surfaces at the elbow. It is important to note that patients who were revised for component loosening were not considered in this study, and therefore the true incidence of component wear may be underestimated.

In a retrieval study of 16 failed Coonrad-Morrey devices together with the periprosthetic tissues, Goldberg et al. reported progressive radiolucencies in all elbows, including both diffuse and focal radiolucencies [58]. UHMWPE particles in the periprosthetic tissues included granules from $<1$ to $10\,\mu m$ in size and flakes measuring up to several hundred micrometers across. The presence of large particles could indicate high contact stresses in the elbow bushings. Titanium-alloy particles ranged from 0.1 to $5\,\mu m$ size and were the most abundant particulate in the tissues. Tissue histology revealed a similar histopathology to that

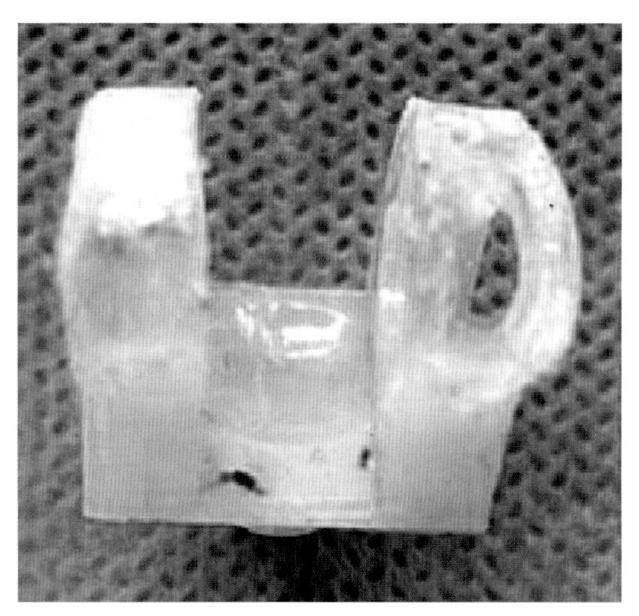

**FIGURE 10.20** Retrieved Osteonics Elbow demonstrating UHMWPE wear. Reprinted with permission from Elsevier [59].

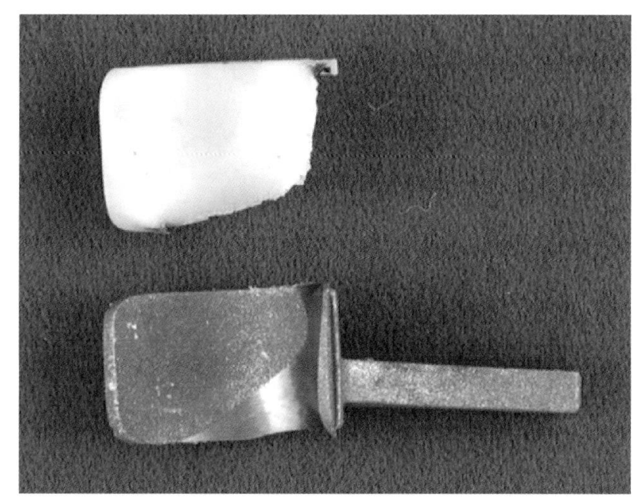

**FIGURE 10.21** Retrieved Kudo (Type 4) Component demonstrating UHMWPE wear and burnishing of the titanium ulnar component. Reprinted with permission from *The Journal of Bone and Joint Surgery, Inc.* [60].

observed in association with osteolysis in hip and knee tissues. The population reported by Goldberg, in contrast to that reported by Lee et al., included only failed prostheses and did not investigate patients who underwent bushing exchange [54, 58]. The observation of extreme metal-on-metal wear of the retrieved components indicates that the revisions were performed for late stage of failure after total wear through of the bushings. The relative contribution of titanium wear particles, as opposed to UHMWPE to osteolysis, is unknown for these cases.

Figgie et al. reported that 5.3% of 170 patients receiving an Osteonics-type linked semiconstrained design had late failure of the locking mechanism at a mean time to failure of 5.5 years [59]. Initial TEA was performed for either rheumatoid arthritis or posttraumatic conditions. At the time of revision, polyethylene debris, metallosis and synovitis were observed. The polyethylene bushings demonstrated evidence of extreme wear through the condylar portion (Figure 10.20). The authors believed that the failure of the locking mechanism was the result of binding of the components after complete UHMWPE wear. Although none of the cases in this study had extensive bone loss at the time of revision surgery, five of the nine later showed mild to severe osteolysis without evidence of implant loosening.

Landor et al. reported a 17% loosening rate in 45 Souter-Strathclyde TEAs from 35 rheumatoid patients with a mean follow-up of 9.5 years [37]. They noted heavy damage or fracture of all of the ulnar components at revision and attributed this to the all UHMWPE design of the ulnar components used at their institution. It had been previously noted that use of metal-backed components led to reduced loosening of the ulnar component [35]. Osteolysis was observed around the metal humeral components but did not result in their migration.

In a study of the unlinked Kudo (Type 4), 26 patients with 32 elbows were followed for an average of 37 months [46]. The average age of the patient population was 56 years. Five elbows demonstrated marked resorption within the condylar portion of the uncemented humeral component followed by subsidence, resulting in revision at an average time of 34 months after implantation. Metallosis of the periprosthetic tissue and burnishing of the outer condylar bearing surface were noted during the revision surgery (Figure 10.21). It has been suggested that the subsidence of the humeral component resulted in edge loading of the component, resulting in destruction of the UHMWPE and impingement of the components on one another [46, 60]. It was recommended against using titanium alloy on UHMWPE as a bearing couple in the elbow. A study by the same authors using the Kudo (Type 5) with a cobalt-chrome bearing surface demonstrated no appreciable osteolysis at an average follow up of 46 months [43]. A further study by Dos Remedios et al. included 13 Kudo Type 4 and 23 Kudo Type 5 TEAs with a mean follow-up of 62 months [48]. Metallosis was observed in five of the seven implants that were revised. Four were titanium alloy Type 4 devices, and one was a cobalt-chrome alloy Type 5. Osteolysis was observed in three of the cases with aseptic loosening, and metallosis was associated with loosening. In one elbow, severe UHMWPE wear was observed with metallosis but without loosening.

The continued refinement of TEA design has improved to the point where aseptic loosening and dislocation are no longer the exclusive reasons for revision surgery. UHMWPE wear has been associated with osteolysis and fracture of the metallic components. Increased wear has been associated with activity levels and the use of titanium on UHMWPE bearing couples. It is expected that future designs will focus on improvements at the bearing couple.

## 10.6 CONCLUSION

TEA presents a proven and successful option for reducing pain and restoring function for an expanding number of clinical indications. While the rates of aseptic loosening and dislocation have been greatly reduced, it is expected that future designs will continue to address these issues. It is also expected that future designs will focus on reducing UHMWPE wear by further improving contact conditions while avoiding in overconstraint of the elbow. At the moment, there does not seem to be a clear consensus regarding the selection of UHMWPE for TEA. Reported designs use conventional UHMWPE with sterilization methods that include ethylene oxide and gamma irradiation in inert gas. There is very little reported in the scientific literature regarding wear testing or retrieval analysis for TEA or the effect of material selection. Future improvements for TEA may be achieved with the use of advanced bearing materials in combination with new implant designs.

## 10.7 ACKNOWLEDGMENTS

This chapter was not written with the financial support of any manufacturer of total elbow arthroplasty systems. However, the manufacturers of some of the products mentioned in this review were contacted by the author and given the opportunity to verify the factual accuracy of the information relating to their products. Institutional support for Judd Day has been received from Zimmer, Inc. I would like to thank Chris Espinosa for many of the illustrations in this chapter, Madeline Olsen for editorial assistance, and Drs. Matthew Ramsey and Bernard Morrey for their insight into elbow arthroplasty.

## REFERENCES

1. Lockard M. Clinical biomechanics of the elbow. *J Hand Ther* 2006 April–June;**19**(2):72–80.
2. Alcid JG, Ahmad CS, Lee TQ. Elbow anatomy and structural biomechanics. *Clin Sports Med* 2004 October;**23**(4):503–17, vii.
3. Bryce CD, Armstrong AD. Anatomy and biomechanics of the elbow. *Orthop Clini North Am* 2008 April;**39**(2):141–54, v.
4. Williams GR, Yamaguchi K, Ramsey ML, Galatz LM. *Shoulder and Elbow Arthroplasty*. Philadelphia: Lippincott Williams & Wilkins; 2005.
5. Morrey BF, An KN, Stormont TJ. Force transmission through the radial head. *J Bone Joint Surg Am* 1988 February;**70**(2):250–6.
6. Li G, Pierce JE, Herndon JH. A global optimization method for prediction of muscle forces of human musculoskeletal system. *J Biomech* 2006;**39**(3):522–9.
7. Amis AA, Dowson D, Wright V. Elbow joint force predictions for some strenuous isometric actions. *J Biomech* 1980;**13**(9): 765–75.
8. Chadwick EK, Nicol AC. Elbow and wrist joint contact forces during occupational pick and place activities. *J Biomech* 2000 May;**33**(5):591–600.
9. Eckstein F, Merz B, Muller-Gerbl M, Holzknecht N, Pleier M, Putz R. Morphomechanics of the humero-ulnar joint: II. Concave incongruity determines the distribution of load and subchondral mineralization. *Anat Rec* 1995 November;**243**(3):327–35.
10. Merz B, Eckstein F, Hillebrand S, Putz R. Mechanical implications of humero-ulnar incongruity—finite element analysis and experiment. *J Biomech* 1997 July;**30**(7):713–21.
11. Markolf KL, Lamey D, Yang S, Meals R, Hotchkiss R. Radioulnar load-sharing in the forearm. A study in cadavera. *J Bone Joint Surg Am* 1998 June;**80**(6):879–88.
12. Markolf KL, Dunbar AM, Hannani K. Mechanisms of load transfer in the cadaver forearm: role of the interosseous membrane. *J Hand Surg [Am]* 2000 July;**25**(4):674–82.
13. Coonrad RW. History of total elbow arthroplasty. In: Inglis AE, editor. *Symposium on total joint replacement of the upper extremity*. St. Louis: Mosby; 1982. p. 75–90.
14. Dee R. Total replacement arthroplasty of the elbow for rheumatoid arthritis. *J Bone Joint Surg Br* 1972 February;**54**(1):88–95.
15. Morrey BF, Bryan RS, Dobyns JH, Linscheid RL. Total elbow arthroplasty. A five-year experience at the Mayo Clinic. *J Bone Joint Surg Am* 1981 September;**63**(7):1050–63.
16. O'Driscoll SW, King GJ. Treatment of instability after total elbow arthroplasty. *Orthop Clini North Am* 2001 October;**32**(4):679–95, ix.
17. Ewald FC. Total elbow replacement. *Orthop Clini North Am* 1975 July;**6**(3):685–96.
18. Kudo H, Iwano K, Watanabe S. Total replacement of the rheumatoid elbow with a hingeless prosthesis. *J Bone Joint Surg Am* 1980 March;**62**(2):277–85.
19. Stevens PS, Street DM. The use of the Stevens-Street elbow prosthesis. *Acta orthopaedica Belgica* 1975 July–August;**41**(4):447–54.
20. Wevers HW, Siu DW, Broekhoven LH, Sorbie C. Resurfacing elbow prosthesis: shape and sizing of the humeral component. *J Biomed Eng* 1985 July;**7**(3):241–6.
21. Inglis AE, Pellicci PM. Total elbow replacement. *J Bone Joint Surg Am* 1980 December;**62**(8):1252–8.
22. Morrey BF, Askew LJ, Chao EY. A biomechanical study of normal functional elbow motion. *J Bone Joint Surg Am* 1981 July;**63**(6):872–7.
23. Schlein AP. Semiconstrained total elbow arthroplasty. *Clini Orthop Relat Res* 1976 November–December;**12**(121):222–9.
24. Trepman E, Vella IM, Ewald FC. Radial head replacement in capitellocondylar total elbow arthroplasty. 2- to 6-year follow-up evaluation in rheumatoid arthritis. *J Arthroplasty* 1991 March; **6**(1):67–77.
25. Ramsey ML, Adams RA, Morrey BF. Instability of the elbow treated with semiconstrained total elbow arthroplasty. *J Bone Joint Surg Am* 1999 January;**81**(1):38–47.
26. Poll RG, Rozing PM. Use of the Souter-Strathclyde total elbow prosthesis in patients who have rheumatoid arthritis. *J Bone Joint Surg Am* 1991 September;**73**(8):1227–33.
27. Shah BM, Trail IA, Nuttall D, Stanley JK. The effect of epidemiologic and intraoperative factors on survival of the standard Souter-Strathclyde total elbow arthroplasty. *J Arthroplasty* 2000 December;**15**(8):994–8.

28. Kelly EW, Coghlan J, Bell S. Five- to thirteen-year follow-up of the GSB III total elbow arthroplasty. *J Shoulder Elbow Surg* 2004 July-August;**13**(4):434–40.
29. Ewald FC, Simmons Jr. ED, Sullivan JA, Thomas WH, Scott RD, Poss R, et al Capitellocondylar total elbow replacement in rheumatoid arthritis. Long-term results. *J Bone Joint Surg Am* 1993 April;**75**(4):498–507.
30. Ljung P, Jonsson K, Rydholm U. Short-term complications of the lateral approach for non-constrained elbow replacement. Follow-up of 50 rheumatoid elbows. *J Bone Joint Surg Br* 1995 November; **77**(6):937–42.
31. Ovesen J, Olsen BS, Johannsen HV, Sojbjerg JO. Capitellocondylar total elbow replacement in late-stage rheumatoid arthritis. *J Shoulder Elbow Surg* 2005 July–August;**14**(4):414–20.
32. Ruth JT, Wilde AH. Capitellocondylar total elbow replacement. A long-term follow-up study. *J Bone Joint Surg Am* 1992 January;**74**(1):95–100.
33. Weiland AJ, Weiss AP, Wills RP, Moore JR. Capitellocondylar total elbow replacement. A long-term follow-up study. *J Bone Joint Surg Am* 1989 February;**71**(2):217–22.
34. Schneeberger AG, King GJ, Song SW, O'Driscoll SW, Morrey BF, An KN. Kinematics and laxity of the Souter-Strathclyde total elbow prosthesis. *J Shoulder Elbow Surg* 2000 March–April;**9**(2):127–34.
35. Ikavalko M, Belt EA, Kautiainen H, Lehto MU. Revisions for aseptic loosening in Souter-Strathclyde elbow arthroplasty: incidence of revisions of different components used in 522 consecutive cases. *Acta Orthop Scand* 2002 June;**73**(3):257–63.
36. Ikavalko M, Lehto MU, Repo A, Kautiainen H, Hamalainen M. The Souter-Strathclyde elbow arthroplasty. A clinical and radiological study of 525 consecutive cases. *J Bone Joint Surg Br* 2002 January;**84**(1):77–82.
37. Landor I, Vavrik P, Jahoda D, Guttler K, Sosna A. Total elbow replace ment with the Souter-Strathclyde prosthesis in rheumatoid arthritis. Long-term follow-up. *J Bone Joint Surg Br* 2006 November; **88**(11):1460–3.
38. Rozing P. Souter-Strathclyde total elbow arthroplasty. *J Bone Joint Surg Br* 2000 November;**82**(8):1129–34.
39. Samijo SK, Van den Berg ME, Verburg AD, Tonino AJ. Souter-Strathclyde total elbow arthroplasty: medium-term results. *Acta Orthop Belg* 2003 December;**69**(6):501–6.
40. Sjoden GO, Lundberg A, Blomgren GA. Late results of the Souter-Strathclyde total elbow prosthesis in rheumatoid arthritis. 6/19 implants loose after 5 years. *Acta Orthop Scand* 1995 October;**66**(5):391–4.
41. Trail LA, Nuttall D, Stanley JK. Comparison of survivorship between standard and long-stem Souter-Strathclyde total elbow arthroplasty. *J Shoulder Elbow Surg* 2002 July–August;**11**(4):373–6.
42. Gill DR, Morrey BF. The Coonrad-Morrey total elbow arthroplasty in patients who have rheumatoid arthritis. A ten to fifteen-year follow-up study. *J Bone Jt Surg* 1998 September;**80**(9):1327–35.
43. Kudo H, Iwano K, Nishino J. Total elbow arthroplasty with use of a nonconstrained humeral component inserted without cement in patients who have rheumatoid arthritis. *J Bone Joint Surg Am* 1999 September;**81**(9):1268–80.
44. Herren DB, O'Driscoll SW, An KN. Role of collateral ligaments in the GSB-linked total elbow prosthesis. *J Shoulder Elbow Surg* 2001 May-June;**10**(3):260–4.
45. Cesar M, Roussanne Y, Bonnel F, Canovas F. GSB III total elbow replacement in rheumatoid arthritis. *J Bone Joint Surg Br* 2007 March;**89**(3):330–4.
46. Kudo H, Iwano K, Nishino J. Cementless or hybrid total elbow arthroplasty with titanium-alloy implants. A study of interim clinical results and specific complications. *J Arthroplasty* 1994 June;**9**(3):269–78.
47. Brinkman JM, de Vos MJ, Eygendaal D. Failure mechanisms in uncemented Kudo type 5 elbow prosthesis in patients with rheumatoid arthritis: 7 of 49 ulnar components revised because of loosening after 2–10 years. *Acta Orthop* 2007 April;**78**(2):263–70.
48. Dos Remedios C, Chantelot C, Giraud F, Migaud H, Fontaine C. Results with Kudo elbow prostheses in non-traumatic indications: a study of 36 cases. *Acta Orthop Belg* 2005 June;**71**(3):273–88.
49. Kudo H, Iwano K. Total elbow arthroplasty with a non-constrained surface-replacement prosthesis in patients who have rheumatoid arthritis. A long-term follow-up study. *J Bone Joint Surg Am* 1990 March;**72**(3):355–62.
50. Rauhaniemi J, Tiusanen H, Kyro A. Kudo total elbow arthroplasty in rheumatoid arthritis. Clinical and radiological results. *J Hand Surg [Br]* 2006 April;**31**(2):162–7.
51. Tanaka N, Kudo H, Iwano K, Sakahashi H, Sato E, Ishii S. Kudo total elbow arthroplasty in patients with rheumatoid arthritis: a long-term follow-up study. *J Bone Joint Surg Am* 2001 October; **83-A**(10):1506–13.
52. Thillemann TM, Olsen BS, Johannsen HV, Sojbjerg JO. Long-term results with the Kudo type 3 total elbow arthroplasty. *J Shoulder Elbow Surg* 2006 July–August;**15**(4):495–9.
53. van der Heide HJ, de Vos MJ, Brinkman JM, Eygendaal D, van den Hoogen FH, de Waal Malefijt MC. Survivorship of the KUDO total elbow prosthesis—comparative study of cemented and uncemented ulnar components: 89 cases followed for an average of 6 years. *Acta Orthop* 2007 April;**78**(2):258–62.
54. Lee BP, Adams RA, Morrey BF. Polyethylene wear after total elbow arthroplasty. *J Bone Joint Surg Am* 2005 May;**87**(5):1080–7.
55. Wright TW, Hastings H. Total elbow arthroplasty failure due to overuse, C-ring failure, and/or bushing wear. *J Shoulder Elbow Surg* 2005 January–February;**14**(1):65–72.
56. Hastings II H, Theng CS. Total elbow replacement for distal humerus fractures and traumatic deformity: results and complications of semiconstrained implants and design rationale for the Discovery Elbow System. *Am J Orthop* 2003 September;**32**(9 Suppl.):20–8, Belle Mead, NJ.
57. Hastings II H. Minimally constrained elbow implant arthroplasty: the discovery elbow system. *Tech Hand Up Extrem Surg* 2004 March;**8**(1):34–50.
58. Goldberg SH, Urban RM, Jacobs JJ, King GJ, O'Driscoll SW, Cohen MS. Modes of wear after semiconstrained total elbow arthroplasty. *J Bone Joint Surg Am* 2008 March;**90**(3):609–19.
59. Figgie MP, Su EP, Kahn B, Lipman J. Locking mechanism failure in semiconstrained total elbow arthroplasty. *J Shoulder Elbow Surg* 2006 January–February;**15**(1):88–93.
60. Little CP, Graham AJ, Karatzas G, Woods DA, Carr AJ. Outcomes of total elbow arthroplasty for rheumatoid arthritis: comparative study of three implants. *J Bone Joint Surg Am* 2005 November;**87**(11):2439–48.
61. Lerch K, Tingart M, Trail I, Grifka J. [Total elbow arthroplasty. Indications, operative technique and results after implantation of an Acclaim elbow prosthesis]. *Orthopade* 2003 August;**32**(8):730–5.
62. Stokdijk M, Nagels J, Garling EH, Rozing PM. The kinematic elbow axis as a parameter to evaluate total elbow replacement: A cadaver study of the iBP elbow system. *J Shoulder Elbow Surg* 2003 January–February;**12**(1):63–8.

Chapter 11

# Applications of UHMWPE in Total Ankle Replacements

Allyson Ianuzzi, PhD and Chimba Mkandawire, PhD, CAISS



## 11.1 INTRODUCTION

The ankle joint was the last joint in which total joint replacement was attempted. For that reason, total ankle replacement (TAR) currently presents more problems than encountered in total hip or knee replacement. In the 1970s, several first-generation TAR designs were introduced. The main distinguishing features among the designs included whether they were fixed versus mobile bearings and whether they had congruent versus incongruent articulating surfaces. Commonalities of the designs included replacement or resurfacing of both the tibial and talar surfaces, the use of cement for bony fixation, and a metal-on-ultrahigh molecular weight polyethylene (hereafter UHMWPE) bearing surface. Based on their clinical performance, many of the early designs were recommended only for older patients with low physical demands, or their use was abandoned entirely.

Many lessons were learned from this first generation of TARs. For example, almost all devices currently are used without cement in clinical practice (although this convention constitutes off-label use in the United States because none of the FDA-cleared devices are cleared for cementless use). One of the more important lessons with respect to the UHMWPE components was highlighted by studies demonstrating high local stresses in the UHMWPE components with incongruent articulating surfaces. All contemporary designs incorporate congruent articulating surfaces. Debate remains as to whether a fixed or mobile bearing is superior. All devices that are currently FDA-cleared in the United States are fixed bearing devices. There are no FDA-approved mobile bearing devices, although approval is pending for one design (STAR).

A query of the Nationwide Inpatient Sample (NIS) indicated that the estimated number of patients discharged from US hospitals after receiving TAR almost doubled between 1997 and 2000 (Figure 11.1), with an estimated 1229 patients discharged after TAR procedures in 2000. Overall the estimated number of patients discharged with TAR rose from 147 patients in 1993 to 870 patients in 2006. By comparison, the estimated number of patients discharged for ankle fusion has fluctuated from 1993 to 2006, where the estimated number of fusions decreased from 7037 patients in 1993 to 6010 patients in 1999, and then it again increased after 1999. These trends could reflect changes in TAR design, the number or types of designs available in the United States over the years, and different recommendations resulting from publications of clinical studies.

The purpose of this book chapter is to provide a historical overview of the TAR. The anatomy and biomechanics of the ankle are briefly reviewed. An in-depth description is provided of early-generation and contemporary TAR designs, with a focus on the UHMWPE component of each design. UHMWPE wear in TAR is also considered. Lastly, available data related to clinical complications and retrieval analysis of the UHMWPE components are reviewed.

## 11.2 ANATOMY

The foot and ankle complex (Figure 11.2) comprises 26 bones within the foot, which are non-sesamoid bones (sesamoid bones vary in quantity and location from person to person), two bones of the lower leg, and approximately 109 ligament and fascicle groups spanning these 28 bones [1].

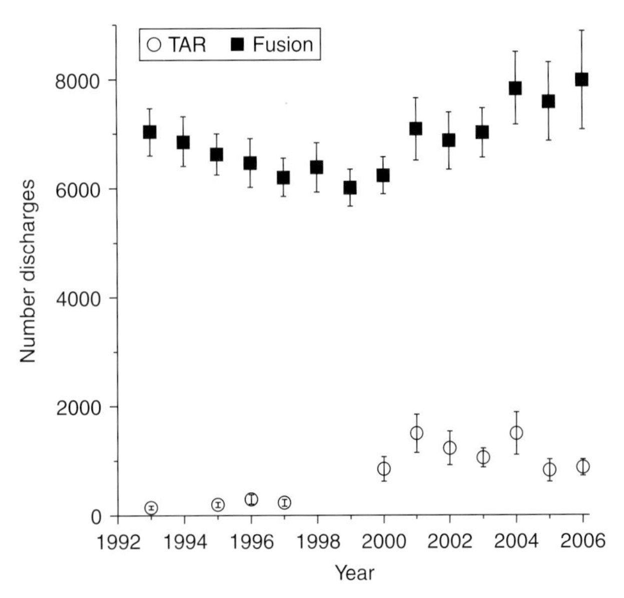


**FIGURE 11.1** Estimated number of discharges for ankle fusion and total ankle replacements (TAR) from 1993 to 2006.

It is noted that accessory bones of the lower leg and foot are developmental abnormalities, which are excluded from this count. There are five functional joint groupings present, the ankle (talocrural joint), subtalar joint, midtarsal or transverse tarsal joint, tarsometatarsal joint, and metatarsophalangeal joints [2, 3].

Due to the detailed nature of the anatomy of this complex, we will only place attention on anatomy of the hindfoot and midfoot. The ankle comprises the articulation between the tibia, fibula and talus, and 10 ligaments and fascicles. The hindfoot (talus and calcaneus) contains the subtalar joint; there are three points of articulation between the inferior aspect (bottom) of the talus and the superior aspect (top) of the calcaneus. Approximately five ligaments span the subtalar joint; however, the cervical ligament (also the strongest of these five ligaments) has four distinct fascicles. There are approximately six ligaments and fascicles that originate on the distal tibia or fibula and insert on the calcaneus or midfoot. The midtarsal joint forms the boundary between the midfoot (navicular, cuboid) and hindfoot, where eight ligaments span.

## 11.3 ANKLE BIOMECHANICS

The motions of the ankle can be described in three orthogonal axes, which include axial rotation (internal/external) in the transverse plane, inversion/eversion in the frontal plane, and dorsiflexion/plantarflexion in the sagittal plane (Figure 11.3). Motion along all three axes is pronation/supination, where pronation is dorsiflexion, eversion and external rotation; and supination is plantarflexion, inversion,

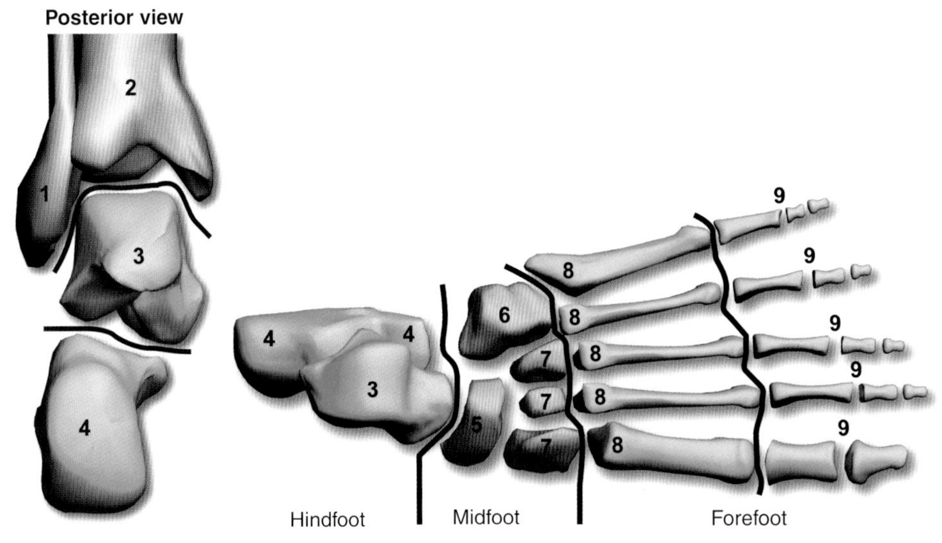


**FIGURE 11.2** The foot and ankle complex. This disarticulated foot is divided into three groupings: the hindfoot, midfoot, and forefoot. Bone nomenclature: 1, fibula; 2, tibia; 3, talus; 4, calcaneus; 5, navicular; 6, cuboid; 7, cuneiforms; 8, metatarsals; 9, phalanges. The following joints are outlined with a solid black line: The ankle is the articulation of the tibia and fibula on the talus, the subtalar joint is the articulation between the calcaneus and talus, the midtarsal joint is between the hindfoot and midfoot, the tarsometatarsal joint is between the midfoot and forefoot, and the metatarsal phalangeal joints are between the metatarsals and the adjoining proximal phalange.

and internal rotation. The sagittal plane range of motion required for walking is approximately 12 degrees of dorsiflexion and 15 degrees of plantarflexion [4]. The majority of this motion occurs at the talocrural joint. During inversion and eversion, the subtalar joint has a greater contribution, with motion at the talocrural joint occurring at the extremes of motion [4–6]. Axial rotation occurs at approximately equal proportions in the two joints [4, 5].

Total passive ankle range of motion within the sagittal plane in cadavers and living subjects is 50 to 70 degrees from maximum dorsiflexion to maximum plantarflexion [2, 7]. A maximum of 35 degrees is required for normal gait, with 10 degrees of dorsiflexion in early stance phase and 25 degrees of plantarflexion at push-off [8]. The average range of sagittal plane motion was 20–27 degrees, with approximately 10 degrees of plantarflexion in early stance and 14–25 degrees dorsiflexion during push-off [9]. Range of sagittal plane motion while ascending stairs was 37 degrees and 56 degrees for descending stairs [7]. Normal gait can be grouped by two phases (Figure 11.4): the swing phase (where the leg swings freely during walking) and stance phase (where the leg is in contact with the ground during walking). The stance phase can further be divided into five subcategories: heel strike or heel contact, foot flat or weight acceptance, midstance, heel-off or push-off, and toe-off [2]. To describe stair ascent or stair descent (as well as abnormal gait), the nomenclature used for the stance phase becomes initial contact, loading response, midstance, terminal stance, and preswing.

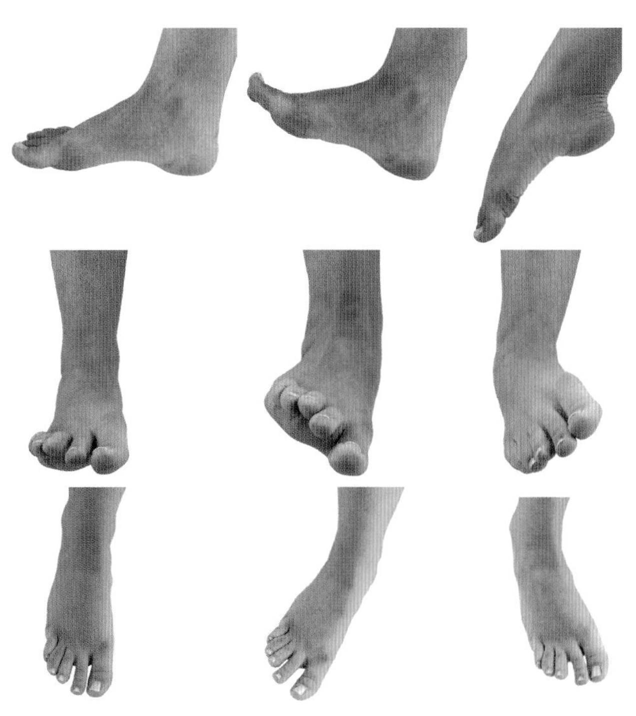

FIGURE 11.3 Motions of the ankle in the sagittal plane (top from left to right: neutral, dorsiflexion, plantarflexion), frontal plane (middle from left to right: neutral, eversion, inversion), and transverse plane (bottom from left to right: neutral, external rotation, internal rotation).

The maximum compressive load through the ankle during normal gait can be greater than five times body weight (compared to the compressive load of three times body weight found in the hip and knee) [9]. Stauffer pointed out that the surface contact area of a normal ankle is approximately $12\,cm^2$ [9]. Maximum shear load was about 80% of body weight [9]. Weight transfer to the fibula can be approximately 17% of body weight [10, 11]. Compressive load studies of the ankle that utilized pressure sensitive film show average contact pressures of 1.84 MPa in the neutral ankle position; 2.16 MPa, 20 degrees of dorsiflexion; 2.14 MPa, 20 degrees of plantarflexion [12]. The Wagner study was performed with each cadaver specimen under 800 N of compression through the tibia and fibula, approximating $1 \times$ body weight [12]. McKinely and others performed similar compression studies, with a piezoresistive elastomer that measured compressive force. That study was performed with 600 N of compressive load (0.87 BW) through quasistatic gait orientations (heel strike, 10% stance phase, 60% stance phase, toe-off) and found a mean compressive load of 2.7 MPa [13]. These studies are conservative estimates of contact pressure because the primary extrinsic muscles of the lower leg were not modeled. Placing tension on the extrinsic leg muscles increases the compressive loads borne at the ankle.

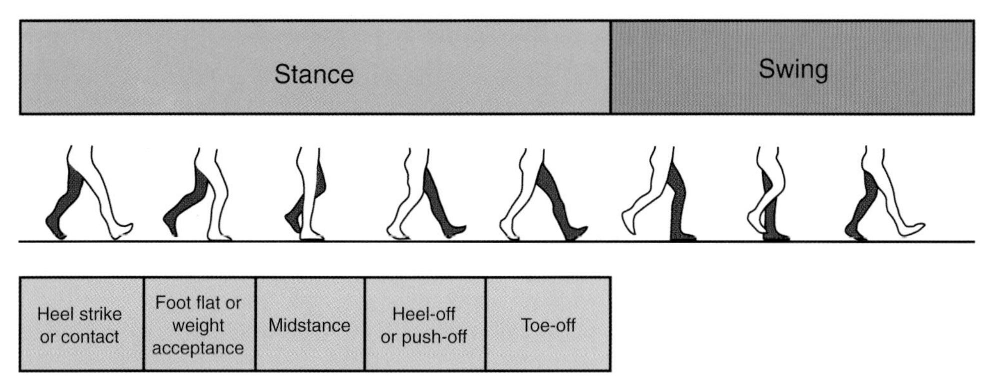


FIGURE 11.4 Schematic of normal gait, demonstrating the stance phase and swing phase.

## 11.4 TOTAL ANKLE REPLACEMENT DESIGN

The goal of a TAR design should be to mimic the ankle joint as closely as possible. Specifically, a suitable range of motion should be available to allow for proper gait patterns and other activities of daily living. TAR designs are expected to have a reproducible surgical technique, minimal bone resection, rapid and adequate bone ingrowth, minimal constraint and replication of physiological ankle motion, and pain relief [14]. Relevant to the UHMWPE components, the design should also offer minimal complications and need for early revision, as well as long-term survivorship [14]. The contact area should be appropriate to distribute the load and avoid high contact stresses. In the next sections, these parameters are considered in discussing historical and modern TAR designs.

The FDA currently has three classifications for TARs, two of which are Class II (metal/polymer or metal/composite semi-constrained cemented prosthesis) and one that is Class III (non-constrained cemented prosthesis). There are currently 20 different TAR designs worldwide that have been released for clinical use or are currently in development [14, 15]. FDA clearance (Class II) is currently limited to the Eclipse Total Ankle Implant (Kinetikos Medical, Inc.), Salto Talaris™ Total Ankle Prosthesis (Tornier), Agility Total Ankle Prosthesis (DePuy Orthopaedics), INBONE Total Ankle (INBONE Technologies, formerly Topez Orthopedics), and the Brigham Total Ankle Prosthesis (Howmedica Corp.). To date, there are no FDA-approved, non-constrained (Class III) TARs, although an FDA panel recommended approval of the STAR device in April 2007, with final approval of the device pending.

### 11.4.1 Early Designs

The first total ankle replacement was designed in 1970 and implanted by Lord and Marotte [16]. The design contained a tibial component with a long stem that was similar to the femoral component of total hip replacements. The talar component was a UHMWPE replacement of almost the entire talar body. After 10 years, only 7 of 25 arthroplasties were considered to be satisfactory, and 12 of the implants failed.

The Richard Smith ankle joint prosthesis (Figure 11.5), designed in 1972, was an unconstrained, ball-and-socket type design with metal tibial "socket" and UHMWPE talar "ball"[17]. It has been described as a multi-axis joint, where rotation of the tibia can occur about any of the major three axes and the rotational motion is not limited [18, 19]. In a clinical study of 24 Smith ankle replacements, with 18 replacements followed for an average of 7 years (implanted between 1975 and 1979), there was an improvement of range of motion in 10 ankles [17]. However, seven patients had loosening of tibial components, with three patients revised at 4+ years [17]. The authors concluded that although some patients were pleased, the Smith prosthesis could not be recommended [17].

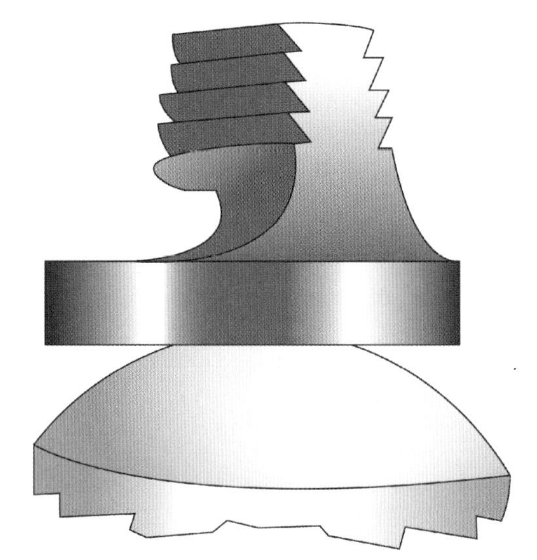

**FIGURE 11.5** Schematic of the Richard Smith ankle joint prosthesis.

The Imperial College of London Hospital (ICLH) implant was a two-part, semi-constrained prosthesis with a metal talar component and UHMWPE tibial component. It was designed in 1972, with subsequent design changes described in 1982 [20]. The design included a dome-shaped metal talar component and a UHMWPE tibial insert with large lateral flanges for fibular articulation; the UHMWPE component was trimmed to size at the time of operation [21] (Figure 11.6). In a clinical study of 24 ankles implanted in 22 patients, the authors concluded that patients were improved compared to their pre-surgery conditions but that the prosthesis was not superior to arthrodesis [22]. After a clinical trial reported in 1982, the procedure was not recommended due to the high complication rates [21].

The ICLH and another prosthesis, the St. Georg [23], can be considered as the basis of the total ankle prostheses that were developed after 1972 [24]. The St. Georg prosthesis was nonconstrained, while the ICLH was semi-constrained [25]. While the ICLH was eventually abandoned, the design of the St. Georg evolved into the "Endo" system, which was a semiconstrained, two-part prosthesis [24].

The Newton ankle prosthesis (designed in 1973) was a two-part, nonconstrained prosthesis that had a UHMWPE tibial component with a flat proximal surface and a curved, cylindrical articulating surface [26, 27] (Figure 11.7). It has been described as a multiaxial joint that allows unrestricted motion about any of the three major axes [19]. The talar component comprised CoCr alloy and had a spherical articulating surface with a distal stem that was implanted into the talus. The result was an incongruent articulating surface that allowed dorsiflexion/plantarflexion, as well as slight inversion, eversion, and rotation [26, 27].

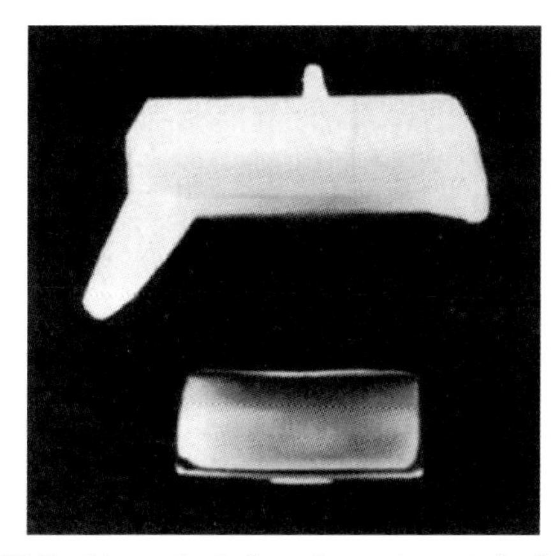
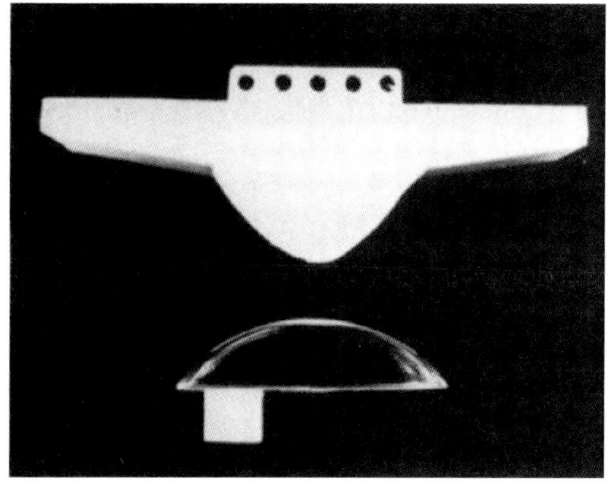

FIGURE 11.6 ICLH ankle prosthesis shown in anterior-posterior (left) and lateral (right) views (reproduced with permission from [21]).

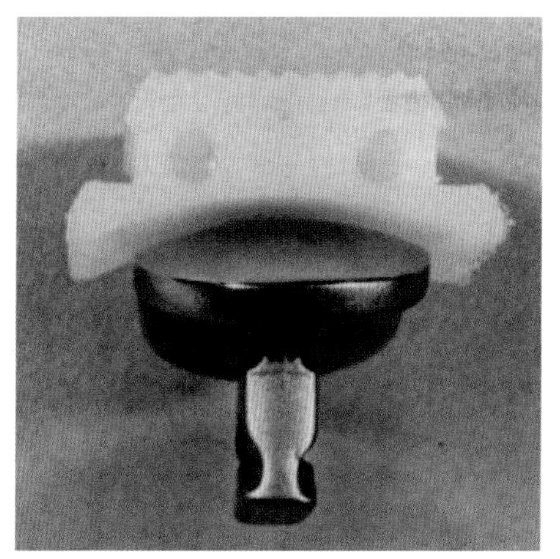
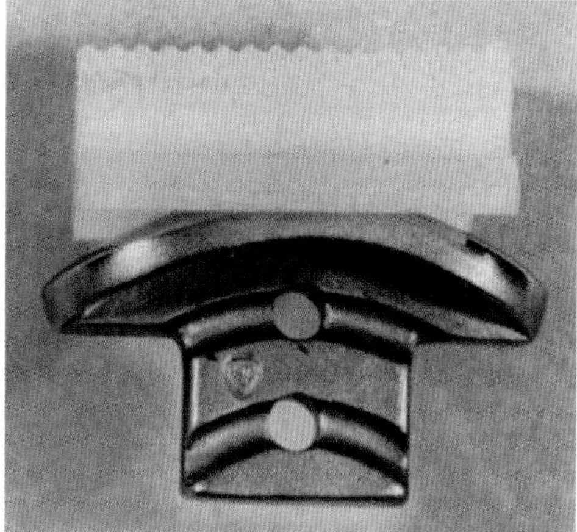

FIGURE 11.7 Newton ankle prosthesis (from [26], with permission).

The Irvine ankle prosthesis (also called the Howmedica total ankle) was a two-part, nonconstrained prosthesis with a UHMWPE tibial component and metal talar component that was designed in 1975 [28, 29] (Figure 11.8). It has been described as a multi-axial joint, which allows tibial motion about any of the three major axes and did not restrict rotational motion [18, 19]. The device was cemented and permitted dorsiflexion/plantarflexion of 114 degrees and abduction/adduction of 40 degrees [29]. Initial clinical results were satisfactory [28]. A biomechanics study in cadaveric human ankle specimens indicated that it did not reproduce normal ankle motion because there was increased coupled motion and hysteresis in ankles with the prosthesis during plantarflexion/dorsiflexion and axial rotation [30].

Designed in 1976, the Thompson Parkridge Richards prosthesis (TPR; Smith & Nephew) was a cemented, semi-constrained device that utilized a convex talar component and UHMWPE tibial component with a concave articular surface [19, 31–36]. The radius of curvature of the tibial component was greater than that of the talar component, resulting in a line contact area of $0.30\,cm^2$ [33]. The device utilized a "lip" on either side that restricted side-to-side movement of the talar component [33]. Patients had some improvement but "disappointing" clinical results due to delayed wound healing and loosening [32]. Subsidence and loosening of the talar component were also observed [31]. Case studies have also been reported where complications of loosening led to revision [37].

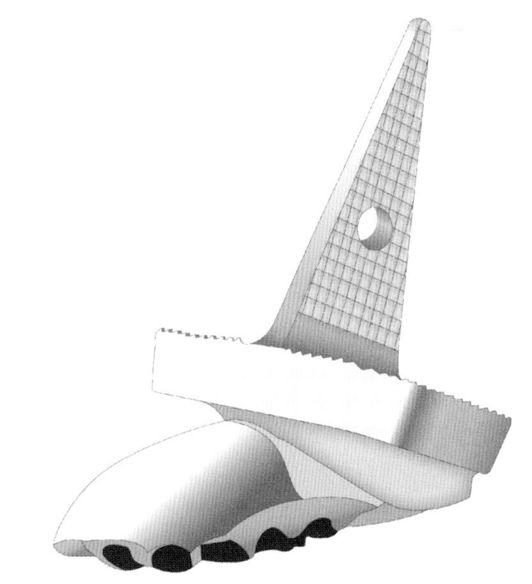

FIGURE 11.8 Schematic of the Irvine ankle prosthesis.

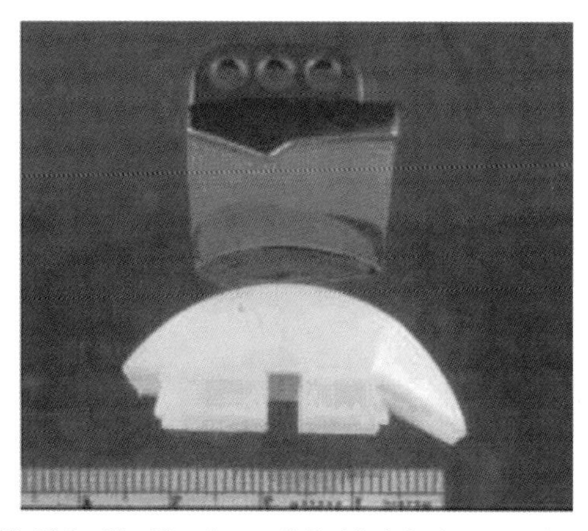

**FIGURE 11.9** The New Jersey Cylindrical Replacement (reproduced from [79] with permission from Endotec, Inc.).

The PCA (also called the Scholz) prosthesis was initially designed in 1976 and implanted in 1984; it was the first porous coated, non-cemented design [25, 38]. This semi-constrained device had a UHMWPE tibial component with a metal-backed tray and a metal talar component.

The New Jersey Cylindrical Replacement design was introduced in 1976 as a two-part prosthesis that had a metal tibial component and a UHMWPE talar dome [39] (Figure 11.9). The congruent spherical surfaces resulted in a contact area of 5.2 $cm^2$ [39]. The prosthesis allowed a total of 65 degrees of motion, with 20 degrees of dorsiflexion and 45 degrees of plantarflexion [39]. This design was subsequently revised and evolved into the contemporary Buechel-Pappas design, which is described later in this chapter.

The Mayo total ankle prosthesis was a two-part, constrained design that consisted of a UHMWPE tibial component and stainless steel talar component with congruent surfaces [40, 41]. The UHMWPE tibial component was concave, while the Vitallium talar component was convex [9]. The resulting articulating area was approximately 9 $cm^2$ [7, 40]. The device theoretically allowed 30 degrees of ankle dorsiflexion/plantarflexion before impingement with bony structures [7, 9]. Clinical results have been described [7, 42–47]. Generally, there were improvements in ROM or functional motion for a large percentage of patients, with some patients experiencing no change or loss of motion [43]. Clinical problems included fracture of the medial malleolus without major long-term effects in the first clinical trials, infection, delayed healing, residual bony impingement, and loosening in younger, active patients [7, 9, 40, 43, 44]. The clinical studies reported that the device should be indicated for less active, older patients [7, 9, 43]. Overall, the procedure was not recommended based on poor short-term and long-term survival rates [46].

Designed in 1984, the Bath and Wessex prosthesis (Howmedica International) was a multi-axial, cemented two-part prosthesis [48, 49]. It had a metal dome for the talus with a congruent UHMWPE tibial component [48, 49]. The correction of malalignment was difficult with this prosthesis, which resulted in undesirable varus/valgus forces [48]. As of a publication in 2001, the prosthesis was no longer manufactured [48].

### 11.4.2 Results of Early Designs

Ultimately, clinical results from early generation designs resulted in recommendations to discontinue use of TAR based on Mayo [43, 45, 46], Conaxial (Beck-Stefee) [50], Bath and Wessex [48], Newton [26], Waugh [18], Smith [18], ICLH [21], and Oregon [18]. Overall, indications for the initial TAR implants were limited due to the high complication rates and failure, with recommendations for use only in elderly patients with limited physical demands [19].

However, valuable lessons were learned from these designs. First, it was recognized that the use of bone cement required a larger resection of bone, which could result in subsidence of the metal components [4]. The community recognized the need to strike a balance between stability of the prosthesis with achieving larger range of motion [4]. Predominant clinical complications included loosening, persistent pain, and delayed wound healing [4]. In a clinical study of 21 patients with varying early-generation designs, signs of loosening (radiolucent lines) and decreased muscle strength occurred in 19/21 patients [18].

Based on objective analysis of gait and electromyography, patients who were implanted with early generation total ankle replacements did not function as well during activities of daily living as had been expected; ranges of motion were within normal limits, although motion patterns were "abnormal" with response to tibiopedal angles during certain aspects of the gait cycle [18] . The authors suggested that this may have been attributed to the weakness of calf and peroneal muscle groups on the side of the TAR.

One of the more valuable lessons learned from early generation designs was related to the UHMWPE component. It was recognized that incongruent articulating surfaces allowed dosiflexion/plantarflexion and axial rotation but resulted in high local stresses on UHMWPE components that could result in high wear rates [4, 39]. Congruent designs were therefore determined to be needed for greater stability and resistance to wear [4].

### 11.4.3 Contemporary Designs

Second-generation and contemporary designs offer improvement in TAR survival rate [36]. Currently, commonly used prostheses include [15]: STAR (Scandinavian Total Ankle Replacement, Waldemar Link), Buechel-Pappas Total Ankle Replacement (Endotec), TNK ankle (Japan Medical Materials), and Agility Total Ankle System (DePuy). Other TAR devices that currently have FDA clearance

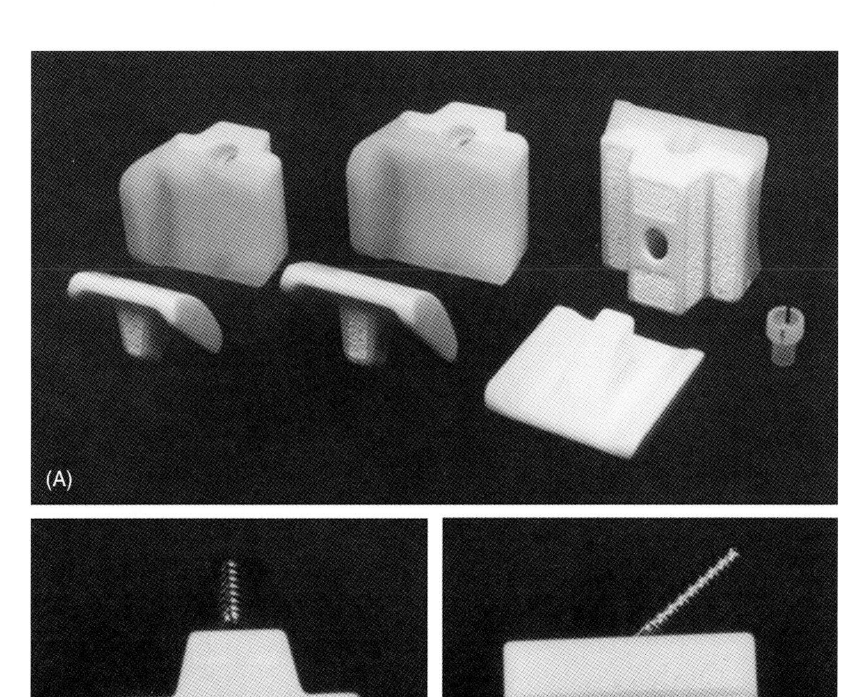

**FIGURE 11.10** Three generations of the ceramic TNK prosthesis with anterior (left) and lateral (right) views; reproduced from [54], with permission.

in the United States include [15]: INBONE Total Ankle (INBONE Technologies, formerly Topez Orthopedics), Salto Talaris™ Total Ankle (Tornier), and Eclipse Total Ankle (Integra Life Sciences Holdings). Although these devices can be used in the United States, little information is available related to the clinical performance of these devices.

### 11.4.3.1 TNK Prosthesis

The TNK prosthesis is unique in that it is a ceramic-on-UHMWPE device that utilizes a UHMWPE tibial articulating surface [51] (Figure 11.10). The device is a cementless, fixed bearing with partially conforming surfaces, having a concave ceramic talus that articulates with a flat UHMWPE tray [52]. The UHMWPE tray is secured to the ceramic tray on the tibial side [52]. Bony fixation methods include hydroxyapatite-coated beads and a tibial screw [52]. Concerns have been raised regarding the articulation and that it may transfer excessive shear and torque to the prosthesis-bone interface, which, along with the ceramic-bone interface, will affect the long-term mechanical fixation [52]. In a clinical study of the TNK, radiolucent lines were prevalent although they did not spread, and there were no instances of talar component sinking [53]. Clinical results also demonstrated a linear decrease in range of motion three years after operation, where range of motion after eight years equaled the range of motion measured preoperatively for both cemented and uncemented implants [51]. The inventors demonstrated that fewer cases of loosening and subsidence were observed in uncemented components compared to cemented components, although this study included a range of ceramic-on-UHMWPE and metal-on-UHMWPE devices [51, 54].

### 11.4.3.2 Agility Total Ankle Prosthesis

The Agility Total Ankle Prosthesis (DePuy) is a semi-constrained, fixed-bearing prosthesis that was designed in 1984 and currently has 510(k) clearance from the FDA [14] (Figure 11.11). In the United States, it is the most commonly used TAR prosthesis and has more than 20 years of clinical experience [55]. The device is currently offered in six sizes of matching tibial and talar components [56]. The device consists of a titanium alloy tibial component, a size-matched and separate UHMWPE insert that slides and locks into the tibial component, and an onlay cobalt-chromium talar component [55, 57, 58]. The tibial articular surface is rotated 20 degrees externally to mimic physiologic ankle anatomy [14, 55]. The semi-constrained

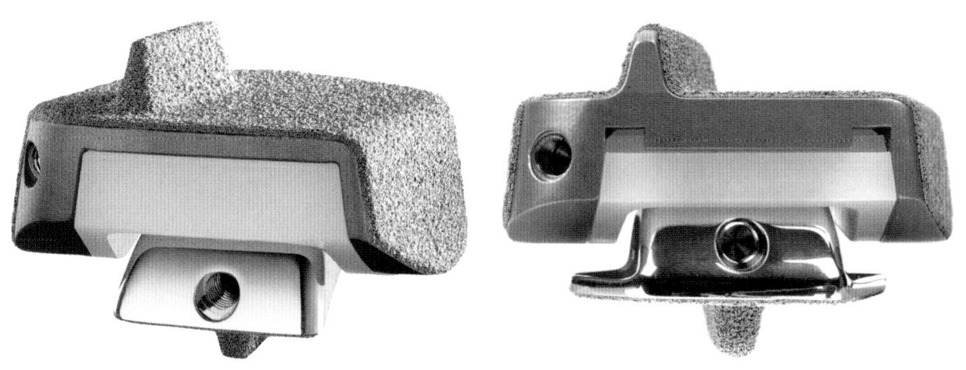

**FIGURE 11.11** The Agility Total Ankle Prosthesis (left) and Agility LP Total Ankle Prosthesis (right) (courtesy of DePuy Orthopaedics).

design allows dorsiflexion/plantarflexion, as well as axial rotation and medial-lateral shift of the talar component within the tibial plane [57–59]. The Agility has unique design attributes that broaden the bony base for the tibial component; this includes an obliquely-rectangular tibial component and a requirement for tibiofibular syndesmosis fusion that allows weight transfer to the fibula [59]. Both the tibial and talar surfaces that interface with bone have porous ongrowth surfaces [57–59]. Due to the patterns of (and larger amounts of) bone resection that are required for the Agility compared to other TARs (e.g., Buchel-Pappas), the Agility is often a potential candidate for revision procedures [60, 61]. Although the Agility is indicated for cemented use only [56], it has been used as an uncemented device off-label in the United States. The Agility permits dorsiflexion/plantarflexion of 60 degrees [57, 59].

The UHMWPE insert of the Agility is sterilized using gas plasma and packaged using a double blister/Tyvek package, which the manufacturer claims "addresses poly insert oxidation and degradation issues" [56]. The resin is either Ticona GUR 1020 or GUR 1050. Thickness of the UHMWPE component varies from 3.73 to 4.7 mm, depending on implant size. Revision components offer thicker "plus 2 mm" UHMWPE inserts that increase the thickness to 5.73 to 6.7 mm [52, 57]. Before 1987, the talar component of the Agility was manufactured from titanium; this was changed to CoCr in 1987 in light of some studies revealing poor wear qualities of titanium against UHMWPE [62].

Complications related to the UHMWPE component have been reported, including bone resorption (lysis), component loosening, and excessive wear resulting in failure of the UHMWPE component. Radiolucent lines have been observed in a clinical study of 95 patients implanted with Agility TARs between 1984 and 1993, but these were not attributed to UHMWPE wear debris due to the short timecourse (two years) and radiographic appearance; the loosening was attributed to high shear stresses between the implant and lateral malleolus [57]. In a continuation study published in 2004 of the first Agility TAR patients with an average follow-up of nine years, 49/117 ankles showed some evidence of mechanical lysis (7/49 progressive), and 18/117 ankles had expansile lysis of late onset (4/18 progressive) [63]. Other clinical studies have also reported lysis surrounding Agility TAR prostheses [62, 64–66]. UHMWPE wear and fracture can occur due to malpositioning, which can generate a region of excessive wear and potentially full-thickness catastrophic fracture of the UHMWPE component [67, 68] (Figure 11.12).

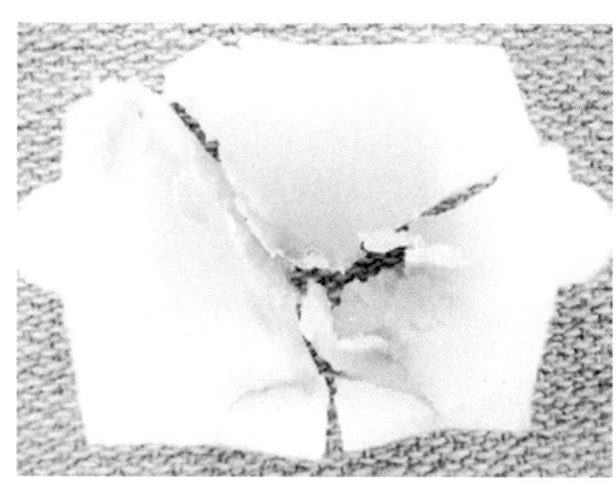

**FIGURE 11.12** Fractured UHMWPE component from an Agility that was retrieved from a patient with a talar component that was malaligned in varus (Copyright © 2008 by the American Orthopaedic Foot and Ankle Society, Inc., originally published in *Foot & Ankle International*, 24(12):901–903, and reproduced here with permission [67]).

Nicholson et al. reported a cadaver study utilizing 10 human ankle specimens, where ankles implanted with Agility TAR prostheses were axially and cyclically preconditioned and then loaded to 700 N for 10 cycles at a frequency of 1 Hz. During cyclic loading, the average contact pressure was 5.6 ± 2.1 MPa distributed over 0.83 $cm^2$, and the average peak pressure was 21.2 ± 5.7 MPa. In a separate phase of the study, the authors observed that contact pressure was larger for smaller component sizes and increased with increasing axial load. The axial loads applied in this study represented approximately 1 × body weight; forces of over 300% body weight could be expected during daily activities such as descending stairs. Therefore, it can be expected that mean peak pressures on the UHMWPE component would exceed those reported in this study.

Cadaveric testing conducted by Valderrabano et al. compared range of motion [69] and talus rotation [70] for normal and fused ankles compared to those implanted with Agility, Hintegra, and STAR prostheses. Compared to the normal condition, the Agility did not show a significant difference in plantarflexion but significantly decreased dorsiflexion [69]. The Agility also showed a significant and substantial increase in inversion/eversion compared to normal ankles, with no significant differences for internal/external rotation [69]. Valderrabano et al. concluded that the increase in overall inversion/eversion range of motion might be explained by the mismatch of size and congruency of the two components of the Agility, which could potentially lead to a tilt of the talus component and result in an overload of ankle ligaments and UHMWPE wear. Cadaveric ankles implanted with the Agility had decreased 60% talus rotation and 80% talus shift compared to normal ankles, while other three-component designs (i.e., Hintegra and STAR) did not differ in talus motion compared to normal ankles [70]. This restriction in talar motion may result in increased stresses within and around the prosthesis, possibly resulting in UHMWPE wear and loosening at the bone-prosthesis interfaces.

### 11.4.3.3 Scandinavian Total Ankle Replacement

The Scandinavian Total Ankle Replacement (STAR) prosthesis is currently manufactured by Link Orthopaedics, but the design was to be acquired by Small Bone Innovations, Inc., in 2008. The device is a three-component, cylindrical, mobile bearing design with CrCoMo tibial and talar components and a mobile, congruent UHMWPE component [33, 71] (Figure 11.13). The contact area is $3.2\,cm^2$ on the talar surface and $6.0\,cm^2$ on the tibial surface [33]. The thickness of the UHMWPE component is 6 to 10 mm [71]. The superior surface of the UHMWPE component is flat, making planar contact with the tibial component to allow translation and medial/lateral and anterior/posterior sliding within the constraints of the surrounding tissue. Dorsiflexion and plantarflexion take place on the curved talar surface. There is a groove in the UHMWPE component that conforms to a crest on the metal talar component; its length is oriented in the anterior-posterior direction to prevent sliding of the prosthesis in the medial-lateral direction [71]. The talar component has wings to replace the medial and lateral talus facets, which protects sculpted surfaces of the talus to allow resection for correcting malalignment, prevents degenerative arthritis of the facets, and allows for retention of blood vessels and ligaments [71]. The UHMWPE components were initially gamma sterilized in air, while more recent components are sterilized with gamma irradiation in a nitrogen-vacuum [72, 73]. Outside of the United States, earlier versions of the metal components have hydroxyapatite coatings to encourage bony ingrowth, while more recent designs available outside the United States have double-coated titanium plasma spray and hydroxyapatite surfaces [72]. Within the United States, CP-Ti plasma spray is available on the bony surfaces of both the tibial and talar components.

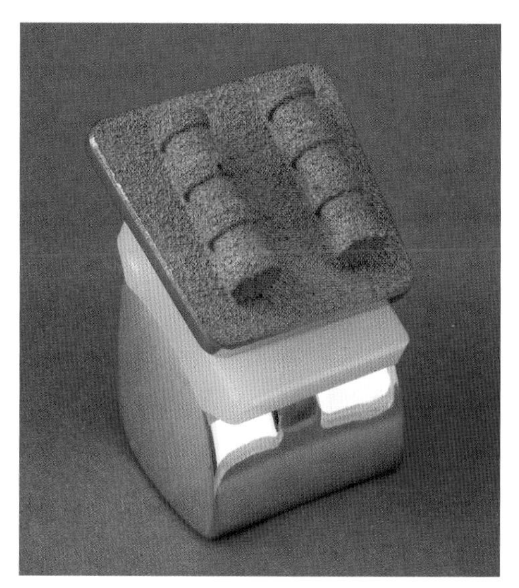

FIGURE 11.13 Scandinavian Total Ankle Replacement (STAR).

Cadaver studies have demonstrated changes in ankle biomechanics in normal ankles versus those implanted with STAR devices [69]. There were no significant differences in inversion/eversion. Dorsiflexion and plantarflexion were significantly reduced in STAR ankles. Significant increases in range of motion were observed during internal tibial rotation, while significant decreases in exterior tibial rotation were observed in the STAR ankles [69].

Biomechanical and clinical studies have emphasized that proper sizing and positioning of the UHMWPE component is essential for function of the STAR device. In a biomechanical study of six cadaveric ankles, anterior lift-off of the UHMWPE component was observed as a result of anterior positioning of the talar component and/or undersizing the UHMWPE component [74]. Anterior positioning could have caused edge loading on the posterior side of the UHMWPE component, where UHMWPE loading is localized to an edge of the component [74]. On the other hand, undersized UHMWPE components could cause subfibular impingement that reduces contact pressure on the anterior edge of the UHMWPE component [74].

Complications related to the UHMWPE component have been discussed in clinical studies. Edge loading of the UHMWPE component creates the potential for increased UHMWPE wear and premature failure. In a study by Wood et al. that investigated 200 STAR ankle implantations, nine patients experienced edge loading of the UHMWPE component [75]. Seven of those patients exhibited preoperative varus or valgus deformity that was greater than 15 degrees [75]. In a case study, a patient developed edge loading of the UHMWPE component due to varus tilting of the talus [76].

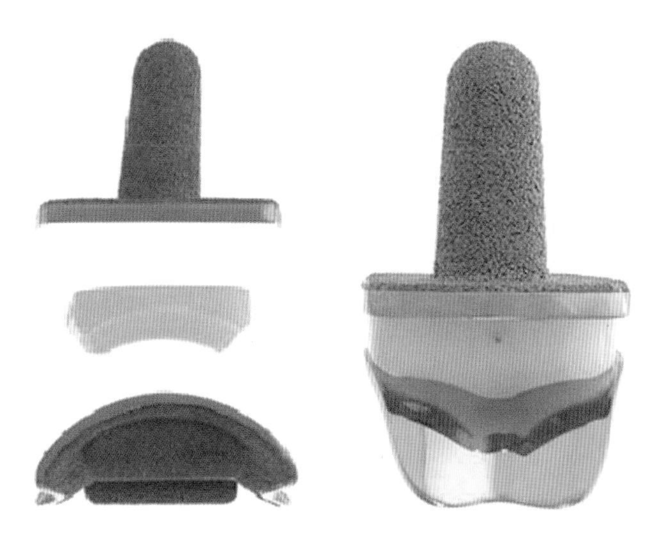

**FIGURE 11.14** The Buechel-Pappas total ankle prosthesis (reproduced from www.endotec.com, with permission).

The patient in this study developed a wear debris cyst in the fibula (confirmed by histology) with subsequent fracture; the UHMWPE component was replaced, and at 3 months post revision the patient's symptoms were improved [76]. Other clinical studies report UHMWPE fractures or revisions related to the UHMWPE component [31, 36, 77]; although these studies did not report a biomechanics rationale for the UHMWPE failure, exaggerated varus/valgus deformity and/or ligamentous instability resulting in accelerated wear could participate in the mechanism for failure of these components as well.

### 11.4.3.4 Buechel-Pappas

The New Jersey Cylindrical Replacement (Figure 11.9) later evolved into a trunion device, which was the first occasion of a three-part mobile bearing system being used in the ankle joint [4]. This later developed into the New Jersey Low Contact Stress total ankle replacement and was described previously [78]. This three-component design was cylindrical and cementless with congruent surfaces, and it was used from 1981 onward by inventors [78] (Figure 11.14). The metal components are manufactured from titanium with a porous coating to promote bony ingrowth, and the mobile core is UHMWPE [79]. A nitride ceramic film is added to the titanium bearing surfaces to improve the wear characteristics [79]. The device enables axial rotation and sliding in mediolateral/anteroposterior directions without constraint at the superior bearing surface [78]. The superior bearing surface is flat and somewhat smaller than the articulating surface of the tibial component, while the inferior surface was congruent to the trochlear groove in the talar component, which allows dorsiflexion/plantarflexion and eversion/inversion [78]. Wear on the superior surface was expected to be relatively low due to the fact that there was little secondary motion (axial rotation) compared to the primary motion (dorsiflexion/plantarflexion) and because the secondary bearing surface was 20% larger than the primary bearing surface [78]. The prosthesis provides 60 degrees dorsiflexion/plantarflexion with congruent contact and 30 degrees axial rotation with congruent contact [78]. Retention of the mobile UHMWPE bearing is achieved via compression of the collateral ligaments and adjacent tissues, as well as gravitational loads [78].

Failure modes of the New Jersey Low Contact Stress design included talar component subsidence, as well as lateral- and medial-bearing subluxation due to ligamentous instability or subsidence [80]. The device was modified to deepen the sulcus in the talar component, two fixation fins were added to the talar component to improve bony fixation, and the UHMWPE meniscal bearing was thickened [80]. Prior to the design change, the UHMWPE component was sterilized using gamma irradiation in air [81]. In the new design, the UHMWPE component is manufactured using 1150 UHMWPE powder, machined from extruded bar and sterilized using ethylene oxide. This design is used today and referred to as the Buechel-Pappas design (Endotec) [4]. Improvements in the design resulted in better survival rates and decrease in talar subsidence [4, 78, 81, 82]. Surgical technique, clinical results, and kinematics have been described extensively [81–84].

The performance of the UHMWPE component of the Buechel-Pappas design is dependent upon proper orientation of the articulating surfaces. Failure of the UHMWPE component has been observed in conjunction with varus or valgus deformity (caused by malleolar fracture or instability) that subsequently leads to edge loading of the UHMWPE component [80, 85, 86]. The existence of a preoperative deformity dramatically influenced the survival of both the Low Contact Stress and Buechel-Pappas designs, where TARs with varus or valgus deformity greater than 10 degrees had an overall survival rate of 48% at eight years, while ankles with neutral alignment preoperatively had an eight-year survival rate of 90% [85]. In one clinical study, osteolysis was suspected radiographically in 3/19 Buechel-Pappas patients at 4.7 years or more follow-up; it is unknown whether edge loading of the UHMWPE component or varus/valgus deformities caused accelerated wear and subsequent osteolysis in these patients [87].

### 11.4.3.5 Salto Talaris™ and Salto™

The Salto Talaris™ Anatomic Ankle (Tornier, Edina, Minnesota, USA) is a metal-on-UHMWPE, semi-constrained cemented TAR (Figure 11.15). It has been in clinical use since 2006 [88]. The device consists of two mating components: a CoCr tibial base in association with a conforming UHMWPE insert and a CoCr talar resurfacing component [89]. The bony interface of the metal components are coated with titanium plasma spray, although the device is intended for cemented use only in the United States [89].

**FIGURE 11.15** The Salto Talaris™ total ankle prosthesis (left) and Salto™ total ankle prosthesis (right), courtesy of Tornier, Inc.

The UHMWPE component is manufactured per ISO 5834-2 [89]. The UHMWPE resin is Type B 4150HP, and it is sterilized using gamma irradiation. Sterilization of the UHMWPE component in a vacuum with double-peel package was introduced at the first launch of the Salto™. The UHMWPE insert combined with the tibial component provides thicknesses of 8, 9, 10, and 11 mm, which is compatible with various levels of instability [88]. The insert is fixed to the tibial component before implantation. The polyethylene insert design allows ±5 degrees of internal-external rotation, ±4 degrees of varus-valgus rotation, and ±2 mm of anterior-posterior translation.

The Salto™ implant (Tornier SAS, Saint Ismier, France) (Figure 11.15) was developed between 1994 and 1996 and has been used clinically since January 1997 [90]. It is distinguished from the Salto Talaris™ Anatomic Ankle in that the Salto™ is a three-component design (unconstrained, possible mobility between the plane top area of the liner and the bottom plane mirror polished tibial component), while the Salto Talaris™ is a two-component design (semi-constrained). The design consists of metal tibial and talar components with a mobile UHMWPE insert. The resin, sterilization, and packaging of the UHMWPE component, as well as the CoCr used in the metal components, are identical to that of the Salto Talaris™. The coating on the bony surface of the Salto is titanium with hydroxyapatite. The Salto™ was intended as an improvement to three-component mobile bearing prostheses to optimize the bearing surfaces, the conical articular surface, accuracy of positioning, and primary fixation of the implant while requiring minimal resurfacing of the bony surfaces [90]. The UHMWPE mobile bearing is available in thicknesses from 4–8 mm (insert only), and it maintains full congruency with the talar component in dorsiflexion and plantarflexion. The UHMWPE component also accommodates as much as 4 degrees varus and valgus deformity or mobility by keeping congruency to avoid edge loading [90]. Coronal plane stability is accomplished via a central, sagittal sulcus on the talar component that conforms to a central ridge on the inferior surface of the UHMWPE component [14]. There is also a medial stop mechanism on the tibial component to avoid impingement with the internal malleolus [14]. While there are few clinical studies published at the time of this writing, one clinical study of the 98 first cases of the prosthesis (93 available at follow-up) showed no evidence of edge loading after a mean follow-up of 35 months (range: 24 to 68 months) [90]. Further clinical follow-up with this patient cohort is pending publication.

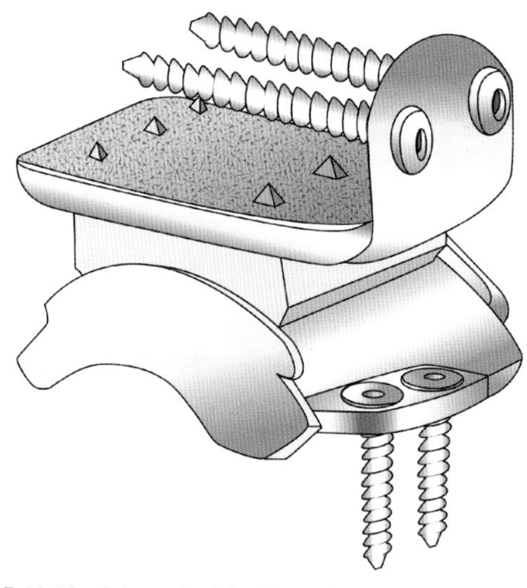

**FIGURE 11.16** Schematic of the HINTEGRA total ankle prosthesis.

### *11.4.3.6 HINTEGRA*

The HINTEGRA prosthesis was first implanted in May 2000 [4]. It is a three-component device (Figure 11.16) and has a unique feature in that the tibial and talar components have screws to reduce the amount of bone resection required at the tibia and talus [4, 91]. The tibial articulating surface is flat, while the talar articulating surface is

**FIGURE 11.17** Mobility Total Ankle System (courtesy of DePuy Orthopaedics).

conically shaped; both surfaces conform to the UHMWPE bearing [91]. The prosthesis provides 50 degrees of congruent contact in dorsiflexion and plantarflexion and 50 degrees of congruent contact in axial rotation [91]. Motion is restricted by the natural soft tissue [91]. The device is also designed with a degree of inclination on the tibial component to provide resistance to the posterior load produced during walking [91]. A biomechanical analysis of cadaveric ankles indicated that ankles implanted with the HINTEGRA had significantly less dorsiflexion, while there was no difference in plantarflexion, inversion/eversion, or internal/external rotation [69]. Clinical results are limited, where one study of 122 HINTEGRA implantations followed for an average of 18.5 months demonstrated a 92% survival rate [91].

### *11.4.3.7 Mobility*

The Mobility Total Ankle System (DePuy) became commercially available in 2004 and is currently being compared to the Agility in a multi-center FDA trial [14] (Figure 11.17). The device is an unconstrained, three-component design with a UHMWPE mobile bearing [52]. The design of the tibial component resembles the Buechel-Pappas because of the conical stem that protrudes from the superior surface, although the sagittal plane dimensions of the Mobility's tibial component are larger than those of the Buechel-Pappas and is tapered posteriorly to avoid overhang and soft tissue impingement [14]. The talar component can be inserted while retaining the medial and lateral aspects of the talus, similar to the Buechel-Pappas [14]. The inferior surface of the UHMWPE component has a ridge that conforms to a

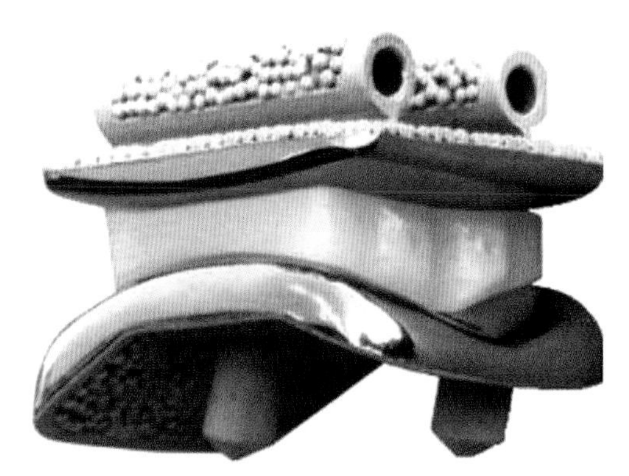

**FIGURE 11.18** BOX prosthesis (reproduced from [4], with permission).

groove on the talar component, and the superior surface is flat and conforms to the flat articulating surface of the tibial component. The constraint offered at the talar articulating surface is therefore dissipated by the tibial articulating surface. The UHMWPE component is manufactured from 1050 UHMWPE powder, which is machined from extruded bar and sterilized using gamma irradiation under vacuum (GVF) [92]. Mechanical and wear testing demonstrated that the Mobility compares favorably with the Buechel-Pappas implant [14]. Similar loads were required to cause dislocation of the UHMWPE component, although the Mobility dislocated less often and generated less UHMWPE wear [14, 92]. Little is known about the clinical performance of this device.

### *11.4.3.8 BOX*

The BOX prosthesis derived its name from its designers from the Instituti Ortopedici Rizzoli, Bologna, Italy, and the laboratory in Oxford, England [14] (Figure 11.18). The design is a three-component device with an UHMWPE mobile bearing. The metal tibial and talar components are manufactured from CoCr with titanium spray coating on the surface [14]. Similar to the Buechel-Pappas design, the BOX resurfaces only the superior dome of the talus, and the sides remain intact [14]. There is a spherical interface between the articular surface of the tibial component and UHMWPE component, allowing rotation about all three axes [93]. There is a concave sulcus in the talar component, and the UHMWPE inferior surface is fully conforming with a bi-concave surface [93]. The size of the UHMWPE component varies in 1-mm steps from 5 mm to 8 mm [93]. The UHMWPE component is manufactured from compression molded sheet from PUR 1020, complies with ISO 5834: Implants for surgery—Ultra-high molecular weight polyethylene Part 1 (Powder form) and Part 2 (Moulded forms), and is sterilized using gamma irradiation in the presence of nitrogen [94].

Similar to all other three-component, mobile bearing designs, the BOX prosthesis is dependent on ligamentous stability to prevent dislocation of the UHMWPE component [93]. Clinical results are currently limited to few patients and short-term follow-up periods [93, 95].

**TABLE 11.1 Contact Pressures for Different TAR Components**

| Device | Type of study | Axial load (N) | Location | Contact pressure (MPa) |
|---|---|---|---|---|
| Agility [98] | FEM | 3330 | Talar surface | Peak: 26–36 |
| Agility [98] | FEM | 3330 | Center | Peak: 20–24 |
| Agility [99] | Cadaver | 700 | Not indicated | Average: 5.6<br>Peak: 21.2 |
| BOX [96] | FEM | 1600 | Tibial surface | Average: 6.4<br>Peak: 10.3 |
| BOX [96] | FEM | 1600 | Talar surface | Average: 10.3<br>Peak: 16.1 |
| STAR [97] | FEM | 3650 | Edges | 20 |
| STAR [97] | FEM | 3650 | Superior/inferior surfaces | 8–10 |

## 11.5 UHMWPE LOADING AND WEAR IN TOTAL ANKLE REPLACEMENTS

Because the ankle joint can experience loading in excess of five times body weight, the TAR needs to tolerate these high loads while maintaining a small contact area. As reviewed by Reggiani et al. [96] and summarized in Table 11.1, contact pressures in TARs have been well studied in cadaveric and finite element studies. Peak contact stresses for ankle replacements are summarized in Table 11.1. In a finite element study of the BOX prosthesis, the tibial and talar articulating surfaces had mean contact pressures of 6.4 and 10.3 MPa, respectively, with peak values of 10.3 and 16.1 MPa, respectively [96]. In a finite element study, the STAR component had larger contact pressures at the anterior, posterior, and internal edges compared to the superior and inferior surfaces (20 MPa versus 8–10 MPa, respectively) under 3650 N of compressive load [97]. In a finite element study of the Agility prosthesis, peak pressures occurred along the talar component edges [98]. Contact pressures in this region ranged from 26 to 36 MPa, with contact pressures in the center ranging from 20 to 24 MPa (a reduction in contact pressure was observed by increasing the width of the talar component) [98].

It has been postulated that the dominant mode of wear for first-generation, incongruent TARs was related to high local stresses occurring approximately 1 mm below the surface [100]. The theory proposed that points of high stress initiate cracks below the surface, and coalescence of these cracks can produce pitting, delamination, and finally propagation of crack(s) throughout the thickness of the UHMWPE component [100]. TARs with congruent surfaces were intended to mitigate these failure mechanisms of the UHMWPE components. However, clinical research of total knee replacements does not support this theory. Conforming and nonconforming total knee replacement designs experience peak stresses that may be sufficiently high to produce localized yielding and deformation of UHMWPE inserts [101]. However, these designs exhibit successful clinical performance [102, 103].

While data related to contact stress is useful, it is not the sole governing entity with respect to failure of UHMWPE components. Factors that influence the oxidation of the UHMWPE (e.g., sterilization, type, packaging), in combination with load distribution, affect its performance. Improvements in TAR UHMWPE component clinical performance likely occurred not only from the development of congruent designs that minimized contact stress but also improvements in UHMWPE sterilization and packaging (e.g., the STAR device, which was gamma sterilized in air and is currently sterilized in a nitrogen-vacuum).

In light of the history of UHMWPE sterilization in other orthopedic devices, as well as cell culture studies that have been conducted, devices that are gamma sterilized in nitrogen (e.g., STAR, Mobility, BOX) benefit from some degree of crosslinking that occurs from the gamma irradiation, which could improve the wear properties of UHMWPE. However, the free radicals formed in the UHMWPE during irradiation increase the component's susceptibility to *in vivo* oxidative degeneration, which subsequently alters the physical, chemical, and mechanical properties of UHMWPE. Barrier packaging and sterilization in an inert environment (e.g., nitrogen-vacuum sterilization of the STAR device, GVF packaging of the Mobility) address the risk of oxidative degeneration during shelf storage. Other sterilization methods (e.g., gas plasma, which is used in the Agility total ankle prosthesis) may improve UHMWPE wear properties without introducing free radicals that can lead to *in vivo* oxidation. While it has been demonstrated that sterilization via ethylene oxide (as is used in the contemporary Buechel-Pappas device) does not alter the physical, chemical, and mechanical properties of UHMWPE, ethylene oxide sterilization is not associated with improved wear rate that is observed with gamma irradiation (refer to Chapter 3).

In a wear simulation study comparing the Buechel-Pappas device with the Mobility, Bell and colleagues utilized a knee wear simulator to evaluate wear in TAR prostheses [92]. The UHMWPE components for the Buechel-Pappas devices were manufactured from 1150 UHMWPE powder, machined from extruded bar, and sterilized using gas plasma. The Mobility UHMWPE insert was manufactured from 1050 UHMWPE powder, machined from extruded bar, and sterilized using gamma irradiation under vacuum (GVF). The components were cyclically loaded under 3.1 kN maximum load using a load profile that mimicked the gait cycle. After the first 5 million cycles (MC), where there was no anterior/posterior displacement, the mean wear rate for the Buechel-Pappas device was (mean ± 95% confidence interval) 10.36 ± 11.8 $mm^3$/MC, and for the Mobility it was (mean ± 95% confidence interval) 3.38 ± 10.0 $mm^3$/MC. After introducing anterior/posterior displacement from 5 MC to 6 MC, wear rate significantly increased for the Buechel-Pappas TAR (16.4 ± 17.4 $mm^3$/MC) and the Mobility (10.4 ± 14.7 $mm^3$/MC). There was no significant difference in wear rate noted between the two designs at any time point, although this may have been due to the small sample sizes (n = 3). The Mobility had wear scars that ranged from 78% to 86% of the surface area, while the wear area for the Buechel-Pappas device was 100% of the contact area. More backside wear occurred on the Mobility design compared to the Buechel-Pappas design. The authors conclude that the observed trends from this study may have been due to differences in UHMWPE formulation, where the 1150 UHMWPE in the Buechel-Pappas design may have had a slightly higher wear rate than the nonstearate containing 1050 UHMWPE in the Mobility. They also suggest that the larger contact area in the Buechel-Pappas design may be susceptible to greater wear via a mechanism of "lubrication starvation."

A similar study was conducted for the BOX prosthesis, where a knee wear simulator was used to evaluate the wear rate of the prosthesis [94]. Loading consisted of plantar-dorsiflexion being applied to the tibial component above, while anterior-posterior displacement and internal-external rotation were imposed on the talar component, while the support of the talar component was self-aligning along the medial-lateral displacement and varus/valgus rotation. The maximum axial load was 1600 N, and the load profile was designed to be similar to walking with −10 to +20 degrees of dorsi-plantarflexion, −2.6 to 7.7 degrees internal/external rotation, and 0 to 8.45 mm of anterior-posterior displacement for 2 million cycles (MC) at 0.9 Hz. Tests were conducted in deionized water at 37°C. The linear penetration rate was 0.0178, 0.0081, and 0.0339 mm per MC for the three specimens. The average wear at 2 MC was 37.2 $mm^3$ (range 15.0–63.3 $mm^3$), corresponding to a wear rate of 18.6 $mm^3$/MC (range: 7.5–31.7 $mm^3$/MC). The variability in wear rate corresponded to the variability seen in wear patterns, where two components demonstrated predominant scratches and polishing in the anterior-posterior direction due to sliding, and the third demonstrated nearly no wear on the most anterior third of the tibial component. The differences in wear pattern may have been due to effects of device positioning or potential damage during the calibration phase.

During an *in vivo* study of the size and shape of wear particles in TARs, synovial fluid was aspirated at least 6 months postoperatively from 15 patients with TAR implants and 11 additional patients with posterior stabilized total knee replacements [104]. The TAR designs included the Agility (n = 4) and STAR (n = 11). The two TAR designs generated UHMWPE wear in similar concentrations, size, and morphology. UHMWPE particles from TARs were significantly rounder compared to total knee replacement. There was also no significant difference between the concentration of UHMWPE particles in TAR patients compared to patients with total knee replacements. The authors acknowledge that the capacity of the ankle joint differs from the knee joint, which may influence the body's response and potential for osteolysis in the presence of similar concentrations of UHMWPE particles.

The few wear studies that have been conducted on TARs highlight the need for more of this type of testing, as well as the need for a standardized test method. The two studies that were published demonstrated comparable wear rates (mean values approximately 10 to 20 $mm^3$/MC), though there was high interspecimen variability. The work of Bell et al. demonstrated how modifications to a loading profile can substantially affect the amount of wear that is generated [92]. A standardized testing protocol would allow for more direct comparisons of wear rates across different devices tested by different research groups. For example, the work of Bell et al. suggests that a larger contact surface does not necessarily result in a reduction in wear; repeatable experimental methods utilizing similar loading profiles, lubricants, and number of loading cycles would allow for a better evaluation of the effects on device design and UHMWPE type and sterilization method.

## 11.6 COMPLICATIONS AND RETRIEVAL ANALYSIS

Intraoperative complications for TARs include injury to neurovascular or soft tissue structures, malpositioning, improper sizing, excessive bone resection, and/or malleolar fractures [15]. Early postoperative complications include infection, impaired wound healing, swelling, stress fractures around the medial malleolus, and/or syndesmotic nonunions (applies to Agility only) [15]. Delayed postoperative complications include deep infection, periprosthetic radiolucencies, aseptic loosening/subsidence, periprosthetic fractures, UHMWPE wear with osteolysis, migration or fracture, heterotopic bone formation, syndesmotic nonunions, and reflex sympathetic dystrophy [15]. UHMWPE wear and aseptic loosening are among the most common complications, as well as wound healing problems and deep infection [15]. A major late complication is failure secondary to loosening [15].

For three-component devices, dislocation of the UHMWPE component may result in altered biomechanics of the prosthesis and accelerated generation of wear particles. As previously reviewed, UHMWPE wear particles in TAR patients were found in similar concentrations as knee replacement patients [104]. This may suggest that the occurrence of osteolysis in TARs could be expected to be similar to that in total knee replacements; however, the effect of the same concentration of UHMWPE particles on the relatively smaller ankle joint remains to be seen. Although the potential for dislocation and subsequent accelerated wear exists for three-component, mobile designs, the added degrees of freedom allows for more physiologic ranges of motion. Ligamentous stability and correction of varus/valgus deformities are important aspects of preventing dislocation of the UHMWPE component. In two-component (fixed-bearing) designs, the partial conformity at the single articulation increases stability but also increases contact stress and wear. These issues, along with considerations regarding the amount of bone resection required for a given design, should be considered when choosing which TAR is appropriate for a particular patient.

## 11.7 CONCLUSION

Many lessons have been learned since first-generation TARs were implanted in the 1970s. Improvements to the first-generation designs have resulted in better loading conditions for the UHMWPE component, and changes to the sterilization methods and packaging have reduced the propensity of UHMWPE oxidative degeneration. Similar to other total joint replacements, a balance between stability and natural joint motion is desired. Improvements to TAR designs and surgical technique will continue to improve the survival and clinical performance of ankle replacements.

## 11.8 ACKNOWLEDGMENTS

This chapter was not written with the financial support of any manufacturer of total ankle replacements. However, some of the manufacturers whose products are mentioned in this review were contacted by the author and given the opportunity to verify the factual accuracy of the information related to their products. Thanks to Chris Espinosa and Michael Drzal of Exponent, Inc.'s Visual Communication Practice for their assistance with figures and photographs. The authors are especially grateful to Hina Patel and Alexis Cohen for their editorial assistance with this chapter.

## REFERENCES

1. Sarrafian SK. *Anatomy of the foot and ankle: descriptive, topographic, functional*. 1st ed. Philadelphia: Lippincott; 1983.
2. Donatelli R, editor. *The biomechanics of the foot and ankle*. 2nd ed. Philadelphia: F.A. Davis Company; 1995.
3. Netter FH, Colacino S. *Atlas of human anatomy*. Summit, N.J.: CIBA-GEIGY Corp., 1989.
4. Vickerstaff JA, Miles AW, Cunningham JL. A brief history of total ankle replacement and a review of the current status. *Med Eng Phys* 2007 December;**29**(10):1056–64.
5. Siegler S, Chen J, Schneck CD. The three-dimensional kinematics and flexibility characteristics of the human ankle and subtalar joints—Part I: kinematics. *J Biomech Eng* 1988 November;**110**(4):364–73.
6. de Asla RJ, Wan L, Rubash HE, Li G. Six DOF in vivo kinematics of the ankle joint complex: application of a combined dual-orthogonal fluoroscopic and magnetic resonance imaging technique. *J Orthop Res* 2006 May;**24**(5):1019–27.
7. Stauffer RN. Total joint arthroplasty. The ankle. *Mayo Clin Proc* 1979 September;**54**(9):570–5.
8. Murray MP, Drought AB, Kory RC. Walking patterns in normal men. *Bone Joint Surg* 1964;**46A**:335–60.
9. Stauffer RN. Total ankle joint replacement. *Arch Surg* 1977 September;**112**(9):1105–9.
10. Wang Q, Whittle M, Cunningham J, Kenwright J. Fibula and its ligaments in load transmission and ankle joint stability. *Clin Orthop Relat Res* 1996;**330**(261–270).
11. Lambert KL. The weight-bearing function of the fibula. A strain gauge study. *J Bone Joint Surg Am* 1971;**53**:507–13.
12. Wagner UA, Sangeorzan BJ, Harrington RM, Tencer AF. Contact characteristics of the subtalar joint: load distribution between the anterior and posterior facets. *J Orthop Res* 1992 July;**10**(4):535–43.
13. McKinley TO, Rudert MJ, Koos DC, Pedersen DR, Baer TE, Tochigi Y, et al. Contact stress transients during functional loading of ankle stepoff incongruities. *J Biomech* 2006;**39**(4):617–26.
14. Deorio JK, Easley ME. Total ankle arthroplasty. *Instr Course Lect* 2008;**57**:383–413.
15. Bestic JM, Peterson JJ, DeOrio JK, Bancroft LW, Berquist TH, Kransdorf MJ. Postoperative evaluation of the total ankle arthroplasty. *AJR Am J Roentgenol* 2008 April;**190**(4):1112–23.
16. Lord G, Marotte JH. [Total ankle prosthesis. Technic and 1st results. Apropos of 12 cases]. *Rev Chir Orthop Reparatrice Appar Mot* 1973 March;**59**(2):139–51.
17. Kirkup J. Richard Smith ankle arthroplasty. *J R Soc Med* 1985 April;**78**(4):301–4.
18. Demottaz JD, Mazur JM, Thomas WH, Sledge CB, Simon SR. Clinical study of total ankle replacement with gait analysis. A preliminary report. *J Bone Joint Surg Am* 1979 October;**61**(7):976–88.
19. Lachiewicz PF. Total ankle arthroplasty. Indications, techniques, and results. *Orthop Rev* 1994 April;**23**(4):315–20.

20. Samuelson KM, Freeman MA, Tuke MA. Development and evolution of the ICLH ankle replacement. *Foot Ankle* 1982 July–August;**3**(1):32–6.
21. Bolton-Maggs BG, Sudlow RA, Freeman MA. Total ankle arthroplasty. A long-term review of the London Hospital experience. *J Bone Joint Surg Br* 1985 November;**67**(5):785–90.
22. Herberts P, Goldie IF, Korner L, Larsson U, Lindborg G, Zachrisson BE. Endoprosthetic arthroplasty of the ankle joint. A clinical and radiological follow-up. *Acta Orthop Scand* 1982 August;**53**(4):687–96.
23. Kaukonen JP, Raunio P. Total ankle replacement in rheumatoid arthritis: a preliminary review of 28 arthroplasties in 24 patients. *Ann Chir Gynaecol* 1983;**72**(4):196–9.
24. Bauer G, Eberhardt O, Rosenbaum D, Claes L. Total ankle replacement. Review and critical analysis of the current status. *Foot and Ankle Surg* 1996;**2**(2):119–26.
25. Gould JS, Alvine FG, Mann RA, Sanders RW, Walling AK. Total ankle replacement: a surgical discussion. Part I. Replacement systems, indications, and contraindications. *Am J Orthop* 2000 August;**29**(8):604–9.
26. Newton III SE. Total ankle arthroplasty. Clinical study of fifty cases. *J Bone Joint Surg Am* 1982 January;**64**(1):104–11.
27. Newton III SE. An artificial ankle joint. *Clin Orthop Relat Res* 2004 July(424):3–5.
28. Evanski PH, Waugh TR. Management of arthritis of the ankle. An alternative of arthrodesis. *Clin Orthop Relat Res* 1977 January–February(122):110–15.
29. Waugh TR, Evanski PM, McMaster WC. Irvine ankle arthroplasty. Prosthetic design and surgical technique. *Clin Orthop Relat Res* 1976 January–February(114):180–4.
30. Michelson JD, Schmidt GR, Mizel MS. Kinematics of a total arthroplasty of the ankle: comparison to normal ankle motion. *Foot Ankle Int* 2000 April;**21**(4):278–84.
31. Schill S, Biehl C, Thabe H. [Ankle prostheses. Mid-term results after Thompson-Richards and STAR prostheses]. *Orthopade* 1998 March;**27**(3):183–7.
32. Jensen NC, Kroner K. Total ankle joint replacement: a clinical follow up. *Orthopedics* 1992 February;**15**(2):236–9.
33. Wood PLR, Clough TM, Jari S. Clinical comparison of two total ankle replacements. *Foot Ankle Int* 2000 July;**21**(7):546–50.
34. De Bastiani G, Vecchini L. Arthroprosthesis of the ankle joint. *Ital J Orthop Traumatol* 1981 April;**7**(1):31–9.
35. Das Jr. AK Total ankle arthroplasty: a review of 37 cases. *J Tenn Med Assoc* 1988 November;**81**(11):682–5.
36. Fevang BT, Lie SA, Havelin LI, Brun JG, Skredderstuen A, Furnes O. 257 ankle arthroplasties performed in Norway between 1994 and 2005. *Acta Orthop* 2007 October;**78**(5):575–83.
37. Schill S. [Ankle arthrodesis with interposition graft as a salvage procedure after failed total ankle replacement]. *Oper Orthop Traumatol* 2007 December;**19**(5–6):547–60.
38. Scholz KC. Total ankle arthroplasty using biological fixation components compared to ankle arthrodesis. *Orthopedics* 1987 January;**10**(1):125–31.
39. Pappas M, Buechel FF, DePalma AF. Cylindrical total ankle joint replacement: surgical and biomechanical rationale. *Clin Orthop Relat Res* 1976 July–August(118):82–92.
40. Stauffer RN. Total ankle joint replacement as an alternative to arthrodesis. *Geriatrics* 1976 March;**31**(3):79–82. 85.
41. Stauffer RN, Chao EY, Brewster RC. Force and motion analysis of the normal, diseased, and prosthetic ankle joint. *Clin Orthop Relat Res* 1977(127):189–96.
42. Lachiewicz PF, Inglis AE, Ranawat CS. Total ankle replacement in rheumatoid arthritis. *J Bone Joint Surg Am* 1984 March;**66**(3):340–3.
43. Stauffer RN, Segal NM. Total ankle arthroplasty: four years' experience. *Clin Orthop Relat Res* 1981 October(160):217–21.
44. Unger III AS, Inglis AE, Mow CS, Figgie HE. Total ankle arthroplasty in rheumatoid arthritis: a long-term follow-up study. *Foot Ankle* 1988 February;**8**(4):173–9.
45. Kitaoka HB, Patzer GL. Clinical results of the Mayo total ankle arthroplasty. *J Bone Joint Surg Am* 1996 November;**78**(11):1658–64.
46. Kitaoka HB, Patzer GL, Ilstrup DM, Wallrichs SL. Survivorship analysis of the Mayo total ankle arthroplasty. *J Bone Joint Surg Am* 1994 July;**76**(7):974–9.
47. Kitaoka HB, Romness DW. Arthrodesis for failed ankle arthroplasty. *J Arthroplasty* 1992 September;**7**(3):277–84.
48. Carlsson A, Henricson A, Linder L, Nilsson JA, Redlund-Johnell I. A 10-year survival analysis of 69 Bath and Wessex ankle replacements. *Foot and Ankle Surg* 2001;**7**(1):39–44.
49. Kirkup J. Rheumatoid arthritis and ankle surgery. *Ann Rheum Dis* 1990 October;**49**(Suppl. 2):837–44.
50. Wynn AH, Wilde AH. Long-term follow-up of the Conaxial (Beck-Steffee) total ankle arthroplasty. *Foot Ankle* 1992 July–August;**13**(6):303–6.
51. Takakura Y, Tanaka Y, Sugimoto K, Tamai S, Masuhara K. Ankle arthroplasty. A comparative study of cemented metal and uncemented ceramic prostheses. *Clin Orthop Relat Res* 1990 March(252):209–16.
52. Easley ME, Vertullo CJ, Urban WC, Nunley JA. Total ankle arthroplasty. *J Am Acad Orthop Surg* 2002 May–June;**10**(3):157–67.
53. Nagashima M, Takahashi H, Kakumoto S, Miyamoto Y, Yoshino S. Total ankle arthroplasty for deformity of the foot in patients with rheumatoid arthritis using the TNK ankle system: clinical results of 21 cases. *Mod Rheumatol* 2004;**14**(1):48–53.
54. Takakura Y, Tanaka Y, Kumai T, Sugimoto K, Ohgushi H. Ankle arthroplasty using three generations of metal and ceramic prostheses. *Clin Orthop Relat Res* 2004 July(424):130–6.
55. Cerrato R, Myerson MS. Total ankle replacement: the agility LP prosthesis. *Foot Ankle Clin* 2008 September;**13**(3):485–94.
56. AGILITY™ Total Ankle System. 2005–2006 [cited 2008; Available from: http://www.depuyorthopaedics.com
57. Pyevich MT, Saltzman CL, Callaghan JJ, Alvine FG. Total ankle arthroplasty: a unique design. Two to twelve-year follow-up. *J Bone Joint Surg Am* 1998 October;**80**(10):1410–20.
58. Saltzman CL, Alvine FG. The agility total ankle replacement. *Instr Course Lect* 2002;**51**:129–33.
59. Alvine FG. Total ankle arthroplasty: new concepts and approaches. *Contemp Orthop* 1991 April;**22**(4):397–403.
60. Assal M, Greisberg J, Hansen Jr ST. Revision total ankle arthroplasty: conversion of New Jersey low contact stress to agility: surgical technique and case report. *Foot Ankle Int* 2004 December;**25**(12): 922–5.
61. Rippstein PF. Clinical experiences with three different designs of ankle prostheses. *Foot Ankle Clin* 2002 December;**7**(4):817–31.
62. Alvine FG. The agility ankle replacement: the good and the bad. *Foot Ankle Clin* 2002 December;**7**(4):737–53, vi.
63. Knecht SI, Estin M, Callaghan JJ, Zimmerman MB, Alliman KJ, Alvine FG, et al. The agility total ankle arthroplasty. Seven to sixteen-year follow-up. *J Bone Joint Surg Am* 2004 June;**86-A**(6):1161–71.
64. Kopp FJ, Patel MM, Deland JT, O'Malley MJ. Total ankle arthroplasty with the agility prosthesis: clinical and radiographic evaluation. *Foot Ankle Int* 2006 February;**27**(2):97–103.

65. Hurowitz EJ, Gould JS, Fleisig GS, Fowler R. Outcome analysis of agility total ankle replacement with prior adjunctive procedures: two to six year followup. *Foot Ankle Int* 2007 March;**28**(3):308–12.
66. Schuberth JM, Patel S, Zarutsky E. Perioperative complications of the agility total ankle replacement in 50 initial, consecutive cases. *J Foot Ankle Surg* 2006 May–June;**45**(3):139–46.
67. Assal M, Al-Shaikh R, Reiber BH, Hansen ST. Fracture of the polyethylene component in an ankle arthroplasty: a case report. *Foot Ankle Int* 2003 December;**24**(12):901–3.
68. Stamatis ED, Myerson MS. How to avoid specific complications of total ankle replacement. *Foot Ankle Clin* 2002 December;**7**(4):765–89.
69. Valderrabano V, Hintermann B, Nigg BM, Stefanyshyn D, Stergiou P. Kinematic changes after fusion and total replacement of the ankle: part 1: range of motion. *Foot Ankle Int* 2003 December;**24**(12):881–7.
70. Valderrabano V, Hintermann B, Nigg BM, Stefanyshyn D, Stergiou P. Kinematic changes after fusion and total replacement of the ankle: part 3: talar movement. *Foot Ankle Int* 2003 December;**24**(12):897–900.
71. Kofoed H. Scandinavian total ankle replacement (STAR). *Clin Orthop Relat Res* 2004 July(424):73–9.
72. Anderson T, Montgomery F, Carlsson A. Uncemented STAR total ankle prostheses. *J Bone Joint Surg Am* 2004 September;**86-A**(Suppl. 1(Pt. 2)):103–11.
73. Carlsson A, Markusson P, Sundberg M. Radiostereometric analysis of the double-coated STAR total ankle prosthesis: a 3–5 year follow-up of 5 cases with rheumatoid arthritis and 5 cases with osteoarthrosis. *Acta Orthop* 2005 August;**76**(4):573–9.
74. Tochigi Y, Rudert MJ, Brown TD, McIff TE, Saltzman CL. The effect of accuracy of implantation on range of movement of the scandinavian total ankle replacement. *J Bone Joint Surg Br* 2005 May;**87**(5):736–40.
75. Wood PL, Deakin S. Total ankle replacement. The results in 200 ankles. *J Bone Joint Surg Br* 2003 April;**85**(3):334–41.
76. Harris NJ, Brooke BT, Sturdee S. A wear debris cyst following STAR total ankle replacement – surgical management. *Foot and Ankle Surg* 2008. online print.
77. Anderson T, Montgomery F, Carlsson A. Uncemented STAR total ankle prostheses. Three to eight-year follow-up of fifty-one consecutive ankles. *J Bone Joint Surg Am* 2003 July;**85-A**(7):1321–9.
78. Buechel FF, Pappas MJ, Iorio LJ. New Jersey low contact stress total ankle replacement: biomechanical rationale and review of 23 cementless cases. *Foot Ankle* 1988 June;**8**(6):279–90.
79. Pappas M, Buechel FF. Biomechanics and design rationale: the Buchel-Pappas ankle replacement system; 1999 [cited; Available from: www.endotec.com/pdf_files/ankle%20bio%20144.pdf
80. Buechel FF, Pappas MJ. Survivorship and clinical evaluation of cementless, meniscal-bearing total ankle replacements. *Semin Arthroplasty* 1992 January;**3**(1):43–50.
81. Buechel Sr. FF, Buechel Jr. FF, Pappas MJ. Twenty-year evaluation of cementless mobile-bearing total ankle replacements. *Clin Orthop Relat Res* 2004 July(424):19–26.
82. Buechel Sr FF, Buechel Jr FF, Pappas MJ. Ten-year evaluation of cementless Buechel-Pappas meniscal bearing total ankle replacement. *Foot Ankle Int* 2003 June;**24**(6):462–72.
83. Komistek RD, Stiehl JB, Buechel FF, Northcut EJ, Hajner ME. A determination of ankle kinematics using fluoroscopy. *Foot Ankle Int* 2000 April;**21**(4):343–50.
84. Ali MS, Higgins GA, Mohamed M. Intermediate results of Buechel Pappas unconstrained uncemented total ankle replacement for osteoarthritis. *J Foot Ankle Surg* 2007 January–February;**46**(1):16–20.
85. Doets HC, Brand R, Nelissen RG. Total ankle arthroplasty in inflammatory joint disease with use of two mobile-bearing designs. *J Bone Joint Surg Am* 2006 June;**88**(6):1272–84.
86. San Giovanni TP, Keblish DJ, Thomas WH, Wilson MG. Eight-year results of a minimally constrained total ankle arthroplasty. *Foot Ankle Int* 2006 June;**27**(6):418–26.
87. Su EP, Kahn B, Figgie MP. Total ankle replacement in patients with rheumatoid arthritis. *Clin Orthop Relat Res* 2004 July(424):32–8.
88. Tornier. Salto Talaris Anatomic Ankle; 2008 [cited 2008; Available from: http://www.tornier-us.com/oes/ankrec004.php
89. Tornier. Summary of safety and effectiveness information 510(k) premarket notification – SALTO talaris total ankle prosthesis., 2006 [cited 2008; Available from: http://www.accessdata.fda.gov/scripts/cdrh/cfdocs/cfPMN/pmn.cfm?ID=20892
90. Bonnin M, Judet T, Colombier JA, Buscayret F, Graveleau N, Piriou P. Midterm results of the salto total ankle prosthesis. *Clin Orthop Relat Res* 2004 July(424):6–18.
91. Hintermann B, Valderrabano V, Dereymaeker G, Dick W. The HINTEGRA ankle: rationale and short-term results of 122 consecutive ankles. *Clin Orthop Relat Res* 2004 July(424):57–68.
92. Bell CJ, Fisher J. Simulation of polyethylene wear in ankle joint prostheses. *J Biomed Mater Res B Appl Biomater* 2007 April;**81**(1):162–7.
93. Leardini A, O'Connor JJ, Catani F, Giannini S. Mobility of the human ankle and the design of total ankle replacement. *Clin Orthop Relat Res* 2004 July(424):39–46.
94. Affatato S, Leardini A, Leardini W, Giannini S, Viceconti M. Meniscal wear at a three-component total ankle prosthesis by a knee joint simulator. *J Biomech* 2007;**40**(8):1871–6.
95. Benedetti MG, Leardini A, Romagnoli M, Berti L, Catani F, Giannini S. Functional outcome of meniscal-bearing total ankle replacement: a gait analysis study. *J Am Podiatr Med Assoc* 2008 January–February;**98**(1):19–26.
96. Reggiani B, Leardini A, Corazza F, Taylor M. Finite element analysis of a total ankle replacement during the stance phase of gait. *J Biomech* 2006;**39**(8):1435–43.
97. McIff TE, Saltzman C, Brown T. Contact pressure and internal stresses in a mobile bearing total ankle replacement. *45th Annual Meeting of the Orthopaedic Research Society*. San Francisco, CA; 2001 February 25–28.
98. Miller MC, Smolinski P, Conti S, Galik K. Stresses in polyethylene liners in a semiconstrained ankle prosthesis. *J Biomech Eng* 2004 October;**126**(5):636–40.
99. Nicholson JJ, Parks BG, Stroud CC, Myerson MS. Joint contact characteristics in agility total ankle arthroplasty. *Clin Orthop Relat Res* 2004 July(424):125–9.
100. Buechel Sr FF, Buechel Jr FF, Pappas MJ Eighteen-year evaluation of cementless meniscal bearing total ankle replacements. *Instr Course Lect* 2002;**51**:143–51.
101. Kurtz SM, Bartel DL, Rimnac CM. Postirradiation aging affects stress and strain in polyethylene components. *Clin Orthop Relat Res* 1998 May(350):209–20.
102. Berger RA, Rosenberg AG, Barden RM, Sheinkop MB, Jacobs JJ, Galante JO. Long-term followup of the Miller-Galante total knee replacement. *Clin Orthop Relat Res* 2001 July(388):58–67.
103. Thadani PJ, Vince KG, Ortaaslan SG, Blackburn DC, Cudiamat CV. Ten- to 12-year followup of the Insall-Burstein I total knee prosthesis. *Clin Orthop Relat Res* 2000 November(380):17–29.
104. Kobayashi A, Minoda Y, Kadoya Y, Ohashi H, Takaoka K, Saltzman CL. Ankle arthroplasties generate wear particles similar to knee arthroplasties. *Clin Orthop Relat Res* 2004 July(424):69–72.

Chapter 12

# The Clinical Performance of UHMWPE in the Spine

Steven M. Kurtz, PhD, Marta L. Villarraga, PhD and Allyson Ianuzzi, PhD



## 12.1 INTRODUCTION

In recent years, UHMWPE has been used in implants to treat chronic neck and back pain, which represent leading causes of disability in the United States and around the world. Neck and back pain may manifest acutely or, more uncommonly, as a chronic condition. Acute back pain typically resolves within weeks, but it is also possible for low-grade symptoms to persist for years following an initial episode [1]. Although less common, chronic and intractable back pain can lead to serious, permanent disability [2]. A complex combination of anatomical, biomechanical, genetic, and psychosocial factors may potentially contribute to chronic neck and back pain, greatly complicating diagnosis and treatment of the condition [2]. As a result, patients presenting with neck or back pain are usually initially treated by conservative therapies, which include antidepressants, massage, exercise, spinal manipulation, combined physical and cognitive behavior training, and corticosteroid injections.

When degenerative disc disease is the underlying cause of pain, conservative therapy may not alleviate the patient's symptoms. In the United States, fusion is the most common surgical procedure for chronic low back pain for patients who do not respond to conservative treatment [1]. In 2005, an estimated 228,000 Americans underwent lumbar spine fusion procedures for intractable lower back pain, and the annual number of lumbar fusions in the United States is projected to increase to 6.5 million by 2030 [3]. However, the clinical success rate of lumbar fusion is variable and reportedly ranges between 16 and 95% [4]. Furthermore, fusion of a painful degenerated disc has been associated with subsequent degeneration at adjacent intervertebral discs (sometimes referred to as "adjacent segment disease" by clinicians) [5, 6].

Starting in the 1960s, total disc arthroplasty was conceived as an alternative treatment to fusion, in an effort to avert adjacent segment degeneration. Researchers adapted successful biomaterials and design principles from hip and knee replacements for total disc replacements (TDR). The first such design, the CHARITÉ artificial disc, consists of two metallic endplates that are fixed to the superior and inferior vertebral bodies of the lumbar spine. A mobile, bioconvex ultra-high molecular weight polyethylene (hereafter, UHMWPE) core articulates against the concave bearing surfaces of the superior and inferior endplates. The CHARITÉ originated in Europe in the 1980s and was approved by the FDA for use in the United States in 2004. Additional TDRs, including the ProDisc, Mobidisc, and Activ-L, were developed within the past decade based on similar design principles and utilize similar biomaterials. The ProDisc was approved by the FDA in 2006; the Mobidisc and Active-L are used in Europe and Asia but are not yet available in the United States.

One of the historical driving factors for the development of disc replacements has been growing dissatisfaction with fusion as a long-term treatment, especially for younger patients. TDR surgery is a promising alternative treatment for advanced degenerative disc disease. The central aims of disc arthroplasty are to restore pain-free motion, as well as the load-carrying capacity of a diseased functional spinal unit. Therefore, disc replacement occupies a unique and important position at one end of the pathological spectrum in the treatment continuum spanning early degenerative disc disease, with spinal fusion at the other end. Proponents of TDR have hypothesized that by maintaining the mobility of the treated disc, long-term adjacent segment degeneration may be forestalled or possibly averted.

The state of knowledge in TDRs is evolving rapidly. Here we provide an overview of TDR designs incorporating metal-on-UHMWPE articulation. This chapter reviews the clinical studies of TDRs containing UHMWPE as well as recent findings from retrieval analysis.

## 12.2 THE CHARITÉ ARTIFICIAL DISC

Total disc arthroplasty can trace its clinical history back to the early 1960s, when Fernström first implanted stainless steel spheres to replace the intervertebral disc [7]. From these modest beginnings, developments in total disc replacement technology have progressed at a slow but exponentially increasing pace. The plethora of patents described by Szpalski et al. attest to engineering creativity over the past 40 years [8]. However, only a small fraction of the innovations described in the body of patent literature has been validated by biomechanical and animal testing, and an even shorter list of implant designs have ever been used clinically in humans.

In this section, we focus on the CHARITÉ Artificial Disc. Earlier artificial disc designs, including Fernström's spheres from the 1960s and the AcroFlex Artificial Disc, which evolved starting in the 1980s, fell out of clinical use soon after their introduction. The early design concepts of Fernström's spheres and the AcroFlex have been reengineered in the 21st century and may yet prove to be viable with the proper choice of biomaterials, surgical instrumentation, and patient indications. In contrast, the CHARITÉ Artificial Disc has been used continuously since its development in the 1980s up to the present day, with only evolutionary changes to its design and biomaterials over the past 20 years. The historical *in vivo* performance of the CHARITÉ thus provides important insight into the complex and demanding environment in which contemporary disc replacements incorporating UHMWPE must function.

With the CHARITÉ, the designers sought to replicate the kinematics, but not the structure, of a vertebral disc. The CHARITÉ design, formally known as the SB CHARITÉ Artificial Disc by its inventors, consists of two metallic endplates fixed to the adjacent vertebral bodies, which articulate against a central core fabricated from UHMWPE (Figure 12.1). This artificial disc was invented by Kurt Shellnack, MD, and Karin Büttner-Janz, MD, PhD,

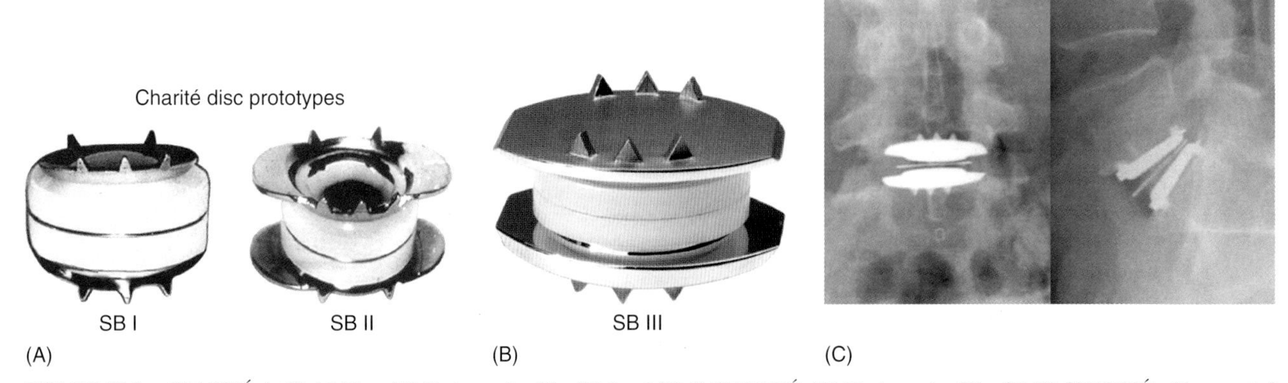


**FIGURE 12.1** CHARITÉ Artificial Disc. (A) Photograph of the SB I and SB II CHARITÉ; (B) Photograph of the SB III CHARITÉ with uncoated endplates; (C) Anterioposterior and lateral radiographs of the SB III CHARITÉ implanted in the lumbosacral spine. Reproduced from [82] with permission.

who were, at the time, affiliated with the Charité Center for Musculoskeletal Surgery at the Medical University of Berlin (Universitätsmedizin Berlin, http://www.charite.de) [9].

In the CHARITÉ design, the polyethylene core has two domed (convex) surfaces, which articulate against the concave metallic endplates. In full flexion or extension, the core is (theoretically) free to translate and rotate, so that the polyethylene rim contacts the metallic rims of the endplates (Figure 12.2). This artificial disc is implanted by an anterior approach. An excellent summary of the design can be found in a recent monograph [10].

The CHARITÉ has gone through four distinct phases in its design history. In this part of the chapter we review the details about the major phases of its design and explain how the CHARITÉ helped to launch the field of modern total disc arthroplasty. In addition to the historical significance of this design, we also briefly highlight some of the controversies surrounding its continued use today.

## 12.2.1 SB CHARITÉ I and II

The first two iterations of the design (i.e., the SB CHARITÉ I and II) occurred between 1984 and 1985 at the Charité Hospital. In this phase of its history, the implants were not commercially available and were implanted in a total of 49 patients [9]. The smooth endplates were fashioned from nonforged stainless steel and only fixed to the vertebral endplates by a number of protruding teeth. A circumferential radiographic wire marker was also included in the polyethylene core. In the SB I design, the endplates were circular and shaped like bottle caps. In the SB II design, the endplates were expanded with lateral wings to provide more coverage of the vertebral endplates (Figure 12.1A).

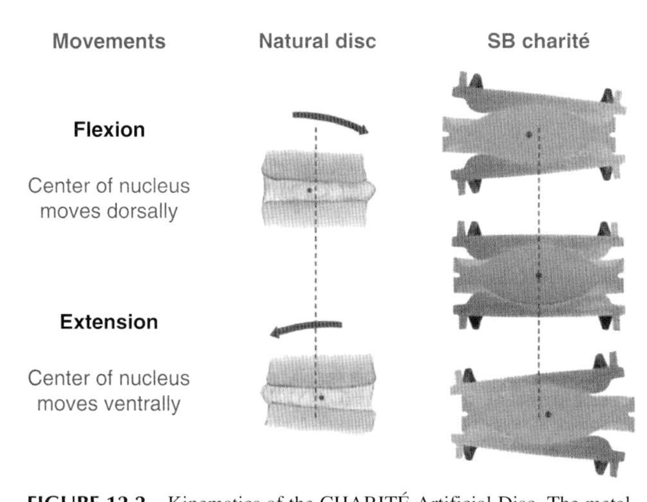


**FIGURE 12.2** Kinematics of the CHARITÉ Artificial Disc. The metallic endplates are designed to contact the rim of the UHMWPE core in flexion and extension. In neutral alignment, load is transferred through the central, domed portion of UHMWPE core. Reproduced with permission Büttner-Janz, Hochschuler, and McAffe [10].

The CHARITÉ's designers adopted the "Low Friction Principle" of polymer-on-metal articulation, employed by Sir John Charnley when he created the first successful hip replacement design [11]. As was covered in earlier chapters of this book, Charnley's hip design evolved to become the dominant joint replacement technology in orthopedics during the 1970s, due largely to the outstanding frictional and fatigue properties of UHMWPE. Until 1997, all the CHARITÉ cores were fabricated from compression molded GUR 412 resin, packaged in air, and gamma sterilized in air.

Only retrospective clinical data are available regarding the SB CHARITÉ I and II designs [10, 12]. McAfee evaluated the early "sub-optimal" results of the Type I and II designs, which he attributed to suboptimal surgical technique (e.g., undersizing, malpositioning) rather than to deficiencies in the design [13]. More recently, investigators from the Charité Hospital have reviewed the performance of 15 patients with the Type I design and 22 patients with the Type II design after an average follow-up of 18.2 and 17.5 years for the two groups, respectively [12] (Table 12.1). Fair to excellent results were found in 87% of patients with Type I implants and in 68% of patients with SB II implants. The higher incidence of poorer results with the Type II implant was due to stainless steel endplate fractures, which occurred in 7/22 patients. No endplate fractures were reported by the authors in the SB I design. Other long-term complications included subsidence, dislocation, and persisting pain (Table 12.1). Interestingly, Putzier et al. reported that heterotopic ossification, leading to spontaneous fusion of the treated level, occurred in 12/15 (80%) of Type I patients and 11/22 (50%) of Type II patients [12].

## 12.2.2 SB CHARITÉ III

The CHARITÉ artificial disc was commercialized by Waldemar Link GmbH & Co. in 1987, resulting in the SB CHARITÉ III (i.e., third iteration design) [10]. As an orthopedic implant manufacturer, Waldemar Link made major changes in the design of the metallic endplates, including changing the material from stainless steel to CoCr alloy. Specifically, Link employed a proprietary CoCr alloy, VACUCAST (0.20–0.25% C, 28.0–30.0% Cr, 5.5–6.5% Mo, maximum 0.5% Ni, maximum 0.5% Fe, 0.4–1.0% Si, 0.4–1.0% Mn, with the remaining 57.1–69% Co) [14]. The backside of the CoCr was "satin finished" by corundum blasting, and fixation was achieved with six sharp teeth [14]. Since the redesign of the endplates by Link, there have been no further reports of component fractures [10, 12]. The geometry and properties of the UHMWPE core remained essentially unchanged by Link at that time. Unlike previous iterations of the design, which were only used at the inventors' institution, this version was launched internationally beginning with France and the Netherlands in 1989 and the United Kingdom in 1990.

**TABLE 12.1** Long-Term Clinical Performance of the SB CHARITÉ (I, II, and III), Adapted from [12]

| CHARITÉ design type | Treatment period (year) | Average follow-up (years) | Number of patients | Number of TDRs implanted | Average age | Fair to excellent patient outcome (%) | Segmental fusion for subsidence | Segmental fusion for dislocation | Segmental fusion for persisting pain | Arthrodesis or spontaneous fusion |
|---|---|---|---|---|---|---|---|---|---|---|
| SB I | 1984–1985 | 18.2 | 15 | 16 | 44 | 13/15 (87%) | 2/15 | 0/15 | 1/15 | 12/15 (80%) |
| SB II | 1985–1987 | 17.5 | 22 | 25 | 45 | 15/22 (68%) | 1/22 | 0/22 | 0/22 | 11/22 (50%) |
| SB III | 1987–1989 | 16.1 | 16 | 22 | 43 | 12/16 (75%) | 0/16 | 1/16 | 0/16 | 9/16 (56%) |
| Total | 1984–1989 | | 53 | 63 | | 40/53 (75%) | 3/53 | 1/53 | 1/53 | 32/53 (60%) |

*No significant difference in long-term clinical performance was noted between the three CHARITÉ designs.*

The clinical performance of the CHARITÉ has been documented in a number of published studies, the majority of which pertain to the Type III design (Table 12.2). Taken together, these studies demonstrate that, in at least some patients, the CHARITÉ TDR has the potential to preserve motion, result in good to excellent clinical outcomes, and survive long-term implantation in the human body. However, these largely retrospective cohort studies also demonstrate broad variations in their complication rates and clinical success rates, raising concerns about the repeatability and reproducibility of the procedure. The largest, and most comprehensive, body of clinical data about the CHARITÉ was collected for the prospective, randomized trial undertaken for the FDA's IDE study in the United States (Table 12.2). Five-year results from the prospective IDE study have recently been presented at national meetings [15].

### 12.2.3 Recent Generations of the CHARITÉ Artificial Disc

There have been a number of changes in the CHARITÉ made by Link, in the late 1990s, and more recently, by DePuy Spine, in 2004. All of these changes have been made on a rolling basis, and it is therefore difficult to appreciate at the present time what effect these will have on the long-term clinical performance.

For example, in 1997, Link changed its packaging of the UHMWPE core, which was initially air permeable ("gamma air"), to two polymeric pouches in a nitrogen/vacuum process (N-VAC) (Figure 12.3) [27]. The UHMWPE material used in the cores was changed to compression molded GUR 1020, converted by Poly Hi Solidur. At the FDA Panel meeting in 2004, it was revealed that the polymeric packaging was also air permeable, resulting in measurable oxidation of the UHMWPE during shelf storage [28]. Based on recently published retrieval analyses, the *in vivo* oxidation behavior of UHMWPE sterilized in N-VAC packaging appears to be indistinguishable from those that were gamma sterilized in air [29]. All of the UHMWPE cores used in the US IDE trial were sterilized in Link's first generation N-VAC packaging.

The current manufacturer of the CHARITÉ (DePuy Spine, Raynhnam, Massachusetts, USA) has made few changes to Link's design for the SB CHARITÉ III. Most notably, DePuy Spine has changed the packaging of the UHMWPE to a foil pouch (GVF). This most recent packaging change was initially implemented in April 2004 and launched worldwide in October 2004 [27]. Traceable retrieval studies will be necessary to establish the natural history of CHARITÉ polyethylene components stored within first-generation (N-VAC) and second-generation (GVF) barrier packaging.

### 12.2.4 Bioengineering Studies of the CHARITÉ

Although much of the literature about the CHARITÉ is related to clinical outcomes, there are an increasing number of studies related to the bioengineering aspects of total disc replacement that have appeared in the literature [30–33]. Among these, the work of Cunningham is perhaps the best known because the data was made available to the FDA during the IDE study and has also been published in the peer-reviewed scientific literature [30–32]. Cunningham's research, therefore, is highly instructive as to the scope and quality of biomechanical testing that is needed to address questions by regulatory bodies, such as the FDA, for the evaluation of total disc replacement technologies. More recently, O'Leary et al. have also published biomechanical data, collected with an *in vitro* cadaveric model, that characterizes the motion patterns for the CHARITÉ [33].

Cunningham's work has been summarized in a recent review article [30] and included both *in vitro* cadaveric spine testing and animal studies. Cunningham's research analyzed the biomechanics and biocompatibility of wear debris for the CHARITÉ. For the *in vitro* biomechanical studies, Cunningham studied control (untreated) lumbar spines, as well as lumbar spines implanted with the CHARITÉ Artificial Disc as compared to the BAK cage and the AcroFlex artificial disc [30, 31]. The lumbar spines were tested with the application of pure moments (±8 Nm) in flexion-extension, lateral bending, and axial rotation. It is noteworthy that Cunningham's tests did not include axial preload and thus quantify the maximum theoretical flexibility of an unloaded spine. Motion was characterized using an OptoTrak 3020 motion system (Northern Digital Inc., Waterloo, Canada) as well as lateral plain radiographs. Under the applied moments, the CHARITÉ exhibited significantly greater axial rotation, but less lateral bending, as compared to the intact spine. In absolute terms, the average axial rotation of the intact spine was measured to be 2.4 degrees as opposed to 3.9 degrees after treatment with the CHARITÉ [31]. There was no significant difference between the intact spine and the CHARITÉ in flexion/extension.

Neural tissue response to wear debris was evaluated by applying 4 mg of particles onto the epidura of New Zealand white rabbits (n = 50, total). Five groups of rabbits were tested (n = 10, each): (1) sham operation (control); (2) stainless steel 316LVM; (3) titanium alloy (Ti-6Al-4V); (4) CoCr alloy; and (5) UHMWPE. The particle sizes ranged between 0.5 and 10 microns in diameter and were verified to be endotoxin free prior to implantation. Animals were sacrificed at 3 and 6 months postoperatively. Even though there was evidence of a chronic inflammatory reaction for all of the particles, it appeared

**TABLE 12.2** Clinical Details of Primary, Peer-Reviewed Studies Involving the SB CHARITÉ (I, II, and III)

| | Büttner-Janz et al. [16] | Griffith et al. [17] | Cinotti et al. [18] | Zeegers et al. [19] | Sott and Harrison [20] | Lemaire et al. [21] | Putzier et al. [12] | FDA IDE Study [22]# | David [23] | Ross [24] |
|---|---|---|---|---|---|---|---|---|---|---|
| Study type | Hcoh | Hcoh | Hcoh | Pcoh | Hcoh | Hcoh | Hcoh | RCT | Hcoh | Hcoh |
| Type of prosthesis | SB I, II, III | SB I, II, III | SB III | SB III | SB III | SB III | SB I, II, III | SB III | SB III | SB III |
| Average age | 43 | 43 | 36 | 43 | 48 | 40 | 44 | 40 | 36 | 46 |
| Age range | 26–59 | 25–59 | 27–44 | 24–59 | 31–61 | 24–51 | 30–59 | 19–60 | 23–50 | 27–73 |
| Number of patients | 62 | 93 | 46 | 50 | 14 | 100 | 53 | 205 | 106 | 160 |
| Number of arthroplasties | 76 | 139 | 56 | 75 | 15 | 147 | 63 | 205 | 106 | 226 |
| Average follow-up in years | 1.3 | 1.0 | 3.2 | 2.0 | 4.0 | 11.3 | 17.3 | 2.0 | 13.2 | 6.6 |
| Good or excellent | 81% | ? | 63% | 70% | 10/14 | 90% | 75% | 57.1%** | 82% (87/106) | 68% |
| Secondary surgery | | | | 17/50 | 1/14 | 5/100 | 12/53 | 11/205 | 11/106 | 12/226 |
| Arthrodesis or spontaneous fusion | | | | | | 2/100 | 32/53 | 2/205 | 3/106 | |
| Complications | 29/76 | 55/139 | 8/46 | 3/50 | 2/15 | 21/100 | 44/53 | 68/205 | 44/106 | 41/123 |
| Motion on flexion-extension radiographs in degrees (range) | | | | | | | | | | |
| Average | 5 | ? | 12.2 | 9 (2–17) | ? | 10.3 (0–16) | ? | 7.1 (6.6–7.7) | 10.1* | |
| L3/L4 | | | – | | | 12.0 (10–16) | | | | 1.4 (0–6) |
| L4/L5 | | | 16 | | | 9.6 (0–15) | | | 12.2 | 4.0 (0–18) |
| L5/S1 | | | 9 | | | 9.2 (0–14) | | | 9.4 | 4.7 (0–16) |

*Legend: Hcoh: Historical cohort study; Pcoh: Prospective cohort study; RCT: Randomized controlled trial; *Average mobility was calculated from the reported data [25]; **Overall clinical success was defined as patients having ≥25% improvement in ODI score at 24 months compared with the preoperative score, no device failure, no major complications, and no neurological deterioration [26]. #Five-year follow-up of this series have been reported in conference presentations [15], but these results have not yet appeared in the peer-reviewed literature.*

**FIGURE 12.3** First-generation, polymeric barrier packaging of the CHARITÉ UHMWPE core by Link, employed between 1998 and 2004. Reproduced from [82] with permission.

localized within the epidural tissue. CoCr particles, in particular, were shown to diffuse from the epidural layers into the cerebrospinal fluid and into spinal cord itself. Although this study provides some insight into the biological response following a massive particle load delivered directly to the epidural layer of the cord, the likelihood of this scenario occurring in the clinical situation is not clear.

O'Leary et al. [33] have studied the motion patterns of the CHARITÉ using an *in vitro* biomechanical testing system that can simultaneously apply pure moments (up to 8 Nm in flexion and 6 Nm in extension) along with a 400N axial preload. The preload is also referred to as a "follower load" because the orientation of the force is such that, by means of cables, it follows the curvature of the spine during flexion-extension testing. Motion was characterized using an OptoTrak 3020 motion system (Northern Digital Inc., Waterloo, Canada) and biaxial motion sensors. Motion of the UHMWPE core was tracked during flexion-extension using digital video-fluoroscopy. Five lumbar spines were used in the testing from patients ranging in age from 39–60 years. Four distinct motion patterns were observed for the UHMWPE core, when tested under compressive preload. With Motion Pattern 1, relative angular motion occurred predominantly between the superior endplate and the core, with little or no core translation (Figure 12.4A). In Motion Pattern 2, lift-off occurred by either the superior endplate from the core, or by the core from the inferior endplate (Figure 12.4B). In Motion Pattern 3, the UHMWPE core "locked" in plane, resulting entrapment, over a portion of the flexion-extension range (Figure 12.4C). In Motion Pattern 4, angular motion occurred between the core and both the superior and inferior endplates (Figure 12.4D). O'Leary et al. concluded that, "CHARITÉ TDR restored near normal quantity of flexion-extension range of motion under a constant physiologic preload; however, the quality of segmental motion differed from the intact case over the flexion-extension range" [33]. Because some of the implants in this study may have been undersized or not optimally placed, the authors further concluded that positioning, changes in lordosis, and the magnitude of the preload would also likely influence the kinematics of the CHARITÉ [33].

## 12.2.5 Controversies Surrounding the CHARITÉ Artificial Disc

The complete history of the CHARITÉ spans 2 decades and numerous iterations in design (1984 to 2004). During this time period, it was implanted in 9000 patients, mostly in Europe. However, the long-term success of the CHARITÉ remains controversial. The broad range of outcomes for the CHARITÉ reported in the literature (Tables 12.1 and 12.2) raises questions about the probability of long-term success of motion preservation with this implant.

In a systematic review of the artificial disc literature published in 2003, the authors concluded that "there is no evidence that disc arthroplasty reliably, reproducibly, and over longer periods of time fulfills the three primary aims of clinical efficacy, continued motion, and few adjacent segment degeneration problems" [25]. Putzier, writing in 2005 from the Charité Hospital in Berlin, where the implant was first used clinically, stated that, "the long-term follow-up study demonstrates dissatisfying results after artificial disc replacement in the majority of the evaluated cases . . . the CHARITÉ artificial disc replacement cannot guarantee long-term near to normal function of spinal motion segment in patients with moderate to severe [degenerative disc disease]" [12]. As a result, the long-term prognosis for patients with a CHARITÉ total disc replacement remains uncertain.

## 12.2.6 The Legacy of the CHARITÉ Artificial Disc

Despite the current controversies surrounding the CHARITÉ, it is nonetheless the icon for contemporary total disc arthroplasty. Other artificial disc designs are currently in clinical use in Europe and may become available in the United States within the coming decade. However, these newer designs build upon the design philosophy established by the CHARITÉ, which adapted the successful bearing concepts from hip and knee replacements for total

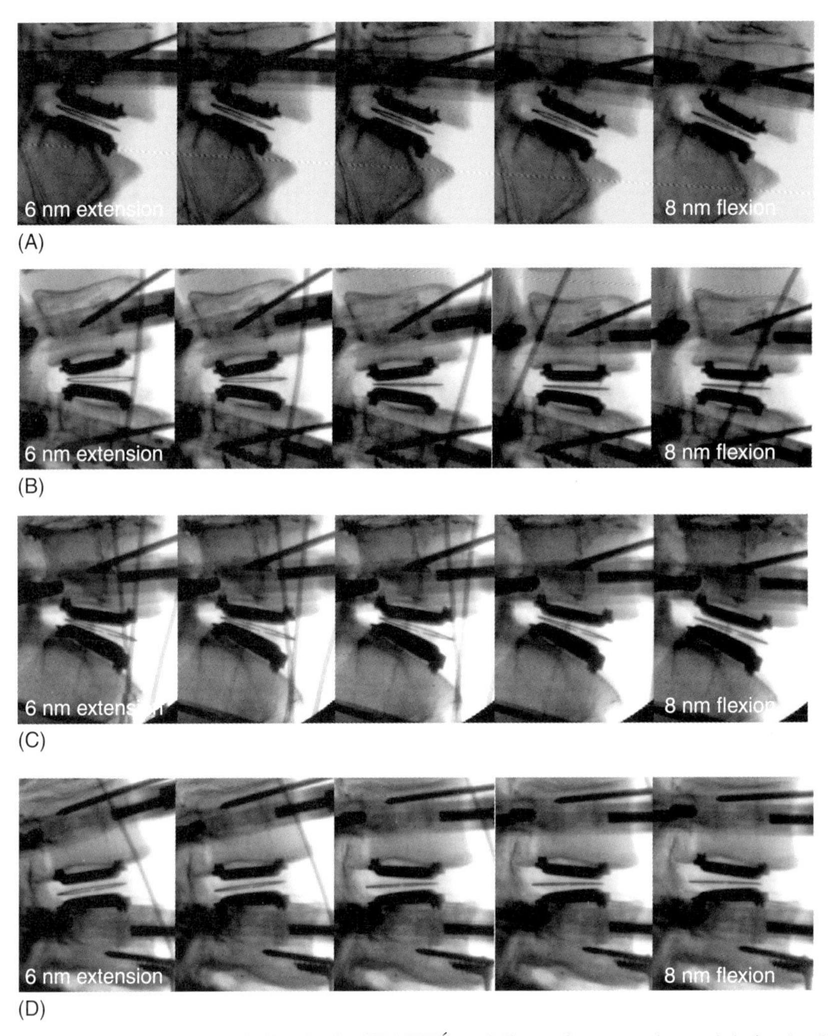


**FIGURE 12.4** Four motion patterns of the UHMWPE Core in the CHARITÉ total disc replacement observed during *in vitro* biomechanical testing. A: With Motion Pattern 1, relative angular motion occurred predominantly between the superior endplate and the core, with little or no core translation. B: In Motion Pattern 2, lift-off occurred by either the superior endplate from the core, or by the core from the inferior endplate. C: In Motion Pattern 3, the UHMWPE Core "locked" in plane, resulting entrapment, over a portion of the flexion-extension range. D: In Motion Pattern 4, angular motion occurred between the core and both the superior and interior endplates. Reproduced from [82] with permission.

disc replacement. The CHARITÉ Artificial Disc is the first design to establish a history of clinical use spanning two decades, with 10-year outcomes available in, albeit limited, retrospective studies. The CHARITÉ is a pioneering design for this reason.

## 12.3 LUMBAR DISC ARTHROPLASTY

As summarized in Table 12.3, there are currently four lumbar total disc replacement designs incorporating UHMWPE in clinical use today: CHARITÉ, ProDisc-L, Mobidisc, and Activ-L. In the previous section, we reviewed the clinical history and bioengineering studies related to the CHARITÉ. The three newer designs differ from the CHARITÉ in several respects, such as the amount of constraint in the bearing and the incorporation of keels into the endplates (Table 12.3).

The amount and quality of available clinical data also varies among the designs (Table 12.3). Although long-term data are available for the ProDisc-L and the CHARITÉ, respectively, the reader should be aware that these long-term studies are limited to retrospective evaluations [12, 21, 23, 34]. In this section, we summarize the design theory and available literature for the ProDisc-L, Mobidisc, and Activ-L artificial discs. Like the CHARITÉ, these devices are based on CoCr-on-UHMWPE articulation and are similarly implanted by an anterior approach. However, the newer designs represent departures from the unconstrained, biconvex mobile bearing philosophy embodied by the CHARITÉ. The three designs incorporate varying degrees of constraint, ranging from a ball-and-socket design of the ProDisc-L to the sliding convex mobile bearing designs of the Mobidisc and Activ-L. It is hoped that these newer designs will be able to overcome some of the

**TABLE 12.3 Summary of Contemporary Lumbar Artificial Disc Designs Incorporating CoCr/UHMWPE Articulations**

| Design | CHARITÉ | ProDisc-L | Mobidisc | Activ-L |
|---|---|---|---|---|
| Current manufacturer | DePuy Spine (Raynham, Massachusetts, USA; http://www.charitedisc.com/) | Synthes (Paoli, Pennsylvania, USA; http://www.synthes.com) | LDR Spine (Troyes, France; http://www.ldrmedical.com/) | Aesculap AG (Tuttlingen, Germany; http://www.aesculap.com/) |
| Number of components for surgeon assembly | Three | Three | Three | Three |
| Number of articulating surfaces | Two | One | Two | Two |
| Bearing design | Mobile bearing | Ball-and-socket | Mobile bearing | Mobile bearing |
| Rotational constraint | Unconstrained | Semiconstrained | Unconstrained | Unconstrained |
| Bone/implant fixation | Teeth with and without CaP/Ti coating | Keel and Ti textured coating | Keel and Ti textured coating | Keel or teeth and Ti textured coating |
| Keel | No | Yes | Yes | Yes* |
| Published clinical studies | Yes (see Table 12.2) | Yes (see Table 12.4) | No | No |
| Longest published average follow-up | 18.2 years [12] | 8.7 years [34] | None | None |
| 2-Year FDA randomized controlled trial | Completed and published [22, 26] | Completed and published (see Table 12.4) | No (Currently in use in Europe and Asia) | In process (currently in use in Europe and Asia) |
| Control Procedure in FDA randomized trial | Anterior interbody fusion with BAK cage | 360 degrees fusion | N/A | CHARITÉ and ProDisc |

***The Activ-L has endplates available with either a keel or teeth.**

long-term complications and controversies that have been identified in recent studies of the CHARITÉ.

### 12.3.1 ProDisc-L

Conceived in 1989 by Thierry Marnay, MD, from Montpellier in France, the ProDisc consists of two metallic endplates and a UHMWPE core (Figure 12.5) [41]. Unlike the CHARITÉ, the UHMWPE core in the ProDisc is firmly attached to the flat, inferior endplate by a locking mechanism (Figure 12.6). The domed (convex) surface of the core articulates against the concave, superior metallic endplate. There have been two versions of the ProDisc (I and II), but only the more recent version of the design (ProDisc II) has been commercially available. The ProDisc II was originally produced by Aesculap AG & Co. (Tuttlingen, Germany) [34] and then commercialized by Spinal Solutions. Today it is manufactured by Synthes (Paoli, Pennsylvania, USA) and the lumbar version of this design is now referred to as the ProDisc-L.

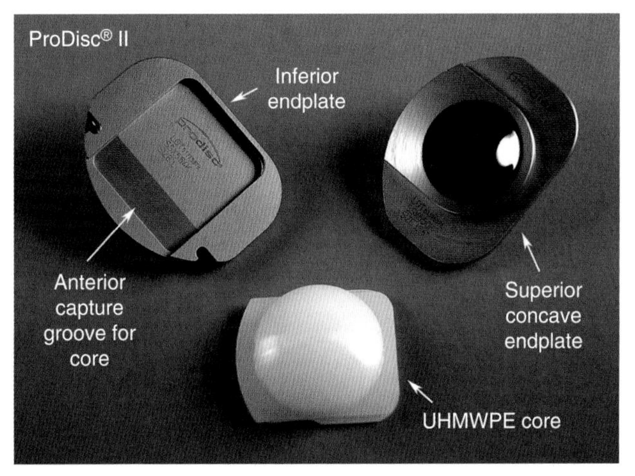


**FIGURE 12.5** Two CoCr alloy endplates and UHMWPE core of the ProDisc II design. Reproduced from [82] with permission.

Besides the CHARITÉ, the ProDisc has the second-longest clinical follow-up of any currently available artificial disc design [34]. Marnay implanted 93 first-generation, ProDisc I artificial discs in 64 patients between March 1990 and September 1993 [34, 35]. The endplates of the ProDisc I were fabricated from Ti alloy (Figure 12.6A). Each endplate featured two short, parallel keels for short-term fixation, and the back surfaces were plasma sprayed with titanium for bone ongrowth.

The intermediate-term clinical results of the ProDisc I patients were published in 2005 [34]. The mean follow-up of this study, which included 55 (86%) patients from Marnay's original cohort, was 8.7 years. Marnay demonstrated significant improvement using nonvalidated outcome scoring methods, which confirmed his hypothesis that the ProDisc I was safe and effective, at least in his hands (Table 12.4). Procedure-related complications were observed in five cases (deep venous thrombosis, iliac vein laceration, transient retrograde ejaculation, and two incisional hernias), but none were implant related. The investigators were not able to detect any measurable UHMWPE wear when comparing the core height immediately after surgery with the core height at the longest follow-up. Subsequent analysis of the range in clinical performance for this cohort by Huang et al. revealed an intriguing correlation between patient outcomes and the range of motion of the operated levels [35]. Patients with more than 5 degrees of flexion-extension range of motion had significantly less postoperative back pain and better clinical outcomes (as measured by postoperative Oswestry Disability Index and Stauffer-Coventry Scores) than patients with less than 5 degrees range of motion.

Marnay followed his patients until 1998 before making further modifications, which resulted in the current ProDisc-L design. Since 1999, more than 5000 ProDisc-L components have been implanted worldwide up to 2005 [41]. In the second-generation design, the endplates are fabricated from CoCr alloy, which has improved tribological properties as compared with Ti alloy. In addition, a single,

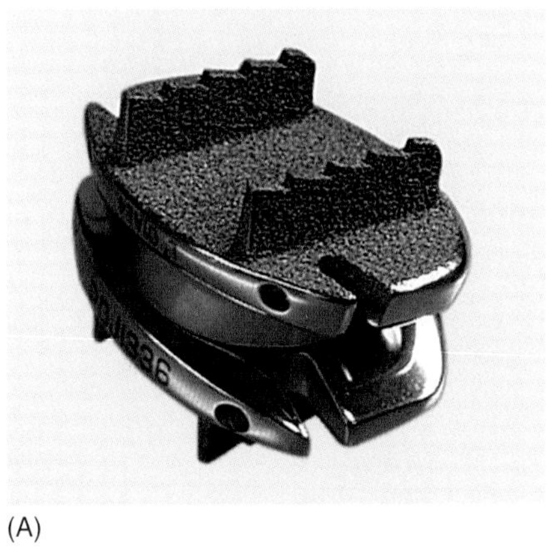

(A)

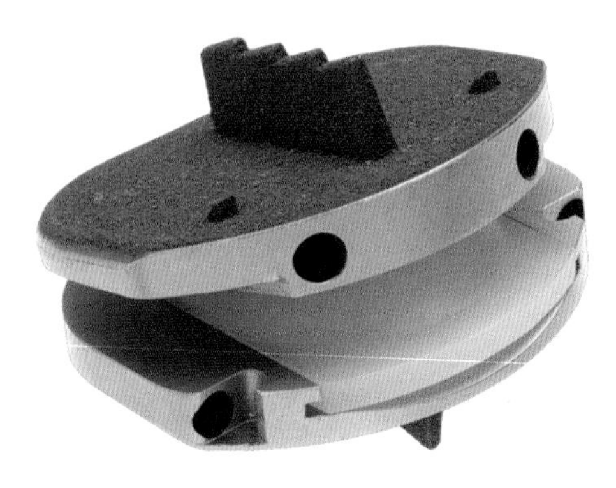

(B)

**FIGURE 12.6** (A) Two titanium alloy endplates and UHMWPE core of the ProDisc I design. (B) Two CoCr alloy endplates and UHMWPE core of the ProDisc II design. Reproduced from [82] with permission.

**TABLE 12.4** Clinical Details of Primary, Peer-Reviewed Studies Involving the ProDisc I and -L

| | Tropiano et al. [34, 35] | Mayer et al. [36] | Tropiano et al. [37] | Zigler [38] | Bertagnoli et al. [39] | Bertagnoli et al. [40] | Delamarter et al. [41] | Zigler et al. [42] |
|---|---|---|---|---|---|---|---|---|
| Study type | Hcoh | Pcoh | Pcoh | RCT | Pcoh | Pcoh | RCT | RCT |
| Type of prosthesis | ProDisc I | ProDisc-L | ProDisc-L | ProDisc-L | ProDisc-L | ProDisc-L | ProDisc-L | ProDisc-L |
| Average age | 46 | 44 | 45 | 38 | 48 (median) | 51 (median) | 40 | 39 |
| Age range | 25–65 | 25–65 | 28–67 | | | 30–60 | 19–59 | |
| Number of patients | 55 | 34 | 53 | 28 | 104 | 25 | 56 | 161 |
| Number of arthroplasties | 78 | 37 | 68 | | 104 | 60 | 91 | 161 |
| Follow-up in months | 104 | 12 | 17 | 12 | 31 (median) | 31 (median) | 18 | 24 |
| Range in follow-up | 85–128 | | 12–24 | | 24–45 | 25–41 | | |
| Good or excellent | 75% | | | | | | | 53.4%* |
| Initial Oswestry Score (%), average ± SD | NA | 19 ± 7 | 56 ± 8 | | 54 | 65 (42–92) | 31 | 63.4 ± 12.6 |
| Oswestry Score (%) at longest follow-up, average ± SD | 18 ± 16 | 7.2 ± 9.6 | 14 ± 7 | Decreased | 29 | 22 (0–48) | 20 | 34.5 ± 24.8 |
| Secondary surgery | 3/55 | 0/34 | 3/53 | | 1/104 | 1/25 | | 6/161 |
| Arthrodesis or spontaneous fusion | 1/55 | 0/34 | 0/53 | | 0/104 | 0/25 | | 0/161 |
| Complications | 10/55 | 3/34 | 5/53 | | 5/104 | 4/25 | | 10/161 |
| Motion on flexion-extension radiographs in degrees (range) | | | | | | | | |
| Average | 4.0 (0–18) | | 9 (2–18) | | 7 | | | 7.7 ± 4.7 |
| L4/L5 | | | 10 (8–18) | | | | 10 | |
| L5/S1 | | | 8 (2–12) | | | | 8 | |

*Legend: Hcoh: Historical cohort study; Pcoh: Prospective cohort study; RCT: Randomized controlled trial; *Overall clinical success was defined by FDA as patients having achieved ≥15% improvement in ODI, SF-36, radiographic success, and neurologic success, all without device failure [42].*

slightly taller keel is used. The textured, back surfaces of the endplates are plasma sprayed with titanium.

The UHMWPE core is machined from compression molded GUR 1020 and is gamma sterilized in nitrogen. Foil packaging is currently used for its oxygen barrier properties (Figure 12.7). The ProDisc-L has a locking detail that engages the UHMWPE core with the inferior endplate. The capture mechanism for the core consists of a sliding, tongue-in-groove along the left and right (laterally), with a raised, triangular-profiled ridge on the inferior surface of the UHMWPE. During implantation, the surgeon slides the tongues on the core backwards into the grooved, capture mechanism of the inferior endplate until the anterior raised ridge of the UHMWPE snaps into a mating slot in the endplate, thereby preventing anterior ejection of the core (Figure 12.5).

The reports of generally positive clinical outcomes of the ProDisc-L are limited to short-term studies (on average, 17–31 months of follow-up), which are summarized in Table 12.4. In addition, published prospective clinical data for the ProDisc-L provide some evidence that the short-term outcomes of multilevel implantation are comparable to single-level disc replacement [40]. In general, the available clinical data for the ProDisc-L is stronger, scientifically speaking, than that for the CHARITÉ, which has historically been documented by mostly retrospective studies in Europe (Table 12.2). With the exception of the FDA trial, none of the CHARITÉ studies included validated outcome measures, such as the Oswestry Disability Index. In contrast, the majority of available studies for the ProDisc-L are prospective and employ validated outcome measures that can be compared with alternative spine procedures (Table 12.4).

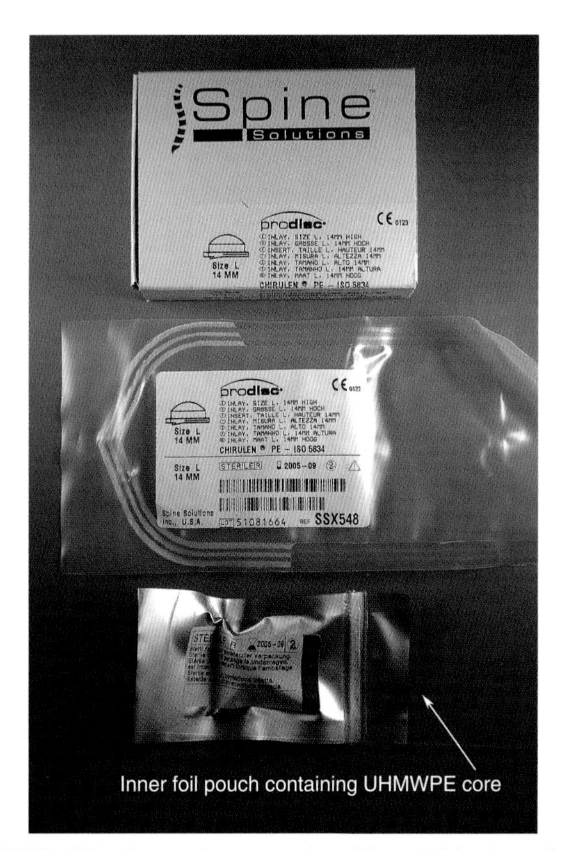


**FIGURE 12.7** Outer polymeric pouch and inner foil barrier packaging of the ProDisc II UHMWPE core developed by Spine Solutions and currently employed by Synthes Spine. Reproduced from [82] with permission.

The prospective, randomized clinical trial for the ProDisc-L, with both single-level and double-level arms, began in the United States in October 2001 and completed its target enrollment by the end of 2003. The reference procedure was an anterior-posterior (360 degrees) fusion; the control patients first underwent an anterior fusion procedure, involving a femoral ring allograft. The patients were then flipped, and a posterior fusion was performed, consisting of bilateral intertransverse fixation with pedicle screws [43]. Although by no means obsolete, a 360-degree procedure represents a highly conservative fusion strategy and arguably a worst-case scenario with which to compare with disc replacement. Two of the 19 clinical sites involved in the pivotal trial have published interim results, with follow-up periods of up to 18 months (Table 12.4) [38, 41, 43, 44]. There was a major difference in the complexity, and hence the operative time, between the disc replacement and the 360-degree fusion (on average, 75.4 versus 218.2 minutes, respectively) [43]. Disc replacement patients recovered faster than controls. Furthermore, the 2-year IDE data indicate that the final outcomes the ProDisc-L, as measured by VAS and Oswestry Disability Scores, are significantly better than 360-degree fusion [42]. Synthes received clearance by the FDA for marketing the ProDisc-L in the United States in August 2006.

Data on short-term complications with the ProDisc II has appeared in the literature thus far [39, 40, 42, 45, 46]. At the 2005 North American Spine Society meeting, for example, Bertagnoli reported an overall complication rate for the ProDisc at 3.0% (15/522 patients) [47]. Although relatively few in number (7/522, 1.4%), many of the complications reported for the ProDisc II, such as wound infections, retrograde ejaculation, epidural or retroperitonial hematoma, appear related to the anterior approach itself [47]. Device-related complications have been reported in 8/522 cases (1.6%) and include partial subsidence of the endplates, anterior migration or partial ejection of the UHMWPE core, and L5 root/motion deficit [47]. Two cases of vertebral split fractures associated with the ProDisc have also been reported in the literature [45]. In both cases, the split fractures were caused by the surgeons when chiseling of the bone grooves to accommodate the keels. Thus, the vertebral split fractures, as well as the majority of the complications reported by Bertagnoli, are likely related to surgical technique. The cases of core ejection may be technique related, particularly if the surgeon fails to properly engage the capture mechanism during installation. Indeed, Bertagnoli opines that "more than 90% of device-related complications are iatrogenic. Poor patient selection, improper implantation, and wrong sizing are the most common examples of surgical errors causing a higher risk of failure" [46].

In the IDE study [42], the 3.7% reoperation rate of the ProDisc-L at 2 years included dislodgement of the UHMWPE inlay due to "extreme trauma" (in two cases) and one case of technical error in which the surgeon inserted the UHMWPE inlay backwards. In one case the inlay dislodged after surgery and was interpreted by the authors to have not been properly inserted. In another case, the entire TDR migrated anteriorly. All of these short-term clinical failures required immediate revision surgery.

Limited test data, related to shock absorption, have been reported for the ProDisc-L in the scientific literature [48]. Dynamic testing at frequency ranges between 0 and 100 Hz, as well as "shock loading" with 0.1 s durations, demonstrated that the ProDisc II exhibits negligible shock-absorption capacity, regardless of the fact that a UHMWPE core is included in the design. Biomechanical range-of-motion evaluations for the ProDisc II were presented at the 2003 meeting of the Orthopedic Research Society [49]. Lipman et al. tested six lumbar spines, with and without a ProDisc II. The specimens were mounted in a 6 degree of freedom spine testing apparatus, attached to a biaxial MTS load frame. The specimens were tested with ±10 Nm flexion/extension, lateral bending, and axial torsion, under applied compressive preloads of 600 N and 1200 N. The investigators found no significant difference in the range of motion of the natural and treated lumbar spine when testing under these conditions.

Wear testing for 5 million cycles of the ProDisc II in comparison to the CHARITÉ was presented at the 2006 ORS meeting [50]. Two different types of duty cycle were employed. In one set of simulations, a curvilinear motion of the endplates was prescribed in flexion/extension and axial rotation. In a second set of simulations, crossing path motion was prescribed in flexion/extension, lateral bending, and axial rotation. The load magnitude varied between 350 and 1750 N; however, no further details about the test conditions (e.g., test temperature and fluid) were provided in the abstract. Simulator kinematics had a significant effect on wear rates, with crossing path motion producing a higher wear rate as compared with curvilinear motion. The wear rates of the ProDisc II and CHARITÉ artificial discs were found to be similar under both sets of simulator conditions.

In summary, there is a growing body of literature supporting the clinical safety and effectiveness for the ProDisc II. The 2-year results of the single-level, prospective randomized trial are now publicly available [42], and 5-year follow-up of this study has been presented at scientific meetings [51]. However, many aspects of the clinical evaluation, in particular long-term complication rates and *in vivo* wear rates, remain yet to be fully explored.

## 12.3.2 Mobidisc

The Mobidisc may be considered a second-generation mobile bearing design and has a clinical track record of less than a decade. Developed by LDR Spine (Troyes, France), clinical use of the Mobidisc began in November 2003. The prosthesis is available in Europe and Asia but not in the United States. The Mobidisc consists of two polished cobalt chromium endplates and a UHMWPE core, which has a spherical domed superior (cranial) surface and a flat inferior (caudal) surface (Figure 12.8). Thus, the Mobidisc is a convex, sliding mobile bearing design. The UHMWPE core has two lateral wings that articulate within lateral capture mechanisms in the inferior endplate. The UHMWPE is gamma-sterilized in a polymeric barrier package.

No peer-reviewed published studies are yet available describing the clinical performance of this design, although the results of an ongoing clinical case series have been reported in a recent book chapter [52]. Steib et al. summarize one surgeon's results of 149 patients that were enrolled and underwent surgery by one of eight surgeons in a prospective clinical case series. After 2 years, back pain improvement on the Visual Analog Scale averaged 60% and improvement on the Oswestry Disability Index was 55%. The mobility of the affected level changed from 4 degrees (range: 0 to 15 degrees) preoperatively to 9.4 degrees (0 to 18 degrees) postoperatively. At the last follow-up, 76% of all prostheses had mobility. A total of 12 complications that resulted in three revision surgeries were reported. Ten cases of subsidence, one laterally malpositioned prosthesis, and one improperly sized prosthesis were reported. While clinical improvements were reported in this series of patients, the results of longer-term, controlled clinical studies have not yet been published.

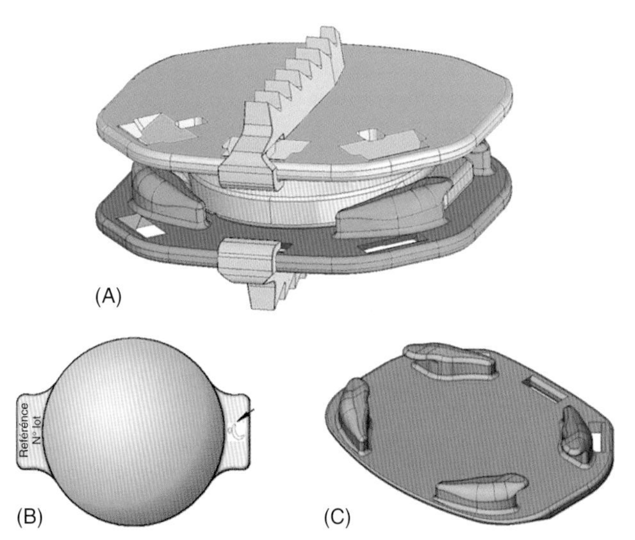


**FIGURE 12.8** (A) Two CoCr alloy endplates and UHMWPE core of the Mobidisc; (B) Design of mobile UHMWPE core, illustrating the lateral wings; and (C) Design of inferior CoCr endplate, illustrating the stops which provide antirotational and translational constraint for the mobile core. Reprinted with permission from LDR Spine.

### 12.3.3 Activ-L

The Activ-L is another second-generation artificial disc and was designed by Aesculap AG (Tuttlingen, Germany) after developing the ProDisc. Unlike the ProDisc, which is a fixed bearing design, the Activ-L is a mobile design consisting of two CoCr endplates and a mobile UHMWPE core (Figure 12.9). The UHMWPE core is flat on the inferior/caudal side to accommodate anterior-posterior translation, and convex on the superior/cranial side to allow rotation of the endplates with respect to one another. The core edge is geometrically constrained in a groove within the inferior endplate (Figure 12.9). According to personal communications with the manufacturer, the UHMWPE core is gamma irradiation sterilized in a barrier package.

The Activ-L received its CE Mark in 2005 and is currently used in Europe and Canada. However, the Activ-L is currently only available for investigational use in the United States by Aesculap Implant Systems (Center Valley, Pennsylvania, USA). An Investigational Device Exception clinical trial for the Food and Drug Administration was initiated in January 2007, and the 24-month follow-up for the target enrollment of 441 patients is expected to be completed by 2010 (http://clinicaltrials.gov/ct2/show/NCT00589797?cond=%22Spinal+Diseases%22&rank=30). The IDE includes both the CHARITÉ and ProDisc as controls. No published studies are available describing the clinical performance of the Activ-L.

## 12.4 CERVICAL DISC ARTHROPLASTY

In several key aspects, total disc arthroplasty is a more attractive technology for the cervical spine than for the lumbar spine. First of all, the biomechanical demands on a cervical prosthesis, which must support the weight of the head and the neck above it, are about an order of magnitude less stringent than in the lumbar region, which instead must support the weight of the entire upper torso. The lower loads of the cervical spine also permit a much broader range of design and biomaterial options, which would not otherwise be feasible in the lumbar spine.

**FIGURE 12.9** Two CoCr alloy endplates and UHMWPE core of the Activ-L design (Aesculap AG, Tuttlingen, Germany).

Secondly, the indications for cervical disc replacement, which include myelopathy and radiculopathy, are broader and more prevalent than for lumbar disc replacement [53]. That is, more patients undergoing cervical fusions today may be candidates for disc arthroplasty, provided certain other contraindications are not met, such as nonintact posterior elements and osteoporosis. In contrast, the list of contraindications in lumbar disc replacement is currently so extensive that it effectively precludes the vast majority of lumbar fusion candidates from receiving an arthroplasty [54].

Thirdly, the surgical access to the anterior cervical spine is easier, and less risky for the patient, than the access to the anterior lumbar spine. Furthermore, the ease of access for cervical disc replacements is preserved during revision, whereas in the lumbar spine, scar tissue and adherent vessels greatly complicate the anterior revision process. For these main reasons, cervical disc arthroplasty appears to be a less risky proposition than lumbar arthroplasty, both in terms of the feasibility of available technology options for the implants, as well as in terms of the risk to the patient during the procedure.

As noted previously, Fernström's stainless steel sphere was the first cervical disc arthroplasty implanted back early 1960s [7]. Although Fernström's 1966 publication acknowledges his use of the spherical prosthesis in the cervical spine, two South African surgeons, Hjalmar Reitz and Mauritius Joubert, are credited with publishing the first paper on cervical disc arthroplasty 2 years earlier [55]. Reitz and Joubert visited with Fernström in Sweden and were "favourably impressed by his results" [55]. After returning home, they implanted 75 spherical, cervical disc replacements in 32 patients who suffered from severe headaches and neck-related arm pain. The surgeons reported generally positive outcomes for their patients who were followed for less than a year. However, Reitz and Joubert were optimistic that Fernström's prosthesis "preserves the mobility of the intervertebral joints, thus obviating the inherent disadvantages of all fusion procedures" [55]. The surgeons acknowledged that longer-term follow-up would be necessary to confirm the viability of their technique. Fernström's pioneering design, along with the experience of Reitz and Joubert, would provide inspiration for future generations of cervical disc replacement technologies.

In this section, we summarize the design features and available literature for contemporary cervical metal-on-UHMWPE artificial discs. Many cervical disc replacements have been proposed over the years, but today five contemporary designs have been documented in the peer-reviewed literature and are currently in clinical trials in the United States: ProDisc-C, PCM, Mobi-C, and Discover (Table 12.5). As detailed in the table, these designs differ from each other in several respects, including the philosophy as well as the constraint in the bearing (Table 12.5).

**TABLE 12.5 Summary of Contemporary UHMWPE/CoCr Cervical Artificial Disc Designs**

| Design | ProDisc-C | PCM | Mobi-C | Activ-C | Discover |
|---|---|---|---|---|---|
| Manufacturer | Synthes (Paoli, Pennsylvania, USA; http://www.synthes.com) | Cervitech (Rockaway, New Jersey; http://www.cervitech.com/) | LDR Spine (Troyes, France, http://www.ldrmedical.com/) | Aesculap AG (Tuttlingen, Germany; http://www.aesculap.com/) | DePuy Spine (Raynham, Massachusetts, USA; http://www.depuyspine.com/) |
| Number of components for surgeon assembly | Two | Two | Three | Three | Two |
| Number of articulating surfaces | One | One | Two | Two | One |
| Bearing design | Ball-and-socket | Surface replacement | Mobile bearing | Mobile bearing | Ball-and-socket |
| Constraint | Semiconstrained | Minimally constrained | Semiconstrained in core translation | Semiconstrained in core translation | Semiconstrained |
| Bone/implant fixation | Keel and Ti textured coating | CaP/Ti coating | Lateral self-retaining teeth and Ti textured coating | Lateral self-retaining teeth and Ti textured coating | Teeth and Ti textured coating |
| Keel | Yes | No | No | No | No |
| Published clinical studies | Yes | No | Yes | No | No |
| Longest published average follow-up | 2 years [56] | None | 7.5 months | None | None |

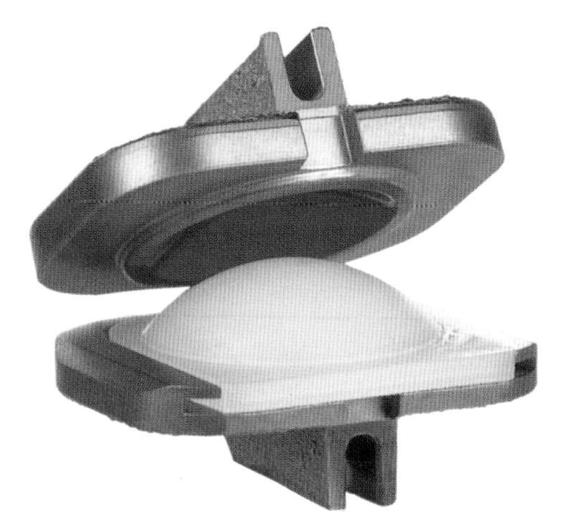

**FIGURE 12.10** The ProDisc-C cervical disc replacement. Implant photograph, reproduced from [82] with permission.

### 12.4.1 ProDisc-C

The ProDisc-C, produced by Synthes, is a cervical version of the lumbar ProDisc II and employs CoCr-on-UHMWPE articulation (Figure 12.10). In contrast to the lumbar design, in the ProDisc-C, the UHMWPE component is provided already inserted into the inferior endplate by the manufacturer. Bone-implant fixation is achieved by a combination of a keel and titanium plasma sprayed surface, similar to the ProDisc II. Naturally, the size and proportion of the keel and articulations are much smaller in the ProDisc-C as compared with the ProDisc-L, but the technology underlying both designs is identical. Synthes completed a prospective randomized clinical trial with the ProDisc-C for the FDA and received clearance to market the device in the United States in 2007.

Bertagnoli and associates have reported promising 1-year clinical follow-up for the ProDisc-C [57, 58]. Starting in December 2002, the ProDisc-C study included 27 single-level patients who were implanted between C4–C5 and C6–C7. The average patient age was 49 years (range: 31–66 years). No device-related or procedure-related complications were reported. After 1 year, the patients' average range of motion (ROM) had improved from 4.2 degrees (preoperative ROM) to 10.2 degrees. The patients' Neck Disability Index (NDI) also improved, on average, from 28.9 points (preoperatively) to 18.8 points at 1 year.

*In vitro* biomechanical testing of the ProDisc-C was performed by DiAngelo et al. [59]. Six cervical spines (C2–C7) were tested sequentially in the intact state, following disc arthroplasty, and then after a simulated single-level fusion. The testing frame was a custom-built load frame attached to a servo-motor load actuator (International Device Corp., Novato, California, USA) with a robotic controller (Adept, Inc., San Jose, California, USA), in line with a load cell. Custom fixtures were then used to drive the cervical spine in displacement control to produce flexion/extension, lateral bending, or axial rotation. The authors justify their displacement control method by suggesting that their procedure is more physiologic than applied pure bending moments and further opine that the use of a follower load "restricts the spine from following its natural path." They measured a significant difference in the extension range of motion of the intact spine (28 degrees ± 6 degrees) as compared with the ProDisc-C (39 degrees ± 2 degrees), but in all other degrees of freedom, the flexibility of the intact spines and the disc replacement were not significantly different. The authors concluded that the ProDisc-C maintains the overall flexibility of the cervical spine, comparable to its intact state.

### 12.4.2 PCM

The Porous Coated Motion (PCM) Artificial Disc is manufactured by Cervitech, Inc. (Rockaway, New Jersey, USA) (Figure 12.11). Like the CHARITÉ, the PCM is a minimally constrained bearing design [60]. The gliding surfaces of the PCM are intended to accommodate smooth translations and rotations. Consequently, the PCM strongly depends upon the competency of the surrounding soft tissues for proper kinematics; in that respect it is more analogous to a cruciate retaining total knee replacement than to a total hip replacement. The inventors compare the PCM with a surface replacement.

The PCM uses the same well-established bearing and coating technologies as the CHARITÉ; the reader is referred to the earlier section for a more detailed description of the bearing materials and coating methods. The cephalad and caudal endplates of the PCM are CoCr alloy. The UHMWPE core is fixed to the caudal endplate, so that motion occurs between the concave, cephalad endplate and the convex UHMWPE surface. The outer surface of the endplates have a serrated profile, for primary fixation, and have the same Ti/CaP coating as the CHARITÉ for long-term bone ongrowth [60]. The porosity of the Ti plasma spray with the PCM is smaller than the CHARITÉ, to account for the finer trabecular architecture in the cervical spine as opposed to the lumbar spine [61].

The anterior profile of the PCM varies with the Low Profile and Fixed designs. The Low Profile design does not protrude beyond the anterior margins of the vertebral column. With the Fixed design, on the other hand, the anterior face of the endplates is flanged for supplemental fixation using cancellous bone screws. The anterior flange protrudes over the surface of the vertebral body. The Low Profile is the standard implant that has been used in the majority of cases, whereas the Fixed design is now only used in extreme cases, such as in revisions and adjacent to multilevel fusions [62].

One-year follow-up data are available for the PCM from a prospective, pilot study conducted by Pimenta and

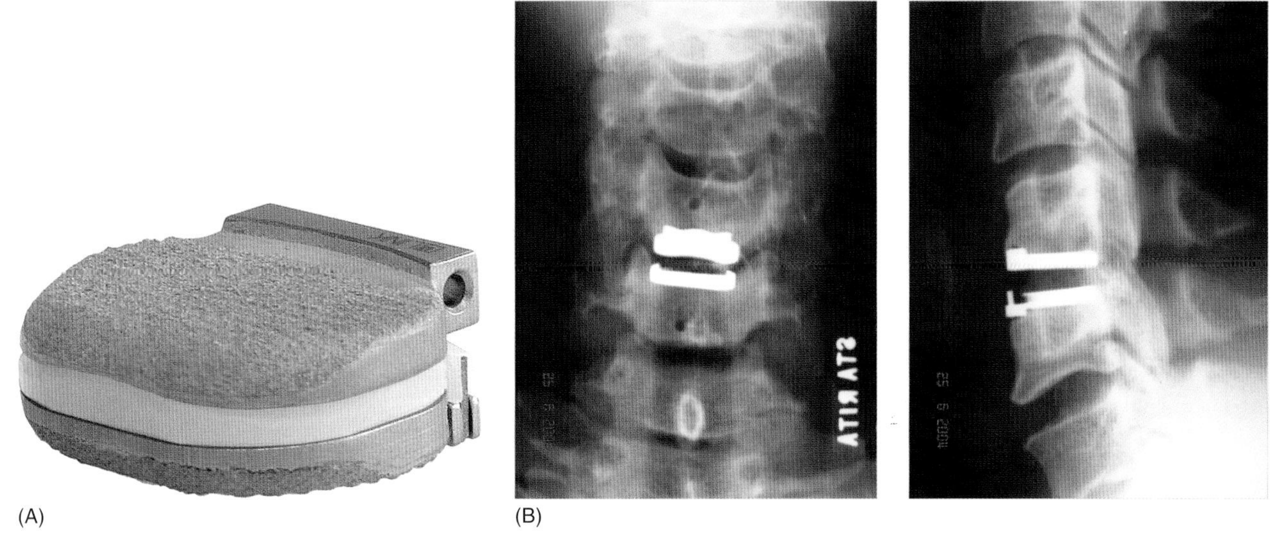

FIGURE 12.11 The PCM cervical disc replacement. (A) Implant photograph; (B) AP and lateral radiographs of the PCM implanted in the cervical spine. Reproduced from [82] with permission.

associates in Brazil. The clinical study began in December 2002 and included 52 patients who were implanted with 81 cervical discs, between C3–C4 and C7–T1. The average patient age was 45 years (range: 28–68 years). Two complications were reported, including one case of anterior migration and one case of heterotopic bone formation. By 1 year, the clinical success was good or excellent in 97% of the cases. The patients' Neck Disability Index (NDI) had improved, on average, to 15 points (range: 40 to 0) following 1 year; preoperatively, the average NDI for the patients was 45 (range: 98 to 18). Interestingly, 10 of the procedures were revisions of previous anterior surgeries [61].

The range of motion of the PCM and role of the posterior longitudinal ligament (PLL) have been evaluated by McAfee and coworkers [63], using the flexibility testing protocol developed by Cunningham et al. [30]. Using this protocol, cervical spine sections between C3–C7 were tested from seven donors under applied pure moments in flexion, extension, left and right lateral bending, and left and right axial torsion. As described previously in the CHARITÉ section of this chapter, Cunningham's protocol does not include axial preload and therefore assesses the maximum theoretical flexibility of the cervical spine column. At C5–C6, no significant difference was observed between the flexion-extension range of motion for the PCM versus the intact spine. In contrast, resecting the PLL had a significant effect on ROM in flexion-extension, axial rotation, and lateral bending, as compared with the intact spine. McAfee et al. also implanted the PCM into the C3–C4 levels of 12 goats, which were followed for 6 months [63]. No significant postoperative complications occurred, and no evidence of wear debris, osteolysis, or particle-related biological reactions was observed.

In a recent study, Dmitriev and coworkers tested the hypothesis that the cervical disc replacement preserves the segmental kinematics and disc pressures of adjacent levels [64]. Ten cervical spines were obtained and treated at the C5–C6 level using the PCM disc replacement, allograft dowel, and allograft dowel with anterior cervical plate. The researchers focused their attention on the motion of adjacent segments, C4–C5 and C6–C7, in addition to the treated level. Disc pressures were monitored using miniature pressure transducers (width = 1.5 mm, height = 0.3 mm, Precision Measurement Co., Ann Arbor, Michigan, USA) inserted into the C4–C5 and C6–C7 discs. The intact spine specimens were first loaded with pure moments (±5 Nm in flexion, extension, left/right lateral bending, and left/right axial rotation) to record the maximum ROM in each degree of the six degrees of freedom. The apparatus for applying the pure moments was similar to that used by Cunningham et al. [30]. After each of the three interventions, the tests were rerun under displacement control at a rate of 3 degrees for all 6 degrees of freedom, using the maximum ROM in that particular degree of freedom as the limit for the test. The investigators observed no significant difference in adjacent level disc pressures or segmental motions between the intact and PCM-treated spines. In general, the greatest difference between treatments was observed in flexion and extension testing. Both fusion treatments resulted in higher disc pressures than the intact or TDR-treated spines during flexion/extension testing. Although limited by the lack of axial preload during the testing, the available data from this study supports the hypothesis that cervical disc replacement preserves adjacent level disc pressures and segmental kinematics.

According to Cervitech's Web site (http://www.cervitech.com), the company has nearly completed enrollment for its prospective randomized clinical trial for the FDA, but no further details are available at the present time.

### 12.4.3 Mobi-C

The Mobi-C is manufactured by LDR Spine (Troyes, France) (Figure 12.12). Like its lumbar analogue, the Mobi-C is a convex sliding mobile bearing design and achieves short-term fixation via lateral teeth and a textured ongrowth surface. The Mobi-C was first implanted in Europe in November 2004, and since that time, over 4300 artificial discs have been implanted, according to the manufacturer. In October 2007, LDR spine announced that it had completed enrollment for its single-level IDE study. Thus, 2-year follow-up is expected to be completed by 2009.

### 12.4.4 Activ-C

The Activ-C is produced by Aesculap AG (Tuttlingen, Germany). The Activ-C is a convex sliding mobile bearing design, similar to its lumbar analogue. Fixation of the CoCr endplates is accomplished by three teeth, along with a Ti spray bone ongrowth surface.

### 12.4.5 Discover

The Discover cervical disc was developed by DePuy Spine (Rayham, Massachusetts, USA) (Figure 12.13). This artificial disc is a fixed bearing, with a convex Ti alloy endplate articulating on a UHMWPE convex inferior core. The core is affixed to the inferior endplates. Bone fixation is achieved by a combination of a textured Ti surface and teeth. The Discover is clinically available in Europe but is not expected to receive clearance by the FDA until 2011.

## 12.5 WEAR AND *IN VIVO* DEGRADATION OF UHMWPE IN THE SPINE

Wear and *in vivo* degradation are key functional aspects of total disc replacements. The majority of *in vivo* clinical data for UHMWPE wear behavior and oxidation in TDRs has been obtained from retrieval studies of the historical CHARITÉ Artificial Disc, manufactured by Link between 1989 and 2004. Limited retrieval data has been published related to the ProDisc-L. *In vivo* wear data for second-generation lumbar artificial discs, as well as for UHMWPE in most cervical disc replacement designs, is not yet available.

Based on the clinical experience of UHMWPE in total hip and knee replacements, the production of wear debris from artificial discs, as well as from other motion-preserving spine implants, is a clinical concern. Wear debris induced osteolysis has been implicated as a potential mechanism for late onset pain following the failure of stainless steel and titanium instrumented fusions [65]. Osteolysis has also been observed around certain total disc replacement designs, such as the Acroflex artificial disc [66], and case studies of osteolysis around CHARITÉ disc replacements have also been reported [67–69]. According to recent conference presentations, the UHMWPE particle load around long-term implanted artificial discs may be comparable to total hip arthroplasty [70], and the periprosthetic particle concentration appears to be correlated with a local inflammatory response [71]. Although the occurrence of osteolysis with metal-on-polyethylene total disc replacements has thus far been relatively rare, the long-term wear behavior of artificial discs remains of clinical importance [69].

Analysis of wear and surface damage in long-term implanted CHARITÉ total disc replacements has been reported in a series of journal publications [72–74]. In the latest update of this multiinstitutional series, 38 CHARITÉ components were retrieved with up to 16 years of implantation [73]. The components were revised for intractable pain and/or facet degeneration. Components were analyzed using optical microscopy, scanning electron microscopy, white light interferometry, and MicroCT. Forty-three percent (15/35) of components analyzed using MicroCT displayed one-sided wear patterns [73]. Significant correlations were observed between implantation time and

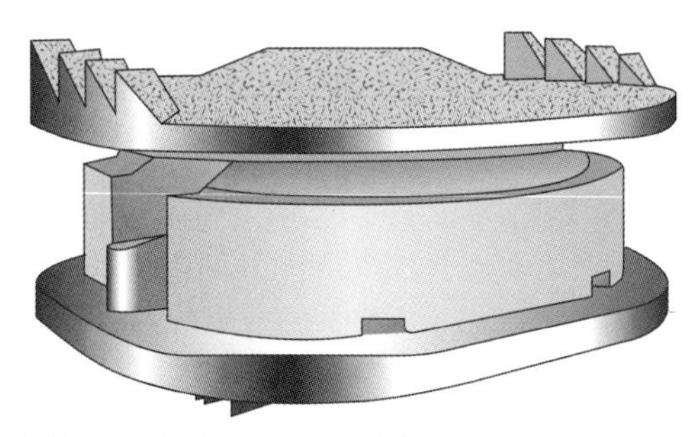

FIGURE 12.12 The Mobi-C cervical disc replacement. Reprinted with permission from LDR Spine.

FIGURE 12.13 The Discover cervical disc replacement. Figure used with permission from DePuy Spine, Inc.

penetration and penetration rate (Figure 12.14). The dome of the components typically exhibited burnishing, which is consistent with the multidirectional wear observed in hip replacements, whereas the rim frequently showed evidence of radial and transverse cracking, often produced by impingement (Figure 12.15). The rim damage modes of plastic deformation, delamination, and cracking were similar to those associated with knee components.

Evidence of dome burnishing, as well as rim impingement, has also been noted in a recent conference poster summarizing a collection of five short-term ProDisc-L prostheses, implanted up to 2.2 years [75]. The bearing surface of the ProDisc-L showed burnishing (3/5 implants), mild scratching, and pitting (3/5 implants). Impingement was noted in 3/5 components and associated with burnishing and plastic deformation. Similar observations of wear, surface damage, and impingement have been noted for the ProDisc-C [75]. The authors of the ProDisc-L retrieval study remarked that, "a potentially worrisome finding is the evidence of impingement. Whether caused by patients achieving a larger range of motion that the implant is designed to accommodate or by component positioning that allows impingement at even smaller range of motion, impingement can be problematic" [75]. A detailed example of a TDR retrieval study for the ProDisc-L, displaying mild anterior impingement, has recently been reported as a case study by Choma and colleagues [76].

Although rim impingement has been observed in retrieved TDRs of different designs, the clinical consequences of chronic rim impingement remain poorly understood. In a study presented at the 2008 Spine Arthroplasty Society meeting [77], a retrieval collection of polyethylene mobile bearing TDRs was analyzed to determine whether rim impingement adversely affected dome penetration. The study showed that 28/40 (70%) of retrieved cores, implanted for 2–16 years (7.9y average), were classified as exhibiting chronic rim impingement based on observations of plastic deformation, burnishing, and/or fracture of the rim. Dome penetration was comparable in chronically impinged cores (average: 0.3, range: 0.1 to 0.9 mm) as compared with nonimpinging cores (average: 0.3, range: 0.1 to 0.5 mm). Rim penetration was significantly greater in chronically impinged cores ($p < 0.05$). Using linear regression, the dome penetration rate for cores with negligible impingement (0.036 mm/y, 95% CI: 0.012 to 0.061 mm/y) appeared slightly higher than in cores with chronic impingement (0.021 mm/y, 95% CI: 0.005 to 0.038 mm/y); however, the difference was not significant.

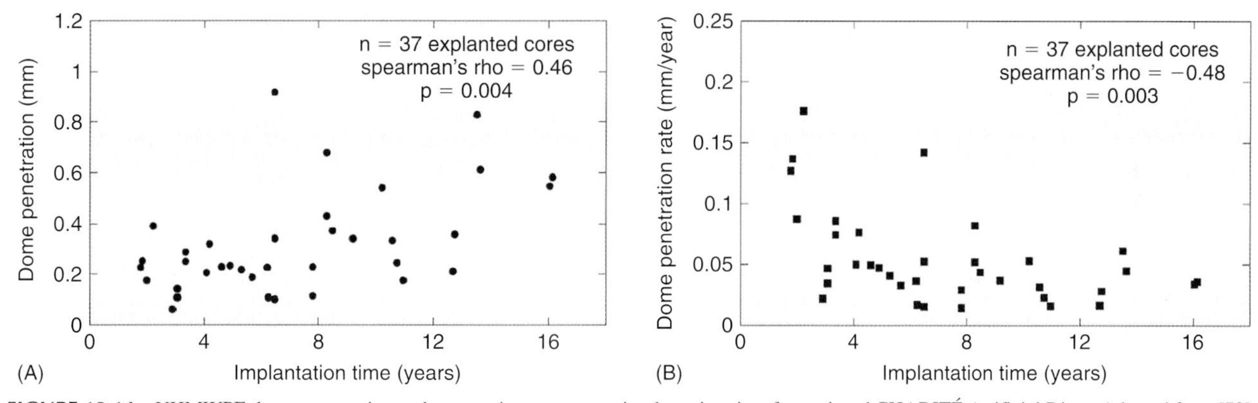


**FIGURE 12.14** UHMWPE dome penetration and penetration rate versus implantation time for retrieved CHARITÉ Artificial Discs. Adapted from [73].

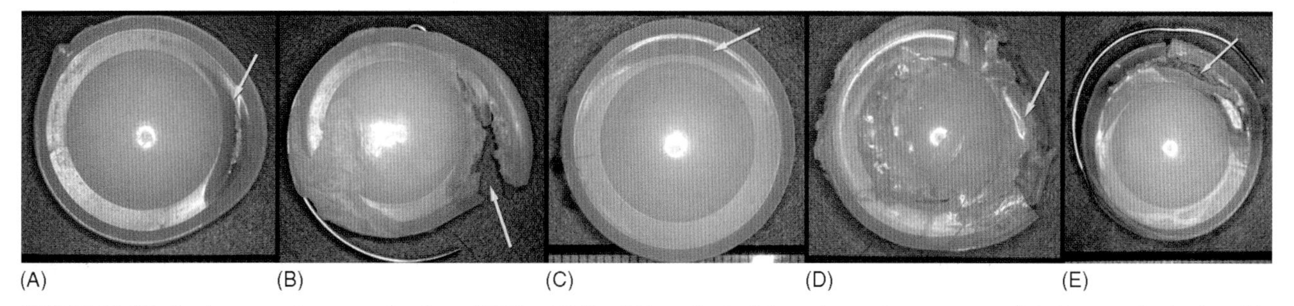


**FIGURE 12.15** Impingement damage to the rims of TDRs: (A) Burnishing, plastic deformation, and transverse crack (white arrow)—implanted in 2000 for 4.9 years (Maa008); (B) Transverse crack, rim fracture (white arrow)—implanted in 1998 for 6.5 years (Maa006); (C) Burnishing, plastic deformation (white arrow)—implanted in 1995 for 10.6 years (Maa018); (D) Rim fracture, transverse cracks (white arrow)—implanted in 1992 for 13.5 years (Maa014); (E) Burnishing, plastic deformation, and transverse crack (white arrow)—implanted in 1989 for 16 years (Maa015). From [73] with permission of Elsevier Science.

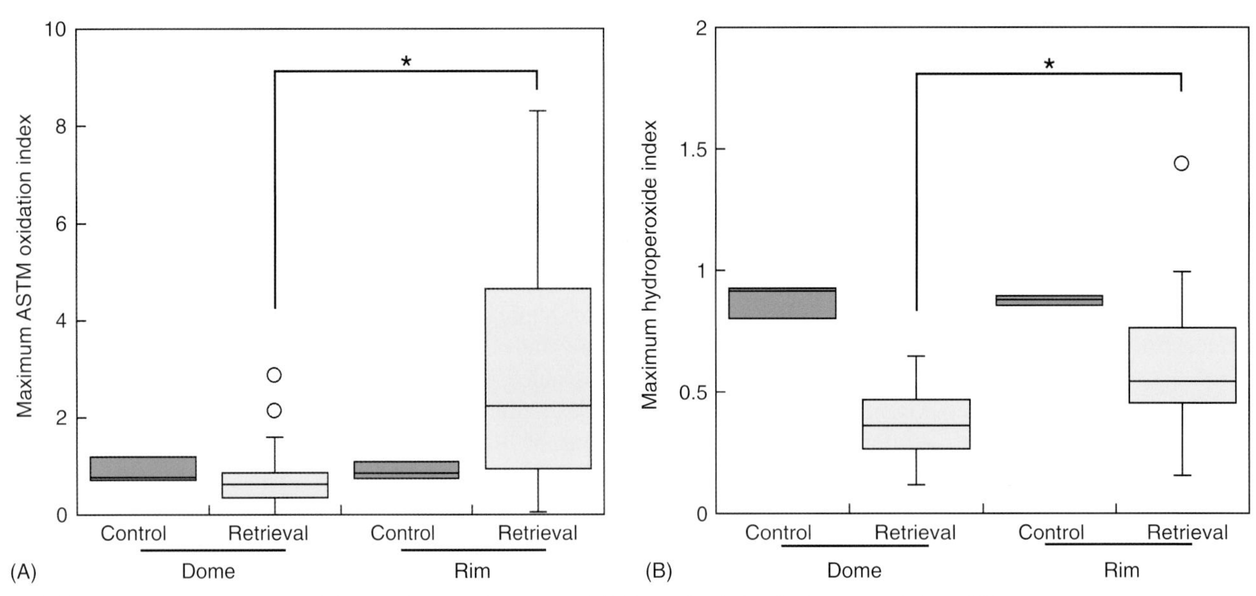


**FIGURE 12.16** Rim versus dome oxidation (A) and oxidation potential (B) in polyethylene TDRs. Adapted from [77].

Thus, the results of this study did not support the hypothesis that chronic rim impingement would be associated with greater dome penetration. However, the findings would suggest that dome wear and impingement are effectively decoupled phenomena and may be studied independently of each other.

In addition to impingement, rim damage observed in polyethylene TDR retrievals has also been associated with postirradiation oxidation [29, 78]. Analysis of explanted CHARITÉ cores using Fourier transform infrared spectroscopy has shown that the exposed rim experiences severe oxidation after 10 or more years [29, 78]. These findings appear consistent for TDRs that were gamma irradiated in air, as well as in first-generation polymeric barrier packaging [29, 78]. However, the central dome appears to be somewhat protected from *in vivo* oxidation due to contact with the metallic endplates (Figure 12.16). No correlation was observed between wear of the central dome and oxidation. These observations are similar to the *in vivo* oxidation patterns noted in artificial hips, which exhibit rim embrittlement after 10 years *in vivo* but show reduced oxidation at the bearing surface where the femoral head contacts the polyethylene [79]. Unlike in hip replacement, the rim of a total disc replacement core may be intended to support chronic loading for the lifetime of the patient. The findings of *in vivo* oxidation in gamma sterilized polyethylene TDR components provide additional motivation for developing *in vitro* mechanical tests that incorporate accelerated aging, or some other oxidative challenge, to simulate changes in the bearing materials that may occur with long-term *in vivo* exposure.

The published CHARITÉ retrieval literature provides crucial long-term *in vivo* wear data for validation of spine wear simulators as well as for *in vitro* biomechanical testing. Multistation, 6-degree of freedom spine wear simulators are now commercially available (Figure 12.17). However, two independent sets of protocols for wear simulation of UHMWPE in the spine, fostered by ASTM International and ISO, have not yet been validated. A recent study presented at the 2008 Spineweek meeting was conducted to better correlate long-term clinical wear rates of the CHARITÉ with simulator wear rates [80]. It was hypothesized that the wear mechanisms of the retrievals would be more accurately simulated by ISO protocols with coupled motion as compared with ASTM-type protocols that resulted in linear motion. Researchers analyzed dome wear rate and surface morphology of 41 CHARITÉ (SB III) explants from 35 patients (71% female). The cores were implanted for 7.5y (range: 1.8–16.3y). Twelve CHARITÉ wear-tested cores and six controls were also examined. Six cores were tested according to an ASTM-type protocol for 10 million cycles [81]; three additional cores were unloaded and soaked. Six cores were tested according to the ISO protocol for 2 million cycles with three loaded and soaked controls. All the cores in this study were produced by the same manufacturer (Link, Germany). The explanted cores typically exhibited burnishing or evidence of adhesive/abrasive dome wear, consistent with multidirectional motion. The wear rate of the explants, obtained by correlation of dome height with implantation time, was 0.023 mm/y. The ASTM-tested cores exhibited unidirectional abrasive wear at a rate of 0.007 mm/Mcycles. The ISO-tested cores exhibited regional burnishing and wear at a rate of 0.124 mm/Mcycles (Figure 12.18). Thus, the ISO protocol generated wear surface morphology that was closer to the retrievals than the ASTM-type protocol, and 1 million cycles of the ISO protocol corresponded, on average, to about 5.6y of clinical wear. The findings from

FIGURE 12.17 Commercially available spine wear simulator for cervical and lumbar artificial discs. Image courtesy of Exponent.

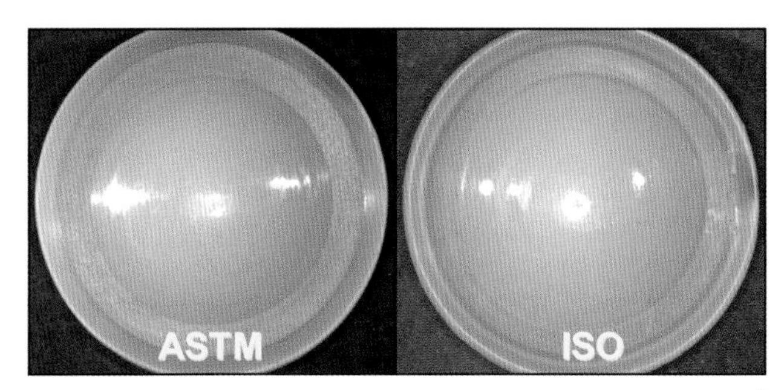


FIGURE 12.18 Comparison of wear tested surfaces of the CHARITÉ Artificial Disc.

this study further suggest that the ISO protocol provides a useful starting point for clinical validation of spine wear simulations incorporating lumbar polyethylene TDRs. Additional research is still needed so that wear simulations of CoCr alloy-on-UHMWPE, let alone other biomaterial combinations, can be considered validated for cervical or lumbar TDRs.

## 12.6 ALTERNATIVES TO UHMWPE IN DISC REPLACEMENT

The current state of the art for UHMWPE in disc replacement is conventional, gamma-sterilized material in barrier packaging. None of the radiation crosslinking and stabilization technologies developed for UHMWPE in hip and knee replacement have yet emerged as clinical candidates for artificial discs. Due in part to controversies we discussed surrounding gamma-air sterilized UHMWPE with the CHARITÉ, the commercial developers of new artificial disc designs in recent years have favored alternative bearing solutions, rather than developing novel UHMWPE formulations specifically for disc replacement. Commercial interest in product differentiation has also fostered a greater willingness to explore alternative bearings in disc replacement than has yet been witnessed in hip and knee arthroplasty.

One technical justification for alternative bearings is to reduce wear and the biological response associated with conventional UHMWPE. Additional technical motivations for alternative bearings are to reduce the thickness of the components, as well as to improve the MRI-compatibility of the implants. For all of these reasons, the field of motion preservation spine implants has witnessed an explosion of different designs with alternative bearing couples.

Until now we have focused on designs incorporating UHMWPE that have reached clinical use. However, the reader should be aware that many more disc replacement designs, as yet unproven and unvalidated, are currently under development or in various stages of animal or human clinical experimentation [82]. All of these disc replacements have the potential to exhibit wear and particulate debris release at some length scale, however microscopic. For materials with longstanding use in total joint replacements, such as UHMWPE, there may be a risk of osteolysis, albeit in a small patient population. With metal-on-metal disc replacements, the long-term biological consequences of the metallic wear debris and their soluble corrosion products in the spine are also unknown at the present time. With other metallic and polymeric bearing materials, the risks associated with their long-term *in vivo* use remain unknown.

In the lumbar spine, metal-on-metal (CoCr/CoCr) is the main alternative to UHMWPE as a bearing surface for TDRs [82]. The Maverick lumbar artificial disc (Medtronic

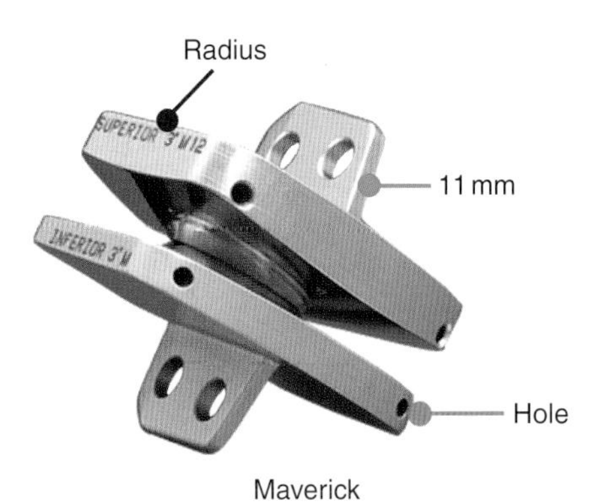


FIGURE 12.19 The Maverick cervical disc replacement. Image courtesy of Medtronic Spinal and Biologics and reproduced from [82] with permission.

FIGURE 12.20 The Prestige cervical disc replacement. Image courtesy of Medtronic Spinal and Biologics and reproduced from [82] with permission.

Spinal and Biologics, Memphis, Tennessee, USA) is a semiconstrained, ball-in-socket articulation (Figure 12.19) [56]. The FlexiCore design (Stryker Spine, Allendale, New Jersey, USA) is a constrained ball-in socket bearing [83]. The Kineflex design (Spinal Motion, Mountain View, California, USA) is mobile bearing metal-on-metal lumbar artificial disc. All three metal-on-metal lumbar disc designs are used in Europe and Asia but are only for investigational use in the United States.

In the cervical design, because of the relatively low service loads, a plethora of alternative bearings are currently being explored [82]. The Prestige artificial disc is a stainless steel ball-in-trough design, with a design history spanning over a decade (Figure 12.20) [84]. The Prestige

FIGURE 12.21 The Bryan cervical disc replacement. Image courtesy of Medtronic Spinal and Biologics and reproduced from [82] with permission.

was approved by the FDA in 2007. The Bryan Artificial Disc, consisting of polyurethane components articulating with titanium endplates, has been used in Europe since the late 1990s (Figure 12.21) [85]. The Bryan is currently for investigational use in the United States. Other artificial disc designs employ polymeric and metallo-ceramic biomaterials without any clinical precedence as long-term orthopedic bearings. Although these novel biomaterials appear to exhibit satisfactory short-term biocompatibility, their biological response over the long term is also simply unknown. At least 10 years of clinical follow-up will be needed to confirm that these novel bearing combinations are comparable to historical bearing materials, which have a clinical history spanning 5 decades of continuous use.

## 12.7 MANY UNANSWERED QUESTIONS REMAIN

Total disc replacement is a new, promising field of spine implant technology that has the potential to revolutionize the treatment of degenerative disc disease. Today, conventional UHMWPE is incorporated in both cervical and lumbar disc arthroplasty. It is clear from both *in vitro* and clinical data that disc replacements can successfully preserve the motion of treated spinal level. Aside from patient satisfaction and the speed of recovery, there are modest clinical benefits with disc replacement that manifest in the short term as compared with fusion. Furthermore, unlike fusion procedures, disc replacements may also need to be revised due to poor implantation technique or failure of the device. On the other hand, over the long term, the primary benefit of disc replacement is expected to be the reduced incidence of adjacent segment degeneration, which will hopefully offset the new, and as yet, poorly quantified risks associated with the technology. It will be many years, probably over a decade, before sufficient data has been generated to evaluate whether the promises inherent in total disc replacement have been fulfilled, and in turn, whether the long-term benefits of the technology justify the additional risks inherent with their implantation and potential revision. The current technologies for disc replacement should be viewed, therefore, as still being in their infancy.

## 12.8 ACKNOWLEDGMENTS

This chapter was supported in part by NIH R01 AR47192. This chapter was not written with the financial support of any manufacturer of total disc replacements. However, all of the producers of disc replacements, whose products are mentioned in this review, were contacted by the author and given the opportunity to verify the factual accuracy of the information related to their products.

The author is indebted to Frank Chan, PhD, Medtronic Spinal and Biologics, for his encouragement, assistance with locating figures, and many helpful discussions related to total disc arthroplasty, without which this chapter would not have been possible. Thanks to Hassan Serhan, PhD, DePuy Spine, Inc., for feedback about the CHARITÉ and Discover designs. Thanks to Noah Bartsch, LDR Spine, for helpful discussions and for providing the images for the Mobidisc and Mobi-C for this chapter.

## REFERENCES

1. Carragee EJ. Clinical practice. Persistent low back pain. *N Engl J Med* 2005 May 5;**352**(18):1891–8.
2. Punnett L, Pruss-Utun A, Nelson DI, Fingerhut MA, Leigh J, Tak S, et al. Estimating the global burden of low back pain attributable to combined occupational exposures. *Am J Ind Med* 2005 December;**48**(6):459–69.
3. Ong K, Lau E, Kurtz SM, Chin KR, Ianuzzi A, Villarraga ML. Projections to 2030 of the prevalence of primary and revision spine fusions in the U.S. *Trans of the 53rd Orthop Res Soc* 2007;**32**:1070.
4. Turner JA, Ersek M, Herron L, Deyo R. Surgery for lumbar spinal stenosis. Attempted meta-analysis of the literature. *Spine* 1992 January;**17**(1):1–8.
5. Hilibrand AS, Robbins M. Adjacent segment degeneration and adjacent segment disease: the consequences of spinal fusion? *Spine J* 2004 November–December;**4**(6 Suppl.):190S–4S.
6. Park P, Garton HJ, Gala VC, Hoff JT, McGillicuddy JE. Adjacent segment disease after lumbar or lumbosacral fusion: review of the literature. *Spine* 2004 September 1;**29**(17):1938–44.
7. Fernström U. Arthroplasty with intercorporal endoprosthesis in herniated disc and in painful disc. *Acta Chir Scand Suppl* 1966;**357**:154–9.
8. Szpalski M, Gunzburg R, Mayer M. Spine arthroplasty: a historical review. *Eur Spine J* 2002 October;**11**(Suppl. 2):S65–84.
9. Büttner-Janz K. History. In: Büttner-Janz K, Hochschuler SH, McAfee PC, editors *The artificial disc*. Berlin: Springer; 2003. p. 1–10.
10. Büttner-Janz K, Hochschuler SH, McAfee PC. *The artificial disc*. Berlin: Springer; 2003.
11. Charnley J. Low friction principle. In: *Low friction arthroplasty of the hip: theory and practice*. Berlin: Springer-Verlag; 1979. p. 3–16.
12. Putzier M, Funk JF, Schneider SV, Gross C, Tohtz SW, Khodadadyan-Klostermann C, et al. Charite total disc replacement—clinical and radiographical results after an average follow-up of 17 years. *Eur Spine J* 2006 February;**15**(2):183–95.
13. McAfee PC. An explanation of early, suboptimal results from Charite Hospital-"Philosophical and metallurgical differences". In: Büttner-Janz K, Hochschuler SH, McAfee PC, editors. *The artificial disc*. Berlin: Springer; 2003.
14. Link HD, Keller A. Biomechanics of total disc replacement. In: Büttner-Janz K, Hochschuler SH, McAfee PC, editors. *The artificial disc*. Berlin: Springer; 2003. p. 33–52.
15. McAfee PC, Banco RJ, Blumenthal SL, Geisler FH, Guyer RD, Holt RT, et al. Prospective, randomized, multicenter FDA IDE study of CHARITÉ artificial disc vs. lumbar fusion: effect at 5-year follow-up of prior surgery on clinical outcomes following lumber arthroplasty. *Transactions of the Spineweek 2008 meeting*. Geneva, Switzerland; 2008 May 26–31. C6.
16. Büttner-Janz K, Schellnack K, Zippel H, Conrad P. Experience and results with the SB Charite lumbar intervertebral endoprosthesis. *Z Klin Med* 1988;**43**:1785–9.
17. Griffith SL, Shelokov AP, Büttner-Janz K, LeMaire JP, Zeegers WS. A multicenter retrospective study of the clinical results of the LINK SB Charite intervertebral prosthesis. The initial European experience. *Spine* 1994 August 15;**19**(16):1842–9.
18. Cinotti G, David T, Postacchini F. Results of disc prosthesis after a minimum follow-up period of 2 years. *Spine* 1996 April 15;**21**(8):995–1000.
19. Zeegers WS, Bohnen LM, Laaper M, Verhaegen MJ. Artificial disc replacement with the modular type SB Charite III: 2-year results in 50 prospectively studied patients. *Eur Spine J* 1999;**8**(3):210–7.
20. Sott AH, Harrison DJ. Increasing age does not affect good outcome after lumbar disc replacement. *Int Orthop* 2000;**24**(1):50–3.
21. Lemaire JP, Carrier H, Ali el HS, Skalli W, Lavaste F. Clinical and radiological outcomes with the CHARITÉ artificial disc: a 10-year minimum follow-up. *J Spin Disord Tech* 2005 August;**18**(4): 353–9.
22. McAfee PC, Cunningham B, Holsapple G, Adams K, Blumenthal S, Guyer RD, et al. A prospective, randomized, multicenter Food and Drug Administration investigational device exemption study of lumbar total disc replacement with the CHARITÉ artificial disc versus lumbar fusion: part II: evaluation of radiographic outcomes and correlation of surgical technique accuracy with clinical outcomes. *Spine* 2005 July 15;**30**(14):1576–1583. Discussion E1388–590.
23. David T. Long-term results of one-level lumbar arthroplasty: minimum 10-year follow-up of the CHARITÉ artificial disc in 106 patients. *Spine* 2007 March 15;**32**(6):661–6.
24. Ross R, Mirza AH, Norris HE, Khatri M. Survival and clinical outcome of SB CHARITÉ III disc replacement for back pain. *J Bone Joint Surg Br* 2007 June;**89**(6):785–9.
25. de Kleuver M, Oner FC, Jacobs WC. Total disc replacement for chronic low back pain: background and a systematic review of the literature. *Eur Spine J* 2003 April;**12**(2):108–16.
26. Blumenthal S, McAfee PC, Guyer RD, Hochschuler SH, Geisler FH, Holt RT, et al. A prospective, randomized, multicenter Food and Drug Administration investigational device exemptions study of lumbar total disc replacement with the CHARITÉ artificial

disc versus lumbar fusion: part I: evaluation of clinical outcomes. *Spine* 2005 July 15;**30**(14):1565–1575. discussion E1387-591.
27. Serhan H. Personal communication. DePuy Spine; 2005.
28. Currier B. CHARITÉ Core Oxidation (Shelf). *CHARITÉ artificial disc FDA panel meeting*. Gaithersburg, MD; 2004 June 2.
29. Kurtz SM, MacDonald D, Ianuzzi A, van Ooij A, Isaza J, Ross ERS. In vivo oxidation and oxidation potential for polyethylene in total disc replacement following gamma sterilization in air and first-generation barrier packaging. *Trans of the 54th Orthop Res Soc* 2008;**33**:1324.
30. Cunningham BW. Basic scientific considerations in total disc arthroplasty. *Spine J* 2004 November–December;**4**(6 Suppl.):219S–230S.
31. Cunningham BW, Dmitriev AE, Hu N, McAfee PC. General principles of total disc replacement arthroplasty: seventeen cases in a nonhuman primate model. *Spine* 2003 October 15;**28**(20):S118–24.
32. Cunningham BW, Gordon JD, Dmitriev AE, Hu N, McAfee PC. Biomechanical evaluation of total disc replacement arthroplasty: an in vitro human cadaveric model. *Spine* 2003 October 15;**28**(20):S110–S117.
33. O'Leary P, Nicolakis M, Lorenz MA, Voronov LI, Zindrick MR, Ghanayem A, et al. Response of CHARITÉ total disc replacement under physiologic loads: prosthesis component motion patterns. *Spine J* 2005 November–December;**5**(6):590–9.
34. Tropiano Jr P, Huang RC, Girardi FP, Cammisa FP, Marnay T Lumbar total disc replacement. Seven to eleven-year follow-up. *J Bone Joint Surg* 2005 March;**87**(3):490–6.
35. Huang RC, Girardi FP, Cammisa Jr FP, Lim MR, Tropiano P, Marnay T Correlation between range of motion and outcome after lumbar total disc replacement: 8.6-year follow-up. *Spine* 2005 June 15;**30**(12):1407–11.
36. Mayer HM, Wiechert K, Korge A, Qose I. Minimally invasive total disc replacement: surgical technique and preliminary clinical results. *Eur Spine J* 2002 October;**11**(Suppl. 2):S124–30.
37. Tropiano P, Huang RC, Girardi FP, Marnay T. Lumbar disc replacement: preliminary results with ProDisc II after a minimum follow-up period of 1 year. *J Spin Disord Tech* 2003 August;**16**(4):362–8.
38. Zigler JE. Lumbar spine arthroplasty using the ProDisc II. *Spine J* 2004 November–December;**4**(6 Suppl.):260S–7S.
39. Bertagnoli R, Yue JJ, Shah RV, Nanieva R, Pfeiffer F, Fenk-Mayer A, et al. The treatment of disabling single-level lumbar discogenic low back pain with total disc arthroplasty utilizing the Prodisc prosthesis: a prospective study with 2-year minimum follow-up. *Spine* 2005 October 1;**30**(19):2230–6.
40. Bertagnoli R, Yue JJ, Shah RV, Nanieva R, Pfeiffer F, Fenk-Mayer A, et al. The treatment of disabling multilevel lumbar discogenic low back pain with total disc arthroplasty utilizing the ProDisc prosthesis: a prospective study with 2-year minimum follow-up. *Spine* 2005 October 1;**30**(19):2192–9.
41. Delamarter RB, Bae HW, Pradhan BB. Clinical results of ProDisc-II lumbar total disc replacement: report from the United States clinical trial. *Orthop Clin North Am* 2005 July;**36**(3):301–313.
42. Zigler J, Delamarter R, Spivak JM, Linovitz RJ, Danielson 3rd GO, Haider TT, et al. Results of the prospective, randomized, multicenter Food and Drug Administration investigational device exemption study of the ProDisc-L total disc replacement versus circumferential fusion for the treatment of 1-level degenerative disc disease. *Spine* 2007 May 15;**32**(11):1155–62. Discussion 1163.
43. Zigler JE. Clinical results with ProDisc: European experience and U.S. investigation device exemption study. *Spine* 2003 October 15;**28**(20):S163–6.
44. Delamarter RB, Fribourg DM, Kanim LE, Bae H. ProDisc artificial total lumbar disc replacement: introduction and early results from the United States clinical trial. *Spine* 2003 October 15;**28**(20):S167–75.
45. Shim CS, Lee S, Maeng DH, Lee SH. Vertical split fracture of the vertebral body following total disc replacement using ProdDisc: report of two cases. *J Spinal Disord Tech* 2005 October;**18**(5):465–9.
46. Bertagnoli R, Zigler J, Karg A, Voigt S. Complications and strategies for revision surgery in total disc replacement. *Orthop Clin North Am* 2005 July;**36**(3):389–95.
47. Bertagnoli R. Complications and rescue strategies in TDR procedures. *Proceedings of the NASS 20th annual meeting*; 2005 April 6–9. p. 19.
48. LeHuec JC, Kiaer T, Friesem T, Mathews H, Liu M, Eisermann L. Shock absorption in lumbar disc prosthesis: a preliminary mechanical study. *J Spinal Disord Tech* 2003 August;**16**(4):346–51.
49. Lipman J, Campbell D, Girardi F, Cammisa F, Myers E, Wright T. Mechanical behavior of the Prodisc II intervertebral disc prosthesis in human cadaveric spines. *49th annual meeting of the orthopedic research society*. New Orleans, LA; 2003. p. 1153.
50. Nechtow W, Hintner M, Bushelow M, Kaddick C. IVD replacement mechanical performance depends strongly on input parameters. *Trans of the 52nd Orthop Res Soc* 2006;**31**:118.
51. Delamarter R, Zigler J, Goldstein J. 5-year results of the prospective, randomized, multicenter FDA invesigational device exemption (IDE) Prodisc-L total disc replacement clinical trial. *Transactions of the Spineweek 2008 meeting*. Geneva, Switzerland; 2008 May 26–31. C3.
52. Steib J, Aubourg L, Beaurain J, Delecrin J, Allain J, Chataigner H, et al. Mobidisc disc prosthesis. In: Yu JJ, Bertagnoli R, McAfee PC, An HS, editors. *Motion preservation surgery of the spine: advanced techniques and controversies*. Philadelphia: Elsevier; 2008. p. 326–9.
53. Auerbach JD, Jones KJ, Fras CI, Balderston JR, Rushton SA, Chin KR. The prevalence of indications and contraindications to cervical total disc replacement. *Spine J* 2007 November 3.
54. Chin KR. Epidemiology of indications and contraindications to total disc replacement in an academic practice. *Spine J* 2007 July–August;**7**(4):392–8.
55. Reitz H, Joubert MJ. Intractable headache and cervico-brachialgia treated by complete replacement of cervical intervertebral discs with a metal prosthesis. *S Afr Med J* 1964 November 7;**38**:881–4.
56. Le Huec JC, Mathews H, Basso Y, Aunoble S, Hoste D, Bley B, et al. Clinical results of Maverick lumbar total disc replacement: two-year prospective follow-up. *Orthop Clin North Am* 2005 July;**36**(3):315–22.
57. Bertagnoli R, Duggal N, Pickett GE, Wigfield CC, Gill SS, Karg A, et al. Cervical total disc replacement, part two: clinical results. *Orthop Clin North Am* 2005 July;**36**(3):355–62.
58. Bertagnoli R, Yue JJ, Pfeiffer F, Fenk-Mayer A, Lawrence JP, Kershaw T, et al. Early results after ProDisc-C cervical disc replacement. *J Neurosurg* 2005 April;**2**(4):403–10.
59. DiAngelo DJ, Foley KT, Morrow BR, Schwab JS, Song J, German JW, et al. In vitro biomechanics of cervical disc arthroplasty with the ProDisc-C total disc implant. *Neurosurg Focus* 2004 September 15;**17**(3):E7.
60. Link HD, McAfee PC, Pimenta L. Choosing a cervical disc replacement. *Spine J* 2004 November–December;**4**(6 Suppl.):294S–302S.
61. Pimenta L, McAfee PC, Cappuccino A, Bellera FP, Link HD. Clinical experience with the new artificial cervical PCM (Cervitech) disc. *Spine J* 2004 November–December;**4**(6 Suppl.):315S–21S.
62. Wefers M. Personal communication. Cervitech, Inc.; 2005.
63. McAfee PC, Cunningham B, Dmitriev A, Hu N, Woo Kim S, Cappuccino A, et al. Cervical disc replacement-porous coated motion

prosthesis: a comparative biomechanical analysis showing the key role of the posterior longitudinal ligament. *Spine* 2003 October 15;**28**(20):S176–85.

64. Dmitriev AE, Cunningham BW, Hu N, Sell G, Vigna F, McAfee PC. Adjacent level intradiscal pressure and segmental kinematics following a cervical total disc arthroplasty: an in vitro human cadaveric model. *Spine* 2005 May 15;**30**(10):1165–72.
65. Hallab NJ, Cunningham BW, Jacobs JJ. Spinal implant debris-induced osteolysis. *Spine* 2003 October 15;**28**(20):S125–38.
66. Fraser RD, Ross ER, Lowery GL, Freeman BJ, Dolan M. AcroFlex design and results. *Spine J* 2004 November–December;**4**(6 Suppl.): 245S–51S.
67. Thorpe PLPJ, Licina P. Osteolysis and complications associated with artificial disc replacement. *2004 annual meeting of the spine society of Australia*. Coolum, Australia; 2004. p. Poster 21.
68. McAfee PC, Geisler FH, Saiedy SS, Moore SV, Regan JJ, Guyer RD, et al. Revisability of the CHARITÉ artificial disc replacement: analysis of 688 patients enrolled in the U.S. IDE study of the CHARITÉ artificial disc. *Spine* 2006 May 15;**31**(11):1217–26.
69. van Ooij A, Kurtz SM, Stessels F, Noten H, van Rhijn L. Polyethylene wear debris and long-term clinical failure of the CHARITÉ disc prosthesis: a study of 4 patients. *Spine* 2007 January 15;**32**(2):223–9.
70. Baxter RM, Steinbeck MJ, Ianuzzi A, van Ooij A, Ross ERS, Isaza J, et al. Periprosthetic polyethylene particle load of retrieved TDR tissue: comparison with THR tissue. *Trans of the 54th Orthop Res Soc* 2008;**33**:389.
71. Punt IM, van Ooij A, de Bruin T, Cleutjens J, Kurtz SM, van Rhijn L. Periprosthetic tissue reactions in revised total disc arthroplasty. *Transactions of the 8th world biomaterials congress*. Amsterdam; 2008 May 28–June 1. P-Sat-A-008.
72. Kurtz S, Siskey R, Ciccarelli L, van Ooij A, Peloza J, Villarraga M. Retrieval analysis of total disc replacements: implications for standardized wear testing. *J ASTM Int* 2006;**3**(6):1–12.
73. Kurtz SM, Patwardhan A, MacDonald D, Ciccarelli L, van Ooij A, Lorenz M, et al. What is the correlation of in vivo wear and damage patterns with in vitro TDR motion response?. *Spine* 2008 March 1;**33**(5):481–9.
74. Kurtz SM, van Ooij A, Ross ERS, de Waal Malefijt J, Peloza J, Ciccarelli L, et al. Polyethylene wear and rim fracture in total disc arthroplasty. *Spine J* 2007;**7**(1):12–21.
75. Wright T, Cottrell J. Retrieval analysis of ProDisc total disc replacements. *7th annual meeting of the Spine Arthroplasty Society*. Berlin, Germany; 2007 May 1–4. P124.
76. Choma TJ, Miranda J, Siskey R, Baxter RM, Steinbeck MJ, Kurtz SM. Retrieval analysis of a ProDisc-L total disc replacement. *J Spin Disord Tech* 2008;**12**. In Press.
77. Kurtz SM, Ianuzzi A, MacDonald D. Does chronic rim impingement influence dome wear in mobile bearing TDR? *Transactions of the 2008 Spine Arthroplasty Society Meeting*. Miami; 2008 May 6–9;8:57.
78. Kurtz SM, van Ooij A, Ross ERS, de Waal J, Isaza J, Peloza J, et al. Clinical significance of polyethylene oxidation for total disc arthroplasty. *Trans of the 53rd Orthop Res Soc* 2007;**32**:1130.
79. Kurtz SM, Hozack WJ, Purtill JJ, Marcolongo M, Kraay MJ, Goldberg VM, et al. Significance of in vivo degradation for polyethylene in total hip arthroplasty. *Clin Orthopaed Relat Res* 2006;**453**:47–57.
80. Kurtz SM, Ianuzzi A, Siskey R, MacDonald D, Cohen A, Dooris AP, et al. Clinical validation of spine wear simulators with long-term TDR wear rates for a mobile bearing polyethylene TDR. *Transactions of the Spineweek 2008 meeting*. Geneva, Switzerland; 2008 May 26–31. C10.
81. Serhan HA, Dooris AP, Parsons ML, Ares PJ, Gabriel SM. In vitro wear assessment of the CHARITÉ Artificial Disc according to ASTM recommendations. *Spine* 2006 August 1;**31**(17):1900–10.
82. Kurtz SM. Total disc arthroplasty. In: Kurtz SM, Edidin AA, editors. *Spine technology handbook*. New York: Academic Press; 2006. p. 303–70.
83. Valdevit A, Errico TJ. Design and evaluation of the FlexiCore metal-on-metal intervertebral disc prosthesis. *Spine J* 2004 November–December;**4**(6 Suppl.):276S–88S.
84. Traynelis VC. The prestige cervical disc. *Neurosurg Clin N Am* 2005 October;**16**(4):621–28.
85. Papadopoulos S. The bryan cervical disc system. *Neurosurg Clin N Am* 2005 October;**16**(4):629–36.

Chapter 13

# Highly Crosslinked and Melted UHMWPE

Orhun K. Muratoglu, PhD



## 13.1 INTRODUCTION

Radiation crosslinking combined with thermal treatment emerged in late 1990s as a technology to improve the wear and oxidation resistance of UHMWPE acetabular components [1–3]. The development of this technology during the past decade led to a series of new alternate UHMWPE bearing materials, including irradiation and melting, irradiation and annealing, sequential irradiation with annealing, irradiation with subsequent mechanical deformation, and irradiation and stabilization with vitamin E. In this section you will find a review for the development of the irradiated and melted UHMWPE technology.

In the late 1980s and early 1990s understanding the wear mechanism of UHMWPE became central to the aim of improving the wear resistance of this polymer. The analysis of acetabular components retrieved during an autopsy or a revision operation showed that the highly worn areas were polished to a glassy finish on gross examination [4]. Upon investigation under a scanning electron microscope, numerous multidirectional scratches along with fine, drawn-out fibrils with a diameter of 1 micrometer or less oriented parallel to each other became apparent on the worn articular surfaces. These fibrils were thought to be the most likely source of the wear particles observed in periprosthetic osteolytic tissue [5]. Thus, wear appeared to occur mostly through the break-up of fibrils formed by large-strain plastic deformation and orientation of the surface. The fibrils were weakened in the transverse direction and subsequently ruptured during multidirectional motion of the hip. These *in vivo* findings were later reproduced in the laboratory by Bragdon et al. [5] and Wang et al. [6, 7] using wear testing devices. These researchers showed that the wear rate of UHMWPE increased markedly when the articulation is switched from a unidirectional to a multidirectional motion. The underlying mechanism is the orientation of the surface fibrils in the direction of motion. This orientation increases the strength of the fibrils in this direction and weakens them in the transverse direction. Hence debris generation occurs easily when the direction of motion is altered. For example, when the articulation comprises crossing motions, like in the hip, the particulate debris is generated by the rupture of the surface fibrils.

Increasing the entanglement density of UHMWPE through melt-annealing was the first proposal to decrease the extent of surface deformation during articulation. However, quenching of the molten UHMWPE was not effective in maintaining what was thought to be a higher entanglement state of the polymer in the melt. This method was later modified by adding radiation crosslinking at above the melt to lock in the entanglements in place. Later it became clear that the increase in wear resistance of the polymer was not necessarily due to the increase in entanglement density but was due to the crosslinking. Hence, crosslinking was advanced as a means to impede the formation of surface fibrils and reduce the wear of UHMWPE. Crosslinking of UHMWPE molecules would decrease the chain mobility needed for the large scale plastic deformation and hence slow down the surface fibril formation.

Crosslinking of UHMWPE can be achieved by subjecting the polymer to ionizing radiation or by using peroxide or silane chemistries [8–10]. So far, silane crosslinking received limited attention. The only study was reported by researchers in the United Kingdom, who developed an acetabular liner fabricated with silane crosslinked polyethylene and reported marked reduction in wear rate in patients up to 10 years *in vivo* [10]. Attempts were also made in using peroxide crosslinking of UHMWPE, which improved the wear resistance, yet resulted in an oxidatively unstable material [11]. The mechanism of oxidation in peroxide-crosslinked UHMWPE has not been studied in detail, and further development of the method for use in orthopedics was abandoned by most researchers in the mid-1990s.

## 13.2 RADIATION CROSSLINKING

UHMWPE crosslinks when exposed to ionizing radiation through free radical recombination reactions. Ionizing radiation cleaves the C-H and C-C bonds, forming free radicals, some of which recombine to form crosslinks (Figure 13.1). The cleavage of C-C bond is chain scission and reduces the molecular weight of the polymer. The reaction of a free radical generated by the cleavage of a C-H and C-C bond results in long-chain branching or Y-linkages. The recombination of two free radicals formed by the cleavage of C-H bonds results in crosslinking or an H-linkage. The rate of formation of H-linkages increases with increasing irradiation temperature [12].

Not all of the free radicals are mobile enough to find each other and recombine. Most of the free radicals formed in the crystalline domains of UHMWPE are thought to be trapped and cause oxidation in the long term. This oxidation was manifested in the form of embrittlement in UHMWPE components that were terminally gamma sterilized [13, 14]. The gamma sterilization generates residual free radicals, which then react with oxygen, forming unstable hydroperoxides. The decay of the hydroperoxides embrittle the polymer through chain scission and recrystallization.

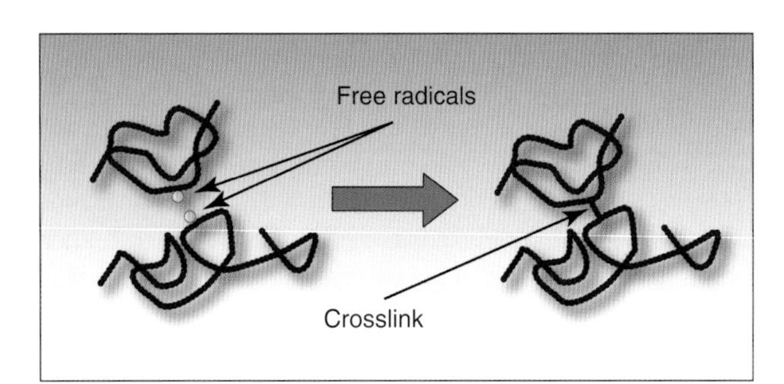


**FIGURE 13.1** During irradiation, carbon-hydrogen bonds are broken, forming free radicals along the backbone of the UHMWPE molecule. The reaction of two free radicals in two separate molecules results in the formation of a crosslink.

Thermal treatment of radiation-crosslinked UHMWPE was proposed to avoid the oxidation that was observed in terminally gamma sterilized UHMWPEs [1–3]. Two main forms of thermal treatment are used: (1) annealing below the melting point of the irradiated polymer or (2) melting after irradiation. The former is discussed in Chapter 14. In this chapter we will review the development of the irradiation and melting method and present the preclinical and clinical evidence supporting the use of this technology.

## 13.3 IRRADIATION AND MELTING

As previously mentioned, the free radicals generated by irradiation are trapped in the crystalline domains. Heating the polymer to above its melting point not only eliminates the crystalline domains or the lamellae releasing the trapped radicals but also imparts them with enough mobility to find and recombine with each other. Thus the residual free radicals are quenched and oxidative stability of the irradiated UHMWPE is improved. Melting of irradiated UHMWPE is carried out in an air convention oven. Typically UHMWPE is irradiated in its bar stock form to be machined into finished implants after the melting step. Any surface oxidation that occurs during the in-air melting is machined away prior to the machining of final components. The irradiated and melted UHMWPEs that are in clinical use are listed in Table 13.1.

Both gamma and electron beam (e-beam) irradiation can be used to crosslink UHMWPE. While the cascade of events that leads to the formation of crosslinks and residual free radicals are similar for both gamma irradiation and electron beam irradiation, there are still important differences between the two methods, specifically in terms of penetration of the effects of radiation and radiation dose rate achieved. Gamma irradiation sources are commonly based on the artificial isotope of cobalt ($^{60}$Co) that generates gamma photons. While the penetration of gamma source into UHMWPE has no practical limitations, the activity level of the gamma source limits the radiation dose rate. With an electron beam the radiation is in the form of accelerated, charged particles. The penetration of the effects of e-beam radiation is limited by the kinetic energy of the electron beam, measured in million electron volts (MeV). With a 10 MeV electron beam incident on a UHMWPE surface, radiation penetrates about 4–4.5 cm. The radiation dose rate that a commercial e-beam accelerator provides is about two orders of magnitude larger than that from a commercial gamma source.

**TABLE 13.1** List of Irradiated and Melted UHMWPEs That Are Currently in Clinical Use

| | Manufacturer | Irradiation temperature | Radiation dose (kGy) | Radiation type | Postirradiation thermal treatment | Sterilization method | Total radiation dose level (kGy) | Residual-free radicals present? | Comments |
|---|---|---|---|---|---|---|---|---|---|
| Longevity | Zimmer | ~40°C | 100 | E-beam | Melted | Gas plasma | 100 | No | Total hips only; also used with large femoral heads |
| Prolong | Zimmer | ~125°C | 65 | E-beam | Melted | Gas plasma | 65 | No | Total knees only with cruciate retaining design |
| Durasul | Zimmer | ~125°C | 95 | E-beam | Melted | EtO | 95 | No | Total hips and cruciate retaining knees; also used with large femoral heads in total hips |
| Marathon | DePuy/JJ | RT | 50 | Gamma | Melted | Gas plasma | 50 | No | Total hips only |
| XLPE | Smith & Nephew | RT | 100 | Gamma | Melted | EtO | 100 | No | Total hips only |

*RT: Room Temperature*

During electron beam irradiation, adiabatic heating of UHMWPE is common. The adiabatic heat generated increases with increasing radiation dose rate. Electron beam radiation offers a greater control of dose rate, which presented a mean to control the temperature rise in UHMWPE during irradiation throughout the development of highly crosslinked UHMWPEs. This also meant that the polymer could be melted during irradiation rather than after irradiation.

Radiation dose is energy per unit mass of the material being irradiated, and it is measured in kGy units, or kJ/g. That is, a 100 kGy radiation dose corresponds to 100 kJ/g of energy that will be converted to heat resulting in a temperature rise, $\Delta T$, in the polymer per the following equation:

$$\text{Dose} = c_p \Delta T + \Delta H_m$$

where $c_p$ (=2.5 kJ/g/C) is the specific heat, and $\Delta H_m$ is the heat of fusion of the polymer. Assuming that the polymer is about 50% crystalline, the heat of fusion would be about 150 kJ/g, and about 430 kGy of radiation dose would be needed to completely melt the UHMWPE from room temperature. This is a rather high dose level and would adversely affect the mechanical properties. The optimum radiation dose level for UHMWPE for total hips is about 100 kGy (discussed later). With a radiation dose of 100 kGy, UHMWPE needs to be heated to at least 120°C to achieve near complete melting of the crystalline domains. Of course the dose rate has to be high and the polymer needs to be insulated to minimize heat loss to the surroundings to achieve near adiabatic conditions. The efforts on using e-beam to melt the UHMWPE led to the development of warm irradiated UHMWPE, now used in the fabrication of total hip and total knee components. In the large scale manufacturing of warm irradiated UHMWPE, a terminal melting step in an air convection oven is used to ensure that complete melting is achieved and that residual free radicals are quenched.

Surprisingly, warm irradiation of UHMWPE resulted in some unusual properties: Warm irradiated UHMWPE displays two distinct peak melting points, one near 115°C and the other at 137°C. This is an indication of a bimodal lamellae thickness distribution. Transmission electron microscopy of warm irradiated UHMWPE corroborated this finding, showing thick lamellae embedded in a matrix comprising smaller and thinner lamellae and the amorphous phase [1]. Warm irradiation and melting of UHMWPE also results in higher ductility and lower strength in comparison with cold irradiation and melting [1].

## 13.4 EFFECT OF RADIATION DOSE, MELTING, AND IRRADIATION TEMPERATURE ON UHMWPE PROPERTIES

Crosslinking, as previously mentioned, occurs primarily in the amorphous phase of UHMWPE. In the crystalline domains the chain segments are not in close enough proximity to allow the recombination of the free radicals—in fact the lattice spacing is smaller than the carbon-carbon bond length. Therefore, for a crosslink to form in the crystalline region, the chain has to kink, which is energetically not favorable. As the absorbed radiation dose increases, so does the crosslink density in a monotonic fashion. Crosslinking starts to saturate at around 100 kGy, beyond which the rate of increase in crosslink density with increasing radiation dose decreases rapidly.

Radiation crosslinking decreases chain mobility and chain stretch; as a result it decreases the ductility of the polymer. This is manifested in a reduction in the elongation at break, toughness, and fatigue crack propagation resistance of the polymer. As the radiation dose increases, there is a monotonic decrease in ductility and toughness, both of which saturate beyond 100 kGy along with crosslink density [2]. The strength of UHMWPE is mainly dependent on its crystallinity, and therefore irradiation by itself does not adversely affect the strength. On the other hand, melting after or during irradiation decreases the crystallinity and decreases the strength. The crystallization kinetics is affected by the presence of crosslinks, which reduce chain mobility and reptation needed for the lamellae to form. Consequently, recrystallization of a radiation crosslinked UHMWPE always results in a lower crystallinity polymer in comparison to the unirradiated UHMWPE. The decrease in crystallinity results in an increase in compliance of the polymer, which is desirable in load bearing applications as it decreases the contact stress [15].

One would expect the wear rate of irradiated and melted UHMWPE to increase given these adverse effects on its mechanical properties. Paradoxically, the wear rate decreases with increasing radiation dose (Figure 13.2) [2, 3]. While the decrease in the mechanical properties would increase the wear rate, the crosslinks decrease the plastic deformation that occurs on the surface during articulation and thus decrease the wear rate. The latter effect is more prominent, and the net effect of radiation crosslinking and melting on wear is favorable. The decrease in the wear rate with increasing radiation dose follows a monotonic function similar to those between mechanical properties and dose, saturating above 100 kGy [2, 3]. Therefore, 100 kGy has been the highest radiation dose used in the fabrication of total joint components made with irradiated and melted UHMWPE.

We previously mentioned that warm irradiation compared to cold irradiation results in higher ductility and lower strength. In addition, the crosslink density of warm irradiated UHMWPE is lower than that of cold irradiated UHMWPE at the same dose level. This could be due to the biphasic morphology developed in UHMWPE after warm irradiation, which is evidenced by the dual peak melting peaks. We speculate that the biphasic morphology is a result of the nonstatistical distribution of crosslinks that is known to occur at high irradiation temperatures with polyethylenes [12, 16, 17]. One of the phases, possibly the continuous one, may be lightly crosslinked and maintain the mechanical properties of the uncrosslinked UHMWPE. The

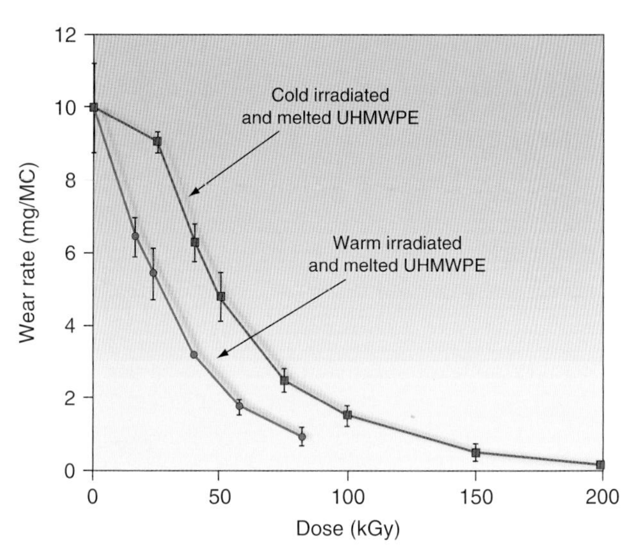


FIGURE 13.2 Average wear rate of cylindrical pins as a function of radiation dose level is shown here. The pins were machined from GUR 1050 UHMWPE blocks that were first irradiated to the indicated radiation dose level and subsequently melted. The wear rate decreases with increasing radiation dose level.

second phase, possibly reinforcing the former continuous phase, may be highly crosslinked and markedly improve the wear resistance of the polymer. As a result the apparent crosslink density would be lower with warm irradiation. In contrast, warm irradiation results in lower wear rates than cold irradiated UHMWPE at the same dose levels (Figure 13.2). This is the reason why warm irradiated UHMWPE components used in total hips (Durasul) are irradiated to 95 kGy, and the cold irradiated ones (Longevity, XLPE) are irradiated to 100 kGy.

## 13.5 EFFECT OF CROSSLINKING ON FATIGUE RESISTANCE

Besides wear, the second most important property change in UHMWPE induced by irradiation is its fatigue resistance. With increasing crosslink density the fatigue crack propagation (FCP) resistance decreases [18, 19]. While irradiation alone decreases the FCP resistance, melting subsequent to irradiation further decreases the FCP resistance [18]. On the other hand, crack initiation resistance increases with crosslinking [20]. The fatigue strength may be decreasing with crosslinking and melting, but understanding the effects of these changes on implant performance is complicated by the design and use of the implants. Fatigue failures reported with irradiated and melted UHMWPE acetabular liners are rare; however, when they occur, these failures are primarily at the rim and almost always occur with steeply vertical shell placement [21]. While some occurrences of fracture were reported with acetabular liners, none were reported with highly crosslinked tibial knee inserts.

## 13.6 OPTIMUM RADIATION DOSE

Orthopedic implant manufacturers chose different radiation dose levels as optimum for irradiation and melting. The debate continues between 100 kGy and 50 kGy as being the optimum dose levels. As we previously explained, the beneficial effects of radiation crosslinking in terms of increasing the wear resistance decrease rapidly above 100 kGy. Thus the optimum radiation dose rate for increasing wear resistance is around 100 kGy. At 50 kGy, some of the mechanical properties, such as toughness and elongation at break, are higher with a lower wear resistance compared to 100 kGy. The argument between 50 and 100 kGy is mainly how much reduction in wear is needed to prevent periprosthetic osteolysis. While the 50 kGy irradiated and melted UHMWPE shows higher wear than 100 kGy, we need long-term (beyond 10 years) clinical studies to determine if the 50 kGy has enough wear resistance to address the problem of periprosthetic osteolysis and loosening.

## 13.7 HIP SIMULATOR DATA

Investigations on the wear behavior of highly crosslinked UHMWPE on functional joint simulators followed the initial preclinical testing using pin-on-disc wear testers. Hip simulator studies of various designs with irradiated and melted UHMWPE all showed improved wear resistance in comparison with conventional UHMWPE. Some studies used gamma in air sterilized UHMWPE and others used ethylene oxide (EtO) or gas plasma (GP) sterilized UHMWPE as controls. Note that the gamma-sterilized controls would have some crosslinking and thus would also have a lower wear rate than the EtO or GP sterilized controls. The hip simulator studies explored the influence of long-term use, artificial aging, presence of third body particles, femoral head size, and, later, component placement on the wear of irradiated and melted UHMWPEs. Some studies were carried out to 30 million cycles, corresponding to 15–30 years of *in vivo* service of an average total hip patient [1, 3, 22–25]. These studies showed that the marked reduction in wear was not affected with the long-term testing and that there were no fatigue type failures. The addition of third body particles, in the form of bone cement, increased the wear of both conventional and highly crosslinked UHMWPEs, yet the latter still exhibited better wear resistance. Artificial aging did not oxidize the irradiated and melted UHMWPEs because of the improved oxidation resistance and as a result did not affect their wear rate.

The femoral head size is an important effect, in that with larger heads the wear rate of conventional UHMWPE acetabular liners increases. Even though there is an increase in the contact area and hence a reduction in contact stress, wear increases with head size due to the dominant increase in the distance traveled at every gait cycle. Historically, smaller femoral heads, first 22 mm then up to 28 and 32 mm diameter, had been used to minimize the wear of the liners. Yet the disadvantage of smaller heads is the decrease in the stability and the range of motion of the joint. With the advent of highly crosslinked UHMWPEs, the use of larger femoral heads became possible. Hip simulator studies showed the wear rate of an acetabular liner with an inner diameter of 46 mm and fabricated with irradiated and melted UHMWPE to be lower than that of a 22 mm conventional UHMWPE acetabular liner (Figure 13.3) [22].

## 13.8 KNEE SIMULATOR DATA

The development of highly crosslinked UHMWPE for use in tibial knee inserts followed the introduction of this material in total hips. Knee simulator studies on the wear behavior of irradiated and melted UHMWPE tibial knee inserts has been widely published [26–33]. These investigations focused on the effect of long-term testing, artificial aging, presence of third body particles, implant design, and most importantly on malpositioning of the implants.

The studies uniformly showed improved adhesive/abrasive wear with irradiated and melted UHMWPE in comparison with conventional gamma-in-air or gamma-in-nitrogen sterilized UHMWPE tibial knee inserts of either cruciate retaining or posterior stabilized designs. Artificial aging of the test components was used to simulate long-term oxidation in the conventional UHMWPE, and the testing of these components on the knee simulators showed delamination-type damage. When the irradiated and melted UHMWPE components were subjected to identical accelerated aging conditions and tested on the knee simulator, they showed no delamination or pitting [32]. In the presence of third body particles, the highly crosslinked UHMWPE components still showed lower wear rate than conventional UHMWPE. Investigations also showed that the highly crosslinked UHMWPE exhibits better wear performance when it is articulated against surgically explanted femoral components that had been extensively damaged during *in vivo* service [31].

In total knees the malpositioning of components can seriously compromise not only the knee function but also the performance of the UHMWPE components used because of the altered stress state [29]. The highly crosslinked UHMWPE tibia inserts were also tested under certain worst-case scenarios. One is the circumstance of a tight posterior cruciate ligament (PCL), which would increase the rollback of the femoral component during flexion and apply high stresses to the posterior edge of the tibial component. Under these conditions accelerated aged conventional UHMWPE tibial inserts showed massive delamination in less than 50,000 cycles of stair climbing, while the aged highly crosslinked components showed no delamination up to 500,000 cycles [29].

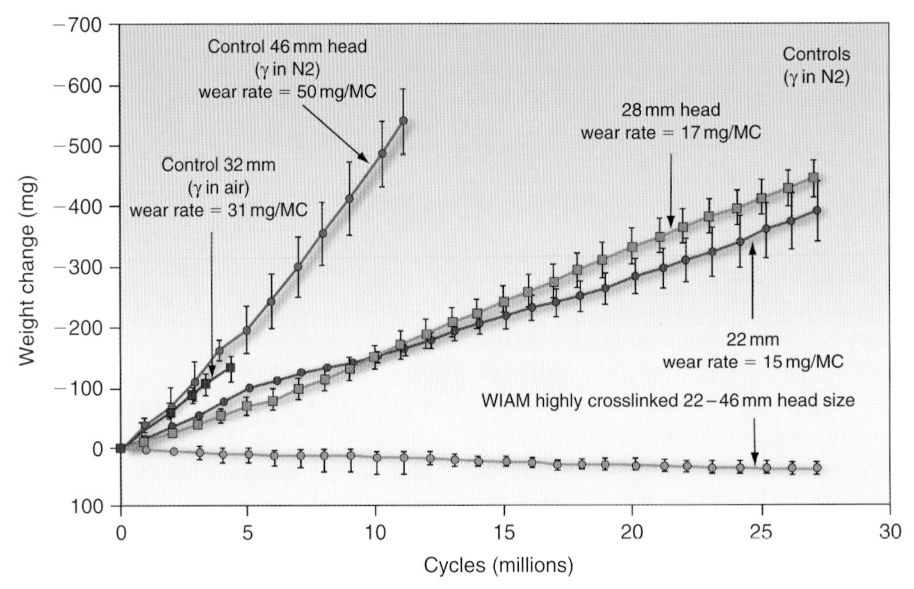


**FIGURE 13.3** Average weight change of acetabular liners as a function of simulated gait cycles is shown here [22]. Control liners were machined from GUR 1050 resin and gamma sterilized in nitrogen. WIAM-95 liners were machined from GUR 1050 resin with an e-beam at a dose level of 95 kilograys and subsequently melted. The WIAM-95 liners were ethylene oxide sterilized. The legend also includes the femoral head sizes used during the hip simulator testing. Note that the wear rate (slope of weight change versus cycles) of the control UHMWPE increases with increasing femoral head size, while the highly crosslinked UHMWPE's wear rate remains unchanged with increasing femoral head size.

Backside wear is another important factor that could affect long-term performance of total knees. Knee simulator studies also showed that the extent of backside wear decreases markedly with irradiated and melted UHMWPE in comparison with that observed in conventional UHMWPE [33].

## 13.9 CLINICAL FOLLOW-UP STUDIES

There are a number of clinical trials comparing the *in vivo* performance of irradiated and melted UHMWPE to that of conventional UHMWPE acetabular liners [34–36]. Radiographic examination of the patients allows the measurement of femoral head penetration into the acetabular liner. The penetration for the first 1–2 years is mostly due to creep and wear, and beyond the second year of follow-up, penetration is attributed to mostly wear alone. The conventional UHMWPEs in these studies are either gas sterilized or sterilized with ionizing radiation. Because the latter introduces some crosslinking, it also results in lower wear rates. Therefore, the comparison of the percentage of wear reduction of the highly crosslinked UHMWPE from different studies is not possible. However, collectively all of these studies show markedly reduced wear rate with highly crosslinked UHMWPE beyond the second year of follow-up, corroborating the laboratory evidence of reduced wear [1, 22]. There are no reports of *in vivo* wear and damage of irradiated and melted UHMWPE tibial knee inserts because of the lack of validated methods of determining the wear and damage rate of these components *in vivo*.

## 13.10 *IN VIVO* CHANGES: EXPLANTS

There has been a number of reports on the *in vivo* performance of irradiated and melted UHMWPE based on changes observed on surgically explanted components [37, 38]. Interestingly, the appearance of the first explants was not very pleasant, in that the articular surfaces were very highly scratched and discolored. This immediately led surgeons to believe that the scratches were due to wear and the discoloration due to oxidation, two things that the material was supposed to address. In closer look the scratching turned out to be plastic deformation from third body particles. We used the shape memory property of UHMWPE and showed by melting of the explants the elimination of the scratches and recovery of the original machining marks [38]. The discoloration was not due to the oxidation of the components but simply due to the absorption of the lipids from the synovial fluid [38]. The initial response from the surgeons was largely affected by their previous experience with conventional UHMWPE, which wears at a rate of about 0.1 mm/year *in vivo*. Wear polishes the articular surface and removes the scratches and discoloration. In contrast with the irradiated and melted UHMWPE, scratches and discoloration accumulate, which points to the lack of the wear of the components. Similar observations were also made with surgically removed tibial knee inserts that were made from irradiated and melted UHMWPE [37].

## 13.11 CONCLUSION

The irradiated and melted UHMWPE has been in clinical use around the globe for the past decade with more than 2 million implantations. Clinical follow-up studies are showing marked reduction in wear rate, and the explanted components are corroborating this observation. The natural progression in research led to the development of second generation technologies, which now combine the improved wear and oxidation resistances with improved strength and fatigue crack propagation resistance. In the coming decade there will be a shift toward the use of these second-generation technologies.

## REFERENCES


1. Muratoglu OK, Bragdon CR, O'Connor DO, Jasty M, Harris WH. 1999 HAP Paul Award. A novel method of crosslinking UHMWPE to improve wear, reduce oxidation and retain mechanical properties. *J Arthroplasty* 2001;**16**(2):149–60.
2. Muratoglu OK, Bragdon CR, O'Connor DO, Jasty M, Harris WH, Gul R, et al. Unified wear model for highly crosslinked ultra-high molecular weight polyethylenes (UHMWPE). *Biomaterials* 1999;**20**(16):1463–70.
3. McKellop H, Shen F-W, Lu B, Campbell P, Salovey R. Development of an extremely wear resistant ultra-high molecular weight polyethylene for total hip replacements. *J Orthop Res* 1999;**17**(2):157–67.
4. Jasty MJ, Goetz DD, Lee KR, Hanson AE, Elder JR, Harris WH. Wear of polyethylene acetabular components in total hip arthroplasty. An analysis of 128 components retrieved at autopsy or revision operation. *JBJS* 1997;**79**(A):349–58.
5. Bragdon CR, O'Connor DO, Lowenstein JD, Jasty M, Syniuta WD. The importance of multidirectional motion on the wear of polyethylene. *Proc Instn Mech Engrs* 1995;**210**:157–65.
6. Wang A, Stark C, Dumbleton JH. Mechanistic and morphological origins of ultra-high molecular weight polyethylene wear debris in total joint replacement prostheses [see comments]. *Proc Inst Mech Eng [H]* 1996;**210**(3):141–55.
7. Wang A. A unified theory of wear for ultra-high molecular weight polyethylene in multi-directional sliding. *Wear* 2001;**248**:38–47.
8. Atkinson JR, Cicek RZ. Silane crosslinked polyethylene for prosthetic applications. Certain physical and mechanical properties related to the nature of the material. *Biomaterials* 1983;**4**:267–75.
9. Atkinson JR, Cicek RZ. Silane crosslinked polyethylene for prosthetic applications. Part II. Creep and wear behavior and a preliminary mouling test. *Biomaterials* 1984;**5**:326–35.
10. Wroblewski B, Siney P, Fleming P. Low-friction arthroplasty of the hip using alumina ceramic and cross-linked polyethylene: a ten-year follow-up report. *J Bone Joint Surg* 1999;**81B**:54–5.
11. Muratoglu OK, Biggs SA, Bragdon CR, O'Connor DO, Merrill EW, Premnath V, et al. Long term stability of radiation and peroxide cross-linked UHMWPE. *23rd Annual Meeting of Society for Biomaterials*. New Orleans, LA; 1997.
12. Horii F, Zhu Q, Kitamaru R, Yamaoka H. C NMR study of radiation-induced cross-linking of linear polyethylene. *Macromolecules* 1990;**23**:977–81.
13. Sutula LC, Collier JP, Saum KA, Currier BH, Currier JH, Sanford WM, et al. Impact of gamma sterilization on clinical performance of polyethylene in the hip. *Clin Orthop* 1995;**319**:28–40.
14. Collier JP, Sperling DK, Currier JH, Sutula LC, Saum KA, Mayor MB. Impact of gamma sterilization on clinical performance of polyethylene in the knee. *J Arthroplasty* 1996;**11**(4):377–89
15. Plank GR, Estok II DM, Muratoglu OK, O'Connor DO, Burroughs BR, Harris WH. Contact stress assessment of conventional and highly crosslinked ultra high molecular weight polyethylene acetabular liners with finite element analysis and pressure sensitive film. *J Biomed Mater Res B Appl Biomater* 2007 January;**80**(1):1–10.
16. Randall JC, Zoepfl FJ, Silverman J. High-resolution solution carbon 13 NMR measurements of irradiated polyethylene. *Radiat Phys Chem* 1983;**22**(1/2):183–92.
17. Kang H, Saito O, Dole M. The radiation chemistry of polyethylene. IX. Temperature coefficient of crosslinking and other effects. *J Am Chem Soc* 1967;**89**(9):1980–6.
18. Oral E, Malhi A, Muratoglu O. Mechanisms of decrease in fatigue crack propagation resistance in irradiated and melted UHMWPE. *Biomaterials* 2006;**27**:917–25.
19. Pruitt LA. Deformation, yielding, fracture and fatigue behavior of conventional and highly cross-linked ultra high molecular weight polyethylene. *Biomaterials* 2005;**26**(8):15.
20. Baker DA, Bellare A, Pruitt L. The effect of degree of crosslinking on the fatigue crack initiation and propagation resistance of orthopedic-grade polyethylene. *J Biomed Mater Res* 2003;**66A**:146–54.
21. Halley D, Glassman A, Crowninshield R. Recurrent dislocation after revision total hip replacement with a large prosthetic femoral head. *J Bone Joint Surg* 2004;**86A**(4):827–30.
22. Muratoglu OK, Bragdon CR, O'Connor DO, Perinchief RS, Estok DM, Jasty M, et al. Larger diameter femoral heads used in conjunction with a highly cross-linked ultra-high molecular weight polyethylene: a new concept. *J Arthroplasty* 2001;**16**(8 Suppl.):24–30.
23. Muratoglu OK, Wannomae K, Christensen S, Rubash HE, Harris WH. Ex vivo wear of conventional and cross-linked polyethylene acetabular liners. *Clin Orthop Relat Res* 2005;**438**:158–64.
24. Endo M, Tipper JL, Barton DC, Stone MH, Ingham E, Fisher J. Comparison of wear, wear debris and functional biological activity of moderately crosslinked and non-crosslinked polyethylenes in hip prostheses. *Proc Inst Mech Eng [H]* 2002;**216**(2):111–22.
25. Wang A. Wear of ultra-high molecular weight polyethylene acetabular cups in a physiological hip joint simulator in the anatomical position using bovine serum as a lubricant. *Proc Inst Mech Eng [H]* 1998;**212**(1):71–2. author reply 72–73.
26. Johnson T, Laurent M, Yao J, Blanchard CR. Comparison of wear of mobile and fixed bearing knees tested in a knee simulator. *Wear* 2003;**255**:1107–12.
27. Muratoglu OK, Bragdon CR, Jasty M, O'Connor DO, Von Knoch R, Harris WH. Knee simulator testing of conventional and crosslinked polyethylene tibial inserts. *J Arthroplasty* 2004;**19**(7):887–97.
28. Asano T, Akagi M, Clarke IC, Masuda S, Ishii T, Nakamura T. Dose effects of cross-linking polyethylene for total knee arthroplasty on wear performance and mechanical properties. *J Biomed Mater Res B Appl Biomater* 2007 November;**83**(2):615–22.
29. Muratoglu OK, Bragdon CR, O'Connor DO, Perinchief R, Jasty M, Harris WH. Aggressive wear testing of a cross-linked polyethylene in total knee arthroplasty. *Clin Orthop* 2002;**404**:89–95.

30. Muratoglu O, Mark A, Vittetoe D, Harris W, Rubash H. Polyethylene damage in total knees and use of highly crosslinked polyethylene. *J Bone Joint Surg* 2003;**85-A**(Suppl. 1):7–13.
31. Muratoglu OK, Burroughs BR, Christensen SD, Lozynsky AJ, Harris WH. In vitro simulator wear of highly crosslinked tibias against explanted rough femoral components. *Proceedings of 50th ORS*. San Francisco; 2004. p. 297.
32. Muratoglu OK, Rubash HE, Bragdon CR, Burroughs BR, Huang A, Harris WH. Simulated normal gait wear testing of a highly cross-linked polyethylene tibial insert. *J Arthroplasty* 2007;**22**(3):435–44.
33. Muratoglu OK, Rubash HE, Bragdon CR, Burroughs BR, Huang A, Harris WH. Simulated normal gait wear testing of a highly cross-linked polyethylene tibial insert. *J Arthroplasty* 2007 April;**22**(3):435–44.
34. Garcia-Rey E, Garcia-Cimbrelo E, Cruz-Pardos A, Ortega-Chamarro J. New polyethylenes in total hip replacement: a prospective, comparative clinical study of two types of liner. *J Bone Joint Surg Br* 2008 February;**90**(2):149–53.
35. Digas G, Karrholm J, Thanner J, Herberts P. 5-year experience of highly crosslinked polyethylene in cemented and uncemented sockets—Two randomized studies using radiostereometric analysis. *Acta Orthop* 2007;**78**(6):746–54.
36. Engh Jr. CA, Stepniewski Jr. A, Ginn SD, Beykirch S, Sychterz-Terefenko C, Hopper RH, et al A randomized prospective evaluation of outcomes after total hip arthroplasty using cross-linked marathon and non-cross-linked ensure polyethylene liners. *J Arthroplasty* 2006;**21**(6, Suppl. 2):17–25.
37. Muratoglu OK, Ruberti J, Melotti S, Spiegelberg S, Greenbaum E, Harris WH. Optical analysis of surface changes on early retrievals of highly cross-linked and conventional polyethylene tibial inserts. *J Arthroplasty* 2003;**18**(7, Suppl. 1):42–7.
38. Muratoglu OK, Greenbaum E, Bragdon C, Jasty M, Freiberg A, Harris WH. Surface analysis of early retrieved acetabular polyethylene liners: A comparison of standard and highly crosslinked polyethylenes. *J Arthroplasty* 2004;**19**(1):68–77.

Chapter 14

# Highly Crosslinked and Annealed UHMWPE

John H. Dumbleton, PhD,DSc, Aiguo Wang, PhD, Kate Sutton, MA,ELS and Michael T. Manley, PhD



## 14.1 INTRODUCTION

Highly crosslinked UHMWPEs were introduced clinically in the 1970s by Oonishi and coworkers [1] and by Grobbelaar and associates [2] and reintroduced by Wroblewski and colleagues [3] in the 1980s. The crosslinking methods used were not based directly on knowledge of UHMWPE structure-property relationships, and clinical use was limited. There was no commercial introduction, but these materials are of historical interest.

In the early 1990s, research focused on improving the understanding between UHMWPE structure, properties, and the relationship of UHMWPE crosslinking to wear reduction. In parallel, there were developments in joint simulation that allowed more realistic determination of wear rates [4]. UHMWPE wear mechanisms were elucidated, and the role of crosslinks in reducing wear was determined [5–7]. Hip simulator studies showed that increased crosslinking resulted in greatly reduced wear [8]. The development of modern highly crosslinked UHMWPE thus has a scientific basis. Crosslinking is achieved by irradiation, and excess free radicals are quenched by heating. Heating above the melting temperature of UHMWPE (remelting) effectively quenches free radicals but changes the microstructure. Free radical quenching below the melting temperature (annealing) preserves the microstructure but is not as effective in quenching free radicals.

Duration Stabilized UHMWPE was the first annealed crosslinked material. Crossfire annealed highly crosslinked UHMWPE (Stryker Orthopaedics, Mahwah, New Jersey, USA) was introduced for total hip components in 1998. In 2005, X3 (Stryker Orthopaedics, Mahwah, New Jersey, USA), a sequentially irradiated and annealed UHMWPE, was introduced. This material was developed based on experience with its predecessors with the intention of using the current data to produce a highly crosslinked UHMWPE available for both the hip and the knee.

## 14.2 DEVELOPMENT OF DURATION STABILIZED UHMWPE

In 1991, John Dumbleton, PhD, in charge of Howmedica Worldwide R&D, wanted to focus on decreasing the incidence of osteolysis and look more closely at UHMWPE oxidation. He believed measurement of wear needed to be more representative of clinical conditions and decided to focus one researcher on understanding of the structure/property relationships of UHMWPE and developing improved materials.

A second researcher would head up a new tribology area that developed hip and knee simulators and simulation techniques for realistically evaluating the improved materials.

D.C. Sun, PhD, introduced many of the techniques used today for UHMWPE evaluation. In 1992, he proposed the stabilization process that led to the development of Duration polyethylene. His work had confirmed that gamma sterilization of UHMWPE resulted in free radicals (and crosslinking), and he believed that residual free radicals could be undesirable because they could react with oxygen and might lead to compromised material properties [9, 10].

During this time, Sulzer introduced UHMWPE gamma sterilized in nitrogen with the goal of preventing oxidation in the package while it was on the shelf. This was a step forward, but free radicals are long-lived and could react with oxygen when the package was opened. Dr. Sun proposed that the Duration components be heated in the package to quench (eliminate) excess free radicals. This stabilization process was carried out at 50°C; the temperature was limited by the package integrity.

During this period in the early 1990s, Aiguo Wang, PhD, from Howmedica worked with MTS Corporation (Minnesota Testing Systems, Eden Prairie, Minnesota, USA) to develop hip simulators. Howmedica received the first hip simulator in 1992, and Dr. Wang spent the next few years developing simulation techniques. In 1994, Dr. Wang evaluated the wear of Duration and found the wear to be lower than that of UHMWPE gamma sterilized in air. In 1995, a 10 million cycle study was completed verifying the earlier test results [11]. While Dr. Wang was conducting his research on Duration, a 510(k) submission for Duration was made to the FDA, and clearance was obtained in August 1995.

The packaging process for Duration was examined closely because the package had to be robust enough to withstand transit conditions (temperature and low pressure) but still be easy to open in the OR. The packages were sealed to prevent oxygen from entering the nitrogen package; the interior of the package was at 15 psi. High altitudes placed considerable stress on the seals. One unique way Howmedica tested package durability was by shipping packages to Petosi, Bolivia and back to Rutherford, New Jersey, after which the packaging integrity was tested. Petosi, located at an elevation of 16,000 feet, is the highest airport in the world. When the package integrity was proven, Duration was launched in late 1996 for hips (ABG II and System 12) and during 1997 for knees (Duracon).

Dr. Sun continued his UHMWPE research and from 1996 to 1998 investigated the effects of radiation dose and quenching temperature on crosslinking and properties. The Howmedica research team realized that crosslinking of the component in the package was overly limiting, so the crosslinking/heating was done on GUR 1150 and later 1050 bar stock. The research team studied the effects of radiation doses up to 100 kGy and used remelting because it more effectively quenched free radicals.

### 14.2.1 The Duration Process

In the Duration process [12], ram extruded GUR 4150 was machined into components (hip or knee), packaged in nitrogen, and gamma sterilized to 30 kGy. Following sterilization the packaged components were heated for 144 hours at 50°C (stabilization). Crosslinks form during sterilization. Excess free radicals are eliminated during the stabilization step. In 1998, ram extruded GUR 1050 replaced GUR 4150 as the starting UHMWPE. This calcium stearate-free grade allowed improved material consolidation during processing.

### 14.2.2 Properties of Duration Stabilized UHMWPE

The free radical content of Duration Stabilized UHMWPE is about one-third that of UHMWPE gamma sterilized in nitrogen but without stabilization annealing. The oxidation resistance was evaluated by heating in air at 80°C [13]. The behavior of Duration was similar to that of virgin UHMWPE. Neither showed oxidation for heating times up to 23 days in contrast to UHMWPE gamma sterilized in air or in nitrogen [14]. Compared to UHMWPE gamma sterilized in air, Duration Stabilized UHMWPE had a lower swell ratio and higher gel content [14, 15]. The level of crosslinking is increased in the surface regions for Duration compared to that of UHMWPE gamma sterilized in air.

The tensile properties of Duration Stabilized UHMWPE were determined using Type IV specimens (ASTM D 638) with testing according to ASTM F 648-96. Impact testing was carried out using double-notched Izod geometry specimens according to ASTM F 648-96. The mechanical properties for Duration Stabilized UHMWPE were as follows: yield strength 23.5 $\pm$ 0.3 MPa; tensile strength 51.8 $\pm$ 2.5 MPa; elongation 336 $\pm$ 11%; impact strength 92 $\pm$ 1 KJ/$m^2$.

The Howmedica investigators knew that the crosslinking level of Duration was higher than that of gamma air UHMWPE, which suggested that the stabilization process eliminated free radicals via crosslink formation. In 1996 Dr. Wang turned his attention to the reasons why crosslinks reduced wear and discovered the impact of crossing path motion at the same time as did Dr. Harris and his group at Massachusetts General Hospital (MGH). Dr. Wang went a step further and published his theory of wear in 1998 [16]. In the 1998 study, wear was evaluated on an MTS hip joint simulator according to a loading-motion protocol for level walking. Testing was carried out with the acetabular component in the inferior position and subsequently in the anatomical position [16]. For the inferior position, the wear of Duration Stabilized UHMWPE was 32% lower than that of UHMWPE gamma sterilized in air. The reduction for Duration was 20% with the cup in the superior position.

### 14.2.3 Clinical Studies

In 1997, Murali Kadaba, PhD, Rama Ramakrishnan, MSc, Dr. Dumbleton at Howmedica, and Ken Greene, MD, at Crystal Clinic (Akron, Ohio, USA) developed a method to clinically evaluate the wear of Duration at the hip. This edge detection method was similar to the Martell method [17], which had not yet been published. A prospective randomized controlled study with the ABG II hip was set up by Dr. Kadaba in the Netherlands. The lead investigator, Dr. Alphons Tonino, published the wear data in 2006 [18], followed by presentations in 2007 [19]. The decreased wear with Duration compared to gamma air UHMWPE was in good agreement with simulator predictions.

The goal of the prospective randomized controlled ABG II hip study begun with Dr. Tonino and other investigators was to compare the clinical performance of Duration Stabilized UHMWPE to that of UHMWPE gamma sterilized in air [18]. The implant was the cementless hydroxyapatite-coated ABG II total hip prosthesis with 28 mm CoCr heads. A total of 133 hips (67 gamma air and 66 Duration) in 127 patients were followed-up for an average of 5 (3–6) years. Wear rates were measured from follow-up radiographs using the edge detection method. The mean wear of the Duration inserts was 0.083 ± 0.056 mm/year, and that of the gamma air inserts was 0.123 ± 0.082 mm/year. The wear reduction for Duration was 32%. Observed femoral osteolysis was 8% for the gamma air inserts and 2% for the Duration inserts at 5 years.

Following-up on the outcomes reported in the 2006 paper [18], Drs. Grimm, Tonino, and Heyligers reported on the 8-year results (mean follow-up, 7–9 years) with 40 patients remaining in the study (23 gamma air and 17 Duration) [19]. The wear for Duration inserts was 0.088 ± 0.03 mm/year, and that of the gamma air inserts was 0.142 ± 0.07 mm/year. The wear reduction with Duration inserts was 38%. Osteolysis was lower in the Duration group.

In a separate study reported on in 2005 at the ORS in Washington, DC by Hirakawa et al., patients were randomized to Duration or gamma air inserts with the same design of hip and with 22 mm femoral heads (28 Duration and 20 gamma air) [20]. At an average of 5 years follow-up, Duration inserts had lower wear (0.112 compared to 0.151 mm/year) than gamma air inserts. In a subset of seven bilateral patients, the wear of Duration inserts was 0.102 compared to 0.168 for gamma air inserts ($p < 0.05$).

### 14.2.4 Summary

Duration Stabilized Polyethylene is significant in that it is the first UHMWPE to employ heat stabilization following irradiation. The processes of free radical elimination included the formation of additional crosslinks. The discovery of reduced wear in hip simulation studies with Duration focused attention on the role of crosslinking in wear reduction. Further investigation into UHMWPE wear led to the identification of the role of crossing-path motion [5, 6]. The mechanism by which crosslinks reduced wear was also determined. All of these findings laid the groundwork for the development of more highly crosslinked UHMWPEs. Hip simulator studies predicted that the wear of Duration inserts would be 20–32% lower than the wear of inserts from UHMWPE gamma sterilized in air. Clinical findings in a prospective randomized controlled study have determined wear reductions of 32% at 5 years and 38% at 8 years follow-up. This finding gives confidence in the predictive value of the wear evaluation protocol using the MTS hip joint simulator.

## 14.3 CROSSFIRE

In 1996, Paul Serekian, director of Materials Research for Osteonics Corporation, began working with raw material manufacturers to investigate the feasibility of producing highly crosslinked UHMWPE slab. In early 1997, he decided to pursue a more in-depth investigation and charged two engineers, Scott Taylor and Kofi Kwarteng, with developing an irradiation-crosslinked UHMWPE for hip bearing applications.

The investigations began by subjecting acetabular inserts to incremental 25–30 kGy doses of gamma irradiation, which was done simply by processing components packaged in nitrogen through successive gamma-irradiation sterilization cycles. The investigators subjected the test articles to cumulative dosages ranging from 0 kGy (non-irradiated) to approximately 130 kGy (five sterilization cycles), and the wear rate was determined by using joint simulation wear testing.

After exploring a range of radiation doses, different radiation sources, different thermal cycles to reduce free radicals, various UHMWPE resins, and packing options, the two investigators found that the process variable with the greatest impact on wear rate was irradiation dosage. They chose gamma irradiation over e-beam because radiation could penetrate greater depths of the UHMWPE material. They chose to crosslink rod stock rather than a packaged product so that higher temperatures could be used for free radical elimination. Annealing rather than remelting was chosen because it would leave the mechanical properties of the UHMWPE unaltered. The investigators were also aware that there were a number of unknowns associated with remelting/recrystallizing the UHMWPE material.

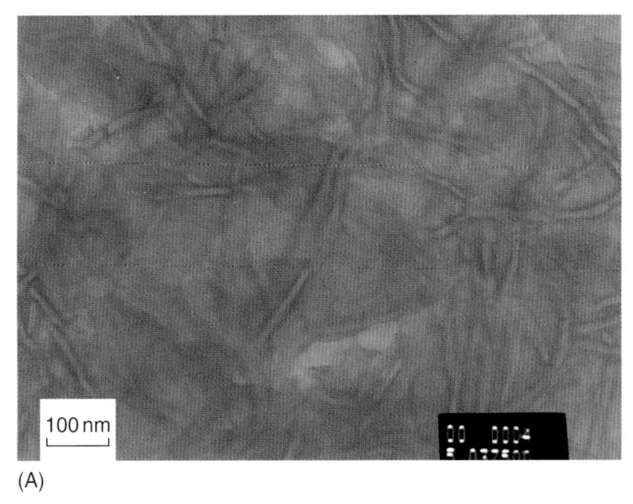


(A)

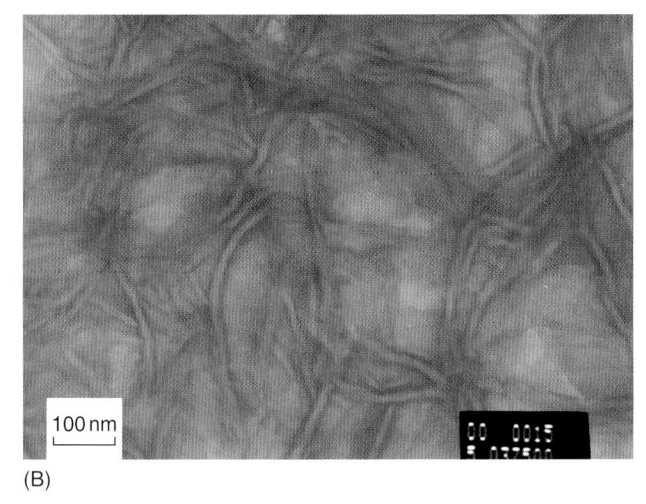


(B)

**FIGURE 14.1** This image shows the transmission electron micrographs of (A) N2/Vac™ and (B) Crossfire® UHMWPE. There is a similar range of lamellar sizes and thicknesses with a tendency for a greater number of thicker crystals in Crossfire®.

By September 1997, Osteonics had developed the new technology and sought clearance from the FDA for the clinical use of Crossfire UHMWPE. Regulatory clearance was secured in September 1998, and the first US clinical implantations of Crossfire occurred throughout the last quarter of that year.

## 14.3.1 Crossfire Process

In the Crossfire process, a GUR 1050 rod is irradiated to 75 kGy in air. The rod is then annealed at 130°C for 8 hours to eliminate free radicals. The outer surface of the rod is machined away, and components are machined from the rod. Components are packaged in nitrogen and gamma irradiated to 30 kGy (N2/Vac). The total radiation dose is 105 kGy. Crossfire acetabular inserts were clinically introduced in late 1998 for use in the hip; Crossfire was not used for knee components.

## 14.3.2 Properties

In a study by Kurtz et al., the free radial level in Crossfire was reported as $1.4 \times 10^{-8}$ mole/cc compared with $3.7 \times 10^{-9}$ mole/cc for N2/Vac [21]. In another study, Collier et al. compared different UHMWPE materials [22]. They reported the free radical content of Crossfire as $7.3 \times 10^{-4}$ mole/dm$^3$ while that of ArCom (Himont 1900 gamma sterilized in argon, Biomet Inc., Warsaw, Indiana, USA) was reported as $3.8 \times 10^{-4}$ mole/dm$^3$. GUR 1050 UHMWPE gamma sterilized in nitrogen at 38 kGy has been shown to have a free radical content of $2.6 \times 10^{15}$ spins/g compared to $4.3 \times 10^{15}$ spins/g for Crossfire [23]. It is difficult to be precise when measuring free radical content because measurements are affected by technique and the time since irradiation. However, even when taking this variability into account, it still appears that the free radical content in Crossfire is roughly double that of UHMWPE gamma sterilized in an inert atmosphere.

The oxidation resistance of Crossfire has been evaluated in a number of studies. Heating for 28 days at 70°C in 3 atmospheres of oxygen resulted in an oxidation index of $0.302 \pm 0.128$ that was statistically higher then that of the starting material of $0.036 \pm 0.03$ ($p < 0.01$) with a subsurface oxidation peak [22]. The crystallinity increased from 58% to 65% during this type of accelerated aging [22]. Both UHMWPE gamma sterilized in nitrogen and Crossfire exhibited subsurface oxidation peaks after heating in air at 80°C for 21 days, but the oxidation was several times higher for Crossfire [23]. *In vivo* oxidation was simulated by placing Crossfire components in a water bath heated to 40°C. After 95 weeks of aging, Crossfire exhibited a subsurface oxidation peak over five times higher than that of UHMWPE gamma sterilized in nitrogen [24].

Figure 14.1 shows transmission electron micrographs of N2/Vac and Crossfire UHMWPE. There is a similar range of lamellar sizes and thicknesses with a tendency for a greater number of thicker crystals in Crossfire. Using differential scanning calorimetry (DSC), N2/Vac and Crossfire displayed a single melt transition at 138°C and 140°C, respectively. The crystallinity of Crossfire is $56 \pm 5\%$, and that of N2/Vac is $57 \pm 3\%$. The difference is not statistically significant [21]. N2/Vac and Crossfire exhibit similar microstructures because stabilization in the Crossfire process was carried out below the melting temperature.

The tensile properties of Crossfire were determined using Type IV specimens (ASTM D 638) with testing according to ASTM F 648-96 [25]. The data are given in Table 14.1. The yield strength and ultimate tensile strength values are similar for N2/Vac and Crossfire. The higher level of crosslinking in Crossfire results in a decrease in elongation and toughness compared to N2/Vac. However, small punch testing indicates

**TABLE 14.1 Mechanical Properties for GUR 1050 Gamma Sterilized in Nitrogen and for Crossfire® [21]**

| Material | Yield Strength (MPa) | Ultimate Tensile Strength (MPa) | Elongation (Per cent) | Tensile Toughness ($J/cm^2$) |
|---|---|---|---|---|
| N2/Vac™ | 24.5 ± 0.4 | 60.0 ± 4.2 | 370 ± 10 | 270 ± 30 |
| Crossfire® | 25.4 ± 0.4 | 61.1 ± 5.0 | 281 ± 1 | 200 ± 20 |

that Crossfire has a notably higher ultimate biaxial strength and toughness compared with the N2/Vac material [21].

The fatigue performance of Crossfire has been compared to that of N2/Vac in a functional test [25]. Acetabular liners with a thickness of 2.5 mm at the rim were mounted in holders that did not provide back surface support. The constructs were mounted on a MTS hip joint simulator and loaded under level walking conditions using fetal bovine calf serum diluted with 50% deionized water as a lubricant. After testing, all liners showed some plastic deformation, but no liners exhibited crack formation up to the test limit of 1 million cycles. Despite being more highly crosslinked than N2/Vac, the behavior of Crossfire was similar in this functional fatigue test because annealing preserved the UHMWPE microstructure.

Hip simulator testing showed a substantial reduction in wear for Crossfire versus N2Vac. Studies comparing the wear of Crossfire to that of N2/Vac were carried out using MTS hip joint simulators under loading conditions for level walking and with bovine serum diluted by 30% to 50% distilled water as the joint lubricant. Wear was measured by weight loss correcting for fluid absorption. Testing was done with CoCr, alumina, and zirconia heads on several designs of acetabular component and with femoral head diameters from 28 to 36 mm.Median absolute wear rates were 51.5 and 3.9 $mm^3$/million cycles for N2/Vac and Crossfire, respectively, corresponding to a 92% wear reduction for Crossfire liners. Further details are given elsewhere [21]. The wear performance of Crossfire was studied under abrasive third-body wear conditions using a Matco hip simulator under level walking with bovine calf serum diluted with 30% distilled water as the joint fluid [26]. Bone cement particles in the size range 10 – 100 μm (50%) and 100 – 400 μm (50%) were mixed with the joint fluid. The Crossfire wear rate was 8.2 ± 3.2 mg/million cycles compared to 37.7 ± 1.1 mg/million cycles for N2/Vac, representing a 72% decrease in wear for Crossfire.

### 14.3.3 Clinical Studies

In a prospective randomized study [27], patients received Secur-Fit HA acetabular components with either Secur-Fit or Secur-Fit Plus HA femoral components (Stryker Orthopaedics, Mahwah, New Jersey, USA). Size 28 mm CoCr femoral heads were used, and randomization was between Crossfire and N2/Vac inserts. There were 24 Crossfire inserts and 22 N2/Vac inserts, with an average follow-up time of 2.3 years (range, 1.8 – 3.2 years). Wear was measured using an edge detection technique [28]. The two-dimensional linear wear rate was 0.12 ± 0.05 mm/year for Crossfire inserts and 0.20 ± 0.10 mm/year for N2/Vac inserts for a statistically significant reduction in wear ($p = 0.001$) of 42% with Crossfire.

A comparison of wear between Crossfire and conventional (N2/Vac) components was made on 56 hips in 47 patients (Crossfire) with 53 case-matched controls [29]. The minimum follow-up was 4 years, with an average of 5 years. Penetration rates were measured using the Martell method [28]. The linear penetration rate was 0.055 ± 0.22 mm/year for Crossfire cups and 0.138 ± 0.073 mm/year for N2/Vac cups. The annual wear was 0.036 mm/year for Crossfire components and 0.131 mm/year for the controls, a reduction of 72%. Radiographic review suggested a reduction in erosive osteolytic lesions of the proximal femur in patients having Crossfire components.

In another study, 27 Crossfire components (25 patients) were matched to 27 components (25 patients) receiving N2/Vac components, and the penetration rates were compared with a follow-up of approximately 6 years [30]. All patients received Secur-Fit acetabular components. All patients had 28 mm CoCr femoral heads. Most received an Omnifit stem (Stryker Orthopaedics, Mahwah, New Jersey, USA) unless bone quality was poor, in which case they received an Eon cemented stem (Stryker Orthopaedics, Mahwah, New Jersey, USA). Penetration rates were measured using the Martell method [28]. The mean total penetration was 0.283 ± 0.253 mm for the Crossfire group and 0.696 ± 0.402 mm for the controls ($p < 0.001$). The wear rate was 0.022 mm/year for the Crossfire group and 0.085 mm/year for the group with N2/Vac components. The decrease in wear was 75%. The comparison of radiographic findings in an active patient with a control, gamma irradiated, barrier-packaged UHMWPE acetabular liner to that of a patient with a Crossfire liner is shown in Figure 14.2. Both films were taken at more than 8 years after index surgery. The Crossfire case demonstrates little head migration and no osteolytic changes. The control shows evidence of visible head migration and wear with osteolytic lesions in the femur and pelvis, although it is still functioning clinically.

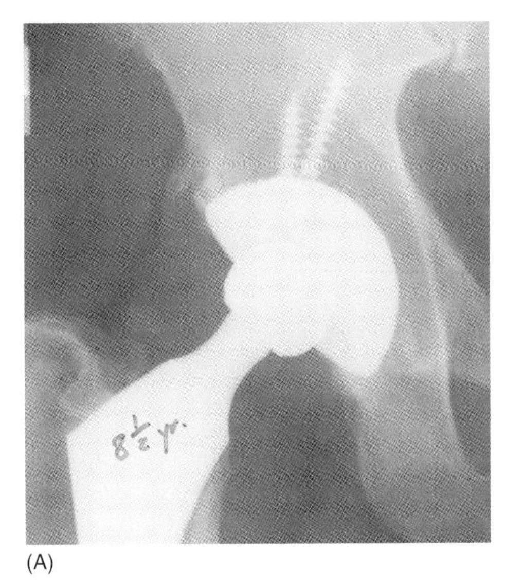

(A)

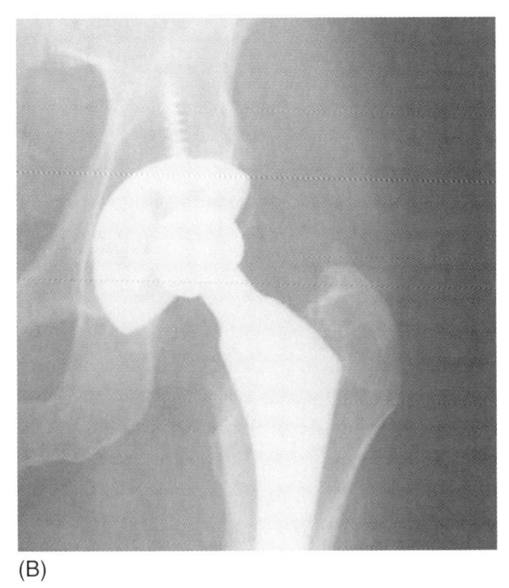
(B)

**FIGURE 14.2** (A) The right hip of a 50 year-old active male with a Crossfire® insert at 8.5 years follow-up. There is little head migration and no osteolysis. (B) The left hip of a 64 year-old active male with an N2/Vac™ insert at 8.5 years follow-up. Head migration is evident as is Zone 2 acetabular osteolysis and proximal femoral osteolysis. *(Images provided courtesy of James A. D'Antonio, M.D.)*

## 14.3.4 Crossfire Retrievals

An analysis of 14 retrieved Crossfire inserts has been reported [31]. The implantation time ranged from 4 to 33 months, and the reasons for removal included infection (3), dislocation (4), heterotopic ossification (2), lysis (2), disability (1), and femoral loosening (1); one of the inserts was removed for an unknown reason. There was no patient information because many of the initial surgeries were carried out at other institutions, and it is not known whether the Crossfire implantations were carried out in a primary or revision setting. The maximum oxidation index was reported for each of the 14 components and ranged from 0.22 to 5.81. A total of 10/14 components had a maximum oxidation index less than 0.57. For the remaining four components, the oxidation indices were 1.23, 2.99, 3.01, and 5.81. There was only a weak correlation between oxidation level and implantation time. Oxidation was highest on the component rim and was low on the articular surface. The maximum oxidation index for a never-implanted component was 0.16. Crystallinity was a maximum at the location of maximum oxidation. For the low oxidation group, the crystallinity values ranged from 64.4 to 69.6%. The high oxidation components had crystallinity values of 74.1, 72.0, 75.0, and 79.2%. The never-implanted component had a crystallinity of 64.9%. The high oxidation group exhibited a subsurface "white band" on sectioning, which is indicative of loss of ductility due to oxidative chain scission.

In a report on 12 retrieved Crossfire components, six of the retrievals exhibited fatigue damage, but five had signs of impingement [32]. None of the components was revised for wear, fatigue, or failure of the locking mechanism. Implantation times ranged from 0.1 to 5.3 years. No other patient data were given. Oxidation was low on the articular surface but was more than an order of magnitude higher at the rim, which is not load bearing in the Crossfire design. Maximum oxidation ranged from 0.4 to 8.8. Rim oxidation correlated with implantation time. A subsurface white band was noted on sectioning at the rim in seven components. One component also showed oxidation at the articular surface. Three components exhibited delamination at the rim in association with impingement, dislocation, or both.

In the preceding studies, Crossfire implants were accepted from different institutions and were accompanied by limited patient data. A different approach was taken by Kurtz et al. [33], who made prior arrangements with certain institutions with high joint volume to acquire patient information to match Crossfire retrievals. In addition, a detailed protocol for implant packaging, shipping, and handling was agreed upon. At the time of writing, analyses were completed on 12 Crossfire inserts from Omnifit Series II designs (Stryker Orthopaedics, Mahwah, New Jersey, USA). All have matching patient data. No component had been removed for wear or osteolysis. Implantation times ranged from 0.02 to 4.8 (mean 1.9) years. Evidence of oxidation was found at the rim of the components where the UHMWPE was exposed to joint fluid. At an average depth of 0.9 ± 0.9 mm (range, 0.2 – 3.0), the maximum oxidation indices ranged from 0.34 to 1.64. Oxidation was lower at the worn area in the bearing dome due to the exclusion of joint fluid by the femoral head. The back surface of the inserts also exhibited low oxidation because these areas also are in contact with the acetabular shell. Mechanical properties (ultimate load) were determined using small punch testing. The mechanical properties at the unworn surface

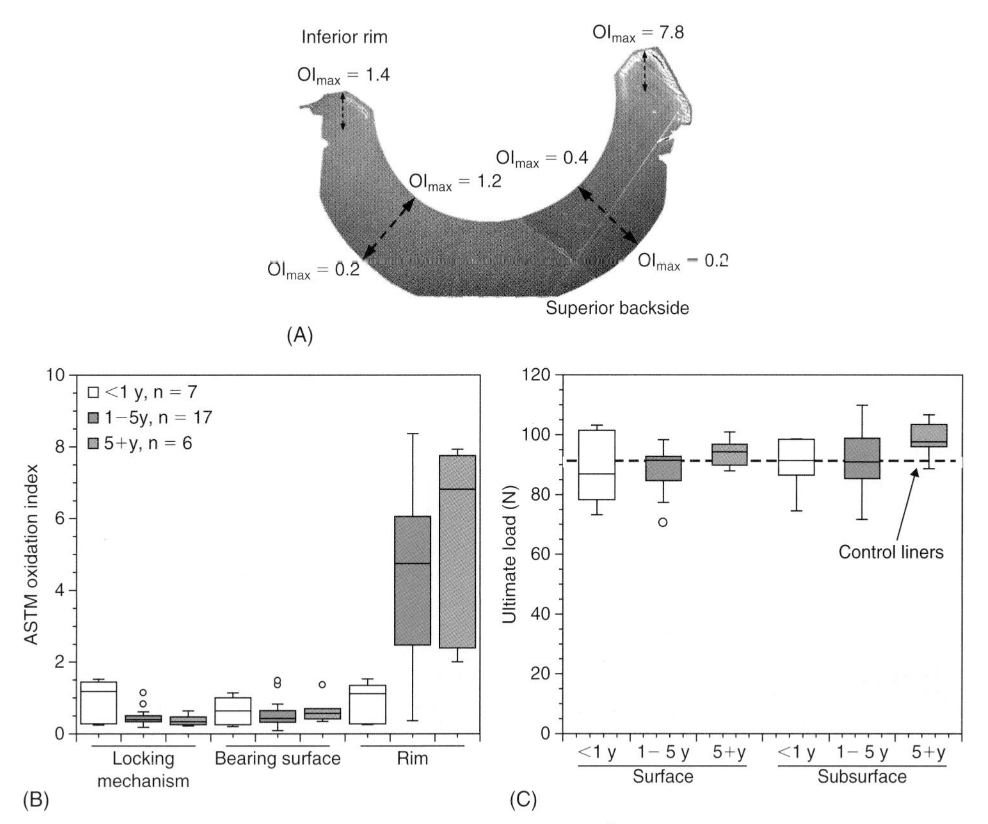


**FIGURE 14.3** (A) Darkfield micrograph of a microtomed section through a Crossfire® insert after 7.4 years *in vivo* (0.1 years of shelf life) showing distribution of oxidation. (B) Oxidation index as a function of the position on the insert for a collection of 30 retrieved Crossfire® components. (C) Ultimate load from small punch testing as a function of the position on the insert for a collection of 30 retrieved Crossfire® components. *(Images provided courtesy of Steven M. Kurtz, Ph.D., Implant Research Center, Drexel University.)*

were lower than those at the worn surface (difference 11.0N, p = 0.01). The properties were also lower (difference 6.0N) in the unworn subsurface compared to the unworn surface (p = 0.02). There were no other significant differences found in property comparison. No correlation between implantation time and either mechanical properties or oxidation could be demonstrated (Figure 14.3).

The Omnifit Series II Crossfire study has been extended to include 21 retrievals with implantation times from 0.02 to 6 years (mean, 3 years) [34]. The implants were revised for reasons unrelated to wear. Considerable regional variation was seen in the oxidation levels with the highest oxidation near the rim (Figure 14.3A). Rim oxidation levels were some five times greater than oxidation at the bearing surface (Figure 14.3B). A slightly lower ultimate load (mean 77N) was observed at the unworn surface compared with the superior bearing surface (mean 93N) (Figure 14.3C). Neither ultimate load nor oxidation level correlated with implantation time. All liners showed evidence of creep and wear. For the 14 components that had been implanted for at least a year, the average rate of penetration was 0.04 ± 0.02mm/year (range, 0.01 – 0.09mm/year).

Fifty-seven Crossfire liners were retrieved during consecutive revision surgeries at a single, high-volume institution [35]. The liners were implanted from 0 to 8 years (average ± SD: 2.8 ± 2.0 years). There were 30 Omnifit Series II and 27 Trident (Stryker Orthopaedics, Mahwah, New Jersey, USA) liners. The implants were revised for reasons unrelated to wear or osteolysis. Ultimate loads at the bearing surface ranged from 70.6 to 103.2N (mean ± SD: 89.7 ± 8.3 N) and from 68.2 to 107.4N (mean ± SD: 95.0 ± 8.6 N) for the Omnifit and Trident designs, respectively. The average subsurface ultimate loads were from 71.7 to 109.8N (mean ± SD: 92.6 ± 9.5N) and from 79.0 to 112.9N (mean ± SD: 98.8 ± 6.8N) for Omnifit and Trident liners, respectively. The ultimate load for unimplanted control liners ranged from 76.6 and 97.9N (mean ± SD: 90.1 ± 8.8N) near the surface and between 67.5 and 100.05N (mean ± SD: 91.0 ± 15.8N) at the subsurface. Omnifit and Trident retrievals had comparable ultimate strength. There was no association between ultimate strength and implantation time.

There was significant regional variation in oxidation. The oxidation range at the rim was 0.3 – 8.4 (mean ± SD: 3.2 ± 2.7). The oxidation at the bearing surface, around the locking mechanism and at the back surface, was much lower than at the rim. There was a strong association between rim oxidation and implantation time (Spearman's

Correlation $\rho = 0.72$, $P < 0.0001$), but a similar association was not noted for bearing oxidation ($P = 0.54$) or oxidation around the locking mechanism ($\rho = -0.29$; $P = 0.03$). Oxidation was observed to be somewhat higher around the locking mechanism for the Omnifit than the Trident liners, presumably due to the different locking mechanism types (groove/locking wire versus groove/equatorial bead).

Retrieved Omnifit and Trident liners showed penetration rates in the range 0.01 – 0.09 mm/year (mean ± SD: 0.04 ± 0.02 mm/year) and in the range 0.00 to 0.1 mm/year (mean ± SD: 0.04 ± 0.03 mm/year), respectively. The penetration rates were lower for implants in place more than 5 years, indicating the influence of initial creep. Only 3/57 liners exhibited fatigue damage at the rim, and that damage was associated with impingement or dislocation.

### 14.3.5 Summary

Crossfire acetabular components have been widely used since late 1998. The manufacturing process preserves the microstructure and strength properties. Functional fatigue properties appear to be superior to those of UHMWPE gamma sterilized in nitrogen. However, gamma sterilization after crosslinking results in a free radical level double that for uncrosslinked UHMWPE gamma sterilized in nitrogen. Crossfire exhibits oxidation during accelerated aging that is higher than that exhibited by UHMWPE gamma sterilized in nitrogen.

Clinical experience with Crossfire components has been good with follow-up results reported out to 6 years. The wear of Crossfire components is some 70% lower than that of N2/Vac. This contrasts with hip simulator predictions of over 90% reduction. There is a suggestion of reduced osteolysis with Crossfire compared to conventional UHMWPE. Retrieval analyses indicate that Crossfire undergoes *in vivo* oxidation. However, high oxidation is only seen at the rim and does not appear to threaten the locking mechanism at 8 years follow-up, which is currently the longest follow-up time for this study. Oxidation at the contact region of the bearing is low due to the protective effect of the femoral head. Wear measurements on retrievals are in good agreement with clinical X-ray measurements. Mean wear rates are in the range 0.01–0.04 mm/year.

## 14.4 X3: SEQUENTIALLY IRRADIATED AND ANNEALED UHMWPE

In April 2002, John Dumbleton agreed to speak about Crossfire UHMWPE at the Harvard Course in Boston, where some presenters commented on Crossfire containing free radicals. Although Crossfire has had, and continues to have, excellent clinical results, the course inspired Dr. Dumbleton to think about the next generation of UHMWPE. He hypothesized with Aiguo Wang and Shi-Shen Yau that it might be possible to create a UHMWPE with little to no free radicals by integrating a sequential irradiation and annealing process in which a low dose of radiation was used followed by annealing. On average, the crosslinks would be far enough apart that there would be sufficient mobility to extinguish the free radicals. The process could then be repeated.

Subsequently, Dr. Wang asked Dr. Yau to empirically determine the best combination of cycles and radiation dose. Dr. Wang initially concentrated on the sequential material for the knee. Dr. Dumbleton's interest was in the sequential material for the hip because he hoped to replace Crossfire. Drs. Dumbleton, Wang, and Yau discussed having a lower radiation dose for the knee material than for the hip, but it turned out that the same process could be used—three sequential doses of 30 kGy for a cumulative total of 90 kGY—with annealing after each cycle and gas plasma sterilization [36, 37].

A patent disclosure for the new UHMWPE material X3 was filed in June 2002 with Aiguo Wang and John Dumbleton as inventors, and the patent application followed shortly thereafter. Subsequently, the names of Shi-Shen Yau and Aaron Essner, who carried out the hip and knee simulator evaluations, were included as inventors, and their names were added to the application in September 2004. X3 was cleared in 2005 for clinical use in the hip and knee.

### 14.4.1 Sequential Crosslinking Process

By 2002, the experience with crosslinked UHMWPEs showed that a radiation dose of 50–100 kGy, and preferably closer to 100 kGy, would be needed to provide sufficient crosslinking for greater wear reduction. It was believed that annealing to remove free radicals had the advantage of retaining the original UHMWPE microstructure and properties, unlike the remelting process. However, the annealing process needed to be made more effective to remove sufficient free radicals to provide oxidation resistance. The sequential process appeared to be the answer rather than a single irradiation and annealing cycle. The researchers concluded that with present crosslinking techniques, terminal sterilization should be by nonionizing means because the steric effects of crosslinking hindered natural elimination of free radicals during heating.

Compression-molded GUR 1020 was chosen over GUR 1050 as the starting material due to its higher ductility and impact strength. Experimental work identified the following sequential irradiation and annealing process as the best balance between cycle time and properties:

- Cycle 1: 30 kGy irradiation followed by annealing for 8 hours at 130°C
- Cycle 2: 30 kGy irradiation followed by annealing for 8 hours at 130°C
- Cycle 3: 30 kGy irradiation followed by annealing for 8 hours at 130°C

The cumulative radiation dose is 90 kGy. Following the sequential irradiation and annealing, the outer 1–3 mm of material was removed, and components were machined and packaged for gas plasma sterilization (Sterrad 200, Advanced Sterilization Produces, Irvine, California, USA). Material produced by the sequential irradiation and annealing process was named X3 and introduced for acetabular components in May 2005. Over the following year, X3 tibial and patella components were introduced for cruciate retaining (CR) and posterior stabilized (PS) TKR. In 2008, total stabilizer (TS) components in X3 became available.

## 14.4.2 X3 Properties

Table 14.2 shows the free radical concentration and crosslink density at stages during the sequential irradiation and annealing process compared with a single irradiation and annealing to the same cumulative dose of 90 kGy [38]. Crosslinks are primarily formed during irradiation. However, there is additional crosslinking during annealing, indicating that one mechanism of free radical elimination is via crosslink formation. The sequential irradiation and annealing process is much more efficient for crosslink formation than is a single irradiation and annealing to the same cumulative dose. Free radicals are largely eliminated during annealing. However, the process becomes less efficient as the crosslinking level increases, which limits the number of useful irradiation and annealing cycles.

The oxidation resistance of X3 has been evaluated in a variety of environments. Figure 14.4 shows the results of aging for 14 days at 70°C in 5 atmospheres of oxygen (ASTM F2003) [38]. Blocks of X3, N2/Vac, virgin UHMWPE (no detectable free radicals), and UHMWPE irradiated to 90 kGy and then annealed (90TS) went through accelerated aging under the preceding conditions. Following aging the blocks were sectioned, and the oxidation index was determined for each section (ASTM F2102). X3 and virgin UHMWPE exhibited minimal oxidation. N2/Vac showed subsurface oxidation peaks, and the sectioned blocks revealed a white band at the subsurface location. The 90TS material showed oxidation intermediate between that of X3 and N2/Vac. Accelerated aging of thin films (180 μm thickness) of material at 90°C in air again indicated that the oxidation resistance of X3 is similar to that of virgin UHMWPE (Figure 14.5). Real-time shelf aging has been carried out (Table 14.3). At 5 years there was no change in mechanical properties and no signs of oxidation. Tensile testing after accelerated aging (ASTM F2003) did not change the tensile properties of X3, but N2/Vac showed lower tensile strength and elongation (Table 14.4) [39]. The differences in the mechanical properties for

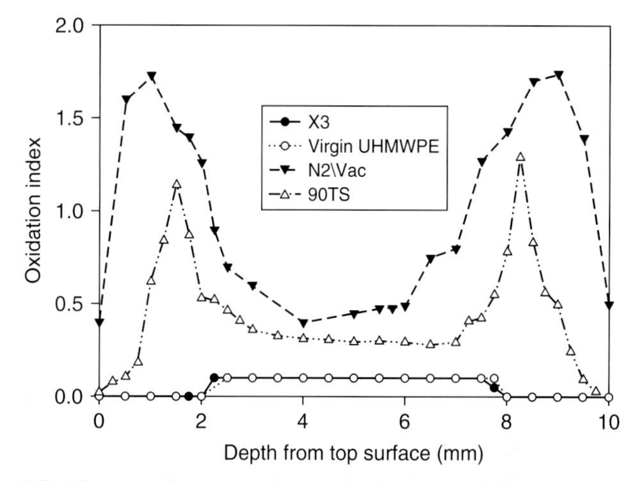


**FIGURE 14.4** The results of aging for 14 days at 70°C in 5 atmospheres of oxygen (ASTM F2003). The oxidation index is shown as a function of depth for X3®, N2/Vac™, 90TS and virgin UHMWPE after accelerated aging.

**TABLE 14.2 Free radical concentration and crosslink density for sequential irradiation and annealing compared to single irradiation and annealing**

| Step | Process | Free radical concentration ($10^{14}$ spins/g) | Crosslink density (mol dm$^{-3}$) |
|---|---|---|---|
| 1 | 30 kGy | 1561 ± 41 | 0.07 ± 0.01 |
| 2 | 30 kGy+annealing | none detected | 0.09 ± 0.01 |
| 3 | Step 2+3 Mrad | 1663 ± 36 | 0.13 ± 0.02 |
| 4 | Step 3+annealing | 7 ± 3 | 0.17 ± 0.02 |
| 5 | Step 4+3 Mrad | 1719 ± 45 | 0.23 ± 0.02 |
| 6 | Step 5+annealing | 9 ± 2 | 0.28 ± 0.03 |
| Single | 90 kGy+annealing | 68 ± 7 | 0.11 ± 0.02 |

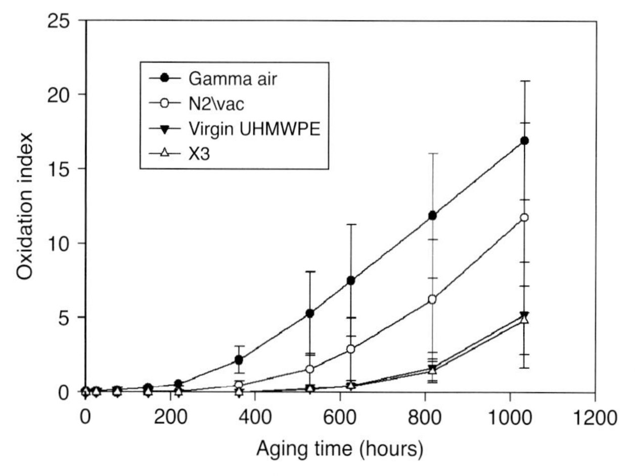


**FIGURE 14.5** Oxidation index as a function of time with thin films in air at 90°C. Accelerated aging of thin films (180 μm thickness) of material at 90°C in air again indicated that the oxidation resistance of X3® is similar to that of virgin UHMWPE.

N2/Vac, reported in Table 14.1 and Table 14.4, result from normal batch-to-batch variation and fall within the acceptance criteria established for that material.

Table 14.5 provides details on the microstructure of X3 compared to that of N2/Vac as measured by small-angle X-ray scattering [38]. There was no significant difference in mean crystal thickness or spacing (p = 0.1). According to DSC measurements, the crystallinity values of X3, N2/Vac, and virgin UHMWPE were 61.7 ± 0.6, 61.5 ± 0.9, and 59.2 ± 0.7%, respectively [38]. There is no significant difference in crystallinity between X3 and conventional UHMWPE, but both materials had significantly higher crystallinity than virgin UHMWPE (p = 0.01). The sequential process preserves the microstructure. The degree of chain scission from irradiation resulting in increased crystallinity is similar for X3 and N2/Vac, despite the former receiving a three times higher radiation dose.

Functional fatigue testing was carried out according to the previously described protocol [25]. Table 14.6 gives the percentage survivorship at 1 million loading cycles. X3 had higher survivorship (89%) than Crossfire (79%) or N2/Vac (80%) [39]. Figure 14.6 shows pictures of retrieved

**TABLE 14.3 Oxidation index and tensile properties for X3® as manufactured and after 5 years shelf aging**

| Properties | At Manufacture | After 5 Years |
|---|---|---|
| Oxidation Index | | |
| Surface | 0.00 ± 0.00 | 0.00 ± 0.00 |
| Bulk | 0.00 ± 0.00 | 0.00 ± 0.00 |
| Tensile Yield Strength (MPa) | 25.3 ± 0.3 | 23.7 ± 0.3 |
| Ultimate Tensile Strength (MPa) | 56.7 ± 1.5 | 56.6 ± 1.7 |
| Elongation (per cent) | 267 ± 7 | 270 ± 6 |

**TABLE 14.5 Long period spacing, crystal thickness and amorphous thickness for X3® and conventional UHMWPE (N2/Vac™)**

| UHMWPE | Long period spacing (nm) | Crystal thickness (nm) | Amorphous thickness (nm) |
|---|---|---|---|
| N2/Vac™ | 38.9 ± 1.0 | 23.0 | 15.9 |
| X3® | 38.2 ± 0.2 | 23.6 | 14.6 |

**TABLE 14.6 Survivorship of X3®, N2/Vac™ and Crossfire® inserts after fatigue testing**

| Material | Number tested | Number failed | Survivorship (percent) |
|---|---|---|---|
| N2/Vac™ | 10 | 2 | 80 |
| Crossfireis® | 14 | 3 | 79 |
| X3® | 9 | 1 | 89 |

**TABLE 14.4 Tensile properties of virgin UHMWPE, N2/Vac™ and X3® before and after aging (ASTM 2003)**

| Material/Property | Aging | Yield strength (MPa) | Tensile strength (MPa) | Elongation (percent) |
|---|---|---|---|---|
| Virgin UHMWPE | No | 22.1 ± 0.5 | 57.0 ± 3.7 | 418 ± 19 |
| | Yes | 22.7 ± 0.6 | 57.5 ± 5.5 | 422 ± 17 |
| N2/Vac™ | No | 23.2 ± 0.4 | 54.8 ± 2.5 | 363 ± 10 |
| | Yes | 27.9 ± 0.4 | 29.9 ± 1.2 | 143 ± 14 |
| X3® | No | 23.5 ± 0.3 | 56.7 ± 2.1 | 267 ± 7 |
| | Yes | 23.6 ± 0.2 | 56.3 ± 2.3 | 266 ± 9 |

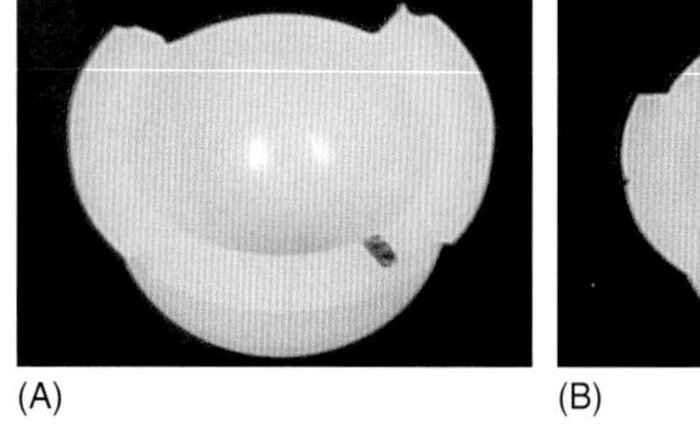
(A)

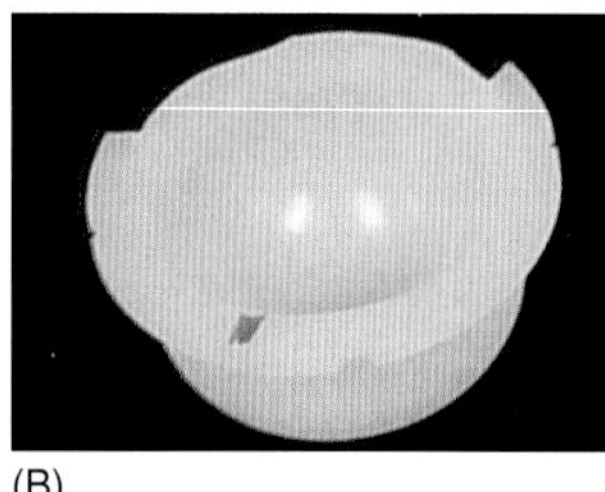
(B)

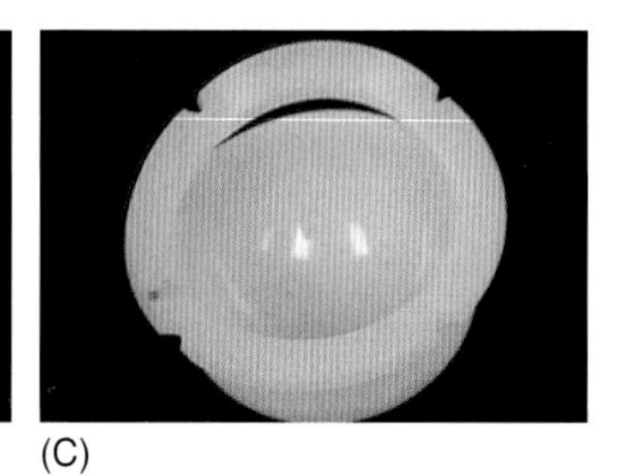
(C)

**FIGURE 14.6** Pictures of compromised components; (A) N2/Vac™, (B) Crossfire®, (C) X3. The X3® component did not separate into two pieces.

components. The X3 component that was removed did not separate into two pieces.

Wear testing was carried out using MTS hip joint simulators with a superiorly applied physiological loading pattern representing level walking at 1 Hz.The maximum and minimum loads were 2450 N and 150 N, respectively. Fetal bovine calf serum (Hyclone Laboratories, Logan, Utah, USA) diluted to 50% concentration with deionized water and stabilized with 20 mmol ethylenediaminetetraacetic acid (EDTA) was the lubricant. Weight loss was measured with correction for fluid absorption. Weight loss increased linearly during the tests until the test conclusion at 10 million cycles. For 32 mm Trident acetabular components (Stryker Orthopaedics, Mahwah, New Jersey, USA) the wear rates were 46.4, 3.6 and 1.3 $mm^3$/million cycles for N2/Vac, Crossfire, and X3, respectively [39]. The wear of X3 was 62% lower than that of Crossfire. Table 14.7 provides a comparison of wear particle dimensions for N2/Vac and X3. There was no significant difference in particle size. Further simulator testing was undertaken with femoral head sizes up to 52 mm and for insert thicknesses from 7.5 to 3.8 mm. The wear rates for these different combinations are shown in Figure 14.7 [40]. Wear was not influenced by either femoral head size or by insert thickness.

Surface (adhesive) wear is the primary type of wear at the hip. Surface wear also occurs at the knee, but delamination wear, caused by a combination of subsurface stress and oxidation, is a major cause of damage at the knee. The resistance to delamination was measured using a ball-on-plate reciprocating test arrangement with bovine calf serum as the lubricant [38]. The initial Herzian contact stress was 61 MPa.UHMWPE gamma sterilized in air, N2/Vac, and X3 were tested following accelerated aging (ASTM F2003). Both gamma air sterilized and N2/Vac showed modest signs of pitting and delamination, but X3 exhibited only mild surface scratching and no signs of pitting or delamination (Figure 14.8).

Knee simulator testing was carried out on MTS knee simulators under level walking conditions (ISO 14243-3) [36]. The Scorpio CR design knee (Stryker Orthopaedics, Mahwah, New Jersey, USA) was used. Weight loss measurements corrected for fluid absorption were made on X3 and N2/Vac tibial inserts. The wear rates were 7.3 ± 0.7 and 34.6 ± 1.5 $mm^3$/million cycles for X3 and N2/Vac, respectively, a decrease of 79% with X3. Accelerated aging (ASTM F2003) did not affect the wear rate with X3

**TABLE 14.7 Hip simulator wear particle dimensions for X3® and N2/Vac™**

| Material | Particle length (μm) | Particle width (μm) | Aspect ratio |
|---|---|---|---|
| N2/Vac™ | 0.409 ± 0.553 | 0.221 ± 0.356 | 2.34 ± 3.48 |
| X3® | 0.471 ± 0.814 | 0.247 ± 0.399 | 1.99 ± 1.76 |

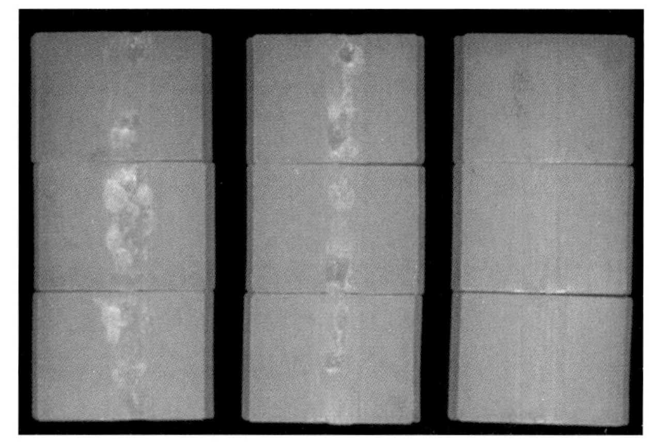

FIGURE 14.8 Bearing surfaces after contact fatigue testing for UHMWPE gamma sterilized in (A) air after $2 \times 10^6$ cycles, (B) nitrogen after $2 \times 10^6$ cycles, and (C) X3® after $5 \times 10^6$ cycles. Both gamma air sterilized and N2/Vac™ showed signs of pitting and delamination, but X3® exhibited only mild surface scratching and no signs of pitting or delamination.

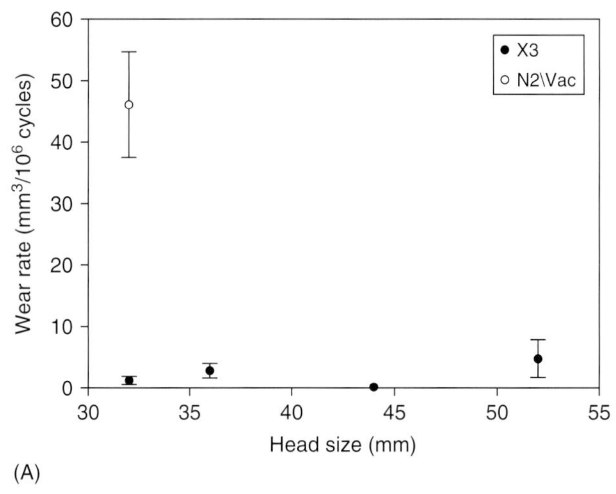


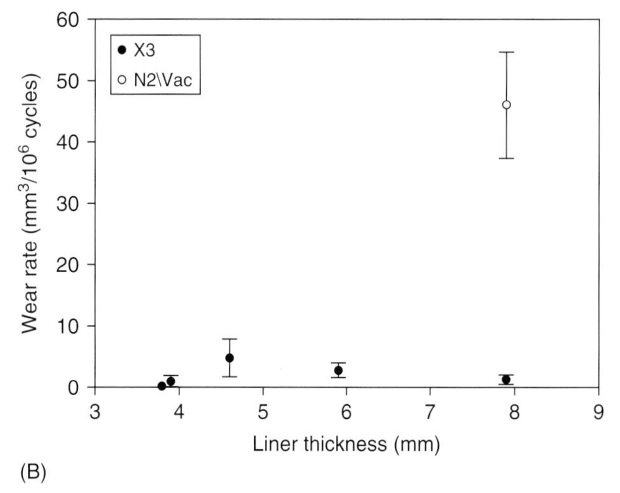


FIGURE 14.7 The wear rate for X3® as a function of head size (32–52 mm) and insert thickness (3.8–7.8 mm) is shown, and the wear of 32 mm N2/Vac™ liners (7.8 mm thickness) is given for comparison. Wear was not influenced by either femoral head size or by insert thickness.

**TABLE 14.8 Wear particle debris length, width and ECD for N2/Vac™ and X3® from Scorpio CR and Scorpio PS knee simulator tests**

| Material | Length (μm) | Width (μm) | ECD* (μm) |
|---|---|---|---|
| Scorpio CR | | | |
| N2/Vac™ | 0.39 ± 0.12 | 0.28 ± 0.07 | 0.37 ± 0.09 |
| X3® | 0.43 ± 0.23 | 0.33 ± 0.17 | 0.42 ± 0.22 |
| Scorpio PS | | | |
| N2/Vac™ | 0.36 ± 0.12 | 0.23 ± 0.06 | 0.32 ± 0.09 |
| X3® | 0.32 ± 0.13 | 0.20 ± 0.05 | 0.28 ± 0.08 |

**Equivalent circle diameter*

tibial inserts in the Scorpio CR design. Scorpio PS design knees (Stryker Orthopaedics, Mahwah, New Jersey, USA) were tested under stair climbing conditions to ensure that there was contact with the post [36]. The wear rates were 8.2 ± 0.7 and 35.8 ± 1.7 $mm^3$/million cycles for X3 and for N2/Vac, respectively, a decrease of 77% with X3. Table 14.8 gives wear particle dimensions for N2/Vac and X3 from the Scorpio CR and Scorpio PS tests.

The wear of X3 has been compared to that of N2/Vac in MTS knee simulator studies with the Triathlon CR TKR (Stryker Orthopaedics, Mahwah, New Jersey, USA) [37]. Level walking conditions were employed (ISO 14243-3). The wear rate for X3 was 5.7 ± 1.5 $mm^3$/million cycles, and for N2/Vac it was 17.7 ± 2.2 $mm^3$/million cycles, a decrease of 68% with X3. Wear particle sizes were not significantly different between N2/Vac and X3. Another study employed Shore-Western knee simulators to evaluate the wear of Triathlon CR knees. The wear with X3 was 86% lower than that of N2/Vac [41]. There was no difference in wear particle morphology between these two materials [42]. No effect of aging or degree of malalignment on wear performance was found in a Triathlon CR study using an AMTI knee simulator [43].

The wear of X3 tibial inserts was compared to that of N2/Vac in MTS knee simulator testing with Triathlon PS TKRs [37]. Stair climbing kinematics and loading were used. The wear rate for X3 was 1.5 ± 0.5 $mm^3$/million cycles and that of N2/Vac was 4.13 ± 0.7 $mm^3$/million cycles, a 64% decrease for X3. The wear with PS components was about one-third that with CR components. This appears to be due to the lower A-P translation exhibited with the Triathlon PS compared to the CR design [37].

### 14.4.3 X3 Clinical Studies

Clinical studies for use of X3 in both the hip and knee are underway. For the hip clinical study, nine centers are currently enrolling patients. In addition to the X3 insert, all patients are receiving a Trident hemispherical cup with an LFIT CoCr 32 mm head and a Secur-fit HA stem. The clinical parameters being studied are wear and osteolysis. Since May 2005, 264 patients have been enrolled. Of those patients, 190 patients have 1 year of follow-up, and 19 patients have 2 years of follow-up.

The performance of X3 in the knee is being examined in two separate studies, one involving the Triathlon CR Total Knee System (Stryker Orthopaedics, Mahwah, New Jersey, USA) and the other involving the Triathlon PS Total Knee System (Stryker Orthopaedics, Mahwah, New Jersey, USA). Both studies have concurrent N2/Vac controls. The Triathlon CR outcomes study involves 12 centers, and 494 knees have been enrolled to date. Of the knees enrolled, 420 (85%) have achieved a minimum follow-up of 1 year. For the X3 arm only, 90 cases have 1-year follow-up; none has yet reached 2-year follow-up. Eleven centers continue to enroll knees for the Triathlon PS outcomes study. To date, 448 knees have been enrolled, and 385 (86%) of those knees have 1 year of follow-up. For the X3 arm only, there are currently 12 cases with 1-year follow-up and none with longer follow-up. Data for both studies continues to be collected and evaluated.

### 14.4.4 X3 Retrievals

A retrieval study conducted by Kurtz et al. [44] examined whether or not highly crosslinked UHMWPE formulations, including X3 sequentially annealed UHMWPE, would have lower oxidation, oxidation potential, and lower head penetration than historical gamma-air and conventional gamma-inert sterilized liners. Of the 139 UHMWPE liners collected from a total of three manufacturers, 91 were highly crosslinked. Forty-three of the 91 were annealed, 41 were remelted, and 7 were sequentially annealed (X3). Thirty-one of the conventional UHMWPE components were gamma sterilized in air, and 17 were gamma sterilized in nitrogen.

Elevated rim oxidation was observed in the gamma-air and annealed groups but not in the remelted and X3 groups. Oxidation and hydroperoxide levels were significantly lower ($p < 0.5$) in the gamma-nitrogen, X3, and remelted groups. This is the first study to report an analysis of retrieved X3 liners. Although the implantation time for the X3 liners was short, the information gathered provides an initial starting point for comparison against longer-term retrievals. The remelted and X3 sequentially annealed liners had zero-to-low levels of oxidation following their brief *in vivo* period but showed measurable oxidation potential.

### 14.4.5 X3 Summary

The sequential irradiation and annealing process has been shown to more effectively eliminate free radicals and form crosslinks than a single cumulative radiation dose of the same magnitude followed by annealing. Steric hindrance due to crosslink formation, which limits free radical quenching, appears to be less of a factor with the sequential process. The sequential and annealing process produces a microstructure that is not significantly different from that of UHMWPE gamma sterilized in nitrogen (N2/Vac).

X3 has oxidation resistance similar to that of virgin UHMWPE with accelerated aging of thin films at 90°C in air or under the conditions in ASTM F2003. Accelerated aging using the latter protocol did not change either the mechanical properties or the wear characteristics of X3. Additionally, X3 had higher survivorship than Crossfire or N2/Vac in functional fatigue testing. In a contact fatigue study comparing the performance of X3 to that of UHMWPE gamma sterilized in air or in nitrogen after aging according to ASTM F2003, X3 did not show any signs of pitting or delamination. Both the gamma air and gamma nitrogen sterilized UHMWPE materials exhibited noticeable delamination.

Hip simulator studies with 32 mm heads showed that the wear rate of X3 was 62% less than that of Crossfire and 97% less than that of N2/Vac. There was no significant difference in wear particle size between X3 and N2/Vac. Further studies demonstrated that the wear rate was unaffected by femoral head size up to 52 mm and for insert thickness down to 3.8 mm.

X3 demonstrates improved wear rate reductions compared to N2/Vac in knee simulator studies with Scorpio and Triathlon designs. The reduction in wear rate is 79% for the Scorpio CR design and 77% for the Scorpio PS design. The wear rate reductions are 68% and 64% for the Triathlon CR and PS designs, respectively. There was no significant difference in particle size between X3 and N2/Vac for measurements made with the CR knees.

X3 was introduced for acetabular components in May 2005 with femoral head sizes up to 36 mm.Subsequently, sizes up to 42 mm were made available. The minimum insert thickness on introduction was 7.8 mm. Insert sizes down to 3.8 mm were made available with the introduction of larger head sizes. X3 was introduced for tibial inserts and patellar components in 2006. X3 is currently available for Scorpio CR, PS, and NRG designs and for Triathlon CR, PS, and TS designs.

Clinical studies are underway for X3 at both the hip and knee. Two-year wear results are expected during late 2008 for the hip. At this short follow-up time the penetration is dominated by creep; it will not be until 5-year results are available that a realistic comparison can be made to the wear of N2/Vac and to Crossfire. Knee clinicals are focused on clinical variables because there is no accepted methodology for *in situ* knee wear measurement. However, wear can be measured using retrievals. Retrieval analysis is ongoing as components become available. Oxidation and mechanical properties are being determined so that the *in vivo* performance of X3 can be compared to that of the earlier generation UHMWPEs.

## 14.5 CONCLUSION

The experience with highly crosslinked UHMWPEs over the last decade shows a progressive development in understanding the relationship between UHMWPE structure (crosslink density and crystallinity) and properties (strength, ductility, and wear). The agreement between clinical results and laboratory predictions for crosslinked UHMWPEs is unusual in orthopedic engineering. When compared to the barrier packaged minimally crosslinked "conventional" materials, annealed highly crosslinked UHMWPE total joint bearings do provide marked reduction in joint wear and reduced radiographic changes in surrounding bone indicative of osteolysis. The clinical literature with Duration and Crossfire components indicates that higher crosslinking can produce a significant wear reduction and that a radiation dose of around 100 kGy provides a good balance between wear and mechanical properties. The use of sequential irradiation and annealing provides the same level of crosslinking at a lower radiation dose with a consequent improvement in mechanical properties that allows X3 to be used for TKR. Simulator testing predicts a 62% wear reduction at the hip for new X3 UHMWPE compared to the earlier generation Crossfire. Based on the excellent predictions for the clinical wear of Duration and Crossfire, we believe that X3 will demonstrate substantially lower wear clinically than these earlier crosslinked UHMWPEs. At the knee, the wear properties of X3 should show notable improvement over those of N2/Vac. However, the efficacy of knee simulation in predicting wear rates has not yet been established.

Retrieval analyses show that oxidation of the earlier generation crosslinked and annealed materials can be identified in those regions of components directly exposed to body fluids (e.g., acetabular component rim) but not at regions protected from oxygenated fluids (e.g., bearing surface and backside). No retrieval of components due to material degradation and failure has been reported as clinical follow-up progresses past 10 years from index operation. It can be expected that the sequentially annealed UHMWPEs will likely resolve any concerns for oxidative degradation in the future.

## REFERENCES


1. Oonishi H, Kadoya Y. Wear of high-dose gamma-irradiated polyethylene in total hip replacements. *J Orthop Sci* 2000;**5**(3):223–8.
2. Grobbelaar CJ, du Plessis TA, Marais F. The radiation improvement of polyethylene prostheses. A preliminary study. *J Bone Joint Surg Br* 1978;**60-B**(3):370–4.
3. Wroblewski BM, Siney PD, Fleming PA. Low-friction arthroplasty of the hip using alumina ceramic and cross-linked polyethylene. A 17-year follow-up report. *J Bone Joint Surg Br* 2005;**87**(9):1220–1.
4. Mejia LC, Brierley TJ. A hip wear simulator for the evaluation of biomaterials in hip arthroplasty components. *Biomed Mater Eng* 1994;**4**:259–71.

5. Wang A, Stark C, Dumbleton JH. Mechanistic and morphological origins of ultra-high molecular weight polyethylene wear debris in total joint replacement prostheses. *Proc Instn Mech Eng [H]* 1996;**210**:157–66.
6. Bragdon CR, O'Connor DO, Lowenstein JD, Jasty M, Syniuta WD. The importance of multidirectional motion on the wear of polyethylene. *Proc Inst Mech Eng [H]* 1996;**210**:157–66.
7. Kurtz SM, Pruitt LA, Jewett CW, Foulds JR, Edidin AA. Radiation and chemical crosslinking promote strain hardening behavior and molecular alignment in ultra high molecular weight polyethylene during multi-axial loading condtions. *Biomaterials* 1999;**20**:449–1462.
8. Wang A, Polineni VK, Essner A, Sun DC, Stark C, Dumbleton JH. Effect of radiation dosage on the wear of stabilized UHMWPE evaluated by hip and knee simulators. *Paper 394*. New Orleans, LA; 1997.
9. Jahan MS, Wang C, Schwartz G, Davidson JA. Combined chemical and mechanical effects on free radicals in UHMWPE joints during implantation. *J Biomed Mater Res* 1991;**25**:1005–17.
10. Premnath V, Harris WH, Jasty M, Merrill EW. Gamma sterilization of UHMWPE articular implants: an analysis of the oxidation problem. Ultra High Molecular Weight Polyethylene. *Biomaterials* 1996;**18**:1741–53.
11. Essner A, Polineni VK, Schmidig G, Wang A, Stark C, Dumbleton JH. Long-term wear simulation of stabilized UHMWPE acetabular cups. *Poster 784*. San Francisco, CA; 1997.
12. Sun DC, Stark C. *U.S. Patent #5,414,049-Non-oxidizing polymeric medical implant;* 1995.
13. Sun DC, Stark C, Dumbleton JH. Development of an accelerated aging method for evaluation of long-term irradiation effects on UHMWPE implants. *Polym Repr* 1994;**35**:969–70.
14. Sun DC, Stark C, Dumbleton JH. Development of stabilized UHMWPE implants with improved oxidation resistance via crosslinking. *Scientific Exhibit S82*. Atlanta, GA; 1996.
15. Sun DC, Schmidig G, Yau S-S, Huffman M, Stark C, Dumbleton JH. Assessment of gel content and crosslinking density in UHMWPE. *Paper 431*. New Orleans, LA; 1997.
16. Polineni VK, Wang A, Essner A, Sun DC, Stark C, Dumbleton JH. Effect of gamma-radiation induced oxidation and crosslinking on the wear performance of Ultra-High Molecular Weight Polyethylene (UHMWPE) Acetabular Cups. In: Gsell RA, Stein HL, Ploskonka JJ, editors. *Characterization and Properties of Ultra-High Molecular Weight Polyethylene*, ASTM STP 1307: American Society for Testing and Materials; 1998. p. 95–108.
17. Martell JM, Berkson E, Berger R, Jacobs J. Comparison of two and three-dimensional computerized polyethylene wear analysis after total hip arthroplasty. *J Bone Joint Surg Am* 2003;**85A**(6):1111–17.
18. Geerdink CH, Grimm B, Ramakrishnan R, Rondhuis J, Verburg AJ, Tonino AJ. Crosslinked polyethylene compared to conventional polyethylene in total hip replacement: pre-clinical evaluation, in-vitro testing and prospective clinical follow-up study. *Acta Orthop* 2006; **77**(5):719–25.
19. Grimm B, Tonino AJ, Heyligers IC. 8-year prospective randomized comparison between a crosslinked and conventional polyethylene in total hip arthroplasty. *p Free Papers-Hip #10*. Florence, Italy; 2007.
20. Hirakawa K, Inaba Y, Murakami K. Randomized prospective comparison between Duration and conventional UHMWPE for primary THA. *Poster P066*. Washington, D.C.; 2005.
21. Kurtz SM, Manley MT, Wang A, Taylor S, Dumbleton JH. Comparison of the properties of annealed crosslinked (Crossfire) and conventional polyethylene as hip bearing materials. *Bulle-Hosp Joint Dis* 2003;**61**(1 & 2):17–26.
22. Collier JP, Currier BH, Kennedy FE, Currier JH, Timmins GS, Jackson SK, Brewer RL. Comparison of cross-linked polyethylene materials for orthopaedic applications. *Clin Orthop Relat Res* 2003;**414**:289–304.
23. Muratoglu OK, Merrill EW, Bragdon CR, O'Connor D, Hoeffel D, Burroughs B, Jasty M, Harris WH. Effect of radiation, heat, and aging on in vitro wear resistance of polyethylene. *Clin Orthop Relat Res Relat Res* 2003;**417**:253–62.
24. Wannomae KK, Christensen SD, Freiberg AA, Bhattacharyya S, Harris WH, Muratoglu OK. The effect of real-time aging on the oxidation and wear of highly cross-lined UHMWPE acetabular liners. *Biomaterials* 2006;**27**(9):1980–7.
25. Wang A, Manley MT, Serekian P; 2003. *Wear and structural fatigue simulation of crosslinked ultra-high molecular weight polyethylene for hip and knee bearing applications*: ASTM International STP 1445.
26. Taylor S, Serekian P, Bruchalski P, Manley MT. The performance of irradiation-crosslinked UHMWPE cups under abrasive conditions throughout hip simulation testing. *Paper 252*. Anaheim, CA; 1999.
27. Martell JM, Verner JJ, Incavo SJ. Clinical performance of a highly cross-linked polyethylene at two years in total hip arthroplasty: a randomized prospective trial. *J Arthroplasty* 2003;**18**(7 Suppl 1):55–9.
28. Martell JM, Berdia S. Determination of polyethylene wear in total hip replacements with use of digital radiographs. *J Bone Joint Surg Am* 1997;**79**(11):1635–41.
29. D'Antonio JA, Manley MT, Capello WN, Bierbaum B, Ramakrishnan R, Naughton M, Sutton K. Five-year experience with crossfire highly crosslinked polyethylene. *Clin Orthop Rel Res* 2005;**441**:143–50.
30. Rajadhyaksha AD, Brotea C, Cheung Y, Kuhn C, Ramakrishnan R, Zelicof SB. Five-year comparative study of highly cross-linked (Crossfire) and traditional polyethylene. *J Arthroplasty* 2009:**24**(2): 161–74. (In Press).
31. Wannomae KK, Bhattacharyya S, Freiberg A, Estok D, Harris WH, Muratoglu O. In vivo oxidation of retrieved cross-linked ultra-high-molecular-weight polyethylene acetabular components with residual free radicals. *J Arthroplasty* 2006;**21**(7):1005–11.
32. Currier BH, Currier JH, Mayor MB, Lyford KA, Collier JP, Van Critters DW. Evaluation of oxidation and fatigue damage of retrieved Crossfire polyethylene acetabular cups. *J Bone Joint Surg Am* 2007;**89**(9):2023–9.
33. Kurtz SM, Hozack WJ, Turner J, Purtill J, MacDonald D, Sharkey P, Parvizi J, Manley MT, Rothman RH. Mechanical properties of retrieved highly crosslinked (Crossfire) liners after short-term implantation. *J Arthroplasty* 2005;**20**(7):840–9.
34. Kurtz SM, Hozack WJ, Purtill JJ, Marcolongo M, Kraay MJ, Goldberg VM, Sharkey PF, Parvizi J, Rimnac CM, Edidin AA. 2006 Otto Aufranc Award Paper: significance of in vivo degradation for polyethylene in total hip arthroplasty. *Clin Orthop Relat Res* 2006;**453**:47–57.
35. Kurtz SM, Austin M, Azzam K, Sharkey P, MacDonald D, Medel FJ, Hozack WJ. Mechanical properties, oxidation and clinical performance of retrieved highly cross-linked crossfire liners after intermediate-term implantation. *J Arthroplasty* 2008. (In Press).
36. Kester MA, Herrera L, Wang A, Essner A. Knee bearing technology. Where is technology taking us?. *J Arthroplasty* 2007;**22**(Suppl 3):16–20.
37. Wang A, Yau SS, Essner A, Herrera L, Manley M, Dumbleton JH. A highly crosslinked UHMWPE for CR and PS total knee arthroplasties. *J Arthroplasty* 2008;**23**(4):559–66.
38. Wang A, Zeng H, Yau SS, Essner A, Manley M, Dumbleton JH. Wear, oxidation, and mechanical properties of a sequentially irradiated and

annealed UHMWPE in total joint replacement. *J Phys D Appl Phys* 2006;**39**:3213–19.
39. Dumbleton JH, D'Antonio JA, Manley MT, Capello WN, Wang A. The basis for a second generation highly crosslinked UHMWPE. *Clin Orthop Relat Res* 2006;**453**:265–71.
40. Herrera L, Lee R, Longaray J, Essner A, Wang A. Hip simulator evaluation of the effect of femoral head size on sequentially crosslinked acetabular liners. *Wear* 2007;**263**:1034–7.
41. Tsukamoto R, Williams PA, Shoji H, Hirakawa K, Yamamoto K, Tsukamoto M, Clarke IC. Wear of sequentially enhanced 9-Mrad polyethylene in 10 million cycle knee simulation study. *J Biomed Mater Res B Appl Biomater* 2008;**86**(1):119–24.
42. Williams P, Tsukamoto R, Clarke I. Wear debris from sequentially crosslinked and crosslinked PE in a knee simulator model. *Paper #218*. San Francisco, CA; 2008.
43. Hermida JC, Fischler A, Bhimji S, Colwell CWJ, D'Lima DD. Effect of component malalignment and aging on sequentially crosslinked polyethylene knee arthroplasty inserts. *Poster #1874*. San Diego, CA: 2007.
44. Kurtz SM, MacDonald D, Brenner E, Medel FJ, Hozack WJ, Purtill JJ, Parvisi J, Austin M, Goldberg VM, Kraay MJ, et al. In vivo oxidation, oxidation potential, and clinical performance of first and second-generation highly crosslinked acetbular bearings for THA. *Poster 1790*. San Francisco, CA; 2008.

Chapter 15

# Highly Crosslinked UHMWPE Doped with Vitamin E

Ebru Oral, PhD and Orhun K. Muratoglu, PhD



## 15.1 INTRODUCTION

Osteolysis as a clinical problem has been the driving force behind the development of crosslinked UHMWPEs (see Chapters 6, 13, and 14). As explained earlier in Chapter 13, postirradiation melting renders crosslinked UHMWPE oxidation resistant by allowing the residual free radicals trapped in crystalline regions to recombine. However, the postirradiation melting step reduces the mechanical properties and fatigue strength of irradiated UHMWPE, which are already decreased by crosslinking due to a decrease in crystallinity that accompanies postirradiation melting.

The development of a second-generation, highly crosslinked UHMWPE was aimed at eliminating the shortcomings of the first-generation, irradiated, and thermally treated UHMWPEs. The clinical effects of these changes in the second-generation material are not yet clear due to the extensive period and large patient populations required to determine clinical success and compare new materials to previously used ones. On the other hand, detailed *in vitro* tests have been developed over the years by researchers in our laboratory and others to predict the causes of *in vivo* damage and failure to the best of our ability. From our perspective, the goal of UHMWPE development for joint arthroplasty is to provide a bearing surface that has enough wear, oxidation, and fatigue resistance such that when it is implanted in a patient, the failure modality will not be the bearing surface material, independent of the surgeon's skill, implant geometry, patient's age, activity level, and weight. Thus, we started working on alternative methods that can simultaneously maximize wear, oxidation, and fatigue resistance, which have surfaced as essential factors in ensuring the longevity of joint implants.

The rationale behind stabilizing the residual free radicals in highly crosslinked UHMWPE with the antioxidant vitamin E (α-tocopherol, Figure 15.1) [1, 2] was to provide oxidation and wear resistance to UHMWPE without

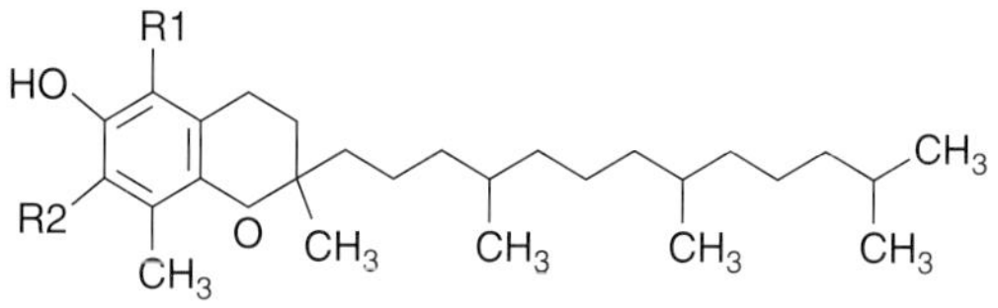

| | R1 | R2 |
|---|---|---|
| α-tocopherol | $CH_3$ | $CH_3$ |
| β-tocopherol | $CH_3$ | H |
| γ-tocopherol | H | $CH_3$ |
| δ-tocopherol | H | H |

**FIGURE 15.1** Tocopherols. Image courtesy of Eric Wysocki, Exponent, Inc.

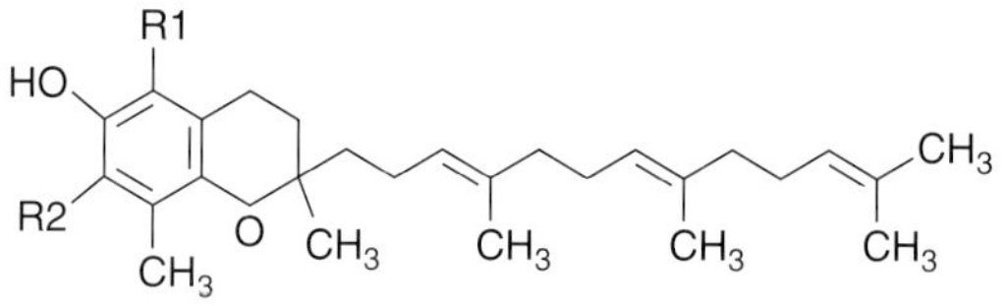

| | R1 | R2 |
|---|---|---|
| α-tocotrienol | $CH_3$ | $CH_3$ |
| β-tocotrienol | $CH_3$ | H |
| γ-tocotrienol | H | $CH_3$ |
| δ-tocotrienol | H | H |

**FIGURE 15.2** Tocotrienols. Image courtesy of Eric Wysocki, Exponent, Inc.

postirradiation melting. Thus, the loss of crystallinity caused specifically by postirradiation melting would be avoided and fatigue strength could be improved. Fatigue strength became a focal point regarding the mostly philosophical concerns, at the time, that the decrease in the *in vitro* fatigue strength of highly crosslinked UHMWPEs would cause premature failure of implants under adverse conditions, especially for use in the knee.

There are two methods of incorporating vitamin E into UHMWPE. One is to blend vitamin E with UHMWPE powder prior to consolidation. When consolidated, the blend can be crosslinked with the use of ionizing radiation. However, the presence of vitamin E in UHMWPE during irradiation reduces the efficiency of crosslinking [3–5]. The properties of vitamin-E-blended UHMWPE are covered in Chapter 16. An alternative method is the diffusion of vitamin E into UHMWPE following radiation crosslinking [1, 6]. The crosslinking efficiency of UHMWPE is not adversely affected in this method because vitamin E is not present during irradiation.

In this chapter, we will focus on the diffusion of vitamin E in radiation crosslinked UHMWPE and the properties of this new generation UHMWPE. It is not yet clear by how much the longevity of the implants will be improved clinically and how much the fatigue strength improvement will allow the use of the new material in younger and more active patients; however, our work continually focuses on the manipulation of the chemistry and semicrystalline morphology of UHMWPE to achieve the goal of obtaining a failure-resistant bearing surface.

## 15.2 FUNCTION OF VITAMIN E

Vitamin E is the most abundant and effective chain-breaking antioxidant present in the human body [7]. Its main function in this regard is to protect cell membranes from oxidative damage. Vitamin E consists of tocopherols and tocotrienols (Figures 15.1 and 15.2), but we use the term "vitamin E" interchangeably in this chapter with "α-tocopherol" because RRR-α-tocopherol has the greatest biological activity and constitutes about 90% of the vitamin in tissues [8]. Moreover, the studies discussed in this chapter have mostly used synthetic vitamin E (D,L-α-tocopherol), which is also often used in dietary supplements along with its acetate or succinate esters. While we discuss the use of vitamin E as a stabilizer for UHMWPE for total joint implant components, the data on the biocompatibility and efficacy of vitamin E is mostly based on oral supplementation. We will discuss the outcome of these studies later in the chapter in the context of the possible local and systemic effects of an intra-articular dose of vitamin E.

The major physiological role of vitamin E is to react with free radicals in cell membranes and protect polyunsaturated fatty acids from degradation due to oxidation [9]. Oxidation of polyunsaturated fatty acids results in active free radicals (LOO•, LO•). The antioxidant activity of α-tocopherol is due to hydrogen donation from the OH group on the chroman ring to a free radical on the oxidized lipid chain. Hydrogen abstraction results in a tocopheryl free radical, which can combine with another free radical. Therefore, tocopherol can theoretically prevent two peroxy free radicals from attacking other fatty acid chains and producing more free radicals [10–12].

α-Tocopherol is an extremely potent antioxidant [13]. Membranes in the body typically contain about one α-tocopherol molecule per 2000 phospholipids to prevent oxidative damage [9]. The reason this works is that the reaction rate between the antioxidant and hydroperoxyl radicals is 10,000 times faster than the reaction between the hydroperoxyl and the hydrocarbon chains [13]. Therefore, the cascading nature of oxidation is stopped, and oxidative damage can be prevented.

Oxidation reactions in polyethylene, which has a lipid-like molecular structure, also follow a similar mechanism of oxidation as lipids do *in vivo*[14–16]. In irradiated UHMWPE, the carbon free radicals resulting from the breakage of the C-H bonds are prevalent [17], and these radicals are alkyl, allyl, and polyenyl free radicals [18], the latter formed at higher radiation doses [19]. Most of these free radicals recombine in the amorphous portion of the polymer [20], with the remaining free radicals trapped in the highly ordered crystalline lamellae [21, 22]. Because the interchain distance is fixed at 4.1 Å in the crystals and the C-C bond distance is 1.52 Å [23], interchain or intrachain crosslinking is highly unlikely. Therefore, free radicals presumably travel along the crystal chains and encounter other free radicals with which to react or encounter the crystalline/amorphous interface. Thus, the decay of free radicals in the crystalline phase is extremely slow [24].

When oxygen is present in irradiated polyethylene, it reacts with the primary free radicals to form peroxy free radicals [25–28]. These peroxy radicals, in the absence of an antioxidant like vitamin E, abstract a hydrogen atom from other polyethylene chains, creating new primary free radicals, which can then react with oxygen to further this chain of reactions [15, 29]. When peroxy free radicals react with hydrogen, they form hydroperoxides, which are not stable and degrade into oxidation products, mainly ketones, esters, and acids [30–33]. The formation of these oxidation products is accompanied by chain scission and a decrease in the molecular weight of polyethylene, deteriorating its mechanical properties [34–36]. In an irradiated polyethylene containing vitamin E, peroxy free radicals presumably abstract a hydrogen atom from vitamin E, forming hydroperoxides without the formation of new free radicals. Vitamin E is also effective in reacting with alkyl radicals [37], but the reaction rate with peroxy free radicals is higher than any other radical [10, 12]. The oxidation cascade in irradiated polyethylene is therefore hindered in the presence of vitamin E. Additional details regarding the radiation chemistry of UHMWPE and the chemical mechanisms of vitamin E stabilization of UHMWPE can be found in Chapter 21.

## 15.3 DIFFUSION OF VITAMIN E IN CROSSLINKED UHMWPE

The main goal of diffusing vitamin E into radiation crosslinked UHMWPE is to obtain enough vitamin E throughout joint implant components to protect the material against long-term oxidation. On the application of this concept, there are several factors to be considered. It is important not to melt the sample during the diffusion process and thereby to avoid the loss of crystallinity. As discussed earlier, postirradiation melting reduces the crystallinity of irradiated UHMWPE, but it is also possible to exacerbate this phenomenon due to a large amount of vitamin E diffusion into newly melted regions, preventing recrystallization. Because the melting range of UHMWPE is wide, starting at 100°C with a peak melting point of about 140°C, the diffusion temperature has to be carefully determined so as not to lose a significant amount of crystallinity. Below the melting point of the crystals, diffusion is limited to the amorphous regions of the polymer because the crystals are impermeable to a molecule as large as vitamin E.

At the same time, increased temperature increases the surface concentration and the saturation concentration of vitamin E at the surface of UHMWPE. Dimensional changes can occur both due to the release of residual stresses from previous processing steps and due to the swelling of the surface with vitamin E.

To circumvent these issues, a two-step diffusion process at elevated temperatures below the melting point was developed involving doping of UHMWPE by soaking in vitamin E with subsequent homogenization at an elevated temperature (Figure 15.3) [6]. The diffusion of vitamin E in radiation crosslinked UHMWPE using doping and subsequent homogenization followed a Fickian model [6], which was used to predict the duration of these steps to obtain a desired vitamin E concentration profile throughout a component of given thickness (Figure 15.3).

The diffusion coefficient of vitamin E in crosslinked UHMWPE is expected to depend upon the amorphous and crystalline morphology of UHMWPE. Below the melting point, crystallinity is not affected significantly as a function of radiation dose, and crosslink density is increased. Surprisingly, there were small differences between the surface vitamin E concentration, penetration depth, and overall amount of diffused vitamin E for 65- and 100-kGy irradiated UHMWPEs doped below the melting point, despite a nearly 40% difference in crosslink density (132 and 182 mol/m$^3$, respectively). All of these outcomes were substantially higher for unirradiated and uncrosslinked UHMWPE, suggesting that crosslinking to 65 kGy does affect diffusion, but the crosslink density differences between 65- and 100-kGy irradiation does not affect diffusion significantly. Also, above the melting point, the diffusion coefficients of vitamin E in uncrosslinked and crosslinked UHMWPEs were similar, showing that the crystalline regions of UHMWPE are a major hindrance to the diffusion of vitamin E in UHMWPE, much more so than crosslinking.

Although there was a significant increase in the amount of diffused vitamin E with increasing temperature, the largest differential increase occurred from 120°C to 130°C, as the temperature approached the peak melting temperature. At 130°C, 24 hours of doping with vitamin E resulted in a 10–15% volumetric change of the irradiated UHMWPE. As previously mentioned, it is preferable to obtain vitamin E throughout UHMWPE joint implants without morphological changes associated with high surface concentrations of vitamin E and without significant distortion of the components due to exposure to high temperature and large amounts of vitamin E. To achieve this goal, the doping step was restricted to 120°C because there were significant volumetric and weight changes above this temperature. When the high surface concentration of vitamin E has been obtained, the homogenization of the surface vitamin E throughout the irradiated component can then be achieved at a higher temperature below the melting point because there is no additional diffusion of vitamin E into the

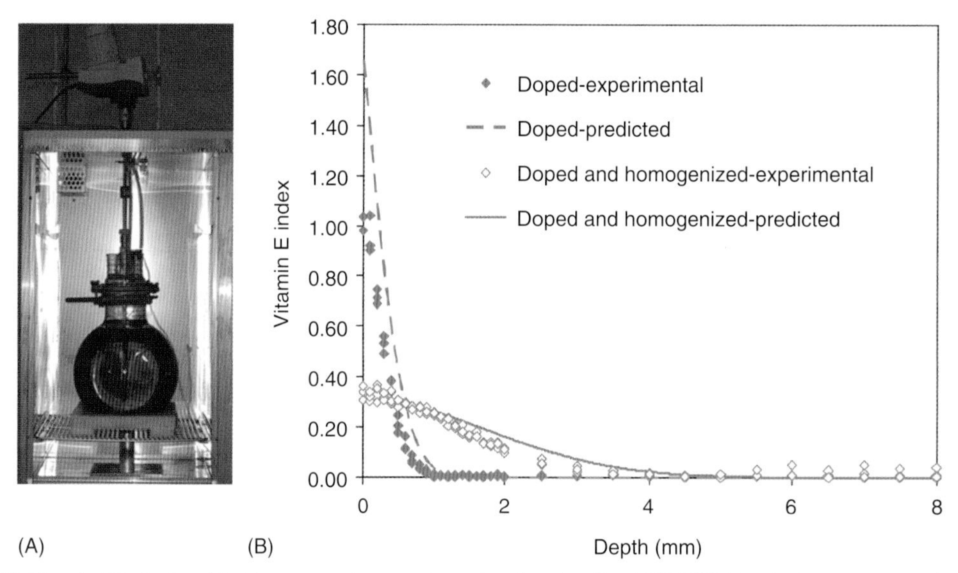


**FIGURE 15.3** (A) Vitamin E bath placed inside a convection oven is used to dope irradiated UHMWPE under inert gas purge; (B) Doping creates a high surface concentration, which is then diffused through the bulk of the sample by subsequent homogenization. A Fickian model can predict doped and homogenized profiles well.

component. The penetration depth is similar in these two methods, namely simple doping versus doping followed by homogenization, but the surface concentration and the diffused vitamin E amount for the doped and homogenized UHMWPE are significantly lower.

By using the model and method previously described, desired concentration profiles can be achieved through radiation crosslinked UHMWPE joint implant components. In the following sections, the wear resistance, mechanical properties, and oxidative stability of vitamin-E-diffused, radiation crosslinked UHMWPE prepared by using the diffusion technique previously discussed will be reported.

Vitamin E diffusion in radiation crosslinked UHMWPE can be performed in air, in inert gas, or in a supercritical fluid medium. Diffusion in air is not preferable due to long exposure times of the components and the vitamin E to air. After contact with vitamin E, the surface of the components are protected against oxidation, and it is desirable to limit the amount of exposure time of the bulk of the material to air until vitamin E diffusion throughout the components is complete. Therefore, inert gas is the environment of choice for the diffusion and homogenization of vitamin E in irradiated UHMWPE. Supercritical carbon dioxide has also been used as a nonreactive medium for doping and homogenization [38, 39]. It is a promising medium also due to its potential to swell UHMWPE and hence enhance the diffusion rate of vitamin E in UHMWPE [38].

## 15.4 WEAR

The premise of vitamin-E-diffused, radiation crosslinked UHMWPE is to improve mechanical and fatigue properties of crosslinked UHMWPEs by avoiding postirradiation melting. At the same time, the wear resistance and oxidative stability should not be sacrificed. In this section, we discuss studies in which the wear resistance of UHMWPE was determined *in vitro* under normal and adverse conditions that are clinically relevant to hip and knee joint replacement bearing surfaces.

### 15.4.1 Hip

While vitamin E interferes with radiation crosslinking of UHMWPE during irradiation, vitamin E diffusion into already crosslinked UHMWPE largely circumvents this issue. A crosslink density equivalent to that of 100-kGy irradiated and melted UHMWPE is desired because this material has been shown *in vitro* and *in vivo* to decrease wear significantly compared to conventional UHMWPE [40, 41]. Therefore, we performed the studies proving the concept that vitamin E prevented oxidation in radiation crosslinked UHMWPE with an initial radiation dose of 85 to 100 kGy with a terminal gamma sterilization dose (~25–40 kGy) after vitamin E incorporation [1, 2].

Wear testing of vitamin-E-diffused, 100-kGy irradiated UHMWPE on a custom-designed pin-on-disc tester in bovine serum in our laboratory showed that the wear rate was comparable to 100-kGy irradiated and melted UHMWPE [1]. In this study, accelerated aging at 80°C for 5 weeks did not change the wear rate of this UHMWPE despite the fact that vitamin E was diffused only 0.5 mm into the surface of this sample and not throughout the entire thickness. This suggested that the wear region was within this surface layer protected by the vitamin E.

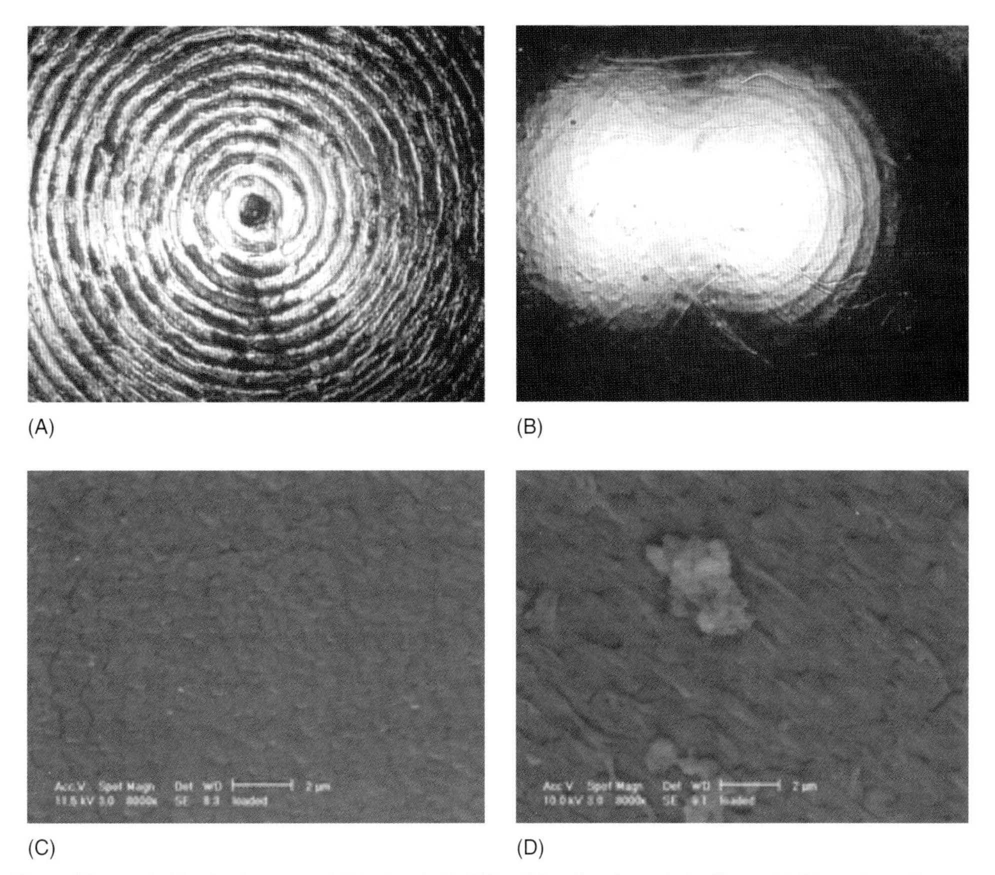

**FIGURE 15.4** Five-million-cycle hip simulator tested (A) vitamin-E-diffused, irradiated acetabular liner with 36mm inner diameter; and (B) conventional liner with 28mm inner diameter; scanning electron microscopy images of the wear surface of (C) vitamin-E-diffused, irradiated acetabular liner with 36mm inner diameter and (D) conventional acetabular liner with 28mm inner diameter after 3 million cycles of hip simulator testing in bovine serum with third body particulate.

After the development of the diffusion method by doping followed by homogenization below the melting point, 5-mm thick acetabular liners were irradiated by 85-kGy gamma irradiation, doped with vitamin E, subsequently homogenized to obtain vitamin E throughout, and terminally gamma sterilized. These acetabular liners (28- and 40-mm inner diameter) were wear tested on a hip simulator against CoCr femoral heads using normal gait load and kinematics at 2Hz in undiluted clean bovine serum for 5 million cycles and compared to conventional UHMWPE gamma sterilized in inert gas [2]. This study showed *in vitro* that under normal conditions, the wear reduction in vitamin-E-diffused, highly crosslinked UHMWPE compared to conventional UHMWPE (92% and 90% reduction in wear for 28- and 40-mm cups, respectively, compared to 28-mm conventional) was comparable to what was observed previously with irradiated and melted UHMWPE compared to conventional UHMWPE [40, 42]. Similar results were observed when these liners were tested in undiluted bovine serum with third body PMMA and barium sulfate particulate (72% and 75% reduction in wear for 28- and 40-mm cups, respectively, compared to 28-mm conventional). The results obtained from these hip simulator studies with vitamin-E-diffused, irradiated highly crosslinked UHMWPE confirmed earlier studies that showed the wear rate of highly crosslinked UHMWPEs were largely independent of femoral head size [42].

The fibrils observed on conventional polyethylene liners under the scanning electron microscope following wear testing have been proposed as precursors for particulate wear debris [43, 44]. The propensity of conventional UHMWPE to form fibrils due to its high plasticity is reflected in its high elongation at break, around 350–400% [45]. The fact that these fibrils were rare on the articular surfaces of vitamin-E-diffused, irradiated UHMWPE acetabular liners indicated that the ability of this polymer to undergo large-scale deformation was reduced by crosslinking, as expected, and not detrimentally affected by vitamin E (Figure 15.4).

Wear debris of conventional UHMWPE has been clearly associated with osteolysis and loosening of implants [46, 47]. Currently, clinical studies up to 5 years have corroborated the *in vitro* simulator testing results of significantly decreased wear in highly crosslinked UHMWPEs compared to conventional UHMWPE [41, 48, 49]. Currently, there is no conclusion on the differences in the biological activity of the wear debris from highly crosslinked

UHMWPE compared to conventional UHMWPE. While the average particle size from highly crosslinked UHMWPEs appears to be smaller than that of conventional UHMWPE, the number of particles is also significantly lower. In addition, conventional UHMWPE has been associated with high oxidation not observed in irradiated and melted UHMWPEs, therefore the activity of the particles may not be the same. To the authors' knowledge, there are currently no published studies comparing the activity of the wear particles from vitamin-E-diffused, irradiated UHMWPE to the prior materials. There is, however, one *in vitro* study that indicates that vitamin-E-containing uncrosslinked UHMWPE increased matrix metalloproteinase 9 (MMP-9) secretion from granulocytes [50]. This may mean that vitamin E has a direct role in regulating immune response when associated with UHMWPE particles.

### 15.4.2 Knee

The use of vitamin-E-stabilized, radiation crosslinked UHMWPE is especially important for use in tibial bearings due to the improved fatigue resistance of this UHMWPE compared to irradiated and melted UHMWPEs while maintaining oxidation resistance. In a knee simulator study, unaged cruciate retaining vitamin-E-stabilized UHMWPE tibial bearings were tested using normal gait load and kinematics for 5 million cycles, then they were accelerated aged and tested for further 3 million cycles. In this study, it was shown that the wear rate of vitamin-E-diffused, irradiated UHMWPE remained the same after the components were aged, while that of gamma sterilized UHMWPE increased significantly[51]. In this study, the wear rate of vitamin-E-diffused, radiation crosslinked UHMWPE was $2.4 \pm 0.5$ mg/million cycles and $2.5 \pm 0.8$ mg/million cycles, respectively, for the unaged 5 million cycles and the subsequent aged 3 million cycles, representing a 91% and a 94% decrease in wear compared to conventional UHMWPE.

In contrast to the conforming hip joint, the motion of the knee joint is in three orthogonal directions with the joint experiencing antero-posterior translation, flexion-extension, and rotation. As a result, in addition to pitting, burnishing, and adhesive/abrasive wear [52], there are compressive forces combined with substantial sliding and shear on the polyethylene surface, making fatigue delamination a major wear mechanism. In a study designed to determine delamination resistance by using unidirectional loading of tibial components [53], we showed that accelerated aged vitamin-E-diffused, irradiated UHMWPE was resistant to delamination in contrast to accelerated aged conventional UHMWPE, which showed severe delamination largely due to severe oxidation and subsurface embrittlement (Figure 15.5).

There is also some evidence that vitamin E protects UHMWPE from defects at grain boundaries, also increasing delamination resistance [54]. This phenomenon is yet to be shown for vitamin-E-diffused UHMWPE after radiation crosslinking, but it could also be a contributing factor in its wear resistance. In summary, vitamin E incorporation after radiation appears not to detrimentally affect the wear resistance of highly crosslinked UHMWPE.

## 15.5 MECHANICAL AND FATIGUE PROPERTIES

The mechanical and fatigue strength of UHMWPE is a strong function of crystallinity and crosslink density. There is a decrease in mechanical and fatigue strength of irradiated and melted UHMWPEs, caused both by crosslinking and the loss of crystallinity during melting, which quenches the residual free radicals [55]. The loss of crystallinity during postirradiation melting was avoided by doping irradiated UHMWPE with vitamin E [2, 45]. Therefore, the ultimate tensile strength, yield strength, and fatigue crack propagation resistance were increased as well. The value for the ultimate tensile strength of 100 kGy irradiated and melted UHMWPE reported in the literature is approximately 35 MPa [45]; the ultimate tensile strength of vitamin-E-diffused, irradiated UHMWPE was measured to be 46 MPa [2].

There are concerns about the lowered fatigue strength of highly crosslinked UHMWPE compared to unaged conventional UHMWPE, especially due to the high stress environment of the knee (Chapter 30). The fatigue strength of surgical-grade UHMWPE is typically quantified by determining the resistance to fatigue crack propagation by cyclically loading a specimen designed to concentrate stresses at a crack tip [56]. The stress range that needs to be applied to propagate the crack at a rate of $10^{-6}$ mm/cycle is reported as a measure of the resistance to crack propagation. The values for unirradiated or gamma-sterilized UHMWPE, which has high fatigue resistance in its unaged state, have been reported as 1.4–2.0 MPa.m$^{1/2}$ [1, 57]. In contrast, irradiated and melted highly crosslinked UHMWPEs had values of 0.55–0.69 MPa.m$^{1/2}$ depending on the radiation dose [1, 57]. The further variation of these values were a result of changes in testing environment (air or liquid), testing temperature, and testing frequency. The fatigue resistance of vitamin-E-diffused, 100-kGy irradiated UHMWPE was reported to be 0.70–0.77 MPa.m$^{1/2}$ [1, 2]. Also importantly, the strength of vitamin-E-diffused, irradiated UHMWPE remained unchanged when exposed to accelerated aging, while that of conventional, gamma sterilized UHMWPE deteriorated significantly [2, 45]. It is crucial to note that while conventional UHMWPE has high fatigue strength in its unaged form before exposure to oxygen, its fatigue resistance is severely deteriorated due to oxidation, to as low as 0.19 MPa.m$^{1/2}$[45]. This is

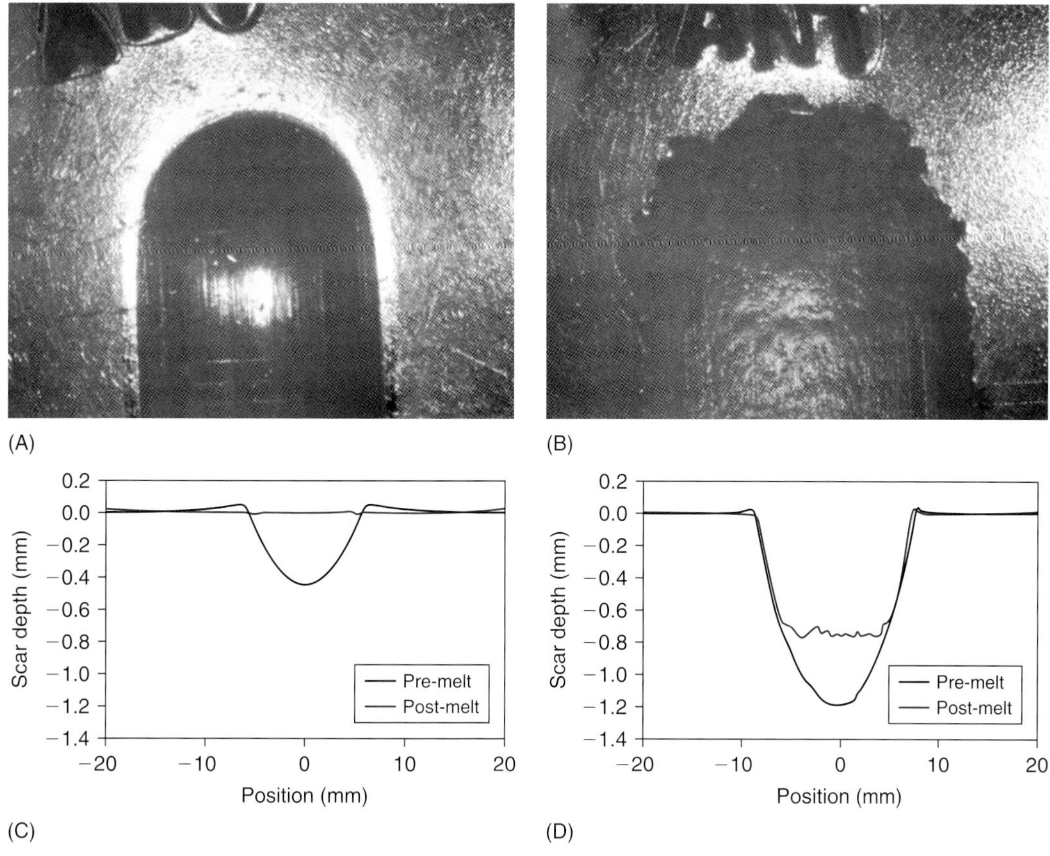


**FIGURE 15.5** The wear scars of (A) vitamin-E-diffused, irradiated UHMWPE and (B) conventional gamma-sterilized (inert) UHMWPE after 2 million cycles of delamination testing. While the deformation was due mostly to creep and was recoverable after melting for vitamin-E-diffused, irradiated UHMWPE (C), there was severe delamination and wear in conventional UHMWPE (D).

undoubtedly the situation *in vivo* as some explanted components can literally crumble upon contact.

It is also important to point out possible effects of vitamin E on UHMWPE without crosslinking. To the authors' knowledge, there have been two studies that looked at the effect of vitamin E addition to UHMWPE without the confounding factor of irradiation [58, 59]. The first determined the effect of vitamin E concentration in uncrosslinked UHMWPE up to 127 wt% on the tensile mechanical properties, while the latter determined the wear properties of a 0.3 wt% vitamin-E-blended UHMWPE using a knee simulator. It was shown that vitamin E did not have a significant effect on the tensile mechanical properties of UHMWPE when vitamin E was blended with UHMWPE or introduced below the melting point of UHMWPE by diffusion, even at very high concentrations. In contrast, the knee simulator study curiously showed that the wear rate of vitamin-E-blended UHMWPE was less than unirradiated UHMWPE. The reason for this remains unclear.

In summary, the fatigue strength of vitamin-E-diffused, irradiated UHMWPE was improved compared to irradiated and melted UHMWPE, and it remained unchanged after accelerated aging, showing improvement compared to oxidation-prone conventional UHMWPE.

## 15.5.1 Adverse Conditions

There are concerns about the ability of highly crosslinked UHMWPEs to withstand especially two clinically adverse situations. One of these is the femoral neck impingement on the rim of an acetabular liner, most often due to the vertical placement of the shell with excessive anteversion [60]. It is likely that gamma-sterilized conventional UHMWPE liners fail under the same circumstances, especially if they have already oxidized to an appreciable degree. Nevertheless, there is increased concern for highly crosslinked UHMWPEs due to their initially lower fatigue strength than unaged conventional UHMWPE and the fact that larger head sizes with thinner polyethylene can now be used, thanks to the independence of the wear rate of highly crosslinked UHMWPEs of femoral head size. In a clinically relevant device fatigue test involving the impingement of the rim of thin polyethylene liners by the neck of the femoral components, simulating the adverse case of vertical malpositioning of the implant, 3.7 mm-thick vitamin-E-diffused, crosslinked acetabular liners with 28- and 40-mm inner diameter were loaded on the rim by the femoral neck for 2 million cycles. There were no fractures in any of the tested liners, including the conventional UHMWPE liners with 28-mm inner diameter used as controls. No apparent differences between unaged conventional UHMWPE and vitamin-E-diffused, highly crosslinked UHMWPE were observed [2].

Another case of concern is the failure of highly crosslinked UHMWPEs due to impingement on the post of a posterior-stabilized tibial knee insert by the cam of the femoral component, either due to component malalignment [61] or by intended design. In a custom-designed test to determine the comparative bending fatigue strength of UHMWPEs under similar conditions, the post of a bending fatigue specimen was subjected to cyclic plastic deformation until failure at peak loads of 400 to 800N [62]. Using statistical analysis, the life expectancy of the components was similar for unaged conventional and vitamin-E-doped, irradiated UHMWPE, but the life expectancy of aged vitamin-E-diffused, irradiated UHMWPE was significantly higher than aged conventional UHMWPE at each applied load. An acceleration factor was calculated as the ratio of the minimum number of cycles at which 90% survivorship is expected for different groups. Over the load range where the analysis was performed, the average acceleration factor for vitamin-E-diffused, irradiated UHMWPE over conventional UHMWPE was 0.92 in their unaged state and 26.4 after aging. This suggested that radiation crosslinking was not a major factor affecting the fatigue strength under these circumstances and that life expectancy was significantly extended for vitamin-E-diffused, irradiated UHMWPE compared to aged conventional UHMWPE (roughly 28 times higher).

In summary, under *in vitro* simulated clinically adverse conditions, vitamin-E-diffused, irradiated UHMWPE behaved similar to unaged conventional UHMWPE and was superior to aged conventional UHMWPE.

## 15.6 OXIDATIVE STABILITY

As mentioned briefly, it is not only important for a joint bearing surface to have the desired properties in its unused form, but it is crucial for these initial properties to be maintained over the expected lifetime of the implant. The major factor limiting the maintenance of the wear and mechanical properties of irradiated UHMWPEs *in vivo* is oxidation (Chapter 22). It has been shown in several types of accelerated aging studies, carried out at elevated temperatures and/or in the presence of pure oxygen under high pressure, that vitamin-E-containing, irradiated UHMWPE is more stable than gamma-sterilized or high-dose irradiated UHMWPE [1, 45]. It is proposed that this is due to the reaction of vitamin E with the primary free radicals on the polyethylene chains and also free radicals resulting from their reaction with oxygen.

While accelerated aging is helpful in comparing the oxidation resistance and oxidation potential of different types of bearing materials, it cannot be used to predict the oxidation timeline or profile of a particular material *in vivo*. The mechanism of oxidation may be different *in vivo* than it is under accelerated aging conditions. Therefore, it is crucial to determine the oxidative stability under more clinically relevant conditions to simulate more closely the aging behavior of this material on the shelf and *in vivo*.

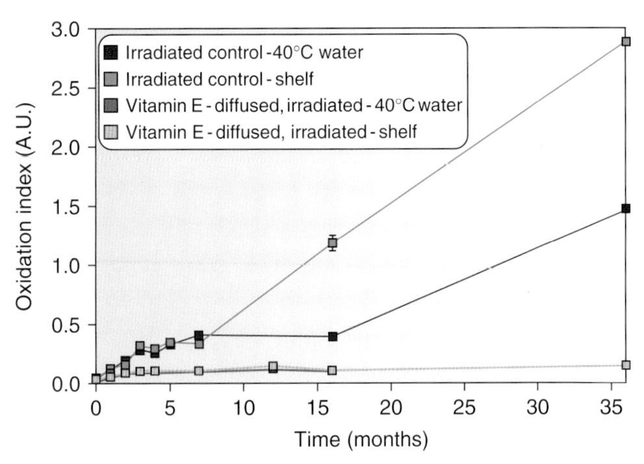


**FIGURE 15.6** Average surface oxidation indices of 100-kGy irradiated control UHMWPE and vitamin-E-diffused, irradiated UHMWPE after real-time aging at room temperature on the shelf and at 40°C in water after 36 months. Image courtesy of Eric Wysocki, Exponent, Inc.

In a study looking at the stabilizing effect of vitamin E on the residual free radicals of irradiated UHMWPE during real-time aging, vitamin-E-diffused, irradiated UHMWPE showed higher oxidative stability than unstabilized, irradiated UHMWPE over 7 months in air at room temperature and in water at 40°C (showing similar trends at 36 months, Figure 15.6). This was accompanied by a faster transformation of the allyl/alkyl free radicals to oxygen-centered free radicals in the former, the reason for which is still under investigation. There are measured free radicals in vitamin-E-diffused, irradiated UHMWPE ($3.5 \times 10^{16}$ spins/g compared to $11.7 \times 10^{16}$ spins/g for 100-kGy irradiated, unstabilized UHMWPE), but the chain breaking antioxidant behavior of vitamin E was evidenced by the early cessation of oxidation in the vitamin-E-diffused, irradiated samples compared to irradiated, unstabilized UHMWPE.

Oxidation is determined by spectroscopic techniques measuring carbonyl moeties on UHMWPE formed as a result of the decay of hydroperoxides. As discussed previously, these hydroperoxides are formed as a result of the first reaction of oxygen with the free radicals on polyethylene and the subsequent hydrogenation of these peroxy free radicals by attacking unreacted polyethylene chains. One mechanism of hydroperoxide production is the reaction of free radicals with the oxygen dissolved in UHMWPE during the initial radiation crosslinking prior to vitamin E incorporation. The other mechanism is the hydrogen abstraction from vitamin E by the peroxy free radicals formed on polyethylene. The fact that there was

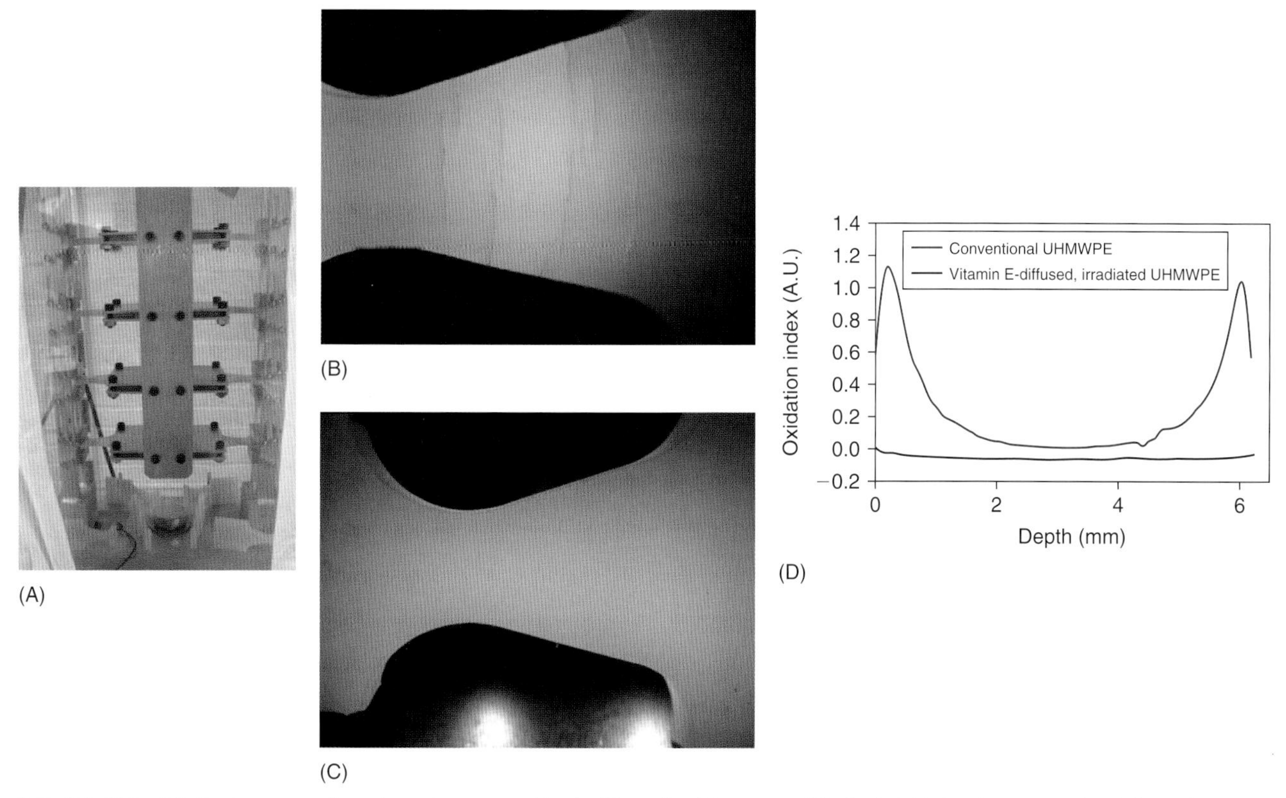


**FIGURE 15.7** (A) Testing setup of accelerated aging combined with cyclic deformation; (B) Crack initiation in a conventional flexural bending sample tested in accelerated aging conditions (80°C, air) in combination with cyclic deformation at an initial stress of 10 MPa; (C) The neck region of a vitamin-E-stabilized, irradiated UHMWPE flexural bending sample tested under the same conditions for 5 weeks showing no crack initiation; (D) Average oxidation profiles of conventional and vitamin-E-diffused, irradiated UHMWPEs that went through accelerated aging in combination with cyclic deformation for 5 weeks.

no further detectable oxidation after 3 months of aging in the vitamin-E-diffused, irradiated UHMWPE (Figure 15.6) indicated that the decay of the hydroperoxides by both mechanisms were likely exhausted and that vitamin E acted as an efficient chain-breaking antioxidant.

Current accelerated aging methods use elevated temperatures to cause an increase in the rate of oxidation in irradiated UHMWPE [63]. Another likely mechanism contributing to *in vivo* oxidation is deformation, which is likely to place the tie-chains between crystallites and the highly immobile crosslinked regions of the polymer under stress. These local stresses can cause chain scission accompanied by new free radicals that can react with oxygen. In a study developed to determine the effect of high-temperature, accelerated aging conditions with deformation [64], flexural fatigue samples were cyclically loaded to generate a maximum stress of 10 MPa in the neck region at 0.5 Hz at 80°C in air (Figure 15.7a), defined as the visible appearance of cracks in the surface of the neck region (Figure 15.7b) or fracture, whichever occurred first. Control materials were aged in the same chamber without loading. The test showed, first, that vitamin-E-diffused, irradiated UHMWPEs did not oxidize with or without loading, and second, that it did not fail (Figures 15.7c and 15.7d) when tested for 5 weeks, in contrast to conventional UHMWPE and sequentially irradiated and annealed UHMWPE, which both failed within 5 weeks. Moreover, the subsurface oxidation peaks in these two oxidation-prone materials were substantially higher than in the unloaded condition (not shown), corroborating the hypothesis that deformation increased oxidation.

In summary, vitamin-E-diffused, irradiated UHMWPE showed high oxidation resistance *in vitro* under accelerated aging and under real-time aging up to 36 months. Aggressive oxidation caused by accelerated aging combined with deformation was also hindered in this material at 5 weeks. Based on this information, it is likely that this material will retain its mechanical properties *in vivo* as well.

## 15.7 BIOCOMPATIBILITY

Vitamin E is found in foods such as vegetable oils, nuts, seeds, whole grains, leafy green vegetables, and avocados [65]. The daily recommended oral intake of vitamin E is generally 15–30 IU, and more specifically approximately 10 mg for males and 8 mg for females [66]. An IU, or International Unit, is a measurement of the biological activity of a given vitamin or drug. In this case, 1 IU is equivalent to the activity of 1 mg of synthetic α-tocopherol acetate. For synthetic α-tocopherol, 1.1 IU is equivalent to 1 mg. The vitamin E used for the stabilization of radiation crosslinked UHMWPEs in the diffused UHMWPEs discussed in this chapter was produced synthetically and was certified to be at least 98% pure α-tocopherol. Natural vitamin E sold for human consumption is required to be at least 96% pure; the impurities in the product are identified as "plant waxes" and are generally accepted to be β-, γ-, and δ-tocopherol. Both synthetic and natural vitamin E are regarded as safe for human consumption by the Food and Drug Administration as additives in prepared food in accordance with current good manufacturing principles [67]. There are no known adverse effects of the consumption of vitamin E naturally occurring in food. Therefore, the evaluation of toxicity of vitamin E is based on data from the intake of vitamin E as a dietary supplement.

Although the local or systemic effects of intra-articular administration of vitamin E are not known, there are numerous studies evaluating the effects of the oral supplementation of this potent antioxidant on the treatment and prevention of various diseases, especially those associated with high oxidative stress. Therefore, the following discussion provides an overview of the effects of vitamin E supplementation on healthy subjects, on patients at risk for cardiovascular disease, and on patients on warfarin anticoagulant therapy. Further discussion on the biocompatibility of vitamin-E-containing UHMWPE can also be found in Chapter 16 (see Section 16.5 and Section 16.6).

### 15.7.1 Vitamin E Toxicity in Clinical Studies with Large Cohorts and Comorbidities

The plethora of studies evaluating the possible benefits and risks of dietary vitamin E supplementation on patients who are suffering from or are at risk of cardiovascular disease is based on the hypothesis that lipid oxidation is active in the mechanisms leading to the formation of arteriosclerotic plaques [68]. It is hypothesized that the antioxidant ability of vitamin E could prevent or reduce lipid peroxidation and hence could slow down the formation of plaques.

Patients on warfarin therapy warrant special consideration because of questions about an increased tendency of bleeding, possibly due to the interference of vitamin E in the action of the vitamin-K-dependent blood clotting factors [69]. It is important to evaluate this risk in healthy populations as well as populations whose vitamin-K-dependent blood clotting activity is already reduced by warfarin anticoagulation.

In a recent meta-analysis of 19 clinical trials comprising 137, 967 subjects with a wide range of comorbidities, Miller et al. [70] have concluded that 400 IU/day and higher doses of oral vitamin E supplementation for a prolonged period of time of at least 1.4 years up to 8.2 years was associated with significant increases in all-cause mortality, and they advised that these doses and durations should be avoided. They also reported that a nonsignificant increase in all-cause mortality was seen starting from a supplementation of 150 IU daily dose for at least a year. Out of the 11 studies with a median dose equal to or above 400 IU/day, two included healthy older adults [71, 72], five included patients with or at risk of heart disease [73–77], one included patients with recent history of large-bowel adenoma [78], one included patients with age-related cataracts [79], one included Alzheimer's disease patients [80], and one included Parkinson disease patients [81]. Two of these studies had patients with more than one indication [76, 77]. It has to be noted that none of these studies included in the meta-analysis showed an adverse outcome in the evaluation of their individual end points, including cardiovascular or all-cause death when reported originally as single studies.

In three other meta-analyses that combined the results of randomized controlled trials for prevention or treatment of cardiovascular disease [82–84], no significant effect of daily vitamin E dietary supplementation up to 800 IU/day (727 mg/day of synthetic α-tocopherol) on cardiovascular deaths was found up to 6.3 years (400 IU/day).

Somewhat in contrast to the previously quoted meta-analysis, in two separate large studies performed on male (39,910) [85] and female (87, 245) [86] health professionals with no prior coronary heart disease, diabetes or hypercholesterolemia, those patients in the quintile receiving the highest dose of vitamin E (above 250 IU/day with a median of 419 IU/day for the men and above 22 IU/day with a median of 208 IU/day for the women) had a significant decrease in the incidence of coronary heart disease compared to the lowest quintile receiving no supplemental vitamin E (median dietary intake 6.4 IU/day and 2.8 IU/day for men and women, respectively). Similarly, Bendich and Machlin [87] reported a review of clinical studies in which vitamin E was supplemented to patients ranging from newborns to elderly in over 15 studies from doses of 250 IU to 2400 IU/day over durations ranging from 28 days to 4.5 years. None of these studies have reported any adverse effects.

In summary, while the data are somewhat conflicting, a recent large meta-analysis reported that vitamin E supplementation with doses equal to and higher than 400 IU/day was associated with an increase in all-cause mortality rates. These effects were associated with prolonged periods of daily oral administration (over 1.4 years) of high doses (≥400 IU/day), while the amount of vitamin E in a single acetabular liner or tibial insert would not exceed approximately 250 mg, corresponding to 275 IU.

### 15.7.2 Toxic Effects of High Dose of Vitamin E in Patients on Warfarin Therapy (Two Cases)

The exact mechanism of the interference of α-tocopherol with anticoagulant therapy is not known; however, it may involve the blocking of the oxidation of vitamin K by vitamin E, thus decreasing the active form of vitamin K. Warfarin, which is a vitamin K antagonist, is also thought to decrease the active form of vitamin K [88].

One case of nose bleeding was reported following rhinoplasty in an adult patient who was recovering with 800–1200 IU/day of vitamin E supplementation [89]. A prospective study was performed on healthy patients who had received vitamin E supplement doses of 2000 IU for 10 days, and no differences in bleeding were found [90, 91]. In the only other reported incidence of a bleeding complication in a patient on vitamin E and warfarin, a 55-year-old male who had self-administered 1200 IU/day of vitamin E experienced coagulopathy with hypoprothrombinemia while receiving both chronic warfarin treatment and clofibrate [92]. The bleeding complication was resolved by discontinuing the vitamin E supplementation. After his clinical and hematological status had been stable for 2 months, vitamin E supplementation was reinstated with 800 IU/day to the same patient while he was still on warfarin therapy, and an increase in the prothrombin time (an indicator of the clotting of the blood) was not observed until 28 days of administration of this high dose.

In light of this case, Kim and White [93] undertook a small double-blind randomized study in which 12 patients on fixed warfarin therapy received 1200 IU/day, 800 IU/day, or placebo doses of vitamin E for 1 month ($n = 4$ for each). None of the patients receiving vitamin E supplements showed a significant change in their prothrombin time, defined by the study authors to be at least 1.4 times the initial range.

Twelve cardiology patients on chronic warfarin treatment (median of 6.2 years) and already showing moderate prolongation of their prothrombin times compared to controls were given daily supplementation of vitamin E at a dose of 100 IU/day (91 mg/day for synthetic α-tocopherol) or 400 IU/day (364 mg/day for synthetic α-tocopherol) for 4 weeks [88]. Prothrombin times were not affected significantly, and there were no clinically observed effects, such as coagulopathy or ecchymoses.

In summary, over the past 30 years, only two case reports have been published that showed possible bleeding complications in high oral vitamin E supplementation in a patient on warfarin. In both cases the vitamin E dose was very high (over 800 IU/day), and one patient was also simultaneously on clofibrate. The small population of patients who were studied prospectively to investigate such a relationship has not indicated that there is likely to be an interaction of clinical significance in that high daily doses were required for 28 days to affect prothrombin time, a circumstance that would be quite unlikely in the case of vitamin E as an additive to UHMWPE for the articulation of a total joint.

Several uncontrolled studies have reported effects such as fatigue, emotional disturbances, thrombophlebitis, breast soreness, creatinuria, altered serum lipid and lipoprotein levels, gastrointestinal disturbances, and thyroid effects. These effects were not consistently observed.

### 15.7.3 Plasma and Tissue Levels of Vitamin E in Healthy Adults with and without Vitamin E Supplementation

Behrens [94] determined the mean serum α-tocopherol concentration for 451 healthy human subjects (both sexes, 19–70 years in age) without dietary supplementation of vitamin E to be 8.82 μg/mL. Given that an average adult has about 5 liters of blood, the total amount of α-tocopherol in the human circulation would be approximately 44 mg.

The relationship between dietary intake and plasma levels in humans is not very clear because the bioavailability of vitamin E is directly related to the absorption efficiency through the intestine and is affected by the amount of food consumed together with the oral vitamin E dose.

In 14 healthy adult male patients, the baseline plasma concentration of α-tocopherol before receiving oral vitamin E supplements was 23 μmol/L (10.1 μg/L), corroborating the study previously discussed. After receiving 1200 IU (800 mg) of RRR-α-tocopherol for 14 days [95], an average plasma concentration was 76 μmol/L (33.5 μg/L). In another study measuring the plasma and tissue levels of α-tocopherol as a result of continuous daily oral supplementation of vitamin E, it was shown that a maximum plateau of plasma concentration was reached at 4–5 days [96].

The tissue concentrations also rise when vitamin E supplementation at high doses is continued over a long time. Tocopherol concentration in tissues is closely related to regulatory mechanisms involved in tissue energy and cholesterol requirements [8] and is not regulated by specific tocopherol requirements. The increase in tissue concentrations is dependent on the fat composition and proximity to systemic circulation of the tissues. For example, while liver tissue has about the same concentration of plasma α-tocopherol, it required 107 days for brain tissue to obtain the same levels of α-tocopherol [97]. Therefore, an increase in the tissue concentrations follows increase in plasma concentrations, but prolonged periods of dosing are required to increase either level significantly.

### 15.7.4 Estimates of Possible Systemic Vitamin Exposure Using the UHMWPE Articulation Components for Total Joint Arthroplasty

Extensive *in vitro* testing was done to assess the possible release of vitamin E from an articular component of a total joint implant under various extraction conditions and under simulated gait [98]. We will discuss this data in light of the normal levels of vitamin E seen in the body as well as the literature reports of toxicity.

A series of tests were performed to quantify the elution of vitamin E out of vitamin-E-containing UHMWPE joint implant components, using environments that would provide different driving forces for vitamin E to elute out of UHMWPE. Extraction in boiling water for 24 hours, an 8-hour soak in 100% isopropyl alcohol, and sonication and extraction in isopropyl alcohol at 50°C for 3 hours were used. Finally, as a very aggressive condition, vitamin E was extracted from irradiated, vitamin-E-diffused, and terminally gamma sterilized UHMWPE by extraction with boiling hexane, which is a very good solvent for vitamin E, to quantify a worst case elution.

When exposed to boiling water or cleaned in isopropyl alcohol (IPA), there was no change in the vitamin E content of vitamin-E-diffused, irradiated acetabular liners as quantified by infrared spectroscopy [98]. When sonicated in IPA at 50°C for 3 hours, only trace amounts of the vitamin E in the liners was eluted. Even after the aggressive extraction for 72 hours in boiling hexane, a clinically irrelevant environment, 20 wt% of vitamin E still remained in the liners [99]. This suggested that under clinically relevant conditions, the complete removal of vitamin E from the components was highly unlikely.

Vitamin E elution out of the vitamin-E-diffused, irradiated UHMWPE acetabular liners could also be influenced by the loading conditions associated with gait, by wear, or by deformation during loading. Simulated normal gait loading of these liners was performed on the hip simulator to quantify wear. Under conditions of simulated gait for 5 million cycles, which corresponds to approximately 5 years worth of gait, the total weight loss of these two sets of liners (corrected by unloaded controls) were approximately 1 mg/million-cycles (Section 15.4). The weight loss was partially due to wear of the UHMWPE, but it may also include some elution of vitamin E. A worst case is considered for possible toxicity, namely, the hypothetical case that all of the measured weight loss was entirely due to vitamin E elution. In this case, the simulated gait would have been associated with a maximum elution of about 5 mg of vitamin E in 5 million cycles. Because 1 million cycles of simulated gait equals approximately 1 year of *in vivo* gait, the measured elution would be approximately 3 μg/day. The assumption is that the release had been uniform over the testing period as the weight loss. Thus, these calculated worst case values of vitamin E elution appear to be very low, substantially less than plasma vitamin E levels in healthy adults without supplementation.

In addition to our work with the elution of vitamin E from irradiated UHMWPE diffused with vitamin E, Wolf et al. [100] have determined that there were no cellular cytotoxic or genotoxic effects of vitamin E extracted from a vitamin-E-blended and terminally gamma sterilized UHMWPE containing 0.8 wt% vitamin E *in vitro*. Based on the studies previously discussed, the elution from the joint components is unlikely to cause significant detrimental effects *in vivo*. It has to be noted also that the earlier version of vitamin-E-diffused, irradiated UHMWPEs contained a high concentration of vitamin E on the surface and lower concentration in the bulk, thus they had a strong driving force out of the component for elution into the joint space. The later versions currently used *in vivo* are incorporated with vitamin E such that the final concentration profile is uniform and is at a lower concentration and thus has a much lower driving force for elution.

### 15.7.5 Animal Studies to Determine the Local Effects of Vitamin E in the Joint Space

Because there were no published reports of the local effects of vitamin E on the periarticular tissue, it is necessary to determine local toxicity using animal studies. Tocopherols that are orally from food or supplements are transported across the gastrointestinal membrane by passive diffusion after solubilization by mixing with chyme and bile salts in the stomach and small intestine [8]. Tocopherols are then incorporated in chylomicrons, which are taken up by the liver. After secretion from the liver, low density lipoproteins (LDLs) and high density lipoproteins (HDLs) are the major carriers of tocopherols in humans [101].

To simulate this mechanism of vitamin E transport conjugated to lipoproteins, we used an emulsifying agent (Tween 80, polysorbate) to solubilize vitamin E in an aqueous solution containing 5 mg/ml vitamin E, 0.25% Tween 80, and 6% ethanol. Two ml of this emulsion was injected to the knee joints of New Zealand white rabbits, and the animals were sacrificed at 2 weeks (n = 3) and at 12 weeks (n = 6). The control knee was injected with the same volume of carrier solution without vitamin E. For all studies, the entire knee joint with its intact capsule was excised and fixed in formalin prior to obtaining sections of the capsule for histological examination. Routine hematoxylin and eosin stained histological sections were prepared and examined by light microscopy to assess the biologic response to the vitamin E injections. The results were interpreted by a pathologist, who was blinded to the experimental conditions. The findings were that regardless of the time *in situ*, the synovial tissue had a normal appearance at harvest and there were no signs of inflammation or sterile puss. Histologically, the synovium had a normal appearance and was lined by one to two cell layers thick of synoviocytes. The tissue of all control knees had the same unremarkable appearance (Jarrett et al., unpublished data). Thus, this study supported the hypothesis that a total of 10 mg of vitamin E introduced by intra-articular injection in a carrier solution would not cause detrimental effects locally.

## 15.8 CONCLUSION AND FUTURE PROSPECTS

Vitamin-E-diffused, radiation crosslinked UHMWPE was given FDA clearance for clinical use in total hips in 2007 and in total knees in 2008. Based on preclinical studies, it shows promise in improving the longevity of highly crosslinked UHMWPE implants due to improved fatigue strength while maintaining the wear and oxidation resistance of irradiated and melted UHMWPE, which has shown good results *in vivo* up to 5 years. As in the case of all joint implants, its clinical performance and whether its fatigue strength improvement compared to first generation crosslinked UHMWPEs will prolong the longevity of joint implants, especially for total knees, will be best determined by prospective randomized clinical studies.

## 15.9 ACKNOWLEDGMENTS

We would like to thank Dr. William H. Harris for his help with the biocompatibility literature survey and for his insights. The work discussed by the authors has been supported by National Institute for Musculoskeletal and Skin Diseases Grant No. AR051142, research grants from Biomet, Inc., Zimmer, Inc., the William H. Harris foundation, and departmental funds.

## REFERENCES

1. Oral E, Wannomae KK, Hawkins NE, Harris WH, Muratoglu OK. α-Tocopherol doped irradiated UHMWPE for high fatigue resistance and low wear. *Biomaterials* 2004;**25**(24):5515–22.
2. Oral E, Christensen S, Malhi A, Wannomae K, Muratoglu O. Wear resistance and mechanical properties of highly crosslinked UHMWPE doped with vitamin E. *J Arthroplasty* 2006;**21**(4):580–91.
3. Parth M, Aust N, Lederer K. Studies on the effect of electron beam radiation on the molecular structure of ultra-high molecular weight polyethylene under the influence of alpha-tocopherol with respect to its application in medical implants. *J Mater Sci-Mater Med* 2002;**13**(10):917–21.
4. Oral E, Greenbaum ES, Malhi AS, Harris WH, Muratoglu OK. Characterization of irradiated blends of alpha-tocopherol and UHMWPE. *Biomaterials* 2005;**26**(33):6657–63.
5. Oral E, Godleski Beckos C, Malhi A, Muratoglu O. The effects of high dose irradiation on the cross-linking of vitamin E-blended UHMWPE. *Biomaterials*. In press.
6. Oral E, Wannomae KK, Rowell SL, Muratoglu OK. Diffusion of vitamin E in ultra-high molecular weight polyethylene. *Biomaterials* 2007;**28**(35):5225–37.
7. Packer L. Protective role of vitamin E in biological systems. *Am J Clin Nutr* 1991;**53**:1050S–55S.
8. Traber MG, Cohn W, Muller DP. Absorption, transport and delivery to tissues. In: Lester P, Jurgen F, editors *Vitamin E in health and disease*. New York: Marcel Dekker, Inc.; 1993. p. 35–52.
9. Packer L, Kagan VE. Vitamin E: the antioxidant harvesting center of membranes and lipoproteins. In: Packer L, Fuchs J, editors *Vitamin E in health and disease*. New York: Marcel Dekker, Inc.; 1993. p. 179–92.
10. Burton G, Ingold K. Autoxidation of biological molecules. 1. The antioxidant activity of vitamin E and related chain-breaking phenolic antioxidants in vitro. *J Am Chem Soc* 1981;**103**:6472–77.
11. Burton GW, Traber MG. Vitamin E: antioxidant activity, biokinetics, and bioavailability. *Annu Rev Nutr* 1990;**10**:357–82.
12. Kamal-Eldin A, Appelqvist L. The chemistry and antioxidant properties of tocopherols and tocotrienols. *Lipids* 1996;**31**(7):671–701.
13. Bowry VW, Ingold K. Extraordinary kinetic behavior of the α-tocopheroxyl (vitamin E) radical. *J Org Chem* 1995;**60**:5456–67.
14. Costa L, Luda MP, Trossarelli L, Brach del Prever EM, Crova M, Gallinaro P. Oxidation in orthopaedic UHMWPE sterilized by gamma radiation and ethylene oxide. *Biomaterials* 1998;**19**:659–68.
15. Al-Malaika S. Autoxidation. In: Scott G, editor. *Atmospheric oxidation and antioxidants*. Amsterdam: Elsevier Science Publishers B.V.; 1993. p. 45–82.
16. Esterbauer H, Gebicki J, Puhl H, Jurgens G. The role of lipid peroxidation and antioxidants in oxidative modification of LDL. *Free Radic Biol Med* 1992;**13**:341–90.
17. Carlsson D, Chmela S, Lacoste J. On the structures and yields of the first peroxyl radicals in γ-irradiated polyolefins. *Macromolecules* 1990;**23**:4934–38.
18. Libby D, Ormerod M, Charlesby A. Electron spin resonance spectra of some polymers irradiated at 77-degrees-K. *Polymer* 1960;**1**:212–18.
19. Lawton E, Balwit J, Powell R. Paramagnetic-resonance studies of irradiated high-density polyethylene. II. Effect of irradiation dose on the radical species trapped at room temperature. *J Chem Phys* 1960;**33**(2):405–12.
20. Dole M. Free radicals in irradiated polyethylene. In: Dole M, editor. *The radiation chemistry of macromolecules*. New York: Academic Press; 1972. p. 335–48.
21. Bhateja S, Duerst R, Aus E, Andrews E. Free radicals trapped in polyethylene crystals. *J Macromol Sci Phys* 1995;**B34**(3): 263–72.
22. Keller A, Ungar G. Radiation effects and crystallinity in polyethylene. *J Phys Chem* 1983;**22**(1/2):155–81.
23. Peacock AJ. *Handbook of polyethylene*. New York: Marcel Dekker Inc.; 2000.
24. Jahan MS, King MC, Haggard WO, Sevo KL, Parr JE. A study of long-lived free radicals in gamma-irradiated medical grade polyethlene. *Radiat Phys Chem* 2001;**62**:141–44.
25. Kuzuya M, Kondo S, Sugito M, Yamashiro T. Peroxy radical formation from plasma-induced surface free radicals of polyethylene as studied by electron spin resonance. *Macromolecules* 1998;**31**: 3230–34.
26. Seguchi T, Tamura N. Mechanism of decay of alkyl radicals in irradiated polyethylene on exposure to air as studied by electron spin resonance. *J Phys Chem* 1973;**77**(1):40–44.

27. Carlsson D, Dobbin C, Wiles D. Direct observations of macroperoxyl radical propagation and termination by electron spin resonance and infrared spectroscopies. *Macromolecules* 1985;**18**:2092–94.
28. Ohnishi S, Sugimoto S, Nitta I. Electron spin resonance study of radiation oxidation of polymers. IIIA. Results for polyethylene and some general remarks. *J Polym Sci: Part A: Polym Chem* 1963;**1**:605–23.
29. Rabek J, Ranby B. Photochemical oxidation reactions of synthetic polymers. In: Kinell P, Ranby B, editors *ESR applications to polymer research*. Stockholm: Almqvist-Wiksell Forlag AB; 1973.
30. Al-Malaika S. Perspectives in stabilisation of polyolefins. *Adv Polym Sci* 2004;**169**:121–50.
31. Assink R, Celina M, Dunbar T, Alam T, Clough R, Gillen K. Analysis of hydroperoxides in solid polyethylene by MAS $^{13}C$ NMR and EPR. *Macromolecules* 2000;**33**:4023–29.
32. Costa L, Luda MP, Trossarelli L. Ultra-high molecular weight polyethylene: I. Mechano-oxidative degradation. *Polym Degrad Stab* 1997;**55**:329–38.
33. Costa L, Luda MP, Trossarelli L. Ultra high molecular weight polyethylene-II. Thermal-and photo-oxidation. *Polym Degrad Stab* 1997;**58**:41–54.
34. Costa L, Jacobson K, Bracco P, Brach del Prever EM. Oxidation of orthopaedic UHMWPE. *Biomaterials* 2002;**23**:1613–24.
35. Kurtz S, Hozack WJ, Marcolongo M, Turner J, Rimnac CM, Edidin A. Degradation of mechanical properties of UHMWPE acetabular liners following long-term implantation. *J Arthroplasty* 2003;**18**(7, Suppl. 1):68–78.
36. Collier J, Sutula L, Currier B, John H, Wooding R, Williams I, et al. Overview of polyethylene as a bearing material: comparison of sterilization methods. *Clin Orthop* 1996;**333**:76–86.
37. Scott G. Antioxidants: chain breaking mechanisms. In: Scott G, editor. *Atmospheric oxidation and antioxidants*. Amsterdam: Elsevier Science Publishers B.V.; 1993. p. 121–60.
38. Godleski CEO, Muratoglu O. Increasing penetration of vitamin E in highly cross-linked UHMWPE for total joint implants by supercritical carbon dioxide. *Transactions, 52nd Annual Meeting of the Orthopaedic Research Society*. Chicago, IL; 2006. 656.
39. Wolf C, Maninger J, Lederer K, Fruhwirth-Smounig H, Gamse T, Marr R. Stabilisation of crosslinked ultra-high molecular weight polyethylene (UHMW-PE)-acetabular components with alpha-tocopherol. *J Mater Sci Mater Med* 2006;**17**:1323–31.
40. Muratoglu OK, Bragdon CR, O'Connor DO, Jasty M, Harris WH. HAP Paul Award. A novel method of crosslinking UHMWPE to improve wear, reduce oxidation and retain mechanical properties. *J Arthroplasty* 1999;**16**(2):149–60.
41. Digas G, Karrholm J, Thanner J, Herberts P. 5-year experience of highly crosslinked polyethylene in cemented and uncemented sockets—Two randomized studies using radiostereometric analysis. *Acta Orthop* 2007;**78**(6):746–54.
42. Muratoglu OK, Bragdon CR, O'Connor DO, Perinchief RS, Estok DM, Jasty M, et al. Larger diameter femoral heads used in conjunction with a highly cross-linked ultra-high molecular weight polyethylene: a new concept. *J Arthroplasty* 2001;**16**(8 Suppl.):24–30.
43. Jasty MJ, Goetz DD, Lee KR, Hanson AE, Elder JR, Harris WH. Wear of polyethylene acetabular components in total hip arthroplasty. An analysis of 128 components retrieved at autopsy or revision operation. *JBJS* 1997;**79**(A):349–58.
44. Edidin AA, Pruitt L, Jewett CW, Crane DJ, Roberts D, Kurtz SM. Plasticity-induced damage layer is a precursor to wear in radiation-cross-linked UHMWPE acetabular components for total hip replacement. Ultra-high-molecular-weight polyethylene. *J Arthroplasty* 1999;**14**(5):616–27.
45. Oral E, Malhi A, Muratoglu O. Mechanisms of decrease in fatigue crack propagation resistance in irradiated and melted UHMWPE. *Biomaterials* 2006;**27**:917–25.
46. Ingham E, Fisher J. Biological reactions to wear debris in total joint replacement. *Proc Instn Mech Engrs* 2000;**214**(H1):21–37.
47. Sochart DH. Relationship of acetabular wear to osteolysis and loosening in total hip arthroplasty. *Clin. Orthop Relat Res* 1999;**363**: 135–50.
48. Olyslaegers C, Defoort K, Simon J, Vandenberghe L. Wear in conventional and highly cross-linked polyethylene cups. *J Arthroplasty* 2008;**23**(4):489–94.
49. Jacobs CA, Christenson CP, Greenwald AS, McKellop H. Clinical performance of highly cross-linked polythylenes in total hip arthroplasty. *J Bone Joint Surg* 2007;**89-A**(12):2779–86.
50. Reno F, Bracco P, Lombardi F, Boccafoschi F, Costa L, Cannas M. The induction of MMP-9 release from granulocytes by vitamin E in UHMWPE. *Biomaterials* 2004;**25**(6):1001.
51. Wannomae K, Micheli B, Lozynsky A, Malhi A, Oral E, Muratoglu O. Vitamin E-stabilized, irradiated UHMWPE for cruciate-retaining knees. *Transactions, 53rd Annual Meeting of the Orthopaedic Research Society*. San Diego, CA; 2007. 1783.
52. Muratoglu OK, Vittetoe D, Rubash H, et al. Damage of implant surfaces in total knee arthroplasty. In: Callaghan J, editor. *The adult knee*. Philadelphia: Lippincott Williams and Wilkins; 2002.
53. Christensen S, Wannomae K, Muratoglu O. Wear and delamination resistance of $\alpha$-tocopherol doped, irradiated UHMWPE. *Transactions, 52nd Annual Meeting of the Orthopaedic Research Society*. Chicago, IL; 2006. 637.
54. Tomita N, Kitakura T, Onmori N, Ikada Y. Aoyama E. Prevention of fatigue cracks in ultrahigh molecular weight polyethylene joint components by the addition of vitamin E. *J Biomed Mater Res* 1999;**48**(4):474–78.
55. Muratoglu OK, Merrill EW, Bragdon CR, O'Connor DO, Hoeffel D, Burroughs B, et al. Effect of radiation, heat, and aging on in vitro wear resistance of polyethylene. *Clin Orthop Relat Res* 2003;**417**:253–62.
56. ASTM E647-08. *Standard test method fro measurement of fatigue crack growth rates*. West Conshohocken, PA: American Society and Testing of Materials International; 2008.
57. Baker DA, Bellare A, Pruitt L. The effect of degree of crosslinking on the fatigue crack initiation and propagation resistance of orthopedic-grade polyethylene. *J Biomed Mater Res* 2003;**66A**:146–54.
58. Oral E, Malhi A, Muratoglu O. Vitamin E does not detrimentally affect properties of UHMWPE at high concentration. *Transactions, 52nd Annual Meeting of the Orthopaedic Research Society*. San Diego, CA; 2007. 1626.
59. Teramura S, Sakoda H, Terao T, Endo MM, Fujiwara K, Tomita N. Reduction of wear volume from ultrahigh molecular weight polyethylene knee components by the addition of vitamin E. *J Orthop Res* 2008;**26**(4):460–64.
60. Halley D, Glassman A, Crowninshield R. Recurrent dislocation after revision total hip replacement with a large prosthetic femoral head. *J Bone and Joint Surg* 2004;**86A**(4):827–30.
61. Chiu YS, Chen WM, Huang CK, Chiang CC, Chen TH. Fracture of the polyethylene tibial post in a NexGen posterior-stabilized knee prosthesis. *J Arthroplasty* 2004;**19**(8):9.

62. Oral E, Malhi A, Wannomae K, Muratoglu O. Highly cross-linked UHMWPE with improved fatigue resistance for total joint arthroplasty. *Journal of Arthroplasty* (Winner of the 2006 "HAP" Paul Award). In press.
63. ASTM F2102-06e1. *Standard guide for evaluating the extent of oxidation in ultra-high-molecular-weight polyethylene fabricated forms intended for surgical implants*. West Conshohocken, PA: American Society for Testing and Materials International; 2006.
64. Nabar S, Wannomae K, Muratoglu O. Environmental stress cracking of contemporary and alpha-tocopherol doped UHMWPEs. *Transactions, 54th Annual Meeting of the Orthopaedic Research Society*. San Francisco, CA; 2008. 1684.
65. Sheppard AJ, Pennington JA, Weihrauch JL. Analysis and distribution of vitamin E in vegetable oils and foods. In: Packer L, Fuchs J, editors *Vitamin E in health and disease*. New York: Marcel Dekker, Inc.; 1993. p. 9–31.
66. Committee on Dietary Allowances FaNB. *Recommended Dietary Allowances*. 9th ed Washington, D.C: National Academy Press; 2000.
67. FDA DoHaHS, Code of Federal Regulations 184.1890:Direct Food Substances Affirmed as Generally Recognized as Safe-[alpha]-tocopherols, Revised 2004 April 1.
68. Diaz MN, Frei B, Vita JA, Keaney JFJ. Antioxidants and artherosclerotic heart disease. *N Engl J Med* 1997;**337**(6):408–16.
69. Corrigan JJ. The effect of vitamin E on warfarin-induced vitamin K deficiency. *Ann N Y Acad Sci* 1982;**292**:361–68.
70. Miller III ER, Pastor-Barriuso R, Dalal D, Riemersma R, Appel LJ, Guallar E. Meta-analysis: High-dosage vitamin E supplementation may increase all-cause mortality. *Ann Intern Med* 2005;**142**(1).
71. Age-Related Eye Disease Study Group. A randomized, placebo-controlled, clinical trial of high-dose supplementation with vitamins C and E and beta carotene for age-related cataract and vision loss: AREDS report no. 9. *Arch Ophthalmol* 2001;**119**(10):1439–52.
72. McNeil JJ, Robman L, Tikellis G, Sinclair MI, McCarthy CA, Taylor HR. Vitamin E supplementation and cataract: Randomized controlled trial. *Opthalmology* 2004;**111**:75–84.
73. Heart Outcomes Prevention Evaluation Study Investigators. Vitamin E supplementation and cardiovascular events in high-risk patients. *N Engl J Med* 2000;**342**:154–60.
74. Stephens NG, Parsons A, Schofiled PM, Kelly F, Mitchinson MJ, Brown MJ. Randomised controlled trial of vitamin E in patients with coronary disease: Cambridge Heart Antioxidant Study (CHAOS). *Lancet* 1996;**347**:781–86.
75. Heart Protection Study Collaborative Group. MRC/BHF heart protection study of antioxidant vitamin supplementation in 20, 536 high-risk individuals. *Lancet* 2002;**360**:23–33.
76. Boaz M, Smetana S, Weinstein T, Matas Z, Gafter U, Iaina A, et al. Secondary prevention with antioxidants of cardiovascular disease in endstage renal disease (SPACE): randomised placebo-controlled trial. *Lancet* 2000;**356**:1213–18.
77. Waters DD, Alderman EL, Hsia J, Howard BV, Cobb FR, Rogers WJ, et al. Effects of hormone replacement therapy and anti-oxidant vitamin supplements on coronary arteriosclerosis in post-menopausal women: a randomized controlled trial. *J Am Med Assoc* 2002;**288**(19):2432–40.
78. The Polyp Prevention Study Group. A clinical trial of antioxidant vitamins to prevent colorectal adenoma. *N Engl J Med* 1994;**331**(3): 141–47.
79. Chylack LJ, Brown N, Bron A, Hurst M, Kopcke W, Thien U, et al. The Roche European American Catarct Trial (REACT): a randomized clinical trial to investigate the efficacy of an oral anti-oxidant micronutrient mixture to slow progression of age-related cataract. *Ophthalmic Epidemiol* 2002;**9**:49–80.
80. Sano M, Ernesto C, Thomas RG, Klauber MR, Schafer K, Grundman M, et al. A controlled trial of selegine, alpha-tocopherol, or both as treatment for Alzheimer's disease. *N Engl J Med* 1997;**336**: 1216–22.
81. Parkinson Study Group. Mortality in DATATOP: a multicenter trial in early Parkinson's disease. *Ann Neurol* 1998;**43**:318–25.
82. Shekelle P, Morton S, Jungvig L. Effect of supplemental vitamin E for the prevention and treatment of cardiovascular disease. *J Gen Intern Med* 2004;**19**(4):380–89.
83. Eidelman R, Hollar D, Hebert P, Lamas G, Hennekens C. Randomized trials of vitamin E in the treatment and prevention of cardiovascular disease. *Arch Intern Med* 2004;**164**(14):1552–56.
84. Vivekanathan DP, Penn MS, Sapp SK, Hsu A, Topol EJ. Use of antioxidant vitamins for the prevention of cardiovascular disease: meta-analysis of randomised trials. *Lancet* 2003;**361**:2017–23.
85. Rimm E, Stampfer M, Ascherio A, Giovanucii E, Colditz G, Willett W. Vitamin E consumption and the risk of coronary heart disease in men. *N Engl J Med* 1993;**328**:1450–56.
86. Stampfer M, Hennekens C, Manson J, Colditz G, Rosner B, Willett W. Vitamin E consumption and the risk of coronary heart disease in women. *N Engl J Med* 1993;**328**:1444–49.
87. Bendich A, Machlin L. The safety of oral intake of vitamin E: data from clinical studies from 1986 to 1991. In: Packer L, Fuchs J, editors *Vitamin E in health and disease*. New York: Marcel Dekker, Inc.; 1993. p. 411–16.
88. Corrigan J, Ulfers L. Effect of vitamin E on prothrombin levels in warfarin-induced vitamin K deficiency. *Am J Clin Nutr* 1981;**34**: 1701–05.
89. Churukian M, Zemplenyi J, Steiner M. Postrhinoplasty epitaxis-Role of vitamin E. *Arch Otolaryngol Head Neck Surg* 1988;**114**(7): 748–50.
90. Hale W, Perkins L, May F, Marks R, Stewart R. Vitamin E effect on symptoms and laboratory values in the elderly. *J Am Diet Assoc* 1986;**86**(5):625–29.
91. Huijgens P, Vandenberg C, Imandt L, Langenhuisen M. Vitamin E and platelet-aggregation. *Acta Haemologica* 1981;**65**(3):217–18.
92. Corrigan J, Marcus F. Coagulopathy associated with vitamin E ingestion. *J Am Med Assoc* 1974;**230**(9):1300–01.
93. Kim J, White R. Effect of vitamin E on the anticoagulant response to warfarin. *Am J Cardiol* 1996;**77**:545–546.
94. Behrens W, Madere R. Alpha- and gamma-tocopherol concentrations in human serum. *J Am Coll Nutr* 1986;**5**(1):91–96.
95. Morinobu T, Yoshikawa S, Hamamura K, Tamai H. Measurement of vitamin E metabolites by high performance liquid chromatography during high dose administration of alpha-tocopherol. *Eur J Clin Nutr* 2003;**57**(3):410–14.
96. Dimitrov NV, Meyer C, Gilliland D, Ruppenthal M, Chenoweth W, Malone W. Plasma tocopherol concentrations in response to supplemental vitamin E. *Am J Clin Nutr* 1991;**53**:723–29.
97. Burton GW, Traber MG, Acuff RV, Walters DN, Kayden H, Hughes L, et al. Human plasma and tissue alpha-tocopherol concentrations in response to supplementation with deuterated natural and synthetic vitamin E. *Am J Clin Nutr* 1998;**67**:669–84.
98. Oral E, Rowell S, Wannomae K, Muratoglu O. Migration stability of alpha-tocop herol in irradiated UHMWPE. *Biomaterials* 2006;**27**(11):2434–39.

99. Oral E, Rowell S, Muratoglu O. Oxidation resistance of Vitamin E-doped, irradiated UHMWPE for total joints following forceful extraction of vitamin E. *Transactions, 52nd Annual Meeting of the Orthopaedic Research Society*. Chicago, IL; 2006. 665.
100. Wolf C, Lederer K, Muller U. Tests of biocompatibility of α-tocopherol with respect to the use as a stabilizer in ultrahigh molecular weight polyethylene for articulating surfaces in joint endoprostheses. *J Mate Sci: Mater Med* 2002;**13**:701–03.
101. Behrens W, Thompson J, Madere R. Distribution of alpha tocopherol in human plasma lipoproteins. *Am J Clin Nutr* 1982;**35**: 691–96.

Chapter 16

# Vitamin-E-Blended UHMWPE Biomaterials

Steven Kurtz, PhD Pierangiola Bracco, PhD and Luigi Costa, PhD



## 16.1 INTRODUCTION

During the past decade, there has been an explosion of international interest in the development of vitamin E as an antioxidant for medical grade UHMWPE. Scientific papers on this topic from Japan [1] and Austria [2–4] began to appear in the scientific literature between 1999 and 2002. However, the recent growth in awareness of vitamin E in orthopedics may be attributed to the efforts of Professor Luigi Costa and collaborators at the University of Torino, who, in September 2003, organized the first international meeting dedicated to improving UHMWPE for total joint replacements (Figure 16.1). In a historical library, before an assembled panel of experts from around the world, Professor Costa reviewed how underlying chemical reactions driving oxidation in gamma sterilized UHMWPE may be effectively blocked with an appropriate antioxidant and recommended vitamin E as "a likely candidate for use in orthopaedic UHMWPE because it is obviously biocompatible and it is approved as antioxidant for food packaging" [5]. Professor Costa called for the routine use of vitamin E in gamma sterilized UHMWPE biomaterials and for international standardization to facilitate its widespread adoption.

Active commercialization of vitamin E in UHMWPE, both by consolidators and by medical device companies, followed soon after. Additional supporting scientific evidence related to vitamin E was provided by international experts in the second UHMWPE meeting, convened by Professor Costa in Torino during September 2005. By the close of 2007, ASTM International published a standard specification for medical grade UHMWPE blended with vitamin E [6], and around the same time, Ticona announced that it would supply medical grade UHMWPE resins containing vitamin E to the orthopedic industry. Medical grade GUR 1020-E and 1050-E, commercially released in 2009, contain 1,000 ppm Vitamin E. In the span of 5 years, vitamin-E-blended UHMWPE has been internationally recognized and commercially released as a new international standard for orthopedic implants [6].

Currently, the primary drawback of vitamin-E-blended UHMWPE is considered to be increased resistance of the material to subsequent radiation crosslinking, which depends upon the method of irradiation as well as concentration level of antioxidant [2, 7, 8]. When trace levels of vitamin E ($\leq$500 ppm, 0.05%) are blended with UHMWPE, the loss of crosslinking efficiency with a 100 kGy gamma irradiation dose is $\leq$10% (Figure 16.2).

**FIGURE 16.1** The first international UHMWPE meeting on September 18, 2003, to discuss vitamin E stabilization. The venue for the meeting was the historic "Giacomo Ponzio" Library of Chemistry at the University of Torino. Professor Costa, standing with a red tie on the right side of the image, can be seen making a point to the speaker, Anuj Bellare, PhD (author of Chapter 19), standing beside the overhead projector. The editor (SMK) is seated at the far end of the table to the left of the projector.

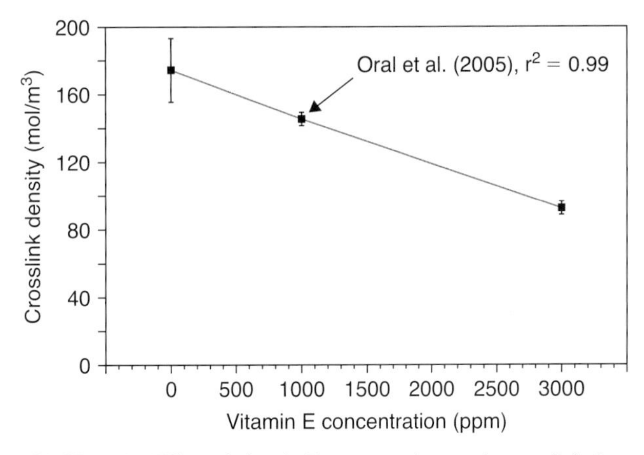


**FIGURE 16.2** Effect of vitamin E concentration on the crosslink density of 100 kGy, gamma irradiated UHMWPE, adapted from [7].

At 1000 ppm (0.1%) of vitamin E, the level in GUR 1020-E and 1050-E resin, the loss in crosslinking efficiency may approach 30% (Figure 16.2). On the other hand, above a concentration level of 3000 ppm (0.3%), elevated levels of crosslinking may not be achievable at all with gamma radiation [8]. Practical considerations for overcoming this limitation, centered mainly around optimization of radiation dosage for blends containing 1000 ppm vitamin E, are covered in this chapter.

In this chapter we also trace the scientific foundation for the acceptance of vitamin-E-blended UHMWPE biomaterials and summarize the development of ideas that led to the commercialization of vitamin-E-blended UHMWPE for use in orthopedics. The reader can find additional information relevant to vitamin E and UHMWPE in many other sections of this *Handbook*. The topic of vitamin E doped, highly crosslinked UHMWPE, as well as the biocompatibility of vitamin E, have already been covered by Oral and Muratoglu in the preceding chapter (Chapter 15). Similarly, the mechanisms of oxidation and vitamin E stabilization (Chapter 21) and the clinical significance of *in vivo* oxidation (Chapter 22) are also covered in separate chapters. Furthermore, a detailed description of electron-spin resonance-based radical measurements in UHMWPE containing vitamin E has been provided by Professor Shah Jahan in Chapter 29 (Section 29.6). As in these other chapters, we will refer to $\alpha$-tocopherol and vitamin E interchangeably for this review.

## 16.2 VITAMIN E AS AN ANTIOXIDANT FOR POLYOLEFINS

The idea of blending vitamin E with UHMWPE for stabilization is not new, and it appears in the patent literature from the 1980s and 1990s. Tocopherol compounds were first proposed as "hygienically safe" stabilizers for polyolefins in the early 1980s [9]. In December 1981, Dolezel and Adamirova from Prague filed a patent on polyolefin stabilization in what was, at the time, the Czechoslovakian Socialist Republic (now the Czech Republic) [9]. The Czech inventors described their motivation to prevent degradation of polyolefins during processing, exposure to weather, food contact, and implant use:

> Polyolefins stabilized with tocopherol have significantly higher resistance to thermooxidative degradation, both at the temperatures at which technical use is made of these materials, and at processing temperatures. They have higher resistance to photooxidative degradation, which is manifest in higher durability when polyolefins stabilized with tocopherol are exposed under natural weather conditions. They are hygienically safe. They can be used in contact with food. They do not cause adverse tissue reactions if polyolefins stabilized in this way are used as implants in the body. On the contrary, they have a positive effect on the processes that take place.

The Czech patent describes a methodology for stabilizing polyolefins by mixture of the polymer with 100 to 50,000 ppm (0.01 to 5%) of "alpha-tocopherol, beta-tocopherol, gamma-tocopherol, delta-tocopherol, or mixtures thereof" [9]. Although this patent covers polyolefins, and UHMWPE is, by definition, a polyolefin, UHMWPE was not specifically mentioned, nor were orthopedic hip and knee components. Furthermore, the actual methodology for mixing the polyolefin with vitamin E was also not disclosed by Dolezel and Adamirova.

Experimental techniques for blending vitamin E with UHMWPE prior to consolidation were first disclosed by Berzen and Luketic in 1994 [10]. Here, the orthopedic application is explicit. Berzen and Luketic were employed by Hoechst (now Ticona) and taught that vitamin E stabilization applies especially to "the UHMW-PE used in artificial hip, shoulder or knee joints" [10]. This European patent covers the manufacture of consolidated forms from UHMWPE powder mixed with antioxidants for use in orthopedic implants.

Although idea of blending of vitamin E with UHMWPE has its origins in the patent literature of the 1980s, the first widespread applications of this technology would actually appear in food packaging, not orthopedics. In the following section, we trace the development of scientific concepts in vitamin E stabilization of polyolefins from a food packaging perspective.

## 16.3 VITAMIN E BLENDS IN FOOD PACKAGING

The food packaging industry started to commercially develop vitamin E as an antioxidant for polyolefins between the late 1980s and early 1990s [11–14]. During that period, vitamin E was commercialized by Hoffman-LaRoche under the name of Ronotec 2001. It was specifically designed for stabilizing polypropylene, which is much more prone to oxidation than polyethylene [12, 14]. Vitamin E was mainly employed as a stabilizer for processing and then added at very low concentrations (100 ppm).

In light of the need to reduce oxidation in orthopedic implants (Chapters 3, 21, and 22), there is considerable motivation to examine the polyolefin melt stabilization literature for clues as to how vitamin E might perform in UHMWPE. On the surface, food packaging may not appear directly relevant to UHMWPE implants. However, the fundamental principles of polymer chemistry apply equally well to polyethylene in both food packaging and orthopedic applications, even if the specifics of these applications differ. If we approach the food packaging literature with the perspective of obtaining better understanding of the fundamental concepts about vitamin E blending with polyethylene, important insight can be obtained in the mechanisms of stabilization; the transformation products of vitamin E; the effects of vitamin E on the color (appearance) of polyolefins; and the levels of vitamin E necessary to stabilize polyethylene. Details about the mechanisms of stabilization and degradation mechanisms of vitamin E in polyethylene are covered in Chapter 21. We will cover the remaining topics in this section.

As early as 1990, vitamin E has been considered an attractive substitute for other hindered phenolic antioxidants used in polyolefin food packaging, such as BHT [12]. The drawback of BHT and other hindered phenols was that oxidation products could leach into packaged food, causing an undesirable aftertaste. Vitamin E was considered a "natural" alternative to synthetic phenolic antioxidants used in food packaging [14]. Many studies of vitamin E

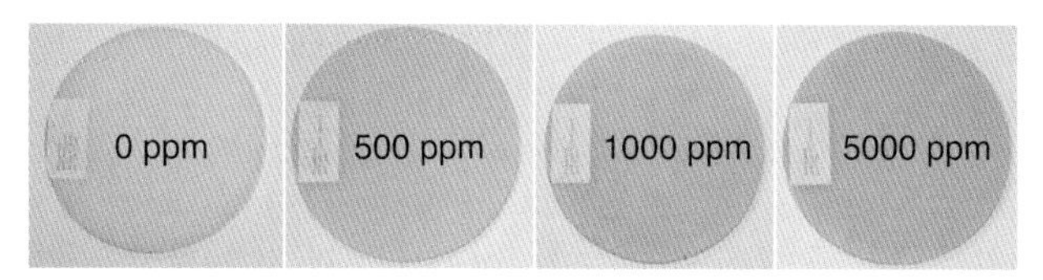


**FIGURE 16.3** Effect of vitamin E concentration on the color of compression molded UHMWPE. The plaques were provided courtesy of Ticona (Oberhausen, Germany).

stabilization of polyolefins, summarized recently by Al-Malaika [15], pertain to preventing degradation during melt processing, which includes methods such as screw extrusion that are not employed with UHMWPE. Nevertheless, as explained in Chapter 21, oxygen cannot be completely excluded from the conversion of UHMWPE resin powder to consolidated form, even in a nitrogen atmosphere. Thus, the protection of UHMWPE from degradation during ram extrusion and compression molding remains a relevant concern.

A recurring theme in the packaging literature related to vitamin E centers around the control of color and appearance of the stabilized polymer [15]. Given the crucial role of packaging for the marketing of food products, considerable research has been performed to understand the reasons underlying color change in vitamin-E-blended polyolefins [15]. Many of the concepts of color change in food packaging are relevant to medical grade UHMWPE. Pharmaceutical grade alpha-tocopherol (>99.9% purity) has the appearance of amber oil, and when mixed with medical grade UHMWPE results in a yellow color in the consolidated polymer (Figure 16.3). As illustrated in Figure 16.3, part of the color change in UHMWPE is clearly related to the concentration of vitamin E blended with the resin. However, any oxidation that occurs during processing, whether by consolidation or by subsequent exposure of the material to ionizing radiation, can also influence color change, because the transformation products of vitamin E are yellow pigments of UHMWPE. In food packaging, it is customary to use secondary antioxidants, such as phosphites, to control color; however, the use of these phosphorous-containing compounds may not be suitable for implant use [15]. If color change is considered undesirable with UHMWPE, the food packaging literature suggests that it may be reduced by minimizing the amount of vitamin E in the material or by reducing the availability of oxygen during processing [15].

To control color, as well as cost, the food packaging industry has interest in minimizing the amount of additives used for stabilization. Indeed, one of the motivations for α-tocopherol over BHT and other phenolic antioxidants is its improved efficiency [12–15]. Early studies by Laermer and colleagues from Hoffman-La Roche suggested that α-tocopherol concentrations as low as 100 ppm could effectively stabilize polyolefins [12,14]. More recent research by Mallegol et al. has also shown that HDPE resin used for water bottles can be effectively stabilized during gamma irradiation using 160 ppm of α-tocopherol [16, 17]. Depending upon the application and processing route, the packaging literature suggests that very low concentrations (100–300 ppm) of vitamin E are sufficient to protect polyethylene from melt processing and shelf oxidation [15]. Overall, the experience with vitamin E in food packaging over the past 2 decades provides a useful starting point for understanding many key parameters associated with stabilization of medical grade UHMWPE.

## 16.4 VITAMIN E STUDIES FROM JAPAN

If the vitamin E movement in UHMWPE biomaterials was born in Torino in 2003, then its conception occurred in Japan by Professor Naohide Tomita from the University of Kyoto. Japan has a history of early adoption of vitamin E both in packaging as well as in orthopedic implants. Vitamin E has been used in Japan as an environmentally friendly additive for polyolefins since the late 1980s [11]. In light of its acceptance as an effective antioxidant, a manufacturer from Japan (Nakashima Medical Co., Ltd., http://www.medical.nakashima.co.jp) was also the first supplier to produce UHMWPE blended with vitamin E for total knee replacements, and these implants have been in clinical use in Japan since 2006 [18] (Figure 16.4). The commercialization of vitamin-E-blended UHMWPE in Japan is largely based on the research efforts of Professor Tomita and Nobuyuki Shibata, who was a doctoral student at the time [1,19–21]. These studies have been recently summarized [18]; consequently the contributions of the Kyoto research group are only briefly reviewed herein.

Professor Tomita and colleagues began work on improving the fatigue wear performance of total knee replacements in the mid-1990s [22–27]. These studies, directed primarily at fatigue damage initiation at grain boundaries following gamma sterilization, led to the incorporation of vitamin E in UHMWPE as a way to prevent delamination by reducing crack formation at grain boundaries [1, 19–21]. In a series of experiments, outlined in the recent review [18], researchers from Kyoto University demonstrated that gamma sterilized UHMWPE components blended with vitamin E exhibited resistance to oxidation and fatigue wear when compared with unstabilized controls. Professor Tomita's research was first published in 1998 by the Japanese Orthopaedic Association [28] but later appeared in *Applied Biomaterials* in 1999 [1]. These initial experiments were performed on Mitsui resin UHMWPE blended with 1000 to 3000 ppm vitamin E and gamma sterilized in air.

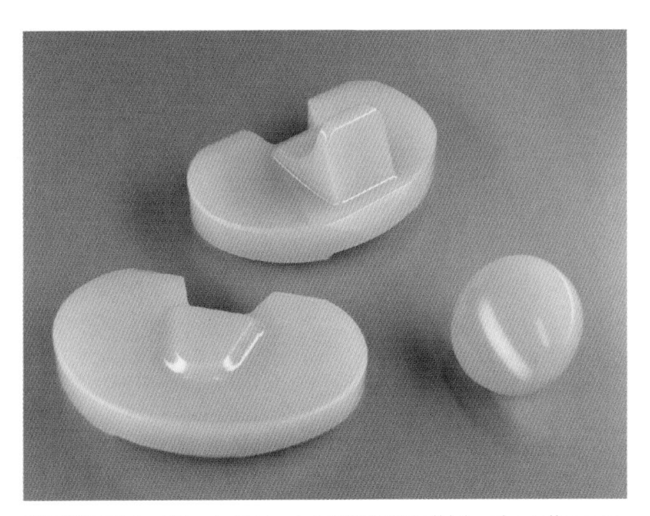

**FIGURE 16.4** Vitamin-E-blended UHMWPE tibial and patellar components of the High Tech II Knee (Nakashima Medical Co., Ltd.). Image provided courtesy of Kunihiko Fujiwara and Yoshio Nakashima.

More recently, knee simulator experiments have been performed with GUR 1050 blended with 3000 ppm vitamin E and likewise observed a reduction in wear as compared with unstabilized controls [21]. These recent experiments were performed without sterilization because the material is ethylene oxide sterilized when used clinically in Japan by Nakashima Medical. Several mechanisms, including improved mechanical properties, reduced oxidation during consolidation, and slight differences in protein adhesion, have been postulated by the authors to explain why, in the absence of gamma sterilization, the addition of vitamin E results in improved wear resistance in a knee simulator.

In summary, UHMWPE blended with vitamin E was first used clinically in Japan following a decade of research activity from the University of Kyoto. A prospective clinical trial of vitamin E has been undertaken of the Hi Tech II Knee (Nakashima Medical) in Japan. The tibial and patellar components are fabricated from compression molded GUR 1050 blended with 3000 ppm vitamin E (Figure 16.4). The UHMWPE insert is ethylene oxide sterilized. The femoral component is Ti-alloy. Although we know from personal communication with the manufacturer, as well as from citations in the literature, that this trial has taken place, the clinical results have not yet been published.

## 16.5 VITASUL AND VITAMIN E STUDIES FROM AUSTRIA

VITASUL is the trade name (currently owned by Zimmer, Inc.) for a gamma nitrogen sterilized, compression molded GUR 1020 material blended with 2000 ppm α-tocopherol. By several unusual twists of fate, and despite an extensive technical development effort spanning over a decade, VITASUL has to date not been released to the orthopedic community. The origins of VITASUL can be traced to the mid-1980s, when researchers from former Sulzer (Winterthur, Switzerland) realized the implications of the oxidation process induced by sterilization with high-energy radiation in air [29]. As a result, the company began to sterilize UHMWPE hip cups in 1986 and knee inserts in 1991 in an inert (nitrogen) atmosphere. During the mid-1990s, Sulzer started to explore the possibility of stabilizing UHMWPE with vitamin E. By 1998, an α-tocopherol-blended UHMWPE material had been developed under the trade name of VITASUL. The approach for fabricating VITASUL is described in a patent by Silvio Schaffner [30]. Concurrently, Sulzer licensed highly crosslinked UHMWPE patents from Massachusetts General Hospital and developed DURASUL (Chapters 13 and 20). In the end, only DURASUL was commercialized. In 2003, Sulzer (which had by then changed its name to Centerpulse) was acquired by Zimmer, Inc. (Warsaw, Indiana, USA). The extensive, international integration of Zimmer and Centerpulse has thus far precluded the launch of VITASUL. In addition, the stabilizing approach has become broader, including highly crosslinked UHMWPE.

In the mid-1990s, Sulzer initiated a collaboration regarding this project with a research group based at the university of Leoben in Austria and coordinated by Professor Klaus Lederer [2–4,31–34]. The Austrian group started a comprehensive research program using UHMWPE GUR 1020 blended with DL-α-tocopherol from Hoffman-La Roche (today known as DSM). The samples were prepared by dissolving the stabilizer in a solvent, which was mixed into the UHMWPE powder drop by drop in a screw cone mixer. The solvent was then evaporated, and the mixture was sintered in an industrial facility [3]. They showed that with this method, reasonable homogeneity of the vitamin E distribution could be achieved (±2% variation from the desired concentration).

Christian Wolf, who was a doctoral student at the time, Professor Lederer, and colleagues explored the oxidative stability of blends for a broad range of concentrations [3], ranging from 1000 to 8000 ppm (0.1% to 0.8% w/w), following gamma sterilization in nitrogen (25 kGy). After extensive accelerated aging in 5 atmospheres of pure oxygen at 70°C (ASTM 2003 methodology), it was demonstrated that the mechanical properties of the stabilized samples (modulus and breaking elongation) stayed constant in the same range of time where those of unstabilized samples dropped. They observed no remarkable difference in the stabilization effect between the samples stabilized with 1000 to 8000 ppm α-tocopherol, but they also found that at 8000 ppm α-tocopherol starts to act as a softening agent for the polymer. Consequently, they concluded that 2000 to 4000 ppm should be suggested as an optimal concentration of stabilizer. A comparison with other well-known stabilizing systems (Irganox 1010 and Irgafos 168 by Ciba-Geigy) showed that α-tocopherol has the same, if not a higher, stabilizing activity.

Wolf and colleagues also performed aging experiments in aqueous hydrogen peroxide solution (10%) and at elevated temperatures (120°C) in air [32]. The material chosen for these experiments was GUR 1020 blended with 4000 ppm (0.4%) and gamma sterilized in nitrogen (25 kGy). The motivation for the hydrogen peroxide aging experiments was to address the concern of *in vivo* oxidation. However, as will be more fully discussed in Chapter 22 (see Section 22.5), the ability of these types of *in vitro*, hydrogen peroxide models to replicate *in vivo* oxidation is questionable. Nevertheless, if we regard these experiments as oxidative challenges, Wolf and colleagues observed that the vitamin-E-blended material exhibited much greater oxidation resistance than its unstabilized control.

Biocompatibility of the vitamin-E-UHMWPE blends were also tested [4]. Based on the observation that the added α-tocopherol may undergo chemical transformations during manufacturing and sterilization by γ-irradiation, the biocompatibility of the transformation products was investigated. Cytotoxicity testing showed no evidence of a cytotoxic behavior of stabilized UHMWPE specimens, nor could genotoxic activity be found. However, it was shown that the adhesion rate and spreading of cells were diminished, indicating that the material was biocompatible but not bioactive. Therefore, the authors suggested that it should not be used as a material for implants on which cells should adhere to build a strong contact between implant and body. This is not the case of orthopedic surgery, where UHMWPE has generally no direct contact to the surrounding bone, being either cemented or used as an inlay in a metal shell or tray.

The same group also investigated the influence of α-tocopherol on the degree of crosslinking induced by electron beam irradiation [2]. They found that a higher radiation dosage in the case of the stabilized samples is necessary to reach the same amount of crosslinking as with the unstabilized samples. However, the influence of the stabilizer concentration turned out to decrease with an increasing radiation dosage and seemed to disappear completely at an absorbed dose of 100 kGy. The authors postulated that at this point α-tocopherol was no longer able to impede crosslinking.

To overcome the disadvantage of hindered crosslinking observed in earlier experiments by Parth and coworkers [2], two alternative methods for doping radiation crosslinked UHMWPE with vitamin E were also investigated [33]. One method involved doping of finished forms by diffusion of pure α-tocopherol under an inert atmosphere at high temperature, while the other used supercritical $CO_2$ to incorporate the additive into the UHMWPE. The subject of doping highly crosslinked UHMWPE with vitamin E is covered extensively in Chapter 15.

One of the few animal studies with vitamin-E-blended UHMWPE was also carried out by this group [31]. Stabilized and unstabilized UHMWPE films were subcutaneously implanted into rats, then morphology and reactivity of surrounding connective tissue were studied at timed intervals (2 weeks, 3 months, and 6 months). In addition, the specimens were examined regarding oxidative degradation with the help of FTIR spectroscopy. No oxidation of either stabilized or unstabilized material could be detected during 6 months *in vivo*. A loss of α-tocopherol was observed, but it was estimated to be in the order of 3% within 10 years, leaving, according to the authors, enough α-tocopherol in the UHMWPE to ensure an adequate lifetime stabilization. Implants had been well tolerated, and a well-defined capsule of coarse connective tissue free of elastic fibers was developed already after 2 weeks of implantation. Dimension and composition of this capsule did not change during the time period studied. Therefore, it was concluded that no *in vivo* consequences would manifest in the presence of α-tocopherol.

To further ensure biocompatibility, *in vitro* tests with human cells were carried out [34]. Two different human cell lines were tested on UHMWPE with and without α-tocopherol with respect to cell viability, proliferation, and morphology. Similar proliferation rates were found with both polyethylene samples. Intact morphology was found in light and electron microscopy on each substrate, while the morphologic characteristics of skin fibroblasts were not changed by any material. Normal adherence and spreading of the fibroblasts was found on both stabilized and unstabilized UHMWPE. These results led the authors to conclude that both materials, UHMWPE and α-tocopherol stabilized UHMWPE, do not show any toxic effects to human cells.

In their series of seven papers published to date [2–4, 31–34], the contributions of Christian Wolf, Professor Lederer, and researchers from Sulzer and the University of Leoben are noteworthy in several respects. The team were the first to clearly establish the oxidative stability as well as the biocompatibility of gamma inert-sterilized vitamin-E-blended, medical grade UHMWPE (in the range of 1000 to 8000 ppm). Although most of their experiments were conducted with GUR 1020, some experiments up to radiation dosage used to produce highly crosslinked UHMWPE were carried out on GUR 1050. Furthermore, the team were the first to recognize the primary drawback of vitamin E blended with UHMWPE, namely its reduction in crosslinking efficiency. Although these concepts would be further refined and more exhaustively explored by subsequent researchers, the VITASUL team developed their idea over the span of a decade, up to the point of commercialization, laying the scientific groundwork for its later adoption by the orthopedic community.

## 16.6 VITAMIN E STUDIES FROM ITALY

Investigators in Japan and Austria were the first to publish their applications of vitamin E blended with UHMWPE in orthopedic implants; however, researchers from Italy have contributed to our understanding of the stabilization chemistry and biologic reactions with this material. As mentioned previously, the international UHMWPE meetings convened in Italy in 2003 and 2005 provided a crucial forum for rapidly disseminating knowledge about vitamin E stabilization to the orthopedic community. This knowledge transfer was made possible by the research group coordinated by Professor Costa from the University of Torino, which in 2003 already had nearly a decade of experience in the stabilization of UHMWPE with vitamin E.

In the early 1990s, researchers from the University of Torino initiated studies on UHMWPE, and, in particular, on radiation-induced oxidation of UHMWPE. As a consequence of their previous experience in the field of oxidation and stabilization of polyolefins, the researchers were immediately oriented toward the search of a suitable stabilizer. By the end of 1994, they filed an Italian patent on vitamin E stabilized UHMWPE for orthopaedic applications. At the same time, they started an initial collaboration with former PolyHiSolidur and Ticona to produce a first experimental UHMWPE sample plate of 20 × 20 × 5 cm containing 5000 ppm (0.5% w/w) of vitamin E. On this sample they started the first experimental studies. In the decade that followed, new samples were produced by blending UHMWPE powder with vitamin E (Ronotec 2001, Hoffman-LaRoche), using a technique similar to that described in the previous section and following by consolidation in an industrial facility. In 2005, additional samples were provided thanks to a renewed collaboration with Ticona, Quadrant, and Orthoplastics.

In this way, a very broad range of vitamin E concentrations, ranging from 250 ppm to 5000 ppm (0.025 to 0.5% w/w), was studied in the 1990s up until the present time by researchers from Torino. FTIR, ESR, and chemiluminescence experiments, along with mechanical testing, solvent expansion, and oxidation induction time measurements were carried out to investigate the oxidative stability of the blends and the influence of the additive on the crosslinking efficiency [35, 36]. The results of these studies allowed researchers to postulate the mechanisms discussed in Chapter 21.

Biocompatibility studies on these materials have been carried out by a collaborating group at the Research Center for Biocompatibility of the University of Eastern Piedmont. Vitamin E has been described as an antinflammatory agent, inhibiting many key events in inflammation, such as release from activated monocytes, monocyte adhesion to endothelial cells, and respiratory burst. Moreover, among the non-antioxidant molecular function of vitamin E, its ability has been reported to activate the protein phosphatase 2 A (PP2A), an enzyme that modulates protein-kinase C activity. Although a broad range of vitamin E effects are known (see Chapter 15, Section 15.7), no reports are available on the effects of vitamin E added UHMWPE on activation of resting granulocytes that represent the first line of biological defence. Because granulocyte activation is an early event in the inflammation process, it has been taken as an indicator for risk assessment of biopolymer-mediated inflammation.

Studies from the Piedmont group indicate that vitamin E-blended UHMWPE was able to induce a PP2A-dependent increase in the matrix metalloproteinase 9 (MMP-9) release from granulocytes in the absence of the typical oxidative stress that is considered a hallmark of granulocytic activation [37]. These data emphasize the possibility that the use of vitamin-E-stabilized UHMWPE could modulate the *in situ* tissue remodeling and immune response through MMP-9 release and growth factor activation at the biomaterial-tissue interface.

Another study was carried out by evaluating the immunoglobulin G (IgG) adsorption onto normal UHMWPE and vitamin E–UHMWPE [38]. Because the cell–biomaterial interactions are mediated mainly from the layer of adsorbed proteins present in a few seconds onto the material surface and, in particular, human macrophages' long-term adhesion seems to be mediated by immunoglobulin G (IgG) and its fragments, then IgG adsorption onto a biomaterial surface could be considered as a potential signal for monocyte–macrophage adhesion and fusion. It was shown that the normal UHMWPE surface was able to adsorb more IgG (especially fragments and single, heavy, chain IgG) compared to vitamin E–UHMWPE. This observation suggests that this surface is less prone to macrophage long-lasting adhesion and, therefore, potentially less prone to foreign-body reaction compared to normal UHMWPE.

## 16.7 VITAMIN E BLENDS AND THRESHOLDS FOR OXIDATIVE STABILITY

Many recent studies, summarized in the previous sections of this chapter, have determined that blending vitamin E with resin powder, in concentrations ranging from 1000 to 8000 ppm prior to consolidation, effectively stabilizes medical grade UHMWPE and prevents oxidation. We have also reviewed the literature of the food packaging industry, in which very low concentrations (100–300 ppm) of vitamin E are used to protect polyethylene from shelf oxidation [15]. However, the effects of trace concentrations of vitamin E on the stability and mechanical behavior of UHMWPE have not been as extensively reported in the orthopaedic literature. The motivation for better understanding the lower threshold limits of stability with vitamin E differ in food packaging and orthopaedics. Cost and appearance are important factors influencing stabilizer choice and dose levels in food packaging. The concerns are different with orthopedic implants. For example, the price of the antioxidant, when factoring all the other costs of delivering an UHMWPE component to market, is not a driving concern. The primary concerns with blending vitamin E in UHMWPE center around the anticipated effectiveness of the stabilization and how long that stabilization can be expected to last *in vivo*. A second important concern, touched upon in the introduction (Figure 16.2), relates to the tendency of vitamin E to reduce the efficiency of crosslinking in UHMWPE. This topic will be more fully discussed in Section 16.9.

A research program was conducted at the first author's institution to determine the minimum vitamin E concentration necessary to protect conventional and highly crosslinked UHMWPE from a severe *in vitro* oxidative challenge [39]. We tested the hypothesis that the oxidation resistance of vitamin-E-blended UHMWPE would be influenced by trace levels of antioxidant (<500 ppm), resin (GUR 1020/1050), and radiation treatment (0, 30, and 75 kGy). Trace concentrations of vitamin E were observed to influence the mechano-oxidative degradation behavior of UHMWPE (Figures 16.5 and 16.6). The minimum concentration of vitamin E needed to stabilize UHMWPE during

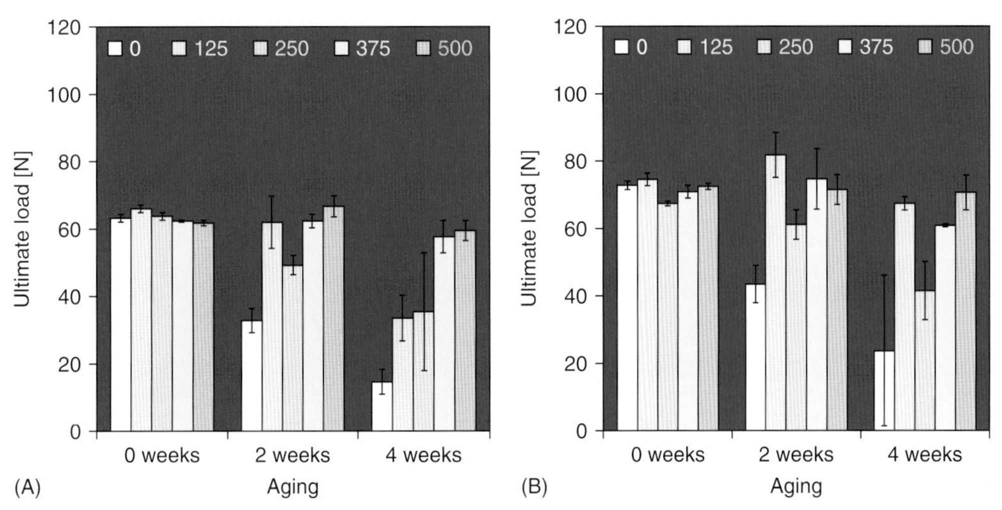


**FIGURE 16.5** Ultimate load in the small punch test for (A) 30 kGy irradiated in nitrogen and (B) 75 kGy irradiated in air GUR 1020 specimens with 0–500 ppm vitamin E and aged 0–4 weeks (ASTM F2003). Similar results were observed for GUR 1050. Adapted from [39].

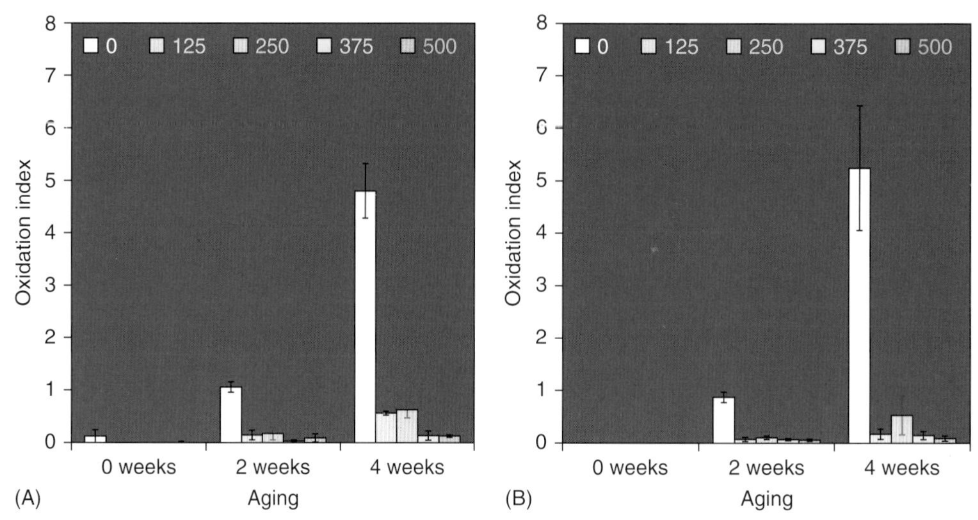


**FIGURE 16.6** ASTM oxidation index for (A) 30 kGy irradiated in nitrogen and (B) 75 kGy irradiated in air GUR 1020 specimens with 0–500 ppm vitamin E and aged 0–4 weeks (ASTM F2003). Similar results were observed for GUR 1050. Adapted from [39].

accelerated aging tests (per ASTM 2003) depended upon the method of radiation processing. For the 30 and 75 kGy irradiated materials, the addition of 125 ppm or more vitamin E was sufficient to maintain baseline mechanical and chemical properties through 2 weeks of accelerated aging. The addition of 375 ppm or 500 ppm, respectively, was necessary to maintain baseline mechanical and chemical properties throughout the 4-week accelerated aging period. Overall, this study demonstrated that elevated concentrations of vitamin E are not necessary for the oxidative stability of conventional and highly crosslinked UHMWPE. The minimum concentration necessary to stabilize irradiated UHMWPE was found to depend on the processing conditions, but not the resin.

The results of this study should not be interpreted as recommending a universal minimum dose level for UHMWPE blended with vitamin E [39]. The research was conducted using pilot scale, vitamin E blending techniques and a standardized compression molding cycle. These processing techniques were judged to be reasonable for the purposes of running a laboratory experiment but were not performed using commercially validated procedures. Because the lower limits of vitamin E stabilization in UHMWPE are process dependent, further research is necessary with large-scale blending techniques, commercial conversion processes, and commercial radiation facilities to establish the lower threshold for a specific manufacturers' materials. Finally, knowledge that the lower limits of stability for UHMWPE can be achieved with less than 500 ppm establishes an important parameter useful in the development of vitamin-E-stabilized conventional and highly crosslinked materials, especially in light of the tradeoff between elevated doses of vitamin E and crosslinked efficiency, as discussed in Section 16.9.

## 16.8 VITAMIN E BLENDS AND MECHANICAL BEHAVIOR

The static mechanical properties of vitamin-E-blended UHMWPE are relatively insensitive to the addition of vitamin E at concentrations up to 4000 ppm [3, 7, 39]. At trace levels (125–500 ppm), vitamin E has negligible effect on the mechanical behavior of both virgin (unsterilized) and gamma irradiated UHMWPE (30 to 75 kGy) [39]. Wolf and coworkers reported on the mechanical properties of GUR 1020 between 1000 and 8000 ppm before and after gamma sterilization in nitrogen (25 kGy). For concentrations of 0 (control), 1000 ppm, 2000 ppm, and 4000 ppm, the elastic modulus, elongation to failure, ultimate tensile strength, and impact strength were indistinguishable for both unsterilized and gamma sterilized materials. For the 8000 ppm blended material, the elastic modulus and ultimate tensile strength were ~20% lower than the other materials tested, whereas the elongation to failure and impact strength remained at baseline values. Either plasticization or differences in crystallization may explain the reduction in modulus and ultimate strength at 8000 ppm observed by Wolf and colleagues [3]. Taken together, previous studies indicate that blending vitamin E with up to 4000 ppm will have minimal impact, either positive or negative, on the static mechanical properties of virgin (unsterilized) medical grade UHMWPE [3, 7, 39]. Indeed, the ASTM standard specification of medical grade, vitamin-E-blended UHMWPE requires that the raw material conform to the physical and mechanical property limits set forth for unstablized material [6].

Previous authors have suggested that vitamin E, at concentrations exceeding 1000 ppm, may improve the fatigue resistance of UHMWPE by acting as a plasticizer [1, 40].

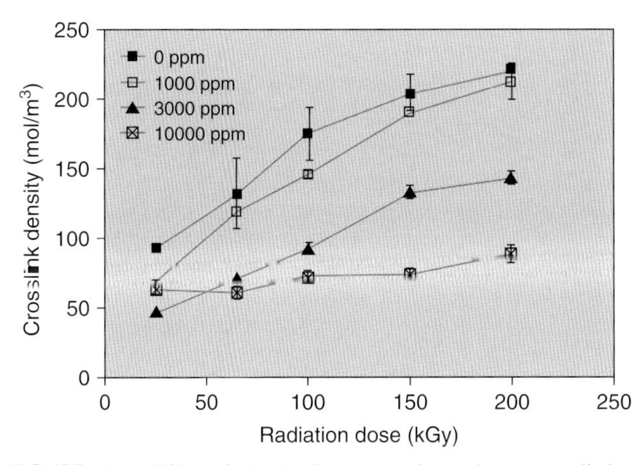


**FIGURE 16.7** Effect of vitamin E concentration and gamma radiation dose on crosslink density of GUR 1050 UHMWPE. Adapted from [8]. Image courtesy of Eric Wysocki, Exponent, Inc.

However, researchers from Berkeley observed no significant difference in the fatigue crack propagation behavior of virgin (unsterilized) UHMWPE when comparing 500 ppm, 1000 ppm, and 5000 ppm concentrations with unstabilized controls [41]. Thus, the precise mechanism for the hypothesized improvement in mechanical properties (if any) for vitamin-E-blended materials below 5000 ppm remains to be confirmed and more fully elucidated in the literature.

## 16.9 VITAMIN E BLENDS AND CROSSLINKING EFFICIENCY

Vitamin E is an effective free radical scavenger, and its ability to stabilize UHMWPE at extremely low concentrations also results in its tendency to reduce the efficiency of crosslinking during irradiation. The ability of vitamin E to reduce crosslinking efficiency has been noted by several previous authors [2, 7, 8]. Because the crosslinking behavior of UHMWPE is related to its wear resistance, the development of vitamin-E-blended materials involves a potential tradeoff in wear resistance versus oxidation stability, as illustrated schematically by Figure 16.2, for concentrations of up to 3000 ppm.

Recently, Oral and coworkers have investigated the efficiency of crosslinking reactions in GUR 1050 UHMWPE blends with up to 10,000 ppm (1%) vitamin E [8] (Figure 16.7). In this study, the specimens were all radiation crosslinked by gamma radiation. At 1000 ppm, Oral and coworkers found that UHMWPE could be highly crosslinked simply by increasing its exposed radiation dosage, consistent with their previous findings [7]. However, the 3000 ppm and 10,000 ppm materials reached a threshold in the maximum crosslinking that could be achieved using this crosslinking technology. In particular, the 10,000 ppm could not be highly crosslinked, even when exposed to 200 kGy gamma radiation (Figure 16.7).

It must be emphasized that these laboratory investigations, using idealized radiation processing steps, should be interpreted as a starting point for commercial development of radiation crosslinked UHMWPE implants containing vitamin E. It is clear, regardless, that simply substituting commercial, off-the-shelf vitamin-E-blended material for use in implants with no other changes in radiation exposure will result in lower crosslink density, and hence lower wear resistance, unless the radiation dosage is increased to compensate for radical scavenging ability of the vitamin E. As a general rule of thumb, when using vitamin-E-containing GUR 1020-E and 1050-E (1000 ppm) and crosslinking using gamma radiation, the absorbed dose may need to be increased by 30% or more to achieve the same crosslinking levels found in unstabilized UHMWPE. In particular, the widely used dose of gamma radiation normally used for sterilization (25–40 kGy) may need to be expanded (e.g., 35–50 kGy) when applied to vitamin-E-containing GUR 1020-E and 1050-E. The ramifications of vitamin E concentration depend entirely on the levels within the material, the radiation processing route, and the joint application for which the stabilized UHMWPE is intended. Ultimately, it is the responsibility of the orthopedic implant manufacturer to determine, in combination with hip and knee wear simulator testing, the appropriate radiation conditions for a particular blend of UHMWPE and vitamin E.

## 16.10 SUMMARY AND CONCLUSION

Within 5 years, UHMWPE blended with vitamin E has emerged as a new biomaterial for use in orthopedic implants. Vitamin E stabilization of polyolefins became widely known for food packaging in the late 1980s, and its use in UHMWPE implants was anticipated in patent filings from the 1980s and 1990s. In 2007, an international standard specification was established by ASTM, followed by commercialization of GUR 1020-E and 1050-E as the first vitamin E-containing UHMWPE resins for use in orthopedics in early 2009. These new developments in stabilized UHMWPE represent the culmination of over a decade of research by academic centers working independently around the world. It is the hope of the authors that the widespread adoption of vitamin-E-containing materials by the orthopedic community will help prevent *in vivo* oxidation in the future, especially in contemporary total knee replacements that continue to be gamma-radiation sterilized.

## 16.11 ACKNOWLEDGMENTS

The authors would like to thank Werner Schneider, Zimmer, Inc., for his feedback regarding the history of VITASUL.

## REFERENCES


1. Tomita N, Kitakura T, Onmori N, Ikada Y, Aoyama E. Prevention of fatigue cracks in ultrahigh molecular weight polyethylene joint components by the addition of vitamin E. *J Biomed Mater Res (Appl Biomater)* 1999;**48**:474–8.
2. Parth M, Aust N, Lederer K. Studies on the effect of electron beam radiation on the molecular structure of ultra-high molecular weight polyethylene under the influence of alpha-tocopherol with respect to its application in medical implants. *J Mater Sci* 2002 October;**13**(10):917–21.
3. Wolf C, Krivec T, Blassnig J, Lederer K, Schneider W. Examination of the suitability of alpha-tocopherol as a stabilizer for ultra-high molecular weight polyethylene used for articulating surfaces in joint endoprostheses. *J Mater Sci* 2002 February;**13**(2):185–9.
4. Wolf C, Lederer K, Muller U. Tests of biocompatibility of alpha-tocopherol with respect to the use as a stabilizer in ultrahigh molecular weight polyethylene for articulating surfaces in joint endoprostheses. *J Mater Sci* 2002 July;**13**(7):701–5.
5. Costa L, Bracco P, Brach del Prever EM, Luda MP. Oxidation and stabilization of UHMWPE. *Trans UHMWPE for arthroplasty: past, present, and future*; 2003. http://www.uhmwpe.unito.it/
6. ASTM F 2695-07. Standard specification for ultra-high molecular weight polyethylene powder blended with alpha-tocopherol (vitamin E) and fabricated forms for surgical implant applications. West Conshohocken, PA: American Society for Testing and Materials; 2007.
7. Oral E, Greenbaum ES, Malhi AS, Harris WH, Muratoglu OK. Characterization of irradiated blends of alpha-tocopherol and UHMWPE. *Biomaterials* 2005 November;**26**(33):6657–63.
8. Oral E, Godleski Beckos C, Malhi AS, Muratoglu OK. The effects of high dose irradiation on the cross-linking of vitamin E-blended ultrahigh molecular weight polyethylene. *Biomaterials* 2008 May 30.
9. Dolezel B, Adamirova L. Method of hygienically safe stabilization of polyolefins against thermooxidative and photoxidative degradation. *Czechoslovakian Socialist Republic Patent 221,403*; 1982.
10. Berzen J, Luketic D. Stabilised polyethylene moulding materials, esp. for implants containing polyethylene with molecular wt. 10^5–10^7, antioxidant, and tocopherol, esp. alpha- tocopherol or vitamin E, as stabiliser. *European Patent 613,923*; 1994.
11. Urata Y. Application of tocopherol for PP resin for food packaging. *Packag Jpn* 1988;**7**:49–54.
12. Laermer SF, Nabholz F. Use of biological antioxidants as polypropylene stabilizers. *Plast Rubber Proc Appl* 1990;**14**(4):235–9.
13. Ho YC, Yam KL, Young SS, Zambetti PF. Comparison of vitamin E, Irganox 1010 and BHT as antioxidants on release of off-flavor from HDPE bottles. *J Plast Film Sheeting* 1994;**10**(3):194–212.
14. Laermer SF, Zambetti PF. Alpha-Tocopherol (vitamin E)—the natural antioxidant for polyolefins. *J Plast Film Sheeting* 1992;**8**(3):228–48.
15. Al-Malaika S. Perspectives in stabilization of polyolefins. *Adv Polym Sci* 2004;**169**:121–50.
16. Mallégol J, Carlsson DJ, Deschêne L. Antioxidant effectiveness of vitamin E in HDPE and tetradecane at 32°C. *Polym Degr Stab* 2001;**73**:269–80.
17. Mallégol J, Carlsson DJ, Deschênes L. A comparison of phenolic antioxidant performance in HDPE at 32–80C. *Polym Degr Stab* 2001:**73**:259–67.
18. Shibata N, Kurtz SM, Tomita N. Advances of mechanical performance and oxidation stability in ultrahigh molecular weight polyethylene for total joint replacement: highly crosslinked and α-tocopherol doped. *J Biomed Sci Eng* 2006;**1**(1):107–23.
19. Shibata N, Tomita N, Onmori N, Kato K, Ikeuchi K. Defect initiation at subsurface grain boundary as a precursor of delamination in ultrahigh molecular weight polyethylene. *J Biomed Mater Res A* 2003 October 1;**67**(1):276–84.
20. Shibata N, Tomita N. The anti-oxidative properties of alpha-tocopherol in gamma-irradiated UHMWPE with respect to fatigue and oxidation resistance. *Biomaterials* 2005 October; **26**(29):5755–62.
21. Teramura S, Sakoda H, Terao T, Endo MM, Fujiwara K, Tomita N. Reduction of wear volume from ultrahigh molecular weight polyethylene knee components by the addition of vitamin E. *J Orthop Res* 2008 April;**26**(4):460–4.
22. Ohashi M, Tomita N, Ikada Y, Ikeuchi K, Motoike T. An observation on subsurface defects of ultra high molecular weight polyethylene due to rolling contact. *Biomed Mater Eng* 1996;**6**(6):441–51.
23. Koizumi M, Tomita N, Tamai S, Oonishi H, Ikada Y. Detection of cracks in polyethylene components of retrieved knee joint prostheses. *J Orthop Sci* 1998;**3**(6):330–5.
24. Todo S, Tomita N, Kitakura T, Yamano Y. Effect of sliding locus on subsurface crack formation in ultra-high-molecular-weight polyethylene knee component. *Biomed Mater Eng* 1999;**9**(1):13–20.
25. Shibata N, Tomita N, Ikeuchi K. Microscopic destruction of ultra-high molecular weight polyethylene (UHMWPE) under uniaxial tension. *Biomed Mater Eng* 2003;**13**(1):47–57.
26. Shibata N, Tomita N, Ikeuchi K. Gamma-irradiation aggravates stress concentration along subsurface grain boundary of ultra-high molecular weight polyethylene (UHMWPE) under sliding fatigue environment. *Biomed Mater Eng* 2003;**13**(1):35–45.
27. Shibata N, Tomita N, Ikeuchi K. Numerical simulations on fatigue destruction of ultra-high molecular weight polyethylene using discrete element analyses. *J Biomed Mater Res A* 2003 March 1;**64**(3):570–82.
28. Tomita N, Kitakura T, Ikada Y, Aoyama E. Fatigue resistance characteristics of polyethlylene containing vitamin E. *J Jpn Orthop Assoc* 1998;**72**(8).
29. Streicher RM. Ionizing irradiation for sterilization and modification of high molecular weight polyethylenes. *Plast Rubber Proc Appl* 1988;**10**(4):221–9.
30. Schaffner S. UHMW polyethylene for implants. *United States Patent No 6,277,390*; 2001.
31. Wolf C, Lederer K, Bergmeister H, Losert U, Bock P. Animal experiments with ultra-high molecular weight polyethylene (UHMW-PE) stabilised with alpha-tocopherol used for articulating surfaces in joint endoprostheses. *J Mater Sci* 2006 December; **17**(12):1341–7.
32. Wolf C, Macho C, Lederer K. Accelerated ageing experiments with crosslinked and conventional ultra-high molecular weight polyethylene (UHMW-PE) stabilised with alpha-tocopherol for total joint arthroplasty. *J Mater Sci* 2006 December;**17**(12):1333–40.
33. Wolf C, Maninger J, Lederer K, Fruhwirth-Smounig H, Gamse T, Marr R. Stabilisation of crosslinked ultra-high molecular weight polyethylene (UHMW-PE)-acetabular components with alpha-tocopherol. *J Mater Sci* 2006 December;**17**(12):1323–31.
34. Wolf C, Lederer K, Pfragner R, Schauenstein K, Ingolic E, Siegl V. Biocompatibility of ultra-high molecular weight polyethylene (UHMW-PE) stabilized with alpha-tocopherol used for joint endoprostheses assessed in vitro. *J Mater Sci* 2007 February 3.

35. Bracco P, Brunella V, Zanetti M, Luda MP, Costa L. Stabilisation of ultra-high molecular weight polyethylene with vitamin E. *Polym Degr Stab* 2007;**92**(12):2155–62.
36. Costa L, Carpentieri I, Bracco P. Post electron-beam irradiation oxidation of orthopaedic UHMWPE blended vith vitamin E. *Polym Degr Stab* 2008. In press.
37. Reno F, Bracco P, Lombardi F, Boccafoschi F, Costa L, Cannas M. The induction of MMP-9 release from granulocytes by vitamin E in UHMWPE. *Biomaterials* 2004 March;**25**(6):995–1001.
38. Reno F, Cannas M. UHMWPE and vitamin E bioactivity: an emerging perspective. *Biomaterials* 2006 June;**27**(16):3039–43.
39. Kurtz SM, Dumbleton J, Siskey RS, Wang A, Manley M. Trace concentrations of vitamin E protect radiation crosslinked UHMWPE from oxidative degradation. *J Biomed Mater Res A* 2008 June 18. In press.
40. Oral E, Wannomae KK, Hawkins N, Harris WH, Muratoglu OK. Alpha-tocopherol-doped irradiated UHMWPE for high fatigue resistance and low wear. *Biomaterials* 2004 November; **25**(24): 5515–22.
41. Kohm A, Furmanski J, Pruitt L. Effect of alpha-tocopherol on UHMWPE fatigue resistance. *Trans of the 52nd Orthop Res Soc* 2006;**31**:662.

Chapter 17

# Composite UHMWPE Biomaterials and Fibers

Steven M. Kurtz, PhD



## 17.1 INTRODUCTION

At a nanometer length scale, UHMWPE is inherently a composite material consisting of an amorphous matrix reinforced by stiffer, crystalline lamellae. UHMWPE composites can also be further engineered at a micro- or nanometer length scale by blending polymer powder resin with micro- or nanoparticles and fibers before consolidation. UHMWPE fibers provide another route to fabricating composites because they can be woven into sheets or plies forming two- and three-dimensional structures with direction-dependent properties. In addition, porous UHMWPE can be created as either an intermediate or final step in the production of open-cell, closed-cell, and reinforced composites. Thus, UHMWPE can play the role of matrix and fiber in a broad range of composite materials, depending upon the application.

UHMWPE composite materials and fibers are widely used in military, industrial, and consumer applications. UHMWPE composite films blended with glass fibers are used as insulative barriers in lead automotive batteries (Figure 17.1), whereas porous UHMWPE materials find applications as water filters. UHMWPE fibers are used in military and personal protection equipment, such as body armor [1] and cut-resistant gloves for kitchen use (Figure 17.2). Overall, the current commercial success for a wide range of UHMWPE fiber and matrix composites with non-medical applications provides motivation for exploring these materials as candidate biomaterials.

In the 1970s, carbon fiber-reinforced (CFR) UHMWPE composites were considered for orthopedic implants and were even commercially introduced (Poly II, Zimmer, Inc., Warsaw, Indiana, USA). However, catastrophic short-term clinical failures [2, 3] eventually led to the abandonment of Poly II and the perception for decades, among surgeons and researchers alike, that UHMWPE composite materials may not be appropriate for orthopedic bearing applications. Today, with the interest in UHMWPE development that was stimulated by radiation and thermal processing techniques, as well as the growing use of UHMWPE composites in nonmedical arenas, there is renewed curiosity about these materials for biomedical applications.

In this chapter, we begin with a historical overview of CFR UHMWPE composites used in orthopedics (Poly II). This chapter also reviews recent advances in UHMWPE

**FIGURE 17.1** Battery separators produced using UHMWPE composites of silica and oil. Image courtesy of Harvey Stein, Ticona, Inc.

matrix and fiber composites for biomedical applications. Peer-reviewed literature has been relied upon wherever possible when compiling this chapter. However, many of the composite and fiber applications are sufficiently novel and proprietary that product information could only be obtained from commercial suppliers or by interviewing historical witnesses. The reader should interpret the nominal values provided by suppliers as such and independently verify key properties that are judged to be essential for biomedical implant performance.

## 17.2 CFR UHMWPE COMPOSITE: POLY II

One carbon-UHMWPE composite, commonly known as Poly II (Zimmer, Inc., Warsaw, Indiana, USA), was developed commercially and used clinically. The CFR-UHMWPE was reinforced by chopped, randomly oriented carbon fibers in a direct compression molded UHMWPE matrix [4]. The carbon fiber reinforcement was initially considered to be responsible for improved wear behavior relative to UHMWPE during initial experimental testing conducted by the manufacturer [5]. However, further studies ultimately revealed that such performance came at the expense of the ductility, decreased crack resistance, and the fiber-matrix interface of the composite [6]. Subsequent wear studies also showed evidence of fiber disruption at the surface and abrasive wear of the metallic counterface [7]. Furthermore, occasional difficulties in the manufacture of the composite material resulted in incomplete consolidation of the powder and carbon fibers in certain implants [8]. Thus, after its clinical introduction, Poly II was found to exhibit wear, fracture, and extensive delamination [2, 3, 9], and, as a result, the material was eventually withdrawn [10].

The clinical failure modes of carbon-UHMWPE composites were not well simulated by the standard testing and analysis methods available at the time the materials were developed. Indeed, initial evaluations of carbon powder- and carbon fiber-reinforced UHMWPE composites provided generally encouraging results. Several studies indicated enhanced tribological and mechanical properties of carbon-UHMWPE composites compared to neat UHMWPE [5, 11–13]. For example, Galante and Rostoker [11] examined a composite containing 25 wt% graphite powder in UHMWPE and, during preliminary pin-on-disc testing in water, found it to have one-seventh to one-thirtieth of the wear of UHMWPE when tested against CoCr alloy. Similarly encouraging results were initially predicted for Poly II. A study by Ainsworth et al. [5] found carbon fiber-reinforced Poly II to have greater compressive and flexural yield strengths and elastic moduli, in addition to a lower wear rate and higher stiffness. Rostoker and Galante observed that carbon fiber reinforcement was responsible for a significant decrease in creep strain in the composite [12]. A study by Wright et al. reported that carbon fiber-reinforced UHMWPE was 88% stiffer than regular UHMWPE and was capable of withstanding 17% greater compressive loads [13].

Biocompatibility studies of CFR-UHMWPE conducted during the 1970s also determined its response to be satisfactory [14]. Poly II was found by Groth et al. to be free of any significant foreign body reactions after clinical implantation of 12 to 15 months [15]. Rushton and Rae investigated the intra-articular response to particulate Poly II in mice and found it to be similar to that of neat UHMWPE [16].

Other studies, conversely, cast doubt on the ultimate viability of carbon-UHMWPE composite usage in hip and knee joint replacement, especially after clinical exposure. After longer-duration pin-on-disc studies using 25% graphite powder-filled UHMWPE, Rostoker and Galante reported that the short-term improvement in wear resistance appeared to be lost [17]. Sclippa and Piekarski

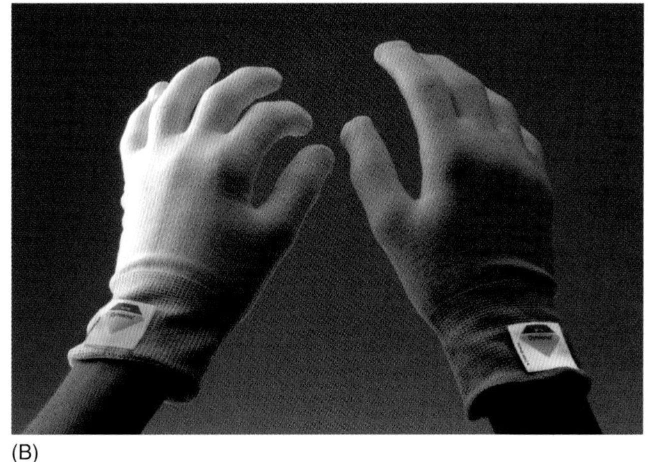

(A) (B)

FIGURE 17.2 Body armor (A) and gloves (B) woven from UHMWPE fibers (Dyneema). Images courtesy of Carina Snijder, DSM Dyneema.

observed poor performance of the reinforced composite in tension due to fiber damage during injection molding as well as generally poor interfacial fiber-matrix bond strength [18]. Others reported issues with polymer molding and incomplete particle fusion. Wright et al. examined clinical results from retrieved components and found that the presence of carbon fibers prevented a firm locking of the polyethylene particles that constituted the matrix of Poly II knee-replacement components [2, 3]. Connolly et al. also found little fiber-matrix adhesion in the composite, such that the reinforced UHMWPE had approximately an eight times greater fatigue crack growth rate compared to plain UHMWPE [6]. An example of poor adhesion of carbon fibers to the UHMWPE matrix is illustrated by scanning electron microscopy of the worn bearing surface of a retrieved patellar component (Figure 17.3). In addition to poor fiber-matrix adhesion, inadequate process control of the direct compression molded components also contributed to incomplete consolidation and short-term clinical failures of Poly II components in some cases [8].

FIGURE 17.3 Exposed carbon fibers at the worn bearing surface of a retrieved Poly II patellar component showing fiber pull-out and poor adhesion to the UHMWPE matrix. The scanning electron microscopy image was taken at 350 times magnification and provided courtesy of Francisco Medel, PhD, Drexel University.

A wear study by McKellop et al. indicated that Poly II exhibited a 10 times greater wear rate than UHMWPE against a variety of common counterfaces, such as 316 stainless steel and CoCr alloy [19]. Because the major mechanism of wear in carbon fiber-reinforced polyethylene is abrasive wear that induces the drawing out of fibers from bearing surfaces [20], the interface strength is of critical importance in the overall performance of Poly II and similar composites.

The clinical use of Poly II persisted into the 1980s until it was eventually abandoned, but no studies have documented its long-term performance. Surprisingly, some of these CFR-UHMWPE tibial components have managed to survive long-term implantation in patients (Figure 17.4). These observations prompted the author and colleagues at Drexel University to revisit the clinical performance of Poly II.

Forty-two Poly II tibial inserts were retrieved at revision surgery from three medical institutions. The retrievals were produced in the Total Condylar, Insall-Burstein, and Miller-Galante designs. Twenty-four conventional UHMWPE components of the same designs served as controls. Patient and clinical information, including implantation time and reason for revision, were available. Surface damage (per Hood et al.) was assessed on all retrievals studied using the methodology described in Chapter 8.

Patient age, weight, and implantation time were statistically comparable for both groups. However, Poly II had a wider implantation range (3.7–32.8y) than UHMWPE (3.5–17.0y) with the same average (11.1y). Patellar complications (n = 22) prevailed in patients with carbon fiber-reinforced components, but the incidence of wear and metalosis was statistically indistinguishable. Poly II inserts had less surface damage within all regions when compared to conventional UHMWPE components. Also, Poly II was

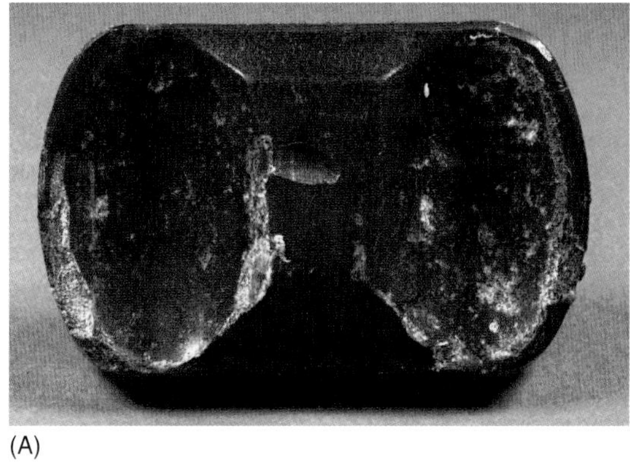

(A)

(B)

FIGURE 17.4 CFR UHMWPE (Poly II) tibial component (A) and patellar component (B) retrieved after 32 years of implantation. Implanted in 1975, this Total Condylar knee replacement was revised in 2006 for instability and patellar dislocation. Image provided courtesy of Francisco Medel, PhD, Drexel University.

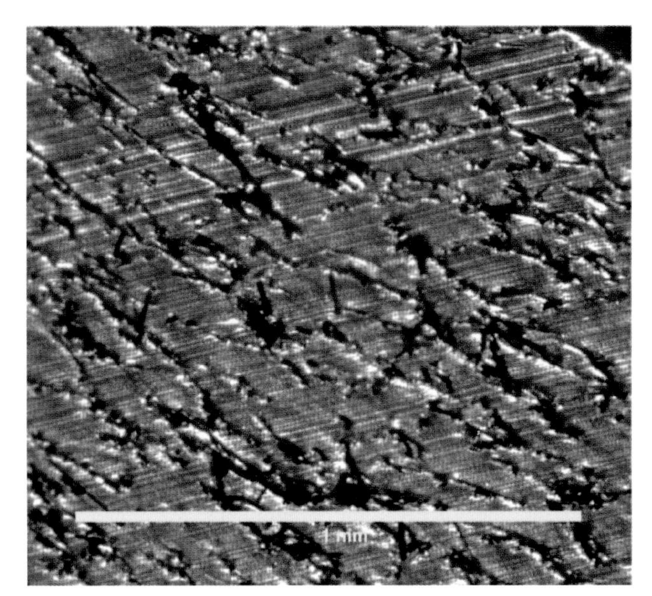


FIGURE 17.5 Dark-field optical micrograph of a 20 μm-thick section of a retrieved Poly II insert. Note the segregation of carbon fibers to the UHMWPE grain boundaries. The yellow scale bar is 1 mm. Image provided courtesy of Dan MacDonald, Drexel University.

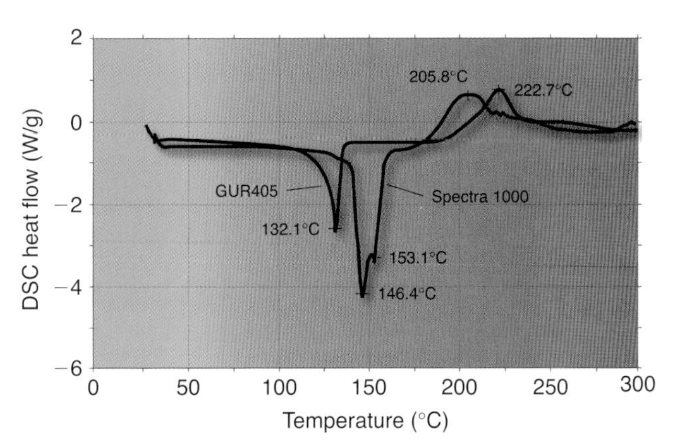


FIGURE 17.6 Differential scanning calorimetry trace of UHMWPE and UHMWPE fibers (Spectra). Adapted from [24]. Image courtesy of Christopher Espinosa, Exponent, Inc.

less sensitive to pitting and delamination but exhibited significantly higher damage scores for abrasion and embedded debris. Dark-field microscopy of long-term implanted Poly II retrievals showed well-consolidated fibers in the UHMWPE matrix (Figure 17.5). In Figure 17.5, the fibers are segregated to the boundaries of the polymer particles.

Contrary to expectations, the recent findings from clinical retrievals document the long-term clinical survivability of Poly II tibial bearings of both cruciate retaining (CR) and posterior stabilizing (PS) designs. Analysis of these long-term retrieved bearings demonstrated complete consolidation through the thickness of the components. Improved understanding of long-term Poly II retrievals may therefore provide motivation to revisit carbon fiber-reinforced polymeric bearings for joint replacement in the 21st century.

## 17.3 UHMWPE HOMOCOMPOSITES

Homocomposites, also referred to as self-reinforced composites, are defined as materials in which the matrix and reinforcement are produced from the same material, in this case UHMWPE. In the 1990s, just prior to the adoption of radiation crosslinking by orthopedics manufacturers, UHMWPE homocomposites were evaluated by multiple research groups for joint replacement applications [21–28]. Self-reinforced UHMWPE materials were developed both by sintering oriented fibers together, as well as by reinforcement of a polymer powder-based matrix with UHMWPE fibers. As we shall see in a subsequent section of this chapter, UHMWPE fibers offer much higher tensile modulus (113 GPa) and ultimate tensile strength (2–4 GPa) when compared with bulk UHMWPE consolidated from resin powder [24, 26]. Furthermore, the crystalline melt transition for UHMWPE fibers is ~10°C higher than powder (Figure 17.6) [24]. As a result, researchers hypothesized that UHMWPE fibers could be compression molded, either by themselves or in combination with resin powder, within a narrow temperature range to retain a "memory" of their initial orientation [22, 23, 25]. The chemical composition and crystallization behavior of the fibers during fabrication of the composites was also hypothesized to promote fiber-matrix compatibility, localized crystallization, and adhesion [24–26].

Polyethylene homocomposites, fabricated by compaction and sintering of continuous HDPE fibers, were investigated by researchers from the University of Leeds in the early 1990s [29–33]. The rationale for fiber compaction processing was to compress and heat only the surfaces of oriented fibers, such that melting and sintering occurred without disrupting the oriented morphology of each fiber. Oriented HDPE composites fabricated using this compaction and sintering technique retained up to 90% of the tensile moduli of the initial fibers [32]. However, the weight average molecular weight of the HDPE fibers was 150,000 g/mole, about an order of magnitude lower than that of UHMWPE.

Researchers at Zimmer, Inc. and Northwestern University successfully adopted this strategy to fabricate oriented UHMWPE composites for orthopedic bearings by sintering compressed continuous Spectra 1000 fibers [21, 27]. Although the modulus and strength of the UHMWPE composites increased by an order of magnitude when tested along the direction of fiber orientation, there was concern that the transverse mechanical properties would be diminished. The wear behavior of these sintered-fiber UHMWPE composites was not reported [34]. However, in their patent application, Price and coworkers taught that

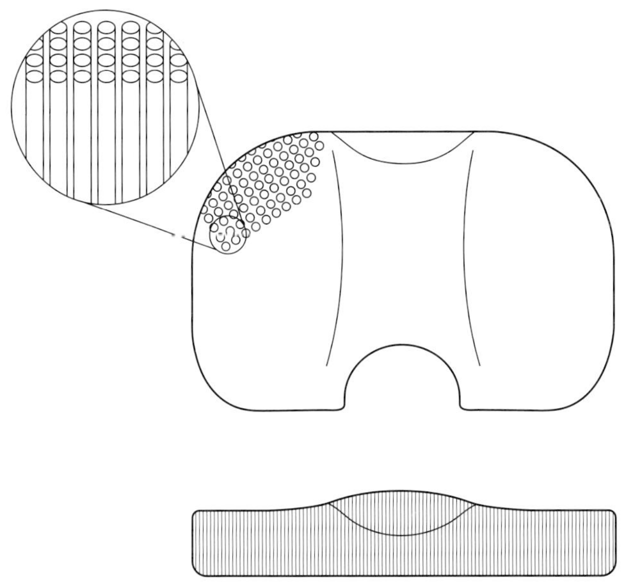

FIGURE 17.7 Schematic of a self-reinforced UHMWPE composite described by Price et al. [21]. In this concept drawing, the sintered UHMWPE fibers are oriented orthogonal to the bearing surface. Image courtesy of Eric Wysocki, Exponent, Inc.

orienting the compacted Spectra 900 fibers perpendicular to the bearing surface would provide the best abrasion resistance (Figure 17.7) [21].

Self-reinforced UHMWPE fiber/matrix composites for orthopedic bearings were evaluated by Deng and Shalaby from Clemson University [22–24]. Their methodology for fabricating the UHMWPE composites, involving both UHMWPE fibers and powder, was taught in United States Patent No. 5,824,411 [22], and supporting data were published in *Biomaterials* [24]. Deng and Shalaby's studies were performed using continuous Spectra 1000 UHMWPE fibers and GUR 405 UHMWPE resin powder (Figure 17.8). For a variety of ply layups, the researchers were able to show improved ultimate strength, impact strength, and creep resistance for continuous fiber-reinforced UHMWPE composites with up to 8% weight fraction of fibers as compared with controls (Figure 17.9). However, no improvement in wear behavior was reported.

Mosleh also described a process of dry blending Spectra 1000 fibers, this time with GUR 415 UHMWPE resin powder prior to compression molding [25, 26]. The Spectra fibers were initially cut to 25 mm length before blending with resin powder in an industrial mixer; the final as-molded fiber lengths were not reported. Below 50% volume fraction of randomly mixed fibers, the tensile and compressive strength of the homocomposite was found to increase with respect to unfilled controls. At high volume fractions (75% fiber) the composite strength decreased, presumably due to difficulties in mixing and bonding [25]. Improved wear resistance was noted under unphysiologic, dry wear test conditions.

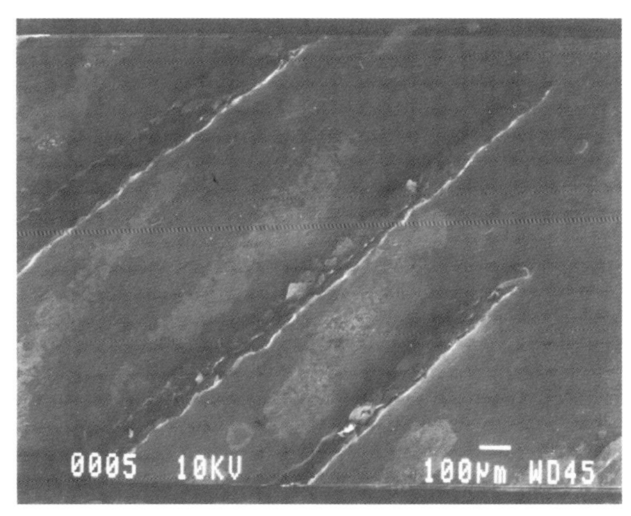


FIGURE 17.8 Scanning electron microscope image of self-reinforced UHMWPE composite microstructure incorporating Spectra UHMWPE fibers, developed by Deng and Shalaby [24]. Image provided courtesy of Meng Deng, PhD, Ethicon, Inc.

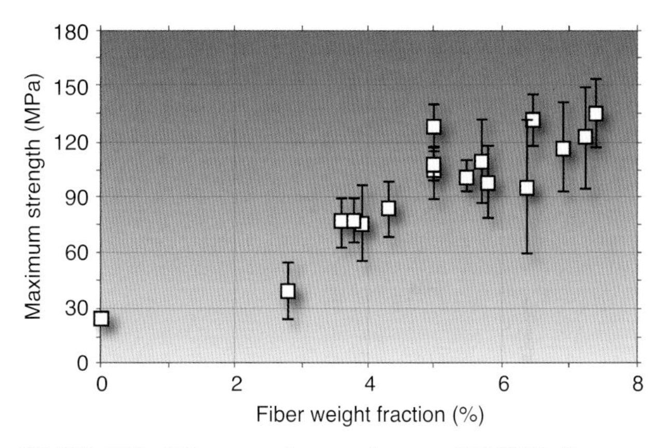


FIGURE 17.9 Ultimate tensile strength versus UHMWPE fiber content, adapted from Deng and Shalaby [24]. Image courtesy of Christopher Espinosa, Exponent, Inc.

UHMWPE fabric-reinforced UHMWPE homocomposites have also been investigated for orthopedic applications, but with mixed results [26, 28, 35]. In a study by Suh et al., Spectra 1000 fabrics were blended with GUR 415, resulting in a 60% fabric volume fraction [26]. The composite layup was produced manually by layering dry fabric sheets and powder in the mold. The authors noted lower wear of the fabric-based composite in dry, but not under bovine-serum lubricated, conditions in a unidirectional reciprocating wear tester [26].

In an effort to further improve the fiber-matrix interfacial strength, Cohen et al. devised a method of developing a prepreg UHMWPE fabric of Spectra 1000 fibers with a solution of GUR 4113 UHMWPE [35]. In the final processing step, the UHMWPE fabric was compression molded into sheets and wear tested in a unidirectional cylinder-on-flat wear tester [28]. However, two of the three compression molded composite samples were not fully consolidated and displayed higher "catastrophic" wear when compared with isotropic controls.

The published literature on self-reinforced UHMWPE homocomposites has not advanced substantially in the past decade. Despite research interest in this topic during the late 1990s, self-reinforced UHMWPE homocomposites were never commercialized for total joint replacement applications. Available data on this topic was generated with Spectra fibers and historical UHMWPE resins, and none of the UHMWPE homocomposites were evaluated with elevated levels of crosslinking. Regardless, depending upon whether chopped or continuous fibers were employed, the improvement in mechanical properties of the UHMWPE homocomposites could be much less than would be expected based on the stiffness of the Spectra fibers, and substantial gains in wear performance could not be convincingly demonstrated in the literature. Furthermore, difficulties encountered by some researchers with processing of the homocomposites [28] likely foreshadowed potential challenges for commercialization when compared with radiation-processed materials.

In retrospect, self-reinforced UHMWPE represents an interesting body of academic research that has yet to reach fruition. Two additional historical factors played a role in the abandonment of this line of scientific inquiry. First, the advent of crosslinked UHMWPE at this time in orthopedic history eclipsed many other contemporary research interests in favor of what was perceived as the "dominant" bearing technology. Second, and perhaps more importantly, all of the pilot research on this topic was performed with Spectra 900 and 1000 fibers produced by Honeywell, which was unwilling to distribute its fibers for use in medical implants at the time. Some of these historical factors may have somewhat abated in the early 21st century. As we shall see in a subsequent section of this chapter, with the recent commercialization of medical grade fibers, one of the historical limitations for proceeding with self-reinforced UHMWPE bearing material research may no longer be a current concern.

## 17.4 UHMWPE MATRIX COMPOSITES FOR ORTHOPEDIC BEARINGS

Although self-reinforced UHMWPE was effectively abandoned for orthopedic bearings, interest has continued in other UHMWPE matrix composites, driven partly by new initiatives in nanocomposites materials science and polymer science. Indeed, UHMWPE matrix composites have been employed industrially for decades, and researchers have continued to explore the effect of historical "standard" fillers on the mechanical and dielectric properties of UHMWPE [36–40]. Solid lubricants, such as glass, graphite, ceramics (e.g., kaolin, $Al_2O_3$), and molybdenum disulfide ($MoS_2$), with particle size ranges on the order of microns, have been evaluated with UHMWPE for improved tribology for industrial applications [36, 41, 42]. In recent years, UHMWPE composite research has expanded to include evaluating the effects of nanoparticles [43, 44], nanofibers [45], and nanotubes [40, 46] on polymer properties, for both industrial and orthopedic applications.

UHMWPE composite research for orthopedic bearings has also recently expanded to include mineral and quasicrystalline fillers [47]. Xie and colleagues have evaluated chemically crosslinked UHMWPE filled with micron-sized quartz particles [47]. Anderson et al. [48] have studied blends of Al-Cu-Fe powders and UHMWPE for orthopedic applications. These new composite UHMWPE bearing materials are still in the early phases of experimental investigation.

At a microscale, untreated glass, aramid, Kevlar, and carbon-based fillers are all insoluble in UHMWPE, and after blending will coat the UHMWPE powder surface [37]. Although fillers can be dry blended with resin prior to compression molding, clumping and aggregation of fillers is a well-known problem [37]. Solvent dispersion of the filler and ultrasonication may be used to improve filler dispersion over the surface of the powder particles. Surface treatment of the filling material, such as by plasma or chemical etching, may also be employed to improve fiber-matrix adhesion [49]. During UHMWPE consolidation, filling materials are typically aggregated at grain boundaries [37], and thus there may be a trade-off in ultimate mechanical properties associated with elevated filler concentrations. On the other hand, elastic modulus and yield strength may be increased by the presence of fiber reinforcement. Thus, a range of processing techniques can be used to optimize the fabrication and properties of fiber- and particle-filled UHMWPE composites.

In a departure from historical approaches to introducing fillers in UHMWPE, researchers from Colorado State University have developed a novel UHMWPE-hyaluronan microcomposite [50, 51]. This microcomposite is detailed in Chapter 18. Hyaluronan is a natural lubricant in articular cartilage. By incorporating this biomolecule into UHMWPE, researchers theorized that wear could be reduced. Because hyaluronan is hydrophilic and UHMWPE is hydrophobic, fabrication of an UHMWPE-hyaluronan microcomposite is by no means a straightforward proposition and is accomplished via a complex and elegant multistep procedure. The process, which is described in greater detail in Chapter 18, begins with the production of a porous UHMWPE substrate, which constitutes the "matrix" of the microcomposite. The hyaluronan is complexed with quaternary ammonium cations (CTA+) and silylated [52]. A solution of silyl hyaluronan-CTA is then diffused into the open celled porous UHMWPE matrix, chemically crosslinked *in situ*, and then hydrolyzed to produce hyaluronan. A similar technique is used to generate a surface layer of hyaluronan on the surface of the UHMWPE. In the final step of the process, the porous UHMWPE-hyaluronan material is compression molded a second time to generate a consolidated microcomposite. By tuning the processing conditions, a microcomposite can be achieved with mechanical properties that exceed those specified in ASTM F648, with modest wear reduction in a multidirectional pin-on-disk wear tester [51].

Improvement in wear is frequently cited as a motivation for current composite research in UHMWPE for orthopedic implants [43, 44, 47–51]. Thus far, the use of fillers alone has not proven effective in reducing the wear rate of UHMWPE by an order of magnitude, as has been observed with extensive radiation crosslinking. Whereas conventional UHMWPE may have been the state of the art when early research on UHMWPE composites was initiated, today UHMWPE matrix composites need to demonstrate superior properties when compared with unfilled radiation crosslinked materials. Because the tribology of UHMWPE in artificial joints is strongly dependent on the kinematics and lubricant, additional research is needed to fully characterize the biotribological behavior of UHMWPE micro- and nanocomposites for specific orthopedic bearing applications.

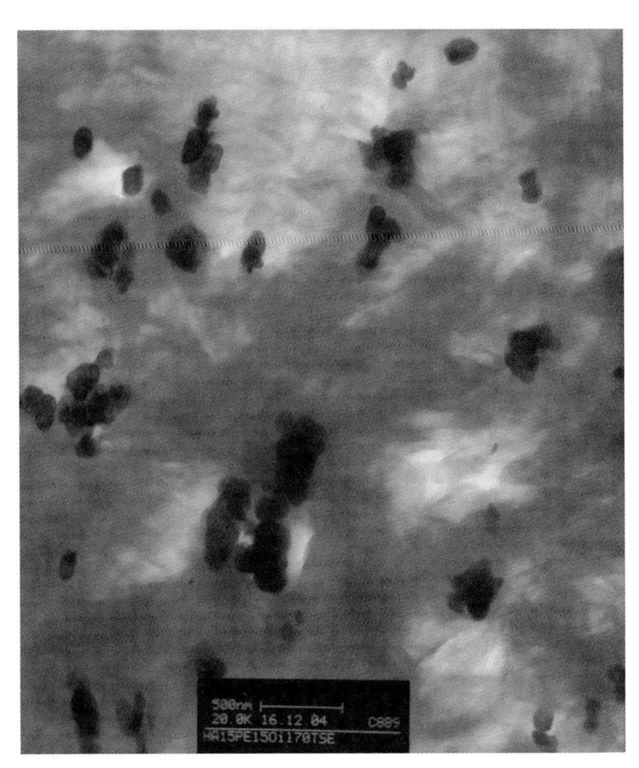


**FIGURE 17.10** Transmission electron microscope image (unstained) of UHMWPE-hydroxyapatite nanocomposite, showing the dispersion of HA nanoparticles in a UHMWPE matrix (HA volume content: 12.9%; hot draw ratio: 0) [64–66]. Figure provided courtesy of Professor Liming Fang, PhD, Yang Leng, and Ping Gao, South China University of Technology.

## 17.5 POLYETHYLENE-HYDROXYAPATITE COMPOSITES

Although most studies of UHMWPE composite biomaterials have focused on improvement in the mechanical properties and wear resistance of the polymer, another line of research has recently emerged in the fabrication of bioactive polyethylene composites. Bonfield et al. initiated research in polymer-based biocomposites for bone replacement [53]. One such material, HA-PEX, is a composite of 40% by volume hydroxyapatite (HA) particles with HDPE [54]. Hydroxyapatite (HA) is a bioactive ceramic with a structure similar to bone mineral. Because of its low molecular weight (less than 1 million), HA-PEX composites have certain practical versatility in processing, including by injection molding. The median particle size of the HA in HA-PEX is on the order of microns, and the weight average molecular weight of the HDPE in HA-PEX is 225,000 [54]. The bioactivity and biocompatibility of HA-PEX has been well documented in the literature [55–63]. Wang and colleagues have shown that substantial improvements in mechanical properties can be achieved by hydrostatic extrusion of HA-PEX [54]. However, the mechanical properties of HA-PEX are not suitable for high load-bearing applications, and its current clinical products include middle ear prostheses (Gyrus AMCI–ENT, Bartlett, Tennessee, USA).

Fang and coworkers have successfully compounded HA nanoparticles with UHMWPE [64–66] with the goal of producing higher strength bone substitutes than HA-HDPE composites (Figure 17.10). To overcome the difficulties associated with dispersion of the HA, Fang et al. developed a wet ball milling and swelling technique to produce particles on the order of 50 nm in diameter. These were mixed with UHMWPE using a variety of techniques, including compression molding and injection molding. The UHMWPE was swollen in pharmaceutical grade paraffin oil to facilitate extrusion [66]. After fabrication, the excess paraffin was extracted using hexane. The resulting 20% HA biocomposite had an elastic modulus of 7 GPa and elongation to failure of 375%. The large deformation mechanical behavior of the biocomposite may be attributed to the fine dispersion of the nanoparticles in the UHMWPE matrix (Figure 17.10). Although UHMWPE-HA composites has passed the proof-of-concept phase, additional research is certainly warranted on the biocompatibility of this material to clinically demonstrate its suitability as a bone replacement material.

## 17.6 UHMWPE FIBERS

As discussed previously, UHMWPE fibers form the building blocks of homocomposites. However, by weaving

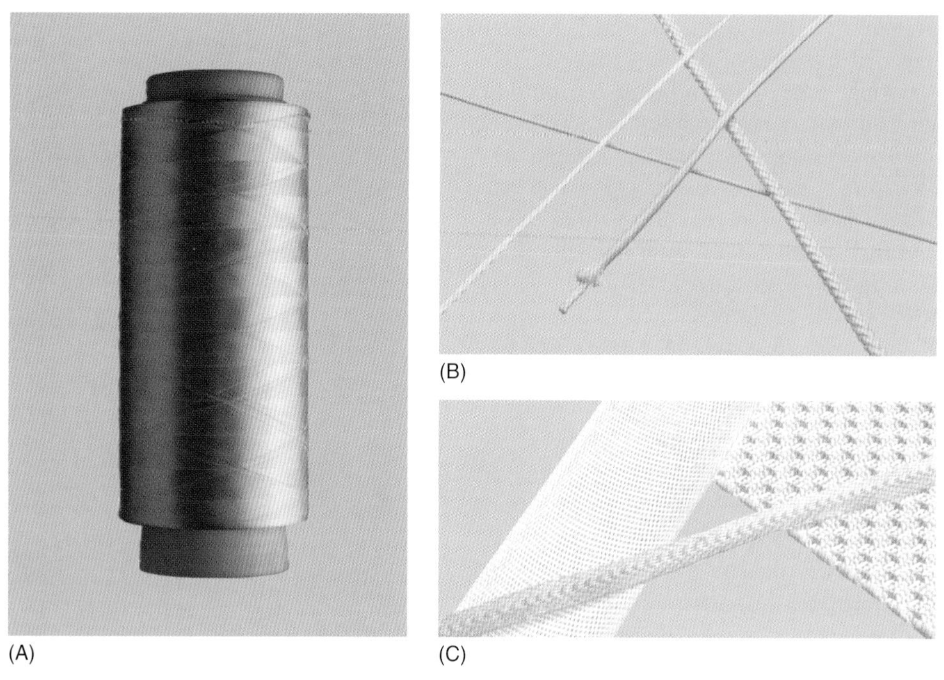

**FIGURE 17.11** Medical grade UHMWPE fibers (Dyneema Purity) (A), braided sutures (B), and two-dimensional weaves (C). Images courtesy of Carina Snijder, DSM Dyneema.

in two or three dimensions, composite structures of UHMWPE fibers can be constructed (Figure 17.11). Cut-resistant gloves are good example of such three-dimensional composites fabricated from UHMWPE fibers (Figure 17.2).

UHMWPE fibers are produced by a gel spinning technique, whereby a UHMWPE solution is extruded through a cone-shaped die (referred to as a *spinnaret*) and then mechanically drawn to orient the molecules parallel to the fiber direction [67]. The uniaxial molecular orientation of the fiber contributes to its high modulus (as high as 222 GPa) and strength (around 8 GPa) [67]. UHMWPE fiber production technology was originally developed by DSM in the Netherlands during the 1980s.

Today, UHMWPE fibers are produced by Honeywell and DSM. Honeywell's UHMWPE fibers are produced under the trade name Spectra and are intended for industrial applications only. Honeywell does not support the use of their UHMWPE fibers for implant applications.

DSM industrial fibers are produced under the trade name Dyneema. In 2006, DSM completed ISO certification of its fiber production for medical applications, which were commercialized under the "Dyneema Purity" trade name. In 2008, DSM announced that it was expanding its UHMWPE fiber products to include three grades (UG, TG, SX), which vary in diameter and strength. In contrast with Honeywell, DSM's UHMWPE fibers are developed specifically for medical device applications, including sutures, for example (Figure 17.11). DSM has conducted biocompatibility and cytotoxicity testing of their fibers to facilitate regulatory burden for starting companies. To date, these fibers have been used primarily in sutures and may be suitable for ligament and tendon repair.

## 17.7 SUMMARY

Over the past 30 years, an extensive body of research has evolved on UHMWPE composites for biomedical applications. Because of its powder size and melt viscosity, UHMWPE poses several serious challenges in processing, which may have been responsible for the short-term clinical failures observed with Poly II during the 1980s. Despite the negative connotations associated with composites from the Poly II experience, researchers have continued to develop innovative techniques to disperse structural and even bioactive fillers in UHMWPE.

The introduction of medical grade UHMWPE fibers is another recent development that will likely make a new generation of composite implants possible. Knowledge in the public domain about UHMWPE fibers for medical devices remains in its infancy. With the degrees of freedom available afforded by fibers and biocompatible fillers, UHMWPE promises to be an expanding platform for innovation by surgeons and biomedical engineers in the coming decade.

## 17.8 ACKNOWLEDGMENTS

The retrieval research covered in this chapter was funded by NIH R01 AR47904. Thanks are due to Clare Rimnac, PhD, Jack Parr, PhD, and Steve Lin, PhD, for insight into the history of Poly II; and Jeremy Gilbert, PhD, for helpful discussions about self-reinforced UHMWPE. The author also thanks Francisco Medel, PhD, Dan MacDonald, and Priyanka Shah, Drexel University, for their collaboration on the retrieval analysis of Poly II; editorial assistance with the chapter; and with the preparation of figures.

## REFERENCES


1. Taylor SA, Carr DJ. Post failure analysis of 0 degrees /90 degrees ultra high molecular weight polyethylene composite after ballistic testing. *J Microsc* 1999 November;**196** (# (Pt 2)):249–56.
2. Wright TM, Rimnac CM, Faris PM, Bansal M. Analysis of surface damage in retrieved carbon fiber-reinforced and plain polyethylene tibial components from posterior stabilized total knee replacements. *J Bone Joint Surg [Am]* 1988;**70**(9):1312–19.
3. Wright TM, Astion DJ, Bansal M, Rimnac CM, Green T, Insall JN, et al. Failure of carbon fiber-reinforced polyethylene total knee-replacement components. A report of two cases. *J Bone Joint Surg [Am]* 1988;**70**(6):926–32.
4. Farling G, inventor; Zimmer Inc., assignee. Human body implant of graphitic carbon fiber reinforced ultra-high molecular weight polyethylene. *United States Patent No. 4,055,862*; 1977.
5. Ainsworth R, Farling G, Bardos D. An improved bearing material for joint replacement prostheses: carbon fiber-reinforced UHMW polyethylene. *Trans 3rd Soc Biomater* 1977;**3**:119.
6. Connelly GM, Rimnac CM, Wright TM, Hertzberg RW, Manson JA. Fatigue crack propagation behavior of ultrahigh molecular weight polyethylene. *J Orthop Res* 1984;**2**(2):119–25.
7. Peterson CD, Hillberry BM, Heck DA. Component wear of total knee prostheses using Ti-6A1-4 V, titanium nitride coated Ti-6A1-4 V, and cobalt-chromium-molybdenum femoral components. *J Biomed Mater Res* 1988;**22**(10):887–903.
8. Parr J. Personal communication. Memphis; 2008.
9. Busanelli L, Squarzoni S, Brizio L, Tigani D, Sudanese A. Wear in carbon fiber-reinforced polyethylene (poly-two) knee prostheses. *La Chirurgia degli organi di movimento* 1996;**81**(3):263–7.
10. Kurtz SM, Muratoglu OK, Evans M, Edidin AA. Advances in the processing, sterilization, and crosslinking of ultra- high molecular weight polyethylene for total joint arthroplasty. *Biomaterials* 1999;**20**(18):1659–88.
11. Galante JO, Rostoker W. Wear in total hip prostheses: an experimental evaluation of candidate biomaterials. *Acta Orthop Scand (Suppl)* 1973;**145**:6–46.
12. Rostoker W, Galante JO. Indentation creep of polymers for human joint applications. *J Biomed Mater Res* 1979 September; **13**(5):825–8.
13. Wright TM, Fukubayashi T, Burstein AH. The effect of carbon fiber reinforcement on contact area, contact pressure, and time-dependent deformation in polyethylene tibial components. *J Biomed Mater Res* 1981 September;**15**(5):719–30.
14. Tetik RD, Galante JO, Rostoker W. A wear resistant material for total joint replacement—tissue biocompatibility of an ultra-high molecular weight (UHMW) polyethylene-graphite composite. *J Biomed Mater Res* 1974 September;**8**(5):231–50.
15. Groth H, et al. Postmortem analysis of carbon-fiber reinforced UHMWPE and UHMWPE prostheses retrieved from a single subject after a service life of 12 to 15 months. *Trans Orthop Res Soc* 1978;**186**.
16. Rushton N, Rae T. The intra-articular response to particulate carbon fibre reinforced high density polyethylene and its constituents: an experimental study in mice. *Biomaterials* 1984 November;**5**(6):352–6.
17. Rostoker W, Galante JO. Some new studies of the wear behavior of ultrahigh molecular weight polyethylene. *J Biomed Mater Res* 1976 March;**10**(2):303–10.
18. Sclippa E, Piekarski K. Carbon fiber reinforced polyethylene for possible orthopedic uses. *J Biomed Mater Res* 1973 January; **7**(1):59–70.
19. McKellop H. Wear of artificial joint materials II. Twelve-channel wear-screening device: correlation of experimental and clinical results. *Eng Med* 1981;**10**(3):123–36.
20. Xiong D. Friction and wear properties of UHMWPE composites reinforced with carbon fiber. *Mater Lett* 2005;**59**:175–9.
21. Price HC, Lin ST, Hawkins ME, Parr JE, inventors; Zimmer, Inc., assignee. Reinforced polyethylene for articular surfaces. *United States Patent No. 5,609,638*; 1997.
22. Shalaby SW, Deng M, inventors; Poly-Med, Inc., assignee. Self-reinforced ultra-high molecular weight polyethylene composites. *United States Patent No. 5,824,411*; 1998.
23. Deng M, Latour RA, Ogale AA, Shalaby SW. Study of creep behavior of ultra-high-molecular-weight polyethylene systems. *J Biomed Mater Res* 1998 May;**40**(2):214–23.
24. Deng M, Shalaby SW. Properties of self-reinforced ultra-high-molecular-weight polyethylene composites. *Biomaterials* 1997 May;**18**(9):645–55.
25. Mosleh M. An UHMWPE homocomposite for joint prostheses. In: Jacobs JJ, Craig TL, editors. *Alternative bearing surfaces in total joint replacement*. West Conshohoken: American Society for Testing and Materials; 1998. p. 256–65.
26. Suh NP, Mosleh M, Arinez J. Tribology of polyethylene homocomposites. *Wear* 1998;**214**:231–6.
27. Megremis SJ, Duray S, Gilbert JL. Self-reinforced composite polyethylene (SCR-PE): a novel material for orthopedic applications. In: Jacobs JJ, Craig TL, editors. *Alternative bearing surfaces in total joint replacement*. West Conshohoken: American Society for Testing and Materials; 1998. p. 235–55.
28. Chang N, Bellare A, Cohen RE, Spector M. Wear behavior of bulk oriented and fiber reinforced UHMWPE. *Wear* 2000;**241**:109–17.
29. Hine PJ, Ward IM, Olley RH, Bassett DC. The hot compaction of high modulus melt-spun polyethylene fibres. *J Mater Sci* 1993;**28**(2):316–24.
30. Olley RH, Bassett DC, Hine PJ, Ward IM. Morphology of compacted polyethylene fibres. *J Mater Sci* 1993;**28**(4):1107–12.
31. Ward IM. New developments in the production of high modulus and high strength flexible polymers. *Progr Colloid Polym Sci* 1993;**92**:103–10.
32. Kabeel MA, Bassett DC, Olley RH, Hine PJ, Ward IM. Compaction of high-modulus melt-spun polyethylene fibres at temperatures above and below the optimum. *J Mater Sci* 1994;**29**(18):4694–9.
33. Kabeel MA, Bassett DC, Olley RH, Hine PJ, Ward IM. Differential melting in compacted high-modulus melt-spun polyethylene fibres. *J Mater Sci* 1995;**30**(3):601–6.
34. Gilbert JL. Personal communication. University of Syracuse; 2008.
35. Cohen Y, Rein DM, Vaykansky L. A novel composite based on ultra-high molecular weight polyethylene. *Compos Sci Technol* 1997;**57**:1149–54.
36. Ramasubramanian N, Krishnamurthy R, Malhotra SK. Tribological characteristics of filled ultrahigh molecular weight high density polyethylene. *Wear* 1993;**162-164**:631–5.

37. Zhang C, Ma C, Wang P, Sumita M. Temperature dependence of electrical resistivity for carbon black filled ultra-high molecular weight polyethylene composites prepared by hot compaction. *Carbon* 2005;**43**(12):2544–53.
38. Xi Y, Ishikawa H, Bin Y, Matsuo M. Positive temperature coefficient effect of LMWPE–UHMWPE blends filled with short carbon fibers. *Carbon* 2004;**42**(8–9):1699–706.
39. Bin Y, Xu C, Zhu D, Matsuo M. Electrical properties of polyethylene and carbon black particle blends prepared by gelation/crystallization from solution. *Carbon* 2002;**40**(2):195–9.
40. Xi Y, Yamanaka A, Bin Y, Matsuo M. Electrical properties of segregated ultrahigh molecular weight polyethylene/multiwalled carbon nanotube composites. *J Appl Polym Sci* 2007;**105**(5):2868–76.
41. Puukilainen E, Saarenpaa H, Pakkanen TA. Compression-molded, lubricant-treated UHMWPE composites. *J Appl Polym Sci* 2007;**104**:1762–8.
42. Guofang G, Huayong Y, Xin F. Tribological properties of kaolin filled UHMWPE composites in unlubricated sliding. *Wear* 2004;**256**(88–94).
43. Xiong DS, Lin JM, Fan DL. Wear properties of nano-Al2O3/UHMWPE composites irradiated by gamma ray against a CoCrMo alloy. *Biomed Mater* 2006;**1**(175–179).
44. Xiong D, Lin J, Fan D, Jin Z. Wear of nano-TiO2/UHMWPE composites radiated by gamma ray under physiological saline water lubrication. *J Mater Sci* 2007 November;**18**(11):2131–5.
45. Ren X, Wang XQ, Sui G, Zhong WH, Fuqua MA, Ulven CA. Effects of carbon nanofibers on crystalline structures and properties of ultrahigh molecular weight polyethylene blend fabricated using twin-screw extrusion. *J Appl Polym Sci* 2008;**107**:2837–45.
46. Gao J, Li Z, Meng Q, Yang Q. CNTs/UHMWPE composites with a two-dimensional conductive network. *Mater Lett* 2008;**62**:3530–2.
47. Xie XL, Tang CY, Chan KY, Wu XC, Tsui CP, Cheung CY. Wear performance of ultrahigh molecular weight polyethylene/quartz composites. *Biomaterials* 2003 May;**24**(11):1889–96.
48. Anderson BC, Bloom PD, Baikerikar KG, Sheares VV, Mallapragada SK. Al-Cu-Fe quasicrystal/ultra-high molecular weight polyethylene composites as biomaterials for acetabular cup prosthetics. *Biomaterials* 2002 April;**23**(8):1761–8.
49. Hofste JM, Schut JA, Pennings AJ. The effect of chromic acid treatment on the mechanical and tribological properties of aramid fibre reinforced ultra-high molecular weight polyethylene composite. *J Mater Sci* 1998 October;**9**(10):561–6.
50. Zhang M, King R, Hanes M, James SP. A novel ultra high molecular weight polyethylene-hyaluronan microcomposite for use in total joint replacements. I. Synthesis and physical/chemical characterization. *J Biomed Mater Res A* 2006 July;**78**(1):86–96.
51. Zhang M, Pare P, King R, James SP. A novel ultra high molecular weight polyethylene-hyaluronan microcomposite for use in total joint replacements. II. Mechanical and tribological property evaluation. *J Biomed Mater Res A* 2007 July;**82**(1):18–26.
52. Zhang M, James SP. Silylation of hyaluronan to improve hydrophobicity and reactivity for improved processing and derivatization. *Polymer* 2005;**46**:3639–48.
53. Bonfield W. Composites for bone replacement. *J Biomed Eng* 1988 November;**10**(6):522–6.
54. Wang M, Ladiezesky NH, Tanner KE, Ward IM, Bonfield W. Hydrostatically extruded HAPEX. *J Mater Sci* 2000;**35**:1023–30.
55. Dalby MJ, Bonfield W, Di Silvio L. Enhanced HAPEX topography: comparison of osteoblast response to established cement. *J Mater Sci* 2003 August;**14**(8):693–7.
56. Dalby MJ, Di Silvio L, Davies GW, Bonfield W. Surface topography and HA filler volume effect on primary human osteoblasts in vitro. *J Mater Sci* 2000 December;**11**(12):805–10.
57. Dalby MJ, Di Silvio L, Gurav N, Annaz B, Kayser MV, Bonfield W. Optimizing HAPEX topography influences osteoblast response. *Tissue Eng* 2002 July;**8**(3):453–67.
58. Dalby MJ, Kayser MV, Bonfield W, Di Silvio L. Initial attachment of osteoblasts to an optimised HAPEX topography. *Biomaterials* 2002 February;**23**(3):681–90.
59. Di Silvio L, Dalby M, Bonfield W. In vitro response of osteoblasts to hydroxyapatite-reinforced polyethylene composites. *J Mater Sci* 1998 December;**9**(12):845–8.
60. Di Silvio L, Dalby MJ, Bonfield W. Osteoblast behaviour on HA/PE composite surfaces with different HA volumes. *Biomaterials* 2002 January;**23**(1):101–7.
61. Huang J, Di Silvio L, Wang M, Tanner KE, Bonfield W. In vitro mechanical and biological assessment of hydroxyapatite-reinforced polyethylene composite. *J Mater Sci* 1997 December;**8**(12):775–9.
62. Rea SM, Best SM, Bonfield W. Bioactivity of ceramic-polymer composites with varied composition and surface topography. *J Mater Sci* 2004 September;**15**(9):997–1005.
63. Rea SM, Brooks RA, Schneider A, Best SM, Bonfield W. Osteoblast-like cell response to bioactive composites-surface-topography and composition effects. *J Biomed Mater Res B Appl Biomater* 2004 August 15;**70**(2):250–61.
64. Fang L, Gao P, Leng Y. High strength and bioactive hydroxyapatite nano-particles reinforced ultrahigh molecular weight polyethylene. *Composites Part B: Eng* 2007;**38**(3):345–51.
65. Fang L, Leng Y, Gao P. Processing of hydroxyapatite reinforced ultrahigh molecular weight polyethylene for biomedical applications. *Biomaterials* 2005 June;**26**(17):3471–8.
66. Fang L, Leng Y, Gao P. Processing and mechanical properties of HA/UHMWPE nanocomposites. *Biomaterials* 2006 July;**27**(20):3701–7.
67. Pennings AJ, van der Hooft RJ, Postema AR, Hoogsteen W, ten Brinke G. High-speed gel-spinning of ultra-high molecular weight polyethlene. *Polym Bull* 1986;**16**:167–74.

Chapter 18

# UHMWPE/Hyaluronan Microcomposite Biomaterials

Susan P. James, PhD, Rachael (Kurkowski) Oldinski, PhD, Min Zhang, PhD and Herb Schwartz, PhD



## 18.1 INTRODUCTION

Other chapters in this *Handbook* detail the various conventional and crosslinked UHMWPE materials that have been used in joint replacement implants since the 1960s. While UHMWPE articulating against a metallic component remains the gold standard for joint replacements, researchers continue to develop modified and improved UHMWPE materials. One such material, BioPoly™, a microcomposite of UHMWPE and hyaluronan (HA) was developed by Professor Sue James and colleagues at Colorado State University. UHMWPE/HA biomaterials are considered microcomposites because a small amount (<5%) of HA is covalently locked into a porous preform of UHMWPE with micron-scale pores before molding the preform to full density.

The vast differences between UHMWPE total joint replacement materials and articular cartilage motivated the development of the UHMWPE/HA biomaterials. UHMWPE is a hydrophobic, "waxy" synthetic polymer, while articular cartilage is a hydrophilic material containing chondrocytes (cells) and an extracellular matrix (ECM) that is comprised of macromolecules, proteoglycans, type II collagen fibers, and water [1]. The proteoglycans are large molecules consisting of a protein core with negatively charged glycosaminoglycan (GAG) side chains. The protein cores of these proteoglycan aggregates are in turn linked to a negatively charged GAG core, HA. When fully hydrated, these large HA-containing molecular complexes have the ability to resist very large compressive loads via hydrostatic pressure [2].

The high concentration of HA (~3 mg/ml) in synovial fluid leads to entanglement and interaction of the GAG chains, imparting viscoelastic gel-like properties to the synovial fluid [3]. While the precise mode of lubrication in articular cartilage depends on sliding speed, joint geometry and loading, it is the interaction of articular cartilage and synovial fluid that plays a key role in the very low friction and wear in synovial joints [4]. There is considerable evidence for the squeeze-film lubrication mode resulting in fluid film separation of the articulating surfaces. The HA-containing synovial fluid forms a gel-like layer that protects articulating surfaces during sliding motion [5]. Thus, the excellent lubrication of natural joints is due to the HA-containing fluid film layer at the surface of the cartilage imparted by the synovial fluid.

This chapter begins with a review of other UHMWPE orthopedic implant surface modifications, the properties of HA, and the medical applications of HA. The synthesis and processing of UHMWPE/HA biomaterials is then described, followed by the chemical, physical, mechanical, and tribological characterization of the UHMWPE/HA biomaterials. Finally, the sterilization, biocompatibility, and commercialization of the UHMWPE/HA biomaterials are covered.

## 18.2 SURFACE MODIFICATION OF UHMWPE

The various bulk material (e.g., crosslinking) and surface structural modifications that are used to enhance the wear resistance of UHMWPE do not take into consideration the surface chemistry of UHMWPE. Indeed, lubrication using the body's own lubricant has not, to date, been a feature of joint replacement implant design. Current total joint replacements operate in the mixed or boundary lubrication regimes, relying on the inherent low friction and wear properties of UHMWPE [6, 7]. UHMWPE implants do not enjoy the lower friction and wear of natural joints because the surface chemistry of UHMWPE is very different from that of articular cartilage. The extreme hydrophobicity of UHMWPE limits the ability of synovial fluid constituents to interact with and lubricate the implant surface. Because the composition of synovial fluid present after postoperative healing following implantation is similar to that present in a healthy joint [8], some efforts have been made to modify the surface chemistry of UHMWPE or to use materials other than UHMWPE that will be better lubricated in the joint.

Widmer and Spencer [9] treated UHMWPE with oxygen plasma. The treated surface was hydrophilic and had a faster and modified protein adsorption. A denser boundary layer of human serum albumin protein was found on the UHMWPE surface. This layer enhanced boundary lubrication, reduced the dynamic friction by 50%, and reduced static friction. Like most plasma surface modifications, this surface modification is relatively short-lived, but it lasts long enough to verify the concept that adsorption of protein from synovial fluid can be influenced by changing surface chemistry and can potentially improve boundary lubrication.

In an effort to take advantage of elastohydrodynamic lubrication and to enhance fluid film lubrication, a thin compliant layer with low elastic modulus was attached to the UHMWPE acetabular surface [10, 11]. The compliant layer acts like a cushion form bearing and maintains an effective lubricant film through elastic surface deformation. Two classes of materials have been investigated for this application: biocompatible elastomers and hydrogel cushions.

## 18.3 POLYURETHANES AND HYDROGELS

The biocompatible elastomers include various polyurethanes [10, 12, 13] and silicone rubbers [11]. Artificial cartilage made of these materials exhibits a significantly lower coefficient of friction than UHMWPE and shows fluid film lubrication under certain conditions. The polyurethane cushions exhibit a coefficient of friction as low as 0.005 under the best conditions [7]. These elastic cushions work well during steady-state sliding due to the resultant full fluid-film lubrication conditions, but the friction increases to an unacceptable level during periods of heavy loading with little movement or at the start-up of motion due to the resultant boundary lubrication conditions [13].

Hydrogel cushions, the second class of materials investigated, were hypothesized to solve the preceding problem by introducing weeping lubrication as a supplemental lubrication mechanism under severe operating conditions. Hydrogels are a class of soft, porous, permeable, crosslinked hydrophilic polymer networks with a structure more like articular cartilage than the previously described elastomers. They readily absorb water and maintain high water contents, typically 40–60% [14, 15]. Upon contact of the articulating surfaces, the water trapped in the hydrogel is squeezed out under pressure, helping to maintain the separation of articular surfaces.

The hydrogels most often used, such as polyvinyl alcohols and poly hydroxyethyl methacrylate, are made from vinyl monomers [15]. The conventional homogeneous hydrogels are not suitable for artificial cartilage materials because of their poor mechanical properties. Strategies to enhance the mechanical properties of synthetic hydrogels include the development of copolymers and the use of semi-interpenetrating polymer networks. Copolymerization between hydrophilic vinyl monomers and hydrophobic vinyl monomers largely increases the strength and modulus of hydrogels, but it also reduces the elongation to break and the water content. Cellulose acetate and cellulose acetate butyrate have higher strengths and moduli and have been used to reinforce hydrogels by forming semi-interpenetrating polymer networks with crosslinked vinyl monomers. While some of these modified hydrogel bearings have moduli and water contents comparable to natural articular cartilage and exhibit lower coefficients of friction (both start-up and steady) than the elastomer cushions, they may not necessarily have good wear resistance and mechanical durability [11, 13, 15]. Inferior tensile and shear properties lead to the degradation or fatigue of cushion materials and debonding of the soft layer from its much stiffer substrate [7, 12].

## 18.4 HYALURONAN

UHMWPE/hyaluronan biomaterials, on the other hand, were designed to exploit the high strength and durability of UHMWPE and the hydrophilic synovial lubricating ability of hyaluronan (HA). A composite approach avoids the weak attachments of plasma surface grafting and the inferior mechanical properties of hydrogel and cushion materials.

HA was chosen to modify UHMWPE joint replacement bearing surfaces because it is a naturally occurring and readily available commercial polymer. It can be produced synthetically via bacteria or harvested from rooster comb or umbilical cords. HA is a naturally occurring polysaccharide with a large unbranched structure consisting of repeating disaccharides of N-acetylglucosamine and glucuronic acid (Figure 18.1) that is synthesized and secreted by various cells throughout the body, including chondrocytes, the cells found in articular cartilage. HA is present in all vertebrate tissues and body fluids with relatively high concentrations in the vitreous humor of the eye, the umbilical cord, synovial joint fluid, and rooster combs [16].

HA was first isolated from bovine vitreous humor in acid form by Karl Meyer and John Palmer of Columbia University in 1934. They named the new polysaccharide "hyaluronic acid," meaning uronic acid from hyaloid (vitreous). The term "hyaluronan" was introduced by Endre Balazs in 1986 to encompass the different forms the molecule can take—for example, the acid form, hyaluronic acid, and the salts, such as sodium hyaluronate, which form at physiological pH [17]. Endre Balazs pioneered the medical use of HA. He derived the main concepts for many applications and prepared the first noninflammatory fraction of sodium hyaluronate, called NIF-NaHA, which was free of impurities that could cause inflammatory reactions [18].

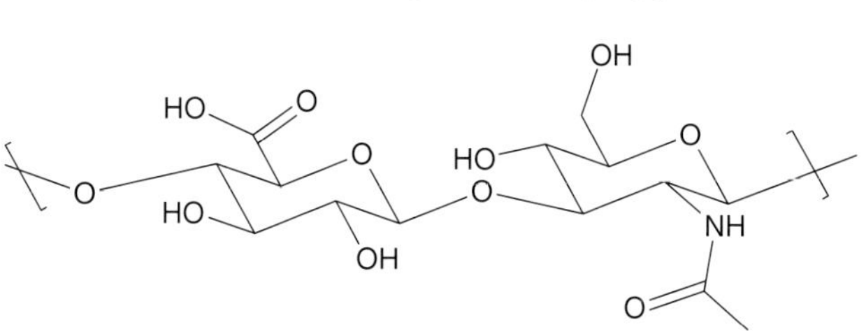

HO O O HO O OH O HO O NH OH O

FIGURE 18.1 Structure of hyaluronan (HA).

HA is utilized for its lubricious and biocompatible properties in the production of various biomaterials and intravenous drug therapies because it has a unique set of physical and biological properties compared with other polysaccharides. These include the following:

- Viscoelasticity: HA carries one carboxyl group (−COOH) per disaccharide unit, which is dissociated at physiological pH thereby conferring a polyanionic characteristic to the compound. The negatively charged flexible chains take on an expanded conformation and entangle with each other at very low concentrations, contributing to the unusual rheological properties and viscoelastic behavior of HA. Solutions made of HA are primarily viscous at low shear rates, but they are primarily elastic at high shear rates [16].
- Hydrophilicity: It has been demonstrated that HA acts as a water-retaining polymer in many tissues [19, 20], including skin [21, 22]. The hydrodynamic volume of HA in solution is 1000 times larger than the space occupied by the unhydrated polysaccharide chain [16]. The large molecular volume forces the overlap of individual HA molecular domains, resulting in extensive chain entanglement and interaction. As a result, the polymer network can hold large amounts of water (up to ~1000 times its weight) like a molecular sponge [23]. This property is unique to HA. Other GAGs, such as heparin, chondroitin sulfates, and keratin sulfate, may form viscous solutions, but they never form viscoelastic polymer networks [24].
- Lubricity: The extraordinary viscoelastic properties also make HA an ideal lubricant. HA in synovial fluid complexes with proteins and penetrates the surface of cartilage, forming a 1–2 μm-thick layer of HA protein complex, which serves as the primary isolating and lubricating layer in joints. Under slow mechanical loading it behaves as a viscous lubricant; however, at higher mechanical loading rates the HA layer changes to a highly deformable elastic system. It absorbs and converts the imposed stress to an elastic deformation and then rebounds to the original condition when the stress is removed [19]. Biological lubrication by HA is not limited to joints. It occurs at most tissue surfaces that slide along each other [23, 25].
- Regulator of cellular activity: HA is not just an inert compound, acting as a vital structural component of connective tissues. It also plays an important role in diverse biological processes, such as cellular migration, mitosis, inflammation, cancer, angiogenesis, and fertilization. HA influences cell behavior by binding specific proteins, termed hyaladherins. Many reports have been published on the role of HA receptors [26–32].

**TABLE 18.1 The Medical Applications of HA and HA Derivatives**

| Medical application | Form and use | References |
|---|---|---|
| Viscosupplementation | Elastoviscous solutions and viscoelastic gels used to replace the pathologic synovial fluid and supplement the endogenous HA to restore normal function to arthritic joints. | [39, 40, 41–44] |
| Viscoseparation | Viscoelastic gels, membranes, and fluids used to separate tissues, prevent adhesions, and decrease scar formation. | [45, 46] |
| Viscosurgery | Elastoviscous solutions and viscoelastic gels used as surgical tools and implants to protect sensitive tissues from mechanical damage during and after surgery. | [47–49] |
| Viscoaugmentation | Viscoelastic gels used to replace or augment the intercellular matrix of dermal and other soft tissues. | [44, 50] |
| Drug delivery | Solutions or gels with chemical modification of various HA functional groups to allow drug attachment are used in drug delivery; targeted drug delivery is facilitated by the binding of HA with cellular receptors. | [51–54] |
| Tissue engineering | Scaffolds cultured with chondrocytes, mesenchymal stem cells, keratinocytes, and endothelial cells are used for cartilage, bone, skin, and vessel repair. | [38, 55–60] |
| Wound healing | Fluids, sponges, solids, gels, slurries, absorbent powders, sheets, nonwoven fleeces, ropes, and meshes have been used to enhance wound healing. | [35, 61–65] |
| Biocompatible and lubricous coatings | Crosslinked or grafted coatings have been used on contact lenses to enhance wetting, on silicone rubber indwelling catheters to prevent occlusion, and on prosthetic devices, such as artificial valves, intraocular lenses, vascular grafts, and pacemakers to improve device biocompatibility. | [66–71] |

However, HA's high solubility, rapid degradation, and short residence time in water still limit the biomedical application of naturally occurring HA, particularly in tissue engineering and viscoseparation applications [33]. Many researchers have modified HA to obtain a more stable, solid material. While the number of modified HA materials used in medicine are far too numerous to describe here, two groups of widely used and commercialized HA derivatives, hylan and HYAFF®, are discussed in the following paragraphs.

Hylan is the generic name for crosslinked HA in which crosslinking occurs at the hydroxyl groups, not affecting carboxyl and acetamide groups [34]. Balazs and his coworkers at Biomatrix have patented a large group of hylans made via various crosslinking methods since the mid-1980s [35]. By adjusting the crosslinking reactions, hylans can be customized to particular medical applications and provided in different forms (fluids, gels, and solids). The derivatives have enhanced rheological properties and longer residence time in the tissue than native HA, and they are just as biocompatible as native HA.

HYAFF is a class of HA esters with the free carboxyl group of glucuronic acid esterified using different types of alcohols. They are the patented products of Fidia Advanced Biopolymer in Italy [36]. Reducing the hydrophilic carboxyl groups and introducing hydrophobic alkylation components increases the stability of the polymer in an aqueous environment. The hydration extent and degradation rate depend on the type of substituent alcohol and degree of esterification. HA and pharmacologically active alcohols are their biodegraded products [37], which are biocompatible, so these materials are intrinsically safe to use [38].

During the past 3 decades, HA and its derivatives have become important therapeutic agents and biocompatible materials due to their unique rheological properties and bioactivity. Table 18.1 provides a brief summary of their medical applications.

Thus, HA has long been used in a variety of medical applications. The UHMWPE/HA biomaterials were developed to make UHMWPE better mimic cartilage by giving UHMWPE an HA-containing surface so that it is better lubricated by synovial fluid. Current joint replacements do not enjoy the low friction and wear of natural joints due to the extreme hydrophobicity of UHMWPE and its surface chemistry, which is very different from that of articular cartilage. UHMWPE materials with improved bulk properties or modified surface microstructures cannot take advantage of the natural joint lubrication mechanisms either, because they maintain hydrophobic surface chemistry similar to conventional UHMWPE. Hydrophilicity obtained via plasma treatment is successful but short-lived. Although the application of soft cushion bearings has achieved desirable lubrication, the degradation and debonding problems block its commercial success. HA is a natural lubricant and has been used for hydrophilic and lubricious coatings of a variety of medical devices, but it has not been used in total joint replacements yet due to the difficulty of fixing it to the articular surface of UHMWPE components in a manner that would survive under severe shear stresses. As shown in the following section, UHMWPE/HA has a lubricous, hydrophilic surface that can withstand cyclic wear, enhanced wear resistance relative to UHMWPE, and excellent biocompatibility.

## 18.5 SYNTHESIS AND PROCESSING OF UHMWPE/HA MICROCOMPOSITES

Mixing extremely hydrophobic UHMWPE and extremely hydrophilic HA is analogous to mixing oil and water because the two polymers are immiscible. Thus, more than a simple blending process must be used to fabricate UHMWPE/HA microcomposites [72]. Furthermore, the goal of processing is to covalently lock the HA within the UHMWPE microstructure so that the HA treatment is long-lived under mechanical stress, while also maintaining the mechanical properties of the host UHMWPE.

Thus, UHMWPE/HA processing involves the following general steps that are later described in greater detail for the different types of UHMWPE/HA biomaterials.

1. Silylation of HA: HA, which is only soluble in water, is made temporarily hydrophobic (by silylation) so that it can be dissolved in xylenes, one of the best solvents for UHMWPE.
2. UHMWPE preform sintering: A porous preform of UHMWPE containing interconnected micron-sized pores is sintered.
3. Solvent infiltration: The UHMWPE porous preform is infiltrated with a xylenes solution containing silylated HA; the hydrophobic UHMWPE structure rapidly wicks up the silylated HA solution.
4. Crosslink silylated HA: The silylated HA is covalently crosslinked to itself, locking it into the porous UHMWPE structure.
5. Hydrolysis of silylated HA: The hydrophobic, crosslinked, silylated HA is transformed to hydrophilic, crosslinked, native HA via hydrolysis. The hydrophilic HA network cannot macroscopically phase separate from the UHMWPE preform because it is crosslinked.
6. HA surface coating: An *optional* surface coating of HA is applied to the "roots" of the crosslinked HA at the surface of the UHMWPE and crosslinked to itself and the HA molecules locked into the UHMWPE preform.
7. Compression molding: The HA-treated UHMWPE preform is molded to full density.

Three different types of UHMWPE/HA microcomposite biomaterials have been manufactured to date: conventional UHMWPE/HA, crosslinked UHMWPE/HA, and crosslinked compatibilized UHMWPE/HA. While all of these biomaterials are considered to be part of the UHMWPE/HA family, the term "UHMWPE/HA" is also used to describe the first type of microcomposite made by Zhang and James in which the UHMWPE portion of the material was not intentionally crosslinked. On the other hand, the UHMWPE is crosslinked in the latter two types of UHMWPE/HA material (crosslinked UHMWPE/HA and crosslinked compatibilized UHMWPE/HA).

The previously listed UHMWPE/HA processing Steps 1–7 are described next. The modifications of those steps for crosslinked and crosslinked compatibilized UHMWPE/HA materials are then explained.

### 18.5.1 UHMWPE/HA

1. Silylation of HA:

The term silylation means the replacement of the active hydrogen in organic compounds, such as H atoms in $-OH$, $-COOH$, and $NH_2$, by a triorganosilyl moiety, especially the trimethylsilyl moiety ($-Si(CH_3)_3$) [73, 74]. The silylation of high molecular weight polysaccharides is performed to create organic-soluble derivatives [75]. Cellulose is the most reported polysaccharide that has been successfully silylated. In the paper and textile industry, cellulose from different sources was silylated to improve its solubility in organic solvents [75, 76] or to produce a melt-processable derivative [77, 78]. Chitin [75] and dextran [79] have also been silylated to improve their hydrophobicity and solubility in organic solvents.

As a polysaccharide, the structure of HA is somewhat similar to that of the previously-mentioned cellulose, chitin, and dextran; yet the methods reported in the literature were not successful in silylating HA. Thus, the James group at Colorado State University developed a new process for the silylation of HA [80] to improve its solubility in organic solvents and its compatibility with UHMWPE. This is the key step to solvent introduction of HA into the UHMWPE porous preforms.

Silylation of HA is particularly difficult because of strong hydrogen bonding within and among HA molecules and its insolubility in any of the common silylation solvents. For these reasons, unlike cellulose, chitin, and dextran, HA cannot be silylated directly. However, Zhang and James were able to overcome this difficulty by first complexing sodium HA with a long aliphatic chain ammonium cation and then silylating this HA complex. The silyl derivative can be regenerated via hydrolysis. The study confirmed that silylation conditions did not cause HA chain scission and further measured the thermal stability of HA and its silyl derivatives to elucidate the compression molding conditions that could be used on UHMWPE/HA materials without causing thermal degradation [80].

High molecular weight sodium HA was precipitated with cetyltrimethylammonium (CTA) bromide from aqueous solution to create HA-CTA (Figure 18.2A). The white precipitate was isolated, dissolved in organic solvent, and

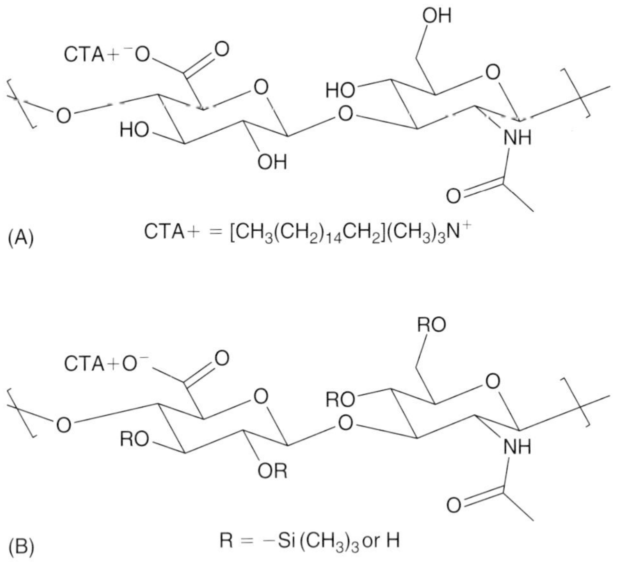

(A) CTA+ = $[CH_3(CH_2)_{14}CH_2](CH_3)_3N^+$

(B) R = $-Si(CH_3)_3$ or H

**FIGURE 18.2** Structure of (A) HA-CTA; (B) silyl HA-CTA.

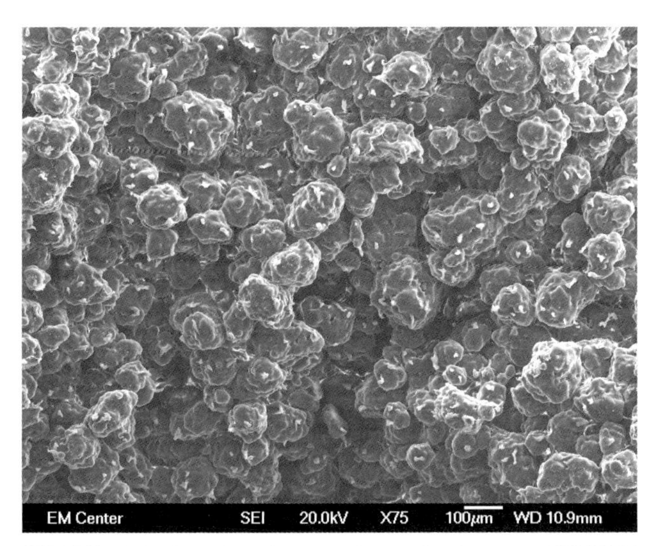


**FIGURE 18.3** Scanning electron micrograph of porous UHMWPE preform.

then silylated with hexamethyldisilazane (HMDS) to create silyl HA-CTA (Figure 18.2B). The HA-CTA intermediate and silyl HA-CTA product were confirmed by Fourier transform infrared (FTIR) spectroscopy, X-ray photoelectron spectroscopy (XPS), and $^{1}$H- and $^{13}$C- nuclear magnetic resonance (NMR) spectroscopy [80].

The silyl HA-CTA with a greater degree of silylation can be dissolved in hexane and xylenes, while those with smaller degrees of silylation are soluble in acetone, tetrahydrofuran, and 1,2-dichloroethane. The time and temperature conditions used to make the xylenes-soluble materials did not cause appreciable degradation (chain scission) of the HA. Silyl HA-CTA is stable in air; polymer samples left in ambient air for 1 year maintained their chemical structure [80].

Thermogravimetric analysis (TGA) demonstrated that the native (and regenerated) HA are more thermally stable than HA-CTA and silyl HA-CTA; furthermore, the crosslinked native HA has a much higher degradation temperature than uncrosslinked native HA. This would indicate that if the crosslinked silyl HA-CTA is hydrolyzed back to crosslinked HA before the final compression molding step, the UHMWPE/HA materials can be molded above the melting point of UHMWPE without degrading the HA. Indeed, work by Zhang et al. [81, 82] and Kurkowski [83] indicate the importance of completely hydrolyzing the crosslinked silyl HA-CTA back to crosslinked HA and removing all moisture before the final compression molding step to prevent HA degradation during compression molding.

2. UHMWPE preform sintering:

The UHMWPE preforms were made by sintering UHMWPE (GUR 1020) in a vacuum oven under relatively low pressures (dead weight), with increases in pressure leading to decreases in porosity. The porous preforms used by Zhang et al., Kurkowski, and Johnson were sintered in a vacuum oven under 313 Pa of nominal pressure at 150°C [81–84]. While the sintering pressure determined the degree of porosity, the sintering time was adjusted for the preform and sintering mold geometry and size, with larger parts requiring longer sintering times to achieve complete sintering throughout the part [84]. Mercury porosimetry was used to confirm the amount, size, and interconnectivity of the porous preforms. A scanning electron micrograph (SEM) of the porous preform structure is shown in Figure 18.3. The preforms have good mechanical integrity and are able to go through the handling required for the remaining processing steps without losing any UHMWPE.

3. Solvent infiltration:

The preforms were thoroughly dried and then placed in a silyl HA-CTA/xylenes solution. While Zhang et al. experimented with various solvent infiltration parameters (solution concentration, number of dips, etc.) [81, 82], it has been determined that the concentration of the silyl HA-CTA solution is the most important parameter for controlling final HA content. Staining treated preforms with the cationic dye toluidine blue O (TBO) confirmed the uniform distribution of the bulk HA, even in relatively large complex shapes such as an acetabular component [84]. The cationic TBO binds to the anionic HA molecules and does not bind to UHMWPE. To date, the UHMWPE/HA biomaterials have been made with a variety of final bulk HA concentrations. The bulk concentrations can be varied depending on the final application (i.e., total joint replacement versus cartilage resurfacing) and whether the optional surface coating of HA is applied (see Step 6).

4. Crosslink silylated HA:

The silyl HA-CTA treated porous preforms are dipped in a polyisocyanate (i.e., HA crosslinker) acetone solution, cured in a vacuum oven to crosslink the silyl HA-CTA in the porous preform, and then washed in acetone to remove residual crosslinker. The optimal concentration of the HA crosslinking was determined by balancing biostability (higher crosslinking levels result in better resistance to hyaluronidase degradation) and hydrophilicity (lower crosslinking levels result in better hydrophilicity as measured by aqueous contact angles) [81–84].

5. Hydrolysis of silylated HA:
   After crosslinking, the treated porous preforms are subjected to a hydrolysis process to revert the crosslinked silyl HA-CTA to hydrophilic, crosslinked HA. The hydrolysis procedure involves placing the treated preforms in a solution of NaCl in ethanol. The precise procedure depends on the size and shape of the sample and is simply performed long enough to ensure full hydrolysis. Zhang and James confirmed that the hydrolysis procedure reverted the silyl HA-CTA back to native HA using FTIR [80], and later studies confirmed full hydrolysis of the preforms using XPS [83].
6. HA surface coating (optional):
   If desired, a surface coating of crosslinked HA can be applied to the UHMWPE/HA biomaterials using an aqueous HA solution [81–84]. HA is applied to the surface in an aqueous solution and then crosslinked into the bulk HA of the porous preform as described in earlier steps. This results in a surface film of HA that is well-adhered to the HA/UHMWPE substrate. Indeed, if plain UHMWPE is painted with a surface layer of HA using an aqueous HA solution and then crosslinked, the result is a surface layer of HA that does not adhere to the UHMWPE but immediately debonds.
7. Compression molding:
   The preform is compression molded to full density. A variety of molding cycles (time, temperature, and pressure profiles) have been used to manufacture the UHMWPE/HA materials depending on the size and geometry of the sample and mold [81–84].

## 18.5.2 Crosslinked UHMWPE/HA

The UHMWPE portion of UHMWPE/HA, manufactured as previously described, is not intentionally crosslinked. Kurkowski studied the effect of moderately crosslinking the UHMWPE portion of UHMWPE/HA materials using dicumyl peroxide (DCP) as a chemical crosslinker [83]. Numerous studies have demonstrated that chemical crosslinking (with silanes or peroxides) of UHMWPE improves *in vitro* and *in vivo* wear resistance, although because the crosslinking occurs during the melt soak stage of compression molding, the percentage crystallinity of the UHMWPE and size of the crystalline lamellae are reduced [85–92].

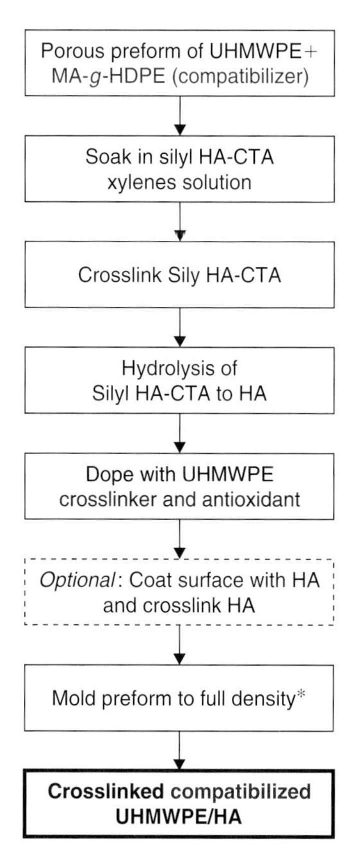


**FIGURE 18.4** UHMWPE/HA materials processing. UHMWPE/HA is made with the black steps, crosslinked UHMWPE/HA is made with black and blue steps, and crosslinked compatibilized UHMWPE/HA is made with the black, blue, and red steps.*UHMWPE crosslinking and UHMWPE/HA compatibilization occur during molding.

The goal of Kurkowski's work was to achieve a level of crosslinking and oxidative stability in crosslinked UHMWPE/HA similar to that in ram extruded bar stock of GUR 1050 that has undergone 50 kGy gamma irradiation followed by remelting, annealing, and gas plasma sterilization. The dry, treated preforms were dipped in a DCP/antioxidant solution after hydrolysis and before final compression molding. The melt soak time and temperature were adjusted to ensure complete decomposition of the DCP. A variety of DCP/antioxidant concentrations were examined, and swell ratio, molecular weight between crosslinks ($M_c$), percentage crystallinity (from differential scanning calorimetry, DSC), electron paramagnetic resonance, oxidative stability (oxidative index after accelerated aging), impact strength, and tensile strength characterization were used to select the concentration that resulted in an oxidatively stable material with a level of crosslinking similar to the 50 kGy gamma irradiated and annealed UHMWPE previously mentioned.

Figure 18.4 provides a schematic of the UHMWPE/HA materials processing steps, indicating in blue the additional steps required to make crosslinked UHMWPE/HA.

### 18.5.3 Crosslinked Compatibilized UHMWPE/HA

Compatibilizers are utilized in polymer chemistry to improve the mechanical properties of multicomponent polymer blends. The process of compatibilization is meant to reduce polymer particle interfacial tension, facilitate chain dispersion, stabilize morphology against severe melt processing conditions, and enhance adhesion between two phases [93].

Kurkowski explored compatibilization of the UHMWPE and HA phases of crosslinked UHMWPE/HA materials using maleic anhydride grafted high density polyethylene (MA-*g*-HDPE) [83]. Maleated polyolefins have been used to compatibilize immiscible polymer blends of polyolefins (e.g., polyethylenes or polypropylenes) and polysaccharides (e.g., cellulose), particularly in the plastic lumber industry [94–97]. The maleated polyolefins go into solution with xylenes, through refluxing and/or dissolving at low concentrations and elevated temperatures [98]. Maleic anhydride has been grafted onto the backbone of different polymers, in the presence of an organic peroxide, either at melt processing temperatures, in solid state, or in solution [98–100]. The maleated polyolefins compatibilize polyolefins and polysaccharides because the anhydride group will react with the hydroxyl groups on the polysaccharides (HA in the case of UHMWPE/HA) to form ester linkages (Figure 18.5), while the hydrophobic polyolefin chain portion of the compatibilizer diffuses (i.e., entangles) with host polyolefin molecules (UHMWPE in the case of UHMWPE/HA).

In addition to the compatibilization previously described, the MA-*g*-HDPE also enhances processing of the UHMWPE/HA material because the HDPE portion of the compatibilizer has a lower melting point than UHMWPE. Thus, molding compatibilized UHMWPE/HA to full consolidation can be achieved with less thermal energy than molding UHMWPE/HA. HDPE and UHMWPE do undergo cocrystallization [101, 102].

To make crosslinked compatibilized UHMWPE/HA, Kurkowski modified the method previously described by using an UHMWPE/MA-*g*-HDPE powder blend to sinter the porous preform [83]. Figure 18.4 provides a schematic of the UHMWPE/HA material processing steps, indicating in blue and red the additional steps required to make crosslinked compatibilized UHMWPE/HA.

## 18.6 CHEMICAL AND PHYSICAL CHARACTERIZATION OF UHMWPE/HA BIOMATERIALS

A variety of characterization techniques have been used to verify the presence of HA in UHMWPE/HA biomaterials, the hydrophilic nature of the UHMWPE/HA surface, the biostability of UHMWPE/HA and the affect of UHMWPE/HA treatment on the crystallinity of UHMWPE [81–84].

### 18.6.1 UHMWPE/HA Composition

FTIR has been used to confirm the presence of HA in the bulk of UHMWPE/HA materials. The UHMWPE/HA spectrum showed both the presence of UHMWPE peaks ($CH_2$, $CH_3$: 2932, 2843, and 1473 $cm^{-1}$) as well as peaks associated with HA ($-OH$: ~3400 $cm^{-1}$, $-COOH$: ~1700 $cm^{-1}$) [83] and not associated with UHMWPE. Note that the carboxylate carbonyl FTIR peaks associated with HA overlap the carbonyl peaks used to identify oxidation in pure UHMWPE. This presents a challenge for assessing potential oxidative degradation in UHMWPE/HA biomaterials.

X-ray photoelectron spectroscopy (XPS), which is a very sensitive surface analysis technique, has also been used to characterize UHMWPE/HA. By measuring the energies of electrons emitted from an X-ray irradiated surface, XPS can provide qualitative and quantitative information on the elemental composition and chemical state of the surface with a sampling depth between 20–100Å [103]. XPS can identify all elements (except H and He) at concentration >0.1 atomic % and identify organic groups via

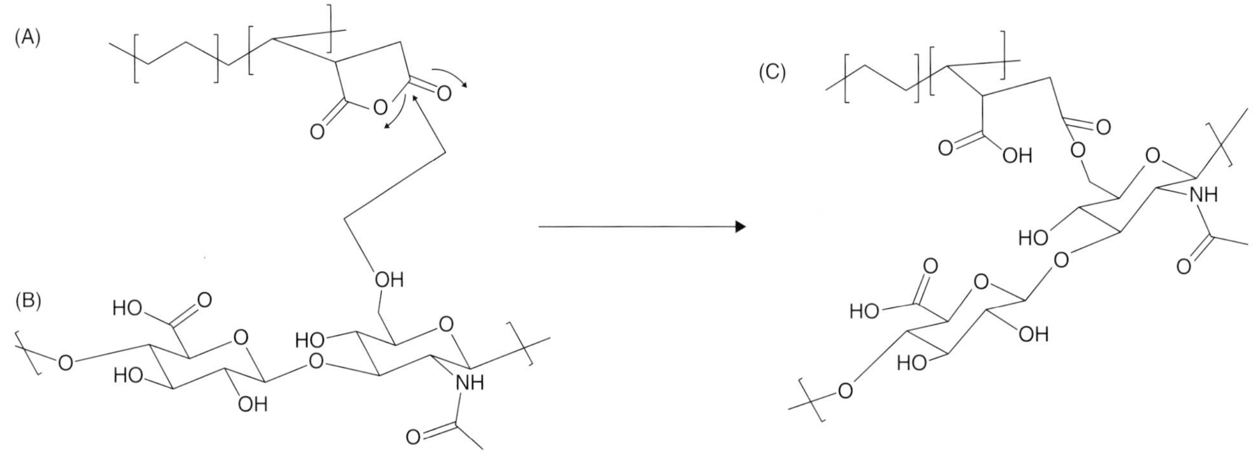

(A) (B) (C)

**FIGURE 18.5** The compatibilization reaction between MA-*g*-HDPE [A] and HA [B] results in an ester linkage [C] between [A] and [B].

high resolution scans. For example, small amounts of nitrogen within HA can be detected with XPS, and XPS can be used to differentiate between nitrogen from −CTA residue and nitrogen on the N-acetyl glucosamine unit of HA.

XPS was used to examine multiple locations on each surface of all control and UHMWPE/HA materials. UHMWPE control samples were either directly molded from GUR 1020 or were molded from GUR 1020 preforms that had not been treated with HA to serve as a sham control for the UHMWPE/HA material processing procedure. UHMWPE/HA samples with and without the optional HA surface coating, as well as crosslinked and crosslinked compatiblized UHMWPE/HA samples have been examined [81, 83].

Zhang et al. found nitrogen, at similar levels to that found in HA, at the surface of UHMWPE/HA, but none was detected in either the UHMWPE preform or molded sham control disk, demonstrating the presence of a thin layer of HA on the microcomposite surface [81]. While Zhang found incomplete hydrolysis of the silyl HA-CTA in some samples, Kurkowski's later work determined hydrolysis conditions that ensured complete hydrolysis of the silyl HA-CTA in the treated preforms and confirmed nitrogen-containing HA (and not residual −CTA) in her hydrolyzed preforms before molding UHMWPE/HA materials [83]. Furthermore, Kurkowski found that the surface of crosslinked compatibilized UHMWPE/HA had a higher concentration of HA than crosslinked UHMWPE/HA and UHMWPE/HA samples indicating that the compatibilizer enhances HA bonding in both the bulk and in the optional surface coating. This result was also confirmed with TBO staining, as described next.

TBO is a cationic dye that binds the negatively charged carboxyl groups on HA with 1:1 ratio under appropriate conditions and does not bind or stain UHMWPE [104]. Thus, TBO staining can be used to qualitatively (i.e., visual surface color) and semiquantitatively determine the amount of HA on the surface of UHMWPE/HA materials. The quantitative assay uses visible absorbance at 632 nm of TBO bound to the surface after elution of the bound dye with an NaCl urea solution and comparison to a standard calibration curve [81]. This process is only deemed semiquantitative because the surface density (nmol/cm$^2$) depends on the precise surface area of the sample, and the surface area is only known nominally (based on macroscopic sample dimensions).

UHMWPE/HA materials will stain from a light to dark blue/purple color with TBO depending on the amount of bulk HA and whether the optional HA surface coating has been applied. For example, the top two UHMWPE/HA samples shown in Figure 18.6 did *not* receive the optional HA surface coating, while the bottom two UHMWPE/HA samples did receive the optional HA surface coating. The samples without the optional surface coating show less HA than those with the coating, as only the "roots" of the bulk HA in the top samples are present at the surface of the implant [83]. As described in the next section, Zhang et al. [81] did find that the surface density of HA does correlate to surface hydrophilicity as measured by aqueous contact angle.

## 18.6.2 UHMWPE/HA Hydrophilicity

Aqueous (distilled water) contact angle measurements are used to characterize the hydrophilicity of solid surfaces with hydrophobic surfaces exhibiting very high contact angles (i.e., water beading up on the surface of a waxed car has a high contact angle) and hydrophilic surfaces exhibiting low contact angles. Water contact angles depend on a number of things including surface roughness, the timing of when the contact angle is measured (some surfaces wet more slowly than others) and the moisture content of the surrounding environment (as water evaporates the contact angle can change).

Zhang et al. [81] measured static water contact angles immediately after application of a water drop on UHMWPE/HA samples that did *not* receive the optional surface coating of HA and *had* been preconditioned in distilled water for 4 hours. The noncoated UHMWPE/HA sample had a much lower water contact angle (~32 degrees) than the UHMWPE control (~91 degrees), indicating the UHMWPE/HA was

**FIGURE 18.6** TBO dyed UHMWPE/HA: The top two UHMWPE/HA samples did not receive the optional HA surface coating (light purple), and the bottom two samples were coated (dark purple).

much more hydrophilic than the UHMWPE. Zhang also found a qualitative correlation between the surface density of bound HA and contact angle, with contact angle decreasing as surface density of HA increased.

UHMWPE/HA samples with the optional HA surface coating exhibited fully hydrated surfaces after preconditioning in distilled water for 4 hours. The water drop immediately spread on these surfaces indicating the formation of a uniform HA layer on these surfaces. These preconditioned surfaces could not be used for the sessile water drop measurements because the surface was too hydrophilic (i.e., the wetting was too rapid). Thus, Zhang examined the kinetics of the aqueous hydration process on *dry* (i.e., no preconditioning in distilled water) UHMWPE/HA samples by measuring the contact angle as a function of time (1 minute intervals) until the water drop evaporated. The water contact angle of the dry UHMWPE/HA samples rapidly decreased from an initial hydrophobic state (~80 degrees) to approximately 50 degrees in 1 minute, to below 20 degrees in approximately 7 minutes. From the trend of the water contact angle versus time curves, it was inferred that within 15–20 min water drops would completely spread in a saturated water-vapor environment, and the surface of the UHMWPE/HA would become completely hydrated.

Figure 18.7 compares a water drop on UHMWPE to a water drop on crosslinked UHMWPE/HA with the optional HA surface coating. Both samples were dry (i.e., no preconditioning in water) before the water drop was placed on the surface and the picture was taken soon after (<1 minute).

### 18.6.3 UHMWPE/HA Biostability

While natural HA has a short residence time in the body, the HA component of the UHMWPE/HA biomaterials should be stable in the body because it is crosslinked and microscopically entangled within hydrophobic UHMWPE, thus limiting the access of enzymes to the HA phase. Zhang et al. [81] examined the biostability of UHMWPE/HA in hyaluronidase (15 or 150 units/ml) phosphate buffered saline (PBS) solutions. She tested both treated (but not fully consolidated) preforms and fully consolidated UHMWPE/HA samples, the former intended to be a more severe test of biostability because the enzyme could more easily gain access to the HA molecules in the nonconsolidated porous preforms. After 10 and 30 days in the enzyme solution at 37°C, TBO dye elution was used to quantify any HA degradation relative to control samples that were not exposed to the enzyme.

Hyaluronidases are enzymes that degrade HA. The presence of hyaluronidases in synovial fluid and its activity against HA have been demonstrated [105], although significant HA degradation in synovial joints has not been observed [106]. Zhang used the same enzyme concentrations in her study as those used by Lowry and Beavers [107] for their HA coatings. The two enzyme levels, 15 and 150 units/mL, were respectively 6 and 60 times the hyaluronidase concentration present in human serum [108]. Thus, Zhang's enzyme degradation environment was much more severe than the human synovial fluid environment, because hyaluronidase activity in synovial fluid is much lower than that in human serum [105].

Zhang used these hyaluronidase studies to determine the minimum level of HA crosslinking with polyisocyanate that resulted in enzymatically stable UHMWPE/HA materials. The UHMWPE/HA samples made with the appropriate concentrations of HA crosslinker exhibited no significant decreases in HA density over the 30 day period and were still lubricious, water-wettable, and exhibited no TBO staining difference when compared to the surfaces before enzyme exposure.

### 18.6.4 UHMWPE Crystallinity in UHMWPE/HA

Chemical crosslinking can reduce crystallinity because the crosslinking occurs in the melt during compression molding. Similarly, remelting irradiation-crosslinked UHMWPE to eliminate free radicals can also reduce crystallinity; in both cases the crosslinked UHMWPE molecules in the melt cannot crystallize as easily as noncrosslinked UHMWPE molecules.

Thus, Kurkowski used DSC to assess the percent crystallinity of some UHMWPE/HA porous preforms and fully consolidated samples to determine the effect of compatibilizer addition, HA addition, and chemical crosslinking on the intrinsic crystallinity of UHMWPE/HA

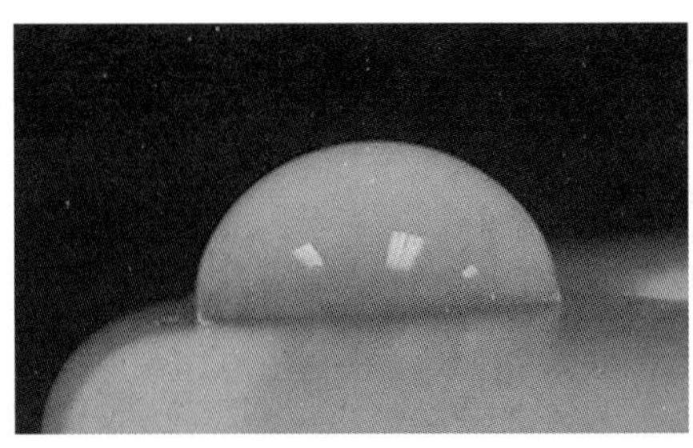

**FIGURE 18.7** A water drop on *dry* UHMWPE (left) and a water drop on *dry* crosslinked UHMWPE/HA with the optional HA surface coating (right).

**TABLE 18.2** Percent Crystallinity Measured by Differential Scanning Calorimetry (Average ± Standard Error of the Mean)

| Sample | % crystallinity (average ± s.e.m.) |
|---|---|
| UHMWPE porous preform | 52.79 ± 0.08 |
| UHMWPE/MA-g-HDPE porous preform | 50.51 ± 0.07 |
| UHMWPE/HA | 47.14 ± 0.04 |
| Crosslinked UHMWPE/HA | 41.56 ± 0.03 |

materials. Her results are shown in Table 18.2. The percent crystallinity was measured in a dry $N_2$ atmosphere per ASTM D3418-03. The $H_f$ was calculated by integrating the DSC endotherm from the second heating curve. The $H_f$ of 100% crystalline UHMWPE was determined to be 288 J/g. The percent crystallinity was calculated by dividing the $H_f$ of the sample by 288 J/g (because the 100% crystalline $H_f$ of UHMWPE/HA is unknown) and multiplying by 100.

The UHMWPE porous preform had a significantly higher percent crystallinity ($p \leq 0.001$) compared to the UHMWPE/HA and crosslinked UHMWPE/HA groups due to the introduction of HA and DCP. While statistically significant, the difference between the UHMWPE and UHMWPE/HA was relatively small. The introduction of the compatibilizer (exhibited by the UHMWPE/MA-*g*-HDPE porous preform) also decreased the percent crystallinity; however, the compatibilized porous preform did not have a significantly different percent crystallinity compared to the UHMWPE/HA treatment group. As expected, the crosslinked UHMWPE/HA displayed the lowest crystallinity percent with a significantly lower percent crystallinity ($p \leq 0.001$) compared to the compatibilized porous preform and the UHMWPE/HA samples.

## 18.7 MECHANICAL AND TRIBOLOGICAL CHARACTERIZATION OF UHMWPE/HA BIOMATERIALS

### 18.7.1 Tensile Properties

As described in Chapter 2 of this *Handbook*, the mechanical properties of compression molded and direct compression molded UHMWPE depend on the molding conditions (time, temperature, pressure) used to consolidate the part. Incomplete consolidation, as evidenced by residual UHMWPE powder particle boundaries in microtome slices of the consolidated product, can result in inferior mechanical properties.

As previously described, one of the challenges in processing UHMWPE/HA materials relates to the fact that HA is less thermally stable than UHMWPE. Furthermore, as explained in Chapter 2, the low thermal conductivity of UHMWPE means that molding cycle conditions required to achieve complete consolidation depend on the particular geometry of the molding press, mold, and UHMWPE part. Finally, in the case of UHMWPE/HA materials with the optional HA surface coating, the HA is concentrated at the surface of the part which receives more thermal energy in the course of a molding cycle than the center of the part. This may practically limit the size of parts that can be compression molded from HA-surface coated UHMWPE/HA materials because the center of excessively large parts may not reach full consolidation before the surface HA begins degrading. The effect of this balance between preventing HA degradation and achieving full UHMWPE consolidation is evident in the mechanical property studies of UHMWPE/HA materials described next.

Zhang et al. [82] performed the first mechanical characterization studies on UHMWPE/HA. Zhang's preforms were molded to full consolidation at a maximum temperature lower than the 205°C often used to consolidate UHMWPE. Her tensile tests were conducted according to ASTM Standard D638-99 (sample type V). She compared the tensile properties of UHMWPE/HA and a sham control UHMWPE sample (i.e., molded from a non-HA-treated preform but using the UHMWPE/HA molding cycle) to ASTM F648-98 specifications and UHMWPE reference materials [109]. All samples were conditioned in distilled water for 24 hours before testing.

The modulus and yield strength values for the sham control and UHMWPE/HA were within the range of values specified by the ASTM standard, but the ultimate strength and elongation-to-failure values were lower than the ASTM specifications and reference materials. Zhang concluded that the reduced properties were not a result of HA introduction into the UHMWPE because the sham control exhibited similarly inferior properties. Microtome slices of both the sham control and UHMWPE/HA samples exhibited some UHMWPE particle boundaries, indicating that the preform process might hinder consolidation during final molding. Therefore, steps were successfully taken to improve the material properties (i.e., preform and molding parameters, crosslinking and compatibilization). It should be noted that Zhang's UHMWPE/HA exhibited better wear resistance than conventional UHMWPE despite its inferior mechanical properties [82].

Kurkowski expanded on Zhang's work, examining the effect of molding parameters on consolidation and mechanical properties of both sham control UHMWPE (i.e., molded from a non-HA-treated preform but using the UHMWPE/HA molding cycle) and on the various types of UHMWPE/HA materials. Her work elucidated a clear "preform effect" in UHMWPE/HA materials. Using more thermal energy than Zhang (i.e., longer melt soak times and/or higher temperatures) Kurkowski found a molding cycle that resulted in full consolidation of UHMWPE when it was molded directly from powder (i.e., no particle memory retention) but inferior consolidation of sham controls. In other words, the process of sintering a preform before molding to full consolidation (with no HA treatment) resulted in inferior properties relative to samples molded directly from powder. The preform process enhanced particle memory retention in the final product.

Kurkowski also examined the effects of crosslinking and compatibilization on the mechanical properties of UHMWPE/HA materials. The mechanical properties of crosslinked and crosslinked compatibilized UHMWPE/HA molded by Kurkowski were superior to those of Zhang's UHMWPE/HA; however, the tensile properties were still inferior compared to compression molded unmodified UHMWPE [83].

Zhang and Kurkowski's experimentation with different molding parameters on different sample geometries (e.g., flat discs from which type V tensile bars were stamped versus thicker cylinders from which wear pins were machined) set the stage for an experienced, industrial converter to develop new, proprietary methods for molding the various UHMWPE/HA formulations into a variety of implant shapes and sizes. Recent collaborative studies by Schwartz Biomedical and Colorado State University have developed proprietary molding conditions that significantly improve the mechanical properties of UHMWPE/HA materials. Under these conditions UHMWPE/HA materials with properties that exceed ASTM standard (type I) requirements can be molded. However, UHMWPE/HA materials molded under these conditions still exhibit a "preform effect." Finally, there is no clear correlation between the amount of crosslinked HA in UHMWPE/HA materials and their mechanical properties. Studies are ongoing to understand and further enhance the mechanical properties of UHMWPE/HA materials.

## 18.7.2 Wear Resistance

Zhang et al. performed pin-on-disk wear testing of UHMWPE/HA pins with the optional HA surface coating [82]. The wear testing was conducted at Zimmer (Warsaw, Indiana, USA) using their custom-built 12-station device, which is capable of crossing motion paths with a static test load (i.e., it was not capable of a typical gait cycle of fluctuating load). A constant load of 445 N (6.9 MPa) was applied to each UHMWPE pin. The square wear path (total length of 60 mm) resulted in UHMWPE cross-shear. The crossing path motion is crucial to the determination of the wear behavior of UHMWPE for use in orthopedic implants [110].

The wear test was performed at room temperature at 1.0 Hz for 1.0 million cycles. The lubricant was undiluted bovine calf serum and wear was measured gravimetrically relative to load-soak controls. Super finished CoCrMo alloy disks were employed as the wear counterface and both conventional UHMWPE (nonsterile) and sham controls (UHMWPE molded from preform with same molding cycle used on UHMWPE/HA) were compared to UHMWPE/HA.

Zhang's wear rate results for UHMWPE/HA are shown in Figure 18.8. The sham control had higher wear rates than the conventional UHMWPE, but the differences between them decreased with increasing test cycles. The wear rate of the UHMWPE/HA was significantly ($p \leq 0.05$) lower than both conventional UHMWPE and the sham control throughout the test, although the differences decreased as the cycles increased. While the wear rates of the conventional UHMWPE and sham control were relatively constant with time, the wear rate of the UHMWPE/HA increased with time. At 1 million cycles

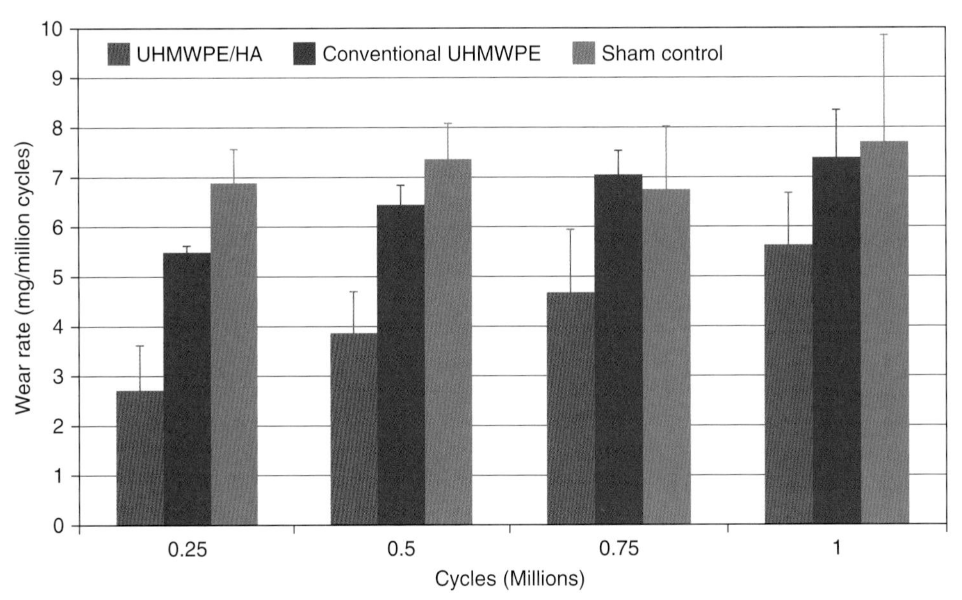


**FIGURE 18.8** Wear rates to 1 million cycles for UHMWPE/HA, conventional UHMWPE and sham control.

the wear rate of UHMWPE/HA was 31% lower than conventional UHMWPE. At 1 million cycles the amount of wear that had occurred on UHMWPE/HA was 56% lower than the amount of wear that had occurred on conventional UHMWPE [82].

The 2007 study by Zhang et al. clearly demonstrated that the hydrophilic HA-containing UHMWPE/HA surface is more wear resistant than untreated UHMWPE. Although Zhang's UHMWPE/HA samples had inferior tensile properties, they still exhibited superior wear properties relative to controls. The incomplete preform consolidation due to Zhang's low-temperature remolding adversely affected the wear properties of the sham control, indicating that improved consolidation should further enhance the UHMWPE/HA's wear resistance.

Kurkowski went on to test the wear resistance of crosslinked and crosslinked compatibilized UHMWPE/HA with the optional HA surface coating to see if they were more wear resistant than UHMWPE/HA and the 50 kGy irradiation crosslinked UHMWPE with a similar level of UHMWPE crosslinking (i.e., $M_c$) as the crosslinked UHMWPE/HA biomaterials [83]. The pin-on-disk wear tests were conducted on an AMTI OrthoPOD machine (Watertown, Massachusetts, USA). The metal counter faces (i.e., disks) were wrought CoCrMo alloy conforming to ASTM F1537 Alloy 2 (UNS 31538) and were highly polished to a surface roughness of less than 10 nm. The wear tests were conducted in 90% bovine calf serum at 37 ± 1°C, following a 10 mm × 10 mm square wear pattern with respect to the disk with no lift. A Paul loading cycle with a peak force of 330 N was applied with a frequency of 1.6 Hz. The duration of the test was $1.98 \times 10^6$ or almost 2 million cycles. At each measurement interval following 330,000 cycles, the amount of wear (i.e., the weight lost during each interval) of the samples was determined using gravimetric measurement with standard soak controls.

Kurkowski's wear rate results for crosslinked and crosslinked compatibilized UHMWPE/HA are shown in Figure 18.9 [83]. For ease of comparison, the results in Figures 18.8 and 18.9 are graphed on the same scale. However, as previously described, Zhang's wear testing on Zimmer's custom-built machine was somewhat different from Kurkowski's wear testing on the OrthoPOD wear tester.

At 1.98 million cycles, the wear rates of crosslinked and crosslinked compatibilized UHMWPE/HA were not significantly different from the control (50 kGy irradiated and stabilized GUR 1050); at 1.98 million cycles the total wear (i.e., mass lost) of crosslinked compatibilized UHMWPE/HA was 5.54 mg, which was not significantly different from the total wear of the control (50 kGy irradiated and stabilized GUR 1050), which was 3.56 mg. The lower wear (mg) and wear rates (mg/million cycles) of crosslinked and crosslinked compatibilized UHMWPE/HA compared to UHMWPE/HA exhibited the positive effect of chemical crosslinking and compatibilization on the wear resistance of UHMWPE/HA materials [83].

The wear rates of the three types of samples shown in Figure 18.9 vary differently as a function of time [83]. The crosslinked UHMWPE/HA wear rate appears to decrease in the first several intervals and then increases after 0.99 million cycles only to begin decreasing again. The wear rate of the crosslinked compatibilized UHMWPE/HA decreases up to 0.99 million cycles and then increases slightly. The 50 kGy irradiated and stabilized GUR 1050 wear rate appears to slowly increase throughout time at least until the last interval.

The fluctuation in wear rate of the UHMWPE/HA samples is likely explained by the fact that after 0.99 million

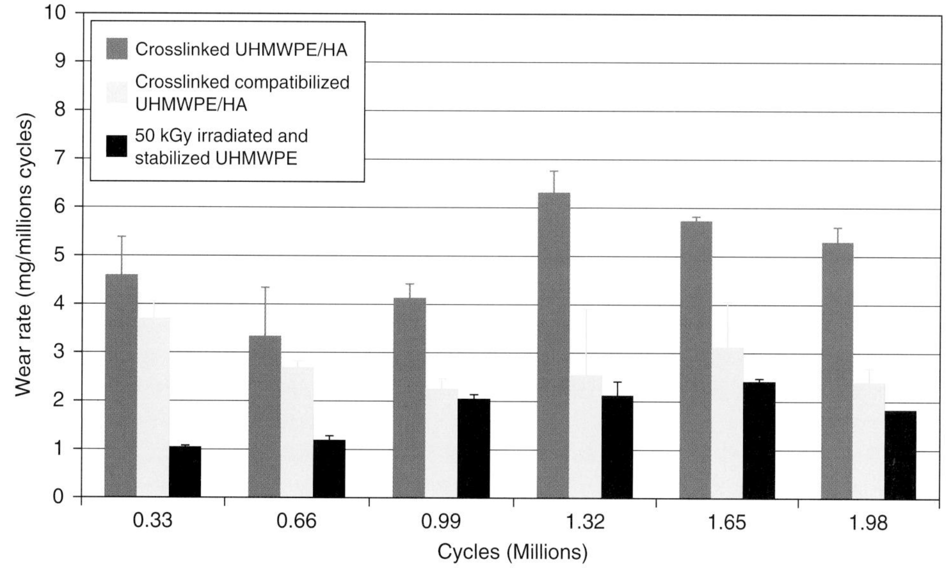


**FIGURE 18.9** Wear rates to 1.98 million cycles for crosslinked UHMWPE/HA, crosslinked compatibilized UHMWPE/HA, and the 50 kGy irradiated and stabilized UHMWPE control.

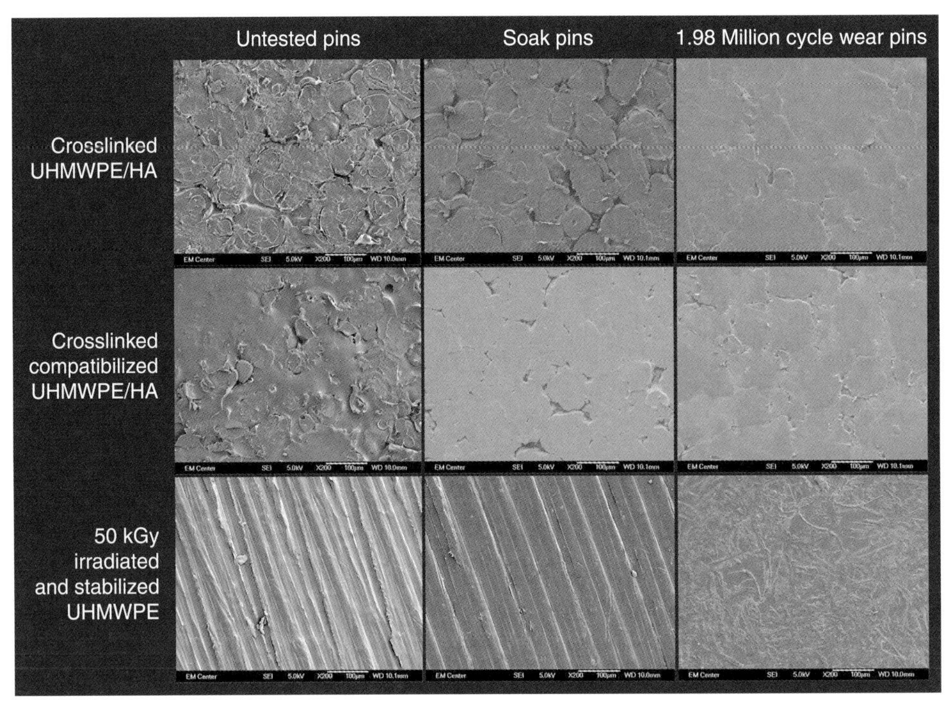


**FIGURE 18.10** Scanning electron micrographs of untested, soak, and 1.98 million cycle wear pins (wear rate data is shown in Figure 18.9). All micrographs were taken at 200 × magnification, and the white micron bar at the bottom of each image is 100 micrometers long.

cycles, the metallic disk counter surfaces were cleaned in an attempt to remove a transfer film that had developed on the disk surfaces as they articulated against the UHMWPE/HA samples. It appears that the removal of this transfer film resulted in an increase in wear rate of the UHMWPE/HA samples.

SEM pictures of untested, soak, and 1.98 million cycle wear pins are shown in Figure 18.10 [83]. The crosslinked and crosslinked compatibilized UHMWPE/HA show minimal signs of wear, with the crosslinked compatibilized UHMWPE/HA surface after 2 million cycles of wear in particular not looking much different than the soak pin. These surfaces exhibit islands of dry (the samples are dried in a vacuum oven for SEM imaging) HA in the optional HA surface coating that was applied to the samples before molding the preform to full density. While the soak pins have changed slightly from the untested pins, perhaps due to loss of unbound HA during soaking, the wear pins look very similar to the soak pins particularly for the crosslinked compatibilized UHMWPE/HA. This may indicate that the 5.54 mg of total wear exhibited by the crosslinked compatibilized UHMWPE/HA pins was only the loss of HA and not UHMWPE. This is supported by the fact that hundreds of mg of HA are in the surface layers of these samples, but less than 10 mg of wear had occurred by 1.98 million cycles. It is hypothesized that HA wear debris are more biocompatible and will not cause the foreign body reaction and resultant osteolysis caused by UHMWPE wear debris. Joint simulator testing and wear debris osteolytic potential will be the subject of future studies.

On the other hand, the 50 kGy irradiated and stabilized GUR 1050 surface after wear testing exhibited signs of abrasive and adhesive wear and the complete loss of machine marks at 1.98 million cycles relative to the soak and untested pins.

## 18.8 STERILIZATION OF UHMWPE/HA BIOMATERIALS

The sterilization methods that are commonly used on medical implants are all being investigated for potential use with UHMWPE/HA biomaterials. Kurkowski performed tensile testing and TBO staining of nonsterile, gas plasma, ethylene oxide, electron beam, and gamma irradiation sterilized UHMWPE/HA samples [83]. TBO staining of the UHMWPE/HA surfaces was not affected by the sterilization process, and she did not find much significant difference between the tensile properties of samples sterilized by the different methods. This may be because her samples were not aged (artificially or naturally) after sterilization. All sterilization methods used on contemporary UHMWPE implants remain under investigation. Future studies will look at the oxidative stability of aged, sterilized samples, but, as previously described, FTIR cannot be used to assess oxidation in UHMWPE/HA materials, so these studies will rely on measurement of oxidation-sensitive properties (i.e., density, mechanical properties) to assess long-term oxidative stability. Ethylene oxide has been used to sterilize the UHMWPE/HA samples used in biocompatibility and animal studies to date.

## 18.9 BIOCOMPATIBILITY OF UHMWPE/HA BIOMATERIALS

*In vitro* biocompatibility tests performed on UHMWPE/HA materials at NAMSA (Northwood, Ohio, USA) included leachable and cytotoxicity testing. Leachables were measured using isopropyl alcohol as a solvent, and cytotoxicity was measured using the ISO Elution Method (ISO 10993 – Biological Evaluation of Medical Devices, Part 5: Test for Cytotoxicity: *in vitro* Methods). HDPE and tin stabilized poly(vinyl chloride) (PVC) were used as negative and positive controls, respectively, in the cytotoxicity tests. Negligible leachables were detected, and there was no evidence of cell lysis or toxicity with the UHMWPE/HA samples. No differences between the UHMWPE/HA material and HDPE were detected (unpublished data) indicating, among other things, that the variety of chemicals used during UHMWPE/HA material processing are removed from the material during processing.

UHMWPE/HA implants for partial knee resurfacing are currently being tested in a long-term large animal model. A short-term (6-week) pilot study of UHMWPE/HA implants in a large animal knee indicated excellent *in vivo* biocompatibility with no adverse reactions to the UHMWPE/HA material (unpublished data).

## 18.10 COMMERCIALIZATION OF UHMWPE/HA BIOMATERIALS

Schwartz Biomedical, LLC (Fort Wayne, Indiana, USA) has licensed the UHMWPE/HA patent family from Colorado State University. Schwartz Biomedical LLC is a tissue engineering company that develops and commercializes novel technologies in the orthopedic industry. Schwartz Biomedical LLC is investigating the use of the UHMWPE/HA biomaterial platform in a variety of applications, including total joint replacements, partial hemiarthroplasty (i.e., resurfacing) and spine applications. Proprietary processing methods are under development with resin converters. UHMWPE/HA materials can be direct compression molded into final implant shape (Figure 18.11) or compression molded into bulk forms from which implants are machined. However, in the latter case if the optional HA surface coating is applied to the articulating surface, care must be taken to protect this surface during subsequent machining operations. Entire implants can be made from UHMWPE/HA, or a layer of UHMWPE/HA can be molded onto an UHMWPE substrate [81–83]. Schwartz Biomedical is currently assembling data to support applications to the FDA for regulatory approval and plans to market the UHMWPE/HA biomaterial under the trademark of BioPoly™.

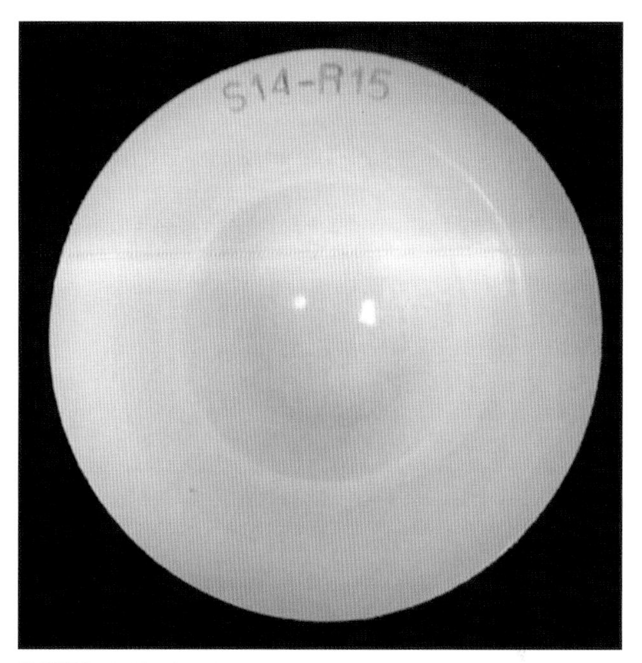

**FIGURE 18.11** Crosslinked UHMWPE/HA acetabular cup.

## 18.11 CONCLUSION

The UHMWPE/HA biomaterials were designed to take advantage of natural synovial lubrication mechanisms. UHMWPE/HA is UHMWPE with a small amount of HA in the bulk that extends HA "roots" at the surface of the UHMWPE to which an optional HA surface coating can be applied. The HA-rich material is hydrophilic, lubricious, and well hydrated. The UHMWPE/HA family of materials includes UHMWPE/HA with nonintentionally crosslinked UHMWPE, crosslinked UHMWPE/HA, and crosslinked compatibilized UHMWPE/HA. UHMWPE/HA wears considerably less than plain UHMWPE, and crosslinked compatibilized UHMWPE/HA has a wear resistance on par with UHMWPE having a similar level of crosslinking. Furthermore, after 2 million cycles of wear, crosslinked compatibilized UHMWPE/HA surfaces look similar to unworn surfaces, indicating little to no UHMWPE wear. Preliminary studies indicate that UHMWPE/HA is noncytotoxic and well tolerated in the knee.

## 18.12 ACKNOWLEDGMENTS

The authors wish to acknowledge all past and present members of the James Group at Colorado State University, particularly Nate Johnson, Guy Beauregard, Jason Marini, Trey Rodgers, and Tim Collins for their help in the development of UHMWPE/HA materials. Lou Matrisciano's help is also greatly appreciated. Past funding by the National Science Foundation and Whitaker Foundation helped develop the scientific platform from which UHMWPE/HA materials were developed. Support from the Indiana 21st Century Research and Technology Fund and Schwartz Biomedical, LLC have helped move the UHMWPE/HA materials down the commercialization pathway.

## REFERENCES

1. Huber M, Trattnig S, Lintner F. Anatomy, biochemistry, and physiology of articular cartilage. *Invest Radiol* 2000;**35**(10):573–80.
2. Mow VC, Fithian DC, Kelly MA, Ewing JW. *Fundamentals of articular cartilage and meniscus biomechanics. Articular cartilage and knee joint function: basic science and arthroscopy*. New York: Raven Press, Ltd.; 1990 p. 1–15.
3. Ghosh P. The role of hyaluronic acid (hyaluronan) in health and disease: interactions with cells, cartilage and components of synovial fluid. *Clin Exp Rheumatol* 1994;**12**:75–82.
4. Dowson D. Modes of lubrication in human joints. *Proc Inst Mech Eng (Part 3J)* 1966–67;**181**:45–55.
5. Hlavdek M. The influence of the acetabular labrum seal, intact articular superficial zone and synovial fluid thixotropy on squeeze-film lubrication of a spherical synovial joint. *J Biomech* 2002;**35**: 1325–35.
6. Unsworth A. The effects of lubrication in hip joint prostheses. *Phys Med Biol* 1978;**23**:253–68.
7. Unsworth A. Recent developments in the tribology of artificial joints. *Tribol Int* 1995;**28**:485–95.
8. Black J. *Handbook of biomaterial properties*. London: Chapman and Hall; 1998.
9. Widmer MR, Spencer ND. Influence of polymer surface chemistry on frictional properties under protein-lubrication conditions: implications for hip-implant design. *Tribol Lett* 2001;**10**:111–17.
10. Auger DD, Dowson D, Fisher J. Friction of cylindrical cushion form bearings for artificial joints. In: Dowson D, editor. *Thin films tribology*. Amsterdam: Elsevier Science Publishers B.V.; 1993. p. 683–92.
11. Sawae Y, Murakami T, Higaki H, Moriyama S. Lubrication property of total knee prostheses with PVA hydrogel layer as artificial cartilage. *JSME Int J (Series C)* 1996;**39**:356–64.
12. Blamey J, Rajan S, Unsworth A, Dawber R. Soft layered prostheses for arthritic hip joints: a study of materials degradation. *J Biomed Eng* 1991;**13**:180–4.
13. Caravia L, Dowson D, Fisher J. A comparison of friction in hydrogel and polyurethane materials for cushion-form joints. *J Mater Sci: Mater Med* 1993;**4**:515–20.
14. Goldsmith AJ, Cliff SE. Investigation into the biphasic properties of a hydrogel for use in a cushion form replacement joint. *Trans of the ASME* 1998;**120**:362–8.
15. Corkhill PH, Trevett AS, Tighe BJ. The potential of hydrogels as synthetic articular cartilage. *Proc Inst Mech Eng (Part H)* 1990;**204**:147–55.
16. Laurent TC. Structure of hyaluronic acid. In: Balazs EA, editor. *Chemistry and molecular biology of the intercellular matrix*. London: Academic Press; 1970. p. 703–28.
17. Laurent TC. Introduction. In: Evered D, Whelan J J, editors *The biology of hyaluronan, Ciba Foundation Symposium*. Chichester: John Wiley & Sons; 1989. p. 1–5.
18. Laurent TC. The tree: hyaluronan research in the 20th century. Glycoforum—Science of Hyaluronan Today 2002 [cited; Available from: http//www.glycoforum.gr.jp/science/hyaluronan/hyaluronanE.html]
19. Balazs EA, Gibbs DA. The rheological properties and biological function of hyaluronic acid. In: Balazs EA, editor. *Chemistry and molecular biology of the intercellular matrix*. London, New York: Academic Press; 1970. p. 1241–53.
20. Balazs EA, Band P. Hyaluronic acid: its structure and use. *Cosmetics and Toiletries* 1984;**99**:65–72.
21. Yates JR. Mechanism of water uptake by skin. In: Elden HR, editor. *Biophysical properties of the skin*. New York: John Wiley and Sons; 1971. p. 485–511.
22. Bettelheim FA. Structure and hydration of mucopolysaccharides. In: Elden HR, editor. *Biophysical properties of the skin*. New York: John Wiley and Sons; 1971. p. 303–27.
23. Hoekstra D. Hyaluronan-modified surfaces for medical devices. *Medical Device & Diagnostic Industry* 1999:51–8.
24. Balazs EA. Viscosurgery in the eye. *Ocul Inflamm Ther* 1983;**1**:91–2.
25. Ogston AG. The biological functions of the glycosaminoglycans. In: Balazs EA, editor. *Chemistry and molecular biology of the intercellular matrix*. London: Academic Press; 1970. p. 1231–9.
26. Hamilton SR, Fard SF, Paiwand FF, Tolg C, Veiseh M, Wang C, et al. The hyaluronan receptors CD44 and Rhamm (CD168) form complexes with ERK1,2 that sustain high basal motility in breast cancer cells. *J Biol Chem* 2007;**282**:16667–80.
27. Garg HG, Hales CA, editors. *Chemistry and biology of Hyaluronan*. 1st ed. Oxford San Diego: Elsevier; 2004.
28. Turley EA, Noble PW, Bourguignon LYW. Signaling properties of hyaluronan receptors. *J Biol Chem* 2002;**277**:4589–92.
29. Toole BP. Hyaluronan-cell interactions in morphogenesis. In: Laurent TC, editor. *The chemistry, biology and medical applications of hyaluronan and its derivatives*. London: Portland Press Ltd; 1998. p. 155–9.
30. Day AJ, Parkar AA. The structure of link module: a hyaluronan-binding domain. In: Laurent TC, editor. *The chemistry, biology and medical applications of hyaluronan and its derivatives*. London: Portland Press Ltd; 1998. p. 141–7.
31. Lesley J. Hyaluronan binding function of CD44. In: Laurent TC, editor. *The chemistry, biology and medical applications of hyaluronan and its derivatives*. London: Portland Press Ltd; 1998. p. 123–33.
32. Heinegard D, Bjornsson S, Morgelin M, Sommarin Y. Hyaluronan-binding matrix proteins. In: Laurent TC, editor. *The chemistry, biology and medical applications of hyaluronan and its derivatives*. London: Portland Press Ltd; 1998. p. 113–21.
33. Coleman PJ, Scott D, Ray J, Mason RM, Levick JR. Hyaluronan secretion into the synovial cavity of rabbit knees and comparison with albumin turnover. *J Physiol* 1997;**503**(3):645–56.
34. Balazs EA, Leshchiner EA. Hyaluronan, its crosslinked derivative—hylan—and their medical applications. In: Inagaki H, Phillips GO, editors *Cellulosics utilization: research and rewards in cellulosics; proceedings of nisshinbo international conference on cellulosics utilization in the near future*. New York: Elsevier; 1989. p. 233–41.
35. Balazs EA, Leshchiner EA, Larsen NE, Band P. Hyaluronan biomaterials: medical applications. In: Wise DL, editor. *Encyclopedic handbook of biomaterials and bioengineering: Part A: materials*. New York: Marcel Dekker; 1995. p. 1693–715.
36. della Valle F, Romeo A, inventors; Fidia, S.p.A., assignee. Esters of hyaluronic acid. *United State of America Patent No. 48551521*; 1989.
37. Rastrelli A, Becaro M, Biviano F, Calderini G, Pastorello A. Hyaluronic acid esters: A new class of semisynthetic

biopolymers—chemical and physico-chemical properties. In: Heimke G, Solteaz U, Lee A, editors *Clinical implant materials: advances in biomaterials*. New York: Elsevier Science Publishers; 1990. p. 199–205.
38. Campoccia D, Doherty P, Radice M, Brun P, Abatangelo G, Williams DF. Semisynthetic resorbable materials from hyaluronan esterification. *Biomaterials* 1998;**19**(23):2101–27.
39. Biomatrix I. Viscoseparation: a historical perspective. In: Kennedy JF, Phillips GO, Williams PA, Hascall VC, editors *Hyaluronan*. Cambridge: Woodhead Publishing Limited; 2000. p. 385–9.
40. Weiss C, Abatangelo G, Weigel PH. Why viscoelasticity is important for the medical uses of hyaluronan and hylans. In: *New frontiers in medical sciences: redefining hyaluronan*. Amsterdam: Elsevier; 2000. p. 89–103. 40.
41. Balazs EA, Denlinger JL. Sodium hyaluronate and joint function. *J Equine Vet Sci* 1985;**5**(4):217–19, 222–228.
42. Balazs EA, Leshchiner A, inventors; Biomatrix, Inc., assignee. Crosslinked-gels of hyaluronic acid and products containing such gels. *United States of America Patent No. 4582865*; 1986.
43. Balazs EA, Leshchiner A, Leshchiner A, Band P, inventors; Biomatrix, Inc., assignee. Chemically modified hyaluronic acid preparation and method of recovery thereof from animal tissues. *United States Patent No. 4713448*; 1987.
44. Biomatrix I. Viscoaugmentation: a historical perspective. In: Kennedy JF, Phillips GO, Williams PA, Hascall VC, editors. *Hyaluronan*. Cambridge: Woodhead Publishing Limited; 2002. p. 41–4.
45. Weiss C. Viscoseparation and viscoprotection as therapeutic modalities in the musculoskeletal system. In: Laurent TC, editor. *The chemistry, biology and medical applications of hyaluronan and its derivatives*. London: Portland Press Ltd; 1998. p. 255–65.
46. Savani RC, Bagli DJ, Harrison RE, Turley EA. The role of hyaluronan-receptor interactions in wound repair. In: Garg HG, Longaker MT, editors *Scarless wound healing: basic and clinical dermatology*. 19th ed. New York: Marcel Dekker; 2000. p. 115–42.
47. Biomatrix I. Viscosurgery: a historical perspective. In: Kennedy JF, Phillips GO, Williams PA, Hascall VC, editors *Hyaluronan*. Cambridge: Woodhead Publishing Ltd; 2002. p. 461–5.
48. Balazs EA, Freeman MI, Kloti R, Meyer-Schwickerath G, Regnault F, Sweeney DB. Hyaluronic acid and replacement of vitreous and aqueous humor. In: Streiff EB, editor. *Mod Probl Ophthalmol*. New York: S. Karger, Basel; 1972. p. 3–21.
49. Balazs EA, inventor; Biotrics, Inc., assignee. Ultrapure hyaluronic acid and the use thereof. *United States of America Patent No. 4141973*; 1979.
50. Denlinger JL. Hyaluronan and its derivatives as viscoelastics in medicine. In: Laurent TC, editor. *The chemistry, biology and medical applications of hyaluronan and its derivatives*. London and Miami: Portland Press Ltd; 1998. p. 235–41.
51. Gustafson S. Hyaluronan in drug delivery. In: Laurent TC, editor. *The chemistry, biology and medical applications of hyaluronan and its derivatives*. London: Portland Press Ltd; 1998. p. 291–303.
52. Pouyani T, Prestwich GD. Functionalized derivatives of hyaluronic acid oligosaccharides: drug carriers and novel biomaterials. *Bioconjug Chem* 1994;**5**(4):339–47.
53. Prestwich GD. Biomaterials from chemically-modified hyaluronan. Glycoforum-Science of Hyaluronan Today 2001 [cited; Available from: http://www.glycoforum.gr.jp/science/hyaluronan/hyaluronanE.html
54. Band PA. Hyaluronan derivatives: chemistry and clinical applications. In: Laurent TC, editor. *The chemistry, biology and medical applications of hyaluronan and its derivatives*. London: Portland Press Ltd; 1998. p. 33–42.
55. Toole BP, Munaim SI, Welles S, Knudson CB. Hyaluronate-cell interactions and growth factor regulation of hyaluronate synthesis during limb development. In: Evered D, Whelan J, editors *The biology of hyaluronan, Ciba Foundation Symposium*: Wiley; 1989.
56. Grigolo B, Roseti L, Fiorini M, Fini M, Giavaresi G, Aldini NN, et al. Transplantation of chrondrocytes seeded on a hyaluronan derivative (HYAFF-11) into cartilage defects in rabbits. *Biomaterials* 2001;**22**(17): 2417–24.
57. Bernard GW, Pilloni A, Kang M, Sison J, Pirnazar P, Hunt D, et al. Osteogenesis in vitro and in vivo with hyaluronan and bone morphogenic protein-2. In: Abatangelo G, Weigel PH, editors *New frontiers in medical science: redefining hyaluronan*. Oxford: Elsevier Science B.V.; 2000. p. 215–31.
58. Solchaga LA, Goldberg VM, Caplan AI. Hyaluronan and tissue engineering. In: Kennedy JF, Phillips GO, Williams PA, Hascall VC, editors *Hyaluronan*. Cambridge: Woodhead Publishing Ltd; 2000. p. 45–54.
59. Hollander DA, Wild M, Konold P, Windolf J. Benzylester hyaluronic acid membranes: a delivery system for autologous keratinocyte cultures in the treatment of complicated chronic and acute wounds. In: Abatangelo G, Weigel PH, editors *New frontiers in medical sciences: redefining hyaluronan*. Amsterdam: Elsevier Science B.V.; 2000. p. 303–11.
60. Brun P, Cortivo R, Radice M, Abatangelo G. Hyaluronan-based biomaterials in tissue engineering. In: Abatangelo G, Weigel PH, editors *New frontiers in medical science: redefining hyaluronan*. Amsterdam: Elsevier Science B.V.; 2000. p. 269–78.
61. Abatangelo G, Martelli M, Vecchia P. Healing of hyaluronic acid-enriched wounds: histological observations. *J Surg Res* 1983;**35**(5):410–16.
62. Breuing K, Eriksson E, Liu P, Miller DR. Healing of partial thickness porcine skin wounds in a liquid environment. *J Surg Res* 1992;**52**(1):50–8.
63. King SR, Hickerson WL, Proctor KG, Newsome AM. Beneficial actions of exogenous hyaluronic acid on wound healing. *Surgery* 1991;**109**:76–84.
64. Murashita T, Nakayama Y, Hirano T, Ohashi S. Acceleration of granulation tissue ingrowth by hyaluronic acid in artificial skin. *Br J Plast Surg* 1996;**49**(1):58–63.
65. Zacchi V, Soranzo C, Cortivo R, Radice M, Brun P, Abatangelo G. In vitro engineering of human skin-like tissue. *J Biomed Mater Res* 1998;**40**(2):187–94.
66. Balazs EA, Leshchiner A, inventors; Biomatrix, Inc., assignee. Hyaluronate modified polymeric articles. *United States of America Patent No. 4500676*; 1985.
67. Beavers EM, inventor; Universal High Technologies, assignee. Lens with hydrophilic coating. *United States of America Patent No. 4663233*; 1987.
68. Beavers EM, Lowry KM, inventors; Beacon Research, Inc., assignee. Non-fogging transparent coatings. *United States of America Patent No. 5148311*; 1992.
69. Halpern G, Campbell C, Beavers EM, Chen HY, inventors. Method of hydrophilic coating of plastics. *United States of America Patent No. 4801475*; 1989.
70. Halpern G, Campbell C, Beavers EM, Chen HY, inventors. Method of hydrophilic coating of plastics. *United States of America Patent No. 4959074*; 1990.
71. DeFife KM, Shive MS, Hagen KM, Clapper DL, Anderson JM. Effects of photochemically immobilized polymer coatings on protein adsorption, cell adhesion, and the foreign body reaction to silicone rubber. *J Biomed Mater Res* 1999;**44**(3):298–307.
72. James SP, Zhang M, Beauregard GP, inventors. Outer layer having entanglement of hydrophobic polymer host and hydrophilic polymer guest. United States of America; 2003.

73. Wang P-C, Renga JM, inventors; The Dow Chemical Company, assignee. Silylation of organic compounds. *United States of America Patent No. 4360686*; 1982.
74. Birkofer L, Ritter A. The use of silylation in organic syntheses. *Angew Chem Int Ed Engl* 1965;**4**(5):417–29.
75. Harmon RE, De KK, Gupta SK. New procedure for preparing trimethylsilyl derivatives of polysaccharides. *Carbohydr Res* 1973; **31**(2):407–9.
76. Mormann W, Demeter J, Wagner T. Partial silylation of cellulose with predictable degree of silylation—stoichiometric silylation with hexamethyldisilazane in ammonia. *Macromol Chem Phys* 1999; **200**(4):693–7.
77. Cooper GK, Sandberg KR, Hinck JF. Trimethylsilyl cellulose as precursor to regenerated cellulose fiber. *J Appl Polym Sci* 1981; **26**(11):3827–36.
78. Hermanutz F, Gahr F, Pirngadi P. Process for producing silylated cellulose products for thermoplastic processing. *Chem Fibers Int* 2001;**51**:271–2.
79. Ydens I, Rutot D, Degee P, Six J-L, Dellacherie E, Dubois P. Controlled synthesis of poly(ε-caprolactone)-grafted dextran copoly mers as potential environmentally friendly surfactants. *Macromolecules* 2000;**33**(18):6713–21.
80. Zhang M, James SP. Silylation of hyaluronan to improve hydrophobicity and reactivity for improved processing and derivatization. *Polymer* 2005;**46**:3639–48.
81. Zhang M, King R, Hanes M, James SP. A novel ultra high molecular weight polyethylene-hyaluronan microcomposite for use in total joint replacements. I. Synthesis and physical/chemical characterization. *J Biomed Mater Res* 2006;**78**(1):86–96.
82. Zhang M, Pare P, King R, James SP. A novel ultra high molecular weight polyethylene-hyaluronan micro-composite for use in total joint replacements, Part II: mechanical and tribological property evaluation. *J Biomed Mater Res Part A* 2007;**82**(1):18–26.
83. Kurkowski RA. *Chemical crosslinking, compatibilization, and direct molding of an ultra high molecular weight polyethylene and hyaluronic acid microcomposite*. Fort Collins: Colorado State University; 2007.
84. Johnson NT. *Manufacturing ultra high molecular weight polyethylene and hyaluronan composite acetabular components*. Fort Collins: Colorado State University; 2007.
85. Kurtz SM, Muratoglu OK, Evans M, Edidin AA. Advances in the processing, sterilization, and crosslinking of ultra-high molecular weight polyethylene for total joint arthroplasty. *Biomaterials* 1999; **20**:1659–88.
86. Shen FW, McKellop HA, Salovey R. Irradiation of chemically crosslinked ultrahigh molecular weight polyethylene. *J Polym Sci: Part B: Polym Phys* 1996;**34**:1063–77.
87. Narkis M, Raiter I, Shkolnik S, Siegmann A, Eyerer P. Structure and tensile behavior of irradiation- and peroxide-crosslinked polyethylenes. *J Macromol Sci—Phys* 1987;**B 26**(1):37–58.
88. Stein HL. *Ultra high molecular weight polyethylene (UHMWPE). Guide to engineering plastics families: thermoplastic resins*. Materials Park: ASM International; 1999 p.167–171.
89. Lewis G. Properties of crosslinked ultra-high-molecular-weight polyethylene. *Biomaterials* 2001;**22**:371–401.
90. Kurtz SM, Pruitt LA, Jewett CW, Foulds JR, Edidin AA. Radiation and chemical crosslinking promote strain hardening behavior and molecular alignment in ultra high molecular weight polyethylene during multi-axial loading conditions. *Biomaterials* 1999;**20**:1449–62.
91. Muratoglu OK, Bragdon CR, O'Conner DO, Jasty M, Harris WH, Gul R, et al. Unified wear model for highly crosslinked ultra-high molecular weight polyethylenes (UHMWPE). *Biomaterials* 1999; **20**:1463–70.
92. Shen FW, McKellop HA, Salovey R. Morphology of chemically crosslinked ultrahigh molecular weight polyethylene. *J Biomed Mater Res* 1998;**41**:71–8.
93. Ajji A, Utracki LA. Interphase and compatibilization of polymer blends. *Polym Eng Sci* 1996;**36**(12):1574–85.
94. Lu B, Chung TC. Synthesis of maleic anhydride grafted polyethylene and polypropylene, with controlled molecular structures. *J Polym Sci: Part A: Polym Chem* 2000;**38**:1337–43.
95. Chandra R, Rustgi R. Biodegradation of maleated linear low-density polyethylene and starch blends. *Polym Degradation Stability* 1997; **56**:185–202.
96. Zhang F, Qiu W, Yang L, Endo T, Hirotsu T. Mechanochemical preparation and properties of a cellulose-polyethylene composite. *J Mater Chem* 2002;**12**:24–6.
97. Zhang J, Sun X. Mechanical properties of poly(lactic acid)/ starch composites compatibilized by maleic anhydride. *Biomacromolecules* 2004;**5**:1446–51.
98. Pompe T, Zschoche S, Herold N, Salchert K, Gouzy M, Sperling C, et al. Maleic anhydride copolymers—a versatile platform for molecular biosurface engineering. *Biomacromolecules* 2003;**4**:1072–9.
99. Ganzeveld KJ, Janssen LPBM. The grafting of maleic anhydride on high density polyethylene in an extruder. *Polym Eng Sci* 1992; **32**(7):467–74.
100. Zhai H, Xu W, Guo H, Zhou Z, Shen S, Song Q. Preparation and characterization of PE and PE-*g*-MAH/montmorillonite nanocomposites. *Eur Polym J* 2004;**40**:2539–45.
101. Boscoletto AB, Franco R, Scapin M, Tavan M. An investigation on rheological and impact behaviour of high density and ultra high molecular weight polyethylene mixtures. *Eur Polym J* 1997;**33**(1):97–105.
102. Lim KLK, Ishak ZAM, Ishiaku US, Fuad AMY, Yusof AH, Czigany T, et al. High-density polyethylene/ultrahigh-molecular-weight polyethylene blend. I. The processing, thermal, and mechanical properties. *J Appl Polym Sci* 2005;**97**:413–25.
103. Ratner BD, Caster DG. Electron spectroscopy for chemical analysis. In: Vickerman JC, editor. *Surface analysis: the principal techniques*. New York: John Wiley and Sons; 1997. p. 43–98.
104. Johnston JB. A simple, nondestructive assay for bound hyaluronan. *J Biomed Mater Res* 2000;**53**:188–91.
105. Bollet AJ, Bonner WMJ, Nance JL. The presence of hyaluronidase in various mammalian tissues. *J Biol Chem* 1963;**238**:3522–7.
106. Fraser JRE, Laurent TC. Turnover and metabolism of hyaluronan. The biology of hyaluronan. *Ciba Found Symp* 1989;**143**:41–53.
107. Lowry KM, Beavers EM. Resistance of hyaluronate coatings to hyaluronidase. *J Biomed Mater Res* 1994;**28**:861–4.
108. Delpech B, Bertrand P, Chauzy C. An indirect enzymoimmunological assay for hyaluronidase. *J Immunol Methods* 1987;**104**:223–9.
109. Bennett AP, Wright TM, Li S. Global reference UHMWPE: characterization and comparison to commercial UHMWPE. *Orthopaedic Research Society*. Atlanta, Georgia; 1996. p. 472.
110. Bragdon CR, O'Conner DO, Lowenstein JD, Jasty M, Syniuta WD. The importance of multidirectional motion on the wear of polyethylene. *Proc Instn Mech Engrs* 1996;**210**:157–65.

Chapter 19

# High Pressure Crystallized UHMWPEs

Anuj Bellare, PhD and Steven M. Kurtz, PhD



## 19.1 INTRODUCTION

High-pressure crystallization refers to a type of crystallization in which a semicrystalline polymer such as UHMWPE is heated to a temperature exceeding its melting temperature, subjected to hydrostatic pressure, usually higher than 350 MPa, and then isobarically cooled to a temperature at which it is able to crystallize. High-pressure crystallization may also proceed isothermally at a temperature above the melting temperature (usually above 210°C) by increasing pressure until it reaches several hundred MPa. Such pressures are generally considered to be high because most common processing conditions, such as compression molding or ram extrusion, are conducted in a range of a few kPa up to a few MPa.

High-pressure crystallization of polyethylene has been widely studied, and several reviews by Wunderlich [1, 2], Bassett [3], Bhateja and Pae [4] and Leute and Dollhopf [5] exist. Various aspects of high-pressure crystallization have been reviewed by Wunderlich et al. in a series of six companion papers [6–11]. However, while these studies have furthered our understanding of crystallization in polymers in general, there haven't been many applications for high-pressure crystallized PE. Medium molecular weight polyethylenes, such as HDPE, are 80% crystalline and have several applications because they are relatively tough. But high-pressure crystallization can increase the crystallinity of HDPE up to 99%, which leads to extremely high brittleness and low ductility [12]. In this respect, ultrahigh molecular weight PE (UHMWPE) is a better candidate for high-pressure crystallization because its high molecular weight (high entanglement density) ensures that the starting polymer is only about 50% crystalline and its high-pressure counterpart does not achieve as high a crystallinity as medium molecular weight PE, resulting in a high-modulus, tough, wear-resistant polymer. An example of such a high-pressure crystallized UHMWPE is Hylamer (Depuy-Dupont, Inc.), which will be discussed in greater detail in Section 19.3.

While in certain cases Hylamer did not perform well clinically due to susceptibility to oxidative degradation, the benefits of high-pressure crystallization were well recognized. Hylamer had a higher resistance to fatigue crack propagation and a higher resistance to creep deformation [13, 14], without a significant change in wear resistance (when it was not oxidized) [15], which are desirable mechanical properties for application in total joint replacement prostheses. Despite the susceptibility of Hylamer to oxidation following gamma sterilization, there has been an interest in revisiting high-pressure crystallization as a processing technique applied to radiation crosslinked UHMWPE. It is well-known that radiation crosslinked UHMWPE exhibits a dose-dependent decrease in both modulus [16] and resistance to fatigue crack propagation

[17] compared to unirradiated UHMWPE. High-pressure crystallization would then have the potential to restore its modulus, resistance to creep deformation, and resistance to fatigue crack propagation by increasing its crystallinity.

In this chapter, we outline the basic concepts of high-pressure crystallization in the context of basic polymer crystallization, discuss the phase diagram of PE, and describe extended-chain crystals followed by a review of Hylamer and recent research on the use of high-pressure crystallization to increase the properties of irradiated UHMWPE.

## 19.2 EXTENDED CHAIN CRYSTALLIZATION

The earliest concepts of crystallization of semicrystalline polymers were obtained by studies of polymers crystallized from very dilute solutions and at atmospheric pressures. Electron microscopy studies revealed that the large macromolecules formed lamellar-shaped single crystals, which had a thickness of 5–50 nm, much less than the length of the extended chain. In 1957, Keller [18] and Fischer [19] proposed that the chains fold back and forth, forming folded chain lamellar crystals, as shown in Figure 19.1. In melt-crystallized lamellae, chain folding is also observed, except there are other conformations, such as loose chain folds and tie molecules, that exit one lamella and enter an adjacent lamella, explained more by a telephone switchboard-like model (see Figure 19.2) compared to an adjacent-reentry model. Chain folding is kinetically driven and thus the chain folded lamellar shaped crystals are not equilibrium crystals but rather metastable. There is a large surface area associated with thin lamellae, and consequently PE lamellae have a high surface free energy associated with them, which explains its metastability. Over the years, the relationship between supercooling and lamellar thickness has been well established using calorimeters and dilatometers. The surface free energy, $\sigma_e$, has been calculated for PE to be equal to $9 \times 10^{-6}$ J/cm$^2$ [20]. Based on the extrapolated equilibrium melting temperature, $T^0_m$, obtained using the Hoffman-Weeks method [21] for PE ($T^0_m = 141.5°C$ [22]), the surface free energy and the heat of fusion, $\delta h_f$, of the PE crystal ($\delta h_f = 293$ J/cm$^3$ [23]), the lamellar thickness can be estimated from the experimental melting temperature, $T_m$, by the Gibbs-Thomson equation [24] :

$$L = 2\sigma_e T^0_m / [\delta h_f (T^0_m - T_m)]$$

(Note: It must be noted that the Gibbs-Thomson equation, when applied to extended-chain crystals, can often lead to erroneous estimates because extended-chain crystals melt very slowly, and this can lead to "superheating" effects, i.e. the melting is slower than conduction of heat into the sample, providing large values of observed melting temperatures, often exceeding the equilibrium melting temperature of PE, which is impossible.)

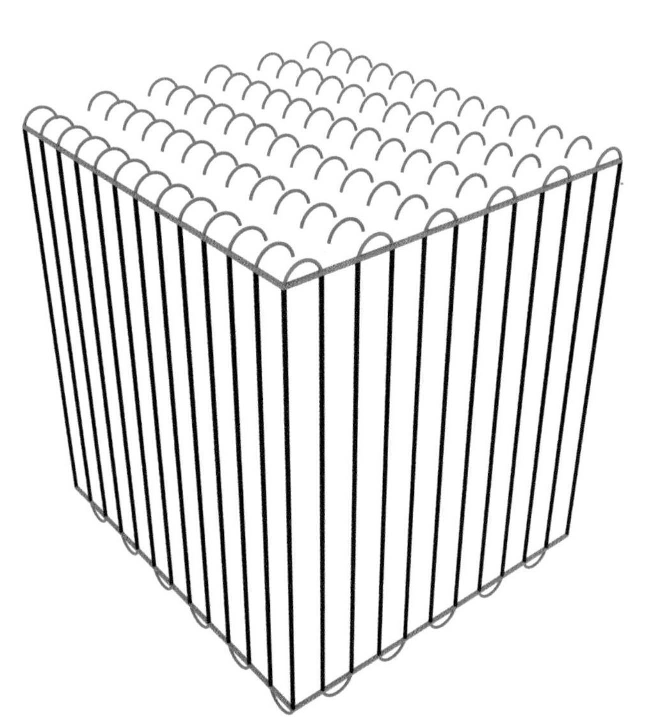

FIGURE 19.1 Polyethylene lamella grown from a dilute solution showing regular chain folds (adjacent reentry model). Image courtesy of Eric Wysocki, Exponent, Inc.

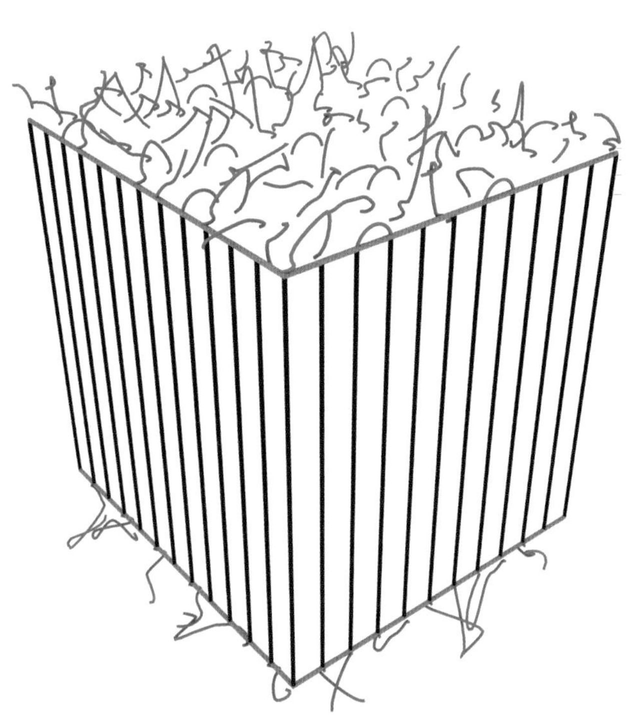

FIGURE 19.2 Polyethylene lamella crystallized from the melt state showing some regular chain folds, loose loops, and tie molecules (switchboard model). Image courtesy of Eric Wysocki, Exponent, Inc.

It was eventually proved experimentally via solution crystallization under pressures up to 600 MPa that the reason for chain folding was kinetic and not thermodynamic [25], i.e. the lamellar thickness was found to change with supercooling (or the difference between melting temperature and crystallization temperature) rather than absolute temperature. An increase in supercooling is also known to increase the rate of crystallization. These crystallization studies of polymer solutions were extended to polymer melts and, in 1964, ultimately led to the discovery of the so-called extended-chain crystals (ECC) of PE [12, 26]. Samples crystallized at 500 MPa had densities as high as 0.994 g/cc (density of the pure PE crystal = 1.0 g/cc) and revealed crystals with an average thickness of 250 nm, with a few as high as 3 μm. Figures 19.3 and 19.4 reveal the large differences in the lamellar thickness between chain-folded lamellae and ECCs. Electron diffraction showed that the c-axis of the PE crystals were parallel to characteristic striations present on the fracture surfaces and were at a moderate angle to the lamellar surfaces similar to chain-folded crystals. Wunderlich defined ECCs as any crystal of a linear high polymer of sufficiently high molecular weight, which has a length of at least 200 nm in the molecular direction [6], which is of the order of magnitude of the accepted lower limit of a high polymer (=20,000 g/mole). Although the low molecular weight fractions of PE can reach full molecular extension, it was discovered that the higher molecular weight fractions, while they crystallized into markedly thicker lamellae than atmospheric crystallized lamellae (the thickness increasing with molecular weight), they could not have achieved full extension, i.e. the longest macromolecules must have folded to some extent, however small [8, 27]. Also, full molecular extension does not limit the size of extended-chain crystals. Hatakeyama et al. [28] crystallized PE of 50,000 g/mole MW for a period of 200 h under 480 MPa. Even though over 98% of the macromolecules were less than 10 μm at full extension, over 20% of the lamellae were 10 μm thick, with an observed maximum thickness of 40 μm, indicating that the macromolecules fit end-to-end inside some of the lamellae. Morphological investigations have shown that, during isothermal crystallization, the extended-chain lamellae increase in thickness with time and were generally thicker at positions farther away from the primary nucleus, giving them a tapered profile. Rees and Bassett [29] explained this in terms of rapid increase in thickness behind the growth front, not dissimilar to chain-folded crystallization. Wunderlich et al. also did not see any major difference between high-pressure crystallization and atmospheric pressure crystallization except that chain extension is substantially more kinetically hindered in the latter case, and they attribute it more to the higher temperature of crystallization at high pressures rather than the effects of pressure alone.

In general, high-pressure crystallized PE contains a mixed morphology of both ECCs and chain-folded crystals, evident from two separate atmospheric melting peaks observed by Wunderlich and Arakawa in PE that was crystallized in a range of 280–380 MPa [26]. The higher melting peak was associated with ECCs, while the lower peak was associated with chain-folded crystals. Bassett et al. showed that that each melting peak was also associated with unique texture [30]. The lower peak of the chain-folded crystals was associated with spherulitic morphology,

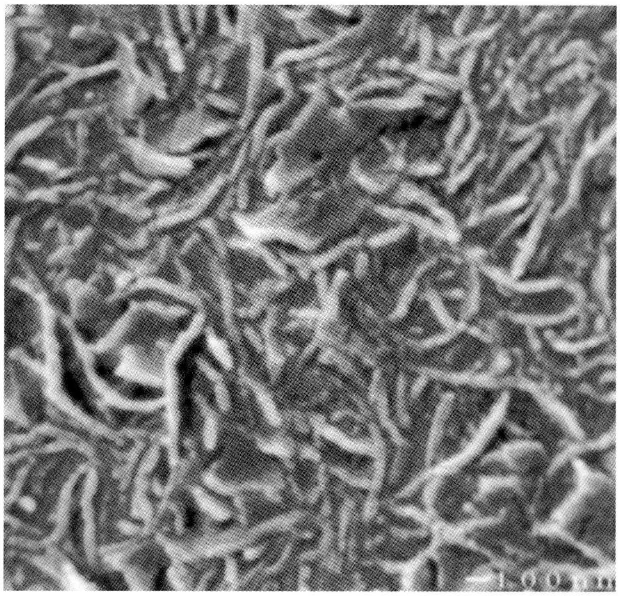

**FIGURE 19.3** Low voltage scanning electron micrograph of permanganic etched freeze fracture surface of UHMWPE.

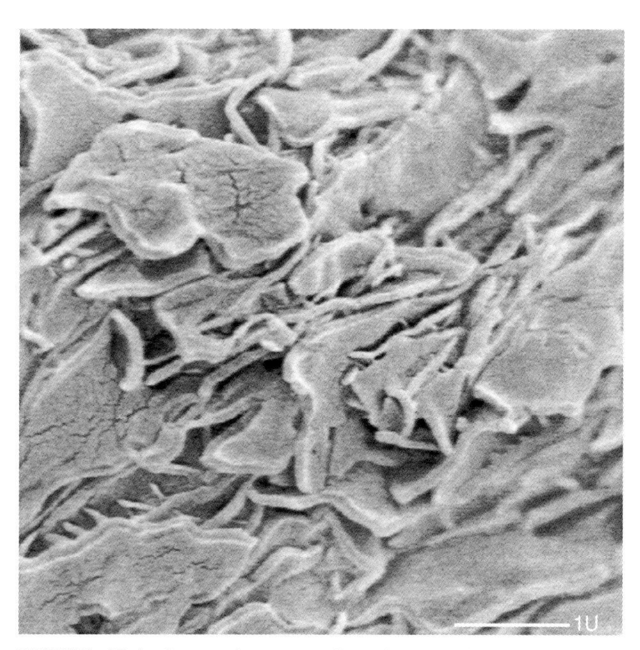


**FIGURE 19.4** Low voltage scanning electron micrograph of permanganic etched freeze fracture surface of high-pressure crystallized UHMWPE.

while the higher peak of the ECCS was associated with sheaf-like structures, which are precursors to spherulites. Bassett [3] notes that the absolute value of thickness alone does not distinguish between ECCs and chain-folded lamellae because the thickness of ECCs is molecular-weight dependent. For example, for very small molecular weights, ECCs of 80 nm thickness were observed for a PE molecular weight of 20,000 g/mole fraction, which is of a similar size as the thickness of chain-folded crystals [31].

Another important discovery was that chain-folded crystallization and ECC growth were independent processes. Under certain isobaric, isothermal conditions, ECC growth occurred before chain-folded lamellae did, while in other conditions they overlapped but the lamellar growth did not transform from one mechanism to another for any single lamella [31]. Kanetsuna and coworkers also showed ECCs forming before chain-folded crystals and that the lamellar morphology suddenly changed from chain-folded to ECC with a small change in isothermal crystallization temperature at 300 MPa [32, 33].

## 19.2.1 Phase Diagram for PE

In 1972, Bassett and Turner melted PE at 410 MPa and higher pressures and showed the appearance of an additional high-temperature melting endotherm, which was reversible [34]. A similar observation was made by Yasuniwa et al. in PE melting at an applied pressure of 400 MPa and crystallization at a pressure of 600 MPa [35]. It was proposed that at high pressures an intermediate phase was present between the melt state and the orthorhombic phase, associated with chain-folded crystallization. The orthorhombic crystal unit cell parameters for PE are: a = 7.4 Å, b = 4.93 Å, and c = 2.534 Å [36] with the "c" axis along the chain direction, as shown in Figure 19.5. Bassett, Block and Piermarini performed X-ray diffraction studies at high pressures using a diamond-anvil device [37], which, is commonly used today to study phase transitions, lamellar thickening, and other nanoscale and microscale morphological alterations associated with elevated pressure and temperatures using small angle and wide angle X-ray diffraction. They discovered that the commonly observed 110 and 200 crystal reflections of the orthorhombic crystal of PE was replaced by a single spacing of 4.23 Å before melting at 600 MPa. This new phase was identified as a hexagonal phase (see Figure 19.6) in which the "a" lattice parameter of the hexagonal crystal increased to give a/b = 1.57, while the "c" direction was preserved during this phase transition. This provided a more open unit cell structure where the PE macromolecules could assume both trans as well as gauche conformation and were therefore more meltlike than in an orthorhombic unit cell.

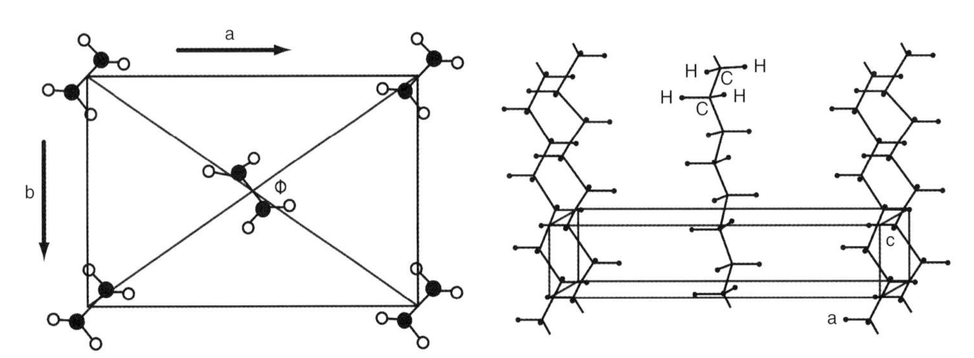


**FIGURE 19.5** Orthorhombic crystal of PE (A) 2-D model and (B) 3-D model (adapted from White JL and Shan H. "Deformation Induces Structural Changes in Crystalline Polyolefins" Polymer-Plastics Technology and Engineering 2006; 45: 317–328). Image courtesy of Eric Wysocki, Exponent, Inc.

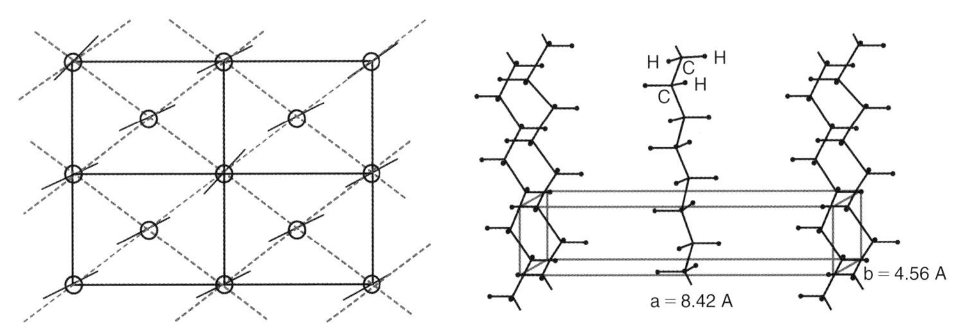


**FIGURE 19.6** Hexagonal crystal of PE (A) 2-D model and (B) 3-D model (adapted from White JL and Shan H. "Deformation Induces Structural Changes in Crystalline Polyolefins" Polymer-Plastics Technology and Engineering 2006; 45: 317–328). Image courtesy of Eric Wysocki, Exponent, Inc.

Based on melting experiments performed at various pressures, a phase diagram for PE is now available, as shown by the schematic of Figure 19.7. At pressures below the triple point, there is melting of the orthorhombic cell at all pressures without transformation into the hexagonal cell. The triple point is defined as the point where the melt, hexagonal, and orthorhombic phases converge. Above the triple point, the orthorhombic unit cell undergoes transformation into a hexagonal unit cell prior to melting. This is a reversible process in that the hexagonal phase is encountered whether the temperature is increased from the orthorhombic to the melt phase or decreased from the melt phase to the orthorhombic phase at elevated pressures. Leute and Dollhopf noted that the triple point cannot be determined accurately and is present above 350 MPa and between 215–220°C for medium molecular weight PE [5]. Bassett estimated that the triple point for very high molecular weight polyethylene was at 270 MPa and above 500 MPa for molecular weights of a few thousand [3]. A study on gamma irradiated solution-grown crystals of PE showed that for an 80 Mrad dose, the triple point dropped from 300 MPa to 200 MPa.

There are two basic processing routes for high-pressure processing of PE: high-pressure crystallization (HPC) and high-pressure annealing (HPA), as shown in Figure 19.7. In HPC, the sample is taken from room temperature into the melt phase and then isothermally pressurized so that it enters the hexagonal phase horizontally at elevated pressures and temperatures. Alternatively, HPC can be performed by entering the melt phase, increasing the pressure above the pressure at the triple point so that the PE is still in the melt phase, and then isobarically cooled down so that HPC occurs in the hexagonal phase. Both these routes are effective in inducing ECCs in PE. When cooling from the melt state at high pressures, crystallization via the hexagonal phase occurs to form lenslike crystals that grow in the direction of the diameter of the lens, up to the order of micrometers in thickness. As the sample is cooled further and passes below the hexagonal-orthorhombic transition line, the gaps between the ECC lamellae are rapidly filled with chain-folded crystals, making the observer feel like they're witnessing a firecracker [35]. It should be noted that ECCs can also be formed below the triple point [31, 37–39]. Rastogi et al. showed upon cooling at elevated pressures below the triple point that crystallization occurred initially via both a hexagonal phase as well as orthorhombic phase. The hexagonal phase transformed into the orthorhombic phase with time, and further crystallization occurred via the orthorhombic phase only. Relatively thick, ECC-type lamellae could be formed by this method. This phenomenon was not reversible, i.e. the transient hexagonal phase was not observed during melting at elevated pressures below the triple point. This is not the case for pressures exceeding the triple point pressure where the hexagonal phase is encountered during both heating and cooling.

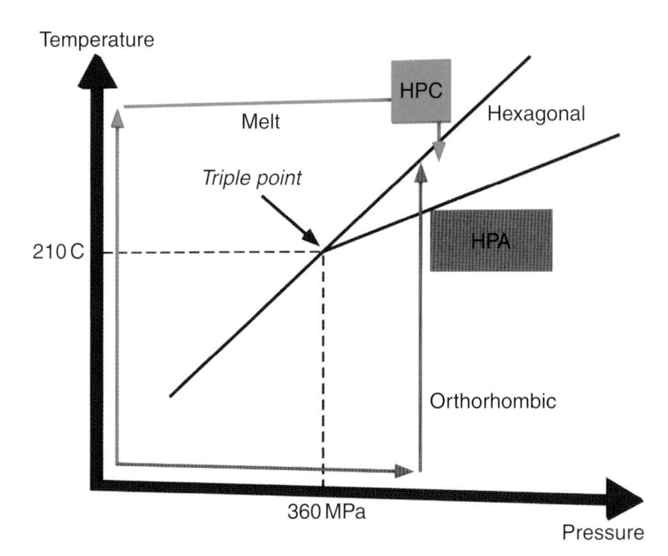


**FIGURE 19.7** Phase diagram for PE showing melt state, orthorhombic state, the triple point, a route for high-pressure crystallization (HPC), and a route for high-pressure annealing (HPA). Image courtesy of Eric Wysocki, Exponent, Inc.

Finally, HPA involves increasing the pressure of PE beyond the triple point pressure and then heating the polymer so that it enters into the hexagonal phase. If the PE is maintained at these elevated pressures and temperatures, the initially chain-folded lamellae begin to increase in thickness via the hexagonal phase. Rees and Bassett showed that chain-folded PE, regardless of forms such as films, fibers, spherulitic, or single-crystal mats, could be converted into ECCs via gradual thickening of lamellae and without loss of original texture (orientation) [29, 40]. Gruner et al. [11] had also conducted high-pressure annealing but interpreted the results in terms of partial melting and recrystallization. Bassett and Carder were the first to report high-pressure annealing via the hexagonal phase, which was performed on oriented sheets of PE [41]. There has been some dispute about the mechanism of thickening. While a strong argument can be made for the case that the annealing process is a sequence of local melting and recrystallization into thick ECCs, this theory would not appear to be plausible because there was evidence of sample history in the final ECC samples, although melting and recrystallization has been shown to also preserve sample history [42, 43]. HPA of ultrahigh molecular weight PE has shown very little increase in lamellar thickening or crystallinity, indicating that large lamellar thickening is confined to the low-entanglement, medium and low molecular weight PEs rather than highly entangled or crosslinked ultrahigh molecular weight PEs.

## 19.3 HYLAMER

Hylamer was the product of research on high-pressure crystallization of UHMWPE in the 1980s. Invented by Li and Howard from E. I. DuPont de Nemours and Company (Wilmington, Delaware, USA), Hylamer was marketed as "Enhanced Polyethylene" by the DePuy-DuPont Orthopaedics joint venture (Newark, Delaware, USA) [13]. Hylamer was produced by the joint venture, which was disbanded in mid-1998. Hylamer was used extensively in the United States, Europe, and the United Kingdom as late as 1999 [44]. Initially gamma sterilized in air prior to 1993, Hylamer was gamma sterilized in nitrogen-purged vacuum packaging between 1993 and 1995 [45]. Starting in 1995, Hylamer was sterilized using gas plasma [45]. It has been estimated that 80,000 Hylamer cups were implanted worldwide [44] before it was replaced by Marathon crosslinked UHMWPE in 1997.

## 19.3.1 Structure and Properties

Hylamer was produced by hot isostatic pressing (HIPping) ram-extruded rods of GUR 415 at elevated temperatures and pressures (approximately 600°F and 65 to 75 ksi; personal communication, Todd Smith, DePuy Orthopedics). These conditions led to the formation of an extended-chain crystallite morphology with thick (200–500 nm) lamellae and high crystallinity (65–71%) [14, 46]. In contrast, conventional low pressure sintered UHMWPE displays a folded-chain crystalline morphology with much thinner lamellae (10–50 nm in thickness) and a crystallinity of 50–55% (Chapter 1). By varying the postconversion heating, pressure, and cooling sequence, two orthopedic biomaterials, Hylamer and Hylamer-M, were developed with varying crystallinities and mechanical properties (Table 19.1). Hylamer was used in acetabular components for hip replacements as well as in glenoid components for total shoulder replacements, whereas Hylamer-M was used in knee bearing components.

## 19.3.2 *In Vitro* Studies of Hylamer

As summarized in Table 19.1, Hylamer had a higher density and crystallinity than GUR 415, which was the conventional UHMWPE of the 1990s. Although the yield and ultimate strength of Hylamer are slightly higher, the most noticeable change occurs in the elastic modulus, which is nearly double for Hylamer as compared to conventional GUR 4150 [14]. Hylamer also exhibited much greater creep resistance than GUR 415 [14].

A finite element analysis performed by Wright et al. [48] shed some light into the structural behavior of acetabular cups manufactured from Hylamer. When there was a clearance between the metal articulating ball and the plastic in a total hip arthroplasty model, the contact stresses for Hylamer were greater than those predicted for conventional UHMWPE. However, depending on the failure criterion used, it was not readily apparent whether or not the material could be predicted to be advantageous for use in acetabular cups.

In 1992, McKellop and colleagues [49] reported on the wear behavior of Hylamer using a hip joint simulator. After 4 million loading cycles, no statistical difference was observed in the wear rates of Hylamer when compared with conventional UHMWPE. The cups in this initial study were produced from the same batch of GUR 415 and gamma irradiated in air, but not subjected to accelerated aging, which at the time had not yet been developed [15]. McKellop later reported that the wear behavior of gas plasma sterilized Hylamer was comparable to never-sterilized GUR 415 acetabular cups throughout a 5 million cycle wear test [15]. The researchers concluded that, whether gamma sterilized or gas plasma sterilized, the wear behavior of Hylamer in a hip simulator was comparable to lot-controlled GUR 415, provided both materials were sterilized and aged in the same manner.

Wang and colleagues also investigated the tribological properties of Hylamer acetabular components after gamma-irradiation in air and in a low oxygen package [50]. The wear rates, as determined using a multidirectional motion hip simulator, were $0.29 \pm 0.08\,mm^3$/year for Hylamer after irradiation in air; $0.17 \pm 0.02\,mm^3$/year for Hylamer after irradiation in a low oxygen package; and $0.12 \pm 0.004\,mm^3$/year for conventional UHMWPE

TABLE 19.1 Summary of Physical and Mechanical Properties of GUR 415, Hylamer, and Hylamer-M, Adapted from [47]

| Property | GUR 415 | Hylamer-M | Hylamer |
|---|---|---|---|
| Degree of crystallinity (%) | 50 | 57 | 68 |
| Density (g/cc) | 0.934 | 0.946 | 0.955 |
| Melt temperature (°C) | 135 | 147 | 149 |
| Yield strength (MPa) | 23.3 | 26.5 | 28.6 |
| Elongation to failure (%) | 339 | 369 | 334 |
| Elastic modulus (GPa) | 1.39 | 2.01 | 2.52 |

after irradiation in air. Thus, in contrast with the wear data reported by McKellop et al. [15, 49], Wang found that gamma-air sterilized Hylamer exhibited substantially higher wear rates than conventional materials. McKellop [15] attributed the discrepancy between his results and Wang's data to differences in the batch of the UHMWPE materials, as well as to differences in the shelf aging, particularly with respect to the materials that were gamma irradiated air.

Work on the change of the bulk mechanical properties of the Hylamer material was performed by Rimnac et al. [51], who exposed both conventional GUR 4150 and Hylamer to a degradative aqueous environment for up to 12 months. They reported that the changes in density and modulus were no greater in the conventional material than in the Hylamer material. However, a contemporaneous report by King et al. [52] suggested that extended-chain morphology material (Hylamer) was more susceptible to oxidative degradation than was a similarly-aged conventional UHMWPE material. This latter experiment used accelerated thermal aging to oxidize both polyethylenes after gamma irradiation and reported a loss in strain energy to break of 68% for Hylamer as opposed to only a 5% change for the GUR 4150 conventional UHMWPE. Both materials showed about the same level of oxidation using FTIR, leading the authors to postulate that the smaller amorphous content in Hylamer led to a greater percentage of tie molecules being broken for a given oxidation index.

## 19.3.3 *In Vitro* Studies of Hylamer-M

Like Hylamer, Hylamer-M exhibited higher crystallinity and yield strength than ram-extruded GUR 415 (Table 19.1). Hylamer-M was also found to have substantially smaller rates of fatigue crack propagation and creep in the laboratory [14]. Because fatigue crack propagation and creep were believed to play important roles in the development of surface damage to knee inserts *in vivo* [53], a material with a slower crack growth rate and improved creep resistance was considered to be advantageous in such inserts.

In the late 1990s, fatigue crack propagation studies were performed at body temperature by Baker and colleagues [54] on GUR 4150HP and Hylamer-M after sterilization using ethylene oxide, gas plasma, gamma radiation in air, and gamma radiation in nitrogen. In contrast with the other sterilization methods, gamma sterilization in air followed by accelerated aging resulted in the initiation of discontinuous surface cracks and the reduction of stress intensity factors at crack inception ($DK_{incept}$) by 40% and 55% for GUR 4150HP and Hylamer-M, respectively.

Small punch test results, as well as transmission electron microscopy, on Hylamer-M retrievals were reported for three retrieved tibial inserts following gamma irradiation in a nitrogen environment and gas plasma sterilization [55]. Beyond the elastic regime, the small punch load-displacement behavior of Hylamer-M displayed a typical response for UHMWPE, including an initial peak load (during initial bending of the disk-shaped specimen), followed by a drawing (stretching) phase under equibiaxial tension. The ultimate load of the Hylamer-M was found to be significantly higher than the gamma air-sterilized GUR 415 retrievals. Minimal lamellar alignment was detected by TEM at the wear surfaces of the retrieved Hylamer-M components (Figure 19.8).

## 19.3.4 Clinical Studies of Hylamer in Hip Arthroplasty

The clinical performance of Hylamer in acetabular inserts has been closely followed by the orthopedic community over the past 2 decades [44, 50, 56–73]. An initial review of a small number of retrieved Hylamer acetabular liners was presented by Muratoglu et al. [58] in which they noted a wear rate of 0.48 mm/year. Similarly, Schmalzried et al. [59] noted elevated wear rates in another series of revised Hylamer cups. It should be noted that because both of these series report measurements based on retrieved, and therefore by definition failed arthroplasties, these studies

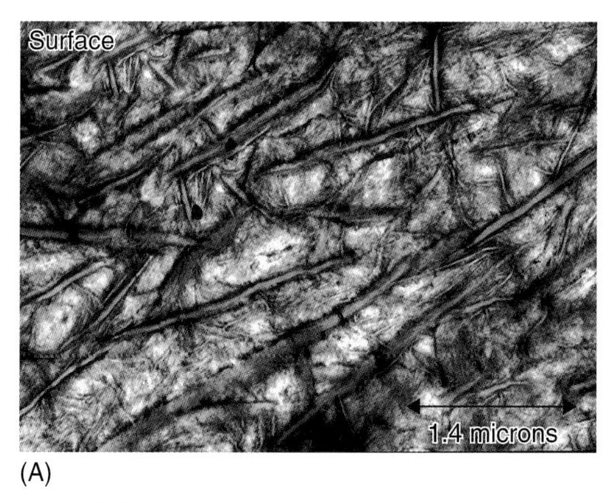

(A)

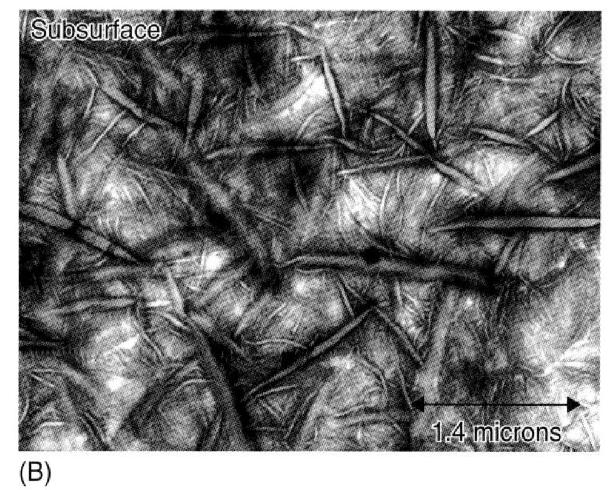

(B)

**FIGURE 19.8** Transmission electron micrograph of a Hylamer-M tibial insert within 2 μm of the surface of a retrieved tibial insert (A), and 4 μm below the articulating surface (B). Images courtesy of Deborah Crane and Lisa Pruitt, University of California at Berkeley.

provide limited insight as to how the polymer might behave in a well functioning arthroplasty.

In 1996, Chmell et al. [57] published a retrospective review of 143 patients implanted with Hylamer acetabular inserts after a follow-up period of at least 2 years. The index surgeries were performed between 1991 and 1992, and thus all of the Hylamer components in this study were gamma irradiated in air. Seventeen patients in the series were implanted with matched femoral components from the same manufacturer as the liner; the remaining patients were implanted with "mismatched" femoral components from a different manufacturer. Overall, the revision rate due to excessive wear was 4.8%, with all six of the revisions occurring within the group of 126 patients with mismatched components. The authors judged the wear-related failure rate to be unacceptable and discontinued use of Hylamer acetabular inserts.

Livingston et al. [56] reported on several series of patients with total hip arthroplasties whose acetabular bearing inserts were made of either Hylamer or conventional UHMWPE. These reviewers were also prompted by isolated incidents of early retrieval of Hylamer bearings, but they extended their study to the measurement of nonrevised patients by measuring the wear rate radiographically. The overall finding was that the wear rate of the Hylamer liners was almost double that exhibited by the conventional liners. Furthermore, they noted that the wear rate of the Hylamer liners rose even more, to about 2.7 times that of the conventional liner, when femoral bearing heads of a different manufacturer than that of the Hylamer product were fitted to the femoral stem. While their report did compare the wear rates of two differing designs as well as differing polyethylenes, it did cast some doubt, when combined with the study by Chmell et al. [57], on the efficacy of using the extended-chain Hylamer material in acetabular cups.

Sychterz et al., in contrast, found no significant difference between the radiographically-determined average wear rates of Hylamer and conventional UHMWPE (Enduron) [62]. Analysis suggested that the wear rates of the two materials appeared to be changing as a function of time. However, only in the third postoperative year were any statistically significant differences observed between the two materials. Consequently, the authors cautioned that further investigation would be necessary to draw conclusions regarding the long-term performance of Hylamer relative to Enduron.

In a multi-institutional review of 1080 patients implanted with Hylamer and conventional UHMWPE (GUR 4150HP), Schmalzried et al. [60] reported that *in vivo* wear rates were highly surgeon dependent, with an up to fivefold variation in wear rate from surgeon to surgeon. Schmalzried's data, along with the results of Sychterz et al. [62], suggested that many surgeons may have used Hylamer preferentially in their younger, more active patients, thereby biasing the results against Hylamer. Nonetheless, some surgeons using Hylamer presented series with wear rates substantially lower than those of colleagues using GUR 4150 based designs.

Collier and coworkers also studied the average wear rate, oxidation, and mechanical properties in retrieved Hylamer and conventional UHMWPE inserts of the same design [61]. Although a statistically significant difference was not observed between the linear wear rates of the two materials, the ultimate tensile strength and elongation to break appeared lower in the Hylamer components for a given level of oxidation, leading the authors to conclude that the impact of oxidation may be more detrimental to Hylamer than to conventional UHMWPE.

Collier's initial hypothesis was further supported by the detailed analysis of 12 Hylamer components that were revised for accelerated wear in a study by Scott et al. [63]. All of the components were gamma sterilized in air and implanted between 1990 and 1992. A total of 11/12 components exhibited pelvic lysis. The implantation time of the liners ranged from 39.2 to 69.4 months (mean ± SD: 50.0 ± 10.5 months), and the average linear wear rate was 0.49 ± 0.13 mm/y (range: 0.33 to 0.72 mm/y). The researchers were able to classify variations in color among three subgroups of retrieved Hylamer liners, ranging from bright white, cream, or light tan, and a darker or mottled tan with a "tiger eye." These color changes were correlated to both the shelf life, which ranged from 2 to 24 months (mean ± SD: 10.5 ± 6.5 months), as well as to the geographic location at which the implants were sterilized. Scott also described a "palpable ridge" between the worn and unworn regions of the Hylamer acetabular bearing surface, which was white in all cases. A representative example of a retrieved Hylamer insert, showing its color variation, is shown in Figure 19.9.

In the past 6 years, a number of reports about Hylamer have emerged from Europe [44, 64–73]. These studies have generally shown elevated wear rates for Hylamer in combination with zirconia femoral heads, as well as stainless steel femoral heads. These elevated wear rates, however, were not always associated with early revision. Wroblewski et al. from Wrightington Hospital found that the clinical results with a mean follow up of 6 years (range: 3 to 8 years) were "excellent" despite an average wear rate of 0.22 mm/y. Despite these "excellent" results, Wroblewski wrote that they had discontinued use of Hylamer at Wrightington because "the high initial rates of penetration did not settle to the expected low levels within the anticipated time"[66].

### 19.3.5 Clinical Studies of Hylamer in the Knee

The performance of Hylamer-M in tibial components has not been extensively documented in the literature. A case

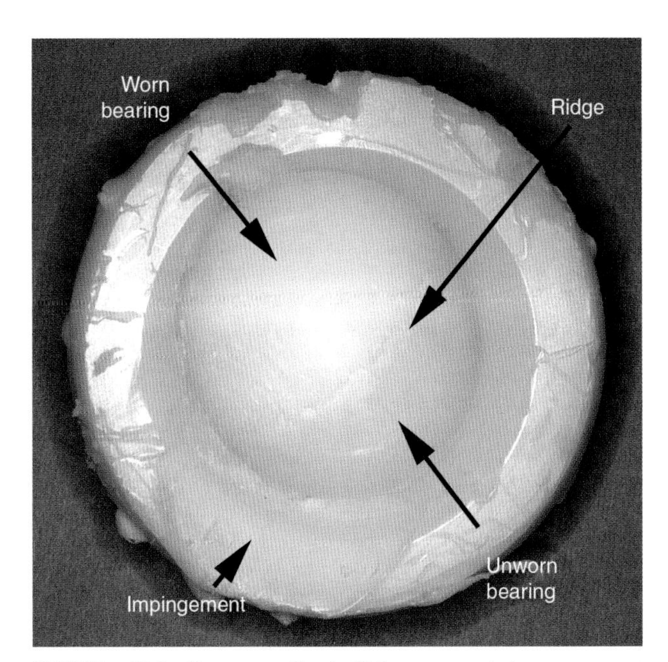


**FIGURE 19.9** Gamma-sterilized Hylamer acetabular component retrieved after 12 years *in vivo* due to wear, osteolysis, and cup loosening. Note that the component has a tan or ivory color, except for worn regions of the bearing surface and a region of rim impingement, which are white. The worn and unworn regions of the bearing surface are separated by a ridge, which also exhibits peripheral delamination. Image courtesy of Dan MacDonald, Drexel University.

report in 1996 by Ries et al. [74] reviewed a Hylamer-M tibial bearing insert in which gross pitting and delamination were seen 2.5 years postoperatively. The report noted that the maximum crystallinity measured in the retrieved implant was 68% at a depth into the plastic that coincided with the plane of delamination. Ahn et al. also reported osteolysis in two cases and delamination in one case of Hylamer-M within 5 years of implantation [75]. Aside from these clinical case reports, data on the performance of Hylamer-M in the knee remains limited.

### 19.3.6 Hylamer: A Current Perspective

A large body of clinical data now indicates that Hylamer is at best comparable to, and more frequently inferior to, conventional UHMWPE in total joint arthroplasty. Scientific evidence would suggest that its short-term wear performance depends upon the oxidation levels at the time Hylamer was implanted. It also must be emphasized that radiographic penetration data is biased in favor of Hylamer because of its greater creep resistance than conventional UHMWPE. Thus, even if Hylamer and conventional UHMWPE displayed the same overall penetration rate, as seen radiographically, it would imply that Hylamer exhibits greater wear, as discussed by Sychterz et al. [62]. Although by the late 1990s, there was some debate as to the reasons underlying variability in wear rates with Hylamer, today these differences appear to be best explained by differences in how the material was sterilized, and if it was gamma sterilized in air, how long the material was exposed to oxygen during shelf aging.

Developed in the decade following Poly II (detailed in Chapter 17), Hylamer marks the second time in orthopedic history in which a major manufacturer attempted, unsuccessfully, to improve UHMWPE for total joint replacements. The circumstances surrounding the evaluation and clinical performance of Hylamer in the 1990s are didactic for the contemporary reader in several respects. Clinical data from Hylamer with various sterilization conditions has been used by researchers at multiple institutions to help validate contemporary hip simulators [15, 50, 76]. Thus, the knowledge gained from the Hylamer experience helped inform the transition in orthopedics to highly crosslinked UHMWPE during the late 1990s. Most importantly, the history of Hylamer provides evidence of the unique vulnerability of high-pressure crystallized UHMWPE to post-irradiation oxidation and clues for how to improve this family of materials for clinical use today.

## 19.4 CROSSLINKING FOLLOWED BY HIGH-PRESSURE CRYSTALLIZATION

There have been several studies on the high-pressure crystallization of high molecular weight polyethylenes [77–81], demonstrating its high modulus combined with ductility, ultimately leading up to the introduction of Hylamer to the orthopedic market. Despite the limited success of Hylamer associated with oxidative degradation following gamma sterilization, it was well acknowledged that high-pressure crystallization increased several mechanical properties, including resistance to fatigue crack growth, resistance to creep deformation, high modulus and high yield stress without compromise to ductility or wear (when it was not oxidized) [13, 14]. In 2002, high-pressure crystallization investigations of UHMWPE began again but this time with the objective of restoring mechanical properties of radiation-crosslinked UHMWPEs, which are known to have lower mechanical properties compared to conventional (uncrosslinked) UHMWPE [82, 83]. It was shown that crosslinking followed by high-pressure crystallization led to smaller and smaller increases in crystallinity with increase in radiation dose. This effect was not surprising because lamellar thickening and growth require "reeling" in or reptation of macromolecules of PE from the melt state, which would be impeded by the chemical connectivity of the crosslinked macromolecules. Thus, higher crosslink density (or radiation dose) would result in a smaller enhancement of crystallinity for the same process conditions. In 2008, Oral et al. also showed that the effectiveness of high-pressure crystallization decreases with increase in radiation dose [84]. In companion studies

[85, 86], it was shown that, for a radiation dose of 5 Mrad followed by high-pressure crystallization, the resistance to fatigue crack propagation, tensile modulus, and yield strength increased without compromising wear resistance.

The effectiveness of high-pressure annealing following radiation crosslinking was also investigated for medical grade UHMWPE [87]. D'Angelo et al. showed that isothermal annealing of 5 Mrad gamma-irradiated UHMWPE at various pressures up to 500 MPa was actually less effective than atmospheric pressure annealing. This was not surprising because annealing is more effective in free radical removal when the annealing temperature is closer to the melting temperature. Annealing at a fixed temperature at higher and higher pressure combined with the fact that the melting temperature increases with increasing pressure (via the phase diagram) means that the samples at high pressures were annealed far below their corresponding melting temperatures than samples annealed at moderate and low pressures. However, high-pressure annealing also means that the annealing temperature can be raised to very high temperatures (such as 220°C at 500 MPa) without melting the PE, where there is moderate gain in crystallinity (in the hexagonal phase), and the high-thermal energy associated with high temperature can also effectively remove free radicals. This observation was independently confirmed by Oral et al., albeit at different irradiation dose and annealing conditions and at high pressures [88] as well as intermediate pressures [89].

## 19.5 HIGH-PRESSURE CRYSTALLIZATION FOLLOWED BY CROSSLINKING

High-pressure crystallization followed by radiation crosslinking has also been studied, primarily as a means to increase the mechanical properties of crosslinked PE without compromising wear [90–93]. Greer et al. showed that that the pre-irradiation crystallinity affected the wear resistance of crosslinked PE. They irradiated a series of PE samples from rapidly quenched PE (low crystallinity) up to Hylamer (high crystallinity, ECC morphology), melted them to quench free radicals and cooled them to room temperature, thereby inducing a new crystalline morphology. Laboratory wear test results showed that a low pre-irradiation crystallinity produced more highly wear resistant PE, probably due to the high amorphous regions available for crosslinking during irradiation, but also due to high entanglement density compared to high-pressure crystallized PE. Oral et al. irradiated high-pressure crystallized PE containing vitamin E or diffused vitamin E into the samples so that no further heat treatments were necessary, which would destroy its ECC morphology [91]. A similar study was later conducted in high-pressure crystallized PE without vitamin E and demonstrated that low rate of radiation dose of electron beam was beneficial in maintaining high wear resistance and mechanical properties compared to high dose rate electron beam treatment [92]. However, the high crystallinity associated with high-pressure crystallization probably results in more trapped free radicals, which need to be removed to prevent long-term oxidation. It was shown by D'Angelo et al. that for an irradiation dose of 5 Mrad with no further heat treatment, the oxidation increase linearly with increase in lamellar thickness of the pre-irradiated material [93]. Therefore, it would be beneficial for such high crystallinity, high-pressure crystallized PEs to have an antioxidant present in them to prevent subsequent oxidative degradation, which occurred in Hylamer.

## 19.6 SUMMARY

In summary, high-pressure crystallization offers a novel processing route to increase the crystallinity of PE and associated mechanical properties, such as resistance to fatigue crack propagation, resistance to creep deformation, tensile modulus, and yield strength without compromising ductility and wear resistance. However, attention must be paid to the effects of radiation on high-pressure crystallized PE because the crystalline regions trap free radicals, which can lead to long-term oxidative degradation. The use of antioxidants or melt quenching prior to high-pressure crystallization may offer a PE with higher mechanical properties for application in joint replacements compared to conventionally processed PE.

## 19.7 ACKNOWLEDGMENTS

Special thanks to Todd Smith, PhD, DePuy Orthopedics, and Ebru Oral, PhD, Massachusetts General Hospital, for providing an editorial review of this chapter.

## REFERENCES

1. Wunderlich B. *Macromolecular physics*. New York, NY: Academic Press; 1973.
2. Wunderlich B. *Macromolecular physics*. New York, NY: Academic Press; 1976.
3. Bassett DC. Chain-extended polyethylene in context—review. *Polymer* 1976;**17**(6):460–70.
4. Bhateja SK, Pae KD. Effects of hydrostatic-pressure on compressibility, crystallization, and melting of polymers. *J Macromol Sci—Rev Macromol Chem Phys* 1975;**C 13**(1):77–133.
5. Leute U, Dollhopf W. A review of experimental data from the high pressure phase in polyethylene. *Colloid Polym Sci* 1980;**258**:353–9.
6. Wunderlich B, Davidson T. Extended-chain crystals. I. General crystallization conditions and review of pressure crystallization of polyethylene. *J Polym Sci Part A-2-Polym Phys* 1969;**7**:2043–50.
7. Davidson T, Wunderlich B. Extended-chain crystals. II. Crystallization of polyethylene under elevated pressure. *J Polym Sci Part A-2-Polym Phys* 1969;**7**:2051–9.

8. Prime RB, Wunderlich B. Extended-chain crystals. III. Size distribution of polyethylene crystals grown under elevated pressure. *J Polym Sci Part A-2-Polym Phys* 1969;**7**:2061–72.
9. Prime RB, Wunderlich B. Extended-chain crystals. IV. Melting under equilibrium conditions. *J Polym Sci Part A-2-Polym Phys* 1969;**7**:2073–89.
10. Prime RB, Wunderlich B, Melillo L. Extended-chain crystals. V. Thermal analysis and electron microscopy of the melting process in polyethylene. *J Polym Sci Part A-2-Polym Phys* 1969;**7**:2081–97.
11. Gruner CL, Wunderlich B, Bopp RC. Extended-chain crystals. VI. Annealing of polyethylene under elevated pressure. *J Polym Sci Part A-2-Polym Phys* 1969;**7**:2099–113.
12. Geil PH, Arakawa T, Anderson FR, Wunderlich B. Morphology of polyethylene crystallized from melt under pressure. *J Polym Sci Part A—General Papers* 1964;**2**:3707.
13. Li S, Howard EG. Process for manufacturing ultra high molecular weight polyethylene shaped articles. *United States Patent, No. 5,037,928*; issued 1991 August 6.
14. Champion AR, Li S, Saum K, Howard E, Simmons W. The effect of crystallinity on the physical properties of UHMWPE. *Transactions of the 40th Orthopedic Research Society*. New Orleans, LA; 1994 February 21–24. p. 585.
15. McKellop H, Shen FW, Lu B, Campbell P, Salovey R. Effect of sterilization method and other modifications on the wear resistance of acetabular cups made of ultra-high molecular weight polyethylene. A hip-simulator study. *J Bone Joint Surg Am* 2000 December;**82-A**(12):1708–25.
16. Gomoll A, Wanich T, Bellare A. J-integral fracture toughness and tearing modulus measurement of radiation cross-linked UHMWPE. *J Orthop Res* 2002 November;**20**(6):1152–6.
17. Baker DA, Bellare A, Pruitt L. The effects of degree of crosslinking on the fatigue crack initiation and propagation resistance of orthopedic-grade polyethylene. *J Biomed Mater Res Part A* 2003 July 1;**66A**(1):146–54.
18. Keller A. A note on single crystals in polymers—evidence for a folded chain configuration. *Philos Mag* 1957;**2**(21):1171.
19. Fischer EW. *Nature* 1957;**12a**:753.
20. Hoffman JD. Role of reptation in the rate of crystallization of polyethylene fractions from the melt. *Polymer* 1982;**23**(5):656–70.
21. Hoffman JD, Weeks JJ. Melting process and equilibrium melting temperature of polychlorotrifluoroethylene. *J Res Natl Bureau of Stand Section A—Phys Chem* 1962;**66**(JAN-F):13.
22. Psarski M, Piorkowska E, Galeski A. Crystallization of polyethylene from melt with lowered chain entanglements. *Macromolecules* 2000 February 8;**33**(3):916–32.
23. Wunderlich B, Dole M. Specific heat of synthetic high polymers. 8. Low pressure polyethylene. *J Polym Sci* 1957;**24**(106):201–13.
24. Hoffman JD. Regime-III crystallization in melt-crystallized polymers—the variable cluster model of chain folding. *Polymer* 1983;**24**(1):3–26.
25. Wunderlich B. Effect of pressure on crystallization of polyethylene from dilute solution. *J Polym Sci Part A—General Papers* 1963;**1**(4):1245.
26. Wunderlich B, Arakawa T. Polyethylene crystallized from melt under elevated pressure. *J Polym Sci Part A—General Papers* 1964;**2**(8PA):3697.
27. Wunderlich B, Melillo L. Morphology and growth of extended chain crystals of polyethylene. *Makromolekulare Chem-Macromol Chem Phys* 1968;**118**(NOV):250.
28. Hatakeyama T, Kanetsuna H, Hashimoto T. Large Polyethylene Crystals Formed under High Pressure. *J Macromol Sci-Phys* 1973;**B7**(2):411–15.
29. Rees DV, Bassett DC. Crystallization of polyethylene at elevated pressures. *J Polym Sci Part A-2-Polym Phys* 1971;**9**(3):385.
30. Bassett DC, Khalifa BA, Turner B. Chain-extended crystallization of polyethylene. *Nat-Phys Sci* 1972;**239**(94):106–8.
31. Bassett DC, Turner B. Phenomenology of chain-extended crystallization in polyethylene. *Philos Mag* 1974;**29**(4):925–55.
32. Kanetsuna H, Mitsuhashi S, Iguchi M, Hatakeyama T, Kyotani M, Maeda Y. Effect of pressure on crystallization of polyethylene. *J Polym Sci Part C—Polym Symp* 1973(42):783–93.
33. Maeda Y, Kanetsuna H. Crystallization and melting of polyethylene under high-pressure. 1. Crystallization by slow cooling from melt. *J Polym Sci Part B-Polym Phys* 1974;**12**(12):2551–65.
34. Bassett DC, Turner B. New high-pressure phase in chain-extended crystallization of polythene. *Nat-Phys Sci* 1972;**240**(103):146.
35. Yasuniwa M, Nakafuku C, Takemura T. Melting and crystallization process of polyethylene under high-pressure. *Polym J* 1973;**4**(5):526–33.
36. Bunn CW. The crystal structure of long-chain normal paraffin hydrocarbons. The "shape" of the >CH2 group. *Trans Faraday Soc* 1939;**35**(1):0482–90.
37. Bassett DC, Block S, Piermarini GJ. High-pressure phase of polyethylene and chain-extended growth. *J Appl Phys* 1974;**45**(10):4146–50.
38. Rastogi S, Kurelec L, Lemstra PJ. Chain mobility in polymer systems: On the borderline between solid and melt. 2. Crystal size influence in phase transition and sintering of ultrahigh molecular weight polyethylene via the mobile hexagonal phase. *Macromolecules* 1998 July 28;**31**(15):5022–31.
39. Rastogi S, Kurelec L. Polymorphism in polymers; its implications for polymer crystallisation. *J Mater Sci* 2000 October;**35**(20):5121–38.
40. Rees DV, Bassett DC. Origin of extended-chain lamellae in polyethylene. *Nature* 1968;**219**(5152):368–70.
41. Bassett DC, Carder DR. Oriented chain-extended polyethylene. 1. Formation and characterization. *Philos Mag* 1973;**28**(3):513–33.
42. Bassett DC, Khalifa BA, Olley RH. Morphological study of chain-extended growth in polyethylene. 2. Annealed bulk polymer. *Polymer* 1976;**17**(4):284–90.
43. Khalifa BA, Bassett DC. Morphological study of chain-extended growth in polyethylene. 3. Annealing of solution-grown lamellae. *Polymer* 1976;**17**(4):291–7.
44. Kiely PD, Harty JA, McElwain JP. Hylamer wear rates and shelf life: a clinical correlation. *Acta Orthop Belg* 2005 August;**71**(4):429–34.
45. Rockwood Jr. CA, Wirth MA Observation on retrieved Hylamer glenoids in shoulder arthroplasty: problems associated with sterilization by gamma irradiation in air. *J Shoulder Elbow Surg* 2002 March-April;**11**(2):191–7.
46. Bellare A, Livinston BJ, Cohen RE, Chmell M, Thomas W, Poss R, et al. Characterization of wear surface and bulk morphology of retrieved Hylamer acetabular cups. *Trans 5th World Biomater Conference*; 1996. p. 982.
47. Li S, Burstein AH. Ultra-high molecular weight polyethylene. The material and its use in total joint implants. *J Bone Joint Surg Am* 1994;**76**(7):1080–90.
48. Wright TM, Gunsallus KL, Rimnac CM, Bartel DL, Klein RW. Design considerations from an acetabular component made from an enhanced form of ultra high molecular weight polyethylene. *Transactions of the 37th Orthopedic Research Society*; 1991. p. 248.

49. McKellop HA, Lu B, Li S. Wear of acetabular cups of conventional and modified UHMW polyethylene compared on a hip joint simulator. *Transactions of the 38th Orthopedic Research Society*. Washington, D.C.; 1992 February 17-20. p. 356.
50. Wang A, Essner A, Polineni VK, Stark C, Dumbleton JH. Lubrication and wear of ultra-high molecular weight polyethylene in total joint replacements. *Tribol Int* 1998;**31**(1-3):17–33.
51. Rimnac CM, Baldini TH, Wright TM, Saum KA, Sanford WM. In vitro chemical and mechanical degradation of Hylamer ultra-high molecular weight polyethylene. *Transactions of the 42nd Orthopedic Research Society*. Atlanta, GA; 1996 February 19-22. p. 481.
52. King R, Kirkpatrick L, Devanathan D, Lin S, Krebs S, Rohr W. Long-term aging behavior of implant grades of polyethylene. *Trans 5th World Biomater Conference*; 1996. p. 196.
53. Wright TM, Bartel DL. The problem of surface damage in polyethylene total knee components. *Clin Orthop* 1986;**205**:67–74.
54. Baker D, Coughlin D, Pruitt L. The effect of accelerated aging and sterilization method on the fatigue crack propagation resistance of Hylamer-M and GUR4150HP at body temperature. *Trans 34th Soc Biomater* 1998;**21**:122.
55. Kurtz SM, Rimnac CM, Pruitt L, Jewett CW, Goldberg V, Edidin AA. The relationship between the clinical performance and large deformation mechanical behavior of retrieved UHMWPE tibial inserts. *Biomaterials* 2000;**21**(3):283–91.
56. Livingston BJ, Chmell MJ, Spector M, Poss R. Complications of total hip arthroplasty associated with the use of an acetabular component with a Hylamer liner. *J Bone Joint Surg Am* 1997;**79**(10):1529–38.
57. Chmell MJ, Poss R, Thomas WH, Sledge CB. Early failure of Hylamer acetabular inserts due to eccentric wear. *J Arthroplasty* 1996;**11**(3):351–3.
58. Muratoglu OK, Imlach H, Estok D, Ramamurti B, Sedlacek R, Jasty M, et al. Analysis of eight retrieved Hylamer acetabular components. *Transactions of the 43rd Orthopedic Research Society*. San Fransisco, CA; 1997 February 9-13. p. 852.
59. Schmalzreid TP, Szuszczewicz ES, Campell PC, McKellop HA. Femoral head surface roughness and patient activity in the wear of Hylamer. *Transactions of the 43rd Orthopedic Research Society*. San Fransisco, CA; 1997 February 9-13. p. 787.
60. Schmalzried TP, Scott DL, Zahiri CA, Dorey FJ, Sanford WM, Kem L, et al. Variables affecting wear in vivo: analysis of 1080 hips with computer-assisted technique. *Transactions of the 44th Orthopedic Research Society*; 1998. p. 275.
61. Collier JP, Bargmann LS, Currier BH, Mayor MB, Currier JH, Bargmann BC. An analysis of Hylamer and polyethylene bearings from retrieved acetabular components. *Orthopaedics* 1998;**21**(8):865–71.
62. Sychterz CJ, Shah N, Engh CA. Examination of wear in duraloc acetabular components: two- to five-year evaluation of hylamer and enduron liners. *J Arthroplasty* 1998;**13**(5):508–14.
63. Scott DL, Campbell PA, McClung CD, Schmalzried TP. Factors contributing to rapid wear and osteolysis in hips with modular acetabular bearings made of hylamer. *J Arthroplasty* 2000;**15**(1):35–46.
64. Norton MR, Yarlagadda R, Anderson GH. Catastrophic failure of the Elite Plus total hip replacement, with a Hylamer acetabulum and Zirconia ceramic femoral head. *J Bone Joint Surg Br* 2002 July;**84**(5):631–5.
65. Iwase T, Warashina H, Yamauchi K, Sugiura S, Hasegawa Y. Early head penetration into cemented Hylamer Ogee socket. *J Arthroplasty* 2003 October;**18**(7):920–4.
66. Wroblewski BM, Siney PD, Fleming PA. Wear of enhanced ultra-high molecular-weight polyethylene (Hylamer) in combination with a 22.225 mm diameter zirconia femoral head. *J Bone Joint Surg Br* 2003;**85**(3):376–9.
67. Cohen J. Catastrophic failure of the Elite Plus total hip replacement, with a Hylamer acetabulum and Zirconia ceramic femoral head. *J Bone Joint Surg Br* 2004 January;**86**(1):148.
68. von Schewelov T, Sanzen L, Onsten I, Carlsson A, Besjakov J. Total hip replacement with a zirconium oxide ceramic femoral head: a randomised roentgen stereophotogrammetric study. *J Bone Joint Surg Br* 2005 December;**87**(12):1631–5.
69. Haines JF, Trail IA, Nuttall D, Birch A, Barrow A. The results of arthroplasty in osteoarthritis of the shoulder. *J Bone Joint Surg Br* 2006 April;**88**(4):496–501.
70. Stea S, Antonietti B, Baruffaldi F, Visentin M, Bordini B, Sudanese A, et al. Behavior of Hylamer polyethylene in hip arthroplasty: comparison of two gamma sterilization techniques. *Int Orthop* 2006 February;**30**(1):35–8.
71. Visentin M, Stea S, De Clerico M, Reggiani M, Fagnano C, Squarzoni S, et al. Determination of crystallinity and crystal structure of Hylamer polyethylene after in vivo wear. *J Biomater Appl* 2006 Ocotober;**21**(2):131–45.
72. Iwakiri K, Iwaki H, Kobayashi A, Minoda Y, Kagiyama H, Kadoya Y, et al. Characteristics of Hylamer polyethylene particles isolated from peri-prosthetic tissues of failed cemented total hip arthroplasties. *J Biomed Mater Res B Appl Biomater* 2008 April;**85**(1):125–9.
73. Skwara A, Stracke S, Tibesku CO, Fuchs-Winkelmann S. Poor mid-term results of total hip arthroplasty with use of a Hylamer liner. *Acta Orthop Belg* 2008 June;**74**(3):337–42.
74. Ries MD, Bellare A, Livingston BJ, Cohen RE, Spector M. Early delamination of a Hylamer-M tibial insert. *J Arthroplasty* 1996;**11**(8):974–6.
75. Ahn NU, Nallamshetty L, Ahn UM, Buchowski JM, Rose PS, Lemma MA, et al. Early failure associated with the use of Hylamer-M spacers in three primary AMK total knee arthroplasties. *J Arthroplasty* 2001 January;**16**(1):136–9.
76. Sychterz CJ, Orishimo KF, Engh CA. Sterilization and polyethylene wear: clinical studies to support laboratory data. *J Bone Joint Surg Am* 2004 May;**86-A**(5):1017–22.
77. Attenburrow GE, Bassett DC. Plastic-deformation of chain-extended polyethylene. *J Mater Sci* 1977;**12**(1):192–200.
78. Ferguson RC, Hoehn HH. Effect of molecular-weight on high-pressure crystallization of linear polyethylene. 2. Physical and chemical characterizations of crystallinity and morphology. *Polym Eng Sci* 1978;**18**(6):466–71.
79. Hoehn HH, Ferguson RC, Hebert RR. Effect of molecular-weight on high-pressure crystallization of linear polyethylene. 1. Kinetics and gross morphology. *Polym Eng Sci* 1978;**18**(6):457–65.
80. Lupton JM, Regester JW. Physical-properties of extended-chain high-density polyethylene. *J Appl Polym Sci* 1974;**18**(8):2407–25.
81. Yasuniwa M, Nakafuku C. High-pressure crystallization of ultra-high molecular-weight polyethylene. *Polym J* 1987;**19**(7):805–13.
82. Turell, M.E., Bellare, A. Effect of crystallization temperature on the morphology of high pressure crystallized ultra-high molecular weight polyethylene. *Abstra Paper Am Chem Soc* 2002 Aug 18;**224**: U508-U508.
83. Turell ME, Scherrer P, Bellare A, Thornhill TS. Effect of high-pressure crystallization on the morphology of crosslinked ultra-high molecular weight polyethylene. *Transactions of the Annual Meeting of the Orthopedic Research Society.* New Orleans, LA; 2003. p. 1409.

84. Oral E, Godleski C, Lozynsky AJ, Muratoglu OK. The Effect of Crosslink Density on Improved Strength in High Pressure Crystallized UHMWPE. *Transactions of the Annual Meeting of the Orthopedic Research Society*. San Francisco, CA; 2008. p. 1681.
85. Simis KS, Bistolfi A, Bellare A, Pruitt LA. The combined effects of crosslinking and high crystallinity on the microstructural and mechanical properties of ultra high molecular weight polyethylene. *Biomaterials* 2006 March;**27**(9):1688–94.
86. Bistolfi A, Turell ME, Lee YL, Bellare A. Tensile and tribological properties of high crystallinity radiation crosslinked UHMWPE. *J Biomed Mater Res Part B: Appl Biomater* 2008;**12**. (in press)
87. D'Angelo F, Conteduca F, Thornhill TS, Bellare A. The effect of pressure-annealing on the oxidation resistance of irradiated UHMWPE. *Transactions of the Annual Meeting of the Orthopedic Research Society*. Chicago, IL; 2006. p. 663.
88. Oral E, Godleski Beckos C, Muratoglu OK. Free radical elimination in irradiated UHMWPE through crystal mobility in phase transition to the hexagonal phase. *Polymer* 2008;**49**:4733–9.
89. Oral E, Godleski C, Lozynsky AJ, Muratoglu OK. Near-melt annealing of irradiated UHMWPE under pressure. *Transactions of the Annual Meeting of the Orthopedic Research Society*. San Francisco, CA; 2008.
90. Greer K, Aust SK, King R. The effect of pre-irradiation crystallinity on the crosslinking and wear of UHMWPE. *Transactions of the Annual Meeting of the Orthopedic Research Society*. Washington DC; 2005. p. 243.
91. Oral E, Wannomae KK, Christensen SD, Muratoglu OK. High pressure crystallized, irradiated and alpha-tocopherol stabilized UHMWPE with high crystallinity, low wear and oxidation. *Transactions of the Annual Meeting of the Orthopedic Research Society*. Washington DC; 2005.
92. Oral E, Godleski C, Muratoglu OK. Highly crosslinked, highly crystalline UHMWPE for total joints. *Transactions of the Annual Meeting of the Orthopedic Research Society*. San Diego, CA; 2007. p. 19.
93. D'Angelo F, Turell ME, Conteduca F, Thornhill TS, Bellare A. Oxidation in UHMWPE without post-thermal treatment. *Transactions of the Annual Meeting of the Orthopedic Research Society*. San Diego, CA; 2007. p. 1781.

Chapter 20

# Compendium of Highly Crosslinked UHMWPEs

Steven M. Kurtz, PhD



## 20.1 INTRODUCTION

The first generation of radiation crosslinking and thermal treatment of UHMWPE aroused intense scientific and commercial interest within the orthopedic community in the late 1990s and led to the commercial release of a second generation of highly crosslinked materials starting in 2005. The proliferation of crosslinking technology into hip and knee replacements has resulted in numerous proprietary UHMWPEs, with trade names like Crossfire, Longevity, and Marathon. Many of these new UHMWPEs are available for hips, whereas others, such as Prolong (Zimmer) and XLK (DePuy), are available exclusively for knees. Certain materials, such as Durasul, E-Poly HXLPE, and X3, can be found in both hip and knee products. The new product nomenclature, as well as the different techniques used in the irradiation, thermal processing, and sterilization of these new materials, may be confusing even to investigators who are accustomed to UHMWPE technology in joint replacement.

This chapter is a compendium of the processing details, packaging information, as well as the most recent research findings related specifically to the clinically available highly crosslinked and thermally treated UHMWPEs provided by the five largest orthopedic manufacturers (DePuy, Zimmer, Biomet, Stryker, Smith & Nephew). The general characteristics of first- and second-generation highly crosslinked UHMWPE materials have already been reviewed in the context of alternative bearings for hip replacement, as well as in the context of knee replacements, within Chapters 6 and 8, respectively. Further details about the development of annealed and remelted UHMWPE, as well as vitamin-E-containing UHMWPEs have also been summarized in Chapters 13–16. This chapter assumes prior knowledge about the basics of crosslinking and thermal treatment technology, which is covered in previous chapters.

Scientific and commercial developments related to radiation crosslinking and thermal treatment of UHMWPE are evolving at a rapid pace. At the time of my first review on this topic, written between 1997 and early 1998, highly crosslinked materials had not yet been clinically launched; the article was published in 1999 after the materials had been released for THA. When I wrote my second review of this subject as a book chapter in collaboration with Orhun Muratoglu in the summer of 2001, highly crosslinked UHMWPEs had just been released for TKA [1]. These previous works, along with a review by Gladius Lewis [2], dealt primarily with the effects of crosslinking on the properties and *in vitro* wear performance of radiation crosslinked UHMWPE. A study by Collier et al. [3] reviewed the physical and tensile properties of highly crosslinked UHMWPE materials for hip replacement before and after accelerated aging.

The first edition of this *Handbook*, published in 2004, contained an overview of first-generation UHMWPE materials available for clinical use at that time (Figure 20.1). A symposium on this topic, convened by the American Society for Testing and Materials (ASTM) in November 2002, provided a wealth of information about first-generation highly crosslinked UHMWPEs. The peer-reviewed proceedings of that symposium [4], printed by ASTM as a Special Technical Publication (STP), were a key reference when compiling the first edition of this chapter.

Since that time, an overview was performed in the context of a combined AAOS/NIH symposium on implant wear and summarized certain characteristics of first- and second-generation UHMWPE materials [5]. This new chapter is intended to update and expand upon the limited information contained in the AAOS/NIH review. In this new chapter we also summarize the clinical experience of first-generation crosslinked UHMWPEs and include overviews of the newer, second-generation materials in clinical use today. In the sections that follow, highly crosslinked materials are reviewed in alphabetic order. Because this chapter deals exclusively with proprietary materials, the contents of this chapter were verified for accuracy with all five of the individual manufacturers by the author.

## 20.2 ALTRX

| | |
|---|---|
| Type of crosslinked material | First generation (remelted) |
| Orthopedic manufacturer | DePuy Orthopedics, Inc. (Warsaw, Indiana, USA), www.depuy.com |
| Joint applications | THA |
| Clinical introduction | 2007 |
| Starting stock material | GUR 1020 ram extruded rod |
| Irradiation crosslinking modality | Gamma |
| Irradiation temperature | Room temperature |
| Irradiation dose (range) | 75 kGy |
| Postirradiation stabilization process | Melted at 155°C for 24 hours and then annealed at 120°C for 24 hours, both in reduced oxygen atmospheres |
| Sterilization modality | Gas plasma |
| Target total irradiation dose | 75 kGy |
| Detectable free radicals? | No |
| Maximum head size available | 48 mm |
| Minimum thickness available | 6 mm |
| Published *in vitro* properties/performance data? | Yes |
| Longest average published follow-up | Not available |

### 20.2.1 Development History and Overview

AltrX is based on technology developed for Marathon (see Section 20.8). AltrX™ was clinically introduced for the Pinnacle cup design in 2007. The AltrX process is similar to that described for Marathon except that ram extruded GUR 1020 is used for AltrX (instead of GUR 1050, which is used for Marathon) and a 75 kGy dose of gamma radiation is used (which is greater than the 50 kGy dose used

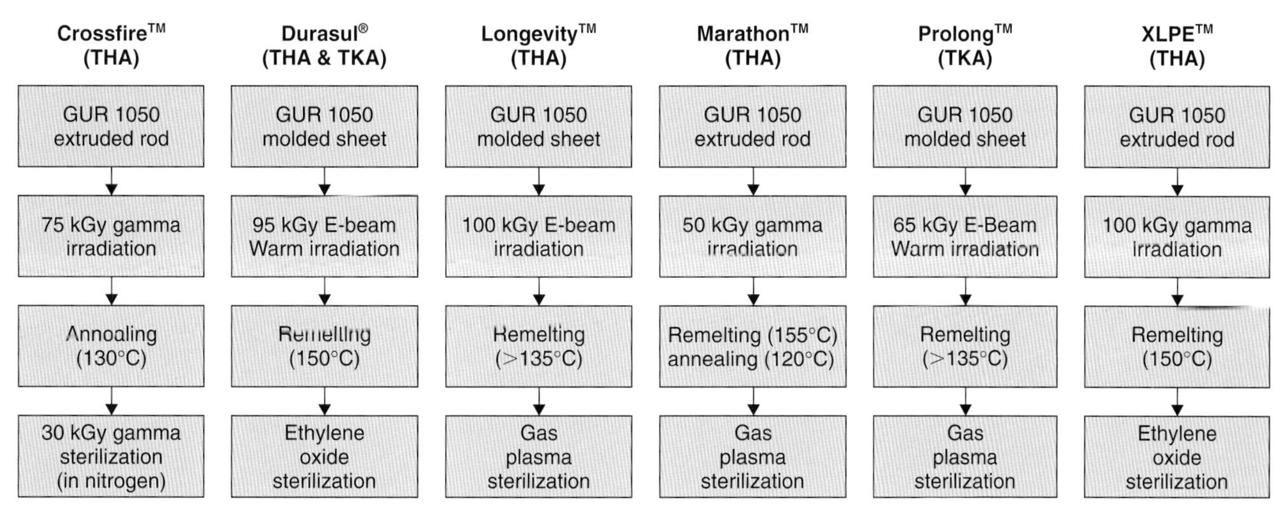


**FIGURE 20.1** Comparison of starting stock materials and processing conditions used to produce first-generation thermally treated and highly crosslinked UHMWPE materials for TKA and THA.

for Marathon). Components are machined from processed AltrX material, enclosed in gas-permeable packaging, and sterilized by gas plasma. This material is not available for knee replacements.

The hood and lip geometry for AltrX liners varies with head size. AltrX 28 mm and 32 mm inner diameter liners are available in Neutral, +4 Neutral, +4 10 Degree, and Lipped configurations. AltrX 36 mm liners are available in Neutral, +4 Neutral, and +4 10 Degree configurations. AltrX 40 mm, 44 mm, and 48 mm liners are available in +4 Neutral and +4 Degree configurations.

### 20.2.2 Properties and *In Vitro* Performance

The mechanical behavior of AltrX has been described by Greer and coworkers [6, 7]. When subjected to identical irradiation and thermal stabilization, ram extruded GUR 1020 exhibits similar wear behavior to GUR 1050 but has improved mechanical properties, such as toughness and ductility. Hip simulator wear testing, also performed by Greer et al., showed that the 75 kGy material would result in less wear relative to the 50 kGy material [6, 7].

The tribological properties for AltrX were recently presented at the 2008 annual meeting of the Orthopedic Research Society [7]. Hip simulator testing was performed using 36 mm diameter, CoCr, and BIOLOX Delta (Ceramtec) femoral heads under smooth and roughened conditions. The lowest *in vitro* wear rates were observed when AltrX was articulated against BIOLOX Delta femoral heads.

### 20.2.3 Clinical Results

Clinical results are not yet available for AltrX.

## 20.3 ARCOM XL POLYETHYLENE

| | |
|---|---|
| Type of crosslinked material | Second generation (mechanically annealed) |
| Orthopedic manufacturer | Biomet, Inc. (Warsaw, Indiana, USA), www.biomet.com |
| Joint applications | THA |
| Clinical introduction | 2005 |
| Starting stock material | GUR 1050 isostatically compression molded |
| Irradiation crosslinking modality | Gamma |
| Irradiation temperature | Room temperature |
| Irradiation dose (range) | 50 kGy |
| Postirradiation stabilization process | Solid state extrusion (mechanical annealing) followed by 130°C thermal annealing |
| Sterilization modality | Gas plasma |
| Target total irradiation dose | 50 kGy |
| Detectable free radicals? | Yes |
| Maximum head size available | 40 mm |
| Minimum thickness available | 4.8 mm |
| Published *in vitro* properties/performance data? | Yes |
| Longest average published follow-up | Not available |

### 20.3.1 Development History and Overview

ArCom XL polyethylene, a second-generation highly crosslinked UHMWPE, was developed by David Schroeder and coworkers at Biomet, Inc. (Warsaw, Indiana, USA). ArCom XL polyethylene was clinically introduced in 2005 and is currently available in the Max-Rom, Hi-Wall, 10 degree, +5 Hi-Wall, and +5 Max-Rom liners in the RingLoc acetabular cup design. The material is not available in total knee replacement designs.

ArCom XL polyethylene is produced in a multistep process that has been described in the literature [8]. Rods of GUR 1050 are isostatically molded and gamma irradiated at room temperature with a nominal dose of 50 kGy. The rods are preheated to 130°C and then hydrostatically extruded with a diametral compression ratio of 1.5. The hydrostatic extrusion produces orientation of the UHMWPE and mechanically anneals the material. The rods are then stress relieved at 130°C prior to machining into components, which are terminally gas plasma sterilized.

### 20.3.2 Properties and *In Vitro* Performance

Details about the *in vitro* test results for ArCom XL polyethylene have been published [8]. Mechanical testing data indicate that the material is anisotropic, with significantly enhanced strength oriented along the long axis of the hydrostatically extruded rod. For the elastic properties, only a 20% difference was noted between the long axis of the rod and the orthogonal, radial direction. The highly crosslinked material was found to contain detectable free radicals, at a concentration that is 90% less than control, gamma inert-sterilized UHMWPE. However, the material was not found to oxidize in accelerated aging tests (ASTM F2003) conducted for up to 4 weeks. Because the material was found to be oxidatively stable, the material is gas sterilized in air-permeable packaging. ArCom XL polyethylene is associated with a 47% improvement in hip simulator wear rate, as compared with conventional gamma-inert sterilized (ArCom polyethylene) material, during *in vitro* testing conducted by the manufacturer.

### 20.3.3 Clinical Results

Clinical results are not yet available for ArCom XL polyethylene.

## 20.4 CROSSFIRE

| Type of crosslinked material | First generation (annealed) |
|---|---|
| Orthopedic manufacturer | Stryker Orthopedics, Inc. (Mahwah, New Jersey, USA), www.stryker.com |
| Joint applications | THA |
| Clinical introduction | 1998 |
| Starting stock material | GUR 1050 extruded rod |
| Irradiation crosslinking modality | Gamma |
| Irradiation temperature | Room temperature |
| Irradiation dose (range) | 75 kGy |
| Postirradiation stabilization process | Annealed at 130°C for unknown duration |
| Sterilization modality | Gamma in nitrogen ($N_2$/Vac), 30 kGy dose |
| Target total irradiation dose | 105 kGy |
| Detectable free radicals? | Yes |
| Maximum head size available | 36 mm |
| Minimum thickness available | 5.9 mm |
| Published *in vitro* properties/ performance data? | Yes |
| Longest average published follow-up | 4.9 years [9] |

### 20.4.1 Development History and Overview

Crossfire, an annealed highly crosslinked UHMWPE, was developed by Scott Taylor and coworkers at Stryker Osteonics Corp. (Allendale, New Jersey, USA). The history of the development of Crossfire is summarized in Chapter 14 (see Section 14.3). Crossfire was clinically introduced in the fall of 1998 for the Series II liner in the Omnifit acetabular cup design. After Osteonics acquired Howmedica in 1999 and formed Strkyer Orthopedics, Crossfire was subsequently extended to the System 12 and Trident acetabular cup designs.

In the Crossfire process, extruded rod bar stock is irradiated with a nominal dose of 75 kGy and then annealed at 130°C [10]. Acetabular components are then machined from the bar stock, barrier packaged in nitrogen, and then gamma sterilized with a nominal dose of 30 kGy. Consequently, components that have been through the Crossfire process have received a total dose of 105 kGy.

An example of the $N_2$/VAC packaging used for Crossfire is shown in Figure 20.2.

### 20.4.2 Properties and *In Vitro* Performance

Details about the *in vitro* test results with Crossfire have been published [10] and are summarized in Chapter 14 (see Section 14.3.2). In comparison with conventional UHMWPE, Crossfire is associated with over 90% improvement in median hip simulator wear rate across

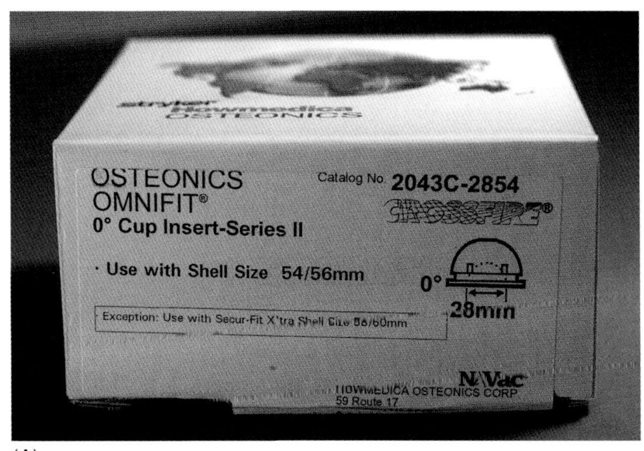

(A)

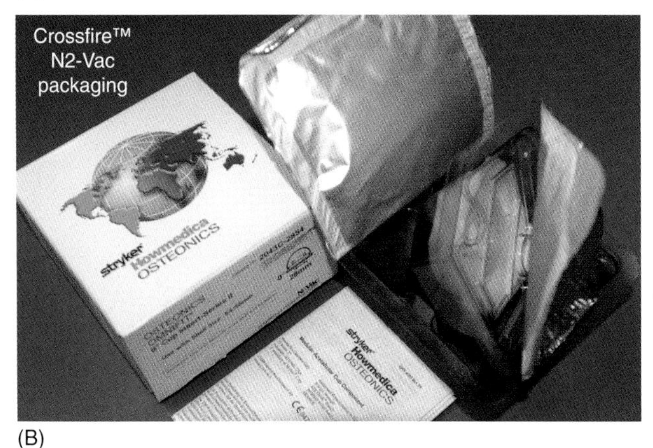

(B)

**FIGURE 20.2** Packaging for Crossfire. (A) Outer packaging box. (B) N2/Vac barrier packaging used for gamma sterilization of Crossfire.

a broad range of implant designs. The crystallinity and microstructure, as well as the yield stress and ultimate strength of Crossfire, were reported to be comparable to conventional, barrier packaged UHMWPE.

Because Crossfire is sterilized by gamma radiation, the material contains free radicals. However, the mechanical performance of the bearing was judged by the designers of Crossfire to be of greater importance than complete elimination of free radicals due to the use of barrier packaging, which minimizes oxidation during shelf storage prior to implantation.

Crossfire is not available for knee applications due to concerns raised during functional fatigue testing [11]. During patella testing, Wang et al. [11] reported that components fabricated from Crossfire were observed to fracture through the support pegs.

### 20.4.3 Clinical Results

Five clinical studies of Crossfire [9, 12–15] have been published to date (Table 20.1), including a 2-year prospective randomized multicenter trial [12]. These five studies report a reduction in head penetration for Crossfire ranging between 42% and 86% (Table 20.1). All of these studies employ 28-mm diameter CoCr femoral heads. See Chapter 14 (Section 14.3.3) for additional information about the clinical studies related to Crossfire.

## 20.5 DURASUL

| | |
|---|---|
| Type of crosslinked material | First generation (remelted) |
| Orthopedic manufacturer | Zimmer, Inc, (Warsaw, Indiana, USA), www.zimmer.com |
| Joint applications | THA and TKA |
| Clinical introduction | 1998 (THA), 2001 (TKA) |
| Starting stock material | GUR 1050 compression molded sheet |
| Irradiation crosslinking modality | Electron beam |
| Irradiation temperature | ~125°C (see Section 20.5.1), WIAM process (see Section 20.5.1) |
| Irradiation dose (range) | 95 kGy |
| Postirradiation stabilization process | Remelted at 150°C for 2 hours |
| Sterilization modality | Ethylene oxide |
| Target total irradiation dose | 95 kGy |
| Detectable free radicals? | No |
| Maximum head size available | 44 mm (USA); 36 mm (Europe and Japan) |
| Minimum thickness available | 5 mm |
| Published *in vitro* properties/performance data? | Yes |
| Longest average published follow-up | 6 years [15] |

### 20.5.1 Development History and Overview

Durasul is based on WIAM (warm irradiation with adiabatic melting) technology developed by a team of researchers, including Orhun Muratoglu, PhD, and William Harris, MD, at Massachusetts General Hospital (MGH, Boston, Massachusetts, USA). The WIAM process technology was licensed to Centerpulse Orthopedics, Inc. (formerly Sulzer, Austin, Texas, USA) and Zimmer, Inc. (see Section 20.7 for Longevity highly crosslinked UHMWPE). Centerpulse clinically introduced Durasul in the Converge acetabular cup design in 1998 (Figure 20.3A). Durasul was later introduced for TKA in the cruciate-retaining Natural Knee design (NK II) at the 2001 American Academy of Orthopedic Surgeons (AAOS). The Durasul all-polyethylene patellar component was introduced in 2002

**TABLE 20.1 Summary of Peer-Reviewed Studies Involving Crossfire (Stryker Orthopedics, Mahwah, New Jersey, USA)**

| | Martell et al. (2003) [12] | Rohrl et al. (2005) [14] | Krushell et al. (2005) [13] | D'Antonio et al. (2005) [9] | Rohrl et al. (2007) [15] |
|---|---|---|---|---|---|
| Study type | RCT | Pcoh | Hcoh | Hcoh | Pcoh |
| Number of institutions | 5 | 1 | 1 | 1 | 1 |
| Cup design | Secur-Fit HA | Osteonics | Microstructured PSL | Microstructured PSL | Osteonics |
| Cup fixation | Noncemented | Cemented | Noncemented | Noncemented | Cemented |
| Head size | 28 mm | 28 mm | 28 mm | 28 mm | 28 mm |
| Head material | CoCr L-Fit | CoCr | CoCr L-Fit | CoCr L-Fit | CoCr |
| Average age | 60 | 58 | 69 | 57.4 | 58 |
| Age range | 28–76 | 49–79 | 45–83 | | 49–79 |
| Number of hips | 46 (24 Crossfire) | 50 (10 Crossfire) | 80 (40 Crossfire) | 109 (56 Crossfire) | 30 (10 Crossfire) |
| Follow-up in years | 2.3 | 3 | 4 | 4.9 | 6 |
| Range in follow-up | 1.8–3.2 | 3 | 2.6–4.7 | 4–5.8 | 6 |
| Number of device failures | None | None | None | None | None |
| Wear methodology | Martell | UmRSA | Ramakrishnan | Ramakrishnan | UmRSA |
| 2-D linear penetration (mm/y)*—crosslinked | 0.12 ± 0.05 | 0.02 | 0.05 ± 0.02 | 0.06 ± 0.02 | 0.01 |
| 2-D linear penetration (mm/y)*—control | 0.20 ± 0.10 | 0.16 | 0.12 ± 0.06 | 0.14 ± 0.07 | 0.07 |
| % reduction | 42 | 85 | 58 | 72 | 86 |
| Radiographic assessment of osteolysis | No | No | No | Yes | No |

*Legend: Hcoh: Retrospective cohort study (Level III); Pcoh: Prospective cohort study; RCT: Randomized controlled trial (Level I); L-Fit: Low friction ion treatment. *The 2-D linear wear is listed for the longest follow-up period and includes the initial bedding-in period.*

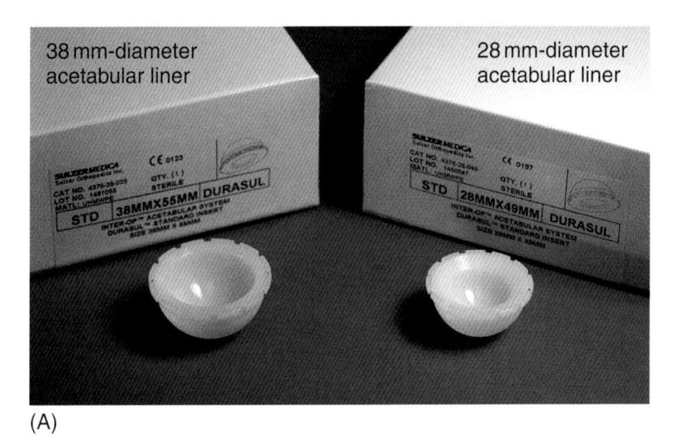


(A)

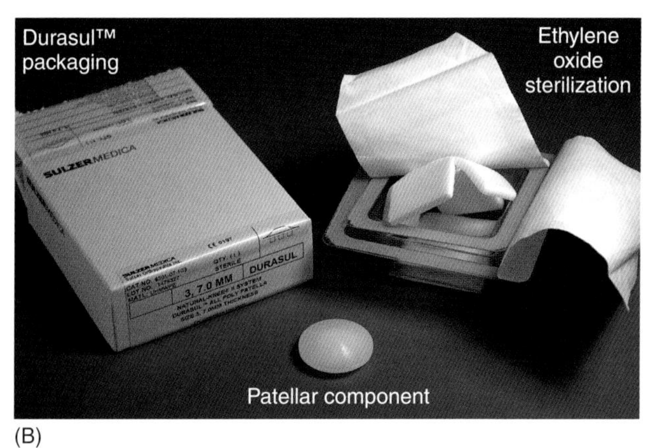


(B)

**FIGURE 20.3** Durasul components for hip and knee arthroplasty with original packaging from 2003 (Centerpulse Orthopedics, formerly Sulzer Orthopedics, currently Zimmer). (A) 28-mm and 38-mm diameter Durasul acetabular liners. (B) Durasul patellar component with packaging used for ethylene oxide sterilization.

(Figure 20.3B). Following the acquisition of Centerpulse by Zimmer, the packaging for Durasul components was updated (Figure 20.4).

In the WIAM process, the UHMWPE sheets or blocks are first machined into preforms (for acetabular liners) or slabs (for tibial inserts) and placed on a conveyor within an oven maintained at a temperature just below the melting temperature of the polymer (around 125°C). The warm UHMWPE pucks are exposed to a 10MeV Rhodotron electron accelerator, which deposits the 95kGy dose within seconds [17]. The dose rate is sufficiently high (~10kGy/s) that the UHMWPE heats up, but not above the melt transition. Because the UHMWPE is about 50% amorphous at the start of the irradiation process, but partially melts to become nearly 100% amorphous by the end of the crosslinking process, Durasul has a spatially varying crosslink density at a microscopic level. Following irradiation, the UHMWPE is maintained at 150°C for stabilization of free radicals (Figure 20.1). Components are then machined from the Durasul material, enclosed in gas-permeable packaging, and sterilized by ethylene oxide gas [17, 18].

## 20.5.2 Properties and *In Vitro* Performance

Because of its processing history, Durasul has a biphasic crystalline microstructure that is manifested in two distinct thermal transitions during a DSC experiment [17]. Overall, Durasul has a lower degree of crystallinity, elastic modulus, yield stress, and ultimate stress as compared with control UHMWPE. Nevertheless, the properties of Durasul are maintained above the recommended guidelines of ASTM and ISO standards for conventional UHMWPE [18]. Muratoglu et al. [18] have verified the oxidative stability of WIAM UHMWPE. ESR studies of Durasul have reported that the material contains no detectable free radicals [18].

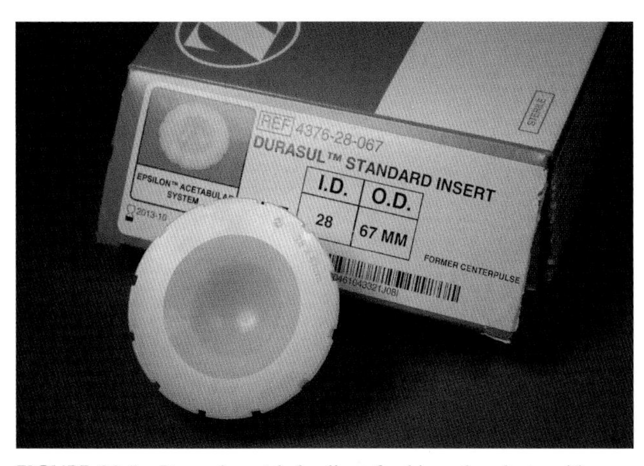


**FIGURE 20.4** Durasul acetabular liner for hip arthroplasty with contemporary packaging from 2008 (Zimmer, Inc., Warsaw Indiana, USA).

Durasul has undergone extensive wear testing in hip and knee simulator studies, under both clean and abrasive lubrication conditions [19, 20]. Hip simulator studies have been conducted with Durasul with femoral heads ranging from 22 to 46mm in diameter for up to 27 million cycles [19]. For all head sizes, Durasul has exhibited significant reductions in wear in the hip simulator. Knee simulator studies also show a significant reduction of wear under normal and aggressive duty cycles [20].

## 20.5.3 Clinical Results

Five clinical studies of Durasul [16, 21–24] have been published (Table 20.2), including the results of a 3-year prospective randomized trial [21]. For three of these studies, which compare Durasul to a control liner, the reduction in head penetration ranged between 20% and 94% (Table 20.2). Three of these studies employ 28-mm CoCr femoral heads, whereas a study by Bragdon [24] included both 28-mm and 32-mm heads, and a study by Geller [23] investigated the clinical performance of 38-mm CoCr heads (Table 20.2).

Dorr and colleagues [16] have reported on the 5-year follow-up for Durasul, the longest series currently available. Although the data from the 37 arthroplasties in the Durasul group were prospectively collected, the control group was retrospectively matched from a group of patients who were implanted 6 months prior to the implantation of Durasul. The implants used in the control group were identical to those in the Durasul group, with the exception of a conventional liner (gamma sterilized in nitrogen) being used for the controls. Dorr [16] reported a 45% reduction in overall 2-D femoral head penetration associated with Durasul liners as compared with control polyethylene liners.

# 20.6 E-POLY HXLPE

| Type of crosslinked material | Second generation (vitamin E doped) |
|---|---|
| Orthopedic manufacturer | Biomet, Inc. (Warsaw, Indiana, USA), www.biomet.com |
| Joint applications | THA and TKA |
| Clinical introduction | 2007 (THA); 2008 (TKA) |
| Starting stock material | GUR 1050 (THA); GUR 1020 (TKA); isostatically compression molded (THA/TKA) |
| Irradiation crosslinking modality | Gamma |

| | |
|---|---|
| Irradiation temperature | Room temperature |
| Irradiation dose (range) | 100kGy |
| Postirradiation stabilization process | Vitamin E doping followed by 130°C annealing to accelerate internal diffusion |
| Sterilization modality | Gamma in argon, 30kGy dose |
| Target total irradiation dose | 130kGy |
| Detectable free radicals? | Yes |
| Maximum head size available | 40mm (THA) |
| Minimum thickness available | 4.8mm (THA) and 10mm (TKA) |
| Published *in vitro* properties/ performance data? | Yes |
| Longest average published follow-up | Not available |

### 20.6.1 Development History and Overview

E-Poly HXLPE is a second-generation highly crosslinked UHMWPE, fabricated using a patented vitamin E doping process developed by researchers at Massachusetts General Hospital (MGH) and commercialized by Biomet, Inc. (Warsaw, Indiana, USA). E-Poly HXLPE was clinically introduced in 2007 for total hip replacements and in 2008 for knee replacements. In the hip, E-Poly is available in Max-Rom, Hi-Wall, 10 degree, +5 Hi-Wall, and +5 Max-Rom liners in the Ring Loc acetabular cup design (Figure 20.5). In the knee, E-Poly HXLPE is available in CR, PS, Anterior Stabilized (AS), and Cruciate Retaining Lipped (CRL) bearings in the Vanguard design. The thinnest configuration is currently 4.8mm in the hip and 10mm in the knee.

The scientific rationale for doping highly crosslinked UHMWPE with vitamin E is explained in Chapter 15. Specifically, E-Poly HXLPE is produced from isostatically molded rods of GUR 1050 (for hips) or GUR 1020 (for knees) that are gamma irradiated at room temperature with a nominal dose of 100kGy. Components are machined from the irradiated rods and doped with a solution of vitamin E. The components are then annealed to equilibrate the diffusion of vitamin E throughout the thickness. After vitamin E doping, the parts are inert packaged in argon and gamma sterilized with a nominal dose of 30kGy. Thus, E-Poly HXLPE components receive a total dose of 130kGy.

### 20.6.2 Properties and *In Vitro* Performance

The scientific data for E-Poly HXLPE is reviewed in Chapter 15. Hip and knee wear testing conducted by the inventors at MGH revealed over 90% reduction in wear in the hip and the knee (Chapter 15), when compared with conventional gamma inert sterilized controls. E-Poly contains free radicals; however, in the presence of vitamin E, the material has been reported to be oxidatively stable. Gamma sterilization is performed, rather than gas sterilization, to ensure sterility of the vitamin E diffused into the UHMWPE, rather than to achieve additional crosslinking.

### 20.6.3 Clinical Results

Clinical results are not yet available for E-Poly HXLPE.

## 20.7 LONGEVITY

| | |
|---|---|
| Type of crosslinked material | First generation (remelted) |
| Orthopedic manufacturer | Zimmer, Inc. (Warsaw, Indiana, USA), www.zimmer.com |
| Joint applications | THA |
| Clinical introduction | 1999 |
| Starting stock material | GUR 1050 compression molded sheet |
| Irradiation crosslinking modality | Electron beam |
| Irradiation temperature | ~40°C |
| Irradiation dose (range) | 100kGy (90–110kGy target dose range) |
| Postirradiation stabilization process | Remelted above the melt transition (>135°C) |
| Sterilization modality | Gas plasma |
| Target total irradiation dose | 100kGy |
| Detectable free radicals? | No |
| Maximum head size available | 36mm |
| Minimum thickness available | 6mm |
| Published *in vitro* properties/ performance data? | Yes |
| Longest average published follow-up | 3.3 years [23] |

### 20.7.1 Development History and Overview

Longevity is based on WIAM (warm irradiation with adiabatic melting) technology patented by Massachusetts General Hospital (MGH, Boston, Massachusetts, USA). The WIAM technology was licensed to Zimmer, Inc. (Warsaw, Indiana, USA). The Longevity process was

**TABLE 20.2** Summary of Peer-Reviewed Studies Involving Durasul (Zimmer, Warsaw, Indiana, USA)

| | Digas et al. (2004) [21] | Manning et al. (2005) [22] | Dorr et al. (2005) [16] | Geller et al. (2006) [23] | Bragdon et al. (2006) [24] |
|---|---|---|---|---|---|
| Study type | RCT | Pcoh | Pcoh | Pcoh | Pcoh |
| Number of institutions | 1 | 2 | 1 | 1 | 1 |
| Cup design | Inter-Op | Inter-Op | Inter-Op | Inter-Op | Inter-Op |
| Cup fixation | Cemented | Cementless | Cementless | Cementless | Cementless |
| Head size | 28 mm | 28 mm | 28 mm | 38 mm | 28 mm (n = 41) and 32 mm (n = 12) |
| Head material | CoCr | CoCr | CoCr | CoCr | CoCr |
| Average age | 55 | 60.9 ± 11.1** | | 62.5** | 60.3 |
| Age range | 42–64 | | | 28–86** | 23–86 |
| Number of hips | 49 (23 Durasul) | 160 (49 Durasul) | 74 (37 Durasul) | 45 (13 Durasul)*** | 123 (53 Durasul)### |
| Follow-up in years | 3 | 2.6** | 5 | 3.3 | 3.8 |
| Range in follow-up | | 2–3.7** | | | 3 years minimum |
| Number of device failures | None | None | None for polyethylene wear# | None for polyethylene wear# | None for polyethylene wear# |
| Wear methodology | RSA | Martell | Martell | Martell | Martell |
| 2-D linear penetration (mm/y)—Crosslinked | 0.08 | 0.010 ± 0.009* | 0.029 ± 0.02 | −0.08 ± 0.26* | 0.03 ± 0.10 (28 mm) 0.01 ± 0.09 (32 mm) |
| 2-D linear penetration (mm/y)—control | 0.10 | 0.176 ± 0.054* | 0.065 ± 0.026 | None | 0.15 ± 0.09 (28 mm) |
| % reduction | 20% | 94% | 55% | N/A | 81% (28 mm only) |
| Radiographic assessment of osteolysis | None | None | No | Yes | No |

*Legend: Hcoh: Retrospective cohort study (Level III); Pcoh: Prospective cohort study; RCT: Randomized controlled trial (Level I); L-Fit: Low friction ion treatment. *The 2-D linear wear is listed for the longest follow-up period and includes the initial bedding-in period. **Data set pooled with Longevity highly crosslinked implants. ***2/15 patients were excluded from radiographic wear analysis due to poor quality radiographs. #Patients were revised from the original series due to an implant recall unrelated to polyethylene wear. ##Supine examination; only 3-D penetration results were reported in this study, which were converted to penetration rates for the table. ###Only 53/146 hips implanted with Durasul had sufficient quality radiographs to be assessed for linear penetration.*

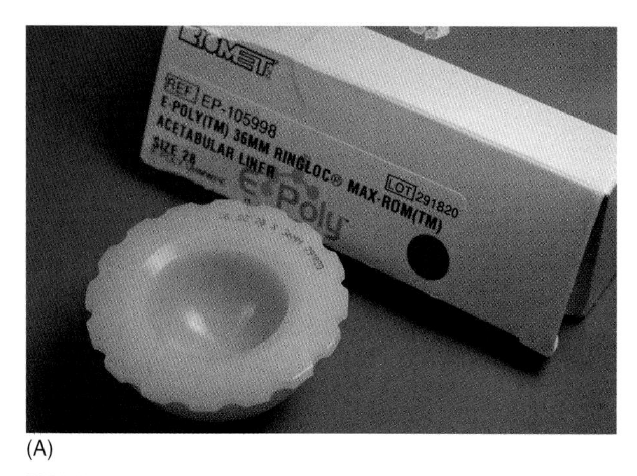


(A)

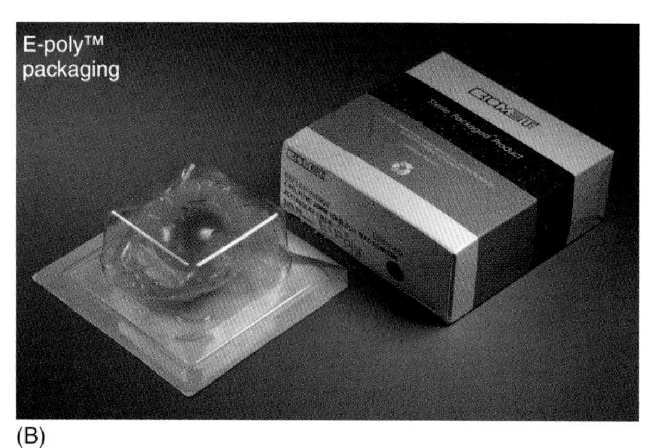


(B)

**FIGURE 20.5** E-Poly component for hip arthroplasty with packaging (Biomet, Inc., Warsaw, Indiana, USA). (A) 36-mm diameter E-Poly acetabular liner. (B) Packaging of E-Poly.

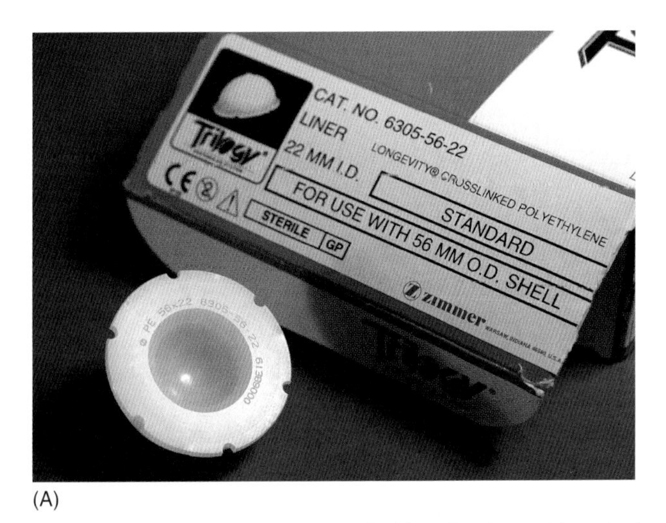


(A)

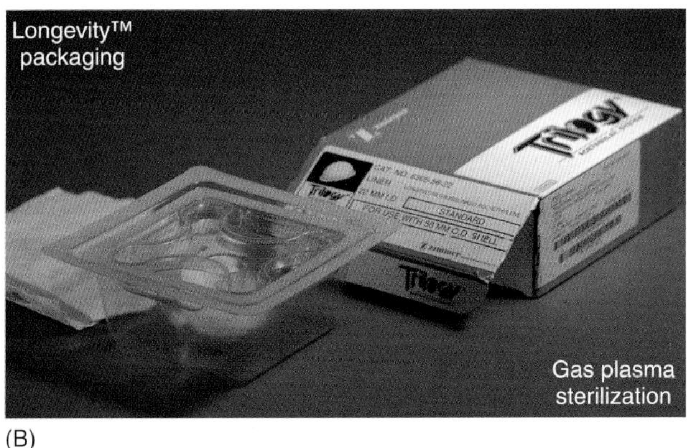


(B)

**FIGURE 20.6** Longevity component for hip arthroplasty with packaging (Zimmer, Inc., Warsaw, Indiana, USA). (A) 22-mm diameter Longevity acetabular liner. (B) Packaging of Longevity.

developed by Zimmer and clinically introduced in the Trilogy acetabular cup design in 1999 (Figure 20.6).

In the Longevity process, the UHMWPE bars are warmed, placed in a carrier on a conveyor, and are exposed to electron beam radiation, with a total dose of 100 kGy. The UHMWPE does not heat above the melt transition during the crosslinking. After irradiation, the UHMWPE is heated above the melt temperature (>135°C) for stabilization of free radicals. Components are then machined from the Longevity material, enclosed in gas-permeable packaging, and sterilized by gas plasma.

### 20.7.2 Properties and *In Vitro* Performance

The crystallinity, tensile properties, and fracture toughness of Longevity have been characterized before and after accelerated aging [25, 26]. In these studies, Longevity was also compared with conventional UHMWPE that was gamma sterilized in nitrogen. The researchers observed that accelerated aging had no significant affect on the oxidation or mechanical properties of Longevity.

The wear properties of Longevity have been studied under normal and abrasive conditions for 22 and 32 mm head sizes [26, 27]. Laurent et al. [27] found that Longevity maintained its superior wear resistance relative to conventional UHMWPE, even under the presence of abrasive particles. Large-head Longevity liners have also been studied and demonstrate that Longevity wear characteristics are independent of head size.

### 20.7.3 Clinical Results

Four clinical studies of Longevity [21–23, 28] are summarized in Table 20.3. In two of these studies [21, 22], head penetration reductions of 31% and 90% were reported (Table 20.3). Two studies employed 28-mm CoCr femoral heads [21, 22]; Geller investigated the use of 36 and 40 mm CoCr heads [23]; and Bragdon investigated 28 mm and 36 mm CoCr heads [28].

Manning et al. [22] have published the longest-term radiographic wear study comparing Longevity to conventional, gamma sterilized polyethylene, with up to 3.7 years follow-up (2.6 years, average). Manning [22] reported a 90% reduction in overall 2-D femoral head penetration associated with Longevity liners as compared with control PE liners.

## 20.8 MARATHON

| Type of crosslinked material | First generation (remelted) |
|---|---|
| Orthopedic manufacturer | DePuy Orthopedics, Inc. (Warsaw, Indiana, USA), www.depuy.com |
| Joint applications | THA |
| Clinical introduction | 1998 |
| Starting stock material | GUR 1050 extruded rod |
| Irradiation crosslinking modality | Gamma |
| Irradiation temperature | Room temperature |
| Irradiation dose (range) | 50 kGy |
| Postirradiation stabilization process | Melted at 155°C for 24 hours and then annealed at 120°C for 24 hours, both in reduced oxygen atmospheres |
| Sterilization modality | Gas plasma |

| | |
|---|---|
| Target total irradiation dose | 50 kGy |
| Detectable free radicals? | No |
| Minimum/maximum head size available | 48 mm |
| Minimum thickness available | 6 mm |
| Published *in vitro* properties/performance data? | Yes |
| Longest average published follow-up | 5.7 years [29] |

### 20.8.1 Development History and Overview

Marathon is based on the collaborative research efforts of Harry McKellop, Fu-wen Shen, and coworkers at the Los Angeles Orthopedic Hospital (Los Angeles, California, USA) [30], and material scientists affiliated with DePuy-DuPont Orthopaedics. The process technology was licensed to DePuy Orthopedics, Inc. (Warsaw, Indiana, USA) and clinically introduced in 1998. Marathon is currently available for the Pinnacle and Duraloc acetabular component systems.

In the Marathon process, extruded rod bar stock is irradiated with a dose of 50 kGy and then remelted at 150°C [30] (Figure 20.1). After remelting, the rods are then annealed at 120°C for 24 h [6]. Acetabular components are machined from the processed bar stock, enclosed in gas-permeable packaging, and then gas plasma sterilized (Figure 20.7).

**TABLE 20.3** Summary of Peer-Reviewed Studies Involving Longevity (Zimmer, Warsaw, Indiana, USA)

| | Digas et al. (2004) [21] | Manning et al. (2005) [22] | Geller et al. (2006) [23] | Bragdon et al. (2007) [28] |
|---|---|---|---|---|
| Study type | RCT | Pcoh | Pcoh | Pcoh |
| Number of institutions | 1 | 2 | 1 | 1 |
| Cup design | Trilogy | Trilogy | Trilogy | Trilogy |
| Cup fixation | Cementless | Cementless | Cementless | Cementless |
| Head size | 28 mm | 28 mm | 36 mm (15) or 40 mm (3) | 28 mm (16) or 36 mm (14) |
| Head material | CoCr | CoCr | CoCr | CoCr |
| Average age | 48 | 60.9 ± 11.1*** | 62.5*** | 56.1 |
| Age range | 29–70 | | 28–86*** | 36–77 |
| Number of hips | 54 (27 Longevity) | 132 (21 Longevity) | 45 (18 Longevity)**** | 30 Longevity |
| Follow-up in years | 2 | 2.6*** | 3.3 | 3 |
| Range in follow-up | | 2–3.7*** | | |
| Number of device failures | None | None | None for polyethylene wear# | None |
| Wear methodology | RSA | Martell | Martell | RSA |
| Linear penetration (mm/y)—crosslinked | 0.11** | 0.018 ± 0.022* | −0.12 ± 0.22* (36 mm) 0.11 ± 0.20* (40 mm) | 0.026 ± 0.024# (28 mm) 0.000 ± 0.058# (32 mm) |
| Linear penetration (mm/y)—control | 0.16** | 0.176 ± 0.054* | None | None |
| % reduction | 31% | 90% | Not available | Not available |
| Radiographic assessment of osteolysis | None | None | None | None |

*Legend: Hcoh: Retrospective cohort study (Level III); Pcoh: Prospective cohort study; RCT: Randomized controlled trial (Level I). *The 2-D linear wear is listed for the longest follow-up period and includes the initial bedding-in period. **Only 3-D linear penetration rates are reported. *** Data set pooled with Durasul highly crosslinked implants. ****12/30 patients were excluded due to lack of radiographs. #Patients were revised from the original series due to an implant recall unrelated to polyethylene wear. #Median steady state wear rate was reported and does not include bedding in prior to 1 year of implantation.*

### 20.8.2 Properties and *In Vitro* Performance

The *in vitro* test results for remelting of crosslinked UHMWPE were originally based on GUR 415 resin [30]. Test results for Marathon, also based on GUR 415, were summarized in an abstract by DiMaio and colleagues [31]. In DiMaio's study, the fatigue behavior, mechanical properties, and hip simulator response of Marathon were evaluated in the aged and unaged condition. The mechanical properties of Marathon were judged to fall within the ASTM specifications for F648.

Greer et al. [6] published the *in vitro* test results for GUR 1020 and GUR 1050, processed with gamma and electron beam radiation, followed by Marathon thermal processing. Although the mechanical properties were found to be sensitive to molecular weight of the resin, the hip simulator wear rate, in contrast, did not appear to be affected by the resin type.

The oxidative stability of Marathon has been assessed by accelerated aging [30, 31]. In both McKellop's and DiMaio's studies, the properties of the crosslinked and remelted UHMWPE were not significantly affected by accelerated aging.

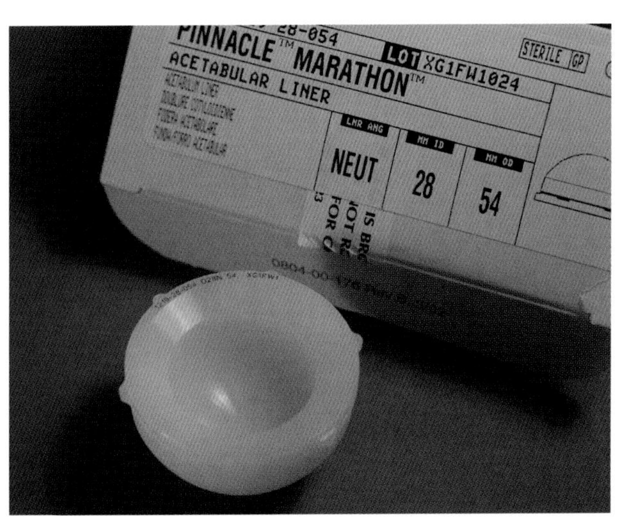


(A)

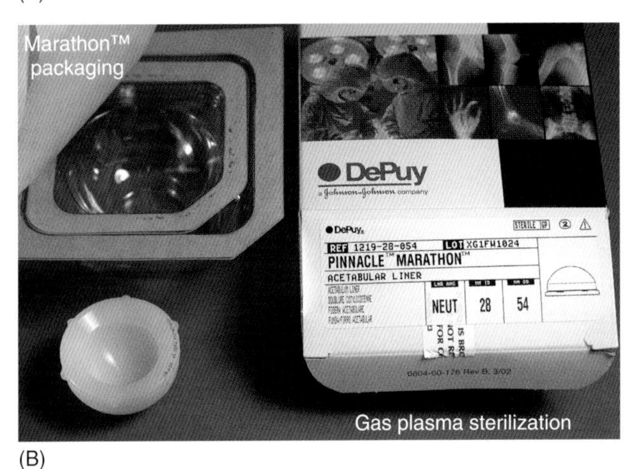


(B)

**FIGURE 20.7** Marathon component for hip arthroplasty with packaging (DePuy Orthopedics, Inc., Warsaw, Indiana, USA). (A) 22-mm diameter Marathon acetabular liner. (B) Packaging of Marathon.

### 20.8.3 Clinical Results

Four clinical studies of Marathon [29, 32–34] are summarized in Table 20.4. The reduction in head penetration for Marathon reported in these studies ranged between 56% and 95% (Table 20.4). All of these studies employ 28-mm CoCr femoral heads. However, a study by Heisel [33] also includes the results for seven Marathon hips with 32-mm head sizes (Table 20.4). In addition, 3/34 (9%) of the femoral heads in Heisel's study [33] were ceramic (the majority, 31/34 (91%), were CoCr).

In perhaps the most comprehensive study yet published for highly crosslinked UHMWPE, Engh and coworkers [29] conducted a prospective randomized trial of Marathon at their institution, with up to 7.2 years follow-up (average, 5.6 years). Nine percent of the original series was lost to follow-up. The control material in this series was uncrosslinked, gas-plasma sterilized polyethylene. Engh [29] reported a 95% reduction in overall 2-D femoral head penetration associated with Marathon liners as compared with control polyethylene liners. However, the use of high wear, uncrosslinked polyethylene as a control complicates the comparison of Engh's findings with other studies of crosslinked polyethylene, which employ gamma sterilized liners as controls. Although there were no revisions or cases of loosening in either the Marathon or control liners, the authors noted a significant difference in the incidence of osteolysis between the two groups. A total of 57.8% of the control patients exhibited radiographic evidence of osteolysis of the pelvis or femur, as compared to an osteolysis incidence of 24.0% in the Marathon patients ($p < 0.001$). It should be noted, however, that patient satisfaction was excellent and indistinguishable between the Marathon (96.2% satisfied with results) and control (99%) groups at the longest follow-up period.

## 20.9 PROLONG

| | |
|---|---|
| Type of crosslinked material | First generation (remelted) |
| Orthopedic manufacturer | Zimmer, Inc. (Warsaw, Indiana, USA), www.zimmer.com |
| Joint applications | TKA |
| Clinical introduction | 2002 |
| Starting stock material | GUR 1050 compression molded sheet |
| Irradiation crosslinking modality | Electron beam |

| | |
|---|---|
| Irradiation temperature | ~120°C |
| Irradiation dose (range) | 65 kGy (58 to 72 target dose range) |
| Postirradiation stabilization process | Remelt above the melting temperature |
| Sterilization modality | Gas plasma |
| Target total irradiation dose | 65 kGy |
| Detectable free radicals? | No |
| PS designs available | Yes |
| Minimum polyethylene thickness available | 6 mm |
| Published *in vitro* properties/performance data? | Yes |
| Longest average published follow-up | Not available |

### 20.9.1 Development History and Overview

Prolong is based on WIAM technology patented by Massachusetts General Hospital (MGH, Boston, Massachusetts, USA) and licensed to Zimmer, Inc. (Warsaw, Indiana, USA). Prolong was clinically introduced for cruciate-sparing, NexGen knee component systems in 2002 (Figure 20.8).

The Prolong process is similar to the previously described Durasul except that an electron beam irradiation dose of 65 kGy is used for Prolong, and continuous compression molded bars are processed as with Longevity. Components are machined from processed Prolong material, enclosed in gas-permeable packaging, and sterilized by gas plasma.

### 20.9.2 Properties and *In Vitro* Performance

Prolong has been subjected to a battery of wear and functional fatigue testing [27, 35]. The wear properties

**TABLE 20.4** Summary of Peer-Reviewed Studies Involving Marathon (DePuy Orthopedics, Warsaw, Indiana, USA)

| | Hopper et al. (2003) [32] | Heisel et al. (2004) [33] | Heisel et al. (2005) [34] | Engh et al. (2006) [29] |
|---|---|---|---|---|
| Study type | Pcoh | Pcoh | Pcoh | RCT |
| Number of institutions | 1 | 1 | 1 | 1 |
| Cup design | Duraloc 100 | Duraloc or Pinnacle | Duraloc | Duraloc 100 |
| Cup fixation | Uncemented | Uncemented | Uncemented | Uncemented |
| Head size | 28 mm | 28 mm (27) or 32 mm (7) | 28 mm | 28 mm |
| Head material | CoCr | CoCr (31) or ceramic (3) | CoCr | CoCr |
| Average age | 60.3 | 60 | 59.7 | 62.5 |
| Age range | 26–80 | 26–83 | 39–79 | 26–87 |
| Number of hips | 98 (48 Marathon) | 58 (34 Marathon) | 6 (3 Marathon) | 209 (105 Marathon) |
| Follow-up in years | 2.9 | 2.8 | 3.2 | 5.7 |
| Range in follow-up | 2.0–3.7 | 2.0–4.4 | 2–4 | 4.1–7.2 |
| Number of device failures | None | None | None | None |
| Wear methodology | Martell | Martell | Martell | Martell |
| 2-D linear penetration (mm/y)*—crosslinked | 0.08 ± 0.24 | 0.02 ± 0.1 | 0.06 ± 0.02 | 0.01 ± 0.07 |
| 2D Linear penetration (mm/y)*—control | 0.18 ± 0.20 | 0.13 ± 0.1 | 0.27 ± 0.02 | 0.20 ± 0.13 |
| % reduction | 56% | 81% | 78% | 95% |
| Radiographic assessment of osteolysis | None | None | None | None |

*Legend: Pcoh: Prospective cohort study; RCT: Randomized controlled trial (Level I); *The 2-D linear wear is listed for the longest follow-up period and includes the initial bedding-in period.*

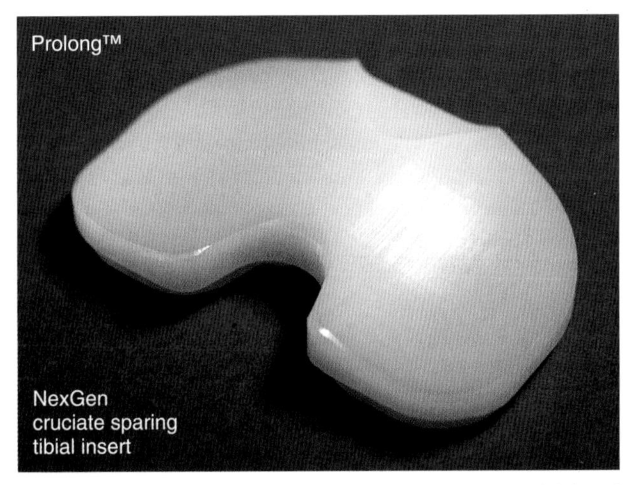


**FIGURE 20.8** NexGen cruciate sparing tibial component fabricated from Prolong (Zimmer, Inc., Warsaw, Indiana, USA).

of Prolong have been studied in a knee simulator under normal and abrasive conditions [27]. Yao et al. have also tested Prolong in delamination and posterior condyle fatigue simulators [35]. The functional fatigue tests were conducted with aged Prolong and compared with conventional UHMWPE that was gamma sterilized in nitrogen. After accelerated aging, Prolong was found to have superior functional fatigue performance relative to conventional UHMWPE. An FDA cleared claim for improved delamination resistance of Prolong over conventional polyethylene has been made based upon these test data.

## 20.9.3 Clinical Results

Clinical results are not yet available for Prolong.

# 20.10 X3

| Type of crosslinked material | Second generation (sequentially annealed) |
|---|---|
| Orthopedic manufacturer | Stryker Orthopedics, Inc. (Mahwah, New Jersey, USA), www.stryker.com |
| Joint applications | THA and TKA |
| Clinical introduction | 2005 |
| Starting stock material | GUR 1020, compression molded sheet |
| Irradiation crosslinking modality | Gamma |
| Irradiation temperature | Room temperature |
| Irradiation dose (range) | 30 kGy in three steps |
| Postirradiation stabilization process | Annealing at 130°C after each gamma sterilization step |
| Sterilization modality | Gas plasma |
| Target total irradiation dose | 90 kGy |
| Detectable free radicals? | Yes |
| Maximum head size available | 44 mm (THA) |
| Minimum thickness available | 3.8 mm (THA) and 6.2 mm (TKA) |
| Published *in vitro* properties/ performance data? | Yes |
| Longest average published follow-up | Not available |

## 20.10.1 Development History and Overview

X3 is a second-generation highly crosslinked UHMWPE developed by Stryker Orthopedics (Mahwah, New Jersey, USA) and is described in detail by its inventors in Chapter 14 (see Section 14.4). X3 was clinically introduced in 2005 for total hip and knee replacements and is available in the Trident and Tritanium acetabular cup designs. For knees, X3 is available in the Scorpio Primary Flex system (CR and PS); the Scorpio NRG Primary system (CR and PS); the Triathlon Primary system (CR, CS lipped, PS), and the Triathlon TS Revision system.

The scientific rationale for sequential irradiation and annealing of highly crosslinked UHMWPE is explained in Section 14.4. Specifically, X3 is produced by gamma irradiating compression molded GUR 1020 with 30 kGy of gamma radiation followed by annealing at 130°C. These irradiation and annealing steps are repeated for a total of three times, resulting in three sequential 30 kGy doses (a total of 90 kGy) that have undergone a total of three annealing steps. Components are machined from the sequentially irradiated and annealed material and terminally gas plasma sterilized.

## 20.10.2 Properties and *In Vitro* Performance

The scientific data for X3 is reviewed in Section 14.4. Hip and knee wear testing conducted by the manufacturer revealed over 90% reduction in wear in the hip and the knee (Section 14.4) when compared with conventional gamma inert sterilized controls. X3 contains free radicals. However, the material has been reported by the manufacturer to be oxidatively stable (Section 14.4).

## 20.10.3 Clinical Results

Clinical results are not yet available for X3.

## 20.11 XLK

| Type of crosslinked material | First generation (remelted) |
|---|---|
| Orthopedic manufacturer | DePuy Orthopedics, Inc. (Warsaw, Indiana, USA), www.depuy.com |
| Joint applications | TKA |
| Clinical introduction | 2005 |
| Starting stock material | GUR 1020 ram extruded rod |
| Irradiation crosslinking modality | Gamma |
| Irradiation temperature | Room temperature |
| Irradiation dose (range) | 50 kGy |
| Postirradiation stabilization process | Melted at 155°C for 24 hours and then annealed at 120°C for 24 hours, both in reduced oxygen atmospheres |
| Sterilization modality | Gas plasma |
| Target total irradiation dose | 50 kGy |
| Detectable free radicals? | No |
| PS designs available | Yes |
| Minimum polyethylene thickness available | 6 mm |
| Published *in vitro* properties/ performance data? | Yes |
| Longest average published follow-up | Not available |

### 20.11.1 Development History and Overview

XLK is based on technology developed for Marathon (see Section 20.8). XLK was clinically introduced for the PFC Sigma total knee component systems in 2005. The XLK process is the same as previously described Marathon except that ram extruded GUR 1020 is used for XLK (instead of GUR 1050 used for Marathon). Components are machined from processed XLK material, enclosed in gas-permeable packaging, and sterilized by gas plasma. XLK is available in the PFC Sigma knee system in PLI, Curved, Curved+, and Stabilized inserts (Figure 20.9).

### 20.11.2 Properties and *In Vitro* Performance

The mechanical test properties of XLK were described by Greer and coworkers [6]. When subjected to identical irradiation and thermal stabilization, ram extruded GUR 1020 exhibits similar wear behavior but improved mechanical properties, such as toughness and ductility. Extensive wear testing of XLK was performed at DePuy [36].

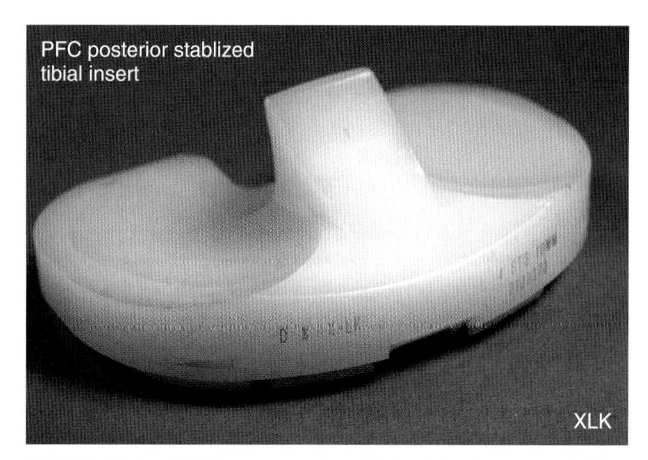


FIGURE 20.9 PFC Sigma posterior stabilized tibial component fabricated from XLK (DePuy, Inc., Warsaw, Indiana, USA).

### 20.11.3 Clinical Results

Clinical results are not yet available for XLK.

## 20.12 XLPE

| Type of crosslinked material | First generation (remelted) |
|---|---|
| Orthopedic manufacturer | Smith & Nephew Orthopedics, Inc. (Memphis, Tennessee, USA), www.smith-nephew.com |
| Joint applications | THA and TKA |
| Clinical introduction | 2001 (THA); 2008 (TKA) |
| Starting stock material | GUR 1050 extruded rod (THA); GUR 1020 Compression molded sheet (TKA) |
| Irradiation crosslinking modality | Gamma |
| Irradiation temperature | Ambient |
| Irradiation dose (range) | 100 kGy (THA); 75 kGy (TKA) |
| Postirradiation stabilization process | Remelted at 147°C for at least 5 hours |
| Sterilization modality | Ethylene oxide |
| Target total irradiation dose | 100 kGy (THA); 75 kGy (TKA) |
| Detectable free radicals? | No |
| Maximum head size available | 36 mm (THA) |
| Minimum thickness available | 6 mm (for THA and TKA) |
| Published *in vitro* properties/ performance data? | Yes |
| Longest average published follow-up | 2 years [37], for THA only |

### 20.12.1 Development History and Overview

Developed by Shilesh Jani, Victoria Good, and coworkers at Smith & Nephew, XLPE was introduced into the US orthopedic market for THA in 2001. XLPE used in the hip is produced from ram extruded GUR 1050, gamma irradiated with a dose of 100 kGy, and remelted (Figure 20.1). XLPE is available in the Reflection and R3 acetabular cup designs.

For the knee, XLPE is fabricated from compression molded GUR 1020 and gamma irradiated with a total dose of 75 kGy. Both the hip and knee formulations of XLPE undergo the same remelting process and are ethylene oxide sterilized. XLPE for the knee was clinically introduced for the Legion High Flex CR and PS designs at the 2008 annual meeting of the American Academy of Orthopedic Surgeons (AAOS).

### 20.12.2 Properties and *In Vitro* Performance

The hip and knee simulator performance of XLPE has been reported against smooth and roughened counterfaces [38–41]. Research published in 2003 suggested that XLPE was more sensitive than conventional UHMWPE to abrasive wear. In the hip simulator, XLPE still exhibited less wear than conventional UHMWPE with a roughened femoral head. This was not the case in the knee simulator, with XLPE exhibiting greater wear than conventional UHMWPE when articulating against a roughened CoCr femoral component. Based on these initial results, Smith & Nephew decided not to release 100 kGy XLPE for knee applications.

In 2007, Parikh and coworkers showed that at 50 and 75 kGy doses, XLPE performed better than conventional UHMWPE in a knee simulator [41]. The lowest wear for either type of XLPE was observed when an OXINIUM femoral component was tested versus a CoCr femoral component in both the roughened and smooth configuration. These tests formed the basis of Smith & Nephew introducing 75 kGy XLPE for total knees in 2008.

### 20.12.3 Clinical Results

Two-year follow up of XLPE in hip arthroplasty was reported at the 2006 annual meeting of the Orthopaedic Research Society [37]. This study also reported findings from an early 50 kGy version of XLPE, prior to the adoption of 100 kGy dose for THA. Using RSA, the average linear penetration rate at 2 years was 0.04 mm/y for 100 kGy XLPE and 0.23 mm/y for the uncrosslinked gas sterilized control (83% reduction). Clinical results are not yet available for XLPE used in total knee applications.

## 20.13 THE FUTURE FOR HIGHLY CROSSLINKED UHMWPE

Innovations and developments in the field of highly crosslinked UHMWPE are still far from over. As we anticipated in the first edition of the *Handbook*, future advances in highly crosslinked UHMWPE for joint arthroplasty should be considered a near certainty. Although the intermediate-term clinical results for highly crosslinked UHMWPEs in THA have thus far been encouraging, further follow-up is still necessary to confirm the long-term patient outcomes with these promising new UHMWPE bearing materials. In 2003, highly crosslinked UHMWPE knee designs were available from two manufacturers (Centerpulse and Zimmer). Today, all of the five major orthopedic manufacturers reviewed in this chapter produce highly crosslinked total knees in both cruciate retaining and posterior stabilized designs. The clinical introduction of highly crosslinked UHMWPE in the knee is still too recent for there to be any consensus based on published clinical data. It is clear, however, based on the proliferation of new highly crosslinked UHMWPE materials for TKA over the past 5 years, that there remains strong interest in applications of this technology for knee replacements. It will be another 5 years until sufficient evidence has been collected to test the hypothesis that first generation materials significantly reduce the incidence of revision in hip replacement, and the outcomes in the knee could still take another decade to reach consensus among orthopedic surgeons and researchers. For all of these reasons, highly crosslinked UHMWPE is expected to be an important clinical research topic in orthopedics throughout the first decades of the 21st century.

## 20.14 ACKNOWLEDGMENTS

This chapter was made possible by the assistance of colleagues at the five orthopedic manufacturers of highly crosslinked UHMWPE, who provided access to representative products and processing information: Jordan Freedman and David Schroeder (Biomet, Inc.); Keith Greer (DePuy Orthopedics); Hallie Brinkerhuff (Zimmer, Inc.); Alissa Sellers (Stryker Orthopedics); and Mark Morrison (Smith & Nephew).

## REFERENCES

1. Muratoglu OK, Kurtz SM. Alternative bearing surfaces in hip replacement. In: Sinha R, editor. *Hip replacement: current trends and controversies*. New York: Marcel Dekker; 2002. p. 1–46.
2. Lewis G. Properties of crosslinked ultra-high-molecular-weight polyethylene. *Biomaterials* 2001;**22**(4):371–401.
3. Collier JP, Currier BH, Kennedy FE, Currier JH, Timmins GS, Jackson SK, et al. Comparison of cross-linked polyethylene materials for orthopaedic applications. *Clin Orthop* 2003 September;**414**(414):289–304.
4. Kurtz SM, Gsell R, Martell J. *Highly crosslinked and thermally treated ultra-high molecular weight polyethylene for joint replacements*. West Conshohoken: American Society for Testing and Materials, ASTM STP 1445; 2004.

5. Crowninshield RD, Muratoglu OK. How have new sterilization techniques and new forms of polyethylene influenced wear in total joint replacement? *J Am Acad Orthop Surg* 2008;**16**(Suppl. 1): S80–5.
6. Greer K, King R, Chan FW. Effects of raw material, irradiation dose, and irradiation source on crosslinking of UHMWPE. In: Kurtz SM, Gsell R, Martell J, editors *Highly crosslinked and thermally treated ultra-high molecular weight polyethylene for joint replacements*. West Conshohoken: American Society for Testing and Materials; 2003.
7. Liao Y-S, Greer K, Alberts A. Effect of head material and roughness on the wear of 7.5 Mrad crosslinked-remelted UHMWPE acetabular inserts. *Trans of the 54th Orthop Res Soc* 2008:1901.
8. Kurtz SM, Mazzucco D, Rimnac CM, Schroeder D. Anisotropy and oxidative resistance of highly crosslinked UHMWPE after deformation processing by solid-state ram extrusion. *Biomaterials* 2006 January;**27**(1):24–34.
9. D'Antonio JA, Manley MT, Capello WN, Bierbaum BE, Ramakrishnan R, Naughton M, et al. Five-year experience with crossfire highly cross-linked polyethylene. *Clin Orthop Relat Res* 2005 December;**441**:143–50.
10. Kurtz SM, Manley M, Wang A, Taylor S, Dumbleton J. Comparison of the properties of annealed crosslinked (Crossfire) and conventional polyethylene as hip bearing materials. *Bull NYU Hosp Joint Dis* 2002–2003;**61**(1–2):17–26.
11. Wang A, Manley M, Serekian P. Wear and structural fatigue simulation of crosslinked ultra-high molecular weight polyethylene for hip and knee bearing applications. In: Kurtz SM, Gsell R, Martell J, editors. *Highly crosslinked and thermally treated ultra-high molecular weight polyethylene for joint replacements*. West Conshohoken: American Society for Testing and Materials; 2003.
12. Martell JM, Verner JJ, Incavo SJ. Clinical performance of a highly cross-linked polyethylene at two years in total hip arthroplasty: a randomized prospective trial. *J Arthroplasty* 2003 October;**18**(7 Suppl. 1):55–9.
13. Krushell RJ, Fingeroth RH, Cushing MC. Early femoral head penetration of a highly crosslinked polyethylene liner vs a conventional polyethlene liner. *J Arthroplasty* 2005;**20**(7 Suppl.):73–6.
14. Rohrl S, Nivbrant B, Mingguo L, Hewitt B. In vivo wear and migration of highly cross-linked polyethylene cups a radiostereometry analysis study. *J Arthroplasty* 2005 June;**20**(4):409–13.
15. Rohrl SM, Li MG, Nilsson KG, Nivbrant B. Very low wear of non-remelted highly cross-linked polyethylene cups: an RSA study lasting up to 6 years. *Acta Orthop* 2007 December;**78**(6):739–45.
16. Dorr LD, Wan Z, Shahrdar C, Sirianni L, Boutary M, Yun A. Clinical performance of a durasul highly cross-linked polyethylene acetabular liner for total hip arthroplasty at five years. *J Bone Joint Surg Am* 2005 August;**87**(8):1816–21.
17. Abt NA, Schneider W. Influence of irradiation on the properties of UHMWPE. In: Kurtz SM, Gsell R, Martell J, editors. *Highly crosslinked and thermally treated ultra-high molecular weight polyethylene for joint replacements*. West Conshohoken: American Society for Testing and Materials; 2003.
18. Muratoglu OK, Bragdon CR, O'Connor DO, Perinchief RS, Travers JT, Jasty M, et al. Markedly improved adhesive wear and delamination resistance with a highly crosslinked UHMWPE for use in total knee arthoplasty. *Transactions of the 27th Annual Meeting of the Society for Biomaterials* 2001;**24**:29.
19. Muratoglu II OK, Bragdon CR, O'Connor D, Perinchief RS, Estok DM, Jasty M, et al. Larger diameter femoral heads used in conjunction with a highly cross- linked ultra-high molecular weight polyethylene: a new concept. *J Arthroplasty* 2001;**16**(8 Suppl. 1):24–30.
20. Muratoglu OK, Bragdon CR, O'Connor DO, Perinchief RS, Jasty M, Harris WH. Aggressive wear testing of a cross-linked polyethylene in total knee arthroplasty. *Clin Orthop* 2002 November (404):89–95.
21. Digas G, Karrholm J, Thanner J, Malchau H, Herberts P. Highly cross-linked polyethylene in total hip arthroplasty: randomized evaluation of penetration rate in cemented and uncemented sockets using radiostereometric analysis. *Clin Orthop* 2004 December;**429**(429):6–16.
22. Manning DW, Chiang PP, Martell JM, Galante JO, Harris WH. In vivo comparative wear study of traditional and highly cross-linked polyethylene in total hip arthroplasty. *J Arthroplasty* 2005 October;**20**(7):880–6.
23. Geller JA, Malchau H, Bragdon C, Greene M, Harris WH, Freiberg AA. Large diameter femoral heads on highly cross-linked polyethylene: minimum 3-year results. *Clin Orthop Relat Res* 2006 June;**447**:53–9.
24. Bragdon CR, Barrett S, Martell JM, Greene ME, Malchau H, Harris WH. Steady-state penetration rates of electron beam-irradiated, highly cross-linked polyethylene at an average 45-month follow-up. *J Arthroplasty* 2006 October;**21**(7):935–43.
25. Bhambri SK, Gsell R, Kirkpatrick L, Swarts D, Blanchard CR, Crowninshield RD. The effect of aging on mechanical properties of melt-annealed highly crosslinked UHMWPE. In: Kurtz SM, Gsell R, Martell J, editors. *Highly crosslinked and thermally treated ultra-high molecular weight polyethylene for joint replacements*. West Conshohoken: American Society for Testing and Materials; 2003.
26. Laurent MP, Johnson TS, Crowninshield RD, Blanchard CR, Bhambri SK, Yao JQ. Characterization of a highly cross-linked ultrahigh molecular-weight polyethylene in clinical use in total hip arthroplasty. *J Arthroplasty* 2008 August;**23**(5):751–61.
27. Laurent MP, Blanchard CR, Yao JQ, Johnson TS, Gilbertson LN, Swarts D, et al. The wear of highly crosslinked UHMWPE in the presence of abrasive particles: hip and knee simulator studies. In: Kurtz SM, Gsell R, Martell J, editors. *Highly crosslinked and thermally treated ultra-high molecular weight polyethylene for joint replacements*. West Conshohoken: American Society for Testing and Materials; 2003.
28. Bragdon CR, Greene ME, Freiberg AA, Harris WH, Malchau H. Radiostereometric analysis comparison of wear of highly cross-linked polyethylene against 36- vs 28-mm femoral heads. *J Arthroplasty* 2007 September;**22**(6 Suppl. 2):125–9.
29. Engh Jr. CA, Stepniewski Jr. AS, Ginn SD, Beykirch SE, Sychterz-Terefenko CJ, Hopper RH, et al A randomized prospective evaluation of outcomes after total hip arthroplasty using cross-linked marathon and non-cross-linked Enduron polyethylene liners. *J Arthroplasty* 2006 September;**21**(6 Suppl. 2):17–25.
30. McKellop H, Shen FW, Salovey R. Extremely low wear of gamma-crosslinked/remelted UHMW polyethylene acetabular cups. *Transactions of the 44th Orthopedic Research Society*. New Orleans, LA; 1998 February 21–24. p. 98.
31. DiMaio WG, Lilly WB, Moore WC, Saum KA. Low wear, low oxidation radiation crosslinked UHMWPE. *Transactions of the 44th Orthopedic Research Society*. New Orleans, LA; 1998 February 21–24. p. 363.
32. Hopper Jr. RH, Young AM, Orishimo KF, McAuley JP Correlation between early and late wear rates in total hip arthroplasty with application to the performance of marathon cross-linked polyethylene liners. *J Arthroplasty* 2003 October;**18**(7 Suppl. 1):60–7.

33. Heisel C, Silva M, dela Rosa MA, Schmalzried TP. Short-term in vivo wear of cross-linked polyethylene. *J Bone Joint Surg Am* 2004 April;**86-A**(4):748–51.
34. Heisel C, Silva M, Schmalzried TP. In vivo wear of bilateral total hip replacements: conventional versus crosslinked polyethylene. *Arch Orthop Trauma Surg* 2005;**125**:555–7.
35. Yao JQ, Lu MP, Johnson TS, Gilbertson LN, Swarts D, Blanchard CR, et al. Improved resistance to wear, delamination and posterior loading fatigue damage of electron beam irradiated, melt-annealed, highly crosslinked UHMWPE knee inserts. In: Kurtz SM, Gsell R, Martell J, editors. *Highly crosslinked and thermally treated ultra-high molecular weight polyethylene for joint replacements*. West Conshohoken: American Society for Testing and Materials; 2003.
36. McNulty D, Swope S. Influence of polyethylene processing, tibial tray surface finish, and modular locking mechanism design on in-vitro wear for total knee arthroplasty. *Trans of the 51st Orthop Res Soc* 2005:0840.
37. Li MG, Zhou ZK, Wood DJ, Rohrl SM, Ioppolo JL, Nivbrant B. Low wear with high-crosslinked polyethylene especially in combination with oxinium heads. A RSA evaluation. *Trans of the 52nd Orthop Res Soc* 2006;**31**:643.
38. Good V, Ries M, Barrack RL, Widding K, Hunter G, Heuer D. Reduced wear with oxidized zirconium femoral heads. *J Bone Joint Surg Am* 2003;**85-A**(Suppl. 4):105–10.
39. Good V, Widding K, Scott M, Jani S. The sensitivity of crosslinked UHMWPE to abrasive wear: hips vs. knees. In: Kurtz SM, Gsell R, Martell J, editors. *Highly crosslinked and thermally treated ultra-high molecular weight polyethylene for joint replacements*. West Conshohoken: American Society for Testing and Materials; 2003.
40. Clarke IC, Green DD, Williams PA, et al. Simulator comparison of XLPE wear with 36 mm CoCr and oxidized zirconium balls in smooth and roughened condition. *Trans of the Seventh World Biomater Congr* 2004:1138.
41. Parikh A, Morrison M, Jani S. Wear testing of crosslinked and conventional UHMWPE against smooth and roughened femoral components. *Trans of the 53rd Orthop Res Soc* 2007;**32**:0021.

Chapter 21

# Mechanisms of Crosslinking, Oxidative Degradation and Stabilization of UHMWPE

Luigi Costa, PhD and Pierangiola Bracco, PhD



## 21.1 INTRODUCTION

Radical reaction pathways govern the crosslinking and degradative reactions of UHMWPE. For these reactions to occur, macroradicals must be induced in the polymer, for example by thermal decomposition of hydroperoxides or by high-energy radiation, which leads to homolytic bond scissions with the production of alkyl macroradicals. In previous chapters, we have often referred to crosslinking and degradation of UHMWPE, but we have not described the chemical mechanisms associated with these pathways in great detail. As noted previously, during irradiation, a cascade of chemical reactions may occur in UHMWPE depending upon the environmental conditions. This chapter describes in detail the chemical reactions that take place in UHMWPE during irradiation and subsequent exposure to oxygen.

## 21.2 MECHANISMS OF CROSSLINKING

Crosslinking of a polymer is defined as the linking of two or more polymeric chains by means of chemical (covalent) bonds. In this way, the molecular mass increases theoretically up to infinity. Thus, crosslinking results in one long, branched molecule with infinite molecular mass. Crosslinking can be achieved by chemical or by radiochemical reactions [1–4].

In chemical crosslinking, one must either extend the polymerization process or add suitable reactants and additives to induce the formation of chemical bonds between adjacent polymeric chains. One example of chemical crosslinking is rubber, which may be chemically crosslinked by addition of sulphur. With UHMWPE, peroxides such as dicumyl peroxide may be mixed with the resin powder to chemically crosslink the polymer during conversion to bulk rod or sheet [4]. However, chemical crosslinking is generally not used to process UHMWPE for medical applications, and for this reason we will focus on radiation crosslinking for the remainder of this chapter.

### 21.2.1 Mechanism of Macroradical Formation During Irradiation

The degradation mechanisms of polyethylene in an inert atmosphere induced by high energy radiation have been widely studied [1–3]. Electron beam and gamma rays, employed for both sterilization and crosslinking processes, have a mean energy some orders of magnitude higher than that of chemical bonds. Their interaction with UHMWPE leads, through a complex energy transfer, to the scission of C-C and C-H bonds, giving $H^{\bullet}$ radicals and primary and secondary macroradicals (Scheme 1) [1, 2, 5]. These macroradicals are dispersed throughout both the amorphous phases of the polymer and likely in the crystalline phase.

In previous studies, researchers have not detected primary alkyl macroradicals in irradiated UHMWPE [6–8]. In addition, NMR and FTIR studies have not revealed an appreciable increase in the concentration of terminal methyl units, which would be associated with radiolytic cleavage of the polymer chains. Because breaking of C-C is a random (stochastic) process, it can be assumed that the primary macroradicals resulting from Reaction 1 undergo recombination, in both amorphous and crystalline phase, giving back a C-C bond, with dissipation of energy in the polymer mass (Scheme 1, Reaction 2).

Orthopedic UHMWPE has a weight average molecular mass of $2 \times 10^{-6}$ amu. The polymer has a high viscosity, even in the molten state. Thus, macroradicals have very low mobility, either in the molten or in the solid state, whereas the $H^{\bullet}$ radical, which has a diameter smaller than 1 Å, can diffuse in the polymer mass, even in the crystalline phase, where distances between C atoms are in the order of 4 Å.

Reaction 5 (Scheme 2) is extremely favored, being exothermal ($\Delta H = -288$ kJ/mol), with a very low entropy variation. The intramolecular process (Reaction 5) is extremely fast and the secondary macroradicals decay, giving vinylene double bonds and molecular hydrogen (a gaseous product), which in turn can diffuse through the polymer mass. Among the vinylene double bonds, the trans-bonds are thermodynamically more stable. However, NMR studies have reported the presence of both cis- and trans-vinylene in the amorphous phase and of trans-vinylene only in the crystalline phase [9–10], formed according to a pseudo-zero order kinetic [11].

$H^{\bullet}$ radicals resulting from Reaction 3 (Scheme 1) are very mobile, and they can extract other H atoms intramolecularly, producing hydrogen. Intermolecular extraction (Reaction 6, Scheme 2) is possible, being exothermal ($\Delta H = -30$ kJ/mol) and associated to a very low entropy variation. The probability of extraction of the H atom decreases according to the following order: allyl, tertiary, secondary, primary, with a reactivity ratio between secondary and tertiary of 1 versus 9 [12]. Then, statistically, the hydrogen atom extracted in Reaction 6 is likely to be a

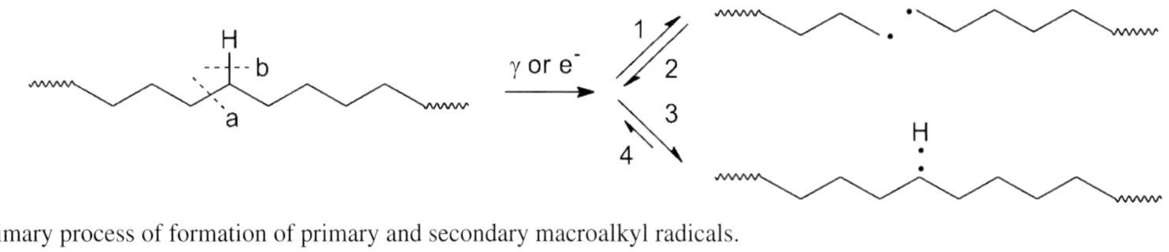

**SCHEME 1** Primary process of formation of primary and secondary macroalkyl radicals.

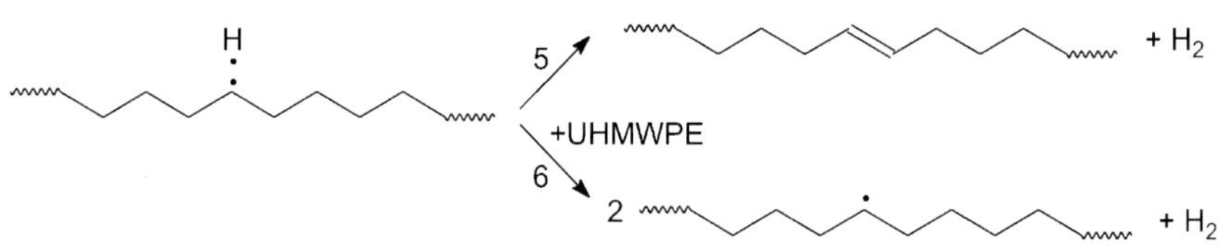

**SCHEME 2** Reactions of hydrogen formation.

secondary hydrogen. Therefore, hydrogen radicals not involved in Reaction 5 decay, giving molecular hydrogen and secondary macroradicals, which in turn can give hydrogen transfer, β scissions, or reactions with other reactive species.

### 21.2.2 Reaction of Isolated Radicals

Reaction 6 (Scheme 2) leads to an increase of isolated macro alkyl radicals in the polymer mass. Secondary alkyl radicals can then migrate along the polymer chain via H transfer. When the hydrogen transfer results in setting of a radical in α to a vinylene double bond, (Scheme 3 Reaction 7), the radicals structure changes from secondary to allyl macroradicals, more stable, which can survive even after a few years in the shelf at room temperature. The H transfer reaction in Scheme 3 is thermodynamically favored because it leads to the formation of more stable allyl macroradicals. The activation energy for this process is only 40 kJ/mol, and thus the reaction can occur even at room temperature at a high rate.

The concentration of surviving macroradicals decreases with time after irradiation [13]. Their living time is on the order of 24 hours in the amorphous phase, but macro alkyl radicals can be found in the polymer bulk even after years, probably trapped in the crystalline phase [14]. The mechanism of decay of macro alkyl radicals is still unknown.

These radical species are responsible for radical reactions (crosslinking, initiation, and propagation of the oxidative cascade). A number of EPR studies investigating the radical species formed in polyethylene following irradiation can be found in the literature. However, the attribution of the EPR signal is often controversial.

The evolution of secondary alkyl macroradicals via β scission, as shown in Scheme 4, rarely occurs at room temperature. Reaction 8 (Scheme 4) is the inverse reaction of polymerization and is endothermic ($\Delta H = 88$ kJ/mol), while Reaction 9 is even more endothermic ($\Delta H = 146$ kJ/mol), due to the higher energy of the C-H bond, compared to the C-C one. Thus, both the reactions are extremely unlikely at room temperature but may occur at higher temperatures (≈200–250°C) in the absence of oxygen.

### 21.2.3 Y-Crosslinking Mechanism

UHMWPE is not just a simple sequence of methylenes -($CH_2$)-, but it also contains small but measurable concentrations of vinyl double bonds, tertiary carbons, and methyl groups. These short and long chain branches, as well as the residues of the catalyst, are incorporated in the less ordered amorphous phase, not in the crystalline lamellae. The amounts of structural irregularities, as well as the degree of crystallinity, depend on the polymer synthesis method and on the subsequent processing conditions of the resin.

Double bonds can react with macroradicals, resulting in the formation of branching and increasing the molecular mass. When a terminal vinyl double bond reacts with the macroradical on an adjacent polymer chain (Reaction 10, Scheme 5), a Y-crosslink is formed [9, 10]. Reaction 10 is exothermal ($\Delta H = -88$ kJ/mol), but it is controlled by inductive and steric effects. In UHMWPE, Reaction 10 occurs only with the terminal vinyl groups at the end of the polymer chains [15]. There is experimental evidence from NMR studies to support the formation of Y-crosslinks from terminal vinyl species, as shown in Scheme 5 [9, 10]. Their consumption has been found to be proportional to the absorbed radiation dose [5]. This crosslinking reaction predominates at room temperature and is the primary mechanism of crosslinking in gamma sterilization, as well as during elevated crosslinking with gamma or electron beam radiation when the UHMWPE is in the solid state [9]. The high efficiency of this reaction also suggests that the rate of H transfer of the macroalkyl radical must be very high.

Due to steric hindrance, the reaction shown in Scheme 5 is not observed with the vinylene double bonds, which are formed in UHMWPE upon irradiation (Reaction 5 in Scheme 2). In lower density polyethylene, this reaction can occur with vinylidene species [16–17]. However, these types of vinylidene bonds are rarely observed in UHMWPE and are mentioned nonetheless for completeness.

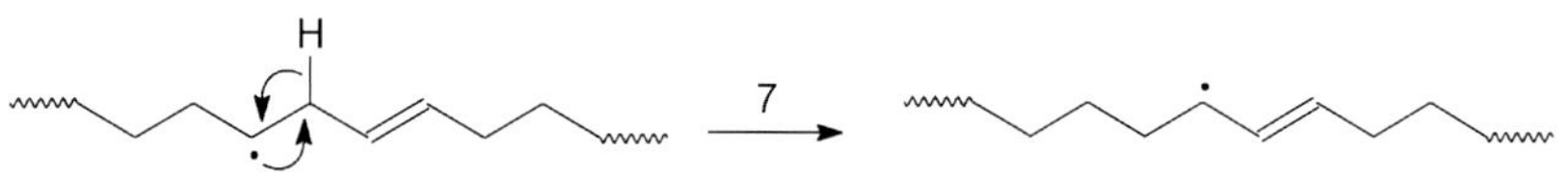

H 7

SCHEME 3 Hydrogen transfer to allyl radical.

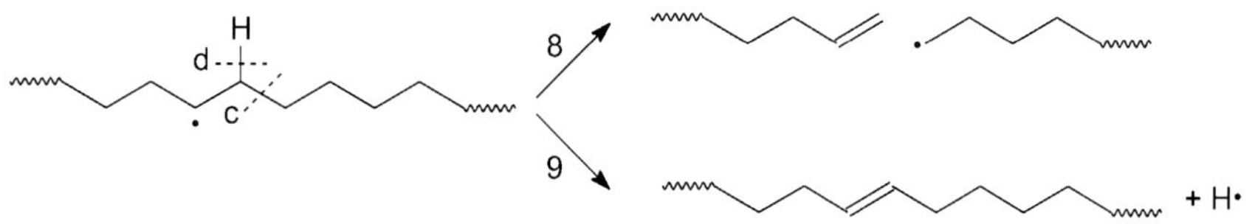

H d c 8 9 + H•

SCHEME 4 β−scission reaction of secondary radicals.

### 21.2.4 H-Crosslinking Mechanism

Another crosslinking mechanism, leading to the formation of H-crosslinks, is shown in Scheme 6. Secondary macroradicals can decay, moving along the polymeric chain, via disproportion (Scheme 6, Reaction 11) or coupling (Reaction 12). Both reactions are exothermal ($\Delta H = -260\,kJ/mol$ and $-313\,kJ/mol$ for Reactions 11 and 12, respectively).

Although thermodynamically feasible at room temperature and experimentally observed with liquid hydrocarbons [18], the formation of H-crosslinking is nonetheless blocked by steric hindrance when UHMWPE is in the solid state. Radicals in the crystalline phase behave differently from radicals in the amorphous phase. Models of the crystalline phase in the solid state show that the minimum distance between two C atoms is 4 Å, thus much higher than the mean C-C bond length (1.54 Å). Therefore, the formation of H-crosslinks (Reaction 11) is sterically hindered in the crystalline phase. The ratio between the kinetic constants of disproportion and coupling for a secondary butyl radical, in the liquid state at room temperature, is about 1.5, increasing by one order of magnitude for the same radical in the solid state [19]. The presence of H-crosslinks in polyethylene irradiated in solid state has been ruled out in previous NMR studies. In addition, given the high rate of transfer of the secondary macroradical, the encounter of the two species involved in Reaction 12 would be statistically favored. Then, if the reaction were easily feasible, it would be a termination pathway for the radical species, which would not be found for long durations after irradiation.

Moreover, experimental evidence shows that the level of crosslinking increases with the radiation dose, but it reaches a plateau for doses higher than 100 kGy, when the majority of the vinyl double bonds have been consumed [13]. A different behavior can take place in the molten state, where the mobility of radicals is higher. Under these conditions, NMR studies confirm the presence of some H-crosslinks [10].

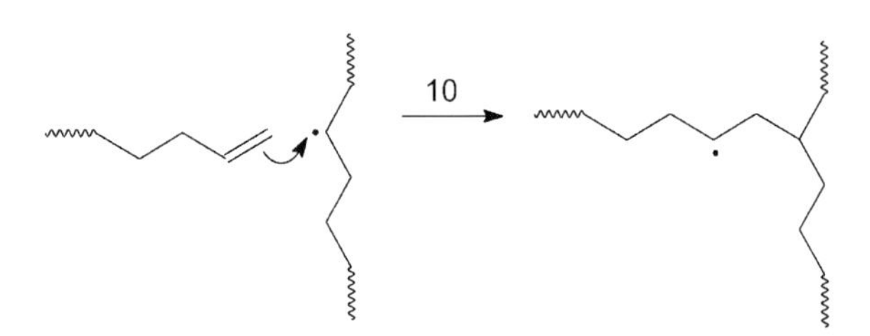


**SCHEME 5** Reaction of vinyl double bonds with secondary radicals.

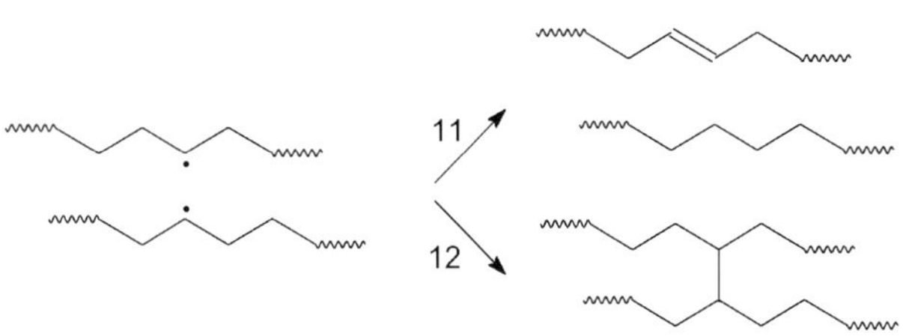

**SCHEME 6** Termination reactions: disproportion (11) and coupling (12).

## 21.3 UHMWPE OXIDATION

### 21.3.1 Introduction

The mechanism of oxidation of short-chain hydrocarbons is widely known under the name of Bolland's cycle (Scheme 7) [20], and it was studied at 70–80°C in the liquid phase, where the hydrocarbon chains and radicals have high mobility. The first step involves the formation of macroradicals, which, in the presence of oxygen, are easily transformed into peroxy radicals (ROO•). These species evolve to hydroperoxides (ROOH) via hydrogen extraction from a hydrocarbon chain, leaving behind a new radical, which reenters the cycle.

The same cyclic process has been proposed for hydrocarbon polymers (polyolefins). In this case, the cycle is initiated by the formation of radicals via homolitic bond scission, occurring when an external event provides enough energy. The most favored scission will involve the weakest bonds, that is, the O-O bonds of peroxides or hydroperoxides created during processing by compression molding or ram extrusion. We must bear in mind that the oxygen is usually present during the conversion of the UHMWPE powder, and it is, in practical terms, nearly impossible to obtain a totally oxygen-free polymer during industrial processing, even when conducted in the presence of an inert gas, such as nitrogen.

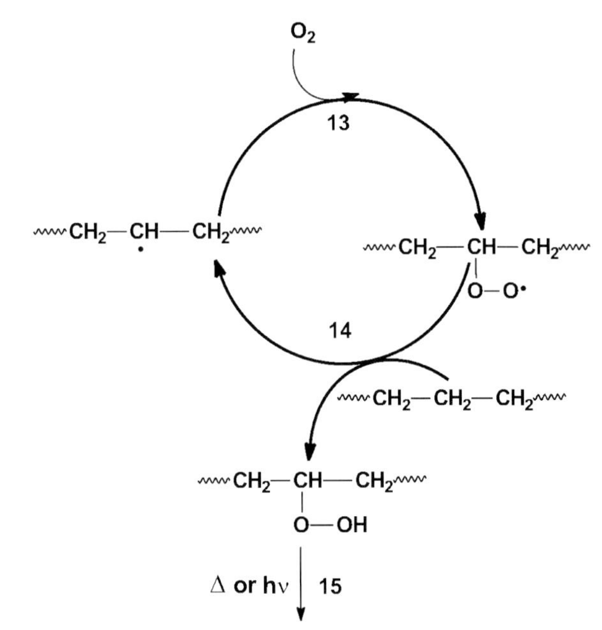


**SCHEME 7** Bolland's cycle: oxidation scheme of hydrocarbons.

The energy needed for bond cleavage can be provided by different forms, such as visible light (in this case the process is called photo-oxidation), heat (thermal-oxidation), mechanical stress (mechano-oxidation), and photon or electron beam (radiation-induced oxidation). During irradiation, a large amount of radicals are formed at a very high rate, creating a favorable scenario for studying the process under controlled conditions. In comparison to the oxidation of liquid hydrocarbons, some differences can be highlighted in the case of UHMWPE whose high molecular mass is responsible for a very low mobility of the polymer chains. Secondary alkyl macroradicals can react with the oxygen solubilized in the polymer bulk and diffused from the external environment, giving peroxy radicals. The peroxy radicals are set in a defined position on the polymer chain, and they cannot migrate as do the alkyl macroradicals [20]. Consequently, the peroxy radicals can abstract H intra- (1–5 extraction, Scheme 8, Reaction 16) or intermolecularly to produce hydroperoxides. The hydroperoxides can be hydrogen bonded or nonhydrogen bonded, but both are thermally stable at room temperature. Once again, this is a cyclic process, thus it continues also after irradiation has finished, but its rate decreases with time and approaches zero on the order of 100 hours, depending on the total amount of radicals, that is, on the radiation dose [8, 21].

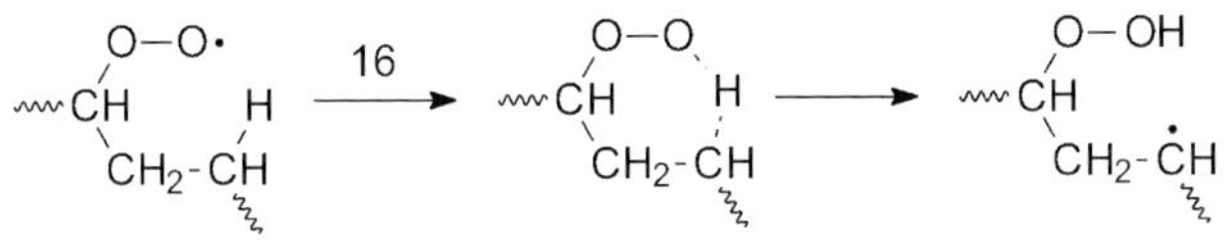

**SCHEME 8** 1–5 H extraction.

## 21.3.2 UHMWPE Post-irradiation Oxidation

Post-irradiation oxidation of UHMWPE at room temperature results in the formation of ketones as main products of the oxidation cycle, together with hydroperoxides and variable amounts of acids, alcohols, esters, and lactones [21]. It has been shown [21] that hydroperoxides in UHMWPE are stable at room temperature and start to decompose at temperature higher than 70°C, therefore ketones and the other oxidized species observed during post-irradiation oxidation cannot result only from the decomposition of hydroperoxides. Considering these results, it can be supposed that ketones are also formed during the first step of oxidation of UHMWPE as a consequence of a direct reaction between macroalkyl radicals and oxygen. A modified Bolland's cycle is proposed in Scheme 9 [21].

In the post-irradiation oxidation process, the rate of oxidation decreases by more than one order of magnitude in the first 100 hours, although alkyl macroradicals are continuously formed along with the formation of hydroperoxides (Scheme 9, Reaction 23). The termination reaction of thermo-oxidative processes is generally described as a Russell reaction between two peroxy species. The relative immobility and the stability of the peroxy radical makes the Russell bimolecular termination strongly disfavored in the solid state at room temperature [22, 23]. More likely,

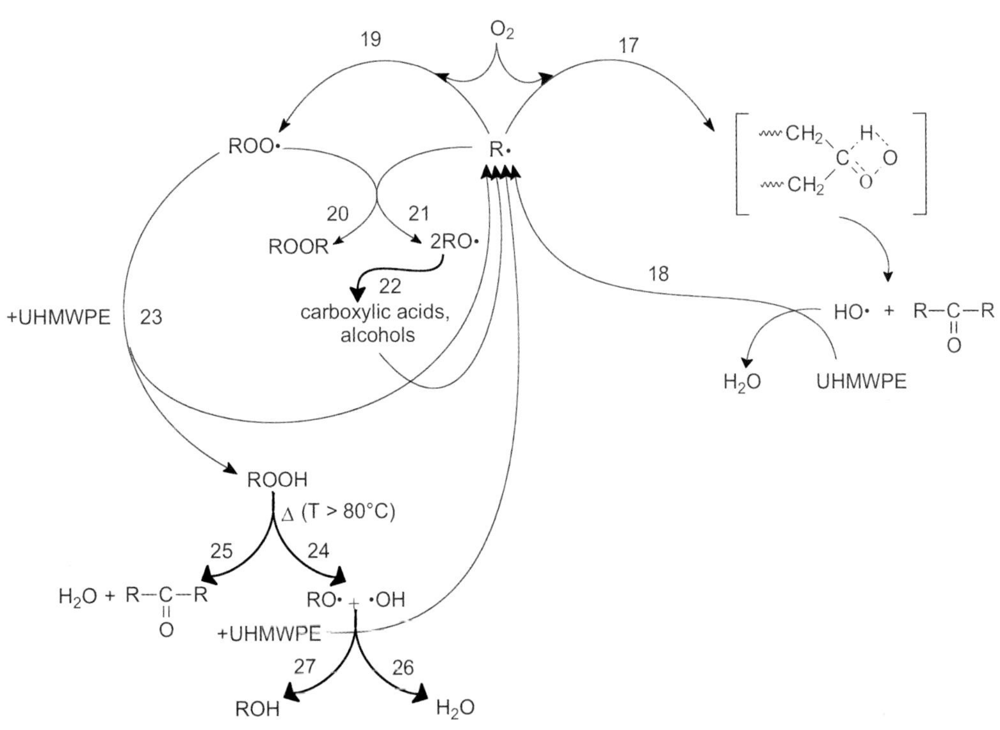


**SCHEME 9** Oxidation scheme of UHMWPE.

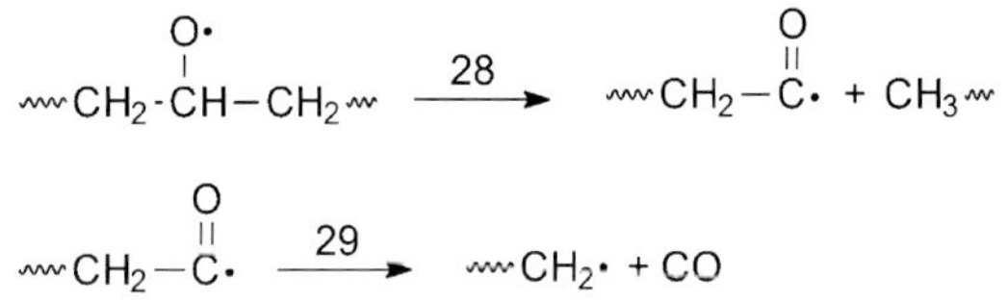

$$\sim CH_2-\overset{O\bullet}{\underset{|}{C}}H-CH_2\sim \xrightarrow{28} \sim CH_2-\overset{O}{\overset{\|}{C}}\bullet + CH_3\sim$$

$$\sim CH_2-\overset{O}{\overset{\|}{C}}\bullet \xrightarrow{29} \sim CH_2\bullet + CO$$

**SCHEME 10** Evolution of the alkoxy radical.

termination occurs between the peroxy radical, fixed on the polymeric chain and the alkyl macroradical, which migrates along the polymeric chain, with formation of peroxides (Scheme 9, Reaction 20). This reaction is not kinetically inhibited, and it is thermodynamic feasible [21]. Nevertheless, the resulting peroxides cannot be revealed with the usual characterization techniques, therefore their presence can only be hypothesized.

## 21.4 CRITICAL PRODUCTS OF THE OXIDATION PROCESS: MACRORADICALS

### 21.4.1 Alkyl Macroradical ($R^\bullet$)

Alkyl macroradicals $R^\bullet$ originate either by direct interaction of the polymer with gamma rays (or e-beam) or by hydrogen abstraction from reactive radicals, such as $ROO^\bullet$, $HOO^\bullet$, $RO^\bullet$, $OH^\bullet$, and $H^\bullet$. The amount of $R^\bullet$ is proportional to the absorbed dose, but their decay is quite fast [13]. Their actual concentration depends on the radiation (e-beam or gamma) dose, dose rate, and time of observation.

Experimental studies have shown that the vinylene concentration is constant with sample thickness [13]. Then we must assume that the alkyl macroradicals produced by gamma radiation are homogeneously distributed within the UHMWPE component because vinylene bonds are the primary product of interaction between polyethylene and high-energy radiation. In electron beam irradiation, the radical distribution is a function of depth due to the cascade effect, which leads to a subsurface maximum in the absorbed dose [2].

### 21.4.2 Peroxy Macroradical ($ROO^\bullet$)

Peroxy macroradicals are formed immediately after irradiation through the reaction between alkyl macroradicals and the oxygen present in the amorphous phase of UHMWPE (Scheme 9, Reaction 19). This reaction is thermodynamically favored. From the kinetic point of view, it requires the encounter of two radical species, which is also favored given the high rate of migration of the alkyl radical. Some $ROO^\bullet$ radicals are stable in UHMWPE, in spite of the large amount of available hydrogen atoms. In fact, hydrogen abstraction has a high activation energy (60kJ/mole) and proceeds slowly at room temperature. $ROO^\bullet$ decay is so slow that they have still been found after many weeks of storage at room temperature in air.

### 21.4.3 Alkoxy Macroradical ($RO^\bullet$)

The alkoxy radical is usually described as a typical product of the thermal decomposition of hydroperoxides. Nevertheless, in the post-irradiation oxidation process at room temperature, it cannot originate from this reaction because all the formed products follow a kinetic similar to that of ketone formation [21]. The reaction between the alkyl macroradical and the peroxy macroradical forms peroxides (Scheme 9, Reaction 20), but we can also hypothesize Reaction 21, Scheme 9. Literature studies demonstrate that the alkoxy radical can give beta-scission (Reaction 28) forming a primary alkyl radical and CO, a product that is found during the irradiation of PE (Scheme 10, Reaction 29) [24]. The activation energy of this reaction is around 50kJ/mole.

## 21.5 CRITICAL PRODUCTS OF THE OXIDATION PROCESS: OXIDIZED PRODUCTS

### 21.5.1 Hydroperoxide (ROOH)

The peroxy radical can extract an H atom from a nearby UHMWPE molecule (activation energy 108kJ/mol) with the formation of another alkyl macroradical in an adjacent position. Then, if more oxygen is available, a second bonded hydroperoxide is formed. On the other hand, if the formed alkyl macroradical migrates along the polymer chain before the reaction with oxygen, the formation of a free hydroperoxide occurs. The ratio of hydrogen-bonded versus non hydrogen-bonded hydroperoxides is constant in post-irradiation oxidation; therefore, it does not depend on the alkyl macroradical concentration nor on the irradiation dose.

The O-O bond in ROOH is thermally unstable at temperatures above 70–80°C (bond energy 170kJ/mol); therefore, it easily decomposes in thermal oxidation, producing very reactive $OH^\bullet$ and $RO^\bullet$ radicals (Scheme 9, Reaction 24). Thus, throughout the thermal oxidation of UHMWPE, a competition between ROOH formation and ROOH consumption occurs. It should be emphasized that an increase of 10°C in temperature generally doubles the rate of chemical reactions. Hydroperoxides determination is the key factor to measure the level and the behavior of the oxidation [25–27].

### 21.5.2 Ketone ($R_2CO$)

Ketones are usually identified as the main product of decomposition of hydroperoxides. However, in post-irradiation oxidation they are also formed at room temperature when the hydroperoxides are thermally stable. Their formation has the same behavior as that of ROOH, including a high rate after irradiation, when the concentration of alkyl macroradicals is high, and when the concentration of hydroperoxides is at a minimum. At the end of the post-irradiation process, when the concentration of hydroperoxides is maximum, the rate of ketones formation is minimum.

A mechanism of reaction justifying the formation of ketones at room temperature is proposed in Scheme 9, Reaction 17. The total products of Reactions 17 and 18 are the same obtained in the hydroperoxides formation and subsequent thermal decomposition (Scheme 9, Reaction 25). The difference is the formation of an intermediate, which immediately turns into ketones and hydroxyl radicals.

Depending on the temperature, we cannot exclude that ketones can also originate from hydroperoxides decomposition, via a "closed" process, which does not create new radicals and then does not spread oxidation. Furthermore, because the process takes place without further chain scissions, it does not induce any variation of the molecular mass and only minimal changes of the mechanical properties of UHMWPE.

Scientific studies and the ASTM oxidation index standard (ASTM F2102) usually give the quantity of ketones and other carboxyl species present as index of the oxidation degree. It must be pointed out that those ketones, though a product of the oxidative process, do not produce polymeric chain scissions and so they do not result in substantial reduction of the UHMWPE mechanical properties. Quantification of ketones is reliable only if the ratio between ketones and carboxylic acids remains constant through the whole oxidative process [21].

### 21.5.3 Carboxylic Acid (RCOOH)

Acids are produced by scission of the polymeric chain, with a mechanism that has not yet been elucidated [21, 28–29]. The alkoxy radical formed through Reaction 21 can decompose via β-scission, according to Reactions 28 and 29 (Scheme 10), with the formation of a methyl chain end, whose increase has been observed in the post-irradiation oxidation process, and of a carbonyl radical, which in turn decomposes giving a primary macroalkyl radical and carbon monoxide (CO), commonly found among the products of irradiation or thermo-oxidation of PE [2, 30]. Primary alkyl macroradicals react with oxygen to form primary hydroperoxides, then the hydroperoxides decomposition results in the formation of acids, as already stated in the literature [28].

Acid formation leads to a decrease in molecular mass and thus to a progressive deterioration of the mechanical properties of the UHMWPE. Acid determination, by using derivatization techniques and IR analyses [25], allows a quantification of the variation of the molecular mass of the oxidized material [31].

$$2ROOH \xrightarrow{30} RO^{\bullet} + ROO^{\bullet} + H_2O$$

**SCHEME 11** Bimolecular thermal decomposition of hydroperoxides.

### 21.5.4 Sec-alcohol ($R_2CHOH$)

Alcohols can be formed via H extraction from the polymer chain by an alkoxy radical (Reaction 27).

### 21.5.5 Ester

Ester formation occurs during oxidative degradation, although the precise mechanism of formation is not completely clear. One hypothesis involves the decomposition of primary peroxides [32].

## 21.6 CONSIDERATIONS ON ACCELERATED AGING METHODS: COMPARISON BETWEEN POST-IRRADIATION OXIDATION AND THERMAL OXIDATION

The thermo-oxidative stability of irradiated UHMWPE at 70°C is related to the concentration of hydroperoxides, as we have reported before. The bimolecular thermal decomposition of hydroperoxides leads to the formation of water and alkoxy and peroxy radicals (Scheme 11).

$RO^{\bullet}$ and $ROO^{\bullet}$, through extraction of H from an UHMWPE molecule, form alcohol and hydroperoxide and two macroalkyl radicals, which continue the oxidation process. Therefore, the process can be accelerated as a function of the temperature [21].

The products formed in the two post-irradiation oxidation processes, at room temperature or in accelerated aging at higher temperature, are substantially the same and, in addition, their relative abundance is the same. Therefore, their formation mechanism also must be the same.

These observations confirm that accelerated oxidation tests carried out above room temperature are still valid in reproducing the in-the-shelf oxidation process of radiation sterilized UHMWPE. A more complex behavior can be expected for *in vivo* oxidation, where other effects (i.e., mechanical stresses) can play a relevant role.

In conclusion, it has been elucidated how hydroperoxides and macroradicals are the key species in determining the oxidation cascade. A standardized measure of these two products could be helpful in evaluation of the "oxidative potential" of a given UHMWPE product.

### 21.6.1 Temperature Effects During Irradiation

Unsterilized UHMWPE components are typically stored at ambient temperature before being sent to a gamma irradiation plant. The temperature of the sterilization cell depends on the external temperature (sterilization in winter or summer results in a different initial cell temperature) and of course on the intrinsic characteristics of the plant itself [33]. Gamma rays deposit energy into the sample, thereby increasing its temperature and the temperature of the cell. Reasonably, the cell temperature is in the range between 25 and 45°C [1].

Orthopedic prostheses are sterilized with a dose of 25 to 40 kGy. Implants receive a dose within this range, typically about 30 kGy (that is, 30 kJ/kg)[34]. The specific heat of UHMWPE is around 1.5 kJ/(K kg). It means that, under adiabatic conditions, the temperature increase of UHMWPE component is about 20°C. However, thermal energy absorbed by UHMWPE during irradiation is partially transferred to the surrounding environment. Like many polymers, UHMWPE has a low linear heat transfer coefficient ($K_{UHMWPE} = 0.33$ W/m K). As a consequence, the temperature of the UHMWPE component will be inhomogeneous and vary between that of the irradiation cell on the surface and that of the adiabatic limit near the center of the component. Based on these considerations, it is estimated that the temperature inside an UHMWPE component during gamma sterilization ranges between 45 to 65°C.

After sterilization, prostheses are stored at ambient temperature and the component has time to reequilibrate. Eventually, after implantation, the temperature of the UHMWPE component rises to body temperature (37°C).

### 21.6.2 Distribution of Oxidized Compounds in the New UHMWPE Prosthetic Components

Fick's law governs the oxygen diffusion into the UHMWPE prosthetic component. Oxygen dissolves in the amorphous phase of polyethylene [35]. As a result, oxidation only occurs in the amorphous domain, where oxygen is available. It has been reported that oxygen solubility in a high density polyethylene, whose degree of crystallinity is similar to that of UHMWPE, is 1 mmol/kg at 25°C [35]. Because UHMWPE components are typically up to 15 mm thick, the implants are saturated with oxygen before radiation sterilization. Following sterilization, oxygen may diffuse into the UHMWPE on the shelf if it is stored in air-permeable packaging.

According to Carlsson [25] and Birkinshaw [36], about 2 mmols of alkyl radical ($R^•$) per kg of UHMWPE are formed for every 10 kGy of irradiation. $R^•$ concentration is proportional to dose, but the constant of proportionality is also a function of the dose rate. However, with a typical sterilization dose of 30 kGy, the $R^•$ concentration is calculated to be 6 mmols/kg. Thus, the reaction between alkyl radicals and oxygen will be oxygen diffusion controlled as soon as the dissolved oxygen is consumed.

The oxidation rate of UHMWPE components, resulting from high-energy radiation, depends on the following factors:

- Dose rate of the irradiation source, which governs the duration of exposure to the environmental conditions at the sterilization facility
- Temperature of the sterilization facility
- Absorbed dose, which controls the generation of sec-alkyl macroradicals
- Sample thickness, which governs the oxygen concentration distribution through the thickness of the implant

Typical conditions, to which gamma sterilized prosthetic components may be subjected, are summarized below:

- Initial temperature of sterilization cell between 15 and 50°C
- Absorbed dose between 28 and 40 kGy
- Dose rate between 10 and 1 kGy/h corresponding to 3–40 hours of irradiation
- Increase of temperature in the sample bulk of more than 20°C

In conclusion, both a dynamic gradient of $[O_2]$ and of temperature will be created through the thickness of an irradiated sample, whereas the dose rate can be considered constant during irradiation. Local oxygen concentration and local temperature control the actual concentration of oxidized species, which varies along the thickness, provided that the $[R^•]$ is constant but dependant on dose rate.

The extent of oxidation within a UHMWPE hip or knee component, at a certain depth beneath the surface, depends on the actual concentration of alkyl radicals and on the amount of available oxygen [37–42]. The actual concentration of alkyl radicals is a function of both the dispensed dose rate and the temperature of the sterilization room because further radicals can result from thermal decomposition of hydroperoxides.

Whereas the distribution profile of the radicals in the bulk cannot be determined, the distribution of the products of their reaction with oxygen can. In other words, the degree and distribution of oxidative degradation in the bulk can be determined by measuring the distribution of the hydroperoxides and of their decomposition products using infrared spectroscopy (FTIR) [43, 44]. In Figure 21.1 the oxidation products after NO (nitrogen monoxide) treatment for a new prosthesis are reported [41–42].

This distribution is a function of the rate at which gamma or e-beam radiation is supplied, the quantity of

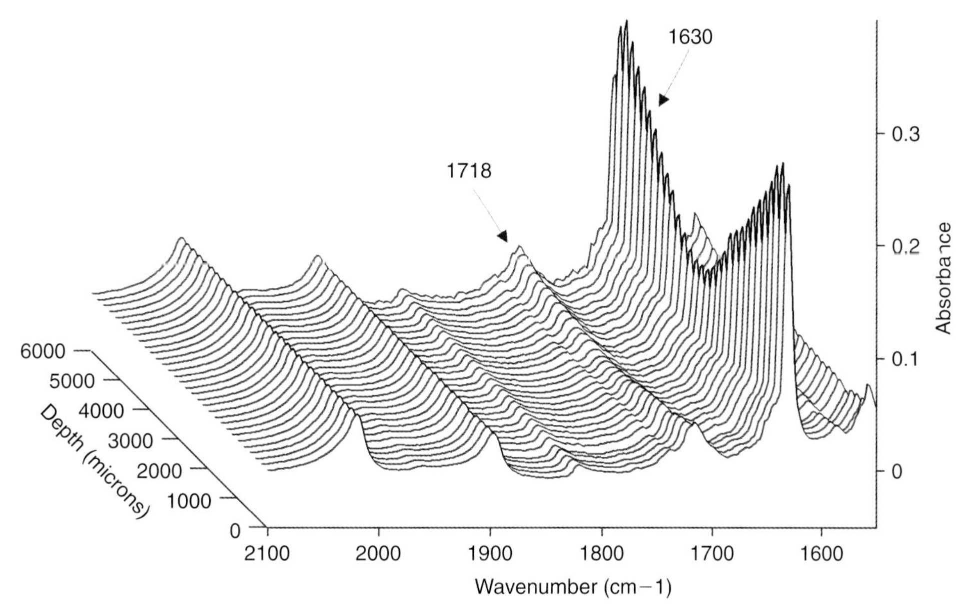


**FIGURE 21.1** FTIR map collected along the cross-section of a gamma sterilized, shelf-aged UHMWPE acetabular cup [41–42].

oxygen present within the polymer at the time of irradiation, the quantity of oxygen diffused afterwards, and the temperature of the sample.

Recent studies have shown that a large variability in the extent and in the distribution of oxidative degradation is observed in shelf-aged UHMWPE components dependent on the previously-mentioned parameters [38–42].

In the case of e-beam irradiation, the observed oxidative processes are qualitatively similar to those occurring with gamma ray irradiation. With e-beam irradiation, the dose rate is orders of magnitude higher than with gamma radiation (10 kGy/s with e-beam versus 1–10 kGy/h with gamma irradiation), and the irradiation time is on the order of seconds with e-beam, as opposed to hours with gamma irradiation. Comparing the two irradiation methods under the same conditions and the same dispensed doses, it has been observed that amount of oxidation was lower after e-beam irradiation than when gamma rays were used, due to the difference in dose rates and the ability of oxygen to diffuse into the UHMWPE during the longer irradiation times used with gamma irradiation [45].

## 21.6.3 Postoxidative Degradation after Implant Manufacture

The oxidative process initiated during sterilization can continue during shelf storage and implantation. The rate and the extent of the process will depend on the storage temperature in the shelf and on the human body temperature, together with the amount of available oxygen *in vivo*. Mechanical stresses developed during *in vivo* use can also play an important role in facilitating the oxidative process.

In addition, oxidative degradation related to poor consolidation has been found in both *in vivo* and shelf-aged prostheses that were either gamma sterilized or ethylene oxide sterilized in the presence of calcium stearate [46]. When oxidation related to consolidation occurs, the oxidation profile through the cups section is inhomogeneous, and the maximum oxidation is observed near the center of the prosthesis. It is worth mentioning that this consolidation-related oxidation mechanism is often accompanied by whitening of the material, visible to the naked eye.

The origins of consolidation-related oxidation are not clear yet, but the processing conditions of the UHMWPE bar or sheet, together with the influence of the machining of the prosthetic component, have been proposed as contributing factors. Poor consolidation of the UHMWPE powder during processing, followed by internal stresses induced by machining, have been hypothesized to lead to the formation of free volumes between individual resin powder particles. These intergranular voids are more easily permeated by oxygen, so that facilitated oxidative degradation can take place at the grain boundaries, even if the initiating radical mechanisms are still unclear.

Figure 21.2 shows a collection of FTIR spectra through the thickness of a shelf-aged, EtO sterilized cup in presence of calcium stearate, where bulk oxidation was evident. Gamma irradiation of the UHMWPE was ruled out by examination of the transvinylene region of the FTIR spectra. The absorption at 1718 cm-1 is attributed to ketones. In this case, the oxidation was associated with poor consolidation of the UHMWPE and has nothing to do with the sterilization process. Again, *in vivo* stresses can have a synergistic effect in enhancing oxidation.

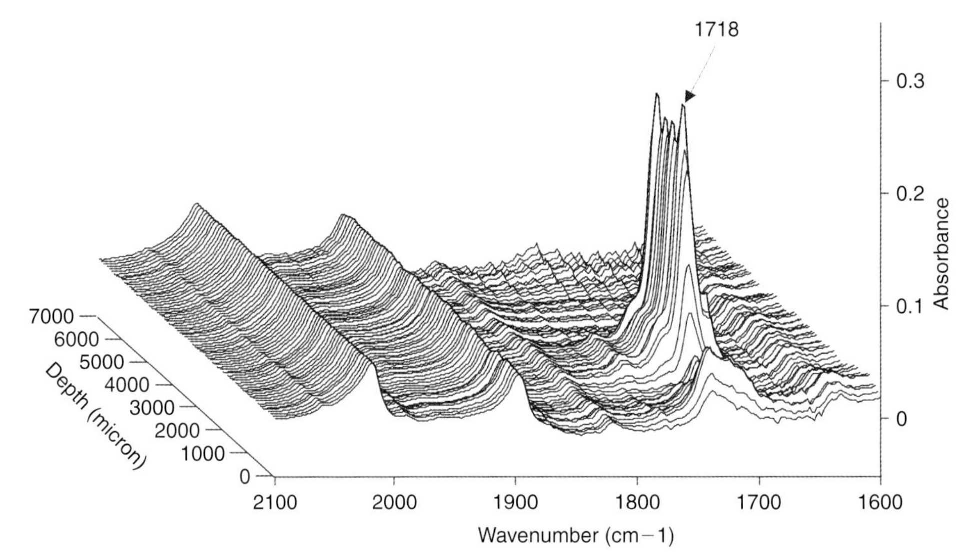


**FIGURE 21.2** FTIR map collected along the cross-section of an EtO sterilized, shelf-aged UHMWPE acetabular cup [46].

# 21.7 STABILIZATION OF UHMWPE

## 21.7.1 Introduction

Most polymers are easily subject to oxidative processes, which strongly compromises their mechanical and aesthetic characteristics. In the first years of industrial production of polymers, one of the main goals of the manufacturers was the complete understanding of the mechanisms of oxidation, of polyolefins in particular, and the development of effective strategies of stabilization [47].

As already discussed in the previous paragraphs, oxidation is basically due to the reaction between macroradicals and oxygen diffused into the polymer from the surrounding atmosphere. Polymers, during their entire lifetime, often undergo many different causes of mechanical, thermal and photo degradation, which create macroradical species, generally in the presence of oxygen. Then it is practically impossible to stop the oxidation process permanently. However, it can be considerably slowed down if the highly reactive polymer macroradical is transformed into a less reactive radical species. This strategy is the basis of polymers stabilization, which results in a longer preservation of the original mechanical characteristics of the polymer.

To stabilize the polyethylene, the addition of a reactive species is required. The additive (ADH) reacts with the polymeric macroradical (alkyl or peroxy), giving a kinetically and thermodynamically stable radical ($AD^{\bullet}$) and a saturated macromolecule, according to the following reaction:

The more stable the $AD^{\bullet}$ radical, the more the oxidation process is slowed down. Therefore, all commercial polyethylene is commonly stabilized to oxidation with specific additives for each application.

$$R\cdot + ADH \longrightarrow RH + AD\cdot$$

**SCHEME 12** Reaction between macroradicals and stabilizer.

The science of stabilization is widely described in a number of comprehensive textbooks [47–49]. The main classes of oxidation stabilizers are hindered phenols and amines. Literature data show that for a radiation-induced process, phenols are generally more effective [3, 49–51].

Despite this evidence, ASTM F-648 and ISO 5834 standards forbid the use of any additive in orthopedic UHMWPE [52]. The ban dates back to the early 1980s, when some of the additives used in polymer stabilization were suspected carcinogenic products (for example, aromatic amines for UV stabilization). The choice to forbid dangerous substances into medical UHMWPE was obviously understandable because the additives can potentially diffuse into the human body during *in vivo* implantation. More questionable was the practice of sterilizing the prosthetic components with gamma radiation in air, while, already in the 1980s, literature studies showed the detrimental effect of this combination [53]. Nevertheless, this remained the preferred method of sterilization until the mid-1980s, when some producers shifted to high energy radiation in inert atmosphere. Today, most sterilization protocols for medical devices using high-energy radiation are validated in air and not in inert atmosphere (ISO 11137 and EN 552) [54].

It is worth mentioning that the first dramatic failures of UHMWPE components in the mid-1980s were attributed to inadequate mechanical properties of the original polymer, despite the evidence that these properties, in pristine UHMWPE, were much better than those required by ASTM F648.

In the 1990s, some researchers showed that a biocompatible stabilizer could have been useful, or even necessary, to stabilize orthopedic UHMWPE [37, 71, 72].

α-tocopherol (synthetic vitamin E, see formula in Scheme 13) is the most widely used among natural stabilizers. Its mechanisms of stabilization have been widely studied in polymers processing, and it has been shown to

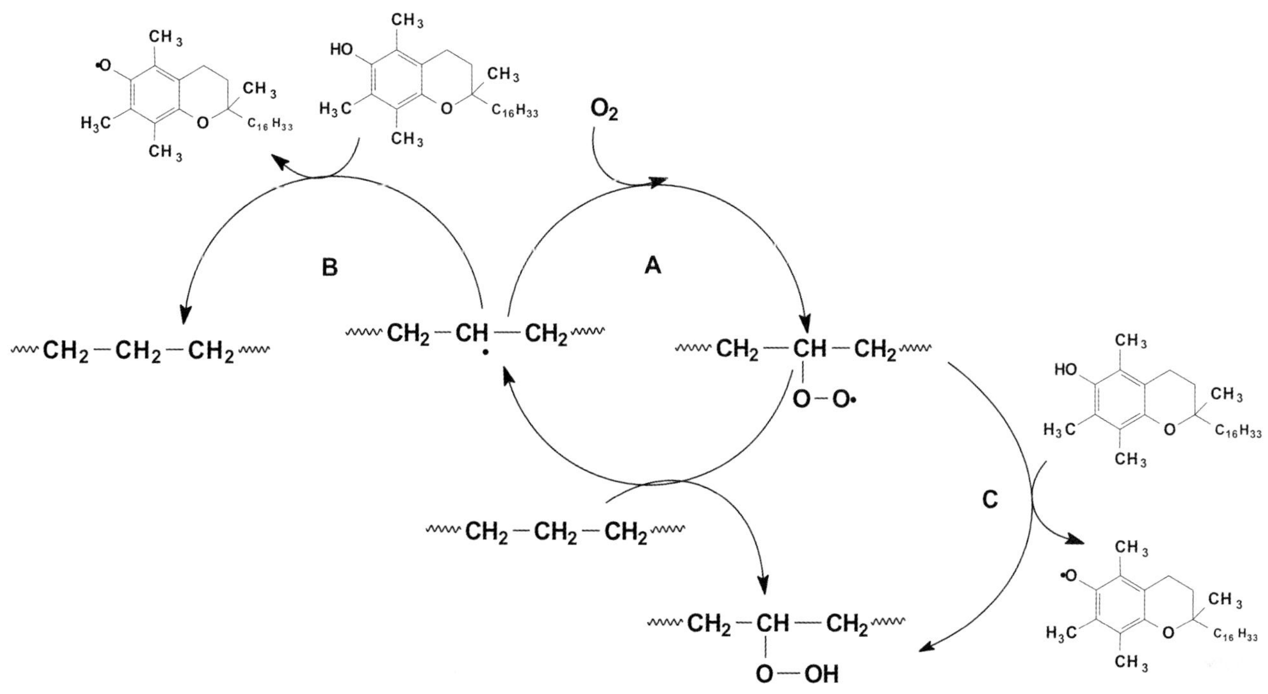


SCHEME 13 Cycle of stabilization of oxidation process by α-tocopherol.

be more effective when compared to other commercial additives [55–60]. Moreover, α-tocopherol is an FDA-approved additive for food and beverages, and it has a commercial turnover of about 30,000 ton/year.

Vitamin E, an apolar molecule that is the most important lipid-soluble antioxidant in humans, is not synthesized in our body, but it is acquired by nourishment. Vitamin E is emulsified together with the fat-soluble components of the food and transported in the body. Lipolysis and emulsification of the formed lipid droplets then lead to the spontaneous formation of mixed micelles [61].

Vitamin E is soluble in UHMWPE (around 3% by weight), while it is insoluble in water and in the synovial fluid, and therefore it cannot be drawn out from the UHMWPE prosthetic components. UHMWPE films with 0.5% of vitamin E irradiated in air at 100 kGy after 60 days in water at 70°C do not show loss of vitamin E or degradation products [62]. This behavior is similar, but in an opposite way, to the diffusion of the apolar products present in the synovial fluid (cholesterol, ester of cholesterol, squalene) [63].

According to the results obtained using norms EN 30993-5, DIN UA 12, and *in vivo* animal tests reported in the literature, vitamin E and its products of oxydegradation are biocompatible and not cytotoxic [64–66].

## 21.7.2 Chemical Mechanisms of Vitamin E Stabilization

The stabilization mechanism of α-tocopherol in UHMWPE have been recently studied in detail [67–69] and partially revised [70]. Phenolic antioxidants are generally known to act with a mechanism that involves the reaction of the additive with the peroxy radical formed during the oxidation process (see Scheme 13, Part C), giving a hydroperoxide and a phenoxy radical, more stable and kinetically less reactive thanks to delocalization of the unpaired electron over the aromatic ring [49].

Recent studies indicate that α-tocopherol can also directly react with alkyl macroradicals, this reaction being more likely due to the higher mobility of the alkyl macroradicals compared to that of peroxy radicals [70]. The alkyl macroradicals react with α-tocopherol, giving a polymeric saturated molecule and a stabilized phenoxy radical (Scheme 13, Part B). The phenoxy radical can in turn interact with another alkyl macroradical leading to the formation of α,β-unsaturated ketones [69, 70]. This mechanism indicates a double efficiency of stabilization of vitamin E: one vitamin E molecule can convert up to two alkyl macroradicals to unreactive species.

Scheme 13 shows the mechanisms of stabilization of α-tocopherol in detail.

Therefore, the α-tocopherol blended with the UHMWPE powder before processing can be consumed by the reaction with the radicals formed during irradiation of the implant either for sterilization or crosslinking purposes [69, 70]. Nevertheless, even if the vitamin E cannot completely avoid the formation of hydroperoxides, it has a clear stabilizing effect against oxidation of UHMWPE. Literature data show that UHMWPE blended with 0.1% of α-tocopherol and irradiated to 30 kGy is much more stable to accelerated aging than original UHMWPE [69].

### 21.7.3 Determination of the Vitamin E Content in UHMWPE

The growing interest of the market for stabilized UHMWPE has raised the need for a quick and easy method to determine the vitamin E content in both untreated and irradiated UHMWPE. Techniques involving extraction and following separation analyses [74, 75] have been ruled out for their complexity, while FTIR and UV spectroscopy were found to provide quite easy and reliable measurements. The best experimental conditions are now under investigation by an ASTM workgroup, but good results are already reported in the literature [69, 70].

Using FTIR spectroscopy, the C-O stretching of the phenol at $1210\,cm^{-1}$ is the most suitable absorption to assess the vitamin E content in virgin, unirradiated UHMWPE. The same absorption can also provide some information regarding the consumption of vitamin E upon irradiation.

The absorption at $1260\,cm^{-1}$, attributed to the Aryl-O-Alkyl ether group, can also provide quantification of vitamin E in untreated UHMWPE [73, 76], but with lower sensitivity. In addition, this absorption does not disappear completely upon irradiation (probably because it is converted into an Alkyl-O-Alkyl ether), being less reliable in evaluating the phenol consumption.

The precise concentration of α-tocopherol in UHMWPE is usually determined by spectral subtraction of a reference antioxidant-free sample [69, 70]. This method permits determination of concentration as low as 0.0375% (375 ppm) of vitamin E [69, 77].

In the study of the stabilization mechanism, importance can assume the separation and the identification of the degradation products induced during the irradiation and oxidation process [69]. In this case it is necessary to extract the products from the polymeric mass with an apolar solvent and to submit them to a separation in liquid chromatography followed by mass spectrometer analysis.

The transformation products during processing in HDPE with vitamin E have been analyzed by Al Malaika [78–79]. The conditions of processing are very different from the conditions of irradiation; therefore, the products can be very different.

## 21.8. *IN VIVO* ABSORPTION OF LIPIDS

UHMWPE, as all the other polyethylenes, is a semicrystalline polymer with a glass transition temperature lower than −70°C, which means that rubbery amorphous regions coexist with crystalline, ordered domains. The crystalline portion is extremely compact; for this reason it is highly unlikely that a molecule may penetrate the space (about 4.5 Å) between the polymeric chains, except for the tiny hydrogen and helium molecules whose diameter is about 2 Å. The rubbery amorphous portion is less compact, also in crosslinked UHMWPE, and has continuous free volume among the flexible polymeric chains. These voids may allow diffusion of not only gaseous molecules but also of molecules of larger size, having flexible linear chains. Obviously, only substances with a certain affinity for polyethylene will be able to diffuse. In other words, apolar, long-chained molecules, which have a fairly good solubility in polyolefines, are the most likely diffusants. The two Fick's laws govern the diffusion process. Diffusion rate is a function of many factors such as temperature, pressure, and molecular mass of the diffusant.

UHMWPE prosthetic components *in vivo* are in intimate contact with joint fluid, basically a filtered component of blood's plasma, rich in lipids and triglycerides.

Figure 21.3 shows a collection of FTIR spectra along the cross section of an EtO sterilized cup, retrieved after 4 years *in vivo*. The absorption at 1740 cm-1 is attributed to ester groups. Note the typical profile, decreasing from the surfaces toward the bulk. Studies revealed the composition of the diffused fraction by extraction and GC/MS analysis [49]. It has been pointed out that the diffused products are mostly squalene (a cholesterol precursor), cholesterol, and cholesteryl esters with fatty acids and triglycerides, whose relative amount is highly variable depending on the patient. Figure 21.4 shows a typical gas chromatogram of the extracted fraction from a UHMWPE component.

One question is whether the diffusion phenomenon can affect the mechanical properties of the polymer and to what extent. Experimental data, obtained *in vitro* with model compounds, suggests that modifications of mechanical properties induced by diffusion are quite limited [80]. *In vivo* diffusion leads to a variation in the surface wetability, then it favors the adhesion to the UHMWPE surface of a monomolecular layer of polar compounds, such as proteins or phospho-lipids [81]. It must be taken into account that *in vitro* tests (i.e., hip or knee simulators) can accelerate the mechanical processes but, even in the presence of a suitable lubricant, do not correctly reproduce this phenomenon, which is strongly time dependent.

## 21.9 CHEMICAL PROPERTIES OF WEAR DEBRIS

Wear debris is formed due to the abrasion and adhesion induced by sliding of a harder body (typically metal or ceramics) on the surface of the UHMWPE component. It follows that the chemical structure of the debris corresponds to that of the surface layer of the UHMWPE component. As previously mentioned, UHMWPE retrieved after *in vivo* implantation usually contains a certain amount of lipids diffused from the synovial fluid. The abundance of these products varies with the clinical situation of the patient, with the time *in vivo*, with loading, etc. Therefore, these compounds will be present in the debris, and their

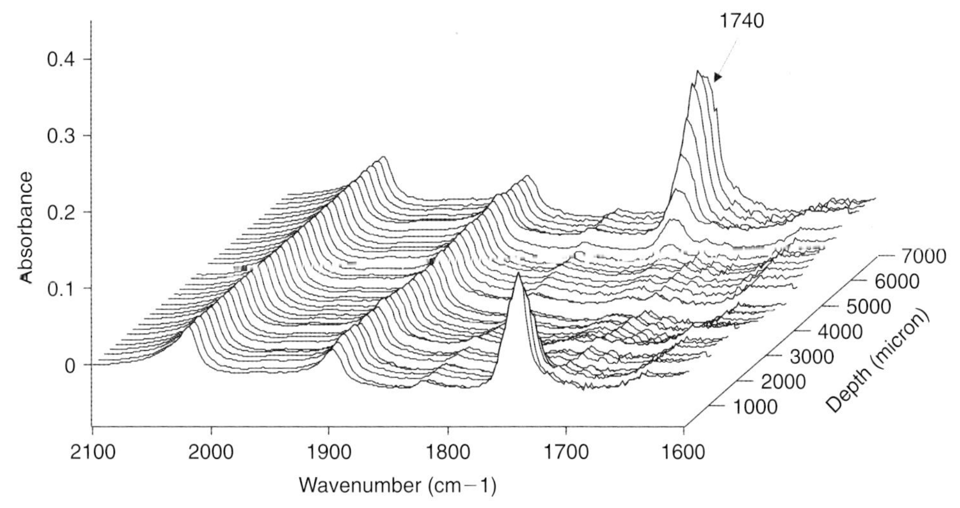


**FIGURE 21.3** FTIR map collected along the cross-section of an EtO sterilized acetabular cup, explanted after 4 years *in vivo*.

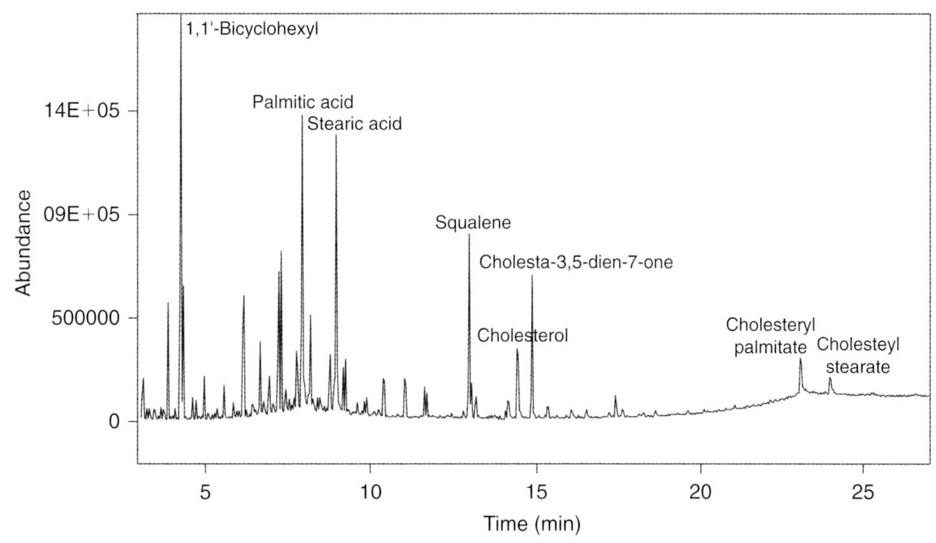


**FIGURE 21.4** Typical GC/MS analysis of the soluble fraction extracted from a retrieved UHMWPE prosthetic component.

concentration will be regulated by the solubility and by the diffusion process.

In addition, if the UHMWPE component is oxidized, the debris will also be oxidized, and this can affect its surface reactivity. Loosening membranes of EtO-sterilized polyethylenes are usually smaller in volume and contain less and smaller particles if compared to those coming from components sterilized by gamma radiation in air and thus oxidized [82]. These observations suggest that together with the shape, size, volume, and number, the chemical characteristic of debris must also be taken into account when studying the biological reaction.

## 21.10 ACKNOWLEDGMENTS

We are very grateful to Professors Elena Brach del Prever and Paolo Gallinaro of the Traumatology and Orthopaedics Hospital (CTO) of Torino, Italy, whose interest in research and continuous encouragements have been the foundation of our work on UHMWPE.

The work discussed by the authors has been supported by research grants from Bioster (Bergamo Italy), Lima Lto., and Zimmer, Inc.

## REFERENCES

1. Clegg D, Collyer A. *Irradiation effects on polymers*. New York: Elsevier Applied Science; 1991.
2. Ivanov VS. *Radiation chemistry of polymers*. The Netherlands: VSP Utrecht; 1992.
3. Carlsson DJ. Degradation and stabilisation of polymers subjected to high energy radiation. In: Scott G, editor. *Atmospheric oxidation and antioxidants*, vol. II. Amsterdam: Elsevier; 1993. p. 495–528.

4. Shen FW, McKellop HA, Salovey R. Irradiation of chemically crosslinked UHMWPE. *J Poly Sci Poly Phys* 1996;**34**:1063–77.
5. Bracco P, Brunella V, Luda MP, Zanetti M, Costa L. Radiation-induced crosslinking of UHMWPE in the presence of co-agents : chemical and mechanical characterization. *Polymer* 2005;**46**: 10648–57.
6. De Vries KL, Smith RH, Franconi BM. Free radicals and new end groups resulting from chain scission: 1. Gamma-irradiation of polyethylene. *Polymer* 1980;**21**:949–56.
7. Igarashi M. Free radical identification by ESR in polyethylene and nylon. *J Poly Sci: Poly Chem Ed* 1983;**21**:2405–25.
8. Lacoste J, Carlsson DJ. Gamma-, photo, and thermally-initiated oxidation of linear low density polyethylene: a quantitative comparison of oxidation products. *J Polym Sci, Part A Poly Chem* 1992;**30**:493–500.
9. Randall JC. Carbon[13] NMR gamma-irradiated polyethylene. In: Guven O, editor. *Crosslinking and scission in Polymers. NATO ASI series C*, vol. 292. Kluwer: Academic Publishers; 1988. p. 57–76.
10. Perez E, Vanderhart DL. A $^{13}C$ Cp-Mas NMR study of irradiated polyethylene. *J Poly Sci: Poly Phys* 1988;**26**:1979–93.
11. Dole M, Milner DC, Williams TF. Irradiation of polyethylene. II kinetics of insaturation effects. *J Am Chem Soc* 1958;**80**:1580–8.
12. Arnaud R, Moisan JY, Lemaire J. Primary hydroperoxidation in low density polyethylene. *Macromolecules* 1984;**17**:332–6.
13. Brunella V, Bracco P, Carpentieri I, Paganini MC, Zanetti M, Costa L. Lifetime of alkyl macroradicals in irradiated UHMWPE. *Polym Degr Stab* 2007;**92**:1498–503.
14. Bhateja SK, Duerst RW, Aus EB, Andrews EH. Free radicals trapped in polyethylene crystals. *J Macromol Sci Phys* 1995;**B34**(3):263–72.
15. Odian G. *Principles of polymerization*. 4th ed Hoboken, NJ: Wiley & Sons; 2004.
16. Shinde A, Salovey R. Irradiation of polyethylene. *J Pol Sci: Polymer Physics Ed* 1985;**23**:1681–9.
17. Smedberg A, Hjertberg T, Gustafsson B. Crosslinking reactions in an unsaturated low density polyethylene. *Polymer* 1997;**38**:4127–38.
18. Woods RJ, Pikaev AK. *Applied Radiation Chemistry. Radiation processing*: Wiley; 1994 p 191.
19. Tilman P, Tilquin B, Claes P. Estimation du rapport Kd/Kc pour des radicuax alkyles en phase solide. *J de Chimie Physique* 1982; **79**:629–32.
20. Bolland JL. Kinetic of olefin oxidation. *Quart Rev Chem Soc* 1949;**3**:1–21.
21. Costa L, Carpentieri I, Bracco P. Post electron-beam irradiation oxidation of orthopaedic UHMWPE. *Polym Degr Stab* 2008;**93**:1695–703.
22. Alam TM, Celina M, Collier JP, Currier BH, Currier JH, Jackson SK, Kuethe DO, Timmins GS. $\gamma$-irradiation of ultrahigh-molecular-weight polythylene: Electron paramagnetic resonance and nuclear magnetic resonance spectroscopy and imaging studies of the mechanism of subsurface oxidation. *J Polym Sci: Part A Polym Chem* 2004;**42**:5929–59.
23. Pratt D, Mills J, Porter N. Role of Hyperconjugation in determing Carbon-Oxygen bond dissociation Enthalpies in alkylperoxyl radicals. *J Am, Chem Soc* 2003;**125**:5801–10.
24. Lazar M, Rychly J, Klimo V, Pelikan P, Valko L. *Free radicals in chemistry and biology*. Boca Raton, FL: CRC Press; 1989 Chapter 3: pg 56.
25. Lacoste J, Carlsson DJ. Gamma-, photo-, and thermally-initiated oxidation of linear low density polyethylene: A quantitative comparison of oxidation products. *J Polym Sci-Polym Chem Ed* 1992;**30**:493–500.
26. Lacoste J, Carlsson DJ, Falicki S, Wiles DM. Polyethylene hydroperoxide decomposition products. *Polym Degr Stab* 1981; **34**:309–23.
27. Carlsson DJ, Brousseau R, Zhang C, Wiles DM. Polyolefin oxidation: quantification of alcohol and hydroperoxide products by nitric oxide reactions. *Polym Degr Stab* 1987;**17**:303–23.
28. Costa L, Luda MP, Trossarelli L. Ultra high molecular weight polyethylene: II. Thermal- and photo-oxidation. *Polym Deg Stab* 1997; **58**:41–54.
29. Zaharadnickova A, Sedlar J, Dastych D. Peroxy acids in photo-oxidised polypropylene. *Polym Deg Stab* 1991;**32**:155–74.
30. Barabas K, Iring M, Kelen T, Tudos F. Study of the thermal oxidation of polyolefines: volatile products in the thermal oxidation of polyethylene. *Degradation and Stab of Polyolefins, IUPAC Microsymp on Macromol, 15th* 1977;**57**:65–71.
31. Costa L, Luda MP, Trossarelli L, Brach del Prever EM, Crova M, Gallinaro P. Oxidation in orthopaedic UHMWPE sterilized by gamma-radiation and ethylene oxide. *Biomaterials* 1998;**19**:659–68.
32. Costa L, Luda MP, Trossarelli L. Ultra high molecular weight polyethylene: 1. Mechano-oxidative degradation. *Polym Degr Stab* 1997;**55**:329–38.
33. Harrison N. Radiation sterilisation and food packaging. In: Clegg D, Collyer A, editors *Irradiation effects on polymers*. New York: Elsevier Applied Science; 1991. p. 319–44.
34. Halls NA. Gamma-irradiation processing. In: Clegg D, Collyer A, editors *Irradiation effects on polymers*. New York: Elsevier Applied Science; 1991. p. 253–96.
35. Billingham NC. Physical phenomena in the oxidation and stabilisation of polymers Chap 6. In: Pospíšil J, Klemchuk PP, editors. *Oxidation Inhibition in organic materials*, vol. II. Boca Raton, FL: CRC Press; 1990. p. 249–98.
36. Birkinshaw C, Buggy M, Daly S, O'Neill M. The effect of $\gamma$-radiation on the physical structure and mechanical properties of ultrahigh molecular weight polyethylene. *J Appl Polym Sci* 1989;**38**:1967–73.
37. Costa L, Luda MP, Trossarelli L, Brach del Prever EM, Crova M, Gallinaro P. In vivo UHMWPE biodegradation of retrieved prosthesis. *Biomaterials* 1998;**19**:1371–85.
38. Yeom B, Yu Y, McKellop HA, Salovey R. Profile of oxidation in irradiated polyethylene. *J Poly Sci Part A Poly Chem* 1998;**36**:329–39.
39. Shen FW, Yu YJ, McKellop H. Potential errors in FTIR measurements of oxidation in ultrahigh molecular weight polyethylene implants. *J Biomed Mater Res* 1999;**48**:203–10.
40. Costa L, Bracco P, Brach del Prever EM, Kurtz SM, Gallinaro P. A survey of oxidation and oxidation potential in contemporary packaging for polyethylene total joint replacement components. *J Biomed Mater Res Part B: Appl Biomater* 2006;**78B**:20–6.
41. Bracco P, Brach del Prever EM, Cannas M, Luda MP, Costa L. Oxidation behaviour in prosthetic UHMWPE components sterilised with high energy radiation in a low oxygen environment. *Polym Degr Stab* 2006;**91**:2030–6.
42. Bracco P, Brunella V, Luda MP, Brach del Prever EM, Zanetti M, Costa L. Oxidation behaviour in prosthetic UHMWPE components sterilised with high energy radiation in the presence of oxygen. *Polym Degr Stab* 2006;**91**:3057–64.
43. Lacoste J, Carlsson DJ. A critical comparison of methods for hydroperoxide measurement in oxidized polyolefins. *Polym Degr Stab* 1991;**32**:377–86.
44. Carlsson DJ, Brousseau R, Zhang C, Wiles DM. Polyolefin oxidation: quantification of alcohol and hydroperoxide products by nitric oxide reactions. *Polym Degr Stab* 1987;**17**:303–23.

45. Ikada Y, Nakamura K, Ogata S, Makino K, Tajima K, Endhoh N, Hayashi T, Fujita S, Fujisawa A, Masuda S, Oonishi H. Characterization of ultrahigh molecular weight polyethylene irradiated with gamma-rays and electron beams to high does. *J Polym Sci Part A: Polym Chem* 1999;**37**:159–68.
46. Costa L, Jacobson K, Bracco P, Brach del Prever EM. Oxidation of orthopaedic UHMWPE. *Biomaterials* 2002;**23**:1613–24.
47. Scott G. *Atmospheric Oxidation and antioxidant*. Amsterdam: Elsevier; 1993.
48. Pospisil J, Klemchuck PP. Boca Raton: CRC Press; 1990.
49. Zweifel H. *Stabilization of polymeric materials*. Berlin: Sprinter – Verlag; 1998.
50. Klemchuk PP. Protecting polymers against damage from gamma radiation. *Radiat Phys Chem* 1993;**41**:165–72.
51. Mallégol J, Carlsson DJ, Deschênes L. A comparison of phenolic antioxidant performance in HDPE at 32–80°C. *Polym Degrad Stab* 2001;**73**:259–67.
52. ASTM F648-07 Standard specification for Ultra-High-Molecular-Weight Polyethylene powder and fabricated form for surgical implants 5. UHMWPE Fabricated Form Requirements. 5.1 Compositional Requirements. 5.1.1 . No stabilizer or processing aids are to be add to the virgin polymer powder during manufacture of a fabricated form.
53. Nusbam HJ, Rose RM. The effects of radiation sterilisation on the properties of UHMWPE. *J Biomed Mat Res* 1979;**13**:557–76.
54. Miller A, Hansen J. Revision of the ISO and EN radiation sterilization standards. *Radiat Phys Chem* 2002;**63**:665–7.
55. Laermer S, Young S, Young T, Zambetti P. Vitamin E: a new primary antioxidant. *Engineering Plast* 1996;**9**:503–26.
56. Al-Malaika S, Goodwin C, Issenhuth S, Burdick D. The antioxidant role of α-tocopherol in polymers II. melt stabilising effect in polypropylene. *Polym Degrad Stab* 1999;**64**:145–56.
57. Al-Malaika S. Vitamin E: an effective biological antioxidant for polymer stabilisation. *Polym Polym Compos* 2000;**8**:537–42.
58. Breese KD, Lamèthe J, DeArmitt C. Improving synthetic hindered phenol antioxidants: learning from vitamin E. *Polym Degrad Stab* 2000;**70**:89–96.
59. Mallégol J, Carlsson DJ, Deschênes L. Post-γ-irradiation reactions in vitamin E stabilised and unstabilised HDPE. *Nucl Inst Meth Phys Res, Sect B* 2001;**85**:283–93.
60. Mallégol J, Carlsson DJ, Deschênes L. Antioxidant effectiveness of vitamin E in HDPE and tetradecane at 32°C. *Polym Degrad Stab* 2001;**73**:269–80.
61. Stocker A. Molecular mechanism of vitamin E transport. *Ann NY Acade Sci* 2004;**1031**:44–59.
62. Results not yet published.
63. Costa L, Bracco P, Brach del Prever EM, Luda MP, Trossarelli L. Analysis of products diffused in UHMWPE prosthesis components *in vivo*. *Biomaterials* 2001;**22**:307–15.
64. Wolf C, Lederer K, Müller U. Tests of biocompatibility of α-tocopherol with respect to the use as a stabilizer in ultrahigh molecular weight polyethylene for articulating surfaces in joint endoprostheses. *J Mater Sci Mater Med* 2002;**13**:701–5.
65. Wolf C, Lederer K, Pfragner R, Schauenstein K, Ingolic E, Siegl V. Biocompatibility of ultra high molecular weight polyethylene (UHMWPE) stabilised with alfa tocopherol used for joint endoprostheses assessed in vitro. *J Mater Sci: Mater Med* 2007;**18**:1247–52.
66. Wolf C, Lederer K, Bergmeister H, Losert U, Bock P. Animal experiments with ultra-high molecular weight polyethylene (UHMW-PE) stabilised with alpha-tocopherol used for articulating surfaces in joint endoprostheses. *J Mater Sci Mater Med* 2006;**17**:1341–7.
67. Wolf C, Maninger J, Lederer K, Fruhwirth-Smounig H, Gamse T, Marr R. Stabilisation of crosslinked ultra-high molecular weight polyethylene (UHMW-PE)-acetabular components with alpha-tocopherol. *J Mater Sci Mater Med* 2006;**17**:1323–31.
68. Wolf C, Macho C, Lederer K. Accelerated ageing experiments with crosslinked and conventional ultra-high molecular weight polyethylene (UHMW PE) stabilised with alpha-tocopherol for total joint arthroplasty. *J Mater Sci Mater Med* 2006;**17**:1333–44.
69. Bracco P, Brunella V, Zanetti M, Luda MP, Costa L. Stabilisation of UHMWPE with vitamin E. *Polym Degr and Stab* 2007;**92**:2155–62.
70. Costa L, Carpentieri I, Bracco P. Post electron-beam irradiation oxidation of orthopaedic UHMWPE stabilized with vitamin E. *Polym Degr Stab*. Submitted.
71. Dolezel B, Adamirova L. Method of hygienically safe stabilization of polyolefines against thermoxidative and photooxidative degradation. *Czechislovakian Socialist Republic Patent 221, 403*; 1982.
72. Berzen J, Luketic D. Stabilised polyethylene moulding materials, for implants containing polyethylene with molecular wt. $10^5$-$10^7$, antioxidant, and tocopherol, α- tocopherol or Vitamin E, as stabiliser. *European Patent 613, 923*; 1994.
73. Oral E, Wannomae KK, Rowell SL, et al. Diffusion of vitamin E in UHMWPE. *Biomaterials* 2007;**28**:5225–37.
74. Strohschein S, Rentel C, Lacker T, Bayer E, Albert K. Separation and identification of tocotrienol isomers by HPLC-MS and HPLC-NMR coupling. *Anal Chem* 1999;**71**:1780–5.
75. Lauridsen C, Leonard SW, Griffin DA, Liebler DC, McClure TD, Traber MG. Quantitative Analysis by Liquid Chromatography—Tandem Mass Spectrometry of Deuterium-Labeled and Unlabeled Vitamin E in Biological Samples. *Anal Biochem* 2001;**289**:89–95.
76. Oral E, Wannomae KK, Hawkins N, et al. α-tocopherol-doped irradiated UHMWPE for high fatigue resistance and low wear. *Biomaterials* 2004;**25**:5515–22.
77. Kurtz SM, Dumbleton J, Siskey RS, Wang A, Manley M. Trace concentrations of vitamin E protect radiation crosslinked UHMWPE from oxidative degradation. *J Biomed Mater Res A* 2008;**12**. Published online.
78. Al-Malaika S, Issenhuth S. Antioxidant role of α-tocopherol in polymers. III. Nature of transformation products during polyolefins extrusion. *Polym Degrad Stab* 1999;**65**(1):143–51.
79. Al-Malaika S, Issenhuth S, Burdick D. The antioxidant role of vitamin E in polymers. V. Separation of stereoisomers and characterisation of other oxidation products of dl-α-tocopherol formed in polyolefins during melt processing. *Polym Degrad Stab* 2001;**73**(3):491–503.
80. Turner, J.L, Kurtz, S.M., Bracco, P., Costa, L The effect of cholesteryl acetate absorption on the mechanical behaviour of crosslinked and conventional UHMWPE. *49th Annual Meeting of Orthopedic Research Society*, New Orleans, LA: 2003 February 2-5. Poster 1423.
81. Pickard JE, Fisher J, Ingham E, Egan J. Investigation into effects of proteins and lipids on frictional properties of articular cartilage. *Biomaterials* 1998;**19**:1807–12.
82. Brach del Prever EM, Costa L, Bistolfi A, Botto Micca F, Bracco P, Linari A, Massè A, Crova M, Gallinaro P. The biological reactivity of polyethylene wear debris is related with sterilisation methods of UHMWPE. *Chir Organi Mov* 2003;**LXXXVIII**:291–303.

Chapter 22

# *In Vivo* Oxidation of UHMWPE

Steven M. Kurtz, PhD



## 22.1 INTRODUCTION

Bodily fluids contain dissolved oxygen as well as reactive oxygen species that have the potential to degrade implanted polymers. Consequently, whether deployed in the hip, knee, shoulder, or spine, UHMWPE components are bathed in body fluids containing potential free radical initiators, as well as molecular oxygen fuel, for *in vivo* oxidation. Today, dissolved molecular oxygen, rather than reactive free radical oxygen species, are considered to be the primary driving force for perpetuating the oxidation reactions in unstabilized, gamma irradiated UHMWPE within the human body. Thus, dissolved oxygen is not considered to initiate *in vivo* oxidation of UHMWPE. Rather, the polymer must already contain unstablized macroradicals (e.g., from gamma sterilization in air or in an low oxygen environment) for the UHMWPE to oxidize *in vivo*.

The aqueous *in vivo* environment for UHMWPE components is in many ways more complex than the gaseous environment to which they are exposed during manufacture, sterilization, and shelf aging. For example, the joint fluid of the hip and knee contains filtered plasma augmented by biomolecules, such as lubricious proteins, that are produced by the synovial membrane [1, 2]. The synovial fluid of patients with arthritis is also known to contain to dissolved oxygen and other gases which originate from filtered plasma [3]. For patients with osteoarthritis, an average (±SD) partial pressure ($pO_2$) of 43 ± 15 mmHg has been measured by Lund-Olesen [4]; in patients with rheumatoid arthritis, the oxygen content may be as low as zero, but on average it has been measured to be 27 ± 19 mmHg [1, 4] (Figure 22.1).

The *in vivo* environment of artificial discs is less well understood than hip and knee replacements. In a healthy avascular disc, the concentrations of oxygen and other biomolecules in the annulus and nucleus are governed by diffusion from vertebral body capillaries that infiltrate the endplate-disc interface [5, 6]. Thus, diffusion-mediated oxygen transport is a primary mechanism of cell nutrition in the spinal disc, and disruption of nutrient supply is hypothesized to contribute to disc degeneration. However, spinal disc tissue changes with degeneration, with a tendency toward increased enervation and vascularization, which will in turn increase oxygen transport [7]. It is into this complex milieu that artificial discs are implanted. Although the biological environment around artificial discs remains the subject of ongoing research, the available data suggests that the biological environment for UHMWPE orthopedic and spine components will typically contain dissolved molecular oxygen [1, 3–6], albeit at a lower partial pressure than 160 mmHg found under standard conditions

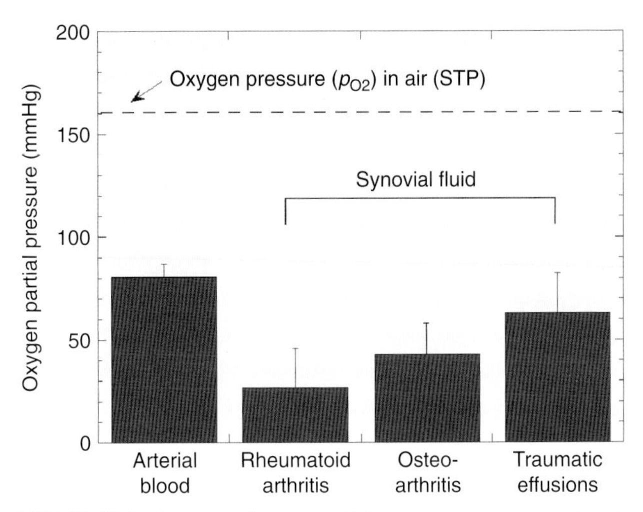


**FIGURE 22.1** Summary of oxygen partial pressure measurements in blood and joints of patients with rheumatoid and osteoarthritis [1,4], compared to air at standard temperature and pressure (STP).

in air (Figure 22.1). Despite its low partial pressure, the molecular oxygen dissolved in body fluids is sufficient to substantially degrade exposed regions of gamma-sterilized (25–40 kGy) UHMWPE after the first decade of service.

This chapter provides an overview of the natural history of *in vivo* degradation in historical and contemporary polyethylene components for hip, knee, and spinal implants. This chapter also summarizes the test methods that are commonly used to evaluate implants for *in vivo* oxidation.

## 22.2 PERSPECTIVE OF *IN VIVO* OXIDATION IN THE 1980s TO THE PRESENT

Although oxidative degradation of gamma sterilized polyethylene during shelf aging in air was a major research theme of the 1990s (Chapter 3), it has not been until the past decade that the orthopedic community has recognized the potential significance of degradation *in vivo*. Researchers began to investigate *in vivo* degradation of polyethylene in the 1980s but were not able to relate the changes they observed with clinical performance [8–10]. Professor Edward Grood and colleagues from the University of Cincinnati were among the first to study crystallinity changes in both *in vitro* aged specimens and retrieved UHMWPE components that were implanted for up to 7 years [8]. Grood observed significantly higher crystallinity (67.2% ± 1.7%) in the loaded regions of implants when compared with unloaded regions (63.3% ± 1.7%). Grood et al. concluded that, "UHMW polyethylene is not a static material but is continually undergoing dynamic changes *in vivo*."

Another useful concept introduced by Grood's research was the notion that the degradation of shelf aged and *in vivo* retrieved components could be compared based on their "total aging time" [8]. In Grood's study, the shelf life of his retrievals was unknown, but he calculated the total aging time of the retrievals by adding the implantation time to the shelf aging time after removal from the body. Although the actual duration of *ex vivo* aging was not explicitly reported, by examination of maximum total aging time for the retrievals (~10 years) and knowing the maximum implantation time (7 years) we can infer that the components were aged up to 3 years prior to evaluation. Thus it was recognized as early as the 1980s that retrieved components continued to degrade after explantation and that this "postremoval shelf aging" (hereafter, *ex vivo* aging) time should be taken into account in the interpretation of data obtained from explants. These early results suggested to Grood and coworkers that *in vivo* aging and *ex vivo* aging of UHMWPE were "of the same order" [8].

Peter Eyerer and colleagues in Germany further studied *in vivo* changes in the density, molecular weight, and oxidation of retrieved cemented acetabular cups, following as much as 11–13 years *in vivo* [9, 10]. Eyerer's retrievals were all fabricated from RCH-1000 (the equivalent of compression molded GUR 412). Although the shelf aging time prior to implantation was unknown, the retrievals were characterized within 1–3 months of revision surgery, thereby averting artifacts due to substantial variations in *ex vivo* aging. Despite considerable variability, Eyerer observed increasing density and decreasing molecular weight with implantation time (Figure 22.2).

Interestingly, Eyerer also observed region-specific differences in properties of the retrieved cups, with the inferior, unloaded bearing surface of the liners showing the greatest changes in density *in vivo* (Figure 22.2). The backside surface of the liners showed lower changes in density (Figure 22.2). Eyerer reported subsurface changes in the properties of UHMWPE, which he attributed to postirradiation aging. Eyerer avoided potential confounding due to *ex vivo* aging by evaluating retrieved components within 1–3 months of removal from the body. In this way, the variability in the UHMWPE could be attributed to variation in radiation conditions, shelf aging, and implantation time.

An important legacy of Peter Eyerer's work was his mentorship of Clare Rimnac, who was at the time a researcher from the Hospital for Special Surgery (HSS) in New York. While visiting Eyerer's laboratory in Stuttgart, Germany, Rimnac developed expertise in density profiling of UHMWPE components and established this capability in her research facilities upon return to the United States. Several studies in the 1990s related to *in vivo* degradation of UHMWPE were based upon density measurement techniques [11, 12]. These studies helped lay a methodological foundation for studying *in vivo* degradation of UHMWPE. Recognizing that substantial variability in retrieved components was due to uncertainty in its initial conditions prior to implantation, Rimnac established the first traceable retrieval

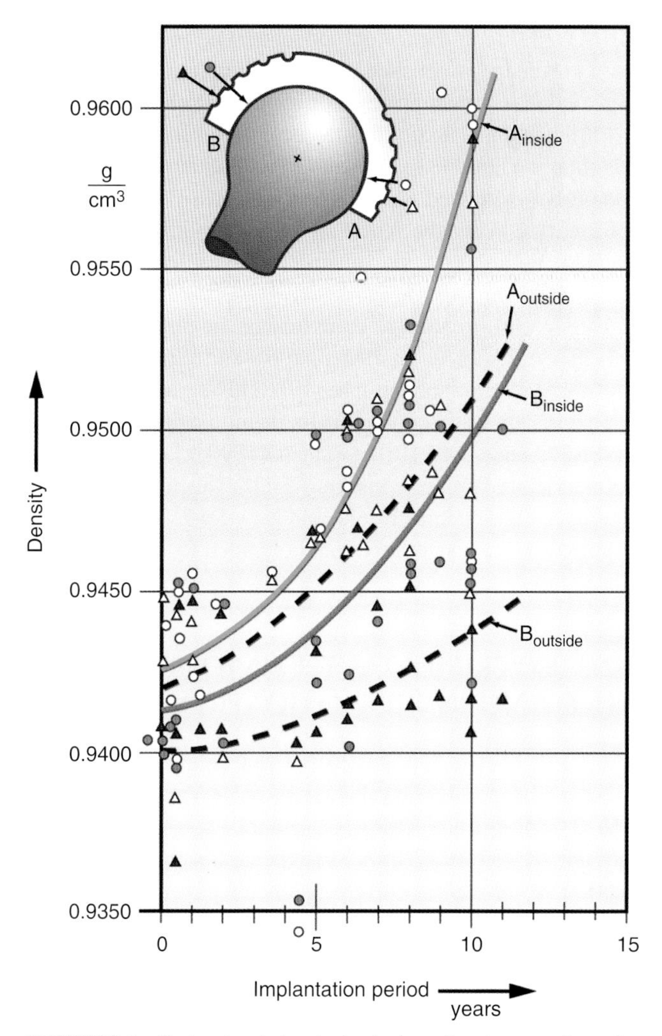


FIGURE 22.2 Regional variations in density in retrieved cemented acetabular components: (A) inferior, unloaded regions; (B) superior, loaded regions. An increasing trend of density with implantation time was observed. Adapted from Eyerer and Ke [9], courtesy of Gil Matityahu, Exponent, Inc.

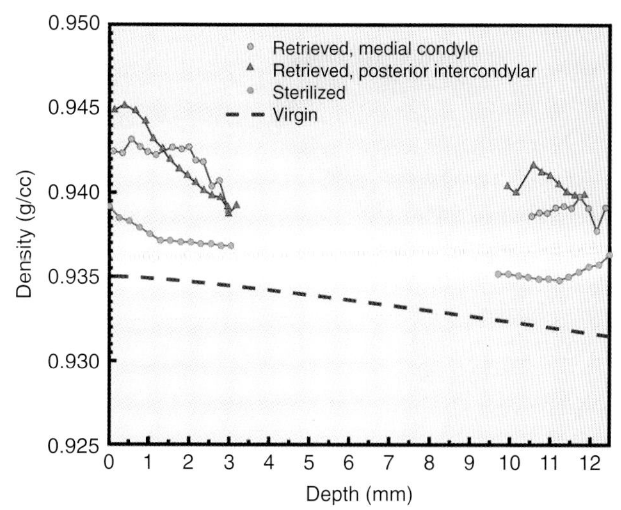


FIGURE 22.3 Subsurface variation in density from as-received ("virgin"), as-sterilized, and as-retrieved conditions for machined GUR 412 Insall Burstein II total knee replacements. Adapted from Bostrom et al. [11], courtesy of Gil Matityahu, Exponent, Inc.

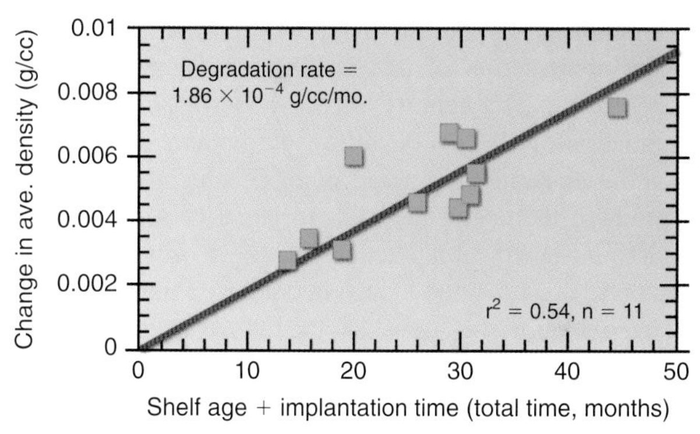


FIGURE 22.4 Change in average density versus total aging time (defined as the sum of shelf life and implantation time) for machined GUR 415 Insall Burstein II total knee replacements. Adapted from Kurtz et al. [12], courtesy of Gil Matityahu, Exponent, Inc.

program at HSS, whereby the density profile in as-received and sterilized components was measured between 1989 and 1992 [11]. Subsequently, retrieved implants at HSS could, upon revision surgery, be traced back to the original lot of material to determine its original density. All of the explants could be traced to determine if they were fabricated from GUR 412 or GUR 415 resin, and they were fabricated with the same Insall Burstein II design. Recognizing the confounding introduced by *ex vivo* aging at room temperature air, researchers at HSS stored traceable implants in a cryogenic freezer at −20°C to minimize degradation after removal from the body [11]. A representative profile from a traceable retrieval is illustrated in Figure 22.3, showing the variation in the density from as-received ("virgin"), as-sterilized, to as-retrieved condition. The retrieval also shows the beginning of a subsurface peak in density approximately 1–2 mm below the surface (Figure 22.3). Using this technique, the independent effects of shelf aging and implantation were investigated.

The HSS retrieval studies provided further support that the total aging time, comprised of the shelf aging and implantation (*in situ*) time, was correlated with progressive, degradative changes in UHMWPE [11, 12]. However, it must be noted that these traceable retrievals studies were exposed to relatively short durations *in vivo* [11, 12]. The longest implantation time of the traceable retrievals was only 31 months. Nevertheless, for this limited collection of retrievals, a significant association between total aging time and average changes in density could be demonstrated (Figure 22.4) [12].

Research on *in vivo* degradation of UHMWPE in the 1980s and 1990s started the gradual accumulation of scientific evidence that later helped usher out the era of gamma sterilization in air for most orthopedic manufacturers in the mid- to late 1990s. However, it must be emphasized that, at the time, none of these early research results could be easily translated to the clinical environment. In 1984, Eyerer and Ke [9] themselves posed this difficult question: “How do property changes influence the medical application of hip cups? We cannot yet give a quantitative answer.” Density changes in UHMWPE could be correlated with changes in mechanical behavior (e.g., higher elastic modulus), which in turn implied higher stresses in implants and greater surface damage [11]; however, these associations were somewhat oblique, circumstantial, and based on measurements from short-term retrievals. In the mid 1990s, differences in the infrared spectra between shelf aged and retrieved components were also observed by Professor Luigi Costa and coworkers from the University of Torino, and in retrospect these observed differences can be attributed to absorbed lipids and biological molecules diffused into UHMWPE [13]. The *in vivo* changes that Grood, Eyerer, Rimnac, Costa, and colleagues measured were viewed as degradative and undesirable, yet researchers at the time could not offer improved alternatives over contemporary UHMWPE, nor could a solution to *in vivo* oxidation be recommended.

Circumstances during the 1990s combined to delay greater appreciation of *in vivo* degradation until the first decade of the 21st century. As we reviewed in Chapter 3, degradation of UHMWPE following gamma irradiation in air and long-term shelf aging in air became a major focus of research inquiry in the mid- to late 1990s, resulting in changes in sterilization practices for major orthopedic manufacturers in the United States. In contrast with shelf aging, *in vivo* oxidation was viewed at the time as a comparatively minor and clinically insignificant issue [14].

The prevailing view in the 1990s that *in vivo* oxidation was less clinically relevant than shelf aging is perhaps best illustrated by a review of the oxidation problem authored by Premnath and coworkers from the Massachusetts Institute of Technology and Harvard Medical School [15]. Although oxygen gas diffusion was identified by Premnath et al. as a potential mechanism for *in vivo* oxidation, this topic received only one paragraph of treatment in a 10-page review on “the oxidation problem” of UHMWPE [15]. Premnath et al. conceded that “it is possible for oxidation reactions to continue within the body, and produce the characteristic oxidation profiles in retrieved samples, although at a much slower rate.” Clearly, many researchers of the mid-1990s were focused on changing the practice of gamma sterilization in air. With the explosion of interest in crosslinking during the late 1990s, radiation and thermal treatment dominated the research topics for UHMWPE for a decade. Because it was judged at the time to have minimal clinical impact, during the 1990s *in vivo* oxidation of polyethylene became relegated to the list of topics of limited practical interest, suitable for academic inquiry.

The author, at Drexel University, and Clare Rimnac, at Case Western Reserve University, began a multi-institutional retrieval program in 2000 to understand *in vivo* oxidation of UHMWPE hip and knee components. The investigations at each university established close collaborations with a local, high volume academic revision center (Rothman Institute in Philadelphia and Case Western Reserve University Hospital in Cleveland) and were able to obtain not only consecutively explanted components but also crucial clinical information about the retrievals, including implant stickers with lot numbers. A key feature of this retrieval program has been obtaining traceable sterilization dates, resin, and processing information to complement the physical, chemical, and mechanical characterization and clinical data for the retrieved components [16–19]. This research program has been supported continuously by the National Institutes of Health since 2001 (R01 Grant AR47904), and today, this collaboration captures consecutive retrievals from five institutions in Ohio, Pennsylvania, and New Jersey. The retrieval examples in this chapter were all obtained from the Drexel-Case retrieval program.

## 22.3 EXPERIMENTAL TECHNIQUES FOR STUDYING *IN VIVO* OXIDATION

Many different techniques have been employed to assess *in vivo* changes to UHMWPE over the past 2 decades. Specialized infrastructural procedures for handling and storage of retrievals have also been developed. In this section we provide a current perspective on methodological issues associated with evaluating *in vivo* oxidation in retrieved UHMWPE components. This section documents not only the contemporary practices at the Drexel and Case retrieval programs, but also alternative approaches at other regional retrieval centers currently investigating *in vivo* oxidation of UHMWPE.

### 22.3.1 Institutional Procedures and Study Design

It is now well appreciated by researchers in the field that retrieved UHMWPE implants containing unstabilized macroradicals continue to oxidize during *ex vivo* storage in air. To address this confounding factor, investigators either (1) analyze the component within a short time frame from removal (preferably less than 1 year) [9]; or (2) store the explanted component in a cryogenic freezer to retard oxidation until it is analyzed [16–18]; and/or (3) store the component in room temperature nitrogen, depriving the material of additional oxygen during storage [20]. Of these methods, we employ a combination of the first two strategies at our institution. UHMWPE retrievals are stored cryogenically at −80°C. The size of our retrieval collection (over 1400 UHMWPE components at present) makes nitrogen storage impractical.

For components that were gamma irradiated in air or in an oxygen-permeable, first-generation barrier package, the

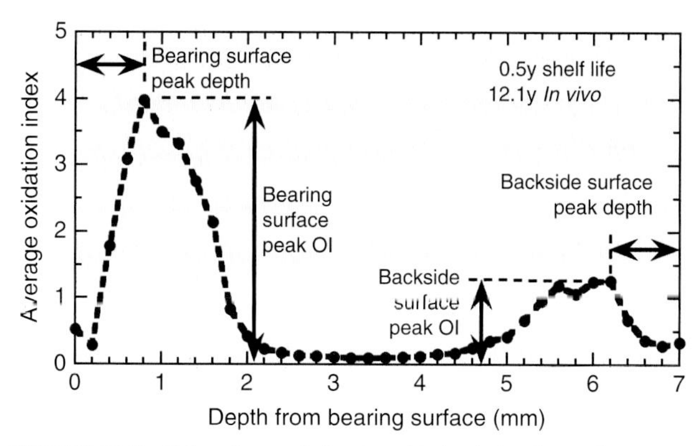


**FIGURE 22.5** Subsurface variation in oxidation index from a long-term implanted GUR 415 acetabular insert, illustrating the difference in magnitude of the subsurface peaks at the bearing surface as compared to the backside surface. Adapted from Kurtz et al. [17].

oxidation that occurs during shelf aging will also contribute variability to the degradation observed after retrieval and potentially obfuscate the interpretation of oxidation measurements. To reduce this initial variability, it is advisable to trace the shelf life of retrieved components. Tracing is typically accomplished by collaboration with the manufacturer. The catalog and serial numbers (sometimes referred to as "lot numbers") of retrieved liners can be obtained from the patients' operative notes or from the inscriptions on the rim of the liners. These serial numbers permit a manufacturer to trace the production history of a specific polyethylene component. With gamma air-sterilized components, the clearest picture of changes that occur *in vivo* will be obtained from retrievals with a shelf life of less than 1 year [16–18]. An example of an oxidation index profile through the thickness of a modular, gamma-air sterilized UHMWPE acetabular liner is shown in Figure 22.5. Because of its short shelf life (0.5 years) and much longer implantation time (12 years), we can unambiguously conclude that the severe oxidation illustrated in Figure 22.5 occurred *in vivo*.

Alternatively, provided the shelf life is known, the degradation prior to implantation can be accounted for in a subsequent statistical data analysis, using a similar strategy to that employed by Bostrom [11]. When investigating contemporary gamma sterilized retrievals with well-validated barrier packaging, the traceability of the retrievals is useful but less important for the purposes of studying *in vivo* oxidation. However, it must be emphasized that barrier packaging is manufacturer specific (Chapter 3); if the specific packaging system for a retrieval is unknown or not yet validated, it is preferable to trace the shelf life of the particular component.

In addition to the sterilization method and date, it is important to trace the UHMWPE resin and consolidation method in a research study of *in vivo* oxidation. Previous studies suggest that the magnitude of oxidation during shelf aging differs between historical GUR and 1900 resins, as well as between historical calcium-stearate-containing

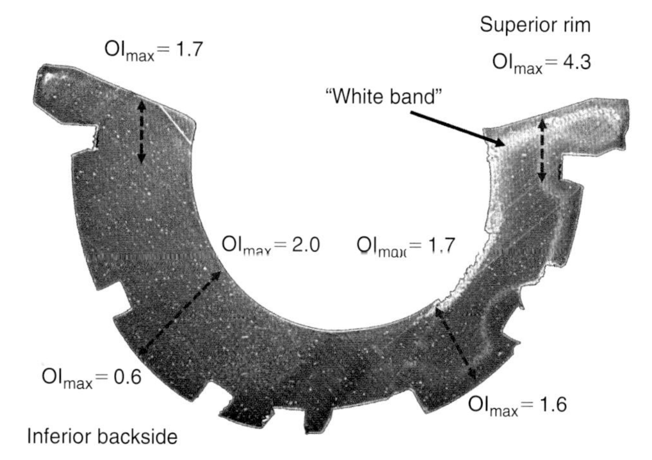


**FIGURE 22.6** Implanted in 1978, this cemented, gamma-air sterilized Charnley-design acetabular component was revised after 29 years *in vivo* due to acetabular loosening and osteolysis. The shelf life and processing details of this explant are unknown. FTIR scans through the thickness were performed along the trajectories illustrated by the dashed blue lines, and the maximum oxidation index ($OI_{max}$) within 3 mm of the surface are shown here for the superior and inferior rim, spherical bearing surface, and backside surfaces.

resins and contemporary resins that contain no added calcium stearate [16–18, 21–24]. It is equally important to trace the manufacturing methodology used to consolidate the resins [21, 23, 24]. All of these factors have been suggested based on findings from shelf aging [21–24]; their influence on *in vivo* oxidation is hypothesized. Until such time as resin and consolidation factors can be decisively ruled out, they should be viewed as potentially introducing variability into the response of a material to *in vivo* oxidation.

## 22.3.2 Experimental Techniques

Fourier transform infrared (FTIR) spectroscopy has superceded density analysis as the preferred method for quantifying *in vivo* oxidation in the 21st century. FTIR offers many advantages over density analysis, including assessment of carbonyl species (the terminal reaction products of oxidation), crystallinity, and double bond (trans-vinylene) content. With density and DSC analysis of crystallinity, it is difficult to discriminate whether physical changes are due to degradation or simply crystalline rearrangement without loss of molecular weight (e.g., annealing), and these crystallinity changes can in any event also be quantified using FTIR [19].

We typically assess the morphology of retrieved UHMWPE components using dark-field microscopy of 200μm-thin transverse sections. These sections are inspected for the presence of and location of white bands, which have been associated with subsurface embrittlement of the polyethylene. A dark-field micrograph of a long-term retrieved cemented acetabular component is illustrated in Figure 22.6.

Because of its age, we can infer that this component was likely fabricated from CHIRULEN (compression molded, GUR 412) material and gamma irradiated in air. The white dots in the section are fusion defects (see Chapter 2), and the superior rim of the component exhibits a pronounced white band consistent with severe oxidation. One can appreciate from Figure 22.6 that both the white band and the distribution of oxidation are nonuniform across the sample. The highest oxidation is found at the rim of the cup, whereas the lowest is found near the back surface. The regional variations illustrated in Figures 22.5 and 22.6 are fairly typical for *in vivo* oxidized acetabular components following gamma irradiation in air and likely reflect the variation in access of the UHMWPE surface to oxygen in the body [17]. These regional variations in oxidation index are consistent with the earlier findings by Eyerer and Ke [9], although it is not entirely clear from the early studies precisely where the density measurements were performed relative to the rim. Because of the regional evolution of oxidation *in vivo*, it is overly simplistic and misleading to conceptualize *in vivo* oxidation as a one- or even two-dimensional problem in time. Instead, the variations of *in vivo* oxidation occur three-dimensionally in space within the component, depending upon its geometry.

Additional experimental details about performing of measurements on UHMWPE using FTIR are covered in Chapter 24; however, it is important to note that within 1 mm of the bearing surface, the spectra collected from retrievals in the vicinity of carbonyl region (1710–1740 $cm^{-1}$) are complicated by the presence of biological contamination, mainly in the form of lipids and proteins [25]. Costa et al. identified the substances extracted from retrieved polyethylene components [26]. Using GC/MS, they found squalene, cholesterol, and cholesteryl derivative to be the main diffused products. These fatty acids can interfere with the standardized assessment of oxidation index using ASTM F2102 [27]; however, these biological contaminants are readily extracted using heptane or hexane [25, 28]. Throughout this chapter, calculations of oxidation index (e.g., Figure 22.5) are reported using the ASTM method following extraction of 200 μm-thin transverse sections in boiling heptane for 6 hours, unless stated otherwise.

Researchers from Dartmouth University currently employ a different method of calculating oxidation index than the ASTM method [22, 29]. Because of the high throughput of their retrieval lab, they do not perform hexane extraction on explants to remove biological contamination. Instead, they define a ketone index using the peak height at 1718 $cm^{-1}$ (associated with ketone band) divided by the peak height of the reference at 1368 $cm^{-1}$ (the same reference band as the ASTM method). The rationale for using this peak is that: (1) the peak height ratio was found to be repeatable and reproducible in a previous interlaboratory study [30]; and (2) lipids have a greater impact on the peak areas under the carbonyl band, especially around 1738 $cm^{-1}$, than at the ketone peak height at 1718 $cm^{-1}$ "minimizes the impact of absorbed species" [22]. The Dartmouth technique, referred to in this chapter as the Dartmouth Oxidation Index (DOI), represents an approximation of the ketone species in the UHMWPE near the surface. Although the impact of absorbed species are minimized, they are not, in fact, eliminated, and the uncertainty associated with this approximation is not clear at the present time. In the electronic supplement to a recent paper, Currier et al. [29] published a formula for converting the DOI values in their studies to the ASTM OI values; both indices are based on the same reference peak ($r^2 = 0.98$):

$$\text{ASTM OI} = 1.91 \times \text{DOI}$$

When referring to data published by Dartmouth University, we have performed this conversion so that all values in this chapter appear in the ASTM scale.

Recently, we have also been interested in examining the hydroperoxide content in retrieved components. Hydroperoxides are precursors to carbonyl formation and thus represent the oxidation potential of polyethylene [31]. The study of the hydroperoxide levels requires the exposure of the sections to nitric oxide (NO) for at least 16 hours to convert hydroperoxides to nitrates, which are easily detected by FTIR spectroscopy [32]. The same regions of interest used for the oxidation analysis are used for the assessment of hydroperoxides. In this chapter, a hydroperoxide index was calculated through the thickness of microtomed sections using the area under the peak between 1600 and 1670 $cm^{-1}$ divided by the area of the reference peak between 1330 and 1396 $cm^{-1}$ (this is the same reference peak used to calculate the oxidation index).

Miniature specimen mechanical testing (e.g., small punch testing) as well as microindentation are also useful techniques for assessing the mechanical behavior of retrieved components. Further details about these mechanical test methodologies can be found in Chapters 32 and 33, respectively. Provided the UHMWPE specimens are relatively flat, as in certain designs of total knee replacements, uniaxial tensile specimens can also be prepared for retrieval analysis [22]. However, the geometry of uniaxial tensile specimens is generally not compatible with the geometry of acetabular inserts, tibial inserts with curved bearing surfaces, or artificial disc components, greatly limiting the applicability of this uniaxial technique across most type of UHMWPE implants. Of these miniature specimen test methods, only the small punch test is available as a standard from the American Society for Testing and Materials.

In contrast with FTIR, biological molecules such as lipids have negligible impact on the results from small punch testing (Figure 22.7). Based on previous extraction studies [26], cholesteryl acetate ($C_{29}H_{48}O_2$, M.M. = 428,8 g/mol) was chosen as a model compound for evaluation the effect of lipids on small punch test results [33]. Miniature disc specimens,

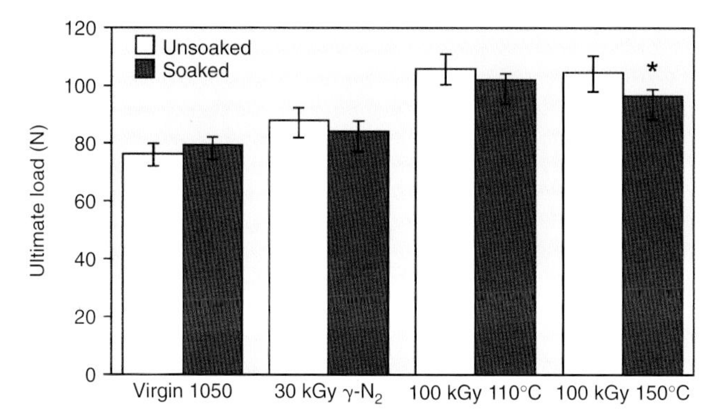


**FIGURE 22.7** Ultimate load comparison of various types of UHMWPE soaked in cholesteryl acetate, a model compound for fatty acids. An * denotes a significant change. These data indicate that lipid extraction prior to small punch testing of retrievals will provide minimal changes in results. Adapted from Turner et al. [33].

6.36 mm wide and 0.5 mm thick, were machined from rods of the following types of conventional and highly crosslinked UHMWPE: (1) conventional GUR 1050, (2) conventional PE $\gamma$-irradiated in nitrogen at 30 kGy (30-$N_2$), (3) crosslinked PE $\gamma$-irradiated in air at 100 kGy and annealed at 110°C (100 kGy-110°C), and (4) crosslinked PE $\gamma$-irradiated in air at 100 kGy and remelted at 150°C (100 kGy-150°C). The small punch specimens were immersed in cholesteryl acetate and maintained in an oil bath at 70°C for 42 days to achieve complete saturation. Soaking significantly ($p = 0.03$) reduced the ultimate load of the remelted, highly crosslinked material (100 kGy-150°C) by 7.9%. All other samples showed no significant change (Figure 22.7). As a result of these studies, it is not currently considered necessary to perform hexane or heptane extraction prior to measuring the mechanical behavior of retrieved components.

### 22.3.3 Correlation of *In Vivo* Oxidation and Mechanical Behavior in Retrievals

The extent of oxidation in UHMWPE is correlated to mechanical behavior; however, the correlation is neither continuous nor linear across the entire range of oxidation values [22]. In the early stages of oxidation, initial reductions in molecular weight of the radiation crosslinked network may at first result in *greater* crystallinity and improved yield strength. However, at elevated levels of oxidation, the ultimate strength and ductility of UHMWPE become compromised [18, 22].

How, then, are we to interpret values in the literature for *in vivo* oxidation in retrievals? Current studies of mechanical behavior in shelf aged components and retrievals, whether based on miniature uniaxial tensile specimens or the small punch test, exhibit considerable variability and scatter, even when oxidation values are measured directly from the miniature specimens themselves. Consequently, published data for the relationship between *in vivo* oxidation and mechanical behavior are necessarily statistical in nature. That is, one may discuss trends for a population of retrievals; however, the available data is not sufficiently precise to make reliable predictions of mechanical properties for a specific region of an individual explant based solely on its oxidation index, except when the measured level is "low" (defined here as OI $< 1$) or when it is beyond a "critical" level (defined here as OI $> 3$). For regions of retrievals falling within the "moderate-to-severe" range of oxidation ($1 \leq$ OI $\leq 3$), the loss of mechanical properties display the greatest variability and lowest correlations with oxidation index.

- Low oxidation (OI $< 1$): For ASTM oxidation index values of less than 1, it is very difficult to demonstrate a negative correlation with mechanical properties from retrievals, regardless of whether they are measured using miniature tension tests or small punch tests [16, 22, 34]. For this reason, oxidation indices of less than 1 may be considered relatively low, and despite evidence of early oxidative changes, it is unlikely that these changes have yet had a substantial negative impact on mechanical behavior.
- Critical oxidation (OI $> 3$): When the measured ASTM OI $> 3$, published data indicate that the ability of the material to withstand long-term mechanical loading *in vivo* has been compromised [16, 22]. In the words of Currier et al. [22], "all mechanical integrity of the polyethylene is lost."

## 22.4 CLINICAL SIGNIFICANCE OF *IN VIVO* OXIDATION

It is important to distinguish between mechanical behavior and clinical performance. Although one may conceptualize different regimes of UHMWPE mechanical behavior based on their oxidation index, the association between oxidation, mechanical behavior, and clinical performance is not straightforward. The clinical significance of *in vivo* oxidation in the hip and knee is further explored in the following sections. Additional details about *in vivo* oxidation in artificial disc replacements can be found in Chapter 12.

### 22.4.1 *In Vivo* Oxidation and Total Hip Arthroplasty

The clinical significance of *in vivo* oxidation continues to be a controversial topic for total hip replacement. *In vivo* oxidation, as a solitary factor, does not govern clinical failure modes of wear, osteolysis, and late implant loosening observed in historical, cemented total hip replacements (Chapter 5). The mechanical behavior of the UHMWPE liner in THA is but one variable in a complex multifactorial problem (Chapter 5). On the other hand, there is clear evidence of critical levels of oxidation in retrieved

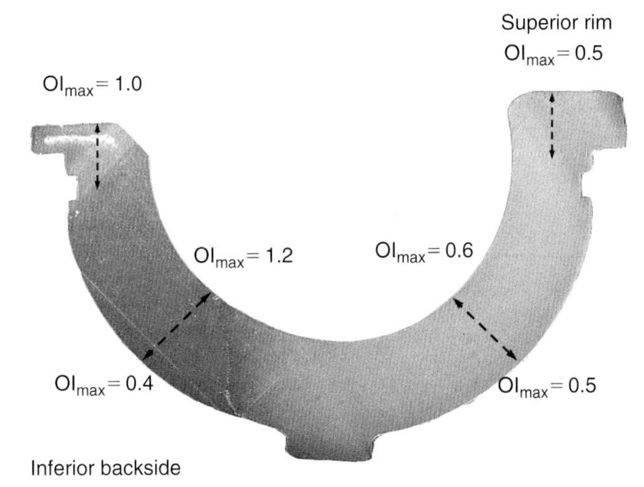


**FIGURE 22.8** Dark-field micrograph of a section through a modular, gamma-inert sterilized acetabular insert fabricated from compression molded GUR 1050. Implanted in 1997, this modular, gamma-nitrogen sterilized acetabular component was revised after 9.3 years *in vivo* due to femoral loosening and UHMWPE wear. FTIR scans were performed through the thickness of the component, as in Figure 22.7. Note the moderate oxidation and subsurface white bands at the inferior rim.

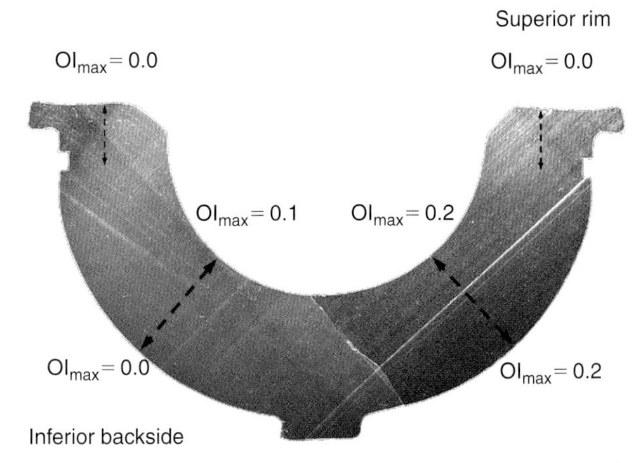


**FIGURE 22.9** Dark-field micrograph of a section through a modular, gas-sterilized acetabular insert fabricated from compression molded GUR 1050 (Longevity, Zimmer, Inc.). Implanted in 2000, this modular, highly crosslinked and remelted acetabular component was revised after 6.7 years *in vivo* due to recurring dislocation. FTIR scans were performed through the thickness of the component, as in Figure 22.7. Note the uniformly low oxidation throughout this section.

cemented liners, especially at the rim (e.g., Figure 22.6), and previous studies with the cemented Charnley prosthesis have shown that rim oxidation increases with implantation time [35]. The key concept that harmonizes these disparate observations is the recent discovery that *in vivo* oxidation is a regional phenomenon, governed by local access of oxygen [18].

There is now strong evidence to support the hypothesis that degradation of radiation-sterilized polyethylene occurs in the body for not only historical gamma air sterilized liners, but also for conventional gamma inert sterilized and annealed highly crosslinked polyethylene (Crossfire) liners as well [18]. Although Crossfire is annealed, following gamma sterilization in nitrogen this material contains between two and three times as many unstabilized macroradicals as conventional gamma inert sterilized UHMWPE [36, 37]. Thus, Crossfire retrievals provide an important group for testing hypotheses underlying both the mechanisms and significance of *in vivo* oxidation for total hip arthroplasty.

Previous studies suggest that a similar mechanism is likely responsible for *in vivo* oxidation in both gamma-sterilized and Crossfire liners [18, 29, 34]. Examples of a conventional gamma inert-sterilized acetabular liner is shown in Figure 22.8; a similar dark-field micrograph for a retrieved Crossfire liner appears in Chapter 14 (Figure 14.3A). A white band and elevated oxidation appears at the rim in both conventional gamma inert and Crossfire retrieved liners. In contrast, gas sterilized acetabular liners fabricated from highly crosslinked and remelted UHMWPE have thus far shown negligible evidence of *in vivo* oxidation either at the bearing or at the rim (Figure 22.9).

In contrast with gamma sterilized liners, studies of shelf aged, gas sterilized liners also show no evidence of oxidation or hydroperoxide levels during exposure to oxygen prior to implantation [38]. Remelted highly crosslinked acetabular liners have also exhibited negligible evidence of *in vivo* oxidation thus far (Chapter 13) [18] (Figure 22.9). Taken together, these general observations support the hypothesis that the initial concentration of unstabilized macroradicals in UHMWPE plays the dominant role in its susceptibility to shelf aging, in the presence of air, as well as to *in vivo* oxidation. The very low levels of oxidation in gas-sterilized and remelted UHMWPE liners also demonstrates that any *in vivo* degradation mechanisms due to reactive oxygen species (e.g., superoxide radicals) are negligible, at least during the first decade of implantation.

*In vivo* oxidation studies of conventional and gamma inert-sterilized retrievals have revealed that the most severe manifestations of *in vivo* oxidation typically occur in regions of the liner experiencing minimal wear, such as the rim of the component, where the UHMWPE has access to oxygen-containing body fluids. The regional distributions of oxidation and oxidation potential in modular, gamma sterilized acetabular liners can be found in Figures 22.10 and 22.11, respectively [42]. When interpreting the magnitude of oxidation among these groups of implants, it is important to appreciate that the implantation times varied, as shown in Table 22.1 [42].

Currier et al. have proposed that *in vivo* oxidation, like shelf aging, follows an exponential relationship with total exposure time for total hip replacements [29]. Even if we restrict our attention to a consistent location of a hip implant (e.g., the rim), fitting of an exponential relationship to *in vivo* oxidation data from both gamma air and

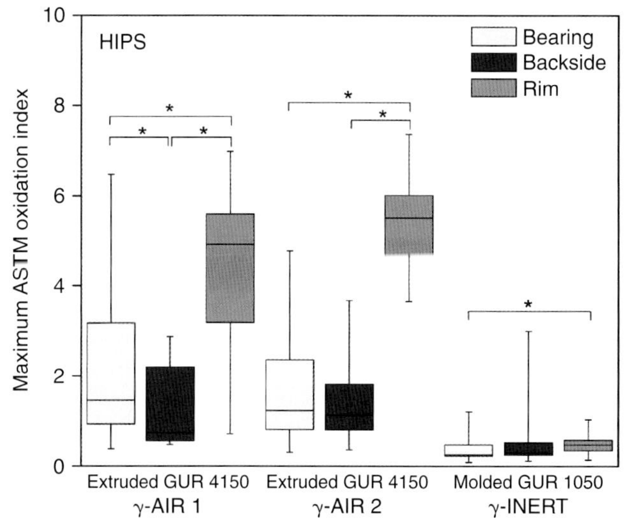


FIGURE 22.10 Average maximum ASTM oxidation indexes for the three categories of historical and conventional acetabular liners, summarized in Table 22.1 [42]. Substantial regional variation was observed in the maximum oxidation of retrieved liners (* significant differences in paired t-tests).

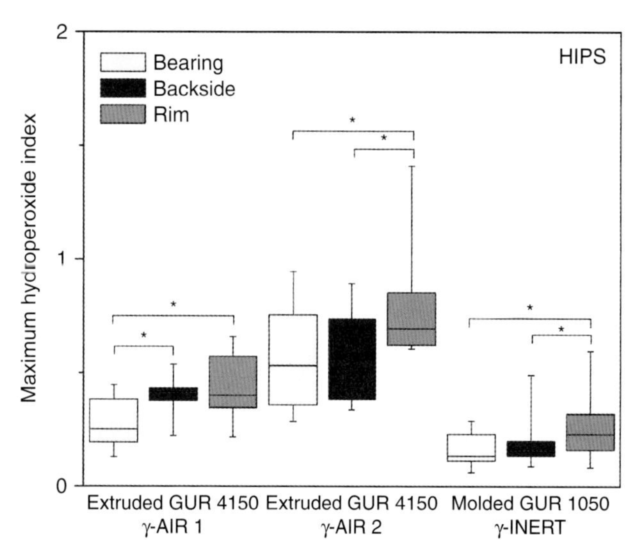


FIGURE 22.11 Average maximum hydroperoxide indexes (oxidation potential) for the three categories of historical and conventional acetabular liners, summarized in Table 22.1 [42]. Substantial regional variation was observed in the maximum oxidation potential of retrieved liners (* significant differences in paired t-tests).

**TABLE 22.1 Shelf Lives and Implantation Time, Average ± Standard Deviation, for Historical and Conventional Hip Retrievals [42]**

| Group | Processing conditions | N | Shelf life (y) | *In vivo* time (y) |
|---|---|---|---|---|
| H-air 1 | γ-air, extruded, GUR 4150 | 10 | 1.1 ± 1.0 | 14.7 ± 4.4 |
| H-air 2 | γ-air, extruded, GUR 4150 | 21 | 0.4 ± 0.3 | 11.2 ± 2.6 |
| H-inert | γ-inert, molded, GUR 1050 | 17 | 0.8 ± 0.7 | 4.0 ± 2.5 |

gamma inert components reveals considerable variability that is not explained by the model ($r^2 = 0.22 - 0.53$, Figure 22.12). The weakest association ($r^2 = 0.22$, Figure 22.12C) is observed at the backside surface. It is clear, therefore, that factors other than the duration of exposure influence the evolution to oxidation *in vivo*. These observations, coupled with the variation in oxygen content in body fluids by disease state (Figure 22.1), suggest that *in vivo* oxidation is patient specific, and population-based retrieval studies cannot be used to reliably predict *in vivo* aging that will be experienced by a particular individual.

Overall, the data from gamma-sterilized liners (Figures 22.10–22.12), combined with evidence from Crossfire retrievals (Chapter 14), all point to a similar scenario in which the femoral head reduces the *in vivo* oxidation of polyethylene at the bearing surface [18]. Consequently, provided rim impingement does not occur and the polyethylene locking mechanisms remain relatively isolated from oxidizing fluid, *in vivo* oxidation does not appear to be clinically important in the first 10 years of implantation for conventional gamma sterilized polyethylene. Recently, evidence of rim delamination has been reported with retrieved Crossfire liners [39]. The clinical circumstances associated with these revisions did not reveal an association between the rim damage and the reasons for revision. However, retrieval studies suggest that *in vivo* degradation is associated with clinically relevant damage modes of rim damage and rim wear-through in specific first-generation acetabular component designs incorporating gamma-air sterilized UHMWPE [18]. Currently, researchers have concluded *in vivo* degradation should be included among the list of long-term failure modes for conventional polyethylene components for THR. Longer-term retrieval studies are needed to determine whether *in vivo* oxidation will result in clinically meaningful damage to Crossfire acetabular liners.

## 22.4.2 *In Vivo* Oxidation and Total Knee Arthroplasty

Although the clinical significance of oxidation in the hip will continue to be debated for many years, there is

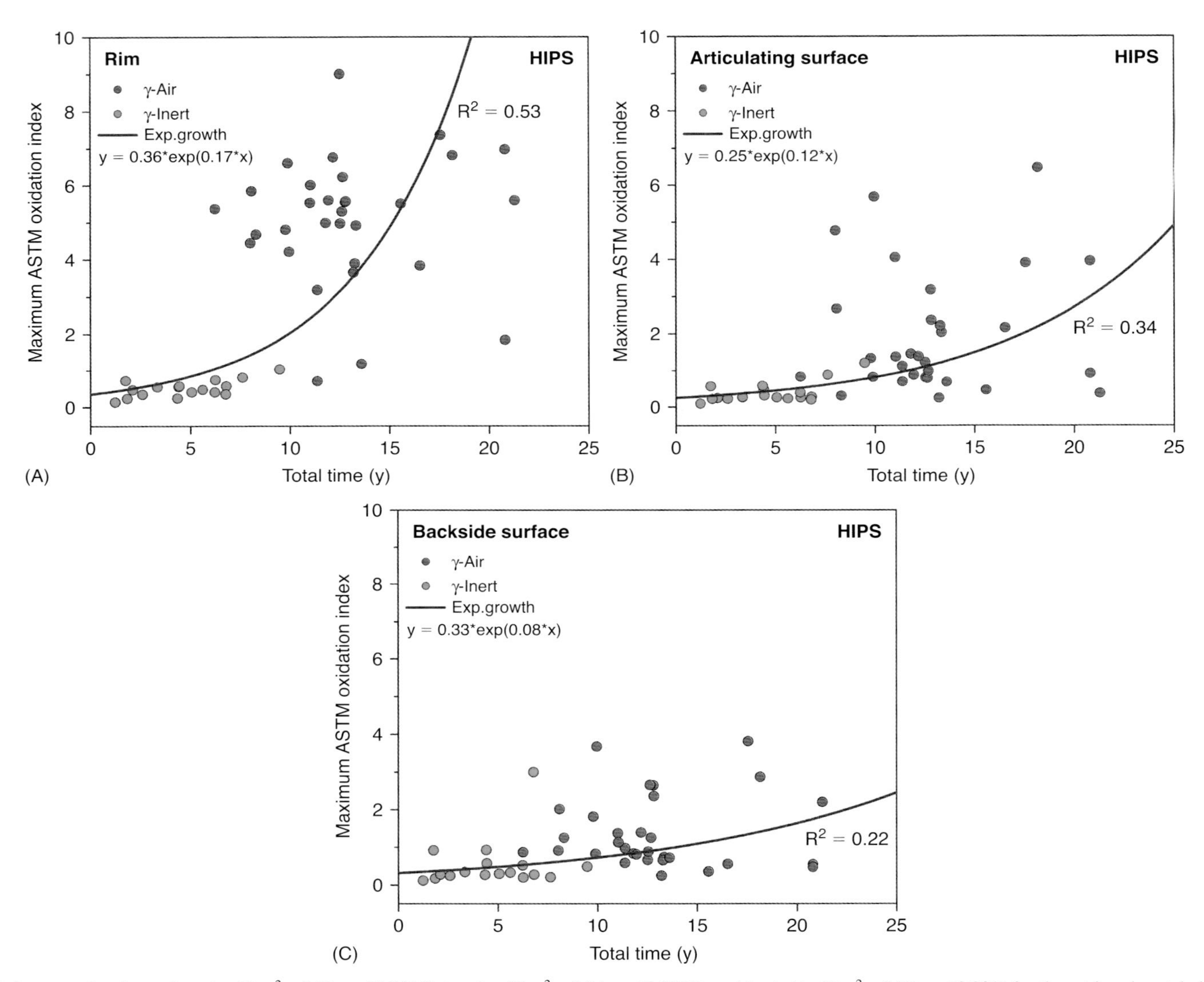


**FIGURE 22.12** Oxidation rate plots by region, rim (A: $r^2 = 0.53$, $p < 0.0006$), bearing (B: $r^2 = 0.34$, $p < 0.0001$), and backside (C: $r^2 = 0.22$, $p < 0.001$) for the retrieved acetabular liners summarized in Table 22.1 [42]. Total time represents the sum of shelf life, implantation time and *ex vivo* life (where appropriate). These models assume that conventional and historical liners oxidize at the same rate, consistent with the model proposed by Currier et al. [29].

greater consensus in the research community that oxidation is related to accelerated wear and delamination of the bearing surface in unicondylar and total knee replacements [22, 40, 41]. Unlike with total hips, the findings from historical gamma-air retrievals would suggest the bearing surface in tibial inserts is not protected to the same extent as it is in total hip arthroplasty (Figure 22.13) [19, 22, 42–45]. At our institution, we have found greater evidence of region-specific oxidation in gamma air sterilized inserts as compared with gamma inert sterilized inserts (Figure 22.14) [19, 42]. The implantation time of these retrievals varied as shown in Table 22.2 [42]. With the collection now available for evaluation, we currently have strong evidence that gamma inert sterilized inserts exhibit *in vivo* oxidation potential that is at least qualitatively similar to those observed following gamma irradiation in air (Figure 22.15). Because of the comparatively short implantation times of inert-sterilized inserts retrieved thus far (Table 22.2), it remains unknown how their rate of *in vivo* oxidation compares with that of historical gamma-air sterilized inserts.

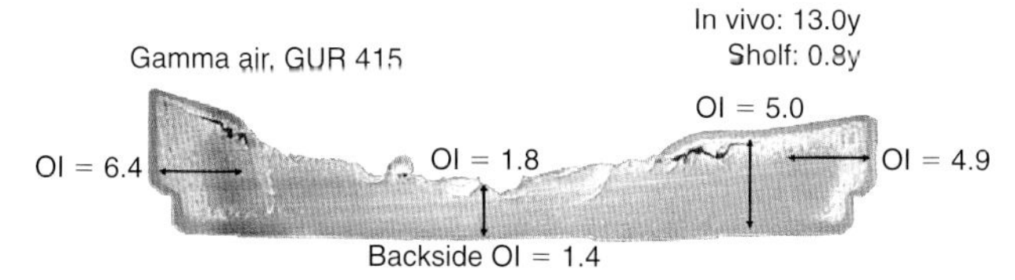


**FIGURE 22.13** Dark-field micrograph of a section through a modular, gamma-air sterilized tibial insert that exhibited severe *in vivo* oxidation and delamination. Note that the backside of the insert was protected from severe oxidation by contact with the metal backing.

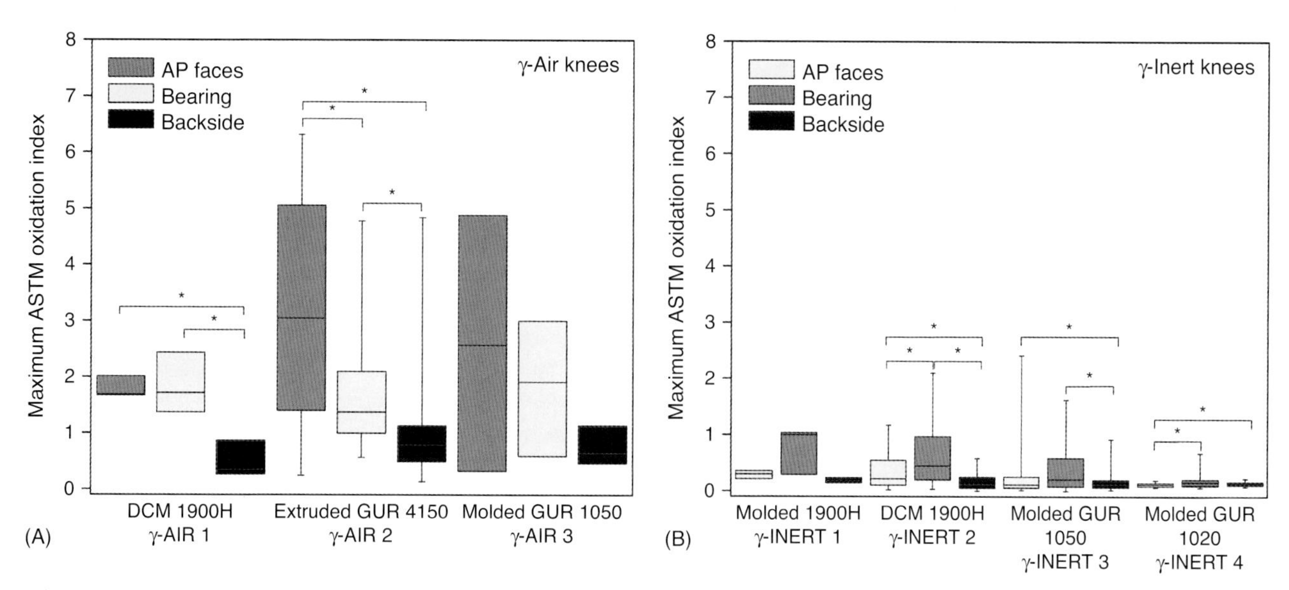


**FIGURE 22.14** Average maximum ASTM oxidation index for the seven categories of historical and conventional tibial inserts, summarized in Table 22.2 [42]. Substantial regional variation was observed in the maximum oxidation of retrieved inserts (* significant differences in paired t-tests).

**TABLE 22.2** Shelf Lives and Implantation Time, Average ± Standard Deviation, for Historical and Conventional Knee Retrievals [42]

| Group | Processing conditions | n | Shelf life (y) | *In vivo* time (y) |
|---|---|---|---|---|
| K-air 1 | γ-air, DCM, 1900H | 3 | 0.6 ± 0.3 | 13.2 ± 0.4 |
| K-air 2 | γ-air, extruded, GUR 4150 | 17 | 1.2 ± 1.6 | 10.4 ± 3.2 |
| K-air 3 | γ-air, molded, GUR 1050 | 3 | 1.2 ± 1.4 | 8.7 ± 2.8 |
| K-inert 1 | γ-inert, molded, 1900H | 4 | 0.4 ± 0.3 | 4.5 ± 2.4 |
| K-inert 2 | γ-inert, DCM, 1900H | 34 | 1.3 ± 1.8 | 3.4 ± 2.3 |
| K-inert 3 | γ-inert, molded, GUR 1050 | 31 | 1.4 ± 1.5 | 2.9 ± 2.3 |
| K-inert 4 | γ-inert, molded, GUR 1020 | 28 | 0.7 ± 0.8 | 1.7 ± 1.4 |

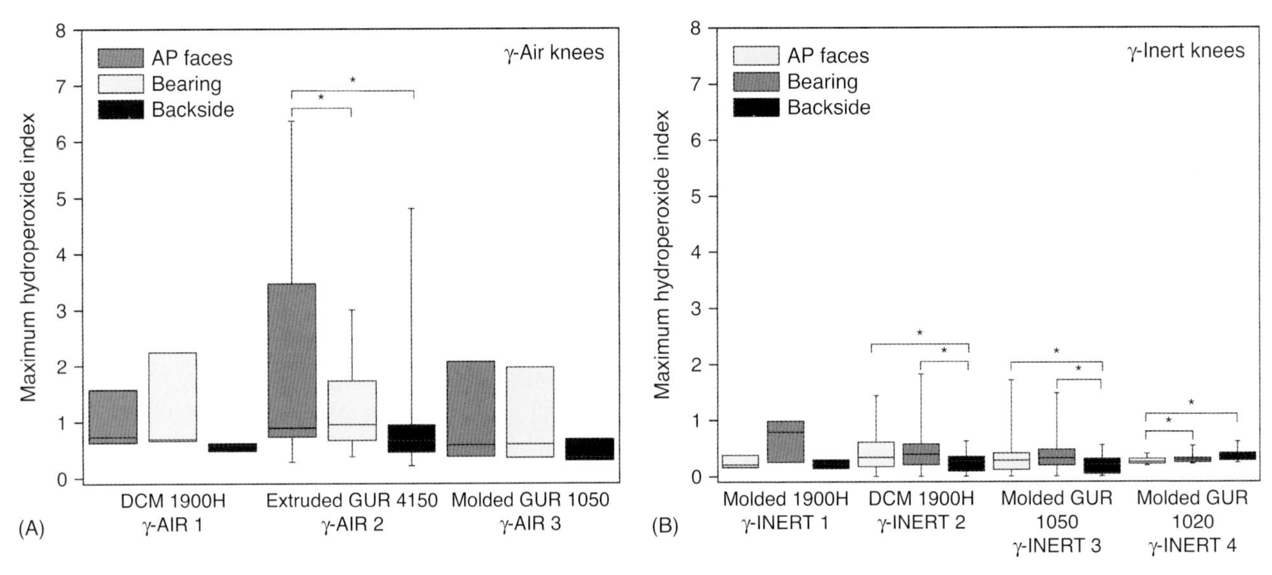


**FIGURE 22.15** Average maximum hydroperoxide index (oxidation potential) for the seven categories of historical and conventional tibial inserts, summarized in Table 22.2 [42]. Substantial regional variation was observed in the maximum oxidation potential of retrieved inserts (* significant differences in paired t-tests).

Gamma-air sterilized tibial inserts with the longest implantation times are associated with the highest oxidation near the bearing surface, as well as with more extensive and prevalent delamination (Figure 22.13). Because of the relatively short shelf lives of traced gamma air sterilized inserts, such as the component illustrated in Figure 22.13, and due to the significant regional variations between the surface and backsides of the insert, oxidation was inferred to have taken place *in vivo*. Consequently, available data support our hypothesis that *in vivo* oxidation is a contributing factor to delamination in TKA [44, 45]. On the other hand, we found no evidence to suggest that *in vivo* oxidation is related to pitting. Wear performance is multifactorial, and current research is focused specifically on material factors and their influence on fatigue damage [19, 22, 42–45].

Our data do not support the hypothesis that resin and conversion techniques substantially influence the resistance of tibial inserts to *in vivo* oxidation [42]. After gamma irradiation in air, oxidation and hydroperoxide levels in components fabricated from direct molded 1900H were similar to those in ram extruded GUR 415 (Figures 22.14 and 22.15). Similarly, after inert sterilization, components fabricated from 1900H were similar to molded GUR 1050 after comparable *in vivo* exposure.

In their study of retrieved knee bearings, Dartmouth University researchers have proposed an exponential model for gamma air and gamma inert components that captures the total time of oxygen exposure following irradiation [22]. As we saw with total hip replacements (Figure 22.12), the fitting of an exponential model to data obtained from knee replacements results in a different quality of fit depending upon the region of study (Figure 22.16). However, across regions, an exponential model only captures between 37–51% of the variation in the combined data set for gamma air and gamma inert retrievals. When we separate out the gamma inert retrievals, an exponential model explains only 16% of the variation in oxidation index with implantation time at the bearing surface (Figure 22.17). These observations reinforce our previous conclusions about the variability observed with oxidation in total hip replacements across different patients.

## 22.5 LABORATORY SIMULATION OF *IN VIVO* OXIDATION

Efforts to simulate *in vivo* oxidation in the laboratory can be traced back to the work of Peter Eyerer from the 1980s [9]. In his 1984 paper, Eyerer introduced the hypothesis that UHMWPE was under chemical attack by the synovial fluid *in vivo*. He postulated that "aggressive synovial fluids" were responsible for *in vivo* degradation and that hydrogen peroxide was a likely culprit [9]. As discussed previously, it is now recognized that oxygen dissolved in the joint fluid, rather than peroxide radicals, are responsible for *in vivo* oxidation. However, Eyerer's early hypothesis inspired a series of additional *in vitro* experiments with hydrogen peroxide during the 1990s [46, 47], which, in hindsight, proved to be overly aggressive and unsuccessful at reproducing *in vivo* oxidation profiles from retrievals.

In the late 1980s, Eyerer also reported on the characterization of hip cups that had been subjected to 1 million cycles of loading in simulators [48]. Although Eyerer noted slight changes in the cups following cyclic loading, the simulator-tested components did not replicate the property distributions observed in retrieved implants. Eyerer

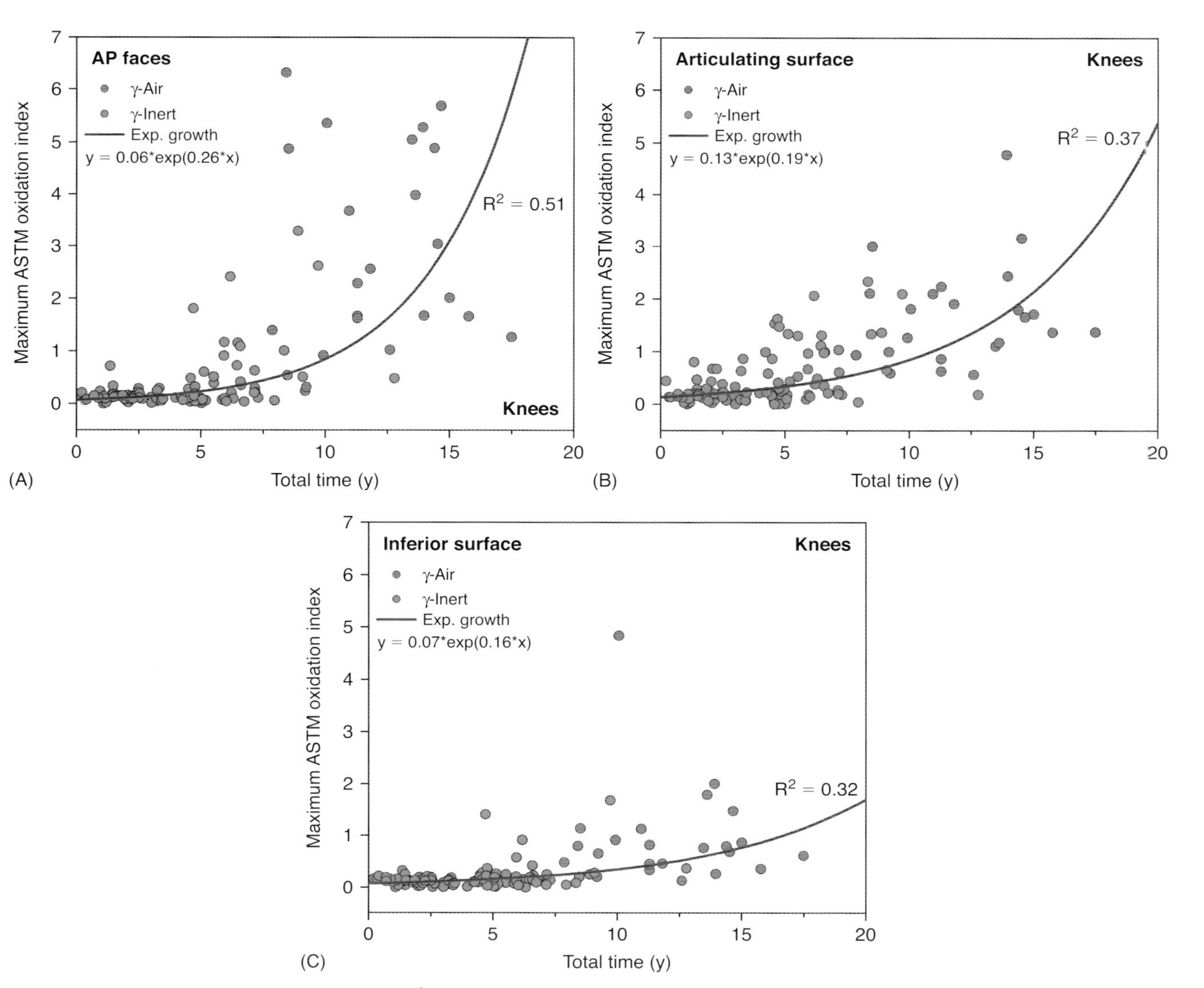


**FIGURE 22.16** Oxidation rate plots by region, anterior-posterior faces (A: $r^2 = 0.51$, $p < 0.0001$), articulating surface (B: $r^2 = 0.37$, $p < 0.0001$), and inferior surface (C: $r^2 = 0.32$, $p < 0.0001$) for the retrieved tibial inserts, summarized in Table 22.2 [42]. These models assume that conventional and historical inserts oxidize at the same rate, consistent with the model proposed by Currier et al. [29].

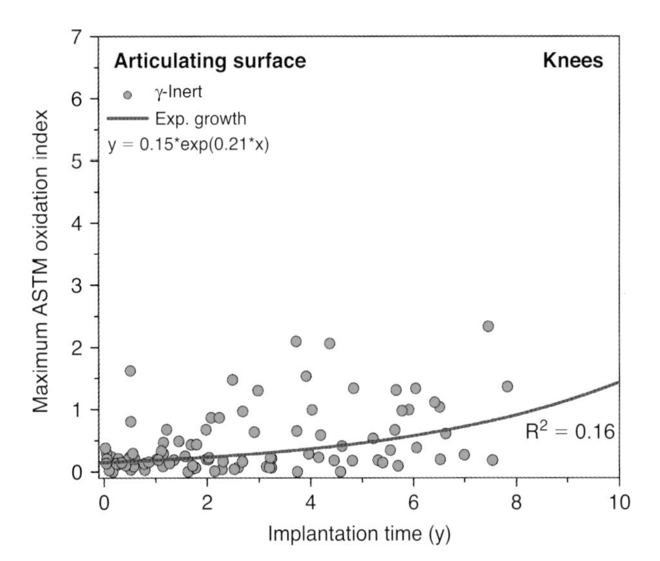


**FIGURE 22.17** Maximum oxidation versus implantation time plot for the conventional tibial inserts at the bearing surface. The exponential growth curve fit predicts on a maximum oxidation index greater than 1.0 after 10 years of implantation for the conventional inserts.

concluded his study prophetically with the prediction that, "as *in vivo* aging processes are slow they cannot be simulated with *in vitro* tests of a much shorter duration" [48].

"Real time" aging of UHMWPE in aqueous environments was reported by Robert Streicher in 1988 [49]. Streicher investigated the effects of aging gamma sterilized UHMWPE in room temperature air and in water at 37°C for up to 9 months. Whether gamma sterilized in air or in nitrogen, Streicher found that the oxidation index (which was referred to as the "normalized extinction C = O," measured by the ratio of the carbonyl band at $1720\,cm^{-1}$ to the reference peak at $1370\,cm^{-1}$) increased over time in water [49]. During the 9-month experiment, Streicher observed greater oxidation in the water-aged samples than in the samples exposed to room temperature air. Streicher attributed the greater oxidation in body-temperature water to "higher molecular mobility at the elevated temperature and the higher oxygen content of the medium" [49].

Even today, the ability to accelerate aqueous aging of gamma sterilized UHMWPE remains highly problematic. At the elevated temperatures necessary to accelerate the degradation process, the oxidation products that are generated in the UHMWPE are unrealistic [50]. However, at body temperature, the time scales that are involved (1+ years) are much too long to be practically useful as a preclinical test method. Even if real time aging is employed, the regional distribution of oxidation will not reproduce the variations observed in hip and knee retrievals. For example, aqueous aging of acetabular liners will produce much higher levels of oxidation at the bearing surface than would be observed clinically if the femoral head reduced access of oxygen to that location. As a result of these difficulties, currently no validated accelerated aging methods exist for reproducing both the mechanism and regional variation of *in vivo* oxidation found in UHMWPE after long-term implantation. At present, researchers interested in realistic and accurate analysis of *in vivo* oxidation in UHMWPE are advised to undertake a carefully controlled retrieval study rather than organize a multiyear real time aging experiment.

## 22.6 SUMMARY AND CONCLUSION

In the mid-1990s, the "oxidation problem" with UHMWPE was mostly thought to occur following gamma irradiation in air and prolonged shelf aging. Although this issue was decisively addressed by impermeable barrier packaging, UHMWPE containing unstabilized macroradicals will oxidize in the human body over time. Under no circumstances can *in vivo* oxidation be considered to be beneficial to the patient, but on the other hand, its link to clinical failure of implant components is not direct and may take over a decade to fully elucidate. Thus, the impact of *in vivo* oxidation continues to be debated in scientific circles, especially with regard to total hip replacements. In the knee, the future implications of *in vivo* oxidation are less ambiguous. As succinctly predicted by Currier et al. [22], "The result of *in vivo* oxidation is expected to be loss of mechanical properties, and susceptibility of polyethylene bearings to eventual fatigue failure."

When the results of shelf aging came to the forefront of orthopedics in the mid- to late 1990s, the clinical significance of oxidation was not fully understood, and researchers could not recommend improved alternatives to contemporary practices of gamma sterilization in air. The situation is different today. A number of well-validated and commercially viable methods are now available to stabilize macroradicals in medical grade UHMWPE, including remelting (Chapter 13), annealing (Chapter 14), and vitamin E stabilization of highly crosslinked (Chapters 15) and conventional UHMWPE (Chapter 16). Hopefully, by the time the next edition of this *Handbook* is published in the future, stabilization of conventional UHMWPE will be widely accepted, and *in vivo* oxidation will be considered a historical concern, suitable only for academic interest.

## 22.7 ACKNOWLEDGMENTS

This research was supported in part by the National Institutes of Health (R01 AR47904), Zimmer, Sulzer, and Stryker Orthopedics. Special thanks are due to Clare Rimnac, Case Western Reserve University, for launching my research on this topic many years ago. The author also thanks Professor Luigi Costa, University of Torino, for his many helpful discussions on this topic over the years and for providing editorial input to this chapter. Thanks are warmly extended to Eric Wysocki, Exponent, Inc., as well as Francisco Medel and Daniel MacDonald, Drexel University, for their assistance with the figures. The author is grateful to the patients, surgeons, and collaborators who made this research possible over the years.

## REFERENCES


1. Regling G, Jessen N, Meister S, Berg R. Intra-articular measurement of resting synovial pO2 (oxygen partial pressure of synovial fluid)—a new point of intersection for clinical research in the areas of arthrosis and pain. In: Regling G, editor. *Wolff's law and connective tissue regulation: modern interdisciplinary comments on Wolff's law of connective tissue regulation and rational understanding of common clinical problems*. Berlin: Walter de Gruyter; 1992. p. 299–320.
2. Ahlqvist J. A physiological approach to the pathogenesis of rheumatoid and other high-protein arthropathies. *Med Hypotheses* 1988 February;**25**(2):77–88.
3. Treuhaft PS, McCarty DJ. Synovial fluid pH, lactate, oxygen and carbon dioxide partial pressure in various joint diseases. *Arthritis Rheum* 1971 July–August;**14**(4):475–84.
4. Lund-Olesen K. Oxygen tension in synovial fluids. *Arthritis Rheum* 1970 November–December;**13**(6):769–76.
5. Grunhagen T, Wilde G, Soukane DM, Shirazi-Adl SA, Urban JP. Nutrient supply and intervertebral disc metabolism. *J Bone Joint Surg Am* 2006 April;**88**(Suppl. 2):30–5.
6. Urban JP, Smith S, Fairbank JC. Nutrition of the intervertebral disc. *Spine* 2004 December 1;**29**(23):2700–9.
7. Roberts S. Disc morphology in health and disease. *Biochem Soc Trans* 2002 November;**30**(Pt 6):864–9.
8. Grood ES, Shastri R, Hopson CN. Analysis of retrieved implants: crystallinity changes in ultrahigh molecular weight polyethylene. *J Biomed Mater Res* 1982;**16**(4):399–405.
9. Eyerer P, Ke YC. Property changes of UHMW polyethylene hip cup endoprostheses during implantation. *J Biomed Mater Res* 1984;**18**(9):1137–51.
10. Kurth M, Eyerer P, Ascherl R, Dittel K, Holz U. An evaluation of retrieved UHMWPE hip joint cups. *J Biomater Appl* 1988;**3**(1):33–51.
11. Bostrom MP, Bennett AP, Rimnac CM, Wright TM. The natural history of ultra high molecular weight polyethylene. *Clin Orthop* 1994;**309**(309):20–8.
12. Kurtz SM, Rimnac CM, Bartel DL. Degradation rate of ultra-high molecular weight polyethylene. *J Orthop Res* 1997;**15**(1):57–61.
13. Brach del Prever E, Crova M, Costa L, Dallera A, Camino G, Gallinaro P. Unacceptable biodegradation of polyethylene in vivo. *Biomaterials* 1996;**17**(9):873–8.
14. Gomez-Barrena E, Li S, Furman BS, Masri BA, Wright TM, Salvati EA. Role of polyethylene oxidation and consolidation defects in cup performance. *Clin Orthop* 1998;**352**(352):105–17.
15. Premnath V, Harris WH, Jasty M, Merrill EW. Gamma sterilization of UHMWPE articular implants: an analysis of the oxidation problem. *Biomaterials* 1996;**17**(18):1741–53.
16. Kurtz SM, Hozack W, Marcolongo M, Turner J, Rimnac C, Edidin A. Degradation of mechanical properties of UHMWPE acetabular liners following long-term implantation. *J Arthroplasty* 2003 October;**18** (7 Suppl. 1):68–78.
17. Kurtz SM, Rimnac CM, Hozack WJ, Turner J, Marcolongo M, Goldberg VM, et al. In vivo degradation of polyethylene liners after gamma sterilization in air. *J Bone Joint Surg Am* 2005 April; **87-A**(4):815–23.
18. Kurtz SM, Hozack WJ, Purtill JJ, Marcolongo M, Kraay MJ, Goldberg VM, et al. Significance of in vivo degradation for polyethylene in total hip arthroplasty. *Clin Orthop Relat Res* 2006; **453**:47–57.
19. Medel FJ, Rimnac CM, Kurtz SM. On the assessment of oxidative and microstructural changes after in vivo degradation of historical UHMWPE knee components by means of vibrational spectroscopies and nanoindentation. *J Biomed Mater Res* 2008. Published online.
20. Wannomae KK, Bhattacharyya S, Freiberg A, Estok D, Harris WH, Muratoglu O. In vivo oxidation of retrieved cross-linked ultra-high-molecular-weight polyethylene acetabular components with residual free radicals. *J Arthroplasty* 2006 October;**21**(7):1005–11.
21. Currier BH, Currier JH, Collier JP, Mayor MB. Effect of fabrication method and resin type on performance of tibial bearings. *J Biomed Mater Res* 2000;**53**(2):143–51.
22. Currier BH, Currier JH, Mayor MB, Lyford KA, Van Citters DW, Collier JP. In vivo oxidation of gamma-barrier-sterilized ultra-high-molecular-weight polyethylene bearings. *J Arthroplasty* 2007 August;**22**(5):721–31.
23. Edidin AA, Villarraga ML, Herr MP, Muth J, Yau SS, Kurtz SM. Accelerated aging studies of UHMWPE. II. Virgin UHMWPE is not immune to oxidative degradation. *J Biomed Mater Res* 2002;**61**(2):323–9.
24. Edidin AA, Herr MP, Villarraga ML, Muth J, Yau SS, Kurtz SM. Accelerated aging studies of UHMWPE. I. Effect of resin, processing, and radiation environment on resistance to mechanical degradation. *J Biomed Mater Res* 2002;**61**(2):312–22.
25. James SP, Blazka S, Merrill EW, Jasty M, Lee KR, Bragdon CR, et al. Challenge to the concept that UHMWPE acetabular components oxidize in vivo. *Biomaterials* 1993 July;**14**(9):643–7.
26. Costa L, Bracco P, del Prever EB, Luda MP, Trossarelli L. Analysis of products diffused into UHMWPE prosthetic components in vivo. *Biomaterials* 2001;**22**(4):307–15.
27. ASTM F 2102-06-e1. *Standard guide for evaluating the extent of oxidation in ultra-high molecular weight polyethylene fabricated forms intended for surgical implants*. West Conshohocken, PA: American Society for Testing and Materials; 2006.
28. Sun DC, Halleck A, Schmidig G, Wang A, Stark C, Dumbleton JH. Fourier transform infrared (FTIR) oxidation analysis of UHMWPE implants: Possible contamination from synovial fluid and serum. In: Gsell RA, Stein HL, Ploskonka JJ, editors *Characterization and properties of ultra-high molecular weight polyethylene*. West Conshohoken: American Society for Testing and Materials; 1998. p. 39–45.
29. Currier BH, Currier JH, Mayor MB, Lyford KA, Collier JP, Van Citters DW. Evaluation of oxidation and fatigue damage of retrieved crossfire polyethylene acetabular cups. *J Bone Joint Surg Am* 2007 September;**89**(9):2023–9.
30. Kurtz SM, Muratoglu OK, Buchanan F, Currier B, Gsell R, Shen FW, et al. Interlaboratory studies to determine optimal analytical methods for measuring the oxidation index of UHMWPE. *Biomaterials* 2001;**22**(21):2875–81.
31. Costa L, Luda MP, Trossarelli L, Brach del Prever EM, Crova M, Gallinaro P. Oxidation in orthopaedic UHMWPE sterilized by gamma-radiation and ethylene oxide. *Biomaterials* 1998 April–May; **19**(7–9):659–68.
32. Bracco P, del Prever EMB, Cannas M, Luda MP, Costa L. Oxidation behaviour in prosthetic UHMWPE components sterilised with high

energy radiation in a low-oxygen environment. *Polym Degrad Stab* 2006;**91**(9):2030–8.

33. Turner J, Kurtz S, Bracco P, Costa L. The effect of cholesteryl acetate absorbtion on the mechanical behavior of crosslinked and conventional UHMWPE. *Trans 53rd Orthop Res Soc* 2003;**28**:1423.
34. Kurtz SM, Hozack W, Turner J, Purtill J, MacDonald D, Sharkey P, et al. Mechanical properties of retrieved highly cross-linked crossfire liners after short-term implantation. *J Arthroplasty* 2005 October;**20**(7):840–9.
35. Hardaker C, Fisher J, Isaac G, Stone M, Older J. Quantification of the effect of shelf and in vivo aging on the in vivo and in vitro wear rates of a series of retrieved charnley acetabular cups. *Transactions of the European Society for Biomaterials*. London, England; 2001. p. T53.
36. Collier JP, Currier BH, Kennedy FE, Currier JH, Timmins GS, Jackson SK, et al. Comparison of cross-linked polyethylene materials for orthopaedic applications. *Clin Orthop* 2003 September;**414**(414):289–304.
37. Kurtz SM, Manley M, Wang A, Taylor S, Dumbleton J. Comparison of the properties of annealed crosslinked (Crossfire) and conventional polyethylene as hip bearing materials. *Bulletin (Hospital for Joint Diseases) New York, NY* 2002–2003;**61**(1–2):17–26.
38. Costa L, Bracco P, Brach del Prever EM, Kurtz SM, Gallinaro P. Oxidation and oxidation potential in contemporary packaging for polyethylene total joint replacement components. *J Biomed Mater Res B Appl Biomater* 2006 July;**78**(1):20–6.
39. Tower SS, Currier JH, Currier BH, Lyford KA, Van Citters DW, Mayor MB. Rim cracking of the cross-linked longevity polyethylene acetabular liner after total hip arthroplasty. *J Bone Joint Surg Am* 2007 October;**89**(10):2212–17.
40. Collier JP, Sperling DK, Currier JH, Sutula LC, Saum KA, Mayor MB. Impact of gamma sterilization on clinical performance of polyethylene in the knee. *J Arthroplasty* 1996;**11**(4):377–89.
41. McGovern TF, Ammeen DJ, Collier JP, Currier BH, Engh GA. Rapid polyethylene failure of unicondylar tibial components sterilized with gamma irradiation in air and implanted after a long shelf life. *J Bone Joint Surg Am* 2002 June;**84-A**(6):901–6.
42. Kurtz SM, Hozack W, Parvizi J, Purtill J, Sharkey P, MacDonald D, et al. Gamma inert sterilization: A solution to polyethylene oxidation? *Transactions of the 2008 AAOS Meeting*. San Francisco; 2008 March 5–8. Scientific Exhibit No. SE06.
43. Kurtz SM, Purtill J, Sharkey P, Topper J, MacDonald D, Kraay M, et al. Clinical significance of in vivo oxidation for gamma sterilized tibial inserts. *Trans 52nd Orthop Res Soc* 2006;**31**:552.
44. Kurtz SM, Parvizi J, Purtill J, Sharkey P, MacDonald D, Shah P, et al. In vivo oxidation contributes to delamination but not pitting in TKA. *Trans 53rd Orthop Res Soc* 2007;**32**:1891.
45. Medel F, Kurtz SM, Brenner E, Sharkey P, Purtill J, Parvizi J, et al. In vivo oxidation and oxidation potential in gamma air, gamma inert, and post-irradiation stabilized polyethylene components. *Trans 54th Orthop Res Soc* 2008;**33**:191.
46. Goldman M, Lee M, Gronsky R, Pruitt L. Oxidation of ultrahigh molecular weight polyethylene characterized by Fourier Transform Infrared Spectrometry. *J Biomed Mater Res* 1997;**37**(1):43–50.
47. Rimnac CM, Burstein AH, Carr JM, Klein RW, Wright TM, Betts F. Chemical and mechanical degradation of UHMWPE: Report of the development of an in vitro test. *J Appl Biomater* 1994;**5**:17–21.
48. Eyerer P, Kurth M, McKellup HA, Mittlmeier T. Characterization of UHMWPE hip cups run on joint simulators. *J Biomed Mater Res* 1987;**21**(3):275–91.
49. Streicher RM. Ionizing irradiation for sterilization and modification of high molecular weight polyethylenes. *Plast Rubber Proc Appl* 1988;**10**(4):221–9.
50. Mazzucco DC, Dumbleton J, Kurtz SM. Can accelerated aqueous aging simulate in vivo oxidation of gamma-sterilized UHMWPE? *J Biomed Mater Res B Appl Biomater* 2006 October;**79**(1):79–85.

Chapter 23

# Pathophysiologic Reactions to UHMWPE Wear Particles

Marla J. Steinbeck, PhD, Ryan M. Baxter, PhD Candidate, and Theresa A. Freeman, PhD



## 23.1 INTRODUCTION

From a clinical perspective, implants for total joint replacement must contribute to a biologically favorable and mechanically stable environment to provide a satisfactory long-term outcome. One of the most important factors determining the success of a total joint replacement is implant wear and the associated adverse biologic reaction elicited by wear debris [1] (Chapter 27). The most obvious gross, clinical manifestation associated with wear debris generation is bone loss (osteolysis; Figure 23.1), which has been identified as a major reason for implant loosening and the need for revision surgery after total hip replacement [2–3]. In 1977, a seminal paper by Hans-Georg Willert demonstrated macrophage activation by UHMWPE wear debris generated by movement of the articulating implant surfaces and by the movement of the implant against surrounding bone [4]. These initial observations led to numerous studies characterizing the chronic inflammatory response in periprosthetic tissue. The challenge in assessing the pathophysiologic response to wear debris associated with joint replacement surgery is to accurately characterize wear particle generation and the ensuing tissue response. This requires complex cellular and molecular analysis of periprosthetic tissues at both early and late implantation times.

The distribution of wear debris is another confounding factor because the cyclic loading of an articulating joint implant may result in intermittent waves of joint fluid pressure. This may distribute wear particles both locally and to more remote sites along the implant interface, as well as other organs. Locally, the dissemination of particles is limited by the density of the surrounding tissue. Large particles are trapped in dense, collagen-rich joint connective tissue, while smaller particles are able to move more freely. A recent cadaveric study showed UHMWPE wear debris was disseminated to the lymph nodes, spleen, and liver after joint replacement surgery [5]. Most disseminated particles were smaller than one micron, but particles as large as 50 microns were identified in abdominal lymph nodes. Lymphatic transport through perivascular lymph channels, as free or phagocytosed particles within macrophages, is the most probable route for wear debris distribution. The nature and ultimate fate of the wear debris,

UHMWPE Biomaterials Handbook

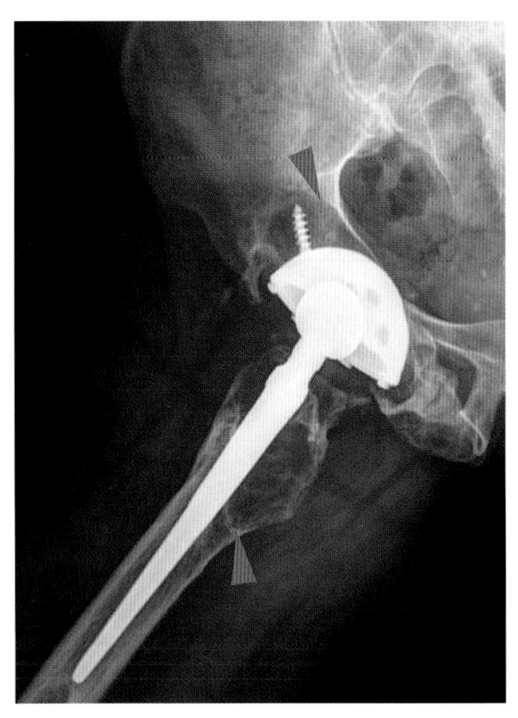

**FIGURE 23.1** Radiograph with pronounced regions of acetabular and femoral lysis (radiolucency indicated with red arrows) from a historical, gamma-air sterilized UHMWPE hip replacement revised 11.4 years post-implantation.

and the implications of long-term systemic exposure, are among the least understood aspects of joint arthroplasty. This is not surprising because the difficulty of particle detection, coupled with the number of tissue locations required to analyze particle distribution, makes this an overwhelming task.

The overall pathophysiologic response to UHMWPE wear debris is a complex process, involving a number of cell types and progressive local and systemic changes with increasing implantation times. Patient-specific factors or responses to wear debris further confound a complete understanding of this process by introducing additional variability of the host response to UHMWPE particles. Potential factors involved in individual-specific responses include genetic polymorphisms in matrix metalloproteinase 1, IL-6, TNF-α, and the vitamin D receptor [6–8]. In general, the frequently undetectable progression of inflammatory, biochemical and morphological changes in tissues surrounding the implant impose a major challenge.

## 23.2 RATIONALE FOR EVALUATING TISSUE RESPONSES

There is an overall need to understand the pathophysiologic responses to wear debris in the hope of developing both early diagnostic tests (serum biomarkers) and treatment modalities to suppress these responses. Insights into the loss of total joint replacement function have been gained by analyzing the cellular and molecular changes in periprosthetic tissues. Cellular and molecular biological techniques, including histology, histochemistry, immunohistochemistry, *in situ* hybridization, polymerase chain reaction (PCR), and Western and Northern blot analysis, have yielded important information concerning the biologic processes in periprosthetic tissues [1, 9–10]. Although gains have been made in understanding the contribution of UHMWPE wear debris volume, size, shape, and biochemical or protein binding characteristics to the tissue response, improvements in methods to achieve a complete understanding of the local and systemic pathophysiologic responses are needed. Improved methods of tissue analysis will aid in developing a standardized approach to evaluate the effects of wear particles for each new type of implant. This information is needed to make improvements in implant design, ultimately resulting in better implant longevity. For those patients with existing implants, this information, as previously mentioned, will aid in the development of diagnostic tests and treatment modalities to reverse the existing host responses.

## 23.3 IMMUNE SYSTEM

Since the seminal work by Hans-Georg Willert, which demonstrated macrophage activation by UHMWPE wear debris, interest in understanding the complete immune response in periprosthetic tissue has exploded [4]. For most nonimmunologists, this complex area is difficult to grasp, but hopefully the following information will provide a basis for a better understanding of the immune system.

The cells that carry out the immune response are called white blood cells or leukocytes. Leukocytes are subdivided into two groups of cells. The first group is the granulocytes, which are cells that contain cytosolic granules and include neutrophils, eosinophils, and basophils. These cells undergo maturation in the bone marrow and then enter the bloodstream. If needed for defense of the host, they are recruited to various tissues in response to infectious agents. During recruitment the leukocytes become tethered, roll, and eventually adhere to activated vascular endothelial cells lining the postcapillary venules of the affected tissue. The cells then migrate through the endothelial cell layer and enter (infiltrate) the tissue [11].

The second group is composed of monocytes and lymphocytes. Monocytic cells mature in the bone marrow, enter the bloodstream and, unlike the other leukocytes, are found in most tissues. After recruitment to the tissues, they are referred to as tissue macrophages, and depending on the specific tissue they may be referred to as histiocytes, dendritic cells (skin), microglial cells (brain), Kupffer cells (liver), etc. These cells make up part of the body's first line of defense against infectious agents or other foreign material, including bacterial products, chemicals, drugs, pollen, food, animal hair and dander, or more specifically to the topic at hand, wear debris generated by joint implant use. Lymphocytes are subdivided into B cells that undergo complete maturation in the bone marrow and T cells that mature in the thymus. B cells develop into plasma cells that release antibodies, while T cells form helper ($T_H$) or cytotoxic ($T_C$) T cells. Only lymphocytes can recognize and differentiate between specific foreign (nonself) and self- molecules (antigens). A separate population of T cells, called regulatory T cells (Treg) specializes in mediating immune suppression. Disruption in the function of these cells is the primary cause of chronic inflammatory and autoimmune diseases.

The immune response is divided into innate immunity and adaptive immunity. The innate immune system is constitutively on guard, reacts immediately, and is responsible for the initial defense against infectious agents and other foreign particles. This response is not specific; therefore, it is not dependant on the type of foreign body or nonhost component. The innate immune system consists of physical (skin and mucosal membranes) and chemical (mucous and antimicrobial peptides) barriers, blood proteins (acute phase proteins and complement), cytokines, phagocytic cells (neutrophils and macrophages), and natural killer cells (NK). Neutrophils and macrophages perform various host defense functions that rely on phagocytic uptake of infectious agents and other foreign particles. Phagocytes are equipped with multiple antimicrobial mechanisms that become activated upon initial contact with foreign particles. NK cells undergo maturation in the bone marrow; most contain granules, but they resemble lymphocytes rather than granulocytes in their appearance. After entering the bloodstream, they accumulate in secondary lymphoid tissues, such as the tonsils, lymph nodes, and spleen. The natural killing by these cells does not require prior host exposure to the target cells and is mediated by granule exocytosis.

### 23.3.1 Adaptive Immune Response

Adaptive, or specific, immunity is controlled by lymphocytes and occurs after exposure to any foreign antigenic substance that is specifically recognized by these cells. Unlike the innate immune response, the adaptive immune response is antigen specific, reacting only with the substance that induces activation, thereby increasing response time as the activated cells undergo clonal expansion. This arm of the immune system also exhibits immunological memory. It "remembers" that it has encountered a foreign antigen and will react more rapidly on subsequent exposure to the same molecule. Memory cells play a major role in both T cell-mediated and B cell, humoral, adaptive immunity. Every adaptive immune response involves the recruitment and activation of not only the effector T and B cells, but also Treg immune suppressor cells. The balance between these cell populations is critical for the appropriate control of the quality and magnitude of the adaptive immune response and for establishing tolerance to self-antigens.

The role of NK cells in the adaptive immune response is to produce cytokines that modulate emerging T and B cell responses and eliminate infected or abnormal host cells prior to T and B cell activation.

### 23.3.2 Cytokines and Chemokines

Cytokines are cellular proteins (chemical messenger proteins) that mediate inflammation and communication between cells of the immune system, but they can affect other local cell types. These proteins are released by cells and affect the behavior of other cells that bear receptors for them. The general categories of cytokines include interleukins (IL), which at the present time include IL-1 to IL-34 and interferons $\alpha$–$\gamma$.

Chemokines (or chemotactic cytokines) are members of a large family of extracellular immunoregulatory proteins that primarily act as chemoattractants for immune cells, recruiting them to areas containing infectious agents or other foreign particles. Differentially expressed by most cell types, chemokines can be loosely divided into two main immunologic groups: homeostatic chemokines, which are expressed constitutively; and inflammatory chemokines, which are induced by infection and injury or in the setting of inappropriate inflammation. Like cytokines, these proteins are produced by immune cells, as well as other cell types, in response to specific stimuli.

## 23.4 IMMUNOLOGIC RESPONSES TO JOINT REPLACEMENT UHMWPE WEAR DEBRIS

Following a large total joint arthroplasty, a newly formed joint capsule or pseudosynovial membrane is typically formed around the implant within the joint space [3]. During this time, capsular tissue becomes established, followed by the formation of an intermediate, highly vascularized fibrous membrane [4]. At the time of revision surgery for aseptic implant loosening, an extended and thickened fibrous membrane with increased numbers of histiocytes and focal areas of decreased vascularization is typically present [12]. The predominant cells found in the intermediate fibrous membrane and other periprosthetic tissues include fibroblasts, histiocytes, infiltrated peripheral blood monocytes, and multinucleated giant cells [10, 13–14] (Figure 23.2). Interspersed throughout the tissue are both micron (typically from gamma air-sterilized UHMWPE components) and submicron UHMWPE debris [15–17]. Most of the UHMWPE particles, ranging in size from 0.1–10$\mu$m (typically less than 2$\mu$m), are ingested and

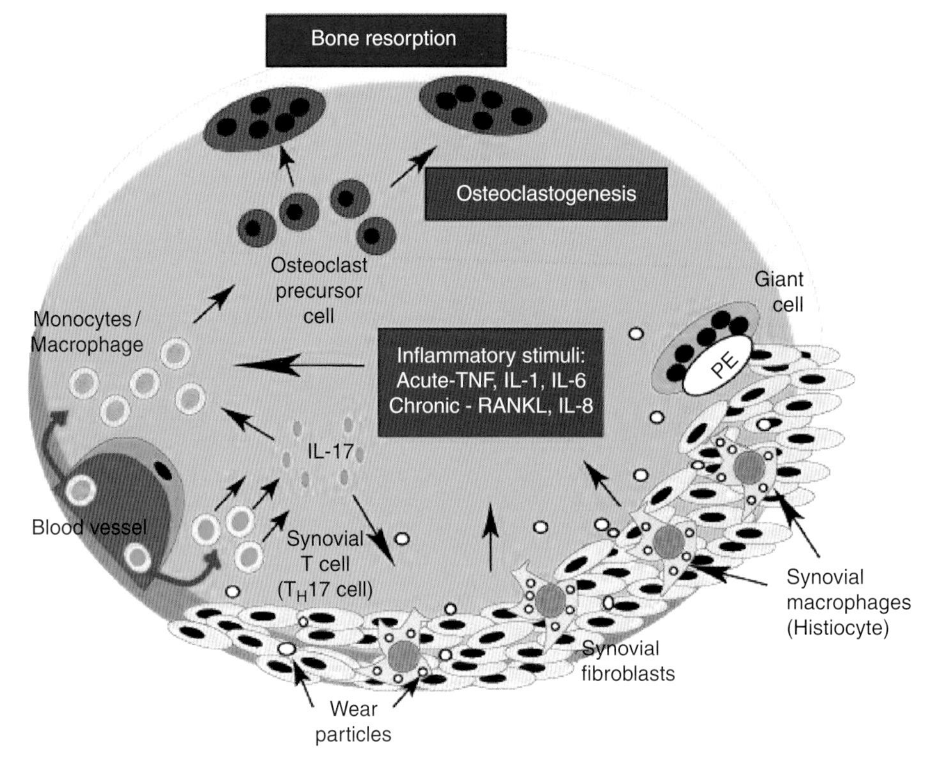


**FIGURE 23.2** Diagram of inflammatory changes preceeding osteolysis. The infiltration of immune cells, including monocytes and T cells, and activation of resident synovial fibroblasts and histiocytes in response to UHMWPE wear debris leads to the production of chemokines, cytokines, and growth factors. Giant cells form in response to larger wear debris. Monocyte/macrophages differentiate into osteoclasts, which are the cells responsible for bone resorption.

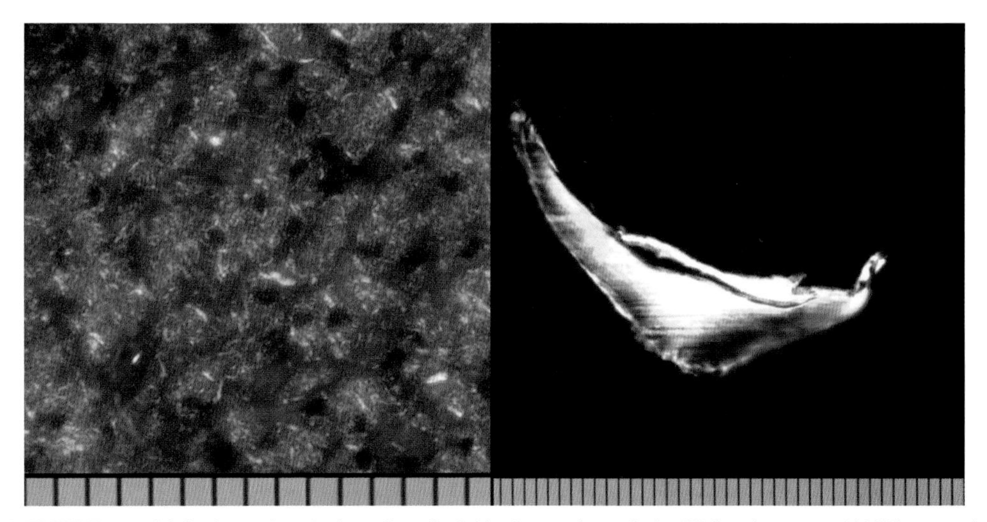

**FIGURE 23.3** UHMWPE wear debris observed *in situ* in periprosthetic hip tissue using polarized light microscopy. (A) Micron- and submicron-sized UHMWPE wear debris (birefringent particles) within phagocytic histiocytes; (B) large UHMWPE wear debris (approximately 300 μm) in periprosthetic tissue from a historical, gamma-air sterilized UHMWPE hip replacement revised 15.6 years post-implantation. Scale interval represents 10 μm in both images.

found within phagosomes of the histiocytes and macrophages (Figure 23.3). The fusion of macrophages to form giant cells occurs in an attempt to ingest the UHMWPE particles, usually >10 μm [18], because 5 μm particles have been observed within these multinucleated giant cells (Figure 23.4) [19]. The actual or attempted ingestion of particles, referred to as phagocytosis or frustrated phagocytosis, respectively, occurs through nonspecific receptors on the cell surfaces and results in activation of these cells. Activation marks the beginning of a chronic inflammatory

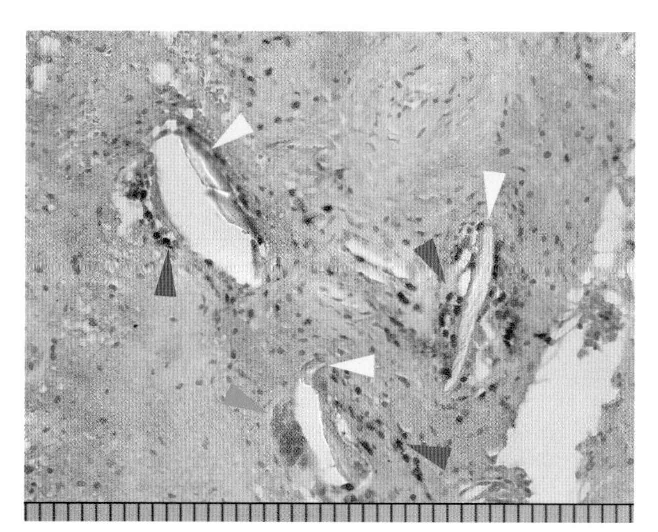

FIGURE 23.4 CD3 immunostained T cells (red arrows) and foreign body giant cell (green arrow) localized with UHMWPE wear debris (yellow arrows) from a historical, gamma-air sterilized UHMWPE hip replacement revised 14.2 years post-implantation. 200× magnification. Scale interval represents 10 μm.

reaction initiated by the accumulation of UHMWPE wear debris and the inability of the phagocytic cells to degrade the ingested or uningested wear debris. During the process of attempted degradation of the UHMWPE wear debris, the cells release protein degrading enzymes and destructive reactive oxygen and nitrogen species (RONS) [20–22]. Thus, without the ability to remove the initiating wear debris stimulus, the inflammatory response continues, and rather than achieving a resolution thereby returning the tissue to a normal functional state, further tissue damage occurs.

The innate immune response induced by wear particles is a chronic inflammatory and nonspecific foreign body reaction. This reaction may also involve mast cells within the periprosthetic tissue [23]. Mast cells are immune cells found in mucosal and connective tissue in relatively low numbers. When activated, the number of mast cells can increase dramatically, and the release of enzymes and other mediators by these cells can stimulate fibroblast proliferation, vascularization, and inflammation.

Based on numerous studies of the periprosthetic tissues, it is the activation of both recruited and resident cells that leads to the production and release of cytokines, chemokines, and growth factors by these cells. The acute response proinflammatory cytokines and chemokines produced by activated fibroblasts, macrophages, and mast cells include IL-1α, IL-1β, IL-6, and TNF-α [24–27]. In most inflammatory reactions, the acute phase is typically a short-lived response, followed by the proinflammatory cytokine and chemokine release that perpetuates a chronic inflammatory response, such as IL-6, IL-8, and monocyte chemoattractant protein (MCP-1), which are also detected in these tissues [6, 28–30]. However, given the continuous introduction of wear particles into the tissue, the acute inflammatory phase may be recurrent, at least during the early implantation years.

While the focus has been on the innate immune response, low numbers of T cells have been observed in periprosthetic membrane tissue [27, 31–32]. The role of T cells is not clear. In a subset of patients, however, there may be a hypersensitivity immune response that is directed at an UHMWPE particle containing bound immunostimulatory proteins [33] or altered, host protein from damaged tissue [34]. A Type II antibody-mediated hypersensitivity reaction is mediated by $T_H$ cells, B cells (plasma cells), and effector cells, typically monocyte/macrophages. The tissue damage associated with this response is mediated by activated macrophages and the release of lysosomal granule contents and the production of RONS. Type IV hypersensitivity or delayed-type hypersensitivity reactions may occur in response to the binding of small proteins (haptens, <1000 daltons) or other host cell material to UHMWPE particles. However, this type of reaction has predominantly been observed when metal debris was present in the periprosthetic tissue (Figure 23.5). Type IV hypersensitivity reactions are mediated by $T_H1$ cells, and the tissue damage associated with this immune reaction is mediated by $T_C$ cells.

An interesting recent development has been the recognition of a new subset of T cells producing IL-17 [35–37]. IL-17 is produced predominantly by memory T cells and acts synergistically with TNF-α to activate synovial fibroblast-like cells. $T_H17$ cells have a distinctive cytokine profile, which includes IL-17, TNF-α and RANKL (receptor activator of nuclear-receptor factor NFκB ligand). These factors affect neutrophil, monocyte/macrophage, and osteoclast mobilization, differentiation, and activation [36]. Osteoclast formation and bone resorption is discussed in Section 23.6. $T_H17$ cells play a role in inflammatory reactions, particularly those involved in autoimmune (self-directed) diseases directed against altered host (self) proteins. These cells are found in the synovial membrane of patients with autoimmune rheumatoid arthritis (RA), and the expression of extracellular matrix proteins and integrin receptors by synovial lining cells are similar in both RA and in tissues from patients with aseptic loosening of THRs [38].

The involvement of NK cells in the removal of damaged cells or cells that have lost "self recognizing" surface receptors in periprosthetic tissues has not been investigated. Like the $T_H17$ cells, NK cells secrete cytokines that regulate T and B cell responses, and they are involved in the development of autoimmune diseases. Finally, the disregulation of Treg cells may also affect the tissue response to wear debris by failure to suppress the activity of stimulated immune cells. The involvement of T cell subsets and NK cells in periprosthetic tissue inflammation associated with UHMWPE wear debris will require additional studies.

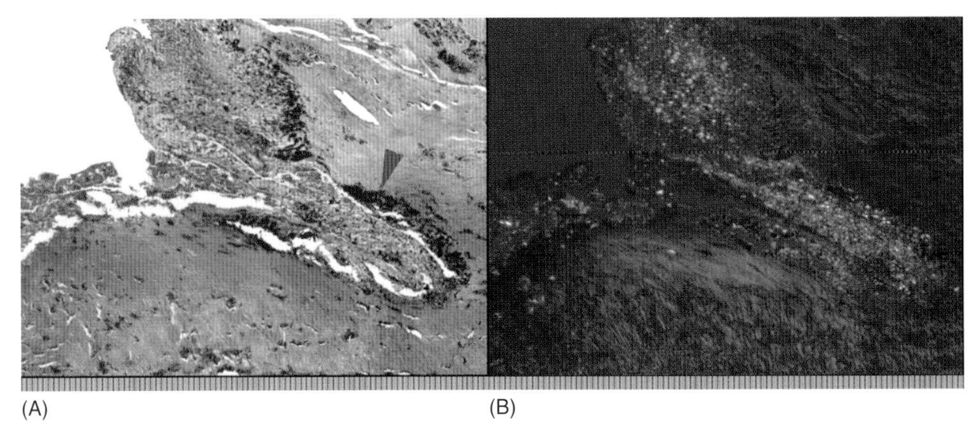


**FIGURE 23.5** Combination of metal (A, red arrow) and UHMWPE wear debris (B) in periprosthetic hip tissue revised 19 years postimplantation. Main image: 100× magnification. Scale interval represents 10 μm.

## 23.5 *IN VITRO* AND *IN VIVO* MODELS USED TO STUDY THE IMMUNE RESPONSE TO UHMWPE WEAR DEBRIS

To study the direct effects of UHMWPE wear debris on monocyte/macrophage activation and to gain insight into the host response, UHMWPE particles with characteristics similar to those found in periprosthetic tissues have been added to cells in culture. Exposure to UHMWPE wear debris within the most clinically relevant size range (0.1–10 μm) results in monocyte/macrophage activation, measured as an increase in the production of IL-1, IL-6, and TNF-α in culture [15, 39–40]. In addition to size-based responses, elevated numbers of particles have been found at all locations in the periprosthetic tissue where chronic foreign-body reactions occur. This suggests that the number of particles may regulate monocyte/macrophage recruitment and subsequent activation. *In vitro* studies have shown that the amount of proinflammatory cytokines released was dependent on the ratio of particles to cells [39]. In studies by Green et al. [39], the volume of particles was related to the cell number at two different concentrations, 10 $\mu m^3$:1 or 100 $\mu m^3$:1. Particles in the 0.24 ± 0.094 μm size range caused the greatest amount of IL-1, IL-6, and prostaglandin $(PG)E_2$ to be released when the ratio of particles to cells was 10 $\mu m^3$:1. At the higher ratio, the larger sized particles caused an increased release of proinflammatory cytokines. Particle composition also effected of the release of proinflammatory cytokines by cultured monocyte/macrophages [41–43]. Specific wear morphologies can also instigate and affect the biological response in periprosthetic tissue. To investigate the effect of particle shape, Fang et al. developed an inverted monocyte/macrophage cell culture system and generated wear particles of varying size and shape [44]. Their findings showed that spherical particles of the same volume were ingested to a greater extent by monocyte/macrophages in culture than the elongated particles. Thus, *in vitro* studies have shown that the biological response of monocyte/macrophages to UHMWPE wear debris is mediated by a variety of specific particle characteristics, including size, number, and shape.

Because the pathophysiologic response to wear debris is not limited to the effects on monocyte/macrophages alone, animal models have been developed to study the *in vivo* response to wear debris. These animal models include rat [45–46], canine [47], and mouse [48–51] models. Particles within the 0.1–10 μm size range isolated from periprosthetic tissues induced a local inflammatory response, which was increased as the particle numbers increased. Interestingly, Yang et al. [52] injected globular and elongated UHMWPE debris into air pouches on the backs of mice and showed that particles with large aspect ratios induced higher levels of TNF-α and IL-1β production, relative to globular particles with similar surface area [53]. This result is interesting because *in vitro* cell studies showed that the globular particles were preferentially ingested by monocyte/macrophages in culture [44].

Mouse models have also been used to evaluate potential treatments for inflammation and subsequent osteolysis [50, 54–59]. The anti-inflammatory treatments tested include TNF-α antagonists and a cyclooxygenase 2 inhibitor ($PGE_2$ source). Treatment targeting osteoclasts and bone resorption include the RANKL receptor inhibitor osteoprotegerin (OPG) and bisphosphonates [29, 58, 60]. Numerous other potential targets have been identified and tested; most of these, however, have not been tested in the patient population because of potential side effects.

For both the *in vitro* and *in vivo* animal studies, acute rather than chronic inflammatory cytokines were elevated in response to particle addition or injection, respectively. The drawback of these studies is that based on the short duration of the experiments, the immune response to UHMWPE wear does not necessarily correlate with the response observed in human tissue after years of implantation and continuous particle generation [29].

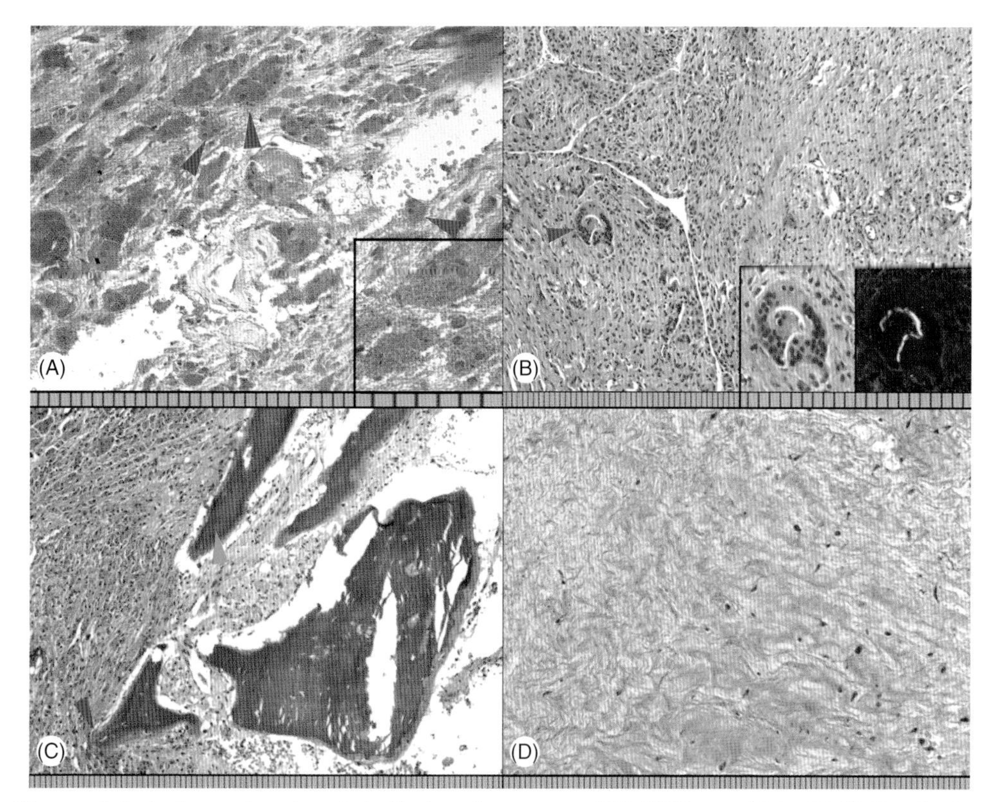


**FIGURE 23.6** Histomorphologic changes in periprosthetic hip tissue in response to UHMWPE wear debris. (A) CD68 antibody stain showing histiocytes (red arrows), 11 years post-implantation (YPIS). Main image: 200× magnification; (B) Foreign body giant cell (red arrow) with internalized UHMWPE wear debris, 15.4 YPIS. Main image: 100× magnification. Left inset of image B shows enlarged giant cell, and right inset shows UHMWPE wear debris that is exhibiting birefringence under polarized light. (C) Representative regions of fibrocartilage transition (red arrow) and tissue calcification (green arrows), 15.6 YPIS. Main image: 100× magnification; (D) Tissue necrosis with a few remaining intact cell nuclei, 19.7 YPIS. Both insets enlarged by 200% relative to original image. Scale interval represents 10 μm.

## 23.6 INFLAMMATORY- AND NONINFLAMMATORY-BASED HISTOMORPHOLOGIC CHANGES IN PERIPROSTHETIC TISSUES

In addition to the inflammatory response observed in periprosthetic tissues, other morphologic changes can take place in tissue and bone surrounding the implant. These changes include tissue fibrosis, necrosis, fibrocartilage formation, heterotopic ossification, and last, but certainly not least, osteolysis of the surrounding bone (Figure 23.6).

To gain insight into the histomorphologic changes in tissue in well-fixed implants, Bos et al. did a cadaveric study of 25 prosthetic hip joints implanted for a mean of 7 (0.25–16) years [61]. Although these were cemented hip implants, and there was a contribution of both cement and UHMWPE wear debris, a timeline of changes within the surrounding tissue and bone was observed as wear accumulation increased. In general, within a year of implantation the initial granulation tissue was more fibrotic and had increased in thickness. After 2.5 years, the number of histiocytes within the fibrous membrane dramatically increased concurrently with the cement wear debris. This process did not plateau; it continued with increasing implant duration and the generation of UHMWPE wear debris. The histiocytes contained large amounts of phagocytosed wear debris and showed degenerative changes in chromatin structure and the disintegration of cell borders. In addition, giant cells were typically found around UHMWPE particles >30 μm. After 4 years of implantation, focal areas of necrosis were observed, which became large, confluent, and sharply delineated areas of necrosis by 5–7 years. A positive correlation with implant duration was found only for necrosis, although the increasing thickness of the fibrous tissue over time correlated with the increasing number of histiocytes and amount of wear debris. Necrosis may be caused by cell injury (e.g., ingestion of wear debris) or inflammation, and in turn necrotic cells release cellular contents that perpetuate and exacerbate inflammation and osteoclast mediated bone resorption. Morphologically, necrotic cells appear swollen and show signs of nuclear and cytoplasmic degeneration. Similar increases in histiocytes, wear debris accumulation, and necrosis were observed in three other cadaveric studies [62–64].

Inflammation, tissue damage, and decreased vascularization [12] can also instigate the conversion of fibrotic tissue to fibrocartilage and heterotopic ossification (bone formation in soft tissue). In addition to inflammation and inflammatory related tissue responses, the formation of fibrocartilage may have a more physical basis, related to micromotion of the implant [13]. Implant movement affects integrin and nonintegrin attachment of cells to extracellular matrix components, and, as a result, adhesion receptor signaling is modified, initiating transformation of fibrous tissue into fibrocartilage (metaplasia). The micro- and macromotion can also lead to ischemia-reperfusion injury of the vascular endothelial cell lining, decreasing vascularization of the tissue, resulting in tissue necrosis or fibrocartilage formation [13]. Formation of fibrocartilage changes the plasticity of the tissue and is an initial step toward heterotopic ossification, which has been observed in tissues around hip, knee, and spine implants. The persistence of local inflammation, regardless of the number of years post initial surgery, and the development of fibrocartilage followed by heterotopic ossification has been observed in a number of other chronic inflammatory conditions [65–68].

The most obvious gross, clinical manifestation associated with wear particle generation is osteolysis, which has been identified as a major reason for implant loosening and the need for revision surgery after total hip replacement [3, 6, 28]. Current scientific theory still maintains that osteolysis around total joint replacements is governed by a critical initiating event of interaction between immune cells specialized to respond to foreign bodies and wear debris accumulation (e.g., monocyte/macrophages and histiocytes) [3]. Evaluation of clinical data showed that patients revised for osteolysis consistently had UHMWPE particle quantities in the order of 10 billion particles per gram weight of tissue, suggesting the existence of a threshold for the infiltration and activation of monocyte/macrophages and the onset of osteolysis [69, 70]. The accumulation of particles may be due to implant interface wear and/or micromotion of implant components with respect to each other (e.g., fretting) [71, 72] (Chapter 27). Both micron and submicron (0.1–10 $\mu$m) debris have been implicated as important contributors to the inflammatory response in periprosthetic tissue and the onset of osteolysis [15–17]. Particle-induced osteolysis, or loss of bone around the implant, can lead to implant loosening, which is the result of a tissue mediated loss of fixation (bone-implant ingrowth). In addition, loosening may occur due to inadequate initial fixation or mechanical loss of fixation over time. As a result, mechanically loose prostheses undergo excessive displacement, subsidence, and/or migration with the application of physiologic loads.

Monocyte/macrophages in periprosthetic tissue have the potential to differentiate into fully functional osteoclasts, capable of bone resorption [73–75]. Soluble products released by activated macrophages and fibroblasts in the periprosthetic tissue control the activity, proliferation, and differentiation of osteoclasts, bone degrading cells, and osteoblasts, bone forming cells (Figure 23.2) [3, 76–78]. Osteoclasts are formed from monocyte/macrophage precursors in response to RANKL and monocyte colony stimulating factor (M-CSF) [81–82]. Osteoclast differentiation, fusion to form multinucleated cells, and activation result from the binding of RANKL to its receptor RANK. The relative expression of RANK, RANKL, and OPG, a molecule that competitively binds RANKL and thus prevents osteoclast activation, determines the amount of bone resorption at a given site. Specifically, the RANKL/OPG ratio is an important determinant of how much bone is resorbed. IL-6 is produced by monocyte/macophages and fibroblasts and enhances the recruitment of osteoclast precursors, promoting osteoclast formation (osteoclastogenesis) [79–80]. In addition, M-CSF and IL-1 are capable of acting directly on osteoclast precursors to promote osteoclast differentiation or indirectly by modulating RANKL and OPG expression. Bone resorption by osteoclasts includes degradation of the organic and inorganic parts of bone after differentiation and fusion by creating a sealing zone between the bone and the osteoclast, into which protons, degradative enzymes (e.g., cathepsin K), and RONS are released. Proton pumps produce an acidic pH, causing hydroxyapatite, the mineral making up the bone, to be dissolved.

## 23.7 CURRENT CONSIDERATIONS BASED ON MORE RECENT FINDINGS AND APPROACHES TO TISSUE ANALYSIS

Recent findings suggest that IL-1$\alpha$, IL-1$\beta$, and TNF-$\alpha$ are not found in freshly retrieved human periprosthetic tissues and that the alternative macrophage activation pathway resulting in IL-8 and macrophage inflammatory protein (MIP)1$\alpha$ release, decreased local osteoprotegerin (OPG) levels, and impaired osteogenesis may be involved in periprosthetic osteolysis [28–29]. Many of the early studies evaluated the production of cytokines by periprosthetic tissues in culture. However, the recent comparative findings of Shanbhag et al. [29] show that the protein profile is affected by placing the tissues in culture and differs considerably from the cytokines, chemokines, and growth factors found in fresh tissue. Using high-throughput protein chips, this group analyzed 29 macrophage inflammation-related cytokines, chemokines, and growth factors. Of the cytokines evaluated, freshly isolated tissue contained IL-6, IL-8, and chemoattractants for activated $T_H1$ cells. The authors suggest that the presence of $T_H1$ cell chemoattractants imply a role for these cells in the inflammatory process. IL-8 is produced by several cell types and is chemotatic for T cells and neutrophils, and it also promotes angiogenesis (blood vessel formation) and osteoclast differentiation [79–80]. Current studies are focused on standardizing the tissue analysis and incorporating new molecular techniques to unravel the complex pathways involved in UHMWPE wear mediated inflammation and subsequent pathophysiologic changes.

## 23.8 EXACERBATION OF THE IMMUNE RESPONSE TO WEAR DEBRIS AS A RESULT OF SUBCLINICAL INFECTION

One of the commonly overlooked reasons for implant loosening and revision surgery for well-fixed implants is subclinical infection [83–84]. The major reason for this oversight is that subclinical infection is very difficult to diagnose. In revision cases when infection is suspected, clinical laboratory tests are performed, which include tissue or joint aspirate bacterial cultures, peripheral blood ESR, WBC and differential, and levels of C-reactive protein [85]. However, all of the current approaches, including bacterial cultures and PCR for bacterial species, in joint implant biology and in other clinical areas, have so far proved to be inadequate [86–88]. Recently, Schroeder, et al. established a histomorphological criteria, in conjunction with polarized light microscopy, to define four types of periprosthetic membranes: periprosthetic membranes with wear debris (type I), periprosthetic membranes with infection (type II), periprosthetic membranes of combined types I and II (type III), and periprosthetic membranes with no obvious wear debris or infection (type IV) [84]. Both the interobserver reproducibility (95%) and the correlation between histopathologic and microbiologic diagnoses (89%, $p < 0.001$) were higher than any of the previous approaches employed. Most notable was that the four types of periprosthetic membranes were observed at significantly different times of revision. This new classification system allows for a standardized diagnostic approach and future studies concerning the etiology and pathogenesis of THR revisions [84].

If infection is present, activation of macrophages can occur in response to several bacterial products, lipopolysacchride (LPS, gram negative bacteria), lipoteichoic acid, and peptidoglycan (gram positive bacteria) [88]. These products tend to adhere to UHMWPE and metal wear debris [89]. *In vitro*, the presence of LPS increased osteoclast formation and bone resorption in response to wear debris by 50–70%. Therefore, the presence of subclinical infection has the potential to greatly increase the host response to the generation of wear debris. In addition, the presence of bacterial products on the wear particles may reduce the threshold required for the initiation of pathophysiologic changes in the periprosthetic tissue and bone.

## 23.9 COMPARATIVE PATHOPHYSIOLOGIC CHANGES IN PERIPROSTHETIC HIP TISSUES FROM HISTORICAL AND HIGHLY CROSSLINKED UHMWPE IMPLANT RETRIEVALS

To date, studies performed on periprosthetic tissue have largely focused on inflammatory factors that contribute to implant loosening. These studies, while informative, have been performed, for the most part, on tissues retrieved from patients receiving historical, gamma air-sterilized UHMWPE components. The current generation of implants, however, are composed of newer, highly crosslinked UHMWPE material. To provide a context of differences between the osteolytic potential of wear debris from historical and highly crosslinked UHMWPE materials, we compared pathophysiologic changes in periprosthetic hip tissues. Unfortunately, because gamma air-sterilized UHMWPE material is no longer used, the retrieved historical components have a significantly higher implantation time as compared to the highly crosslinked components. While this is of concern, we believe that providing a context related to differences in the pathophysiologic tissue responses at the time of revision is very valuable in obtaining a better understanding of implant longeveity.

In this study, bright field microscopic images of H&E stained tissue sections were evaluated for inflammatory cell infiltration, foreign body giant cells (GCs), histiocytes, necrosis, and fibrocartilage. A corresponding polarized light microscopic image of each tissue section was also analyzed to determine the presence of relatively small (0.5–2 μm) and large (>2 μm) UHMWPE particles and to provide an evaluation of the histomorphology in the regions associated with wear debris. We believe this analysis provided a unique comparison of the pathophysiologic tissue responses highlighted in the previous sections for the historical UHMWPE components.

In our comparison of historical and highly crosslinked UHMWPE component tissue histomorphology, we showed more prevalent and extensive inflammation, histiocytosis, necrosis, and wear debris in tissues retrieved from gamma air-sterilized UHMWPE hip implants. The presence of inflammation and histiocytes correlated with the accumulation of small UHMWPE wear debris (0.5–2 μm), and GCs were consistently found near large UHMWPE particles (>2 μm). In addition, the presence of histiocytes and small wear debris correlated with focal regions of tissue necrosis. Our finding that the presence of histiocytes with ingested small wear debris and GCs associated with large wear debris from historical implant components are in agreement with those reported by others [91–93]. These findings are consistent with inflammatory-mediated histomorphic changes of periprosthetic tissue associated with phagocytosis of a continuous influx of UHMWPE wear particles, cellular activation, and production of cytokines and chemokines, ultimately resulting in cell death and focal necrosis of the tissues [61]. Collectively, these factors can contribute to tissue dysfunction, pain, osteolysis, and implant loosening.

For the highly crosslinked cohort of patients, inflammation and necrosis were the most consistent findings, although histiocytes were observed in the majority of these patient tissues. The presence of histiocytes, without measureable wear debris, suggested that wear debris below the 0.5 μm detection limit might be present. Tissue from several patients also contained metal and fibrocartilage, suggesting that micromotion of the femoral stem may contribute to metal wear generation and fibrocartilage formation.

In general, our observations imply that the overall tissue response to wear debris and/or mechanical stimuli directly contribute to implant loosening of both gamma air-sterilized and highly crosslinked UHMWPE component hip implants, and these factors may ultimately result in the need for revision surgery. As the highly crosslinked UHMWPE components are retrieved with increasing times of implantation, we will have the opportunity to perform a more direct comparison of the pathophysiologic tissue response between the historical and highly crosslinked UHMWPE component implants, providing more conclusive answers to the question of biological compatibility.

## 23.10 CONCLUSION

This chapter summarizes the current understanding of both the morphologic and inflammatory-related tissue changes that result from the generation of UHMWPE wear debris. The overall tissue response to wear debris is both complex and patient specific. However, the limitations of earlier methods of tissue preservation and *in vitro* and *in vivo* model systems are being increasingly recognized. In addition, technological methods for tissue analysis are continually being updated, such as the use of protein microarrays to measure specific profiles of proteins and cell signaling factors. With new technological advances and increased implantation time comparisons for highy crosslinked UHMWPE components, an improved understanding of regional and systemic immune responses to UHMWPE wear debris will continue to evolve. Ultimately, a better understanding of these processes will provide insight into the early diagnosis and treatment of inflammatory mechanisms involved in the loss of implant function.

## 23.11 ACKNOWLEDGMENTS

The research outlined in this chapter was supported by NIH R01 AR47904.

## REFERENECES

1. Goodman S. Wear particulate and osteolysis. *Orthop Clin North Am* 2005 January;**36**(1):41–8.
2. Goldring SR, Clark CR, Wright TM. The problem in total joint arthroplasty: aseptic loosening. [comment]. *J Bone Joint Surg Am* 1993 June;**75**(6):799–801.
3. Purdue PE, Koulouvaris P, Potter HG, Nestor BJ, Sculco TP. The cellular and molecular biology of periprosthetic osteolysis. *Clin Orthop* 2007 January;**454**:251–61.
4. Willert HG. Reactions of the articular capsule to wear products of artificial joint prostheses. *J Biomed Mater Res* 1977 March;**11**(2): 157–64.
5. Urban RM, Jacobs JJ, Tomlinson MJ, Gavrilovic J, Black J, Peoc'h M. Dissemination of wear particles to the liver, spleen, and abdominal lymph nodes of patients with hip or knee replacement. [see comment]. *J Bone Joint Surg Am* 2000 April;**82**(4):457–76.
6. Looney RJ, Schwarz EM, Boyd A, O'Keefe RJ. Periprosthetic osteolysis: an immunologist's update. *Curr Opin Rheumatol* 2006 January;**18**(1):80–7.
7. Malik MHA, Jury F, Bayat A, Ollier WER, Kay PR. Genetic susceptibility to total hip arthroplasty failure: a preliminary study on the influence of matrix metalloproteinase 1, interleukin 6 polymorphisms and vitamin D receptor. *Ann Rheum Dis* 2007 August;**66**(8): 1116–20.
8. Wilkinson JM, Wilson AG, Stockley I, Scott IR, Macdonald DA, Hamer AJ, et al. Variation in the TNF gene promoter and risk of osteolysis after total hip arthroplasty. *J Bone Miner Res* 2003 November;**18**(11):1995–2001.
9. Harris WH. Osteolysis and particle disease in hip replacement. A review. *Acta Orthop Scand* 1994 February;**65**(1):113–23.
10. Hicks DG, Judkins AR, Sickel JZ, Rosier RN, Puzas JE, O'Keefe RJ. Granular histiocytosis of pelvic lymph nodes following total hip arthroplasty. The presence of wear debris, cytokine production, and immunologically activated macrophages. [see comment]. *J Bone Joint Surg Am* 1996 April;**78**(4):482–96.
11. *Fundamental immunology*. 6th ed. Lippincott: Williams & Wilkins; 2008.
12. Gallo J, Kaminek P, Ticha V, Rihakova P, Ditmar R. Particle disease. A comprehensive theory of periprosthetic osteolysis: a review. *Biomed Pap Med Fac Palacky Univ Olomouc, Czech Repub* 2002 December;**146**(2):21–8.
13. Konttinen YT, Zhao D, Beklen A, Ma G, Takagi M, Kivela-Rajamaki M, et al. The microenvironment around total hip replacement prostheses. *Clin Orthop* 2005 January(430):28–38.
14. Willert HG, Bertram H, Buchhorn GH. Osteolysis in alloarthroplasty of the hip. The role of ultra-high molecular weight polyethylene wear particles. *Clin Orthop* 1990 September(258):95–107.
15. Schmalzried T, Campbell P, Schmitt A, Brown I, Amstutz H. Shapes and dimensional characteristics of polyethylene wear particles generated in vivo by total knee replacements compared to total hip replacements. *J Biomed Mater Res: Appl Biomater* 1997;**38**:203–10.
16. Shanbhag AS, Bailey HO, Hwang DS, Cha CW, Eror NG, Rubash HE. Quantitative analysis of ultrahigh molecular weight polyethylene (UHMWPE) wear debris associated with total knee replacements. *J Biomed Mater Res* 2000;**53**(1):100–10.
17. Tipper J, Galvin A, Williams S, McEwen H, Stone M, Ingham E, et al. Isolation and characterization of UHMWPE wear particles down to ten nanometers in size from in vitro hip and knee joint simulators. *J Biomed Mater Res* 2006;**78A**:473–80.

18. Elfick APD, Green SM, Krikler S, Unsworth A. The nature and dissemination of UHMWPE wear debris retrieved from periprosthetic tissue of THR. *J Viomws Mater Res* 2003;**65A**:95–108.
19. Drees P, Eckardt A, Gay RE, Gay S, Huber LC. Mechanisms of disease: molecular insights into aseptic loosening of orthopedic implants. *Nat Clin Pract Rheumatol* 2007 March;**3**(3):165–71.
20. Kobayashi A, Bonfield W, Kadoya Y, Yamac T, Freeman MA, Scott G, et al. The size and shape of particulate polyethylene wear debris in total joint replacements. *Proc Inst Mech Eng [H]* 1997;**211**(1):11–15.
21. Wang ML, Sharkey PF, Tuan RS. Particle bioreactivity and wear-mediated osteolysis. *J Arthroplasty* 2004 December;**19**(8): 1028–38.
22. Takagi M, Santavirta S, Ida H, Ishii M, Mandelin J, Konttinen YT. Matrix metalloproteinases and tissue inhibitors of metalloproteinases in loose artificial hip joints. *Clin Orthop* 1998 July(352):35–45.
23. Solovieva SA, Ceponis A, Konttinen YT, Takagi M, Suda A, Eklund KK, et al. Mast cells in loosening of totally replaced hips. *Clin Orthop* 1996 January(322):158–65.
24. Chiba Jr J, Schwendeman LJ, Booth RE, Crossett LS, Rubash HE A biochemical, histologic, and immunohistologic analysis of membranes obtained from failed cemented and cementless total knee arthroplasty. [see comment]. *Clin Orthop* 1994 February(299):114–24.
25. Jiranek WA, Machado M, Jasty M, Jevsevar D, Wolfe HJ, Goldring SR, et al. Production of cytokines around loosened cemented acetabular components. Analysis with immunohistochemical techniques and in situ hybridization. [see comment]. *J Bone Joint Surg Am* 1993 June;**75**(6):863–79.
26. Shanbhag AS, Jacobs JJ, Black J, Galante JO, Glant TT. Cellular mediators secreted by interfacial membranes obtained at revision total hip arthroplasty. *J Arthroplasty* 1995 August;**10**(4):498–506.
27. Goldring SR, Jasty M, Roelke MS, Rourke CM, Bringhurst FR, Harris WH. Formation of a synovial-like membrane at the bone-cement interface. Its role in bone resorption and implant loosening after total hip replacement. *Arthritis Rheum* 1986 July;**29**(7):836–42.
28. Koulouvaris P, Ly K, Ivashkiv LB, Bostrom MP, Nestor BJ, Sculco TP, et al. Expression profiling reveals alternative macrophage activation and impaired osteogenesis in periprosthetic osteolysis. *J Orthop Res* 2008 January;**26**(1):106–16.
29. Shanbhag AS, Kaufman AM, Hayata K, Rubash HE. Assessing osteolysis with use of high-throughput protein chips. *J Bone Joint Surg Am* 2007 May;**89**(5):1081–9.
30. Koreny T, Tunyogi-Csapo M, Gal I, Vermes C, Jacobs JJ, Glant TT. The role of fibroblasts and fibroblast-derived factors in periprosthetic osteolysis. *Arthritis Rheum* 2006 October;**54**(10):3221–32.
31. Goldring SR, Schiller AL, Roelke M, Rourke CM, O'Neil DA, Harris WH. The synovial-like membrane at the bone-cement interface in loose total hip replacements and its proposed role in bone lysis. *J Bone Joint Surg Am* 1983 June;**65**(5):575–84.
32. Clohisy JC, Calvert G, Tull F, McDonald D, Maloney WJ. Reasons for revision hip surgery: a retrospective review. *Clin Orthop* 2004 December(429):188–92.
33. DeHeer DH, Engels JA, DeVries AS, Knapp RH, Beebe JD. In situ complement activation by polyethylene wear debris. *J Biomed Mater Res* 2001 January;**54**(1):12–19.
34. Wooley PH, Fitzgerald Jr RH, Song Z, Davis P, Whalen JD, Trumble S, et al. Proteins bound to polyethylene components in patients who have aseptic loosening after total joint arthroplasty. A preliminary report. [see comment]. *J Bone Joint Surg Am* 1999 May;**81**(5):616–23.
35. Bettelli E, Kuchroo VK. IL-12- and IL-23-induced T helper cell subsets: birds of the same feather flock together. *J Exp Med* 2005 January 17;**201**(2):169–71.
36. Kolls JK, Linden A. Interleukin-17 family members and inflammation. *Immunity* 2004 October;**21**(4):467–76.
37. Miossec P. Interleukin-17 in rheumatoid arthritis: if T cells were to contribute to inflammation and destruction through synergy. *Arthritis Rheum* 2003 March;**48**(3):594–601.
38. Konttinen YT, Li TF, Xu JW, Takagi M, Pirllä L, Silvennoinen T, et al. Expression of laminins and their integrin receptors in different conditions of synovial membrane and synovial membrane-like interface tissue. *Ann Rheum Dis* 1999;**58**:683–90.
39. Green T, Fisher J, Matthew J, Stone M, Ingham E. Effect of size and dose on bone resorption activity of macrophages by in vitro clinically relevant ultra high molecular weight polyethylene particles. *J Biomed Mater Res: Appl Biomater* 2000;**53**:8.
40. Green T, Fisher J, Stone M, Wroblewski B, Ingham E. Polyethylene particles of a 'critical size' are necessary fo the induction of cytokines by macrophages in vitro. *Biomaterials* 1998;**19**:6.
41. Haynes DR, Boyle SJ, Rogers SD, Howie DW, Vernon-Roberts B. Variation in cytokines induced by particles from different prosthetic materials. *Clin Orthop* 1998 July(352):223–30.
42. Sethi RK, Neavyn MJ, Rubash HE, Shanbhag AS. Macrophage response to cross-linked and conventional UHMWPE. *Biomaterials* 2003 July;**24**(15):2561–73.
43. Shanbhag AS, Jacobs JJ, Black J, Galante JO, Glant TT. Macrophage/particle interactions: effect of size, composition and surface area. *J Biomed Mater Res* 1994 January;**28**(1):81–90.
44. Fang H-W, Ho Y-C, Yang C-B, Liu H-L, Ho F-Y, Lu Y-C, et al. Preparation of UHMWPE particles and establishment of inverted macrophage cell model to investigate wear particles induced bioactivites. *J Biochem Biophys Methods* 2006 October 31;**68**(3): 175–87.
45. Howie DW, Vernon-Roberts B, Oakeshott R, Manthey B. A rat model of resorption of bone at the cement-bone interface in the presence of polyethylene wear particles. *J Bone Joint Surg Am* 1988 February;**70**(2):257–63.
46. Pap G, Machner A, Rinnert T, Horler D, Gay RE, Schwarzberg H, et al. Development and characteristics of a synovial-like interface membrane around cemented tibial hemiarthroplasties in a novel rat model of aseptic prosthesis loosening. *Arthritis Rheum* 2001 April;**44**(4):956–63.
47. Shanbhag AS, Hasselman CT, Rubash HE. The John Charnley Award. Inhibition of wear debris mediated osteolysis in a canine total hip arthroplasty model. *Clin Orthop* 1997 November(344):33–43.
48. Kaar SG, Ragab AA, Kaye SJ, Kilic BA, Jinno T, Goldberg VM, et al. Rapid repair of titanium particle-induced osteolysis is dramatically reduced in aged mice. *J Orthop Res* 2001 March;**19**(2):171–8.
49. Merkel KD, Erdmann JM, McHugh KP, Abu-Amer Y, Ross FP, Teitelbaum SL. Tumor necrosis factor-alpha mediates orthopedic implant osteolysis. *Am J Pathol* 1999 January;**154**(1):203–10.
50. Schwarz EM, Benz EB, Lu AP, Goater JJ, Mollano AV, Rosier RN, et al. Quantitative small-animal surrogate to evaluate drug efficacy in preventing wear debris-induced osteolysis. *J Orthop Res* 2000 November;**18**(6):849–55.
51. Wooley PH, Morren R, Andary J, Sud S, Yang S-Y, Mayton L, et al. Inflammatory responses to orthopaedic biomaterials in the murine air pouch. *Biomaterials* 2002 January;**23**(2):517–26.
52. Yang SY, Mayton L, Wu B, Goater JJ, Schwarz EM, Wooley PH. Adeno-associated virus-mediated osteoprotegerin gene transfer

protects against particulate polyethylene-induced osteolysis in a murine model. *Arthritis Rheum* 2002 September;**46**(9):2514–23.

53. Yang S, Ren W, Park Y, Sieving A, Hsu S, Nasser S, et al. Diverse cellular and apoptotic responses to variant shapes of UHMWPE particles in a murine model of inflammation. *Biomaterials* 2002;**23**:3535–43.
54. Yang S, Wu B, Mayton L, Evans CH, Robbins PD, Wooley PH. IL-1Ra and vIL-10 gene transfer using retroviral vectors ameliorates particle-associated inflammation in the murine air pouch model. *Inflamm Res* 2002 July;**51**(7):342–50.
55. Yang S-Y, Mayton L, Wu B, Goater JJ, Schwarz EM, Wooley PH. Adeno-associated virus-mediated osteoprotegerin gene transfer protects against particulate polyethylene-induced osteolysis in a murine model. *Arthritis Rheum* 2002 September;**46**(9):2514–23.
56. Ren W, Wu B, Peng X, Hua J, Hao H-N, Wooley PH. Implant wear induces inflammation, but not osteoclastic bone resorption, in RANK(-/-) mice. *J Orthop Res* 2006 August;**24**(8):1575–86.
57. Ren W, Wu B, Peng X, Mayton L, Yu D, Ren J, et al. Erythromycin inhibits wear debris-induced inflammatory osteolysis in a murine model. *J Orthop Res* 2006 February;**24**(2):280–90.
58. Ren W, Zhang R, Markel DC, Wu B, Peng X, Hawkins M, et al. Blockade of vascular endothelial growth factor activity suppresses wear debris-induced inflammatory osteolysis. *J Rheumatol* 2007 January;**34**(1):27–35.
59. von Knoch F, Heckelei A, Wedemeyer C, Saxler G, Hilken G, Brankamp J, et al. Suppression of polyethylene particle-induced osteolysis by exogenous osteoprotegerin. *J Biomed Mater Res Part A* 2005 November 1;**75**(2):288–94.
60. Abu-Amer Y, Darwech I, Clohisy JC. Aseptic loosening of total joint replacements: mechanisms underlying osteolysis and potential therapies. *Arthritis Res Ther* 2007;**9**(Suppl. 1):S6.
61. Bos I, Fredebold D, Diebold J, Lohrs U. Tissue reactions to cemented hip sockets. Histologic and morphometric autopsy study of 25 acetabula. *Acta Orthop Scand* 1995 February;**66**(1):1–8.
62. Fornasier V, Wright J, Seligman J. The histomorphologic and morphometric study of asymptomatic hip arthroplasty. A postmortem study. *Clin Orthop* 1991 October(271):272–82.
63. Santavirta S, Konttinen YT, Bergroth V, Eskola A, Tallroth K, Lindholm TS. Aggressive granulomatous lesions associated with hip arthroplasty. Immunopathological studies. *J Bone Joint Surg Am* 1990 February;**72**(2):252–8.
64. Schmalzried TP, Kwong LM, Jasty M, Sedlacek RC, Haire TC, O'Connor DO, et al. The mechanism of loosening of cemented acetabular components in total hip arthroplasty. Analysis of specimens retrieved at autopsy. *Clin Orthop* 1992 January(274):60–78.
65. Kaplan FS, Glaser DL, Hebela N, Shore EM. Heterotopic ossification. *J Am Acad Orthop Surg* 2004 March–April;**12**(2):116–25.
66. Liu K, Tripp S, Layfield LJ. Heterotopic ossification: review of histologic findings and tissue distribution in a 10-year experience. *Pathol Res Pract* 2007;**203**(9):633–40.
67. Rifas L. T-cell cytokine induction of BMP-2 regulates human mesenchymal stromal cell differentiation and mineralization. *J Cell Biochem* 2006 July 1;**98**(4):706–14.
68. Steiner DRS, Gonzalez NC, Wood JG. Mast cells mediate the microvascular inflammatory response to systemic hypoxia. *J Appl Physiol* 2003 January;**94**(1):325–34.
69. Kobayashi A, Freeman MA, Bonfield W, Kadoya Y, Yamac T, Al-Saffar N, et al. Number of polyethylene particles and osteolysis in total joint replacements. A quantitative study using a tissue-digestion method. *J Bone Joint Surg Br* 1997 September;**79**(5):844–8.
70. Kadoya Y, Revell PA, Kobayashi A, al-Saffar N, Scott G, Freeman MA. Wear particulate species and bone loss in failed total joint arthroplasties. *Clin Orthop* 1997 July(340):118–29.
71. Kadoya Y, Kobayashi A, Ohashi H. Wear and osteolysis in total joint replacements. [see comment]. *Acta Orthop Scand Suppl* 1998 February;**278**:1–16.
72. Maloney WJ, Smith RL. Periprosthetic osteolysis in total hip arthroplasty: the role of particulate wear debris. *Instr Course Lect* 1996;**45**:171–82.
73. Greenfield EM, Bi Y, Ragab AA, Goldberg VM, Van De Motter RR. The role of osteoclast differentiation in aseptic loosening. *J Orthop Res* 2002 January;**20**(1):1–8.
74. Sabokbar A, Fujikawa Y, Neale S, Murrary D, Athanasou N. Human arthroplasty derived macrophages differentiate into osteoclastic bone resorbing cells. *Ann Rheum Dis* 1997;**56**:414–20.
75. Haynes DR, Crotti TN, Potter AE, Loric M, Atkins GJ, Howie DW, et al. The osteoclastogenic molecules RANKL and RANK are associated with periprosthetic osteolysis. *J Bone Joint Surg Br* 2001 August;**83**(6):902–11.
76. Xu JW, Konttinen YT, Waris V, Patiala H, Sorsa T, Santavirta S. Macrophage-colony stimulating factor (M-CSF) is increased in the synovial-like membrane of the periprosthetic tissues in the aseptic loosening of total hip replacement (THR). *Clin Rheumatol* 1997 May;**16**(3):243–8.
77. Gehrke T, Sers C, Morawietz L, Fernahl G, Neidel J, Frommelt L, et al. Receptor activator of nuclear factor kappaB ligand is expressed in resident and inflammatory cells in aseptic and septic prosthesis loosening. *Scand J Rheumatol* 2003;**32**(5):287–94.
78. Crotti TN, Smith MD, Findlay DM, Zreiqat H, Ahern MJ, Weedon H, et al. Factors regulating osteoclast formation in human tissues adjacent to peri-implant bone loss: expression of receptor activator NFkappaB, RANK ligand and osteoprotegerin. *Biomaterials* 2004 February;**25**(4):565–73.
79. Devlin RD, Reddy SV, Savino R, Ciliberto G, Roodman GD. IL-6 mediates the effects of IL-1 or TNF, but not PTHrP or 1,25(OH)2D3, on osteoclast-like cell formation in normal human bone marrow cultures. *J Bone Mine Res* 1998 March;**13**(3):393–9.
80. Manolagas SC, Jilka RL. Bone marrow, cytokines, and bone remodeling. Emerging insights into the pathophysiology of osteoporosis. *N Engl J Med* 1995 February 2;**332**(5):305–11.
81. Boyle WJ, Simonet WS, Lacey DL. Osteoclast differentiation and activation. *Nature* 2003 May 15;**423**(6937):337–42.
82. Suda T, Kobayashi K, Jimi E, Udagawa N, Takahashi N. The molecular basis of osteoclast differentiation and activation. *Novartis Found Sym* 2001;**232**:235–47. discussion 247-250.
83. Della Valle CJ, Zuckerman JD, Di Cesare PE. Periprosthetic sepsis. *Clin Orthop* 2004 March(420):26–31.
84. Schroeder JH, Morawietz L, Matziolis G, Leutloff D, Gehrke T, Krenn V, et al. Loosening of joint arthroplasty-the potential of the periprosthetic membrane. *J Bone Joint Surg-Br* 2008;**88-B**(SUPP_I, 64).
85. Robbins GM, Masri BA, Garbuz DS. Duncan CP. Evaluation of pain in patients with apparently solidly fixed total hip arthroplasty components. *J Am Acad Orthop Surg* 2002 March–April;**10**(2):86–94.
86. Cazzavillan S, Ratanarat R, Segala C, Corradi V, de Cal M, Cruz D, et al. Inflammation and subclinical infection in chronic kidney disease: a molecular approach. *Blood Purif* 2007;**25**(1):69–76.
87. Pajkos A, Deva AK, Vickery K, Cope C, Chang L, Cossart YE. Detection of subclinical infection in significant breast implant capsules. [see comment]. *Plast Reconstr Surg* 2003 April;**111**(5): 1605–11.

88. Savarino L, Baldini N, Tarabusi C, Pellacani A, Giunti A. Diagnosis of infection after total hip replacement. *J Biomed Mater Res Part B, Appl Biomater* 2004 July 15;**70**(1):139–45.
89. Bi Y, Seabold JM, Kaar SG, Ragab AA, Goldberg VM, Anderson JM, et al. Adherent endotoxin on orthopedic wear particles stimulates cytokine production and osteoclast differentiation. *J Bone Mine Res* 2001 November;**16**(11):2082–91.
90. Goodman SB, Chin RC, Chiou SS, Schurman DJ, Woolson ST, Masada MP. A clinical pathologic biochemical study of the membrane surrounding loosened and nonloosened total hip arthroplasties. *Clin Orthop* 1989 July(244):182–7.
91. Athanasou NA. The pathology of joint replacement. *Curr Diagn Pathol* 2002;**8**(1):26–32.
92. Ito S, Matsumoto T, Enomoto H, Shindo H. Histological analysis and biological effects of granulation tissue around loosened hip prostheses in the development of osteolysis. *J Orthop Sci* 2004;**9**(5):478–87.
93. Kim KJ, Chiba J, Rubash HE. In vivo and in vitro analysis of membranes from hip prostheses inserted without cement. *J Bone Joint Surg Am* 1994 February;**76**(2):172–80.

Chapter 24

# Characterization of Physical, Chemical, and Mechanical Properties of UHMWPE

Stephen Spiegelberg, PhD



## 24.1 INTRODUCTION

There are many properties to consider when developing a new UHMWPE material for implant applications. The characteristics of the UHMWPE powder may be measured to verify that there are not variations in different manufacturing lots. After consolidation, when the polyethylene powder is compressed into a solid slab of material via ram extrusion, compression molding, or other techniques including isostatic pressing, manufacturers will test the solid polyethylene slab to determine the quality of the consolidation and to determine if any deleterious effects have occurred. Highly crosslinked UHMWPE is commonly used now, whereby the polyethylene slab is exposed to radiation, which forms chemical bonds between polymer chains. Several analytical techniques have been developed to assess if the material has been exposed to enough radiation to improve the desirable properties of the UHMWPE, while at the same time ensuring that other properties are not diminished by the radiation.

This chapter summarizes the common test techniques used in the orthopedic community today for analyzing UHMWPE during the various stages of processing. The techniques described in this chapter can also be used to characterize explanted (retrieved) components in an effort to better understand the reasons for revision. The tests can be applied to UHMWPE powder, consolidated stock, or irradiated stock, shown in Figure 24.1.

## 24.2 WHAT DOES THE FDA REQUIRE?

The Food and Drug Administration (FDA) has set guidelines for the battery of tests a researcher should conduct on a new UHMWPE material prior to its gaining approval for implantation into humans. This list represents the minimum number of tests that should be conducted; individuals may

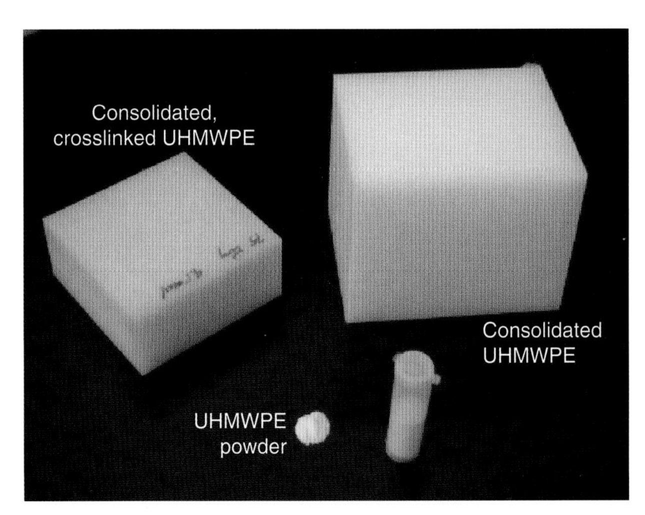


**FIGURE 24.1** UHMWPE powder, consolidated stock, and irradiated stock. The irradiated stock has not been thermally treated and shows the characteristic yellow color caused by free radicals generated from electron beam.

want to conduct more tests on their own, and the FDA may require other tests depending on the nature of the material being considered for FDA clearance.

Table 24.1 summarizes some of the tests discussed in the FDA guidance document *Data Requirements for Ultra High Molecular Weight Polyethylene (UHMWPE) Used in Orthopedic Devices* [1]. If the new polyethylene has almost identical properties to a polyethylene already on the market after stage 1 testing, no further testing may be required. If it does differ in properties from an existing FDA-approved polyethylene, stage 2 testing will be necessary. If the stage 2 test results are also different from existing FDA-approved polyethylenes, stage 3 tests may be required.

The American Standards for Testing and Materials (ASTM) has compiled a list of the recommended properties that UHMWPE components must meet or exceed if they are to be used in orthopedic devices. This list is summarized in ASTM F648 and D4020-01a. The FDA usually refers to these standards when considering a new UHMWPE for orthopedic implants. A summary of all possible tests, including their ASTM standard numbers, where applicable, are summarized in Table 24.2 and indicate where the individual tests are appropriate for UHMWPE powder, consolidated stock, or irradiated stock.

**TABLE 24.1 Tests Required by the FDA for a New UHMWPE Submission**

| Stage 1 tests | Stage 2 tests | Stage 3 tests |
|---|---|---|
| Ultimate tensile strength | Creep | Biocompatibility |
| Yield strength | Wear | |
| Young's modulus | Fatigue | |
| Poisson's ratio | Crack propagation | |
| % elongation | J-Integral | |
| Molecular weight | Thin sectioned photomicrograph | |
| Density and porosity | IR spectra and chemical structure | |
| % crystallinity | | |
| Glass transition temperature, $T_g$ | | |
| Crystallization temperature range, $T_c$ | | |
| Melting temperature, $T_m$ | | |
| Oxidation temperature, $T_o$ | | |

## 24.3 PHYSICAL PROPERTY CHARACTERIZATION

The physical properties of UHMWPE include its morphology, which encompasses both microscopic and macroscopic analysis of the material, as well as thermal properties and molecular weight.

### 24.3.1 Differential Scanning Calorimetry

Differential scanning calorimetry, or DSC, is the most common technique used to measure the thermal properties of the UHMWPE and provides the melting point and degree of crystallinity, as briefly summarized in Chapter 1. Given that the glass transition of UHMWPE is around −160°C, most researchers will not measure this parameter. A picture of a DSC is shown in Figure 24.2. This unit is equipped with an autoloader, which facilitates testing of multiple samples. In DSC, a small amount of the UHMWPE, approximately 5–10 mg, is measured on an accurate microbalance and sealed in an aluminum sample pan. The sample pan, along with an empty reference pan, is placed in the DSC chamber, which houses two heaters, one for each pan. The DSC heats the two pans at a known rate, usually 10°C/min, from 0 to 200°C, and monitors the heat flow. A DSC trace for GUR 1050 powder is shown in Figure 24.3. The melting endotherm at 141°C is clearly visible. When the endotherm is integrated from 50 to 160°C ($\Delta H_{endotherm}$) and normalized with the heat of fusion of pure UHMWPE ($\Delta H_f$ = 289 J/g), the calculated degree of crystallinity is $X$ = 77%.

**TABLE 24.2 Common Test Techniques Used to Assess the Various Properties of UHMWPE**

| Physical | Chemical | Mechanical | *In Vitro* |
|---|---|---|---|
| Transmission electron microscopy[2,3] | Fourier transform infrared spectroscopy (ASTM F2102, F2381)[1,2,3] | Small punch (ASTM F2183) [2][2,3] | Accelerated aging (ASTM F2003-02)[2,3] |
| Scanning electron microscopy[1,2,3] | Electron spin resonance spectroscopy[3] | Compression (ASTM D2990)[2,3] | Wear testing (ASTM F732-00)[2,3] |
| Density (ASTM D1505)[2,3] | Gel permeation chromatography (ASTM D6474)[1,2] | Tensile (ASTM D2990, D638)[2,3] | |
| Differential scanning calorimetry (ASTM F2625)[1,2,3] | Dilute solution viscometry (ASTM D2857, F4020)[1,2] | Fatigue (ASTM E647)[2,3] | |
| Oxidation induction time, oxidation induction temperature (ASTM D3895, D2009)[2,3] | Swelling analysis (ASTM D2765, F2214)[3] | J-integral (ASTM D6068)[2,3] | |
| Fusion assessment[2] | Sol-gel (ASTM D2765)[2,3] | Creep (ASTM 2990)[2,3] | |
| | Trace element (ASTM F648)[1] | Izod impact (ASTM D256)[2,3] | |

*The superscripts indicate which tests are suitable for the UHMWPE in a powder state, after consolidation, and after crosslinking via irradiation: [1]Powder state; [2]consolidated; [3]postirradiation.*

**FIGURE 24.2** DSC with autoloader.

$$\%X = \frac{\Delta H_{endotherm}}{\Delta H_f}\%$$

The degree of crystallinity will depend on the integration range used and the heat of fusion of pure UHMWPE. For this reason, an integration range of 50–160°C and a heat of fusion of 289.3 J/g was standardized for UHMWPE in ASTM F2625 in 2007. Researchers should report the integration range and degree of fusion used when reporting crystallinity of UHMWPE.

After consolidation, the melting point and the degree of crystallinity decrease, usually down to 140°C and 60%, respectively. This fact is illustrated in the second heat of the GUR 1050 powder shown in Figure 24.3. For a DSC trace of consolidated UHMWPE, see Figure 1.8 in Chapter 1. The crystallinity further decreases after radiation if melt annealing is performed following radiation. Consequently, DSC should be performed after each of these processing steps.

Oxidation induction time (OIT) measurements are less commonly performed, but they do provide some information about the oxidative stability of the UHMWPE. In this test, according to ASTM D3895, the DSC is heated to 200°C then held isothermally. The purge gas, which is normally nitrogen, is then switched to pure oxygen. The time to the start of the degradation exotherm caused by oxidation is reported as the OIT. Samples with a longer OIT are more oxidation stable under these conditions than those with a shorter OIT. Open sample pans are used in this test technique. An example of an OIT measurement is shown in Figure 24.4 for UHMWPE. An alternative technique is the oxidation onset temperature (OOT) as described in ASTM D2009. Similar to oxidation induction time, UHMWPE samples are placed in open pans and heated in

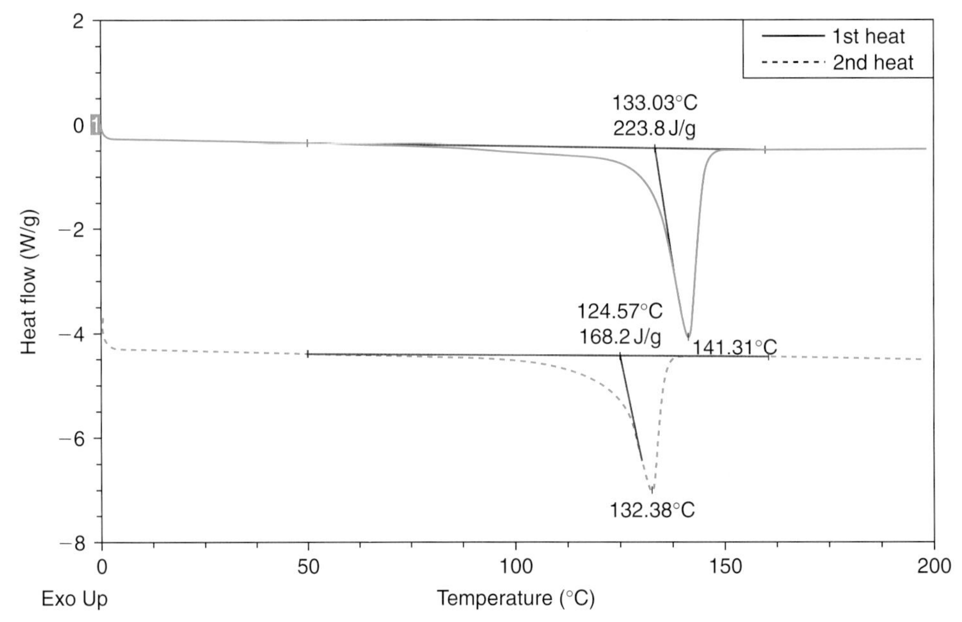


**FIGURE 24.3** DSC trace of GUR 1050 powder (first heat and second heat). The crystallinity of the first heat, based on the melting endotherm and a heat of fusion of 289 J/g, was 77% based on an integration range of 50–160°C. The crystallinity of the second heat dropped to 58%.

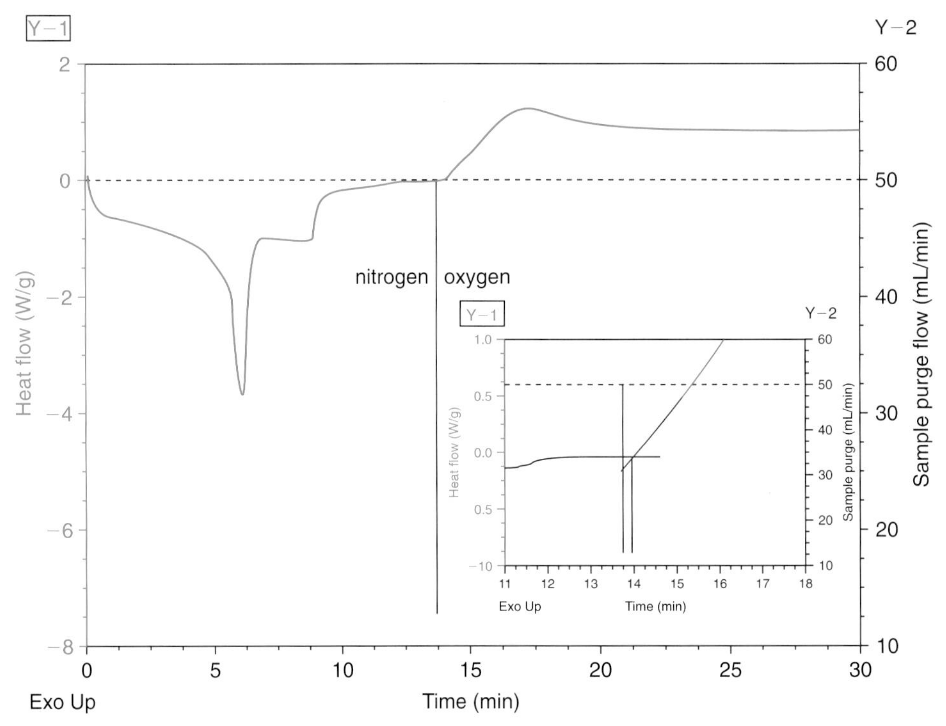


**FIGURE 24.4** Oxidation induction time trace of UHMWPE. After achieving 200°C, the nitrogen purge is switched over to pure oxygen. The oxidation exotherm occurs shortly after. The time difference between the intersection of the extrapolated lines before and during the exotherm (14.1 minutes) and the switch over to oxygen (13.75 minutes) is the OIT, which is 0.35 minutes in this example. This calculation is shown in the inset plot.

a DSC. However, pure oxygen is used for the entire heating run, and the temperature is increased until an exotherm appears. The onset temperature of this exotherm is reported as the OOT and again provides an indication of the oxidation resistance of the material.

## 24.3.2 Scanning Electron Microscopy

Scanning electron microscopy (SEM) is used to examine the morphology of the powder before consolidation. The procedure is straightforward: Two-sided carbon tape is

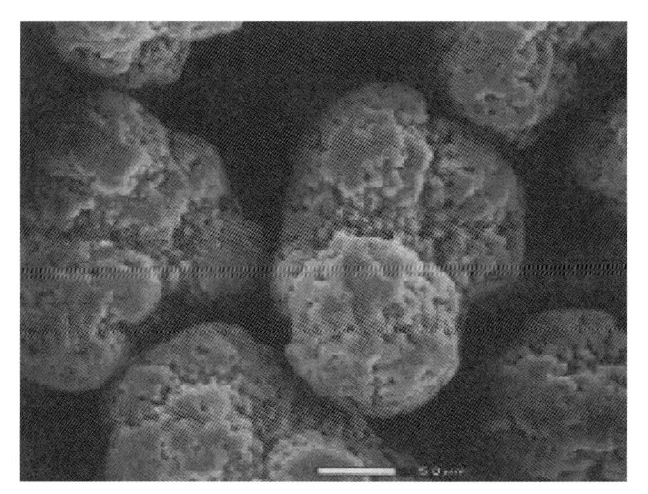

FIGURE 24.5 Scanning electron micrograph of UHMWPE powder.

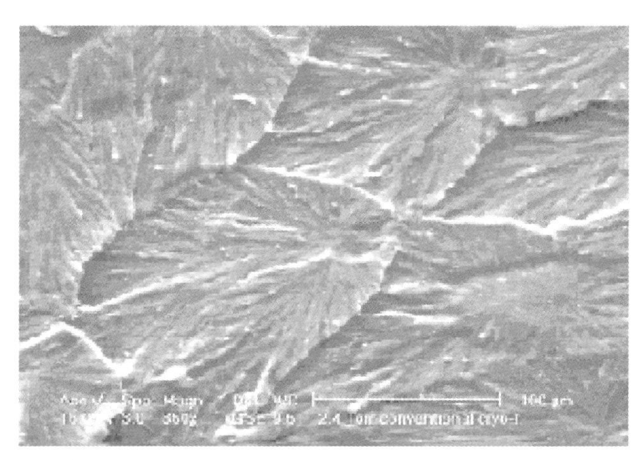

FIGURE 24.6 Scanning electron micrograph of consolidated UHMWPE, showing freeze-fractured surface.

fixed to a SEM sample stub, and the UHMWPE powder is sprinkled onto the surface. A light gold, platinum, or carbon coating is applied (≈100 Å), and the sample is examined in an SEM chamber. A typical SEM micrograph of UHMWPE powder is shown in Figure 24.5. The flakes are 50–100 μm in diameter. SEM can also be used to examine the consolidated resin, as shown in Figure 24.6. This sample was freeze fractured and clearly shows the UHMWPE flakes on the fracture surface. Voids in the sample due to poor consolidation would be readily visible.

### 24.3.3 Intrinsic Viscosity

The intrinsic viscosity (IV) is related to the molecular weight of the polymer through the Mark-Houwink relationship [3]. The technique used to perform IV measurements is described in ASTM D2857 and D4020. A Ubbelohde No. 1 viscometer, as shown in Figure 24.7, is placed in an oil bath at 135°C. The polymer powder is

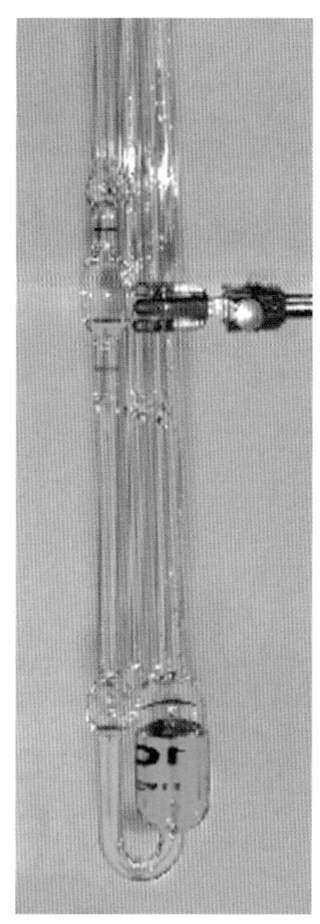

FIGURE 24.7 Ubbelohde viscometer.

dissolved in decahydronaphthalene at 150°C with a volume of solvent in milliliters equal to 1.8 times the mass of UHMWPE in milligrams. After mixing for 1 hr, the solution is transferred to the viscometer and allowed to equilibrate at 135°C. The elution time through the capillary of the viscometer is measured $t_s$ and compared with the elution time of the pure solvent $t_o$ and the constant for the viscometer, $k$. The relative viscosity $\eta_r$, specific viscosity $\eta_{sp}$, intrinsic viscosity $[\eta]$, and viscosity-average molecular weight are computed as follows:

$$\eta_r = \frac{t_s - k/t_s}{t_o - k/t_o} = \eta_{sp} + 1$$

$$[\eta] = (2\eta_{sp} - 2\ln\eta_r)^{1/2}/c$$

$$M_v = 5.37 \times 10^4[\eta]^{1.37}$$

The latter equation is termed the Margolies equation and is based on the Mark-Houwink parameters for polymer-solvent systems. It is only relevant for UHMWPE at 135°C in decahydronaphthalene. Other values have been reported for the power value, which will obviously affect the reported molecular weight [4]. Consequently, researchers should describe which expression is used when reporting a viscosity-averaged molecular weight.

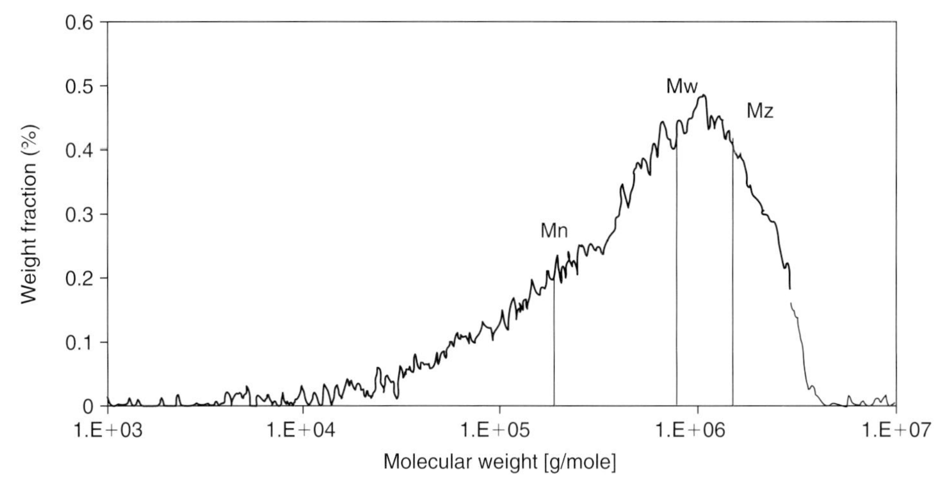


**FIGURE 24.8** Molecular weight distribution for UHMWPE, relative to UHMWPE standards. The number-average, weight-average, and z-average molecular weights are shown ($M_n$, $M_w$, $M_z$).

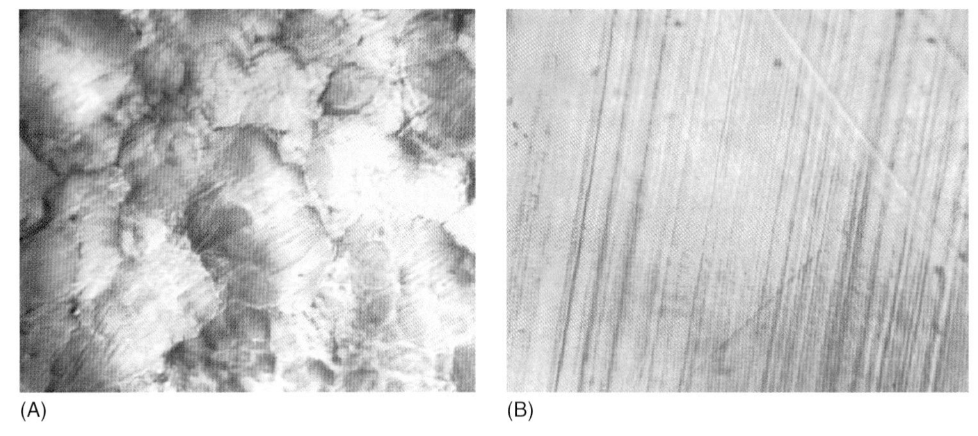


**FIGURE 24.9** Optical micrograph of microtomed UHMWPE films prepared by compression molding. The image on the left shows clear boundaries between the original UHMWPE flakes, indicating poor consolidation. The image on the right shows good consolidation. The parallel lines in this micrograph are knife marks caused during microtoming (magnification: 15×).

#### *24.3.3.1 Gel Permeation Chromatography*

An alternative technique to measure the molecular weight is via gel permeation chromatography (GPC). In GPC, the polymer is again dissolved in a suitable solvent, which for UHMWPE is usually 1,2,4-trichlorobenzene at 135°C, in relation to ASTM D6474. The dilute solution is injected in a flow stream that passes through columns containing a rigid gel with different porosities. The smaller polymer chains are temporarily entrained by the pores in the gel, while the larger polymer chains pass through the column with less pore interaction. As a result, the chains become separated according to size. As they leave the gel columns, a refractometer detector measures the relative concentration of material. By calibrating the columns with polymers of known molecular weight, the molecular weight distribution of the UHMWPE can be determined. A typical distribution for UHMWPE is shown in Figure 24.8, where UHMWPE standards were used. The moments of molecular weight were calculated from the distribution and are plotted on the curve [5]. This technique, like any technique that requires solvating the UHMWPE, will only measure the molecular weight of the soluble portion of the material.

### 24.3.4 Fusion Assessment

The morphology of the consolidated UHMWPE should be analyzed to ensure that proper fusion has occurred during the molding process. Transmission optical microscopy is an easy approach to examining this morphology. A microtomed slice of material is placed under a microscope and examined at 10–40×. According to the morphology evaluation protocol described in ASTM F648 (Annex 2), UHMWPE films of at least 100 μm thickness are viewed under transmitted light or dark field illumination at no less than 40× magnification. The images shown in Figure 24.9 show a sample with poor consolidation,

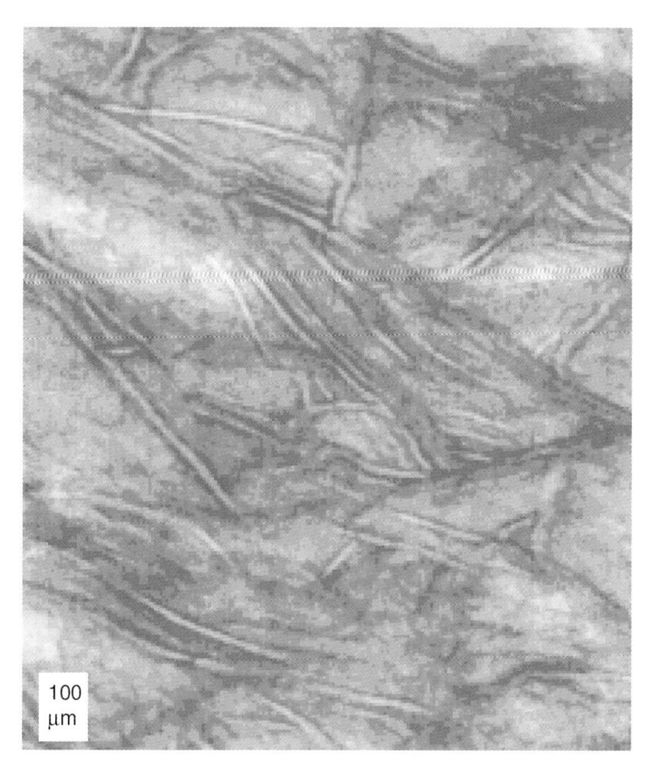


FIGURE 24.10 Transmission electron micrograph of consolidated UHMWPE, showing lamellae.

indicated by pronounced grain boundaries, and a sample with good consolidation.

## 24.3.5 Transmission Electron Microscopy

The morphology of the consolidated UHMWPE can be examined with transmission electron microscopy (TEM) to a very high degree of magnification. Structures down to 100 nm can be readily viewed with this technique. Changes in crystalline structure with consolidation and radiation treatment can also be viewed. As opposed to the surface analysis technique of SEM, TEM takes a spatial average of the polymer morphology, relying on differences in electron density to yield structure. Consequently, microtoming is required to yield 50–200 nm thick slices, which are then stained with chlorosulfonic acid to enhance contrast. An example of the ribbon-like lamellae found in UHMWPE is shown in Figure 24.10. This micrograph shows primary large lamellae, with smaller lamellae interspersed.

## 24.3.6 Density Measurements

The density of the UHMWPE is related to both its crystallinity and any porosity that may be present. One of the more accurate techniques for measuring density involves the use of a density gradient column, as described in ASTM D1505-99e1. In this technique, a vertical cylindrical tube is filled with two or more miscible liquids that have a density range that span the density of the material to be investigated. The tube is temperature controlled because the density of the fluids will change with temperature. The resulting liquid mixture will have a linearly varying density with respect to height, with the lowest density at the top of the column. Glass floats of accurately known densities are placed at the top of the column and then fall to the height in the column where their density matches that of the fluid. These floats are used to calibrate the column. The density of the test sample is determined by placing a piece in the column, measuring the height in the column where it settles, and comparing this height with the calibration curve. For UHMWPE, an ethanol-water column will provide the necessary range of 0.72–1 g/ml density range.

Alternatively, the material can be determined by the water displacement technique. The sample is weighed carefully in air with a sensitive analytical balance. The same sample is then suspended from a wire attached to the balance then placed in water (or any liquid that will cause it to sink). The weight of the sample in the liquid, less the weight of the wire, is used to determine the mass of liquid displaced by the sample, which provides the volume of the sample. The density can then be calculated.

# 24.4 CHEMICAL PROPERTY CHARACTERIZATION

Chemical characterization of the UHMWPE powder is typically performed at the powder manufacturing site as part of their quality control protocol. However, researchers may want to periodically verify the purity and composition of the powder. ASTM 648 outlines several tests that can be used to analyze the powder. With regards to chemical characterization, trace element analysis and Fourier transform infrared spectroscopy (FTIR) are the two common approaches.

## 24.4.1 Trace Element Analysis

In trace element analysis, one is typically looking for elements that could have been inadvertently introduced during the powder manufacturing process. The most common trace elements found in UHMWPE powder include titanium, calcium, chlorine, and aluminum. Quantification of these elements is usually carried out by weighing known amounts of the powder, pyrolizing the UHMWPE through heat or microwave digestion, and performing mass spectroscopy, such as inductively-coupled plasma mass spectroscopy (ICP-MS), to quantify the yields of these materials, reporting results in parts per million. Sample preparation is important to get good quantitative results. Samples must be sealed in containers prior to pyrolization or microwave digestion of the material; if not, erroneously low concentrations can be measured due to airborne loss of materials. Chlorine has to be measured via titrimetric methods. Semiquantitative and qualitative elemental analysis on the surface of consolidated parts can be measured with energy dispersive spectroscopy (EDS), which is conducted with scanning electron microscopy.

## 24.4.2 Fourier Transform Infrared Spectroscopy

FTIR is performed to measure the chemical structure of the UHMWPE. FTIR can be performed on either the UHMWPE powder using the potassium bromide (KBr) pellet technique or on the consolidated UHMWPE. An FTIR with a microscope attachment, which is useful for examining the chemical structure as a function of position in a sample, is shown in Figure 24.11. The UHMWPE powder is blended in a 1:20 ratio with KBr powder that has been carefully dried in an oven to remove all absorbed water. The two powders should be ground together with a mortar and pestle and then placed in a KBr pellet anvil. One should place enough powder to cover the bottom of the mold, then press. If done correctly, the KBr pellet should be transparent, with semiopaque regions where the UHMWPE powder is sitting. One should collect and average at least 32 scans in transmission mode with an FTIR bench. The background spectrum should be collected on a pure KBr pellet. Any unusual peaks may indicate the presence of contaminants or oxidation. This technique is semiquantitative.

After consolidation of the UHMWPE powder into solid stock, few if any chemical changes should have occurred. Degradation, or chain scission, caused by heat or mechanical deformation, may sometimes appear after consolidation in isolated cases. FTIR can be used to see if bulk degradation has occurred in the sample.

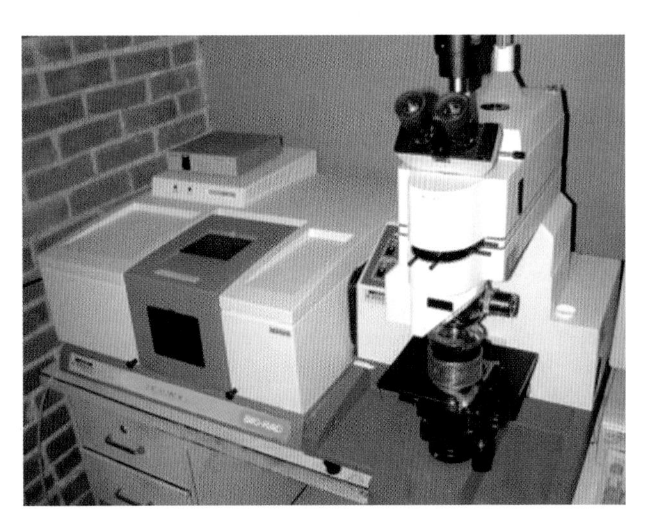

FIGURE 24.11 FTIR spectrometer with microscope attachment.

To perform FTIR on consolidated material, researchers will typically prepare a thin film of the material and then perform the infrared analysis in transmission mode. A thin film can be prepared by microtoming. A researcher will cut the consolidated UHMWPE slab with a band saw into a piece small enough to fit into the chuck of a microtome, shown in Figure 24.12. They will then microtome several millimeters away from the cut surface to avoid sampling material than may have degraded in the band saw. A 100–200 μm thick sample is then microtomed and placed in the FTIR bench or on the microscope stage of a micro-FTIR system. A minimum of 24 scans should be conducted and ratioed against a blank background. Typical spectra are shown in Figure 24.13 and Figure 24.14.

To look for evidence of degradation, researchers will usually examine the terminal vinyl group at 910 $cm^{-1}$, or

FIGURE 24.12 Sliding microtome, used for preparing thin films of UHMWPE.

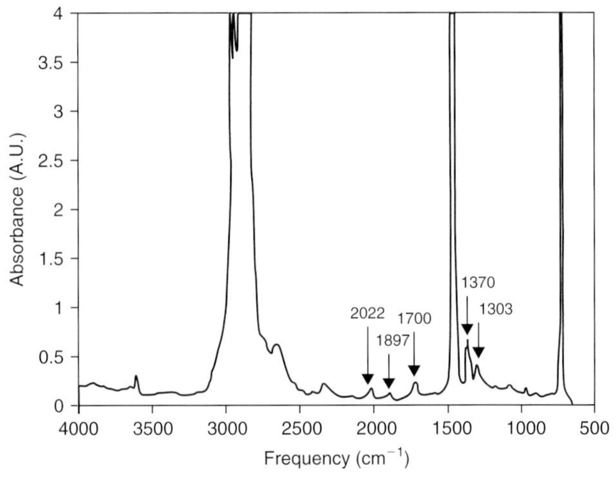


FIGURE 24.13 FTIR trace of UHMWPE irradiated to 15 Mrads and gamma sterilized in air, with key peak locations indicated.

a double bond at the end of a polymer chain, as shown below:

$$—CH{=}CH_2$$

The presence of a terminal vinyl group usually means that the polymer chain has broken, which will leave behind a vinyl group on each end.

Oxidation can also sometimes occur, which leads to carbonyl formation. Carbonyls, which include ketones, esters, and ethers, have a principal absorption peak around $1700\,cm^{-1}$. ASTM F2102 describes the technique to measure the *oxidation index*, or the area of the carbonyl absorption peak ratioed against the methylene stretch at $1396\,cm^{-1}$ [6–8].

In addition to the previously mentioned peaks, the uniformity of the received radiation dose can be determined by examining the trans-vinylene peak at $965\,cm^{-1}$ [9, 10]. Trans-vinylene groups appear in UHMWPE as a radiolytic product during exposure to ionizing radiation with a yield related to the number of crosslinks formed. Monitoring this group as a function of location in the sample allows one to determine the degree of uniformity of the received radiation dose.

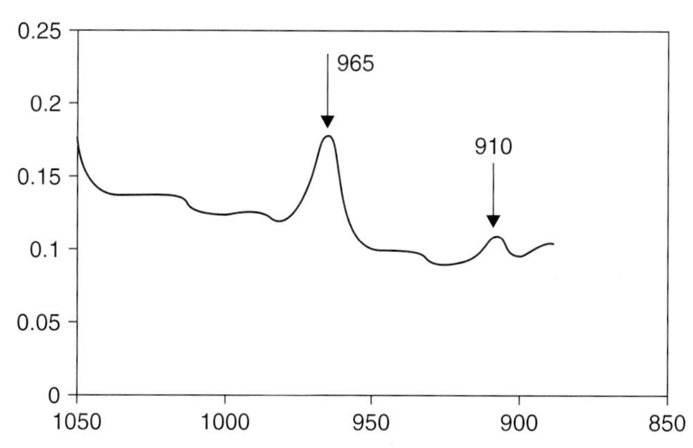


**FIGURE 24.14** Peaks of interest in the footprint region of the spectra shown in Figure 24.12.

The predecessors to terminal oxidation products, namely primary and secondary peroxides, can be identified via FTIR through a derivatization process described by Costa and coworkers [6, 11]. In this method, films of UHMWPE are placed in a vessel with nitric oxide, which reacts with peroxide groups to form primary and secondary nitrates that are visible at 1642 and $1633\,cm^{-1}$ when viewed with infrared spectroscopy.

## 24.4.3 Electron Spin Resonance

Residual free radicals trapped in the UHMWPE are known to cause long-term stability problems in implanted UHMWPE prosthesis. The free radicals can react with oxygen to form unstable hydroperoxides, which can decompose to carbonyls. The latter are readily susceptible to chain scission and subsequent loss of mechanical properties. The only known method of directly measuring residual free radicals is via electron spin resonance spectroscopy, or ESR (also known as electron pair resonance, or EPR) [12]. Figure 24.15 shows an ESR trace of unirradiated and irradiated, or sterilized, UHMWPE. The multiple peaks in the irradiated sample indicate the presence of residual free radicals. The number of free radicals per gram can be determined by performing a double

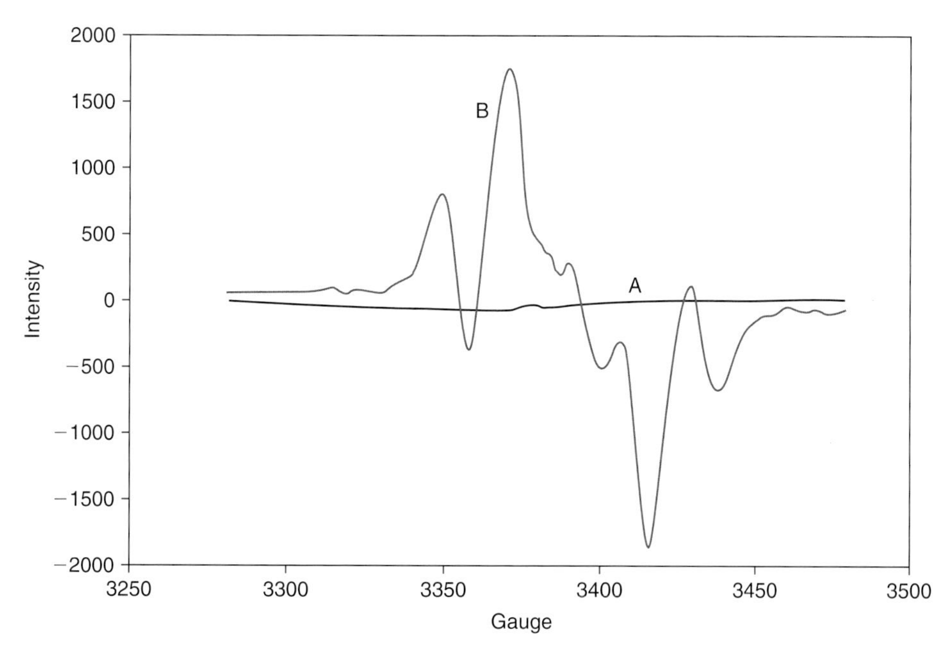


**FIGURE 24.15** ESR trace of unsterilized (A) and gamma-sterilized (in air) (B) UHMWPE.

integration on this curve. ESR is discussed in greater detail in Chapter 29.

## 24.4.4 Swell Ratio Testing

When a crosslinked UHMWPE is placed in a good solvent, it cannot dissolve, but it will swell to a certain degree. The degree of crosslinking can be indirectly determined by measuring the swell ratio of the crosslinked UHMWPE. In ASTM D2765, a crosslinked specimen is weighed, allowed to swell in a chamber containing xylene at 130°C, and is then reweighed. The swell ratio $q$ is determined from the weight gain and the densities of the UHMWPE and xylene.

An alternate approach is discussed in ASTM 2214. In this technique, the change in sample height is monitored with a probe that is in light contact with the sample. A swelling apparatus is shown in Figure 24.16, which uses a laser micrometer to monitor the height change of the sample. An example of the transient response of the swelling of crosslinked UHMWPE is shown in Figure 24.17. Based on the steady state swell ratio $q$, the interaction parameter $\chi_1$ for UHMWPE and xylene at 130°C, and the molar volume of xylene $V_1$, the crosslink density $\nu_x$ (moles of crosslinks/unit volume) can be calculated from Flory network theory [13] from the following expression:

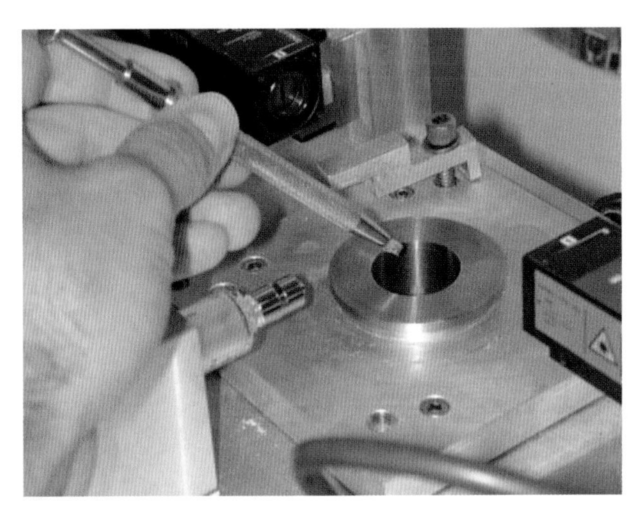

FIGURE 24.16 UHMWPE cube (4 × 4 × 4 mm) being inserted into test chamber of swelling apparatus. The laser micrometer is visible.

$$\nu_x = -\frac{\ln(1-q^{-1}) + q^{-1} + \chi_1 q^{-2}}{V_1(q^{-1/3} - q^{-1}/2)}$$

This expression assumes a network with tetrafunctional crosslinking (i.e., no branching). This technique is particularly useful for samples than do not swell isotropically due to residual stress from molding or oriented structure.

Similar to the ASTM method for swelling analysis, sol-gel measurements involve weighing, or gravimetric, methods to determine the degree of crosslinking in polymer systems. However, instead of weighing the solvent uptake in the polymer matrix, one measures the amount of material extracted (i.e., soluble) from the network. In this manner, one measures the percentage of gel, or fully crosslinked, material in the sample. UHMWPE exposed to sterilization levels of radiation (25–35 kGy) typically exhibits greater than 75% gel, with higher radiation doses approaching 100% gel. ASTM D2765 describes several methods for making these measurements. In one approach, the sample is carefully weighed, packaged in a preweighed porous stainless steel mesh bag, then placed in a flask containing boiling xylene or decahydronaphthalene and equipped with a condensor. After 24 hours, the bag and remaining sample are dried in a vacuum oven and reweighed. In another approach, the weighed samples are swelled in xylene at 110°C (the melting point of the UHMWPE is reduced by about 30°C due to the

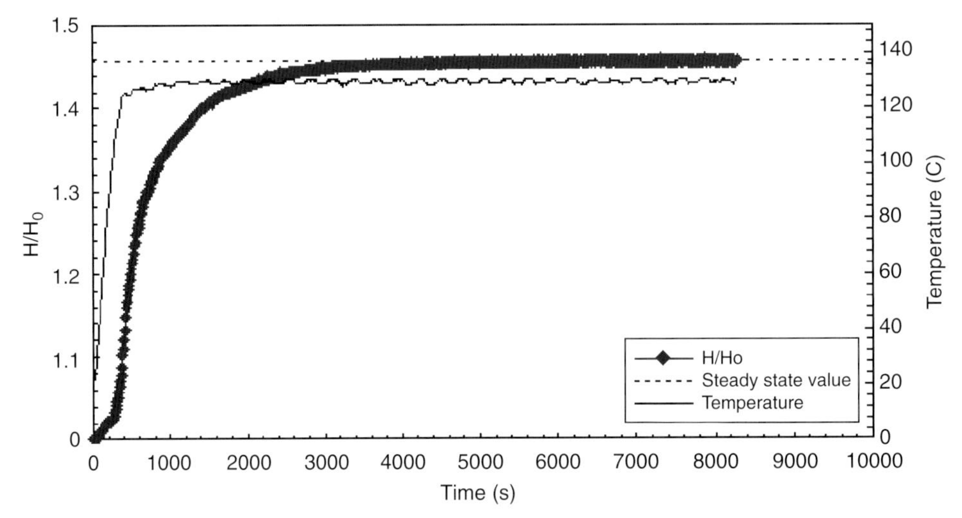


FIGURE 24.17 Transient swelling data for UHMWPE in xylene at 130°C. Steady state was achieved after approximately 1 hour.

presence of the solvent) then weighed when fully swollen. The gravimetric approach can be used effectively for polymer films and unusual shaped structures and for polymers with low crosslink density. However, they may be less effective in computing a crosslink density as the swelling method previously described because it averages the swelling over all the dimensions, and solvent loss during weighing can compromise the results. An antioxidant must be used in both of these methods to prevent thermally-induced chain scission.

Finally, some researchers have used trichlorobenzene (TCB) extraction to examine soluble UHMWPE content. In this method, 200 mg samples of UHMWPE are placed in 100 ml of TCB with some antioxidant (N-phenyl-2-naphthylamine, 0.1 wt%) and heated to 170°C for 6 hours. The solution is then hot filtered at 170°C, and the nonsoluble portion is washed in dichloromethane and dried at 105°C for 6 hours. The difference in weight before and after TCB extraction provides a measure of the nonsoluble portion.

## 24.5 MECHANICAL PROPERTY CHARACTERIZATION

Mechanical properties are critical for UHMWPE that is to be used in an orthopedic device. The following tests are commonly used by researchers in preparing an FDA application.

### 24.5.1 Poisson's Ratio

Poisson's ratio $\nu$ indicates the variation of the strain ($\varepsilon = \Delta L/L_0$) of a sample in the transverse direction $\varepsilon_t$ and the longitudinal direction $\varepsilon_L$ when the sample is subjected to either tension or compression, and is defined as:

$$\nu = -\frac{\varepsilon_t}{\varepsilon_L}$$

To measure the Poisson's ratio, a sample must be instrumented with strain gauges in both transverse and longitudinal directions, which are monitored while the sample is deformed below its elastic limit in either tension or compression.

### 24.5.2 J-Integral Testing

J-integral tests indicate the resistance of a material to crack propagation under a steady tensile deformation and are a means of reporting the toughness of the material in geometries that contain notches or flaws. ASTM D6068 describes the test technique for monitoring crack propagation in plastics. In this test, a test specimen, usually in the form of a compact tensile specimen or a three-point bend specimen, is notched with an initial flaw. It is pulled apart at a constant crosshead speed, monitoring the tensile load along with the crack length $\Delta a$, which is usually measured optically. ASTM D6068 recommends the testing of a minimum of seven test specimens per formulation to generate a J-R curve. Each specimen is loaded to a different displacement, removed from the test jig, and rapidly fractured. It has been found that the best method for rapid fracture is to freeze the tested specimens in liquid nitrogen and then break then with a blunt impact [14]. The crack length $\Delta a$ is usually measured optically. The $J$ parameter is computed from the load-displacement curves and the sample geometry to generate a plot like that shown in Figure 24.18. Researchers report $J_Q$, which is the intersection of the $J$-$\Delta a$ line with the 0.2 mm offset line. This technique provides criteria to determine if the test was conducted in plane strain conditions (i.e., the test results are independent of sample geometry). If the criteria determine the test results were in plane stress conditions, researchers should be cautious in reporting data.

### 24.5.3 Tensile Testing

Tensile testing is perhaps the most common of the mechanical tests conducted on UHMWPE and is described in detail in ASTM D638. In this test, a *dogbone* specimen is machined or punched out, as shown in Figure 24.19A, and pulled at a user-specified crosshead speed in a tensile load frame. There are several standard size dogbones described in this standard. It should be noted that the use of microdogbones can lead to specimen-size-dependent tensile properties, and thus caution should be used when comparing these values with those reported in ASTM F648. The load $F$ and displacement $\Delta L$ are monitored and are converted to engineering stress and strain as shown in Figure 24.19B. The area $A$ is the initial cross-sectional area of the sample. Young's modulus $E$, the yield stress $\sigma_Y$,

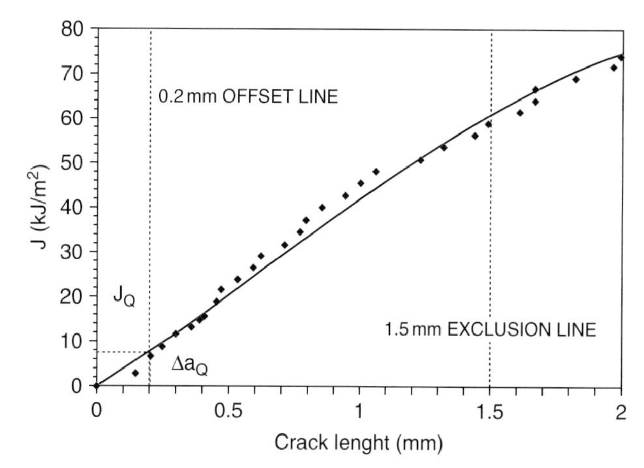


**FIGURE 24.18** J-integral plot for UHMWPE compact tensile specimen.

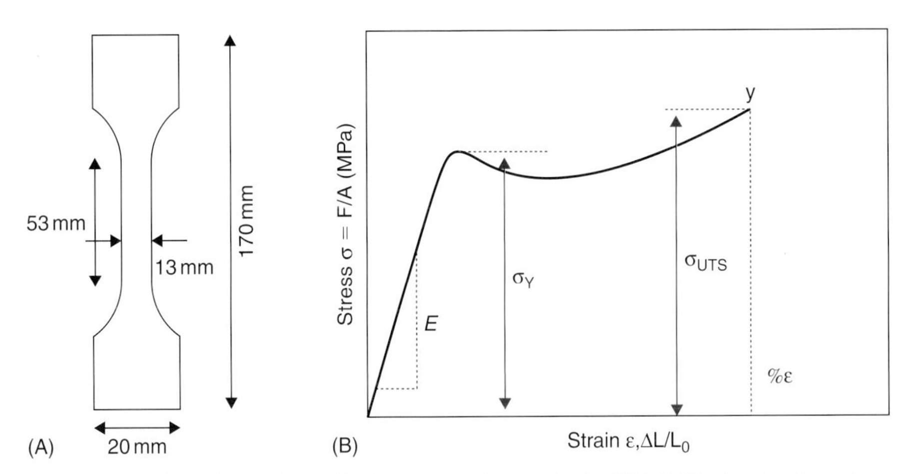


FIGURE 24.19 (A) Type I "dogbone" tensile specimen. (B) A typical tensile test plot for UHMWPE, showing the various parameters that can be determined.

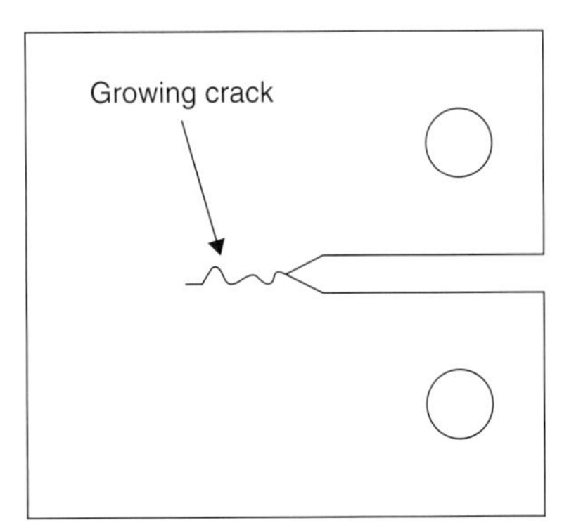


FIGURE 24.20 Compact tensile specimen for fatigue crack growth studies.

ultimate tensile stress $\sigma_{UTS}$, and percent elongation % $\varepsilon$ are all determined from this curve. Because of the potential for displacement at the grips, researchers usually use an extensometer to monitor axial strain of the sample in the gauge region. Optical extensometers, which monitor displacement of targets on the sample, usually work the best, given their noncontact nature. However, it is often difficult to maintain good adhesion of the targets to UHMWPE during testing.

## 24.5.4 Fatigue Testing

Fatigue testing, as described in ASTM E647, monitors the crack propagation in a specimen as it is subjected to an oscillating tensile load. The number of oscillations $N$ used in these tests are typically in the millions of cycles. The purpose of the test is to monitor the crack propagation resistance under cyclical loading, a type of deformation to which most implants will be subjected [15].

In this test, a specimen, most often a compact tensile specimen, as used in the J-integral tests and shown in Figure 24.20, is placed in a hydraulic load frame and then oscillated at either a fixed displacement (displacement control) or between fixed loads (load control). The choice of test depends in part on the application for the material. During the test, the crack length $\Delta a$ is periodically measured. The stress intensity factor $K$, which depends on the load range used, sample geometry, and crack length, is computed from the raw data ($a$ versus $N$), which is then used to prepare a curve of log d$a$/d$N$ versus log $\Delta K$.

## 24.5.5 Creep

Creep, or room-temperature flow upon application of a stress such as body weight, is an issue in orthopedic devices containing a UHMWPE [16]. These components are usually machined to tight tolerances, and long-term stability of the implant depends on maintaining these tolerance dimensions in the body. If an acetabular liner or tibial tray undergoes excessive creep, the leg can shorten or dislocations can occur. Minimization of creep is thus desired. Creep can occur when the UHMWPE is subjected to tensile or compressive stresses exceeding the yield stress of the material for a sufficient period of time, which depends on sample shape and constraining environment. The bedding-in period in the first 2 years, which is often discussed in clinical assessment of wear, is attributed to creep of the UHMWPE. Creep can often be reduced through crosslinking.

Creep is measured by application of a constant load to a sample, then measuring the height displacement of the load as a function of time, and reporting the data in terms of strain $\varepsilon$ versus time, as shown in Figure 24.21. As described in ASTM D2990, a stress on the order of 8 MPa is applied for up to 60 minutes, during which time an extensometer monitors displacement of the sample. The data is often converted to compliance $J(t)$, which is inversely proportional to the modulus, as expressed relative to the applied compressive stress $\sigma(t)$ as:

$$\varepsilon(t) = J(t)\sigma(t)$$

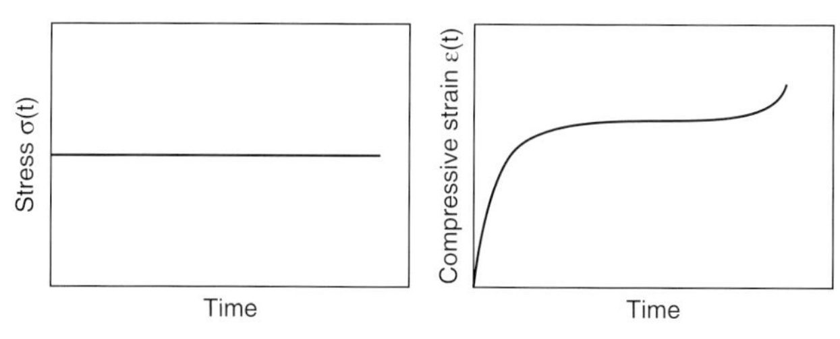


**FIGURE 24.21** Applied load history and resultant strain history for creep analysis of polyethylene.

## 24.5.6 Impact Testing

ASTM F648 calls for impact characterization via Izod-type specimens. According to ASTM D256, a pendulum impact machine contacts a double-notched test specimen, and the breaking energy is evaluated by the terminal swing height of the pendulum after fracturing the sample. Extensive round-robin testing in ASTM has demonstrated that the impact results are highly sensitive to the prenotching technique, which is performed with a fresh razor blade of specific size, as well as the notching technique itself. Orthopedic manufacturers using this technique are standardizing on a single notcher design.

## 24.5.7 Small Punch

A newer mechanical analysis has emerged in recent years that makes use of the biaxial deformation of small discs of UHMWPE, yielding a load-displacement curve [2]. Researchers have demonstrated a dependence on the area under the load-displacement curve to wear testing and suggest that the biaxial nature of wearing is the reason for the good comparison [17]. The obvious benefit of this technique is the small sample size, allowing spatial measurement of mechanical properties on retrieved components. This technique is described in further detail in ASTM F2183 and Chapter 32.

## 24.6 OTHER TESTING

Finally, two standard methods have emerged to anticipate the response of implanted UHMWPE components prior to and during implantation. Accelerated aging, through increased temperatures and oxygen content, is used to try to predict oxidation levels over many years of shelf storage [18, 19]. Hip simulators are used to monitor wear behavior under a physiologically-accurate wearing pattern [20–22]. These two approaches are discussed extensively in other manuscripts and are beyond the scope of this chapter.

## 24.7 CONCLUSION

There is a large battery of tests available to the researcher to examine the effects of radiation on the properties of UHMWPE. These effects include evolution of morphological structure, production of chemical species, and changes in mechanical properties. Caution must be used in sample preparation, test conditions, and data interpretation. This list of analytical tests is not all encompassing. As noted in previous sections, some regulatory agencies may require additional tests, depending on the nature of the material and the final application. Additionally, researchers may want to perform other tests to explore additional benefits of their material.

## REFERENCES

1. Data requirements for ultra high molecular weight polyethylene (UHMWPE) used in orthopedic devices, FDA; 1995.
2. Kurtz SM, Jewett CW, Foulds JR, Edidin AA. Miniature specimen mechanical testing technique scaled to articulating surface of polyethylene components for total joint arthroplasty. *J Biomed Mater Res* 1999;**48**(1):75–81.
3. Fox TG, Flory PJ. Viscosity-molecular weight and viscosity-temperature relationships for polystyrene and polyisobutylene. *J Am Chem Soc* 1948;**70**:2384–95.
4. ISO. Plastics—determination of the viscosity of polymers in dilute solution using capillary viscometers—Part 3: polyethylenes and polypropylenes; 2003.
5. Rempp P, Merrill EW. *Polymer synthesis*. Heidelberg: Huethig & Wepf; 1986.
6. Costa L, Luda MP, Trossarelli L, Brach del Prever EM, Crova M, Gallinaro P. Oxidation in orthopaedic UHMWPE sterilized by gamma radiation and ethylene oxide. *Biomaterials* 1998;**19**:659–68.
7. Spiegelberg SH, Schaffner SR. Oxidation profiles in shelf-stored ultra high molecular weight polyethylene samples. *Society for Biomaterials, 25th Annual Meeting*. Providence, RI; 1999.
8. Edidin AA, Jewett CW, Kalinowski A, Kwarteng K, Kurtz SM. Degradation of mechanical behavior in UHMWPE after natural and accelerated aging. *Biomaterials* 2000;**21**:1451–60.
9. Lyons BJ, Johnson WC. Radiolytic formation and decay of trans-vinylene unsaturation in polyethylene. In: Reichmanis E, Frank CW, O'Donnell JH, editors. *Irradiation of polymeric materials: processes, mechanisms, and applications*. Washington, DC: American Chemical Society; 1993. p. 62–73.
10. Johnson WC, Lyons BJ. Radiolytic formation and decay of trans-vinylene unsaturation in polyethylene: fournier transform infra-red measurements. *Radiat Phys Chem* 1995;**46**(4–6):829–32.
11. Costa L, Luda MP, Trossarelli L. Utra high molecular weight polyethylene-II. thermal- and photo-oxidation. *Polym Degrad Stab* 1997;**58**:41–54.
12. Jahan MS, Wang C-H, Schwartz G, Davidson JA. Combined chemical and mechanical effects on free radicals in UHMWPE joints during implantation. *J Biomed Mat Res* 1991;**25**:1005–17.
13. Flory PJ. *Principles of polymer chemistry*. Ithaca and London: Cornell University Press; 1953.
14. Cole JC, Lemons JE, Eberhardt AW. Gamma irradiation alters fatigue-crack behavior and fracture toughness in 1900H and GUR 1050 UHMWPE. *J Biomed Mater Res (Appl Biomater)* 2002;**63**:559–66.
15. Baker DA, Bellare A, Pruitt L. Effects of degree of crosslinking on the fatigue crack initiation and propagation resistance of orthopedic-grade polyethylene. *J Biomed Mater Res Part A* 2003;**66A**(1):146–54.

16. Kang PH, Nho YC. Effect of gamma-irradiation on ultra-high molecular weight polyethylene recrystallized under different cooling conditions. *Radiat Phys Chem* 2001;**60**(1–2):79–87.
17. Edidin AA, Kurtz SM. Validation of a modern hip simulator using four clinically-applied polymeric biomaterials. *Society for Biomaterials, 25th Annual meeting*. Providence, RI; 1999.
18. Kurtz SM, Muratoglu OK, Buchanan F, Currier B, Gsell R, Greer K, Gualtieri G, Johnson R, Schaffner S, Sevo K, Spiegelberg S, Shen FW, Yau SS. Interlaboratory repoducibility of standard accelerated aging methods for oxidation of UHMWPE. *Biomaterials* 2001;**22**:1731–7.
19. Sun DC.e.a. Simple accelerated aging method for long-term post-radiation effects in UHMWPE implants. *Fifth World Biomaterials Congress*. Toronto, Canada; 1996.
20. Schroeder DW, Pozorski KM. Hip simulator wear testing of isostatically molded UHMWPE effect of EtO and gamma irradiation. *42nd Annual Meeting, Orthopaedic Research Society*. Atlanta, GA; 1996.
21. McKellop H. Wear characteristics of UHMW polyethylene: a method for accurately measuring extremely low wear rates. *J Biomed Mater Res* 1978;**12**:895–927.
22. Ramamurti BS, Estok DM, Jasty M, Harris WH. Analysis of the kinematics of different hip simulators used to study wear of candidate materials for the articulation of total hip arthroplasties. *J Bone Joint Surg* 1998;**16**:365–9.

Chapter 25

# Tribological Assessment of UHMWPE in the Hip

Aaron Essner, MS and Aiguo Wang, PhD



## 25.1 INTRODUCTION

The modern era has seen the introduction or resurgence of ostensibly low wearing, high performance bearing constructs. Prior to the introduction of any of these bearing technologies into the marketplace, hip simulation data is almost always required by the relevant regulatory body. Beyond regulatory clearance, hip simulator data is necessary to the development of those technologies as well. The need for and role of hip simulation is well established in orthopedics.

There is, however, a healthy skepticism that exists regarding hip simulator data. Most surgeons believe that the data typically presented to them is in fact real and a precise representation of *in vitro* outcomes. But the question of reliability, that is, clinical correlation or validity, is a basic litmus test that hip simulation deservedly must face. This doubt is firmly rooted in historical clinical experience with bearing technologies that did not perform in the field as expected based on initial laboratory predictions (e.g., Poly II and Hylamer, covered in Chapters 17 and 19, respectively).

A basic fact that cannot be overlooked is that not all hip simulator data is equivalent. There are many variables that play a role in hip simulation, and the precise way these variables are addressed can affect the clinical relevance of the data. The details of the particular test methodology employed are of critical importance. There are numerous variables that can influence the tribological performance of hip bearing devices *in vivo*. It is impossible to simulate all of these known variables *in vitro* let alone the unknown ones. However, broadly speaking, three key categories of factors are well recognized to affect the clinical performance of the bearing devices: (1) implant design including bearing materials, (2) surgical techniques, and (3) patient factors. In engineering terms, these three categories of factors can be translated into: (1) biomechanical factors (design, surgical, patient), (2) biological factors (patient), and (3) biomaterial factors (design). It is the interactions between these three categories of factors that determine the tribological performance of implant bearing materials and devices.

## 25.2 BIOMECHANICAL FACTORS

Hip simulation is not a new science. The literature traces hip simulation back to the 1960s, with pioneers like Sir John Charnley emphasizing the importance of *in vitro* testing [1–3], as discussed in Chapter 4. This early work on *in vitro* evaluation of hip bearing materials was mainly based on trial and error, but it successfully predicted the superior wear performance of UHMWPE over PTFE that opened up the modern era of total hip arthroplasty. In Charnley's original study using a pin-on-plate wear test machine, UHMWPE outperformed PTFE by a factor of 1600 [4]. In his later clinical follow-ups, UHMWPE outperformed PTFE by a factor of 20 [2, 5]. Little progress was made during the 1970s and 1980s in the development of more wear-resistant polymeric bearing materials. This period coincided with the widespread use of pin-on-plate or pin-on-disk type wear devices in the screening of new bearing materials. It would take another decade before Charnley's clinical wear rate ratio between PTFE and UHMWPE was finally reproduced in the laboratory with the help of modern joint simulators [6–9].

One of the most widely used hip simulators is the orbital bearing machine (OBM) conceived and designed by McKellop and Clarke [10]. Three key features of this orbital simulator are multidirectional motion, physiologic loading, and serum lubrication [11–15]. In addition, the McKellop-Clarke simulator consisted of multiple stations that were capable of testing a candidate bearing material against a control material with statistical significance in a single test that normally lasted for months. Although the OBM is not the only possible configuration among all modern hip simulator designs, it is the one that has the longest history of use in hip simulation and has now been manufactured or replicated by multiple groups [10, 13, 15, 16–18]. A schematic of one station on an OBM machine is shown in Figure 25.1. One component of a ball-and-socket hip replacement (either the femoral head or ball, or the acetabular cup or socket) is mounted on a baseplate and allowed to interface with the remaining component that is mounted vertically above. In the original work by McKellop & Clarke, the configuration was inverted, and use of the machine focused on simple material comparison or "wear test" screening [19–21]. The baseplate is mounted on a bearing block inclined at 23 degrees, and an enclosure or specimen chamber is mounted to the baseplate. The vertical mount for the counter component usually incorporates allowance for alignment (e.g., a floating bearing) so that any eccentricity is eliminated or minimized. The baseplate is also constrained from axial rotation, allowing only oscillatory motion. Because the bearing between the inclined block and baseplate is in motion, the machine configuration has become known as an orbital bearing machine (OBM).

By mounting the inclined block assembly on a rotating axial actuator, both motion and loading can be applied. As the inclined block rotates, the baseplate and chamber assembly experience oscillatory motion of ±23 degrees. This motion occurs in the physiologically equivalent saggital and coronal planes, thereby applying flexion-extension and abduction-adduction of ±23 degrees. At first glance internal-external rotation may appear to be absent or forbidden by constraint, but detailed analysis of the interface motion experienced between a head and cup shows that the pole of the ball-and-socket assembly experiences a sinusoidal motion [11, 22, 23]. The specific magnitude of this motion depends on geometric details of the baseplate assembly and ranges from around ±7 degrees for small baseplates to ±10 degrees for larger ones. The composite

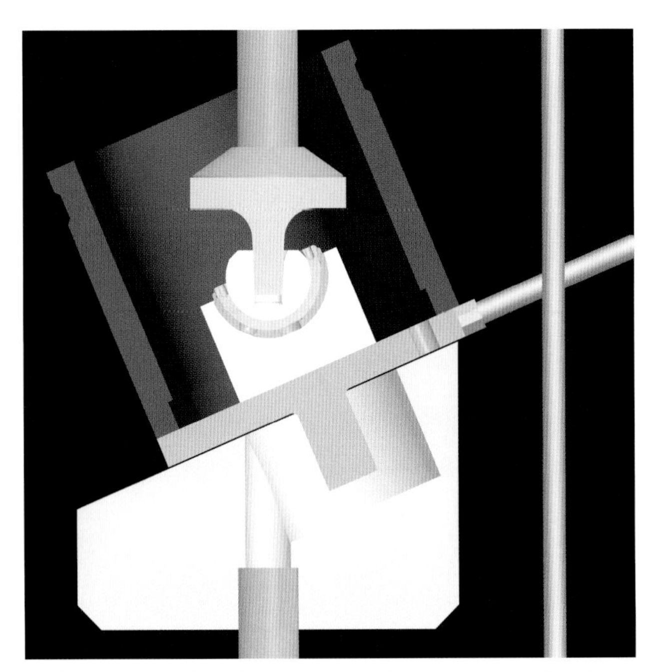

**FIGURE 25.1** Schematic of an orbital bearing machine (OBM) hip simulator in the inverted configuration.

motion of the OBM results in a cross-path or cross-shear motion at the bearing interface. This signature stress pattern of the hip joint is critical for reproducing the mechanism and magnitude of polyethylene (PE) wear [11–15, 24]. Historically, the OBM was used to assess bearing surface or Mode 1 wear as described by McKellop [25, 26].

While motion on an OBM is prescribed by the inclined block, there is still considerable flexibility in the remaining details. The hip device may be positioned anatomically (cup superior, head inferior) as is commonly done by contemporary researchers [11, 27–30] and as shown in Figure 25.2. This positioning creates the same "tunneling" effect of a femoral head penetrating into an acetabular component as reported by Charnley [2, 6]. An additional benefit of anatomical positioning is the possibility of addressing additional biomechanical factors in hip simulation. This allows factors such as acetabular cup abduction angle, head-liner subluxation, and neck to rim impingement to be incorporated into hip simulation models. The component mounting method may use rigid or compliant materials to retain the hip device to the baseplate but must allow repeatable positioning if the device is removed and remounted during testing. The specific load profile and magnitude applied may also vary depending on specific test goals. This dynamic profile may reflect activities such as stumbling or running.

One of the most significant advancements in hip joint simulation is the realization of the fundamental importance of multidirectional motion to reproduce the clinically relevant wear mechanisms, wear debris, and wear magnitudes of UHMWPE. The necessity to incorporate these multidirectional motions was discovered in the analysis of

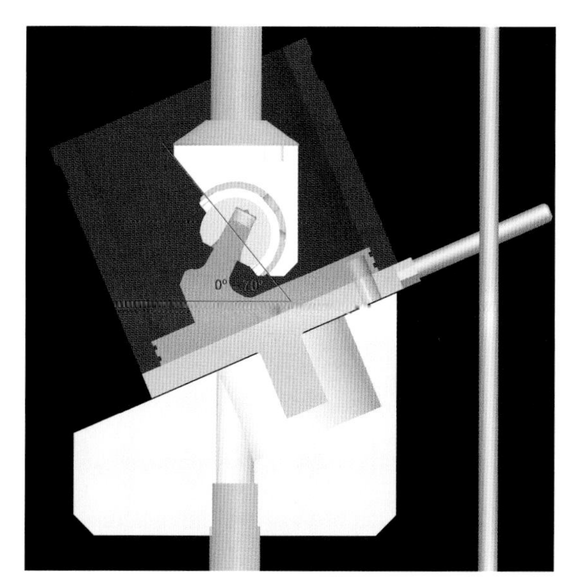

**FIGURE 25.2** Schematic and photograph of an OBM simulator set up with anatomical positioning.

the etiology of wear for UHMWPE bearings [11–15]. This work identified the fact that PE molecules would preferentially orient or align in the direction of the primary stress vector applied to an acetabular surface by a femoral head. Stress vectors applied orthogonally to this primary orientation direction would potentially split these molecular strands, thereby creating wear debris and thus wear. This understanding would lead to development of improved UHMWPE bearings as will be discussed later. However, the orbital bearing machine that serendipitously incorporated the essential element of cross-path motion also played a key role in the development of modern crosslinked UHMWPEs that are widely used in THR today.

## 25.3 BIOLOGICAL FACTORS

It is well recognized that human articular joints are lubricated by synovial fluid. This fluid provides not only lubrication to the moving joint, but it also provides for nutrition and biological support of articular cartilage. It is excreted by the synovial membrane and joint capsule tissue surrounding the joint and is essentially an ultrafiltrate of blood [7, 8, 31–36]. Its specific constituents depend on many factors, including disease condition, but it likely varies, perhaps considerably, from patient to patient [7, 8, 31–36]. This biologic fluid changes when a patient undergoes a joint replacement operation. This is due to the fact that the native tissue and structures are disturbed from the surgical intervention, with capsular tissue sometimes being removed. The fluid lubricating joint replacements is therefore not exactly synovial fluid. A better description is "joint fluid." In fact, there are substantial differences in chemical composition and viscosity. Joint fluid after arthroplasty typically contains more proteins with more variability in specific protein types than synovial fluid [7, 8, 31–36]. It also usually loses up to 90% of its hyaluronic acid content, the high molecular weight constituent that is the primary booster of viscosity in a natural joint [7, 8, 31–36].

A review of the literature regarding orthopedic wear testing, be it simple screening or hip simulation, reveals that the biological nature of the fluid environment surrounding a joint replacement has not always been fully appreciated. Water and saline have been used as fluid test media for a variety of hip bearing materials [37–41]. Some of this work was conducted using simple screening machines, but some data was generated using simulator configurations [37–41]. The data generated and reported was both variable and substantially different than clinical reporting of like materials [2, 42–44]. Beyond the biomechanical factors that may rule out simple screening test configurations, the biological nature of joint fluid requires consideration to establish a clinical link for *in vitro* data.

In their description of the OBM hip simulator, McKellop and Clarke discussed the use of bovine serum as a lubricant [10]. This fluid seems to be an appropriate choice in that it is biological in nature and contains proteins. Work has shown that it is these proteins that are critical to the role of lubrication for joint replacements, and the fact that they are absent in water or saline leads to artifact when using either as a lubricant [8]. The direction of error introduced (too much or too little wear) is also a function of specific material. For example, PTFE (Teflon) and UHMWPE behave differently in this manner [6].

Simple inclusion of proteins by use of bovine or other animal sera may not be sufficient to mimic the lubricating properties of joint fluid. In a controlled study the influence of protein concentration on the wear magnitude of both PTFE and UHMWPE was determined [6, 8]. The ratio of wear rates of both materials was compared to clinical data reported by Charnley [8]. This allowed determination of a serum protein concentration range that yielded wear rate ratios in the clinical range reported by Charnley. Too much total protein yielded irrelevant wear rates, but so did no protein.

The specific protein also plays a role [8]. Observations of test data on a specific material and device generated over time on an OBM using the same total protein concentration showed varying performance. Stepwise analysis eliminated test methodology or specimen variability as factors and ended up identifying serum as the responsible factor. Assays of the serum revealed that the albumin: globulin ratio for the media changed over time, sometimes appreciably. Even when diluted to the same clinically relevant total concentration (as determined by the Charnley data), the specific albumin:globulin makeup affected wear. This ratio must be accounted for and should be matched to joint fluid as closely as possible.

A potential factor in the specific protein role regarding wear is thermal stability. A hip joint simulator, even when applying anatomical motions and loading and using serum lubricant, is still not alive. This means that biological processes are not present, including homeostasis, metabolism, and catabolism. Frictional heat that is generated at a bearing interface is dissipated to and through the specific materials surrounding that bearing. This heat has been both measured as well as calculated, with flash interface temperatures as high as 60°C or more [45–49]. The significance of this is thermal degradation of proteins. At elevated temperatures, proteins precipitate out of solution. This not only changes the protein concentration of the serum but also provides an artificial solid lubricant that protects bearing surfaces from wear [45, 47, 49]. The thermal degradation behavior of albumin and globulin differs, with globulins somewhat more stable. Gamma globulin in particular, with its high molecular weight, seems to be important. This may at least partly explain the importance of A:G ratio, at least in polymer bearing wear.

It is reasonable to expect that protein degradation occurs in humans with hip joint replacements, just as has been observed in simulators. The difference is in the processes of homeostasis, metabolism, and catabolism. These processes focus on maintaining the joint space in a relatively consistent condition. That is, temperature is maintained, with anything more than the very local surfaces prevented from overheating. In the simulator environment this means providing an extensive heat sink to prevent steady increase and overheating of test environment temperature. Overheating would yield excessive protein denaturing as well as thermal softening of test materials [45–49]. While a cursory viewpoint would consider heating the environment to body temperature (37°C), an understanding of the bearing interface and the potential for thermal artifact to artificially decrease wear overrides body temperature concerns. Frictional heating accounts for this. The metabolism and catabolism processes provide for the biological removal of waste products in the body. That means degraded proteins that have been generated can be removed. The hip simulator environment must also provide for this to prevent these degradation products from creating artifact. This can be accounted for through turnover or supply of fresh proteins and removal of degraded ones. Frequent lubricant changes or a large enough lubricant volume can account for this.

## 25.4 BIOMATERIAL FACTORS

There have been several biomaterials used as bearings in hip replacement over the years. These materials include PTFE (Teflon), polyester (Delrin), UHMWPE, polyacetal, CoCr, Ti, alumina, zirconia, and more recently, oxidized zirconium-niobium (Oxinium) and zirconia toughened alumina (Biolox Delta) [2, 6, 50–55]. All of these materials have the potential to wear when used as bearings because wear occurs as a function of opposing surfaces subject to contact, motion, and loading. As a consequence of wear, microscopic and, indeed, sometimes macroscopic particles of a bearing material can be produced. It is the biologic immune system response to wear debris particles that has been identified as a cause of osteolysis and aseptic loosening [56–61]. This immune system response is patient specific, with contributing factors including the amount, size, and shape of the particles as well as the material itself. The literature has indicated that when wear is below a threshold value, the relative incidence of osteolysis is low [58–61]. This information therefore yields a wear resistance goal for the development of bearing materials.

The materials mentioned earlier have all been attempted to mitigate wear and aseptic loosening and thereby increase the longevity of a hip bearing. The technologies have met with varied success, but some "improvements" were clinical failures [2, 44, 58, 61–82]. It is these failures that are actually most useful to engineers and researchers. This is because they represent known clinical failure models that an *in vitro* hip simulation must be able to reproduce. If a given test model cannot replicate a known failure, how could the predictive value of such a model be trusted? Replication of known failures is a prerequisite to be able to establish clinical validity.

A good starting point is PTFE compared to UHMWPE, or the so-called Charnley model mentioned earlier [6, 8]. This subject has been the focus of lubricant optimization, but it was also the basis for an extensive hip simulation round-robin test. This work allowed multiple laboratories to generate data on both materials using the particular lab's standard test methodology. Variables included component orientation (inverted or anatomical), load profile (specific waveform, maximum load), and lubricant details. What the work revealed was the effect of these methodology details on duplicating the clinical Charnley ratio. Test details matter.

Another material with clinical history that has shown varied simulator performance is Hylamer. This material is a form of UHMWPE that was subjected to a high pressure remelting process, resulting in a highly crystalline chain extended microstructure [55], as was discussed in Chapter 19. Published wear data includes both pin-on-disk as well as simulator data [83, 84]. In consideration of the biomechanical factors discussed earlier, the pin-on-disk data can be disregarded. However, the simulator data, which presumably used inverted positioning, compliant component mounting, and a small volume (approximately 100 ml) of high protein serum (100% bovine serum diluted only with the addition of EDTA) showed similar wear performance of Hylamer and ordinary UHMWPE (both gamma-air sterilized). Clinical experience did not establish equivalent wear performance for these two materials, with Hylamer showing much higher wear and associated incidence of osteolysis [62–65, 80–82]. As summarized in Chapter 19, variation in sterilization and aging conditions (i.e., Hylamer oxidation) between the simulator and clinical situation is now thought to explain at least part of the apparent discrepancy.

Although the exact role of oxidation in either this laboratory or clinical data is not precisely known, a later simulator study retrospectively duplicated this relative performance [22, 97]. The major difference between these two OBM studies lies in the test methodology details. This includes anatomical positioning, more rigid component mounting, and, perhaps most significant, use of a large volume (≈450 ml) of serum lubricant (≈20 g/l total protein, 2:1 A:G) optimized from the Charnley model. It is interesting to note that not only was the wear rate ratio duplicated, but also were the magnitudes of Hylamer and UHMWPE wear reported in the clinical literature reproduced in the later simulator test.

This work is retrospective in nature and does not necessarily guarantee prospective predictive capability. To speak to that, OBM hip simulation utilizing this clinically representative methodology was applied in the development of several current generation UHMWPE materials. One of the first of these was intentionally crosslinked Duration Stabilized UHMWPE [98]. Details about the Duration process are covered in Chapter 14. This material involved irradiating nitrogen packaged UHMWPE components to generate free radicals and then subjecting them to heat treatment for stabilization. Other crosslinked PEs follow this same principle. In fact, this radiation-heat treating principle formed the basis for all first generation crosslinked PE materials on the market today (see Chapters 13 and 14).

Preclinical hip simulator data for Duration material predicted an approximate 30–35% reduction in wear for this technology compared to gamma-air sterilized UHMWPE [11, 12, 87–88]. A prospective randomized clinical study was initiated on these same two offerings in 1996. At a recent UHMWPE symposium in Madrid (September 2007), Grimm et al. reported 10-year data on these materials showing a 42% reduction for Duration ($p<0.01$, Duration 0.08 mm/yr, gamma-air 0.14 mm/yr) [89]. This 10-year clinical data confirmed the prediction of the hip simulator testing. It should be noted that this was for the clinically correlated simulator methodology previously described. Not surprisingly, OBM simulator data from that same clinical research group, using their specific methodology, showed a predicted performance of roughly 60% lower wear. This again shows the influence of test details on the outcome of OBM data, even when generated on nearly identical machines.

An understanding of the importance of multidirectional motion in the wear of UHMWPE was a key not only for focusing on hip simulators over wear screening devices, but it also was critical to the development of the Duration Stabilized technology mentioned. This work highlighted the role of crosslinking in the wear of UHMWPE and led to proliferation of intentionally crosslinked UHMWPEs beyond the Duration Stabilized material. Subsequent research indicated a nonlinear effect of crosslinking to resist crossing shear motion [11, 12, 90–92]. Polyethylene wear is a function of dose, with the effect becoming asymptotic around 100 kGy (10 Mrad). This is shown in Figure 25.3.

One of the first generation highly crosslinked UHMWPEs is Crossfire, which involves gamma-irradiating UHMWPE to 75 kGy followed by below melt annealing (130°C) [93] (Chapter 14). Acetabular liners are then fabricated, packaged in $N_2$, and then gamma-sterilized at 30 kGy. OBM hip simulator evaluation using the optimized methodology discussed here yielded approximately 90% less wear than conventional nitrogen packaged, gamma-sterilized material [9]. Clinical studies of this material have shown *in vivo* performance in this 90% range [68].

Other PE materials using this radiation-heat treating principle include Longevity, Durasul, XLPE, and Marathon, among others [53, 94, 95] (Chapter 20). Radiation source,

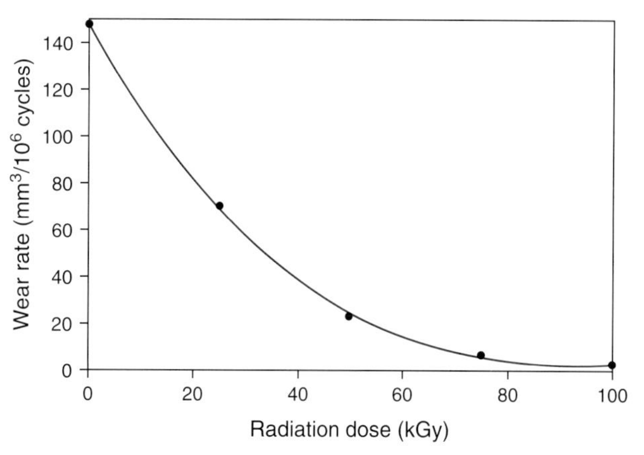


**FIGURE 25.3** Hip wear reduction dose (crosslinking) effect curve.

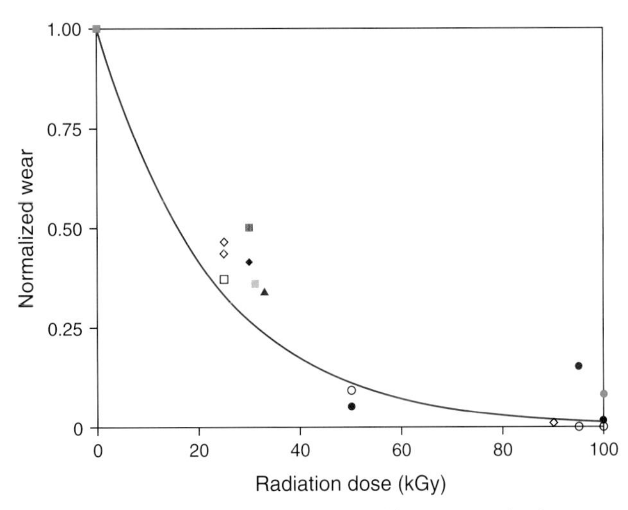


**FIGURE 25.4** Normalized dose (crosslinking) wear reduction curve including *in vitro* master curve, *in vitro* data points (open symbols), and clinical data points (closed symbols). All data points normalized to percentage reductions of respective control materials. This neutralizes the effect of specific design details (head size, thickness, etc.).

total dose, and heat treatment details vary, but all are crosslinked more than respective control materials. All have indicated a wear reduction benefit from crosslinking, which has now also been confirmed by clinical data [70, 71, 75–82, 96, 97]. Specific percentage reductions in wear depend on crosslinking process details and the specific control material used for comparison, but the data supports the dose effect observation mentioned earlier.

In fact, there has been consistent global consensus on the wear-reducing benefit of crosslinking to UHMWPE. When clinical data is superimposed on laboratory wear reduction curves of hip wear versus crosslinking, a remarkable overlay is achieved. This is shown in Figure 25.4. Data was normalized to allow comparison across design details (head size, thickness, etc.). This is perhaps the first time in orthopedic history that predictions made by engineers and researchers in the laboratory across the globe have been confirmed by clinical studies conducted by surgeons worldwide.

## 25.5 WEAR RANKING AND MAGNITUDE

As the previous section discussed, it is possible to use differing test methods yet yield the same relative ranking of UHMWPE materials with clinical experience. This is true for OBM machines, but it is also true when comparing OBM and three-axis machines. In a three-axis machine flexion-extension, abduction-adduction, and internal-external rotation are independently controlled. As mentioned earlier, these motions are coupled in an OBM and not separable. A comparison between these two simulator machines shows that both ranked the same three materials nearly identically [106, 111]. However, the study also showed that overall magnitudes were different by nearly five times.

While *in vitro* determination of hip component lifetime has not and may never be fully established, it is useful to evaluate and develop bearings under aggressive wear conditions. Without sufficient magnitudes, subtle performance questions cannot be answered. As described, details of the test methodology influence this fact, and from an engineering perspective it is almost always better to err on the side of aggressive evaluation. In that way, additional factors of safety can be built into technologies and designs. If a model is to be aggressive and clinically relevant, additional biomechanical, biological, and biomaterial factors may need to be considered.

## 25.6 OBM DEVELOPMENTS

In engineering terms, the performance of a given device is a function of both the material and the design. The ability of a simulator to incorporate an actual device design is a tremendous advantage over screening machines. The former allows a functional fatigue evaluation of the device because both the material and design are subject to motion and loading. This fatigue aspect should not be overlooked because rim fracture of acetabular components has been reported in the recent literature [99–105]. To have true clinical value, an OBM model should be able to reproduce this failure mechanism, and this has been reported [106, 107].

At first glance it may appear that due to the somewhat fixed nature of how motion is achieved with an OBM-type simulator, biomechanical evaluation of devices beyond bearing surface wear may not be possible or practical. But recent work has shown that OBM machines can, in fact, be very versatile [108–112].

There has been much attention focused on hard-hard bearings (CoC, MoM) in recent times, including observations on short-term retrievals. One such observation has been the presence of wear stripes on both ceramic-ceramic and metal-metal bearings [43, 113–115]. Component positioning has been identified as critical for hard-hard bearing performance *in vivo*, and it may play a role in stripe formation [43, 116–119]. The ability to incorporate this in a hip simulation is important. Biomechanical analysis has indicated that stripes may form as the femoral head subluxes from the acetabular cup [113, 120]. This phenomenon can occur under a variety of activities and may be due to either soft tissue laxity or impingement. Regardless of the specific cause, the event can bring the femoral head into contact with the edge of the acetabular cup as it separates. This can cause a relatively high contact stress to develop in the region of both the head and cup that are in contact when any physiological force is applied. This high stress may cause local damage that alters the surface finish in the contact regions, allowing them to be visually identified as "stripes." McKellop refers to this as Mode 2 wear [25, 26].

In the development or evaluation of any bearing technology, it is important to account for these types of clinical events. There have now been multiple separation models developed and discussed for hip simulator application, two differing ones utilizing OBM simulators [108–112]. Similar to the principles discussed regarding hip simulator output overall, it is critical for any *in vitro* model to establish a clinical correlation, at minimum, through comparison to retrievals showing the same etiology. This has been done for these separation models.

An additional surgical concern relates to rim impingement. This speaks to Mode 4 wear as reported by McKellop and involves two nonbearing surfaces [25, 26]. The ability to address this concern shifts an OBM from a hip "wear" simulator to a hip simulator. The former has in general been regarded as a material tester, while the latter is far more comprehensive and is a tool for functional device evaluation. Once again, clinical observations on retrieved devices drive this need. Up to 70% of retrieved liners show evidence of rim or impingement damage [121–125].

Recently a model allowing an OBM simulator to create clinically evidenced impingement has been developed [108–112]. Unlike separate or standalone test models used to evaluate this aspect of a construct, this OBM model applies normal gait kinematics as well as neck to rim impingement, allowing the net effect of this occurrence on device performance to be evaluated. While this particular model incorporates force control, others have discussed inclusion of displacement-driven impingement in OBM test models [111]. A possible limitation of a displacement-driven impingement model is a reduction in impingement contact and force over test duration due to rim creep or geometry change.

## 25.7 STANDARDIZATION

The two major global standards organizations are ASTM and ISO. The approach taken within these organizations regarding hip simulation is somewhat different. ASTM efforts generally predate those within ISO, dating back to a hip simulator guide (F1714). This guide focused on general recommendations regarding points to consider when setting up a hip simulation. These points were generally broad, discussing topics like applying dynamic force and kinematic motions to hip components mounted in a simulator. Lubricant was no more detailed than "bovine serum," and no specifics were given regarding component orientation or mounting arrangements. This guide only highlighted major points to consider. Either OBM or three-axis machines were allowed. In fact, nearly any machine configuration would be allowed if reasonable justification was provided.

The ISO approach is somewhat different in that it is very specific [126]. ISO 14242 details everything from component orientation to the particular loads and motions to apply. It is essentially a step-by-step set of instructions for conducting a hip simulator test. Interestingly, at the time this standard was approved, there had not been a single data point reported in the literature generated using the method. The logic given was that by taking a hip device, mounting it in a machine the same way it is mounted in the body, and applying physiological loads and motions the results will be clinically valid. While this approach seems perfectly logical, it does not take into consideration the potential for laboratory-induced artifact. This type of artifact was highlighted in the biological factors section and points to the importance of making accommodation to minimize error introduced by the fact that a simulator is not a living system. The ISO standard also does not allow an OBM-type simulator to be used. This has been regarded as a rigid oversight of the clinical relevance that has been established for some OBM methodologies [70, 71, 75–82, 96, 97]. Approval of the standard was obtained through international voting, with member countries exercising votes. Because the majority of international hip simulator experience exists within the United States, the negative vote it logged was overruled by the international community, which included countries with no known hip simulator capacity or experts—OBM, three-axis, or otherwise.

To rectify this situation, the ASTM hip wear committee has withdrawn the hip guide and drafted a proposed subset of ISO 14242 that covers the OBM-type simulator. Approval of this will allow adoption of the overall ISO hip standard within ASTM and therefore a unified global approach to hip wear simulation. This is only a starting point, however. Neither the standard nor the working document covers the separation or impingement models mentioned in the previous section and may still need further refinement regarding lubricant details, for example.

## 25.8 SUMMARY

Hip wear simulation using an OBM has come a long way in the nearly 30 years it has existed. An understanding of the test method details determined through a series of experiments has allowed development of a clinically relevant model. Contemporary OBM models are now complex enough to include the simulation of surgical and patient risk factors in the evaluation of new bearing materials and design concepts. Some of the major risk factors, such as vertical acetabular cup positioning, head-cup separation, and neck-cup impingement, have been successfully incorporated in modern OBM. This allows additional safety to be built into designs and technologies that are developed from them. Surgeons must note that not all OBM models are identical. It is the details of how an experiment is run that influence the clinical correlation. While an understanding of the variables may seem cumbersome, surgeons should at minimum ask to see proof of *in vivo* correlation for a given simulator model. By doing so they can have confidence in selecting a bearing technology that offers their patients the most benefit.

## 25.9 ACKNOWLEDGMENT

The authors would like to thank Lizeth Herrera, Laryssa Rodriguez, and Reginald Lee for their help with references and figures.

## REFERENCES


1. Charnley J. Factors in design of an artificial hip joint. In: *Lubrication and wear in living and artificial human joints. Proc. Inst. Mech. Eng* 1966;**181**(3J):104.
2. Charnley J, Kamangar A. The optimum size of prosthetic heads in relation to the wear of plastic sockets in total hip replacement of the hip. *Med Biol Eng* 1969;**7**:31–9.
3. Charnley J. The wear of plastic materials in the hip. In: *Plastics in medicine and surgery*. London: The Plastics and Rubber Institute; 1975. 3.
4. Waugh HG. The plan fulfilled 1959–1969. In: *John Charnley: the man and the hip*. London: Springer-Verlag; 1990. 4.
5. Charnley J, Cupic Z. The nine and ten year results of the low-friction arthroplasty of the hip. *Clin Orthop Rel Res* 1973;**25**:95–9.
6. Clarke IC, Good VD, Anissian L, et al. Charnley wear model for validation of hip simulators-ball diameter versus polytetrafluoroethylene and polyethylene wear. *Proc Inst Mech Eng* 1997;**211**:25–36.
7. Clarke IC, Chan FW, Essner A, Good V, Kaddick C, Lappalainen R, et al. Multi-laboratory simulator studies on effects of serum proteins on PTFE cup wear. *Wear* 2001;**250**(1–12):188–98.
8. Wang A, Essner A, Schmidig G. The effects of lubricant composition on in vitro wear testing of polymeric acetabular components. *J Biomed Mater Res Part B: Appl Biomater* 2004;**68B**:45–52.
9. Wang A, Essner A, Copper J. The clinical relevance of hip simulator testing of high performance implants. *Semin Arthro* 2006;**17**:49–55.
10. McKellop HA, Clarke IC. Evolution and evaluation of materials-screening machines and joint simulators in predicting in vivo wear phenomena. In: *Functional Behavior of Orthopedic Biomaterials*, Vol. II. Boca Raton: CRC Press; 1984.
11. Wang A, Essner A, Polineni VK, Stark C, Dumbleton JH. Lubrication and wear of ultra-high molecular weight polyethylene in total joint replacements. *Tribol Int* 1998;**31**(1–3):17–33.
12. Wang A, Sun DC, Yau S-S, Edwards B, Sokol M, Essner A, Polineni VK, Stark C, Dumbleton JH. Orientation softening in the deformation and wear of ultra-high molecular weight polyethylene. *Wear* 1997;**203–204**:230–41.
13. Wang A. A unified theory of wear for ultra-high molecular weight polyethylene in multi-directional sliding. *Wear* 2001;**248**:38–47.
14. Muratoglu OK, Bragdon CR, O'Connor DO, Jasty WH, Gul MR, McGarry F. Unified wear model for highly crosslinked ultra-high molecular weight polyethylenes (UHMWPE). *Biomaterials* 1999;**20**:1463–70.
15. Bragdon CR, O'Connor DO, Lowenstein JD, Jasty M, Syniuta WD. The importance of multidirectional motion on the wear of polyethylene. *Proc Instn Mech Engrs [H]* 1996;**210**:157–66.
16. Mejia LC, Brierley TJ. A hip wear simulator for the evaluation of biomaterials in hip arthroplasty components. *Biomed Mater Eng* 1994;**4**(4):259–71.
17. Saikko V. A 12-station anatomic hip joint simulator. *Proc Instn Mech Engrs [H]* 2005;**219**:437–48.
18. Clarke IC, Good V, Williams P, Schroeder D, Anissian L, Stark A, et al. Ultra-low wear rates for rigid-on-rigid bearings in total hip replacements. *Proc Inst Mech Eng [H]* 2000;**214**(4):331–47.
19. Clarke IC. Wear of artificial joint materials IV—hip joint simulator studies. *Eng Med* 1981;**10**(4):189–98.
20. Clarke IC. Wear-screening and joint simulation studies vs. materials selection and prosthesis design. *CRC Crit Rev Biomed Eng* 1982;**8**:29–91.
21. Clarke IC, McKellop HA, Okuda R, McGuire P, Young R, Hull D. Materials and prosthesis design criteria—hip simulator studies. *Trans Orthop Res Soc* 1982.
22. Saikko V, Calonius O. Slide track analysis of the relative motion between femoral head and acetabular cup in walking and in hip simulators. *J Biomech* 2002;**35**:455–64.
23. Calonius O, Saikko V. Slide track analysis of eight contemporary hip simulator designs. *J Biomech* 2002;**35**:1439–50.
24. Paul J. Loading on normal hip and knee joints and on joint replacements. In: Schaldach M, Hohman D, editors *Engineering in medicine, Vol. 2, advances in artificial hip and knee joint technology*. New York: Springler-Verlag; 1976.
25. McKellop HA. Wear modes, mechanisms, damage, and debris. Separating cause from effect in the wear of total hip replacements. In: Galante JO, Rosenberg AG, Callaghan JJ, editors *Total hip revision surgery*. New York: Rave Press, Ltd.; 1995.
26. McKellop HA. The lexicon of polyethylene wear in artificial joints. *Biomaterials* 2007;**28**:5049–57.
27. Saikko V. A 12-station anatomic hip joint simulator. *Proc Inst Mech Eng [H]* 2005;**219**(6):437–48.
28. McKellop HA, Campbell P, Park SH, Schmalzried TP, Grigoris P, Amstutz HC, Sarmiento A. The origin of submicron polyethylene wear debris in total hip arthroplasty. *Clin Orthop* 1995;**311**:3–20.
29. Bigsby RJA, Hardaker CS, Fisher J. Wear of ultra-high molecular weight polyethylene acetabular cups in a physiological hip joint simulator in the anatomical position using bovine serum as a lubricant. *Proc Inst Mech Eng [H]* 1997;**211**:265–9.
30. Polineni VK, Essner A, Wang A, Sun DC, Stark C, Dumbleton JH. Effects of radiation induced oxidation and crosslinking on the wear of UHMWPE acetabular cups. *ASTM Symp on characterization and properties of UHMWPE*; 1996.
31. Dumbleton J. *Tribology of natural and artificial joints*. Amsterdam: Elsevier; 1981.
32. Bole GG. Synovial fluid lipids in normal individuals and patients with rheumatoid arthritis. *Arthr Rheum* 1962;**6**:589–601.
33. Levick JR, et al. Synovial fluid: determinants of volume turnover and material concentration. In: Kuettner K, editor. *Articular cartilage and osteoarthritis*. New York: Raven Press; 1992. p. 529–41.
34. Saari H, Santavirta S, Nordstrom D, Paavolaien P, Konttinen YT. Hyaluronate in total hip replacement. *J Rheum* 1993;**20**(1):87–90.
35. Lentner C, editor. *Geigy scientific tables. Vol. 1. Units of measurements, body fluids, composition of the body, nutrition*. West Caldwell: Ciba-Geigy; 1981:161.
36. Binette JP, Schmid K. The proteins of synovial fluid: a study of the alpha 1-alpha 2 globulin ratio. *Arth Rheum* 1965;**46**:14–28.
37. Calonius O, Saikko V. Analysis of polyethylene particles produced in different wear conditions in vitro. *Clin Orthop Relat Res* 2002;**399**: 219–30.
38. Saikko V. Wear of polyethylene acetabular cups against alumina femoral heads: 5 prostheses compared in a hip simulator for 35 million walking cycles. *Acta Orthop Scan* 1993;**64**(5):507–12.
39. Clarke IC, Dai QC, Brahm A, Gustafson A. Effects of serum and water lubrication on PTFE/CoCr implants. *Trans Combined Orthop Res Soc* 1995:204.
40. Wang A, Essner A, Stark C, Dumbleton JH. Comparison of the size and morphology of UHMWPE wear debris produced by a hip joint

simulator under serum and water lubricated conditions. *Biomaterials* 1995;**17**:865–71.
41. Derbyshire B, Fisher J, Dowson D, Hardaker C, Brummitt K. Comparative study of the wear of UHMWPE with zirconia ceramic and stainless steel femoral heads in artificial hip joints. *Med Eng Phys* 1994;**16**:229–302.
42. Campbell P, Ma S, Yeom B, McKellop HA, Schmalzried T, Amstutz HC. Isolation of predominantly submicron-sized UHMWPE wear particles from periprosthetic tissues. *J Biomed Mater Res* 1995;**29**:127–31.
43. Nevelos JE, Ingham E, Doyle C, Fisher J, Nevelos AB. Analysis of retrieved alumina ceramic components from Mittelmeier total hip prostheses. *Biomaterials* 1999;**20**(19):1833–40.
44. Orishimo KF, Hopper RH, Engh CA. Long-term in vivo wear performance of porous-coated acetabular components sterilized with gamma irradiation in air and ethylene oxide. *J Arthroplasty* 2003; **18**:546–52.
45. Liao YS, McKellop H, Lu Z, Campbell P, Benya P. The effect of frictional heating and forced cooling on the serum lubricant and wear of UHMW polyethylene cups against cobalt-chromium and zirconia balls. *Biomaterials* 2003;**24**(18):3047–59.
46. Liao YS, Benya PD, Lu B, McKellop H. Stability of serum as a lubricant in wear simulator test of prosthetic joints. *Proc World Biomater Congr* 1996.
47. Wang A, Polineni VK, Essner A, Stark C, Dumbleton JH. Quantitative analysis of serum degradation and its effect on the outcome of hip joint simulator wear testing of UHMWPE. *Trans Orthop Res Soc* 1999;**24**:73.
48. Lu Z, McKellop H. Frictional heating of bearing materials tested in a hip joint wear simulator. *Proc Inst Mech Eng [H]* 1997;**211**(1):101–8.
49. Lu Z, McKellop H, Liao P, Benya P. Potential thermal artifacts in hip joint wear simulators. *J Biomed Mat Res Part B: App Biomater* 1999;**48**(4):458–64.
50. McKellop H, Clarke IC, Markolf KL, Amstutz HC. Friction and wear properties of polymer, metal, and ceramic prosthetic joint materials evaluated on a multichannel screening device. *J Biomed Mater Res* 1981;**15**(4):619–53.
51. Saikko V. Wear and friction properties of prosthetic joint materials evaluated on a reciprocating pin-on-flat apparatus. *Wear* 1993;**166**(2):169–78.
52. Walter A. Wear-screening of ceramic-to-ceramic components for total hip replacements by ring-on-disc device and joint simulator test. In: *High tech ceramics*. Amsterdam: Elsevier Science Publishers B.V.; 1987.
53. Kurtz SM. *The UHMWPE Handbook: ultra-high molecular weight polyethylene in total joint replacement*. San Diego: Elsevier Academic Press; 2004.
54. Park JB, Lakes RS. *Biomaterials: an introduction*. 2nd ed. New York: Plenum Press; 1992 p 141–65.
55. Li S, Burstein AH. Ultra-high molecular weight polyethylene: the material and its use in total joint implants. *J Bone Joint Surg Am* 1994;**76**:1080–90.
56. Amstutz H, Campbell P, Kossovsky N, Clarke IC. Mechanism and clinical significance of wear debris-induced osteolysis. *Clin Orthop Relat Res* 1992;**276**:7–18.
57. St. John KR, editor. *Debris from medical implants mechanisms of formation and biological consequences*: ASTM STP 1144; 1992.
58. Dumbleton J. A literature review of the association between wear rate and osteolysis in total hip arthroplasty. *J Arthroplasty* 2002; **17**(5):649–61.
59. Sochart DH. Relationship of acetabular wear to osteolysis and loosening in total hip arthroplasty. *Clin Orthop Relat Res* 1999; **363**:135–50.
60. Oparaugo PC, Clarke IC, Malchau H, Herberts P. Correlation of wear debris-induced osteolysis and revision with volumetric wear-rates of polyethylene: a survey of 8 reports in the literature. *Acta Orthop Scand* 2001;**72**(1):22–8.
61. Orishimo KF, Claus AM, Sychterz CJ, Engh CA. Relationship between polyethylene wear and osteolysis in hips with a second-generation porous-coated cementless cup after seven years of follow-up. *J Bone Joint Surg Am* 2003;**85**:1095–9.
62. Livingston BJ, Chmell MJ, Spector M, Poss R. Complications of total hip arthroplasty associated with the use of an acetabular component with a Hylamer liner. *J Bone Joint Surg Am* 1997;**79-A**:1529–38.
63. Chmell MJ, Poss R, Thomas WH, Sledge CB. Early failure of Hylamer acetabular inserts due to eccentric wear. *J Arthroplasty* 1996;**11**:351–3.
64. Muratoglu OK, Imlach H, Estok D, Ramamurti B, Sedlacek R, Jasty M, Harris WH. Analysis of eight retrieved Hylamer acetabular components. *Trans Orthop Res Soc* 1997;**22**:852.
65. Schmalzreid TP, Szuszczewicz ES, Campbell PC, McKellop HA. Femoral head surface roughness and patient activity in the wear of Hylamer. *Trans Orthop Res Soc* 1997;**22**:787.
66. Kurtz SM, Muratoglu OK, Evans M, Edidin AA. Advances in the processing, sterilization and crosslinking of ultra-high molecular weight polyethylene for total joint arthroplasty. *Biomaterials* 1999;**20**:1659–88.
67. Charnley J. Tissue reactions to polytetrafluoroethylene (letter). *Lancet* 1963:1379.
68. Rohrl SM, Li MG, Nilsson K, Nivbrant B. Very low wear of non-remelted highly cross-linked polyethylene cups an RSA study lasting up to 6 years. *Acta Orthopaedica* 2007;**78**(6):739–45.
69. Nivbrant B, Roerhl S, Hewitt BJ, Li MG. In vivo wear and migration of high crosslinked poly cups: a RSA study. *Trans Orthop Res Soc* 2003;**28**:358.
70. Bragdon CR, Digas G, Karrholm J, et al. RSA evaluation of wear of conventional vs. highly crosslinked polyethylene acetabular component in vivo. *Trans Am Assoc Hip Knee Surg* 2002;**12**:23.
71. Hopper RH, Young AM, Orishimo KF, McAuley JP. Correlation between early and late wear rates in total hip arthroplasty with application to the performance of highly crosslinked polyethylene liners. *J Arthroplasty* 2003;**18**(7 Suppl. 1):60–7.
72. Wright TM, Rimnac CM, Faris PM, Bansal M. Analysis of surface damage in retrieved carbon fiber-reinforced and plane polyethylene tibial components from posterior stabilized total knee replacements. *J Bone Joint Surg* 1988;**70**(9):1312–19.
73. Elke R. Particle disease: status and today's solutions. *Proc 7th Biolox Symp* 2002.
74. D'Antonio JA, Manley MT, Capello WN, Bierbaum BE, Ramakrishnan R, Naughton M, Sutton K. Five-year experience with crossfire highly cross-linked polyethyelene. *Clin Orthop Rel Res* **441**:143–450.
75. Digas G, Karrholm J, Malchau H, Bragdon CR, Herberts P, Thanner J, Estok D, Plank G, Harris WH. RSA evaluation of wear of conventional vs. highly cross-linked polyethylene acetabular components in vivo. *Trans Orthop Res Soc* 2003;**28**:1430.
76. Digas G, Karrholm J, Thanner J, Malchau H, Herberts P. Highly cross-linked polyethylene in total hip arthroplasty: randomized evaluation of penetration rate in cemented and uncemented sockets using radiostereometric analysis. *Clin Orthop* 2004;**429**:6–16.
77. Rabenseifner L. In vivo results wit highly cross-linked polyethylene in the hip. *Trans 7th EFORT Congr* 2003;**7**:1137/380.
78. Manning DW, Chiang PP, Martell JM, Galante JO, Harris WH. In vivo wear of traditional vs. highly crosslinked polyethylene. *Trans Orthop Res Soc* 2004;**29**:1478.

79. Digas G. New polymer materials in total hip arthroplasty. *Acta Orthop* 2005;**76**(Suppl. 315):1–84.
80. Schmalzreid TP, Scott DL, Zahiri CA, Dorey FJ, Sanford WM, Kem L, Humphrey W. Variables affecting wear *in vivo*: analysis of 1080 hips with computer-assisted technique. *Trans Soc Biomater* 1998;**21**:216.
81. Sychterz CJ, Shah N, Engh CA. Examination of wear in Duraloc acetabular components: two- to five-year evaluation of Hylamer and Enduron liners. *J Arthroplasty* 1998;**13**:508–14.
82. Collier JP, Bargmann LS, Currier BH, Mayor MB, Currier JH, Bargmann BC. An analysis of Hylamer and polyethylene bearings from retrieved acetabular components. *Orthopaedics* 1998;**21**: 865–71.
83. Huber J, Plitz W, Walter A, Refior HJ. Comparative tribological studies of Chirulen, Hylamer, and Enduron combined with A1203. *Orthopade* 1997;**26**:125–8.
84. McKellop HA, Lu B, Li S. Wear of acetabular cups of conventional and modified UHMW polyethylenes compared on a hip joint simulator. *Trans Orthop Res Soc* 1992;**17**:356.
85. Essner A, Polineni VK, Wang A, Stark C, Dumbleton JH. Hip simulator wear of "enhanced" UHMWPE acetabular inserts. *Trans Orthop Res Soc* 1998;**23**:774.
86. Sun DC, Stark C, Dumbleton JH. Development of an accelerated aging method for evaluation of long term irradiation effects on UHMWPE. In: *ACS symposium series, irradiation of polymers*: American Chemical Society; 1996. p. 340–9. 86.
87. Essner A, Polineni VK, Schmidig G, Wang A, Stark C, Dumbleton J. Long term wear simulation of stabilized UHMWPE acetabular cups. *Trans Orthop Res Soc* 1997;**22**:784.
88. Essner A, Wang A, Martell J, Edidin A. *In-vitro* and *in-vivo* acetabular cup wear corroboration. *Trans Orthop Res Soc* 2001;**26**:1007.
89. Grimm B. 1st generation results at the hip. Prediction for 2nd generation. *Trans. of the 3rd UHMWPE International Meeting*. Madrid, Spain; 2007 September 14–15.
90. Martell JM, Verner JJ, Incavo SJ. Clinical performance of a highly cross-linked polyethylene at two years in total hip arthroplasty: a randomized prospective trial. *J Arthroplasty* 2003;**18**(7 Suppl. 1):55–9.
91. Endo M, Tipper JL, Barton DC, et al. Comparison of wear, wear debris and functional biological activity of moderately crosslinked and non-crosslinked polyethylenes in hip prostheses. *Proc Inst Mech Eng* 2002;**216**(2):111.
92. McKellop H, Shen F, Lu B, Campbell P, Salovey R. Effect of sterilization method and other modifications on the wear resistance of acetabular cups made of ultra-high molecular weight polyethylene: a hip simulator study. *J Bone Joint Surg Am* 2000;**82**:1708–25.
93. Kurtz SM, Manley M, Wang A, Taylor S, Dumbleton J. Comparison of the properties of annealed crosslinked (crossfire) and conventional polyethylene as hip bearing materials. *Bull Hosp Joint Dis* 2002–2003;**61**:17–26.
94. McKellop H, Shen F, Lu B, Campbell P, Salovey R. Development of an extremely wear-resistant ultra high molecular weight polyethylene for total hip replacements. *J Orthop Res* 1999;**17**:157–67.
95. Muratoglu OK, Bragdon CR, O'Connor DO, Jasty M, Harris WH. A novel method of cross-linking ultra-high molecular-weight polyethylene to improve wear, reduce oxidation, and retain mechanical properties. *J Arthroplasty* 2001;**16**(2):149–60.
96. Bragdon CR, Kwon YM, Geller JA, Greene ME, Freiberg AA, Harris WH, Malchau H. Minimum 6-year followup of highly cross-linked polyethylene in THA. *Clin Orthop Relat Res* 2007;**465**:122–7.
97. Bitsch RG, Loidolt T, Heisel C, Ball S, Schmalzried TM. Reduction of osteolysis with use of Marathon cross-linked polyethylene. A concise follow-up, at a minimum of five years, of a previous report. *J Bone Joint Surg Am* 2008;**90**:1487–91.
98. Essner A, Wang A. Comparison of two different hip wear simulators. *Trans Orthop Res Soc* 2000:217.
99. Bono JV, Sanford L, Toussaint JT. Severe polyethylene wear in total hip arthroplasty. Observations from retrieved AML plus hip implants with an ACS polyethylene liner. *J Arthroplasty* 1994;**9**:119–25.
100. Berry DJ, Barnes CL, Scott RD, Cabanela ME, Poss R. Catastrophic failure of the polyethyhlene liner of uncemented acetabular components. *J Bone Joint Surg Br* 1994;**76**:575–8.
101. Hamadouche M, Witvoet J, Porcher R, Meunier A, Sedel L, Nizard R. Hydroxyapatite-coated versus grit blasted femoral stems. A prospective, randomised study using EBRA-FCA. *J Bone Joint Surg Br* 2000;**83**:979–87.
102. Garcia-Rey E, Garcia-Cimbrelo E. Long-term results of uncemented acetabular cups with an ACS polyethylene liner. A 14–16-year follow-up study. *Int Orthop* 2007;**31**(2):205–10.
103. Tower SS, Currier JH, Currier BH, Lyford KA, Van Citters DW, Mayor MB. Rim cracking of the cross-linked longevity polyethylene acetabular liner after total hip arthroplasty. *J Bone Joint Surg Am* 2007;**89**:2212–17.
104. Halley D, Glassman A, Crowninshield RD. Recurrent dislocation after revision total hip replacement with a large prosthetic femoral head. *J Bone Joint Surg Am* 2004;**86**:827–30.
105. Patel J, Scott JE, Radford WJ. Severe polyethylene wear in uncemented acetabular cup system components: a report of 5 cases. *J Arthroplasty* 1999;**14**(5):635–6.
106. Essner A, Yau S-S, Schmidig G, Wang A, Dumbleton JH, Manley MT. Acetabular liner functional fatigue performance of crosslinked UHMWPE. *Trans Orthop Res Soc* 2005:245.
107. Wang A, Manley M, Serekian P. Wear and structural fatigue simulation of crosslinked ultra-high molecular weight polyethylene for hip and knee bearing applications. *J ASTM Int* 2004;**1**(1).
108. Nevelos J, Ingham E, Doyle C, Streicher R, Nevelos A, et al. Microseparation of the centers of alumina-alumina artificial hip joints during simulator testing produces clinically relevant wear rates and patterns. *J Arthroplasty* 2000;**15**(6):793–5.
109. Manaka M, Clarke IC, Yamamoto K, Shishido T, Gustafson A, Imakiire A. Stripe wear rates in alumina THR—comparison of microseparation simulator study with retrieved implants. *J Biomed Mater Res* 2004;**69B**(2):149–57.
110. Lee R, Longaray J, Cardinale M, Essner A, Wang A, Walter WL. Controllable separation model for hard on hard hip wear simulation. *Trans Orthop Res Soc* 2008:33.
111. Holley K, Furman B, Babalola O, Lipman J, Padgett D, Wright T. Impingement of acetabular cups in a hip simulator comparison of highly cross-linked and conventional polyethylene. *J Arthroplasty* 2005;**20**:77–87.
112. Lee R, Longaray J, Essner A, Wang A, Capello W, D'Antonio J. A neck to rim impingement model for hip wear. *Trans Combined Orthop Res Soc* 2007.
113. Walter W, Insley G, Walter W, Tuke M. Edge loading in third generation alumina ceramic-on-ceramic bearings. *J Arthroplasty* 2004;**19**(4):402–13.
114. Shishido T, Clarke IC, Williams P, Boehler M, Asano T, Shoji H, et al. A clinical and simulator wear study of alumina ceramic THR to 17 years and beyond. *J Biomed Mater Res* 2003;**67B**(1):638–47.
115. Morlock M, Delling G, Rüther W, Hahn M, Bishop N. In-vivo wear of hip surface replacements—a comparison of retrievals and simulator results. *J Biomech* 2006;**39**:S121.

116. Brodner W, Grübl A, Jankovsky R, Meisinger V, Lehr S, Gottsauner-Wolf F. Cup inclination and serum concentration of cobalt and chromium after metal-on-metal total hip arthroplasty. *J Arthroplasty* 2004;**19**(8 Suppl. 3):66–70.
117. De Haan R, Pattyn C, Gill HS, Murray W, Campbell PA, De Smet KA. Correlation between inclination of the acetabular component and metal ion levels in metal-on-metal hip resurfacing replacement. *J Bone Joint Surg Br* 2008;**90-B**:1291–7.
118. Campbell P, Beaulé PE, Ebramzadeh E, et al. A study of implant failure in metal-on-metal surface arthroplasties. *Clin Orthop* 2006;**453**:35–46.
119. De Haan R, Campbell PA, Su EP, De Smet KA. Revision of metal-on-metal resurfacing arthroplasty of the hip: the influence of malpositioning of the components. *J Bone Joint Surg Br* 2008;**90-B**:1158–63.
120. Taylor S, Manley M, Sutton K. The role of stripe wear in causing acoustic emissions from alumina ceramic-on-ceramic bearings. *J Arthroplasty* 2007;**22**(7):47–51.
121. Shon WY, Baldini T, Peterson MG, Wright TM, Salvati EA. Impingement in total hip arthroplasty. *J Arthroplasty* 2005;**20**(4):427–34.
122. Isaac GH, Wroblewski BM, Atkinson JR, Dowson DA. Tribological study of retrieved hip prostheses. *Clin Orthop Res* 1992;**276**:115–25.
123. Yamaguchi M, Akisue T, Bauer TW, Hashimoto Y. The spatial location of impingement in total hip arthroplasty. *J Arthroplasty* 2000;**15**(3):305–13.
124. Barrack RL, Schmalzried TP. Impingement and rim wear associated with early osteolysis after a total hip replacement: a case report. *J Bone Joint Surg Am* 2002;**84**:1218–20.
125. Bosco JA, Benjamin JB. Loosening of a femoral stem associated with the use of an extended-lip acetabular cup liner: a case report. *J Arthroplasty* 1993;**8**(1):91–3.

Chapter 26

# Tribological Assessment of UHMWPE in the Knee

Hani Haider, PhD



## 26.1 INTRODUCTION

The era of mass users of total knee replacement (TKR) is quickly approaching, and virtually all current TKR designs use UHMWPE as a bearing component. With rising demand for TKR from a significant part of the aging population, more manufacturers and more types of prostheses continue to emerge, and the number of surgeons who carry out TKR and revision operations is probably increasing fastest of all. Testing of such variety of implants poses a challenge. Testing itself needed to be standardized to help the industrial design process and to help the regulatory bodies (e.g., FDA in the United States and CE in Europe) in the screening of implants prior to their use clinically.

Testing of implants can be generally categorized into *in vivo* and *in vitro* testing. The former is described in Chapters 23 and 28. In this chapter we are concerned with *in vitro* tests. The obvious main benefit of such tests is that the evaluation can be made at the TKR design and prototyping stage and thus can not only screen risks prior to clinical use but can aid with useful feedback to the design process. Also, most variables can be more easily, systematically, and precisely controlled *in vitro*, almost to the level of elimination, for highly useful comparisons between implants prior, during, and after their design. Uncontrolled variables can, on the other hand, greatly complicate *in vivo* testing and results interpretation (see Section 26.4.1 for examples of such errors).

Durability testing is only a subset of a wide and growing range of possible other *in vitro* TKR motion tests [1]. These include early lower limb simulators for testing knee implants in cadaveric specimens [2], the Oxford Rig [3, 4], variations of it for 3-D motion of the knee to investigate ligament length patterns [5], and continuing upgrades such as with the Purdue Simulator [6, 7], and even later versions to assess deep knee flexion [8]. There are also special simulators with multiple muscle actuators to measure cruciate ligament strain [9], multiaxial freedoms and actuators to assess TKR constraint [10], or the robotic type of simulators for studying the role of ligaments in cadaveric knees [11] and lately robotic simulators [12, 13] for studying high knee flexion too. Wear testing of UHMWPE in a TKR is typically done on what is commonly referred to as a "knee simulator," but wear testing should be viewed in the correct perspective of all those other types of tests, on machines that are also called sometimes called by the same name. This chapter contrasts and critiques the most common methods and simulators for wear testing of UHMWPE bearings in TKR.

It is clear to those who regularly perform knee wear testing, and especially those who try hard to do testing well, that the "devil is in the detail". With the exception of some specialist workshops (e.g., by ASTM) even scientific conferences appear lately to be more focused on results; they leave little room for detailed discussion, especially on methodological detail where most of the variability in results emanate, in the author's view. Nevertheless, this chapter is still not a detailed how-to manual. International standards (where available) are cited instead. Published international standards typically describe such methods in sufficient detail. Nor is this text a review of the results of TKR tests from the literature, as useful as that may be. The results are too many and varied; only some are cited here to illustrate specific points. The main emphasis here is on matters of rationale and know-how to avoid pitfalls in the choice of test methods, in their implementation, and in the interpretation of results. This emphasis should help engineers in the orthopedic industry and in the research community who hunger for the type of detail they cannot usually find in published research papers, usually for lack of space in such result-oriented research articles.

Standardized test methods and international standards are neither yet sufficiently mature nor are they rigorously or uniformly applied. The evolution of standards is naturally slow because of their almost consensus-based nature, having to be unambiguously clear, very doable, and inclusive of sometimes conflicting styles and constraints.

Four categories of durability wear tests are typically done for UHMWPE destined for a TKR application: (1) pin-on-disk material screening tests, (2) force-control knee simulator tests, (3) displacement-control simulator tests, and (4) special damage mode simulation tests. The other (first three) categories will be described at varying levels of detail in this chapter. The second and third test methods will be preceded by describing some general considerations of the biomechanics of the knee, coordinate systems, and the physical variables involved. This will set the scene for a brief portrayal of the history of knee wear simulator development, not surprisingly setting the scene also for the much debated force- and displacement-control testing approaches. The difference between those will be critically highlighted. The final category of testing wear of UHMWPE in TKR relates to variances of these methods to screen for special failure modes, such as on TKR stabilizing posts, high flexion, and those specific for mobile bearing TKRs. This category of testing is not dealt with here because it is typically custom designed to address suspected particular design-related failure modes. Reference to standardized testing techniques is frequently made in the text, elaborating wherever possible on the rationale behind such methods.

## 26.2 PIN-ON-DISC TESTING OF UHMWPE DESTINED FOR KNEE REPLACEMENTS

Pin-on-disk (POD) is essentially a materials tribology test that is frequently used to screen materials rather than whole TKR systems. Because it is only marginally within the scope of the chapter, this category of tests will be only briefly visited here to highlight special considerations required for TKR testing applications.

In POD testing of UHMWPE for TKR systems, a single station or multistation machine is configured to articulate the end of a pin onto a flat disk to measure frictional and wear characteristics of the given material pair. The setup usually provides for a constant or dynamically varying compressive force on the pin to induce a simulated stress level that would occur in the TKR *in vivo*. POD in general engineering testing (outside orthopedics) typically provides a reciprocating linear motion or curvilinear motion where a disk is rotated under a stationary pin with some compression applied. Resisting force to the pin is measured as well as the compressive force to estimate the coefficient of friction. In orthopedics, conventional UHMWPE or one of the many improved alternatives are tested in a POD configuration as a bearing material for TKR, or what is tested is an alternative metal, ceramic, or coated metal versus UHMWPE (e.g., a nanostructured diamond on a titanium alloy [14]). Among the many complications required here is whether to simulate just sliding, as in simply reciprocating pin or disk articular motion; or gliding [15], such as a metal pin translating relative to a UHMWPE disk, thus producing fatigue type loading and wear as well as adhesive wear of the polymer; or rolling, which is more like the latter but with less friction; or any combination of those effects.

Which one or combination of these motions are actually simulated (or occur) depends on whether the UHMWPE acts as a pin or disc, the shape of each, and whether either pin and/or the disk moves and how. A wide spectrum of POD test machines have therefore been used in a multitude of configurations from the very simple (e.g., linear reciprocal) to the highly sophisticated and flexible, such as with two-axis computer driven motions for programmable cross paths (Figures 26.1 and 26.2).

What drove POD testing of UHMWPE for TKR to become exceedingly more complex is that UHMWPE wear

**FIGURE 26.1** The AMTI OrthoPOD. An example of a six-station computer controlled pin-on-disk test machine, programmable for two dimensional paths, environmentally controlled specimen chambers and with friction force measuring capability. This machine has variable dynamic compressive force and planetary gear pin rotations over a rotating disc platform with six specimen disk holders. Both motions are servo motor controlled and programmable for any two-dimensional paths, with frictional measuring capability.

**FIGURE 26.2** MAX-Shear wear experimental testing system. Another example of a six-station computer controlled pin-on-disk test machine. Here a computer driven X-Y stage motor is used with static compressive force on each station [22].

is highly accelerated with multidirectional articular paths and cross-shear [16–19], as would occur in knee replacement implants. In one experimental pin-on-disk study [19], only 15 degrees of crossing motion in sliding increased wear over 10 times. Some studies suggested little reason to exaggerate angles of motion crossings beyond this angle. A recent study on *in vivo* kinematics of one TKR design and thorough postprocessing analysis of the crossing motion found the included angle of bidirectional crossing motion not to exceed 10 degrees [20].

There is a standard test method, ASTM F732 [21], for POD testing, which is comprehensive enough to provide guidance to delineate the motions previously described and attempt to map them to suitable UHMWPE applications across knee, hip, and other implants. This ASTM standard also gives sufficiently detailed procedures to simulate, actuate, measure, and document such POD testing.

The following are some technical considerations that are useful to remember when screening UHMWPE for a TKR application:

- Surface finish: It is obvious, but worth remembering, that when comparing alternatives of bearing couple materials (e.g., against UHMWPE), the surface finish, especially of the metallic material, has to be kept constant as a controlled variable. Sometimes the different surface finish is inherent in the alternative material (e.g., a surface that is difficult to polish, or much easier/cheaper to polish to achieve a superior smoother finish), and therefore the surface finish becomes one of the surrogate variables of interest. In POD tests of materials simulating orthopedic joint replacement conditions, the wear factor was found to rapidly increase for Ra>50nm values [23]. Considering this level is about the highest (roughest) used in polished metallic articulating surfaces for implants, it is wise to have surfaces polished well below this value in POD tests.
- Edge effects: It is unlikely that full alignment can be achieved and maintained for good control of contact area (and stress) with a flat metal pin articulating on a flat UHMWPE surface. The edge effects would dominate and can be highly complicated with reversing and crossing motions. Metal pins need to be spherical or highly convex, with the contact areas then approximated using "Hertzian" contact [24]. UHMWPE pins can be flat; with reasonable alignment and some deformation, good predictable contact area can be maintained.
- Wear versus wear rate or wear factor: Attention should be given to expressing or interpreting the results from POD tests. In some cases, wear is presented as a wear factor, typically the volumetric wear per unit load and (per unit) distance of articular motion. This implies some dependence on stress, which is itself disputed and adds confusion (discussed later).
- Stress level: Care should be taken in selecting a combination of load and pin contact area to reproduce the

stress levels expected in the knee. TKR stresses are typically higher than for hips and span a wider range (e.g., 2–50 MPa) depending on the TKR design [25]. The dependence of wear of UHMWPE on contact stress and apparent contact area is inconsistent in the published literature [26–31]. Wear was found not to (necessarily) rise with stress in the loading regime of TKR according to some of these studies. Wear of UHMWPE in the stress range for orthopedic implant applications is as (or even more) sensitive to the contact area [28]. The wear factor in one study was even found to decrease with rising stress at the low stress levels in the range above [31]. At low stress, higher contact areas engage more of the surface asperities, giving more opportunities for material removal. Therefore, the combination as well as the contact area and stress individually need to be physiologically realistic or take account of the worst case scenarios in the design of a POD UHMWPE screening test for TKR.

- Constant or dynamic load: To simulate the loading of a convex TKR femoral component surface with sliding, gliding, and rolling against UHMWPE, the compressive load in a POD test needs to be dynamic and intermittent to induce dynamic UHMWPE stresses with reasonable representation of lubrication replenishment of the articulating surfaces. This is typically achieved through dynamic varying of the load on the pin (for example in the system of Figure 26.1). With a UHMWPE pin, articulation on a smaller area of the poly is maximized, and wear measurements of UHMWPE weight loss are relatively more accurate given small weight pins. Alternatively, the UHMWPE material can be configured as the larger area (disk) bearing material, with a metallic pin having a spherical articulating surface, mimicking more directly TKR bearing systems (e.g., the system of Figure 26.2 [22]). In the latter situation, the actual contact area has frequently been approximated using a simple Hertzian ball-on-flat contact mechanics model [24]. Such estimates have, in the past, provided reasonable approximations when cross-validated with finite element analysis computations and static pressure sensitive film experiments [32]. Care should still be taken because the TKR loading is complex in that it is dynamic, noting that Hertzian contact theory relies on several assumptions, among which are small, frictionless, and continuous contact surfaces, which both deform [24]. A typically metallic femoral component is almost infinitely rigid in most designs compared to a UHMWPE tibial bearing insert.
- Direct volumetric measurement of UHMWPE wear: Wear estimates that rely on thickness (or surface scanning) measurements are plagued with creep and recovery errors associated with viscoelastic deformation and some shape memory properties in UHMWPE [33]. Where possible, such techniques should be avoided, in the author's opinion, or the creep and recovery errors should be comprehensively accounted for, with great difficulty.

## 26.3 TESTING UHMWPE WITHIN WHOLE TKR SYSTEMS

### 26.3.1 Degrees of Freedom and a Suitable Joint Coordinate System

The intact knee joint, or one with an installed TKR implant, has 6 degrees of freedom. Because there is no rigid linkage between the bones, the femur exercises three displacements and three rotations (relative to the tibia), and vice versa, to varying levels. Any structure with components capable of rotation and translation in three dimensions can be greatly complicated without a suitable choice of axes for these motions. Typically, the final position/orientation of the structure is influenced by the order (sequence) of motions along and rotations about these axes. A Cartesian coordinate system fixed in space would obviously not be a good choice for the knee. It is much more efficient to fix such an axis system to one bony segment (typically the tibia, based on defined anatomical landmarks) and have the other bone(s), defined similarly, move relative to the first. With simple Cartesian axes, the kinematic position would depend on the order in which the rotations (usually termed Eulerian angles) occur, and it would therefore be complex to describe the knee motions, let alone simulate, actuate, and control them in a wear testing machine. The sequence of rotations, which in reality need to occur simultaneously, would influence the desired kinematic state, compromising both accuracy and speed of response of a simulator.

Various schemes have been published to characterize the natural knee motion, including estimates of the effective axis of flexion by relating it to known anatomical landmarks, such as the transepicondylar line [34]. Various medial pivoting themes [34, 35] have attempted to model the apparent coupling between AP translation with IE rotation and flexion. The posterior femoral rollback onto the proximal tibia, combined with the internal rotation of the tibia with higher flexion, was approximated to a simpler rotation around a proximal-distal axis fixed at the center of the medial compartment of the tibia. Some themes went further to ascribe a helical motion [36] (more widely known as screw motion) to the knee. These comprehensive studies with astute observations have contributed to our understanding of how the knee moves. It is however important to remember that they mostly represented unloaded (passive) knee motion, and it was not clear how they could provide quantitative and unambiguous coordinate axes around which practical knee wear simulators could be built.

The main desire is to have a system that precisely and unambiguously describes the relative position between the femur and tibia and their relative motions in time. The coordinate

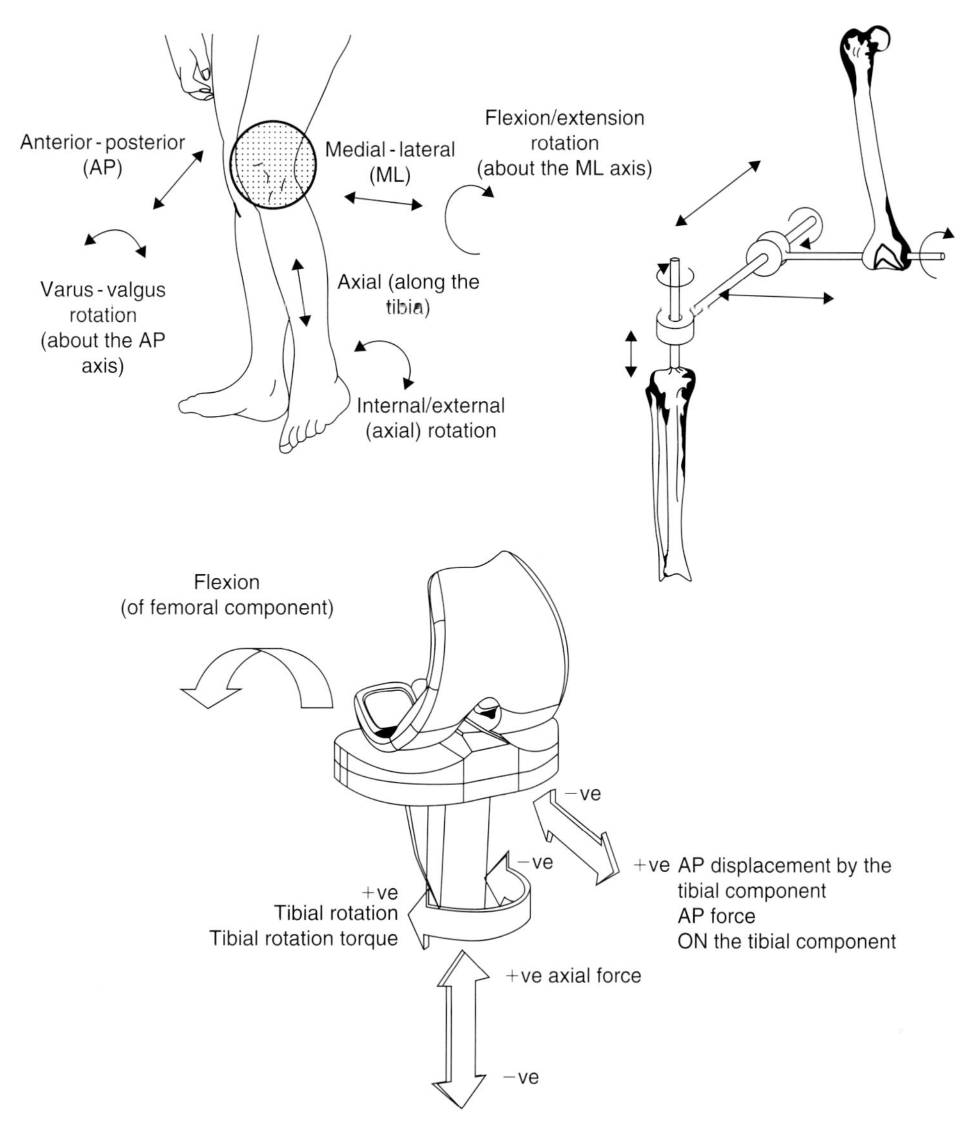


**FIGURE 26.3** Top left: Anatomically-based axis system and degrees of freedom of the human knee joint. Top right: Schematic depicting the Grood and Suntay [37] knee coordinate system with a ML (flexion) axis, which floats in the AP direction. Bottom: Application to a left TKR. Note: External forces and torques ON which components, and motion BY which component. (This sign convention is consistent with ISO 14243-1 [38], and a version of this diagram has been suggested for insertion into a revision of this standard.)

system most typically approximated in knee simulators, to varying degrees of success, was proposed by Groot and Suntay [37], illustrated in Figure 26.3. It conceptualizes a four-link kinematic chain, of which the tibia and femur are the first and last, each along an axis fixed to the tibia and femur respectively (as body-fixed axes). Two imaginary links are placed between the body fixed axes; these are not fixed to either bone. Abduction or varus-valgus occurs around one, and flexion around the other, the latter being a "floating" axis which is allowed to translate in the anterior-posterior direction. In the way it was defined, this system provided a set of independent generalized coordinates that harnessed commonly employed clinical terminology and permitted precise definitions of clinical descriptions of knee motion. It allows unambiguous measurements and control of flexion-extension (FE), varus-valgus (VV), and internal-external (IE) (sometimes called axial) rotations, and anterior-posterior (AP), medial-lateral (ML), and proximal-distal (axial) displacements. Note that in this system, within realistic physiological ranges of knee motion, the rotations actually appear to commute. This means that the net final position of the femur relative to the tibia becomes independent of the sequence of the individual rotations. When understood and applied well, this system brought a much needed simplification for knee simulators.

Literally (loosely!) resembling a mechanical hinge, the knee joint mainly provides FE rotation with an angle range of approximately −5 degrees to 140 degrees. Very moderate ranges of axial (lift-off), medial-lateral displacement, and varus-valgus rotation occur. These three degrees of freedom are highly constrained by a mixture of some surface congruency of the tibia-femoral interfaces, the cruciate and collateral ligaments, and other soft tissue. While also moderately constrained by the same mechanisms, the knee is relatively more lax on the remaining two degrees of freedom. The AP translation (approximately ±10mm) helps adjust the effective torque arm length of the quadriceps muscles to balance the flexion/extension moment due to gravity of the body, and it prevents bony impingement at high flexion angles. While the knee joint is flexed, the center of mass of the body above the knee acts with a moment arm (about the ML axis) of many inches in length, whilst the moment arm of either muscle (quadriceps or hamstrings) group is a fraction of this distance. Large muscle forces and joint reactions result, which reach multiples of human body weight.

The flexion moments themselves are of central interest in many biomechanic texts and clinical gait studies [39]. However, our interest in them in TKR testing is indirect only, as the flexional frictional resistance is modest in simulators when testing very low friction TKR bearings. Nevertheless, those large muscle forces also produce large joint reaction forces in a TKR, which have major components axially and significant components in the AP direction [25, 40]. These axial and AP knee joint reaction forces need to be reproduced in TKR and UHMWPE wear testing, and they are usually reproduced by actuating them as forces or reproducing the expected motions from them.

A simple qualitative observation of a human while walking shows the body swinging left and right a little as well undulating up and down a little (typically 50mm amplitude). The continuous acceleration/deceleration of the body center of mass adds modest perturbations to the joint reaction forces (typically reaching 15%). The slight twisting of the body over the load bearing foot necessitates transmission of a time varying axial torque (and rotation) between the tibia and femur, with an angular range of up to approximately ±30 degrees. The AP translation and IE rotation motions, superimposed on flexion extension, while the knee doubles in function as a hinge and a thrust bearing, are crucial factors on wear and durability of UHMWPE in TKR implants.

A wide range of TKR designs has emerged over the years (see Chapters 7 and 8), all aiming to cater for this complexity. Some are generically different from the anatomically shaped bicondylar type. In one extreme, various early implants were simple hinge designs with essentially one degree of freedom. On the other extreme, mobile bearing TKRs are extremely lax; they have different translational and rotational degrees of freedom provided by separate bearing mechanisms within. Rotating hinge revision TKRs are an example where a simple hinge design rotates on a secondary thrust bearing so the whole TKR assembly offers unconstrained relative rotation about two axes and effectively infinite stiffness rotationally in varus-valgus and translationally in AP. Some mobile bearing TKRs offer all six degrees of freedom with highly nonlinear and highly discontinuous stiffness because of unconstrained axial rotation and/or AP translation limited sometimes by rigid mechanical stops.

## 26.3.2 Brief History of Knee Wear Simulators

The earliest wear simulator that the author has read about was a 1973 knee and hip machine that provided flexion with axial force only, in which knees were tested in Ringer solution [41]. The first machine designed to specifically simulate and measure TKR wear was developed in Leeds in the mid- to late-1970s [42]. It was a single station hydraulic machine through which axial compressive and anterior-posterior forces were applied in synchrony with flexion, using drip-fed distilled water as lubricant. The AP was force controlled between 66N in the anterior and 220N in the posterior directions (respectively) in a (busy!) waveform that reversed the force direction seven times. The VV and IE were allowed to freely rotate. In a later study [43] published on this simulator, wear (and of course creep) was estimated from the penetration of the metallic femoral component into the polyethylene as measured optically by reflective dual-index holography. The UHMWPE bearing insert was coated with a 0.1nm thick pure aluminum coating (by vacuum vapor deposition) to improve the quality of the optical contouring process, before and after the test (the coating was removed during the actual testing!). That simulator was reported to capture both adhesive and abrasive wear, and the wear factors that resulted were similar to those of pin-on-disk studies (with water also).

Motivated by the need to test the (then) revolutionary new concepts in rotating platform mobile bearing knee designs, the New Jersey simulator provided flexional rotation, axial compressive force, AP motion, and IE rotation [44, 45].

Both the Leeds and the New Jersey knee machines must have been pioneers of their time in wear simulators because, even as late as 1992, knee wear simulator designs were being published, which provided for only axial force varying with flexion [46]. Such simulators had IE free to rotate, and ML and VV were locked but adjustable during setup for alignment. AP displacement was coupled mechanically with flexion to provide "partial constraint of the centroid path" [46].

The preceding were all single (test) station machines. The need for multiple stations in a knee simulator was soon felt. It is important to note that wear is a gradual damage accumulation process that results from highly repetitive gait activities, such as walking. TKR implants *in vivo* need to survive tens of years in service, and thus millions of cycles of walking at least. Considering the polymeric nature of UHMWPE and its viscoplastic behavior (see Chapter 35), there would be a speed limit on experimental simulation of loading, and this is typically chosen to be about 1Hz and should not exceed 1.5–2Hz at most. Considering that a typical TKR patient would walk 1–2 million cycles (MC) per year, reasonable simulations should therefore last at least 5 million cycles or more to represent a reasonable duty period *in vivo*. With those low frequencies, a typical TKR wear test with overheads of setup time, breakdown for maintenance, and wear measurements would last a minimum of 2.5 months but typically up to 6–9 months. Therefore, a knee simulating machine for wear testing has to have many stations to be practical and usefully test, compare, and screen TKR designs. The earliest multichannel knee simulator was a 12-station machine that provided flexion and axial force only for testing tibial fixation [47].

In 1997 work on the Durham simulator was published, which was specifically designed as a TKR wear simulator, with six test stations [48]. This actuated and controlled flexion, axial force, and AP motion under displacement control (range ±15.6 mm), with fixed ML position and passive rotations allowed for IE and VV. This study ran an 8 million cycle test on one of this simulator's stations and highlighted the importance of long wear tests. Diluted bovine serum was used for lubricant. Only a very modest wear rate of 2.62 mg/MC resulted in that study, presumably because of the absence of actuated IE or torque or rotation, and therefore cross-paths of UHMWPE must have been undersimulated.

Almost at the same time, 1997, early work to develop the Stanmore knee simulator was published in preliminary concept on one prototype station [51]. This simulator allowed dynamic flexion and axial force variation like all the others, with ML and VV freedoms, but also provided for dynamic servo control of AP *force* and IE *torque* as inputs and physically simulated some restraint to the AP and IE motions with simple elastomeric bumpers. However, besides being single station, soft elastomeric bumpers of course could not provide a consistent stiffness between tests or even a constant stiffness within a test for millions of simulator cycles. A full-fledged later implementation of the same concept from Stanmore/England was the four-station force-controlled knee simulator (Figure 26.4 and Figures 26.8–26.10) with springs instead of bumpers [49, 50]. It also featured AP force and IE torque actuations that were more decoupled and provided more precise freedoms in ML and VV. This configuration provided the basis for a TKR wear testing methodology [38, 49, 50, 52–55] known as force control, which continues to be refined even at the present day, and more of it will be described later in this chapter (see Section 26.3.4).

Meanwhile, displacement-controlled knee simulators continued to be developed in a variety of configurations and used for TKR testing. Examples of those machines are the early version of the AMTI knee simulator (Figure 26.5), the Shore Western knee simulator (Figure 26.15), and the ProSim knee simulator (Figure 26.16). The test method used on those machines will be described in detail later in Section 26.3.8.

In capping the above brief summary of the development history of knee simulators, we note the emergence of two general trends, force-control and displacement-control knee simulators. The control refers to the way AP force/translation and IE torque/rotation were imposed on the TKR. These two trends have been much debated and

**FIGURE 26.4** Lower left and right: First version of the Stanmore Knee Simulator [49, 50]. The electronics of this simulator have been upgraded by the Instron Corporation, and the mechanics have been upgraded in Nebraska as shown in Figure 26.8.

**FIGURE 26.5** Hydraulic six-station displacement-control Knee Wear Simulator by AMTI, Watertown, Massachusetts, USA.

continue to be the subject of exciting discussions to the present day. The two approaches have themselves evolved and gradually converged into two well-known standard test methods; the first is the force-control method, described in detail in ISO 14243-1 [38], and the second is displacement control, described in ISO 14243-3 [56]. They both specify the relative flexional angular movement between the articulating femoral and tibial components of a TKR, the pattern of the applied axial force synchronized with flexion, speed and duration of testing, sample configuration, and test environment to be used for the wear testing of UHMWPE. The two methods differ in how AP forces and displacement are controlled and how tibial torques or rotation angles are controlled. In choosing which method to use, or in using either, it is important to realize the subtle but important differences and their implications. The two approaches are therefore described in more detail and more critically contrasted in the following sections.

### 26.3.3 Contemporary Knee Wear Simulators and the Much-Debated Force- versus Displacement-Control Paradigms

In durability testing of UHMWPE bearings in whole TKR systems, a simulating machine must inherently energize as many of the failure modes as can result from continual use of the TKR *in vivo*. To reproduce the wear mechanisms the correct combinations of loading and relative motion between the parts of the TKR must be actuated. Therefore, knee simulator machines, as well as provide multitest stations, correct temperature and lubrication environments; allow disassembly during testing to measure wear; perform repetitive cyclic actuations which best approximate to a realistic human activity such as walking; should importantly provide passive ML and VV freedoms; and actuate the important motions of flexion/extension, AP translation and rotation while the joint is fully loaded and undergoing

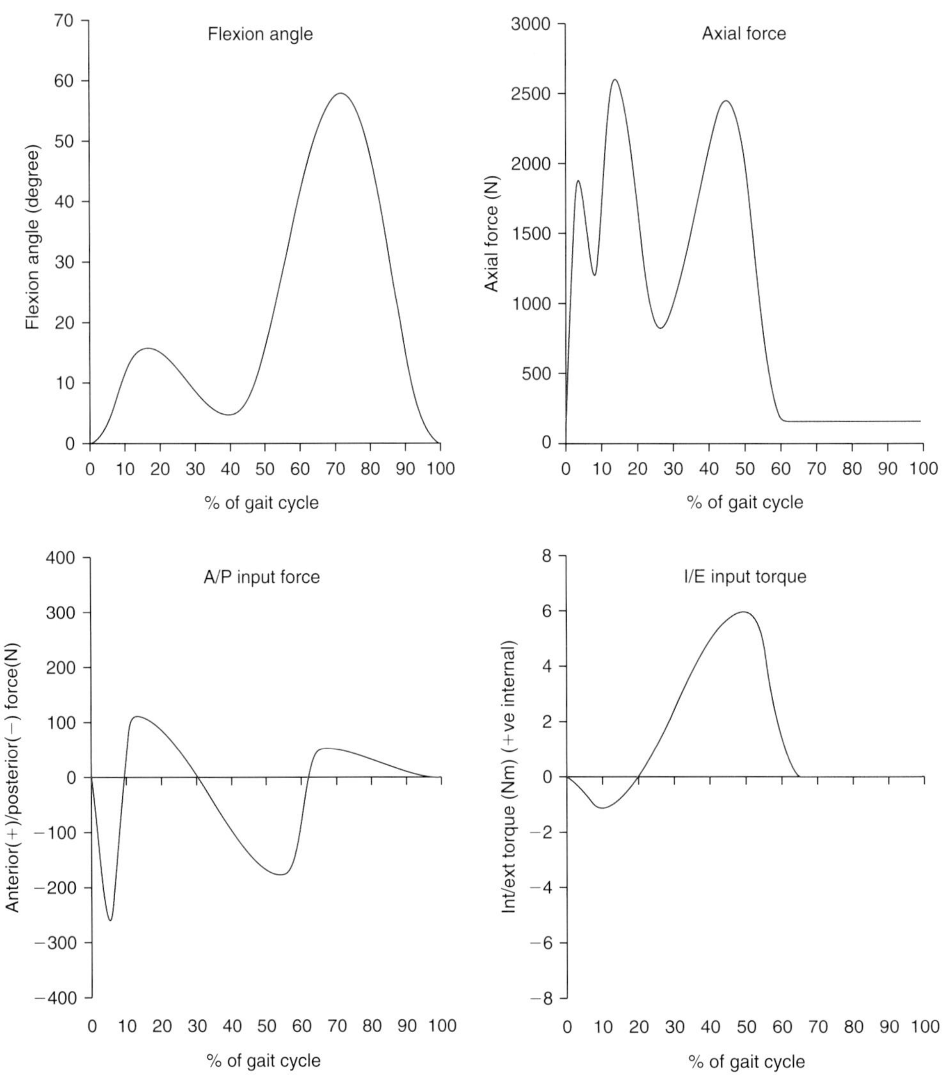


**FIGURE 26.6** Input waveforms of ISO 14243-1 (force control) [38]. The flexion and axial force (top two plots) are also the same for ISO 14243-3 (displacement control) [56].

physiologically realistic joint reaction contact forces. It is almost a prerequisite, therefore, for any standardized TKR wear testing method that it should have predefined, clearly communicated, and easily implemented test conditions to ensure that an appropriate, physiologically realistic movement of the TKR parts in the AP plane occur throughout the test. As will be seen in the following discussion, this latter requirement presents the major difference between the two standardized knee wear test methods [38, 56].

Most current knee simulators treat the flexion angle as an independent variable to which all other variables are synchronized in time in a simulated dynamic gait activity (e.g., walking). Flexion/extension is the kinematic variable with the largest dynamic range within the knee and is easiest to measure externally on a moving human. Flexion has also been abundantly and accurately characterized for different human activities, so it yields itself as efficient in characterizing the phases of any cyclic human activity (or actuation waveform) on a simulator. Figure 26.6 (top left) shows the familiar two-peaked shape curve of flexion angle during walking used in both ISO test methods [38, 56].

Most TKR simulator studies apply the axial compressive component of the joint contact force in synchrony with flexion/extension. This is the largest external force in amplitude resulting from the body weight and its inertia, and compounded upon both are the significant quadriceps and hamstring muscle forces (depending on stance) at all flexion angles above zero. Figure 26.6 (top right) shows the force waveform typically used as input. These waveforms were based on the results of quasi-static analyses of a geometric model of the knee, with electro-myographic (EMG) measured data, ground-to-foot force, and kinematic data [57, 58].

Note the familiar triple peaks of force reaching multiple body weights during the stance phase (0–60% of the walking gate cycle between heel strike and toe off). Notice the modest flexion peak of only 15 degrees reached during the stance phase and the highest flexion peak reached at 58 degrees in the swing phase with nominal (but not zero) forces during the latter.

FIGURE 26.7 The EndoLab knee simulator. This four-station machine is servo-hydraulically actuated for full force-control testing with soft tissue simulation. One of the test stations can be optionally used as a loaded soak control station.

TKR wear is a complex function of the integral of the joint reaction forces and the rolling/sliding and crossing path motions between the implant articulating surfaces. The ratio of distances and velocities between rolling and sliding and the degree to which the motion paths cross are largely influenced by the anterior-posterior (AP) motion and internal-external (IE) rotation kinematics. Displacement-controlled simulation directly actuates input waveforms for the anterior posterior (AP) motion and internal external (IE) rotation of the tibial component relative to the femoral component in synchrony with axial force and flexion angle during walking. This type of dynamic position control allows for robust simulator mechanisms with prolonged wear simulation runs without excessive maintenance. It does, however, require prior and precise knowledge of the actual AP translations and IE rotations expected with ideal insertion *in vivo* for that particular TKR. For example, in a displacement-controlled simulator wear study [59] on a widely-used TKR design with a long clinical history, a normal walking gait was found to have a reasonable wear rate of 17 mg/Mc with a realistic AP displacement range, peaking at under 5.2 mm and a modest IE rotation range of 8 degrees. This wear rate dropped by a staggering 90% and 95% in the absence of AP motion and IE rotation in turn, respectively.

Even in the same (theoretically) assumed average patient with a (theoretically) ideal surgical installation, with a well defined/simulated repetitive gait activity, such motions would primarily depend on the detailed design (shape, materials, and surface finish) of the TKR's articular bearing surfaces, which actually govern its AP and IE laxity.

For example, there would be less AP displacement and IE rotation in highly constrained designs than for low constrained designs, and those kinematics may be much curtailed, or one could become zero for hinge-type revision TKR. Therefore, for simulation in a displacement-controlled regime, the precise kinematics need to be known beforehand, which has not been possible with new designs, except recently with unconfirmed degrees of accuracy using finite element analysis computation [60–62].

Importantly and as briefly explained elsewhere [63], even with a known clinically used design with presumably accurate kinematic data from *in vivo* tests (e.g., using fluoroscopy), displacement-control wear testing would be prone to a major flaw, in the opinion of the author, for the following reason: Some wear, or in some cases most of the wear, may occur in those regions of high compressive or shear stress interaction or impacts between the harder metallic femoral component and the softer UHMWPE bearing insert. Examples of those are impingements onto the posterior stabilizing posts or climbs over steep anterior or posterior ridges of the UHMWPE. The polyethylene deforms viscoelastically

FIGURE 26.8 Upgraded design of the Instron-Stanmore Knee Simulator. Three machines are located at the University of Nebraska Medical Center. Lower: Upgraded designs of flexion brackets, femoral component holders, and tibial component dishes, which are now stainless steel, round, and much larger to provide higher range of TKR motion and can accommodate optional serum bags to increase the amount of lubricant per station and extend the simulator for use with larger and different TKR designs. Middle left: Custom made femoral component holders as one-piece stainless steel jigs, which are precisely manufactured to fix to the main simulator femoral brackets and are directly cemented with an exact fit to a femoral component. Middle: Holder with femoral component cemented. Top left: Stations running without serum bags. Middle right: One specimen station viewed from the proximal anterior perspective.

in those regions under high stress contact. In force-control testing, the stresses in each cycle would continue to be physiologically realistic as the motions would self-adjust (minutely but automatically) to keep the forces and torque reactions as desired/assumed. In displacement-control testing, the motions are repeated by the simulator actuators precisely without any such adjustments. Imperfect alignment causes either exaggerated stresses from slight (e.g., a few micron) overimpingement, or it causes superficially attenuated stresses in those regions where the motions run minutely short of the physiologically realistic/assumed values. The viscoelastic deformation of an anterior ridge or stabilizing post in the early cycles also grossly alters the impact for later cycles. This would cause the wear during the majority of the test to be undersimulated.

Until 2004, there was no published standard nor agreed inputs for the displacement-control TKR testing. The nearest such document was a set of gradually evolving guidelines in an ASTM draft standard document, which has since been discontinued. The current ISO standard displacement-control test method [56] was based on the earlier developed force-control method [38] but prescribed new displacement-control inputs instead, and these inputs will be described later (see Section 26.3.7).

## 26.3.4 Force-Control TKR Simulation for UHMWPE Wear Testing

In force-controlled simulation, the knee is treated as a generalized joint required to transmit a deterministic AP force and IE torque across the knee to allow quasi-static equilibrium at each phase/instant of a slow ambulatory gait activity, such as walking. It is assumed here that for such equilibrium to occur, on an average simulated patient, at each posture within a repeated walking gait activity, the external forces acting on the knee joint as a full system (i.e., treated like a black box or surrounded by its own free body diagram) are independent of the design of the TKR implanted. With good estimates based on comprehensive studies, these forces and torques can then be treated as standard inputs and applied to any knee design during simulation of walking. The force-controlled simulation was thus intended to automatically replicate (*in vitro*) the kinematics and kinetics appropriate for that particular TKR design to simulate its wear.

The input waveforms for the force-control method (Figure 26.6) were originally adapted for the (then draft, now published) ISO 14243-1 standard [38] from the rich studies presented in the early 1970s by Morrison [58] and later results from other studies [57]. Morrison derived a comprehensive "quasi-static" theoretical model of the knee joint and computed knee forces based on ground-to-foot reaction force and kinematic data from gait studies and electro-myographic data with assumed/simplified anatomic geometric muscle insertion points. He also assumed zero coefficient of friction of the articulating surfaces, among many other assumptions. The force-control ISO standard itself [38] did not explain the process of how the input curves were devised. The curves were devised through a collaborative research study [50] that proposed and used the input waveforms without elaboration of how the curves were interpreted from the original cited studies.

Many investigators have had to check this validity themselves by reverting to the original studies cited because some have questioned the accuracy of the AP and IE torques and even their sign convention. Those attempts, to the knowledge of the author, verified those interpretations to be good approximations, including the sometimes disputed sign/sense of the AP force. Further, the waveforms continued to prove harmonious with other findings since. They were also verified to be physiologically realistic

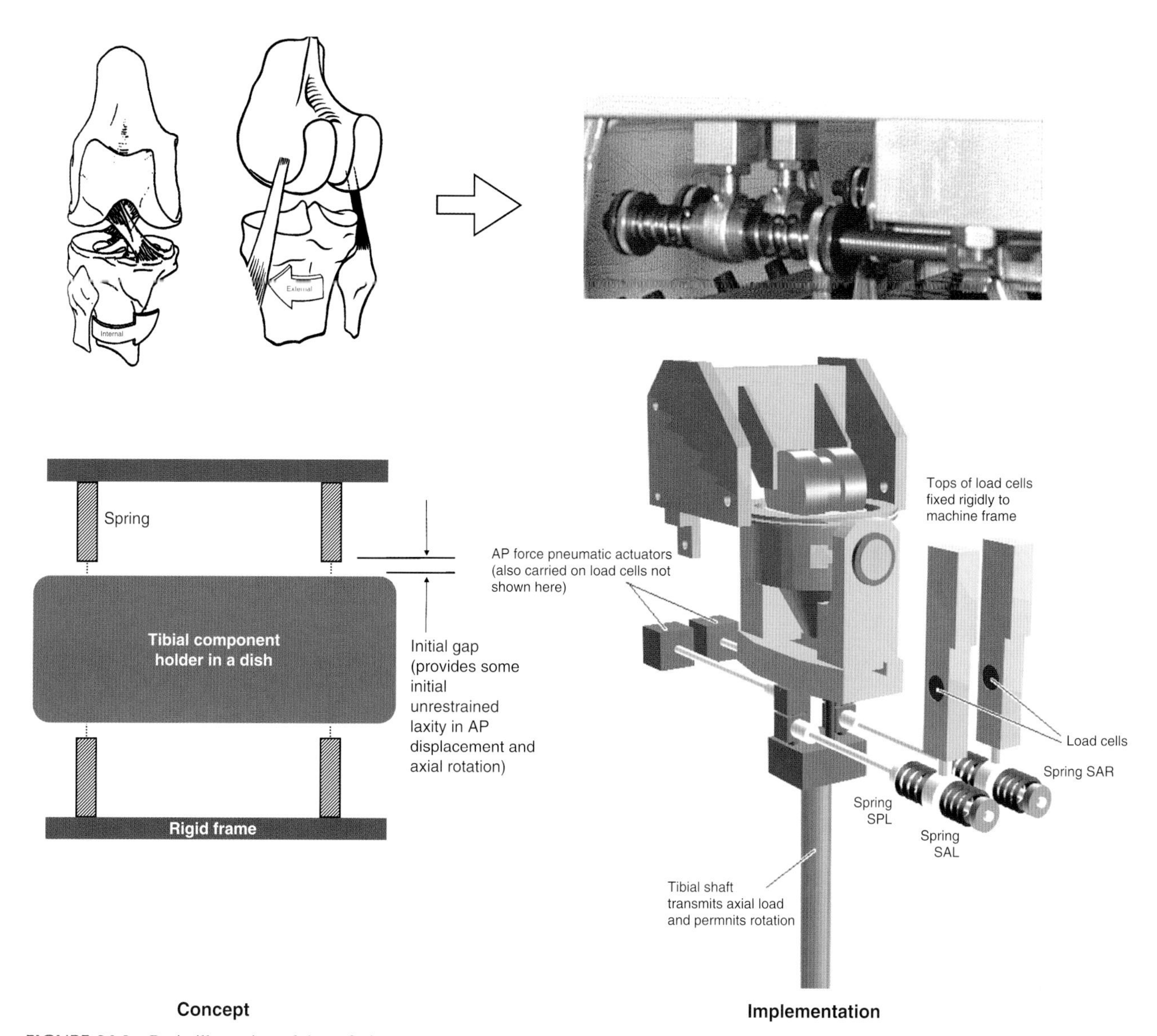


FIGURE 26.9 Basic illustration of the soft tissue restraint simulation system concept and implementation in the Instron-Stanmore Knee Simulator. Top right: The actual spring assembly. Bottom right: The configuration of the main AP force actuators and soft-tissue restraint springs on one station of the first design of the Instron-Stanmore Knee Simulator.

with direct telemetry data from a distal femoral knee replacement [64, 65] and were found to be compatible with and within the range of more recent theoretically modeled/predicted forces [66] and more recent *in vivo* telemetry data from studies with an instrumented TKR [67].

## 26.3.5 Soft Tissue Simulation in the Force-Control Test Method

For physiologically realistic motions to follow in a simulation where the external (to the knee) AP force and IE torque are (force) controlled, the system requires additional simulation of the ligament and other soft tissue constraints that would occur *in vivo*. This soft tissue restraint would act to attenuate/limit the motions *in vivo* in response to those forces. The restraint of ligaments after TKR surgery and other stabilizing soft tissue depends on the instantaneous AP position and IE angle. In the force-control method, these were initially intended to be simulated with springs, which were desired to modulate the restraint forces and torques based on the actual instantaneous distances and angles moved. Figure 26.9 shows schematically how such a spring configuration was implemented in the force-controlled Instron-Stanmore Knee Simulator.

The initial design, which was adopted in the ISO standard [38], was for the springs to be precompressed such that they would never become loose. Alternatively, the springs can be loose with a gap, such that compression only starts (or ends) when the motion exceeds the gap. It can be shown that the stiffness of this whole assembly with given

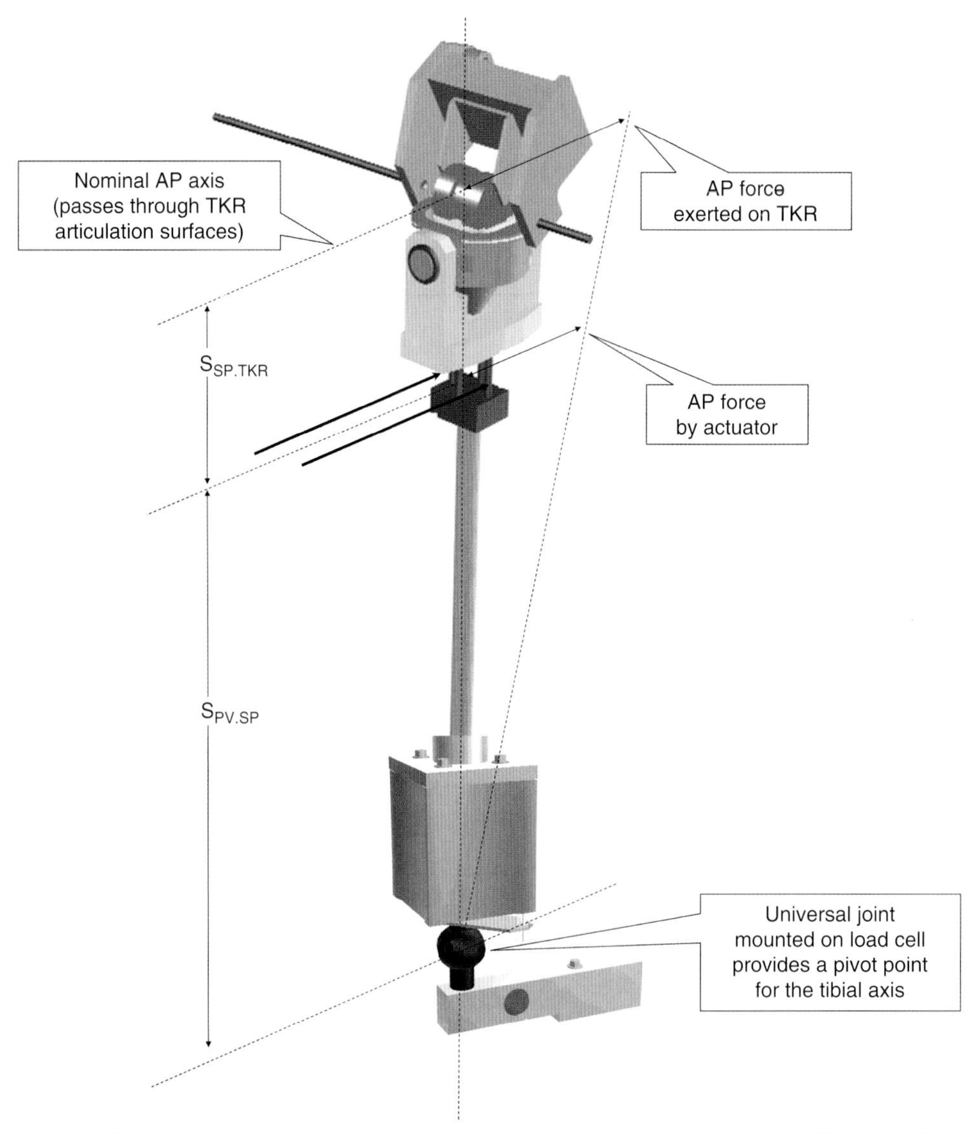


**FIGURE 26.10** Schematic illustrating the leverage between where the AP actuators exert forces versus the position of the tibial component on the Instron-Stanmore Knee Simulator.

symmetrical springs set with two-sided precompression drops by 50% when any gap occurs on either side (i.e., two-sided compression becomes one-sided). This situation can easily occur even with moderate motion of the TKR specimen and should preferably be prevented by avoiding precompression altogether.

The stiffness of the springs used and their detailed configuration and settings (i.e., precompression or gap and gap distance) all affect the restraint and thus affect the TKR specimen kinematics [55], which in turn influence the wear. As will be seen later, the *non*-precompressed springs (i.e., with gaps) would also more realistically simulate the physiological soft tissue situation.

In descriptions and results from the force control simulation method [49, 50], initial kinematic studies [53] and wear studies using this simulator, and even in the version of the ISO standard [38] published at the time of this edition of this book, there has not been a definite specification of the details of how the soft tissue restraint springs were set up, all of which would strongly influence the restraint. The standard specified only overall net linear and rotational stiffness to be achieved. However, the detailed spring configuration that affects the stiffness greatly (e.g., precompressed springs double the stiffness, and intermittent freeing of one compressive spring by it reaching its natural length abruptly drops the stiffness) has limited the ability for consistency within each lab and limited the opportunity for reliable data comparison between different laboratories. The way this soft tissue simulation was specified even shed doubts on the credibility of the methodology itself. For example, no differences in spring stiffness nor their configuration were specified to distinguish testing posterior cruciate ligament sparing from sacrificing TKRs. A series of later studies [54, 68] had, however, been conducted to address these issues and recommended the stiffness and gap settings for the two most common situations, a resected ACL with retained PCL and a TKR with both cruciates resected, which would approximate how the soft-tissue restraint acts *in vivo*. Those findings are summarized here because they have formed the technical basis for a revision effort to update the force-control ISO 14243-1 [38] standard (being carried out at the time of writing of this chapter).

A mathematical model was developed using the stiffness values of each of the four springs, their settings (level of precompression or gap), and the detailed geometry of the simulator to calculate AP restraint force curves versus AP displacement and how these change with any IE rotation superimposed. Various coupling effects between AP motion and IE rotation and discontinuities resulted from mathematically simulating the behavior of the configuration/assembly of Figure 26.9. These coupling effects affected the net overall AP translational and IE rotational stiffness of the whole assembly and the ranges of motion. For example, the range of restrained linear translation was found to be curtailed by a sudden effectively infinite increase in stiffness when a spring reaches its minimum length when its coils are fully compressed and touching or when a simulator actuator reaches its limit of range of motion. This range of motion is curtailed further if there is some simultaneous IE rotation that eliminates part or all of a spring gap or starts compressing one spring before any net linear translation occurs for the whole assembly (see Figure 26.9).

The same analysis was done for rotational restraint with simultaneous linear displacement superimposed. The most optimum spring values, settings, and gaps were chosen in those studies [54, 68] to approximate to published data on the contribution of ligaments to the laxity of the natural knee [69]. The results further confirmed that the behavior of the system was as sensitive to the gap or level of precompression as to the stiffness of the springs [54, 68]. The configuration with the best match between the spring characteristics and the experimentally measured laxity curves for knee joints [69] for the preserved PCL case recommended, for the Instron-Stanmore Knee Simulator, soft (7.24 N/mm) springs on the ACL side and harder (33.8 N/mm) springs on the PCL side, with a 2.5 mm gap on each to simulate the flat horizontal section of the typical anatomical S-shaped laxity curve [54]. For both resected ACL and PCL, the soft (7.24 N/mm) springs for both sides were found to be best, provided that a 2.5 mm gap was set on each side [68]. These values are dependent on the simulator design, and extreme care needs to be taken when extrapolating the preceding values to other types of simulators.

In the Instron-Stanmore Knee Simulator, the location at which these spring forces were applied to the shaft assembly carrying the tibial component was subject to leverage, as shown in Figure 26.10. This simulator-specific leverage was accounted for based on the machine's exact geometry. A similar distance magnification correction was also made to relate the AP distance moved in the plane of the spring system to the actual distance moved by the TKR specimen.

When these restraint settings are generalized (for any simulator) into one linear net restraint (and not two parallel springs), and when the leverage issue is corrected for, the desired net stiffness values recommended for soft tissue restraint are as detailed in Table 26.1. These generalized settings have been proposed as part of a major revision of the force-control ISO standard 14243-1 [38].

How these recommended settings compare to the current ISO standard settings, and how both compare to the physiological functions of the soft tissue restraint from the literature, are shown in Figures 26.11 and 26.12. Note in Figure 26.11 how the physiological soft tissue restraint [69] also shows some coupling between AP linear and IE rotational laxities. The AP laxity was curtailed if no rotation was allowed and vice versa [69].

It is important to note that the physiological knee restraint as measured from the cadaveric study [69] included the restraint from the shape of the anatomical articulating surfaces (joint conformity restraint) as well as the contribution

**TABLE 26.1 The Current and Planned Revisions of Soft Tissue Restraint Characteristics for Wear Testing under Force Control**

| | | |
|---|---|---|
| **Linear AP restraint** | | |
| Current (2005 edition) ISO setting (no gap, and same for PCL retained and resected) | 30 N/mm | |
| **Proposed revision and settings used in Nebraska since 2000** | Within ± 2.5 mm of the neutral position | Beyond |
| PCL sacrificing TKR | 0 | 9.3 N/mm |
| PCL retaining TKR | 0 | 44 N/mm |
| **Rotational IE restraint** | | |
| Current (2005 edition) ISO setting (no gap, and same for PCL retained and resected) | 0.6 Nm/degree | |
| **Proposed revision and settings used in Nebraska since 2000** | Within ± 6 degrees of the neutral position | Beyond |
| PCL sacrificing TKR | 0 | 0.13 Nm per degree |
| PCL retaining TKR | 0 | 0.36 Nm per degree |

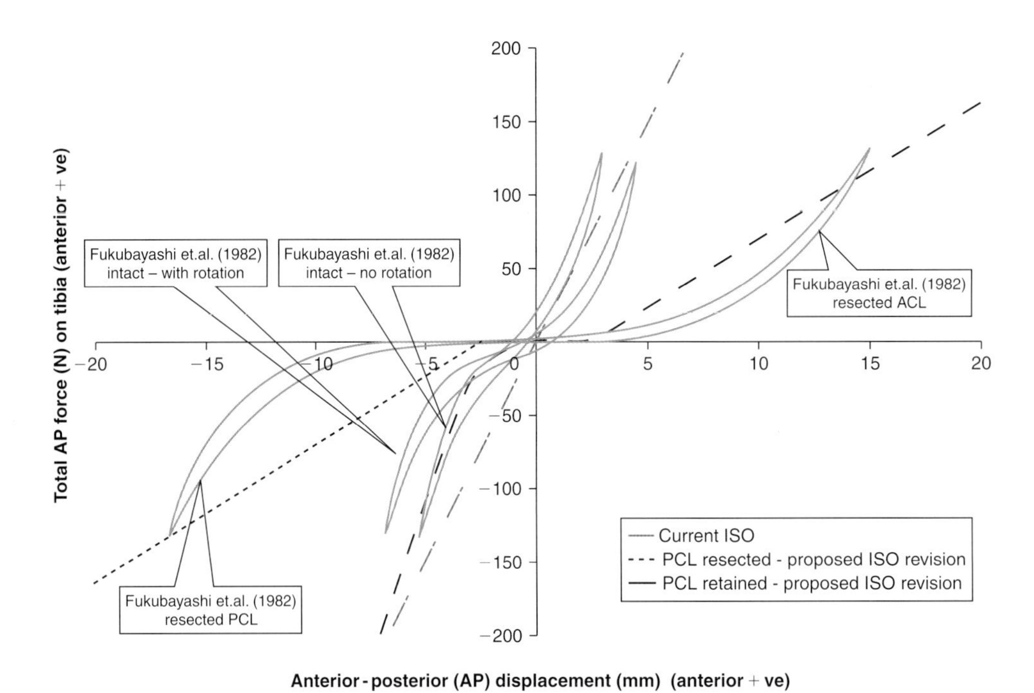


**FIGURE 26.11** Anterior-posterior (linear) soft tissue restraint stiffness settings of the current ISO 14243-1 [38] standard and the proposed stiffness revisions, all superimposed on the physiological restraint of an intact knee or with PCL or ACL resected (from the work of Fukubayashi et al., 1982 [69]).

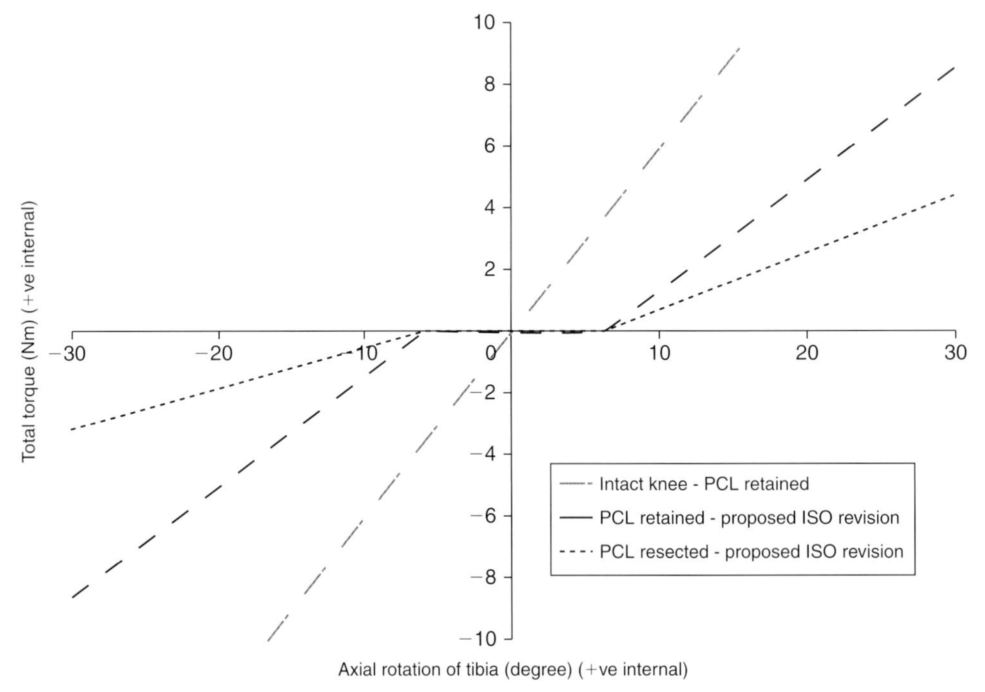


**FIGURE 26.12** Internal-external (rotational) soft tissue restraint stiffness settings of the current ISO 14243-1 [38] standard and the proposed stiffness revisions.

from the soft tissue. Therefore, the actual pure soft tissue contribution must be even less *in vivo*. Now observe in Figure 26.11 the straight line simulated restraint currently in the 14243-1 ISO standard [38]. It far exceeds any physiological value when the ACL is resected (which is virtually every time in TKR surgery), and it certainly exceeds the *in vivo* PCL retaining constraint, let alone the PCL resected case from [69]. This means that all tests done strictly according to the current ISO settings have excessively restrained TKR motion, and therefore it is not surprising that they typically produce little wear. Since 2000 in the author's laboratory in Nebraska, the soft tissue restraint springs on the knee simulator have *on all tests* been set according to the new proposed ISO revision version shown in Table 26.1 and as shown in Figures 26.11 and 26.12.

There was some initial fear that the gaps would cause impact loading and the springs would fatigue with multimillion cycle use of the machine. With specially designed and made plastic protective collars/bushes, these configurations have been tried on over 10 separate 5-million cycle tests and the spring assemblies proved durable, with the gaps remaining accurate when they were properly set and locked into those settings. It is clear from Figure 26.11 that the stiffness and settings found from the analysis of [54, 68] for the two surgical/ligament situations are obviously very different from each other and are different from the tests with this simulator, which have been published thus far by several laboratories [50–53, 70, 71]. As mentioned earlier, the use of consistent as well as physiologically realistic settings are a precursor for valid comparison of wear and kinematics between TKRs and laboratories.

The 2.5 mm gap in the springs attenuates the simulated ligamentous constraint very close to the neutral position, which has a zero value *in vivo* too near the knees' neutral position. This gap, and the stiffer ligamentous restraint with a PCL-retaining TKR, compared to the sacrificing have made TKR kinematics under force control more like what is expected from their designs [55] and has produced higher and more discriminating wear results.

Regardless of the desired spring settings, it has proven to be (doable but) difficult to accurately implement the restraint physically on a simulator, which requires intricate setting, tuning, and checks throughout a testing period. The nonlinearity and coupling effects between linear and rotational restraint add to those difficulties. A new concept of virtual soft tissue simulation was envisaged, which is simpler and more accurate in implementation. The soft tissue restraint could be modeled in software utilizing the real-time feedback signals of the kinematics and soft-tissue restraint look-up tables, incorporating their effects within the AP force and IE torque actuators. This allows any non-linear and more accurate representation.

This concept takes advantage of measured real-time feedback of AP and IE rotation from sensors on board the knee simulator. Based on those instantaneous AP and IE implant positions/motions, a real-time (or strictly, semi-real-time) computation is made of the influence/contribution of the desired soft tissue springs or any virtual mathematical restraint models. The computed AP force and IE torque contributions are added (vectorially, mathematically) to the electronic command signals for the externally actuated AP and IE torques of the simulator.

The bandwidth of TKR dynamics is well contained in less than 50 Hz (i.e., no physiologically meaningful force spikes or drastic kinematic position changes can occur in less than 20 ms in time, or strictly 40 ms, when considering aliasing sampling errors). Even modest modern electronic data acquisition and control systems can reach at least 1 kHz data sampling, processing, and control loop closure frequency. Therefore, the process of incorporating a soft tissue control correction to the external simulator actuators can be made to occur at least 10–20 times before a TKR changes its kinematic position, therefore, the correction can be considered virtually instantaneous.

This concept is now termed "virtual soft tissue" simulation and has been elegantly implemented in the latest models of the AMTI knee simulator (Figure 26.18). This implementation was combined with adaptive control in this simulator, allowing the electronics to gradually exaggerate or attenuate the cyclic actuation waveforms until the resulting measured forces and torques are closely, cycle by cycle, made to match the desired input forces and torques with the desired soft tissue model. Comprehensive quantitative studies and formal presentations of this novel system are yet to be published.

## 26.3.6 TKR Kinematics Measured from the Force-Control Wear Testing Method

With the same standardized repeatable external forces, and approximated soft tissue restraint physically simulated as described in the previous section, different TKR designs under the force-control test method would experience different kinematics of motion as governed by their individual design details.

Depending on the simulator, with suitable calibrated sensors (Figure 26.13), this force-control concept could therefore facilitate assessment of the dynamic performance of a TKR implant by measuring its kinematics *in vitro* during a simulated walking activity.

Various studies comparing a wide range of measured TKR kinematics under identical standardized conditions had been carried out in this way. An early one of these studies [53] tested a set of generically different TKR designs of highly different constraint, including fixed, mobile-bearing, and hinged designs. Those TKRs indeed produced expectedly different kinematics; some moved less than half of what others did in the stance phase of walking.

Another study compared four fixed and three mobile bearing TKRs of different designs, all PCL cruciate-retaining (CR), but those produced different kinematics, too, and again reflected the expected variation from their difference in constraint [72]. For example, rotating platform bearings rotated more than fixed bearings, and highly dished fixed bearings translated less than shallower fixed bearings.

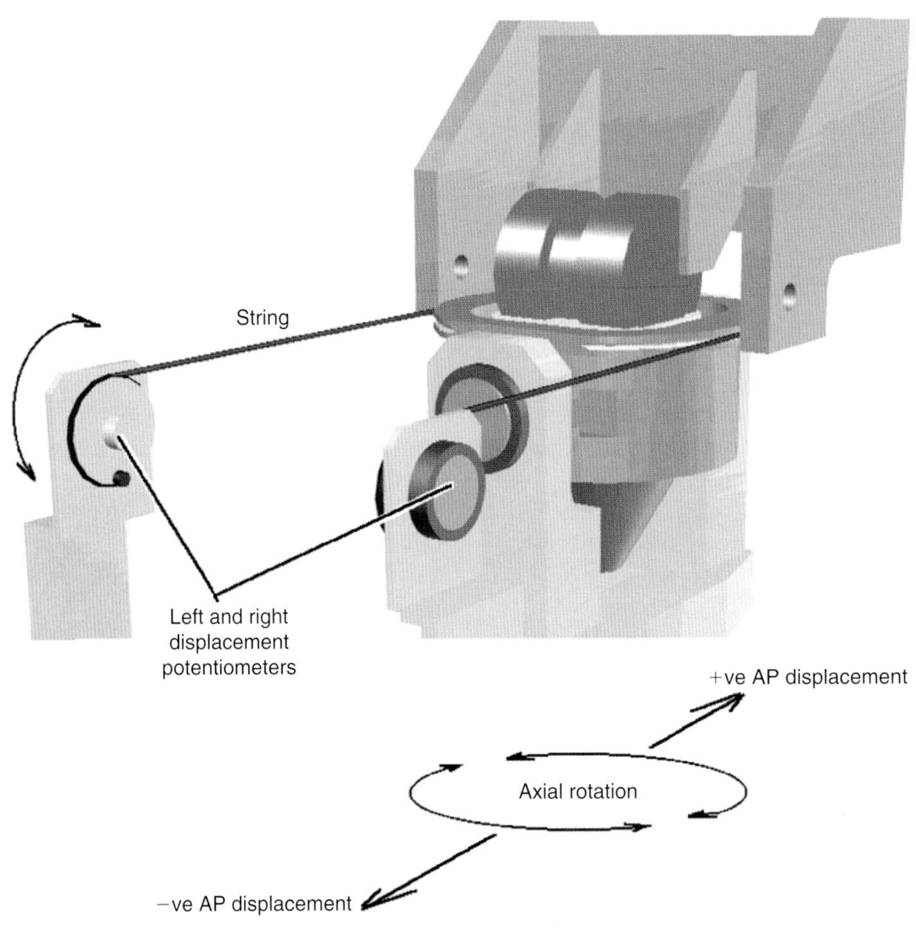


**FIGURE 26.13** The configuration of two potentiometer sensors with strings that allow measurement of two parallel motions from which a net AP displacement and a net IE rotation angle were calculated. This configuration allows very accurate measurement of the TKR kinematics, yet it minimizes influence of the sensors or their strings on the freedom of the tibial component to translate proximally-distally and in AP, ML, or rotate in IE and VV. The TKR flexion occurs through rotation of the femoral component holder, so it does not affect and is unaffected by these AP and IE motion sensor measurements.

If the small frictional and inertial forces of the simulator implant holders were neglected, then the external actuated (and measured) simulator input forces, then the soft tissue (spring) restraint forces (also measured on the simulator) and the TKR articulating surface (joint reaction) forces should all balance in a quasi-steady-state equilibrium at each phase of the walking gait cycle. Therefore the latter (TKR joint reaction forces) were computed from the force measurements at each phase of the simulator cycles in both of the previously mentioned studies [53, 72]. As expected, some (the highly constrained) TKRs relied mainly on their inherent conforming surfaces or constraint in limiting their motions, yet the motion of other TKRs were limited mostly by the simulated soft tissue (springs).

It is interesting to note that the TKRs of the [72] study were supposedly indicated for the same type of patient conditions (all CR), yet their function through their measured kinematics varied markedly between the TKR designs, even between fixed-bearing types. The fact that the TKRs' kinematic behavior varied was not surprising considering their slightly different sagittal profile/radii and design philosophies. The wise conclusion from that study [72] was not to expect different TKR clinical products to behave the same way even if they were (through device manufacturer marketing) indicated for the same patient conditions. This conclusion should further extend to the use and durability of UHMWPE in TKR. The real wear and durability of UHMWPE will strongly depend on the detailed design of the TKR system, as well as the generic type of TKR and to which patient indications it is intended.

The same different fixed and mobile cruciate retaining TKR designs in [72] were also tested under the same force-control conditions but with various patient and surgical variability imposed as could occur clinically to establish how sensitive the test method was and how forgiving each TKR design was for such misalignments [55]. An enhanced duty cycle was also applied to simulate a heavier patient or one with more vigorous activities. It was found that the differences in kinematics between TKR became higher, and in some cases, the relative values between designs changed.

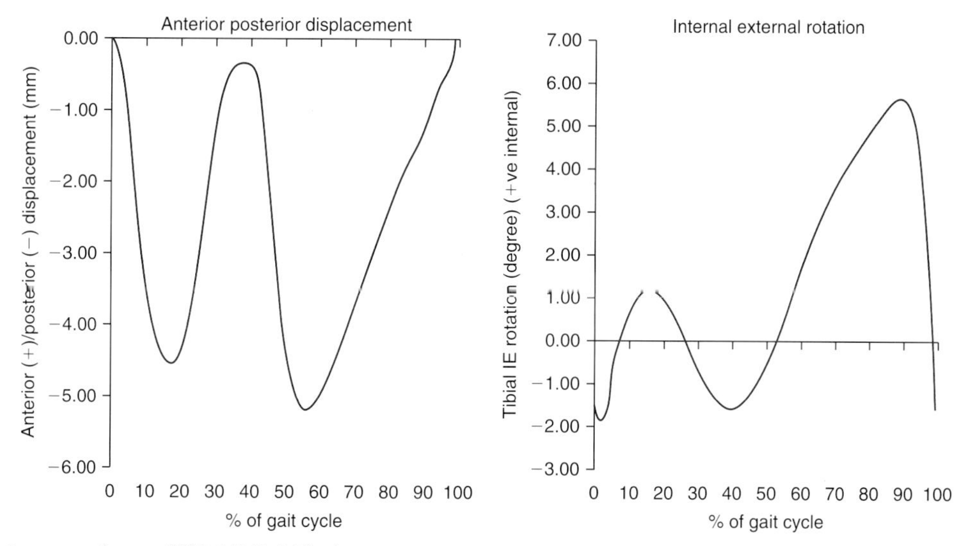


**FIGURE 26.14** Input waveforms of ISO 14243-3 (displacement control) [56]. The flexion and axial force (not shown here) are the same as for force control (ISO 14243-1 [38], shown in Figure 26.7).

Different setups were tested to simulate variations in surgical placement equally imposed on all the TKR designs.

Changing the line of action of the compressive force in the frontal plane, equivalent to a change in the frontal varus-valgus angle, did not change the kinematics. Rotating the tibial component about a vertical axis in either direction shifted the internal and external rotational curves (offset them) in that same direction of misalignment. The largest affect on kinematics was that of changing the tightness of the PCL. For a tight PCL, the displacements and rotations were reduced greatly; a loose PCL had the opposite affect.

Interesting to note was that in all of the previously mentioned tests [55], the resulting displacements and rotations of the mobile-bearing knee replacements were in the same general range as for the fixed-bearing designs, or even less in some cases. Good insight was gained from the preceding study about the relative kinematic behavior of different TKR designs and their sensitivity to patient factors and the alignment of the components at surgery.

## 26.3.7 Standardization of the Displacement-Control TKR Simulation Method for Wear Testing

The kinematics used as inputs in the ISO 14243-3 displacement-control standard [56] are shown in Figure 26.14. The origin of these AP and IE input motion curves in this later standard has been frequently enquired about. Whilst the current author had not contributed to the emergence of this (the displacement-control version) standard, various verbal communications with the originators [73] of this ISO standard and the proposers of its input curves [74] clarified the history of these kinematic waveforms. They originated from a gait laboratory PhD study [75] of *in vivo* measured kinematics extracted using the Point Cluster Technique [76]. Those very curves in ISO 14243-3 [56] were the results from one male patient with a cruciate-retaining Zimmer NexGen CR total knee replacement system.

One related publication that referred to the input curves was also from a Point Cluster Technique study on 10 patients with TKRs of similar design but these were posterior stabilized (PS), where the posterior–cruciate ligament had been substituted [74]. As mentioned earlier, the ISO 14243-3 kinematic curves were from the single patient CR data. It is interesting to note from the study of [74] that the data processing to calculate the surface contact slip velocity profiles had shown marked differences, especially on the knee's medial side between CR and PS knees. Indeed, the conclusion of that study wisely alluded to the need to vary the slip velocities and kinematics when testing different TKR designs. Regardless of how representative these input curves were, they have proven to be useful as part of an ISO standard taking tests with displacement control some way toward conversion of methodology, especially on many simulators still operating today that are not capable of the force-control scheme.

The preceding detailed discussion should render obvious some of the pros and cons of both the force- and displacement-control methods. As mentioned earlier, the displacement control is simpler in concept and more robust in implementation on simulators. However, what displacement inputs should be used for each TKR? One encouraging study demonstrated that if the standard forces in a force-control test were used and suitably chosen displacements in a displacement-control test were also employed, similar wear rates would result for the same implant design

[77]. In that particular study [77], the knee implant tested in both methods was the very cruciate-retaining TKR design for which the displacement-control inputs (kinematics) were originally devised [74, 75].

How wise is it to use the same AP and IE rotations for different TKR designs in a standardized test? Even in the same TKR design, how would UHMWPE deformations relieve or accentuate stresses under identical displacements and rotations in later cycles of the same test? In the previously mentioned study [77], not only the *in vivo* kinematics of that very TKR design were used, but also the TKR was a relatively lax cruciate-retaining design, so similar wear resulted. It is for the preceding reasons that an important limitation was included, perhaps understated, in the opening "Scope" section of the ISO 14243-3 [56] (the displacement-control standard). It brought attention to its questionable applicability to knee designs with a high degree of constraint, which could result in damage to the articulating components in the early stages of the test that would not be representative of clinical service. The author questions further here: What is a high degree of constraint, and how much damage would be considered unrealistic? Constraint is a function of the detailed design of a TKR and thus varies within a continuum ranging from zero laxity of fully constrained hinged knee systems to the highly lax mobile bearing knee systems [10]. TKR kinematics were found to vary even with slight differences among generically similar TKR designs that had only subtle changes in the sagittal and frontal profiles/radii reflecting various design philosophies [55, 72].

In turn, there have been a variety of criticisms of the force-control method in both concept and in implementation. Among the most profound relate to the complexity and difficulty of implementation in a simulator. Force control is naturally a challenge with fast, dynamically changing forces and torques (lower plots of Figure 26.6). It becomes especially more difficult to physically implement when the target load (usually the tibial component of a TKR in this case) is itself moving fast in complex ways against fast-changing and wide-ranging resistance to motion, coupled in a way such that IE rotation somewhat affects the resistance to AP motion and vice versa. Added to this are the intricacies required in the soft-tissue simulation system. None of the simulators to date have managed to implement force control accurately enough and independently across multiple stations with sufficient ease to make this technique and its technology conveniently transferable across TKR test laboratories. It has been noticed, however, that due to the highly cyclic and repetitive nature of simulating a walking gait up to millions of cycles, a thorough (sometimes seemingly exaggerated) investment of tuning and retuning at the start of a test does bring adequate rewards in sufficient accuracy.

Another fundamental criticism goes to the heart of the concept of force-control and soft-tissue simulation. "In order to specify the correct spring stiffness" (for soft tissue simulation), early critics argued that "some estimate of sensible anterior-posterior translation would be required. Hence, since an estimate of deflection is needed to implement load control, why not control displacement in the first place?" [48]. The answer to this should be clear from the earlier description of the force-control concept (Section 26.3.4) and the soft-tissue simulation (Section 26.3.5). In summary, in force control the simulated soft tissue will not be chosen to suit individual TKR designs but would be the same and is fixed for all designs, except for whether to simulate the PCL action or not for CR and PS TKR designs, respectively. Unless the soft-tissue simulation is widely out of the physiological range, the effects of a wrongly standardized simulated restraint model would be of a secondary level in its influence, especially in comparing the wear of varying TKR designs. If the displacement and rotation are made constant (displacement control) in testing all designs, those kinematics are of a primary order in affecting wear, and they do need to be customized for each TKR design. Further, in Section 26.3.3, it was argued that even with the appropriate kimematics for a given TKR design, the impacts/impingements on UHMWPE posts and viscoelastic stresses will be self-tuned/adjusted in force control but might not be in rigidly implemented displacement control.

## 26.4 CONSIDERATIONS AND PITFALLS IN KNEE WEAR TESTING OF UHMWPE

### 26.4.1 Variable Control and the Number of TKR/UHMWPE Samples to Test

#### *26.4.1.1 Variable Control*

Consider the requirement for testing a TKR system with a different (say, improved) version of UHMWPE material. Typically, TKR samples with the new material are compared with others of the same TKR design, size, and everything else but with a standard or previous version of UHMWPE as a control. It is obvious that the best control would be one with good clinical history and success. Testing on a knee simulator takes time (e.g., typically a 5 million cycle test takes 6–9 months, including setup and downtime) and is highly costly. In this situation, it is frequently asked how many TKR samples of each material should be tested. The general answer, of course, is that if a significant difference (or absence of it) is to be proven statistically, then the level of that difference should be chosen sensibly such that it would be of clinical significance too. Also, the study/test should secure sufficient "statistical power" [78]. Both of these will depend on the variability (sometimes expressed as variance or standard deviation in the variable of interest) inherent in the results of any testing process, so it is worth examining the potential sources of variability in some detail.

*In vitro* TKR wear tests can, in principle at least, be much more variable controlled than *in vivo* studies and therefore require less sample numbers. Clinical TKR wear or retrieval studies are subject to variability from:

1. Patient anatomy
2. Surgical installation, implant alignment, and soft tissue/ligament tuning
3. Capsule and lubricant conditions
4. Patient daily activities and type of long-term use
5. Duration of duty *in vivo*
6. Limitations in the method of measurements of *in vivo* wear
7. Other factors, including the ones listed in the following discussion for *in vitro* wear tests

Some of the preceding sources of variability are serious in their affects on UHMWPE wear, which is influenced by the TKR kinematics and stresses. An interesting study [61] developed a probabilistic finite element (FE) analysis TKR model to evaluate the effects of two levels of parameter variability with standard deviations of TKR component alignment. The study predicted modest variability of less than 10% (1.66 MPa) in peak contact pressure but up to 226% (3.44 mm) in resulting anterior-posterior (AP) translation and up to 169% (4.30 degrees) in internal-external (IE) rotation. The critical alignment parameters were the tilt of the tibial insert and the IE rotational alignment of the femoral component. The authors commented that the observed variability in kinematics and, to a lesser extent, contact pressure, obviously had the potential to impact UHMWPE wear.

All of the surgical and alignment variables previously listed can be more precisely controlled or eliminated with hardware and detailed experimental procedure in knee simulator testing of UHMWPE wear. Many factors can be more precisely set during installation, alignment, and tuning in a simulator, especially with the relative ease of monitoring and retuning of simulator conditions (for example, see Section 26.4.3). Having eliminated *in vitro* the vast majority of the typically uncontrollable variables *in vivo*, a typical knee simulator test would typically leave the following three remaining or new sources of error and variability:

- Manufacturing tolerances between specimens: These are typically minute because testing standards generally prescribe that TKR test samples should all have passed quality control checks by the manufacturer. Manufacturing tolerances are of the order of microns in dimensions. These are orders of magnitude less in influence than any of the *in vivo* variables previously listed.
- Errors in gravimetric measurements of UHMWPE wear: At each stage/interval of the test, gravimetric weight measurements of UHMWPE should be carried out with advanced instruments (scales) with preferably 10 millionths of a gram resolution or better. The gravitational approach to measure wear of UHMWPE is fundamentally weak, if not flawed. In experimentation and instrumentation theory, subtracting two similar large quantities to determine the required variable being the small difference between them is riddled with errors, especially when the two larger quantities are analog and not digital in their nature and measurement. UHMWPE wear at each testing interval may be as low as a fraction of a milligram or may reach 50 mg at most, yet the UHMWPE component weight is hundreds of times heavier.
An example is given here to illustrate this. A typical UHMWPE TKR bearing insert component may weigh 22 g with some small *absolute* experimental uncertainty in each weighing measurement (of, say, ±0.1 mg, thus <0.0005% relative error). The weight measurements are subtracted to determine a net interim weight loss and thus wear of typically a few milligrams (say, weight loss due to an interim wear interval = 1.5 mg). This result is naturally subject to the sum of the absolute errors of the two measurement constituents, therefore the net interval wear result is (1.5 ± 0.2 mg), hence the uncertainty progresses up to a ±13% *relative* error in the final result [79]. Until methods evolve to track and quantitatively measure the amount (weight or volume) of the actual wear particles, the gravimetric technique remains the most accurate and offers the best compromise.
In the author's own laboratory, the wear of each TKR sample, at each interval of the test, is measured hundreds of times (quickly and automatically by computer) and averaged until statistical stability is reached each time. With a running average continuously updated, more data continue to be taken until the running average ceases to change beyond the 10 μg resolution of the instrument. Such techniques added to the stringent protocol requirements of testing standards (ISO 14243-2) [80] can minimize those errors. The gravitational weighing errors naturally become more significant in screening implant wear as the wear becomes smaller with improved materials. It is therefore fortunate that at some future stage wear may become so small that it loses clinical relevance, and then those tiny errors in wear measurements *in vitro* become unimportant. Some argue that this is the situation now with highly crosslinked UHMWPE whose wear may not exceed the uncertainty level in gravimetric wear measurements and liquid absorption levels.
- Temporal variability in UHMWPE wear during a test: A change in the wear rate may result at different stages/intervals of a prolonged multimillion cycle TKR wear test: In a typical study, measurements are taken at 11–12 stages of (say) a 5 million cycle test, representing 10–11 different intervals of wear for each specimen. A linear regression analysis (least square error method) is typically used to average the results from all those intervals resulting with the (mathematically) best fitting linear wear rate for each test sample. Therefore, the 11

interval/stage measurements of wear average out this third source of variability/error, which happens to be the most serious, in the author's view. Sometimes the temporal variability may stem from liquid absorption rate changes, and, of course, they have to be corrected. Some temporal variability relates to the environment (temperature and humidity) or a change/drift in testing conditions. Ambiguity may arise in recognizing such phenomena because some temporal variability may actually represent a true physical result. For example, "bedding in," "wearing in," "run in," and "run-away" wear are terms frequently used in total hip replacements (see Chapter 5), but similar effects can actually occur in TKR tests, too. For example, when a lax TKR wears, the conformity of the articulating surfaces may gradually increase, and this may slightly diminish the kinematics range and also the stress, and thus could alter the wear rates between the start and end of a long test. This phenomenon was reported in [50] where AP and especially the rotational laxities were observed to reduce with time over a long wear test. This was attributed to the formation of a more congruent wear track because of wear and deformation. Another possible explanation was that the coefficient of friction increased with time because of embedded particles or slight roughening of the articulating metallic surfaces [50].

### 26.4.1.2 Number of TKR/UHMWPE Samples to Test

Because the variables *in vitro* can be more controlled as previously outlined, sometimes as few as two TKR samples versus another two would suffice in comparing UHMWPE wear in TKR systems. In the author's laboratory, two TKR samples were compared to another two, with a slightly different (incrementally changed) design to improve the geometry to reduce wear, where retrievals showed it had locally occurred. The results showed a threefold improvement (reduction) in wear, which was statistically significant ($p < 0.05$). In another example from the same lab, the wear was tested of two different TKR mobile bearing designs. Four samples of each TKR were tested, one design type tested to 5 million cycles, the other to 7 million cycles, but both had identical test methods. Although the wear rates of each showed the usual scatter, a statistically significant difference in wear rates resulted ($p < 0.05$) between these two mobile bearing TKR designs.

The same has been echoed in TKR wear studies from other labs. In a recently published study [81], highly crosslinked UHMWPE for TKR was proven in a displacement-controlled test of three TKR samples of each kind to wear statistically significantly less ($P = 0.034$) than conventional UHMWPE. In a force-controlled TKR wear study [70], a fivefold increase in wear rate ($p = 0.002$) was found under different simulator input conditions comparing pure walking to a 70:1 combination of walking with stair climbing. This test compared four samples of a clinically used TKR in each loading condition, with the TKR designs being slightly different. Finally, in an another force-controlled knee simulator test, two different fixed bearing TKR designs were compared, where four samples of each were tested at a time, and the wear rates for one of them was shown to be 10 times less than the other [62]. In that study, the two implant designs were tested in separate setups at different times, at different frequencies of walking cycle (1 Hz versus 1.4 Hz), and the results of wear were retrospectively compared.

The preceding examples showed statistically significant differences in wear rates between designs when such differences existed. Sometimes such differences are smaller or do not exist, and it becomes more difficult to choose how many samples to test and whether to demonstrate the differences or to verify their absence. A good *in vitro* TKR wear study, where possible, should be well designed in terms of sample numbers and criteria for assessing the results prior to the actual simulator testing. The choice of sample numbers to test should ensure sufficient statistical power [78] for the study. The following is an example as a recently published case study [63].

To compare the UHMWPE wear of rotating platform (RP) TKR to a fixed bearing (FB) one of the same design, a recently published study compared the wear of mobile to fixed bearing total knees of identical design in other aspects [63]. The knees had the same femoral components and were tested with identical force-control inputs. Valid justification was needed for choosing a sample number of TKRs to test from each group. The sample numbers chosen were justified based on a previously published study [82] (by a different lab), which compared the wear rates of the same two implant designs with the displacement-control method. That study [82] reported a mean wear rate of $4.86 \pm 3.55$ mg/million cycles (Mc) for the RP versus a mean of $21.3 \pm 5.52$ mg/Mc for the FB. These results showed over 75% reduction in wear with the RP, and therefore such reduction was of high clinical interest. Both wear values were in the range of what TKRs produce with conventional UHMWPE, and the difference (75% reduction in wear) was highly worthy for an alternative design to reduce the risk of osteolysis and increase longevity *in vivo*. Those results from the cited study [82] were used as the effect size in a statistical power analysis [78] for the later study [63]. A two-sided t-test was performed with a two-sided alpha level of $p = 0.05$. Those parameters yielded 40% power (probability of correctly rejecting the null hypothesis of no difference between the two groups [78]) by testing two implants in each group, 85% with three implants in each group, and 97% power with four implants in each group. In the case study [63], four samples of each group were chosen and thus approximated to 97% statistical power based on the aforementioned assumptions and the effect size published in the study that was cited [82]. Typically, 80% or higher statistical power is deemed acceptable in this field, and so three samples could have also sufficed. Please see any good statistical text [78] for a comprehensive explanation of how a type II (β) error (statistical power = 1−β) interacts with a type I (α) error (level of statistical significance, or p value), and their implications on the required sample numbers (n) in trying to detect a meaningful difference between two groups. Remember, a meaningful difference in the UHMWPE wear in the context of TKR testing is that of clinical significance, and this invariably relates to the potential onset of osteolysis and risk of TKR failure by loosening (see Chapter 8).

## 26.4.2 Soaking UHMWPE Prior to Testing and Soak Controls during Testing

In the previous section, the principal weakness of estimating UHMWPE gravimetrically was highlighted. Relatively large errors are unfortunately inherent in having to subtract an UHMWPE bearing component weight before and after wear. The example given showed how an interim/interval weight loss in UHMWPE of, say, 1.5 mg may be subject to an overall uncertainty of over 10% due to error progression alone in simple weight measurements, each made to better than 0.001% resolution. Therefore, any other seemingly negligible sources of uncertainty in gravimetric UHMWPE wear measurements can have disproportionate effects on the final wear results and need to be avoided or corrected for.

Soaking of liquid into the UHMWPE material is the most important source of such uncertainty. UHMWPE is essentially porous, and both *in vivo* and during testing *in vitro*, it is submerged and saturated with liquid lubricant. Wear testing measurements are done when the UHMWPE is supposedly, at least nominally, dry, to reduce as far as possible the weight of residual liquid as a source of uncertainty. The first measure to reduce the soaking effect is to presoak the UHMWPE specimens prior to the test for 2–3 weeks in deionized water or in the same diluted serum lubricant used for later testing. This would ensure that the UHMWPE has stabilized and saturated with liquid as it would later be during testing. In the author's own laboratory, a minimum of 2 weeks pretest soaking is continued until the weight rise on any specimen in any extra 24 hour period becomes less than 10% of the overall amount soaked up to that point.

Various cleaning protocols have been used prior to wear measurements, usually ending in some rigorous drying process. Some involve drying with modest heating for long hours and some under vacuum (desiccation) or a combination of both. In the author's lab, the rigorous cleaning and drying protocol described in ISO 14243-2 [80] is usually followed. This involves a well-defined multi-time-step sequence of cleaning and rinsing with detergent and deionized water in an ultrasonic bath, then drying for at least 30 min in a vacuum chamber to a pressure of below 13.3 Pa (approximately 100 mTorr) under ambient temperature.

To correct for any remaining liquid absorption, one or more control UHMWPE specimens are included in the test as prescribed in ISO 14243-1 [38]. These are placed in an identical diluted serum lubricant medium as the actual test specimens so they soak similarly and are cleaned, dried, and weighed similarly. Any changes in the control specimens weight are considered to have occurred only due to the soaking process under the same conditions in that test interval. Those weight changes are used as corrections to the actual weight losses due to wear. Such control specimens are commonly referred to as passive soak controls.

Alternatively, loaded or active soak control specimens can be used. These are not only placed in the same lubricant medium but are also subjected to the same dynamic compressive loads (axial forces of a TKR) during testing without any other motions. The idea here is that the compression/decompression may accelerate or in some way alter the soaking process, and this is suspected to occur more with the highly crosslinked UHMWPE materials. In a prospective revised version of the ISO standards, 14243-1 and 2 [38, 80], whether to use passive or loaded soak controls is expected to be made optional.

How much soaking is expected and typically corrected during a typical TKR wear test? In the author's own laboratory, passive soak controls are more frequently used for conventional UHMWPE, and active controls are used for highly-crosslinked UHMWPE unless there are special requirements otherwise. Each test has its own controls, but if the results from over 20 tests with different UNMWPE lots, different TKR designs, and different levels of crosslinking are averaged, the overall correction from passive soak controls has averaged 0.89 ± 0.85 mg/Mc. From active soak controls, the overall average was higher, 1.65 ± 0.88 mg/Mc. In both cases, the soaking correction increased the net wear rate from the apparent gravimetric measurements. Obviously, the significance of these amounts depends on the wear rate of the actual tested TKR specimens and also depends on the initial presoaking process. The intricacy of those corrections assume a larger importance with the very modern very low wearing highly-crosslinked UHMWPE in TKRs, which explains the early reports of negative wear rates. In screening tests of large contact area TKR implants with conventional UHMWPE with high wear rates reaching tens of mg/Mc, whether a passive or active soak control is used does not really affect the relative ranking between different TKR designs or materials under test.

In the author's opinion, this whole liquid soaking and correction process is riddled with mystery and has evolved in a pseudo-scientific fashion. The reader is reminded that the drying and desiccation under vacuum process is supposedly leaving the specimen consistently dry before each weight measurement. Therefore, it is rather unknown how dry should be dry, and therefore liquid soaking data results are inherently noisy to say the least, but as usual, these methods are used until better methods are developed and are demonstrated to improve reliability.

FIGURE 26.15 Hydraulic six-station displacement-control Knee Wear Simulator by Shore Western Manufacturing Inc., Monrovia, California, USA.

## 26.4.3 Symmetry of Applied Inputs and Simulator Tuning

Useful results have begun to flow from different research laboratories on the kinematics and wear of TKR implants using the force-control method. A crucial ingredient in the reliability of results is the consistency of the input force waveforms. The standard ISO TKR simulation test of UHMWPE wear (of, say, 14243-1 [38]) is based on defined inputs for simulating a walking cycle originating as mentioned (Section 26.3.4) from theoretical, empirical, and *in vivo* studies. However good or inexact, the waveform inputs are largely assumed to be (actually) input in the same way across simulator stations, throughout long tests, across simulators, and across laboratories. How closely these standard control inputs are physically actuated on the simulator is very important. Comparison of durability can only be valid between implants and studies of consistent inputs. In this section, natural discrepancies of this kind are shown in a case study and shows how they could be reduced from the unacceptable to the negligible.

In this case study, the TKRs were two rotating and translating mobile bearing knees and two fixed-condylar types by the same manufacturer, both with conventional UHMWPE. After appropriate installation and initial mechanical alignment and intricate setting of the soft-tissue restraint springs on a force-controlled Instron-Stanmore Knee Simulator, the measured input forces were symmetrical between stations of identical TKR designs. The solid curves of Figure 26.17 show the intended (at the time, an early 1999 draft) ISO force input waveforms in synchrony with flexion angle for level walking. These inputs produced the (symbol) traces on the same graphs (Figure 26.17) as actuated by the simulator and measured by its calibrated and well-zeroed sensors. The output kinematics were also found to be consistent between each pair of identical specimens. However, for wear testing, the load-bearing stance phase is very important, and the measured curves did not follow the desired command waveforms. For example, during the stance phase (i.e., less than 60% of the cycle) there should have been two distinct peaks of axial force, a distinct peak of positive AP force, and virtually no rotational torque.

Even at 1 Hz, the most popular testing frequency, the results in Figure 26.17 showed a mismatch in magnitude and a more serious one in phase lag (approximately 18 degrees, equivalent to 50 ms) due to the inertia and slower motor-driven flexion actuator compared to the lighter and faster pneumatic force actuators. At the load bearing stance, where any wear mechanism would be strongest, the forces were sometimes 50% less due to the phase mismatch. Note also from Figure 26.17 the flexion did not reach below 7–8

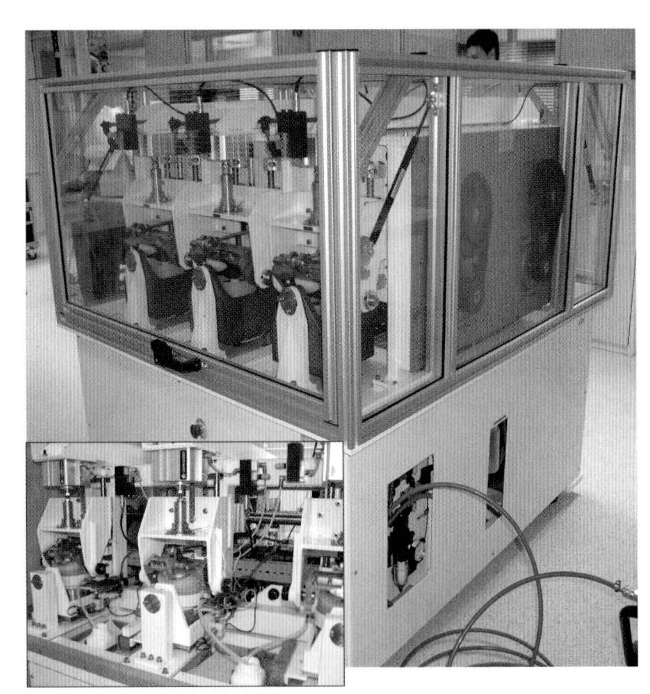

FIGURE 26.16 Six-station displacement-controlled Knee Wear Simulator by ProSim, United Kingdom. This machine has force-control capability for AP and IE torque but no soft tissue simulation.

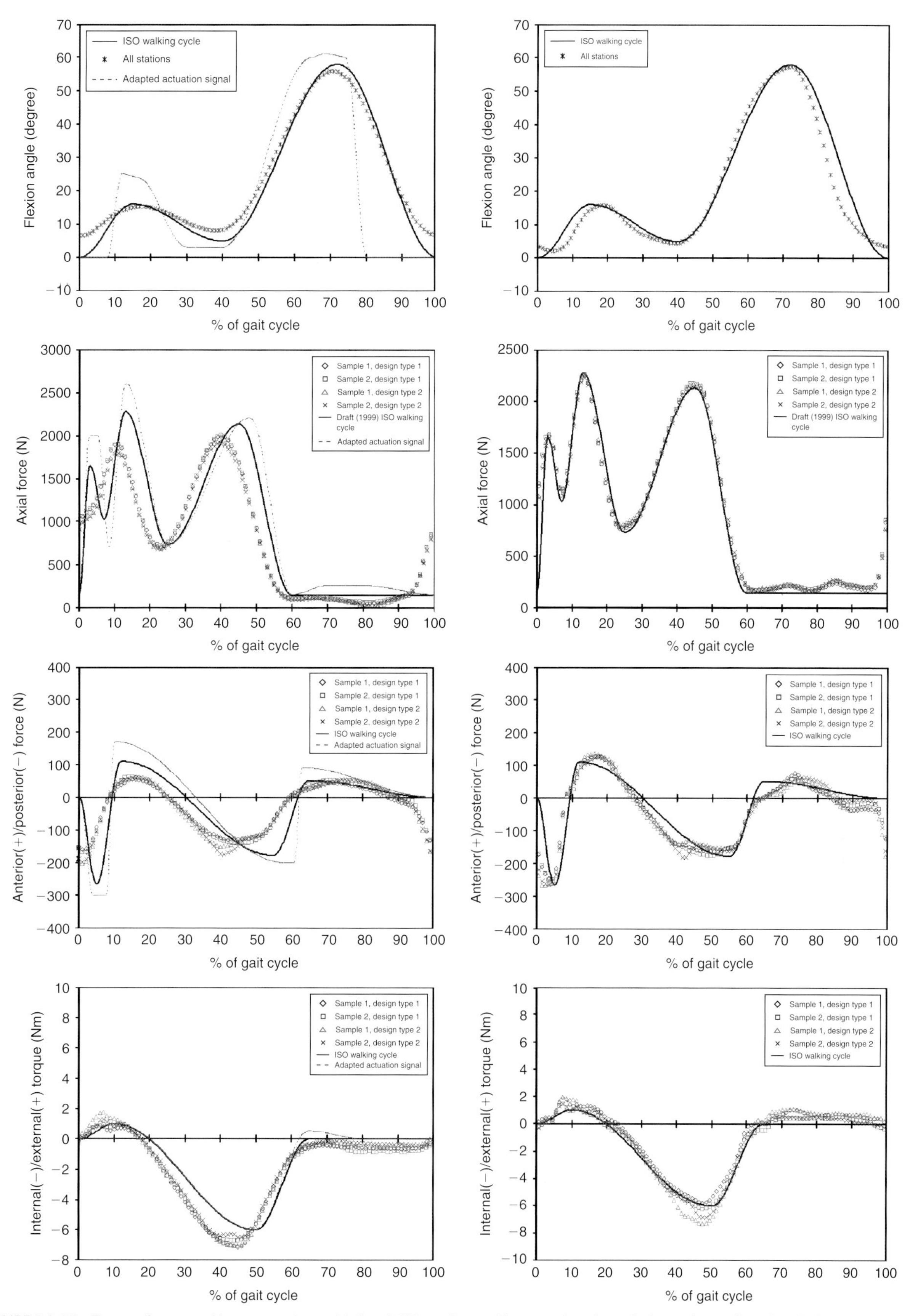


**FIGURE 26.17** Traces of measured input actuations with four TKR specimens. Two samples of one design and two of another design were tested on a four-station Instron-Stanmore simulator. The left column shows the input waveforms as measured without adaptation of the command waveforms. The right column shows the measured signals after the command waveforms have been adapted to approximate to the desired waveforms. The right column plots show an example after a set of iterative tuning steps (adaptations of the command signals), which could be improved upon further, such as further tuning of the flexion angle. (Note that the desired waveform of axial force shown here was an early 1999 version of the then-draft ISO standard, with axial forces peaking at about 2300 N. The forces were elevated slightly in the actual published standard to peak at about 2600 N.)

FIGURE 26.18 AMTI Six-Station knee simulator capable of displacement control and force control with two sets of three identical TKR specimens in each. Each bank of three runs independently with its own programmable virtual soft tissue restraint system.

degrees when it should have been zero. Note lastly the absence of the first of the three axial load peaks at heel strike. All this pointed to the need of applying a correction to each input waveform.

Eleven iterative tuning processes were carried out until the inputs shown in the dotted lines of the left column of Figure 26.17 produced the response of the right column plots.

Measurement with sensors of the (actually) actuated input waveforms, thorough data processing, critical observation of the results, and repeated adaptation of those waveforms until they are at least symmetrical on all stations and follow the desired waveforms in phase and magnitude are therefore recommended as crucial but admittedly arduous steps for accurate wear simulation and testing of UHMWPE wear in TKR. It is the force-motion combinations at the load bearing stance that mostly govern wear, and so any substantial mismatches in that phase must be minimized. The corrections presented here are naturally frequency dependent and to some extent machine and specimen dependent. The verification and correction procedure to the inputs in wear testing of TKR implants with UHMWPE should be implemented to make results comparison within and across different laboratories more reliable. Indeed, the process of verification of the actuated input signals by measurement should be formally reported in the test results.

Only a machine with independent closed-loop control on each station would (attempt to) prevent this effect, resulting in double the complexity and cost of the simulator machine itself. Some modern and advanced simulators feature the so-called "adaptive control" in their driver software. In those simulators, adaptation is made automatically during running tests.

## 26.4.4 Choice of Lubricant in Testing UHMWPE Wear in Knee Replacements

In TKR simulation and wear testing, a substitute for the synovial fluid is needed as a lubricant. The primary functions of the knee's synovial fluid are lubrication and a medium to carry nutrition to the articular cartilage and waste products back to the blood vessels. Obviously, the lubrication and cooling functions and the stability against deterioration in the lubricant with time are needed in the *in vitro* testing situation. Therefore, however complex synovial fluid may be biologically, its simulant *in vitro* need not be that complex if it provides cooling and lubrication steady enough for test intervals long enough between convenient and economical lubricant changes. The purpose of this section is to attempt to draw from the literature and wealth of experience of others what is thought to be important and how to invite convergence in the types and dilution levels of lubricants used in wear testing.

As with most body fluids, the synovial fluid consists primarily of a high percentage of water. It is, however, a complex lubricant with protein content and hyaluronic acid, among other constituents. After knee joint replacement, the synovial fluid is expected to continue to be secreted into the joint space to lubricate the articulation of a typical CoCr metallic alloy femoral component with a UHMWPE tibial bearing and patellar component. However, its nature and content would vary between patients, their ages, activities, and the state of their joints. In wear testing, the lubricant is treated as one of the (supposedly!) controlled variables, assuming a simulated lubricant of an average TKR patient. Historically and for very good economical and practical reasons, there appears to be agreement in most of the published literature that bovine blood serum diluted with deionized water has more or less similar biological content, especially in terms of proteins, and is thus a good stimulant in TKR wear testing.

Nevertheless, the source, detailed composition and dilution level of the bovine serum has been considered an important factor influencing *in vitro* tested wear results [83–86]. The most consistently discussed, important, and mostly agreed upon of those factors is the concentration of proteins, which act as boundary lubricants on the UHMWPE. The interplay between the source/initial composition of the serum, its dilution level with deionized water, and the heat generation/removal from the articulated surfaces affects protein precipitation onto the articulating surfaces, degradation of the proteins, and even calcification, which all affect the resulting wear processes.

One common pitfall in writing and interpreting published studies on this topic is to quote a percentage dilution of the protein in bovine serum. The percentages of dilution in published studies have varied widely. Among the reasons is the initial concentration of proteins in the serums used themselves, which depends on the source, processing, supplier, and even individual lot/batch used. Another difficulty in interpreting the literature is the emphasis of early studies on hip implants, assuming applicability on knee and general pin-on-disc studies. This may be so, but there is the extra difficulty in that the literature findings can be inconsistent or even incoherent sometimes.

From various published studies (such as [83–86] and others cited therein and elsewhere), the protein concentration in human synovial fluid centers around 20 g/l for healthy joints and rises to 30 g/l for some joint replacement patients and can reach even 55 g/l for some osteoarthritic knees. The ISO standard 14243-1 [38] (at the time of writing this chapter) had specified a percentage dilution (which is not useful for the reasons previously described), and it also specified for the protein concentration to *exceed* 17 g/l. Having not specified an exact value, various laboratories gathered data for 3–4 years with their different chosen protein concentrations. In an effort to target one value for the prospective revision of the ISO standard, and without much expertise on serum lubricant content, the author surveyed many orthopedic company and academic laboratory experts in an ASTM forum about what they thought and used and found the average to be around 20 g/l. This also happens to coincide with the protein concentration of normal knee synovial fluid, and it coincides with many test results that already exist on many implants across some laboratories. One popular other concentration used in TKR testing on many implant designs was actually 30 g/l from a very active lab in Germany. The two values were not that different in the author's view, and the literature [83–86] tends to suggest that 20 g/l may be expected to produce higher wear in testing than 30 g/l (see the end of this section for a specific citation of [87] to this end). Knee simulator testing is frequently accused of not being harsh enough to produce physiologically realistic wear levels that are seen *in vivo* or from UHMWPE bearing retrievals. Also, vitamin E doped or sequentially processed modern highly crosslinked UHMWPEs are increasingly advocated for TKR use. Their wear is so small that it makes them hardly measurable in simulator tests. Therefore, in the author's own laboratory, a protein concentration of 20 g/l has been used consistently throughout, and this is the value proposed after long discussion for the forthcoming ISO standard revision.

More variability may perhaps occur in UHMWPE wear testing results due to inconsistent volumes (per station) of diluted serum lubricant used, its cleanliness, consistency of serum replacement, and regular replenishment with deionized water and the onset of deterioration, protein deposits, and calcification. Attention to those areas may be more rewarding in achieving more consistent results. For this reason, additives are typically used to attempt to overcome such problems or delay their influence until a serum batch is due to be replaced (e.g., after nearly 1 week of, say, 0.5 millions walking cycles at 1 Hz simulator frequency). In the author's laboratory and many others, 2 g/l of sodium azide is usually added to the lubricant solution to slow bacterial growth, and 20 mM (7.45 g/l) of ethylenediaminetetraacetic acid (EDTA) is added to reduce calcification and protein precipitation.

The debate about lubricant serum type, content, and dilution for UHMWPE wear tests in knee simulators continues in earnest. More and more clinically relevant recipes for serum lubricant continue to emerge, partly divergent and sometimes convergent, in conclusions. For example, the importance of adding hylauronic acid was stressed and found to multiply simulator wear rates by nearly seven times [86]. Another recent study compared sera of different protein subconstituents from alpha calf and newborn and bovine sources, which were all found to be different and produced different TKR wear rates [87]. In the latter study, wear was found to increase with protein degradation and with more water dilution and decrease particularly with more albumin and $\alpha$-1 globulin proteins in combination. Such studies continue to be useful and add to the much-needed knowledge in this field. However, caution is advised against prematurely standardizing methodologies based on early results. For example, sodium azide as well as hylauronic acid were used in the [86] study, where they were added to the lubricant in an anodized aluminum tibial dish holder in the knee simulator. That anodized aluminum surface is known to corrode, adding unaccounted-for third body debris particles (a possible serious multiplier of wear based on the author's own experience with this simulator and that very tibial dish holder that had to be upgraded with a stainless steel version very early on). Although statistical significance ($P<0.05$) was quoted for the findings in [87], only three TKR samples were tested for each effect for only 1.5–3 million cycle test stages. The differences in resulting wear rates in those intervals were approximately 20–50% only, which brings into question the statistical power of the findings. More test samples, more prolonged studies, and more variable control can only help extend such efforts to bring more consistent *in vitro* test lubricants with better isolated and more understood variable effects. Until then, in the humble opinion of the author, the more consistent and simply defined the serum lubricant for UHMWPE wear testing is, the more data comparison it will facilitate across laboratories.

## 26.5 CONCLUDING REMARKS AND FUTURE DIRECTIONS IN UHMWPE AND TKR LONGEVITY TEST METHODS

As a recap, this tour of tribological testing of UHMWPE destined for TKR started with pin-on-disk (POD) testing, where preliminary screening of *materials* is needed. POD testing was described and could not take any consideration of the TKR design in the planning of the test, except for choices of articular contact area of the pin and the stress levels through suitable choices of loading. POD is relatively simple, economical, and fast (typically $<2$ million cycles) and can be highly informative whenever new materials are to be assessed. Edge effects and other pitfalls were discussed, and wear was shown to be highly sensitive to the motion paths.

An abridged version of the history of knee simulators was outlined, as well as their use to evaluate whole TKR systems for functional performance and longevity. Knee *wear* simulators were shown to be a subset of other types of TKR simulators. They facilitate the testing of UHMWPE and other materials in combination with the individual TKR design and in combination with the implant installation/alignment factors. The reader should now be aware and hopefully able to contribute to the seemingly perpetual debate about the merits of force-control versus displacement-control testing methodologies, which were delved into in some detail here. Various pitfalls and areas of focus and attention in UHMWPE wear testing were discussed, hopefully convincing the reader that the process of TKR wear testing is typically highly costly and time consuming, especially if done well and with a critical eye for detail.

The future will carry exciting further developments into more sophisticated virtual soft-tissue simulation for the force-control methodology. More daily activities beyond walking will be simulated and in programmed sequential combinations with strenuous and rest periods. Testing will take the TKR systems to extremes of their motion and possibly reveal more UHMWPE failure modes in regions previously unconsidered *in vitro*. A new debate emerges as to whether there should be UHMWPE test methods dedicated especially for mobile bearing knee replacements. The answer is obvious for potential special damage modes that relate to the mobility, deformation, and potential subluxation of the UHMWPE bearing insert. Thus a stream of new test methods continue to be proposed and discussed (e.g., at ASTM). It is less clear what modifications are needed in testing UHMWPE wear during simulated walking with a mobile bearing knee.

There are other pressing needs for standardized testing methods, for example, testing of the UHMWPE in the patella. The forces and stresses there are equal if not higher than in the tibial bearings. Also, patella failures have been rife in the historical literature and do need the testing community's attention. The future may also carry exciting possibilities for new simulators to emerge or for current simulators and methods to be adapted to test implants for partial resurfacing of joints that share loads and contribute to the native articular surfaces of the knee.

*In vitro* testing of full artificial joint *systems* will remain a serious and costly endeavor, especially for total knee replacement. Perhaps *the* most important thing to stress in conclusion is the *importance of testing* itself as an integral part of the implant design process. The rest of the messages in this chapter pertain to how to do the testing well.

## REFERENCES


1. Walker PS, Haider H. Characterizing the motion of total knee replacements in laboratory tests. *J Clin Orthop Relat Res* 2003;**410**:54–68.
2. Shaw JA, Murray MD. Knee joint simulator. *Clin Orthop Relat Res* 1973;**94**:15–23.
3. Harding ML, Harding L, Goodfellow JW. A preliminary report of a simple rig to aid study of the functional anatomy of the cadaver human knee joint. *J Biomech* 1977;**10**:517–21.
4. Zavatsky AB. A kinematic-freedom analysis of a flexed-knee-stance testing rig. *J Biomech* 1997;**30**:277–80.
5. Rovick JS, Reuben JD, Schrager RJ, Walker PS. Relation between knee motion and ligament length patterns. *J Clin Biomech* 1991;**6**:213–20.
6. Zachman NJ, Hillbery BM, Kettlekamp DB. Design of a load simulator for the dynamic evaluation of prosthetic knee joints. ASME 1978:78-DET-59.
7. Maletsky LP, Hillberry BM. Simulating dynamic activities using a five-axis knee simulator. *J Biomech Eng* 2005;**127**(1):123–33.
8. Maletsky LP, Clary CW, Guess TM, Mane AM, Reeve AN. In vitro simulation of deep knee flexion using static and dynamic machines. *Proc. ASME 2008; Summer Bioengineering Conference (SBC2008)*. Marco Island, Florida, USA.
9. Dürselen L, Cales L, Kiefer H. The influence of muscle forces and external loads on cruciate ligament strain. *Am J Sports Med* 1995; **23**(1):129–36.
10. Haider H, Walker PS. Measurements of constraint of total knee replacement. *J Biomech* 2005;**38**(2):341–8.
11. Darcy SP, Kilger RHP, Woo SL-Y, Debski RE. Estimation of ACL forces by reproducing kinematics between sets of knees, a novel non-invasive methodology. *J Biomech* 2006;**39**(13):2371–7.
12. Li G, Most E, Sultan PG, Schule S, Zayontz S, Park SE, Rubash HE. Knee kinematics with a high-flexion posterior stabilized total knee prosthesis: an in vitro robotic experimental investigation. *J Bone Joint Surg [Am]* 2004;**86-A**(8):1721–9.
13. Most E, Li G, Sultan PG, Park SE, Rubash HE. Kinematic analysis of conventional and high-flexion cruciate-retaining total knee arthroplasties: an in vitro investigation. *J Arthroplasty* 2005;**20**(4):529–35.
14. Hill MR, Catledge SA, Konovalov V, Clem WC, Chowdhury SA, Etheridge BS, Stanishevsky A, Lemons JE, Vohra YK, Eberhardt AW. Preliminary tribological evaluation of nanostructured diamond coatings against ultra-high molecular weight polyethylene. *J Biomed Mater Res Part B, Appl Biomater* 2008;**85B**:140–8.
15. Cornwall GB, Bryant JT, Hansson CM. The effect of kinematic conditions on the wear of ultrahigh molecular weight polyethylene (UHMWPE) in orthopaedic bearing applications. *Proc Inst Mech Eng, Part H* 2000;**215**:95–106.
16. Bragdon CR, O'Connor DO, Lowenstein JD, Jasty M, Syniuta WD. The importance of multidirectional motion on the wear of polyethylene. *Proc Inst Mech Engrs, Part H: J Eng Med* 1996;**210**(H3):157–65.

17. Muratoglu OK, Bragdon CR, O'Conner DO, Jasty M, Harris WH, Gul R, McGarry F. Unified wear model for highly crosslinked ultra-high molecular weight polyethylenes (UHMWPE). *J Biomater* 1999;**20**:1463–70.
18. Burroughs BR, Blanchet TA. Factors affecting the wear of irradiated UHMWPE. *Tribol Trans* 2001;**44**:215–23.
19. Wang A. A unified theory of wear for ultra-high molecular weight polyethylene in multi-directional sliding. *J Wear* 2001;**248**:38–47.
20. Hamilton MA, Sucec MC, Fregly BJ, Banks SA, Sawyer WG. Quantifying multidirectional sliding motions in total knee replacements. *J Tribol* 2005;**127**:280–6.
21. ASTM F732-00. Standard test method for wear testing of polymeric materials used in total joint prostheses. *ASTM Int* 2006. West Conshohocken, PA.
22. Gevaert MR, LaBerge M, Gordon JM, DesJardins JD. The quantification of physiologically relevant cross shear wear phenomena on orthopaedic bearing materials using a novel wear testing machine. *J Tribol* 2005;**127**(4):740–9.
23. Fisher J, Dowson D, Hamdzah H, Lee HL. The effect of sliding velocity on the friction and wear of UHMWPE for use in total artificial joints. *J Wear* 1994;**175**:219–25.
24. Lewis G. Contact stress at articular surfaces in total joint replacements. Part II: analytical and numerical methods. *Bio-Med Mater Eng* 1998;**8**:259–78.
25. Walker PS. Biomechanics of total knee replacement designs, chapter 15. In: Mow V, Huiskes R, editors. *Basic orthopaedic biomechanics and mechano-biology*. 3rd ed: Lippincott, Williams and Wilkins; 2004. p. 657–702.
26. Walker PS, Blunn GW, Lilley PA. Wear testing of materials and surface for total knee replacement. *J Biomed Mater Res (Appl Biomater)* 1996;**33**:159–75.
27. Sathasivam S, Walker PS, Campbell PA, Rayner K. The effect of contact area on wear in relation to fixed bearing and mobile bearing knee replacements. *J Biomed Mater Res* 2001;**58**(3):382–90.
28. Mazzucco D, Spector M. Effects of contact area and stress on the volumetric wear of ultrahigh molecular weight polyethylene. *J Wear* 2003;**254**(5-6):514–22.
29. Mazzucco D, Spector M. Contact area as a critical determinant in the tribology of metal-on-polyethylene total joint arthroplasty. *J Tribol* 2006;**128**(1):113–21.
30. Saikko V. Effect of contact pressure on wear and friction of ultra-high molecular weight polyethylene in multidirectional sliding. *Proc Inst Mech Eng, Part H: J Eng Med* 2006;**220**(7):723–31.
31. Vassiliou K, Unsworth A. Is the wear factor in total joint replacements dependent on the nominal contact stress in ultra-high molecular weight polyethylene contacts? *Proc Inst Mech Eng, Part H: J Eng Med* 2004;**218**:101–7.
32. Bartel DL, Bicknell VL, Wright TM. The effect of conformity, thickness and material on stresses in ultra-high molecular weight components for total joint replacement. *J Bone Joint Surg (A)* 1986;**68-A**:1041–52.
33. Lee K-Y, Pienkowski D. Viscoelastic recovery of creep-deformed ultra-high molecular weight polyethylene (UHMWPE). In: Gsell RA, editor. Chapter in *Characterization and properties of ultra-high molecular weight polyethylene*. STP1307-EB ASTM Publications.
34. Churchill DL, Incavo SJ, Johnson CC, Beynnon BD. The transepicondylar axis approximates the optimal flexion axis of the knee. *J Clin Orthop* 1998;**356**:111–8.
35. Iwaki H, Pinskerova V, Freeman MAR. Tibiofemoral motion 1: the shapes and relative movements of the femur and tibia in the unloaded cadaver knee. *J Bone Joint Surg (B)* 2000;**82B**:1189–95.
36. Blankevoort L, Huiskes R, De Lange A. Helical axes of passive knee joint motions. *J Biomech* 1990;**23**:1219–29.
37. Grood ES, Suntay WJ. A joint coordinate system for the clinical description of three-dimensional motions: application to the knee. *J Biomech Eng* 1983;**105**:136–44.
38. ISO 14243-1. Implants for surgery—wear of total knee-joint prostheses—Part 1: loading and displacement parameters for wear-testing machines with load control and corresponding environmental conditions for test, International Organization for Standardization; 2002.
39. Andriacchi TP, Johnson TS, Hurwitz DE, Natarajan RN. Musculosleletal dynamics, locomotion and clinical applications, chapter 3. In: Mow V, Huiskes R, editors. *Basic orthopaedic biomechanics and mechano-biology*. 3rd ed. Lippincott, Williams and Wilkins; 2004. p. 91–121.
40. Nordin M, Frankel VH. Biomechanics of the knee, chapter 7. In: Nordin M, Frankel VH, editors. *Basic biomechanics of the musculoskeletal system*. 3rd ed. Lippincott, Williams & Wilkins; 2001.
41. Swanson SAV, Freeman MAR, Heath JC. Laboratory tests on total joint replacement prosthesis. *J Bone Joint Surg [B]* 1973;**55B**:759–73.
42. Dowson D, Jobbins B, O'Kelly J, Wright V. A knee joint simulator, chapter 7. *Evaluation of artificial hip joints*. UK: Biological Engineering Society; 1977. p. 79–90.
43. Dowson D, Gillis BJ, Atkinson JR. Penetration of metallic femoral components into polymeric tibial components observed in a knee joint simulator. American Chemical Society Symposium Series No. 287, Polymer Wear Contr 1985; p. 215–28.
44. Pappas MJ, Buechel FF. New Jersey knee simulator. *Trans of the 11th Annu Int Biomater Symp* 1979;**3**:101.
45. Pappas MJ, Buechel FF. On the use of a constraint radius femoral component in meniscal bearing knee replacement. *J Orthop Rheum* 1994;**7**:27–9.
46. DiAngelo DJ, Harrington IJ. Design of a dynamic multi-purpose joint simulator. *Adv Bioeng* 1992;**22**:107–11. ASME Bioeng Div Publ BED.
47. Walker PS, Hsieh HH. Conformity in condylar replacement knee prostheses. *J Bone Joint Surg* 1977;**59B**:222–8.
48. Burgess IC, Kolar M, Cunningham JL, Unsworth A. Development of a six station knee wear simulator and preliminary wear results. *Proc Inst Mech Eng, Part H: J Eng Med* 1997;**211**:37–47.
49. Haider H, Walker PS, Blunn GW, Perry J, DesJardins J. A four channel force control knee simulator: from concept to production. *Proceedings of the 11th Annual Symposium, International Society for Technology in Arthroplasty [ISTA]*, Marseille, France; 1998. p. 213–14.
50. Walker PS, Blunn GW, Perry JP, Bell CJ, Sathasivam S, Andriacchi TP, Paul JP, Haider H, Campbell PA. Methodology for long-term wear testing of total knee replacements. *Clin Orthop Relat Res* 2000;**372**:290–301.
51. Walker PS, Blunn GW, Broome DR, Perry J, Watkins A, Sathasivam S, Dewar ME, Paul JP. A knee simulating machine for performance evaluation of total knee replacements. *J Biomech Tech Note* 1997;**30**:83–9.
52. Blunn GW, Bell CJ, Walker PS, Chaterjee S, Perry J, Cambell P, Haider H, Paul JP. Simulator testing of total knee replacements, chapter 9. In: Hutchings IM, editor. *Friction, lubrication and wear of artificial joints*. Professional Engineering Publishing; 2003. p. 113–25. ISBN 1 86058 363 6.
53. DesJardins JD, Walker PS, Haider H, Perry J. The use of a force-controlled dynamic knee simulator to quantify the mechanical performance of total knee replacement designs. *J Biomech* 2000;**33**:1231–42.

54. Haider H, Walker PS. Analysis and recommendations for the optimum spring configurations for soft tissue restraint in force-control knee simulator testing. *Trans Orthop Res Society, 48th Annual Meeting*; 2002.
55. Haider H, Walker P, DesJardins J, Blunn G. Effects of patient and surgical alignment variables on kinematics in TKR simulation under force-control. *J ASTM Int (JAI)* 2006;**3**(10).
56. ISO 14243-3. Implants for surgery—wear of total knee-joint prostheses—Part 3: loading and displacement parameters for wear-testing machines with displacement control and corresponding environmental conditions for test, International Organization for Standardization; 2004.
57. Mikosz RP, Andriacchi TP, Andersson GBJ. Model analysis of factors influencing the prediction of muscle forces at the knee. *J Orthop Res* 1988;**6**:205–14.
58. Morrison JB. The mechanics of the knee joint in relation to normal walking. *J Biomech* 1970;**3**:51–61.
59. Johnson TS, Laurent MP, Yao JQ, Gilbertson LN. The effect of displacement control input parameters on tibiofemoral prosthetic knee wear. *Wear* 2001;**250**:222–6.
60. Knight LA, Pal S, Coleman JC, Bronson F, Haider H, Levine DL, Taylor M, Rullkoetter PJ. Comparison of long-term numerical and experimental total knee replacement wear during simulated gait loading. *J Biomech* 2007;**40**(7):1550–8.
61. Laz P, Pal S, Halloran J, Petrella J, Rullkoetter PJ. Probabilistic finite element prediction of knee wear simulator mechanics. *J Biomech* 2006;**39**(12):2303–10.
62. Rawlinson JJ, Furman BD, Li S, Wright TM, Bartel DL. Retrieval, experimental, and computational assessment of the performance of total knee replacements. *J Orthop Res* 2006;**24**(7):1384–94.
63. Haider H, Garvin K. Rotating platform versus fixed-bearing total knees—an in vitro study of wear. *Clin Orthop Relat Res* 2008;**466**:2677–85.
64. Taylor SJG, Walker PS, Perry JS, Cannon SR, Woledge R. The forces in the distal femur and the knee during walking and other activities measured by telemetry. *J Arthroplasty* 1998;**13**:428–37.
65. Taylor SJG, Walker PS. Force and moments telemetered from two distal femoral replacements during various activities. *J Biomech* 2001;**34**:839–48.
66. Taylor WR, Heller MO, Bergmann G, Duda GN. Tibio-femoral loading during human gait and stair climbing. *J Orthop Res* 2004;**22**:625–32.
67. D'Lima DD, Patil S, Steklov N, Chien S, Colwell Jr. C. In vivo knee moments and shear after total knee arthroplasty. *J Biomech* 2007;**40**:S11–7.
68. Haider H, Sekundiak TD, Garvin KL. Simulation of the spring-based soft tissue restraint in testing knee wear under force control. *Proceedings of the First International Conference on Mechanics of Biomaterials & Tissues*. Hawaii, USA; 2005 December.
69. Fukubayashi T, Torzilli PA, Sherman MF, Warren RF. An in-vitro biomechanical evaluation of anterior-posterior motion of the knee. *JBJS* 1982;**64-A**:258–64.
70. Benson LC, DesJardins JD, Harman MK, LaBerge M. Effect of stair descent loading on UHMWPE wear in a force-controlled knee simulator. *Proc Inst Mech Engrs (Part. H, J Eng Med)* 2002;**216**(6): 409–18.
71. DesJardins JD, Banks SA, Benson LC, Pace T, Laberge M. A direct comparison of patient and force-controlled simulator total knee replacement kinematics. *J Biomech* 2007;**40**(15):3458–66.
72. Haider H, Walker PS, Blunn GW. Are the kinematics of different TKR designs targeted for the same patient the same? *Trans Orthop Res Society, 48th Annual Meeting*. Dallas, TX; 2002.
73. Verbal communication with Drs. Todd. S. Johnson and Les Gilbertson of Zimmer Inc. between 2001 and 2008.
74. Johnson TS, Andreacchi T, Laurent M. Development of a knee wear test method based on prosthetic in vivo slip velocity profiles. *Trans Orthop Res Society, 46th Annual Meeting*. FL; 2000.
75. Johnson TS. *PhD Thesis*. University of Minnesota; 1999.
76. Andriacchi TP, Alexander EJ, Toney MK, Dyrby CO, Sum J. A point cluster method for in vivo motion analysis: applied to a study of knee kinematics. *J Biomech Eng* 1998;**120**:743–9.
77. Haider H, Alberts LR, Laurent MP, Johnson TS, Yao J, Gilbertson LN, Walker PS, Neff JR, Garvin KL. Comparison between force-controlled and displacement-controlled in-vitro wear testing on a widely used TKR implant. *Trans Orthop Res Society, 48th Annual Meeting*. Dallas, TX; 2002 February.
78. Anderson DR, Sweeney DJ, Williams TA. Chapter 8. In: *Introduction to Statistics, Concepts and Applications*. 3rd ed. St. Paul, MN: West Publishing Co. ISBN: 0-314-02813-7.
79. Taylor JR. *An introduction to error analysis—the study of uncertainties in physical measurements*: University Science Books, Oxford University Press; 1982 ISBN 0-935702-10-5.
80. ISO 14243-2. Implants for surgery—wear of total knee-joint prostheses—Part 2: methods of measurement, International Organization for Standardization; 2006.
81. Muratoglu OK, Rubash HE, Bragdon CR, Burroughs BR, Huang A, Harris WH. Simulated normal gait wear testing of a highly cross-linked polyethylene tibial insert. *J Arthroplasty* 2007;**22**(3):435–44.
82. Fisher J, McEwen HMJ, Tipper JL, Jennings LM, Farrar R, Stone MH, Ingham E. Wear-simulation analysis of rotating-platform mobile bearing knees. *Orthop* 2006;**29**:36–41.
83. Clarke C, Chan FW, Essner A, Good V, Kaddick C, Lappalainen R, Laurent M, McKellop H, McGarry W, Schroeder D, Selenius M, Shen MC, Ueno M, Wang A, Yao JQ. Multi-laboratory simulator studies on effects of serum proteins on PTFE cup wear. *J Wear* 2001;**250**:188–98.
84. Liao YS, Benya PD, McKellop HA. Effect of protein lubrication on the wear properties of materials for prosthetic joints. *J Biomed Mater Res, Appl Biomater* 1999;**48**:465–73.
85. Yao JQ, Laurent MP, Johnson TS, Blanchard CR, Crowninshield RD. The influences of lubricant and material on polymer/CoCr sliding friction. *J Wear* 2003;**255**:780–4.
86. DesJardins JD, Aurora A, Tanner SL, Pace TB, LaBerge DM. Increased total knee arthroplasty ultra-high molecular weight polyethylene wear using a clinically relevant hyaluronic acid simulator lubricant. *Proc Inst Mech Eng, Part. H, J Eng Med* 2006;**220**(5): 609–23.
87. Brandt JM, Charron K, Zhao L, Medley JB, MacDonald SJ, Bourne RB, McCalden, RW. Development of a clinically relevant lubricant for in-vitro wear testing of total knee replacements. *Proc. 75th Annual Meeting AAOS, Poster* 137;477.

Chapter 27

# Characterization of UHMWPE Wear Particles

Joanne L. Tipper, PhD, Laura Richards, PhD, Eileen Ingham, PhD and John Fisher, PhD



## 27.1 INTRODUCTION

Although most total joint prostheses remain stable for many years, a significant proportion of patients experience loosening, with about 10% requiring revision 10 years after primary arthroplasty [1–3]. It is generally accepted that osteolysis is initiated by the biological response to wear particles, which are released from ultrahigh molecular weight polyethylene (UHMWPE) acetabular liners and tibial trays [4–6]. Wear particles up to 10 $\mu$m in size are predicted to be phagocytosed by resident macrophages and cause the release of inflammatory cytokines and chemokines, including tumour necrosis factor alpha (TNF-$\alpha$), interleukin-1beta (IL-1$\beta$), IL-6, and IL-8 [7–10]. These cytokines stimulate the release of other mediators and the ensuing inflammatory cascade results in the formation of a periprosthetic granulomatous tissue reaction [3, 5]. Histopathological studies have revealed that these inflammatory cytokines facilitate osteoclastogenesis, which results in resorption of bone and a painfully loose implant. Revision surgery is the only option for patient and surgeon [11].

Insights into failure modes of total joint prostheses have been gained by analyzing the particulate wear debris present in periprosthetic tissues from around failed joint replacements [12–19]. In addition, the response of various cell types, including macrophages [7–9, 20–22], fibroblasts [23, 24], and osteoblasts [25, 26], has been investigated *in vitro*. Animal models of particle-induced osteolysis have been developed [27, 28] and the molecular mechanisms of osteolysis have been explored [29].

## 27.2 RATIONALE FOR WEAR PARTICLE ISOLATION

Particle size, volume, morphology, and composition have been shown to be critical determinants of the biological response to wear particles [7, 8, 20, 21]. In recent years orthopedic manufacturers have focused on improving the wear properties of UHMWPE and improving the design of joint prostheses, which has resulted in an increase in the number of new devices coming to the market. Preclinical testing of new devices and materials has become more stringent recently, and it is recognized that quantification of the wear particles and the ability to predict osteolytic potential of the wear debris released is a valuable marketing tool. It is therefore important that investigations into the *in vivo* and *in vitro* cellular response to particles should be underpinned by accurate and comprehensive particle characterization techniques.

It is essential that proteins are removed from tissues and simulator lubricants so that only purified particles are recovered for analysis. Any lipids and proteins that are not removed may mask the true particle sizes and morphologies. Many of the particle isolation techniques reported to date utilize strong acids [30, 31], alkalis [13, 17, 32, 33], or enzymes [19] to digest tissues or protein-containing lubricants from *in vitro* joint simulators, followed by filtration to isolate the particles from the digested suspension. Individual particle information is then gained from scanning electron microscopy analysis, which can be time consuming and expensive. More recently, techniques using automated particle sizers, for example, the Malvern Mastersizer and Zetasizer [34, 35] and the Nanosight particle sizer [36], or automated analysis of scanning electron micrographs [37] have been reported, which attempt to speed up the process of particle characterization while removing observer bias. Details of the different procedures for the isolation and characterization of UHMWPE particles from periprosthetic tissues and joint simulator lubricants are outlined in the following discussion.

## 27.3 DELIPIDIZATION OF SAMPLES

Delipidization of tissues and simulator lubricants is necessary to achieve recovery of purified particles. Contaminating lipids from tissues or those produced by the breakdown of bacteria in serum lubricants of simulator samples, if not removed, will form a layer over the particles during the subsequent filtration steps. Complete removal of lipids is therefore important. This is generally achieved by incubation with chloroform:methanol (2:1) at room temperature for 12–72 h, either before [12, 37] or after [13, 17] digestion with strong alkali, concentrated acids, or enzymes. Tissues delipidized before digestion can be removed from the solvents directly into the chosen digestion solution. Tipper et al. [13] found more complete lipid removal was possible when repeated extraction with chloroform:methanol was performed after the tissues had been digested. However, because the combination of strong alkali with these solvents evokes an exothermic reaction, care should be taken to ensure that tissue digests are cooled to 4°C prior to combining with chloroform:methanol. Chlorofom: methanol extraction results in contaminating lipids and proteins partitioning at the interface between the digested tissue solution (upper layer) and the solvent layer (lower layer). When extraction is combined with centrifugation at 2000 g, the upper layer containing UHMWPE particles can be removed to a clean tube, leaving the unwanted cellular debris behind at the interface. The process is usually repeated until the upper layer containing the UHMWPE particles has visibly cleared of lipids.

## 27.4 ALKALI DIGESTION OF PERIPROSTHETIC TISSUES AND SIMULATOR LUBRICANTS

Several methods for the digestion of tissues and simulator lubricants have been described, which usually involve incubation with either potassium or sodium hydroxide at concentrations between 4 M and 12M, at temperatures above 55°C, for 1 to 7 days. Shanbhag et al. [18] digested periprosthetic tissues with 4 M potassium hydroxide (KOH) at 56°C for 48 h, while Campbell et al. [12] digested periprosthetic tissues using 5 M sodium hydroxide (NaOH) for 1–5 h at 65°C in a shaking water bath. Many of the subsequently published protocols are based upon the latter method. Tipper et al. [13] found that the higher concentration of 12M KOH was more effective for the complete digestion of formalin-fixed retrieval tissues at 60°C over 5 to 7 days. More recently, Baxter et al. [38] compared digestion of porcine and human capsular tissue with sodium hydroxide and potassium hydroxide at concentrations of 5 M, 10M, and 15M over 24 h at 65°C. These authors found that 5 M KOH followed by 5 M NaOH produced the best results in terms of most complete tissue digestion.

## 27.5 ACID DIGESTION OF PERIPROSTHETIC TISSUES AND SIMULATOR LUBRICANTS

Margevicius et al. [30] found that digestion of periprosthetic tissue samples with concentrated nitric acid at room temperature over 48 h was more effective than digestion with hydrochloric acid for the isolation of UHMWPE, cobalt chromium, and titanium alloy wear particles. In contrast, Niedzwiecki et al. [39] found that using concentrated hydrochloric acid (37% by volume) to digest simulator serum at a 5:1 volume ratio at 60°C for 45 minutes removed serum proteins effectively and produced UHMWPE particles that were comparable in size and shape to those produced by digestion with 5 M NaOH. Slouf et al. [37] also reported on a method that used 65% nitric acid to digest periprosthetic tissues at room temperature over 24 h. This technique incorporated repeated washing of the upper layer of the digest with nitric acid, neutralization with 12M KOH, and washing with water until the sample was purified.

While strong alkalis and acids do not appear to have an adverse affect on the size and shape of UHMWPE wear particles [39], it is not recommended that these reagents are used when isolating metal wear particles, particularly cobalt chrome alloy particles, because it has been shown that these reagents have an affect on the size and composition of cobalt chromium particles [40]. Gentler enzymatic methods of digestion are recommended for isolation of metal wear particles from periprosthetic tissues and simulator lubricants [41, 42].

## 27.6 ENZYME DIGESTION OF PERIPROSTHETIC TISSUES AND SIMULATOR LUBRICANTS

Campbell et al. [32] evaluated the ability of various enzymes, including crude and pure papain, collagenase, and protease, to digest fresh and formalin fixed capsular tissues compared to sodium hydroxide. This study revealed that crude papain introduced particulates that masked UHMWPE particles during scanning electron microscopy analysis and that the protease and collagenase enzymes were ineffective against fixed tissues. Pure papain was able to digest fresh tissues effectively but was sensitive to changes in pH brought about by the formalin preservative in the fixed tissues. This could be overcome by extensive washing of the fixed tissues prior to digestion. There may also be issues of cost associated with pure papain. In contrast, concentrations of sodium hydroxide above 5 M were effective in digesting both fresh and fixed tissue samples. Maloney et al. [19] utilized papain to digest periprosthetic tissues at 65°C over 24 h to recover UHMWPE, titanium, and cobalt chrome alloy particles. If the tissue was not completely digested, then a further aliquot of papain was added, and it was subjected to an additional 24 h digestion time. Other enzymes used to digest periprosthetic tissues include pronase used for 72 h at 37°C followed by collagenase at 37°C for 72 h [37]. However, both enzymes were subject to self-digestion and had to be replenished regularly over the digestion period. In addition, Proteinase K has been used to digest hip simulator lubricants at 37°C for 45 minutes with stirring at 350 rpm [39].

## 27.7 CENTRIFUGATION OF SAMPLES

Whether alkali or acid digestion is used, various centrifuge steps or separation by density gradient centrifugation are the preferred methods used to remove biological contaminants and limit the agglomeration of particles. UHMWPE has a density of around 0.94 g/cc and will easily separate from heavier elements, such as metal and ceramic, when centrifuged in the correct medium; however, removing all biological material is often very challenging. In addition, some authors have suggested that centrifugation or ultracentrifugation causes particle loss and deformation [43]. It has been recognized that loss of UHMWPE particles occurs at each stage of the digestion and centrifugation protocol, and Visentin et al. [43] remarked that the average particle retrieval rate using the method of Campbell et al. [12] was just 40%.

Campbell et al. [12] employed sucrose density gradient centrifugation at 40,000 rpm over 3 h to separate UHMWPE particles from contaminating cellular debris and metal wear particles. Using this method the UHMWPE particles remain at the top of the gradient, the cellular debris will collect toward the middle, and any metal or ceramic particles that may be present in the tissue sample will sediment to the bottom of the gradient. Thus the particles of interest are spatially removed from contaminating proteins and may be harvested, washed, and purified using isopropanol density gradients before undergoing further analysis. However, this method relies on collecting particles banded diffusely at interfaces or particular locations in the centrifuge tube used for the gradient, making it difficult to ensure that all isolated particles have been collected. This may account for the poor particle retrieval rate of this method [43].

Other methods utilize ethanol precipitation combined with high-speed centrifugation (up to 20,000 g) for 2 h to remove contaminating proteins before filtration of the whole tissue or simulator lubricant digest [13, 17, 44]; particle recovery rate of this method has been determined to be in the region of 70–75% [45].

## 27.8 FILTERING TO RECOVER PARTICLES

Initially filter membranes with a pore size of 0.1–0.2 μm were utilized to collect all the particles isolated [12]. However, the diverse size range of UHMWPE particles generated both *in vivo* and *in vitro* quickly became apparent, and samples have been sequentially filtered through filter membranes with progressively smaller pore sizes, such as 10 μm, 1 μm, and 0.1 μm [13].

More recently, as UHMWPE particles in the nanoscale have been identified [15, 17, 46, 47] additional filters with pore sizes in the range of 0.01–0.015 μm are now being utilized.

## 27.9 POLARIZED LIGHT MICROSCOPY OF TISSUE SAMPLES

Initial attempts to characterize UHMWPE wear particles released into the periprosthetic tissues around failing implants used histological staining and polarized light microscopy to identify UHMWPE [48, 49]. When observed using polarized light, particles of UHMWPE have a characteristic birefringent "glow" [49, 50]. However, it was soon recognized that submicron-sized individual particles could not be observed due to the limited resolution of the light microscope (approximately 0.5–1 μm). Intracellular particles were described in association with macrophages and giant cells, and particles were identified in the range of "less than one micrometre" to 100 μm. Most of the intracellular particles were $<10\,\mu m$ in length. Particle shape and morphology were described as shard or needle-shaped, granular, and flakelike [19, 49]. This technique could be used to semiquantitatively grade the amount of wear debris present within the tissue sections but did not attempt to estimate actual numbers or volumes of particles. Schmalzried et al. [6] investigated the use of oil red O staining to identify UHMWPE quantitatively within tissue sections; however, two "no particle" control specimens stained positive for UHMWPE. Neither of these techniques distinguishes agglomerates of particles, and therefore the size ranges of particles identified by polarized light microscopy are likely to be overestimated.

## 27.10 SCANNING ELECTRON MICROSCOPY ANALYSIS

Scanning electron microscopy (SEM) has been used extensively to analyze particles isolated from periprosthetic tissue samples or simulator lubricants that have been digested and isolated onto filter membranes. The resolution of a standard SEM is approximately 0.1 μm. Campbell et al. [12] used magnification ranges up to 10,000× to observe particles from tissues down to 0.07 μm in size. Using this higher resolution analysis, more detailed particle shape descriptors were reported, and particle size was measured more accurately using digital image analysis systems. Campbell reported on two distinct morphologies of particles: rounded and elongated particles; the elongated particles were further separated into fibrils (short and thin, <3 μm long and <0.25 μm thick) and shreds (up to 12 μm long and up to 1 μm thick). Also around this time the first particle size distributions were published, which revealed that the majority of the particles (>90%) were less than one micrometer in size [19]. However, particle frequency distributions were similar for all patients regardless of implant type, with the mode of the distribution typically falling in the submicron size range [22, 30, 51].

Tipper et al. [13] digested 18 samples of periprosthetic tissue that had been isolated from failed Charnley hip prostheses. These authors reported particles ranging from 0.1 μm to 1000 μm in size, using magnification ranges up to 25,000×. In addition to particle frequency distributions, which showed the mode of the distribution to be in the 0.1–0.5 μm size range, these authors reported mass distributions for all patients. Using the mass distributions, the wear particles from different patients were successfully differentiated, with each patient demonstrating an individual distribution of particles. This study was the first to recognize the importance of the particles above 10 μm in size. Up until this time many studies had concentrated on the smaller submicron-sized particles because they had been shown to be the most biologically active in terms of causing the release of osteolytic cytokines from macrophages [21]. Indeed, several studies showed that particles >10 μm in size exhibited minimal biological activity. However, the study by Tipper et al. [13] recognized that a relatively small number of larger particles could account for a large proportion of the total wear volume, and consequently the total wear volume of particles would have lower biological activity.

More recently, mass distributions have been replaced by volume [35] or area distributions [15, 17]; the latter approach takes the two-dimensional area of the wear particles and converts it to volume assuming a constant particle thickness [45].

Scott et al. [31, 47] analyzed hip simulator debris from conventional, 50 kGy, and 100 kGy crosslinked UHMWPE at magnifications of 10,000× and 20,000×. Particles with equivalent circle diameters (ECD) of 0.19, 0.11, and 0.08 were observed for the conventional, 50 kGy, and 100 kGy UHMWPE samples, respectively. Scott et al. [31, 47] was one of the first authors to report on the isolation of particles in the nanometer size range from *in vitro* hip simulator lubricants.

Galvin et al. [46] and Tipper et al. [15] used high-resolution field emission gun-scanning electron microscopy (FEG-SEM) to analyze particles of UHMWPE from hip and knee simulator lubricants. FEG-SEM can achieve magnifications over 200,000×. With these advances in technology, these authors identified particles as small as 10 nm in size, with the mode of particles isolated from the hip simulator lubricant lying in the nanometer size range (<0.1 μm). More recently, Richards et al. [17] used alkaline digestion, filtration, and FEG-SEM analysis to isolate and analyze UHMWPE from periprosthetic tissues from around failed Charnley hip prostheses. Using this high-powered technique, *in vivo* generated UHMWPE particles in the nanometer size range were observed for the first time in all tissue samples examined. Other authors have also recently reported the presence of nanometer-sized UHMWPE particles in some periprosthetic tissue samples [52].

## 27.11 IMAGE ANALYSIS OF UHMWPE WEAR PARTICLES

Digital image analysis software represents a useful tool to assist with particle size analysis. Particles can be sized automatically by the software, with multiple measurements of maximum width, maximum length, area, and perimeter being recorded simultaneously. Alternatively, particles can be sized manually, with each individual particle being traced with a cursor. There are drawbacks associated with both methods. One of the limitations of automatic sizing is that the system can count clumped or agglomerated particles as one large particle. This can be overcome by manual manipulation of the images prior to automatic sizing. However, particle isolation can be subject to contamination, either with biological or environmental contaminants. Biological contaminants might be lipids or proteins that originate from the tissue or simulator lubricant sample or salt crystals that originate from the isolation procedure. Common environmental contaminants encountered include silicon dust from the atmosphere. Automatic sizing procedures do not discriminate between particles of UHMWPE and contaminating particles, leaving this technique open to overestimation of the number of particles. However, this method is quickly executed and lends itself to accurately sizing large numbers of particles.

Manual sizing allows each individual particle to be assessed for inclusion into the size distribution, making it simple to identify and exclude contaminants. However, manual sizing is very time consuming and tedious and may be less accurate because it relies on the operator's accuracy to trace large numbers of particles.

However, both methods have the following further limitations associated with SEM observation: observer bias toward a particular particle size (may be overcome with random sampling techniques; see Richards et al. [17]); only a small number of sample micrographs are taken from across the filter membrane representing analysis of a very small area of the filter, which introduces the possibility of misrepresentative sampling; and small numbers of particles analyzed can introduce systematic and observer bias, which can be compounded by sampling errors. Elfick et al. [34] reasoned that image analysis techniques improved upon the sampling statistics of these methods by increasing the numbers of particles analyzed, but the accuracy of the size distribution is dependent upon resolution of discrete particles.

## 27.12 AUTOMATED PARTICLE ANALYSIS

Given the problems associated with image analysis of wear particles previously described, automated particle size analysis by Coulter counter seemed like an obvious way forward. Maloney et al. [19] and Margevicius et al. [30] analyzed tissue membranes using a Coulter Multisizer, which employed electrical impedance technology. These authors reported the mean diameter of particles to be 0.63 μm and 0.40 μm, respectively. This method meets a number of pitfalls; firstly, particles are assumed to be spheroids. Secondly, the minimum detection range is 0.4 μm to 0.5 μm; and lastly, agglomerated particles would result in particle size overestimation, which would be included in the analysis. Various studies have utilized automated sizing equipment. Elfick et al. [34, 35] used laser diffraction particle analysis (LDPA) to analyze particles from both hip simulator samples [34] and periprosthetic tissue samples [34, 35]. LDPA uses Mie theory to predict the volume distribution of the debris in suspension by the scattering of light from two laser beam sources, where the smallest particles diffract light by a larger angle. The lower detection limit of the Malvern Multisizer 2000 used in this study was 0.02 μm according to the manufacturers; however, the minimum particle size detected in tissue samples was 0.113 μm, almost tenfold larger than the smallest particle detected by FEG-SEM analysis of periprosthetic hip tissues [17]. The authors concluded that particles smaller than 0.113 μm were not produced *in vivo*, but they go on to speculate on the relationship between small particle size and the increased propensity for particle transport away from the implant site.

More recently, automated particle sizers with improved lower detection thresholds have become available commercially, such as the Malvern Zetasizer (detection range 0.006–6 μm) and the Nanosight particle sizer (detection range 0.01–800 μm). These may be useful for sizing particles with a limited size range in the nanoscale, such as metal or ceramic wear particles, but have limited applicability to sizing UHMWPE particles due to the large size range of particles (0.01–1000 μm). However, all automated sizing suffers from the inability to visualize the particles directly and determine the degree of agglomeration accurately.

Slouf et al. [37] reported a novel automated quantitative analysis technique, SEMq, which was based on a light scattering theory. In this method automated image analysis of electron micrographs is used to determine the total amount of wear particles released into the tissues around total joint replacements. This technique was shown to provide rapid, reliable, and reproducible results, which indicated that the distribution of wear particles around TJRs is nonhomogeneous [37]. However, this technique does not calculate absolute numbers of wear particles but rather determines relative numbers of wear particles in one sample compared to another. The technique may also be sensitive to errors introduced by particle clumping and protein contamination of filters used for SEM analysis.

## 27.13 STANDARDS

It has been recognized that a standardized method of particle retrieval from the tissues and simulator lubricants, followed by debris characterization and quantification, is required for a uniform approach to debris response investigations. However, there are a number of different standards available from the British Standards Institution (BSI), International Organization for Standardization (ISO), and the American Society for Testing and Materials (ASTM), which currently offer conflicting advice for isolation of wear particles from periprosthetic tissues and simulator lubricant samples.

The British standard (BS ISO 17853:2003 Wear of implant materials—Polymer and metal wear particles—Isolation, characterization and quantification) [53], and the ISO standard (ISO 17853 Wear of implant materials—Polymer and metal wear particles—Isolation, characterization and quantification) [53] are identical documents. Both current standards, issued in 2003, recommend the use of 5 M sodium hydroxide to digest tissue and simulator lubricants to recover both polymer and metal wear particles. As discussed in Section 27.5, strong acids and alkalis are not suitable reagents for use in the recovery of metal wear particles from tissue or simulator lubricants due to adverse effects on particle size and shape. Enzymatic digestion protocols are recommended for use to recover metal wear particles to avoid these effects. The BSI and ISO standards recommend purification of particles by sucrose density gradient centrifugation followed by filtration through filter membranes with pore sizes of 0.1 μm, followed by observation of particles using SEM analysis at magnifications between 1000× and 5000×. Since the publication of this standard, a number of studies have identified UHMWPE particles in the nanometer size range [15, 17, 46, 47, 52], and filters with pore sizes in the range of 0.01–0.05 μm are commonly employed to isolate particles in this size range. In addition, UHMWPE fibrils up to 1 mm in length are generated both *in vivo* (13) and *in vitro* (15) and magnifications of 1000× would be too high to observe particles of this size. Magnifications between 150× and 100,000× may be more appropriate to capture images of all sizes of particles after sequential filtration through 10, 1, and 0.015 μm filters or similar [17]. The ISO standard is currently being updated to incorporate recent advances in technology, that is, increased magnification range achievable with FEG-SEM and to ensure that particles in the nanometer size range are isolated and observed.

The ASTM standard (F561-05a Standard Practice for Retrieval and Analysis of Medical Devices, and Associated Tissues and Fluids) [54] is also currently being revised to reflect similar changes. The current standard is very detailed, giving four recommended methods of tissue digestion (papain, 4 M KOH and pronase, concentrated nitric acid and 5 M sodium hydroxide) and two recommended methods for digestion of protein containing simulator lubricants (12M sodium hydroxide and concentrated hydrochloric acid). However, there is no indication of which methods are suited to particular types of materials, and the overall effect is one of confusion rather than guidance. Filtration is the recommended method for the recovery of wear particles and sequential filtration, and the need for smaller pore size filters to recover particles in the nanometer size range are discussed. All the standards recommend that chemical analysis by energy dispersive X-ray analysis (EDX) or Fourier transform infrared spectroscopy (FTIR) analysis are performed to eliminate contaminating particles or verify the identification of UHMWPE particles, respectively.

In addition, all the particle isolation standards previously referred to recommend that particle morphology should be characterized using a series of predefined descriptions, such as length, width, equivalent circle diameter, area, perimeter, aspect ratio, form factor, elongation, and roundness, as described in detail in ASTM F1877-05 (Standard Practice for Characterization of Particles) [55]. These parameters are described in more detail in Section 27.14.

## 27.14 PARTICLE MEASUREMENTS (SIZE/SHAPE DESCRIPTORS)

Many different shape descriptors have been employed over the past 15 years; these include spherical, spheroidal, granular, angulated, globular, floret, flakes, shards, needle-like, fibrillar, and ribbons. Particles of UHMWPE generated *in vitro* and *in vivo* tend to exhibit many different morphologies over a number of size ranges, that is, fibrils isolated from periprosthetic tissues have been described up to 1 mm in length, but also smaller fibrils of around 1 μm were also observed by SEM at higher magnifications [13]. A detailed description of the nomenclature used for particle morphology description can be found in the ASTM Standard Practice for Characterization of Particles (F1877-05) [55].

Simple measurements are shown in Table 27.1.

**TABLE 27.1 Simple Measurements**

| | |
|---|---|
| Area of particle: This does not include any holes within the object. | |
| Perimeter of particle: The outline of the object. | 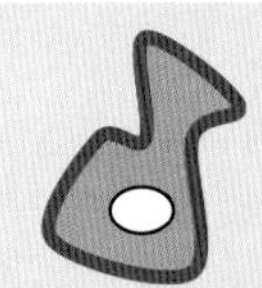 |
| Maximum diameter ($d_{max}$): Defined as the length of the longest line joining two points of the object outline and passing through the centroid. | 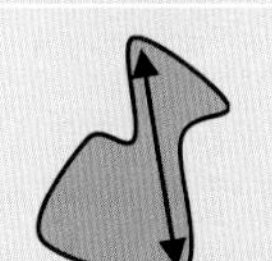 |
| Minimum diameter ($d_{min}$): Defined as the length of the shortest line joining two points of the object outline and passing through the centroid. | 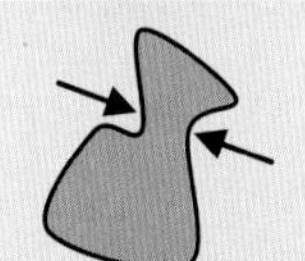 |
| Size (length, L): Defined as the Feret diameter (i.e., caliper length) along major axis of the object. |  |
| Size (width, W): Defined as the Feret diameter (i.e., caliper length) along minor axis of the object. | 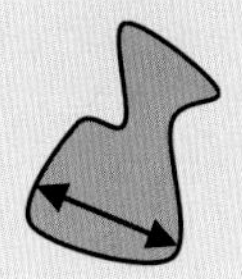 |

In addition, more detailed measurements of equivalent circle diameter (ECD), which is defined as the diameter of a circle with an area equivalent to the area of the particle, and has units of length:

$$ECD = (4 \times A/\pi)^{1/2}$$

Aspect ratio (AR) is defined as the ratio of maximum diameter to minimum diameter:

$$AR = d_{max}/d_{min}$$

Elongation (E) is similar to the aspect ratio but is a more suitable measure for long fibril-type particles, where the major axis line does not stay within the boundaries of the particle. It is given as a ratio of particle length to width:

$$E = L/W$$

Roundness (R) is a measure of how closely a particle resembles a circle. A perfect circle has a value of one:

$$R = (4A)/(\pi\, d_{max}^{\ 2})$$

Form factor (FF) is similar to roundness but is based on the perimeter (p) of the particle outline rather than the major diameter:

$$FF = (4\,\pi A)/p^2$$

These measurements provide a more complex analysis of particle shape and morphology.

## 27.15 PREDICTING FUNCTIONAL BIOLOGICAL ACTIVITY

Traditionally, prediction of the clinical performance of different materials or devices has been assessed by wear testing using laboratory simulators prior to clinical implantation, which allows measurement of the total wear volume produced in $mm^3$ and wear rate of a particular type of material/device in $mm^3$/million cycles of simulation. This enables materials or devices to be ranked using their wear performance. Such measurements are useful when assessing the performance of new materials against existing ones and have led to the introduction of newer materials with improved wear resistance into clinical practice, for example, crosslinked UHMWPEs have been shown to have significantly lower wear rates compared to noncrosslinked UHMWPEs [56, 57]. However, differences in the wear particle characteristics can lead to altered biological responses and a different potential for the development of osteolysis. It has been shown that the biological response to different types of UHMWPE wear debris will vary depending on the kinematic conditions [58], oxidative state [45], and level of crosslinking [59, 60]. Therefore, it is desirable to combine information on both the wear volume of a material with its wear particle characteristics, in terms of size, volume, and biological activity, to give a clearer indication of the osteolytic potential of the material or joint replacement device.

Fisher et al. [61] devised a novel method for predicting the osteolytic potential of UHMWPE wear particles generated in hip or knee simulators from different UHMWPEs or different device designs, using functional biological activity (FBA). This approach integrates the volumetric concentration of wear particles in three different size ranges (<0.1–1.0 μm, 1.0–10 μm, >10 μm), with the biological activity index for wear particles in each of the three size ranges (as determined in previous studies) [7]. The percentage volumetric concentration of wear as a function of particle size, $C(r)$, can be determined by isolating the particles from the bovine serum lubricant from laboratory wear simulations using one of the methods previously described. Using the particle area distributions, and assuming a constant particle thickness, a volume distribution can be determined, giving the volumetric concentration of wear particles in each of the three size ranges: <0.1–1 μm, 1–10 μm, >10 μm.

The biological activity function $B(r)$ for UHMWPE wear debris was determined as a function of particle size, $r$, at fixed volumetric concentrations. The wear particles were used to stimulate primary human macrophages obtained from six human donors in triplicate [7]. Tumor necrosis factor-alpha (TNF-α) was used as a marker of biological activity because it has previously been shown to correlate with bone resorption [62]. Biological activity, $B(r)$, of the particles in the different size ranges was normalized with respect to particle in the smallest size range, which was assigned a value of unity (Figure 27.1). Normalization was carried out on each donor with the mean from at least three replicates to eliminate donor-to-donor variability. The index of specific biological activity (SBA) of UHMWPE wear debris is defined as the relative biological activity per unit volume. It is calculated by integrating the biological activity function, $B(r)$, and the volumetric concentration function, $C(r)$, over the size range of the wear particles:

$$SBA = \int C(r)B(r)$$

The total size range studied was <0.1–100 μm.

Determining the volumetric concentration of wear particles in each size range for each material or device and then integrating these values with the biological activity function for UHMWPE allows the SBA for each material or device tested to be determined. The SBA can then be integrated with the volumetric wear rate (in $mm^3$/million cycles) to give a measure of functional biological activity (FBA),

$$FBA = V \times SBA$$

where $V$ has units of $mm^3/10^6$ cycles or $mm^3$/year. The values of SBA and FBA can be compared for the wear

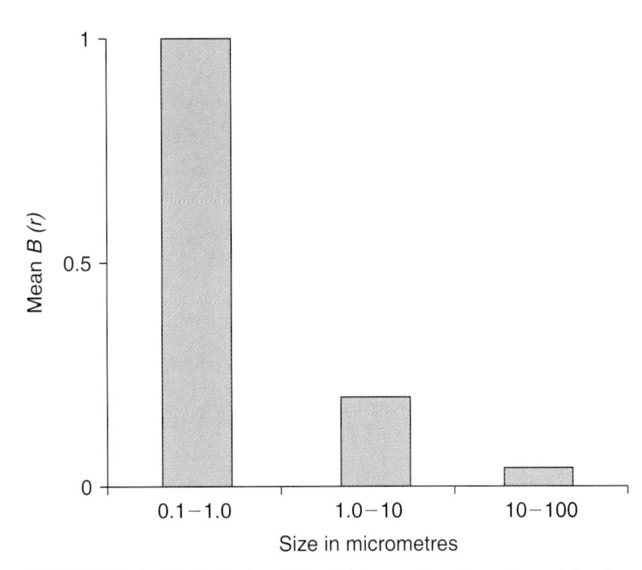


**FIGURE 27.1** Biological activity $B(r)$ as a function of particle size derived from *in vitro* cell culture studies.

particles from all materials or devices. To illustrate the use of this predictive methodology, the following example compares the wear, wear debris, and specific and functional biological activity of two different UHMWPEs, GUR 1020 GVF and GUR 1050 (sterilized with 25 and 50 kGy gamma radiation) in the hip and the knee.

## 27.15.1 Materials

GUR 1020 was irradiated with 25–40 kGy of gamma radiation in a vacuum and then foil packed (GVF), and hence was moderately crosslinked. GUR 1050 was either gamma irradiated in air with 25 kGy of radiation or was moderately crosslinked with 50 kGy of gamma radiation, followed by remelting above 155°C for 24 h to eliminate residual free radicals. The GUR 1020 40 kGy GVF UHMWPE acetabular cups and tibial trays were off-the-shelf clinical products supplied by DePuy International, a Johnson & Johnson company. All other UHMWPE acetabular cups and tibial trays were manufactured specifically for these studies.

In the hip GUR 1020 25–40 kGy (GVF) UHMWPE acetabular cups were compared to GUR 1050 25 kGy UHMWPE acetabular cups in a 10-station ProSim hip simulator against 28 mm diameter cobalt chromium femoral heads, which were highly polished, with $R_a$ values of $<0.02\,\mu m$. Fixed bearing and mobile bearing total knee replacements (TKRs) were investigated in the Leeds knee simulator [63]. Three press fit condylar (PFC) Sigma Rotating Platform (DePuy International, UK) mobile bearing knees (GUR 1020 25–40 kGy GVF) were compared to three PFC Sigma fixed bearing knees (GUR 1020 25–40 kGy GVF; DePuy International, UK). In addition, three PFC Sigma fixed bearing TKRs with UHMWPE inserts manufactured from GUR 1050 50 kGy were investigated.

## 27.15.2 Methods

The methodologies for both the hip and knee simulator tests are described below together with details of the particle isolation technique employed in this study.

### *27.15.2.1 Hip Simulator*

A minimum of three 28 mm prosthesis pairs of each UHMWPE material were studied under standard simulation conditions in a 10-station Leeds ProSim hip joint simulator for 5 million cycles [64]. The cups were positioned superiorly and inclined at 35 degrees to the horizontal plane in an anatomical configuration. The simulator applied two independently controlled motions. The head underwent flexion extension (+30 degrees and −15 degrees), and the acetabular insert underwent ±10 degrees rotation. A twin peak time-dependent loading curve was applied with a peak load of 3 kN at heel strike and toe off, and a swing phase load of 50N was also applied.

Tests were carried out in 25% (v/v) newborn calf serum (Harlan Seralabs, UK) with 0.1% (w/v) sodium azide to retard bacterial growth. The lubricant was changed every 330,000 cycles and stored frozen at −20°C until required for wear debris analysis. Wear was determined volumetrically every million cycles. The three-dimensional geometry of the cups was determined before the test and every million cycles using a coordinate measuring machine (Kemco 400 3D, Keeley Measurement Co., UK).

### *27.15.2.2 Knee Simulator*

In the fixed bearing PFC Sigma TKR, 10 mm thick tibial inserts were assembled by snap fit into size three titanium alloy (Ti-6Al-4 V) tibial trays. The 10 mm thick PFC Sigma RP bearings freely rotated within size three, polished cobalt chrome alloy (CoCrMo) tibial trays. All PFC Sigma and PFC Sigma RP components articulated with size three posterior cruciate retaining, right cobalt chrome alloy femoral components. All devices were tested in the Leeds six-station ProSim force/displacement controlled knee simulator under ISO conditions. All components were tested using “high” kinematic inputs, which closely reproduce those found in the natural knee and were chosen because they represent a normal walking gait cycle [65]. The knee prostheses were subjected to ±5 degrees internal/external (IE) rotation and 0–10 mm anterior-posterior translation. Fixed bearing knees were simulated under displacement control, whereas the rotating platform mobile bearing knees were simulated under force control (ISO 14243) in the AP direction due to RP design constraints, which restrict movement in this axis. Testing was performed at 1 Hz using 25% (v/v) newborn calf serum plus 0.1% (w/v) sodium azide to retard bacterial growth. The lubricant was changed every 330,000 cycles. Knee components were tested to 5 million cycles.

Wear measurements were determined gravimetrically every million cycles using unloaded soak controls as a comparison. Serum lubricants were stored frozen at −20°C until required for wear debris analysis.

UHMWPE wear particles were isolated according to the methods of Tipper et al. [15]. Three replicates were analyzed for each prosthesis type. Particles were isolated from each lubricant by alkaline digestion and sequential filtration onto 10, 1.0, and 0.1 μm filters. Filter samples were coated with 5 nm of platinum for analysis by field emission gun-scanning electron microscopy (FEG-SEM) and image analysis using Image Pro Plus (Media Cybernetics, USA). Maximum diameter and area measurements were taken for a minimum of 150 particles in each size range to generate size and area distributions for the particles from the different materials. Percentage data were arcsine transformed, and the transformed data were analyzed by two-way ANOVA. The minimum significant difference ($p < 0.05$) between group means was calculated using the T-method.

The osteolytic potential of the wear particles generated in the hip and knee simulators was predicted using the method of Fisher et al. [61] to calculate functional biological activity (FBA) as previously described. The values of SBA and FBA were compared for the wear particles from the two UHMWPEs in both the hip and knee simulations.

## 27.15.3 Results

The wear rates, sizes and morphologies of the wear particles together with the predicted biological activity indices for both the hip and knee prostheses are reported below.

### *27.15.3.1 Volumetric Wear Rates*

The wear rates for the two UHMWPEs in the hip and knee are shown in Figure 27.2. In the hip, the GUR 1020 GVF material exhibited a lower wear rate (35 ± 9 mm$^3$/million cycles) compared to the GUR 1050 25 kGy (46 ± 9 mm$^3$/million cycles); however, this difference was not significant. In the knee in the PFC Sigma fixed bearing, the GUR 1020 had a significantly higher ($p < 0.05$) wear rate than the moderately crosslinked GUR 1050 50 kGy UHMWPE, at 22.8 ± 6 mm$^3$/million cycles compared to 13 ± 4 mm$^3$/million cycles [66]. Thus, the introduction of moderate levels of crosslinking into the UHMWPE component improved the wear resistance of the UHMWPE and the performance of the TKR device. However, the PFC Sigma RP mobile bearing knee (GUR 1020 GVF) exhibited a much lower wear rate of 5.2 ± 4 mm$^3$/million cycles. This represented a fourfold reduction in wear compared to the PFC Sigma fixed bearing with GUR 1020 UHMWPE and a 2.5-fold reduction in wear compared to the PFC Sigma fixed bearing with GUR 1050 50 kGy UHMWPE.

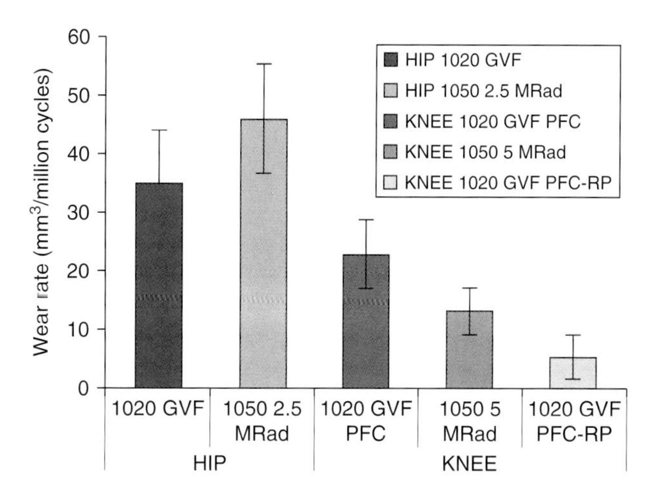


**FIGURE 27.2** Mean volumetric wear rates (±95% confidence limits) for GUR 1020 and GUR 1050 UHMWPEs in the hip and the knee.

### *27.15.3.2 Wear Particles*

The wear particles isolated from the lubricants of both the TKR and the THR prostheses were predominantly less than 1.0 μm in size (Figure 27.3). However, a small number of larger particles >1 μm in size were also observed (Figure 27.4). The mode of the frequency distribution for the particles from both the THR and TKR prostheses was in the <0.1–1 μm size range, indicating that the majority of the wear particles were submicrometer in size (Figure 27.5). There were no significant differences in the particle size distributions between the fixed bearing and the Rotating Platform TKR designs or between the TKR and THR prostheses (Figure 27.5). The mode of the distribution for the volumetric concentration of particles as a function of size was in the <0.1–1.0 μm size range for the two different UHMWPEs in the THRs (Figure 27.6); however, the GUR 1020 GVF UHMWPE had approximately 40% of the wear volume comprised of submicrometer-sized wear particles, compared to almost 90% for the GUR 1050 25 kGy UHMWPE. In the knee prostheses the mode of the volume distribution for the GUR 1020 GVF UHMWPE components, fixed and mobile bearing designs, was in the 1.0–10 μm (Figure 27.6). However, for the PFC Sigma fixed bearing with the GUR 1050 50 kGy UHMWPE components, the mode of the volume distribution was in the smaller particle size range (<0.1–1 μm), with over 60% of the wear volume comprised of submicrometer wear particles.

The volumetric concentrations of wear particles in the different size ranges had a marked effect on the biological activity indices for the different UHMWPEs and the prostheses. In the hip, the SBA of the GUR 1050 25 kGy UHMWPE was markedly higher than the GUR 1020 GVF UHMWPE, at 0.89 compared to 0.51 (Table 27.2). This increase in SBA can be accounted for by the much higher volumetric concentration of wear particles in the more

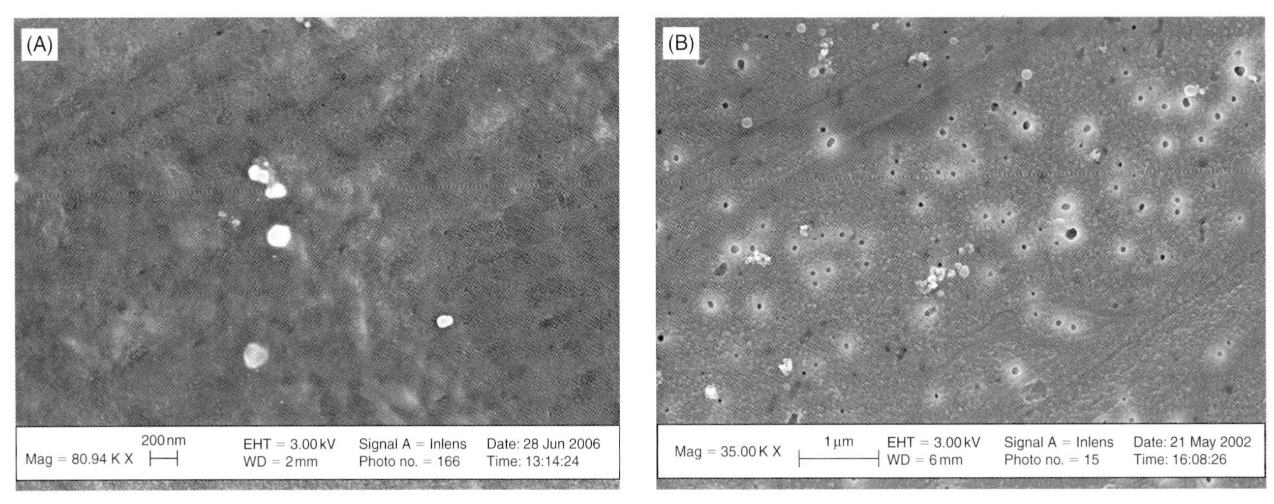


**FIGURE 27.3** Scanning electron micrographs of (A) submicrometer-sized UHMWPE particles from the hip (magnification 80,940x, size bar 200 nm) and (B) submicrometer-sized UHMWPE particles from the knee (magnification 35,000x, size bar 1 μm).

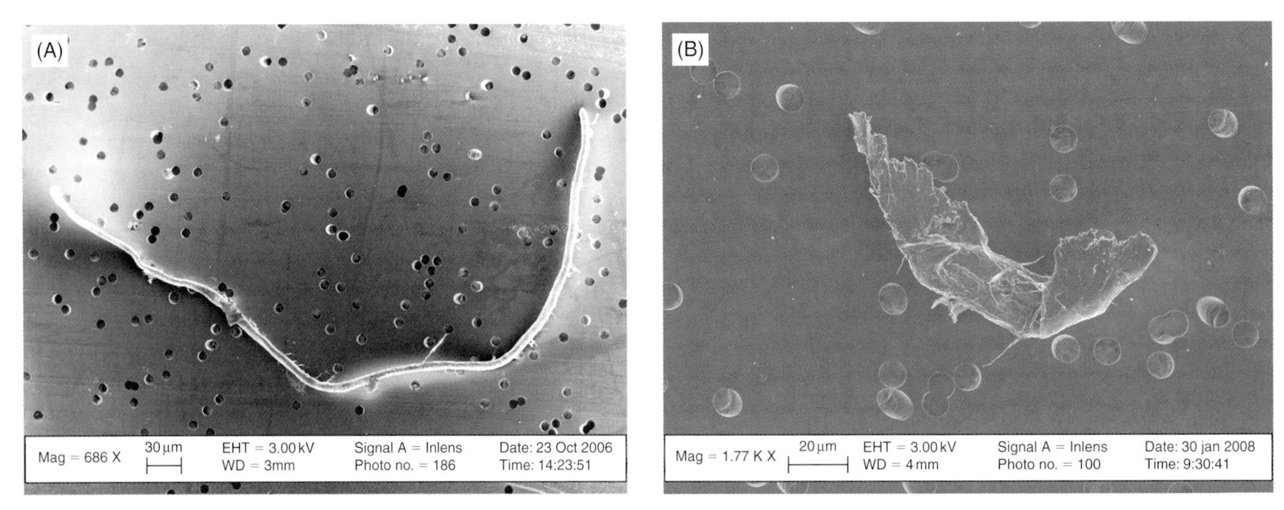


**FIGURE 27.4** Scanning electron micrographs of (A) larger UHMWPE wear particles from the hip (magnification 686x, size bar 30 μm) and (B) larger UHMWPE wear particles from the knee (magnification 1770x, size bar 20 μm).

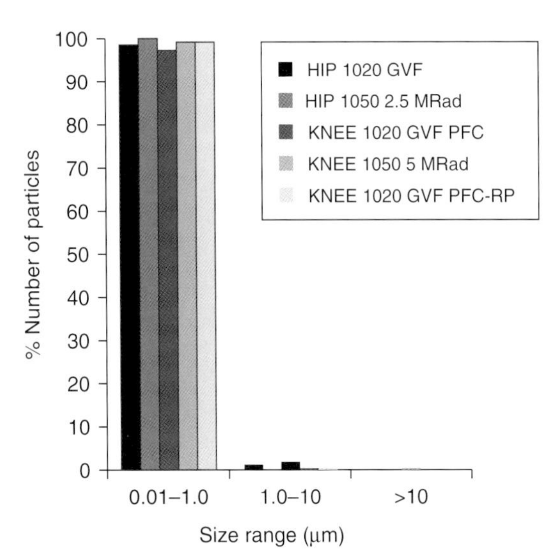


**FIGURE 27.5** Frequency distributions as a function of particle size for GUR 1020 and GUR 1050 UHMWPEs in the hip and knee.

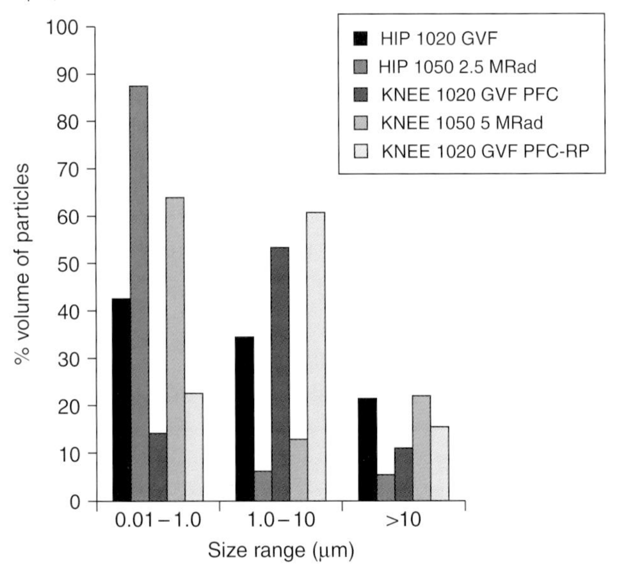


**FIGURE 27.6** Volume distributions as function of particle size for GUR 1020 and GUR 1050 UHMWPEs in the hip and knee.

**TABLE 27.2 Specific Biological Activity and Functional Biological Activity Values for GUR 1020 and GUR 1050 under Different Simulation Conditions**

| Material/implant design | Specific biological activity (SBA) | Functional biological activity (FBA) |
|---|---|---|
| HIP GUR 1020 GVF | 0.51 | 17.81 |
| HIP GUR 1050 25 kGy | 0.89 | 41.1 |
| KNEE PFC Sigma fixed GUR 1020 GVF | 0.26 | 5.9 |
| KNEE PFC Sigma RP GUR 1020 GVF | 0.36 | 1.86 |
| KNEE PFC Sigma fixed GUR 1050 50 kGy | 0.68 | 8.83 |

biologically active size range (<0.1–1 μm) for the GUR 1050 material. In the fixed bearing knee design, again the GUR 1020 GVF UHMWPE had a lower SBA value compared the GUR 1050 50 kGy UHMWPE, at 0.26 compared to 0.68 (Table 27.2). Once again, this increase in SBA for the GUR 1050 material can be explained by a much higher volumetric concentration of particles in the smaller and more biologically active size range (<0.1–1 μm). The Rotating Platform mobile bearing TKR had an SBA somewhere between the two (Table 27.2). In the hip, high SBA coupled with a higher wear rate for the GUR 1050 25 kGy UHMWPE meant that the FBA for this material was more than twice as high as the GUR 1020 GVF UHMWPE material (Table 27.2). In the fixed bearing PFC Sigma knee, although the GUR 1050 50 kGy material had a higher SBA than the GUR 1020 GVF UHMWPE, the reduction in wear rate conferred by the introduction of moderate levels of crosslinking led to similar FBA values. The PFC Sigma rotating platform TKR design had the lowest FBA or osteolytic potential, and in general the TKR designs exhibited lower FBA values or osteolytic potential than the THRs (Table 27.2).

### *27.15.3.3 Discussion*

In the hip there was no significant difference in the wear rates of the two UHMWPE materials; however, in the PFC Sigma fixed bearing knee the introduction of moderate levels of crosslinking into the GUR 1050 UHMWPE caused a significant reduction in the wear rate. Changing the design to a mobile bearing rotating platform knee, but keeping the UHMWPE constant (GUR 1020 GVF) caused a fourfold, highly significant reduction in the wear. Rotating Platform mobile bearing TKR devices have been shown to redistribute the motions between the femoral-insert and tray-insert articulating surfaces and translate complex input motions into more unidirectional motions. The majority of the rotation occurs at the tibial-articulating interface, and as a result the motion at the femoral-insert interface is predominantly unidirectional. This results in reduced cross-shear forces and, hence, reduced wear rates compared with fixed bearing TKR devices [67]. In contrast, rotation of the knee with the PFC Sigma fixed bearing occurs entirely at the femoral-insert articulation. The resulting multidirectional wear path at this interface increases the amount of cross-shear on the UHMWPE bearing and therefore produces a greater polymer wear rate when subjected to high kinematic inputs [67,68].

Consideration of the wear debris size and morphology and its resulting specific biological activity is important, and this is illustrated by the wide range of values determined for the two materials under different kinematic conditions and with different methods of sterilization. In general, the knee prostheses had lower specific biological activity values than the hip, and the lower molecular weight GUR 1020 GVF had lower values of SBA than the GUR 1050 UHMWPE in both the knee and the hip. In the hip, the SBA value for the GUR 1020 UHMWPE was 40% lower than the GUR 1050 UHMWPE, and in the fixed bearing knee it was 60% lower. This is due to the high volumetric concentrations of smaller, more biologically active wear particles generated by the GUR 1050 material in both the hip and the knee. These results were confirmed when the wear rates were taken into consideration, where the GUR 1020 GVF material had lower FBA values in the hip and knee compared to the GUR 1050 UHMWPE materials. However, the limitations of this study were that the materials chosen differed in both their molecular weight and methods of sterilization and therefore were not strictly comparable, that is, GUR 1050 25 kGy UHMWPE was studied in the hip, whereas in the knee the GUR 1050 was irradiated with 50 kGy gamma radiation.

In the studies reported here, lower wear rates were recorded in the knee for rotating platform mobile bearing knee designs and moderately crosslinked UHMWPE. The increase in the biological activity of the wear particles from crosslinked UHMWPE was predicted from analysis of debris size, volume, and morphology. Of equal importance is the observation that GUR 1020 GVF (irradiated in an inert atmosphere), which has become standard, produces wear particles with lower biological activity in both the knee and the hip compared to the GUR 1050 UHMWPEs investigated in this study. Relatively low wear rates in TKR prostheses, coupled with the low biological activity of particles produced indicates that current designs and materials could provide adequate osteolysis-free *in vivo* implant service times for the majority of patients. In the hip, the introduction of highly crosslinked GUR 1050 100 kGy UHMWPE has been shown to cause an 80% reduction in wear rate (46). Although the wear particles have a high

SBA, the reduction in wear rate is so great that the FBA value for this material is reduced almost sixfold [69]. Highly crosslinked UHMWPE has not yet been introduced clinically in the knee, and therefore the effects on wear and particle distributions remain to be determined.

## 27.16 CONCLUSION

This chapter clearly shows that there are many published methods suitable for separating and characterizing UHMWPE wear particles from tissues and simulator lubricants. Many of these methods successfully remove proteins and lipids from particles, preserve particle morphology, and yield similar particle size distributions. Clearly, advances in technology are continually being employed, for example the use of high-resolution FEG-SEM has allowed increasingly smaller particles to be observed and characterized. In addition, the limitations of each method are now widely recognized. Particle characterization, in terms of size and volume distributions and particle contribution to the osteolytic potential of new and existing UHMWPE materials, has been recognized by academic and industrial stakeholders as being an important step in the preclinical assessment of total joint replacements. However, there is no one standard method for isolation or characterization of UHMWPE particles that can be applied to both tissues and simulator lubricant samples, and this would clearly be advantageous.

## REFERENCES


1. Amstutz HC, Campbell P, Kossovsky N, Clarke IC. Mechanism and clinical significance of wear debris-induced osteolysis. *Clin Orthop Relat Res* 1992 March;**1**(276):7–18.
2. Ingham E, Fisher J. Biological reactions to wear debris in total joint replacement. *Proc Inst Mech Eng Part H, J Eng Med* 2000 January 1;**214**(1):21–37.
3. Jacobs JJ, Roebuck KA, Archibeck M, Hallab NJ, Glant TT. Osteolysis: basic science. *Clin Orthop Relat Res* 2001 December; **1**(393):71–7.
4. Goldring SR, Schiller AL, Roelke M, Rourke CM, O'Neil DA, Harris WH. The synovial-like membrane at the bone-cement interface in loose total hip replacements and its proposed role in bone lysis. *J Bone Joint Surg Am Vol* 1983 June 1;**65**(5):575–84.
5. Ingham E, Fisher J. The role of macrophages in osteolysis of total joint replacement. *Biomaterials* 2005 April 1;**26**(11):1271–86.
6. Schmalzried TP, Jasty M, Harris WH. Periprosthetic bone loss in total hip arthroplasty. Polyethylene wear debris and the concept of the effective joint space. *J Bone Joint Surg Am Vol* 1992 July 1;**74**(6):849–63.
7. Matthews JB, Besong AA, Green TR, Stone MH, Wroblewski BM, Fisher J, Ingham E. Evaluation of the response of primary human peripheral blood mononuclear phagocytes to challenge with in vitro generated clinically relevant UHMWPE particles of known size and dose. *J Biomed Mater Res* 2000 November 1;**52**(2):296–307.
8. Matthews JB, Green TR, Stone MH, Wroblewski BM, Fisher J, Ingham E. Comparison of the response of primary human peripheral blood mononuclear phagocytes from different donors to challenge with model polyethylene particles of known size and dose. *Biomaterials* 2000 October 1;**21**(20):2033–44.
9. Horowitz SM, Gonzales JB. Effects of polyethylene on macrophages. *J Orthop Res* 1997 January 1;**15**(1):50–6.
10. Shanbhag AS, Jacobs JJ, Black J, Galante JO, Glant TT. Human monocyte response to particulate biomaterials generated in vivo and in vitro. *J Orthop Res* 1995 September 1;**13**(5):792–801.
11. Willert HG. Reactions of the articular capsule to wear products of artificial joint prostheses. *J Biomed Mater Res* 1977 March 1;**11**(2):157–64.
12. Campbell P, Ma S, Yeom B, McKellop H, Schmalzried TP, Amstutz HC. Isolation of predominantly submicron-sized UHMWPE wear particles from periprosthetic tissues. *J Biomed Mater Res* 1995 January;**29**(1):127–31.
13. Tipper JL, Ingham E, Hailey JL, Besong AA, Fisher J, Wroblewski BM, Stone MH. Quantitative analysis of polyethylene wear debris, wear rate and head damage in retrieved Charnley hip prostheses. *J Mater Sci Mater Med* 2000 February;**11**(2):117–24.
14. Tipper JL, Firkins PJ, Ingham E, Fisher J, Stone MH, Farrar R. Quantitative analysis of the wear and wear debris from low and high carbon content cobalt chrome alloys used in metal on metal total hip replacements. *J Mater Sci Mater Med* 1999 June;**10**(6):353–62.
15. Tipper JL, Galvin AL, Williams S, McEwen HM, Stone MH, Ingham E, Fisher J. Isolation and characterization of UHMWPE wear particles down to ten nanometers in size from in vitro hip and knee joint simulators. *J Biomed Mater Res A* 2006 September 1;**78**(3):473–80.
16. Tipper JL, Hatton A, Nevelos JE, Ingham E, Doyle C, Streicher R, et al. Alumina-Alumina artificial hip joints. Part II: characterisation of the wear debris from in vitro hip joint simulations. *Biomaterials* 2002 August;**23**(16):3441–8.
17. Richards L, Brown C, Stone MH, Fisher J, Ingham E, Tipper JL. Identification of nanometre-sized ultra-high molecular weight polyethylene wear particles in samples retrieved in vivo. *J Bone Joint Surg Br* 2008 August;**90**(8):1106–13.
18. Shanbhag AS, Jacobs JJ, Glant TT, Gilbert JL, Black J, Galante JO. Composition and morphology of wear debris in failed uncemented total hip replacement. *J Bone Joint Surg Br* 1994 January;**76**(1):60–7.
19. Maloney WJ, Smith RL, Schmalzried TP, Chiba J, Huene D, Rubash H. Isolation and characterization of wear particles generated in patients who have had failure of a hip arthroplasty without cement. *J Bone Joint Surg Am* 1995 September;**77**(9):1301–10.
20. Ingram JH, Stone M, Fisher J, Ingham E. The influence of molecular weight, crosslinking and counterface roughness on tnf-alpha production by macrophages in response to ultra high molecular weight polyethylene particles. *Biomaterials* 2004 August;**25**(17):3511–22.
21. Green TR, Fisher J, Stone M, Wroblewski BM, Ingham E. Polyethylene particles of a 'critical size' are necessary for the induction of cytokines by macrophages in vitro. *Biomaterials* 1998, December; **19**(24):2297–302.
22. Shanbhag AS, Jacobs JJ, Black J, Galante JO, Glant TT. Macrophage/particle interactions: effect of size, composition and surface area. *J Biomed Mater Res* 1994 January;**28**(1):81–90.
23. Yao J, Glant TT, Lark MW, Mikecz K, Jacobs JJ, Hutchinson NI, et al. The potential role of fibroblasts in periprosthetic osteolysis: fibroblast response to titanium particles. *J Bone Miner Res* 1995 September;**10**(9):1417–27.

24. Maloney WJ, Smith RL, Castro F, Schurman DJ. Fibroblast response to metallic debris in vitro. Enzyme induction cell proliferation, and toxicity. *J Bone Joint Surg Am* 1993 June;**75**(6):835–44.
25. Dean DD, Schwartz Z, Blanchard CR, Liu Y, Agrawal CM, Lohmann CH, et al. Ultrahigh molecular weight polyethylene particles have direct effects on proliferation, differentiation, and local factor production of MG63 osteoblast-like cells. *J Orthop Res* 1999 January;**17**(1):9–17.
26. Vermes C, Chandrasekaran R, Jacobs JJ, Galante JO, Roebuck KA, Glant TT. The effects of particulate wear debris, cytokines, and growth factors on the functions of MG-63 osteoblasts. *J Bone Joint Surg Am* 2001 February;**83-A**(2):201–11.
27. Dowd JE, Schwendeman LJ, Macaulay W, Doyle JS, Shanbhag AS, Wilson S, et al. Aseptic loosening in uncemented total hip arthroplasty in a canine model. *Clin Orthop Relat Res* 1995 October(319):106–21.
28. Howie DW, Vernon-Roberts B, Oakeshott R, Manthey B. A rat model of resorption of bone at the cement-bone interface in the presence of polyethylene wear particles. *J Bone Joint Surg Am* 1988 February;**70**(2):257–63.
29. Clohisy JC, Hirayama T, Frazier E, Han SK, Abu-Amer Y. Nf-Kb signaling blockade abolishes implant particle-induced osteoclastogenesis. *J Orthop Res* 2004 January;**22**(1):13–20.
30. Margevicius KJ, Bauer TW, McMahon JT, Brown SA, Merritt K. Isolation and characterization of debris in membranes around total joint prostheses. *J Bone Joint Surg Am* 1994 November;**76**(11):1664–75.
31. Scott M, Widding K, Jani S. Do current wear particle isolation procedures underestimate the number of particles generated by prosthetic bearing components?. *Wear* 2001 October;**251**:1213–17.
32. Campbell P, Ma S, Schmalzried T, Amstutz HC. Tissue digestion for wear debris particle isolation. *J Biomed Mater Res* 1994 April;**28**(4):523–6.
33. Mabrey JD, Afsar-Keshmiri A, Engh GA, Sychterz CJ, Wirth MA, Rockwood CA, Agrawal CM. Standardized analysis of UHMWPE wear particles from failed total joint arthroplasties. *J Biomed Mater Res* 2002;**63**(5):475–83.
34. Elfick AP, Green SM, Pinder IM, Unsworth A. A novel technique for the detailed size characterization of wear debris. *J Mater Sci Mater Med* 2000 May;**11**(5):267–71.
35. Elfick AP, Green SM, Krikler S, Unsworth A. The nature and dissemination of UHMWPE wear debris retrieved from periprosthetic tissue of THR. *J Biomed Mater Res A* 2003 April 1;**65**(1):95–108.
36. Kinbrum A, Unsworth A, Kamali A. The wear of high carbon metal on metal bearings. *Transactions of the 54th Annual Meeting of the Orthopeadic Research Society*; 2008:1905.
37. Slouf M, Eklova S, Kumstatova J, Berger S, Synkova H, Sosna A, et al. Isolation, characterization and quantification of polyethylene wear debris from periprosthetic tissues around total joint replacements. *Wear* 2007 April 10;**262**:1171–81.
38. Baxter R, Steinbeck M, Tipper J, Rimnac C, Parvvizi J, Marcolongo M, Kurtz S. Evaluation of methods for periprosthetic tissue digestion and polyethylene wear particle analysis. *Transactions of the 54Th Annual Meeting of the Orthopaedic Research Society*; 2008:58.
39. Niedzwiecki S, Klapperich C, Short J, Jani S, Ries M, Pruitt L. Comparison of three joint simulator wear debris isolation techniques: acid digestion, base digestion, and enzyme cleavage. *J Biomed Mater Res* 2001 August;**56**(2):245–9.
40. Catelas I, Bobyn JD, Medley JB, Krygier JJ, Zukor DJ, Petit A, Huk OL. Effects of digestion protocols on the isolation and characterization of metal-metal wear particles. I. Analysis of particle size and shape. *J Biomed Mater Res* 2001 June 5;**55**(3):320–9.
41. Doorn PF, Campbell PA, Worrall J, Benya PD, McKellop HA, Amstutz HC. Metal wear particle characterization from metal on metal total hip replacements: transmission electron microscopy study of periprosthetic tissues and isolated particles. *J Biomed Mater Res* 1998 October;**42**(1):103–11.
42. Brown C, Williams S, Tipper JL, Fisher J, Ingham E. Characterisation of wear particles produced by metal on metal and ceramic on metal hip prostheses under standard and microseparation simulation. *J Mater Sci Mater Med* 2007 May;**18**(5):819–27.
43. Visentin M, Stea S, Squarzoni S, Antonietti B, Reggiani M, Toni A. A new method for isolation of polyethylene wear debris from tissue and synovial fluid. *Biomaterials* 2004 November;**25**(24):5531–7.
44. Galvin AL, Tipper JL, Jennings LM, Stone MH, Jin ZM, Ingham E, Fisher I. Wear and biological activity of highly crosslinked polyethylene in the hip under low serum protein concentrations. *Proc Inst Mech Eng [H]* 2007 January;**221**(1):1–10.
45. Besong AA, Tipper JL, Ingham E, Stone MH, Wroblewski BM, Fisher J. Quantitative comparison of wear debris from UHMWPE that has and has not been sterilised by gamma irradiation. *J Bone Joint Surg Br* 1998 March;**80**(2):340–4.
46. Galvin AL, Williams S, Hatto P, Thompson J, Isaac G, Stone M, et al. Comparison of wear of ultra high molecular weight polyethylene acetabular cups against alumina ceramic and chromium nitride coated femoral heads: 15th international conference on wear of materials. *Wear* 2005;**259**:972–6.
47. Scott M, Morrison M, Mishra SR, Jani S. Particle analysis for the determination of UHMWPE wear. *J Biomed Mater Res B Appl Biomater* 2005 May;**73**(2):325–37.
48. Gruen TA, Amstutz HC. A failed vitallium/stainless steel total hip replacement: a case report with histological and metallurgical examination. *J Biomed Mater Res* 1975 September;**9**(5):465–77.
49. Schmalzried TP, Callaghan JJ. Wear in total hip and knee replacements. *J Bone Joint Surg Am* 1999 January;**81**(1):115–36.
50. Savio JA, Overcamp LM, Black J. Size and shape of biomaterial wear debris. *Clin Mater* 1994;**15**(2):101–47.
51. McKellop HA, Campbell P, Park SH, Schmalzried TP, Grigoris P, Amstutz HC, Sarmiento A. The origin of submicron polyethylene wear debris in total hip arthroplasty. *Clin Orthop Relat Res* 1995 February(311):3–20.
52. Lapcikova M, Slouf M, Dybal J, Zolotarevova E, Entlicher G, Pokorny D, et al. Nanometer size wear debris generated from ultra high molecular weight polyethylene in vivo. *Wear* 2009 January(266):349–55.
53. Wear of implant materials, polymer and metal wear particles. Isolation, characterisation and quantification. ISO Standard 17853; 2003.
54. Standard practice for retrieval and analysis of medical devices and associated tissues and fluids. ASTM Standards Committee F561-05A; 2005.
55. Standard practice for characterisation of particles. ASTM Standards Committee F1877-05; 2005.
56. Endo M, Tipper JL, Barton DC, Stone MH, Ingham E, Fisher J. Comparison of wear, wear debris and functional biological activity of moderately crosslinked and non-crosslinked polyethylenes in hip prostheses. *Proc Inst Mech Eng [H]* 2002;**216**(2):111–22.
57. Galvin A, Kang L, Tipper J, Stone M, Ingham E, Jin Z, Fisher J. Wear of crosslinked polyethylene under different tribological conditions. *J Mater Sci Mater Med* 2006 March;**17**(3):235–43.
58. Besong AA, Tipper JL, Mathews BJ, Ingham E, Stone MH, Fisher J. The influence of lubricant on the morphology of ultra-high molecular

weight polyethylene wear debris generated in laboratory tests. *Proc Inst Mech Eng [H]* 1999;**213**(2):155–8.

59. Endo MM, Barbour PS, Barton DC, Fisher J, Tipper JL, Ingham E, Stone MH. Comparative wear and wear debris under three different counterface conditions of crosslinked and non-crosslinked ultra high molecular weight polyethylene. *Biomed Mater Eng* 2001;**11**(1):23–35.
60. Yamamoto K, Clarke IC, Masaoka T, Oonishi H, Williams PA, Good VD, Imakiire A. Microwear phenomena of ultrahigh molecular weight polyethylene cups and debris morphology related to gamma radiation dose in simulator study. *J Biomed Mater Res* 2001 July;**56**(1):65–73.
61. Fisher J, Bell J, Barbour PS, Tipper JL, Matthews JB, Besong AA, et al. A novel method for the prediction of functional biological activity of polyethylene wear debris. *Proc Inst Mech Eng [H]* 2001;**215**(2):127–32.
62. Green TR, Fisher J, Matthews JB, Stone MH, Ingham E. Effect of size and dose on bone resorption activity of macrophages by in vitro clinically relevant ultra high molecular weight polyethylene particles. *J Biomed Mater Res* 2000 September;**53**(5):490–7.
63. Barnett PI, McEwen HM, Auger DD, Stone MH, Ingham E, Fisher J. Investigation of wear of knee prostheses in a new displacement/force-controlled simulator. *Proc Inst Mech Eng [H]* 2002;**216**(1):51–61.
64. Barbour PS, Stone MH, Fisher J. A hip joint simulator study using simplified loading and motion cycles generating physiological wear paths and rates. *Proc Inst Mech Eng [H]* 1999;**213**(6):455–67.
65. Lafortune MA, Cavanagh PR, Sommer HJ, Kalenak A. Three-Dimensional kinematics of the human knee during walking. *J Biomech* 1992 April;**25**(4):347–57.
66. Fisher J, McEwen HM, Tipper JL, Galvin AL, Ingram J, Kamali A, et al. Wear, debris, and biologic activity of cross-linked polyethylene in the knee: benefits and potential concerns. *Clin Orthop Relat Res* 2004 November(428):114–19.
67. McEwen HM, Barnett PI, Bell CJ, Farrar R, Auger DD, Stone MH, Fisher J. The influence of design, materials and kinematics on the in vitro wear of total knee replacements. *J Biomech* 2005 February; **38**(2):357–65.
68. Wang A, Stark C, Dumbleton JH. Mechanistic and morphological origins of ultra-high molecular weight polyethylene wear debris in total joint replacement prostheses. *Proc Inst Mech Eng [H]* 1996; **210**(3):141–55.
69. Tipper JL, Galvin AL, Ingham E, Fisher J. Estimation of the osteolytic potential of noncrosslinked and crosslinked polyethylenes and ceramic total hip prostheses. *J Am Soc Test Mater Int* 2006;**3**:1–17.

Chapter 28

# Clinical Surveillance of UHMWPE Using Radiographic Methods

Charles R. Bragdon, PhD



## 28.1 INTRODUCTION

As recounted in Chapter 4, ultra-high molecular weight polyethylene (UHMWPE) has been the primary bearing material used in total hip arthroplasty since its introduction by Charnley in 1962 [1]. UHMWPE is used to create the acetabular component of the reconstruction and in most cases is coupled with a femoral head made of a cobalt chromium alloy, though other materials such as ceramic femoral heads have also been used [2–5]. The original UHMWPE acetabular components were cemented into the acetabular bed of the pelvis. This design is still in use today along with a modular design consisting of a metal shell, which can be fixed to the skeleton with either bone cement or by biological fixation [6–8].

Wear of the articular surfaces has been a concern throughout the inception of joint arthroplasty. The early work by Charnley [9] using PTFE as the bearing surface in the acetabular component was a failure due to the extremely high rate of wear of this material *in vivo* (Chapter 4). It is now understood that periprosthetic osteolysis was associated with this high wear material but was not recognized or investigated at that time. However, periprosthetic osteolysis continued to occur after the bearing surface was changed to UHMWPE but was believed to be either associated with sepsis or changes in the mechanical environment [9, 10]. It is now well established that periprosthetic osteolysis is secondary to particulate debris liberated from the implants [11–18]. At first, it was assumed that this process was solely related to the acrylic cement used for implant fixation, so called "cement disease." However, later investigations indicated that the process could occur in the absence of polymethylmethacrylate and that a number of different types of particles, if of the appropriate size and number, could elicit an osteolytic reaction [19–31]. Though particle-induced osteolysis can be caused by many different particles associated with the total hip reconstruction (various metals, bone cement, UHMWPE, and ceramics), due to its prevalence, UHMWPE wear debris is the primary cause of particle-induced periprosthetic osteolysis around the contemporary THR [32–37]. Periprosthetic osteolysis does not occur in all THR patients, and the response of individual patients having comparable amounts of UHMWPE wear can be quite variable. Current investigations are underway to better understand this variable response. When it does occur, periprosthetic osteolysis can result in a dramatic loss of bone, failure of implant fixation, and the eventual need for revision surgery (Figure 28.1).

Taking advantage of improvements in computer technology, a number of computer-assisted techniques have been developed for making radiographic measurements [46–53]. While these techniques use a variety of image analysis and computational techniques, they are all designed to more quickly and reliably measure the changes in femoral head penetration into the acetabular component from clinical radiographs. Few, and to some extent, inconsistent results exist in the literature concerning accuracy and precision of the various contemporary methods of performing radiographic measurements of femoral head displacement [46, 54–58]. One reason for this is the continuing development of the methods associated with the introduction of new and improved calibration instruments, new mathematical algorithms, and the recent introduction of digital radiographic images [59].

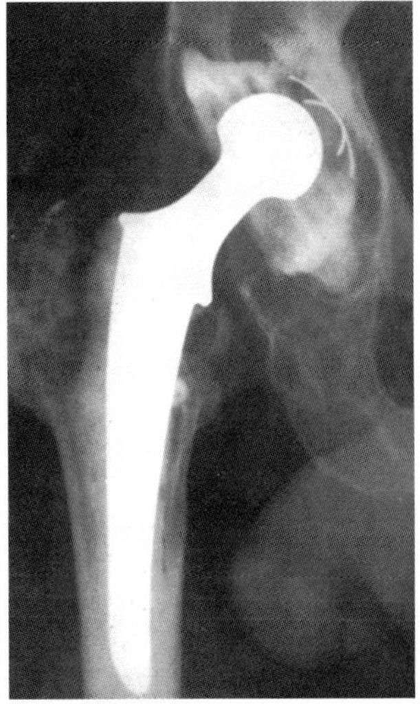
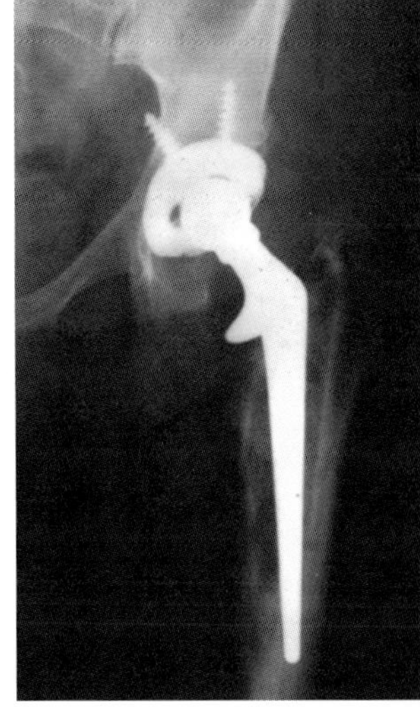

**FIGURE 28.1** Periprosthetic osteolysis, caused by UHMWPE wear debris, can result in massive bone loss around the acetabular and femoral components, necessitating difficult revision arthroplasty.

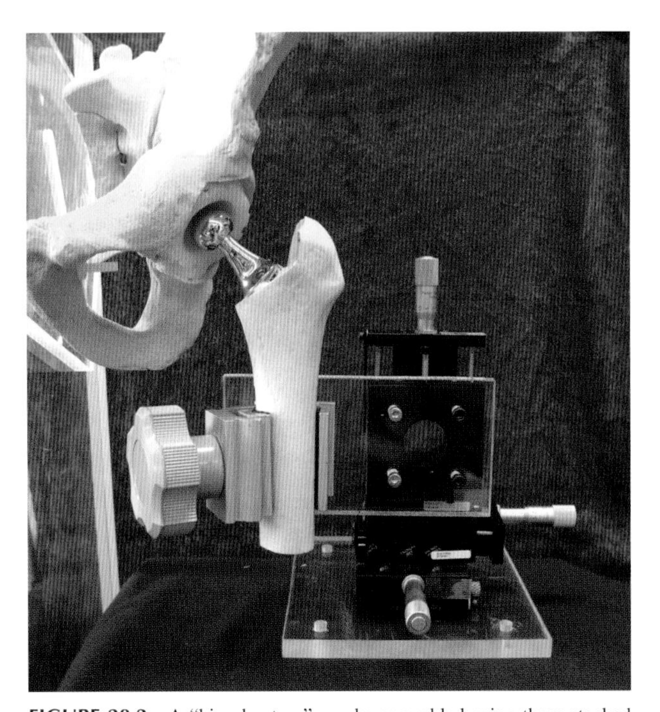

**FIGURE 28.2** A "hip phantom" can be assembled using three stacked linear micrometers (Edmund Optics, Barrington, New Jersey, USA, PN: NT 37-982) and a Plexiglas frame. Pelvic and femoral sawbone models (Pacific Research Laboratories, Vashon, Washington, USA) are used to hold the total hip replacement components in an anatomical position. The ID of the socket has been enlarged to enable displacement of the femoral head.

Bragdon et al. developed a hip phantom capable of simulating the magnitude and direction of femoral head penetration over time [60]. This physical model was useful in evaluating various parameters that affect wear measurements as well as in comparing the accuracy and precision of difference software packages [60–62]. In addition to evaluating different methodologies, a phantom model is used for training and validating new users of a system. A similar phantom can be fabricated using readily available materials (Figure 28.2).

While there are many computer-assisted methods available for measuring femoral head penetration from sequential radiographs, these methods can be divided into two general categories. Radiostereometric analysis (RSA) is a specialized technique that requires a specific type of radiograph with the use of additional methods to perform precise measurements. All other measurement techniques currently employed use standard radiograph projections of the hip and are extensions of the early manual techniques. Each category of techniques has advantages and disadvantages, which will be discussed in this chapter.

## 28.2 EARLY MANUAL METHODS FOR RADIOGRAPHIC MEASUREMENT

Since Charnley's early experience with using PTFE as a bearing material, the importance of measuring wear *in vivo* over time has been recognized. In fact, Charnley placed a wire marker in his first UHMWPE components for this exact purpose [38] (Chapter 5, Figure 5.4). The measurement of femoral head penetration into the polyethylene liner of total hip replacements over time from sequential radiographs has become a standard way of assessing wear of the UHMWPE liner over time. The earlier manual techniques [39–43] relied on the visual determination of the border of the femoral head, the visual determination of the edge of the UHMWPE on the radiograph, and the manual measurement of the changes in the position of the center of the femoral head in relation to the acetabular component using a ruler, digital micrometer, or template overlays. Though these techniques have been used successfully in studies of long duration where there has been a relatively large displacement of the femoral head, these techniques can result in a high variability among different users, and they lack the precision to yield useful information in shorter *in vivo* time periods or in cases where relatively small amounts of femoral head displacement has occurred [44]. An overview of the history of these manual techniques was recently presented by McCalden et al. [45].

## 28.3 RADIOSTEREOMETRIC ANALYSIS

Radiostereometric analysis (RSA), developed by Selvik et al. [51], is considered to be the most accurate method of determining the magnitude of relative displacements from radiographic images for a variety of applications. In its original form, this technique requires that a number of small tantalum beads be placed in each region of interest so that each component can be considered mathematically as a rigid body segment. Tantalum beads of 0.8 or 1.0 mm are commonly used due to their high radio density and biocompatibility. A pair of stereo radiographs is taken with the patient positioned in front of a calibration cage at the center of intersection of the two X-ray beams. This allows for the reconstruction of a three-dimensional coordinate system (Figure 28.3). Using automated image analysis methods, relative displacements of two rigid bodies can be calculated from sequential pairs of stereo radiographs. Among the many applications of this technique, this method has been used to evaluate growth plate integrity, joint kinematics, implant stability, spinal fusion integrity, and wear [63–70].

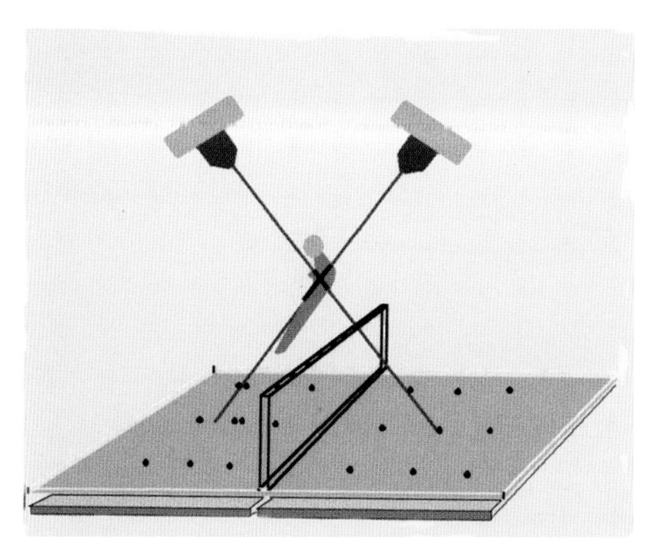

**FIGURE 28.3** The RSA calibration cage contains two groups of tantalum beads, which are used to create a three-dimensional coordinate system. The hip is placed at the intersection of the two X-ray beams. The three-dimensional location of the *in vivo* markers and the center of the femoral head can then be calculated (Drawing courtesy of Dr. Johan Kärrholm).

To measure femoral head displacement into a UHMWPE acetabular component, the penetration of the center of the femoral head into the UHMWPE is calculated relative to a group of tantalum beads that have been inserted into the acetabular component (Figure 28.4).

However, the requirement to have tantalum beads placed *in vivo* limits the use of this method to a relatively small group of patients. Several commercially available software packages have been developed to facilitate the RSA analysis. UmRSA [55], RSA-CMS [53], and the Oxford system [48] are the most developed. Though they are all based on the same mathematical principles, they vary in the level of

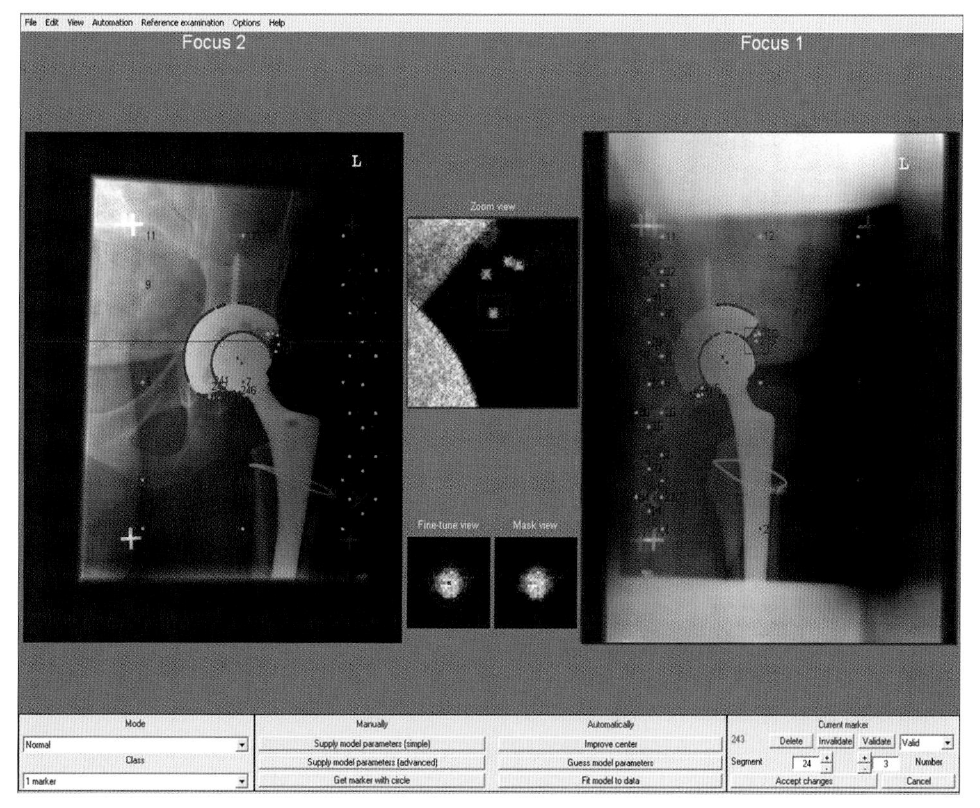


**FIGURE 28.4** An example of a pair of stereo radiographs of a hip with the beads of the calibration cage and the UHMWPE marked in green (UmRSA v 6.0, RSA Biomedical, Sweden). The displacement of the center of the femoral head over time is measured against the group of tantalum markers placed in the acetabular component at the time of surgery.

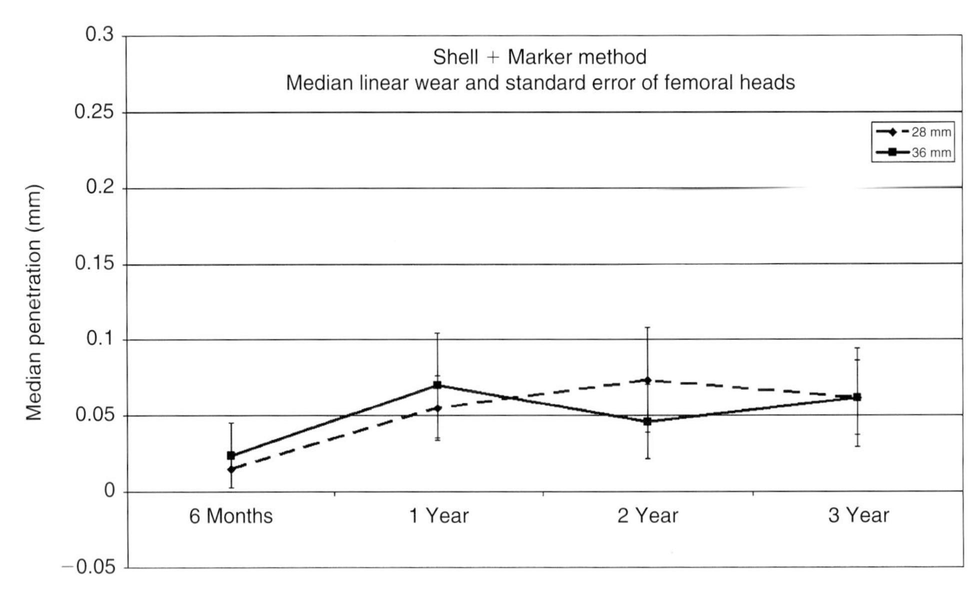


**FIGURE 28.5** Graph of the median superior displacement of the femoral head relative to the acetabular bead segment over time for patients having primary THR using highly crosslinked UHMWPE and either 28 or 36 mm diameter femoral heads. The early penetration up to the first year represents predominantly creep of the UHMWPE. Thereafter, there is no further statistically significant migration of the femoral head, indicating negligible wear of the bearing surface. (Reprinted from J Arthroplasty, 22:125-9, 2007 with permission from Elsevier [73].)

automation, algorithms used for automatic edge detection, and the resolution of the images that can be analyzed.

RSA methods are being developed to eliminate the need for tantalum markers in the UHMWPE either by using CAD/CAM models of the implants provided by the manufacturer or by using edge detection of the outer curvature of the acetabular component [71, 72]. Bragdon et al., using the UmRSA software, have shown in a comparative analysis using the same clinical films that similar results were obtained by either using the tantalum markers in the UHMWPE or by just using the edge detection of the acetabular component. However, the variance of the data was the lowest when the information of the bead position and the edge detection was combined [73].

The precision of the RSA method is dependant on the number of tantalum beads, the degree to which they are distributed three dimensionally, as well as their relative distance to the other object or segment. While in theory each patient would have a unique measurement precision, in reality good precision can be achieved by maximizing the number of beads (a minimum of three) and by having them sufficiently disbursed so they do not fall in a single line or plane. The precision of an RSA study can be determined by the use of a double examination where two pairs of films are acquired on the same day [74]. The measured motion between these double examinations represents the error. A standard deviation of the error of the double examinations for all patients in a study can then be used to calculate a 95% or 99% confidence interval, which represents the precision of the study group.

For RSA studies, femoral head penetration data is usually presented in a line graph format plotting penetration versus time (Figure 28.5). Due to the relatively small number of patients enrolled in an RSA study, the median femoral head penetration and standard error are commonly reported, and appropriate statistical methods of comparison, such as the Mann-Whitney test, are used.

## 28.4 NON-RSA METHODS

The two most commonly used non-RSA methods for determining femoral head penetration from clinical radiographs are the Hip Analysis Suite and PolyWare software packages, due to the fact that they have the most well-developed programs and documentation. While other software programs have been developed for this purpose, a description of the approach taken by these two techniques provides an overview of these non-RSA techniques. Neither the Hip Analysis Suite nor PolyWare require either *in vivo* markers or a calibration cage because they use conventional A/P and lateral projections of the hip, and therefore they can be applied to a larger group of THR patients.

### 28.4.1 Hip Analysis Suite

The Hip Analysis Suite is a software package developed by Martell and introduced in 1997 [49] (Figure 28.6). This method is commonly referred to as the Martell method. Precision of this technique under ideal experimental conditions has been shown to be ±25 microns with a 95% confidence interval [62]. While not as accurate as the RSA methods, the long-term femoral head penetration measurements using this technique in a series of THR patients

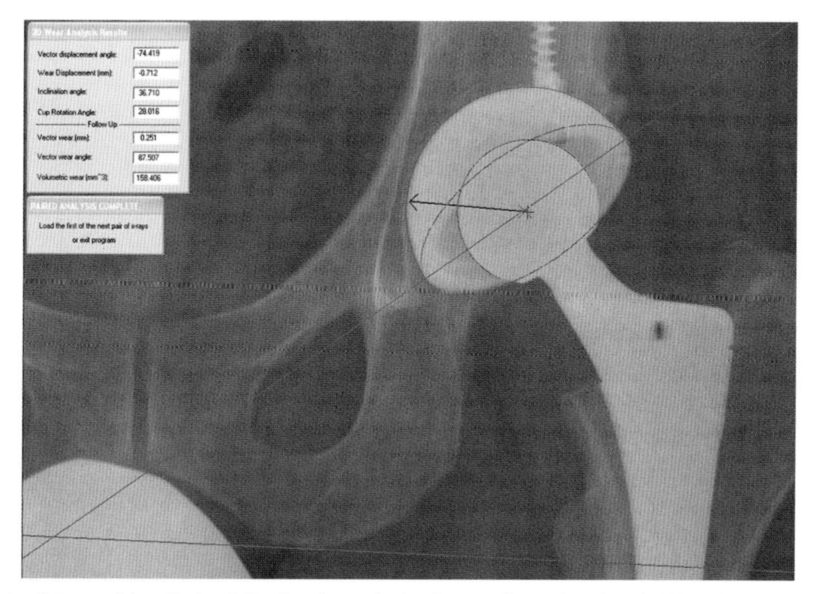


**FIGURE 28.6** An example of the resulting display following the analysis of two radiographs using the Martell method indicates the magnitude and direction of the resulting displacement vector. Postprocessing macros supplied by Dr. John Martell facilitate tabulation and graphing of a large number of penetration measurements.

having conventional UHMWPE components has shown similar trends in the pattern of femoral head penetration to an RSA analysis, though the actual magnitude of penetration measured by the two methods was different [75]. Hip Analysis Suite can be used to analyze two-dimensional or three-dimensional changes in the femoral head center relative to the acetabular component using either A/P or A/P and lateral radiographs. The sign convention used for the vector direction is based on the original manual method of Livermore [41]. Because the femoral head penetration is reported as a vector having both magnitude and direction, negative values of femoral head penetration can result. In studies of conventional UHMWPE, negative values most commonly occur when measuring short intervals of time when relatively little true penetration has occurred [76]. When measuring the femoral head penetration of new wear-resistant, highly crosslinked UHMWPE components, positive and negative values are common, even beyond 5 years postoperatively [76–78] (Figure 28.7). Early negative femoral head penetration values can also occur due to postsurgical muscle laxity, which allows the femoral head to sit slightly out of the full depth of the acetabular recess when the patient is lying supine. Having the patient internally rotate his or her legs while lying supine can minimize this subluxation of the hip. In addition, it is advisable to eliminate early postoperative films from the analysis, using films taken more than 6 weeks after surgery and with the time interval between radiographs greater than 6 months. With measurement of highly crosslinked UHMWPE components, the persistence of negative femoral head penetration values represents the limitation of the measurement technique in discerning very small changes in the position of the femoral head over time and indicates the persistence of the low wear of the bearing material *in vivo*.

Reliable femoral head penetration measurements are extremely dependant on film quality. Underexposed and overexposed radiographs should not be included in film series because the edge detection algorithm will be unable to track the true edge of the components. While prospective studies of femoral head penetration allow for attempts to improve the consistency of radiographic quality, many studies that used these techniques are retrospective studies in which image analysis was not originally contemplated. However, in either case, it is likely that some radiographs will not be appropriate for analysis. As a general rule, films where less than two-thirds of the outline of the femoral head can be seen by visual inspection should be excluded. Similarly, radiographs having low contract between the bone and acetabular shell should be excluded.

The position of the patient in relation to the film plate can also create projection artifacts that can adversely affect the femoral head penetration measurements. The ideal position of the patient on the radiograph results in the center of the femoral head lying in the middle third of the image. When the patient is positioned high or low in relation to the center of the X-ray source, elliptical distortion occurs, which compromises the circle-fitting algorithms. While elliptical fitting can be performed, its use has not been shown to improve the quality of the data over circle fitting. For this reason, as a general rule, films where the center of the femoral head falls outside of the middle third of the image should not be used.

While cross-table lateral films are useful for clinical evaluation, the quality of the images are often not ideal for

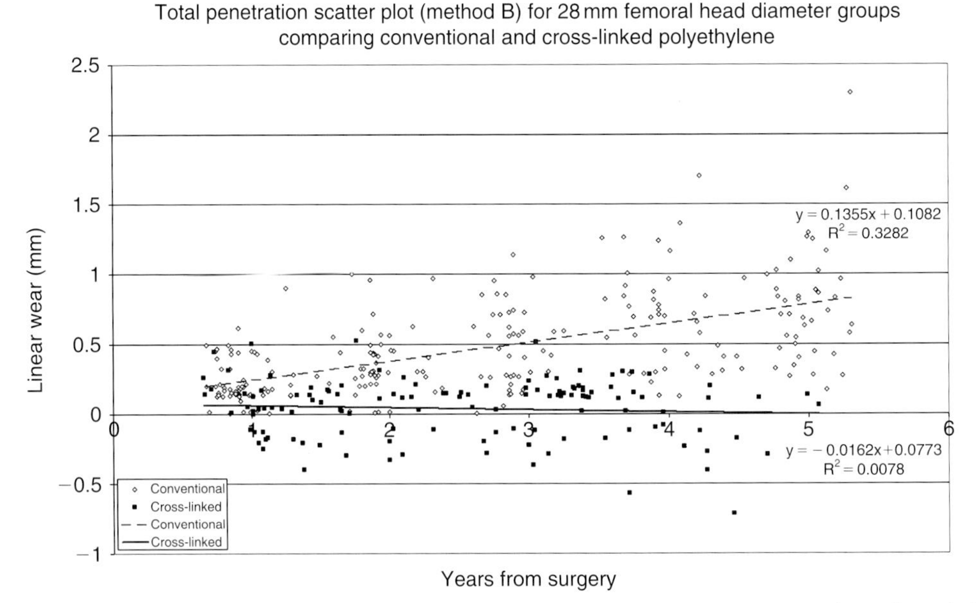


**FIGURE 28.7** A scatter plot of all femoral head penetration measurements made relative to the early postoperative radiograph for two groups of patients, one having a primary THR using conventional UHMWPE and the other using highly crosslinked UHMWPE. The linear regression line represents the femoral head penetration rates. While both groups have data distributed evenly on either side of the regression line, negative values persist throughout the follow-up period due to the fact that very little penetration occurs. There is no correlation between femoral head penetration and time using this form of highly crosslinked UHMWPE. (Reprinted from J Arthroplasty, 21:935-943, 2006 with permission from Elsevier [76].)

visualizing the borders of the femoral head and acetabular shell. Martell et al. found that because the majority of the femoral head penetration occurs superiorly and medially, which is in the plane of the A/P projection, the 3-D determination of femoral head penetration was comparable to the 2-D penetration measurements with the associated disadvantage of increased scatter of the data due to the use of the cross-table lateral projection [58].

As described by Bragdon et al. [76], there are several ways to display and analyze femoral head penetration data when a series of patients have multiple radiographs for analysis over time. One way is to report the average of the total femoral head penetration by using the earliest postoperative and the latest follow-up film. While ignoring all changes that occur in the intervening films, this method provides a snapshot of the femoral head penetration at a single time, and comparisons can be made using a student's t-test. Penetration values must be normalized by dividing each value by the total time between films to arrive at an average femoral head penetration rate. While this is a valid approach for reporting penetration data and is useful in patient series where intervening radiographs are not available, it has several limitations. The results for each patient are dependant on one observation so that image quality is even more important. Also, it does not allow monitoring of changes in the penetration rate over time. As shown in Figure 28.7, when multiple films for each patient are available for analysis, multiple observations over time for each patient can be made based on an appropriate early postoperative baseline film. Femoral head penetration rates for a population or for an individual patient can be calculated by a linear regression analysis of a scatter plot of the data. By taking multiple observations for each patient, the effect of the variability of film quality is minimized. In addition, by using different films within the series as the baseline film, changes in the wear behavior of the bearing surface over time, which may not be otherwise appreciated, may be detected.

## 28.4.2 PolyWare

Similar in principle to the Hip Analysis System, the PolyWare software package was developed by Peter Devane and was presented in 1994 in the Frank Stinchfield Award Paper [46, 79]. This method is commonly referred to as the Devane method. It was designed specifically for measuring three-dimensional changes in the position of the center of the femoral head relative to the center of the acetabular component by using serial radiographs of both the anterior/posterior (A/P) and lateral radiographic views of the reconstructed hip. Edge detection algorithms are used to to determine the centers of the femoral head and acetabular component. Correction for magnification is made by using the known diameter of the femoral head. There are several versions of this software available with the newer version, PolyWare Auto version 6.01, requiring only the A/P radiograph and minimal user interaction. This is made possible by incorporating known dimensions of the acetabular implants within a reference table of the software. This software reports the magnitude of vector wear but not direction, essentially the absolute value of the vector displacement. Greene et al. [80] performed a comparative study between PolyWare Auto and Hip Analysis Suite software results using a group of clinical radiographs of THR patients having highly crosslinked UHMWPE acetabular components. While the formats of the output values are dissimilar, they were able to perform direct statistical comparisons by comparing the absolute values of the Martell method to the output of the Devane method. In addition, they converted the output of the Devane method to a vector output having magnitude and direction based on the sign convention used by Martell [49]. They report significant differences in the femoral head migration output values. A similar study using the most recent versions of both programs compared the head penetration vector in terms of absolute value of the magnitude of the vector as well as the magnitude with its directional component of sign value. This comparison showed no significant differences between the wear and wear rates observed by the programs when the sign convention was included, but it showed a significant difference when the absolute value was taken. These important differences in the data output of these two programs as well as other programs must be considered when attempting to compare femoral head penetration results from different studies.

## 28.5 OTHER FACTORS TO CONSIDER

Femoral head penetration into UHMWPE acetabular components is a result of both removal of UHMWPE from the bearing surface, true wear, and creep of the UHMWPE by plastic and elastic deformation. It is generally understood that creep occurs early as a result of loading of the articulation and is not a concern after the first year of *in vivo* use [73, 76, 81–88]. As seen in Figure 28.5, creep seems to account for roughly 0.1 mm of femoral head penetration in the first year. In an attempt to eliminate the component of creep for femoral head displacement measurements, some studies have ignored the early radiographs and started the analysis using a film after the first postoperative year as the baseline film [76–78]. While in theory creep does not stop, the rate of femoral head penetration due to creep decreases substantially after the first year to a nearly undetectable level. It may be, however, that this continued creep could accumulate over time to an amount that is discernable in long-term studies that have a sufficient number of follow-up radiographs. It may be that this accumulative effect of continued creep may be more apparent in studies of new, low wear UHMWPE. For this reason, the affect of continued creep on the results of long-term clinical studies of femoral head displacement needs to be considered.

A key assumption of any femoral head penetration measurement is that the femoral head is seated into the deepest portion of the acetabular component at each radiographic examination. However, subluxation of the hip does occur under certain circumstances and is affected by the position of the leg and the degree of muscle tone. Several studies have been conducted to determine if radiographs taken while the patient is standing would significantly improve the resulting data but have reached conflicting conclusions [89–91]. Bragdon et al. performed a comparative study on a large group of patients who had both standing and supine radiographs taken on the same day as part of an RSA study. As part of the radiographic protocol, all patients were asked to slightly internally rotate their legs during the supine examination. The use of this technique increases the muscle load across the hip and, in the majority of cases studied in their paper, seemed to have helped minimize the difference between the results of using standing or supine radiographs. However, in a number of patients, the difference in the position of the femoral head between a standing and a supine examination that were taken on the same day was greater than could be explained by measurement error alone. It is likely that some of the patient-to-patient variability that is reported in such radiographic studies is due in part to the variability in seating of the femoral head during the supine examination. From the analysis of the 117 hips in their paper, the use of supine radiographs in an RSA evaluation of wear appeared to be a reasonable approach provided that the number of patients enrolled in the study is sufficient to accommodate the added patient-to-patient variability and that care is taken during the examination to ensure that this source of variation was minimized.

The techniques for measuring femoral head penetration from radiographs described here are primarily used in evaluating femoral head penetration in total hip replacements that use UHMWPE as the acetabular bearing surface articulating against a hard cobalt chrome or ceramic femoral head. While wear is also a concern in other bearing couples, such as metal on metal and ceramic on ceramic, these alternative bearings are not ideal candidates for these wear measurement techniques. Not only does the radiodensity of these materials create difficulties in visualizing the borders of the femoral heads, but substantial wear can occur without significant displacement of the femoral head. While wear resulting in 0.2 to 2.0 mm of wear can often occur with conventional UHMWPE bearings with no significant impact on function, substantial wear of hard-on-hard bearings results in an early failure of the reconstruction, necessitating revision surgery long before this wear can be visualized on plain radiographs. The utility of using these measurement techniques on long-term studies of hard-on-hard bearings has not yet been demonstrated.

The goal of using radiographic analysis techniques to measure wear in total hip arthroplasty has changed over time. Originally, they were used to correlate the amount of femoral head penetration with the incidence of periprosthetic osteolysis. This association is now well established and irrefutable. New bearing materials, such as hard-on-hard and highly crosslinked UHMWPEs, have been developed and are currently in use to minimize articular wear. These radiographic analysis techniques are now being used to monitor the *in vivo* wear performance of these new bearings, primarily highly crosslinked UHMWPE, as a form of an early warning method to ensure that the low wear performance demonstrated in *in vitro* testing translates to the clinical setting and that these low wear properties do not degrade over time. The year 2008 will mark the first decade since the introduction of highly crosslinked UHMWPE in total hip arthroplasty. While long-term studies continue to monitor patients with these existing highly crosslinked UHMWPEs, new formulations continue to be developed and introduced to the market, making the continued use of these techniques an important part of clinical follow-up studies. However, the desired end result of low-wear bearings is the demonstrable decrease, if not elimination, of periprosthetic osteolysis. Several recent studies have reported on the use of CT scans as a means for early detection of osteolysis before it is manifested on plain radiographs [92–98]. With the continued advances in imaging technology and metal artifact suppression, it is likely that CT or another type of imaging technique will be added as a tool for the clinical surveillance of UHMWPE. These techniques will be useful not only for evaluating total hip replacements, but also other joints such as total knee replacements where no reliable techniques for evaluating wear are currently available.

## REFERENCES


1. Waugh W, Charnley J. *The man and the hip*. New York: Springer; 1990.
2. D'Antonio J, Capello W, Manley M. Alumina ceramic bearings for total hip arthroplasty. *Orthopedics* 2003;**26**(1):39–46.
3. Tanaka K, Tamura J, Kawanabe K, Shimizu M, Nakamura T. Effect of alumina femoral heads on polyethylene wear in cemented total hip arthroplasty. Old versus current alumina. *J Bone Joint Surg Br* 2003;**85**(5):655–60.
4. Inzerillo VC, Garino JP. Alternative bearing surfaces in total hip arthroplasty. *J South Orthop Assoc* 2003;**12**(2):106–11.
5. Urban JA, Garvin KL, Boese CK, Bryson L, Pedersen DR, Callaghan JJ, Miller RK. Ceramic-on-polyethylene bearing surfaces in total hip arthroplasty. Seventeen to twenty-one-year results. *J Bone Joint Surg—Am Vol* 2001;**83-A**(11):1688–94.
6. Callaghan JJ, Kim YS, Brown TD, Pedersen DR, Johnston RC. Concerns and improvements with cementless metal-backed acetabular components. *Clin Orthop Relat Res* 1995(311):76–84.
7. Chmell MJ, Rispler D, Poss R. The impact of modularity in total hip arthroplasty. *Clin Orthop Relat Res* 1995(319):77–84.
8. Torchia ME, Klassen RA, Bianco AJ. Total hip arthroplasty with cement in patients less than twenty years old. Long-term results. *J Bone Joint Surg (Am)* 1996;**78**(7):995–1003.
9. Charnley J, Follacci F, Hammond B. The long-term reaction of bone to self-curing acrylic cement. *J Bone Joint Surg* 1968;**50B**:822–9.
10. McKee G, Watson-Farrar J. Replacement of arthritic hips by the McKee-Farrar prosthesis. *J Bone Joint Surg* 1966;**48-B**:245–59.
11. Harris W, Schiller A, Scholler J, Freiberg R, Scott R. Extensive localized bone resorption in the femur following total hip replacement. *J Bone Joint Surg* 1976;**58A**:612–18.
12. Willert H. Reactions of the articular capsule to wear products of artificial joint prostheses. *J Biomed Mater Res* 1977;**11**:157–64.
13. Goldring S, Jasty M, Paiement G. Tissue response to bulk and particulate biopolymers in a rabbit wound chamber model. *Ortho Trans*; 1986. p. 288.
14. Goldring SR, Schiller AL, Roelke M, Rourke CM, O'Neill DA, Harris WH. The synovial-like membrane at the bone-cement interface in loose total hip replacements and its proposed role in bone lysis. *J Bone Joint Surg—Am Vol* 1983;**65**(5):575–84.
15. Glowacki J, Jasty M, Goldring S. Comparison of multinucleated cells elicited in rats by particulate bone, polyethylene, or polymethylmethacrylate. *J Bone Miner Res* 1986;**1**:327–31.
16. Goldring S, Jasty M, Roelke M, Rourke C, Bringhurst F, Harris W. Formation of a synovial-like membrane at the bone-cement interface. Its role in bone resorption and implant loosening after total hip replacement. *Arthritis Rheum* 1986;**29**:836–42.
17. Goldring M, Goldring S. Skeletal tissue response to cytokines. *Clin Orthop* 1990;**258**:245–78.
18. Jiranek WA, Machado M, Jasty M, Jevsevar D, Wolfe HJ, Goldring SR, Goldberg MJ, Harris WH. Production of cytokines around loosened cemented acetabular components. Analysis with immunohistochemical techniques and in situ hybridization [see comments]. *J Bone Joint Surg—Am Vol* 1993;**75**(6):863–79.
19. Amstutz H, Campbell P, Kossovsky N, Clarke I. Mechanism and clinical significance of wear debris-induced osteolysis. *Clin Orthop* 1992;**276**:7–18.
20. Bal BS, Jiranek WA, Jasty M, Harris WH. Periprosthetic femoral osteolysis around an uncemented nonmodular Moore prosthesis. *J Arthroplasty* 1997;**12**(3):346–9.
21. Bauer TW. Severe osteolysis after third-body wear due to hydroxyapatite particles from acetabular cup coating. [comment]. *J Bone Joint Surg—Br Vol* 1998;**80**(4):745.
22. Borssen B, Karrholm J, Snorrason F. Osteolysis after ceramic-on-ceramic hip arthroplasty. A case report. *Acta Orthop Scand* 1991;**62**:73–5.
23. Brown IW, Ring PA. Osteolytic changes in the upper femoral shaft following porous-coated hip replacement. *J Bone Joint Surg—Br Vol* 1985;**67**(2):218–21.
24. Callaghan JJ. Failure mechanisms of hip arthroplasty. *Orthopedics* 1995;**18**(9):849–50.
25. Doorn PF, Mirra JM, Campbell PA, Amstutz HC. Tissue reaction to metal on metal total hip prostheses. *Clin Orthop Relat Res* 1996(329 Suppl.):S187–205.
26. Harris W. Osteolysis and particle disease in hip replacement. A review. *Acta Orthop Scand* 1994;**65**:113–23.
27. Harris W. The problem is osteolysis. *J Biomed Mater Res* 1996;**31**:19–26.
28. Jasty M, Floyd WEI, Schiller AL, Goldring SR, Harris WH. Localized osteolysis in stable, non-septic total hip replacement. *J Bone and Joint Surg* 1986;**68A**:912–19.
29. Jasty M, Smith E. Wear particles of total joint replacements and their role in periprosthetic osteolysis. *Curr Opin Rheumatol* 1992;**4**:204–9.
30. Maloney WJ, Jasty M, Rosenberg A, Harris WH. Bone lysis in well-fixed cemented femoral components. *J Bone Joint Surg—Br Vol* 1990;**72**(6):966–70.
31. Wroblewski BM. Wear and loosening of the socket in the Charnley low-friction arthroplasty. *Orthop Clin North Am* 1988;**19**:627–30.

32. Schmalzried T, Jasty M, Rosenberg A, Harris W. Histologic identification of intracellular polyethylene wear debris using oil red O stain. *J Appl Biomater* 1993;**4**:119–25.
33. Jasty M, Bragdon C, Jiranek W, Chandler H, Maloney W, Harris WH. Etiology of osteolysis around porous-coated cementless total hip arthroplasties. *Clin Orthop Relat Res* 1994(308):111–26.
34. Schmalzried T, Jasty M, Rosenberg A, Harris W. Polyethylene wear debris and tissue reactions in knee as compared to hip replacement prostheses. *J Appl Biomater* 1994;**5**:185–90.
35. McKellop HA, Campbell P, Park SH, Schmalzried TP, Grigoris P, Amstutz HC, Sarmiento A. The origin of submicron polyethylene wear debris in total hip arthroplasty. *Clin Orthop Relat Res* 1995(311):3–20.
36. Bragdon C, O'Connor D, Lowenstein J, Jasty M, Syniuta W. The importance of multidirectional motion on the wear of polyethylene. *Proc Inst Mech Eng [H]* 1996;**210**:157–65.
37. Jasty M, Goetz DD, Bragdon CR, Lee KR, Hanson AE, Elder JR, Harris WH. Wear of polyethylene acetabular components in total hip arthroplasty. An analysis of one hundred and twenty-eight components retrieved at autopsy or revision operations. *J Bone Joint Surg—Am Vol* 1997;**79**(3):349–58.
38. Charnley J. Total hip replacement by low-friction arthroplasty. *Clin Orthop* 1970;**72**:7–21.
39. Charnley J, Cupic Z. The nine and ten year results of the low-friction arthroplasty of the hip. *Clin Orthop* 1973;**95**:9–25.
40. Charnley J, Halley DK. Rate of wear in total hip replacement. *Clin Orthop Relat Res* 1975(112):170–9.
41. Livermore J, Ilstrup D, Morrey B. Effect of femoral head size on wear of the polyethylene acetabular component. *J Bone Joint Surg* 1990;**72-A**:518–28.
42. Kang JS, Park SR, Ebramzadeh E, Dorr L. Measurement of polyethylene wear in total hip arthroplasty—accuracy versus ease of use. *Yonsei Med J* 2003;**44**(3):473–8.
43. Pollock Sr. D, Sychterz CJ, Engh CA. A clinically practical method of manually assessing polyethylene liner thickness. *J Bone Joint Surg—Am Vol* 2001;**83-A**(12):1803–9.
44. Clarke IC, Gruen T, Matos M, Amstutz HC. Improved methods for quantitative radiographic evaluation with particular reference to total-hip arthroplasty. *Clin Orthop Relat Res* 1976(121):83–91.
45. McCalden RW, Naudie DD, Yuan X, Bourne RB. Radiographic methods for the assessment of polyethylene wear after total hip arthroplasty. *J Bone Joint Surg Am* 2005;**87**(10):2323–34.
46. Devane PA, Bourne RB, Rorabeck CH, Hardie RM, Horne JG. Measurement of polyethylene wear in metal-backed acetabular cups. I. Three-dimensional technique. *Clin Orthop Relat Res* 1995(319):303–16.
47. Hardinge K, Porter ML, Jones PR, Hukins DW, Taylor CJ. Measurement of hip prostheses using image analysis. The maxima hip technique. *J Bone Joint Surg—Br Vol* 1991;**73**(5):724–8.
48. Kiss J, Murray DW, Turner-Smith AR, Bulstrode CJ. Roentgen stereophotogrammetric analysis for assessing migration of total hip replacement femoral components. *Proc Inst Mech Eng. Part H-J Eng Med* 1995;**209**(3):169–75.
49. Martell JM, Berdia S. Determination of polyethylene wear in total hip replacements with use of digital radiographs. *J Bone Joint Surg—Am Vol* 1997;**79**(11):1635–41.
50. Pedersen DR, Brown TD, Hillis SL, Callaghan JJ. Prediction of long-term polyethylene wear in total hip arthroplasty, based on early wear measurements made using digital image analysis. *J Orthop Res* 1998;**16**(5):557–63.
51. Selvik G. A stereophotogrammetric system for the study of human movements. *Scand-J-Rehabil-Med-Suppl* 1978;**6**:16–20.
52. Shaver SM, Brown TD, Hillis SL, Callaghan JJ. Digital edge-detection measurement of polyethylene wear after total hip arthroplasty. *J Bone Joint Surg—Am Vol* 1997;**79**(5):690–700.
53. Vrooman HA, Valstar ER, Brand GJ, Admiraal DR, Rozing PM, Reiber JH. Fast and accurate automated measurements in digitized stereophotogrammetric radiographs. *J-Biomech* 1998;**31**(5):491–8.
54. Alfaro-Adrian J, Gill HS, Murray DW. Cement migration after THR. A comparison of charnley elite and exeter femoral stems using RSA. *J Bone Joint Surg—Br Vol* 1999;**81**(1):130–4.
55. Borlin N, Thien T, Karrholm J. The precision of radiostereometric measurements. Manual vs. digital measurements. *J Biomech* 2002; **35**(1):69–79.
56. Kärrholm J, Herberts P, Hultmark P, Malchau H, Nivbrant B, Thanner J. Radiostereometry of hip prostheses. Review of methodology and clinical results. *Clin Orthop* 1997(344):94–110.
57. Valstar ER. *Digital roentgen stereophotogrammetry. Development, validation, and clinical application*. Thesis, Leiden, Holland: University of Leiden; 2001. ISBN 90-9014397-1, NUGI 743.
58. Martell JM, Berkson E, Berger R, Jacobs J. Comparison of two and three-dimensional computerized polyethylene wear analysis after total hip arthroplasty. *J Bone Joint Surg—Am Vol* 2003;**85-A**(6):1111–17.
59. Börlin N. *Model-based Measurements in Digital Radiographs*. PhD Thesis, Umeå University, Sweden: Department of Computing Science; 2000. (ISBN91-7191-843-4).
60. Bragdon CR, Malchau H, Yuan X, Perinchief R, Karrholm J, Borlin N, Estok DM, Harris WH. Experimental assessment of precision and accuracy of radiostereometric analysis for the determination of polyethylene wear in a total hip replacement model. *J Orthop Res* 2002;**20**(4):688–95.
61. Bragdon CR, Estok DM, Malchau H, Karrholm J, Yuan X, Bourne R, Veldhoven J, Harris WH. Comparison of two digital radiostereometric analysis (RSA) methods in the determination of femoral head penetration in a total hip replacement phantom. *J Orthop Res* 2004;**22**(3):659–64.
62. Bragdon II CR, Martell JM, Estok DM, Greene ME, Malchau H, Harris WH. A new approach for the Martell 3-D method of measuring polyethylene wear without requiring the cross-table lateral films. *J Orthop Res* 2005;**23**(4):720–5.
63. Mjöberg B. Loosening of the cemented hip prosthesis. The importance of heat injury. *Acta-Orthop-Scand-Suppl* 1986;**221**:1–40.
64. Malchau H, Kärrholm J, Wang YX, Herberts P. Accuracy of migration analysis in hip arthroplasty. Digitized and conventional radiography, compared to radiostereometry in 51 patients. *Acta Orthop Scand* 1995;**66**(5):418–24.
65. Kärrholm J, Snorrason F. Subsidence, tip, and hump micromovements of noncoated ribbed femoral prostheses. *Clin-Orthop* 1993(287):50–60.
66. Hildebrand H, Aronson S, Kullendorff CM, Selvik G. Roentgen stereophotogrammetric short-term analysis of growth rate in children operated for Crohn's disease. *Acta-Paediatr-Scand* 1991; **80**(10):917–23.
67. Lundberg A. Kinematics of the ankle and foot. In vivo roentgen stereophotogrammetry. *Acta-Orthop-Scand-Suppl* 1989;**233**:1–24.
68. Söderqvist I, Wedin P-Å. Determining the movements of the skeleton using well-configured markers. *J Biomech* 1993;**26**. p. Technical note 1473-77.
69. Uvehammer J, Karrholm J, Brandsson S, Herberts P, Carlsson L, Karlsson J, Regner L. In vivo kinematics of total knee arthroplasty: flat

compared with concave tibial joint surface. *J Orthop Res* 2000; **18**(6):856–64.
70. Kärrholm J, Hansson LI, Selvik G. Longitudinal growth rate of the distal tibia and fibula in children. *Clin Orthop Relat Res* 1984(191):121–8.
71. Borlin N, Rohrl SM, Bragdon CR. RSA wear measurements with or without markers in total hip arthroplasty. *J Biomech* 2006; **39**(9):1641–50.
72. Kaptein BL, Valstar ER, Stoel BC, Rozing PM, Reiber JH. A new model-based RSA method validated using CAD models and models from reversed engineering. *J Biomech* 2003;**36**(6):873–82.
73. Bragdon CR, Greene ME, Freiberg AA, Harris WH, Malchau H. Radiostereometric analysis comparison of wear of highly cross-linked polyethylene against 36- vs 28-mm femoral heads. *J Arthroplasty* 2007;**22**(6 Suppl. 2):125–9.
74. Karrholm J, Herberts P, Hultmark P, Malchau H, Nivbrant B, Thanner J. Radiostereometry of hip prostheses. Review of methodology and clinical results. *Clin Orthop Relat Res* 1997(344):94–110.
75. Bragdon II CR, Martell JM, Greene ME, Estok DM, Thanner J, Karrholm J, Harris WH, Malchau H. Comparison of femoral head penetration using RSA and the Martell method. *Clin Orthop Relat Res* 2006;**448**:52–7.
76. Bragdon CR, Barrett S, Martell JM, Greene ME, Malchau H, Harris WH. Steady state penetration rates of electron beam-irradiated, highly cross-linked polyethylene at an average 45-month follow-up. *J Arthroplasty* 2006;**21**(7):935–43.
77. Bragdon CR, Kwon YM, Geller JA, Greene ME, Freiberg AA, Harris WH, Malchau H. Minimum 6-year followup of highly cross-linked polyethylene in THA. *Clin Orthop Relat Res* 2007;**465**:122–7.
78. Geller JA, Malchau H, Bragdon C, Greene M, Harris WH, Freiberg AA. Large diameter femoral heads on highly cross-linked polyethylene: minimum 3-year results. *Clin Orthop Relat Res* 2006;**447**:53–9.
79. Devane PA, Bourne RB, Rorabeck CH, MacDonald S, Robinson EJ. Measurement of polyethylene wear in metal-backed acetabular cups. II. Clinical application. *Clin Orthop Relat Res* 1995(319):317–26.
80. Greene ME, Bragdon CR, Pitcairn SB, Malchau H. A comparison of the devane and martell methods for in vivo measurement of polyethylene wear of patients with highly cross-linked polyethylene THR components. *54th Annual Meeting of the Orthopaedic Research Society*. San Francisco, CA; 2008.
81. Penning J, Pras H, Pennings A. Influence of chemical crosslinking on the creep behavior of ultra-high molecular weight polyethylene fibers. *Colloid Polym Sci* 1994;**272**:664–76.
82. Ramamurti BS, Estok DM, Bragdon CR, Weinberg EA, Jasty M, Harris WH. Dimensional changes in metal-backed polyethylene acetabular cups under cyclic loading. *24th Annual Meeting of the Society for Biomaterials*. San Diego, CA; 1998.
83. Treharne R, Brown N. Factors influencing the creep behavior of poly(methyl methacrylate) cements. *J Biomed Mat Res* 1975;**9**:81–8.
84. Rostoker W, Galante JO. Indentation creep of polymers for human joint applications. *J Biomed Mat Res* 1979;**13**:825–8.
85. Estok II DM, Bragdon CR, Plank GR, Huang A, Muratoglu OK, Harris WH. The measurement of creep in ultrahigh molecular weight polyethylene: a comparison of conventional versus highly cross-linked polyethylene. *J Arthroplasty* 2005;**20**(2):239–43.
86. Davidson JA, Schwartz G. Wear, creep and frictional heat of femoral implant articulating surfaces and the effect on long-term performance-Part I, a review. *J Biomed Mater Res* 1987;**21**:261–85.
87. Bragdon C, O'Connor D, Weinberg E, Skehan H, Muratoglu O, Lowenstein J, Harris W. The role of head size on creep and wear of conventional vs. highly cross-linked polyethylene acetabular components. *Society for Biomaterials 25th Annual Meeting Transactions*. Providence, RI; 1999.
88. Atkinson JR, Cicek RZ. Silane crosslinked polyethylene for prosthetic applications. Part II. Creep and wear behavior and a preliminary molding test. *Biomaterials* 1984;**5**:326–35.
89. Martell JM, Leopold SS, Liu X. The effect of joint loading on acetabular wear measurement in total hip arthroplasty. *J Arthroplasty* 2000;**15**(4):512–18.
90. Smith PN, Ling RS, Taylor R. The influence of weight-bearing on the measurement of polyethylene wear in THA. *J Bone Joint Surg Br* 1999;**81**(2):259–65.
91. Moore KD, Barrack RL, Sychterz CJ, Sawhney J, Yang AM, Engh CA. The effect of weight-bearing on the radiographic measurement of the position of the femoral head after total hip arthroplasty. *J Bone Joint Surg—Am Vol* 2000;**82**(1):62–9.
92. Engh Jr. CA, Sychterz Sr. CJ, Young AM, Pollock DC, Toomey SD, Engh CA Interobserver and intraobserver variability in radiographic assessment of osteolysis. *J Arthroplasty* 2002;**17**(6):752–9.
93. Puri L, Wixson RL, Stern SH, Kohli J, Hendrix RW, Stulberg SD. Use of helical computed tomography for the assessment of acetabular osteolysis after total hip arthroplasty. *J Bone Joint Surg Am* 2002;**84-A**(4):609–14.
94. Claus Sr. AM, Totterman SM, Sychterz CJ, Tamez-Pena JG, Looney RJ, Engh CA Computed tomography to assess pelvic lysis after total hip replacement. *Clin Orthop Relat Res* 2004(422):167–74.
95. Puri L, Lapinski B, Wixson RL, Lynch J, Hendrix R, Stulberg SD. Computed tomographic follow-up evaluation of operative intervention for periacetabular lysis. *J Arthroplasty* 2006;**21**(6 Suppl. 2):78–82.
96. Garcia-Cimbrelo E, Tapia M, Martin-Hervas C. Multislice computed tomography for evaluating acetabular defects in revision THA. *Clin Orthop Relat Res* 2007;**463**:138–43.
97. Leigh W, O'Grady P, Lawson EM, Hung NA, Theis JC, Matheson J. Pelvic pseudotumor an unusual presentation of an extra-articular granuloma in a well-fixed total hip arthroplasty. *J Arthroplasty* 2008.
98. Reish TG, Clarke HD, Scuderi GR, Math KR, Scott WN. Use of multi-detector computed tomography for the detection of periprosthetic osteolysis in total knee arthroplasty. *J Knee Surg* 2006;**19**(4):259–64.

# Chapter 29

# ESR Insights into Macroradicals in UHMWPE

M. Shah Jahan, PhD



## 29.1 INTRODUCTION

Electron spin resonance or ESR is a technique that can directly detect and quantify unpaired or odd electrons in atomic or molecular systems. The materials that contain unpaired electrons are known as paramagnetic materials because they exhibit a net magnetic moment in an external magnetic field. For this reason, ESR is also known as EPR or electron paramagnetic resonance. While both names are used in practice, a third name has been introduced to replace them. In line with NMR or nuclear magnetic resonance, the new name is EMR or electron magnetic resonance because the magnetic resonance in ESR/EPR results from the electron magnetic moment. In this chapter, the term ESR is used.

ESR is the only technique that is particularly suitable, not only for direct detection, but also for quantitative analyses of free radicals in a solid or fluid. In polymers, free radicals are generally formed as a result of the unpairing of electrons from a particular molecular site (chain, side, or ring) by mechanical or chemical means, or by irradiation with ionizing radiation. The possession of an unpaired electron makes the free radical a good candidate for ESR study. Soon after the ESR instruments became available in mid-1950, free radical measurements in polymers, including polyethylene (PE), began. Most of the early works on PE radicals are compiled by Malcolm Dole in two volumes entitled *The Radiation Chemistry of Macromolecules* [1]. Readers are referred to Chapters 9, 13, and 14 in Volume I, and Chapters 1 and 13 through 16 in Volume II. In 1995, Jan F. Rabek also published a large volume of works, including ESR results, on polymer degradation [2]. In the latter publication, particular attention should be given to Chapters 3 and 10 because of their relevance to the subject of this chapter.

In low density PE, radicals are unstable at room temperature. This low density PE, linear or stretched, was the primary material under study in the aforementioned works of Dole and Rabek. ESR measurements in these early works were therefore conducted at liquid nitrogen temperature. Nevertheless, the results of these measurements on relatively simple PE can help in understanding the macroradical processes in more complex matrix of ultrahigh molecular weight polyethylene (UHMWPE).

UHMWPE has gained its importance in orthopedic science and technology for more than 5 decades as a material of choice for total hip- and knee-joint components. However, the physical and chemical properties of UHMWPE are adversely affected by free radicals that are generated in the polymer matrix as a result of radiation sterilization or crosslinking [3–13]. Consequently, the

ESR technique serves as an important diagnostic/research tool for testing irradiated UHMWPE.

Keeping in mind that ESR knowledge can be useful for the orthopedic or biomaterials community, this chapter starts with the basic principle of ESR and provides ESR data on the radicals generated, propagated, and terminated in UHMWPE. Emphasis is given on the short- and long-term oxidation results and on identification of the peroxy and oxygen-induced radicals. ESR results on vitamin E-doped UHMWPE and quantitative ESR are also presented.

## 29.2 BASIC PRINCIPLE OF ESR

ESR spectroscopy is based on the fundamental properties of electrons. They are mass ($m_e = 9.11 \times 10^{-31}$ kg), charge ($e = -1.6 \times 10^{-19}$ coulomb), and spin(s). Therefore, an unpaired electron in atoms or molecules (for example, hydrogen atoms, transition metal ions, free radicals), possesses both magnetic moment and angular momentum. The ratio between magnetic moment and the angular momentum is known as gyromagnetic ratio. The net magnetic moment of a material (solid, liquid, or gas) can be zero because of the random nature of the spinning electrons. In an external magnetic field (H) the spin magnetic moment aligns parallel or antiparallel to the field, and the spinning electrons are split or divided into high (+) and low (−) energy states. The energy separation ΔE between the states is known as Zeeman energy and is given by $\Delta E = g\beta H$, where H is the external steady magnetic field, β is the electron Bohr magneton ($\beta = 9.27 \times 10^{-24}$ J/T), and g is the spectral splitting factor, commonly known as the g-value. Transitions between these states can be stimulated by an oscillating magnetic field of frequency f when and only when its energy equals ΔE; that is, $hf = \Delta E = g\beta H$, where h is Planck's constant ($h = 6.63 \times 10^{-34}$ J.s). The condition $hf = g\beta H_r$ is known as resonance condition, where $H_r$ is the external magnetic field at resonance. The g-value is related to gyromagnetic ratio (γ) as $\gamma = g\beta/\hbar$, where β is the Bohr magneton (previously given), and $\hbar = h/2\pi$. *Free electrons have a g-value of 2.0023. The g-value of a radical species can be used as an identifying mechanism of the radical type.* The ESR resonance condition is similar to that of the well-known NMR (nuclear magnetic resonance) where g is replaced by $g_N$ (nuclear g-value) and β is replaced by $\beta_N$ (nuclear magneton). While ESR operates at microwave frequency (GHz), NMR operates at radio frequency (MHz).

The g-value, therefore, is expressed as

$$g = hf/\beta H_r \tag{1}$$

and the resonance magnetic field position is expressed by

$$H_r = hf/g\beta. \tag{2}$$

For electrons, the frequency f falls within the microwave regions. Because g is a constant, or a characteristic value, of a species (free radical, for example), resonance field is directly proportional to microwave frequency. The ESR spectrometer operates at the frequency band $\sim 9 \times 10^9$ Hz, or 9 GHz, and is known as an X-band spectrometer; it is most commonly used to study organic or polymeric materials, including biomaterials. The g-value for organic or polymeric radicals varies only slightly from the free electron value of $g = 2.0023$. The corresponding magnetic field for $g = 2.0$ is 3000 Gauss (G) or 0.3 Tesla (T).

NMR (nuclear magnetic resonance) signal intensity is recorded as a function of frequency at a constant magnetic field. Unlike NMR spectrometers, an ESR spectrometer requires a resonator (a microwave cavity) that operates at a fixed frequency, and it records the resonance absorption signal as a function of external magnetic field. The resonance magnetic field of an L-band (f = 1 GHz) spectrometer is 357 G, an S-band (f = 4 GHz) is 1430 G, a K-band (f = 24 GHz) is 8570 G, and a Q-band (f = 35 GHz) is 12,500 G. There are yet other special purpose ESR spectrometers that operate at very low or high frequencies.

At resonance, absorption of microwave energy by the lower energy state (−) spins produces an absorption signal that is generally very weak and may not be directly recorded or displayed. Therefore, an ESR spectrometer, unlike NMR, FTIR (Fourier transformed infrared) or optical spectrometers, records a first derivative of the absorption signal, which results from the modulation of the external magnetic field at 100 kHz (at X-band). A spectrometer may also have the option to produce a second-derivative absorption signal. An unpaired electron formed as a result of a broken covalent bond in a molecule, for example, creates a radical (by definition), and the ESR will most likely produce a single line, or a singlet (see Figure 29.1), provided there is no net nuclear magnetic moment coupled to the unpaired spin or no resolvable hyperfine coupling. Figure 29.1A shows the energy states of an unpaired electron in a magnetic field H with separation $\Delta E = g\beta H$. The transition between the (+) and (−) states is detected as the first-derivative of absorption at resonance $\Delta E = g\beta H_r = hf$, where $H_r$ is the resonance magnetic field and f is the resonance frequency. Figure 29.1B shows an experimentally recorded ESR spectrum due to a stable organic radical, 2, 2-Diphenyl-1-picrylhydrazyl (DPPH). The corresponding absorption curve, shown in Figure 29.1C, is obtained by integrating the first-derive curve or spectrum; the shaded area under the curve is proportional to radical concentration. For computational purposes, resonance magnetic field position $H_r$ (the crossover point in the first-derivative curve, or the peak of the absorption curve, $H_r = 3515.824$ G) and the line width, $\Delta H_{pp} = 1.33$ G, can be measured visually by zooming into the experimental spectrum. As shown in the Figure 29.1C, the magnetic field position of the peak equals $H_r$-value. The resonance frequency is automatically

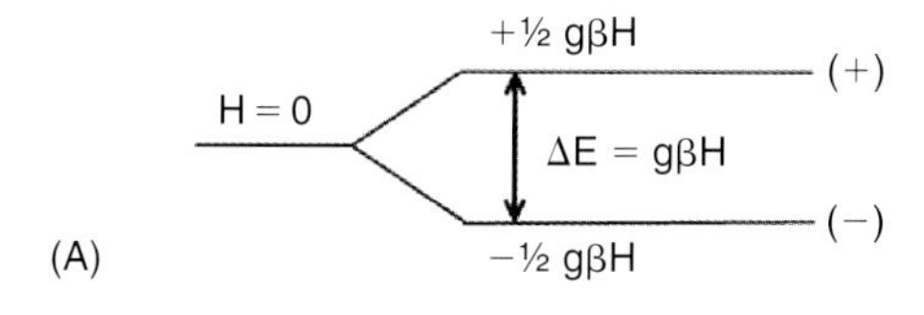


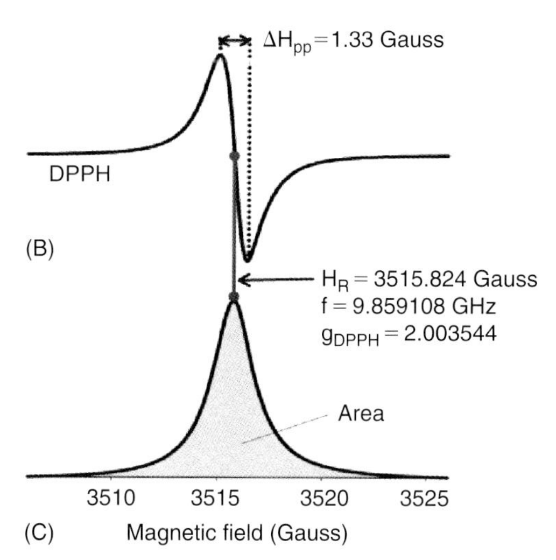


**FIGURE 29.1** (A) Zeeman energy $\Delta E$ between the high (+) and low (−) energy states of an unpaired electron spin in an externally-applied magnetic field H. (B) Experimentally recorded first-derivative ESR spectrum of 2, 2-Diphenyl-1-picrylhydrazyl (DPPH). (C) absorption curve obtained by integrating the first-derivative spectrum.

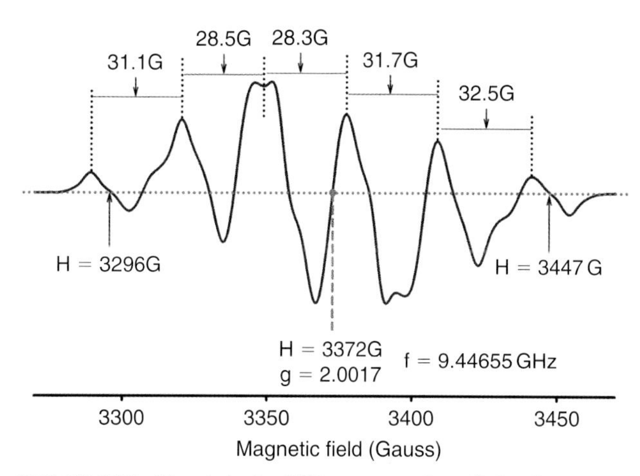


**FIGURE 29.2** First derivative ESR spectrum of an alkyl radical —$CH_2$ —$^{\bullet}$CH—$CH_2$—. Also shown in the figure are the separation between the lines, magnetic field positions of the first line (H = 3296 G) and the sixth line (H = 3447 G), the center of resonance H = 3372 G, frequency f = 9.4465 GHz, and the g-value for the radical g = 2.0017.

recorded at the time of measurement (Bruker EMX). Using f = 9.859108 GHz in the resonance formula (1), the g-value is calculated as g = 2.003544. A more accurate g-value was obtained by taking the average of five measurements. The calculated g-value was g = 2.00360 ± 0.00005, which is same as the textbook value of 2.0036 ± 0.0006. DPPH is frequently used as a standard for quantitative ESR or as a g-marker. The shaded area under the absorption curve is proportional to the radical concentration.

Although structural information about the radical may not be found from a singlet, ESR parameters such as g-value and the line-width or peak-to-peak separation $\Delta H_{pp}$ can provide a basic idea about the radical. Additional discussion about the singlet is presented later.

If a proton is situated near the unpaired spin (at the broken site of a molecule), each electron Zeeman energy state will split into two (for nuclear spin I = 1/2) or three (for nuclear spin I = 1) states. For each nuclear spin, electron spin energy level splits into 2I + 1 levels. The magnetic field positions of the resonance lines for I = 1/2 are given by

$$H = H_r \pm 1/2\, a \tag{3}$$

where $H_r$ is the resonance field position for the radical and 'a' is the hyperfine coupling constant, or hyperfine constant. In this case, two lines are obtained: one at $H_1 = H_r - 1/2$ a; the other at $H_2 = H_r + ½$ a. Therefore, the separation $(H_2 - H_1 = (H_r + 1/2\ a) - (H_r - 1/2\ a) = a)$ between the two lines is 'a.' In a multiline spectrum, the hyperfine constant can be determined by measuring the hyperfine separation or, simply, the separation between the lines. For g-value calculation, the value of the magnetic field position $H_r$ is used, although there may not be any real signal at that position.

For a radical containing five equivalent protons in its neighborhood, each of I = 1/2, six resonance lines are observed in its ESR spectrum (2 × 5/2 + 1 = 6); the six-line spectrum of the alkyl radical in UHMWPE is a good example (see Figure 29.2). One α-proton and four β-protons in the alkyl radical (—$CH_{2,\beta}$ —$^{\bullet}$CHα —$CH_{2,\beta}$) effectively act like five equivalent protons because of random orientation of the molecule in a heterogeneous polymer matrix. Intensity ratios (theoretical) of the lines are 1:5:10:10:5:1. The intensity of, and the separation between, the first derivative lines in Figure 29.2 cannot provide accurate results because of the overlapping side lines (or doublets). However, for g-value calculation, the magnetic field positions of the first line (H = 3296 G), the sixth line (H = 3447 G), and the center of resonance $H_r$ = 3372 G are recorded. Using the operational frequency f = 9.4465 GHz in the formula (3), the g-value for the radical was calculated to be g = 2.0017, approximately. Measuring the separation between the lines, the approximate value for the hyperfine constant was found to be $a_{alkyl}$ = 30.4 G. If, however, nonequivalent or different protons of nuclear spin I = 1/2 or 1, are coupled to the same unpaired electron of the radical, line positions and/or intensity ratios can get complicated because of the combined or overlapping effects of different hyperfine couplings. For background of ESR, readers are referred to Chapter 6, *Basic Principle of Spectroscopy* by Chang [14] and Chapter 1, *Electron Paramagnetic Resonance* by Weil et al. [15].

## 29.3 FREE RADICALS IN UHMWPE

The main or primary radicals frequently reported in the literature are: the alkyl —$CH_2$ —•CH —$CH_2$—, the allyl —$CH_2$—CH = CH— •CH $CH_2$— or —$CH_2$— •CH—CH = CH—$CH_2$—, and the polyenyl —HC•—H(CH = CH)$_n$—. The secondary or tertiary radicals are: peroxy, —$CHO_2$•—, alkoxy, —CHO•—, vinyl >CH=•$CH_2$—, and oxygen-induced radicals (OIR). In simpler terms, a main polymer radical is denoted as P• or R•, the peroxy radical as $PO_2$•, POO•, $RO_2$•, or ROO•, and the alkoxy radical as PO• or RO•. A brief discussion of each one of these radicals is given in the following sections. Note that the presence of an unpaired electron at the broken site of a molecule is conventionally denoted with a superlative dot symbol (•), such as —•CH—.

### 29.3.1 Alkyl Radical

The alkyl radical is formed as a result of abstraction of a hydrogen atom from the —$CH_2$— group in the PE chain. The site of the molecule or group (—•CHα—, for example) with one less hydrogen has, therefore, one unpaired electron. The hydrogen atom at the unpaired electron site is known as α-hydrogen, and the four nearest (two on the left and two on the right) are known as β-hydrogen atoms. The molecular structure of the alkyl radical is expressed as —$CH_{2,\beta}$ —•CHα — $CH_{2,\beta}$—. These β-hydrogen atoms are considered as equivalent atoms insofar as their interaction with the unpaired electron is concerned. Of course, orientation difference of each β-hydrogen atom with respect to the molecular chain can make them nonequivalent. The next-nearest hydrogen atoms in the chain of —$CH_2$—s are known as γ-hydrogens, and they can have an effect on an ESR signal as well. As previously presented, one unpaired electron produces one single line in the ESR signal (or ESR spectrum). The single resonance line splits into two equally intense lines (known as hyperfine lines) with a separation "a" (known as hyperfine separation or coupling constant) when it is coupled with a nearby unpaired proton, such as hydrogen nucleus. Further splitting of the hyperfine lines, with different coupling constant, can occur if the same unpaired electron is also coupled to additional nearby protons (β-hydrogen atoms in polyethylene (PE) chains, for example). In UHMWPE, however, the effects of α-hydrogen and β-hydrogen atoms on the ESR spectrum cannot be distinguished because of random orientation of the molecular chains within the polymer matrix. The typical six-line pattern in the ESR spectrum of the alkyl radical is due to an average coupling of the five hydrogen atoms, one α- and four β-hydrogen atoms (see Figure 29.2). The line separation "a" varies between 23 and 30 Gauss. Each of the lines in Figure 29.2 are broad and exhibit additional features, such as side or satellite lines, or doublets. These additional features in the spectrum may arise from the complex nature of UHMWPE, possessing crystalline, amorphous and interfacial regions in any given matrix. Additional features in the ESR spectrum may arise due to different operating parameters as well such as microwave power, modulation amplitude, temperature, time constant, etc. As a result, attribution of an ESR signal to a particular radical species in UHMWPE becomes nontrivial. One needs to pay careful attention to all of these variables before attributing a signal or spectrum to a particular species. In the early 1960s a number of investigators made very thorough investigations of the irradiated PE and made a precise structural assignment to each radical species based on the ESR spectrum. For example, Dole (see Chapter 14 [1]) observed five hyperfine lines with a separation of 28 G between them ($a_\beta$=28 G) due to four equivalent β-hydrogen atoms in the alkyl radical in oriented molecules in a stretched PE sample at 77 K. Additionally, each hyperfine line was found to split in two with a separation of 13 G (aα=13 G) produced by one α-hydrogen atom.

### 29.3.2 Allyl Radical

Shown in Figure 29.3 is an approximate molecular diagram of the allyl radical —$CH_2$—CH = CH—•CH—$CH_2$—↔—$CH_2$—•CH—CH = CH—$CH_2$—. As previously mentioned, precise structural analyses of the PE radicals were conducted in the early 1960s (see Chapter 14 [1]). For example, using oriented linear PE at temperature 77–293 K, Ohnishi identified a seven-line ESR spectrum due to four β-hydrogen and two α-hydrogen atoms [16, 17]. The separation between the lines was found to be approximately 21 G. Effects of $C_1$-Hα and $C_3$-Hβ bond angles (see Figure 29.3) on the ESR spectrum were also measured at different temperatures. Concentration of the allyl radical at low temperatures was determined by subtracting the alkyl radical concentration at that temperature, or at room temperature when the alkyl radical had decayed. In UHMWPE, a clean seven-line (ESR) fingerprint of the allyl radical is obtained at room temperature following gamma irradiation at the same temperature in nitrogen or vacuum. Figure 29.4A shows a first-derivative

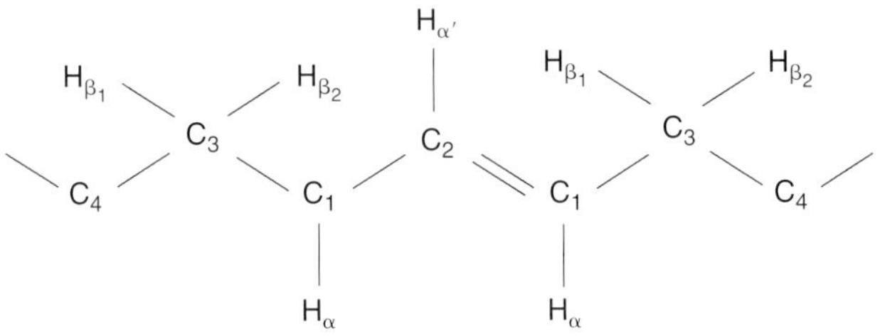

**FIGURE 29.3** Molecular structure of the allyl radical showing approximate orientation of the α- and β-hydrogen atoms.

ESR spectrum of the radical with hyperfine coupling constant $a_{allyl}$ = 13.6G, and the g-value, g = 2.0016, approximately. Note that the g-value for the radical corresponds to the resonance magnetic field ($H_r$ = 3372G) of the entire spectrum, which is the crossover point of the middle line. If alkyl radicals are also present, side lines will appear on both ends (right and left) of the spectrum. The weak broad lines, one on the left and one on the right, shown on the magnified y-scale in Figure 29.4B, are the members of the six-line spectrum due to the alkyl radical; the other four lines are overlapped by the strong lines due to the allyl radical. The center or the resonance magnetic field ($H_r$ = 3508G) of the alkyl radical can be approximately calculated by measuring the magnetic field positions (3433G and 3583G) of the two lines. The g-value for the radical was calculated as 2.0025, which is slightly greater than the g-value

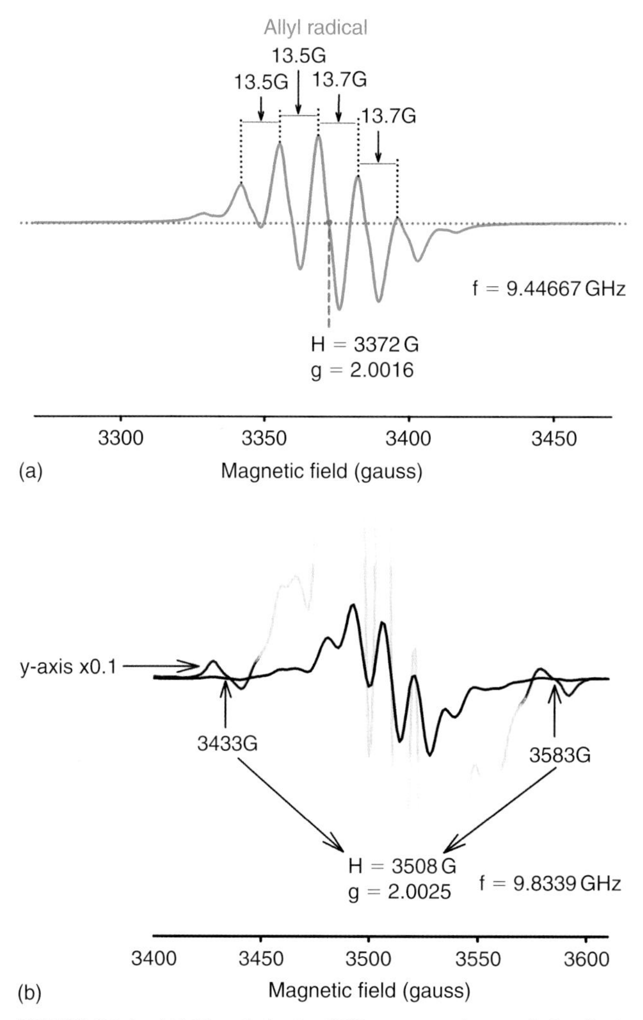


**FIGURE 29.4** (A) First derivative ESR spectrum due to allyl radical. (B) First-derivative ESR spectrum due to the alkyl and allyl radicals present in the same sample (GUR 1050, γ-irradiated in nitrogen and then exposed to air before measurements. The presence of the alkyl radical is shown by the weak lines (magnified) at the extreme left and right ends of the spectrum. The seven-line pattern of the allyl radical appears at the central part of the spectrum.

(g = 2.0017) obtained from the predominantly alkyl ESR spectrum (Figure 29.2). Although each radical has its own characteristic a- and g-value, differences like this can occur due to experimental errors. In practice, therefore, any g-value between 2.001 and 2.003 is acceptable for the alkyl, allyl, and polyenyl radicals. However, to obtain statistically significant results, one needs to make a large number of measurements and compare with the theoretical values.

Two important observations are made at this point: (1) a spectrum like this can be a good indicator of the presence of both alkyl and allyl radicals, and (2) by following any changes in one of these two lines, at extreme right or left, and one of the center lines (due to allyl) one could follow the reaction process of one radical in presence of the other. Subtraction methods may also be used to find the concentration of the alkyl and allyl radicals; however, the presence of di-, tri-, or polyenyl, or vinyl radicals, should be considered as well. To get a good estimate of the constituent radicals, one could deconvolute the experimental spectrum by applying a simulation method (see Section 29.3.3).

## 29.3.3 Polyenyl Radical

Ohnishi et al. provides a good discussion on polyenyl radicals, $CH_2$—$^{\bullet}CH$(—CH = CH)$_n$—$CH_2$— [17], produced as a result of the migration of the alkyl radical. The polyenyl radicals are quite stable due to the formation of conjugated double bonds. According to Ohnishi et al., polyenyl radicals are produced at high doses (>1000kGy), while the allyl radical is formed at intermediate doses (100–1000kGy) and the alkyl is formed at low doses (<10kGy). Nonetheless, low-level polyenyl radicals are formed at low doses as well but may not be directly observed because of the presence of strong resonance lines of other radicals having very similar g-values. Using the WinSim-2002 (Bruker) simulation program, we found concentration of the allyl (n = 1), dienyl (n = 2), trienyl (n = 3), and higher order polyenyl (n > 3) radicals. A simulation of an ESR spectrum due predominantly to the allyl is shown in Figure 29.5A and to the alkyl in Figure 29.5B. Table 29.1 lists the abundance and hyperfine coupling constants of these split radicals. One should note that the simulation of the allyl spectrum may not include alkyl and that the alkyl may not include allyl. Therefore, these simulations are not perfect, but they provide a means to get a general estimate.

## 29.3.4 Peroxy Radical

The peroxy radical (—$H_2C$—$HCO_2^{\bullet}$—$CH_2$—) is the precursor to the oxidation process in UHMWPE. It leads to the production of hydroperoxide and carbonyl specie (>CO), including esters, aldehydes, alcohols, etc. These nonradical or neutral by-products or end products in UHMWPE can be detected or quantified by FTIR technique. In practice, the FTIR absorption band near 1680–1720$cm^{-1}$ due to

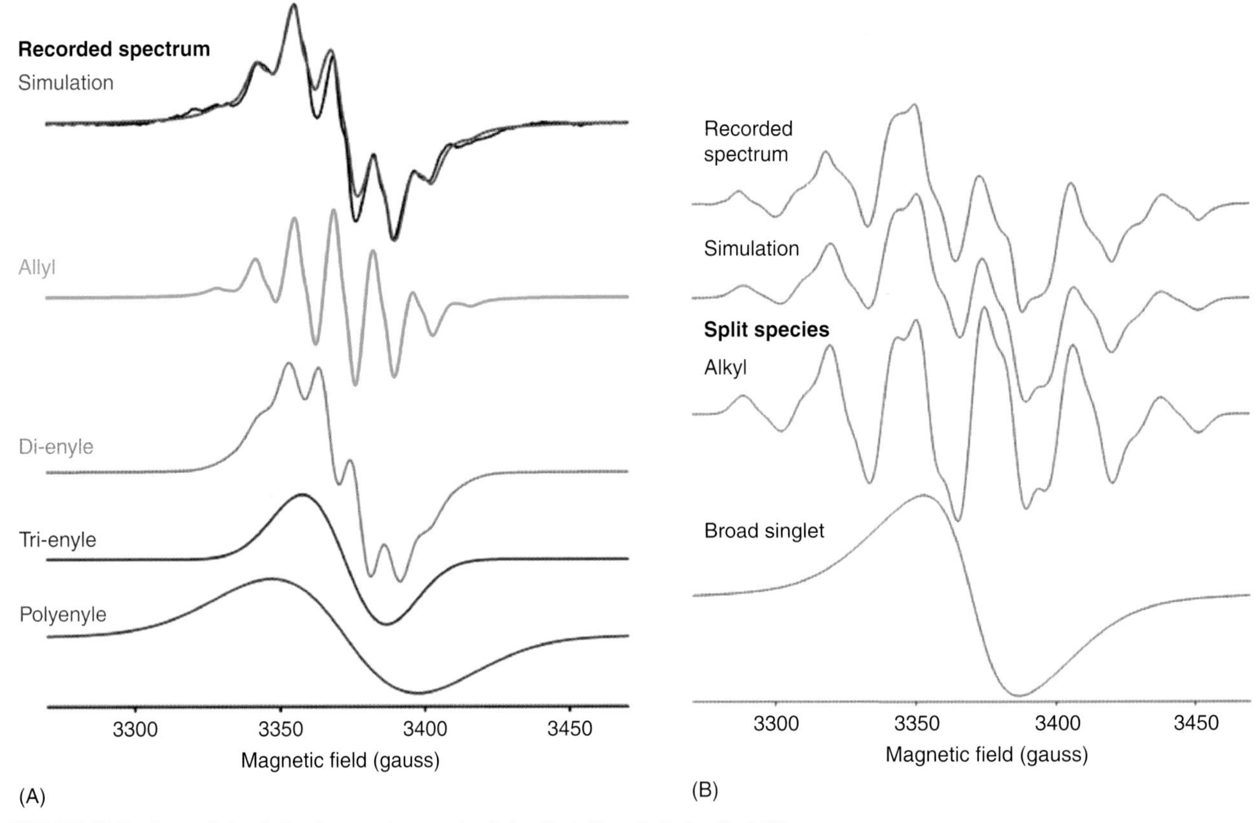


FIGURE 29.5 Spectral simulation for a predominantly allyl radical (A) and alkyl radical (B).

**TABLE 29.1 Abundance, Spin 1/2 Hyperfine Coupling Constants, Percentage Lorentzian and Line Width Used for Spectral Simulations in Simulations 1 and 2**

| | Simulation 1 | | | | Simulation 2 | |
|---|---|---|---|---|---|---|
| | Allyl | Dienyl | Trienyl | Singlet | Alkyl | Singlet |
| Abundance | 23% | 17.70% | 26.70% | 32.60% | 25.70% | 74.30% |
| a-α | 2@13.3 G | 3@8.9 G | 4@7.7 G | NA | 1@22.1 G | NA |
| a-β1 | 1@4.0 G | 2@1.9 G | 3@0.7 G | NA | 4@31.6 G | NA |
| a-β2 | 4@13.3 G | 4@8.9 G | 4@7.7 G | NA | NA | NA |
| a-anisotropy | NA | NA | NA | NA | 1@7.2 G | NA |
| % Lorentzian | 87.4 | 100 | 88.3 | 29 | 0 | 59.2 |
| Line width (G) | 4.3 | 6 | 7.4 | 18.8 | 4.28 | 23.4 |

the carbonyl group is used to measure oxidation levels in UHMWPE and is known as the oxidation index (OI). ESR is the only technique that can directly detect peroxy radicals in solid or liquid, but such detection is challenging when applied to UHMWPE due to the overlapping signals from different radicals present in the complex polymer matrix. Furthermore, each individual signal changes due to the ongoing reactions of the radicals from the surface to the core of a given test sample in the presence or absence of oxygen.

Discussions and speculations based on experimental data have been made about the peroxy radical in UHMWPE, but to our knowledge, no clear ESR evidence has been reported. Many investigators have attributed the so-called single-line spectrum to peroxy [8, 17–20]. However, as presented in the next section, the single line was found to be made of two separate overlapping spectra, each due to a different radical, neither of which was a peroxy. In this section, we present an ESR method for successful detection of the peroxy radical in UHMWPE made recently in our laboratory.

## 29.3.5 ESR Evidence of the Peroxy Radical in UHMWPE

Because of the unique orientation of the oxygen molecule-with-unpaired spin (—$O_2^{\bullet}$—) with respect to the PE molecular axis, the peroxy radical produces (theoretically) asymmetric lines in the ESR spectrum with two g-values, $g_{\parallel}$ and $g_{\perp}$. Ranby and Rabek presented a discussion on the dependence of the ESR spectrum on the orientation of the peroxy end in polytetrafluoroethylene (PTFE) radical [21]. "Rabek provides an ESR spectrum of the peroxy radical in polypropylene and also in polyethylene (Chapter 3 [2]). The polypropylene-peroxy radical presented is a clear ESR spectrum, which can be considered as the fingerprint for the peroxy radical; the ESR spectrum of the polyethylene-peroxy (low density polyethylene) is not as well defined."For a better understanding, readers are referred to a survey of the mechanisms for production, propagation, and termination of the peroxy radicals in polyolefins, including PE, compiled by J. F. Rabek (Chapter 3 [2]) and by Dole (Chapter 13 [1]).

Direct detection of the primary peroxy radical in UHMWPE was made in our laboratory in 2004 [22]. We used 200-μm thick sample (GUR 1050) and irradiated it with X-rays in nitrogen or vacuum. Immediately after irradiation the sample was transferred to the ESR cavity, which was continuously purged with nitrogen. In conjunction with microwave power saturation technique (PST), ESR spectra were recorded at and below room temperature. At different time intervals, the flow of nitrogen was stopped to allow oxygen (air) to react with the radicals. As shown in Figure 29.6, when the sample was exposed to air for 1 minute and the microwave power increased to 32 mW (Figure 29.6B), the resonance lines due to the alkyl and the allyl were suppressed (saturated), and a broad asymmetric signal (one line) was formed (Figure 29.6C). Although this is not a fingerprint of a peroxy radical, it was an indication of the formation of one. The fingerprint was obtained when the temperature of the sample was decreased to 118 K by maintaining a flow of cold nitrogen gas through the cavity (airflow was stopped). The unambiguous peroxy structure in the ESR signal is evidenced by its well-defined $g_{\parallel}$ and $g_{\perp}$ positions ($g_{\parallel} = 2.032$ and $g_{\perp} = 2.003$) as shown in Figure 29.6D. When the temperature was increased to

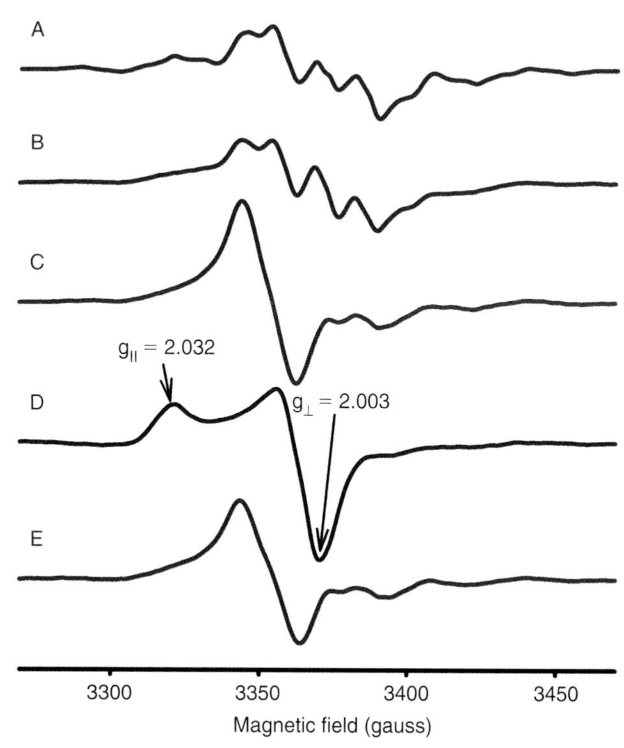


**FIGURE 29.6** Various ESR spectra of one GUR 1050 film sample, x-irradiated for 1 hr in a vacuum at 293 K (room temperature). The sample was transported in nitrogen to the ESR cavity, which had nitrogen flowing through it. The film did get exposed to air for a couple of seconds during the transfer. (A) and (B) were recorded at 8 mW and 32 mW of power, respectively. The sample was then exposed to air for 1 min, and (C) was recorded at 32 mW. The temperature was then lowered to 118 K, and (D) was recorded. (E) was recorded after the sample was brought back to room temperature (293 K) again.

293 K (room temperature) and the nitrogen flow was maintained, the spectrum (Figure 29.6E) returned to its original structure (Figure 29.6C), as expected. The result of these measurements suggests that the asymmetric broad single line at room temperature can serve as an indicator of the presence of peroxy radical; however, confirmation should be made by recording ESR signal at low temperature.

Shown in Figure 29.7 is a similar result obtained when the test was performed on a sample containing predominantly alkyl radical (1−mW scan). At 100 mW microwave power, the alkyl radical signal was saturated and the peroxy signal was revealed. As indicated by the decrease in intensity of the signal, recorded at 15 min, 30 min, and 76 min (1 h and 16 min), the peroxy radical concentration decreased in a nitrogen environment at room temperature by recombination. When the sample was opened to air after 76 min in nitrogen, the peroxy-signal intensity increased, indicating the production of the new peroxy radical. Keep in mind that the alkyl radical may not produce an ESR signal at a high microwave power, but it can react with oxygen that is present in the system and form a peroxy radical.

As shown in Figure 29.8, production of the peroxy radical by the allyl was demonstrated by first allowing the alkyl radical to decay to allyl in 124 hours in nitrogen and then allowing the latter to react with oxygen (air). The corresponding ESR spectra are shown in Figure 29.8.

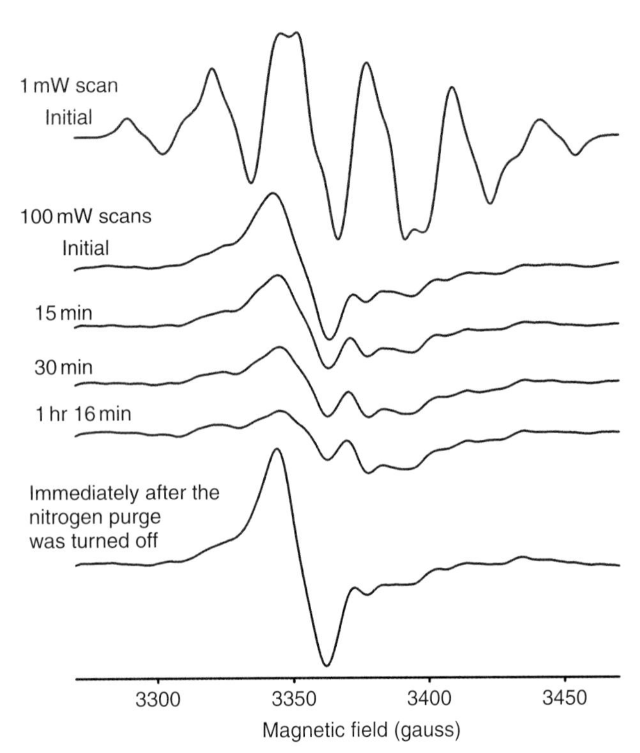


**FIGURE 29.7** ESR spectra of the peroxy radical in a 200-μm GUR 1050 film, x-irradiated for 2 hours in a vacuum ($10^{-6}$ torr) and subsequently opened to air. The first spectrum shown was taken 6 minutes after quickly being opened to air and subsequently put in a nitrogen flow. The nitrogen flow was turned back off after 1 hr 16 min.

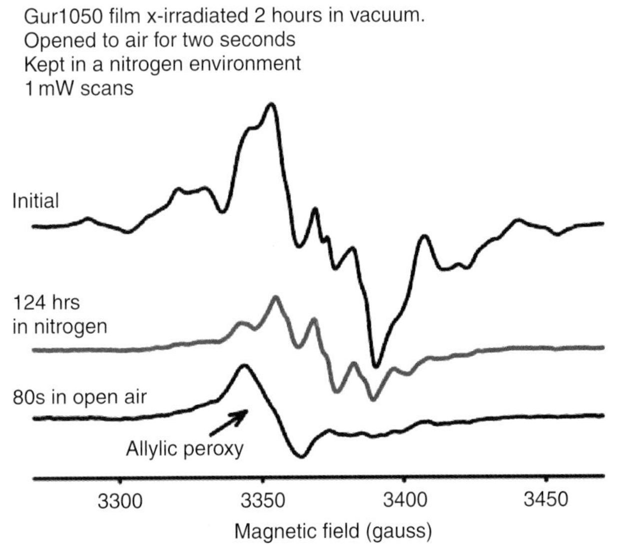


**FIGURE 29.8** ESR spectra of GUR 1050, x-irradiated for 2 hr in vacuum and subsequently transported to the ESR cavity, which had a continuous flow of nitrogen running through it. Measurements were taken at 1 mW of microwave power and 1 G modulation amplitude.

### 29.3.6 Half-Life of the Peroxy Radical in UHMWPE

The decay (in nitrogen) and production (in air) of the peroxy radical was monitored by measuring the normalized peak-to-peak height of the corresponding ESR spectrum.

The decay in nitrogen was greater than that in the open air, possibly due to existing peroxy radicals continuing to decay while little or no generation of new peroxy radicals occurred in the absence of oxygen. In open air, however, generation of new peroxy radicals can occur while others decay. When most of the peroxy radicals decayed in the time interval 80–100 min, subsequent exposure of the same sample to oxygen (air) resulted in the production of a few more peroxy radicals, as indicated by the positive trend in the intensity-time curve. The peroxy radical in UHMWPE is, therefore, very short lived with a half-life of 10–20 min. From the work of Auerbach and Sanders [23], Dole concluded that the half-life of the peroxy radical in linear PE at room temperature was 20 min, which is very close to our result in UHMWPE [1].

## 29.4 LONG-LIVED OXYGEN-INDUCED RADICAL IN UHMWPE

In addition to the short-lived peroxy radical (previously discussed), growth of a long-lived radical becomes evident in the ESR spectrum of an irradiated UHMWPE when the primary radicals decay. Because this long-lived radical is produced only in the presence of oxygen, it is known as an oxygen-induced radical (OIR), for lack of a better term. While we have been observing the OIR in a large number of samples for more than 20 years, its lifetime is yet to be known. Being a terminal species in the polymer system, the OIR is also known as a "terminal" or "residual" radical. However, distinction must be made between the long-lived OIR and the "residual primary" radicals found in UHMWPE following a long-term postirradiation aging in an inert environment (>10 years in this laboratory). Only a small fraction of the primary radicals survive the long-term aging in an inert environment; the longer the aging time, the lower the concentration (see Section 29.4.1). They are also labeled as "residual" or "residual primary" radicals.

### 29.4.1 ESR Evidence of the Long-Lived Radicals in UHMWPE

The growth of the OIR following the decay of the primary radicals in the presence of oxygen is illustrated in Figure 29.10 in which the ESR spectrum is recorded as a function of time at room temperature in air for two different samples, A (GUR 4150) and B (GUR 4120). Sample A was irradiated with X-rays at room temperature in air, and its

ESR tests began immediately after irradiation. Notice that the initial ESR spectrum of the sample A resembles that of the alkyl radical shown in Figure 29.2. Sample B was aged at 75°C for 6 years in vacuum following gamma irradiation in vacuum at room temperature and was subsequently opened to air for oxidation to occur. The ESR spectrum of the sample B, recorded before the sample was open to air, is very weak, but it exhibits the characteristic features of that of the primary alkyl/allyl radicals (see Figures 29.4A and 29.4B). It is important to note that the primary radicals in the sample B decayed significantly, most likely via radical-radical recombination as evidenced by the weak ESR spectrum, but did not transform to other species during the aging period. In the presence of oxygen, these long-lived primary radicals ("residual primary" radicals) in the sample B decayed, and the OIR grew in the same manner as those in the sample A; the growth time for the OIR is approximately 35 days. The growth of the OIR is evidenced by the appearance of the broad line in the ESR spectrum near 3500 G and disappearance of the initial resonance lines as shown in Figure 29.10. From these and similar results obtained from the ESR measurements on a large number of UHMWPE samples, GUR 4120, GUR 4150, GUR 1020, GUR 1050, and Himont 1900, following x-, gamma-, or e-beam irradiation, we make the following observations:

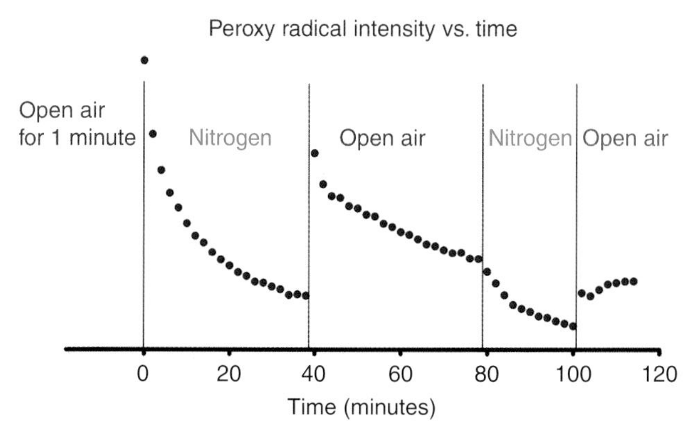


**FIGURE 29.9** Relative peak-to-peak height of the ESR signal due to the peroxy radical in a 200-μm GUR 1050 film, x-irradiated for 1 hr in a vacuum ($10^{-6}$ torr) and subsequently opened to air. The spectra were recorded at 100 mW. When opened to air, the nitrogen flow was turned on and off a couple of times.

1. OIR can form in the presence of oxygen in UHMWPE resin following irradiation with X-rays, gamma rays, or electron beam.
2. OIR is detectable by ESR in about 35 days after an irradiated sample is exposed to oxygen (air).
3. The concentration of the OIR is directly proportional to that of the primary radicals present at the time of initial oxidation.

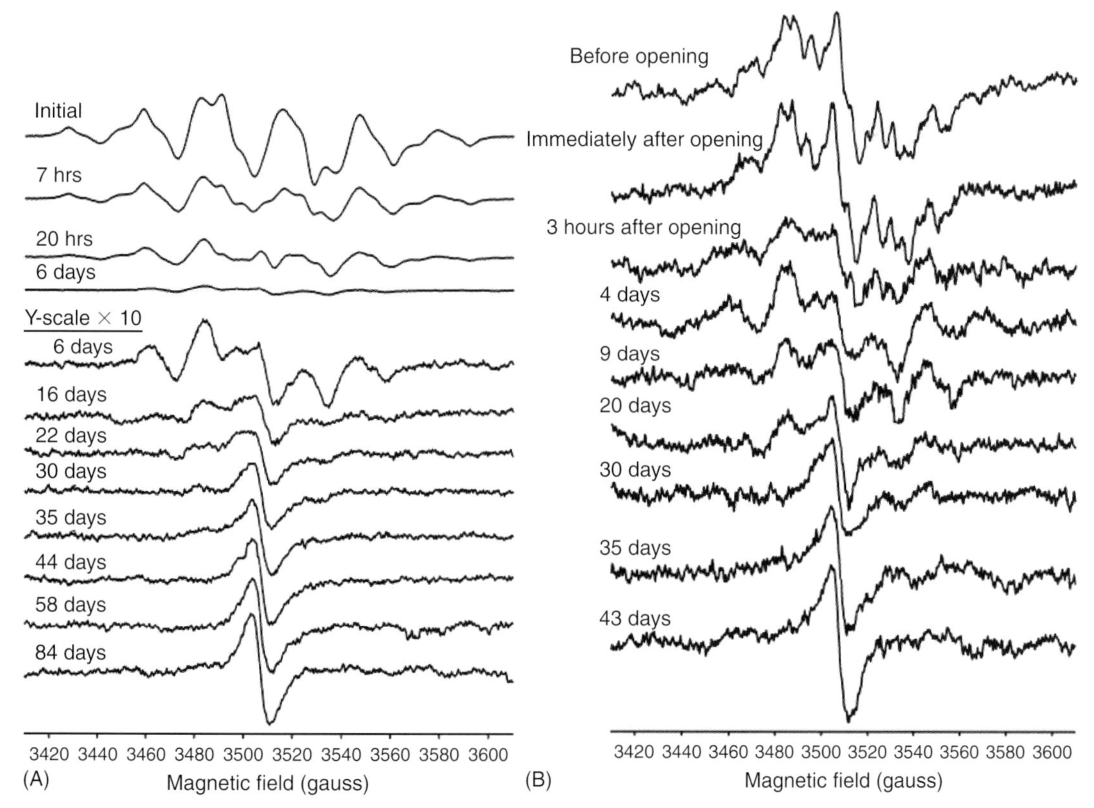


**FIGURE 29.10** ESR spectrum recorded as a function of time at room temperature to monitor the growth of the oxygen-induced radical. (A) GUR 4150 bar irradiated with X-ray for 1.3 hours in air and stored in air at room temperature. (B) GUR 4120 was gamma irradiated in an evacuated quartz tube and then stored for 6 years at 75°C in a vacuum. The sample was then removed from the tube to allow oxidation to occur at room temperature.

## 29.4.2 Growth and Decay of the Oxygen-induced Radicals

Figure 29.11 shows the growth pattern of the OIR for three samples, GUR 4150, GUR 4120 and Himont 1900, following gamma irradiation in vacuum or nitrogen at room temperature and subsequently exposing them to air. In this figure, peak-to-peak height of the ESR line due to the OIR is plotted as a function of time. As shown in the figure, the OIR reaches its maximum in about 200 days and then decays very slowly, and thus it has a very long life. It is not clear why GUR 4150 has consistently higher concentration (intensity of the ESR line) than GUR 4120 or Himont 1900.

The presence of the long-lived OIR in the shelf-aged or retrieved hip- or knee-joint components can be demonstrated by ESR. Figure 29.12 shows ESR data on the gamma-irradiated tibial plateau inserts (TPI). Figure 29.12A shows images of two TPIs (GUR 4120): a clinically retrieved TPI (GCR) on the left and a shelf-aged TPI (GSS) on the right. The holes in each TPI are the locations from where samples were punched out, and they were numbered for identification (not shown). Figure 29.12B shows a representative ESR spectrum of a clinically retrieved sample from position 5 (GCR5). As shown in the inset, the weak line on the shoulder of the main spectrum is an indication of the presence of a trace amount of primary radical. The bar graph in Figure 29.12C shows the distribution of radical concentration (OIR) as a function of the position in a TPI. The average concentration and the standard deviation (error bar) were determined by measuring 3–5 samples in a given region. For comparison, a gamma-irradiated control (GRM), a nonirradiated control (URM), and an ethylene-oxide-sterilized shelf-aged (ESS) TPI were also included in this study. The ESS sample did not show the presence of any radicals, and the gamma-irradiated control (GRM) had a concentration of an order of magnitude higher. In both GCR and GSS samples, random distribution of the radicals is evident, though the concentration in the GCR samples is, on the average, slightly less than that in the GSS samples. The lower concentration in the GCR samples could possibly be a result of annealing of the radicals due to a rise in temperature during articulation. Temperature rise on the articulating surfaces of the TPIs and acetabular cups was discussed by Davidson et al. [24].

In another example, the OIR were detected in a retrieved acetabular cup following 6–8 years of use *in vivo*. The cup (GUR 4150) was gamma-sterilized (30 kGy, 60 Co) in air before implantation. As shown in Figure 29.13A, samples were prepared from around the periphery of the articulating surface of the cup. Figure 29.13B shows the radical concentration (peak-to-peak height of the ESR line due to the OIR) as a function of the sample position at retrieval and after 13 years of shelf age following the retrieval. Two observations are made from this study: (1) OIR concentration was reduced by an order of magnitude in 13 years of postretrieval aging in air, and (2) the distribution pattern of the concentration over the articulating surface, showing the low- and high-wear areas, did not change. It is reported by Jahan et al. [5] that the low-radical concentration region corresponds to high-wear areas and the high-radical concentration corresponds to low-wear areas.

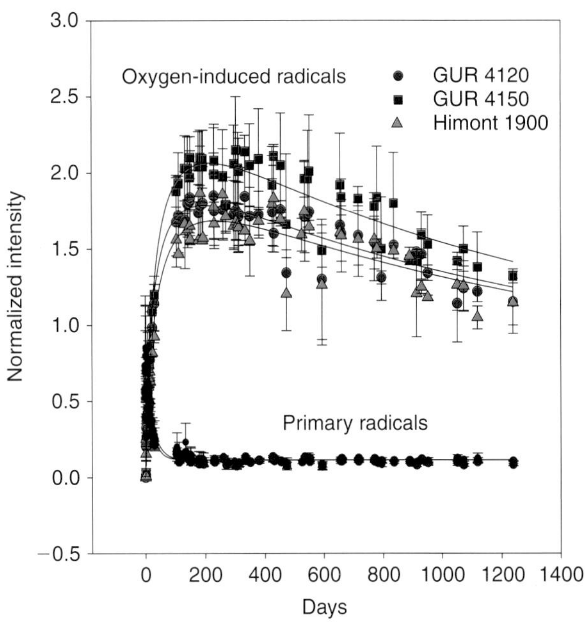


**FIGURE 29.11** Growth of the OIR as the primary radical decays in UHMWPE, GUR 4120, GUR 4150, and Himont 1900.

## 29.4.3 Identification of the Oxygen-induced Radicals by ESR

As shown in Figures 29.10 and 29.12, the ESR spectrum due to OIR is a broad single line, $\Delta H_{pp}$, is 9 G, and its g-value varies between 2.001 and 2.005, approximately. Many reports can be found in the literature about the OIR [8, 17–20]. Most of the reports suggest that it is a peroxy ($POO^{\bullet}$), but others say that it is an alkoxy ($PO^{\bullet}$) or a polyenyl ($P^{\bullet}$) species. Some of these reported works were conducted on low density polyethylene, and the oxidation reaction was investigated by exposing the test samples to oxygen at less than room temperature.

Since 2002, our laboratory (Biomaterials Research Laboratory at the University of Memphis) has been exploring the possibility of determining the structure of the OIR in UHMWPE [22, 25, 26]. Using a high-sensitive ESR spectrometer (Bruker EMX300), microwave power saturation technique (PST) was applied at low temperature (118 K) to demonstrate that the single line was made of two overlapping spectra due to two different radical species, R1 and R2. Further discrimination between these species was made by thermal treatment at 130–140°C at very low (for R1) or high (for R2) microwave power. For this study a high-dose (1000 kGy, 60 Co) sample was used to obtain a strong ESR signal. The sample (GUR 4150) was irradiated in 1988 in room temperature air, and this particular study was conducted from 2004 to 2006. Figure 29.14B shows a typical single-line ESR spectrum detected at room temperature (23°C) under standard operating conditions (1.0 mW microwave power and 1.0 G modulation amplitude). This figure further illustrates that the room-temperature signal becomes a well-defined symmetric single line at microwave power 0.01 mW (see Figure 29.14A). The peak-to-peak width of the line $\Delta H_{pp} = 5.5$ G and the g-value, $g = 2.0044 \pm 0.0003$ were also determined. This resonance

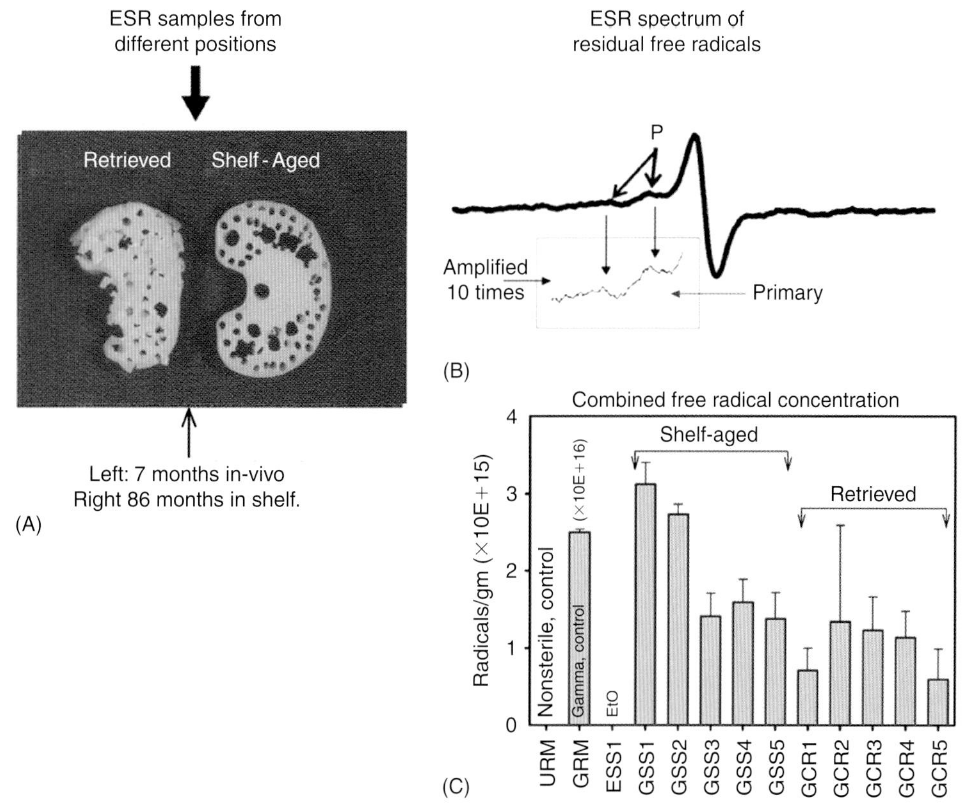


FIGURE 29.12 (A) Retrieved (left) and shelf-aged (right) TPIs. (B) Representative ESR spectrum of one sample showing a strong signal due to OIR and the presence of a trace amount of the primary radical (P). (C) Radical concentration in all TPI samples. Note that the concentration in the gamma-irradiated control (GRM) is an order of magnitude higher.

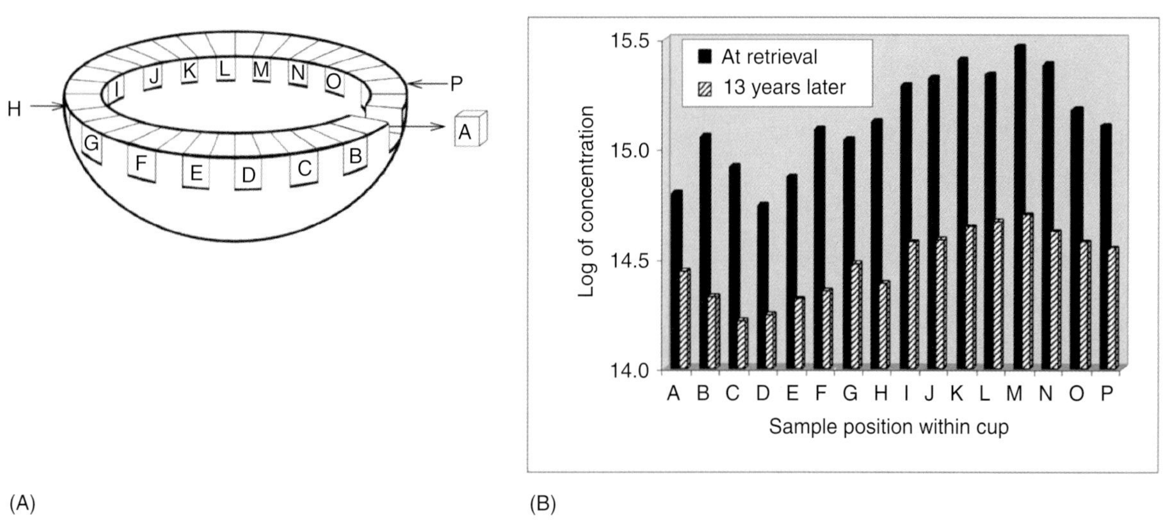


FIGURE 29.13 (A) Schematic diagram of an acetabular cup showing ESR sample locations (periphery of the articulating surface). (B) Concentration of the OIR plotted as a function of the sample position.

signal was attributed to a carbon-centered radical R1 (discussed later). When microwave power was increased to 160 mW, the signal due to R1 was reduced, and the weak shoulder shown in Figure 29.14B increased (see Figure 29.14C). The broad signal exhibited six hyperfine lines, and the signal due to R1 was further reduced to a nondetectable level at low temperature (−133°C) and 10 mW microwave power (see Figure 29.14D). The second signal (first-derivative spectrum with the six hyperfine lines) is attributed to a second type radical R2. Microwave power saturation tests suggest that R2 is an oxygen-centered radical and R1 is a carbon-centered radical because the former saturates at a higher microwave power than the latter (see Figure 29.15).

Discussion about the dependency of the ESR signal intensity of oxygen-centered radicals on microwave power can be found in [10]. Further discrimination of these radicals was performed by heating the sample at 75°C, 100°C, 130°C, and 140°C. Figures 29.16A and 29.16B show the ESR spectra due to the radicals R1 and R2, respectively, recorded as a function of time at 130°C. At this temperature, R2 suffered more loss (decayed) than did R1, and when heated for 90 min, the signal due to R2 fell below the detection limit of the spectrometer. At 140°C both R1 and R2 decayed completely in about 18 min. This result suggests that both R1 and R2 may have been trapped within crystalline regions; R1 resides completely and R2 is partially inside crystalline regions—a dangling radical. Therefore, the most likely species for R1 is a polyenyl radical (—$^{\bullet}$CH—[CH = CH—]m—), a radical having a large number (m) of conjugated double bonds. A radical of this type can produce a single-line ESR spectrum with no resolvable proton splitting due to delocalization of the unpaired spin. Radical 2 could be a similar radical, but it is stretching out of, or dangling at, the crystalline regions because it can be annealed at slightly lower than melting temperature.

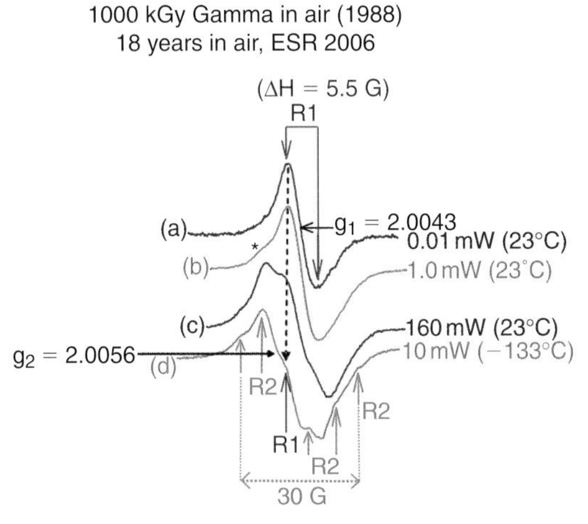


**FIGURE 29.14** First derivative ESR spectrum recorded as a function of microwave power at different temperatures. UHMWPE (GUR 4150) was gamma-irradiated (1000 kGy, 60 Co) in 1988 and spectra recorded in 2006. The spectral features due to the radicals R1 and R2 are shown with arrows.

The ESR signal due to R2 exhibits proton coupling (~6 lines, $a_H$ = 4.8 G), which is more pronounced when the delocalization is frozen at a lower temperature (−133°C), suggesting that the number of conjugated bonds for R2 is less than that for R1. That means R2 could be a dienyl or trienyl (m = 2 or 3) radical, whereas R1 could be a polyenyl (m > 3). Table 29.2 lists the ESR parameters of R1 and R2.

While the details of this investigation can be found in the literature [22, 25, 26], the important findings about the OIR can be highlighted as follows:

1. In presence of oxygen, two radicals (R1 and R2) remain trapped in the irradiated UHMWPE for a very long time (>20 years in this laboratory).
2. One of them (R1) is a polyenyl P$^{\bullet}$ (—$^{\bullet}$CH— [CH = CH—]$_m$— with m>3).

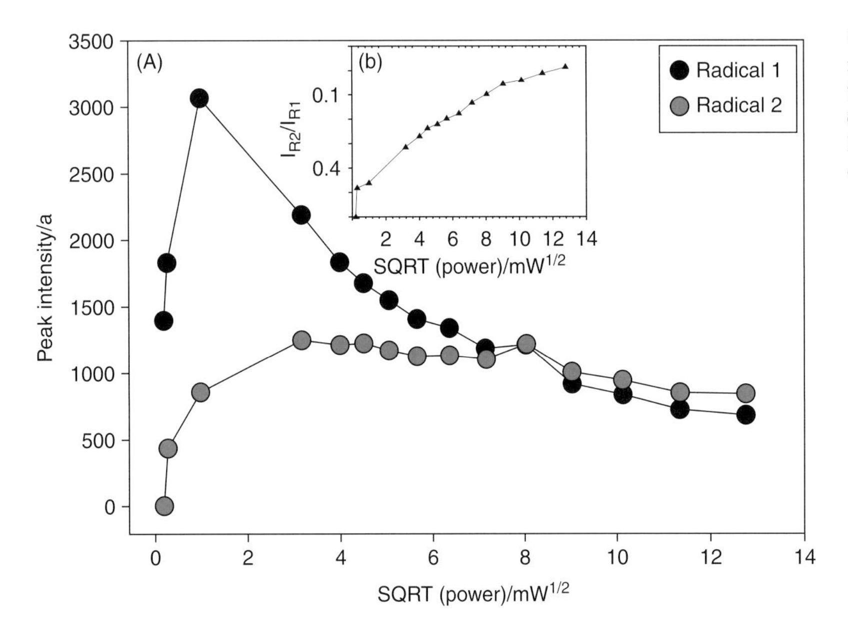


**FIGURE 29.15** Microwave power saturation technique (PST). (A) Signal intensity as a function of the square root of the microwave power for the radicals R1 and R2. (B) Ratio of the signal intensities of R1 and R2 plotted as a function of the square root of the microwave power (see inset).

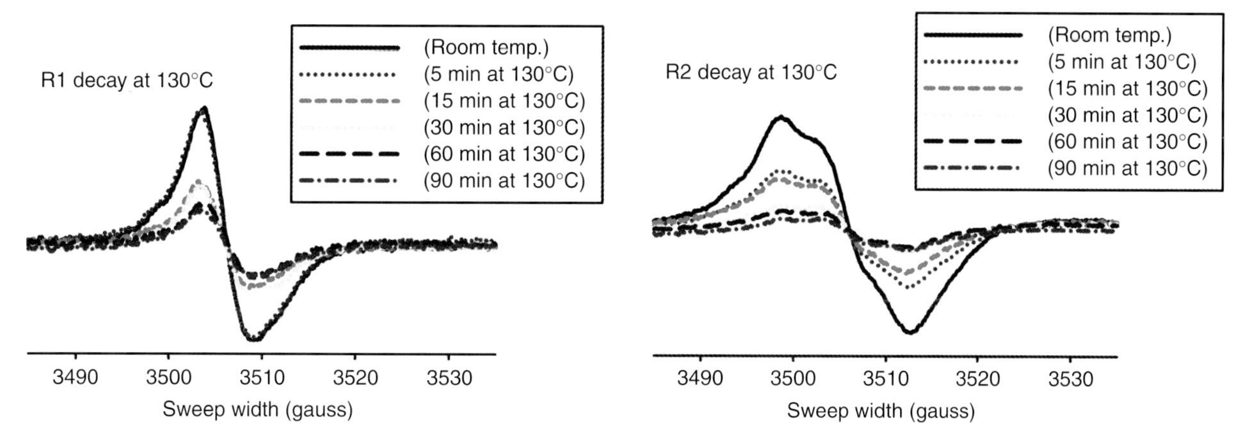


**FIGURE 29.16** First-derivative ESR spectrum of radical R1 (A) and R2 (B) are recorded as a function of time at 130°C.

**TABLE 29.2 ESR Parameters of Radical-1 (R1) and Radical-2 (R2)**

| Radical | g-value | Number of lines | Line width | Hyperfine constants, aH |
|---|---|---|---|---|
| R1 | 2.0044 ± 0.0003 | 1 | 5.5 G | - |
| R2 | ~2.0056 | 6 | 9.4 G | 4.8 G |

3. The second radical R2 is an oxygen-centered polyenyl PO$^{•}$(—$^{•}$OCH— [CH = CH—]$_m$— with m = 2 or 3).
4. R2 is a dangling radical; it resides in a less crystalline region than R1.
5. There is no ESR evidence of the presence of a long-lived *peroxy* radical.

## 29.5 INTERMEDIATE RADICALS

In recent years, ESR work on UHMWPE has been limited to investigations of the peroxy (POO$^{•}$) and the long-lived OIR because of their potential damaging effects (oxidative degradation). Consequently, intermediate radical processes remain significantly unexplored. Of course, chemical models or schemes describing the processes for generation, propagation, and termination of the PE radicals can be found in the literature [1, 2, 27, 28]. This section presents ESR evidence of certain intermediate radical reaction in irradiated UHMWPE.

As shown in Figure 29.10, it takes 30–45 days for the OIR to grow as the primary radicals decay in air. Interestingly, the "primary residual" radicals present in a vacuum-sealed sample following heating at 75°C for 6 years also produces OIR within the same length of time when opened to air. The signal intensity in the latter case is, however, very weak because of the low primary radical concentration at the time of reaction with oxygen (air). This scenario suggests that the oxidation potential of the irradiated UHMWPE components can be reduced, but not eliminated, by heating at a temperature of 75°C in a vacuum or inert environment. Heating or annealing at a near-melting temperature may eliminate all radicals or reduce their concentration to a nondetectable limit of ESR. Another important observation is that the polyenyl radicals that constitute the terminal OIR (previously discussed) are not the primary or initial polyenyl; rather, they are generated as a result of the reaction of the primary radicals in the presence of oxygen.

During the initial 30–45 days when the primary radicals decay in the presence of oxygen, the ESR spectrum changes as a function of time and becomes very complex. The complexity in the ESR spectrum arises from the fact that the radicals at or near the surface (OIR polyenyl and/or peroxy), in the core (primary alkyl/allyl), and the other unknown secondary or tertiary radicals in between the surface and the core produce their own individual spectrum within the same region of the resonance magnetic field. Therefore, unambiguous identification of the radical type or types becomes very difficult. In one of our studies, we recorded an ESR signal and attributed it to a vinyl-type radical. The vinyl feature in the spectrum became clear right before the appearance of the OIR signal, at which time the signal due to the alkyl/allyl decayed or disappeared. For verification, a computer simulation of the spectrum was performed (see Figure 29.17).

In another study, we measured radical concentration (RC) in a series of samples (1.20 ± 0.05 mm thick) prepared from the surface to the core of a 1-inch cube of

UHMWPE (GUR 4150). The cubes were aged for 1 year in Ringer solution, bovine serum, ethyl alcohol, or air at room temperature following gamma irradiation (30kGy, 60 Co) in air at room temperature. Each surface sample showed the presence of the surface radical (OIR), and the core sample showed the primary radicals. The remaining samples showed the presence of a combination of different radicals. At about 5-mm depth at that particular time of measurements, RC was minimal or none (see Figure 29.18). Following these measurements, the samples were kept in air at room temperature and measured again after 10 months. To our expectation, each sample showed the presence of identical OIRs. Figure 29.18 also shows the representative ESR spectra. This study concluded that a nonradical region could be found between the surface and the core of an irradiated UHMWPE solid during the time of decay of the primary and growth of the OIR. It was further suggested that the *nonradical region* could possibly be the region of hydroperoxy species, which cannot be detected by ESR [29].

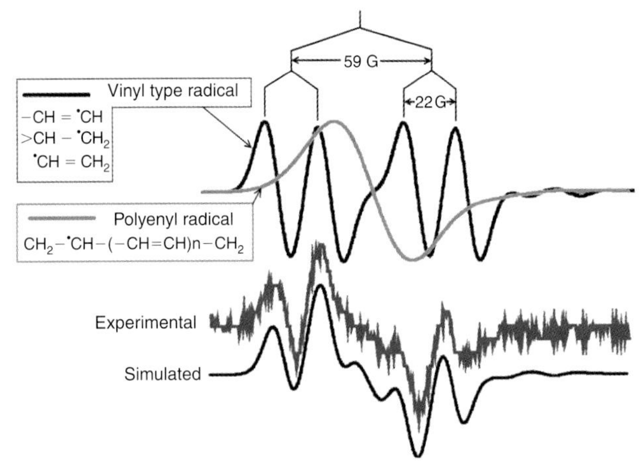


**FIGURE 29.17** First-derivative ESR spectrum of gamma-irradiated UHMWPE (GUR 4150) recorded during the decay of the primary radical but before the appearance of the OIR. Experimental, simulated, and the split spectra of the vinyl and polyenyl radicals are also shown.

## 29.6 ESR OF VITAMIN E-DOPED UHMWPE

Vitamin E (α-tocopherol, α-T) has been found to improve oxidation resistance of UHMWPE [30–32] by scavenging free radicals. Most of the reports on the antioxidation behavior of α-T are, however, based on the FTIR data or mechanical test results; ESR data are minimal at best, to

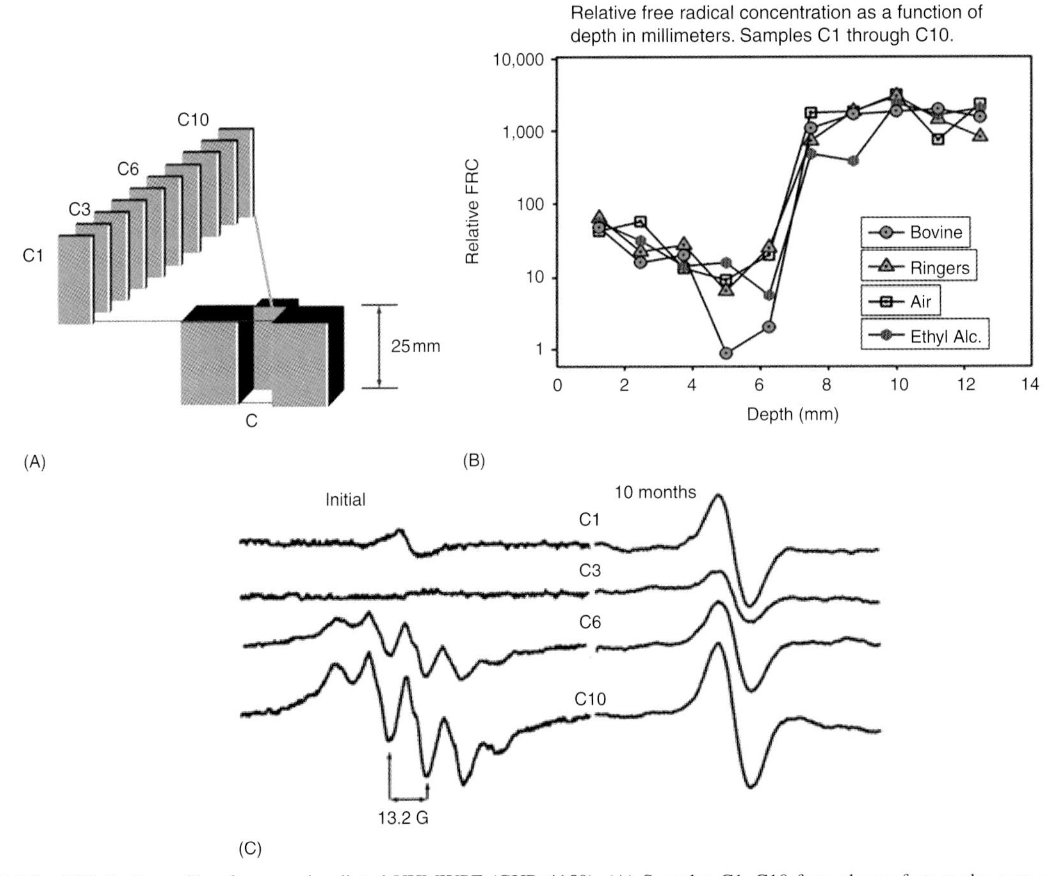


**FIGURE 29.18** ESR depth profile of gamma-irradiated UHMWPE (GUR 4150). (A) Samples C1–C10 from the surface to the core were prepared following postgamma aging for 120 days in air, bovine serum, Ringer solution, or ethyl alcohol. (B) Relative free radical concentration as a function of depth. (C) ESR spectrum of representative samples recorded immediately after they were prepared (left) and 10 months later (right).

our knowledge. While a short summary of the recent ESR works is provided in this section, details can be found in the publications by Ridley and Jahan [33, 34]. More knowledge about the ESR of the antioxidants (tocopherols or Irganox®) can be gained from the reports on the reactions of vitamin E with lipids by Chen et al. [35] and Walker et al. [36] and on low density polyethylene (LDPE) by Jaworska et al. [37] and Yamazaki and Seguchi [38].

As previously mentioned, we reported ESR results on α-T-doped UHMWPE powder resin (labeled as "α-T-resin") and compression molded solid [33]. In practice, less than 1.0 wt% α-T is blended with UHMWPE to avoid plasticization [39], but we used concentrations up to 25% v/v to allow detection of α-T radicals in the presence of strong resonance signals of the primary or secondary PE radicals. Jaworska et al. [37] used 10% Irganox® in low density PE (LDPE) for detection of the antioxidant radicals at liquid nitrogen temperature (77 K).

Vitamin E produces its own radical (tocopheroxyl, α—T— O•) when exposed to room light or ionizing radiation. The molecular structure of vitamin E, and of its radical, are shown in Figure 29.19A. The ESR spectrum of the vitamin E radical is shown in Figure 29.19B (as received) and 29.19C (after gamma irradiation). As shown in Figure 29.19C, the first-derivative ESR spectrum consists of seven resonance lines with separation between the lines $\Delta H = 5\,G$, the line width $\Delta W = 2\,G$, and the intensity ratios 1:6:15:20:15:6:1 (shown by the stick diagram in Figure 29.19D). The characteristic spectral splitting factor, g-value, of the radical is $g = 2.0044$ (shown by an arrow). These data are consistent with the reports found in the literature [35, 40, 41].

According to our recent report [33], vitamin E (α-tocopherol) was found to quench the primary PE radicals in the powder sample (GUR 1050) when gamma irradiation (30 kGy, 60 Co) was performed in air, but it did not quench any radicals in the nitrogen environment. When the irradiated powder was removed from nitrogen and placed in air, radical reactions occurred, and the OIR formed as detected by ESR. So far as the decay of the radicals in the presence of oxygen (air) is concerned, no detectable difference was found between the powder samples with or without vitamin E. The good fit between the experimental and the computer-generated ESR spectra shown in Figure 29.20A for the initial measurements, and in Figure 29.20B for the final measurements (71 days later), confirm the fact that the vitamin E (αTO•) and the resin (primary or OIR) radicals decay independently.

Three observations are made in this study.

1. So far as the postirradiation ESR in air is concerned, no detectable difference is found between the nitrogen-packaged and the open-air α-T-resin samples.
2. A high concentration of α-T allows vitamin E radicals (ESR signal of αTO•) to be monitored in the presence of the relatively strong signal due to PE radicals. This is potentially useful for comparison of the radical-radical reaction [37].
3. Although UHMWPE resin powder has no clinical relevance, a powder sample provides a better means for investigation of free radical reaction in presence of oxygen (air) because it has a large effective surface area, which is easily accessible to oxygen.

Similar postirradiation oxidation behavior was also observed in compression-molded solids (UHMWPE, GUR 1050) containing 0.0% (control), 1.0%, or 10.0% α-T. No detectable difference was found in the subsequent decay pattern between the control (0% α-T) and vitamin-E containing UHMWPE. However, one important observation was made in this study. Compared to the control (0% α-T), all UHMWPE-α-T samples were found to contain about an order of magnitude less primary radicals (see Figure 29.21).

Because of lack of initial data points, it is speculated that the quenching of radicals by α-T must occur during

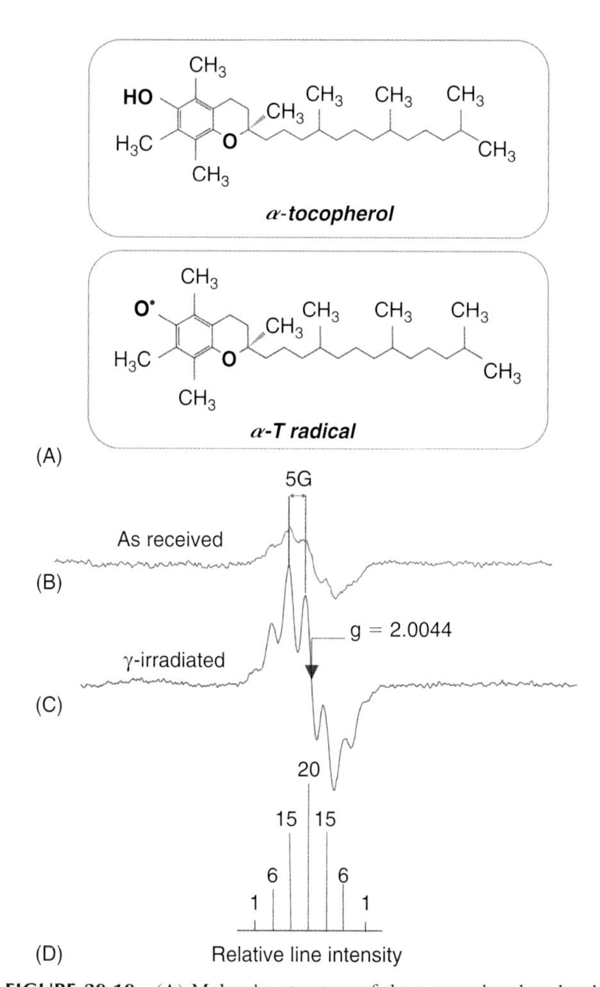


FIGURE 29.19 (A) Molecular structure of the α-tocopherol molecule and the α-tocopheroxyl radical (αTO•). (B) ESR spectrum of the αTO• radical present in as-received vitamin E. (C) Fingerprint of the αTO• with g-value $g = 2.0044$ and line separation $\Delta H = 5\,G$. (D) The relative intensities of the seven lines are shown by stick diagram.

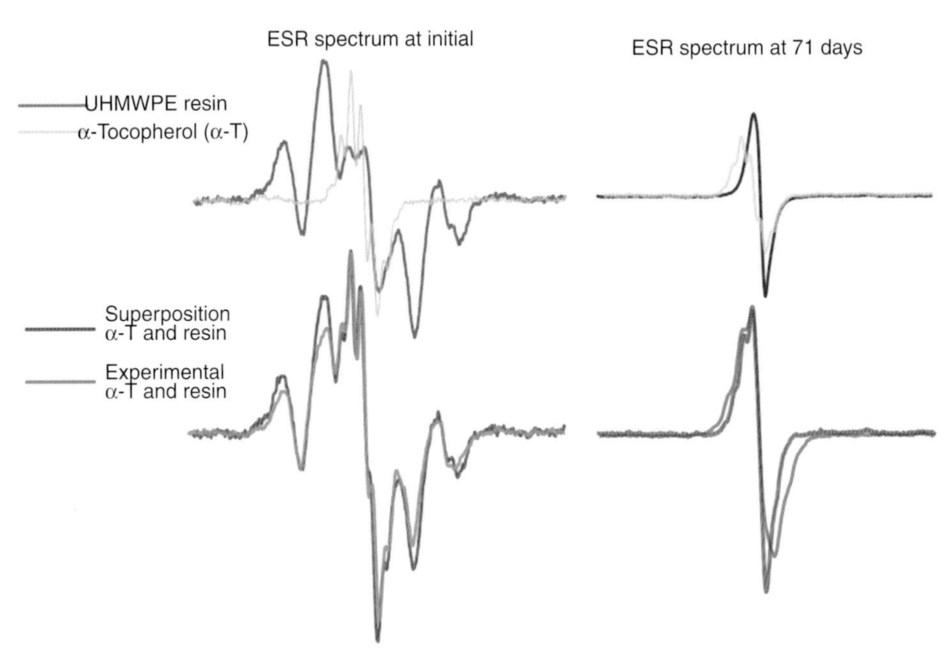


**FIGURE 29.20** (A) The ESR spectrum due to the primary/initial radical in vitamin E (αTO˙) and UHMWPE (top). Bottom figure illustrates the fit between the experimental and the computer-generated spectra. (B) The ESR spectrum due to the OIR and the radical in vitamin E (αTO˙) and UHMWPE (top). Bottom figure illustrates the fit between the experimental and the computer-generated spectra.

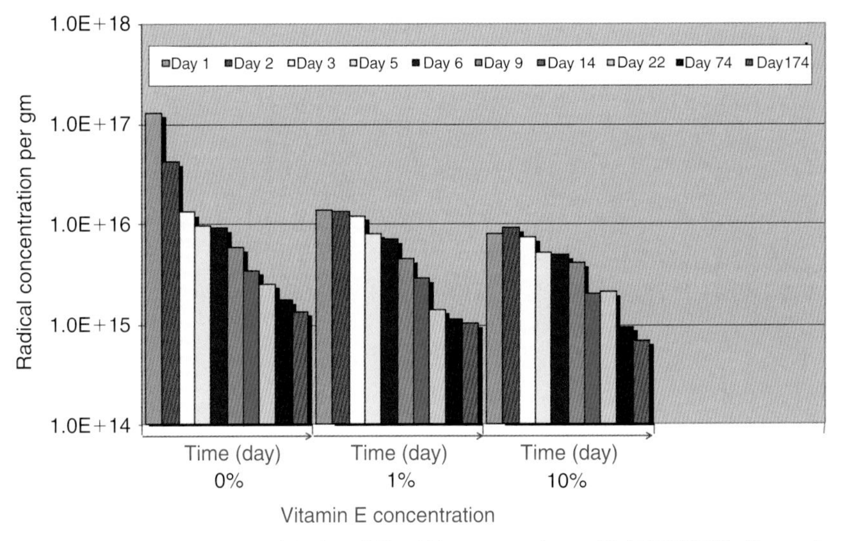


**FIGURE 29.21** Free radical concentration plotted as a function of % α-T in compression molded UHMWPE. Gamma irradiation (30 kGy, 60 Co) was performed in air or nitrogen.

or immediately after irradiation. The time taken (3–5 days) in receiving the irradiated samples from the vendor is the main delay between the irradiation and the ESR measurements. Further study is needed to address this issue.

## 29.7 QUANTITATIVE ESR IN UHMWPE

ESR has been in use for more than 5 decades. During that time, high-sensitive ESR spectrometers have been designed and built to detect low level paramagnetic ions, ionic impurities, and free radicals in a host material. Still, quantitative measurement to determine the "absolute" number of spins or radicals remains a challenge to the ESR community. As pointed out by Mazur [42], careful operational procedures might reduce potential errors to within 2–5% within a given laboratory; however, quantitative analyses of the same sample in different laboratories may vary between 100 and 400%. He listed a number of sources of errors that could lead to such a wide variation in the results. The following list shows the ones that are particularly important for UHMWPE work:

- Sample related: Sample volume and shape, sample tube, sample positioning into the microwave cavity
- Microwave cavity: Microwave field profile, cavity Q-factor, modulation field profile
- Spectrometer: Spectrometer calibration, signal-to-noise ratio, and mechanical and electronic stability of spectrometer
- Reference standard: Selection of a standard, absolute spin concentration, storage, and handling of the standard
- Environment: Laboratory temperature, temperature variation inside the cavity, relative humidity, moisture or condensation inside the cavity, and moisture in the "sample" and the "standard"
- Computational: Limits of integration, base-line correction, noise correction, double integration, and formula for final number

For reliable test results, it is essential to have a standard protocol that addresses most or all of the previously mentioned sources of errors and to demonstrate consistency and reproducibility of the results for a long period of time. The most common standards used for quantitative ESR are: (1) the stable organic free radical 2-2-diphenyl-1-picrylhydrazyl (DPPH); (2) Mn2+ in MgO; (3) ruby; (4) vanadyl (V) in K2SO4 or NaCl; and (5) K3CrO8 in K3NbO8. The fifth standard, K3CrO8 in K3NbO8, was recently employed by Cage et al. for g-factor, spin concentration, and field calibration at high field [43]. Ruby is Cr3+-doped aluminum oxide (Al2O3). It is suitable for use with UHMWPE because its anisotropic resonance lines do not interfere with those of the PE radicals. For "alanine dosimetry," Nagy et al. [44] used ruby as an "adjacent reference" by permanently fixing it inside the cavity wall.

The importance of the cavity Q-factor for sample and standard has been discussed by Blakley et al. [45]. Q-factor is given by: $Q=f_0/\Delta f_0$, where $h\Delta f_0$ is the amount of dissipative energy at resonance frequency $f_0$ (h is Planck's constant). Less dissipation, or loss in microwave energy, results in a higher Q-factor and better sensitivity of the system. The Bruker EMX spectrometer, equipped with a high sensitive cavity (ER4119HS), registers the Q-factor at the time of measurement.

ESR theory for determination of the number of spins or radicals in a given sample is described in detail by Poole [46]. The theory, however, does not fall within the scope of this chapter. Nonetheless, to gain knowledge about the selection of standard reference materials and experimental variables, additional references are cited [47–50].

## REFERENCES


1. Dole M. Radiation chemistry of macromolecules Vol. 1. New York; 1972. Vol. 2. New York, 1973.
2. Rabek J. *Polymer photodegradation: mechanisms and experimental methods*: Springer; 2007.
3. Eyerer P. Property changes of UHMW polyethylene during implantation—first hints for the development of an alternative polyethylene. In: Lee SM, editor. *Advances in biomaterials Vol. 1*: CRC Press; 1987.
4. Igarashi M. Free radical identification by ESR in polyethylene and nylon. *J Polym Sci: Polym Chem* 1983;**21A**:2405–25.
5. Jahan MS, Wang C, Schwartz G, Davidson JA. Combined chemical and mechanical effects on free radicals in UHMWPE joints during implantation. *J Biomed Mater Res* 1990;**25**(8):1005–17.
6. Jahan MS, Thomas DE, Banerjee K, Trieu H, Haggard W, Parr JE. Effects of radiation-sterilization on medical implants. *Radiat Phys Chem* 1998;**51**:93–594.
7. Nausbaum HJ, Rose RM. The effects of radiation sterilization on the properties of ultra-high molecular weight polyethylene. *J Biomed Mater Res* 1979;**13**:557–76.
8. O'Neill P, Birkinshaw C, Leahy J, Barklie R. The role of long lived free radicals in the ageing of irradiated ultra high molecular weight polyethylene. *Polym Degrad Stab* 1999;**63**:31–9.
9. Streicher RM. Influence of ionizing radiation in air and nitrogen for sterilization of surgical grade polyethylene for implants. *Radiat Phys Chem* 1988;**31**:693–8.
10. Seguchi T, Tamura N. Mechanism of decay of alkyl radicals in irradiated polyethylene on exposure to air as studied by electron spin resonance. *J Phys Chem* 1973;**77**:40–4.
11. Sun DC, Stark C. *U.S. Patent No. 414,049*; 1995.
12. Kurtz SM, Muratoglu OK, Evans M, Edidin AA. Advances in the processing, sterilization, and crosslinking of UHMWPE for total joint arthroplasty. *Biomaterials* 2000;**20**:1659.
13. Buchanan FJ, White JR. The influence of gamma irradiation and ageing on degradation mechanisms of ultra-high molecular weight polyethylene (UHMWPE). *J Mater Sci Mater Med* 2001;**12**(1):29.
14. Chang R. *Basic principles of spectroscopy*. New York: Robert Krieger Publishing Company; 1978. 104–129.
15. Weil JA, Bolton JR, Wertz JE. *Electron paramagnetic resonance*. New York: John Wiley and Sons, Inc.; 1994. 1–32.
16. Ohnishi S, Sugimoto S, Nitta I. Electron spin resonance study of radiation oxidation of polymers. Results for polyethylene and some general remarks. *Polymer* 1962;**2**:119.
17. Ohnishi S, Sugimoto S, Nitta I. Electron spin resonance study of radiation oxidation of polymers. IIIA. Results for polyethylene and some general remarks. *J Polym Sci* 1963;**1A**:605–23.
18. Nakamura K, Ogata S, Ikada Y. Assessment of heat and storage conditions on γ-ray and electron beam irradiated UHMWPE by electron spin resonance. *Biomaterials* 1998;**19**:2341–6.
19. Costa L, Luda MP, Trossarelli L, Brach del Prever EM, Crova M, Gallinaro P. Oxidation in orthopaedic UHMWPE sterilized by gamma-radiation and ethylene oxide. *Biomaterials* 1998;**19**: 659–68.
20. Ohnishi S, Ikeda Y, Sugimoto S, Nitta I. On the ESR singlet spectra frequently observed in irradiated polymers at a large dose. *J Poly Sci* 1960;**47**(149):503–7.
21. Ranby B, Rabek JF. *Photodegradation, photooxidation and photostabilization of polymers, principles and applications*. New York: John Wiley; 1975 60–71.

22. Jahan MS, Durant J. Investigation of the oxygen-induced radicals in ultra-high molecular weight polyethylene. *Nucl Instr Meth Phys Res* 2005;**B 236**:166–71.
23. Auerbach I, Sanders LH. *Polymer* 1969;**10**:579.
24. Davidson JA, Schwartz G, Lynch G, Gir S. Wear, creep and frictional heating of femoral implant articulating surfaces and the effect of long-term performance- Part II, friction, heating and torque. *J Biomed Mater Res: Appl Biomater* 1988;**22**:69–91.
25. Durant J, Jahan MS. EPR power saturation techniques and spectral differentiation are used to isolate and simulate radical species in UHMWPE. *Nucl Instr Meth Phys Res* 2005;**B 236**:160–5.
26. Jahan MS, Fuzail M. Examination of the long-lived oxygen-induced radicals in irradiated ultra-high molecular weight polyethylene. *Nucl Instr Meth Phys Res* 2007;**B 265**:67–71.
27. Costa L, Bracco P. Mechanism of croslinking and oxidative degradation of UHMWPE. In: Kurtz SM, editor. *The UHMWPE handbook: ultra-high molecular weight polyethylene in total joint replacement*. New York: Academic Press; 2004 chapter 11. p. 235–50.
28. Khelidj N, Colin X, Audouin L, Verdu J, Monchy-Leroy C, Prunier V. Oxidation of polyethylene under irradiation at low temperature and low dose rate. Part I: The case of pure radiochemical initiation. *Polym Degrad Stab* 2006;**91**:1593–7. Part II. Low temperature thermal oxidation. *Polymer Degradation Stability*; 2006, 91, 1598–605.
29. Jahan MS, Stovall JC, King MC. Observation of a non-radical intermediate in the oxidation pathway of free radicals in gamma-irradiated medical grade polyethylene. *Nucl Instr Meth Phys Res* 2001;**B 185**:323–7.
30. Oral E, Wannomae KK, Hawkins N, Harris WH, Muratoglu OK. α-Tocopherol-doped irradiated UHMWPE for high fatigue resistance and low wear. *Biomaterials* 2004;**25**:5515–22.
31. Shibata N, Tomita N. The anti-oxidative properties of α-tocopherol in α-irradiated UHMWPE with respect to fatigue and oxidation resistance. *Biomaterials* 2005;**26**:5755–62.
32. Tomita N, Kitakura T, Onmori N, Ikada Y, Aoyama E. Prevention of fatigue cracks in ultra high molecular weight polyethylene joint components by the addition of vitamin E. *J Biomed Mater Res (Appl Biomater)* 1999;**48**(1999):474–8.
33. Ridley MD, Jahan MS. Measurements of free radical in vitamin E-doped ultra-high molecular weight polyethylene: dependence of materials processing and irradiation environments. *Nucl Instr Meth Phys Res* 2007;**B 265**:62–6.
34. Ridley MD, Jahan MS. Effects of packaging environments on free radicals in γ-irradiated UHMWPE resin powder blend with vitamin E. *J Biomed Mater Res A* 2009 March 15;**88**(4):1097–103.
35. Chen Z. Biology and life sciences. In: Kawamori A, Yamauchi J, Ohta H, editors. EPR in the 21st Century; 2002.
36. Walker M, Beckert D, Lash J. Interaction of UV light-induced α-tocopherol radicals with lipids detected by an electron spin resonance prooxidation effect. *Photochem Photobiol* 1998;**68**(4):502–10.
37. Jaworska E, Kaluska I, Strzelczak-Burlinkska G, Michalik J. Irradiation of polyethylene in the presence of antioxidants. *Radiat Phys Chem* 1991;**37**(2):285–90.
38. Yamazaki T, Seguchi T. ESR study on chemical crosslinking reaction mechanisms of polyethylene using a chemical agent-II. The effect of phenolic antioxidants. *J Polym Sci, Part A: Polym Chem* 1997;**35**(12):2431–9.
39. Oral E, Greenbaum ES, Malhia AS, Harris WM, Muratoglu OK. Characterization of irradiated blends of α-tocopherol and UHMWPE. *Biomaterials* 2005;**26**:6657–63.
40. Boguth W, Nieman H. Electron spin resonance of chromanoxy free radicals from $\alpha-,\varepsilon^2-,\beta-,\gamma-,\delta-$tocopherol and Toco*. *Biochimica et Biophysica Acta* 1997 April 27;**BBA 55931**(248):121–30.
41. Maguire JJ, Wilson DS, Packer L. Mitochondrial electron transport-linked tocopheroxyl radical reduction. *J Biol Chem Commun* 1989;**264**(36):21462–5.
42. Mazur M, Valco M. Error sources in quantitative EPR spectroscopy. *EPR Newsletter* 2003;**13**:27.
43. Cage B, Weekley A, Brunel LC, Dalal N. $K_3CrO_8$ in $K_3NbO_8$ as a proposed standard for g-factor, spin concentration, and field calibration in high-field EPR spectroscopy. *Anal Chem* 1999;**71**:1951–7.
44. Nagy V, Sleptchonok OF, Desrosiers MF, Weber RT, Heiss AH. Advancements in accuracy of the alanine EPR dosimetery sysytem, Part III: usefulness of an adjacent reference sample. *Rad Phy Chem* 2000;**59**:429–41.
45. Blakley RL, Henry DD, Morgan WT, Clapp WL, Smith CJ, Barr D. Quantitative electron paramagnetic resonance: the importance of matching the Q-factor of standards and samples. *Appl Spectrosc* 2001;**55**(10):1375–81.
46. Poole CP. *In Electron spin resonance, a comprehensive treatise on experimental techniques*. 2nd ed.: John Wiley & Sons, Inc.; 1983 chapter 11D, 400–409.
47. Nagy V. Choosing reference samples for EPR concentration measurements: Part 1, general introduction and systems of S = 1/2. *Anal Chim Acta* 1997;**339**(1–2):1–11.
48. Madej A, Dyrek K, Mattusch J. Preparation and evaluation of the quality of standards for quantitative EPR measurements of spin concentration. *Fresenius J Anal Chem* 1991;**341**:707–8.
49. Burns D, Flockhart BD. Application of quantitative EPR. *Phil Trans R Soc London* 1990;**333A**:37–48.
50. Chang T, Kahn AH. Standard reference materials: electron paramagnetic resonance intensity standards: SRM 2601, description and use. NIST's special publication 1978: 260-59, 38–45.

Chapter 30

# Fatigue and Fracture of UHMWPE

Francisco J. Medel, PhD, and Jevan Furmanski, PhD



## 30.1 INTRODUCTION

In recent years, the mechanical performance of UHMWPE has attracted a great deal of interest [1–12]. Fatigue crack growth resistance and fracture strength are desirable properties used to guide the development of first- and second-generation highly crosslinked UHMWPE formulations [2, 13–17]. The clinical significance of the fatigue and fracture properties of UHMWPE materials depends on the prosthetic device. Abrasive and adhesive wear mechanisms usually prevail in retrieved acetabular liners, whereas fatigue-damaged surfaces are often observed in explanted tibial inserts [2, 18]. Therefore, wear resistance is of principal importance in acetabular components, whereas the typical lower congruency and higher contact pressures of the knee joint make fatigue and fracture properties more relevant [16]. Nevertheless, acetabular component cracking and fractures at the rim of acetabular liners have been reported recently [19, 20], typically associated with abnormal loading or impingement at the nonarticulating region [19]. Furthermore, if the UHMWPE formulation has not been stabilized, *in vivo* oxidation may alter its initial properties [21, 22], potentially influencing the wear and mechanical performance of prosthetic components in the long term. It can be thus concluded that medical grade UHMWPE formulations should represent a balance between design considerations, wear resistance, oxidation stability, and fatigue and fracture properties as well.

The development of both first- and second-generation highly crosslinked UHMWPEs has motivated numerous studies on fatigue and fracture properties of these materials, as well as of conventional UHMWPE, but they differ as far as the philosophical and experimental approaches are concerned. The characteristics of highly crosslinked UHMWPE have already been summarized in detail in Chapters 6, 13–15, and 20, and the reader is referred to these earlier sections if the terminology in this chapter regarding these materials seems unfamiliar. The aim of this chapter is to provide a theoretical background on fatigue and fracture concepts, as well as the different philosophical approaches and experimental techniques available with special regard to UHMWPE. We also outline the main findings on fatigue and fracture properties of contemporary medical UHMWPEs.

## 30.2 FATIGUE RESISTANCE

### 30.2.1 Basic Concepts of Fatigue Resistance

Suresh's classic definition states that fatigue is the progressive and localized structural damage that occurs when a material is subjected to cyclic loading [23]. Coherently, fatigue life is defined as the number of cycles needed to reach catastrophic failure of the studied material. From a structural point of view, the fatigue damage progression initiates microstructure changes that subsequently lead to the nucleation of microscopic cracks. Further on, the microscopic cracks grow and coalesce to form a dominant macroscopic crack that propagates in a stable manner until structural instability or complete fracture is reached. Therefore, the fatigue life of a specific material or component comprises five stages, namely development of permanent damage via microstructural changes, initiation of microscopic cracks, growth and coalescence of microscopic flaws into dominant cracks, stable propagation of the dominant crack, and final fracture [23]. Thus, the total life is a composite of events, associated with the initiation and propagation of a critical fatigue crack. Based on these definitions, two main philosophical approaches may be adopted to analyze the fatigue strength of materials: the total life and defect tolerant approaches. The total life approach aims to predict the service life of a component presupposing that the starting material and laboratory specimens are free of defects, that is, uncracked or unflawed, and comparing laboratory data of the fatigue life as a function of stress to predicted service conditions. In contrast, the defect tolerant approach presumes that the component and the laboratory specimen are both flawed to some degree, by at least the maximum size not detectable by the inspection method employed. In this case, the initiation time is ignored because it is presumed that initiation has either occurred or is irrelevant. Thus, the defect tolerant approach is only concerned about the growth of the presumed flaws or the prevention of their growth through design. Further, the defect tolerant approach is more conservative than the total life approach because it presumes that there is no initiation time, when, in general, some time is needed to generate a coherent flaw in the material. The time predicted to propagate a flaw to failure by the defect tolerant approach is, therefore, a lower bound estimate. This approach makes use of fracture mechanics and fatigue crack propagation experiments and analyses to predict the behavior of flaws in components, compared to the more traditional stress analysis employed in the total life approach to fatigue analysis.

### 30.2.2 Fatigue Analysis: Total Life Approach

To predict the total life of a component under fatigue conditions, simple experiments are run on unnotched, nominally smooth specimens of the component material, where the time (or number of load cycles) to failure is recorded as a function of the applied stress. Stress analysis allows translation between the laboratory data and predicted stresses in the component, thus allowing estimation of the service life of the component. The time to failure of the laboratory specimen is termed the total life because it proceeds from a presumably unflawed initial condition, where part of the life is expended in causing the damage that can coalesce into a coherent flaw, and the rest is spent propagating that flaw to failure of the component. Fatigue laboratory specimens are typically subjected to cyclic load or strain, describing a custom-defined waveform from a minimum stress, $\sigma_{min}$, to a maximum stress, $\sigma_{max}$, or from a minimum strain, $\varepsilon_{min}$, to a maximum strain, $\varepsilon_{max}$. Accordingly, total life experiments may be classified as stress-based or strain-based depending on what parameter was controlled during testing. Various parameters are usually given to describe a stress-based total life fatigue test, namely, stress range, $\Delta\sigma = \sigma_{max} - \sigma_{min}$, stress amplitude, $\sigma_a = 1/2(\sigma_{max} - \sigma_{min})$, mean stress, $\sigma_m$, stress ratio, $R = \sigma_{min}/\sigma_{max}$, frequency, $\nu$, number of cycles to failure, $N_f$, and the waveform (e.g., sine, triangle) [9]. Similar parameters are defined for strain-based total fatigue life experiments. Classically, total fatigue life experiments in which failure is reached in less than 1000 cycles are termed low-cycle fatigue, and laboratory specimens typically experience cyclic strains within the plastic range. In contrast, cyclic elastic strains prevail in high-cycle fatigue experiments.

The characterization of the material (stress-based) fatigue strength utilizes the construction of the so-called stress-life, or S-$N_f$, plot (Figure 30.1), where S represents the stress amplitude and $N_f$ the cycles to failure (or some critical state) on a semilogarithmic scale. S-$N_f$ plots account for the number of cycles needed to reach failure at decreasing stress amplitudes, and they are mathematically expressed

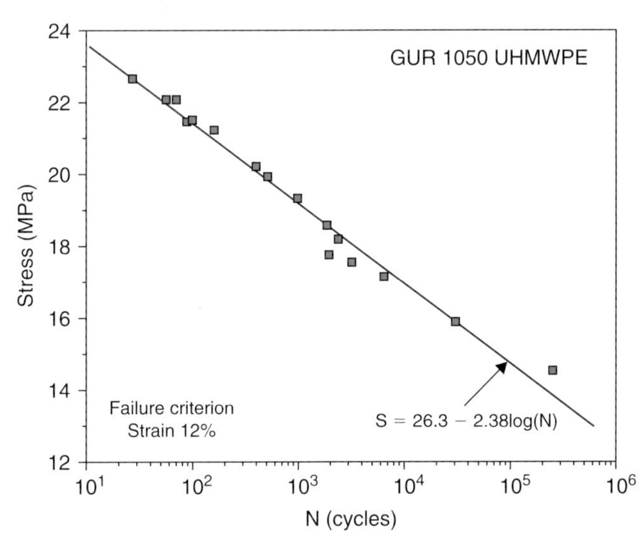


**FIGURE 30.1** Typical stress-yield life, S-N, plot for as-received compression molded GUR 1050 UHMWPE.

by means of the Basquin relation, $S = A\ N_f^{\ B}$. The highest stress level that results in run-out (survival at 10 million cycles, or "infinite life") in the S-$N_f$ plot is termed the endurance limit [9].

### 30.2.3 Total Life Fatigue Testing in UHMWPE

As opposed to metals [23, 24], the question of fatigue flaw nucleation remains poorly understood for polymers, UHMWPE among them. Nevertheless, Williams and DeVries [25] have demonstrated nucleation of defects monitoring the number of free electrons in an amorphous polymer sample under cyclic load and found that the number of broken bonds in the material continually increased and matched the waveform of the load. Thus, chemical bonds in the polymer break every load cycle of the experiment, and damage accumulates over time, providing a mechanism for damage accumulation and flaw nucleation. However, the main limitation of total fatigue life experiments in the case of UHMWPE is that they rarely, if ever, result in the actual rupture of the specimen below its ultimate tensile strength, not even at substantial cyclic plastic strains [26]. This might lead to the erroneous conclusion that if a UHMWPE part is devoid of flaws that it will not fail in fatigue. To the contrary, cracks have been seen to initiate at and propagate from notches and stress concentrations in UHMWPE components [27–29]. As a consequence, the formation and propagation of fatigue cracks in UHMWPE has been predominantly tackled from a defect tolerant point of view, and thus far there is a paucity of studies that report total life testing to catastrophic failure of UHMWPE [12, 30, 31]. Moreover, only one has done so in a traditional high-cycle fatigue manner [32].

Total life studies can differ in loading conditions, shape of specimens, temperature, and environment. While Weightman and Light employed a rotating-bending fatigue setup [32], the works by Villarraga and coworkers [12, 31] represent an adaptation of the small punch technique for evaluation of the total life to fracture of various UHMWPE formulations under biaxial loading conditions. On the other hand, O'Connor and colleagues built a multi-station flexural fatigue tester to load conventional and warm irradiated adiabatically melted (WIAM) crosslinked UHMWPE fatigue specimens by bending them for 20 million cycles. Despite the high number of cycles in this study, WIAM crosslinked specimens did not fail [30]. Because total fatigue life testing requires extremely long experimental times, making machinery complications very frequent, Pruitt and coworkers [33] and Puértolas and colleagues [8, 34, 35] have used an alternative criterion (instead of complete fracture) for prematurely halting the fatigue test. They consider that microdamage has been initiated when UHMWPE reaches a critical quantity of plastic deformation (~12%) because evidence of microscopic fatigue damage has been reported based on ultrasmall angle X-ray scattering assessments [36]. Employing the mentioned criterion, Baker et al. found that the yield lives (i.e., number of cycles up to 12% strain) of gamma irradiated UHMWPEs subsequently treated with consecutive remelting and annealing processes were inferior to that of the virgin material, except for the material irradiated at the highest dose (200 kGy) [33]. The reason for the preserved yield life at the highest radiation dose in this study was attributed to strengthening of the amorphous regions, although the separate effects of irradiation and thermal treatments were not fully discriminated. On the other hand, electron-beam irradiated but not thermally stabilized UHMWPEs demonstrated longer yield lives than virgin material as the radiation dose increased, according to Urries et al. [34]. In further studies from the same group [8, 35], post-irradiation annealing and remelting treatments were found to be detrimental to the yield life of as-irradiated UHMWPEs, with remelting reducing yield lives below the levels of the virgin material regardless of the radiation dose. The yield life improvement in as-irradiated materials was correlated with the increase in lamellar thickness introduced by irradiation, whereas the reduction in yield life caused by annealing and remelting was associated with a decrease in lamellar thickness.

While the foregoing yield life results indicate relative resistance to the accumulation of inelastic strain in a fatigue scenario, they should not be used interchangeably with true total life data because definitive one-to-one correspondence between this metric and ultimate failure under static or cyclic loading conditions remains to be confirmed. Thus, there is still a need to verify these trends with similar UHMWPE formulations tested under cyclic loading up to catastrophic rupture. The accumulation of strain under sustained cyclic loading may be more indicative of the creep resistance of the material, which is indeed pertinent to its fatigue and fracture resistance, as discussed later in the viscoelastic crack propagation section.

### 30.2.4 Fatigue Analysis: Defect Tolerant Approach

#### *30.2.4.1 Linear Elastic Fracture Mechanics*

The defect tolerant approach assumes that materials or structural components are intrinsically flawed, as previously mentioned. In this scenario the fatigue life is based on the propagation of an initial crack or notch, which grows to a critical size driven by cyclic loading [9, 37, 38]. Linear elastic fracture mechanics (LEFM) is typically adopted to study fatigue crack propagation phenomena. LEFM assumes that materials behave in a purely elastic manner in the region close to a moving crack tip, and it provides a simple framework in which a single parameter, the stress intensity factor, K, adequately describes the stress and strain environment. The disadvantage is that no material behaves in a purely elastic manner under the severe conditions that take place at a moving crack tip. Nevertheless, for a wide range of problems and intrinsic material behavior, LEFM adequately describes the stress and strain environment during fatigue crack propagation and fracture.

In LEFM, the stress near a sharp crack tip increases with the power $1/\sqrt{r}$ as the distance from the crack, $r$, approaches zero, regardless of the mode of loading or geometry of the crack. There are three fatigue crack propagation modes depending on the loading conditions: Mode I is the crack-opening mode (applied stress perpendicular to crack plane and crack front), Mode II is the shearing mode, with a shearing type displacement parallel to the crack plane and perpendicular to the crack front, and Mode III is the twisting mode, where the displacement is a shear parallel to crack face and front. The stress intensity factor (in Mode I) can then be defined as:

$$\lim_{r \to 0} \sigma_{ij}^{(I)} = \frac{K_I}{\sqrt{2\pi r}} \left( f_{ij}^{(I)}(\theta) \right) \quad (1)$$

where the local stress field, $\sigma_{ij}$, is dependent on the stress intensity factor, K, and $f_{ij}^{(I)}$ is a function describing the mode-dependent stress distribution with the angle $\theta$ from the crack plane.

K is uniquely related to the length of the crack, a, the geometry of the system (through a geometric correction Y), and the remote loading, $\sigma_\infty$, on the crack by:

$$K = Y\sigma_\infty \sqrt{\pi a} \quad (2)$$

Figure 30.2 schematically depicts the Mode I stress intensity factor and its relationship to local and global stress. The opening mode is typically the most critical, and thus the most studied. The region near the crack tip where K, and therefore Equations 1 and 2, are applicable is called the singularity-dominated zone. As long as nonlinear, plastic, or time-dependent effects are negligible in the singularity-dominated zone, LEFM provides an excellent means to quantify the crack tip condition. Otherwise, LEFM must be supplemented or replaced with other approaches.

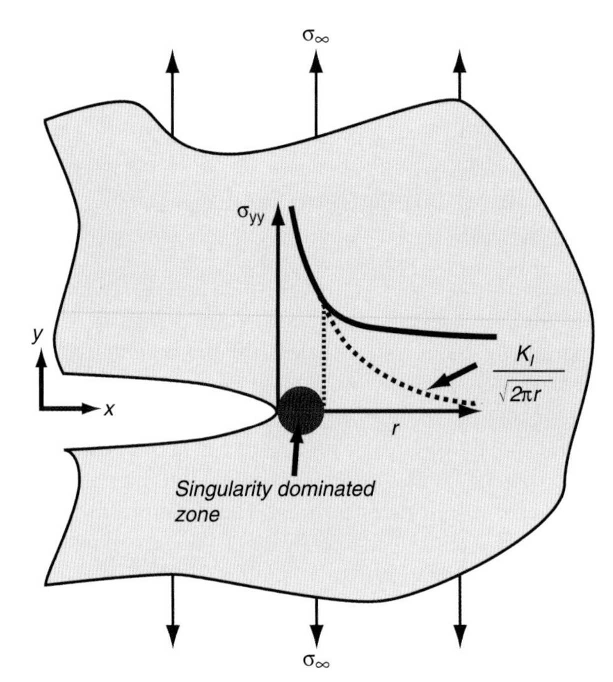


FIGURE 30.2 The LEFM stress intensity factor, K, describes the singular region of stress near a sharp crack tip. So long as a non-linear or inelastic zone of behavior (such as plastic deformation) is small compared to the singularity-dominated zone, K is an apt metric of crack tip effects.

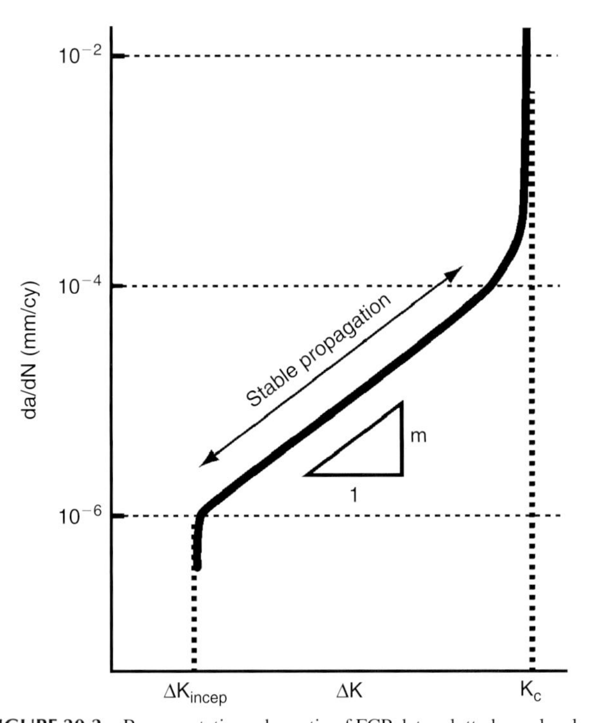


FIGURE 30.3 Representative schematic of FCP data, plotted on a log-log scale. The stable propagation region is shown as conforming to Equation 3, that is, the Paris equation. Crack propagation is often stable between $10^{-6}$ and $10^{-4}$ mm/cycle.

### 30.2.4.2 Fundamentals of Fatigue Crack Propagation

From an empirical standpoint, three regimes can be distinguished into the fatigue crack propagation (FCP): Regime I, where no significant propagation occurs, and the crack is essentially inactive; stable propagation or Regime II; and unstable propagation or Regime III (Figure 30.3). The onset or threshold of Regime II (at around $10^{-6}$ mm/cycle) is given by the stress intensity factor at crack inception, $\Delta K_{incep}$, which is the minimum stress intensity factor range needed to drive crack propagation, while $K_c$ is the critical stress intensity factor for unstable crack propagation and fast fracture. In Regime II, the rate of the stable crack growth under cyclic loading conditions follows the Paris equation for many engineering materials:

$$\frac{da}{dN} = C\,(K_{\max} - K_{\min})^m = C\Delta K^m \quad (3)$$

where $a$ is the crack length, $N$ represents the number of load cycles, $\Delta K$ is the stress intensity factor range, and C and m are empirical fit parameters that are treated as material properties. Note that, for a constant magnitude applied cyclic stress, a growing crack experiences a constantly increasing stress intensity factor and so eventually reaches the critical state and fractures.

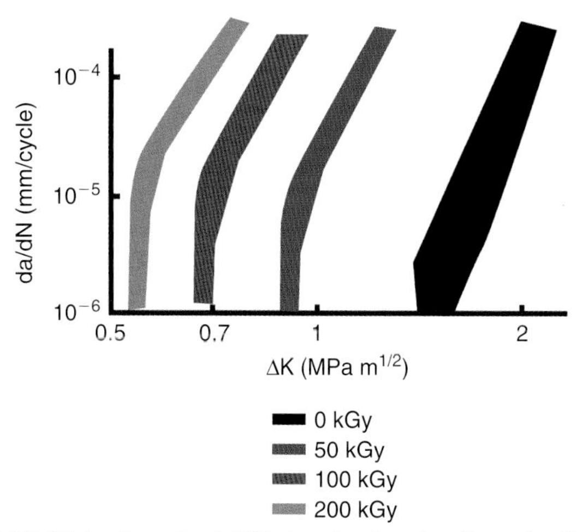


**FIGURE 30.4** Scatterband FCP data showing the effect of radiation dose on the resistance to crack inception and propagation in UHMWPE. Adapted from [33].

## 30.2.5 Fatigue Crack Propagation Testing in UHMWPE

Fatigue crack propagation experiments are performed on standard fracture specimens, such as notched compact tension specimens, typically using servohydraulic load frames. Before testing, the notch tip is sharpened with a razor blade and then subjected to fatigue precracking to obtain an initial crack as sharp as possible. When the initial crack has been created, a load waveform with a fixed stress ratio, R, is applied to the specimen and the crack growth is monitored by means of optical microscopy techniques. As with the total life fatigue approach, temperature effects are also relevant in crack propagation experiments of UHMWPE because its fatigue resistance has been shown to be substantially lower at body temperature than at room temperature [3]. A complete description of fatigue crack growth specimens and experiments can be found in ASTM E647-08 [39].

Research on fatigue crack propagation of conventional and highly crosslinked UHMWPEs has confirmed a detrimental effect of increasing radiation-crosslinking levels on the resistance to crack inception and crack growth [5, 33, 35, 40–42]. Figure 30.4 contains scatterbands of data from Baker et al. [33], indicating a continuous decrease in fatigue crack propagation performance with increasing radiation dose. Post-irradiation remelting has been demonstrated to further decrease fatigue crack propagation properties, according to several studies [5, 8, 10, 35, 40]. On one hand, a reduced capacity for plastic deformation due to a high crosslink density has been posited as the main cause for reduced fatigue crack growth performances of as-irradiated highly crosslinked UHMWPEs. On the other hand, the drop in crystalline content associated with post-irradiation remelting apparently accounts for a portion of the further reduction in the fatigue crack propagation behavior of highly crosslinked UHMWPEs. A positive effect of increasing crystallinity on this property has also been demonstrated for post-irradiation annealed highly crosslinked UHMWPEs and for highly crystalline and highly crosslinked UHMWPEs [5, 8, 41].

## 30.2.6 Viscoelastic Fatigue Crack Propagation

### *30.2.6.1 Viscoelastic Crack Propagation Models and Predictions of FCP Phenomena*

Work by Schapery, Saxena, Williams, and others details the analysis of cracks in creeping, strain rate dependent materials, and provides a predictive basis for the apparently brittle nature of FCP in UHMWPE [43–48]. Particularly useful are the models developed by Schapery and Williams, which directly link the intrinsic, constitutive viscoelastic relaxation behavior of the material to the advance of a stable crack tip [46, 48]. The power of these models is the predictive nature of the mechanics in relating FCP dynamics to the material's viscoelastic behavior that is easily measured in a simple one-dimensional creep test. The elementary consequences of the models result in the static mode fatigue crack propagation behavior that is observed in UHMWPE, and thus potentially provide a first-principles explanation of the fatigue and fracture behavior by the material.

Following the development by Williams [48], we consider a viscoelastic solid that relaxes homogeneously according to a simple power law (greater than time t = 0):

$$C_{(t)} = \frac{1}{E_0}\left(\frac{t}{\tau_0}\right)^d \quad (4)$$

where C is the material compliance, $\tau_0$ is a time constant, $d$ is the relaxation power law exponent, and $E_0$ is the instantaneous (t = 0) elastic modulus. For a fixed load (P = constant), the solution for the J-integral, which represents a measure of the fracture toughness of the material (see the following section on fracture resistance), is:

$$\left(\frac{J}{J_0}\right)_P = \left(\frac{t}{\tau_0}\right)^d \quad (5)$$

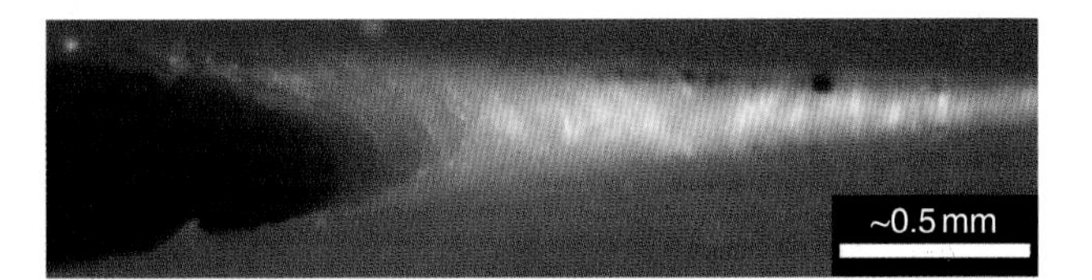


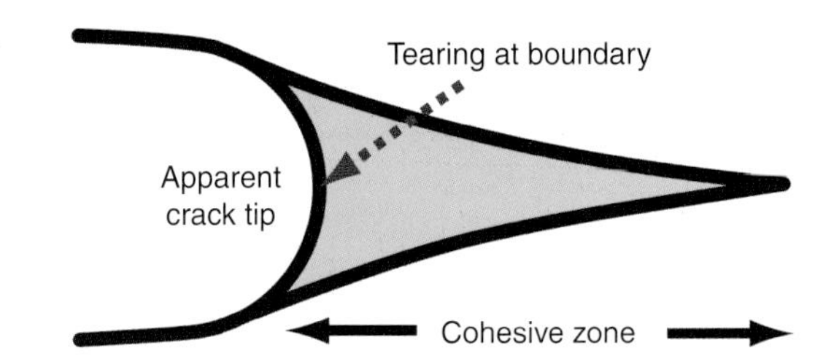


**FIGURE 30.5** The cohesive zone, also known as the process zone or failure zone, is apparent ahead of the crack tip in UHMWPE. This image was taken using an optical microscope on a virgin UHMWPE side grooved compact tension specimen. The material in the cohesive zone is highly strained but apparently lacks voids or craze structures. The boundaries of the cohesive zone show tearing of the failing material, producing the familiar "diamond" tearing lines observed on fracture specimens.

where $J_0$ is the static solution for the J-integral at t = 0. Clearly, $J$ is increasing continuously at the same rate as the compliance. The consequence of this is that if the load is continuously applied (in a static or cyclic fashion), the threshold for crack propagation, $J_c$, is eventually overcome, thus rendering fracture inevitable. This prediction for the initiation of cracks in UHMWPE structures has not yet been verified, but it may shed light on the origin of the recently reported fractures initiating at stress concentrations in crosslinked UHMWPE hip bearings [20, 27, 29].

It can be seen from the foregoing that the deformation and failure of the viscoplastic material adjacent to the crack front (i.e., the cohesive zone) dominates the material resistance to crack propagation in UHMWPE (see Figure 30.5). Thus, the creep resistance of the material in the cohesive zone should play a dominant role in the growth of cracks [49]. Few studies have analyzed the creep behavior of UHMWPE at large strains, which is clearly of import in this case [50]. This literature implies that small strain creep experiments are unlikely to be relevant to the behavior of a cohesive zone comprised of highly strained post-yield material because the creep rate was seen to dramatically reduce at large strains. Therefore, the creep and relaxation behavior of highly strained bulk UHMWPE deserve more attention, leading toward a better mechanistic understanding of crack growth under dynamic loading conditions.

### 30.2.6.2 Fatigue Crack Propagation Relations for Cyclic and Static Mode Behavior

The Paris equation describes the rate of stable crack propagation in materials known to grow cracks due to cyclic mode phenomena (Equation 3). On the other hand, crack growth in static or creeping mode may be described by:

$$\frac{da}{dt} = A\,K_{(t)}^{\;n} \qquad (6)$$

which is analogous to the Paris equation, except that the process depends on the instantaneous K rather than some excursion or cyclic component of it [51]. This may be designated as static mode crack propagation because the process is not sensitive to cyclic phenomena per se. Stable crack propagation in materials that behave in this manner also approximately correlates to the maximum of the applied cyclic stress intensity, that is,

$$\frac{da}{dN} = BK_{\max}^{\;q} \qquad (7)$$

Equation 7 does not in general agree with the Paris equation. A basic approach to account for both static and cyclic mode components of crack propagation involves superposition of the two processes [52]:

$$da = \left[\frac{\partial a}{\partial N}\right]_t dN + \left[\frac{\partial a}{\partial t}\right]_N dt \qquad (8)$$

where each bracketed differential has the non-differentiated parameter fixed. In Equation 8, therefore, crack advance is the sum of a component only sensitive to loading and unloading events and a second component that is independent of cycling but reflects temporal (static mode) effects. These bracketed expressions correspond to the preceding Equations 3 and 6 and can be rewritten (with f = dN/dt):

$$\frac{da}{dN} = \left[\frac{da}{dN}\right]_{fatigue} + \frac{1}{f}\left[\frac{da}{dt}\right]_{creep} \qquad (9)$$

where a material may be more sensitive to the so-called fatigue (cyclic damage) or creep (static mode; independent of cycling) components of the total crack propagation, and this equation represents how creep contributions to FCP are practically considered. This formalism simply superimposes the effect of each mechanism, which excludes synergistic effects of combined static and cyclic mode FCP.

### 30.2.6.3 Static Mode Fatigue Crack Propagation in UHMWPE

Crack propagation in the static mode is not universally termed a fatigue process. Ritchie refers only to cyclic processes as specifically related to fatigue [53], while static mode growth of cracks in polymers is often instead called slow crack growth [54–57]. It is important to note, however, that while cyclic mode growth will not occur under static loading conditions, static mode crack propagation occurs under cyclic loading conditions.

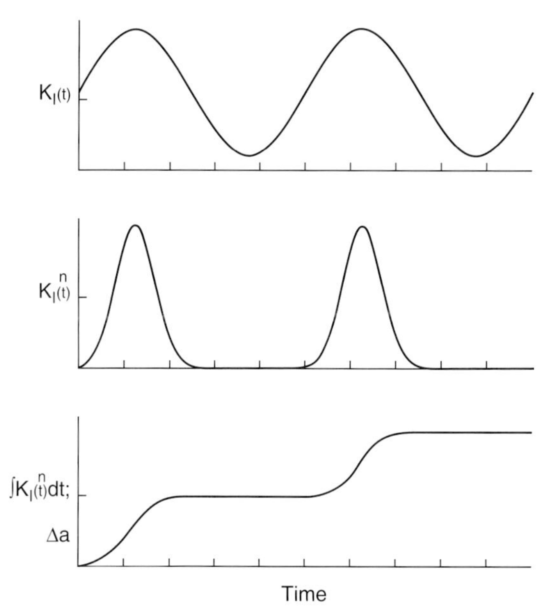


FIGURE 30.6 Graphical representation of static mode FCP—Equation 6. A sinusoidal stress intensity (top) results in intermittent, pulsed crack velocity (middle). The time integral of Equation 6 (bottom) shows that the growth is stepwise periodic with the waveform.

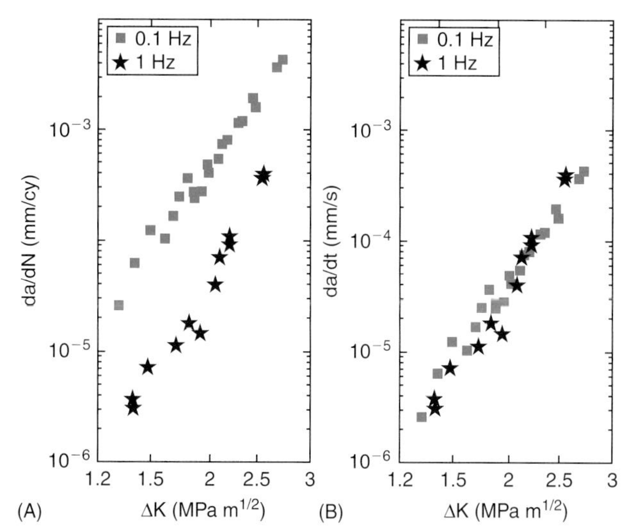


FIGURE 30.7 Crack propagation in UHMWPE is frequency independent in time. Part (A) shows an apparently low rate of propagation *per cycle* for higher frequency testing, while (B) demonstrates that the true propagation rate *per unit time* collapses over a decade of frequency.

Consider a sinusoidal load-controlled test and the associated stress intensity factor, K, as shown at the top of Figure 30.6. The rate of crack propagation, da/dt, is then found by carrying out Equation 6 (see middle plot in Figure 30.6 with n = 8). If the time integral of Equation 6 is taken, the total amount of crack advance over a given period is found, as shown at the bottom of Figure 30.6. It is evident here that so long as crack growth is in the stable regime, incremental crack growth occurs under cyclic loading conditions.

The predictable nature of crack advance under these conditions admits two simple consequences for cyclic loading scenarios. First, for a given waveform, the number of cycles should be irrelevant because the time integral of the power law is insensitive to frequency. Second, for two different waveforms (e.g., sine and square) of the same stress range and load ratio, the time integral of the FCP power law will not be identical, and thus some waveforms will appear more detrimental than others in a mathematically predictable manner. Furmanski and Pruitt tested these two consequences of static mode FCP behavior in conventional UHMWPE to evaluate the extent of the applicability of the static mode model [58].

Medical grade GUR 1050 ram extruded UHMWPE, unirradiated, was machined into compact tension specimens. Two experiments were conducted on a servohydraulic load frame to test the static mode nature of FCP in the material. In the first experiment, two sinusoidal load-controlled waveforms, load ratio R = 0.1, were run on separate specimens. A 1 Hz or 0.1 Hz sinusoidal waveform was repeatedly run for 33.4 minutes, and between these tests the specimen was removed and crack advance recorded using an optical microscope. Thus, the duration of the experiment, *T*, and the time integral of the power law was held constant, while the number of cycles and frequency were varied. A second experiment used identical specimens to the first, and the effects of square and sine waveforms at 1 Hz for the same duration were compared. In this case, test duration, number of cycles, and frequency were held constant, while the time integral of the FCP power law was varied.

The results of the frequency effect showed that the crack growth *per cycle* was approximately an order of magnitude greater in the 0.1 Hz test, which is expected for static mode FCP because there are tenfold more cycles in the 1 Hz test (Figure 30.7A). However, when crack growth is plotted *per unit time*, the data collapse (Figure 30.7B), showing that the process of FCP depends fundamentally on duration but not on the number of cycles.

The crack growth rate of the square waveform was consistently greater than for the sine waveform, by a constant multiple factor (Figure 30.8A). Thus, the stable crack growth regimes appear parallel on the log-log axes. Such a factor is expected of static mode FCP, upon inspection of Equation 10, as the integral of the FCP power law is greater for the square waveform than for the sine waveform. This multiplicative factor, *Q*, is merely the ratio of the predicted crack growth from Equation 10 during an experiment of duration λ:

$$Q = \frac{\Delta a_{square}}{\Delta a_{\sin e}} = \frac{\int_0^\lambda \left(K_{(t)square}\right)^n dt}{\int_0^\lambda \left(K_{(t)\sin e}\right)^n dt} \approx 2.1 \qquad (10)$$

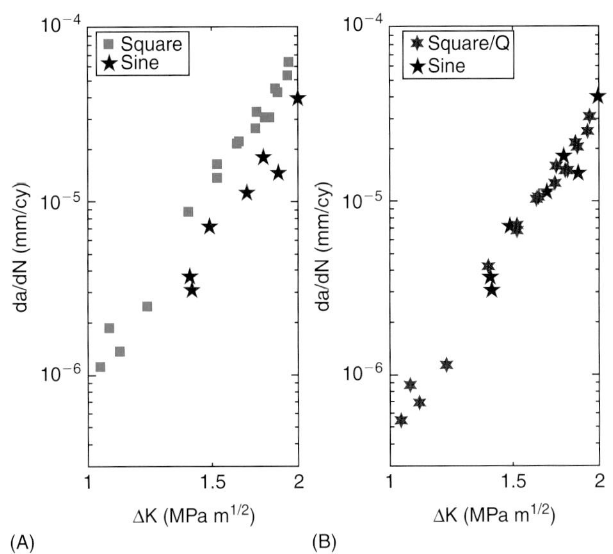


**FIGURE 30.8** (A) Sine and square waveform results in accelerated crack growth. This effect is well predicted from the basic static mode FCP power law (Equation 6), and when the square waveform data are normalized by the predicted difference in FCP velocity, with that of the associated sine waveform, the data collapse (B).

When the square waveform data are normalized by Q, all the data collapse to one set (Figure 30.8B). The prediction of static mode crack propagation rate, using only the FCP power law exponent and Equation 6, shows excellent agreement with the experimentally observed difference in crack velocity for square and sine waveforms. Aggregated, the results demonstrate that under cyclic, purely tensile loading conditions, FCP in conventional UHMWPE occurs in the static mode and does not undergo substantial cyclic mode damage.

### 30.2.6.4 $K_{max}$ Dominated Fatigue Crack Propagation in UHMWPE

Another hallmark of static mode FCP is the correlation of the process to the peak stress intensity factor ($K_{max}$), as opposed to $\Delta K$ that is used in the Paris equation. Furmanski and coworkers recently examined the effect on fatigue crack propagation under various R conditions and demonstrated that FCP in UHMWPE correlates to $K_{max}$ and not $\Delta K$ [37] (also see Table 30.1). Furmanski and Pruitt have recently shown that $K_{max}$ dependence (Equation 7) can be derived directly from Equation 6 [59].

In this work, three experiments were performed, each with a different load range sequence. The first experiment employed a fixed load ratio R = 0.1. A second experiment with fixed load ratio R = 0.5 was run in similar fashion to assess the effect of increased mean stress on FCP in this material. A third experiment with $P_{max}$ held constant probed the overall effect of varying load ratio.

The results for the R = 0.1 experiment show both a prominent crack inception stress intensity range (Figure 30.9), and a stable crack propagation regime that fits to the classic Paris equation (Equation 3), with a power law exponent *m* = 9.5. The data from the R = 0.5 experiment also demonstrate a stable propagation region throughout the tested range, however, without subinception propagation. These data apparently demonstrate a deleterious effect of a superimposed additional static mean stress applied in concert with the cyclic stress. The black star data in Figure 30.9 correspond to the fixed $P_{max}$ (variable R) experiments. At high stress intensity ranges (low R), the $P_{max}$ data in Figure 30.9 collapse to the R = 0.1 data. However, at lower stress intensity ranges, the data appear to converge to a constant rate of crack propagation. The low stress intensity (high R) range waveforms employed, with $P_{max}$ fixed, approach a steady state applied load $P_{max}$, with the cyclic component negligible compared to the static component of the load. The

**TABLE 30.1** Fatigue Crack Propagation Characteristics of Brittle and Ductile Materials (UHMWPE meets the requirements to be considered an essentially brittle material from a fracture mechanics perspective.)

| | UHMWPE | Brittle | Ductile |
|---|---|---|---|
| Propagation mode | Static | Static | Cyclic |
| Correlating parameter | $K_{max}$ | $K_{max}$ | $\Delta K$ |
| Power law exponent | 8–20 | 8–50 | 2–4 |
| Fracture toughness $K_c$ (Mpa$\sqrt{}$m) | 0 kGy 4<br>100 kGy 3 | SiC 3<br>Si 1 | Ti-6Al-4V 65–100<br>4340 Stl 50–100 |

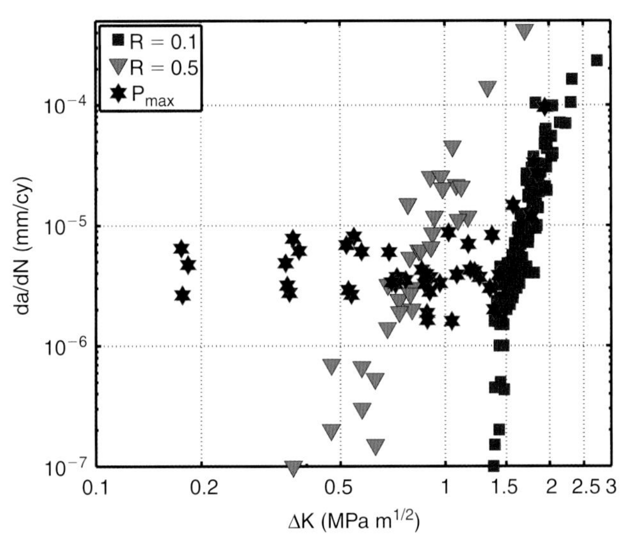


**FIGURE 30.9** FCP data from R = 0.1, R = 0.5, and $P_{max}$ constant experiments. Disagreement between these data indicates that $\Delta K$ is not a sufficient parameter to predict crack propagation behavior.

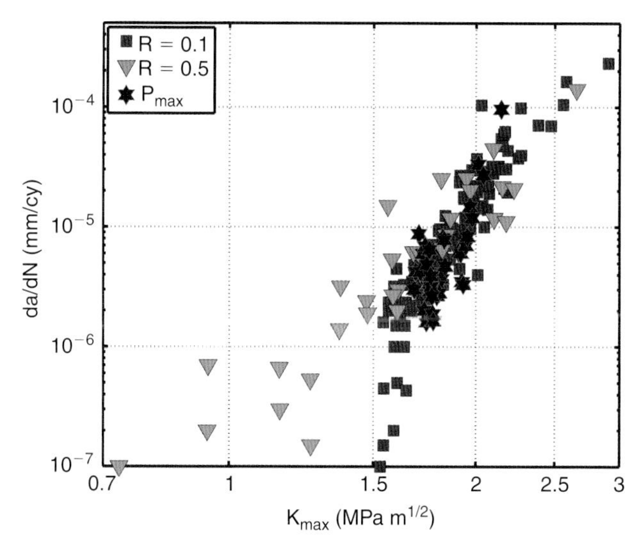


**FIGURE 30.10** Data from all experiments already shown in Figure 30.9, but plotted with $K_{max}$ as the presumed FCP parameter. All data in the stable crack propagation regime collapse to the same curve, confirming $K_{max}$ as the FCP parameter for this regime of crack growth.

disagreement between the data from these experiments indicates that $\Delta K$ is not a generally adequate correlating parameter for fatigue crack propagation in UHMWPE.

When the experimental data in Figure 30.9 are plotted using $K_{max}$ as the FCP correlating parameter (Figure 30.10), the stable propagation results of all three experiments collapse to one common trend. This indicates that $K_{max}$, rather than $\Delta K$, is the dominant FCP correlating parameter for conventional UHMWPE. This result demonstrates that stable crack growth in UHMWPE, under these experimental conditions, is described by Equation 7, and that cracks can propagate in this material without significant cyclic loading.

Finally, we observe that crack propagation in UHMWPE occurs reliably in the static mode, and the toughness and FCP power law exponent match the expectations for classically brittle engineering materials. Table 30.1 summarizes the comparison between ceramics, metals, and UHMWPE in terms of fatigue crack propagation and fracture behavior. This evidence shows that UHMWPE meets the requirements to be considered an essentially brittle material from a fracture mechanics perspective [53, 60] and deserves the consideration in design and application that is normally given to brittle, damage-intolerant engineering materials.

## 30.2.7 Non-Conventional Fatigue Experimental Approaches

Apart from the traditional fatigue methods, other experimental approaches have been used to reproduce fatigue conditions mechanically similar to those in clinical service of total joint replacements. As an example, we present here two of the studies present in the UHMWPE literature, which simulate more precisely the environment of knee joints, and worst-case clinical scenarios for artificial hip and knee bearings, respectively.

### *30.2.7.1 Sliding Fatigue Approach in UHMWPE and Vitamin-E-Doped UHMWPE*

Tomita and colleagues developed a bidirectional sliding fatigue test in an attempt to simulate conditions closer to those found in total knee UHMWPE components [61]. In this test, a titanium alloy pin (hemispheric tip shape 3 mm in radius) slid against a smooth UHMWPE sample with simple reciprocating or U-shaped pathways. The specimens were immersed in water at 37°C, and the load applied was 196 N. After the sliding fatigue test, scanning acoustic tomography allowed for detection of subsurface cracks in the fatigued UHMWPE specimens. Tomita's group first recognized the positive effect of the addition of vitamin E (0.1%) on the inhibition of subsurface crack formation and eventual delamination in conventional UHMWPE, based on sliding fatigue results [61]. In a subsequent study, the same authors reported a deleterious acceleration of crack initiation and propagation at the subsurface grain boundaries of gamma irradiated UHMWPE specimens, which evolved to surface asperities and final delamination of the specimen [62]. Again, the manufacture of compression molded UHMWPE with the addition of vitamin E appeared to be efficacious in preventing formation of subsurface cracks.

### *30.2.7.2 Structural Fatigue Resistance Approach*

Wang and coworkers have employed a structural fatigue experimental approach to evaluate the fatigue performance of acetabular liners and tibial inserts made of various crosslinked UHMWPEs working in conditions close to worst-case clinical scenarios [63]. Thin acetabular liners (2.5 mm thickness; 32 mm diameter) and cemented patellar components were tested in hip and patello-femoral simulators, respectively. Acetabular liners were placed with a 4 mm gap between the dome of the liner and the metal shell so that cups were abnormally supported at the rim. Synchronized physiological loading pattern (150–2450 N) and cross-path motion were applied for 1 million cycles. Cobalt chrome femoral heads and diluted calf serum lubrication were used. The structural fatigue test was stopped four times every 250,000 cycles to allow for inspection of rim fracture and wear assessment of non-fractured liners. Similarly, patellar components were articulated against well-aligned and misaligned (6 degrees internally rotated) femoral components. Patellar components were subjected to a constant 2224 N load, and they rotated from 60 degrees to 120 degrees at 1.33 Hz with serum lubrication for 1 million cycles. In contrast to the aligned test, the patellar component was not centered in the femoral track, and edge loading occurred at different angles (60 degrees and 120 degrees flexion).

Most remelted crosslinked UHMWPE liners tested in Wang's study exhibited open cracks right below the rim, regardless of the radiation dose (5 and 10 Mrads). On the contrary, all annealed crosslinked UHMWPE cups (total radiation dose 10.5 Mrads) survived 1 million cycles without signs of cracking. With regard to patellar components, fracture or cement debonding did not occur in the normal alignment test after 1 million cycles. However, all the remelted highly crosslinked (10 Mrad) UHMWPE patellae fractured at the peg before the end of the test, whereas patellae produced from control material (3 Mrad) experienced no fractures. Therefore, post-irradiation thermal treatments were concluded to have a more pronounced effect than the total radiation dose on structural fatigue resistance. The authors further correlated the structural fatigue performance with ultimate tensile strength and yield stress, concluding that material strength, rather than ductility, dictates structural fatigue resistance.

## 30.3 FRACTURE RESISTANCE

### 30.3.1 Introduction

It is well known that UHMWPE is a tough (highly resistant to fracture) polymer due to its extremely high molecular weight and resulting microstructure. In fact, UHMWPE has been often used as a toughening agent in the development of polymer blends [64–68]. Because UHMWPE is an energetically tough and highly ductile polymer, the wear of UHMWPE components in total joint replacements, especially in TKA, is not caused by purely fatigue mechanisms but by a combined fatigue-fracture process [7, 69]. Uniaxial tensile tests, impact experiments, and multiple methods based on fracture mechanics approaches have been applied to characterize the resistance to fracture of polymers. In particular, several studies on both comparative and absolute estimations of the toughness of UHMWPE, as well as the characterization of changes in this property caused by manufacturing processes (i.e., consolidation, irradiation, and stabilization), have been conducted since the late 1980s. The aim of this section is to summarize the experimental details concerning each one of the techniques previously mentioned and to briefly review their application to the characterization of various medical grade UHMWPEs.

### 30.3.2 Uniaxial Tensile Failure: Work to Fracture and Estimated $K_c$

Even though the uniaxial tensile test is one of the simplest experimental methods, it provides essential information about the mechanical behavior and properties of materials. In this experiment, a nominally smooth dogbone specimen is subjected to uniaxial deformation at a constant strain rate until failure (see ASTM D638-08) [70]. The result of the tensile test is typically displayed as the well-known stress (load per unit area)-strain curve (see Chapters 6 and 35), which should properly be converted to true stress and strain when analyzing large deformation behavior of UHMWPE. The area under the true stress-strain curve gives an estimate of the total strain energy needed to deform and eventually rupture the tensile specimen. This strain energy at specimen rupture is termed the energetic toughness, or sometimes the work to fracture, of the material, and often serves as an estimation of the overall toughness of the material. It must be emphasized, however, that this is not an actual measure of its fracture toughness. For example, compared to engineering alloys, UHMWPE is energetically tough but exhibits low fracture toughness and resistance to crack propagation.

When specimens rupture in tensile experiments, the failure typically initiates at an identifiable crack or flaw within the specimen. A method to obtain estimations of the material toughness from this result is based on the inspection of fracture surfaces and the application of the linear elastic fracture mechanics. Immediately prior to material fracture, and provided plane-strain conditions hold, the stress intensity factor reaches a critical value, which is considered a measure of the fracture toughness, $K_c$:

$$K_c = Y\left(\frac{a}{W}\right)\sigma_u\sqrt{\pi a} \qquad (11)$$

where $\sigma_u$ is the ultimate true stress. As an example, the fracture surface of a 100 kGy electron-beam irradiated UHMWPE tensile specimen with an initial microvoid or flaw in the center of the exposed surface is shown in Figures 30.11A–B. Detailed inspection of the fracture surface revealed the so-called stable crack propagation region surrounding the initial flaw and the fast fracture region. For this particular case, the ultimate stress and the features of the fracture surface gave a $K_c$ of 1.7 MPam$^{1/2}$. This estimated $K_c$ did not exactly agree with that obtained in traditional fatigue crack propagation analyses, but it verified that the trend relating ultimate stress to flaw size match the predictions of LEFM [4, 5]. Similarly, Gencur et al. reported $K_c$ results for as-received, radiation sterilized, and post-irradiation stabilized highly crosslinked UHMWPEs. They observed a decreased $K_c$ for both annealed and remelted highly crosslinked UHMWPEs ($2.8 \pm 0.4$ and $3.0 \pm 0.6$ MPa$\sqrt{\text{m}}$, respectively) with respect to as-received and radiation sterilized UHMWPE ($4.0 \pm 0.5$ and $4.5 \pm 1.1$ MPa$\sqrt{\text{m}}$, respectively), which was attributed to the higher crosslink density of the former [4].

### 30.3.3 Izod and Charpy Impact Tests

While the uniaxial tensile test may be considered as a quasi-static measure of the fracture resistance of materials,

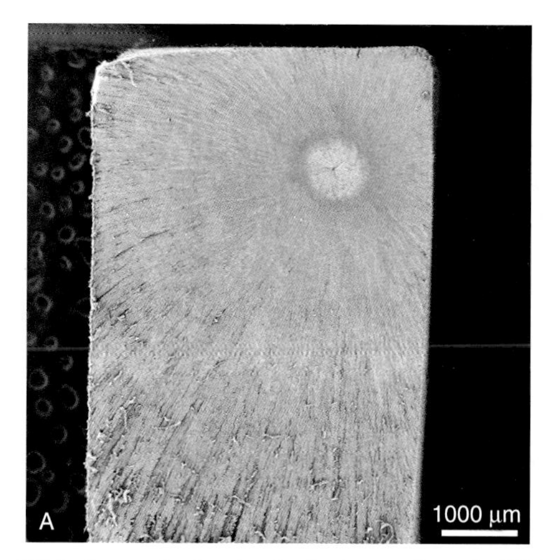

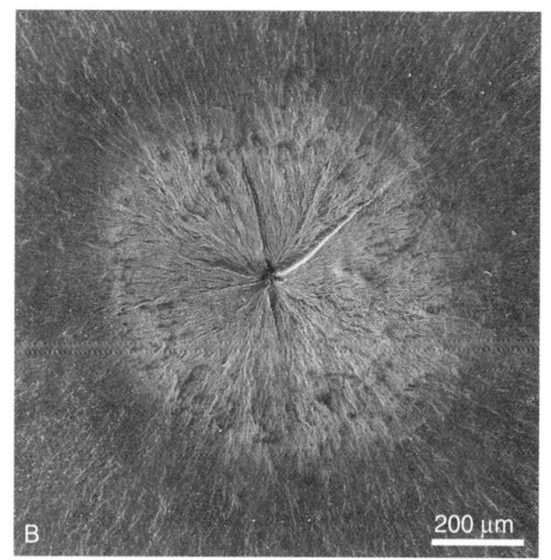


**FIGURE 30.11** SEM micrographs showing the fracture surface of a 100 kGy electron-beam irradiated UHMWPE tensile specimen (A), and detail of the stable crack propagation region with initial microvoid (B).

the impact test represents a dynamic scenario, where laboratory specimens accommodate deformation and fracture processes in a much shorter time scale (much less than a second). The classic Izod and Charpy tests are the most widespread methods to evaluate the impact strength of materials. In both, a pendulum with a dead weight at the end of its arm is swung down, striking and typically fracturing a notched specimen. The specimen fracture absorbs part of the energy of the pendulum, and this fact is reflected in the maximum height reached by the pendulum after the strike. The main difference between the Izod and Charpy tests lies on the position of the notched specimen, which is vertical for the former and horizontal for the latter. In addition, while the notch is held facing the pendulum in the Izod test, the position is the opposite in the Charpy test. Descriptions of the Izod and Charpy impact tests, and their specific modifications for UHMWPE, can be found in ASTM F 648 and ISO 5834 [71]. The energy absorbed during a notched impact test is often compared to the results of fracture toughness or fatigue crack propagation tests, but the link between them is only qualitative. Impact tests involve high strain rates and dynamic crack propagation, both of which alter the behavior of the material and crack tip zone, so use of impact properties should be undertaken judiciously.

It has to be mentioned that double-notched Izod specimens are needed for testing of UHMWPE because complete fracture would not be achieved with single-notched specimens. The minimum required impact toughness strength for UHMWPE to be used in biomedical implants ranges from 25 to 126 $kJ/m^2$ (Izod method) or from 30 to 180 $kJ/m^2$ (Charpy method), depending on the viscosity number of the UHMWPE powder (ASTM F648). Despite being a key property according to ASTM F648, there are only few reports on the impact strength of medical UHMWPE in the orthopedic literature [8, 72, 73]. All these studies confirm decreasing impact strength with increasing radiation doses for UHMWPE, with drops ranging from 35% to 65% after irradiation at doses between 110 and 150 kGy. This effect, however, may be altered depending on other variables, such as radiation type, irradiation temperature, or subsequent thermal treatment. In addition, Greer et al. reported significantly higher impact strengths for lower molecular weight UHMWPE resins at identical radiation dose, little effect of the fabrication method, and higher toughness for electron-beam irradiated specimens compared to those irradiated with gamma rays at the same dose [72]. These facts point out that the ability of UHMWPE to sustain high-energy impacts is progressively lost with increasing radiation doses (i.e., increasing crosslink density). The features present in the fracture surfaces of non-irradiated and 150-kGy irradiated UHMWPE impact specimens also support this conclusion (see Figures 30.12A–B). Note the presence of bands with signs of extensive deformation in the virgin polymer (Figure 30.12C–D) in contrast to the smooth fracture surface in the irradiated and remelted condition.

## 30.3.4 Elastic–Plastic Fracture Mechanics and the J-integral Concept

As explained in the fatigue crack propagation section, the stress intensity factor, K, describes the state of stress near a crack tip. However, K is not exactly a "driving force" for fracture. The energy release rate, $\boldsymbol{g}$, is defined as the change in potential energy, Π, for an increment of crack area, A:

$$\boldsymbol{g} = -\frac{d\Pi}{dA} \tag{12}$$

where $\boldsymbol{g}$ is a "driving force" for crack propagation because it is the gradient of the potential energy associated with crack extension, and it possesses units of energy/area. When this

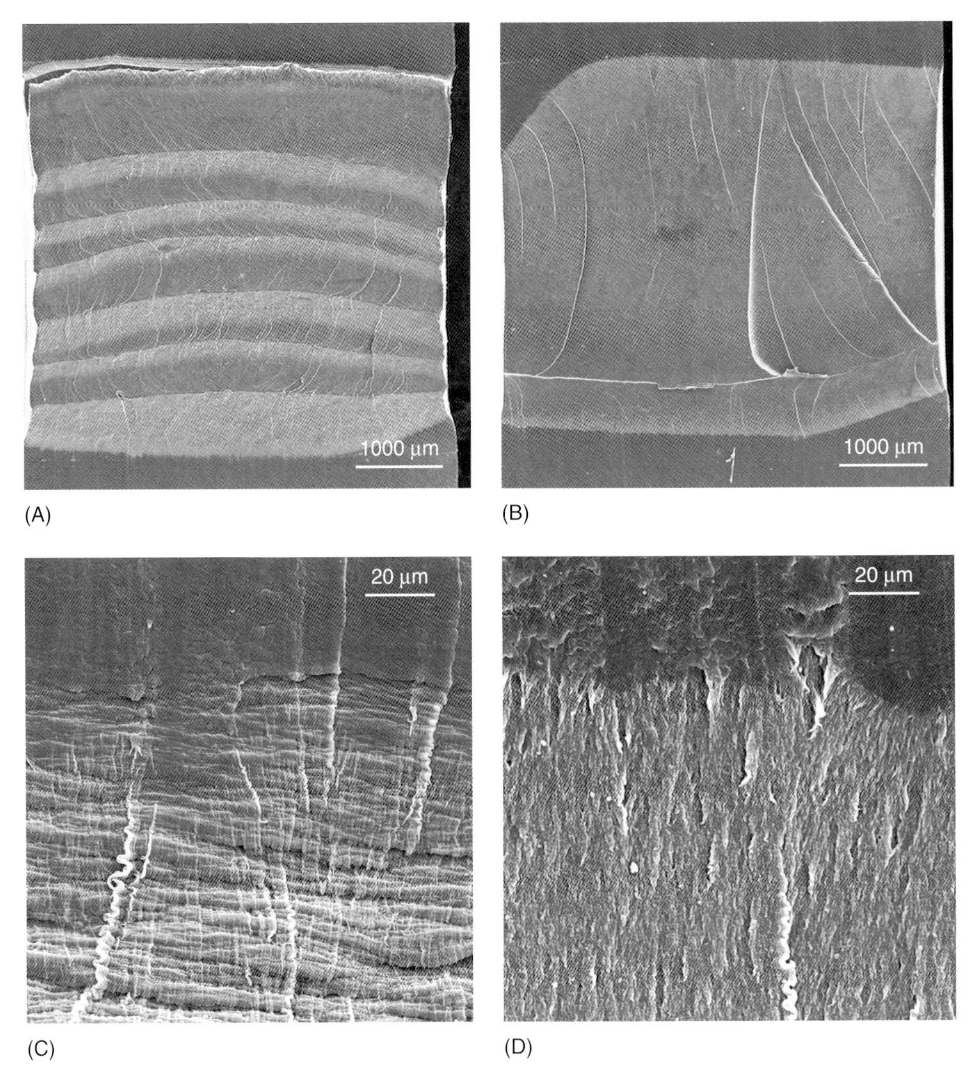


**FIGURE 30.12** SEM micrographs of fracture surfaces corresponding to as-received (A) and 150 kGy electron-beam irradiated and remelted (B) UHMWPE Izod impact specimens. Virgin UHMWPE fracture surfaces exhibit alternate regions of ductile and brittle fracture (C), whereas ductile features are absent in highly crosslinked UHMWPE specimens (D).

energy release rate exceeds a critical value, where the resistance to crack propagation cannot keep up with energy lost due to crack growth, the system is unstable and fast fracture occurs. Thus, $\boldsymbol{g}$ describes conditions of subcritical behavior and fast fracture. In a linear elastic, isotropic material there is a direct correspondence between $\boldsymbol{g}$ and $K$,

$$\boldsymbol{g} = \frac{K^2}{E'} \tag{13}$$

where $E'$ is the elastic modulus, $E$, in plane stress or $E/(1-\nu^2)$ for plane strain. The concept of energy dissipation is an important one because it is able to abstract the details of the crack tip conditions during crack extension without losing any information about the conditions there. Dealing with energy concepts, such as the energy dissipated due to crack growth, is an effective approach for overcoming the limitations of LEFM when non-linear, plastic, and time-dependent effects come into play.

A successful means for analyzing fracture when LEFM is not accurate involves the J contour integral, also called J-integral or just $J$. It is the path-independent value of a closed contour integral around the crack tip that represents the total energy being consumed to extend the crack [74] during a monotonic loading cycle:

$$J = \int_{\Gamma} \left( w dy - \mathbf{T} \frac{\partial \mathbf{u}}{\partial x} ds \right) \tag{14}$$

where $\Gamma$ is the path or contour, $w$ is the strain energy density of the material, $y$ and $x$ are the coordinate directions, $\mathbf{T}$ is the traction vector, $\mathbf{u}$ is the displacement vector, and s is the arc length along the contour.

The essential conclusion of this and related work is that although the material behavior near a crack may be inelastic, the monotonic stress-strain curve of an elastic–plastic material is not distinguishable from that of a non-linear elastic material. This fact renders the analysis of elastic–plastic fracture tractable (neglecting cyclic effects) by remotely computing the total energy associated with crack extension, including deformation processes that occur fairly remote from the crack tip.

The J contour integral computes the change in potential energy with crack extension for a non-linear elastic material in the same way that ***g*** does for linear elastic materials. Thus for a linear elastic material, $\boldsymbol{g} = J$. Furthermore, $J$ has been used to describe the stress intensity at the crack tip for power-law hardening elastic–plastic materials. A common uniaxial stress-strain relation for such materials, the Ramberg-Osgood relation, is:

$$\frac{\varepsilon}{\varepsilon_y} = \frac{\sigma}{\sigma_y} + \alpha\left(\frac{\sigma}{\sigma_y}\right)^n \tag{15}$$

where $\varepsilon$ and $\varepsilon_y$ are the strain and yield strain, respectively, whereas $\sigma$ and $\sigma_y$ are the stress and the yield stress, and n and $\alpha$ are material fit parameters. The stress near the crack tip in a power-law hardening material is described by the so-called HRR equation [75], named after the authors, Hutchinson, Rice, and Rosengren, who introduced it:

$$\sigma_{ij} = k_1\left(\frac{J}{r}\right)^{\left(\frac{1}{n+1}\right)} \tag{16}$$

In the HRR equation, $k_I$ is a proportionality factor that is a function of material properties (i.e., $\alpha$, E, and $\sigma_y$), geometry, $\theta$, r (the distance from the crack tip), and $n$ (the plastic power-law hardening exponent from the Ramberg–Osgood relation). Note that the strength of the singularity depends on $n$, while the relative strength of the field is governed by the stiffness and yield strength. If $n$ is 1, then Equation 16 mathematically reduces to an analog of LEFM because the Ramberg-Osgood relation becomes, in this case, a linear equation.

Another useful relationship connects $J$ to the crack tip opening displacement (CTOD), $\delta$, which defines the distance that the crack faces are pulled apart near the crack tip and is a measure of toughness and ductility. This relationship, shown as Equation 17, holds well beyond the limits of LEFM [27], where $\sigma_f$ is the flow stress of the material at the crack tip.

$$J = \sigma_f \delta \tag{17}$$

This relationship is useful for quantifying energy dissipated in non-linear elastic materials and thus the driving force, particularly where $\delta$ is either easily observed or a convenient physical parameter to track.

## 30.3.5 J-integral Based Fracture Toughness Testing

### 30.3.5.1 Multi-specimen versus Single-Specimen Methods

Single edge-notched three point bend, compact, and disk-shaped compact specimens are typically recommended for J-integral based fracture toughness evaluations. Prior to testing, UHMWPE specimens are typically precracked either under fatigue conditions or by other notching techniques to induce the sharpest pre-existing crack possible at the root of the machined notch. After precracking, the specimen (or a series in the case of multi-specimen methods) is loaded at a constant rate to a selected displacement level to provoke stable crack extension or tearing. According to ASTM E1820 [76], a standard originally developed for metals, two different methods may be utilized depending on how crack extension is measured. The basic (multi-specimen) procedure relies on the use of optical methods to measure crack lengths, whereas elastic compliance techniques provide crack length estimations in the resistance curve (single specimen) procedure. Even though a valid J-integral versus crack extension, J-R, curve cannot be constructed from the multi-specimen procedure, the computation of J values and crack lengths for all the specimens is required to obtain an initiation toughness value, $J_{IC}$. J is calculated as the summation of the elastic, $J_e$, and plastic, $J_{pl}$, components according to the following equations:

$$J = J_e + J_{pl} = \frac{K^2(1-\nu^2)}{E} + \frac{\eta A_{pl}}{B_N b_o} \tag{18}$$

$$K = \frac{P}{(BB_N W)^{1/2}} f(a_o/W) \tag{19}$$

where K is the stress intensity factor, $f(a_o/W)$ is a function of the original crack length, $a_o$, and the width, W, of the specimen, $\nu$ is the Poisson ratio, E is the Young's modulus, $b_o$ is the original uncracked ligament, that is W minus $a_o$, $B_N$ is the net thickness (B if no side grooves are present in the specimen), $\eta$ is a geometrical factor that accounts for the ratio between the uncracked ligament and the width of the specimen ($\eta = 2+0.522b_o/W$), and finally, $A_{pl}$ is the area beneath the load-displacement curve corresponding to the plastic deformation of the specimen. When the selected displacement is reached, the specimen is unloaded and the crack may be marked to help with further length assessment. Then, complete breakage of the specimen is needed to expose the crack for optical or SEM measurements of both the original crack size, $a_o$, and the final physical crack size, $a_p$, at nine points, two of them near the surface (at 0.005 W from the edges) and seven evenly distributed through the fracture surface. The average of the two near-surface measurements is combined with the remaining seven crack length measurements to calculate final averages of the crack lengths. After plotting J values versus crack extension, $\Delta a = a_p - a_o$, the value of $J$ for incipient stable crack growth, $J_{IC}$, is found. $J_{IC}$ estimations require determining and drawing a construction (or blunting) line in accordance with the following equation:

$$J = M\sigma_f \Delta a \tag{20}$$

where M = 2 or otherwise determined as the slope of the linear regression line of at least six data points in the region $0.2J_Q \leq J_i \leq 0.6J_Q$. As previously shown, $\sigma_f$ is the flow stress in the elastic–plastic material ahead of the crack tip. Then, two exclusion lines with the same slope as the construction line are plotted so that they intersect the x-axis at 0.15 and 1.5 mm. In addition, an offset line parallel to the construction and exclusion lines has to be plotted intersecting the abscissa at 0.2 mm (Figure 30.13). Finally, a regression line of the data is determined according the following power law expression:

$$\ln J = \ln C_1 + C_2 \ln\left(\frac{\Delta a}{k}\right) \tag{21}$$

The intersection of this regression line with the offset line (see Figure 30.13) defines interim values of J, $J_Q$, and crack extension, $\Delta a_Q$. $J_Q$ is an appropriate measure of $J_{IC}$ as long as the coefficient $C_2$ is less than 1, the thickness and the initial uncracked ligament of the specimen are higher than $25J_Q/\sigma_f$, and the slope of the power law regression line at $\Delta a_Q$ is less than $\sigma_f$.

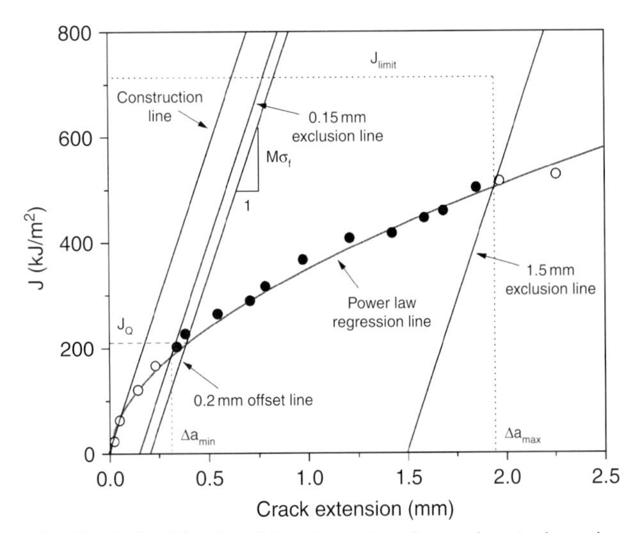


**FIGURE 30.13** Plot describing the analytical procedure to determine $J_{IC}$ from the multi-specimen method. Full circles are used to obtain the best-fitted resistance curve (red line).

It bears mentioning that the assigned value to the flow stress, $\sigma_f$, is somewhat in dispute for UHMWPE. It is classically specified, in ASTM E1820, as the so-called effective yield strength, which is the average of the 0.2% offset yield strength, $\sigma_y$, and the ultimate tensile strength, UTS. However, the flow stress has also been specified for UHMWPE as either the yield strength, $\sigma_y$, or the ultimate tensile strength [11, 77–79]. The choice of flow stress clearly impacts the resultant value of $J$, and this ambiguity remains a difficulty in the field. Recent work has addressed the ambiguity of the flow stress [11], which is detailed in the following sections.

Because compliance measurements do not require complete fracture of test specimens to assess crack extension, a single specimen is subjected to consecutive unloading/loading sequences in the resistance curve procedure to promote longer crack extensions at each step. The construction of a J-R curve relies on the computation of $J_i$ and crack length, $a_i$, at a load $P_i$ for every step i from crack opening displacements measured on the load line (i.e., compliance measurements). The equations used for side-grooved compact specimens are the following:

$$J_i = \frac{(K_i)^2(1-\nu^2)}{E} + J_{pl(i)} \tag{22}$$

$$U_x = \frac{1}{[EB_e V/P_i]^{1/2} + 1} \tag{23}$$

$$a_i/W = [C_0 + C_1U_x + C_2U_x^2 + C_3U_x^3 + C_4U_x^4 + C_5U_x^5] \tag{24}$$

where $K_i$ is the stress intensity factor at load $P_i$, $\nu$ is the Poisson ratio, E is the elastic modulus, $J_{pl(i)}$ represents the correction accounting for plastic deformation of the specimen, $U_X$ is the normalized compliance, V is the displacement between measurement points, $V/P_i$ is the compliance, $B_e = B-(B-B_N)^2/B$ where B is the thickness of the specimen and $B_N$ is the net thickness, and W is the width of the specimen. In general, the compliance method assumes that for a given specimen geometry, the relationship between normalized compliance, $U_x$, and the normalized crack length, a/W, is unique [10]. This implies that any linear elastic, isotropic, and homogeneous material should follow the same U versus a/W curve, thus defining the standard compliance coefficients in Equation 24. These standard coefficients, however, are not valid for conventional UHMWPE or highly crosslinked UHMWPE formulations, as shown by Varadajaran and Rimnac [10], and they have to be experimentally determined for every distinct formulation. Finally, the thickness, B, of the specimen, as well as the original, $b_o$, and final, b, uncracked ligaments, impose a maximum J-integral capacity and maximum crack extension, which are given by $J_{max} = b\sigma_f/20$ or $J_{max} = B\sigma_f/20$, whichever is smaller, and $\Delta a_{max} = 0.25b_o$, respectively. An initiation value of toughness may be also calculated following a similar procedure as in the multi-specimen procedure.

### 30.3.5.2 J-integral and the Single Specimen Normalization Method

Both multi-specimen and single specimen methods described in ASTM E 1820 have obvious drawbacks. The basic multi-specimen procedure is material and time intensive, and a strictly valid J-R curve cannot be obtained. On the other hand, the single specimen or resistance curve procedure relies on elastic compliance methods to assess crack extension. While compliance methods are well suited for metals, materials exhibiting time-dependent properties, that is a viscoelastic behavior, make such crack extension measurements complicated and inaccurate for these materials, polymers among them. The method of normalization can be used to measure crack extension, overcoming some of these limitations and still using a single specimen. Descriptions of this procedure can be found in Landes et al., Varadarajan et. al., and in ASTM E 1820 as well [11, 76, 80].

The normalization method is based on the assumption that the load can be seen as the product of separable functions of crack length and displacement. Therefore, if load and displacement are known, it is possible to calculate crack length by means of this relationship:

$$P = G(a/W)H(\nu_{pl}/W) \tag{25}$$

where a, and W, are again the crack length and the width of the specimen, respectively, and $\nu_{pl}$ is the plastic component of the displacement. G is function only of the crack length normalized by the width, whereas H is function only of the normalized plastic displacement. Then, the load can vary due to both changes in crack length and in plastic displacement for a stationary crack. The G and H functions are given by the following relationships:

$$G(a/W) = WB(b/W)^{\eta_{pl}} \tag{26}$$

$$H(\nu_{pl}/W) = \frac{(L + M\,\nu_{pl}/W)(\nu_{pl}/W)}{N + \nu_{pl}/W} \tag{27}$$

where B is specimen thickness, b is the uncracked ligament, W−a, $\eta_{pl}$ stems from the calibration of J, and L, M, and N are unknown fitting constants. It has to be noted that during the test the total displacement, $\nu$, is registered, instead of the plastic displacement, $\nu_{pl}$. The total displacement consists of elastic and plastic terms:

$$\nu = \nu_{el} + \nu_{pl} \tag{28}$$

$$\nu_{el} = PC(a/W) \tag{29}$$

where C(a/W) is a compliance function, which depends on both the crack length and the load. Because P is a function of crack length and plastic displacement, and the latter is, in turn, a function of crack length via the compliance function, the set of Equations 25–29 has to be solved in an iterative way. On the other hand, the H function accounts for the stress-strain behavior of the material (actually, sometimes H is termed the hardening function), and therefore the L, M, and N constants are not assumed a priori because variations from specimen to specimen are expected. These constants are determined for each specimen during testing by choosing three calibration points, typically at the beginning and end of the test, as well as at near maximum load. For the initial and final calibration points the physical crack length needs to be measured on the exposed fracture surface, whereas for the third point the calibration values are assumed and varied to optimize the fitting functions. To measure the initial crack length point, however, may be problematic, and a set of calibration points are taken within a particular displacement range, the so-called separable blunting zone, in which apparent crack extension is assumed to be exclusively due to blunting (see Varadarajan et al. [11] for more details). After L, M, and N have been determined, the set of Equations 25, 26 and 27 can be solved for crack length results given load and displacement from the test. Finally, *J* values can be obtained by applying Equations 18 and 22.

Varadarajan and colleagues have recently validated the single-specimen normalization method for J-integral fracture toughness testing of UHMWPE [11]. Moreover, they have addressed the flow stress controversy because there is no overall agreement on what the appropriate flow stress at the tip of the crack is for the computation of *J* from either blunting line or crack tip opening displacement based techniques. As previously mentioned, yield strength, ultimate tensile strength, and effective yield strength have been used as the flow stress to obtain J-integral based fracture toughness measurements of UHMWPE materials [11, 76–79]. Note that the use of ultimate tensile strength or yield stress as the flow stress has critical implications for the computed value of the J-integral (previously discussed) because conventional and highly crosslinked UHMWPE formulations have similar yield strengths but substantially different ultimate strengths [6]. The work by Varadarajan and Rimnac gives an overview of some of the various approaches to the specification of the flow stress [11]. They also conducted experiments using a novel crack tip opening displacement approach to demonstrate that specifying the ultimate tensile strength as the flow stress was appropriate. Nevertheless, Varadarajan and Rimnac recommended avoiding the blunting line approach due to the inherent ambiguity of the flow stress. Instead they recommended specifying $J_{Ic}$ at the condition where the crack tip opening displacement, $\delta$, abruptly reduces its growth subsequent to the application of the load. Their investigation of the discontinuous increase in apparent

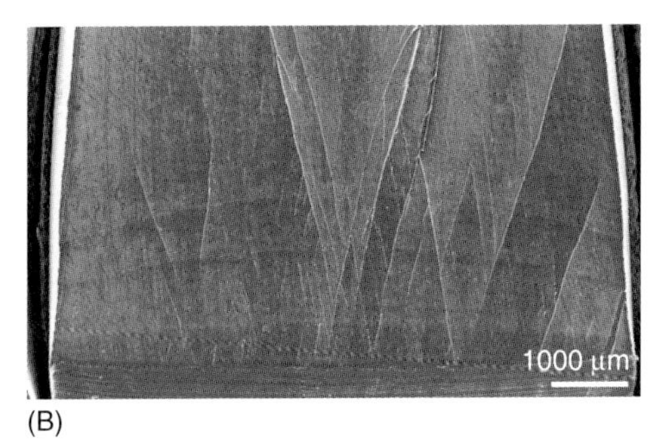


**FIGURE 30.14** Typical SEM micrographs of the fracture surface of virgin UHMWPE (A) and 100 kGy electron-beam irradiated and remelted (B) compact tension J-integral specimens showing the physical cracks. The "diamond" tearing lines appeared as a consequence of fatigue pre-cracking and post-experiment rupture. The crack extension was measured using the five-point average method recommended by ASTM D6068.

crack advance and decrease in the growth of $\delta$ coinciding with $J_c$ also conceptually agree with similar observations of the dissipated work at the crack tip in cyclic loading, referred to as the hysteresis method (discussed later).

### 30.3.5.3 J-integral Based Methods Developed for Polymers

The methods described thus far were originally developed for testing of metallic materials, although they have been applied to study polymers and other materials. To date, there exists only one J-integral based standard protocol specifically developed for polymers, which can be found in ASTM D6068 [81]. Other non-conventional protocols, such as the hysteresis method developed by Lu and colleagues for toughened acrylonitrile-butadiene-styrene, are also present in the polymer literature [69, 82, 83], but they are based on viscoelastic properties of polymers.

ASTM D6068 describes a multi-specimen procedure similar to the one described in ASTM E1820. Prenotched of fatigue precracked three point bend and compact specimens are suggested for testing. Again, the test specimen is loaded up to a selected displacement believed to induce certain crack extension and then is broken to expose the fracture surface. Apart from registering the load versus displacement curve, optical evaluations of the original and final crack length for each specimen has also to be conducted to obtain five-points averaged crack lengths. As an example, typical SEM micrographs of as-manufactured, and 100 kGy electron-beam irradiated and remelted UHMWPEs used for crack length evaluations are shown in Figures 30.14A–B. The data points that eventually will compose the J-R curve must fulfill some qualification requirements to be considered valid data points, similar to those included in ASTM E1820. Specifically, a minimum of seven data points are required to build a J-R curve, which have also to be evenly spaced throughout the region defined by the 0.05 mm exclusion line and the $\Delta a_{max}=0.1b_o$ exclusion line, with at least three data points in the first quadrant (small crack extensions). J is corrected for the energy of indentation, $U_i$, and calculated by means of the following relationships:

$$J = \frac{\eta U}{B(W - a_o)} \tag{30}$$

$$U_T = U + U_i \tag{31}$$

where B, W, $a_o$, are again the thickness, width, and original crack length of the specimen, respectively, $\eta$ equals to $2+0.522b_o/W$ for compact specimens, and $U_T$ and U are the total energy and the energy required to extend the crack, respectively. Finally, power law regression fits of the data are obtained, and the J-R curves are determined. J-R curves from this method can serve to rank different polymers or a polymer subjected to different treatments or processing conditions.

The hysteresis method is based on the behavior of the load versus displacement curve of certain pre-cracked polymeric specimens while subjected to cyclic loading and unloading (see Figure 30.15) [82, 83]. There is an obvious difference between the input energy, the area under the loading portion of the load-displacement curve, $A_o$, and the recovery energy, which is the area under the unloading portion, $A_r$. The hysteresis energy can then be defined as the difference between the input energy and the recovery energy. From a physical standpoint, the hysteresis energy is thought to account primarily for the energy consumed in enlarging the crack tip plastic zone before the onset of crack extension. However, when the crack extension begins the release of strain energy will contribute to the hysteresis energy as well, defining a transition in this property between crack blunting and crack extension and thus yielding an estimation of $J_{IC}$. The hysteresis energy, $H_E$, can be calculated by defining a hysteresis ratio, $H_R$, which is given by the following equation:

$$H_R = (A_o - A_r)/A_o \tag{32}$$

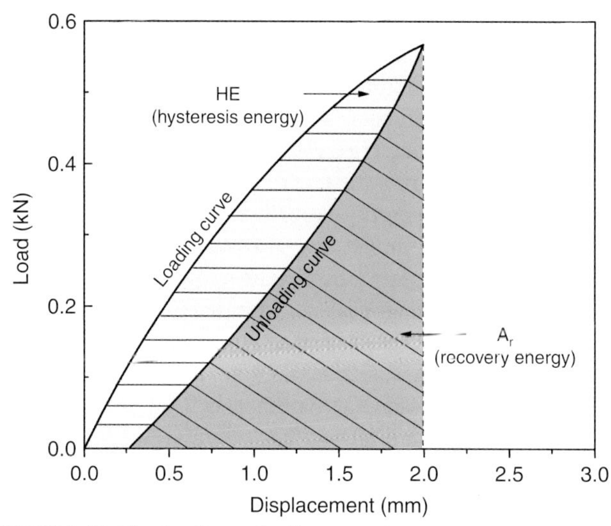


**FIGURE 30.15** Loading–unloading curve in the J-integral hysteresis method. The hysteresis energy (in light yellow) accounts for the energy consumed in enlarging the crack tip plastic zone before the onset of crack extension. When the crack extension begins, the release of strain energy also contributes to the hysteresis energy.

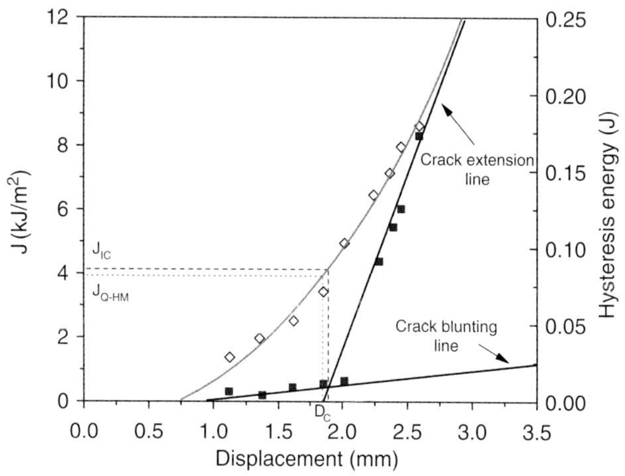


**FIGURE 30.16** Determination of $J_{IC}$ based on the hysteresis method. J values and hysteresis energy values versus displacement are plotted for every specimen. The linear regression of the hysteresis energy data before and after the onset of crack extension defined the crack blunting line and crack extension line, respectively. $J_{IC}$ can be estimated as the *J* value that corresponds to the critical displacement, $D_c$, where the crack blunting line and the crack extension line intersect. The red dotted lines define $D_c$ and $J_{Q-HM}$ (tentative J value) according to the modified hysteresis method introduced by Lewis and Nyman [69].

and then HE is given by:

$$H_E = H_R U \tag{33}$$

where U is the energy under the loading curve at a selected displacement. By representing together the usual *J* and $H_E$ results, it is possible to determine a $J_{IC}$ estimation as the *J* value that corresponds to the critical displacement, $D_c$, where the crack blunting line and the crack extension line intersect (see Figure 30.16). The linear regression of the hysteresis energy data before and after the onset of crack extension defines the transition between the crack blunting line and crack extension line. Clearly, the main advantage of the hysteresis method is that no crack length measurements are needed, avoiding the controversy regarding crack blunting and the flow stress. It has to be borne in mind, however, that constant crack speed conditions held in Lu's work, and so they considered it appropriate to use the hysteresis ratio for a full loading/unloading cycle. In general, the hysteresis ratio will be a function of the local strain, strain rate, temperature, and the amount of crack growth [83].

Lewis and Nyman applied a variant of the hysteresis method to study the fracture resistance of UHMWPE after sterilization by different techniques and accelerated aging [69]. In this study, a single specimen was loaded in a three-point bending configuration, recording the load, P, versus displacement, D, curve until the latter reached 8 mm. The areas under the P-D curve up to six selected displacements, $D_i$ were calculated and used to obtain the corresponding $J_i$ values. The $J_i$-$D_i$ pairs were plotted and a curve best fitted to the data points was also obtained. Then, the asymptotic crack extension line was drawn and its intersection with the displacement axis defined a critical displacement, $D_c$ (see Figure 30.16). The tentative fracture toughness of the material, $J_{Q-HM}$, was given as the J value that corresponded to $D_c$. Finally, the following equation had to be fulfilled to consider $J_{Q-HM}$ as a valid $J_{IC}$ value:

$$B, b_o > 25\,(J_{Q-HM}/\sigma_f) \tag{34}$$

where B is the thickness of the specimen, $b_o$ is the original uncracked ligament, and the flow stress, $\sigma_f$, was specified as $0.5(\sigma_y + \mathrm{UTS})$, that is, the effective yield strength. The main findings on fracture properties of UHMWPE obtained by Lewis and Nyman using this method are summarized in the following section.

## 30.3.6 J-integral Fracture Toughness of UHMWPE

The scientific reports on the characterization of the fracture toughness of UHMWPE reflect the multitude of variants of the J-integral approach introduced in the previous sections. Multiple specimen methods, such as those described for metals in ASTM E1820 and in the now obsolete ASTM E813, or the specifically developed for plastics in ASTM D6068, have been adapted and applied to UHMWPE [8, 77–79, 84, 85]. On the other hand, alternative single specimen methods, namely the normalization and the hysteresis methods, have also provided estimations of the resistance to stable crack growth of UHMWPE [11, 69, 80]. In the next paragraphs, a general overview of the main findings on the fracture toughness of UHMWPE is presented.

Rimnac et al. estimated a $J_{IC}$ value of 99.5 kJ/m$^2$ for ram extruded UHMWPE based on the multi-specimen procedure described in ASTM E813-87 [79, 86]. In this work, the data was qualified by specifying the flow stress as the ultimate tensile strength to obtain the blunting line, and the usual 0.15 and 1.5 mm exclusion lines. Pascaud and coworkers employed the same multi-specimen method to determine $J_{IC}$ of ram extruded GUR 4150 UHMWPE, but they considered five different analytical methods to calculate $J_{IC}$ from J versus crack extension plots in an attempt to find the optimal method [78]. $J_{IC}$ results ranging from 27 kJ/m$^2$ and 238 kJ/m$^2$ were recorded among the various methods included in that study. However, the authors proposed a modified analytical method consisting of linear fitting of the J-R data between $\Delta a/b_o = 0.6\%$ and $\omega = (b_o/J)(dJ/d\Delta a)>3$. $J_{IC}$ was obtained as the intersection between the R curve and the blunting line, which was calculated using ultimate tensile strength as the flow stress. According to this modified method, GUR 4150 UHMWPE specimens machined parallel to the extrusion direction exhibited a $J_{IC}$ of 66.5 kJ/m$^2$. In a further study, the same authors found that gamma sterilization (25–40 kGy) in air diminished the toughness of UHMWPE by 50% [87]. Lewis and Nyman [69] also reported a decrease in the fracture toughness of UHMWPE from 92 kJ/m$^2$ to 44.3 kJ/m$^2$ after gamma radiation sterilization based on the modified hysteresis method. Further, accelerated aging at elevated temperature (70°C) and oxygen pressure (507 kPa) for 2 weeks caused severe drop in the fracture toughness of gamma sterilized UHMWPE, lowering toughness down to 1.9 kJ/m$^2$. Although less affected than gamma sterilized UHMWPE, unsterilized, EtO, and gas plasma sterilized UHMWPEs were not completely immune to artificial aging.

As far as highly crosslinked UHMWPE are concerned, Gillis and colleagues utilized ASTM E813-81 to evaluate fracture resistance properties of highly crosslinked UHMWPEs in two consecutive studies [84, 85]. These studies confirmed decreasing toughness with increasing radiation doses and similar resistance to fracture after both gamma and electron beam irradiation at identical doses. In addition, they found higher fracture toughness for GUR 1020 resins with respect to GUR 1050 and GUR 4150 resins and a slight influence of the consolidation method [84]. Gomoll and coworkers also confirmed a deleterious effect of radiation crosslinking on the steady state J-integral fracture toughness of UHMWPE, in which they specified the flow stress as the tensile yield strength for calculating *J* values based on the Rice and Sorensen model [77]. Cole and colleagues based their $J_Q$ assessments of highly crosslinked UHMWPEs on the multi-specimen procedure and the construction line proposed by Pascaud et al. They found that radiation doses ranging from 33 to 100 kGy had a detrimental effect on the fracture toughness of both 1900H and GUR 1050 UHMWPE samples, the only exception being GUR 1050 specimens irradiated at the lowest dose. It has to be mentioned that all the specimens in Cole's study underwent post-irradiation annealing at 110°C for 24 hours, but the separate effects of irradiation and thermal stabilization were not discriminated. Recently, the separate effect of electron beam irradiation and subsequent remelting and annealing on the J-integral fracture toughness of UHMWPE has been evaluated per ASTM D6068 [8]. Apart from the decrease due to irradiation (see Figure 30.17A), the toughness of UHMWPE was found to further diminish after subsequent remelting at 150°C for 2 hours. In contrast, annealing at 130°C for 2 hours slightly increased toughness with respect to the as-irradiated UHMWPE (see Figure 30.17B). Finally, the findings of the works by Varadajaran and Rimnac addressed some of the issues associated with the single specimen normalization method and the use of the blunting

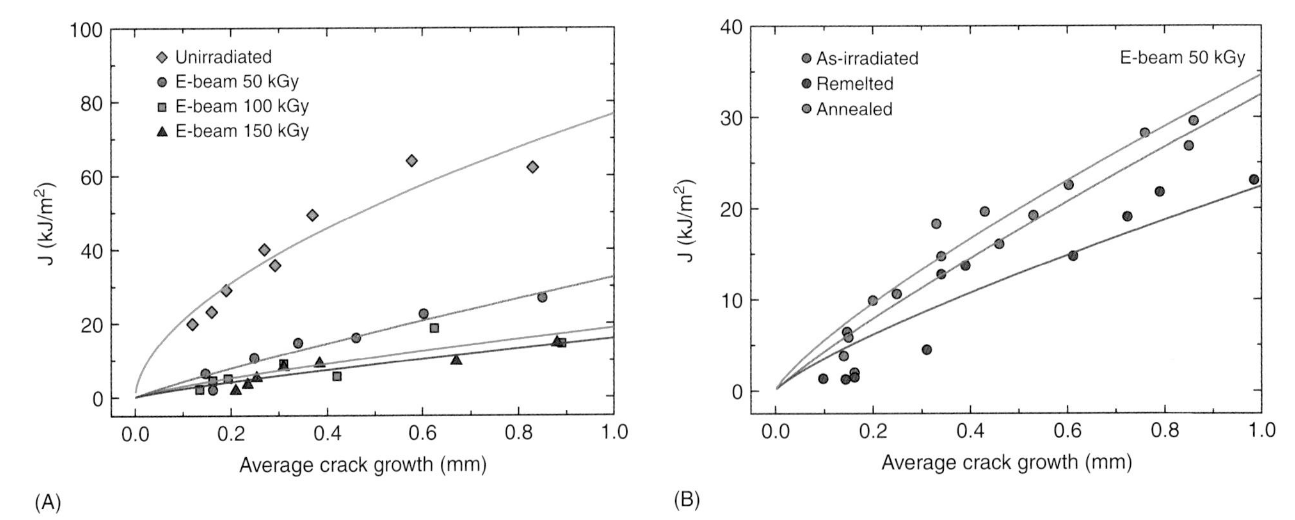


**FIGURE 30.17** J-R curves obtained according to ASTM D6068 for electron-beam irradiated (A) and post-irradiation thermally stabilized UHMWPE materials (B). The fracture toughness of UHMWPE noticeably dropped with increasing radiation doses. Post-irradiation remelting had a detrimental effect on the resistance to fracture of e-beam irradiated UHMWPE, whereas annealing slightly improved its fracture toughness.

line, while studying fracture properties of conventional and highly crosslinked UHMWPE materials [11, 88]. Moreover, they confirmed that crack tip opening displacement methods provide accurate assessments of $J_Q$ for those materials, as mentioned before. Their results are consistent with the previously reported on the effects of irradiation and thermal stabilization, with highly crosslinked remelted UHMWPE showing the least resistance to fracture compared to gamma inert sterilized, and highly crosslinked and annealed UHMWPEs. They also showed a further reduction in fracture toughness at 37°C (body temperature) for the previous materials in comparison with room temperature resistance to fracture.

Again, it is worth noting the existing controversy on the specification of the flow stress to obtain quantitative J-integral based toughness assessments. As noted by Varadarajan and Rimnac, the blunting line equation recommended by ASTM E1820 is strictly applicable only for metals, and it has been shown to be unsatisfactory for other polymers. Moreover, the validation of blunting line equations for UHMWPE materials may be problematic if it is based on observations of the deformed zone caused by blunting of the crack tip. In words of Gomoll and colleagues [77]: "ASTM standard D813-81 (and ASTM D6068-81) choose to avoid the issue of the determination of a flow stress for rate dependent viscoelastic–plastic materials due to uncertainty in its measurement in viscoelastic–plastic materials." The practical implications of the selection of the flow stress as the yield strength, effective yield strength, or ultimate tensile strength may alter qualitatively and quantitatively toughness estimations, especially when dealing with conventional and crosslinked UHMWPEs. Therefore, the pending challenge is to gain more knowledge about the mechanisms of damage and failure during crack extension in the case of UHMWPE materials, as the roles of yielding, strain hardening, and rupture during stable crack growth have yet to be ascertained. The recently developed crack tip opening displacement method (CTOD) for UHMWPE materials by Varadarajan and Rimnac [11], and future refinements of the hysteresis method, appear to be promising steps forward, which may address the current limitations of the extant J-integral methods, giving deeper insight into the underlying fracture mechanisms of UHMWPE.

## 30.4 SUMMARY AND CONCLUSION

The study of the fracture and fatigue properties of UHMWPE materials has received considerable attention over the last 2 decades. The fatigue resistance of UHMWPE is now widely accepted as critical to the structural performance of orthopedic implant bearings, and such data is now commonplace in regulatory approval submissions. With regard to fatigue properties, both the total life (stress analysis based) and defect tolerant (fracture mechanics based) approaches have been employed to characterize conventional and highly crosslinked UHMWPE materials, although the limitations associated with total life experiments have rendered fatigue crack propagation experiments prevalent.

The fatigue crack propagation resistance of UHMWPE has been shown to decrease with increasing radiation doses by multiple investigators. A further reduction in fatigue crack growth performance has also been confirmed due to remelting of the polymer subsequent to irradiation, motivating so-called second generation formulations that are not remelted after crosslinking. Despite the increasing understanding of the fatigue performance of UHMWPE, its behavior under fatigue loading conditions is still under active research. An emergent topic is the static mode, viscoelastic fatigue crack propagation of UHMWPE because recent studies in this area have demonstrated static mode (i.e., non-cyclic) crack propagation in an uncrosslinked formulation.

As far as the fracture resistance of UHMWPE is concerned, uniaxial tensile and impact test methods have been used to provide qualitative estimations of the fracture properties of the polymer. Stricter toughness assessments under stable crack extension conditions have been also conducted based on J-integral methods. In fact, numerous variations of J-integral methods have been used to determine the resistance to fracture of UHMWPE. The development of a standard J-integral based experimental technique, along with sound analytical qualification of the data, for UHMWPE is still a pending challenge for the orthopedic and scientific communities. In this sense, methods relying the crack tip-opening displacement and the specimen hysteresis are promising alternatives that may address the current issues with quantitative J-integral based toughness evaluations, providing deeper insight into the damage and fracture mechanisms during stable crack extension of UHMWPE.

Despite the present limitations of fracture toughness evaluation methods, the fracture properties of conventional and highly crosslinked UHMWPE formulations have been extensively characterized, identifying a detrimental effect of high radiation doses and subsequent remelting treatments, and a less noticeable impact of crystallinity changes on the fracture resistance behavior of UHMWPE. The fatigue and fracture performance of crosslinked UHMWPE is expected to remain an important design and clinical concern for total joint replacements for years to come.

## REFERENCES

1. Rimnac C, Pruitt L, WIWSE. How do material properties influence wear and fracture mechanisms? *J Am Acad Orthop Surg* 2008; **16**:S94–S100.
2. Kurtz SM. *The UHMWPE handbook: ultra-high molecular weight polyethylene in total joint replacement*. New York: Academic Press; 2004.

3. Baker DA, Hastings RS, Pruitt L. Compression and tension fatigue resistance of medical grade ultra high molecular weight polyethylene: the effect of morphology, sterilization, aging and temperature. *Polymer* 2000 January;**41**(2):795–808.
4. Gencur SJ, Rimnac CM, Kurtz SM. Failure micromechanisms during uniaxial tensile fracture of conventional and highly crosslinked ultra-high molecular weight polyethylenes used in total joint replacements. *Biomaterials* 2003 October;**24**(22):3947–54.
5. Gencur SJ, Rimnac CM, Kurtz SM. Fatigue crack propagation resistance of virgin and highly crosslinked, thermally treated ultra-high molecular weight polyethylene. *Biomaterials* 2006 March; **27**(8):1550–7.
6. Kurtz SM, Villarraga ML, Herr MP, Bergstrom JS, Rimnac CM, Edidin AA. Thermomechanical behavior of virgin and highly crosslinked ultra-high molecular weight polyethylene used in total joint replacements. *Biomaterials* 2002 September;**23**(17):3681–97.
7. Lewis G. Properties of crosslinked ultra-high-molecular-weight polyethylene. *Biomaterials* 2001 February;**22**(4):371–401.
8. Medel FJ, Peña P, Cegoñino J, Gómez-Barrena E, Puértolas JA. Comparative fatigue behavior and toughness of remelted and annealed highly crosslinked polyethylenes. *J Biomed Mater Res Part B-Appl Biomater* 2007 November;**83B**(2):380–90.
9. Pruitt LA. Deformation, yielding, fracture and fatigue behavior of conventional and highly cross-linked ultra high molecular weight polyethylene. *Biomaterials* 2005 March;**26**(8):905–15.
10. Varadarajan R, Rimnac CM. Compliance calibration for fatigue crack propagation testing of ultra high molecular weight polyethylene. *Biomaterials* 2006 September;**27**(27):4693–7.
11. Varadarajan R, Rimnac CM. Evaluation of J-initiation fracture toughness of ultra-high-molecular-weight polyethylene used in total joint replacements. *Polym Test* 2008 August;**27**(5):616–20.
12. Villarraga ML, Kurtz SM, Herr MP, Edidin AA. Multiaxial fatigue behavior of conventional and highly crosslinked UHMWPE during cyclic small punch testing. *J Biomed Mater Res Part A* 2003 August 1;**66A**(2):298–309.
13. Wright TM. Polyethylene in knee arthroplasty—What is the future? *Clin Orthop Relat Res* 2005 November(440):141–8.
14. Rimnac CM, Kurtz SM. Ionizing radiation and orthopaedic prostheses. *Nucl Instrum Methods Phys Res Section B-Beam Interact Mater Atoms* 2005 July;**236**:30–7.
15. Jasty M, Rubash HE, Muratoglu O. Highly cross-linked polyethylene—The debate is over—In the affirmative. *J Arthroplasty* 2005 June;**20**(4):55–8.
16. Ries MD. Highly cross-linked polyethylene—The debate is over—In opposition. *J Arthroplasty* 2005 June;**20**(4):59–62.
17. Kurtz SM, Muratoglu OK, Evans M, Edidin AA. Advances in the processing, sterilization, and crosslinking of ultra-high molecular weight polyethylene for total joint arthroplasty. *Biomaterials* 1999 September;**20**(18):1659–88.
18. McKellop HA. The lexicon of polyethylene wear in artificial joints. *Biomaterials* 2007 December;**28**(34):5049–57.
19. Birman MV, Noble PC, Conditt MA, Li S, Mathis KB. Cracking and impingement in ultra-high-molecular-weight polyethylene acetabular liners. *J Arthroplasty* 2005 October;**20**(7):87–92.
20. Tower SS, Currier JH, Currier BH, Lyford KA, Van Citters DW, Mayor MB. Rim cracking of the cross-linked longevity polyethylene acetabular liner after total hip arthroplasty. *J Bone Joint Surg-Am Vol* 2007 October;**89A**(10):2212–17.
21. Medel FJ, Rimnac CM, Kurtz SM. On the assessment of oxidative and microstructural changes after in vivo degradation of historical UHMWPE knee components by means of vibrational spectroscopies and nanoindentation. *J Biomed Mater Res* 2009 May; 89A(2):530–538.
22. Kurtz SM, Hozack WJ, Purtill JJ, Marcolongo M, Kraay MJ, Goldberg VM, et al. Otto Aufranc Award paper—Significance of in vivo degradation for polyethylene in total hip arthroplasty. *Clin Orthop Relat Res* 2006 December(453):47–57.
23. Suresh S. *Fatigue of materials*. 2nd ed. New York: Cambridge University Press; 2001.
24. Hertzberg RW, Manson JA. *Fatigue of engineering polymers*: Academic Press (London) LTD; 1980.
25. Williams ML, DeVries KL. *Proc 5th Cong Rheol*; 1970. p. 139.
26. Baker DA. *Macro- and microscopic evaluation of fatigue in medical grade ultrahigh molecular weight polyethylene*. Berkeley, CA: UC Berkeley; 2001.
27. Furmanski J, Atwood SA, Bal BS, Anderson MR, Penenberg BL, Halley DK, et al. *Fracture of Highly Crosslinked UHMWPE Acetabular Liners (Sci. Exhibit)*. San Francisco: American Academy Orthopaedic Surgeons; 2008.
28. Furmanski J, Gupta S, Chawan A, Kohm A, Lannutti J, Jewett B, et al. Aspherical femoral head with highly cross-linked ultra-high molecular weight polyethylene surface cracking—A case report. *J Bone Joint Surg-Am Vol* 2007 October;**89A**(10):2266–70.
29. Furmanski J, Lehman RL, Schnaser E, Goldberg VM, Kraay MJ, Pruitt LA, et al. In vivo crack initiation in retrieved cross-linked UHMWPE acetabular liners. *Transactions of the 54th Orthopaedic Research Society*; 2009.
30. O'Connor DO, Muratoglu OK, Bragdon CR, Lowenstein J, Jasty M, Harris WH. Wear and high cycle fatigue of a highly cross-linked UHMWPE. *Trans 45th Orthop Res Soc* 1999;**24**:816.
31. Villarraga ML, Edidin AA, Herr MP, Kurtz SM. Multiaxial fatigue behavior of oxidized and unoxidized UHMWPE during cyclic small punch testing at body temperature. In: Kurtz SM, Gsell RA, Martell JA, editors. *Crosslinked and thermally treated ultra-high molecular weight polyethylene for joint replacements*. West Conshohocken, Pa.: ASTM International; 2004.
32. Weightman B, Light D. A comparison of Rch 1000 and Hi-Fax 1900 ultrahigh molecular-weight polyethylenes. *Biomaterials* 1985; **6**(3):177–83.
33. Baker DA, Bellare A, Pruitt L. The effects of degree of crosslinking on the fatigue crack initiation and propagation resistance of orthopedic-grade polyethylene. *J Biomed Mater Res Part A* 2003 July 1;**66A**(1):146–54.
34. Urriés I, Medel FJ, Ríos R, Gómez-Barrena E, Puértolas JA. Comparative cyclic stress-strain and fatigue resistance behavior of electron-beam- and gamma-irradiated ultrahigh molecular weight polyethylene. *J Biomed Mater Res Part B-Appl Biomater* 2004 July 15;**70B**(1):152–60.
35. Puértolas JA, Medel FJ, Cegónino J, Gómez-Barrena E, Ríos R. Influence of the remelting process on the fatigue behavior of electron beam irradiated UHMWPE. *J Biomed Mater Res Part B-Appl Biomater* 2006 February;**76B**(2):346–53.
36. Baker DA, Pruitt L, Bellare A. Ultra-small angle X-ray scattering to detect fatigue damage in polymers. *J Mater Sci Lett* 2001; **20**(12):1163–4.
37. Furmanski J, Pruitt LA. Peak stress intensity dictates fatigue crack propagation in UHMWPE. *Polymer* 2007 June 4;**48**(12):3512–19.
38. Suresh S. *Fatigue of materials*. 2nd ed. Cambridge: Cambridge University Press; 1998.
39. ASTM E647-08. Standard Test Method for Measurement of Fatigue Crack Growth Rates.

40. Oral E, Malhi AS, Muratoglu OK. Mechanisms of decrease in fatigue crack propagation resistance in irradiated and melted UHMWPE. *Biomaterials* 2006 February;**27**(6):917–25.
41. Simis KS, Bistolfi A, Bellare A, Pruitt LA. The combined effects of crosslinking and high crystallinity on the microstructural and mechanical properties of ultra high molecular weight polyethylene. *Biomaterials* 2006 March;**27**(9):1688–94.
42. Cole JC, Lemons JE, Eberhardt AW. Gamma irradiation alters fatigue-crack behavior and fracture toughness in 1900H and GUR 1050 UHMWPE. *J Biomed Mater Res* 2002 October;**63**(5):559–66.
43. Saxena A. *Nonlinear fracture mechanics for engineers*: CRC Press; 1997.
44. Schapery RA. Correspondence principles and a generalized J integral for large deformation and fracture-analysis of viscoelastic media. *Int J Fract* 1984;**25**(3):195–223.
45. Schapery RA. Theory of crack initiation and growth in viscoelastic media .1. Theoretical development. *Int J Fract* 1975;**11**(1):141–59.
46. Schapery RA. Theory of crack initiation and growth in viscoelastic media.2. Approximate methods of analysis. *Int J Fract* 1975; **11**(3):369–88.
47. Schapery RA. Theory of crack initiation and growth in viscoelastic media.3. Analysis of continuous growth. *Int J Fract* 1975; **11**(4):549–62.
48. Williams JG. *Fracture mechanics of polymers*. Chichester: Ellis Horwood Ltd; 1984.
49. Furmanski J, Gupta S, Atwood SA, Pruitt LA. Creep behavior of cross-linked UHMWPE and its influence on clinical fracture failure of orthopaedic implants. *Proceedings of the 53rd Orthopaedic Research Society*; 2008.
50. Crissman JM, Khoury FA, McKenna GB. Relationship between morphology and mechanical properties of ultra high molecular weight polyethylene; 1982.
51. Evans AG. Fatigue in Ceramics. *Int J Fract* 1980;**16**(6):485–98.
52. Dumpleton P, Bucknall CB. Comparison of static and dynamic fatigue crack-growth rates in high-density polyethylene. *Int J Fatigue* 1987 July;**9**(3):151–5.
53. Ritchie RO. Mechanisms of fatigue-crack propagation in ductile and brittle solids. *Int J Fract* 1999;**100**(1):55–83.
54. Bassani JL, Brown N, Lu X. J-Integral correlation of the initiation of slow crack-growth in linear polyethylene. *Int J Fract* 1988 September;**38**(1):43–59.
55. Brown N, Bhattacharya SK. The initiation of slow crack-growth in linear polyethylene under single edge notch tension and plane-strain. *J Mater Sci* 1985;**20**(12):4553–60.
56. Brown N, Donofrio J, Lu X. The transition between ductile and slow-crack-growth failure in polyethylene. *Polymer* 1987 July; **28**(8):1326–30.
57. Brown N, Lu X, Huang Y, Harrison IP, Ishkawa N. The fundamental material parameters that govern slow crack-growth in linear polyethylenes. *Plast Rubber Compos Process Appl* 1992;**17**(4):255–8.
58. Furmanski J, Pruitt L. Static mode fatigue crack propagation of UHMWPE (manuscript in preparation). 2008.
59. Furmanski J, Pruitt L. Origin and precision of Kmax correlations for static mode crack propagation (manuscript in preparation). 2008.
60. Ritchie RO, Gilbert CJ, McNaney JM. Mechanics and mechanisms of fatigue damage and crack growth in advanced materials. *Int J Solids Struct* 2000 January;**37**(1–2):311–29.
61. Tomita N, Kitakura T, Onmori N, Ikada Y, Aoyama E. Prevention of fatigue cracks in ultrahigh molecular weight polyethylene joint components by the addition of vitamin E. *J Biomed Mater Res* 1999 August;**48**(4):474–8.
62. Shibata N, Tomita N, Onmori N, Kato K, Ikeuchi K. Defect initiation at subsurface grain boundary as a precursor of delamination in ultrahigh molecular weight polyethylene. *J Biomed Mater Res Part A* 2003 October 1;**67A**(1):276–84.
63. Wang A, Manley MT, Serekian P. Wear and structural fatigue simulation of crosslinked ultra-high molecular weight polyethylene for hip and knee bearing applications. In: Kurtz SM, Gsell RA, Martell JA, editors. *Crosslinked and thermally treated ultra-high molecular weight polyethylene for joint replacements*. West Conshohocken, Pa.: ASTM International; 2004.
64. Boscoletto AB, Franco R, Scapin M, Tavan M. An investigation on rheological and impact behaviour of high density and ultra high molecular weight polyethylene mixtures. *Eur Polym J* 1997 January;**33**(1):97–105.
65. Huang DH, Yang YM, Li BY. Impact behavior of phenolphthalein poly(ether sulfone) ultrahigh molecular weight polyethylene blends. *J Appl Polym Sci* 1998 January 3;**67**(1):113–18.
66. Lim KLK, Ishak ZAM, Ishiaku US, Fuad AMY, Yusof AH, Czigany T, et al. High-density polyethylene/ultrahigh-molecular-weight polyethylene blend. I. The processing, thermal, and mechanical properties. *J Appl Polym Sci* 2005 July 5;**97**(1):413–25.
67. Lim KLK, Ishak ZAM, Ishiaku US, Fuad AMY, Yusof AH, Czigany T, et al. High density polyethylene/ultra high molecular weight polyethylene blend. II. Effect of hydroxyapatite on processing, thermal, and mechanical properties. *J Appl Polym Sci* 2006 June 5;**100**(5):3931–42.
68. Ranade RA, Wunder SL, Baran GR. Toughening of dimethacrylate resins by addition of ultra high molecular weight polyethylene (UHMWPE) particles. *Polymer* 2006 May 31;**47**(12):4318–27.
69. Lewis G, Nyman JS. A new method of determining the J-integral fracture toughness of very tough polymers: application to ultra high molecular weight polyethylene. *J Long-Term Eff Med Implants* 1999;**9**(4):289–301.
70. ASTM D638-08 Standard Test Method for Tensile Properties of Plastics.
71. ASTM F648-07e1 Standard Specification for Ultra-High-Molecular-Weight Polyethylene Powder and Fabricated Form for Surgical Implants.
72. Greer KW, King RS, Chan FW. The effects of raw material, irradiation dose, and irradiation source on crosslinking of UHMWPE. In: Kurtz SM, Gsell RA, Martell J, editors. *Crosslinked and thermally treated ultra-high molecular weight polyethylene for joint replacements*. West Conshohocken, Pa.: ASTM International; 2004.
73. Abt NA, Schneider W, Schon R, Rieker CB. Influence of electron beam irradiation dose on the properties of crosslinked UHMWPE. In: Kurtz SM, Gsell RA, Martell J, editors. *Crosslinked and thermally treated ultra-high molecular weight polyethylene for joint replacements*. West Conshohocken, Pa.: ASTM International; 2004.
74. Rice JR. A path independent integral and the approximate analysis of strain concentration by notches and cracks. *J Appl Mech* 1968;**35**:379–86.
75. Anderson TL. *Fracture Mechanics*. 2nd ed: CRC Press; 2004.
76. ASTM E1820-08 Standard Test Method for Measurement of Fracture Toughness.
77. Gomoll A, Wanich T, Bellare A. J-integral fracture toughness and tearing modulus measurement of radiation cross-linked UHMWPE. *J Orthop Res* 2002 November;**20**(6):1152–6.
78. Pascaud RS, Evans WT, McCullagh PJJ, Fitzpatrick D. Critical assessment of methods for evaluating J(Ic) for a medical grade

ultra high molecular weight polyethylene. *Polym Eng Sci* 1997 January;**37**(1):11–17.

79. Rimnac CM, Wright TM, Klein RW. J-integral measurements of ultra high molecular-weight polyethylene. *Polym Eng Sci* 1988 December;**28**(24):1586–9.
80. Landes JD, Bhambri SK, Lee K. Fracture toughness testing of polymers using small compact specimens and normalization. *J Test Eval* 2003 March;**31**(2):126–32.
81. ASTM D6068-96(2002)e1 Standard Test Method for Determining J-R Curves of Plastic Materials.
82. Lu ML, Lee CB, Chang FC. Fracture-Toughness of Acrylonitrile-Butadiene-Styrene by J-Integral Methods. *Polym Eng Sci* 1995 September;**35**(18):1433–9.
83. Lu ML, Chiou KC, Chang FC. Fracture toughness characterization of a PC/ABS blend under different strain rates by various J-integral methods. *Polym Eng Sci* 1996 September;**36**(18):2289–95.
84. Duus LC, Walsh HA, Gillis AM, Noisiez E, Li S. The effect of resin grade, manufacturing method and crosslinking on the fracture toughness of commercially available UHMWPE. *Transactions of the 46th Orthopedic Research Society*; 2000. p. 544.
85. Gillis AM, Schmieg JJ, Bhattacharyya S, Li S. An independent evaluation of the mechanical chemical and fracture properties of UHMWPE crosslinked by 34 different conditions. *Transactions of the 45th Orthopedic Research Society*; 1999. p. 908.
86. ASTM E813-87 Test Method for JIC, A Measure of Fracture Toughness (Withdrawn).
87. Pascaud RS, Evans WT, McCullagh PJJ, FitzPatrick DP. Influence of gamma-irradiation sterilization and temperature on the fracture toughness of ultra-high-molecular-weight polyethylene. *Biomaterials* 1997 May;**18**(10):727–35.
88. Varadarajan R, Dapp EK, Rimnac CM. Static fracture resistance of ultra high molecular weight polyethylene using the single specimen normalization method. *Polym Test* 2008 April;**27**(2):260–8.

Chapter 31

# Development and Application of the Notched Tensile Test to UHMWPE

Michael C. Sobieraj, PhD and Clare M. Rimnac, PhD



## 31.1 INTRODUCTION

In engineering design it is well known that a preponderance of failures emanate from stress concentrations such as microstructural or geometrical discontinuities (geometric stress concentrations are generically termed as notches) [8–11]. The field of total joint replacements is not an exception to this rule. In the absence of severe wear, gross fracture of total joint replacement components tends to originate from stress concentrations such as fillets, undercuts, and grooves (Figure 31.1). There have been reports in the literature of cracks and/or gross fracture along acetabular rims [13–16], the stabilizing posts in noncruciate sparing tibial components [6, 17–20], and along the rims of total disc replacements [5]. Notably, there have been reports in the literature of rim fracture of highly crosslinked remelted UHMWPE components after short-term implantation (7 months to 3.8 years) [21–23]. Therefore, the assessment of stress raisers, and a better understanding of their influence on the behavior of UHMWPE, is of keen interest to the orthopedic community. These stress concentrations are inherent and necessary design features, and the behavior of UHMWPE in their vicinity is important to the survival of the implant. To begin to understand the effects of notches on the behavior of UHMWPE components *in vivo*, our group developed a novel system for the assessment of the response of UHMWPE to geometric stress concentrations.

## 31.2 OVERVIEW OF NOTCH BEHAVIOR

The presence of a notch alters the stress state near the notch and may increase the tendency for brittle fracture (as opposed to gross yielding) in potentially four ways: by introducing high local stresses; by creating a triaxial stress state ahead of the notch; by causing high local strain hardening; and by magnifying the strain rate locally [24]. As long as no local yielding at the notch root occurs, the stress at the notch root is defined as:

$$\sigma = K_t S \quad (1)$$

UHMWPE Biomaterials Handbook

where $K_t$ is the elastic stress concentration factor (which is only a function of geometry), and S is the nominal stress across the area [8]. When this calculated stress is greater than the yield stress of the material, local yielding occurs and Equation (1) no longer applies. When local yielding occurs, the magnitude of the stress concentration factor is lowered due to the redistribution of stresses caused by plastic flow [8].

Depending on the ability of the material to undergo plastic deformation in the vicinity of the notch, the material can either demonstrate a tendency toward brittle fracture (notch sensitivity) or failure after yielding at a stress greater than the unnotched yield stress (notch strengthening) [24]. Notch strengthening of ductile materials can be explained using a cylindrical specimen with a circumferential groove (Figure 31.2). When the specimen is loaded in tension, there is a

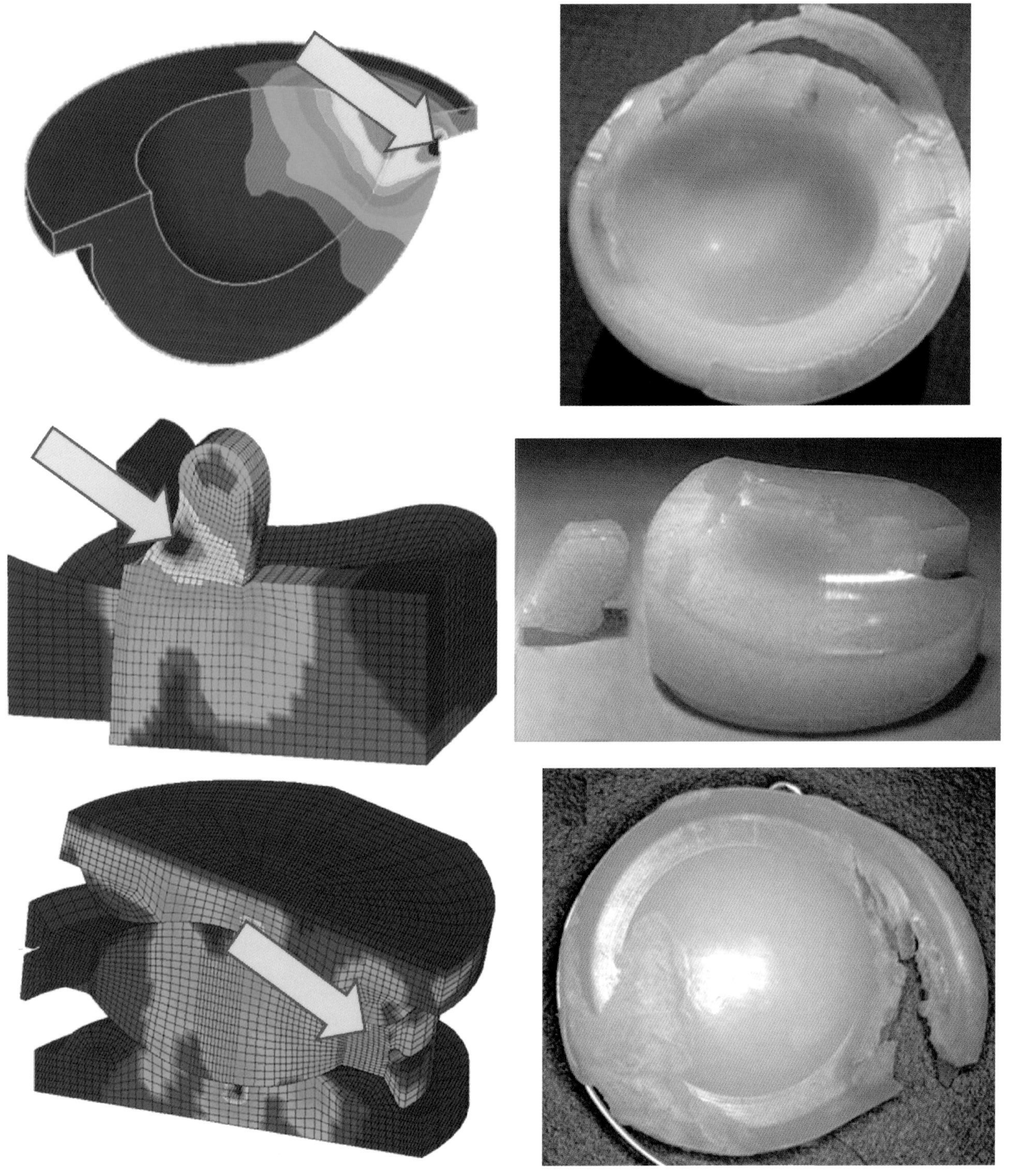

**FIGURE 31.1** FEA simulations of UHMWPE components with areas of potential stress concentration in the design noted by the arrows and retrieved components with fractures originating from similar locations [4–7].

triaxial stress state at the notch root, which is due to the stress concentration effect of the notch. In ductile materials, when the net section stress (the average stress applied to the reduced cross section) reaches the material yield point, the material in this section attempts to deform plastically. However, the notch root material also seeks to contract (due to incompressibility in plastic flow), but it is constrained by the bulk of the material that is still experiencing elastic stresses. The resulting tensile stresses in the tangential and radial directions make it necessary to increase the axial stress to initiate plastic flow. This leads to the apparent notch strengthening [25].

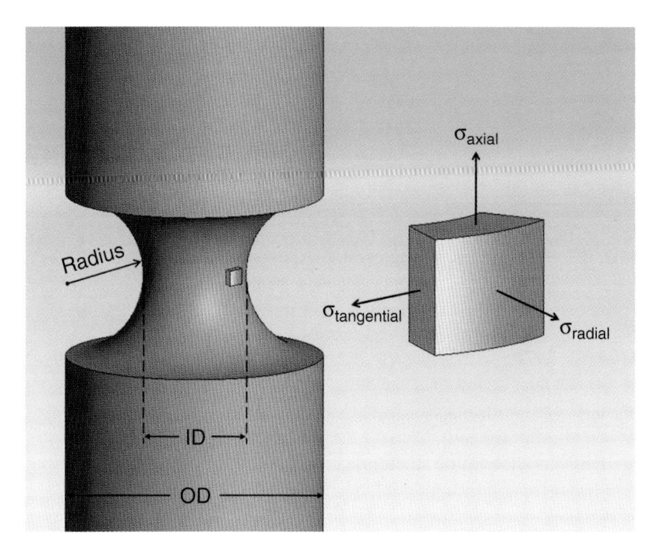


**FIGURE 31.2** A cylindrical bar with a circumferential groove. The enlarged element shows the directions of the principal stresses generated in the notch under an applied tensile load.

Notched monotonic tensile specimens can be a useful approach to begin to understand the effect of structural notches on the behavior of UHMWPE total joint replacement components. Therefore, we developed a testing methodology to characterize the stress–strain and fracture behavior of a notched tensile specimen. Additionally, the stress–strain behavior of notched tensile specimens can be used to challenge the Hybrid Constitutive Model for UHMWPE (see Chapter 35) with a multiaxial stress state. Accurate prediction of the behavior of a notched specimen by a simulation utilizing the Hybrid Model would be one validation of its accuracy in describing the mechanical behavior of UHMWPEs.

This chapter describes our development of the notched tensile test and findings for three studies [1–3] in which we evaluated two to five formulations of conventional and highly crosslinked UHMWPEs (Table 31.1).

**TABLE 31.1 Summary of the UHMWPE formulations, specimen geometries, displacement rates, and environmental conditions for the three studies [1–3] reviewed in this chapter. All dimensions (OD, ID, Radius) are in mm**

| Study | Materials | Geometries | | | | | Rates | Environment | |
|---|---|---|---|---|---|---|---|---|---|
| | Name | Name | OD | ID | Radius | $K_t$ | mm/min | Presoak | Testing |
| I | Virgin | Smooth | 10 | NA | NA | NA | 30 | None | Ambient air |
| | 25 kGy Radiation Sterilized in $N_2$ | Shallow | 10 | 8 | 0.45 | 2.9 | | | |
| | 100 kGy 110°C Annealed | | | | | | | | |
| | 100 kGy 150°C Remelted | Deep | 10 | 6 | 0.45 | 2.9 | | | |
| II | 30 kGy Radiation Sterilized in $N_2$ | Smooth | 8 | NA | NA | NA | 30 | 37°C PBS bath for 8 weeks | 37°C air |
| | 90 kGy Sequentially Annealed | Notched | 8 | 6 | 0.45 | 2.7 | 150 | | |
| III | 30 kGy Radiation Sterilized in $N_2$ | Smooth | 8 | NA | NA | NA | 30 | 37°C PBS bath for 8 weeks | 37°C air |
| | 65 kGy 130°C Annealed | Moderate | 8 | 6 | 0.90 | 2.1 | | | |
| | 65 kGy 150°C Remelted | | | | | | 150 | | |
| | 100 kGy 130°C Annealed | Deep | 8 | 6 | 0.45 | 2.7 | | | |
| | 100 kGy 150°C Remelted | | | | | | | | |

## 31.3 OVERVIEW OF THE NOTCHED TENSILE TEST METHOD

### 31.3.1 Specimen Geometry and Loading

There are multiple geometries and loading modes that can be used to examine notch effects on materials [8 11]. However, we elected to use a circumferentially grooved cylindrical dogbone specimen because it is experimentally and computationally attractive due to its axisymmetric design (Figure 31.2). As a control to the notched specimen, a cylindrical dogbone specimen with no notch (smooth) is used.

### 31.3.2 Data Acquisition and Stress–Strain Determination

Prior investigations into the behavior of notched specimens of materials have used the approach of recording a load and displacement and then photographing the fracture surface immediately after the test to obtain the ultimate axial true stress [26]. This method is acceptable in metals, but UHMWPE shows substantial strain relaxation upon fracture (Figure 31.3), which leads to inaccuracies in the calculated true ultimate stress. Also, this method does not provide information as to what is the deformation behavior of the notch during the test itself. Additionally, any method that uses a form of measurement that involves contacting extensometry risks premature fracture of a UHMWPE specimen due to the creation of a stress riser at the point of contact of the extensometer. Therefore, we developed a video-based system that could capture the stress–strain behavior throughout the duration of the test and avoid specimen contact [3].

The system relies on a CCD video camera with a 50 mm macro lens to acquire the diametral strain of the notched specimens. The camera's output is fed directly into a PC, via a video-to-USB converter, and the images are recorded for future analysis at a frame rate of 30 Hz and a resolution of 320 × 240 pixels in uncompressed AVI format. The camera is mounted on a moving stage, which allows the user to keep the notch in the center of the field of view. Simultaneously, load data is recorded. The load and the video data are then combined using MATLAB code (the algorithm is summarized in Figure 31.4). This code assumes incompressibility of UHMWPE in both the elastic and plastic ranges, which is an acceptable approximation because Bartel et al. [27] have reported that the Poisson's ratio for UHMWPE is 0.46.

The smooth specimens are tested using standard methods in which a noncontacting video extensometer is used to gather strain data in addition to the load cell data [3]. The same incompressibility assumptions are also made in the analysis of the smooth specimens.

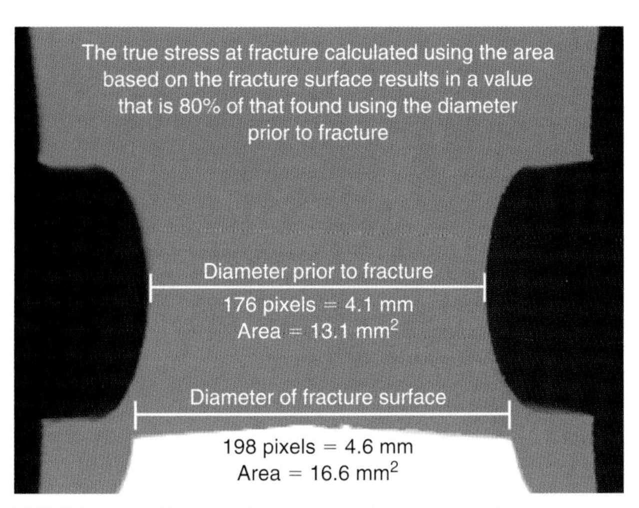


**FIGURE 31.3** Close-up view (a frame from a recorded testing video) of the notched region of a specimen immediately prior to fracture (gray region). The white region displays the image of the fracture specimen 4 seconds after fracture (from the same video recording). There is already a significant amount of strain relaxation at this point, and the resulting area has increased by approximately 25%.

### 31.3.3 Mechanical Characterization of Notch Effects

For both the notched and the smooth specimens, the engineering and true yield stress and strain and ultimate stress and strain can be gathered (with respect to the average axial values across the cross-section). Additionally, four ratios, two that describe the notch effect on yield behavior and two that describe the notch effect on postyield behavior, can be calculated. For each of the specimens, a notch strengthening ratio with respect to stress and a notch strengthening ratio with respect to strain can be calculated:

$$\phi_{\sigma\, or\, \varepsilon} = \frac{X_y}{X_{y_{smooth}}} \tag{2}$$

where $\phi$ is the notch strengthening ratio, and $X_y$ and $\mathbf{X}_{\mathbf{y}_{smooth}}$ are either the true yield stress based on the of the individual specimen and the mean true yield stress of the smooth specimens, respectively, or the corresponding strains. This ratio provides an estimate as to the degree of inhibition on plastic flow caused by the geometric constraint of the notch.

A hardening ratio with respect to stress and a hardening ratio with respect to strain can also be calculated for each of the specimens:

$$\psi_{\sigma\, or\, \varepsilon} = \frac{X_y}{X_u} \tag{3}$$

where$\psi$ is the hardening ratio, $X_u$ is either the true ultimate stress or strain for the specimen, and $X_y$ is the true yield stress or strain for the specimen. This ratio provides some insight into the degree of orientation hardening, and chain alignment, that the material undergoes during the test.

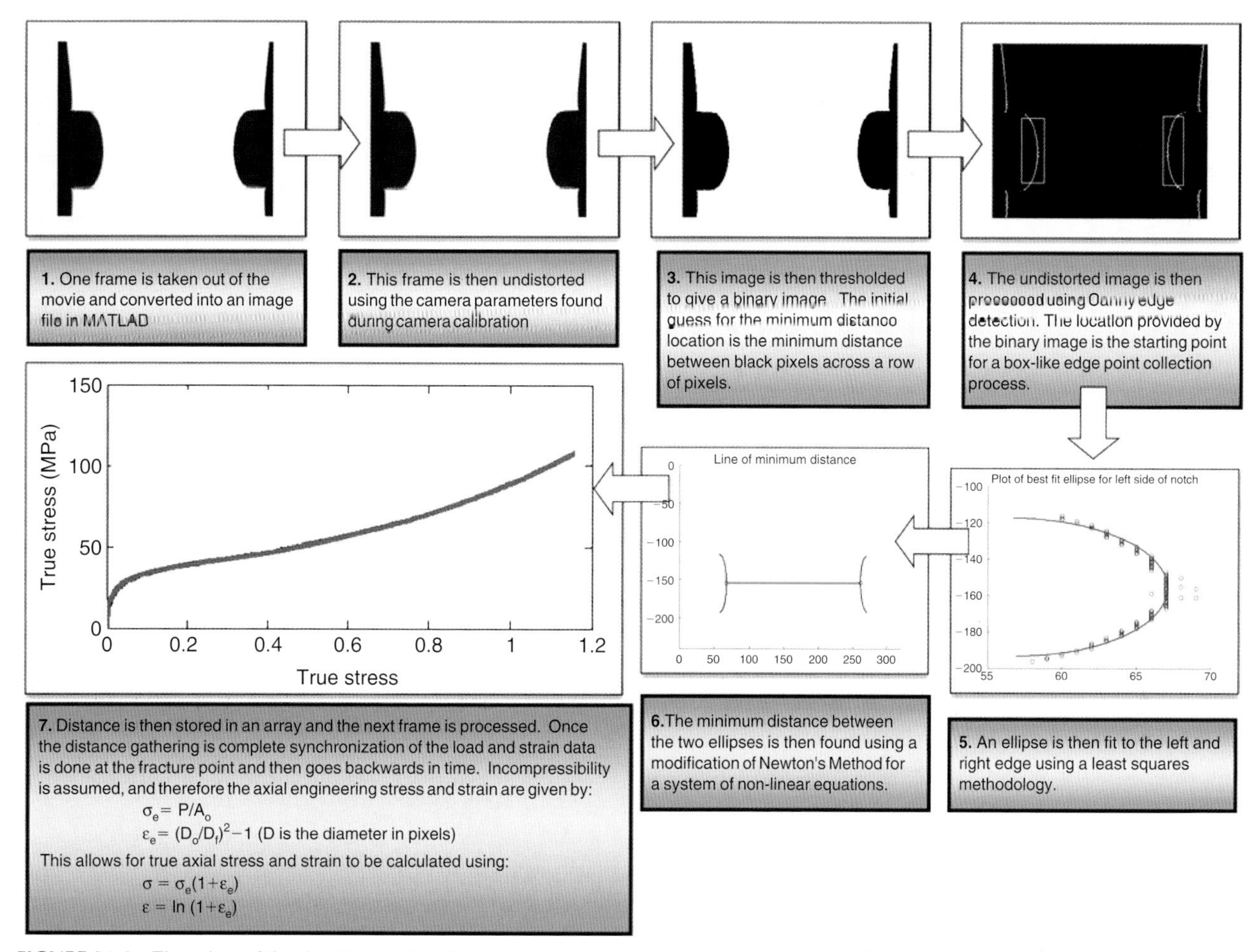


**FIGURE 31.4** Flow chart of the algorithm used to find axial true stress and axial true strain for notched specimens.

## 31.3.4 Material Characterization and Fracture Micromechanism

It has been shown in both experimental and theoretical works that crystallinity and lamellar thickness can be related to several mechanical properties and that the morphology of UHMWPE changes with loading [28]. Therefore, we have found that tracking these changes during unnotched and notched tensile tests provides insight into the interplay of morphological evolution and stress state [2]. Standard differential scanning calorimetry (DSC) methods can be used to examine the evolution of bulk crystallinity in both the smooth and notched conditions compared to untested (undeformed) controls. Changes in lamellar thickness distributions can also be examined via DSC [29, 30].

Scanning electron microscopy (SEM) can be utilized to examine changes in fracture micromechanism with notching. For comparison, we have found that smooth specimens in uniaxial tension show a three-part micromechanism of fracture consisting of void coalescence, slow crack growth, and fast fracture [1–3, 31] (Figure 31.5).

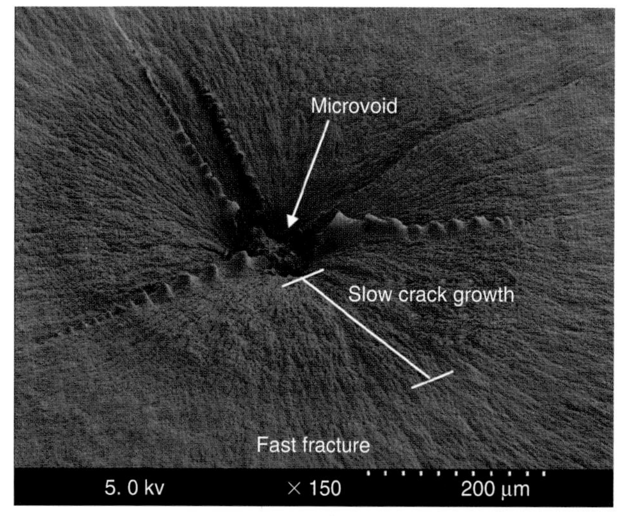


**FIGURE 31.5** SEM fractograph of a tensile smooth specimen (study II, SA material, 30mm/min) fracture surface [2]. This illustrates the three part micromechanism of fracture consisting of void coalescence, slow crack growth, and fast fracture seen in UHMWPE specimens tested in uniaxial tension.

## 31.4 RECENT STUDIES OF CONVENTIONAL AND HIGHLY CROSSLINKED UHMWPE MATERIALS

To date, we have conducted three studies (I [3], II [2], and III [1]) utilizing this tensile notch testing methodology on UHMWPE. The materials, geometries, displacement rates, and testing conditions for these studies are summarized in Table 31.1.

### 31.4.1 Mechanical Characterization of Notch Effects

We found that the true axial stress–strain curves in the three studies showed similar trends when comparing notched specimens to smooth specimen controls (Figure 31.6). All of the notched specimens showed an elevation of the yield stress and a reduction of the ultimate stress. Also, the stress–strain curves of the notched specimens appear truncated when compared to the smooth specimens in that they lack the steep final portion of the curve that is typically associated with orientation hardening. In all of the studies and notched conditions tested, the crosslinked materials showed decreased ductility when compared to the conventional material.

Statistical analyses (ANCOVA) showed that UHMWPE material (e.g., conventional versus crosslinked), displacement rate, and specimen geometry (e.g., smooth versus notched) all significantly influence the axial monotonic properties (yield stress and strain and ultimate stress and strain) [1–3]. In some cases, the *interactions* between these terms were also found to be significant for some of properties. The most important and influential of these was the interaction between specimen geometry and UHMWPE material. This implies that different UHMWPE formulations respond differently to the presence of a notch.

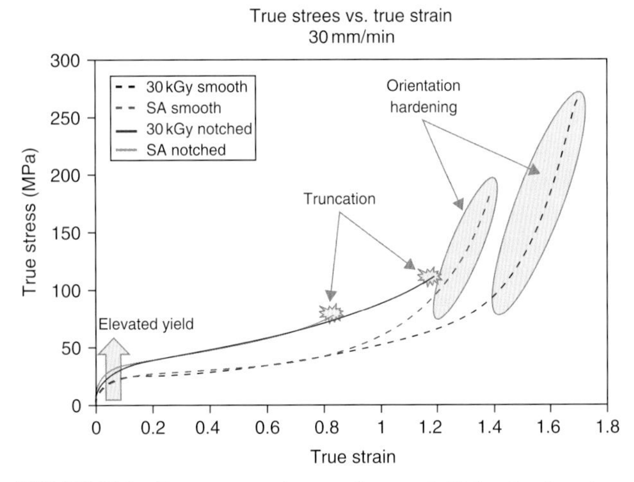


**FIGURE 31.6** True stress-strain curve from study II showing the orientation hardening region in smooth specimens and showing the truncation of this region and the elevated yield stress in notched specimens.

We also found that both increasing the depth of the notch (decreasing the ID/OD ratio) while keeping $K_t$ constant (study I), and increasing the $K_t$ while keeping the ID/OD ratio constant (study III) resulted in an increase in the notch strengthening ratio (Figure 31.7). Additionally, we found that the notch strengthening ratio is dependent on the formulation of UHMWPE being tested and that there is a significant interaction between UHMWPE material and notch geometry. Therefore, notch depth and $K_t$, and their interaction with a UHMWPE formulation, affect the magnitude of plastic flow inhibition during a test.

Finally, all of the UHMWPE materials in the three studies showed a significant decrease in the hardening ratio upon the introduction of a notch. In study I we found that deepening the notch did not change the stress-based hardening ratio (left side of Figure 31.8). In study III, we found that increasing $K_t$ significantly decreased the hardening ratio for all of the materials studied at both rates (right side of Figure 31.8). In all of the studies, the hardening ratio was significantly affected by the UHMWPE material. The results suggest that $K_t$ and the UHMWPE formulation are more important than the notch depth in determining the effect of a notch on the stress-based hardening ratio under monotonic loading.

### 31.4.2 Material Characterization and Fracture Micromechanism

In studies II and III, we found that smooth specimens taken to failure exhibited a decrease in bulk crystallinity compared with undeformed controls (Figure 31.9). Also, depending on the UHMWPE formulation, smooth specimens generally exhibited a greater decrease in bulk crystallinity than did notched specimens taken to failure (Figure 31.9). We also observed that the lamellar thickness distributions qualitatively showed a greater change for smooth samples than did the notched specimens when compared to the control distributions (Figure 31.10).

Perhaps the most dramatic change in material behavior upon the introduction of a notch was in the micromechanism of fracture. In study II, we first noticed that notched specimens showed two different fracture patterns (referred to as one-zone and two-zone). Fractures showing the one-zone pattern all appeared to have fracture initiation from a point on the notch surface, with fast fracture first propagating inward and then radiating outward (Figure 31.11A). The two-zone fractures appeared to have initiated circumferentially at the notch surface, followed by a region of stable crack growth, and then followed by fast fracture (Figure31.11B). In study II, we also observed that the tendency toward a one-zone versus two-zone fracture pattern was affected by material and displacement rate. All of the 90 kGy sequentially annealed specimens, at both displacement rates, failed by a one-zone fracture mechanism. For the 30 kGy irradiated in $N_2$ material, all of the specimens

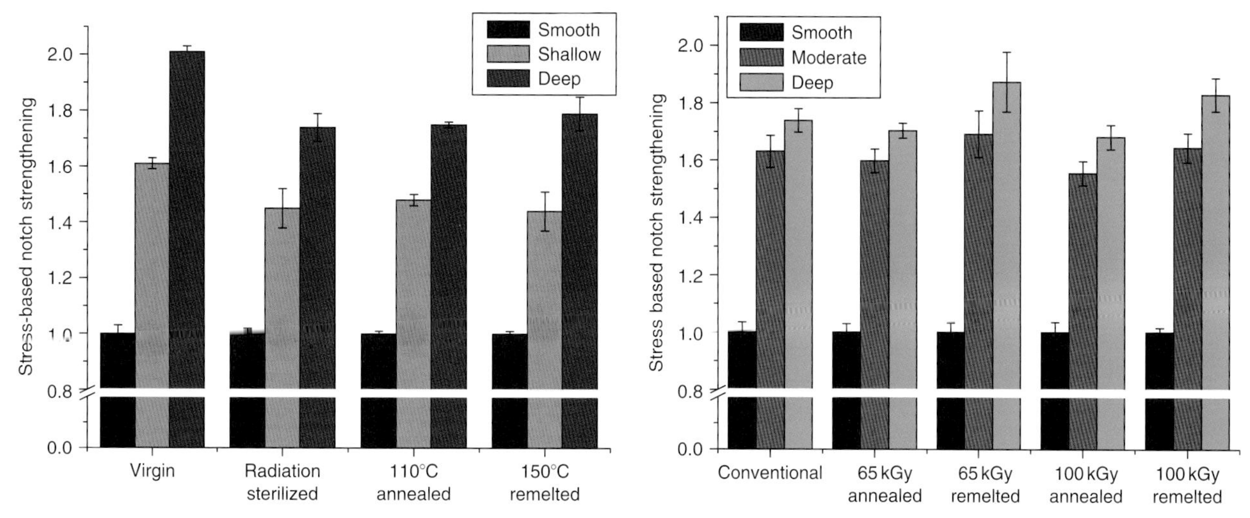


**FIGURE 31.7** Bar charts of the stress-based notch strengthening ratios for study I (left) and for study III (150 mm/min, right) [1, 2].

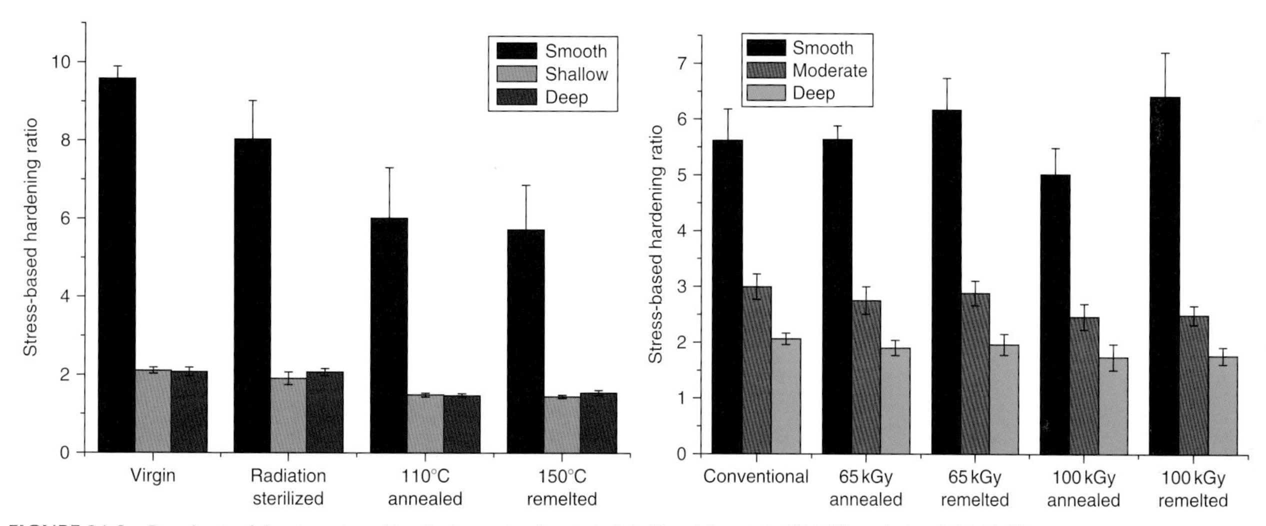


**FIGURE 31.8** Bar charts of the stress-based hardening ratios for study I (left) and for study III (150 mm/min, right) [1, 3].

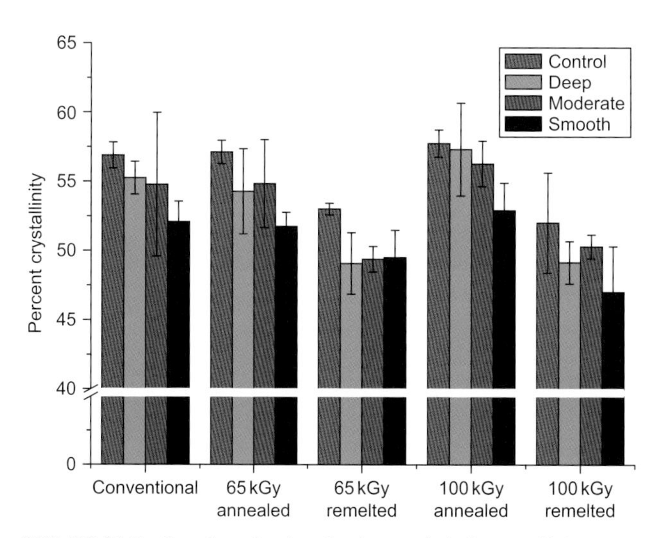


**FIGURE 31.9** Bar chart showing the changes in bulk crystallinity upon deformation [1]. Control is each material in the undeformed condition.

tested at the lower displacement rate failed by the two-zone fracture mechanism, whereas at the higher rate approximately 50% failed by the one-zone fracture mechanism. In study III we observed that some of the highly crosslinked materials did show evidence of a two-zone fracture mechanism in both the deep and the moderate geometries. However, consistent with study II, the conventional material had the greatest tendency toward a two-zone mechanism, and increasing the rate appeared to increase the percentage of one-zone failures (Figure 31.12).

### 31.4.3 Hybrid Model Finite Element Analysis

Investigators at Exponent Inc. performed finite element analyses (FEA) of the smooth and notched samples. The results from the smooth specimen tension tests of studies

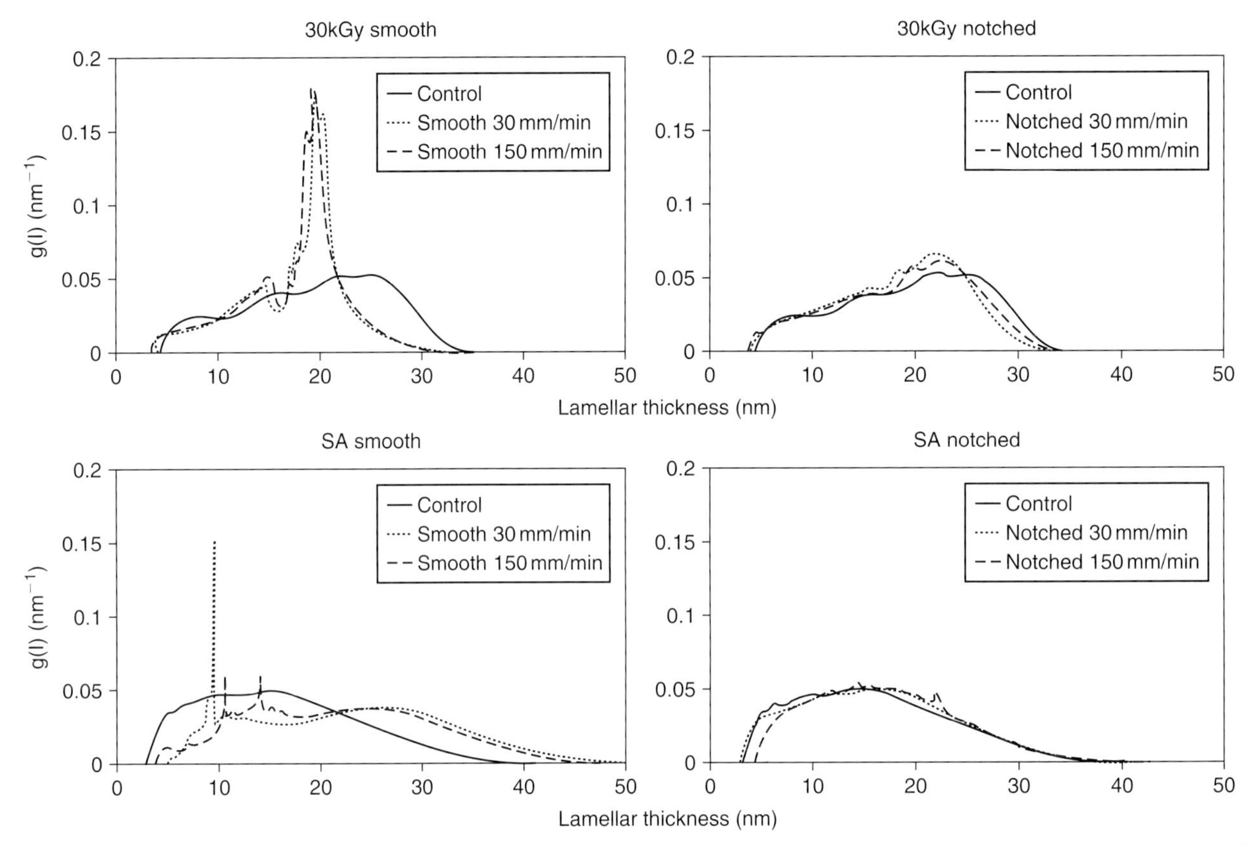


**FIGURE 31.10** Lamellar thickness distributions for all the material-geometry-rate combinations in study II with their respective control (undeformed) distributions [2].

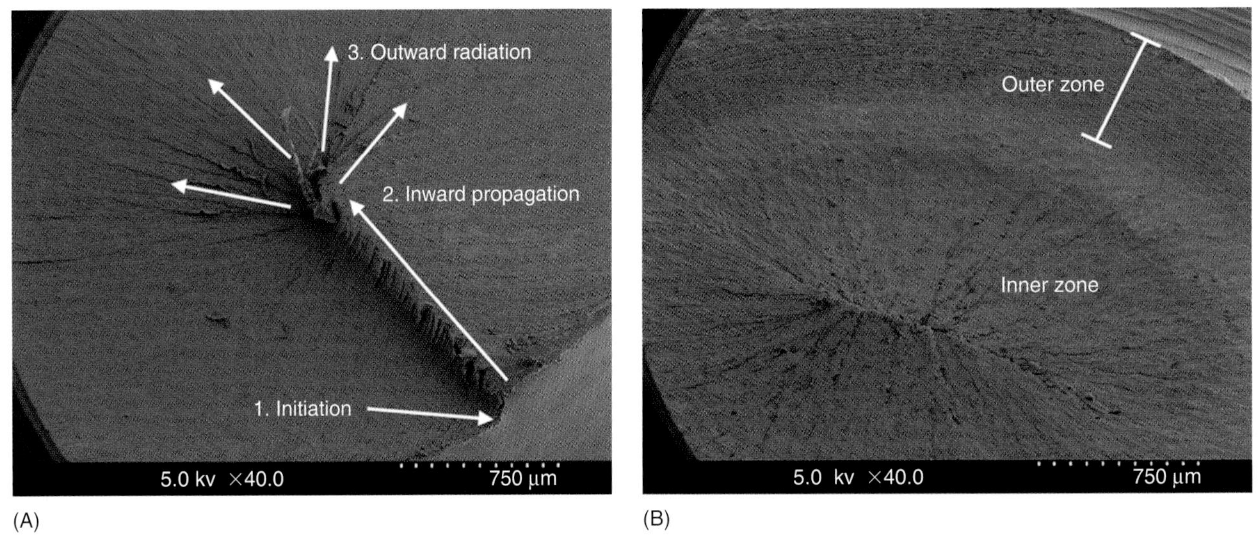


**FIGURE 31.11** Representative SEM fractographs showing the two fracture patterns seen in the notched specimens in study II. (A) one-zone pattern (Sequentially Annealed, 30 mm/min); (B) two-zone pattern (30 kGy, 30 mm/min).

I and II were used to calibrate the Hybrid Model (see Chapter 35) and obtain the critical chain stretch at fracture:

$$\lambda_{chain} = \sqrt{\frac{e^{2\varepsilon} + 2e^{-\varepsilon}}{3}} \tag{4}$$

where $\varepsilon$ is the uniaxial true strain at fracture [32, 33]. (Compared with other stress and strain criteria, it has been found that critical chain stretch is the most accurate failure criterion for UHMWPE [33]). The calibrated Hybrid Model was implemented as a user-material in LS-DYNA

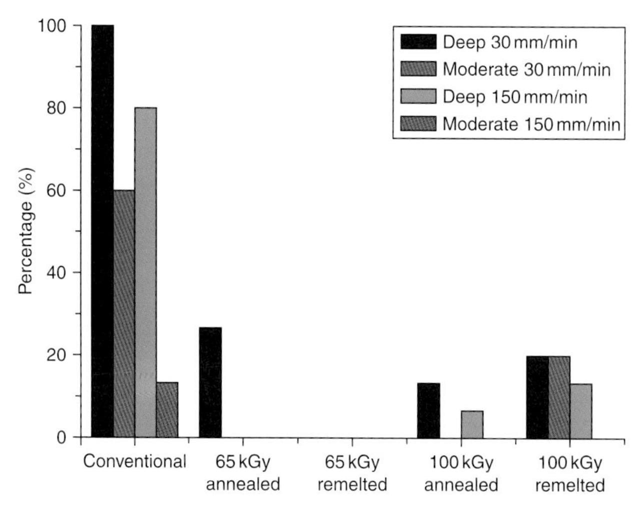


**FIGURE 31.12** Percentage of failed notched specimens showing a two-zone fracture pattern in study III. The remainder showed a one-zone pattern [1].

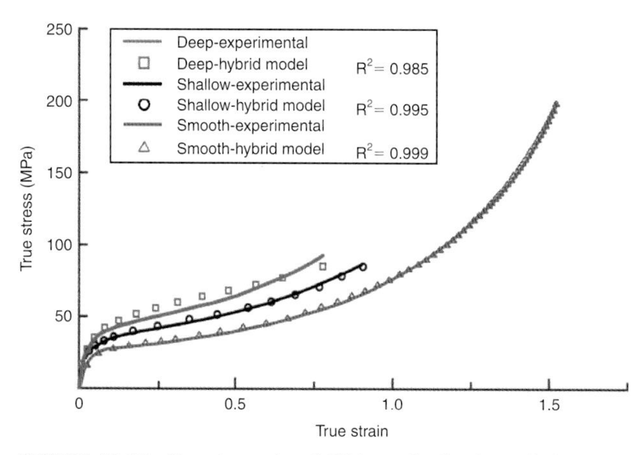


**FIGURE 31.13** Experimental and FEA results for the radiation sterilized material in study I [12].

(LSTC, Livermore, California, USA), which was then used to simulate the notched specimen tests. Validation of the Hybrid Model was performed by direct comparison of the effective stress–strain response from the FEA and the notched experimental tests and by calculating the coefficient of determination, $R^2$.

The results of FEAs performed using study I show that Hybrid- Model-based FEAs predict the monotonic behavior of the notched specimens quite accurately (Figure 31.13) [12]. This was also shown to be true for study II [34]. Taken together, this is strong validation that the model works well under monotonic conditions in describing the constitutive behavior of the different UHMWPE formulations, even when challenged with multiaxial loading. The FEA conducted on the geometries and materials used in study I (Figure 31.14), and also study II, showed that both the stress and the molecular chain stretch are greatest near the surface of the notch. In Figure 31.14, following 5 mm of extension, both of the notch geometries demonstrate that the stress in the notch exceeds the yield stress. In addition, both stress and chain stretch gradients are present, showing that the material is in a triaxial stress state even well after the onset of plastic flow. Note the qualitative similarity in the FEA stress contours of the notched specimens and the FEA contours in the vicinity of rims and other design-induced stress concentrations shown in Figure 31.1.

## 31.5 THE EFFECTS OF NOTCHES ON DEFORMATION AND FRACTURE MECHANISMS

The monotonic behavior of UHMWPE materials in the presence of a notch can be interpreted by consideration of uniaxial tensile deformation processes for semicrystalline polymers. In the smooth condition, the uniaxial tensile deformation process for a semicrystalline polymer is a several stage process involving tie molecule stretching; crystalline lamellar rotation and alignment; fine slip, progressing to coarse slip, and then fragmentation of lamellae; and finally, chain alignment with further degradation of lamellae [35]. The onset of coarse slip in the crystals is the physical reason for the onset of yielding [36, 37]. This several-stage process can result in a decrease in crystallinity, as we observed in our studies of UHMWPE. The occurrence of notch strengthening and the decrease in the hardening ratios upon notching that we found in our studies suggest that these mechanisms are affected by the triaxial state of stress in the vicinity of the notch. Further, the elevation of the yield stress with notching suggests that the lamellar rotation and/or the slip mechanisms in the lamellar deformation process are inhibited. This is supported by our crystallinity findings in studies II and III that showed smaller changes in the bulk crystallinity for specimens exposed to a triaxial stress state as compared to the uniaxial condition. The greater changes in the lamellar thickness distributions of the smooth specimens compared with the notched specimens taken to failure also support that there is an inhibitory effect of the triaxial stress state on these deformation processes.

The decrease in hardening ratio with notching points to a decrease in chain alignment during deformation because the chains are no longer exposed to a single axis of pull. Void formation and coalescence in uniaxial tests of the generic class of polyethylenes (including UHMWPE) appears to be related to chain alignment [38]; therefore, a triaxial stress state (like that in the vicinity of the notch) should inhibit chain alignment and suppress the coalescence of such voids. Complementary to this suppression of voids, fracture should initiate near the notch because that is where the highest local stresses and strains are predicted to be. Taken together, these conditions likely explain the altered mechanism of fracture from the smooth to the notched conditions that we observed in our studies.

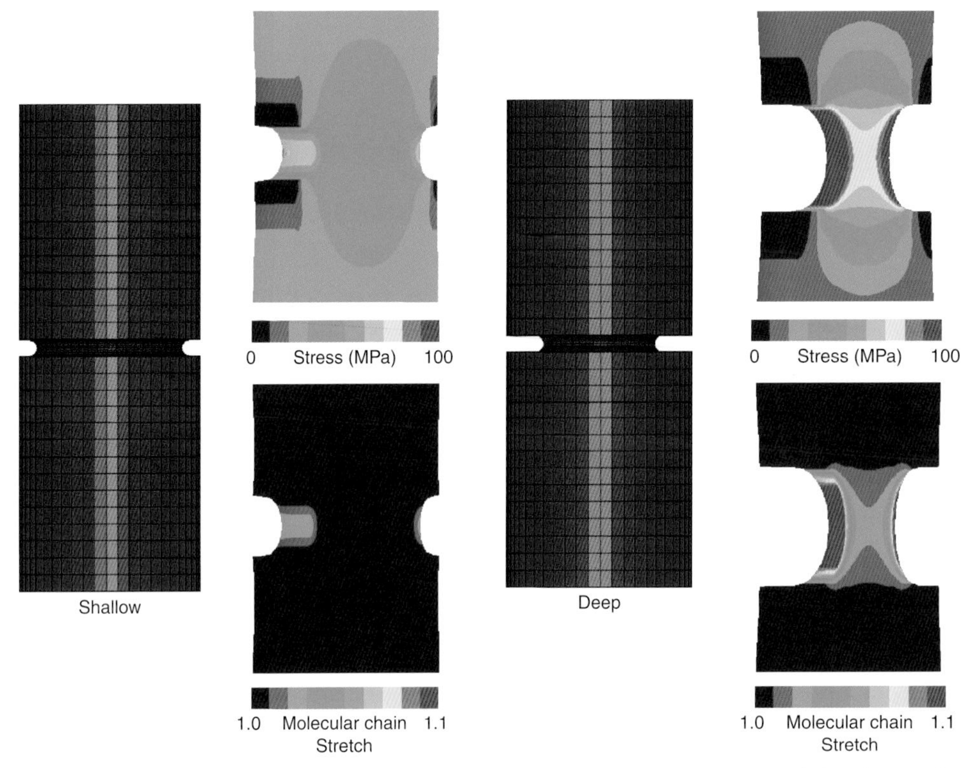


**FIGURE 31.14** FEA results for the radiation sterilized material in study I using the Hybrid Model. The unloaded meshes are shown in red and they represent an overall length of 25 mm, OD = 10 mm, ID = 8 mm or 6 mm (shallow and deep, respectively), and a notch radius of 0.45mm. These were then subjected to a 5 mm extension, and the slightly angled contours of the resulting effective stress and molecular chain are shown to their right [12].

We found that these general trends and fracture micromechanisms occurred regardless of whether the UHMWPE formulation was radiation sterilized or crosslinked. However, the magnitude of the changes upon notching is dependent on the specific material formulation.

## 31.6 CONCLUSION AND FUTURE DIRECTIONS

The novel video-based method used for the evaluation of axial stress–strain behavior of notched cylindrical specimens is a valuable tool. Experimentally, this method is simple to implement and has the advantage of providing strain behavior up to failure using a noncontacting extensometry approach. The notched tensile test is also an attractive experimental method with which to check the validity of the development of constitutive models of UHMWPE (see Chapter 35). The methodology has the advantage of creating a multiaxial stress state without contact, and the specimen can be modeled as axisymmetric in finite element analyses.

The studies that have been conducted using this method thus far have provided insight into the similarities and differences in stress–strain behavior, morphological evolution, and fracture micromechanisms between uniaxial and triaxial loading conditions for different UHMWPE formulations. Because we have shown that UHMWPE formulation significantly affects the response to a notch, we believe that UHMWPE formulation must be considered in the design of the stress riser features of components, including undercuts, fillets, and locking mechanisms. For example, a locking mechanism that is acceptable for a conventional UHMWPE acetabular liner may not be appropriate for a highly crosslinked one.

The success of the finite element simulations in predicting the stress–strain response of the notched tensile tests provides support for the Hybrid Model and insight into the effects of the notch on the stress state. The noted qualitative similarity in the FEA contours of the simulation of design-induced stress concentrations in UHMWPE joint replacement components (Figure 31.1) compared to monotonic notch testing suggests that this approach is a useful tool for evaluating UHMWPE formulation response to design notches.

We have primarily examined the effect of notching on the monotonic behavior of UHMWPE. However, *in vivo*, UHMWPE joint replacement components are also subjected to cyclic loading. Thus, the effect of notching on the fatigue behavior of different UHMWPE formulations is also of interest. We are currently working on a modification of this approach for fatigue life testing [39].

## 31.7 ACKNOWLEDGEMENTS

This research was supported by NIH (NIAMS) R01 AR 047192 and Stryker. Some of the UHMWPE materials and specimens were donated by Orthoplastics, Ltd.

## REFERENCES


1. Murphy J, et al. *Unpublished data*. 2008, Musculoskeletal, Mechanics, and Materials Lab; Departments of Mechanical and Aerospace Engineering and of Orthopaedics; Case Western Reserve University and University Hospitals Case Medical Center: Cleveland, OH.
2. Sobieraj M, et al. Notched stress-strain behavior of a conventional and a sequentially annealed highly crosslinked UHMWPE. *Biomaterials* 2008. Accepted.
3. Sobieraj MC, Kurtz SM, Rimnac CM. Notch strengthening and hardening behavior of conventional and highly crosslinked UHMWPE under applied tensile loading. *Biomaterials* 2005;**26**:3411–26.
4. Bergstrom J, et al. Development and Implementation of an advanced user material model for UHMWPE. *9th International LS-DYNA Users Conference*. Dearborn, MI: 2006.
5. Kurtz S, et al. Polyethylene wear and rim fracture in total disc arthroplasty. *Spine J* 2006;**7**(1):12–21.
6. Mauerhan D. Fracture of the polyethylene tibial post in a posterior cruciate–substituting total knee arthroplasty mimicking patellar clunk syndrome. *J Arthroplasty* 2003;**18**(7):942–5.
7. Kolb E. A finite element analysis of design factors affecting performance of acetabular components in total hip arthroplasty. In: *Department of Mechanical and Aerospace Engineering*. Cleveland: Case Western Reserve University; 2000.
8. Collins J. *Failure of materials in mechanical design*. 2nd ed New York: John Wiley & Sons; 1993.
9. Dowling NE. *Mechanical behavior of materials: engineering methods for deformation, fracture, and fatigue*. 2nd ed Upper Saddle River, NJ: Prentice Hall; 1999.
10. Pluvinage G. *Fracture and fatigue emanating from stress concentrations*. Boston: Kluwer Academic Publishers; 2003.
11. Pluvinage G, Gjonaj M, editors. *Notch effects in fatigue and fracture. Nato Science Series*. Boston: Kluwer Academic Publishers; 2001.
12. Bergstrom J, Bowden A. *Unpublished Data on the Finite Element Simulation of Notched Specimens of Conventional and Highly Crosslinked UHMWPE*. 2006.
13. Astion DJ, et al. The porous-coated anatomic total hip prosthesis: failure of the metal-backed acetabular component. *J Bone Joint Surg Am* 1996;**78**(5):755–66.
14. Birman M, et al. Cracking and impingement in ultra–high-molecular-weight polyethylene acetabular liners. *J Arthroplasty* 2005;**20** (Suppl. 3):87–92.
15. Collier JP, et al. Mechanisms of failure of modular prostheses. *Clin Orthop* 1992(285):129–39.
16. Berry DJ, et al. Catastrophic failure of the polyethylene liner of uncemented acetabular components. *J Bone Joint Surg Br* 1994; **76**(4):575–8.
17. Chiu Y, et al. Fracture of the polyethylene tibial post in a NexGen posterior-stabilized knee prosthesis. *J Arthroplasty* 2004;**19**(8): 1045–9.
18. Hendel D, Garti A, Weisbort M. Fracture of the central polyethylene tibial spine in posterior stabilized total knee arthroplasty. *J Arthroplasty* 2003;**18**(5):672–4.
19. Mariconda M, Lotti G, Milano C. Fracture of posterior-stablized tibial insert in a Genesis knee prosthesis. *J Arthroplasty* 2000; **15**(4):529–30.
20. Mestha P, Shenava Y, D'Arcy J. Fracture of the polyethylene tibial post in posterior stabilized (Insall Burstein II) total knee arthroplasty. *J Arthroplasty* 2000:814–18.
21. Beaule S, et al. Jumbo femoral head for the treatment of recurrent dislocation following total hip replacement. *J Bone Joint Surg* 2002;**84**:256–63.
22. Halley D, Glassman A, Crowninshield R. Recurrent dislocation after revision total hip replacement with a large prosthetic femoral head—A case report. *J Bone Joint Surg* 2004;**86**:827–30.
23. Tower S, et al. Rim cracking of the cross-linked longevity polyethylene acetabular liner after total hip arthroplasty. *J Bone Joint Surg* 2007;**89**:2212–17.
24. Goolsby R, Chatterjee A. Notch sensitivity and fractography of polyolefins. *Polym Eng Sci* 1983;**23**(3):117–24.
25. Hertzberg R. *Deformation and fracture mechanics of engineering materials*. 4nd ed New York: John Wiley & Sons; 1996.
26. Mourad A-HI, et al. Ultra high molecular weight polyethylene deformation and fracture behaviour as a function of high strain rate and triaxial state of stress. *Int J Fract* 2003;**120**(3):501–15.
27. Bartel DL, Bicknell VL, Wright TM. The effect of conformity, thickness, and material on stresses in ultra-high molecular weight components for total joint replacement. *J Bone Joint Surg [Am]* 1986;**68**(7):1041–51.
28. Sobieraj M, Rimnac C. Ultra high molecular weight polyethylene: mechanics, morphology, and clinical behavior. *J Mech Behav Biomed Mater* 2008. Submitted.
29. Alberola N, Cavaille J, Perez J. Mechanical spectrometry of alpha-relaxations of high-density polyethylene. *J Polym Sci Part B: Polym Phys* 1990;**28**:569–86.
30. Crist B, Mirabella F. Crystal thickness distributions from melting homopolymers or random copolymers. *J Polym Sci Part B: Polym Phys* 1999;**37**:3131–40.
31. Gencur SJ, Rimnac CM, Kurtz SM. Failure micromechanisms during uniaxial tensile fracture of conventional and highly crosslinked ultra-high molecular weight polyethylenes used in total joint replacement. *Biomaterials* 2003;**24**(22):3947–54.
32. Bergstrom J, Rimnac C, Kurtz S. An augmented hybrid constitutive model for simulation of unloading and cyclic loading behavior of conventional and highly crosslinked UHMWPE. *Biomaterials* 2004;**25**:2171–8.
33. Bergstrom J, Rimnac CM, Kurtz SM. Molecular chain stretch is a multiaxial failure criterion for conventional and highly crosslinked UHMWPE. *J Orthop Res* 2005;**23**(2):367–75.
34. Kurtz S, et al, Validation of hybrid model and ultimate chain stretch criterion for a second-generation highly crosslinked UHMWPE. *6th Combined Meeting of the ORS. 2007, Orthopaedic Research Society*. Honolulu, HI. p. Poster 498.
35. Courtney TH. *Mechanical behavior of materials*. 2nd ed Boston: McGraw Hill; 2000.
36. Sirotkin RO, Brooks NW. The effects of morphology on the yield behaviour of polyethylene copolymers. *Polymer* 2001;**42**:3791–7.
37. Young RJ. A dislocation model for yield in polyethylene. *Philos Mag* 1974;**30**:85–94.
38. Butler MF, Donald AM, Ryan AJ. Time resolved simultaneous small- and wide-angle x-ray scattering during polyethylene deformation-II. cold drawing of linear polyethylene. *Polymer* 1998; **39**(1):39–52.
39. Sobieraj M, et al, Multi-axial fatigue of a conventional and sequentially annealed highly crosslinked UHMWPE. *6th Combined Meeting of the Orthopaedic Research Societies*. Honolulu, HI: Orthopaedic Research Society; 2007.

# Chapter 32

# Development and Application of the Small Punch Test to UHMWPE

Avram A. Edidin, PhD



## 32.1 INTRODUCTION

Mechanical testing of miniature specimens—as opposed to bulk large specimens—is advantageous when the need arises to directly probe the properties of a specific component rather than a generic material. Testing of miniature specimens obtained directly from a specific structure becomes paramount when evidence suggests that the properties of the material under consideration are known to change over time. In such cases, there can be no assurance that assays of supposedly equivalent material, aged in a similar manner, will have any fidelity to individual components of interest.

Nowhere does this dilemma between a direct testing of a component and testing of an analog appear more forcibly in biomechanics than in the determination of mechanical properties from UHMWPE orthopedic bearings. These components are found in total hip, knee, and shoulder arthroplasties and contain between 40 and 200 g of material, and the designs typically have characteristic length scales of between 4 and 60 mm. It may readily be appreciated that fabrication of any mechanical test specimen, even one greatly reduced in size, will be quite challenging and that obtaining replicates from a single specimen will be even more so.

Unfortunately, the failure modes of UHMWPE bearings are predominantly mechanical at one or multiple length scales, making accurate determination of mechanical properties through time paramount. Furthermore, the properties of UHMWPE are known to vary with starting resin grade, processing methodology, irradiation regime, and storage environment [1]. The number of interdependent variables suggests that attempts to replicate the eventual condition of a particular UHMWPE component will be an inexact approximation at best.

This chapter reviews the development and application of miniature specimen mechanical testing techniques, based on the small punch test, to the characterization of UHMWPE components for total joint replacement. The development of the small punch test as applied to UHMWPE was motivated by two clinically relevant and related problems: the relationship between process and system variables and polymer degradation, and the relationship between mechanical properties and the ensuing wear of the arthroplasty bearing. To a first order approximation, the polymer degradation problem was motivated by mechanical failure of tibial plateau bearings, and the wear problem was motivated by the clinical need to reduce the prevalence of small particle mediated late-onset osteolysis in total hip arthroplasty systems.

## 32.2 OVERVIEW AND METRICS OF THE SMALL PUNCH TEST

The small punch test methodology (also referred to as the "disk bend test") was originally proposed by Manahan et al. [2] for the characterization of metallic components from the power industry. Small punch testing was adapted for UHMWPE by Kurtz et al. [3, 4] for direct characterization of the mechanical behavior of UHMWPE components. In the small punch test for UHMWPE, a disc of material 0.0200-inch thick by 0.250-inch diameter is indented by a hemispherical punch, thereby creating a biaxial state of tension in the material. Furthermore, due to the enclosure of the specimen in the punch-and-die apparatus, the thickness tolerance is only 0.0005-inch. Specimen preparation may be performed by any conventional machining operations performed so as to minimize anisotropy in the finished specimens. A schematic of specimen preparation and testing apparatus is shown in Figure 32.1. Additional details of the small punch test for UHMWPE are provided in ASTM F2183 [5].

Naturally, specimens of this size are particularly attractive to the study of UHMWPE bearing components because multiple specimens can be obtained from almost any single component. Further suitability is noted in that multiple specimens may often be prepared through the *thickness* of the bearing, which is of interest in the study of oxidative degradation, wear induced chain alignment, or any other mechanism that leads to a material inhomogeneity through the thickness of a bearing.

Four basic metrics are obtained from a load-displacement curve of a material subjected to the small punch test. These include the initial stiffness, peak and ultimate load, and the maximum displacement at failure. A fifth derived metric, integral work-to-failure, hereafter WTF, is found by integrating the area under the load-displacement curve and represents the total energy required to fail the specimen. Many, but not all, tests have distinct membrane bend and membrane stretch portions, roughly represented by the portion of the curve before and after the peak load is reached, respectively. A typical load-displacement curve for GUR 1020 UHMWPE is shown in Figure 32.2.

While quantitative metrics form the basic for performing statistical comparisons, a fair amount of information about a material may be gleaned by observing the shape of load-displacement curves. In particular, the degree of geometric strain hardening, so called because it is observed only on the load-displacement curve rather than a true stress–strain curve, often differentiates pristine from degraded UHMWPE or ordinary from crosslinked UHMWPE. An example comparing virgin (unirradiated) to crosslinked UHMWPE is shown in Figure 32.3.

Linear polyethylenes with molecular weights below about 1,000,000 typically display geometric strain softening. Degradation of UHMWPE is often first observed through a reduction or roll-off of the load-displacement curve as compared to the nondegraded material. As will be discussed shortly, this observation represents a mechanical means of detecting chain scission arising from oxidation.

During the development of the small punch test for UHMWPE, a validation analysis was performed comparing the results obtained from small punch testing with properties measured during conventional, large-specimen

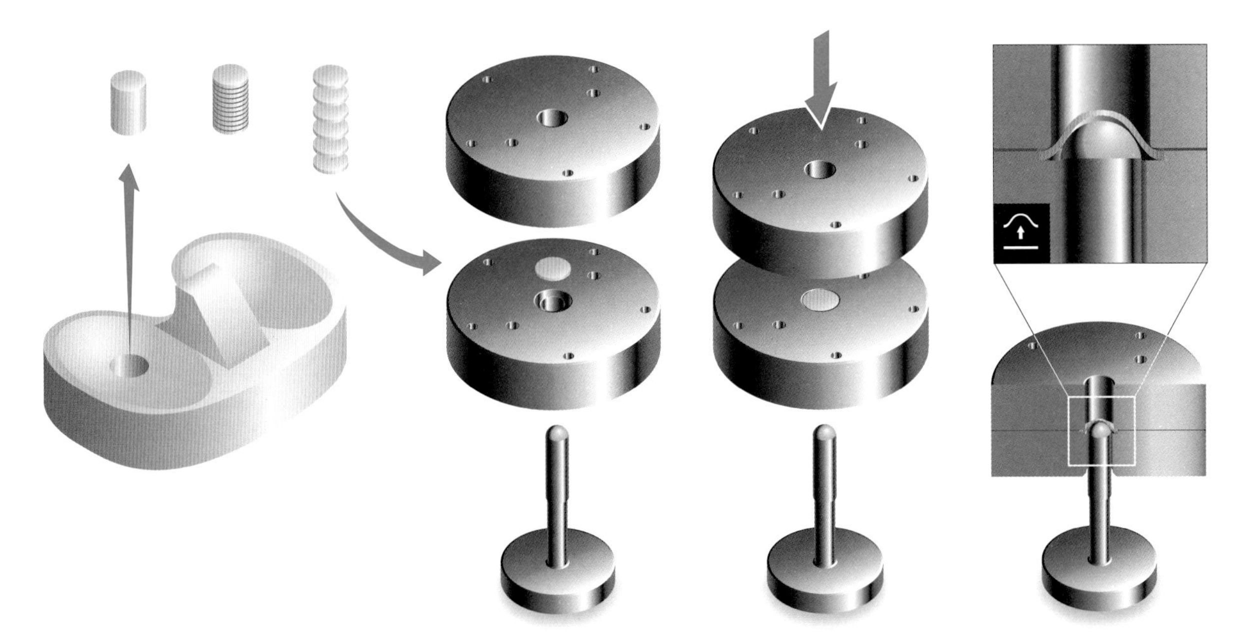

**FIGURE 32.1** Schematic of specimen preparation and small punch test. Typically cores of net diameter are taken with a custom coring tool set to 0.250 inch ± 0.001 inch and then sliced into specimens on a lathe. Following facing operations, the final specimens are 19.7–20.0 mils thick. Specimens are then inserted into a capturing die apparatus through which the hemispherical punch passes. The closed die is mounted to the crosshead of the testing machining and forced over the fixed punch. Dimensions provided here in English units were originally developed in United States.

uniaxial mechanical tests [3]. Historically, mechanical tests of UHMWPE were performed in uniaxial tension using specimens such as those described in ASTM D648. Kurtz et al. describe an experiment comparing the uniaxial response of 4150 and 1120 based UHMWPE specimens to the response of the same materials observed under equibiaxial tension [3]. The researchers availed themselves of the ability to test the same exact material in both arms of their experiment due to the small specimen size required for the small punch testing. They observed, via an inverse approach using a finite element model of the small punch setup to simulate the test itself, that the initial stiffness in the small punch test could be related linearly to the material's modulus via the relationship:

$$E = 13.5k$$

where $k$ is the stiffness (in N/mm) observed in the membrane bend portion of the test and E is the conventional Young's modulus (in MPa).

Work by Asano et al. [6] used the small punch test metrics to investigate the optimum irradiation dose suitable for crosslinking of knee bearings. They measured the toughness, believed to be related to fatigue damage accumulation, in a range of inserts crosslinked via gamma irradiation to cumulative doses of 0 to 200 kGy. Their data showed increasing toughness through 75 kGy but decreasing toughness thereafter. Further investigation of the small punch metrics suggested that while elongation to failure decreased with increasing dose, toughness nonetheless increased up to 75 kGy cumulative dose because the ultimate small punch peak load increased sufficiently to compensate up to 75 kGy cumulative dose, but not at greater doses.

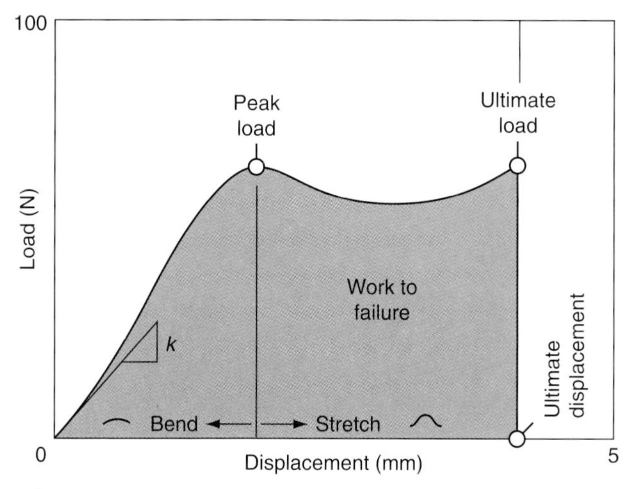


**FIGURE 32.2** Representative load displacement curve developed by testing of a small punch specimen in equibiaxial tension; the curve here reflects typical behavior of GUR 1020. Primary metrics include initial stiffness, peak load, ultimate load, and ultimate displacement. Work-to-failure is shown in gray. Unirradiated (virgin) UHMWPE exhibits a bend and a stretch region as shown.

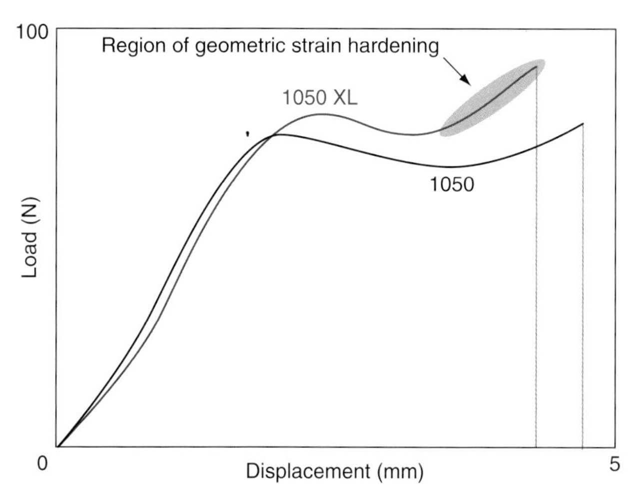


**FIGURE 32.3** Load displacement curves developed from representative testing of Hoechst 1050 resin in both as-converted, labeled 1050, and crosslinked, labeled 1050 XL condition. Crosslinking increases the amount of geometric strain hardening in the region shown in gray. In general, substantial information can be gleaned from the overall form of the load-displacement curve as well as from the quantitative metrics themselves.

## 32.3 ACCELERATED AND NATURAL AGING OF UHMWPE

Following the validation of the small punch test for virgin (unirradiated) UHMWPE and the reconciliation of the reported metrics between the conventional large-specimen uniaxial and the novel miniature specimen biaxial tests, investigators began to apply the small punch test to characterize the effects of oxidative degradation that occurs during accelerated and natural aging on the mechanical behavior of UHMWPE [4]. Kurtz et al. [4] used the small punch test to investigate the *mechanical* signature of oxidative degradation by comparing the mechanical behavior of GUR 4150 and 1120 small punch specimens before and after aging in different atmospheres. They observed that the miniature small punch specimens were resistant to subsequent degradation following accelerated aging in an air oven if the specimens had been irradiated in a nitrogen atmosphere. This finding mirrored that of Clough and Gillen [7], who observed similar "protective" benefits of nitrogen atmosphere irradiation when similar experiments were performed in linear polyethylenes with lower molecular weights. Representative load-displacement curves from the experiment are shown in Figure 32.4.

Later work by Edidin et al. [8] suggests that the protective benefit of irradiating UHMWPE in an inert environment such as nitrogen gas is not as apparent when artificial aging is performed on the bulk specimens as opposed to the prepared small punch test specimens. More specifically, the degree to which changing the irradiation atmosphere protects from future degradation appears to be a function of heating rate, whereby very fast rates quench

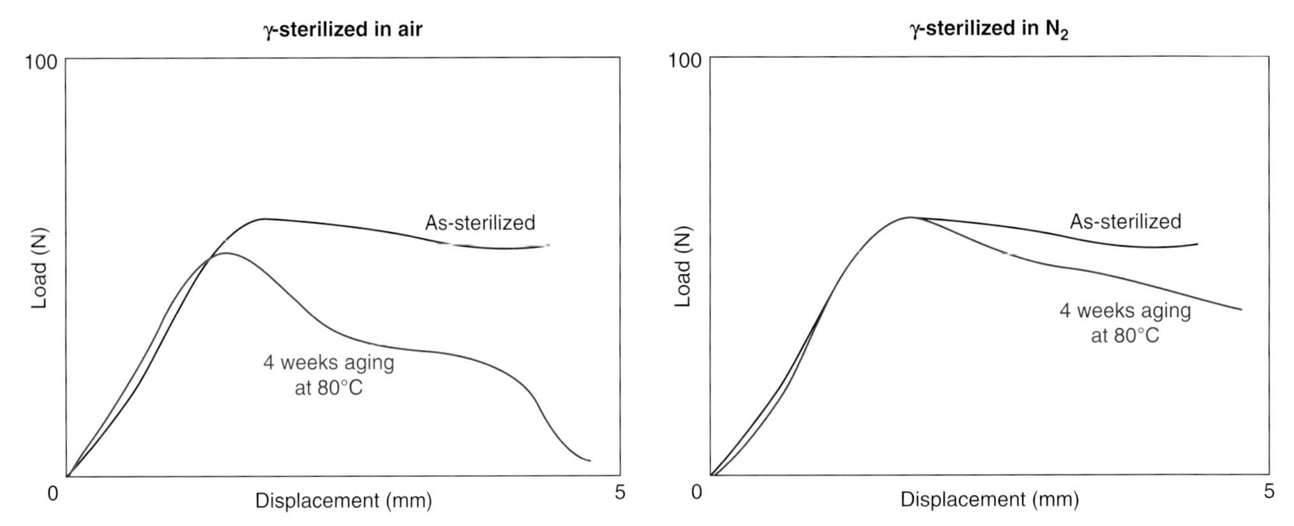


**FIGURE 32.4** Load displacement curves observed by testing specimens prepared from GUR 4150 resin. At left, the specimens irradiated in air and aged in an air atmosphere oven are seen to have degraded mechanical properties. At right, specimens irradiated in nitrogen but also aged in an air atmosphere oven are seen to degrade to a markedly lower degree. Adapted from Kurtz et al. [4].

the surface layer, thereby stopping further free-radical mediated oxidative degradation [9]. How this interplay of variables might affect *in vivo* oxidation of irradiated UHMWPE is discussed further on in this review.

As previously noted, an important motivator in the development of the small punch test for UHMWPE was its ability to assess the mechanical performance of retrieved arthroplasty components, whether retrieved from the shelf or from the body. During the mid-1990s, manufacturers began to switch the irradiation and storage atmosphere from air to either nitrogen or argon. By 2000 some initial data describing the mechanical performance of inert atmosphere irradiated components could be obtained using the small punch test to assay the components. These data were then compared to similar components irradiated in air stored for the same period of time [10]. The results of this comparison were the first to show that insert gas packaging systems were capable of preventing oxidation during shelf storage. Figure 32.5 compares specimens made from GUR 1020 components 10 years after irradiation and storage in either air or nitrogen atmosphere packaging.

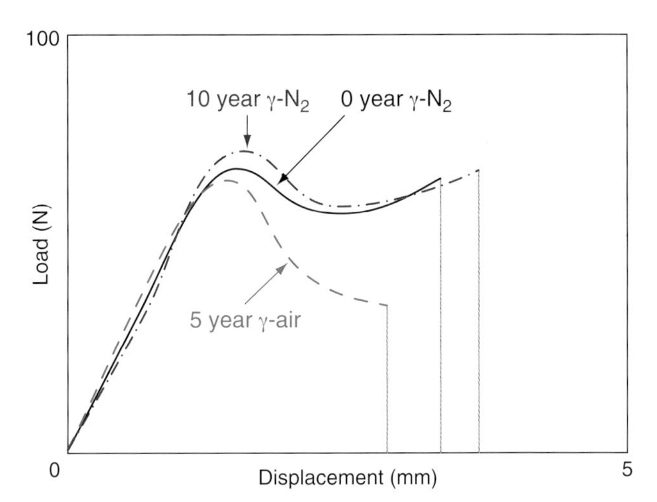


**FIGURE 32.5** Representative load-displacement curve obtained from bearing surfaces of three acetabular components irradiated and stored in either nitrogen or air. After 10 years storage in nitrogen, the load-displacement curve is nearly identical to that obtained immediately post-sterilization in nitrogen. The component irradiated and stored in air is substantially degraded after only 5 years. The base resin was GUR 1020 compression molded in all cases. Adapted from Edidin et al. [10].

Testing of naturally and artificially aged UHMWPE arthroplasty components repeatedly showed that the first mechanical signature of degradation was a decrease in the load-carrying capability of the specimens at displacements beyond that recorded at peak load. Oxidative degradation has been previously shown to reduce effective molecular weight of polyethylene by cleaving the long backbone chains of the semicrystalline polymer matrix. Edidin et al. proposed the hypothesis that the load-displacement curve of degraded UHMWPE should have certain similarity to the load-displacement response of nondegraded linear PE with intrinsically lower molecular weight. The hypothesis was tested by comparing the load-displacement curves and resultant metrics obtained from GUR 1050, 1020, GHR 8110, and LM 6007.00 HDPE. These materials have molecular weights of approximately $3–5 \times 10^6$, $1 \times 10^6$, 500,000, and 120,000, respectively. As may be seen in Figure 32.6, lower molecular weight polyethylenes exhibit load-displacement behaviors akin to those exhibited by degraded UHMWPE [1, 11].

Clearly oxidative chain scission has an effect on mechanical performance similar to reduction of molecular weight ab initio. This experiment provided the strongest mechanical evidence of oxidation to date, in that a

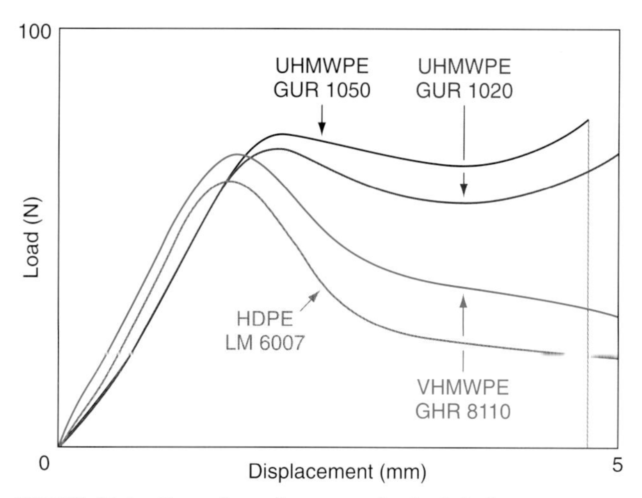


**FIGURE 32.6** Comparison of representative load-displacement curves obtained from testing of four linear polyethylenes of varying molecular weights. The roll-off in load sustained by the specimen in the lower molecular weight materials is similar to that exhibited by higher molecular weight materials that have undergone oxidative degradation (see the example in Figure 32.5). Adapted from Kurtz et al. [1] and Edidin et al. [11].

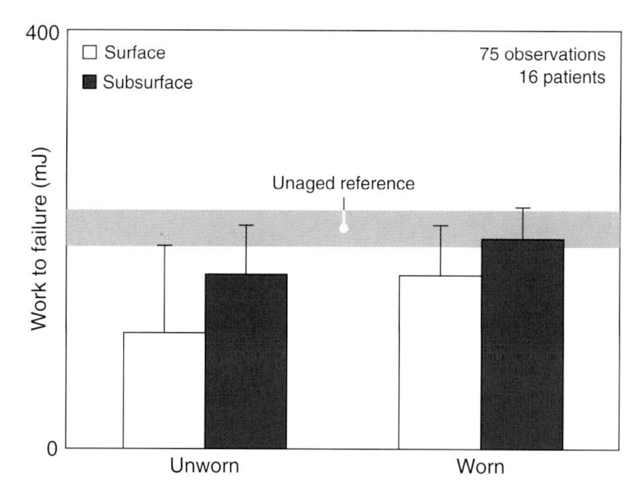


**FIGURE 32.7** Work-to-failure versus implantation time for an entire cohort of 16 retrieved acetabular components. Data was obtained from specimens machined from the surface and subsurface of the worn and unworn region of the articulating surface. Adapted from Kurtz et al. [14].

mechanistic relationship between oxidation and molecular weight was displayed.

## 32.3.1 *In Vivo* Changes in Mechanical Behavior of UHMWPE

The efficacy of barrier inert atmosphere packaging, described in the previous section, gives some assurance that a just-opened UHMWPE bearing component will have minimal oxidative degradation regardless of time spent on the shelf. However, the question remains as to whether or not oxidative degradation occurs *in vivo*. Premnath et al. [12] suggest that *in vivo* oxidation is likely to be a very slow process based on an argument that oxygen dissolved in water saturates at 1/8 the concentration in air. An earlier examination of retrieved components had indirectly suggested that in-service time and oxidation level were not linked [13], but this study and others suffered from unknown shelf and postretrieval aging times; only *in vivo* times were known.

Recently Kurtz and colleagues presented data from retrieval analyses based on examination of well-characterized components obtained from previously consented patients [14]. By judicious use of patient records, coupled with the use of manufacturing records, these authors were able to track each bearing from the time of manufacture, through implantation, and eventually to the research laboratory. Chemical and mechanical small punch testing analysis of the retrieved acetabular bearings was performed in both the articulating and nonarticulating regions. The study was designed such that the time interval between explantation and testing was minimized, thereby decreasing the opportunity for degradation post-vivo. Furthermore, components were stored in nitrogen during this interval. The authors reported on 16 retrieved components originally irradiated and stored in air, with a mean shelf aging time of 0.4 years, a mean implantation time of 11.5 years, and a mean postimplantation time of 0.7 years whilst stored at $-5°C$. Due to the relatively short *ex vivo* aging times compared to the lengthy *in vivo* times, a case was made that any degradation observed would be attributable to *in vivo* oxidative degradation.

Small punch testing of the retrieved specimens showed that the unworn region of the articulating surface was significantly degraded compared to the worn region of the articulating surface. A summary of the differences observed is shown in Figure 32.7 for all 75 small punch tests performed on the acetabular liners retrieved from 16 patients. The finding of maximal degradation in the unworn region was novel; the experiment was not designed to determine whether the difference in degradation rates between the worn and unworn regions reflected a loss of material at the articulating surface or whether loading affected the degradation pathway itself. These data were among the first to conclusively show that *in vivo* degradation occurs in the absence of confounding variables associated with some earlier retrieval studies, such as unknown shelf and postexplantation times. The data also suggest that the assumptions of Premnath et al. [12], related to the concentration in water at room temperature, do not necessarily pertain to the conditions *in vivo*, in which the joint fluids are maintained at body temperature.

Sugano et al. [15] had the opportunity to analyze retrieved all-poly acetabular components following long-term implantation. Two components were mechanically and chemically assayed: a conventional cup irradiated to 25 kGy and a crosslinked component irradiated to 1000 kGy, both in air. They found that while both components exhibited chemical signatures of oxidative chain scission, the mechanical properties of the 1000 kGy crosslinked component were maintained as demonstrated through comparison with a never-implanted 1000 kGy crosslinked control. More specifically, the authors observed the 1000 kGy components to have ultimate load of about 100 N in the equibiaxial small punch test, whereas the conventional 25 kGy retrieved components exhibited ultimate load of about 70 N with geometric strain softening observed with continuing displacement. The data were used to support a hypothesis that crosslinking and its induced higher ultimate load is beneficial in the hip-wear environment even in the presence of oxidation.

## 32.4 EFFECT OF CROSSLINKING ON MECHANICAL BEHAVIOR AND WEAR

In the late 1990s, problems associated with submicron UHMWPE wear debris, such as increased incidence of osteolysis with increasing wear [16], grew to the point that concerted efforts to reduce the flux of particulate emanating from the bearing couples *in vivo* were made. The outcome of these research efforts was the development of various highly crosslinked materials that exhibited markedly reduced wear in laboratory simulators. If, in fact, the primary wear mechanism in a hip arthroplasty is in some way related to the mechanical performance of the bearing, then the small punch test would likely be able to detect differences between the mechanical signature of conventional and highly crosslinked UHMWPEs. Mechanical assessment of several proprietary highly crosslinked UHMWPEs was performed in 2001 by Edidin et al. [17], who repeatedly observed that highly crosslinked UHMWPE components exhibited marked geometric strain hardening following attainment of the initial maximum load on the load-displacement curve. Representative material curves are shown in Figure 32.8.

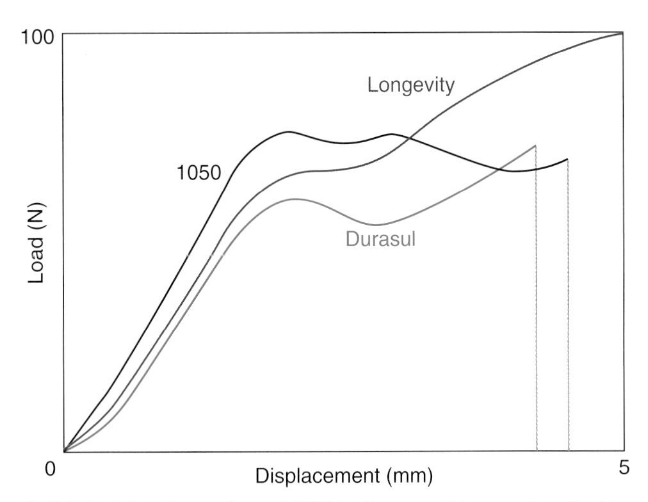


FIGURE 32.8 Comparison of 1050 in three conditions as described by the respective load-displacement curves. Longevity and Durasul are commercially available formulations of highly crosslinked UHMWPE, which exhibit the characteristic geometric strain-hardening in the drawing phase of the small punch test. Free radical quenching by remelting UHMWPE above its peak melt transition typically reduces the observed peak load as exhibited here [17].

When assayed using the small punch test, the load-displacement curve for gamma irradiated 1050 exhibits greater geometric strain hardening and greater change from its nonirradiated condition than do the same data obtained using virgin compared to irradiated 1020, suggesting that all other things being equal, the efficiency of crosslinking is lower in lower Mw resins. Figure 32.9 illustrates this finding by comparing the load-displacement curve of GUR 1020 and 1050 resins converted in the same manner (compression molding) before and after irradiation. The increased efficiency of crosslinking observed in higher molecular weight resins has been corroborated in a chemical manner by Spiegelberg et al. [18], who noted that a greater crosslink density was observed in UHMWPEs with lower polydispersion indices, with 1050 having the lowest polydispersion index of the materials tested.

An alternative explanation for the increased wear resistance of 1050 is based upon an energy argument arising from the observation that crosslinked materials typically exhibit higher integral work-to-failure (WTF) than noncrosslinked UHMWPEs. However, because crosslinking may also influence other metrics that change wear rates, such as hindering chain mobility and/or changing surface lubricity, Edidin and Kurtz designed an experiment to investigate solely the effect of changing WTF on observed wear rates [11]. They used a spectrum of semicrystalline polymers with previous clinical application to test the hypothesis that wear rate was proportional to the WTF of a material when tested under equibiaxial tension. The materials chosen were UHMWPE, HDPE, polyacetal, and polytetrafluoroethylene (PTFE). The basic design of the experiment consisted of fabricating identical acetabular liners on production tooling from the four materials, measuring gravimetric wear on a hip simulator using commonly accepted protocols [19] and measuring the articulating surface WTF using the small punch test. All materials were tested in the native, nonsterile and nonirradiated condition to limit confounding variables. ANOVA analysis revealed correlations between gravimetric wear rate and several of the SP metrics, but the best correlates, with $r^2$ of 0.96, were found between the WTF and wear rate. This correlation, shown in Figure 32.10, displayed a highly linear relationship between each material's WTF and the ensuing wear rate, suggesting that intrinsic equibiaxial toughness or WTF plays a fundamental role in the generation of wear.

Crosslinked UHMWPEs are typically tougher than their noncrosslinked equivalents, as may be observed in Figure 32.9. However, the data shown in Figure 32.9 were obtained using materials with complicated and proprietary thermal annealing and/or quenching steps following

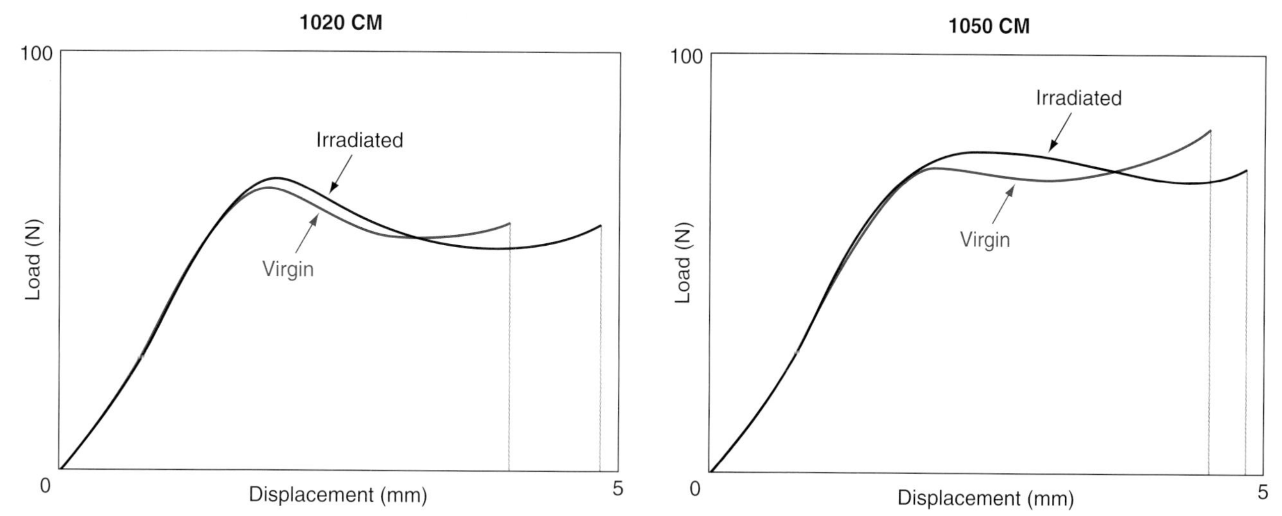


**FIGURE 32.9** Comparison of representative load-displacement curves obtained using GUR 1020 and 1050 resins converted via compression molding before and after irradiation. A more substantial change following irradiation of the degree of geometric strain hardening is observed for the GUR 1050 material than for the GUR 1020 material, suggesting that the higher molecular weight based resin is more efficiently crosslinked. A single irradiation dose of 25 to 40 kGy was absorbed by both materials, which were irradiated simultaneously. Adapted from data in Edidin et al. [8].

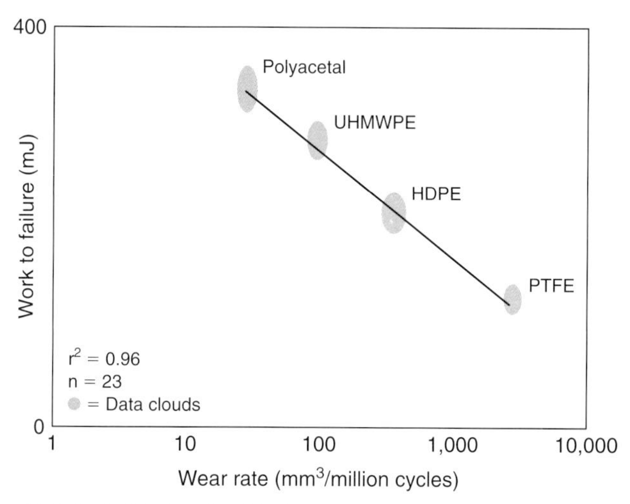


**FIGURE 32.10** Relationship between work-to-failure as calculated from equibiaxial loading to failure for materials with historical clinical application. Data suggest that adhesive–abrasive wear rates are correlated with the innate WTF of the bearing material [11].

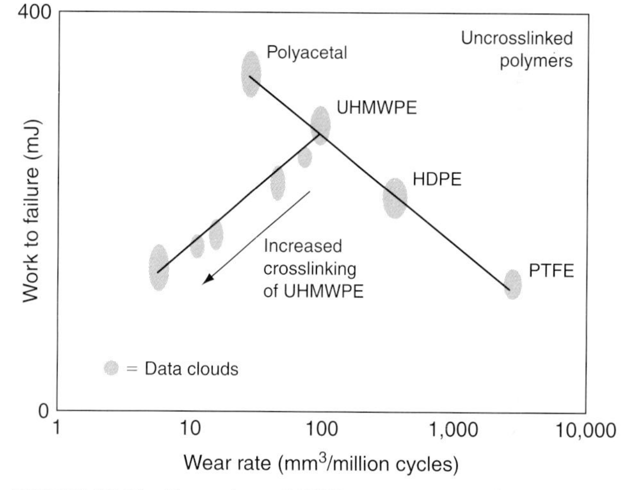


**FIGURE 32.11** Comparison of WTF versus wear rate for crosslinked and noncrosslinked acetabular components fabricated from a variety of polyolefins tested in a multistation wear simulator. Increasing the crosslink density of UHMWPE without thermal annealing still leads to a substantial decrease in wear rate. Adapted from Edidin et al. [11, 20].

gamma irradiation. Because annealing tends to substantially increase the elongation failure, it is not guaranteed that the low wear exhibited by highly crosslinked UHMWPE in the simulator is necessarily a continuation of the curve in Figure 32.10 extrapolated out to even tougher materials. In a thought experiment, Kurtz and Edidin combined the data in Figure 32.10 with data obtained from sequentially irradiated acetabular components with no thermal treatment postirradiation to produce the curve shown in Figure 32.11 [11, 20].

Akagi et al. [21] used metrics from the small punch test to investigate the behavior of candidate crosslinked UHMWPE knee bearings. The authors tested the candidates following wear simulation, using nontested bearings as controls. They further reasoned that the best knee bearing material was the one that accumulated the least damage under load because knee failure modes arise from fatigue damage accumulation leading, in extreme cases, to frank bearing failure. They found that while overall toughness increased with increasing dose, the difference between control and wear-tested bearings was minimized at 50 kGy, even though wear rate continued to decrease monotonically with increasing dose out to 200 kGy. The authors concluded that if accumulated damage is the appropriate design criterion, then the optimal irradiation dose for crosslinked knee bearings is both different and lower than that suggested for hip bearings.

In sum, the experiments relating integral toughness to gravimetric wear metrics suggest that while marked geometric strain hardening is observed in all crosslinked UHMWPEs tested to date, the causality of integral toughness to the wear performance of such components remains unclear and may vary in importance as a function of intended use.

## 32.5 SHEAR PUNCH TESTING OF UHMWPE

The prominence of geometric strain hardening in highly crosslinked UHMWPE, coupled with uncertainty as to the overall role of integral toughness on wear resistance, begged the question of whether or not the presumed resistance to chain mobility in equibiaxial tension evidenced by geometric strain hardening is solely responsible for decreased wear. Wear at the acetabular bearing surface, in particular, has been attributed to the "cross-shear" phenomenon wherein debris is liberated following two orthogonal passes of the femoral head's contact point across some portion of the contact patch of the bearing [19]. If these orthogonal passes induce some sort of shear related damage accumulation, then modification of the mechanical test apparatus to permit shear loading might be able to differentiate wear resistance better than the tensile loading apparatus.

Kurtz et al. [22] developed a modified punch that loaded the specimen in shear for a substantial portion of the range of travel of the punch in a typical test. Rather than a domed head, the modified punch used a flat head and was mated to a die without a radius at the pinch point between the specimen and the upper die face. In keeping with recommendations in the literature, the annular clearance between the punch and the die was set at 0.033 mm, yielding a length to width ratio greater than 15, previously shown to minimize bending during shear testing [23, 24]. No particular thickness to width ratios were designed into the apparatus because it was assumed that the disk-shaped specimens would provide sufficient lateral restraint to eliminate out-of-plane deformations previously associated with shear tests. Correlation between *in vitro* data and finite element analyses was demanded to yield a relationship between the initial stiffness of the UHMWPE and its Young's modulus as:

$$E = 0.283k$$

where $k$ is the initial stiffness (in N/mm) observed on the load-displacement curve and E is the Young's modulus (in MPa). The consistently observed features of the shear punch test are shown in Figure 32.12. Complete shear-through of the specimen occurs when the punch has traversed the initial thickness of the specimen (0.5 mm), but pure shear conditions do not exist beyond the displacement resulting in the peak load.

Testing using the shear punch apparatus of conventional and highly crosslinked UHMWPE strongly

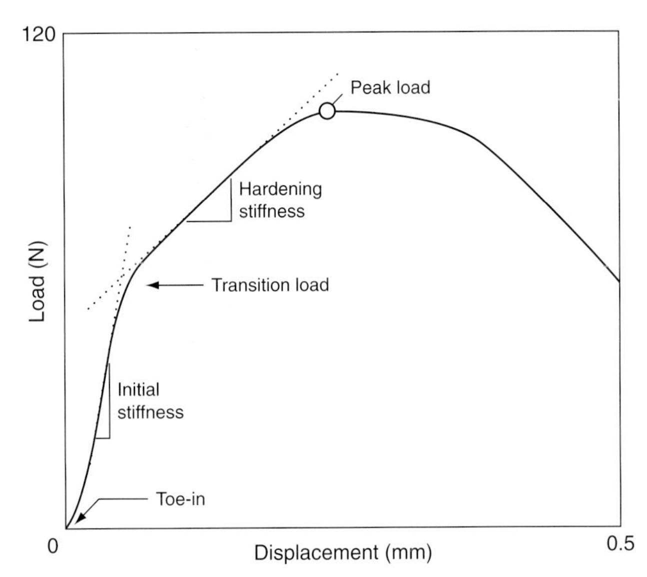


**FIGURE 32.12** Characteristic features observed on load-displacement curves obtained using the shear punch apparatus. An initial and hardening stiffness in addition to the peak load are the primary metrics used to perform analytical comparisons. Like the conventional small punch test, the shear punch test also may be interpreted qualitatively. Adapted from Kurtz et al. [22].

suggested the representative behaviors in shear were, in fact, quite similar as shown in Figure 32.13. This finding suggests that the "cross-shear" motion pattern by the femoral head on the bearing surface likely still results in tensile final rupture of the wear debris fibrils and that therefore the term "cross-shear" is a misnomer.

Kurtz et al. also examined the shear behavior of shelf-aged tibial bearing components to investigate the role played by shear failure in the breakdown of degraded components [22]. The surface and subsurface mechanics of components sterilized and stored in air were measured. Because degradation occurs in a spatially highly nonuniform manner, comparisons were made between the heavily degraded subsurface region and the substantially less degraded surface region. This comparison revealed substantial differences on the overall appearance of the two curves, with the more heavily degraded subsurface specimens exhibiting a complete lack of geometric shear strain hardening and a higher initial stiffness as shown in Figure 32.14. Bartel et al. [25] suggested that the peak stresses in a loaded tibial bearing occur approximately 1 mm below the articulating surface, or in the same region as the subsurface samples tested by Kurtz et al. [22]. Thus, clinical failure modes of UHMWPE bearings observed in retrieved tibial bearings, consisting primarily of pitting and delamination, may arise from the substantially lower resistance to both tensile and shear strain hardening postyield. Lowered integral toughness may also play a role, as suggested by Kurtz et al. [26], who observed a trend relating lower toughness with higher qualitative damage scores in a pilot retrieval study. Overall, the tensile and shear small punch

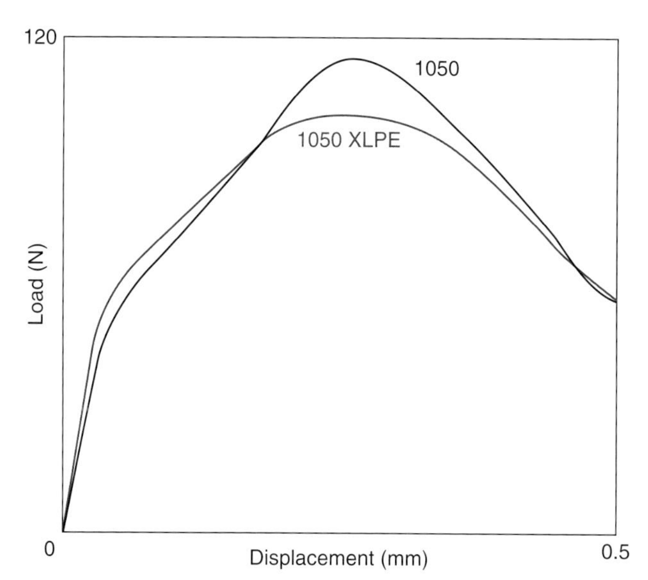


**FIGURE 32.13** Load-displacement curves observed by testing virgin and crosslinked UHMWPE in a shear punch apparatus. The curves are quite similar, suggesting that changes in shear resistance are unlikely to explain the increased resistance to wear exhibited by crosslinked UHMWPEs. Materials tested were virgin GUR 1050 and GUR 1050 irradiated to 75 kGy and annealed at temperatures below the melt point. Adapted from Kurtz et al. [22].

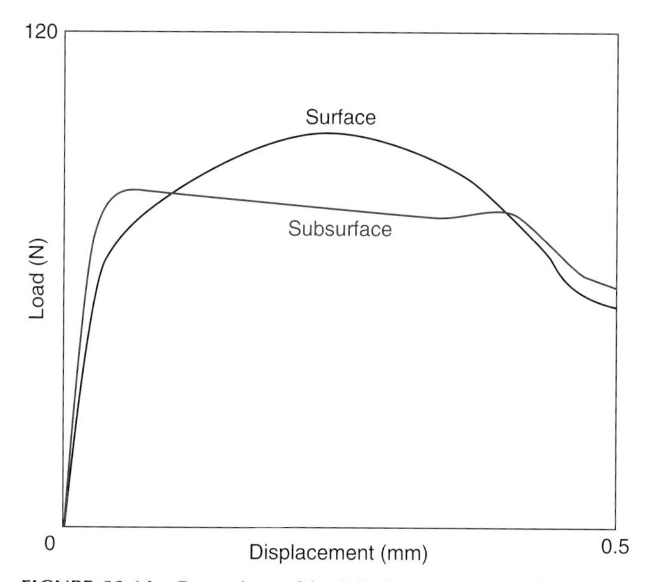


**FIGURE 32.14** Comparison of load-displacement curves from specimens obtained at the surface and at approximately 1500 microns subsurface from a tibial bearing component originally made from GUR 1120 sheet, irradiated with 25 to 40 kGy of gamma irradiation and aged on the shelf in air for 8.5 years. Oxidative degradation in shelf-aged components is known to peak subsurface, and these data suggest that resistance to shear flow is highly compromised by degradation. Adapted from Kurtz et al. [22].

test findings have proven to be key tools in understanding the relationship between the mechanics of UHMWPE to both adhesive/abrasive wear and frank mechanical breakdown of arthroplasty bearing components.

## 32.6 FATIGUE PUNCH TESTING OF UHMWPE

Seminal work first applied the equibiaxial tension small punch test to the investigation of degradation and wear mechanisms because these problems were manifest at the time of the test's development. However, as previously discussed briefly in the review of the shear punch test development, bearing components fail not just from continual adhesive/abrasive wear mechanisms but also from fatigue mechanisms, such as delamination and frank failure associated with crack initiation and propagation. The means of determining resistance to failure in fatigue of aged or crosslinked materials could provide useful information on the suitability of various bearing materials to their proposed application. For example, while some highly crosslinked materials are remelted, others are annealed, and thus the relative fatigue performance of such materials may be as important as their resistance to *in vitro* or *in vivo* oxidation. In addition, the suitability of crosslinked materials in general to applications undergoing cyclic loading, such as tibial plateau bearings, requires the development of new testing methods. Therefore, methodological changes were made to the tensile small punch apparatus, as described by Villarraga et al. [27], to permit cyclic loading of the test specimen at body temperature. These changes included the addition of a boroscope of some sort, typically a medical arthroscope, to permit direct visualization of the specimen during the crack initiation and propagation phases of specimen failure. An environmental chamber was used to encase the punch and die apparatus to permit testing of specimens at the body temperature. Lastly, the fatigue punch test was conducted in a servo-hydraulic testing machine as opposed to the electro-mechanical testing machine typically used under displacement control in the previously described quasi-static punch tests.

Testing of materials under cyclic loading to determine resistance to cracking may be performed with or without a preexisting flaw in the test specimen. The fatigue small punch test is performed using the same specimen as the static test. Failure is presumed to occur when a critical level of damage accumulation has occurred in the absence of a created flaw or defect; this methodology follow a "total-life" or damage accumulation model. Because damage accumulation models are primarily useful to make comparisons among the performance of different materials, as opposed to providing quantitative estimates of a material's absolute performance in the presence of an existing flaw, it is desirable to have specimens fail in a reasonable (fairly low) number of cycles. To this end, Villarraga et al. [27] describe first testing each material in a conventional static small punch test to determine the mean peak load ($P_{max}$) exhibited by that material when tested at a constant loading rate of 200 N/s. Because the peak load represents the inflection between membrane bending and membrane

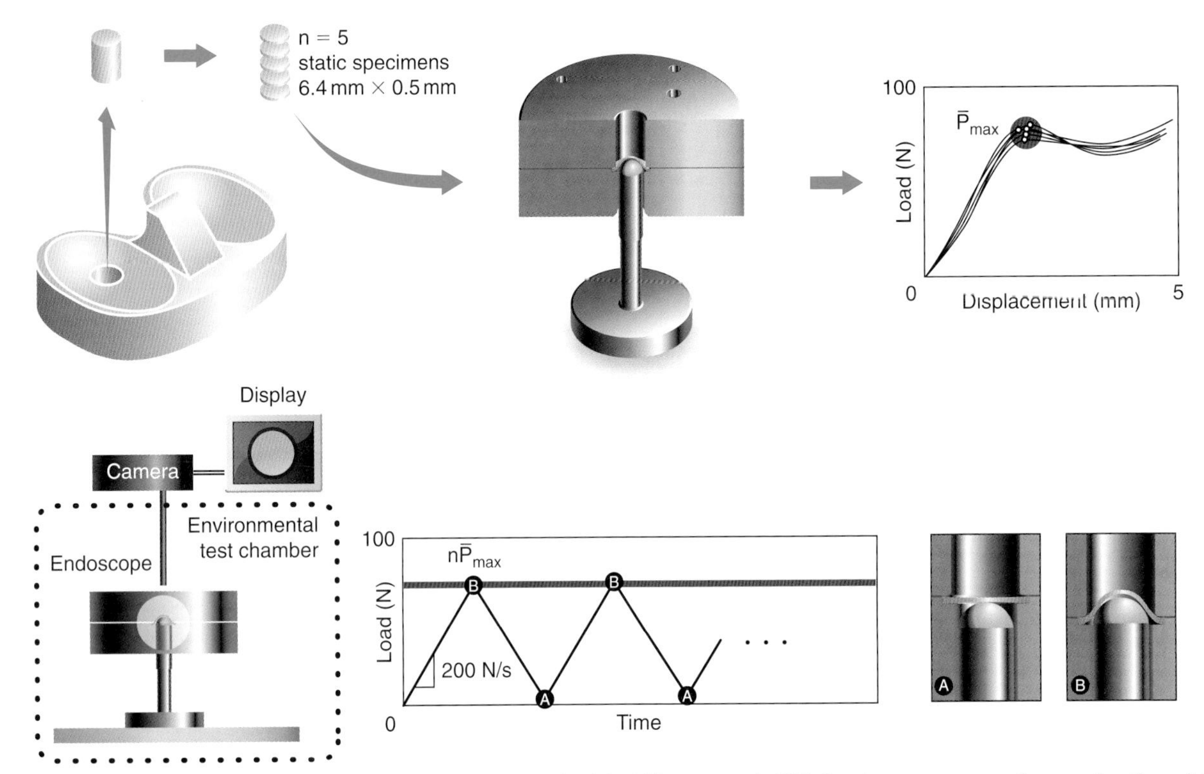


**FIGURE 32.15** Schematic of fatigue punch methodology as described by Villarraga et al. [27]. Specimens are prepared conventionally and then tested to failure at a constant load rate of 200 N/s at 37°C to determine peak load. Subsequently additional specimens from the same material are tested under cyclic loading using a triangular waveform ramping, again at 200 N/s to a percentage of the peak load previously determined. The percentage is varied across the specimen group to cause failure under a greater or lower number of cycles.

stretch deformation modes with its associated chain realignment, it is a natural choice for the limit load in the fatigue punch test. The fatigue punch test itself is then performed under cyclic loading conditions to a maximum load of 0.55 to 0.90 $P_{max}$, depending on the number of cycles to failure expected or desired, with higher percentages shortening the test. Loading rate during the cyclic portion of the test is also 200 N/s. The overall experimental schematic is shown in Figure 32.15.

In their work to develop the fatigue punch test, Villarraga et al. [27] tested four conditions of GUR 1050 UHMWPE used in contemporaneous clinical applications. These conditions included virgin (representing EtO or gas plasma sterilized bearings), gamma sterilized (2.5 to 4.0 Mrad in nitrogen), annealed highly crosslinked UHMWPE (100 kGy and 110°C), and remelted highly crosslinked UHMWPE (100 kGy and 150°C). All testing was conducted at 37°C using an environmental chamber. Two metrics comparing the relative fatigue resistance of a given material were developed. These were a classic S-N curve relating maximum peak load to survival and a hysteresis metric giving some indication of the relative energy retained in the test specimen during each cyclic excursion.

Substantial differences were observed between the control (virgin GUR 1050) material and both crosslinked materials in their displacement to failure, their hysteresis curves, and in the overall appearance of the specimens at failure. The control material exhibited a dramatic increase in displacement just prior to failure not exhibited by either of the crosslinked materials. Lower loads led to greater number of cycles prior to failure at larger displacements in all materials tested. Representative examples of the displacement versus cycles relationships obtained at both relatively low and somewhat higher number of cycles to failure are shown in Figure 32.16. The corresponding appearance of the kinds of specimens tested to generate the curves in Figure 32.16 are shown following fracture in Figure 32.17.

The maximum hysteresis behavior was relatively similar across the materials, with the exception of that exhibited by the XLPE annealed material, which appeared to require lower loads to generate the same amount of stored energy in the specimen. The cumulative hysteresis plot for each material is shown in Figure 32.18. How the ability of a candidate bearing material to dissipate energy during a load–unload cycle relates to its suitability as a bearing component material will require further study in conjunction with carefully monitored clinical application.

## 32.7 CONCLUSION

In summary, over the past decade the small punch test has been extensively developed to permit direct assay of the mechanical properties of orthopedic UHMWPE bearings.

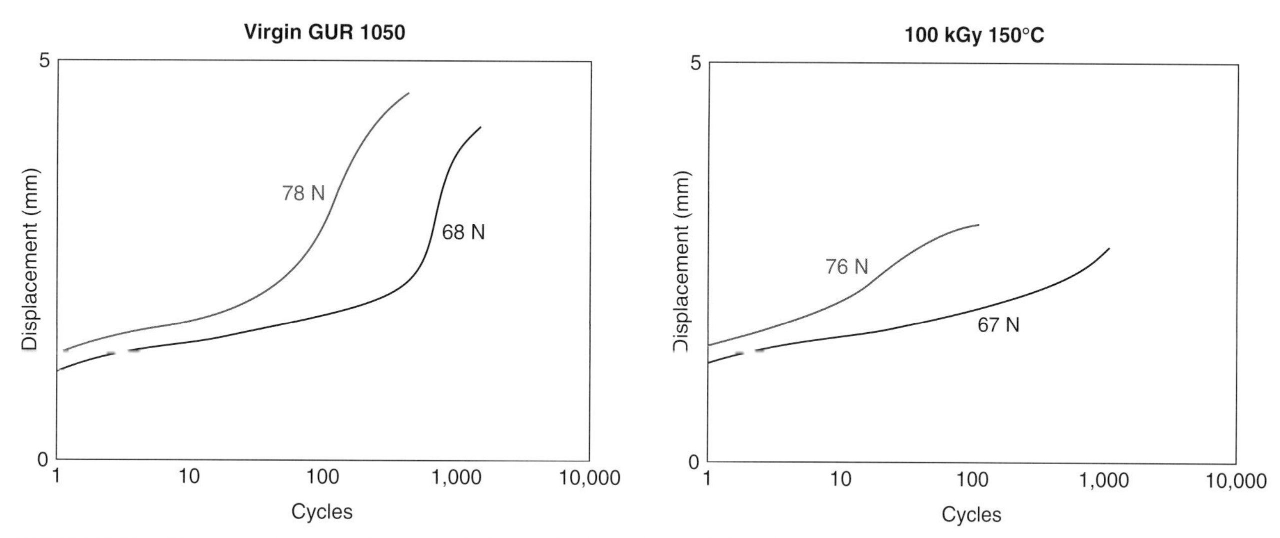


**FIGURE 32.16** Displacement versus cycles curves obtained following testing of virgin and remelted highly crosslinked GUR 1050 specimens at differing peak loads. Crosslinked specimens exhibited substantially lower displacements at failure than control specimens. Adapted from Villarraga et al. [27].

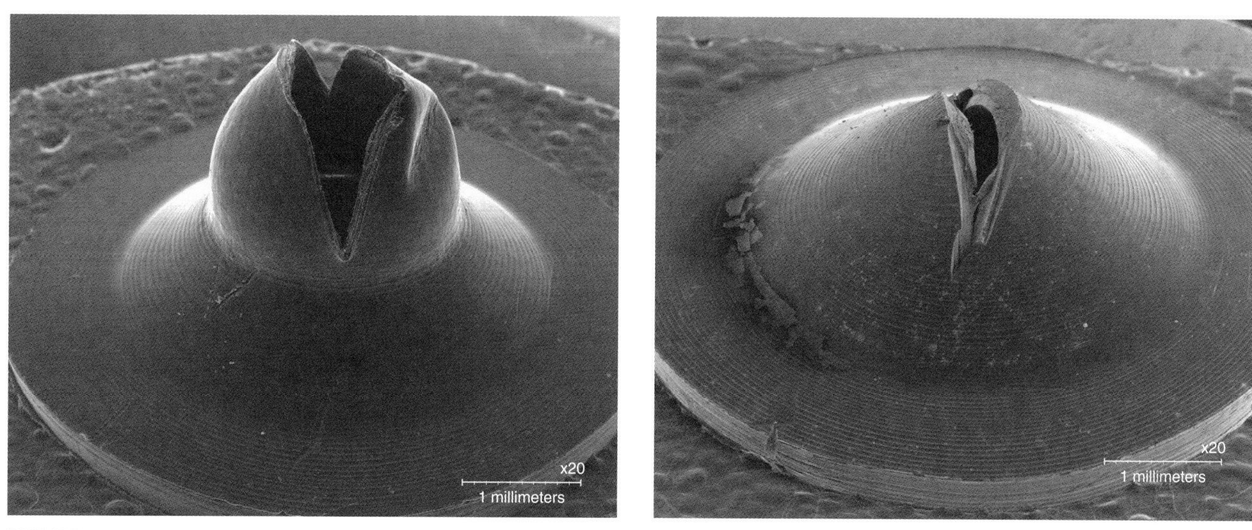


**FIGURE 32.17** Representative low-power magnification SEM photographs of samples of the same materials tested in Figure 32.16. At left, virgin GUR 1050 specimen; at right, a specimen of remelted (150°C) highly crosslinked GUR 1050. The specimen at left failed at 7289 cycles, that on the right failed at 7994 cycles. Adapted from Villarraga et al. [27].

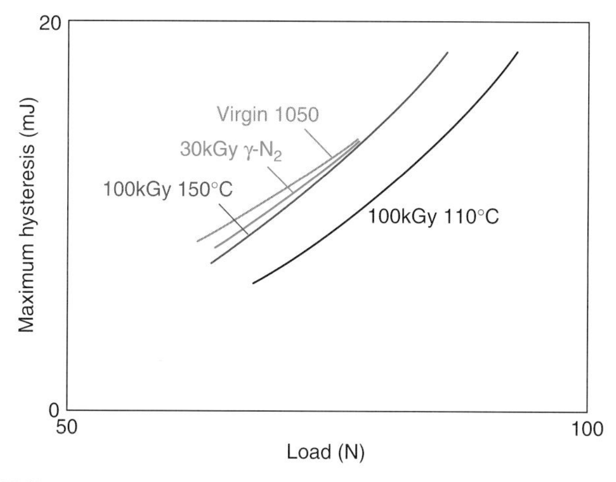


**FIGURE 32.18** Maximum hysteresis exhibited as a function of load for each of the materials tested by Villarraga et al. during the development of cyclic small punch test. Adapted from Villarraga et al. [27].

Originally adapted strictly as a miniature-specimen, biaxial tension test, it has since been extended to permit testing in both shear and cyclic modalities. As such, it has proven to be extremely useful in the detection of the mechanical signature of oxidative degradation and the signature associated with elevated levels of irradiation-induced crosslinking. Further testing has enabled associations between wear and mechanical properties to be made from the selfsame components, although a quandary remains as to which mechanical property drives wear reduction. Most recently, application of the small punch specimen geometry to a cyclic loading regime has permitted initial investigation into the suitability of highly crosslinked UHMWPE materials to applications wherein they will be subjected to cyclic loading. The greatest strength of the small punch testing methodology is its minimal requirement for material, thereby permitting mechanical assays to be made from almost any experimental specimen configuration, and one would expect this strength to continue to be leveraged across the mechanical, chemical, and tribological spectra that describe the orthopedic bearing world.

## REFERENCES


1. Kurtz SM, Muratoglu OK, Evans M, Edidin AA. Advances in the processing, sterilization, and crosslinking of ultra- high molecular weight polyethylene for total joint arthroplasty. *Biomaterials* 1999; **20**(18):1659–88.
2. Manahan MP, Argon AS, Harling OK. The development of a miniaturized disk bend test for the determination of postirradiation mechanical properties. *J Nuclear Materials* 1981;**104**(1-3):1545–50.
3. Kurtz SM, Foulds JR, Jewett CW, Srivastav S, Edidin AA. Validation of a small punch testing technique to characterize the mechanical behavior of ultra-high molecular weight polyethylene. *Biomaterials* 1997;**18**(24):1659–63.
4. Kurtz SM, Jewett CW, Foulds JR, Edidin AA. A miniature-specimen mechanical testing technique scaled to the articulating surface of polyethylene components for total joint arthroplasty. *J Biomed Mater Res (Appl Biomater)* 1999;**48**(1):75–81.
5. ASTM F 2183-02. *Standard test method for small punch testing of ultra-high molecular weight polyethylene used in surgical implants*. West Conshohocken, PA: American Society for Testing and Materials; 2008.
6. Asano T, Akagi M, Clarke IC, Masuda S, Ishii T, Nakamura T. Dose effects of cross-linking polyethylene for total knee arthroplasty on wear performance and mechanical properties. *J Biomed Mater Res B Appl Biomater* 2007;**83**(2):615–22.
7. Clough RL, Gillen KT. Radiation-thermal degradation of PE and PVC: Mechanism of synergism and dose rate effects. *Radiat Phys Chem* 1981;**18**(3-4):661–9.
8. Edidin AA, Herr MP, Villarraga ML, Muth J, Yau SS, Kurtz SM. Accelerated aging studies of UHMWPE. I. Effect of resin, processing, and radiation environment on resistance to mechanical degradation. *J Biomed Mater Res* 2002;**61**(2):312–22.
9. Kurtz SM, Pruitt LA, Crane DJ, Edidin AA. Evolution of morphology in UHMWPE following accelerated aging: the effect of heating rates. *J Biomed Mater Res* 1999;**46**(1):112–20.
10. Edidin AA, Muth J, Spiegelberg S, Schaffner SR. Sterilization of UHMWPE in nitrogen prevents oxidative degradation for more than ten years. *Transactions of the 46th Orthopedic Research Society*, 2000 February 9–13, Orlando, FL: 2000. p. 1.
11. Edidin AA, Kurtz SM. The influence of mechanical behavior on the wear of four clinically relevant polymeric biomaterials in a hip simulator. *J Arthroplasty* 2000;**15**(3):321–31.
12. Premnath V, Harris WH, Jasty M, Merrill EW. Gamma sterilization of UHMWPE articular implants: an analysis of the oxidation problem. *Biomaterials* 1996;**17**(18):1741–53.
13. Gomez-Barrena E, Li S, Furman BS, Masri BA, Wright TM, Salvati EA. Role of polyethylene oxidation and consolidation defects in cup performance. *Clin Orthop* 1998;**352**(352):105–17.
14. Kurtz SM, Hozack W, Marcolongo M, Turner J, Rimnac C, Edidin A. Degradation of mechanical properties of UHMWPE acetabular liners following long-term implantation. *J Arthroplasty* 2003;**18**(7 Suppl. 1):68–78.
15. Sugano N, Saito M, Yamamoto T, Nishii T, Yau SS, Wang A. Analysis of a retrieved UHMWPE acetabular cup crosslinked in air with 1000 kGy of gamma radiation. *J Orthop Res* 2004; **22**(4):828–31.
16. Dumbleton JH, Manley MT, Edidin AA. A literature review of the association between wear rate and osteolysis in total hip arthroplasty. *J Arthroplasty* 2002;**17**(5):649–61.
17. Edidin AA, Kurtz SM. Development and validation of the small punch test for UHMWPE used in total joint replacements. In: Katsube N, Soboyejo W, Sacks M, editors. *Functional Biomaterials*. Winterthur, Switzerland: Trans Tech Publications Ltd; 2001. p. 1–40.
18. Spiegelberg SH, Kurtz SM, Edidin A. Effects of molecular weight distribution on the network properties of radiation- and chemically crosslinked ultra-high molecular weight polyethylene. *Trans 25th Society for Biomaterials*; 2003. 215.
19. Wang A, Essner A, Polineni VK, Stark C, Dumbleton JH. Lubrication and wear of ultra-high molecular weight polyethylene in total joint replacements. *Tribology International* 1998; **31**(1-3):17–33.
20. Edidin AA, Pruitt L, Jewett CW, Crane DJ, Roberts D, Kurtz SM. Plasticity-induced damage layer is a precursor to wear in radiation-cross-linked UHMWPE acetabular components for total hip replacement. Ultra-high-molecular-weight polyethylene. *J Arthroplasty* 1999;**14**(5):616–27.
21. Akagi M, Asano T, Clarke IC, Niiyama N, Kyomoto M, Nakamura T, et al. Wear and toughness of crosslinked polyethylene for total knee replacements: a study using a simulator and small-punch testing. *J Orthop Res* 2006;**24**(10):2021–7.
22. Kurtz SM, Jewett CW, Bergstrom JS, Foulds JR, Edidin AA. Miniature specimen shear punch test for UHMWPE used in total joint replacements. *Biomaterials* 2002;**23**(9):1907–19.
23. G'Sell C, Boni S, Shrivastava S. Application of the plane simple shear test for determination of the plastic deformation of solid polymers at large strains. *J Materials Sci* 1983;**18**:903–18.
24. Gul R. Improved UHMWPE for use in total joint replacement. Boston: Ph.D. *Dissertation, Massachusetts Institute of Technology*; 1997.
25. Bartel DL, Bicknell VL, Wright TM. The effect of conformity, thickness, and material on stresses in ultra-high molecular weight components for total joint replacement. *J Bone Joint Surg [Am]* 1986;**68**(7):1041–51.
26. Kurtz SM, Rimnac CM, Pruitt L, Jewett CW, Goldberg V, Edidin AA. The relationship between the clinical performance and large deformation mechanical behavior of retrieved UHMWPE tibial inserts. *Biomaterials* 2000;**21**(3):283–91.
27. Villarraga ML, Kurtz SM, Herr MP, Edidin AA. Multiaxial fatigue behavior of conventional and highly crosslinked UHMWPE during cyclic small punch testing. *J Biomed Mater Res* 2003; **66A**(2):298–309.

Chapter 33

# Nano- and Microindentation Testing of UHMWPE

Jeremy L. Gilbert, PhD and James D. Wernle, PhD



## 33.1 INTRODUCTION

Previous chapters in this book have focused on engineering aspects of tribological testing of UHMWPE for hip and knee implants (Chapters 25 and 26), and the methods used to characterize wear particles have been reviewed (Chapter 27). One area that has been lacking in our overall analysis of the process of wear in UHMWPE is in the study of the fundamental aspects of wear behavior by way of the structure, properties, and performance of UHMWPE on the scale of both the deformation events that lead to wear and the particles that are formed from the wear of UHMWPE. Wear processes of UHMWPE include abrasive (both two and three body), adhesive, and fatigue wear, and each of these results in the development of particles of UHMWPE that are on the scale of microns or smaller. It is only logical, therefore, to be concerned with the mechanical behavior of UHMWPE on the scale of particles generated *in vivo* during normal wear generation. That is, because the process of wear particle generation takes place on the scale of microns and smaller, the structure, properties, and performance of UHMWPE on the scale of microns or smaller are all critical in understanding how UHMWPE wears. Therefore, the goal of this chapter is to describe the work that has been done to explore the structure, properties, and performance of UHMWPE on the scale of microns or smaller. The chapter will focus on work that has involved the use of mechanical test methods like nanoindentation, microindentation, or, in general, depth-sensing indentation (DSI) testing along with techniques that require the atomic force microscope (AFM) and/or other high magnification imaging methods (e.g., scanning electron microscopy, SEM). Thus, this work will describe indentation testing and SEM and AFM characterization of UHMWPE and will review various testing approaches and the effects of geometry, processing, degradation, and deformation. Briefly, the geometric relationship between the tip and sample plays a role in the analysis of indentation data. Therefore, to obtain reproducible results, correction factors and limitations have been elucidated. In addition, processing and degradation effects have focused on gamma irradiation-induced oxidation and crosslinking, as well as changes in the structure of UHMWPE using various pressure–temperature control methods (e.g., large crystal, highly crystalline UHMWPE—so-called Hylamer). The viscoelastic behavior of UHMWPE is discussed, as are approaches to study viscoelasticity in indentation testing.

This chapter will also describe new methods for exploring the micro- and nanomechanical processes of wear. Here, a surface deformation mechanics analysis will be shown where the surface strains associated with single asperity wear behavior can be determined and related to both polymer structural changes (e.g., orientation of crystallites) and to changes in mechanical properties of the surface deformed UHMWPE (orientation softening) when these deformations take place on the scale of the particles that are formed (e.g., microns or smaller).

## 33.2 DEPTH-SENSING INDENTATION TESTING METHODS

Indentation testing has been a part of materials characterization for decades. However, indentation testing had been primarily a hardness measurement technique that did not focus on the specific load-deflection behavior of the indenter and the sample surface, only the residual indent and its relationship to the applied load. It was not until the 1990s and the widespread use of the atomic force microscope that people became interested in directly measuring the load-deflection characteristics during indentation testing. This desire was partially brought about by the need to estimate the area of contact of the indenter during loading. Indenting on the small scale makes visual verification of residual indents (which allows for hardness calculations) more difficult. Therefore, work based on elastic theory was proposed to describe contact area, given the known indenter cross-sectional area with depth. This specifically has become known as "depth-sensing indentation." In 1992, Oliver and Pharr [1] adapted the theoretical work of Ian Sneddon [2] and developed an experimental approach to the measurement of the surface elastic modulus of materials from the unloading portion of the load-deflection curve. Sneddon's solution to the indentation of an elastic half space by way of a spherical (and other geometries) indenter resulted in the basic equations used in all depth-sensing indentation (DSI) tests today.

In a typical DSI test, a hard indenter of known tip geometry is driven into a sample surface of sufficient thickness. The indenter engages the sample during loading and induces deformations that can be elastic, plastic, and viscoelastic in character depending on the nature of the material under investigation. A schematic of the test method and surface geometry is shown in Figure 33.1.

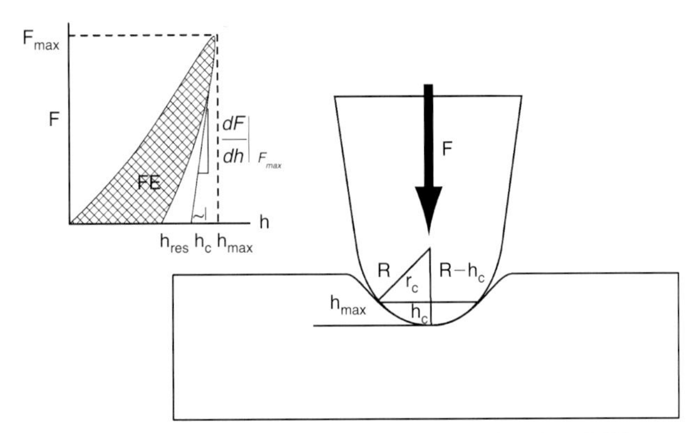


**FIGURE 33.1** Schematic of a depth-sensing indentation (DSI) test. An indenter is pushed with force, F, into the surface an amount, h. The load-depth curve is generated. A schematic for a spherical indenter is shown. Corrections for the contact area calculation are made by finding $h_c = (h_{max}+h_r)/2$. The parameters, $h_{max}$, $F_{max}$, HE, and dF/dh are used to find the hardness, modulus, and energy dissipation resulting from indentation.

The basic approach to DSI testing is to determine the load-depth (F-h) relationship by direct measurement. Load-indent tests can have a pause at the peak load to allow some transient relaxation to occur in viscoelastic materials. Then, several parameters can be determined from these data. The elastic modulus is determined from Sneddon's equations for stiffness and effective elastic modulus. The stiffness relation is defined as:

$$\frac{1}{S} = \frac{\sqrt{\pi}}{2\sqrt{A}}\frac{1}{E_{eff}} \tag{1}$$

where $E_{eff}$ is the effective modulus and the stiffness is the slope of the unloading curve at maximum load. The effective modulus is then calculated as:

$$\frac{1}{E_{eff}} = \frac{1-\nu^2}{E} + \frac{1-\nu_i^2}{E_i} \tag{2}$$

where $\nu_i$ is the Poisson's ratio of the indenter, and $E_i$ is the elastic modulus of the indenter. If the difference between the indenter modulus and sample modulus are greatly different, the elastic modulus can be simplified as follows:

$$E = \frac{(1-\nu^2)\sqrt{\pi}}{2\sqrt{A}}\left.\frac{dF}{dh}\right|_{F_{max}} \tag{3}$$

where E is the sample modulus to be determined, A is the projected area of the indenter, which is a function of the contact depth, $h_c$, $\nu$ is the Poisson's ratio (0.43 for UHMWPE [3]), and dF/dh (S) is the stiffness or the slope of the unloading curve at the maximum force. The slope is most often calculated by fitting a power law function to approximately the top 75% of the unloading data and taking a derivative of that function at the maximum load.

The area-depth function depends on the shape of the indenter used, and there are standard relationships for specific indenters. For example, a Berkovich indenter (typical of a diamond tip indenter in many AFM systems) has $A = 24.5h_c^2$, while for a spherical indenter, $A = \pi(2Rh_c + h_c^2)$, where R is the indenter radius. There are corrections made for the depth, h, to become $h_c$ to account for the actual contact area because A is less than what would be calculated for h (see Figure 33.1). The following equation finds $h_c$:

$$h_c = h_{end} - \varepsilon * \frac{F_{max}}{\partial F/\partial h} \tag{4}$$

where $\varepsilon$ is a correction factor based on tip shape ($\varepsilon = 0.72$ for a cone, 0.75 for a sphere, 1 for a cylindrical indenter).

Other parameters, besides E, that can be determined from DSI tests include the hardness of the material and the energy dissipated during an indent cycle. Continuous stiffness measurement (CSM) is another method that can be used to find complex modulus values for viscoelastic analysis of indentation testing. In continuous stiffness measurements, the indenter, while being indented into the sample, is also cycled over a small amplitude, and the relative amplitude and phase lag between displacement and load gives the viscoelastic response of the material in terms of in-phase and out-of-phase response [4,5].

The hardness of the material can be found by:

$$H = \frac{F_{max}}{A(h_c)} \quad (5)$$

Given that the hardness is inversely proportional to the contact area, and modulus is proportional to the square root of the contact area, accurate knowledge of the tip geometry is of paramount importance. Reliance on common formulas for indenter tips often leads to incorrect property determination. Therefore, imaging to map the change of tip area with height, using either SEM or AFM techniques, is beneficial. Two examples of this are shown in Figure 33.2. A microindentation tip (Figure 33.2(A), made of diamond, exhibits a conical shape with a hemispherical end (r = 15.65 μm). Using geometry, an area function can be determined for indents below and above a threshold level (where the geometry changes from hemispherical to conical). In Figure 33.2B, a three-dimensional depiction of a diamond nanoindenter, based on an AFM height image, is shown. The image was acquired by reverse imaging the tip using a TGT1 grid of very sharp silicon spikes (NT-MDT, Moscow, Russia) with radius of curvature less than 10 μm. Information from AFM height data can be reconstructed and, cross-sectional areas for individual heights can be numerically calculated using a thresholding technique in an imaging program (Image J). The change of area with tip height can be fit with a function. In this indenter's case, a sixth order polynomial captures the function well (Figure 33.3). Upon close inspection, two distinct regions of the curve are noted: one that corresponds to the blunted (rounded) tip end (inset) and another that describes the pyramidal shape of the indenter (paraboloid).

Gilbert et al. [3,6,7] showed that the hysteresis energy (HE, see Figure 33.1) associated with microindentation of UHMWPE varied with the maximum indent depth squared for a spherical indenter, as can be seen in Figure 33.4 for three different UHMWPE materials [3]. Thus, it was

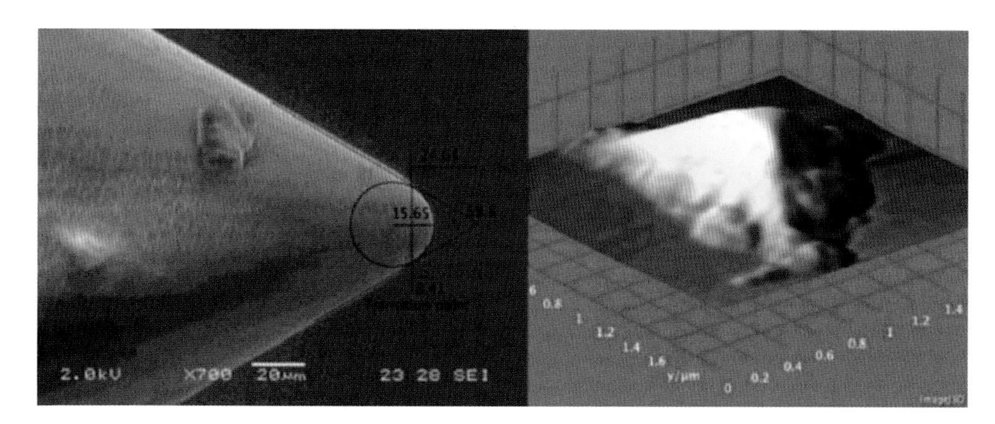


FIGURE 33.2 Geometry of micro- and nanoindenter tips. The height of the nanotip is 500 μm.

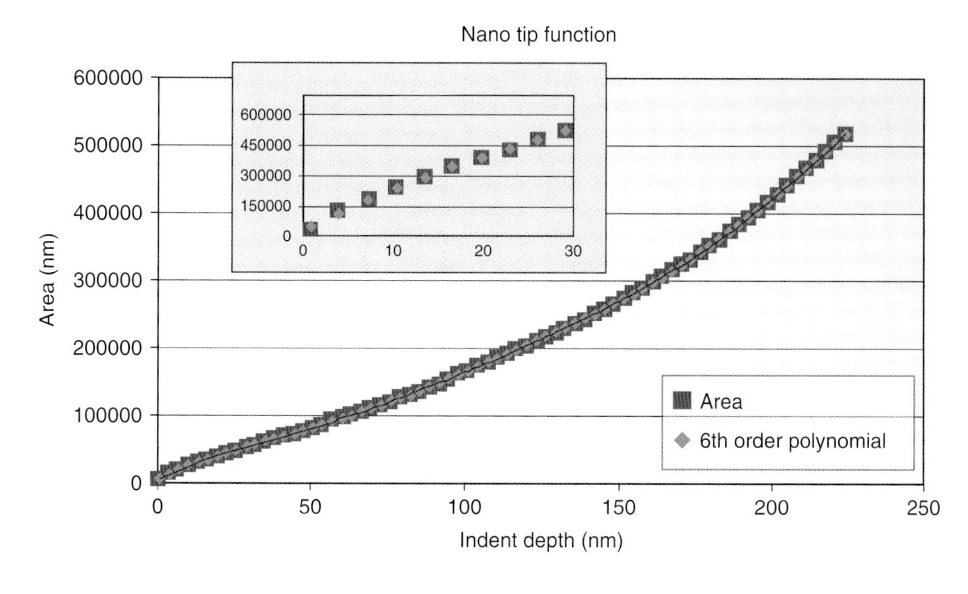


FIGURE 33.3 Area function for nanoindenting tip showing the accuracy of fit of a sixth order polynomial.

determined that a parameter, called the energy dissipation factor (EDF), defined as:

$$EDF = \frac{HE}{h_{\max}^2} \tag{6}$$

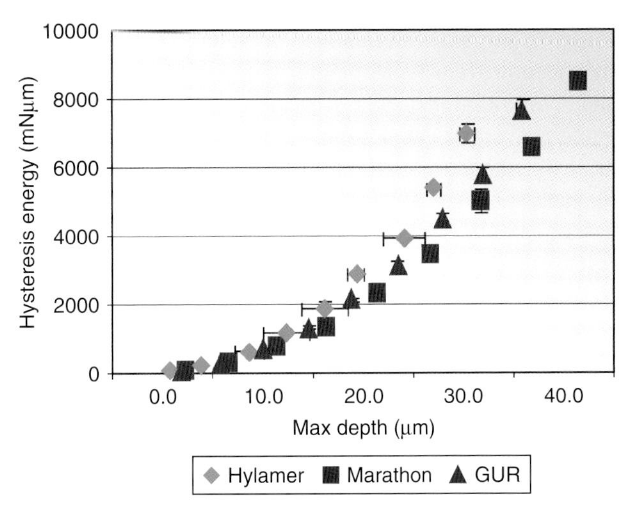


**FIGURE 33.4** Plot of the hysteresis energy associated with an indentation test as a function of the maximum indentation depth. HE varies with indent depth squared for all three different UHMWPEs tested, showing the parabolic relationship (i.e., second order behavior). Dividing HE by the square of the maximum indent depth yields the EDF for a material. Each of these materials has different EDFs with Hylamer having the largest and Marathon the lowest.

is independent of the depth of the indentation and is only a function of the specific material and the geometry of the indenter (i.e., it is a property of the material being indented for a specific indenter). In some respects, this indentation parameter is similar to the loss modulus of a viscoelastic material in that the hysteresis energy associated with cyclic loading of a viscoelastic material depends on the loss modulus and the square of the cyclic strain [8, 9].

The energy dissipation factor, therefore, is specific to the material being tested (for a particular indenter geometry), and changes in this property reflect changes in the material's energy dissipation during indentation.

These parameters, arising solely from analysis of the indenter geometry and load-depth curves, while providing important information about material behavior, do not reveal the underlying deformation state of the material undergoing indentation. To gain an insight into this deformational state, a simulation of a two-dimensional, ideally elastic, material undergoing indentation by a hemispherical indenter is shown in Figure 33.5. The concentric rings, Figure 33.5A, in the material beneath the indenter show the elastic strain energy distribution in the material as a function of location beneath the indenter. Figures 33.5B and 33.5C schematically show the hydrostatic strain and the deviatoric strain distributions as a function location beneath the indenter. It is clear from this figure that the strains (and stresses) in the material are complex and variable with location and fall away roughly spherically

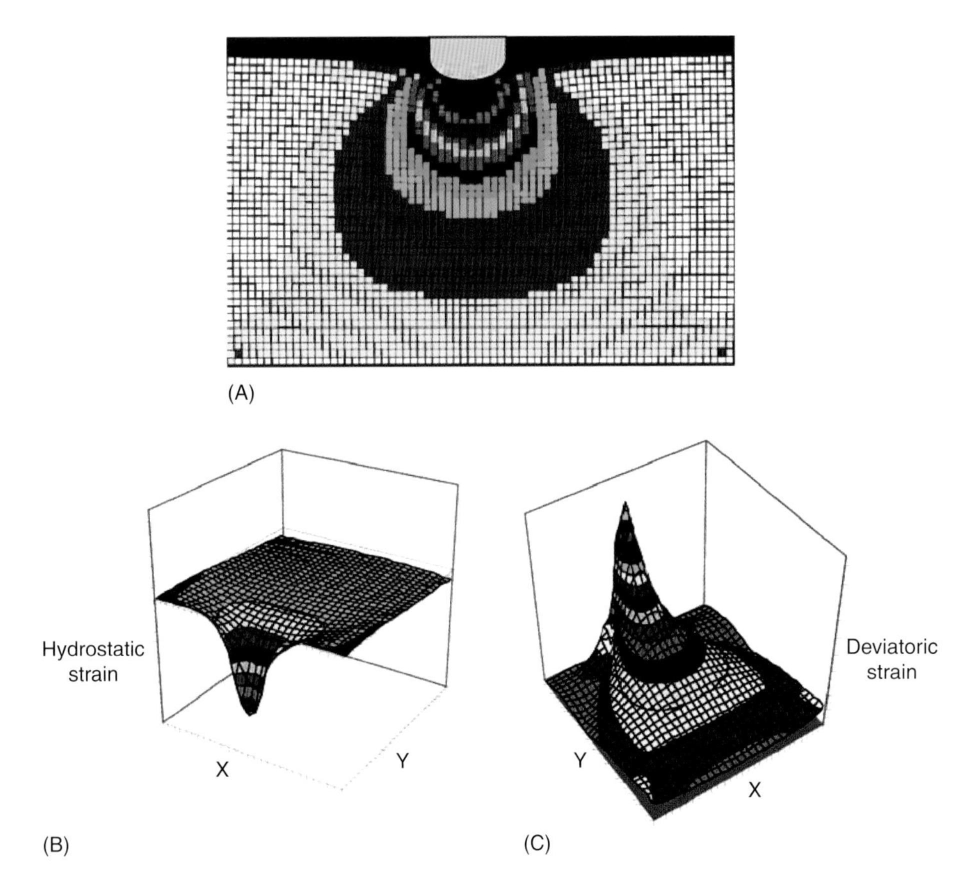


**FIGURE 33.5** (A) Two-dimensional simulation of an indentation test using a multiply connected spring-mass model where the top-middle seven nodes are loaded by a simulated spherical indenter. The model was allowed to relax to an equilibrium condition. The bands in (A) represent elastic strain energy and can be seen to be highly heterogeneous and variable as one moves away from the loading location. Also shown are the schematic representations of the hydrostatic strain (B) and the deviatoric strain (C) that arise from indentations of surfaces. In this example, the model thickness is seen to be affecting the strain distribution under the indenter.

into the material. This figure points out the importance of having an adequately thick sample relative to the depth of indentation and indenter radius. If the sample is too thin, the measured properties will be affected by sensing the underlying support. Wernle and Gilbert [7], as well as others, have shown that typically a thickness of about 10 to 20 times the indent depth is needed for valid measurements of hardness and modulus. This figure also shows how much material is affected during an indentation test and that it is greater than the material in direct contact with the indenter.

Many investigators utilize commercially available indentation test systems (e.g., Hysitron, Inc. and Veeco Instruments' AFM), while others have designed and built custom indentation systems. The advantage of custom systems (of which the authors have developed one) is the ability to design a system capable of a wide range of operations under a wide range of environments and loading conditions. For example, being able to accurately indent a sample sequentially at many locations (e.g., at 100 μm intervals over centimeters) in a profiling or mapping process allows for determination of the spatial distribution of properties. Also, postindentation analysis of the material with additional characterization methods like FTIR and SEM allow for direct correlations between, for example, deformation state and properties or oxidation state and properties. Commercial systems have advantages of being designed for very high sensitivity and small indentation volume interactions. However, testing small regions can also give rise to highly variable results due to surface deformation and structural morphology (e.g., crystalline versus amorphous regions).

## 33.3 INDENTATION TESTS ON UHMWPE: STRUCTURE-PROPERTY TESTING

The literature on indentation testing of UHMWPE has covered a variety of aspects of the behavior of UHMWPE. These include effects of processing, oxidation, prior deformation, extent of crystallinity, frequency of loading, and several other issues related to UHMWPE. Indentation tests have spanned a range of sizes from nanoindentation up to indentation tests in the scale of tens of microns. The advantages of these nano- and microscale tests include the ability to sample very small regions of the material, an ability to profile mechanical properties in a single sample at small intervals (hundreds of nanometers and tens of microns, respectively), and an ability to correlate the indentation response with other measurements of material structure and properties.

Indentation tests of UHMWPE reported in the literature are discussed in the following sections. They have been broken down into general categories based on what was being investigated. Indentation testing of UHMWPE has focused on the effects of material processing and preparation of the surfaces for testing, the effects of radiation induced crosslinking, the effects of oxidation, and the viscoelastic behavior of the material during indentation testing. Each of these areas will be taken in turn.

### 33.3.1 DSI Testing: Effects of Processing, Surface Preparation, and Prior Deformation

DSI testing of the effects of processing (other than crosslinking) and surface preparation on UHMWPE have been carried out by Gilbert [3, 6], Wernle [7, 10], and Ho [11, 12]. Processing effects included the assessment of large crystal, highly crystalline UHMWPE (Hylamer), which was shown to have a significantly higher modulus and hardness and a higher energy dissipation factor than either GUR 1020 or GUR 1050 [3]. That is, highly crystalline UHMWPE tended to have higher mechanical properties and more energy loss during indentation than the other two standard, non-crosslinked resins in use (GUR 1020 and GUR 1050).

Examples of the differences in indentation response for these three materials as well as a crosslinked UHMWPE are shown in Figure 33.6. The indentation response, obtained by indenting a 39 μm spherical alumina indenter up to about 30 μm into the surface, shows a dependence on the starting material. Hylamer shows higher loading slopes and lower extent of residual indentation, and it reflects a higher modulus for this material (about 1.3 GPa) than GUR 1020

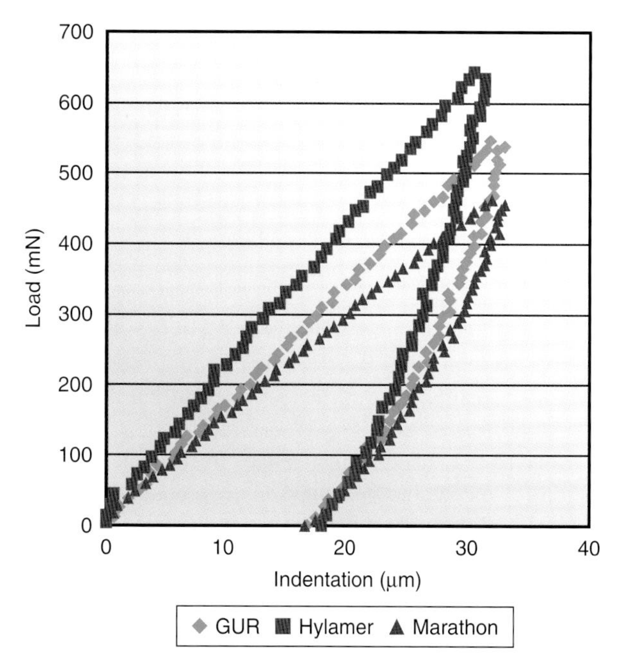


**FIGURE 33.6** Indentation tests of three different UHMWPE materials, (GUR 1020, Hylamer, and Marathon). Note the differences in loading slope and maximum load for approximately the same loading deflection. Moduli, hardnesses, and EDFs can be determined from these indents.

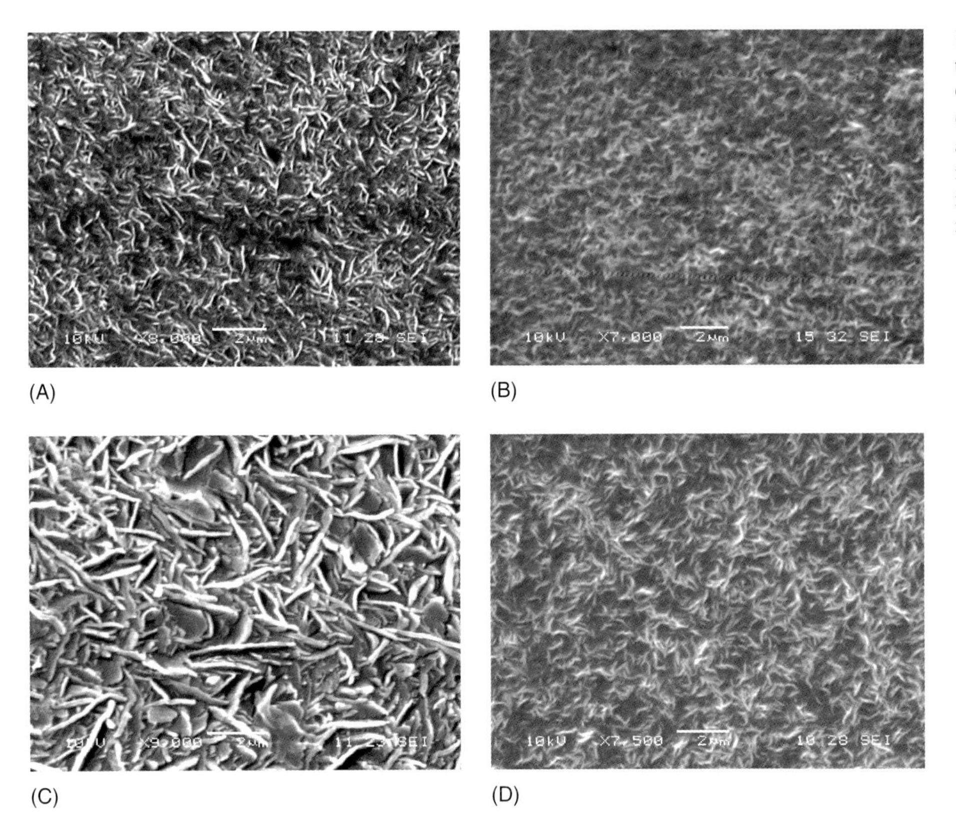


**FIGURE 33.7** SEM comparisons of the crystal structure of UHMWPE for (A) GUR 1050, (B) GUR 1020, (C) Hylamer (a large-crystal UHMWPE material), and (D) Marathon (a 5 Mrad gamma irradiation crosslinked UHMWPE. The image is ca. 7000 to 9000× original magnification.

(E = 0.8 GPa) or Marathon. Hardness and energy dissipation factors measured in DSI testing showed differences with material as well. Hylamer had higher hardness (120 MPa versus 90 MPa for GUR 1020) and higher EDF (5.8 mN/$\mu$m$^2$ versus GUR1020 at 4.8 mN/$\mu$m$^2$).

The structure of these four different materials are shown in Figure 33.7, which are high magnification SEM images of GUR 1050, GUR 1020, Hylamer, and Marathon (a 5 Mrad gamma irradiation crosslinked material) UHMWPE materials after etching with a sulfuric acid–phosphoric acid–hydrogen peroxide process [13]. Figure 33.8 shows the same materials after etching, imaged using AFM techniques. It is clear that the crystallites in these materials vary in terms of size. The etching has removed most of the amorphous polymer between crystals, but the normal structure would include amorphous polymer tying the crystals together and coupling their reorienting and other changes to the deformation field of the polymer during indentation or wear processes.

Ho et al. [11, 12] investigated the effects of surface preparation for indentation testing. They explored compression molding and different microtoming and etching conditions (cryo-microtoming at different temperatures and etching for different times). They found that both hardness and modulus decreased with increasing penetration depth. Modulus ranged from 1.5 to 2.8 GPa, and hardness ranged from 40 to 110 MPa. These tests were at very small nanoindentation areas and depths and had variations that were scaled to the morphology (crystal versus amorphous regions) because of the small volume interrogated.

It is important to note that microtoming of UHMWPE will result (as does wear deformation) in surface alignment of polymer due to the deformation associated with microtome cutting. An example of crosslinked UHMWPE polymer (Marathon) alignment resulting from microtoming at room temperature with a glass knife can be seen in Figures 33.9 and 33.10. Figure 33.9 shows two magnifications of crosslinked UHMWPE (Marathon), revealing the surface tabs that consist of crystals tied together by oriented amorphous polymer in a so-called shish-kebab morphology [14–17].

Figure 33.10 is a scanning electron micrograph of a similar region as Figure 33.9, showing the pulled tabs and the shish-kebab-like structure within the tabs.

In the work by Gilbert et al. [3, 6] and Wernle [7], microindentation, rather than nanoindentation, was used. The samples of UHMWPE were microtomed at room temperature to obtain flat, smooth surfaces for testing. As previously mentioned, there are some significant advantages to the microindentation approach compared to the use of nanoindentation methodology. First, in using a larger indenter, larger loads, and larger indent depths, a greater volume of the material is investigated. Thus, while it is still a very localized measure of properties, microindentation will provide a more volume-averaged response of the material so that local variations in topography and

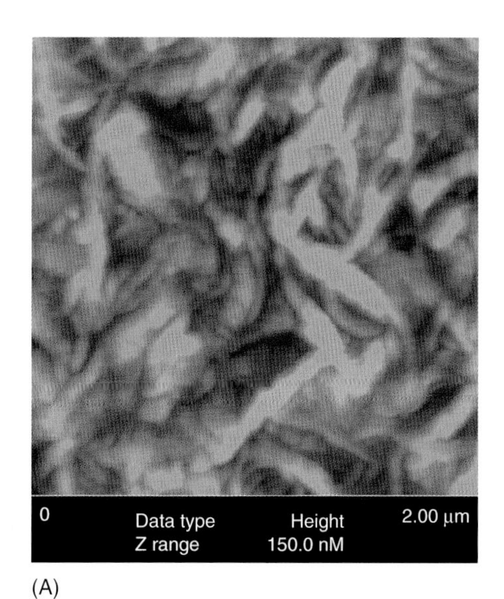


(A)

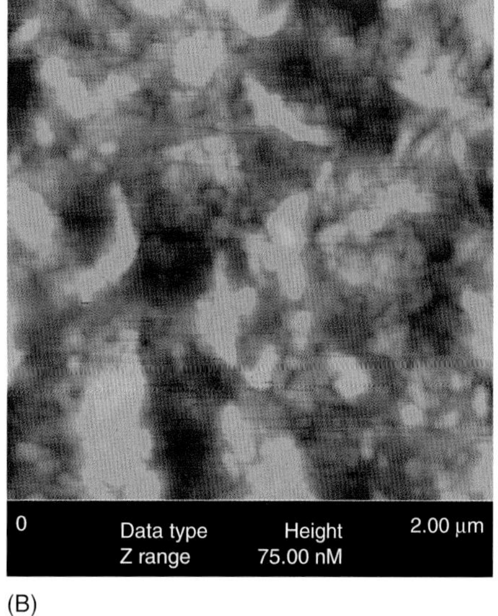


(B)

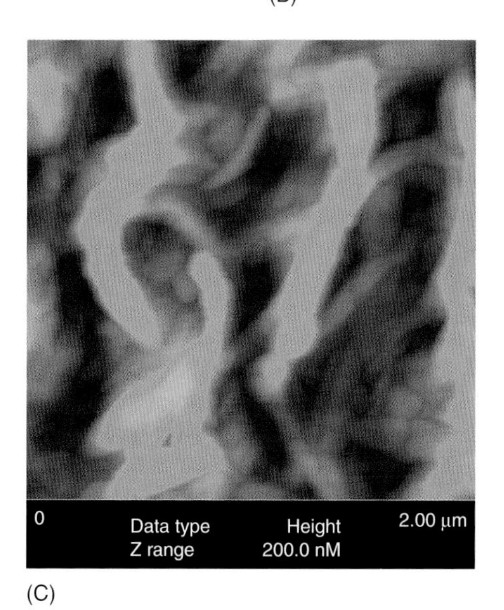


(C)

**FIGURE 33.8** Atomic force microscope images (2 µm on a side and between 75 and 200 ηm tall) of UHMWPE samples. (A) GUR 1050, (B) GUR 1020, (C) Marathon. Note the very different crystallite morphologies for each material. Hylamer is not shown because of the difficulty imaging these large crystals in etched samples in the AFM.

crystalline/amorphous content do not add to the variability of the measured response. This helps, for example, in minimizing surface orientation effects from the microtome or other surface preparation effects, whereas nanoindentation, because it is performed in the very top surface layer (only a micron or less deep), will be susceptible to changes in the surface due to microtoming. Second, the microindentation approach allows for micron-resolution controlled motion of the surface relative to the indenter and up to centimeters of lateral translation so that precision-indent profiling at regular spacings can be performed and the results can be directly correlated with ancillary measurements, such as FTIR microscopy/spectroscopy. This is in contradistinction to nanoindentation systems, where it is often difficult to test larger indenters and larger regions and then perform correlations with other local structure/property measurements.

### 33.3.2 DSI Testing: Effects of Oxidation and Crosslinking

Several authors have investigated the effects of both oxidation and radiation-induced crosslinking of UHMWPE and the effects of these conditions on the DSI properties.

Wernle and Gilbert [7] and Gilbert and Merkhan [6] investigated the indentation behavior of retrieved and shelf-aged UHMWPE tibial inserts that had significant oxidation-induced damage. These samples had developed white bands as could be seen in thin sections of the materials, present due to the combination of oxidized species in the UHMWPE and the microtoming of the sample. These investigations directly correlated to FTIR measurements of the oxidation index (see Figure 33.11) with the measured indentation properties. Figure 33.12 shows the profile of both the oxidation index (the carbonyl peak [1720 cm−1]

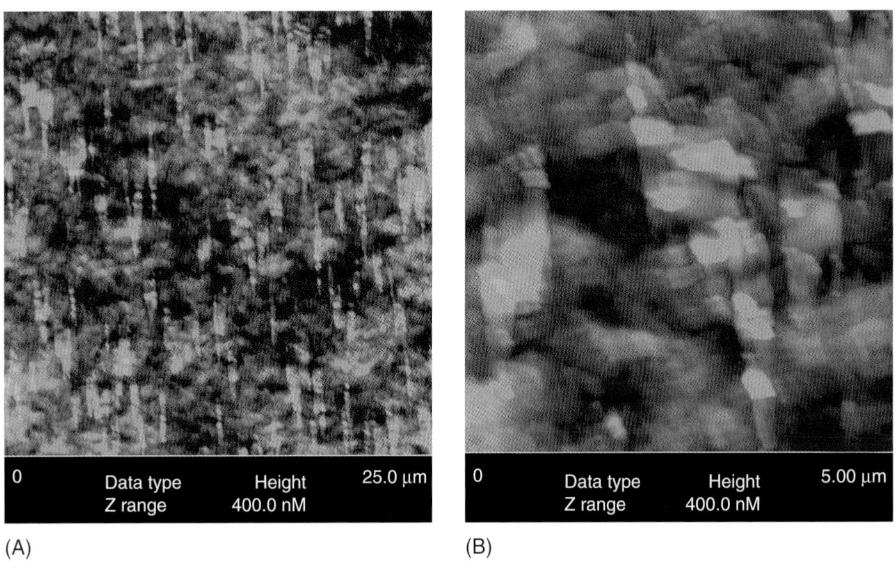


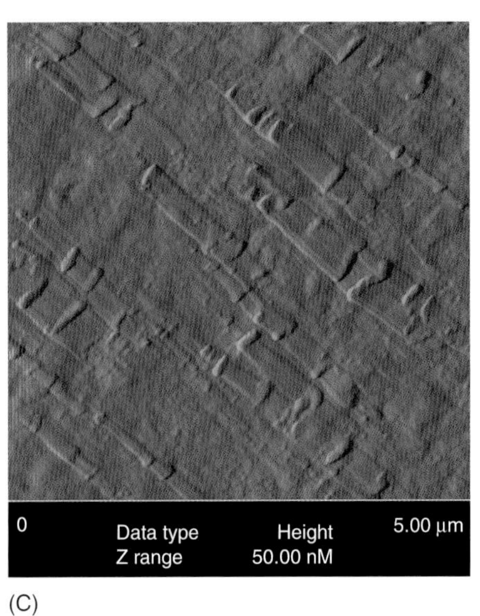


**FIGURE 33.9** AFM height images (A and B) and deflection image (C) of Marathon after microtoming. Note the pulled tabs and oriented polymer structure. This structure is known as a shish-kabob structure where crystals of UHMWPE are the "kabobs" and the amorphous polymer is the "shish" [24].

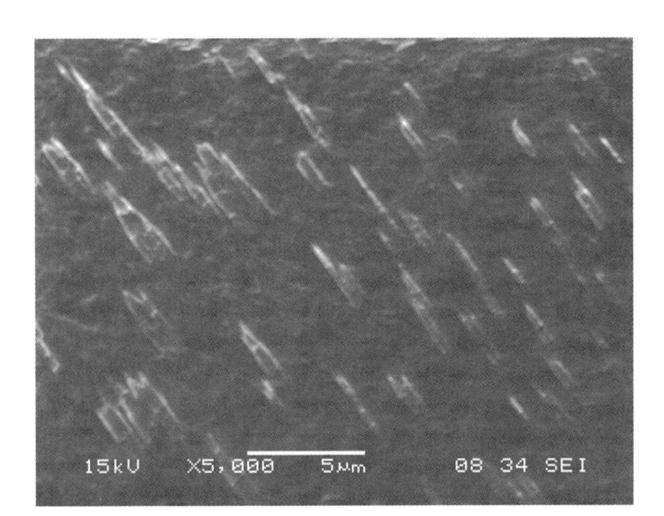


**FIGURE 33.10** SEM micrograph of microtromed Marathon confirming the shish-kabob morphology of the pulled tabs resulting from interaction with the glass knife.

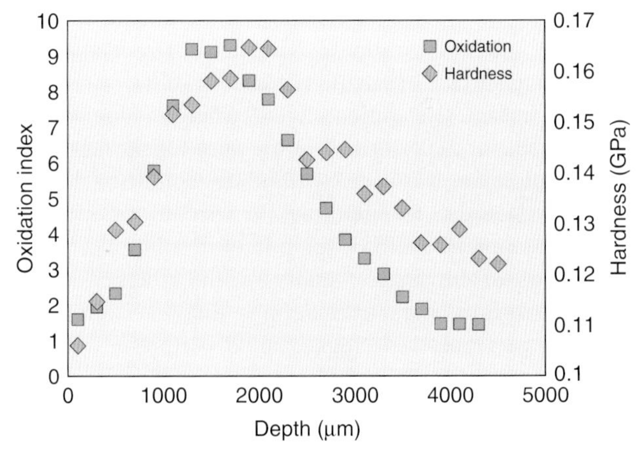


**FIGURE 33.11** Comparison of hardness measurements and the oxidation index (OI) for retrieved UHMWPE tibial bearing samples. Hardness was profiled with microindentation (after: [19]) while the OI was obtained by FTIR spectroscopy at the sites of the indentation testing so that direct correlations could be made.

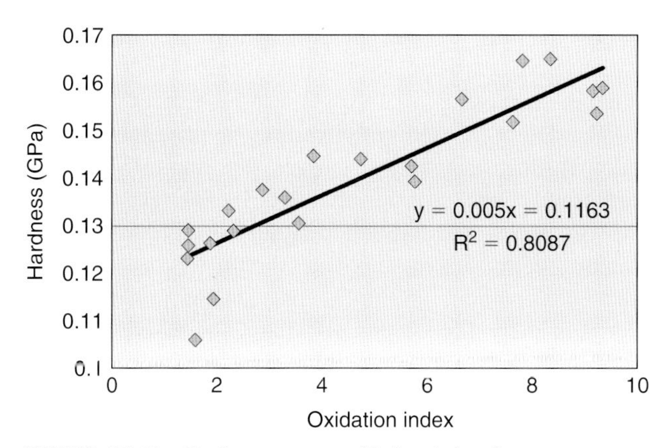


**FIGURE 33.12** Hardness versus oxidation index for measurements made in the same locations on a retrieved UHMWPE tibial component. Note the linear correlation that results between properties and OI.

divided by a reference methylene peak at $2022\,cm^{-1}$) and hardness across a retrieved GUR 1050 UHMWPE component. The results of this work showed that the profile of hardness, modulus, and EDF into the depth of these samples mirror the oxidation index (OI) profile. In fact, when the indentation properties were plotted against OI, an approximately linear correlation was observed [7] (see Figure 33.10). That is, hardness, modulus, and EDF all increased linearly with OI. These results indicated that oxidation processes result in local structural changes (e.g., increases in crystallinity) that result in increased mechanical properties. Of course, indentation response does not characterize residual ductility in the material, so the loss of ductility with oxidation was not observed.

Both Medel et al. [18] and Roy [19] found that oxidation profiles could be observed based on modulus variation through an oxidized UHMWPE. Indeed, the general consensus is that oxidation increases mechanical properties measured by the DSI test.

Crosslinking, on the other hand, is seen to decrease the indentation behavior of UHMWPE. Studies by Gilbert [3], Roy [19], and Zhou [20] each show that mechanical properties measured by indentation are decreased with crosslinking. Park et al. [5], on the other hand, using continuous stiffness measurements, showed increases in modulus for crosslinked materials, which is contrary to the consensus view. This difference could be due to variance in the methods used.

## 33.3.3 DSI Testing: Viscoelastic Behavior

There are relatively few papers that have explored the viscoelastic response of UHMWPE to indentation. Typically, indentation tests of viscoelastic materials have included a pause in the load–unload curves at the maximum load to allow creeplike deformation to occur [21]. Researchers have typically inserted a hold time of between 10 and 100 seconds before unloading in the hopes of reducing any viscoelastic response impacting the determination of the modulus of the material. A limitation of this approach is that there is no assessment of whether the hold time is sufficient to allow complete relaxation (which is unlikely). Viscoelastic responses in UHMWPE will typically last for minutes or longer at room or body temperatures, and the simple inclusion of a dwell time of some arbitrary length is not adequate to address this issue. Also, by holding at the peak for some arbitrary time, the subsequent modulus measurement will then just be the time-dependent relaxation modulus at some location in the overall relaxation modulus curve with little knowledge as to where in that curve one is measuring. To find an appropriate hold time, a convergence analysis is required as done in nanoindentation of other materials [22].

Gilbert and Merkhan [6] explored the effects of rate of loading on the response of UHMWPE to indentation testing. They found that there were modest increases in measured modulus, hardness, and EDF with increasing frequency over the range tested. These authors also provided a methodology, based on heredity integrals, whereby the stress-strain response in indentation testing may be modeled using linear viscoelastic elements. A Maxwell model approximation was used in this publication, which is an oversimplification; however, it should be pointed out that this approach is suitable for a wide range of viscoelastic models and behaviors if the spring and dashpot elements are appropriately described.

An example of the basic approach to using a heredity integral to find expressions for the nominal stress and strain response of the surface is described in the following discussion. First, a viscoelastic model for the material of interest needs to be chosen. In this example, a standard linear model is used consisting of a spring and dashpot in series connected in parallel to another spring (see Figure 33.13). The first spring can be thought of as the difference between the unrelaxed and relaxed modulus, while the second spring is the relaxed modulus. If we assume a constant indentation rate ($dh/dt$ = constant) and that the nominal indent strain is directly related to indent depth and indenter radius ($e = kh/R$, where k is a constant), then the load response over time to a specific indenter strain history can be determined with the appropriate viscoelastic relaxation modulus function (see Figure 33.13). Figure 33.13 shows a comparison between an actual microindentation test and a curve fit to the response using the heredity integral shown. The load is just the nominal stress times the instantaneous indent area. The constant, k, can be estimated by taking the stress at the peak load and divide it by the modulus found experimentally (in this problem, the indenter is spherical, R = 38 um, and k = 10, with a 1 sec indent test 0.5 s loading and 0.5 s unloading). The parameters for the result shown is: $E_1 = 2.4$ MPa, $E_2 = 270$ MPa, and $\eta = 450$ MPa. The relaxation function (E(t)) chosen for this analysis, again, is an

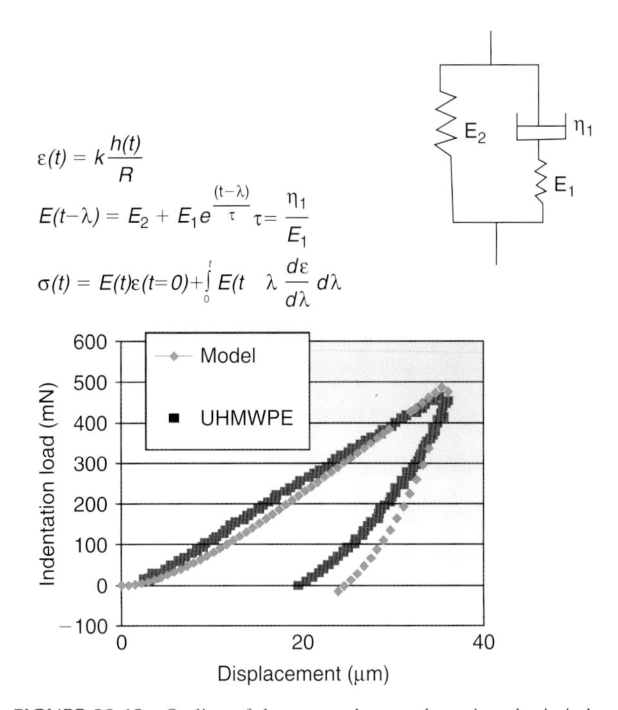


**FIGURE 33.13** Outline of the approach to explore viscoelastic indentation test results. A standard linear model approximation of indentation testing of a viscoelastic material is used in conjunction with the heredity integral and relationships between indent depth and strain, and indent load and stress, to arrive at an approximation of the indentation behavior.

oversimplification. Work is ongoing to make as realistic a relaxation function as possible that relates to the actual deformation behavior of UHMWPE, which can then be used to describe the viscoplastic deformation behavior of UHMWPE.

## 33.4 NANOSCRATCH SINGLE ASPERITY WEAR TESTS AND THEIR EFFECTS ON INDENTATION BEHAVIOR

Recent efforts in the area of nanotribology and nanowear of UHMWPE have included using AFM-based tips to place single-asperity wear scratches into surfaces and to observe the microstructural changes that result. Schmid et al. [23] investigated plowing of AFM tips across UHMWPE. They noted that with cross-plowing, the surface polymer could first be deformed and piled up and then pulled into a fibril. This, they proposed, shows a cross-shear mechanism for particulate generation in UHMWPE. This research was verified by Wong [24], who noted a wall formation created by orthogonally intersecting scratches in AFM nanoscratching, a possible initiation site for particulate debris. Ho et al. applied nanoscale asperties of varying contact pressure to determine changes in dynamic coefficients of friction [25]. They noted that at high contact pressures (greater than 190 MPa), transients in friction exist and may contribute to the mechanism of particle generation. With sufficiently high contact pressure, plowing of the surface was observed, a wear deformation mechanism frequently noted in retrievals.

The authors are currently working on combining indentation testing, quantitative surface deformation measurements, and characterization techniques for surface structural changes due to wearlike deformation to further our understanding of the interrelationships between local structure, wear deformation, and local mechanical properties. In fact, it is likely that the process of asperity-based wear will result in a feedback process whereby surface deformation will depend on structure and properties but also that the deformation process itself will alter these structures and properties and these changes will affect the subsequent wear behavior. Feedback interactions between wear, structure, and properties are not well understood for UHMWPE; however, these relationships will likely be important if we are to gain a clear understanding of how UHMWPE wears.

Recent work by the authors has sought to explore the details of this complex micron-scale wear behavior [26]. In these studies, model single-asperity wear events are imparted to prepared UHMWPE surfaces under controlled loads (stresses), scratch length, and asperity geometry. The surface and near surface strain states are then quantified by quantitative image analysis of the displacement field of the surface based on the movement of grid marks scribed on the surface. When the near surface strain is quantified, both the local structural reorganization (crystallite orientation) and the changes in local mechanical properties are measured with indentation testing. This combination of local indentation testing and quantitative strain analysis of the wear process provides a clear and detailed relationship of how structure and properties affect wear and how wear affects structure and properties. The effects of surface deformation, resulting from single asperity wearlike deformation, on material properties can then be explored with indentation testing.

To perform either nanoscale single asperity wear strain analysis or micron-scale analysis, a freshly microtomed UHMWPE surface is placed into an AFM (Nanoscope IIIa, Digital Instruments, Santa Barbara, California, USA) and a diamond tip for nanoindentation testing is used to place an X–Y grid of scratches into the surface (1 μm or 20 μm intervals for nano- and microscale testing, respectively). The depth of these grids ranges from less than 100 nm for nano- to approximately 2 μm for microtesting. Afterward, a single asperity of known geometry (e.g., the diamond tip for nanowear, and a hemispherical tip for larger wear measurements) is brought into contact with the surface, and a known load is applied while it is scratched across the surface. By tracing the deformation of the grid due to the scratch, a plot of displacement field (i.e., the vectors that describe the motion from the initial grid location to the final grid location) can be obtained and plotted as a function of location. With the appropriate coordinate system, oriented to the scratch direction and perpendicular to

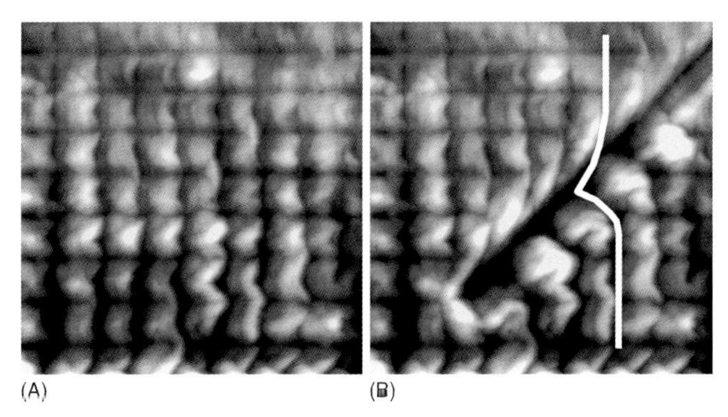


**FIGURE 33.14** (A) Atomic force microscope images of an 8 μm × 8 μm region of UHMWPE in which a 1 μm grid pattern was scribed into the surface. (B) The same identical region after scratching the nanoindenter into the UHMWPE (Hylamer) with a much higher load. The white line overlay roughly approximates the deformation field for the surface strain. The deformation displacement field across the scratch can be measured and used to find the strain field (from the gradient of the displacement field) associated with the asperity-surface wear interaction.

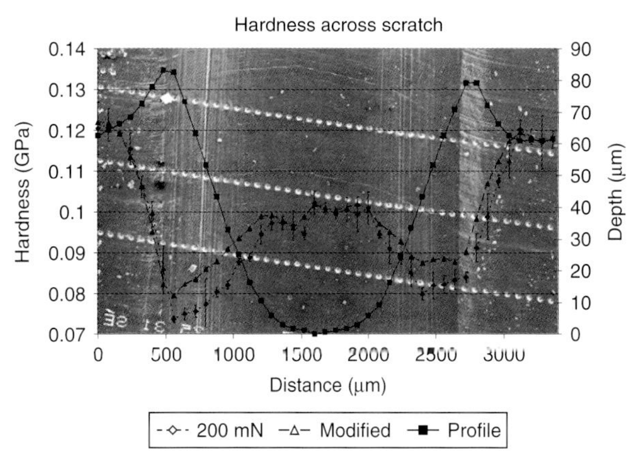


**FIGURE 33.15** Composite image of a scanning electron micrograph of a larger single asperity wear, indent marks from indentation testing across the wear track, and profiles of in-plane (solid white line) and out-of plane (dashed white line) deformation fields. Also shown are the uncorrected and corrected hardness profiles. Note the significant drop in hardness with location in the asperity wear track. Corrections for nonvertical surface testing was performed on the hardness results. Hardness varies linearly with changes in overall strain state.

it (Y and X, respectively), the slope of the displacement field with location can be found and the surface strains determined [26].

The strains into the depth of the sample can also be determined with the appropriate sectioning of the sample and performing a similar set of steps on the vertically oriented face. Thus, local strain states resulting from known asperity geometries and loads can be determined.

This method of wear asperity deformation measurement can be combined with other measurements to more fully describe local interactions. For example, DSI testing can be performed over the span of the asperity wear track, and the local mechanical properties can be directly correlated with the measured strain. Finally, surface structural characterization can be performed by a combination of methods from etching, FTIR, and other techniques to relate surface strains with reorientation of the crystals, for example.

An example of a nanoscratch surface strain experiment is shown in Figure 33.14. This is a Hylamer surface before and after the scratch. A 1 μm-spaced square grid was scribed into the surface with low loads (Figure 33.14A) and then a high load scratch was imparted (Figure 33.14B). The displacement of the grid can be used to find the surface strain as a function of location. Also, at this size scale, the interaction of microstructural elements (e.g., crystals) with the scratch deformation can be visualized. With Hylamer, a large crystal UHMWPE described in Chapter 19, the surface polymer becomes very drawn out, much more than either standard GUR 1050 or crosslinked UHMWPE [26].

This approach can also be applied to larger asperity-surface interactions where more in-depth correlation between surface wear events, imparted strain, and changes in indentation properties can be directly correlated. In these experiments, a larger spherical asperity (radius of about 2 mm) is dragged across a microtomed UHMWPE surface into which linear markings are present to track the surface deformation. The displacement of these lines across the scratch are then used to find the strain. With larger wear tracks, it is then possible to go into the deformed region with microindentation testing and investigate the properties of the surface as a function of the strain state of the surface. Modulus, hardness, and energy dissipation factor can all be determined across the scratch and related to the surface strain present. An example of the composite results from such an experiment is shown in Figure 33.15, an SEM micrograph with a single asperity wear track running about vertically in the image. Also present are the residual indents from three indentation profile experiments across the length of the indent. Overlaid on this micrograph are white lines that represent the in-plane deformation and the depth profile due to the wear asperity (white lines) and a graph of the hardness measurements (both uncorrected and corrected for apparent softening due to loading of a sloped surface) resulting from the indentation testing. From both the in-plane and normal-to-the-plane displacement profiles (see Figure 33.16A), the strains ($\varepsilon_{xy}$ and $\varepsilon_{xz}$, Figure 33.16B) can be found using the Equation 7. It is important to note that the hardness of the material decreases significantly (upwards of 30% decrease) with increasing surface strain. This is true as well with the other materials measured by indentation testing. Thus, wear softening of the surface material takes place in a strain-dependent way [10].

$$\mathbf{U}(x,y) = U_x(x,y,z)\mathbf{i} + U_y(x,y,z)\mathbf{j} + U_z(x,y,z)\mathbf{k}$$

$$\varepsilon_{ij} = \frac{1}{2}\left(\frac{\partial U_i}{\partial x_j} + \frac{\partial U_j}{\partial x_i}\right) \quad (7)$$

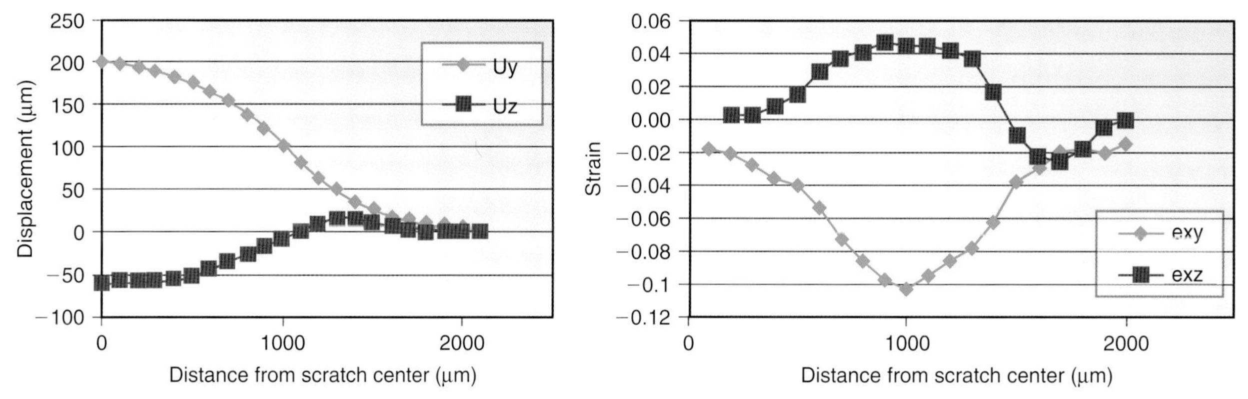


**FIGURE 33.16** (A) In-plane and out-of plane displacement profiles for the single asperity wear scratch in Figure 33.9. (B) The strains that result from the gradients of the displacement fields.

It can be seen that the peak in the wear-based strain arises partway between the outer contact point and the center of the asperity wear track (i.e., where the gradient in the displacement field is a maximum). The surface strain arising from the asperity-wear interaction results in a decrease in the hardness (see Figure 33.15) as well as the modulus and EDF (data not shown). That is, there is a strain softening process whereby the mechanical properties of modulus, hardness, and EDF decrease substantially (upwards of 35% in some UHMWPEs) as a function of the strain imparted by the asperity.

This combination of single asperity wear testing, surface deformation mapping of the resultant deformation field, and the subsequent indentation test measurements of the changes in mechanical properties due to the wear process will provide detailed and significant information of the interrelationships between wear, structure, and property evolutions. These observations may also help discern the fundamental deformation mechanisms that result in wear particle formation and how different UHMWPE starting materials evolve with wear deformation.

## 33.5 SUMMARY AND CONCLUSION

Depth-sensing indentation testing of UHMWPE has provided important information concerning surface mechanical properties and the effects of processing and oxidation. These techniques, when combined with high-magnification imaging using AFM, SEM, and FTIR can now provide detailed relationships between structure, properties, and wear performance, as well as the feedback interactions between these. Wear has been shown to induce local surface property softening, and factors such as crosslinking, oxidation, and crystal size variations have all been shown to influence surface mechanical properties.

DSI methods are versatile and capable of helping to model viscoelastic and viscoplastic behavior. These techniques will continue to provide important new information concerning the fundamental surface behavior of UHMWPE.

## 33.6 ACKNOWLEDGMENTS

Elements of this work have been financially supported by DePuy Orthopedics, Inc.

## REFERENCES

1. Oliver WC, Pharr GM. An improved technique for determining hardness and elastic modulus using load and displacement sensing indentation experiments. *J Mater Res* 1992;**7**(6):1564–83.
2. Sneddon I. The relationship between load and penetration in the axisymmetric Boussinesque problem for a punch of arbitrary profile. *Int J Eng Sci* 1965;**4**(3):47–57.
3. Gilbert JL, Cumber J, Butterfield A. Surface micromechanics of ultrahigh molecular weight polyethylene: microindentation testing, crosslinking, and material behavior. *J Biomed Mater Res* 2002;**61**(2):270–81.
4. Higgins JE, et al. *Nanoindentation Results from Direct Molded vs. Machined UHMWPE Tibial Bearings*. United States; 1999.
5. Park K, et al. Quasi-static and dynamic nanoindentation studies on highly crosslinked ultra-high-molecular-weight polyethylene. *Biomaterials* 2003;**25**(12):2427–36.
6. Gilbert JL, Merkhan I. Rate effects on the microindentation-based mechanical properties of oxidized, crosslinked, and highly crystalline ultrahigh-molecular-weight polyethylene. *J Biomed Mater Res* 2004;**71A**(3):549–58.
7. Wernle JD, Gilbert JL. Micromechanics of shelf-aged and retrieved UHMWPE tibial inserts: indentation testing, oxidative profiling, and thickness effects. *J Biomed Mater Res, Part B: Appl Biomater* 2005;**75B**(1):113–21.
8. Rosen SL. *Fundamental principles of polymeric materials*. NY: J. Wiley; 1993.
9. Ward IM. *Mechanical properties of solid polymers*: John Wiley and sons; 1983; 492.
10. Wernle J, Gilbert JL. Three dimensional strain mapping of single asperity wear in UHMWPE: effects of load and material on surface mechanical properties. *Biomaterials* 2008. In Press.
11. Ho SP, et al. Nanoindentation properties of compression-moulded ultra-high molecular weight polyethylene. *Proc Inst Mech Eng [H]* 2003;**217**(5):357–66.
12. Ho SP, et al. Effects of the sample preparation temperature on the nanostructure of compression molded ultrahigh molecular weight polyethylene. *Proc Inst Mech Eng [H]* 2002;**216**(H2):123–33.

13. Olley RH, Bassett DC. An improved permanganic etchant for polyolefins. *Polymer* 1982;**23**(12):1707–10.
14. Hobbs JK, Miles MJ. Direct observation of polyethylene Shish-Kebab crystallization using in-situ atomic force microscopy. *Macromolecules* 2001;**34**(3):353–5.
15. Matsuba G, et al. Structural analysis of shish-kebab with X-ray and neutron scattering measurements. *PMSE Prepr* 2005;**93**:173–4.
16. Hsiao BS, et al. Unexpected Shish-Kebab structure in a sheared polyethylene melt. *Phys Rev Lett* 2005;**94**(11):117802/1–117802/4/.
17. Ania F, et al. Comparative study of size and distribution of lamellar thicknesses and long periods in polyethylene with a shish-kebab structure. *J Mater Sci* 1996;**31**(16):4199–206.
18. Medel F, Rimnac C, Kurtz S. On the assessment and microstructure changes after in vivo degradation of historical UHMWPE knee components by means of vibrational spectroscopies and nanoindentation. *J Biomed Mater Res* 2008. in press.
19. Roy M. Influence of cross linking and oxidation on the microstructural mechanical properties of UHMWPE. *Mat. Res. Soc Symp Proc* 2005;**847**:1541–6.
20. Zhou J, et al. Tribological and nanomechanical properties of unmodified and crosslinked ultra high molecular weight polyethylene for total joint replacements. *Trans. ASME* 2004;**126**:386–94.
21. Klapperich C, Komvopoulos K, Pruitt L. Nanomechanical properties of polymers determined from nanoindentation experiments. *J Tribol* 2001;**123**(3):624–31.
22. Bushby AJ, Ferguson VL, Boyde A. Nanoindentation of bone: comparison of specimens tested in liquid and embedded in polymethylmethacrylate. *J Mater. Res.* 2004;**19**(1):249–59.
23. Schmid S, et al. Single asperity plowing of metallic and polymeric surfaces in an atomic force microscope: an overview of recent developments. *Mat. Res. Soc. Proc.* 1998;**522**:391–7.
24. Wong BKP, et al. Nano-wear mechanism for ultra-high molecular weight polyethylene (UHMWPE) sliding against a model hard asperity. *Tribol Lett* 2004;**17**(3):613–22.
25. Ho SP, et al. Nanotribology of CoCr-UHMWPE TJR prosthesis using atomic force microscopy. *Wear* 2002;**253**(11–12):1145–55.
26. Wernle J, Gilbert JL. Microscale and nanoscale surface strain mapping of single asperity wear in ultra high molecular weight polyethylene: effects of materials, load, and asperity geometry. *J Biomed Mater Res* 2008. In Press.

Chapter 34

# MicroCT Analysis of Wear and Damage in UHMWPE

Dan MacDonald, PhD Candidate, Anton Bowden, PhD and Steven M. Kurtz, PhD



## 34.1 INTRODUCTION

UHMWPE wear continues to be an important factor in the long-term survivability of orthopedic implants [1–3]. As established in Chapter 23, periprosthetic osteolysis occurs secondary to the generation of UHMWPE wear debris in total hip arthroplasty (THA) [1, 2, 4]. In the last decade, efforts have been made to increase the wear resistance of UHMWPE [1, 5, 6]. Highly crosslinked UHMWPE (Chapters 13 and 14) was introduced in THA 10 years ago and more recently in total knee arthroplasty. UHMWPE wear volume and wear patterns must be fully evaluated to understand historic UHMWPE wear processes; to evaluate the efficacy of highly crosslinked UHMWPE; and to analyze future UHMWPE formulations that attempt to improve the clinical performance of UHMWPE.

There are currently many methods to evaluate UHMWPE wear, including direct [7, 8], gravimetric [9, 10], radiographic [11–13], optical [14], fluid displacement [15], and more recently, microcomputed tomography (microCT) based [16, 17]. All of these techniques have their own unique advantages and disadvantages. The reader is referred to Chapter 28 for a comprehensive overview of radiographic methods. Optical, fluid-displacement, and gravimetric methods have been shown to provide accurate results; however, they are generally time consuming and have limited value in that they do not identify the regions of the implant that have lost material [9, 10, 14, 15].

MicroCT methods provide reliable, accurate, and relatively fast assessment of wear and damage of UHMWPE components [16, 17]. Like all current methods of assessing wear, a drawback of these methods is that actual wear cannot be easily distinguished from creep. For that reason, the term penetration is used throughout this chapter and refers to the cumulative effect of material removal (true wear) and creep deformation. Some of the microCT techniques discussed require *a priori* knowledge of the nominal implant geometry, while others use measured geometric parameters obtained from unworn portions of the components. The purpose of this chapter is to describe the advantages and drawbacks of microCT methods for evaluating wear and damage in UHMWPE.

## 34.2 MICROCT SCANNING

MicroCT is an established technique that has impacted many different fields, including medicine, geology, engineering, and manufacturing. This powerful technology uses X-rays to nondestructively generate three-dimensional geometries and can provide information on the internal features of an object.

A microCT apparatus consists of three major components: a microfocus X-ray source, a calibrated X-ray detector, and a dedicated computer for data acquisition and processing. Tomography refers to a mathematical technique that reconstructs cross-sectional images of a solid based on observing the attenuation of source signals (in this case X-rays) originating in many different directions [18]. The mathematical technique is referred to as filtered backprojection and represents the explicit numerical inverse Radon transform. With enough stacked, consecutive images, the entire volume of an object can be digitally described both internally and externally in three dimensions.

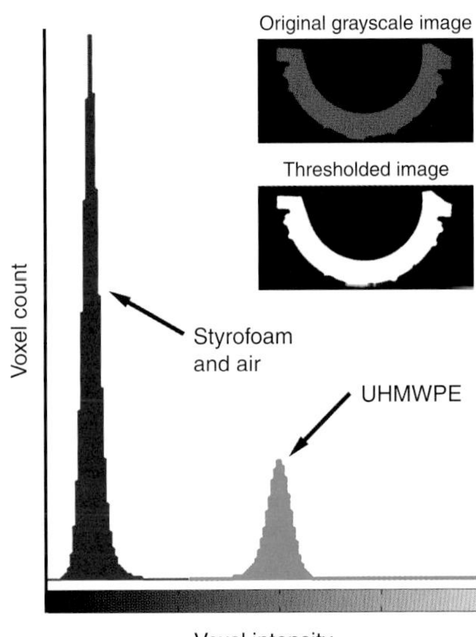


**FIGURE 34.1** Typical histogram of an UHMWPE component. The red peak on the left represents the Styrofoam and air, while the green peak on the right represents the UHMWPE component. A grayscale image of one slice of a hip component is in the upper right. Directly below is the thresholded image of the same hip component.

### 34.2.1 Practical Considerations in Scanning

One of the major considerations for any microCT scan is resolution versus file size. Although higher resolution is highly desirable for enhanced accuracy, as the resolution is increased, the file size can increase quickly. At the present time, it is common for a single microCT scan to require 2 or more gigabytes of storage space. A typical active research scanning site can quickly fill multiple terabytes of storage space. Fortunately, as resolution capabilities continue to increase, costs for computer storage continue to decrease.

### 34.2.2 Motion Artifacts During Scanning

When scanning UHMWPE, it is important to consider artifacts initiated by any movement of the component. As the microCT scans components, the specimen holder rotates throughout the course of the scan. This movement may be sufficient to allow portions of a loose, scanned component to "shift," thereby making it impossible to analyze using the techniques described throughout this chapter. To eliminate this artifact, components should be packed firmly in a specimen holder using Styrofoam. The very low density of Styrofoam translates into a much lower X-ray attenuation coefficient (and correspondingly lower voxel intensity) for the material. The packing materials can thus be easily separated from the UHMWPE components in the microCT images, yet provide sufficient constraint to hold the components firmly throughout the scanning process.

### 34.2.3 Image Segmentation

In the techniques that follow, a fundamental procedure in each technique is object segmentation. This is the process of identifying the regions of the image set that represent the actual device. For an UHMWPE component that is scanned with sufficient image quality (i.e., sufficient contrast and signal to noise), segmentation can be achieved using a simple threshold gray value. In this procedure, voxels above, or brighter than, a certain threshold value are assigned a value of 1, and voxels below that threshold are assigned a value of 0. The threshold value is easily identifiable for UHMWPE as the minimum between two distinct peaks in the image grayscale histogram (Figure 34.1). The peak to the left represents the air and packing foam, while the peak on the right represents the UHMWPE component. In Figure 34.1, the grayscale image of one slice of hip component is shown in the upper right, while the thresholded image is directly beneath it. By segmenting the dataset, it is possible to isolate and measure the volume of the UHMWPE.

## 34.3 EVALUATION OF PENETRATION IN THA USING GEOMETRIC PRIMITIVES

The unworn bearing surfaces of most modern hip arthroplasty implants are geometric primitives (usually hemispheres). Using this knowledge, manual, rigid, three-dimensional registration techniques can be utilized to isolate and measure the penetration volume. This method was first described by Bowden et al. [16]. The overall procedure is to manually align computer-generated geometric primitives corresponding to the nominal size of the component in the three-dimensional image space of the microCT from the retrieved component. Unfilled voxels between the primitives and the component correspond to component penetration and can be examined not only to identify

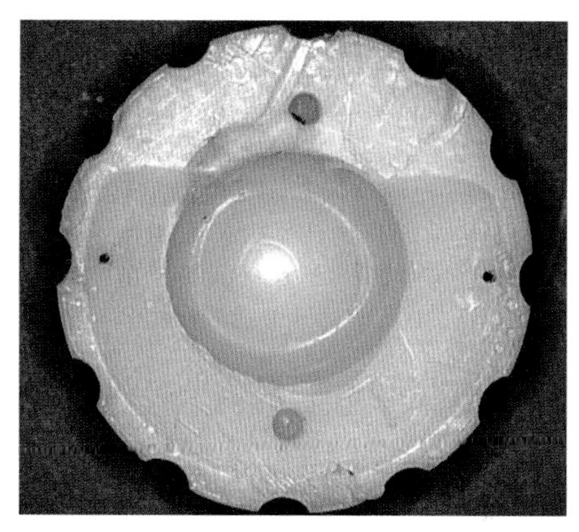
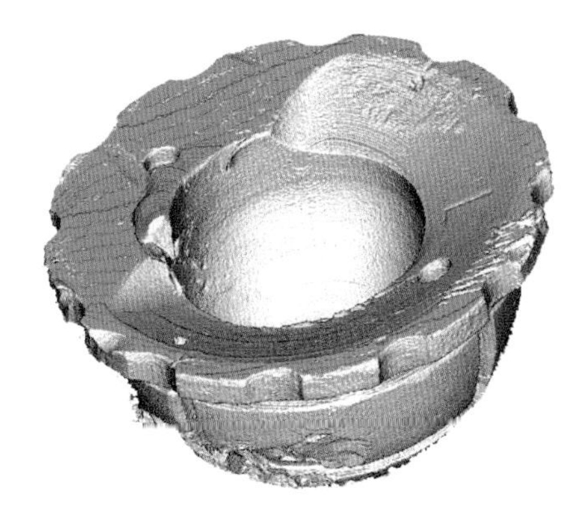

FIGURE 34.2 Stereomicrophotograph (left) and 3-D microCT reconstruction (right) of a total hip replacement that was implanted for 29 years.

penetration volumes but also to quantitatively measure the magnitude of the scalar penetration field across the bearing surface of the implant.

An example of scanned hip component is seen in Figure 34.2. This component was scanned using a high-resolution desktop, cone-beam scanner (Scanco μCT 80, Switzerland). The scanner energy was 45 kVp (177 μA), and the component was scanned at a uniform 74 μm resolution (16-bit precision) using a 1024 × 1024 in plane image matrix.

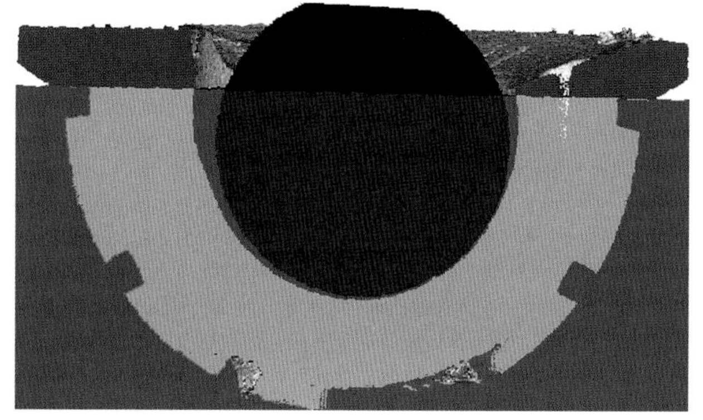

FIGURE 34.3 Registration of geometric primitives. After registration of the primitives (spherical head and rectilinear box), the penetration can be easily isolated and separated from the rest of the dataset by thresholding.

## 34.3.1 Three-Dimensional Alignment

To isolate, visualize, and measure the UHMWPE penetration from the microCT data, two manual alignment steps are used. A spherical binarized image of the appropriate nominal size is aligned inside the microCT image data of the UHMWPE liner cavity using manual rigid image registration techniques. The alignment is performed by aligning the surface of the sphere with the unworn portions of the acetabular component. Nominal size can be determined by reviewing operative reports, from direct measurement of the retrieved femoral head or from measurement of unworn portions of the implant.

After the nominal geometry has been aligned, a second image alignment is used to align a rectilinear, binarized image block with the entire interior cup region. The purpose of this block is to establish a distinct grayscale value for the penetration region of the acetabular cup. The cumulative image consisting of the original microCT image data with the aligned geometric primitives can then be quantitatively examined. An example of a properly registered hip cup can be seen in Figure 34.3. For this example, image registration was performed using commercial image analysis software (Analyze, AnalyzeDirect, Rochester, Minnesota, USA).

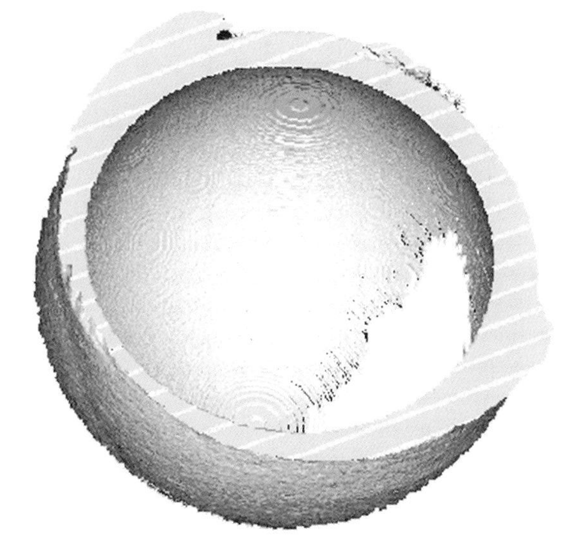

FIGURE 34.4 Penetration volume of an acetabular hip component that was implanted for 29 years.

## 34.3.2 Volume Measurement

Following registration, the penetration volume can be identified and isolated by thresholding the cumulative image dataset. The volume is then calculated by totaling the number of voxels associated with the particular histogram peak that corresponds to the penetration volume. An example of the penetration volume can be seen in Figure 34.4. By dividing the magnitude of the penetration volume by the implantation time, it is possible to estimate an average penetration rate.

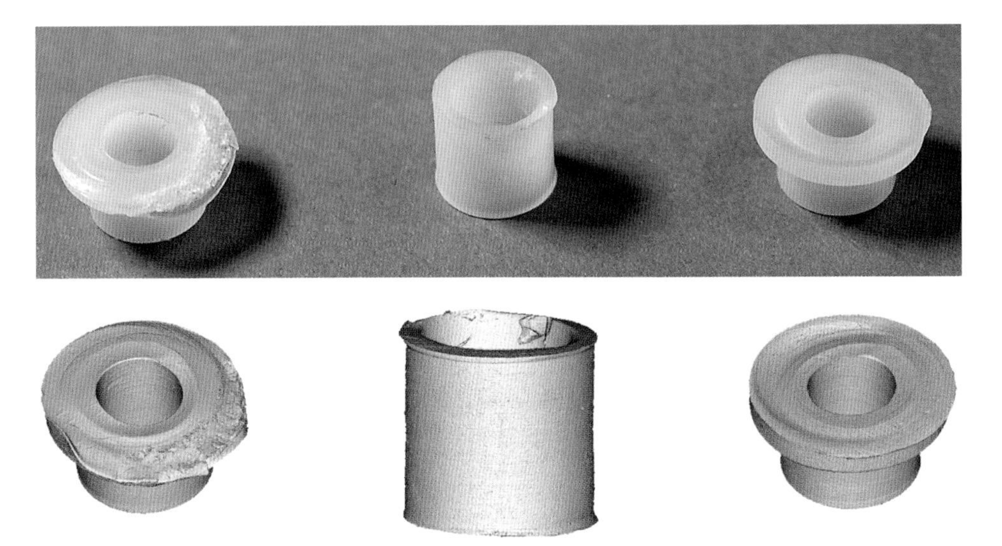

FIGURE 34.5 Photographs and microCT renderings of the three main bearing components in total elbow arthroplasty.

### 34.3.3 Uncertainty Analysis

The uncertainty associated from this technique arises from three different sources: scanner-associated uncertainty (systematic accuracy), interobserver uncertainty, and intraobserver uncertainty. Scanner uncertainty was estimated by comparing gravimetric measurements of wear simulator control specimens with those obtained using the outlined technique and resulted in an absolute wear volume variation of less than 0.6% [16]. The observer uncertainty is a result of manual registration and segmentation. Observer uncertainty was assessed using the techniques outlined in ASTM E691 (Standard Practice for Conducting an Interlaboratory Study to Determine the Precision of a Test Method). Repeatability ($s_r$) is a measure of the uncertainty associated with measurements taken by a single observer. Reproducibility ($s_R$) is a measure of the uncertainty associated with measurements taken by multiple observers. It was found that the observer uncertainty was higher in components that had very small amounts of penetration ($s_r = 48.4\%$ and $s_R = 53.5\%$) as compared to components with greater penetration values ($s_r < 10\%$ and $s_R < 12\%$) [16]. Reduction in uncertainty can be obtained with increased voxel resolution. Nevertheless, the repeatability and reproducibility of the technique for components with large amounts of penetration compared favorably with radiographic techniques ($s_r = 4-17\%$ and $s_R = 0-20\%$) [19–21] and other techniques such as fluid displacement ($s_r < 10\%$ and $s_R < 10\%$) [22, 23].

## 34.4 EVALUATION OF PENETRATION IN NONREGULARLY SHAPED COMPONENTS

Although the method described in Section 34.3 is very useful in determining the amount of penetration in total hip replacement, the technique is not readily transferred to components that are not defined by geometric primitives. Total elbow replacement (Figure 34.5) and total disc replacement (Figure 34.6) components are examples. Additionally, in some components such as those used in total elbow replacement, a significant portion of the penetration occurs on the inner surface, making it difficult or impossible to use direct measurement techniques (Figure 34.7).

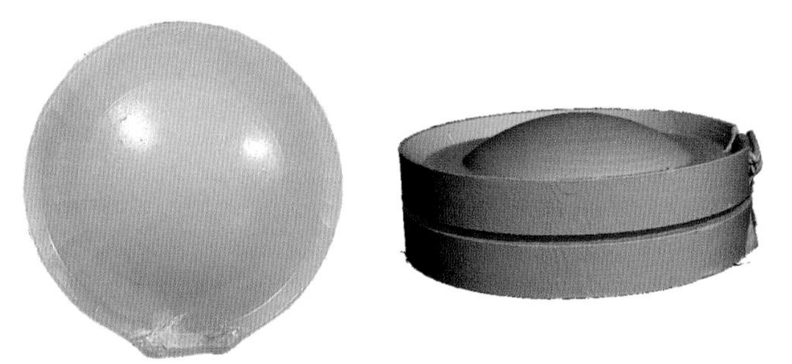

FIGURE 34.6 Photograph and microCT rendering of a mobile bearing total disc replacement. This component was implanted for 11 years.

FIGURE 34.7 MicroCT renderings of the ulnar component in total elbow replacement. Note that penetration is observed on the interior surface of the component, making it difficult to accurately measure penetration using other techniques. The component on the left demonstrates clear pitting that could not be visualized nondestructively otherwise.

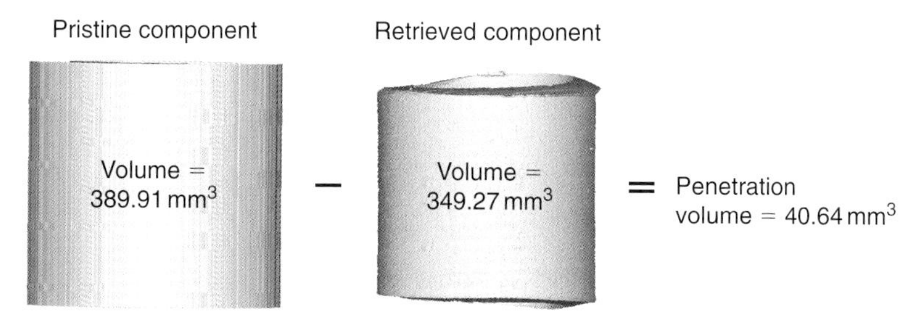


FIGURE 34.8 Equation used for calculating the penetration of elbow UHMWPE components. The retrieved component volume is simply subtracted from the undamaged component volume.

Therefore, newer techniques have been developed that utilize knowledge of the original component geometry as obtained from manufacturers' drawings or from microCT scans of unimplanted, pristine components.

## 34.4.1 Volumetric Penetration in Total Elbow Replacement

This technique requires that the user have access to the manufacturer's drawings or exemplar specimens of the components that are to be analyzed. Using a commercial computer aided design (CAD) program (AutoCAD 2004, San Rafael, California, USA), three-dimensional models were rendered according to the appropriate drawings. Within the CAD program, the volume of the component can be calculated from the rendered model.

A retrieved component was scanned in the microCT using the same scanner and energy settings as for the hip component mentioned previously but with a 20 μm uniform resolution (Figure 34.4). The three-dimensional dataset is subsequently thresholded. The volume of the component can be calculated by summing the connected voxels of the appropriate histogram peak. The penetration volume can then be directly calculated by subtracting the retrieved volume from the pristine volume as obtained from CAD software (Figure 34.8). To obtain a volumetric penetration rate, the magnitude of the penetration volume is simply divided by the implantation time.

## 34.4.2 Spatial Visualization of Penetration in Total Elbow Replacement Using Manual Registration

While the methodology described in Section 34.4.1 is useful for quantifying the volume of penetration seen in retrieved components, it does not provide insight into the patterns of penetration displayed in the retrievals. Analysis of retrieval penetration patterns has shown that implants with similar quantities of penetration can result from significantly different loading conditions. Certain retrieved penetration patterns can be indicative of design or implantation

FIGURE 34.9 An example of the registration of a retrieved humeral bushing and the nominal surface. The green area represents the area of the retrieved component that was damaged due to penetration (creep and wear), while the red area represents outward plastic deformation. The yellow area represents regions in common between the components.

factors that influence the long-term performance of UHMWPE components. Spatial visualization techniques have been developed to separate regions of component penetration (via wear or creep) from regions that have been plastically deformed outwardly from the space of the pristine component.

The procedure requires a baseline exemplar geometry, which can be obtained from solid modeling CAD software or, alternatively, from scanning unimplanted exemplars using the same protocol as for the damaged components. Both the exemplar and retrieved components are isolated using the same threshold value. After scanning and thresholding, the images of the components are then registered with each other using manual or semiautomated techniques. Typically, the backside penetration of the retrieved implants is substantially less than that observed on the primary bearing surfaces. This is often evidenced by the presence of machining marks from the original manufacturing process. Therefore, these surfaces are preferentially used as a reference to properly align the worn and exemplar components. An example demonstrating this technique is shown in Figure 34.9. The green area represents the penetration of the retrieved component into the bearing space, while the red area represents outward deformation of the damaged component from the pristine condition. The yellow area shows the regions in common between the registered pristine and damaged components. The quality of the registration can be quantified using standard image registration metrics, such as the sum of the squared difference or a mutual information metric. However, because these metrics represent an overall comparison of image alignment, rather than a comparison of unworn portions of components, registration quality metrics should be interpreted with caution.

### 34.4.3 Quantitative, Spatial Visualization of Penetration in Total Disc Replacements Using Automated Registration

Despite the fact that the method described in Section 34.4.2 is useful for determining the areas of the implant that undergo penetration or outward deformation, it does not provide information about the magnitude of penetration. Additionally, because the registration is performed manually, there is an element of user-related uncertainty associated with the procedure. To address these issues, an algorithm has been developed by Shkolnikov, Bowden, et al. to automatically align retrieved components with surface models and to display a quantitative penetration map for the entire component [17].

To assess penetration, retrieved components are scanned in the microCT at high resolutions. Three-dimensional models of the nominal surface geometry are needed and may be obtained from solid modeling CAD software or, alternatively, from microCT scans of pristine, unimplanted components. The accuracy of the technique is governed in part by the level of discretization of the nominal surface geometry, so it is important that the surface models are highly discretized (typically 1 to 2 million triangles).

The automatic registration algorithm consists of three distinct phases: initial alignment, position optimization, and penetration depth calculation. The first phase initially aligns the volumetric microCT data from the retrieved component with the discretized surface geometry. A preliminary alignment is obtained by aligning the principle axes and center of mass for the nominal surface with that of the microCT dataset associated with the retrieved component. Using an iterative optimization technique, the contact area between the nominal surface and the surface of the retrieved component is maximized. In cases of severe penetration or iatrogenic damage, the initial alignment can be corrected using a built-in graphical user interface (GUI).

In the second phase, position optimization and penetration is calculated for the retrieved implant at each triangle of the highly discretized nominal geometry. Linear penetration is calculated by moving along the local normal of each triangular surface into and out of the nominal surface. For the purposes of optimal alignment, this distance is limited to much less than the size of the implant (0.5% of the longest implant dimension). This is done to prevent large areas of damage or penetration from biasing the automatic alignment. When the normal surface intersects a component voxel, the signed distance from the surface is defined as

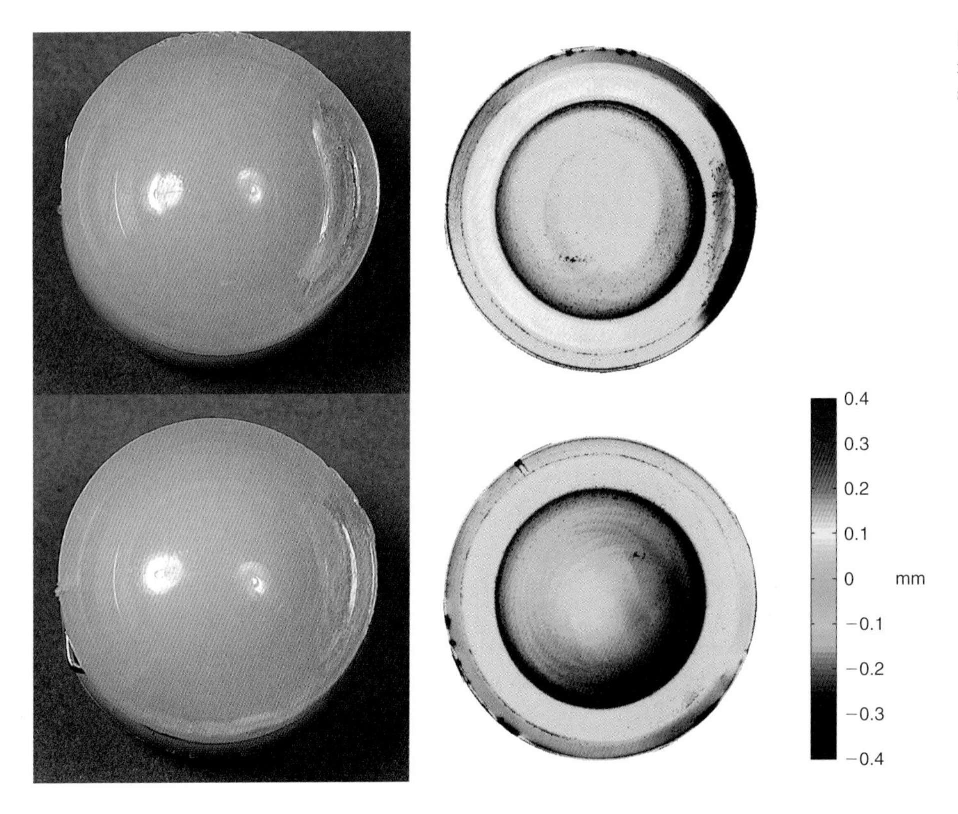


**FIGURE 34.10** Photographs and corresponding penetration maps of both sides of a mobile bearing TDR.

penetration. The composite penetration value for a particular implant orientation/position pair is calculated by summing the penetrations for each triangle of the surface. A Nelder-Mead simplex optimization algorithm was used to obtain the orientation and position of the nominal implant surface that minimizes the composite penetration value.

During the final phase of the technique, the linear penetration value is once again calculated at each individual triangle as described for the optimization procedure, except that the tracing distance is not limited to a small size. The penetration at each triangle is then displayed as a color map on the highly discretized nominal surface (Figure 34.10). The magnitude of component penetration is easily obtained from the composite penetration function.

## 34.5 ASSESSING SUBSURFACE CRACKING USING MICROCT

As shown in Chapter 22, UHMWPE components can undergo oxidative degradation both on the shelf and *in vivo*. One of the consequences of severe UHMWPE oxidation is subsurface cracking that ultimately leads to delamination [24].

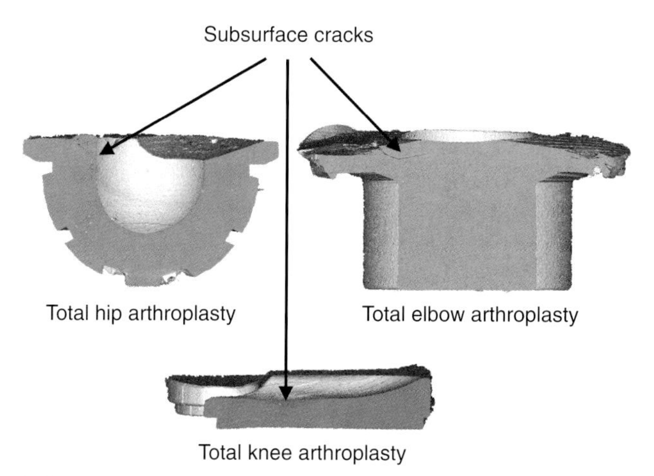


**FIGURE 34.11** Examples of subsurface cracking in total hip arthroplasty, total elbow arthroplasty, and total knee replacement.

This damage mode is prevalent in historic total knee replacements; however, it also occurs in contemporary UHMWPE components that have been sterilized in an inert environment. Currier et al. have hypothesized that severe oxidation can lead to delamination of the rim of highly crosslinked UHMWPE acetabular liners [25]. Therefore, it is important to be able to investigate the subsurface cracking and delamination that can occur in UHMWPE.

MicroCT has a distinct advantage over many other wear analysis techniques in that it is possible to analyze damage to the internal portions of the UHMWPE. The same 3-D microCT datasets that were used to analyze penetration can be used to examine subsurface cracks using an image-processing program (e.g., Analyze, AnalyzeDirect, Rochester, Minnesota, USA). Users can investigate each individual slice of data for the presence or absence of subsurface cracking. Additionally, crack length and width can be measured if desired. Examples of subsurface cracking can be seen in Figure 34.11.

## 34.6 USING MICROCT TO VISUALIZE THIRD-BODY WEAR

Total joint replacements may exhibit third-body wear because bone, bone cement, ceramic, or metal particles may be present in the joint space. Metallic particles can create beam scattering artifacts in microCT image datasets and therefore make it difficult to properly analyze the UHMWPE components. However, bone cement particles and ceramic debris do not exhibit these artifacts and can be readily visualized in the microCT scans.

UHMWPE has a distinct difference in X-ray attenuation as compared to bone cement and ceramics, allowing microCT image datasets to be thresholded to differentiate third-body particles from the UHMWPE component (Figure 34.12). Using the same tools that were used to calculate the penetration in earlier sections, the volume of third-body particles can be calculated within an image processing software package.

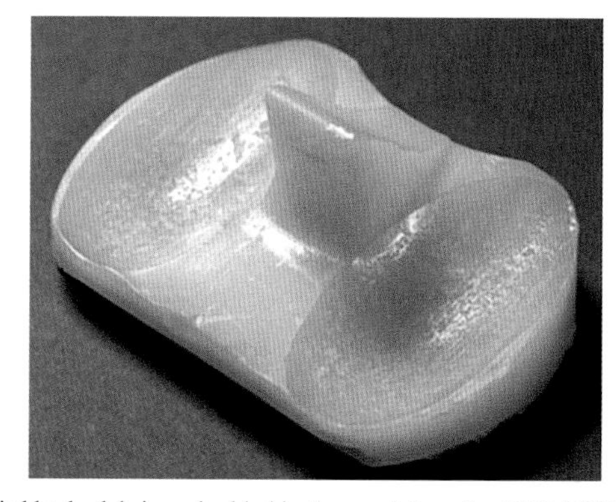
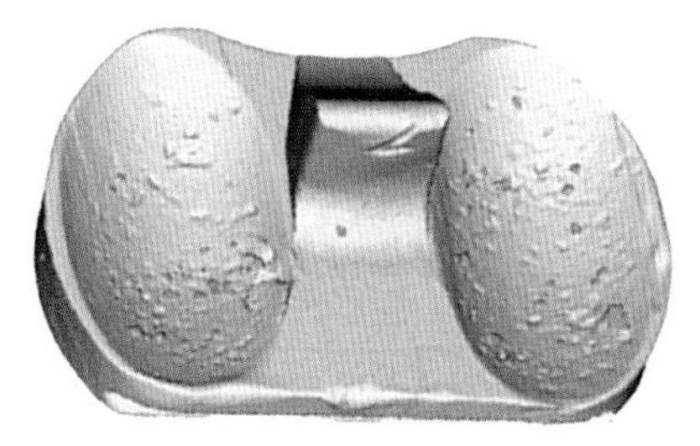

**FIGURE 34.12** Third body debris embedded in the condyles of an UHMWPE component. In the microCT dataset, the third body particles are in green.

## 34.7 CONCLUSION

MicroCT imaging technology provides a way to accurately assess the wear performance of retrieved UHMWPE components. Unlike other radiographic methods, all microCT methods are *ex vivo* in nature, limiting their use in a clinical environment. However, because microCT methods provide three-dimensional quantification of the wear and damage patterns of UHMWPE components, they are extremely valuable in increasing our understanding of the relationships between UHMWPE formulation, loading conditions, mechanical design, and long-term performance.

## 34.8 ACKNOWLEDGMENTS

Special thanks are due to Judd Day, PhD, Exponent, Inc., who developed the techniques used in total elbow replacement and provided editorial assistance with this chapter. The development of these techniques was supported in part by National Institutes of Health (NIAMS) NIH R01 AR47904.

## REFERENCES

1. Dumbleton JH, Manley MT, Edidin AA. A literature review of the association between wear rate and osteolysis in total hip arthroplasty. *J Arthroplasty* 2002 August;**17**(5):649–61.
2. Harris WH. The problem is osteolysis. *Clin Orthop Relat Res* 1995 Febuary(311):46–53.
3. Maloney WJ, Smith RL. Periprosthetic osteolysis in total hip arthroplasty: the role of particulate wear debris. *Instr Course Lect* 1996;**45**:171–82.
4. Willert HG, Bertram H, Buchhorn GH. Osteolysis in alloarthroplasty of the hip. The role of ultra-high molecular weight polyethylene wear particles. *Clin Orthop Relat Res* 1990 September(258):95–107.
5. Manning DW, Chiang PP, Martell JM, Galante JO, Harris WH. *In vivo* comparative wear study of traditional and highly cross-linked polyethylene in total hip arthroplasty. *J Arthroplasty* 2005 October;**20**(7):880–6.
6. Marshall A, Ries MD, Paprosky W. How prevalent are implant wear and osteolysis, and how has the scope of osteolysis changed since 2000?. *J Am Acad Orthop Surg* 2008;**16**(Suppl. 1):S1–6.
7. Pollock D, Sychterz CJ, Engh CA. A clinically practical method of manually assessing polyethylene liner thickness. *J Bone Joint Surg* 2001 December;**83-A**(12):1803–9.
8. Sugano N, Saito M, Yamamoto T, Nishii T, Yau SS, Wang A. Analysis of a retrieved UHMWPE acetabular cup crosslinked in air with 1000kGy of gamma radiation. *J Orthop Res* 2004 July;**22**(4):828–31.
9. Affatato S, Bordini B, Fagnano C, Taddei P, Tinti A, Toni A. Effects of the sterilisation method on the wear of UHMWPE acetabular cups tested in a hip joint simulator. *Biomaterials* 2002 March;**23**(6):1439–46.
10. D'Lima DD, Hermida JC, Chen PC, Colwell CW Polyethylene cross-linking by two different methods reduces acetabular liner wear in a hip joint wear simulator. *J Orthop Res* 2003 September;**21**(5):761–6.
11. Devane PA, Bourne RB, Rorabeck CH, Hardie RM, Horne JG. Measurement of polyethylene wear in metal-backed acetabular cups. I. Three-dimensional technique. *Clin Orthop Relat Res* 1995 October(319):303–16.
12. Martell JM, Berdia S. Determination of polyethylene wear in total hip replacements with use of digital radiographs. *J Bone Joint Surg* 1997 November;**79**(11):1635–41.
13. Selvik G. A stereophotogrammetric system for the study of human movements. *Scand J Rehabil Med* 1978;**6**:16–20.
14. Thomsen M, Burandt C, Gortz A, Dathe H, Meesenburg DK, Spiering S, et al. An optical method for determining deformation and form changes in polyethylene surfaces of explanted acetabular cups of hip endoprostheses. *Biomed Tech* 1999 September;**44**(9):247–54.
15. Yamamoto K, Imakiire A, Masaoka T, Shishido T, Mizoue T, Clarke IC, et al. Wear mode and wear mechanism of retrieved acetabular cups. *Int Orthop* 2003;**27**(5):286–90.
16. Bowden AE, Kurtz SM, Edidin AA. Validation of a micro-CT technique for measuring volumetric wear in retrieved acetabular liners. *J Biomed Mater Res* 2005 October;**75**(1):205–9.
17. Bowden AE, Shkolnikov Y, MacDonald D, Kurtz SM. Development and validation of an automated microCT-based technique for mapping damage of explanted polymeric components for TDR. Poster No. 196. *7th Annual Meeting of the Spine Arthroplasty Society,* May 1–4, Berlin, Germany; 2007.
18. Kak CA, Slaney M. *Principles of computerized tomographic imaging (Classics in Applied Mathematics)*: Society for Industrial Mathematics; 2001 p. 1–3.
19. Collier MB, Kraay MJ, Rimnac CM, Goldberg VM. Evaluation of contemporary software methods used to quantify polyethylene wear after total hip arthroplasty. *J Bone Joint Surg* 2003 December; **85-A**(12):2410–18.
20. Devane PA, Bourne RB, Rorabeck CH, MacDonald S, Robinson EJ. Measurement of polyethylene wear in metal-backed acetabular cups. II. Clinical application. *Clin Orthop Relat Res* 1995 October(319):317–26.
21. Hui AJ, McCalden RW, Martell JM, MacDonald SJ, Bourne RB, Rorabeck CH. Validation of two and three-dimensional radiographic techniques for measuring polyethylene wear after total hip arthroplasty. *J Bone Joint Surg* 2003 March;**85-A**(3):505–11.
22. Jasty M, Goetz DD, Bragdon CR, Lee KR, Hanson AE, Elder JR, et al. Wear of polyethylene acetabular components in total hip arthroplasty. An analysis of one hundred and twenty-eight components retrieved at autopsy or revision operations. *J Bone Joint Surg* 1997 March;**79**(3):349–58.
23. Smith SL, Unsworth A. A comparison between gravimetric and volumetric techniques of wear measurement of UHMWPE acetabular cups against zirconia and cobalt-chromium-molybdenum femoral heads in a hip simulator. *Proc Inst Mech Eng* 1999;**213**(6):475–83.
24. Kurtz SM, Parvizi J, Purtill J, Sharkey P, MacDonald D, Brenner E, et al. *In vivo* oxidation is a primary contributor to delamination in TKA. Poster No. 185. *75th Annual Meeting of the American Academy of Orthoopaedic Surgeons,* March 5–9, San Francisco, CA; 2008.
25. Currier BH, Currier JH, Mayor MB, Lyford KA, Collier JP, Van Citters DW. Evaluation of oxidation and fatigue damage of retrieved crossfire polyethylene acetabular cups. *J Bone Joint Surg* 2007 September;**89**(9):2023–9.

Chapter 35

# Computer Modeling and Simulation of UHMWPE

Anton E. Bowden, PhD, Erin Oneida, PhD Candidate and Jorgen Bergström, PhD



## 35.1 INTRODUCTION

Computer models used to simulate the mechanical behavior of UHMWPE in joint arthroplasty components continue to evolve and increase in sophistication. The strategies for predicting the mechanical response of UHMWPE are typically phenomenologically or physically based, and a particular implementation can vary from very simple to extremely complex. This chapter introduces existing methodologies for predicting the mechanical response of UHMWPE, with the goal of outlining the value and range of application of available tools for modeling UHMWPE.

A thorough understanding of the mechanics of UHMWPE is important for efforts to improve the performance of orthopedic components. Elastic properties, resistance to plastic deformation, stress and strain at failure, fatigue behavior, and wear resistance of UHMWPE are all believed to play roles in the life expectancy of a UHMWPE bearing. There exists a fundamental relationship between a material's intrinsic mechanical properties and how a structure made of the material will respond to mechanical stimuli. This material-specific, fundamental relationship is referred to as a constitutive model, and it is of great importance for simulating the behavior of a UHMWPE component.

In this chapter, we review different candidate constitutive models for UHMWPE, list the strengths and limitations of each model, and show how they can be used in simulations of UHMWPE subjected to loading. One of the most important uses of a validated constitutive model is to predict failure of the structure at hand on either a macroscopic or microscopic length scale. Macroscopic failure modes (e.g., visible deformation or large-scale fracture) are associated with overall component failure. Microscopic failure modes (e.g., wear particle generation) are also important and represent failure of microstructural components within the implant material that is not necessarily visible. Although there has been substantial effort put forth in the area of microstructural modeling of UHMWPE [1, 2], this chapter does not attempt to review this class of failure theories. Instead, it focuses on recent constitutive modeling advances for conventional and highly crosslinked UHMWPE that are appropriate for examining the macroscopic behavior of orthopedic components.

## 35.2 OVERVIEW OF AVAILABLE MODELING AND SIMULATION APPROACHES

Many approaches can be taken to model the mechanical behavior of UHMWPE. The two main classes consist of analytical and computational techniques, each of which can employ material models of varying sophistication. Table 35.1 provides a brief description of different modeling approaches with their associated advantages and disadvantages. The analytical closed-form solution methods by necessity require greatly simplified material model descriptions. Using software packages, such as MATLAB (The Mathworks, Natick, Massachusetts) or Maple (Maplesoft, Waterloo, Ontario, Canada), for numerical analysis calculations can allow for the evaluation of more advanced constitutive models, but only for very simple geometries. Finite element (FE)-based simulation methods are more flexible in that they can allow for the use of sophisticated models for polymer behavior, ranging from linear elastic to very advanced viscoplastic constitutive models, and complex geometries. Though finite element methods are computationally relatively expensive to implement, the decrease in the cost of computing power has enabled them to become practical tools for validating material models and analyzing the response of mechanical systems.

Finite element analysis involves the discretization of a bounding geometry into smaller, discrete elements, typically triangles or rectangles and their three-dimensional analogs. This discretization enables the solution of complicated boundary value problems, involving both complex geometries and complex material behavior, using a weak solution of the ensuing Raleigh-Ritz problem. The size and scope of the simulations performed today have increased tremendously following a number of numerical performance improvements, including faster computers that permit the execution of more complicated analyses encompassing large deformations and complex constitutive models; the development of robust FE solvers that can reliably accommodate geometric and material nonlinearity, contacting surfaces, and time dependence; improvements in automated meshing software that allow for the meshing of difficult initial geometries and, if required for numerical stability, a remeshing of deformed geometries; and, most importantly from the perspective of the present chapter, the development of physically-based constitutive models that accurately capture the experimentally determined behavior of UHMWPE.

Any finite element analysis requires the input of appropriate geometry, loading conditions, and a constitutive model. A satisfactory material model may not be available for the material of interest, requiring the development of a new model. Key to the development of such a model is

**TABLE 35.1** Summary of Available Modeling Approaches for Predicting the Behavior of UHMWPE

| Modeling approach | Advantages | Disadvantages |
|---|---|---|
| Handbook solution (e.g., using a linear elastic material model) [3] | + Fast and easy to perform | – Simple geometries only<br>– Does not consider nonlinear material behavior<br>– Only for very small strains<br>– May give inaccurate results |
| Closed-form analytical solution (e.g., using an isotropic plastic material model) [4] | + Fast and easy to perform<br>+ More accurate than a linear elastic solution | – Handbook solutions are available for very few geometries and loading conditions<br>– May give inaccurate results |
| Computer-assisted analytical solution for simple geometries (e.g., isotropic plasticity, linear viscoelasticity, hyperelasticity) | + Allows comparison of different constitutive models<br>+ Relatively easy to perform | – Solutions can be found for very few geometries and loading conditions<br>– May give inaccurate results |
| Finite element analysis using a simple material model (e.g., linear elasticity, hyperelasticity, linear viscoelasticity, isotropic plasticity) | + Can account for complex geometries<br>+ Relatively easy to perform | – Does not consider the true material behavior in general deformation states<br>– Typically valid only for small to intermediate deformations<br>– May give inaccurate results |
| Finite element analysis using an advanced material model for polymers (e.g., augmented Hybrid Model) | + Can account for complex geometries<br>+ Can account for the experimentally determined, characteristic response<br>+ Enables simulations of complex, thermomechanical, composite responses | – Advanced material models are currently only available as user-supplied subroutines for finite element packages<br>– Requires expertise both to calibrate and use |

good experimental data describing the candidate material's behavior under some set of known loading conditions. After a plausible material model is created, validation of the model is performed both by comparison to the original data and, more importantly, by comparison to additional data obtained from a different loading condition. The material model could then be used, along with a supplied geometry and loading conditions, to simulate a particular event.

## 35.3 CHARACTERISTIC MATERIAL BEHAVIOR OF UHMWPE

Conventional and highly crosslinked UHMWPE materials exhibit similar qualitative behavior due to the underlying similarities of the material microstructures. In this section, we will examine the behavior of four different UHMWPE materials to illustrate the characteristic material response of UHMWPE. These data (from experiments also detailed in [5]) will be used to study various material models in the rest of this review. Two of the four materials were "conventional" UHMWPE and two were highly crosslinked UHMWPE. All materials were created from the same lot of ram-extruded GUR 1050 (Figure 35.1).

The first conventional UHMWPE material was a control, virgin, unirradiated material. The second conventional UHMWPE material was gamma radiation sterilized in nitrogen with a dose of 30 kGy (referred to as "30 kGy, $\gamma$-$N_2$"). The two highly crosslinked UHMWPEs were both gamma irradiated in air with a dose of 100 kGy. One of the crosslinked materials was heat treated until the specimen center reached 110°C for 2 hours (referred to as "100 kGy, 110°C"). The second was heat treated at 150°C for 2 hours (referred to as "100 kGy, 150°C"). The degree of crystallinity of the four material types was determined by differential scanning calorimetry (DSC), see Table 35.2.

The uniaxial tension response of the four materials was examined at room temperature using specimens with a diameter of 10 mm and a gauge length of 25 mm. The specimens were monotonically loaded to failure at a grip displacement rate of 75 mm/min. A noncontacting video extensometer was used to measure the axial strain in the gauge region. Representative true stress–true strain curves for the four types of UHMWPE are shown in Figure 35.2. It is clear that the highly crosslinked materials fail at lower stress and strain values than do the two conventional materials. The figure also shows that the four materials behave rather similarly up to a strain of about 0.8 and that the remelted crosslinked material (100 kGy, 150°C) has a slightly lower flow stress.

The uniaxial compressive response of the four materials was examined using cylindrical specimens with a diameter of 10 mm and a height of 15 mm. The compression testing was

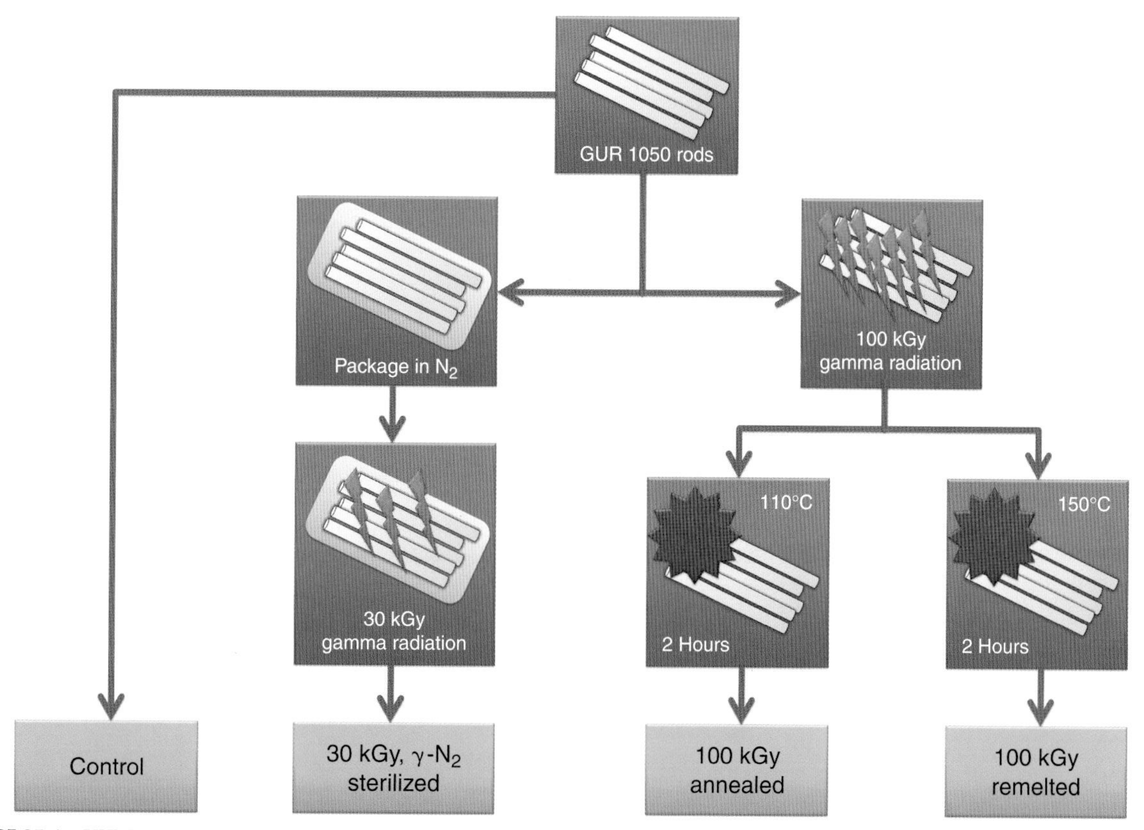


**FIGURE 35.1** UHMWPE materials used in the experimental investigation [5].

**TABLE 35.2** Degree of Crystallinity of the Four Different Types of GUR 1050 [5]

| Material | Degree of crystallinity |
|---|---|
| Unirradiated | 0.50 |
| 30 kGy, $\gamma$-$N_2$ | 0.51 |
| 100 kGy, 110°C | 0.61 |
| 100 kGy, 150°C | 0.46 |

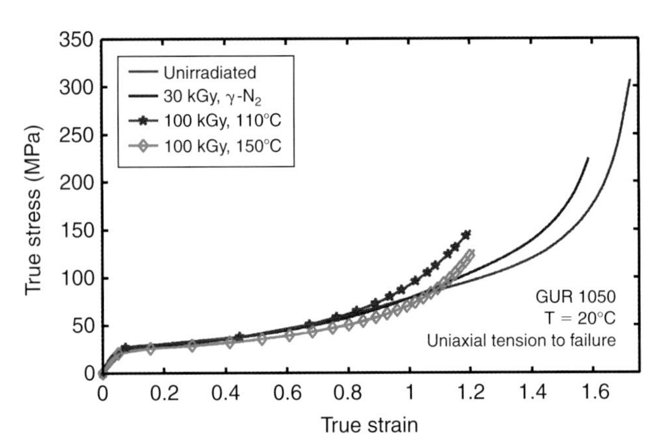


**FIGURE 35.2** Representative experimental uniaxial tension data for the four UHMWPE materials. The tension experiments were performed with a grip displacement rate of 75 mm/min.

performed at room temperature using an electromechanical load frame, and the displacement of the load platens was used to measure the applied deformation. The resulting true stress–true strain response (Figure 35.3) was calculated from the measured force-displacement response assuming constant volume deformation. The compressive behavior of the four materials follows the same trends as observed in the tension experiments: there are only small differences among the observed behavior for all of the materials, and the crosslinked material that was remelted (100 kGy, 150°C) has a slightly lower yield stress. The influence of different strain rates on the compressive behavior of the highly crosslinked GUR 1050 that was heat treated at 110°C (100 kGy, 110°C) is shown in Figure 35.4. As shown in the figure, the yield strength is strongly dependent on the applied strain rate, and the Young's modulus is weakly dependent on the applied strain rate. The significant amount of nonlinear recovery during unloading is also noteworthy.

The influence of temperature on the uniaxial compressive behavior of the annealed material (100 kGy, 110°C) is shown in Figure 35.5. The figure shows that the yield stress and flow behavior are highly dependent on the applied temperature.

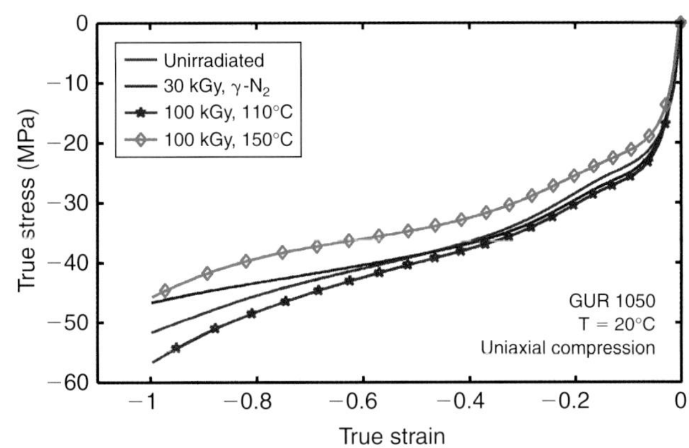


**FIGURE 35.3** Representative experimental uniaxial compression data for the four UHMWPE materials. The compression experiments were performed at an engineering strain rate of −0.02/s.

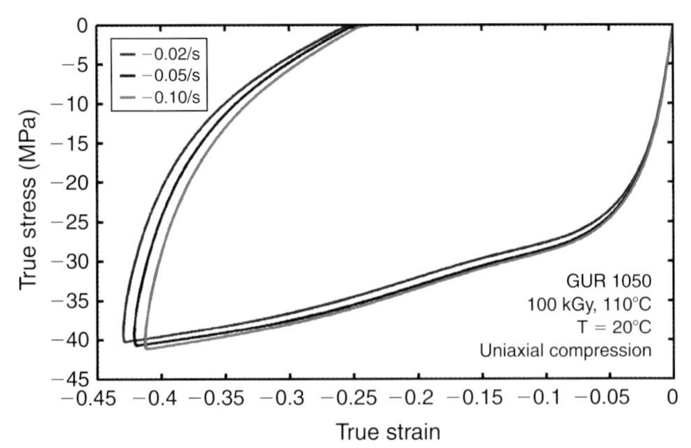


**FIGURE 35.4** Influence of different engineering strain rates on the uniaxial compression behavior of GUR 1050 (100 kGy, 110°C).

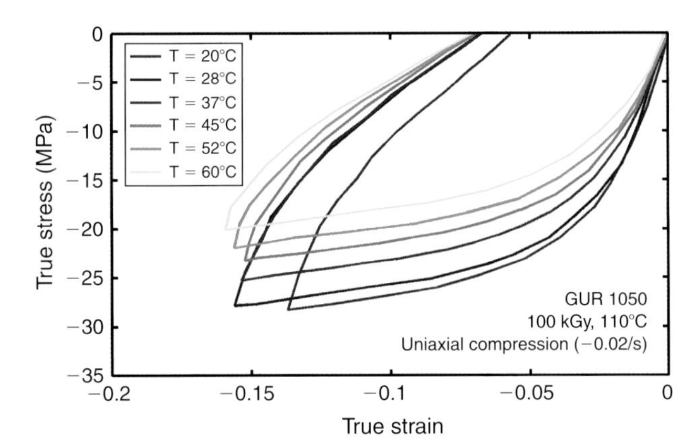


**FIGURE 35.5** Influence of temperature on the compressive stress–strain response (engineering strain rate = −0.02/s) of GUR 1050 (100 kGy, 110°C).

To evaluate the performance of different constitutive models, it is important to perform multiaxial tests in addition to uniaxial tension and compression experiments. The multiaxial tests are necessary for material model validation studies because it is known that some constitutive models

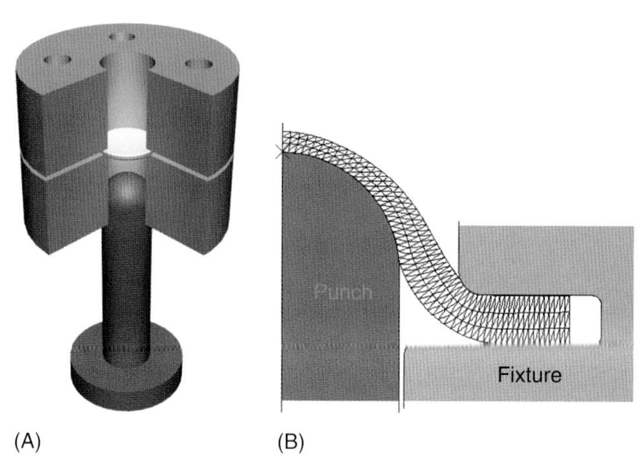


FIGURE 35.6 (A) Schematic representation of the small punch setup. (B) Depiction of the finite element mesh used for simulations [6].

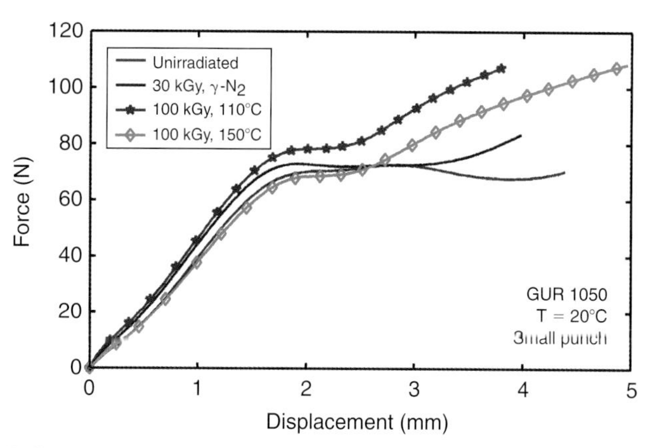


FIGURE 35.7 Representative experimental equibiaxial small punch data for the four UHMWPE materials. The punch test was performed at a displacement rate of 0.5 mm/min.

can be made to agree well with one simple test but are not good predictors of general, multiaxial deformation histories. In this work, we have probed the multiaxial response by using a set of experiments [6] performed using miniaturized disc specimens that had diameters measuring 6.4 mm and were 0.5 mm thick. A closed-loop servohydraulic test system was used for the experiments. The specimens were tested by indentation with a custom-built, hemispherical punch head at a constant displacement rate of 0.5 mm/min at room temperature following ASTM 2183 (2008) [7]. A schematic of the experimental setup is shown in Figure 35.6 A, and the resulting force-displacement results are shown in Figure 35.7. The punch force response is, in general, initially roughly linear with the applied punch displacement, then reaches a plateau, and finally starts to increase until final failure. It is also interesting to note that in the small punch tests, the highly crosslinked materials fail at higher force levels than do the conventional materials, which differs from their characteristic behavior in uniaxial tension (Figure 35.2).

## 35.4 MATERIAL MODELS FOR UHMWPE

Like other thermoplastics, UHMWPE exhibits a complicated nonlinear response when subjected to external loads. As demonstrated in the previous section, its mechanical response is characterized by linear viscoelasticity at small deformations, followed by distributed yielding, viscoplastic flow, and material stiffening at large deformations until ultimate failure occurs. The response is further complicated by a dependence on strain rate and temperature; higher deformation rates and lower temperatures increase the stiffness of the material.

It is clear that thermoplastics are very different from and typically exhibit a broader range of behavior than other structural materials, such as metals. The observed behavior is a manifestation of the different microstructures of the two types of materials and the different micromechanisms controlling the deformation resistance. It is therefore not surprising that different material models should be used when simulating UHMWPE compared to metals. Specifically, and as will be shown in Section 35.4.4, traditional material models for metals (e.g., the $J_2$-plasticity model [4]), although convenient and often familiar to the simulation engineer, should be used with great caution. They are rarely, if ever, a good choice for thermoplastics such as UHMWPE. The magnitudes of the true stress in tension and compression for a given magnitude of applied strain are often very different for UHMWPE. Stress predictions based on most constitutive models developed for metals (including the $J_2$-plasticity model) are symmetrical in tension/compression. As will be discussed later in this chapter, there are models that can more accurately capture the difference between tensile and compressive behavior of polymers.

Several candidate material models are available to predict the behavior of UHMWPE. Because the models have varying degrees of complexity, computational expense, and difficulty in extracting the material parameters from experimental data, it is a good idea to use the simplest material model that captures the necessary material characteristics for the application and situation at hand. Unfortunately, it is often difficult to determine *a priori* the required conditions on the material model. Hence, it is recommended that in general, a more advanced model be used to ensure accuracy and reliability of the predicted data. At a later stage, a less advanced model can be attempted if the computational expense of repeated analyses is too great. At that time, the accuracy of the less advanced model can also be evaluated, and the predictions made can be compared to those generated using the more advanced model.

Given their obvious advantages, the development of advanced constitutive models for UHMWPE and other thermoplastics is an active area of research that is continuously evolving and improving. In the last few years, models

**TABLE 35.3 Representative Values of Young's Modulus at Room Temperature for a Few Different Types of UHMWPE [5]**

| Materials | Young's modulus (MPa) |
|---|---|
| Unirradiated | 830 |
| 30 kGy, γ-$N_2$ | 930 |
| 100 kGy, 110°C | 990 |
| 100 kGy, 150°C | 780 |

attempting to predict the majority of the experimentally observed characteristics of thermoplastics have been developed [6, 8–16]. A number of traditional material models can also be used to predict different aspects of UHMWPE behavior. These models are often easier to use but have a limited domain of applicability. The three main models of this category are: linear elasticity, hyperelasticity, and linear viscoelasticity. These models also have the added benefit of being directly available in commercial finite element packages.

The next few sections present these traditional models and how they apply to UHMWPE. Then, as an example of a more advanced material model, the augmented Hybrid Model [6, 10–12] is presented.

## 35.4.1 Linear Elasticity

Linear elasticity is the most basic of all material models. Only two material parameters need to be determined experimentally: Young's modulus ($E$) and Poisson's ratio ($\nu$). Young's modulus can be obtained directly from uniaxial tension or compression experiments; typical values [5] for a few select UHMWPEs at room temperature are presented in Table 35.3.

The Poisson's ratio can be determined by measuring the transverse strain during uniaxial tension or compression experiments. Due to the small magnitude of the transverse strain, it is often difficult to accurately determine the Poisson's ratio. Instead, it is often sufficient to assume a value for Poisson's ratio of about 0.4. Note that unless the UHMWPE component is highly confined, the Poisson's ratio has a very weak influence on the predicted material response.

Only in special circumstances can UHMWPE be represented with a high degree of accuracy using linear elasticity. As shown in Figure 35.8, linear elasticity is a reasonable model only if the strain in the material is less than about 1.0%. Although a linear elastic representation of UHMWPE is only accurate for small strains, such a

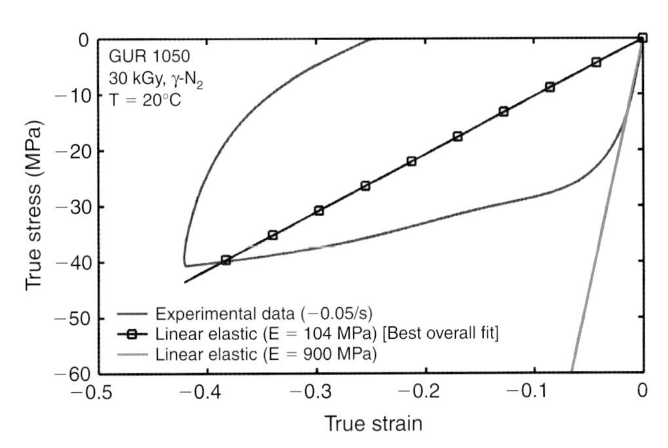


**FIGURE 35.8** Comparison between experimental data (obtained from a uniaxial compression test with an engineering strain rate of −0.05/s) for UHMWPE (GUR 1050, 30 kGy γ-$N_2$) and predictions made using linear elasticity theory.

model can be of value if the UHMWPE component is part of a larger system where the response of the UHMWPE component is not the focus of the study and has little influence on the overall response.

## 35.4.2 Hyperelasticity

A natural extension of linear elasticity is hyperelasticity [17]. Hyperelasticity is a collective term for a family of models that have a strain energy density that only depends on the currently applied deformation state (and not on the history of deformations). This class of material models is characterized by a nonlinear elastic response and does not capture yielding, viscoplasticity, or time dependence. Strain energy density is the energy that is stored in the material as it is deformed and is typically represented either in terms of the strain invariants: $\hat{I}_1, \hat{I}_2$, and $J$, where

$$\hat{I}_1 = \hat{\lambda}_1^2 + \hat{\lambda}_2^2 + \hat{\lambda}_3^2 \tag{1}$$

$$\hat{I}_2 = \hat{\lambda}_1^{-2} + \hat{\lambda}_2^{-2} + \hat{\lambda}_3^{-2} \tag{2}$$

$$J = \det[\mathbf{F}] = \lambda_1\lambda_2\lambda_3 \tag{3}$$

or expressed directly in terms of the distortional principal stretches: $\hat{\lambda}_1$, $\hat{\lambda}_2$, and $\hat{\lambda}_3$. The distortional stretches can be obtained from the applied principal stretches by $\hat{\lambda}_i = J^{-1/3}\lambda_i$, $i=1, 2, 3$. The two main types of hyperelastic models are the polynomial model and the Ogden model [18]. In the polynomial model, the strain energy density is given by

$$W = \sum_{i+j=1}^{N} C_{ij}(\hat{I}_1 - 3)^i(\hat{I}_2 - 3)^j + \sum_{i=1}^{N}\frac{1}{D_i}(J-1)^{2i} \tag{4}$$

and in the Ogden model, the strain energy density is given by:

$$W = \sum_{i=1}^{N} \frac{2\mu_i}{\alpha_i^2}(\hat{\lambda}_1^{\alpha_i} + \hat{\lambda}_2^{\alpha_i} + \hat{\lambda}_3^{\alpha_i} - 3) + \sum_{i=1}^{N} \frac{1}{D_i}(J-1)^{2i} \quad (5)$$

where $C_i$ and $D_i$ are scalar material parameters and $\alpha_i$ are constants. If $N=2$, $\alpha_1=2$, and $\alpha_2=-2$, the Ogden model reduces to a Mooney-Rivlin model, which is often used for modeling incompressible rubber materials. Commercial finite element packages typically contain a number of other hyperelastic representations as well.

Hyperelastic models are often used to represent the behavior of crosslinked elastomers where the viscoelastic response can sometimes be neglected compared to the nonlinear elastic response. Because UHMWPE behaves differently than do elastomers, there are only a few specific cases where a hyperelastic representation is appropriate for simulations involving UHMWPE. One such case is when the loading is purely monotonic and at a single loading rate. Under these conditions, it is not possible to distinguish between nonlinear elastic and viscoelastic behavior, and a hyperelastic representation might be considered. Note that if a hyperelastic model is used in an attempt to capture the stress–strain response that is observed at large strains, including yielding, then there is a significant risk that the model will not be unconditionally- or Drucker-stable [4]. A lack of Drucker-stability implies that even if a hyperelastic model fits the uniaxial experimental data, there is a risk that predictions of multiaxial deformation states will be in error. Attempting to use an Ogden model to capture the behavior of data from a cyclic test could lead to poor results. Figure 35.9 shows the best possible fit of an Ogden 1-term hyperelastic model to the experimental data for GUR 1050 (30 kGy, $\gamma$-$N_2$) over the strain range of the experimental data. From the figure, it is clear that this model is not able to reproduce the observed experimental data. In general, it is often safer to use a more sophisticated model capable of capturing yielding and flow behavior in a more robust and accurate way.

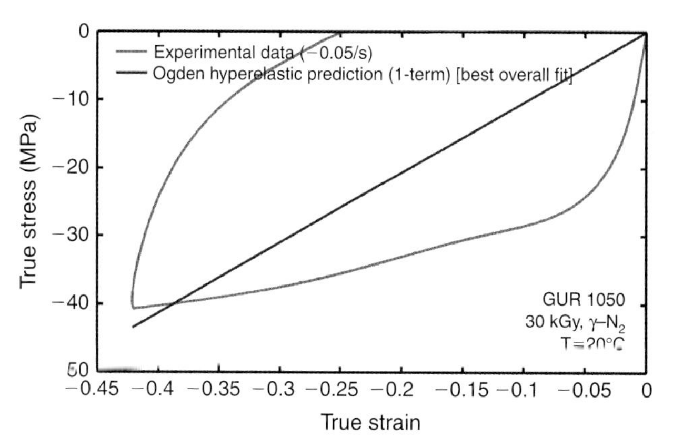


**FIGURE 35.9** Comparison between experimental data (obtained from a uniaxial compression test with an engineering strain rate of −0.05/s) for UHMWPE (GUR 1050, 30 kGy $\gamma$-$N_2$) and predictions made using the Ogden hyperelasticity model.

## 35.4.3 Linear Viscoelasticity

Linear viscoelasticity is an extension of linear elasticity and hyperelasticity that enables predictions of time-dependence and viscoelastic flow. Linear viscoelasticity has been extensively studied both mathematically [19] and experimentally [20] and can be very useful when applied under the appropriate conditions. Linear viscoelastic models are available in all major commercial finite element packages and are relatively easy to use.

The basic foundation of linear viscoelasticity theory is Boltzmann's superposition principle, which states: "Every loading step makes an independent contribution to the final state." This idea can be used to formulate an integral representation of linear viscoelasticity. The strategy is to perform a thought experiment in which a step function of strain is applied, $\varepsilon(t) = \varepsilon_0 \cdot H(t)$ where $H(t)$ is the Heaviside step function, and the stress response, $\sigma(t)$, is measured. Then a stress relaxation modulus can be defined by $E(t)=\sigma(t)/\varepsilon_0$. Note that $\varepsilon_0$ does not have to be infinitesimal due to the assumed superposition principle. To develop a model capable of predicting the stress response from an arbitrary strain history, one begins by decomposing the strain history into a sum of infinitesimal strain increments, $\Delta\varepsilon_i$, at $\tau_i$:

$$\varepsilon(t) = \sum_{i=1}^{N} \Delta\varepsilon_i H(t-\tau_i) \quad (6)$$

The stress response can then be written:

$$\sigma(t) = \sum_{i=1}^{N} \Delta\varepsilon_i H(t-\tau_i) E(t-\tau_i) \quad (7)$$

In the limit as the number of strain increments goes to infinity, the stress response becomes:

$$\sigma(t) = \int_{-\infty}^{t} E(t-\tau)d\varepsilon(t) = \int_{-\infty}^{t} E(t-\tau)\frac{d\varepsilon(\tau)}{d\tau}d\tau \quad (8)$$

This equation can be generalized to a three-dimensional deformation state for an isotropic material as follows:

$$\boldsymbol{T}(t) = \int_{-\infty}^{t} 2G(t-\tau)\dot{\mathbf{e}}d\tau + \mathrm{I}\int_{-\infty}^{t} K(t-\tau)\dot{\phi}d\tau \quad (9)$$

where $G(t)$ is the stress relaxation shear modulus, $\dot{e}$ the rate of change of deviatoric strains, $K(t)$ the stress relaxation bulk modulus, and $\dot{\phi}$ the rate of change of volumetric strains. Only two relaxation moduli need to be determined to predict any arbitrary deformation. The relaxation moduli can be determined from stress relaxation tests and are typically specified in finite element packages as a power series of exponential functions (Prony series):

$$G(t) = G_{\infty} + \sum_{i=1}^{N} G_i e^{-t/\tau_i} \tag{10}$$

where $G_{\infty}$, $G_i$, and $\tau_i$ are material parameters.

The theory behind linear viscoelasticity is simple and appealing. It is important to realize, however, that the applicability of the model to UHMWPE is restricted to strains below the yield strain. Examples comparing predictions based on linear viscoelasticity for UHMWPE specimens tested past yield to the actual experimental data are shown in Figures 35.10 and 35.11. Figure 35.10 shows the best predictive response of a linear viscoelastic model based on a one-term Prony series, and Figure 35.11 shows the best response of a two-term Prony series viscoelasticity model. The figures show that adding terms to the Prony series typically improves the predictive response.

Although the predictions in Figure 35.11 are relatively good, these results are deceiving. The danger of using linear viscoelasticity for strains past yielding is illustrated in Figure 35.12. This figure shows that, although linear viscoelasticity theory can be made to fit one experimental tension or compression test at small to moderate strains, the fit parameters give very poor predictions at slightly different deformation rates.

Another interesting aspect of linear viscoelasticity is that it can be extended to enable predictions at different temperatures. Further, the long-term, time-dependent response at a given temperature can be predicted based on material characterization at several temperatures. The approach to developing these predictions is based on a time–temperature superposition principle [20], which has been shown to work well in a restricted temperature range as long as the requirement of infinitesimal strains is satisfied.

## 35.4.4. Isotropic $J_2$-Plasticity

An example of a material model based on the physics of material behavior is classical metals plasticity theory. This theory, often referred to as $J_2$-flow theory, is based on a Mises yield surface with an associated flow rule and involves rate-independent isotropic hardening [21]. Physically, plastic flow in metals is a result of dislocation motion, a mechanism known to be driven by shear stresses and to be insensitive to hydrostatic pressure. Consequently,

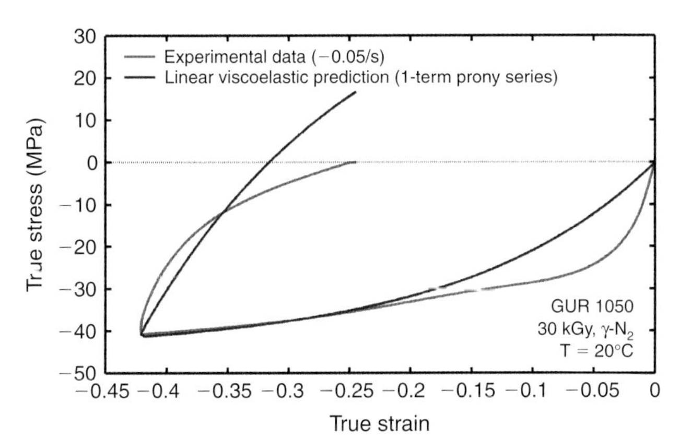


**FIGURE 35.10** Comparison between experimental data (obtained from a uniaxial compression test with an engineering strain rate of −0.05/s) for UHMWPE (GUR 1050, 30 kGy $\gamma$-$N_2$) and predictions made using linear viscoelasticity theory (one-term Prony series).

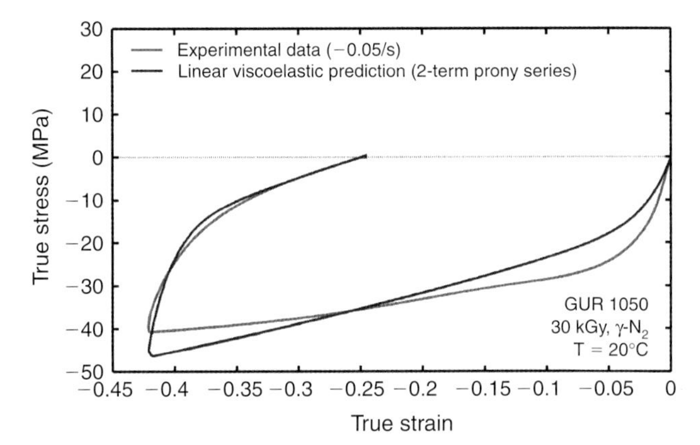


**FIGURE 35.11** Comparison between experimental data (obtained from a uniaxial compression test with an engineering strain rate of −0.05/s) for UHMWPE (GUR 1050, 30 kGy $\gamma$-$N_2$) and predictions made using linear viscoelasticity theory (two-term Prony series).

metals are observed to be nearly incompressible during plastic deformation.

For materials where the basic physics of deformation is different from that previously described, the $J_2$-plasticity model often performs quite poorly. For example, material systems that develop texture or other forms of anisotropy during deformation and materials that behave differently in tension and compression are poor candidates for a $J_2$-plasticity model. Because the mechanisms governing plastic deformation in UHMWPE are quite different from those in metals, a more robust material model is often more appropriate.

The deformation resistance of amorphous polymers in the glassy regime is determined by a combination of thermally activated segmental rearrangements [22] (primarily responsible for the plastic behavior at small to moderate strains) and orientation induced strain hardening due to polymer chain stretching and alignment (primarily responsible for the deformation resistance at large strains [23]). In addition, for semicrystalline polymers, such as UHMWPE,

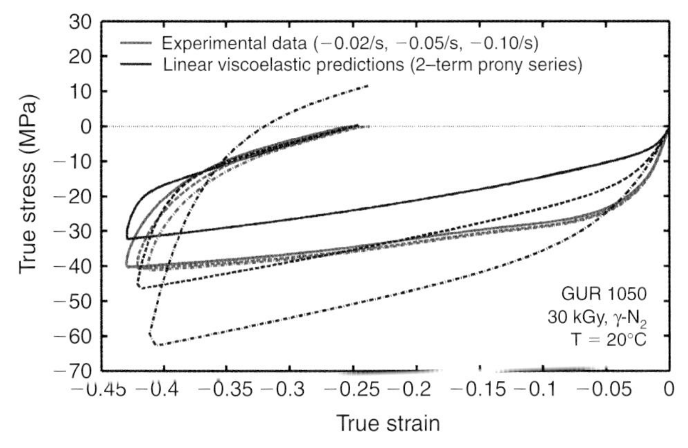


**FIGURE 35.12** Comparison between experimental data (obtained from a uniaxial compression test with engineering strain rates of −0.02/s, −0.05/s, and −0.10/s) for UHMWPE (GUR 1050, 30 kGy $\gamma$-$N_2$) and predictions made using linear viscoelasticity theory (two-term Prony series).

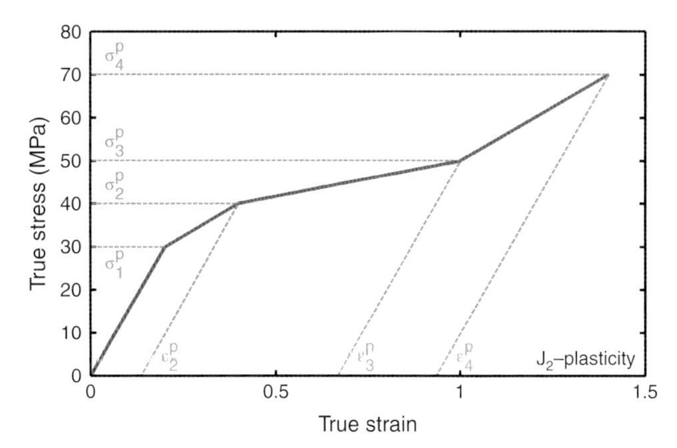


**FIGURE 35.13** Definition of material parameters used in $J_2$-plasticity theory.

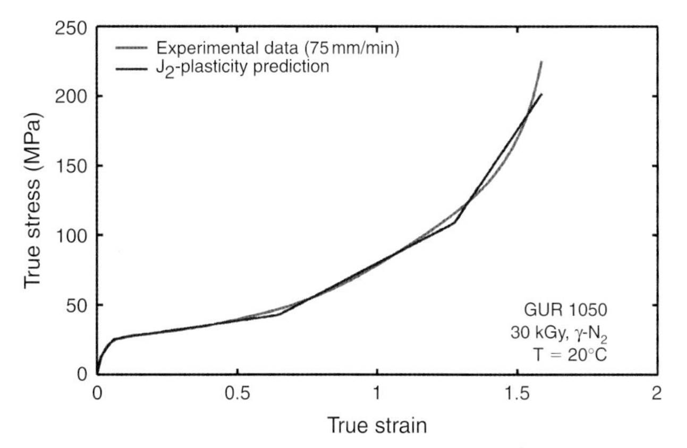


**FIGURE 35.14** Comparison between experimental data (obtained from a uniaxial tension test with a grip displacement rate of 75 mm/min) for UHMWPE (GUR 1050, 30 kGy $\gamma$-$N_2$) and predictions made using the $J_2$-plasticity model.

mechanisms associated with crystallographic slip, twinning, martensitic transformations, and crystallite/lamellar rotation also play a role in resisting deformation [24].

Finding the material parameters for $J_2$-plasticity theory is straightforward and can be obtained from a simple uniaxial tension experiment. The $J_2$-plasticity material model is a piecewise linear model (Figure 35.13), where the material parameters specify the vertices of the stress–strain curve. The $J_2$-plasticity model can be made to fit monotonic, constant-strain rate, constant-temperature test data well (Figure 35.14). The basic model is insensitive to strain rate; however, some FE implementations (e.g., LS-DYNA, LSTC, Livermore, California, USA) have utilized a Cowper-Symonds approach [25] to specifying limited variation in material behavior with strain rate. A significant limitation of $J_2$-plasticity theory is that it always predicts linear unloading behavior. As shown in Figure 35.15, this makes the model inappropriate for certain loading situations, such as cyclic loading of UHMWPE. This is a serious limitation because UHMWPE joint components undergo large deformations at the articulating surface while subjected to cyclically applied loads. In general, the $J_2$-plasticity model is not a robust tool for predicting the large-deformation-to-failure behavior of UHMWPE. Despite its limitations, the $J_2$-plasticity model has been the most widely used approach in the orthopedic research community for simulating the behavior of UHMWPE.

To address the limitations of the existing material models, a new constitutive model was recently developed for conventional and highly crosslinked UHMWPEs [6]. This new model, which is inspired by the physical micromechanisms governing the deformation resistance of polymeric materials, is an extension of specialized constitutive theories for glassy polymers that have been developed during the last 15 years, and it will be discussed in the next section.

## 35.4.5 The Hybrid Model

A number of more advanced models capable of predicting the yielding, viscoplastic flow, time-dependence, and large strain behavior of UHMWPE and other thermoplastics have recently been developed [6, 9–13]. One model formulation that was specifically developed for UHMWPE, the augmented Hybrid Model (which has been recently updated), will be thoroughly detailed (based on information largely contained in [11]) in the remainder of this section.

The basic framework for the Hybrid Model (HM) has evolved from its initial formulation [10] into a more complex constitutive model, the augmented Hybrid Model [11], which has been shown to predict, with great accuracy, the behavior of UHMWPE specimens subjected to several different loading scenarios. The model was developed based on the micromechanical behavior of semicrystalline polymers at the molecular level. At the core of the augmented HM is the multiplicative decomposition of the applied deformation gradient (**F**) into two separate components—elastic

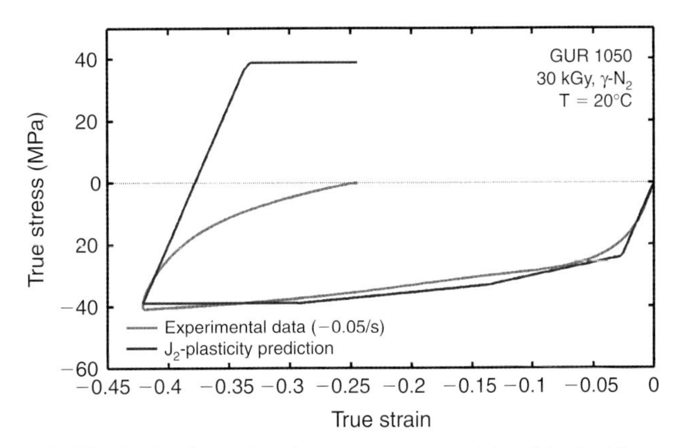


**FIGURE 35.15** Comparison between experimental data (obtained from a uniaxial compression test with an engineering strain rate of −0.05/s) for UHMWPE (GUR 1050, 30 kGy $\gamma$-$N_2$) and predictions made using the $J_2$-plasticity model.

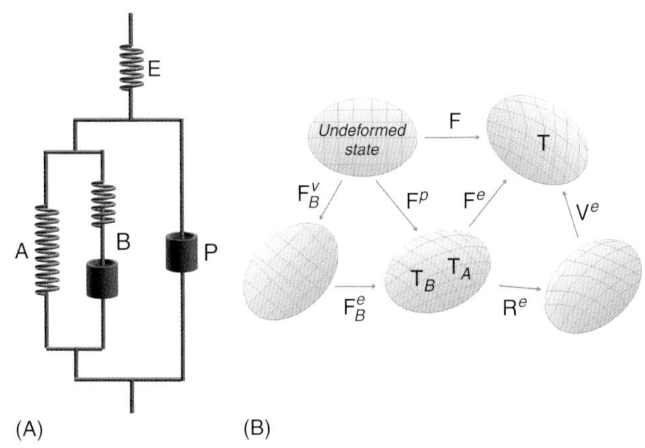


**FIGURE 35.16** (A) Rheological representation of the augmented HM. (B) Deformation map showing the kinematics and stress tensors used in the augmented HM. These figures illustrate how the model represents the viscoplastic flow and how the deformation state is generalized into three dimensions [11].

($\mathbf{F}^e$) and viscoplastic ($\mathbf{F}^p$) deformation gradients—according to $\mathbf{F}=\mathbf{F}^e\,\mathbf{F}^p$. The rheological representation depicting the one-dimensional embodiment of this constitutive model (Figure 35.16A) clearly identifies the elastic (E), which possesses the only linear element in the representation, and viscoplastic (P) components, and also contains backstress network components (A and B). A map that depicts the relationships between the deformation gradients and stress tensors can be seen in Figure 35.16B. A careful study of each component in the augmented HM will now be undertaken.

The first component of the augmented Hybrid Model relates to the elastic behavior of UHMWPE, which is captured by the elastic spring (E) in the rheological representation and the elastic deformation gradient $\mathbf{F}^e$. The polar decomposition [26] of $\mathbf{F}^e$ involves a left stretch tensor, $\mathbf{V}^e$, and a rotation tensor, $\mathbf{R}^e$. Using $\mathbf{V}^e$, the logarithmic true strain ($\mathbf{E}^e$) can be calculated via $\mathrm{E}^e = \ln[\mathbf{V}^e]$, and finally, the Cauchy stress for the system can be determined by:

$$\mathbf{T} = \frac{1}{J^e}(2\mu_e \mathbf{E}^e + \lambda_e\,\mathrm{tr}[\mathbf{E}^e]\mathbf{1}) \tag{11}$$

where $\mu_e$ and $\lambda_e$ are Lamé constants and can be calculated using Young's modulus ($E_e$), Poisson's ratio ($\nu_e$), and the following elastic relationships: $\mu_e = E_e/(2(1+\nu_e))$ and $\lambda_e = E_e\nu_e/((1+\nu_e)(1-2\nu_e))$. The elastic Jacobian represents the relative volume change of the elastic deformation and may be calculated from the elastic deformation gradient: $J^e = \det[\mathbf{F}^e]$.

The first component of the backstress network, A, controls the equilibrium behavior of the viscoplastic response and is dependent on $\mathbf{F}^p$. The stress, $\mathbf{T}_A$, acting on this component can be calculated by:

$$\mathbf{T}_A = \frac{1}{1+q_A}[\mathbf{T}_{8chain}(\mathbf{F}^p;\mu_A,\lambda_A^{lock},\kappa_A) + q_A\mathbf{T}_I(\mathbf{F}^p;\mu_A)] \tag{12}$$

$$\mathbf{T}_{8chain}(\mathbf{F}^p;\mu_A,\lambda_A^{lock},\kappa_A) = \frac{\mu_A}{J^p\bar{\lambda}_A}\cdot\frac{L^{-1}(\bar{\lambda}_A/\lambda_A^{lock})}{L^{-1}(1/\lambda_A^{lock})}\,\mathrm{dev}[\mathbf{B}^{p*}] + \frac{\kappa_A\ln(J^p)}{J^p}\mathbf{1} \tag{13}$$

$$\mathbf{T}_I(\mathbf{F}^p;\mu_A) = \frac{\mu_A}{J^p}\left[I_1^{p*}\mathbf{B}^{p*} - \frac{2I_2^{p*}}{3}\mathbf{1} - (\mathbf{B}^{p*})^2\right] \tag{14}$$

where $\mathbf{F}^p$ is the current viscoplastic deformation gradient; $\mu_A$, $\lambda_A^{lock}$, and $\kappa_A$ are backstress network material parameters equal to the shear modulus, locking stretch, and bulk modulus of the network, respectively; $q_A$ is a material parameter controlling the influence of $\mathbf{T}_{8chain}$ versus $\mathbf{T}_I$ on $\mathbf{T}_A$; $J^p = \det[\mathbf{F}^p]$; $\mathbf{B}^{p*} = (J^p)^{-2/3}\mathbf{F}^p(\mathbf{F}^p)^T$ is a left Cauchy-Green deformation tensor; $I_1^{p*} = \mathrm{tr}[\mathbf{B}^{p*}]$ and $I_2^{p*} = \mathrm{tr}[(\mathbf{B}^{p*})^{-1}]$ are the invariants of $\mathbf{B}^{p*}$; $\bar{\lambda}_A = \sqrt{\mathrm{tr}[\mathbf{B}^{p*}]/3}$; and $L^{-1}(\bullet)$ is the inverse Langevin function for which the following approximation exists [27]:

$$L^{-1}(x) = \begin{cases} 1.31446\tan(1.58956x) + 0.91209x, & \text{if } |x| < 0.84136 \\ 1/(\mathrm{sign}(x) - x), & \text{if } 0.84136 \le |x| < 1. \end{cases} \tag{15}$$

Component B in Figure 35.16A was added to the model to incorporate time-dependent viscoplasticity in the backstress network. The decomposition of the deformation gradient associated with this component, $\mathbf{F}^p$, into elastic and viscoelastic deformation gradients ($\mathrm{F}^p = \mathrm{F}_B^e\mathrm{F}_B^v$), allows for the calculation of the stress, $\mathbf{T}_B$, that is leading to the backstress network viscoelastic flow. Due to the structure of the augmented HM, this stress can be calculated using the elastic deformation gradient of component B ($\mathbf{F}_B^e$). The formulation for $\mathbf{T}_B$ is similar to that of $\mathbf{T}_A$, with the only differences being the addition of a $s_B$ term and the use of $\mathbf{F}_B^e$ instead of $\mathbf{F}^p$ as follows:

$$\mathbf{T}_B = \frac{1}{1+q_A}\left[\mathbf{T}_{8chain}\left(\mathbf{F}_B^e; s_B\mu_A, \lambda_A^{lock}, \kappa_A\right) + q_A s_B \mathbf{T}_I\left(\mathbf{F}_B^e;\mu_A\right)\right] \quad (16)$$

$$\mathbf{T}_{8chain}\left(\mathbf{F}_B^e; s_B\mu_A, \lambda_A^{lock}, \kappa_A\right) = \frac{s_B\mu_A}{J_B^e\bar{\lambda}_B}\cdot\frac{L^{-1}\left(\bar{\lambda}_B/\lambda_A^{lock}\right)}{L^{-1}\left(1/\lambda_A^{lock}\right)} \text{dev}\left[\mathbf{B}_B^{e*}\right] + \frac{\kappa_A \ln\left(J_B^e\right)\mathbf{1}}{J_B^e} \quad (17)$$

$$\mathbf{T}_I\left(\mathbf{F}_B^e;\mu_A\right) = \frac{\mu_A}{J_B^e}\left[I_1^{Be*}\mathbf{B}_B^{e*} - \frac{2I_2^{Be*}}{3}\mathbf{1} - \left(\mathbf{B}_B^{e*}\right)^2\right] \quad (18)$$

where $\mu_A$, $\lambda_A^{lock}$, $\kappa_A$, and $L^{-1}$ (•) are as previously defined; $q_A$ is a material parameter controlling the influence of $\mathbf{T}_{8chain}$ versus $\mathbf{T}_I$ on $\mathbf{T}_B$; $J_B^e = \det\left[\mathbf{F}_B^e\right]$; $\mathbf{B}_B^{e*} = (J_B^e)^{-2/3}\mathbf{F}_B^e(\mathbf{F}_B^e)^T$ is a left Cauchy-Green deformation tensor; $I_1^{Be*} = \text{tr}\left[\mathbf{B}_B^{e*}\right]$ and $I_2^{Be*} = \text{tr}\left[\left(\mathbf{B}_B^{e*}\right)^{-1}\right]$ are the invariants of $\mathbf{B}_B^{e*}$; $\bar{\lambda}_B = \sqrt{\text{tr}\left[\mathbf{B}_B^{e*}\right]/3}$; and $s_B$ dictates the relative stiffness of the main components (A and B) of the backstress network and varies as the material is plastically deformed according to:

$$\dot{s}_B = -p_B\cdot\left(s_B - s_{Bf}\right)\cdot\dot{\gamma}_P \quad (19)$$

where $p_B$ is a material parameter that controls the transition rate associated with the distributed yielding event; $s_{Bf}$ is a material parameter specifying the final value that $s_B$ can reach at fully developed plastic flow (when implementing the model an initial value, $s_{Bi}$, will also be required); and $\dot{\gamma}_P$ is equal to the magnitude of the viscoplastic flow rate:

$$\dot{\gamma}_P = \dot{\gamma}_0\cdot(\tau_P/\tau_P^{base})^{m_P} \quad (20)$$

where $\dot{\gamma}_0$ serves to adjust the units and has a constant value of 1/s; $\tau_P^{base}$ and $m_P$ are material parameters; and $\tau_P$ is equal to the effective shear stress that drives the viscoplastic flow: $\tau_P = \|\text{dev}[\mathbf{T}_P]\|_F$, where F implies that the Frobenius norm is taken and $\mathbf{T}_P$ is defined in Equation 21.

The final component of the augmented HM, as clearly seen in the rheological representation (Figure 35.16A), is the P component. The stress in this component can be easily calculated by:

$$\mathbf{T}_P = \mathbf{T} - [\mathbf{F}^e(\mathbf{T}_A + \mathbf{T}_B)\left(\mathbf{F}^e\right)^T]/J^e \quad (21)$$

Where $J^e = \det[\mathbf{F}^e]$.

Working with the augmented HM requires the use of 13 material parameters (the acquisition of these parameters will be fully detailed later) and knowledge of $\mathbf{F}$, $\mathbf{F}^p$, $\mathbf{F}_B^v$, and $s_B$ at different time points. $\mathbf{F}$ can be determined for the deforming object, and finding the values of $s_B$, $\mathbf{F}^p$, and $\mathbf{F}_B^v$ at new time points will require the use of Equation 19, $\dot{\mathbf{F}}^p = \mathbf{L}^p\mathbf{F}^p$, and $\dot{\mathbf{F}}_B^v = \mathbf{L}_B^v\mathbf{F}_B^v$, where:

$$\mathbf{L}_B^v = \dot{\gamma}_B^v\left(\mathbf{F}_B^e\right)^{-1}\frac{\text{dev}\left[\mathbf{T}_B\right]}{\tau_B}\mathbf{F}_B^e \quad (22)$$

and

$$\mathbf{L}^p = \dot{\gamma}_P\left[\left(\mathbf{R}^e\right)^T\frac{\text{dev}[\mathbf{T}_P]}{\tau_P}\mathbf{R}^e\right] \quad (23)$$

The previously undefined symbols in Equation 22 will now be defined as follows: $\tau_B = \|\text{dev}\left[\mathbf{T}_B\right]\|_F$, where F refers to the Frobenius norm, and:

$$\dot{\gamma}_B^v = \dot{\gamma}_0\left(\frac{\tau_B}{\tau_B^{base}}\right)^{m_B} \quad (24)$$

where $\tau_B^{base}$ and $m_B$ are material parameters.

While the starting costs associated with the initial implementation of the augmented HM are high, subsequent use of the model is fairly straightforward and involves the identification of and use of the following 13 material parameters: $E_e$ and $\nu_e$ (small strain elastic constants); $\mu_A$, $\lambda_A^{lock}$, $\kappa_A$, and $q_A$ (backstress network hyperelastic constants); $s_{Bi}$, $s_{Bf}$, $p_B$, $\tau_B^{base}$, and $m_B$ (backstress network flow constants); and $\tau_P^{base}$ and $m_P$ (viscoplastic constants). To obtain most, or all, of these parameters, an iterative procedure can be used to calibrate the model to existing data obtained from simple experimental tests performed on specimens of the UHMWPE material of interest. To obtain the sample set of material parameters corresponding to the augmented Hybrid Model formulation used in Bergström, Rimnac, and Kurtz [11] and shown in Table 35.4, an iterative computer program, which was based on the Nelder-Mead simplex minimization algorithm, was used to identify parameters leading to predictions that correlated well with the experimental data (both uniaxial tensile and cyclic). The use of such a program required estimates of initial parameter values, and for this particular case they were based on values included in Bergström, Rimnac, and

Kurtz [6]. Coefficient of determination values ($r^2$) were calculated throughout the iterative procedure, and, for each of three chosen GUR 1050 materials (30 kGy, $\gamma$-$N_2$; 100 kGy, 110°C; and 100 kGy, 150°C), the parameter set that led to the highest $r^2$ value was determined to be the optimum set for that specific type of UHMWPE. By comparing the three different optimum parameter sets, it became apparent that the values for nine of the parameters were very similar for all three materials, meaning that these values were not dependent on crosslinking density or thermal treatment. These nine values were then fixed such that they were equivalent in all three of the parameter sets, and the iterative procedure was then performed again with only four parameters allowed to vary for each material. The final, optimum parameter sets identified through this procedure are shown in Table 35.4.

The parameter sets in Table 35.4 were used to simulate tests involving the three materials under study [11]. Simulations of the behavior of UHMWPE (GUR 1050, 30 kGy $\gamma$-$N_2$) when subjected to a uniaxial tension test (Figure 35.17), a cyclic uniaxial fully reversed tension-compression test (Figure 35.18), and a small punch test (Figure 35.19) were in close agreement with the experimental data and illustrate how a single set of parameters can be used in the augmented HM to simulate several different loading scenarios. The small punch test data illustrates the ability of the augmented HM, with parameters obtained from uniaxial tests, to predict a multiaxial deformation history. The ability to predict a multiaxial deformation history sets the augmented HM apart from other constitutive models such as the $J_2$-plasticity model, which has been shown to accurately predict monotonic, uniaxial deformation histories but has difficulties when it comes to simulating multiaxial deformation tests [6].

**TABLE 35.4 Augmented Hybrid Model Material Parameters for GUR 1050 (30 kGy, $\gamma$-$N_2$; 100 kGy, 110°C; and 100 kGy, 150°C)**

| Material parameter | 30 kGy $\gamma$-$N_2$ | 100 kGy $\gamma$ 110°C | 100 kGy $\gamma$ 150°C |
|---|---|---|---|
| $E_e$ (MPa) | 2020 | 2009 | 1270 |
| $\nu_e$ | 0.46 | 0.46 | 0.46 |
| $\mu_A$ (MPa) | 8.22 | 10.15 | 8.14 |
| $\boldsymbol{\lambda_A^{lock}}$ | 4.40 | 2.80 | 2.52 |
| $\kappa_A$ (MPa) | 2000 | 2000 | 2000 |
| $q_A$ | 0.20 | 0.20 | 0.20 |
| $s_{Bi}$ | 40.0 | 40.0 | 40.0 |
| $s_{Bf}$ | 10.0 | 10.0 | 10.0 |
| $p_B$ | 27.0 | 27.0 | 27.0 |
| $\boldsymbol{\tau_B^{base}}$ (MPa) | 25.0 | 26.2 | 20.7 |
| $m_B$ | 9.50 | 9.50 | 9.50 |
| $\tau_P^{base}$ (MPa) | 8.00 | 8.00 | 8.00 |
| $m_P$ | 3.30 | 3.30 | 3.30 |

*Small strain elastic constants: $E_e$ is Young's modulus and $\nu_e$ is Poisson's ratio (note that the Young's modulus and Poisson's ratio values were calculated during the overall optimization procedure and may not be appropriate for more general application in other material models); backstress network hyperelastic constants: $\mu_A$ is the shear modulus, $\lambda_A^{lock}$ is the locking stretch, $\kappa_A$ is the bulk modulus, and $q_A$ is a parameter specifying the asymmetry between tension and compression; backstress network flow constants: $s_{Bi}$ controls the initial relative stiffness between $\mathbf{T}_A$ and $\mathbf{T}_B$, while $s_{Bf}$ controls the final relative stiffness, $p_B$ controls the distributed yielding transition rate, $\tau_B^{base}$ controls the yield strength of network B, and $m_B$ controls the rate-dependence of network B; viscoplastic constants: $\tau_P^{base}$ controls the yield strength of network P, and $m_P$ is a parameter that controls the rate-dependence of network P. The bold text highlights the four parameters that vary among the materials [11].*

## 35.5 DISCUSSION

UHMWPE exhibits a complicated, nonlinear mechanical response that is dependent upon initial processing (radiation and/or thermal treatment), as well as the strain rate and temperature. Due to UHMWPE's complex mechanical behavior, finding an appropriate constitutive model for simulating the behavior of a UHMWPE component can be challenging. For example, the use of a simple elastic or viscoelastic model may yield accurate results for a particular component geometry and material formulation in a specific loading scenario. Small changes to any of these parameters, however, may render this choice of constitutive model inappropriate.

This chapter explored the behavior of linear elastic, hyperelastic, linear viscoelastic, and isotropic $J_2$-plasticity models, as well as the use of the augmented Hybrid Model. Linear elasticity theory proved to only be suitable for modeling UHMWPE specimens subjected to very small strains, while exploration of hyperelasticity theory led to the conclusion that it is often safer to use a more sophisticated constitutive model when modeling UHMWPE. The use of linear viscoelasticity theory led to a reasonable prediction for the response of the material during a uniaxial compression test; however, even small changes to the strain rate rendered the previously identified material parameters unsatisfactory. Isotropic $J_2$-plasticity theory provided excellent predictions under monotonic, uniaxial, constant-strain rate, constant-temperature conditions, but it was unable to predict reasonable results for a cyclic test. The augmented Hybrid Model was capable of predicting the behavior of UHMWPE during a uniaxial tension test, a cyclic uniaxial fully

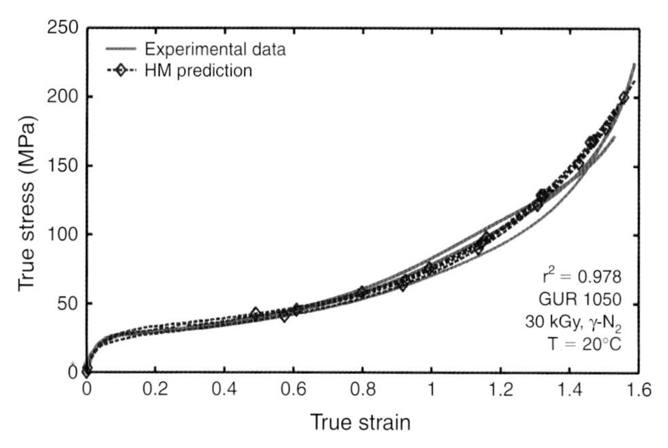


**FIGURE 35.17** Comparison between experimental data (obtained from uniaxial tension tests with grip displacement rates of 30, 75, and 150 mm/min) for UHMWPE (GUR 1050, 30 kGy $\gamma$-$N_2$) and predictions made using the augmented Hybrid Model [11].

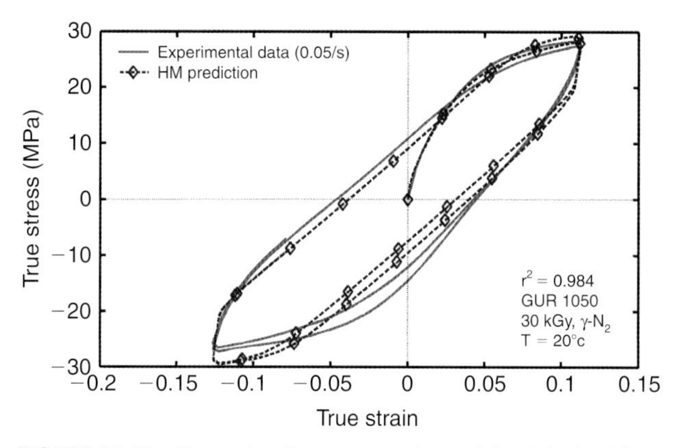


**FIGURE 35.18** Comparison between experimental data (obtained from a cyclic uniaxial fully reversed tension-compression test) for UHMWPE (GUR 1050, 30 kGy $\gamma$-$N_2$) and predictions made using the augmented Hybrid Model [11].

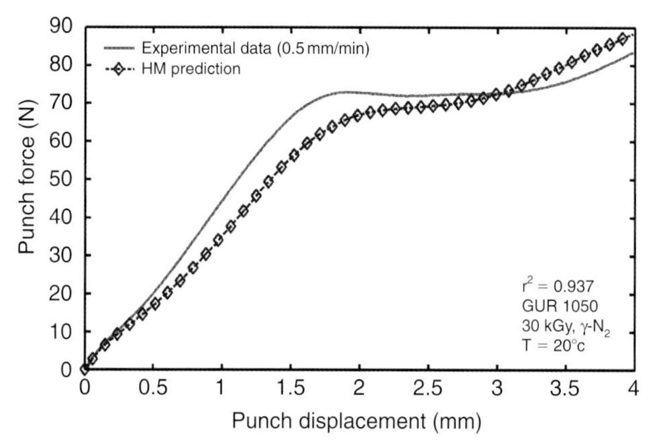


**FIGURE 35.19** Comparison between experimental small punch data (punch rate=0.5 mm/min) for UHMWPE (GUR 1050, 30 kGy $\gamma$-$N_2$) and predictions made using the augmented Hybrid Model (adapted from [11]).

reversed tension-compression test, and a small punch test. Although it is the most complicated to implement, when used in the context of finite element modeling, the augmented Hybrid Model was shown to produce satisfactory predictions across the widest variety of tests for both conventional and highly crosslinked UHMWPEs.

Overall, a number of different approaches can be taken toward modeling UHMWPE components. The analyst must carefully consider what type of constitutive model is most appropriate for a particular situation. Presently, the most versatile and accurate material model for predicting the macroscale behavior of both conventional and highly crosslinked UHMWPE specimens is the augmented HM.

Modeling of damage processes such as fatigue, fracture, and wear have not been explored in this chapter but could be important to consider depending on the particular research objectives. Constitutive models for UHMWPE will continue to evolve as our understanding of the micromechanics and damage behavior of this versatile material are more fully understood and as the availability of computational resources continues to increase.

## ACKNOWLEDGMENTS

Special thanks to Steven Kurtz, Exponent, Inc., and Clare Rimnac, Case Western University, for the experimental test characterization and for the collaborative efforts resulting in the development of the augmented Hybrid Model (HM). Research on the Hybrid Model was supported by NIH R01 47192.

## REFERENCES

1. Baudriller H, Chabrand P, Moukoko D. Modeling UHMWPE wear debris generation. *J Biomed Mater Res B Appl Biomater* 2007 Febuary;**80B**(2):479–85.
2. Medhekar V, Thompson RW, Wang A, McGimpsey WG. Modeling the oxidative degradation of ultra-high-molecular-weight polyethylene. *J Appl Polym Sci* 2003;**87**(5):814–26.
3. Roark RJ, Budynas RG, Young WC. *Formulas for stress and strain*. New York, NY: McGraw-Hill; 2001.
4. Lubliner J. *Plasticity theory*. New York, NY: Pearson Education POD; 1998.
5. Kurtz SM, Villarraga ML, Herr MP, Bergstrom JS, Rimnac CM, Edidin AA. Thermomechanical behavior of virgin and highly crosslinked ultra-high molecular weight polyethylene used in total joint replacements. *Biomaterials* 2002 September;**23**(17):3681–97.
6. Bergstrom JS, Rimnac CM, Kurtz SM. Prediction of multiaxial mechanical behavior for conventional and highly crosslinked UHMWPE using a hybrid constitutive model. *Biomaterials* 2003 April;**24**(8):1365–80.
7. ASTM. F2183. *Standard test method for small punch testing of ultra-high molecular weight polyethylene used in surgical implants*. West Conshohocken, PA: American Society for Testing and Materials; 2008.
8. Arruda EM, Boyce MC, Jayachandran R. Effects of strain-rate, temperature and thermomechanical coupling on the finite strain deformation of glassy-polymers. *Mech Mater* 1995 January;**19**(2–3):193–212.
9. Avanzini A. Mechanical characterization and finite element modelling of cyclic stress-strain behaviour of ultra high molecular weight polyethylene. *Mater Des* 2008;**29**(2):330–43.

10. Bergstrom JS, Kurtz SM, Rimnac CM, Edidin AA. Constitutive modeling of ultra-high molecular weight polyethylene under large-deformation and cyclic loading conditions. *Biomaterials* 2002 June;**23**(11):2329–43.
11. Bergstrom JS, Rimnac CM, Kurtz SM. An augmented hybrid constitutive model for simulation of unloading and cyclic loading behavior of conventional and highly crosslinked UHMWPE. *Biomaterials* 2004 May;**25**(11):2171–8.
12. Bergstrom JS, Rimnac CM, Kurtz SM. Molecular chain stretch is a multiaxial failure criterion for conventional and highly crosslinked UHMWPE. *J Orthop Res* 2005 March;**23**(2):367–75.
13. Dusunceli N, Colak OU. Modelling effects of degree of crystallinity on mechanical behavior of semicrystalline polymers. *Int J Plast* 2008;**24**(7):1224–42.
14. Hasan OA, Boyce MC. A constitutive model for the nonlinear viscoelastic viscoplastic behavior of glassy-polymers. *Polym Eng Sci* 1995 Febuary;**35**(4):331–44.
15. Khan A, Zhang HY. Finite deformation of a polymer: experiments and modeling. *Int J Plast* 2001;**17**(9):1167–88.
16. Kletschkowski T, Schomburg U, Bertram A. Endochronic viscoplastic material models for filled PTFE. *Mech Mater* 2002 December; **34**(12):795–808.
17. Ogden RW. *Non-linear elastic deformations*. Mineola, NY: Dover Publications; 1997.
18. Dorfman A, Muhr A. *Constitutive models for rubber*. Philadelphia, PA: Taylor & Francis; 1999.
19. Christensen RM. *Theory of viscoelasticity*. Mineola, NY: Dover Publications; 2003.
20. Ward IM, Sweeney J. *An introduction to the mechanical properties of solid polymers*. 2nd ed West Sussex, England: Wiley; 2004.
21. Khan AS, Huang S. *Continuum theory of plasticity*. New York, NY: Wiley; 1995.
22. Argon AS. A theory for the low temperature plastic deformation of glassy polymers. *Philos Mag* 1973;**28**:839–65.
23. Haward RN. *Physics of glassy polymers*. Boston, MA: Kluwer Academic Publishing; 1973.
24. Lin L, Argon AS. Structure and Plastic-deformation of polyethylene. *J Mater Sci* 1994 January, 15;**29**(2):294–323.
25. Cowper GR, Symonds PS. *Strain hardening and strain rate effects in the impact loading of cantilever beams*. Providence, RI: Brown University, Applied Mathematics Report; 1958.
26. Gurtin ME. *An Introduction to continuum mechanics*. New York, NY: Academic Press; 1981.
27. Bergstrom JS, Boyce MC. Mechanical behavior of particle filled elastomers. *Rubber Chem Technol* 1999 September–October;**72**(4): 633–56.

# Index

## A



## B



## C

# D

## G



## H



## I

## J



## K



## L

## M



## N



## O

## P

## Q



## R



## S

## Y



## Z